Petroleum Geology of Northwest Europe: Proceedings of the 4th Conference

Volume 2

Petroleum Geology of Northwest Europe: Proceedings of the 4th Conference

held at the Barbican Centre, London 29 March–1 April 1992

Volume 2

edited by

J. R. Parker

Shell UK Exploration and Production, London

with

I. D. Bartholomew	Oryx UK Energy Company, Uxbridge
W. G. Cordey	Shell UK Exploration and Production, London
R. E. Dunay	Mobil North Sea Limited, London
O. Eldholm	University of Oslo
A. J. Fleet	BP Research, Sunbury
A. J. Fraser	BP Exploration, Glasgow
K. W. Glennie	Consultant, Ballater
J. H. Martin	Imperial College, London
M. L. B. Miller	Petroleum Science and Technology Institute, Edinburgh
C. D. Oakman	Reservoir Research Limited, Glasgow
A. M. Spencer	Statoil, Stavanger
M. A. Stephenson	Enterprise Oil, London
B. A. Vining	Esso Exploration and Production UK Limited, Leatherhead
T. J. Wheatley	Total Oil Marine plc, Aberdeen

1993
Published by
The Geological Society
London

THE GEOLOGICAL SOCIETY

The Society was founded in 1807 as The Geological Society of London and is the oldest geological society in the world. It received its Royal Charter in 1825 for the purpose of 'investigating the mineral structure of the Earth'. The Society is Britain's national learned society for geology with a membership of 7500 (1992). It has countrywide coverage and approximately 1000 members reside overseas. The Society is responsible for all aspects of the geological sciences including professional matters. The Society has its own publishing house which produces the Society's international journals, books and maps, and which acts as the European distributor for publications of the American Association of Petroleum Geologists and the Geological Society of America.

Fellowship is open to those holding a recognized honours degree in geology or a cognate subject and who have at least two years relevant postgraduate experience, or have not less than six years relevant experience in geology or a cognate subject. A Fellow who has not less than five years relevant postgraduate experience in the practice of geology may apply for validation and, subject to approval, may be able to use the designatory letters C. Geol (Chartered Geologist).

Further information about the Society is available from the Membership Manager, The Geological Society, Burlington House, Piccadilly, London W1V 0JU, UK.

Published by The Geological Society from:
The Geological Society Publishing House
Unit 7
Brassmill Enterprise Centre
Brassmill Lane
Bath BA1 3JN
UK
(*Orders:* Tel. 0225 445046
Fax 0225 442836)

First published 1993

British Library Cataloguing in Publication Data
A catalogue record for this book is available from the British Library

ISBN 0-903317-85-0

Distributors
USA
AAPG Bookstore
PO Box 979
Tulsa
Oklahoma 74101-0979
USA
(*Orders*: Tel. (918) 584-2555
Fax (918) 584-0469)

Australia
Australian Mineral Foundation
63 Conyngham St
Glenside
South Australia 5065
Australia
(*Orders*: Tel. (08) 379-0444
Fax (08) 379-4634)

India
Affiliated East–West Press PVT Ltd
G-1/16 Ansari Road
New Delhi 110 002
India
(*Orders*: Tel. (11) 327-9113
Fax (11) 331-2830)

Japan
Kanda Book Trading Co.
Tanikawa Building
3-2 Kand Surugadai
Chiyoda-Ku
Tokyo 101
Japan
(*Orders*: Tel. (03) 3255-3497
Fax (03) 3255-3495)

Typeset by Bath Typesetting, Bath, Avon

Printed on acid-free paper at The Universities Press (Belfast) Ltd, Alanbrooke Road, Belfast BT6 9HF, Northern Ireland, UK

Contents

Structural styles and their evolution in the North Sea area

Fluids: migration, overpressure and diagenesis

Field management

Core workshop

VOLUME 1

New plays in established areas: the Tertiary

The Cretaceous

The Jurassic: from regional models to field development; the impact of sequence stratigraphy on hydrocarbon geology

The Triassic: recent advances, discoveries and developments

Permo-Carboniferous and older plays

Irish Sea basins

Irish Sea basins

Introduction and review

R. E. DUNAY

Mobil North Sea Limited, 3 Clements Inn, London WC2A 2EB, UK

The onshore Triassic sandstones adjacent to the East Irish Sea Basin (EISB) have been recognized as potential hydrocarbon reservoir since the discovery of the Formby Oil Field, first drilled in 1939 on an active seep by D'Arcy Exploration Company, a predecessor of BP (Kent 1985). From this small shallow onshore oil accumulation, in which the Triassic reservoir is sealed by Quaternary Boulder Clay, a total in excess of 70 000 barrels of oil was recovered by the cessation of production in 1965.

The first commercial offshore gas discovery, the South Morecambe Field, was made in 1974 by a subsidiary of British Gas plc. This field, which lies within Blocks 110/2a, 110/3a and 110/8a, is the largest accumulation found to date in the EISB, with recoverable reserves in excess of 4 TCF, contained within the Triassic Sherwood Sandstone Group (Bushell 1986). This success was followed in 1976 by the discovery and subsequent delineation of the North Morecambe Field, which occurs in the northern portion of Block 110/2a. This accumulation contains proven gas reserves of 1.08 TCF, also within Triassic Sherwood Sandstone reservoirs (Stuart 1993).

The discoveries in Block 110/13, made in 1990 and 1991 by Hamilton Brothers Oil and Gas Ltd, are extremely significant. The Douglas Field, with oil reserves in excess of 100 MMBBL contained in Triassic sandstones, is the first commercial oil accumulation reported in this area since the discovery of the Formby Field. The adjacent Hamilton and Hamilton North gas accumulations are reported to have total reserves of approximately 800 BCF, also in Triassic sandstones. In 1992, Hamilton reported a significant Triassic oil and gas discovery on Block 110/15, a 12th Round Block adjacent to the Formby Field. These recent successes, coupled with other non-commercial, as well as rumoured, discoveries indicate that the ultimate hydrocarbon potential of the East Irish Sea Basin is yet to be realized.

Eight papers are presented in this section. Three of the papers, by **Jackson and Mulholland**, by **Hardman et al.** and by **Meadows and Beach** address, respectively, regional aspects of the tectonic regime, geochemical modelling, and Triassic reservoir quality within the East Irish Sea Basin. The papers by **Arter and Fagin**, and by **Naylor et al.** concern exploration evaluation of the northern portion of the EISB and the neighbouring Kish Bank Basin, offshore Eire, respectively. The three remaining papers, by **Knipe et al.**, by **Cowan et al.** and by **Stuart** are devoted to studies on the two Morecambe gas accumulations.

Jackson and Mulholland conclude that the Carboniferous/Lower Jurassic succession in the EISB and associated basins exhibits four distinct structural styles. The Morecambe gas fields are considered to occur in a structural style characterized by large tilted fault blocks with stacked halite detachments. The Douglas and Hamilton discoveries are considered to occur in a structural style distinguished by small step-faulted tilted blocks without thick halite deposition. The authors correlate the Permo-Triassic succession of the EISB and adjacent basins with that of the Southern North Sea, and postulate that permanent physical barriers between these basins were absent during the Permo-Triassic. **Hardman et al.** indicate that

Namurian (E2-R1) shales comprise the major source rock for both gas and oil accumulations in the EISB. They suggest that maximum burial occurred during early Tertiary time, with a subsequent two-stage inversion event. The present distribution of hydrocarbon type throughout the EISB is thought to be controlled by the maturity and timing of migration in the various areas coupled with the effects of the Tertiary tectonic event. **Meadows and Beach** address the reservoir quality of the Triassic Sherwood Sandstone. The variability in reservoir quality between the fluvial, sheetflood, and aeolian facies associations is considered to be controlled by the distribution of quartz cement, with the better reservoir quality of the aeolian facies association being due to a relative lack of quartz cement. The authors suggest that these differences in quartz content may be caused by differences in sand provenance.

Arter and Fagin discuss aspects of exploration in the northern EISB, where Tertiary dolerites are intruded into the Triassic succession. These intrusions can result in velocity pull-ups, creating false structures on the time maps of the Triassic reservoir. Subsequent drilling also revealed halite cement within the Triassic reservoir. The authors postulate that dyke intrusion may have mobilized the halite, causing the observed porosity reduction within the Triassic sandstone objective. **Naylor et al.** report on the exploration results within the related Kish Bank Basin, offshore Eire. Two wells tested the Permo-Triassic objective, neither discovering hydrocarbons. Valid structures were drilled, and the reservoir quality of the Permo-Triassic sandstones was good, as was the seal. The authors suggest that the lack of exploration success is due to the absence in this basin of good quality Westphalian and other Carboniferous source rocks.

In regard to the Morecambe papers, **Knipe et al.** address the tectonic history of the Morecambe area. The authors produce a new model for the structural evolution of the South Morecambe and North Morecambe Fields. They have related the growth of platy illite to the structural model, which has facilitated mapping of the Top Platy Illite surface outside well control. **Cowan et al.**, in their dipmeter study of Morecambe wells, conclude that the distribution of the Triassic reservoir facies associations has been controlled by the structural evolution of the area. The major fluvial channel associations display dominantly westerly palaeocurrent orientations. In the final paper in this section, **Stuart** describes the geology of the North Morecambe Field. Although the geology here is similar in many respects to that of South Morecambe, significant differences are reported. For example, in contrast to the South Morecambe structure, which was breached after initial hydrocarbon emplacement and subsequently re-charged with gas, the North Morecambe structure remained intact and contains early emplaced hydrocarbon. The author also demonstrates the separation of the North Morecambe accumulation from that of South Morecambe.

References

BUSHELL, T. P. 1986. Reservoir geology of the Morecambe Field. *In*: BROOKS, J., GOFF, J. C. AND VAN HOORN, B. (eds) *Habitat of*

Palaeozoic Gas in NW Europe. Geological Society, London, Special Publication **23**, 189–207.

KENT, P. E. 1985. UK onshore oil exploration, 1930–1964. *Marine and Petroleum Geology*, **2**, 56–64.

STUART, I. A. 1993. The geology of the North Morecambe Gas Field, East Irish Sea Basin. *In*: PARKER, J. R. (ed.) *Petroleum Geology of Northwest Europe: Proceedings of the 4th Conference*. Geological Society, London, 883–895.

Tectonic and stratigraphic aspects of the East Irish Sea Basin and adjacent areas: contrasts in their post-Carboniferous structural styles

D. I. JACKSON and P. MULHOLLAND[1]

British Geological Survey, 19 Grange Terrace, Edinburgh EH9 2LF, Scotland, UK
[1] *Present address: BP Research, Sunbury Research Centre, Chertsey Road, Sunbury-on-Thames, Middlesex TW16 7LN, UK*

Abstract: A revised geological map and lithostratigraphic correlation chart are presented for Carboniferous–Lower Jurassic strata in the East Irish Sea Basin (EISB) and contiguous basins of the NW–SE- to NNW–SSE-trending Clyde Belt. The significance of the exceptionally thick Triassic succession (4375 m) is documented in relation to other Triassic rifts of NW Europe. Basin margin faults within the Clyde Belt are basement-controlled and located either along the margins of exposed and concealed Ordovician–early Devonian buoyant granites, or orthogonal to earlier Paleozoic structures.

Four basic and markedly different structural styles have been identified from seismic interpretation. The styles comprise: solitary Caledonoid-trending folds (typically elongate centroclines) with minimal NW–SE faulting (style 1); synthetic and antithetic fault systems in single NW–SE half-graben on the fringes of the Clyde Belt (style 2); linked half-graben with large tilted fault blocks and stacked Permo-Triassic halite detachments in the centre and north of the EISB (style 3); and small step-faulted tilt blocks without thick Permo-Triassic halites, but again forming linked half-graben, in the south and on the fringes of the EISB (style 4). The styles are largely independent of the age of the basin-fill, except style 3 which is restricted to Permian–Lower Jurassic successions.

The mixed oil and gas province of the East Irish Sea Basin (EISB) (Bott 1964; Bott and Young 1971; Wright *et al.* 1971; Colter and Barr 1975; Colter 1978; Jackson *et al.* 1987; Ziegler 1987; Stuart and Cowan 1991) currently contains the greatest amount of discovered commercial hydrocarbons on the UKCS excluding the North Sea, with 21 gas and 3 oil discoveries reported from 49 drilled exploration and appraisal wells. The numerous half-graben exhibit complex inter-relationships and are infilled by an exceptionally thick (aggregate maximum 5.7 km) Permian–Lower Jurassic sequence, including probably the thickest Triassic (4375 m) on the UKCS (Figs 1, 2 and 3).

Seismic interpretation of Carboniferous–Jurassic successions covering a wider area than an earlier stratigraphical account (Jackson *et al.* 1987) has revealed a remarkable range of contrasting structural styles within the restricted region of the EISB and contiguous areas (Fig. 2). Geographical variation in both Permian and Triassic play type is strongly influenced by structural style, which consequently determines the exploration strategy.

The EISB is the largest basin within the necklace of deep Triassic troughs (Fig. 1), which stretches NW–SE from the Highland Boundary Fault to the Cheshire Basin and which was termed the Clyde Belt by McLean (1978). Permian–early Jurassic subsidence along this rift took place both in sags with slight internal fault control, and in graben and half-graben dominated by syn-depositional growth faulting. Permo-Triassic basins in the Clyde Belt developed during a phase of NE–SW to ENE–WSW extension. In the EISB the sub-basins are orthogonal to the trend of the Carboniferous basins which developed under dominantly NNW–SSE extension and which were folded along Caledonoid lines during the Variscan orogeny (Lee 1988; BGS 1991).

Post-Devonian reactivation of basement-controlled faults occurred repeatedly. Pre-existing folds and Lower Paleozoic basement ridges (especially where cored by buoyant Caledonian granites) also exerted a major influence on thickness and facies during the Carboniferous and Permian–early Jurassic phases of extension and subsidence. Post-depositional block faulting (and mild compression) took place along the same fabric during the Variscan and post-early Jurassic phases of uplift, erosion and inversion.

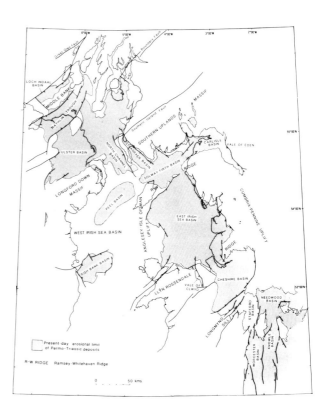

Fig. 1. Location of Permo-Triassic basins and basinal divides within the Clyde Belt. Modified from McLean (1978) and BGS maps.

Throughout the Permian–early Jurassic interval the EISB expanded in area, and subsidence occurred in a series of depositional pulses separated by non-sequences at major lithostratigraphic boundaries (Fig. 3). Upper Permian deposits overlap the Lower Permian along parts of the West Crop and east of the Lancashire Coalfield Boundary Fault (Figs 2 and 9a). The Sherwood Sandstone Group (Lower Triassic) cuts out the Upper Permian south of the Leyland Basin near Chorley (Fig. 2) and in Furness (Rose and Dunham 1977, p. 88).

From *Petroleum Geology of Northwest Europe: Proceedings of the 4th Conference* (edited by J. R. Parker).

Overlap of the Sherwood Sandstone Group by the Mercia Mudstone Group (Middle to Upper Triassic) is seen at the southern end of the Pennines outside the present area. The Permo-Triassic axial successions within the Clyde Belt have been partly exhumed, but preserved by lower relative uplift of the basins compared to the bounding highs.

This contribution uses map examination and seismic interpretation to discuss pre-Permian and later structural influences on the tectonic history of the EISB and environs. An analysis of the post-Carboniferous fault displacement history attempts to quantify the respective amounts of fault movement occurring during initial growth-related subsidence and during later uplift. In a concluding section, the area is subdivided on the basis of structural style in order to demonstrate the influence of basin orientation, basin location and nature of sedimentary fill upon coevally-formed traps.

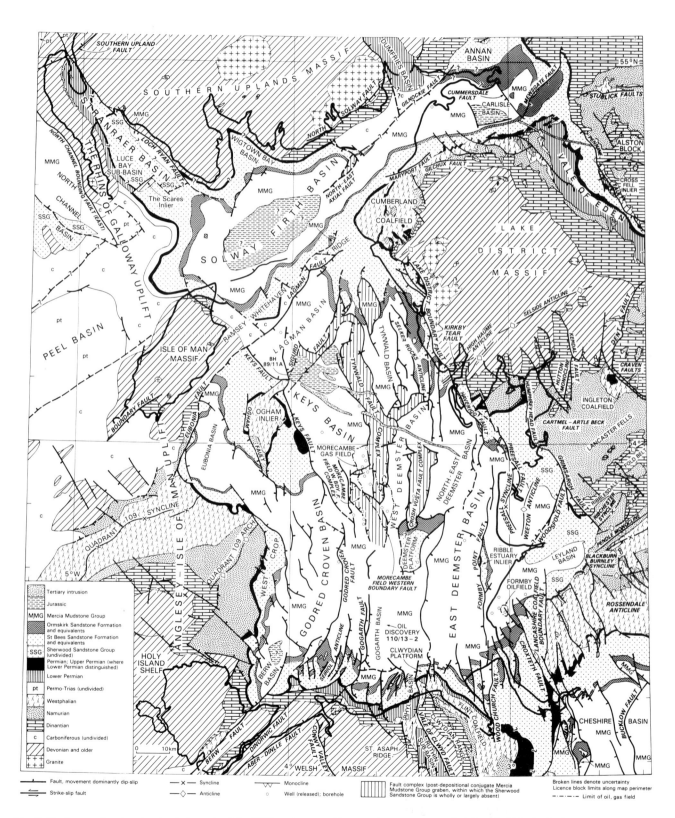

Fig. 2. Geology and major structural elements of the EISB and surroundings; onshore areas compiled from British Geological Survey maps. Minor faults omitted except at designated outcrops; only major faults shown onshore. Boundaries conjectural in certain areas.

Stratigraphical prologue

Recent seismic interpretation, well correlation and facies analysis has established the stratigraphic age equivalence of internal Permian–early Jurassic events throughout the Solway Firth Basin and EISB. Correlation with the Cheshire Basin, Ulster Basin and Southern North Sea (Fig. 3) also demonstrates a shared event-stratigraphy, and is taken to indicate the absence of permanent physical barriers between the various basins. In consequence, certain stratigraphical observations made in Jackson *et al.* (1987) have been reinterpreted and are discussed below.

1. An exceptionally thick Collyhurst Sandstone (Lower Permian) succession, estimated to attain 1150 m in Block 110/17, has been mapped by the authors on seismic data (Figs 9a, 9d) in an ENE–WSW belt extending from the Berw Basin to the Formby Oilfield (Fig. 2). The sandstones are overstepped abruptly by transgressive Manchester Marls to the west (Fig. 9a) indicating firstly, the infill of a palaeo-topographic low by Lower Permian deposits and secondly, a contemporaneous unfaulted western margin to the EISB. Data from the Formby area on the east supports a broadly similar configuration, with Collyhurst Sandstone proved to 715 m in Formby 1 (Kent 1948; Wray and Cope 1948) and absent in Formby 4 (Falcon and Kent 1960, figs 18 and 19).

   Offshore, the sandstones thin northwards against the southern flanks of two cross-cutting and discontinuous pre-Permian ridges; thick sandstones are generally absent north of *c.* 53° 38′ N. One ridge is coincident with a regional Caledonoid upfold extending from the Quadrant 109 Arch in the west (Figs 2 and 4) to its likely continuation on the EISB eastern margin as the High Haume Anticline (Dunham and Rose 1949). The second ridge is a broad NW–SE zone running from the Isle of Man through the structural highs of the Ogham Inlier, the Morecambe Field pericline, the Deemster Platform and the Ribble Estuary Inlier to the Formby area and the Rossendale Block (hereafter the Ogham–Formby line, Fig. 4). To the south of the ridge intersection, the combined ridge crests approximately delineate the change in facies between two differing Permian successions. The Manchester Marls overlie above thick Collyhurst Sandstone to the south of the crest line, and the St Bees Evaporites rest on thin Basal Breccia (Brockram) to the north (Fig. 3; Smith *et al.* 1974).

2. In areas of thick Collyhurst Sandstone, the reflector designated Top Carboniferous in Jackson *et al.* (1987) is now believed to correspond to a horizon at or near the base of the Upper Permian (termed the Near Base Upper Permian reflector on Figs 9 and 10). The event varies laterally from a single reflector to a thin high-amplitude package, and relates to acoustic impedance and tuning interference effects in carbonates and anhydrites within or at the base of the Upper Permian rather than a Top Carboniferous event.

3. Detailed log correlations indicate that the Silicified Zone of the St Bees Sandstone Formation (Colter and Barr 1975; Colter 1978; Burley 1984; i.e. the 'lower unit' of Jackson *et al.* 1987, figs 2 and 8) correlates with the Bunter Shale Formation of the Southern North Sea (cf. Well 49/21-2, Rhys 1974; Fisher and Mudge 1990). Similarly, the 'upper unit' of the St Bees Sandstone Formation (Wilmslow Sandstone Formation equivalent, Jackson *et al.* 1987) equates with the Bunter Sandstone Formation of the Southern North Sea (Fig. 3). The gamma-log res-

ponse of the Bröckelschiefer of the Southern North Sea (Whittaker *et al.* 1985, p. 33 and fig. 24; Fisher and Mudge 1990) can be matched with a slightly reduced radioactivity in the basal unit (20–30 m thick) of the St Bees Sandstone Formation (Wells 110/8-1, 110/8-2 and 110/9-1).

4. The authors regard the so-called 'Keuper Waterstones' of the Morecambe Field (Colter and Barr 1975; Stuart and Cowan 1991) as argillaceous sandstones of floodplain facies deposited by high sinuosity streams within the Budleighensis River (Wills 1970, fig. 5). Thus they are analogous to the Nether Alderley Red Sandstone Member deposited in the Eastern River of the Cheshire Basin (Thompson 1970, fig. 7). The beds comprise the topmost Ormskirk Sandstone Formation and are not viewed as an arenaceous equivalent of the Tarporley Siltstone Formation (Fig. 3; see discussion in Jackson *et al.* 1987). The anhydrite-dolomite marker unit (Colter 1978) together with the Rossall Salts (lower leaf) and most of the Unit A mudstones (Jackson *et al.* 1987) are correlated jointly with the Hambleton Mudstones of the Blackpool area (Fig. 3; Wilson and Evans 1990).

5. The correlation (Wilson 1990) of the thick halite in Wells 110/3-2 and 110/7-1 with the Preesall Salt (Northwich Halite Formation of the Cheshire Basin) is accepted, together with the revised nomenclature of mudstones in the Mercia Mudstone Group where recognition permits in the offshore area (Fig. 3). The yellow miospores upon which a Carnian age and correlation with the Wilkesley Halite Formation of the Cheshire Basin were proposed (Jackson *et al.* 1987) are now regarded as Triassic contaminants introduced as caving, probably from the overlying Pleistocene or from drilling mud.

6. *Arnioceras, Caenisites*? and bivalves were recovered from grey silty Lower Lias mudstones in BGS shallow borehole 89/11A in the Keys Basin (Fig. 2) and indicate a Sinemurian age (*Arnioceras semicostatum* to *Asteroceras obtusum* zones, probably *Caenisites turneri* Zone, H.C. Ivimey-Cook, pers. comm., 1990). They lie at a revised estimate of 600 m above the Near Top Triassic disconformity (Jackson *et al.* 1987, fig. 3a). A similar disconformity overlain by presumed Lower Jurassic rocks has been observed on seismic data in the Solway Firth and Berw basins (Fig. 2) and possibly marks the base of the Penarth Group. If the Berw Basin outlier exists, it would support the prediction by Greenly (1919, p. 778) of a Jurassic and Upper Cretaceous outlier in the EISB, based upon an analysis of the glacial erratics of North Anglesey.

Tectonic setting of the North Irish Sea and Clyde Belt

The North Irish Sea, loosely defined as the area between the North Channel and a line joining the north coast of Anglesey to Wicklow Head, is bisected by a N–S post-early Jurassic axis of uplift (the Anglesey–Isle of Man–Rhins of Galloway Uplift, Fig. 2). Carboniferous and subordinate Lower Paleozoic rocks occupy the West Irish Sea Basin to the west of the axis, in contrast with the area to the east where Permo-Triassic rocks crop out in the EISB and satellite basins such as the Solway Firth Basin (BGS 1991). Continuity of Triassic sedimentation over the axis, albeit slightly condensed over a major high, is provided by a comparison of Triassic thicknesses along the erosional eastern margins of the Kish Bank, ?Peel and North Channel basins with those of the erosional western margins of the EISB and Solway Firth basins.

An account of the tectonic evolution of the North Irish Sea

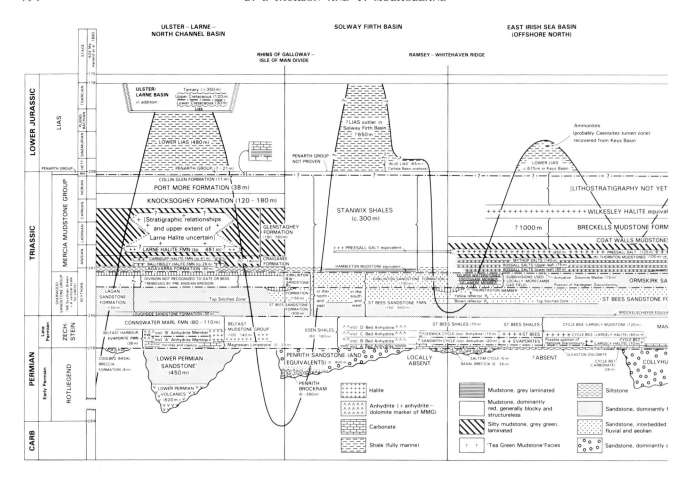

Fig. 3. Time-stratigraphic cross-section (for the interval early Permian to end-early Jurassic) through the Ulster, Solway Firth, East Irish Sea and Cheshire basins to show facies relationships, lithostratigraphic nomenclature and suggested equivalents in the Southern North Sea. Vertical scale as indicated and based on Harland *et al.* 1990. (NB For reasons of clarity, the Sherwood Sandstone Group interval is drawn at four times the vertical scale). Compiled from Rhys (1974), Smith *et al.* (1974), Warrington *et al.* (1980), Cope *et al.* (1980), Smith (1986) and other references cited, modified where necessary.

from the Lower Paleozoic to the present is beyond the scope of this contribution, but the main findings of a larger study (in preparation) are summarized below.

Relationship of Clyde Belt to Triassic of NW Europe

West of the Pennine–Dent–Malvern Line (Fraser *et al.* 1990), NW England and the Clyde Belt of the North Irish Sea are characterized by post-Carboniferous normal dip-slip faults of large vertical displacement but short length (Smith 1985, map 2; Kent 1949). This corrugated Base Permo-Triassic surface contrasts markedly with the much smoother form over the Southern North Sea and Eastern England Shelf (Smith 1985, map 2) and suggests a prolonged difference in crustal behaviour sustained through several depositional and erosional episodes.

Many of the bounding highs of the North Irish Sea are cored by buoyant Ordovician to early Devonian (Acadian) granites (Fig. 2; Bott 1978) which form a discontinuous ring of plutons. The role of these granites has been crucial in providing crustal instability and isostatic readjustment during both subsidence and uplift.

The thick Permo-Triassic succession of the EISB can be divided into four depositional episodes. Following Stephanian–Autunian faulting (Ziegler 1990), isolated intermontane basins were filled with thick aeolian Lower Permian sandstones (Collyhurst Sandstone and equivalents). Ingress for the Late Permian Bakevellia transgression from the Boreal Ocean (Smith *et al.* 1974) lay through the outer Marginal Belt and the

northern part of the Clyde Belt of McLean (1978). The Clyde Belt was subsequently the site of a linear rift with axial sagging occupied by the Budleighensis River in early Triassic times (Wills 1970), followed by further rapid subsidence and halite precipitation in epicontinental seas during late Triassic times (Warrington *et al.* 1980). Superposition of these sequences and their preservation during regional uplift, allied to the buoyancy of onshore Caledonian granites, is responsible for the EISB containing one of the thickest Permo-Triassic successions in NW Europe (Ziegler 1990).

Regional compilations of Triassic isopachs for NW Europe (Ziegler 1990, enclosure 38) indicate that the line of maximum Triassic subsidence along the Clyde Belt can be followed through the Worcester Graben to the Pays de Bray Fault and Paris Basin and ultimately to Tethys. This trend is parallel to other major Triassic troughs of NW Europe (e.g. the Bay of Biscay Rift and the North Danish–Polish Trough). The line is part of a reticulate network of NW–SE and NE–SW rifts related to the Boreal Ocean and North Atlantic. The Clyde Belt–Pays de Bray line is thus viewed as a failed break-up axis or short-cut linking the Boreal Ocean to Tethys during the suturing of Pangaea and the early opening of the North Atlantic.

Structural influences on basin location and trend in the North Irish Sea

The larger basins of the Clyde Belt are separated by transverse Caledonoid-trending Lower Paleozoic ridges or schwelle

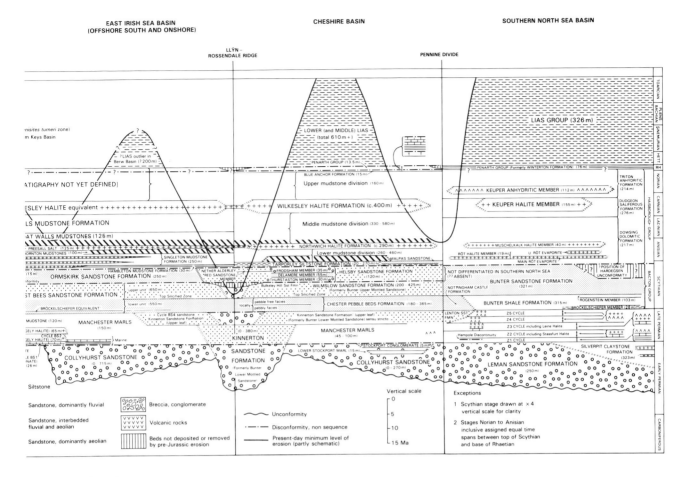

Fig. 3. (continued)

(Fig. 1). Throughout the Permian and Triassic, the ridges experienced condensed deposition and contrasting facies developments compared to the adjoining basins (Figs 2 and 5; Jackson et al. 1987, fig. 7). Three examples from the study area are the Southern Uplands–Longford Down Massif, the Ramsey–Whitehaven Ridge (Bott 1964; Bott 1978), and a ridge in the south, here termed the Llŷn–Rossendale Ridge, which extends from the St Tudwal's Arch through the Llŷn Peninsula, Clwydian Range and Wirral Peninsula to the Rossendale Block (Figs 2 and 5). Throughout the Triassic, but especially during deposition of the Mercia Mudstone Group, the EISB was essentially a giant half-graben with regional northward thickening towards the active margin at the Ramsey–Whitehaven Ridge (Lagman Fault) and a trailing margin located at the Llŷn–Rossendale Ridge.

Faults and folds of post-Devonian age are known to have a long history of movement in onshore areas bordering the North Irish Sea (Moseley 1972; Soper and Moseley 1978; Warren et al. 1984, p. 135; Earp and Taylor 1986; Arthurton and Wadge 1981; Lee 1988). The same relationship holds for the offshore area. Pre-existing Caledonoid (NE–SW), Charnoid (NW–SE) and Malvernoid (N–S) structural trends were reactivated repeatedly in space and time. Caledonoid trends are particularly associated with fold elements (e.g. Quadrant 109 folds, Ribblesdale Fold Belt, Solway Firth Syncline).

These trends are evident during Permian–early Jurassic deposition and post-early Jurassic uplift and indicate the constant interplay between sagging and fault-related subsidence. A trellised network of Caledonoid, Malvernoid and Charnoid axes of uplift and subsidence is found throughout the EISB. The present-day occurrence of Permian–Lower Jurassic outliers in topographically low areas around the North Irish Sea

margins testifies to the continued post-Jurassic influence of the Permo-Triassic growth faults and subsidence patterns.

Basement-controlled faulting (Johnson 1967; Bott 1968) of Carboniferous and Permian–Lower Jurassic successions is particularly evident along the margins of the Clyde Belt (McLean 1978) giving rise to hangingwall depocentres. These dip-slip faults are limited in length but of large vertical displacement and occur in two contexts. One suite is located along the margins of Ordovician–early Devonian granites with maximum displacement and heave opposite the roof crest (e.g. Lake District Boundary Fault, North Solway Fault, Pennine Fault). Fault displacement decreases abruptly to fault tips beyond the granite outcrop (Fig. 4). The second group was initiated at or near the axial crest of orthogonal Lower Paleozoic ridges (e.g. the Vale of Clwyd, Loch Ryan and Lancashire Coalfield Boundary faults, Fig. 2; BGS 1990, 1991). Displacement declines from a maximum at the point of initiation (Walsh and Watterson 1991) to fault tips outside the confines of the ridge. Exceptionally, the Formby Point Fault appears to straddle and to be bound by the limits of an orthogonal Carboniferous/Variscan structure rather than a Lower Paleozoic ridge. The projected extensions of the Bowland Line and Pendle Lineament, marking the limits of the Ribblesdale Fold Belt (Arthurton et al. 1988) appear to coincide with the fault tips of the Formby Point Fault.

Following sinistral Acadian (late Caledonian) transpression (Soper et al. 1987), post-Devonian strike-slip movement is believed to be relatively insignificant, with small movements probably taken up as transfer elements on ENE–WSW faults. The left-stepping en échelon arrangement of dip-slip faults and strike ramps along the eastern boundary of the EISB may be further evidence for transtension (Fig. 2). Permo-Triassic

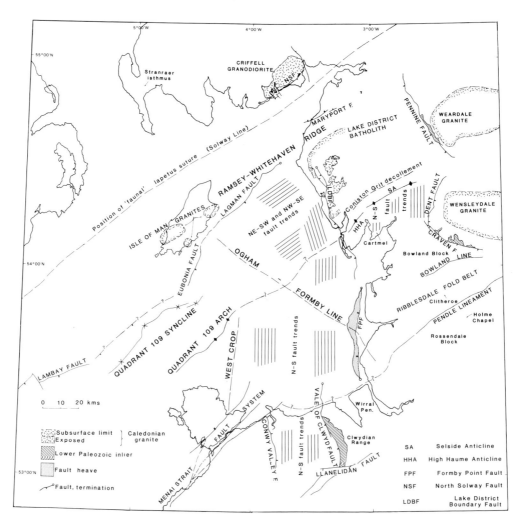

Fig. 4. Crustal and pre-Permian structures in relation to subsequent basin development in the North Irish Sea. Data included from Moseley (1972), Bott (1978), Gibbons (1987), Arthurton *et al.* (1988) and BGS maps.

hangingwall depocentres and Carboniferous hangingwall out-liers are juxtaposed with present-day footwall inliers across the major NW–SE dip-slip growth faults. Condensed successions (Jackson *et al.* 1987, figs 4 and 7) were laid down over footwall blocks, partly due to footwall uplift and elastic rebound. The inliers arise from unroofing during Variscan and post-early Jurassic uplift, accelerated by isostatic readjustments where the footwall is cored by a granite. Striking examples of this juxtaposition from both granite and orthogonal basement ridge settings are provided by the Vale of Clwyd, Formby Point, Keys and Lake District Boundary faults, and by the Pennine and Craven faults further to the east (Fig. 2).

Strike-slip movement is believed to be concentrated on Caledonoid trends (Jackson *et al.* 1987), possibly along reacti-vated Lower Paleozoic lineaments, e.g. the Skipton Rock Fault (Arthurton 1983), the Kirkby Tear Fault (Rose and Dunham 1977, p. 71) and the Morecambe Field offset (Ebbern 1981).

The Iapetus suture and Menai Strait–Pendle line are two Caledonoid-trending crustal lineaments which traverse the area and have been the subject of much recent debate (Fig. 4).

The Solway Firth Basin and Peel Basin overlie the position of the 'faunal' Iapetus Suture along the Solway Line (see review by Todd *et al.* 1991). On the basis of earlier studies using deep seismic profiles (Hall *et al.* 1984) and geoelectric data (Beamish and Smythe 1986), Chadwick and Holliday (1991) have proposed that the Solway Basin and Northumber-land Trough were formed by early Carboniferous extension on

steeply dipping faults, rooting on a N-dipping shear zone believed to represent the mid and upper crustal trace of the Iapetus suture.

The Maryport Fault and its en échelon continuation to the east was identified by Chadwick and Holliday (1991) as an important fault on the south side of the Carlisle/Solway Basin. Westwards the line of the Maryport Fault offshore is traced on present evidence by a number of discontinuous en échelon N-dipping faults rather than a single continuous fracture. These faults run along the northern edge of the Ramsey–Whitehaven Ridge, which is also picked out by a strong gravity lineament. In post-Carboniferous times, the northern boundary of the Ramsey–Whitehaven Ridge appears to have behaved more as a hinge zone.

The Llŷn–Rossendale Ridge has been identified in this account as the post-Carboniferous divide between the EISB and Cheshire Basin. The Permo-Triassic seismo-stratigraphic units show progressive southward thinning onto the hinge-like northern margin. Although no fault has been mapped along this northern boundary in Permo-Triassic deposits, the pos-ition appears to be coincident with the presumed trace of the Menai Strait–Pendle line at deeper crustal levels. It is identified by a weak gravity lineament running close to the present coastline of North Wales and NW Wirral.

The nature of the Menai Strait–Pendle line is seen where it emerges from beneath the Permo-Triassic deposits, a charac-teristic which it shares with its counterpart, the Llanelidan–

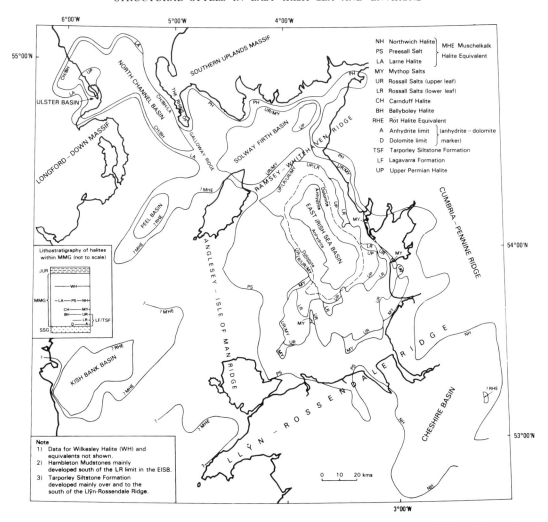

Fig. 5. Depositional limits of halites in the Mercia Mudstone Group and Upper Permian within the North Irish Sea; limits are largely conjectural outside the EISB, in undrilled areas and where removed by post-Triassic erosion. Halite identified on basinward side of limit. Incorporates data from Warrington et al. (1980), Wilson (1990) and other sources quoted in the text.

Bala Fault, on the western margin of the Cheshire Basin (Earp and Taylor 1986). In the west, the Menai Strait Fault System is a transpressive sinistral shear of late Precambrian–early Cambrian age, but with mostly dip-slip movement in post-Arenig times (Gibbons 1987). In the east, the Pendle Monocline is a south-verging fold formed by Variscan inversion of a Dinantian depocentre adjacent to the Pendle Fault (Gawthorpe 1987). The Ribblesdale Fold Belt has been traced to the north of the Formby Oilfield (Falcon and Kent 1960, p. 35) but to date has not been identified to the west of the Formby Point Fault. There is evidence for an ENE–WSW (?strike-slip) fault offshore but this lies some way to the north of the Menai Strait–Pendle line.

North and south structural domains of the East Irish Sea Basin

Four criteria can be used to distinguish a 'northern' from a 'southern' structural domain within the EISB (Jackson et al. 1987), though the geographical limits are different for each case.

First, the EISB is broadly a NW–SE asymmetric graben (or even a graben-within-graben) in the north where it is fault-bounded against the Lower Paleozoic massifs (Fig. 2). Triassic deposits thicken slightly towards the Eubonia Fault, and markedly towards the Lake District Boundary Fault (Jackson et al. 1987, fig. 3a). This configuration contrasts with the south where the basin is an asymmetric N–S sag in cross-section (BGS 1991; Jackson et al. 1987, fig. 3b) with Permian and

Triassic sequences thinning by onlap onto Carboniferous margins.

Second, the basin is subdivided by an elbow-shaped bend in fault orientation (Fig. 2; Jackson et al. 1987, pp. 192–3) swinging around a line which joins the Quadrant 109 Arch to the High Haume Anticline (Dunham and Rose 1949, pp. 15, 28). North–south-trending faults predominate to the south, and contrast with intersecting NW–SE and NE–SW trends in the north. This line is believed to reflect the offshore continuation of folds and dislocations identified in the Caledonian basement of the southern Lake District, in particular the Selside Anticline and décollements at the base and top of the Coniston Grits (Moseley 1972; Soper and Moseley 1978).

In addition, the line would appear to reflect variation in the pre-Carboniferous subcrop offshore. An Ordovician–Llandovery subcrop with Caledonoid trends is predicted to the north of the line as a continuation of the Caledonoid grain of the northern Lake District and Isle of Man. In contrast, a Wenlock–Přídolí subcrop with Malvernoid trends is suggested in the south joining the known outcrops in North Wales (Warren et al. 1984) and the Cartmel–Kendal areas (Moseley 1972). These trends may be inherited from the tectonic grain of underlying Precambrian rocks (Moseley 1972). A similar elbow-shaped change in fault trend is noted in the Cheshire Basin across the Llanelidan Fault and its buried continuation (Earp and Taylor 1986), and which is also believed to separate Ordovician and Silurian rocks. The influence of higher Permo-Triassic depositional rates along Caledonoid-trending sags on each side of the Quadrant 109 Arch–High Haume Anticline

line is reflected in the shape of the EISB western margin. A northern sag runs through the Eubonia and Lagman basins and a southern sag through the Berw and Leyland basins to the Clitheroe outliers (Fig. 2).

The West Crop of the EISB marks the western limit of N–S trends in the southern domain (Figs 2 and 4). In North Wales, the N–NNW-trending Conwy Valley Fault (1800 m downthrow to the east) similarly denotes the western limit of N–S trends in the Wenlock–Ludlow rocks of the Denbigh Moors (Warren *et al.* 1984). It is postulated that the northwards continuation of the Conwy Valley Fault, displaced sinistrally across the Menai Straits Fault System, may act as a deep-seated control beneath the line of the West Crop. The West Crop also indents the southern margin of and constricts the line of the Craven–Dublin Basin. The Craven Basin, the Quadrant 109 Syncline and the Dublin Basin may represent three individual Caledonoid-trending depocentres oblique to the overall ENE–WSW trend of the Craven–Dublin Basin (Ramsbottom *et al.* 1978).

Third, the Rossall Salts (lower leaf) and the halites within the St Bees Evaporites are confined to the north and central parts of the EISB (Figs 3 and 5; Jackson *et al.* 1987, figs 7b and 9). Regional detachments and bedding-parallel zones of high ductile strain are closely associated with these halites, which also occur at major competence contrasts. In contrast, in the southern domain halite detachments are believed to be absent or less effective. There, the Rossall and Mythop salts are thin or absent (Fig. 5; Wilson 1990), the Preesall Salt is much reduced in thickness, the Mercia Mudstone Group contains thin interbedded carbonate (Well 110/7-2), and the Manchester Marls lack halite.

Permian and Triassic halites consistently highlight areas of maximum contemporary subsidence and have exerted a crucial role on a distinctive structural style within the EISB (style 3, see below; Jackson *et al.* 1987). The Permo-Triassic halite distribution is shown in Fig. 5.

Finally, a discontinuous palaeo-topographic ridge, the Ogham–Formby line, trending NW–SE across the EISB is interpreted as an intra-rift transfer zone (Rosendahl *et al.* 1986; Morley *et al.* 1990). It underlies a series of small horsts and is slightly oblique to the axis of the Clyde Belt rift (Fig. 2). A polarity switch is observed over the line with predominantly westerly-tilted half-graben to the northeast of the ridge, and mainly easterly-tilted half-graben to the southwest. As discussed earlier, the line separates successions of contrasting facies in both Lower and Upper Permian rocks.

The superimposition of these four criteria acting throughout Carboniferous and Permian–early Jurassic extension and during Variscan and post-early Jurassic compression heightens the contrast between the northern and southern domains.

Fault displacement history

The displacement history for two widely separated, large syn-depositional faults is summarized on Fig. 6. Data for the Keys Fault (end-Carboniferous to end-early Jurassic) and for the Formby Point Fault (end-early Permian to end-Triassic) is taken from seismic lines and supplemented with nearby well control.

Displacements arising purely from growth faulting are plotted and summed for the major seismo-stratigraphic units, and compared with the present-day displacement of the Base Permo-Triassic horizon. Although simplistic (e.g. by discounting reversal of fault movement during inversion and by ignoring differential compaction effects between the hangingwall and footwall), the analysis demonstrates that only *c.* 36% of the post-Carboniferous (290 Ma, Harland *et al.* 1990) fault displacement across the Keys Fault has occurred in post-early Jurassic times (178 Ma). Similarly only *c.* 24% of the post-early

Permian displacement on the Formby Point Fault is post-Triassic (208 Ma) in age.

Thus some 64–76% of the displacement of Permian outcrops across these two faults (Fig. 2) arises from intra-Permo-Triassic syn-depositional movement of 102 Ma duration. Other faults are presumed to possess comparable histories (e.g. Godred Croven Fault, Jackson *et al.* 1987, fig. 3b). Therefore, the widely held assumption (e.g. Dunham and Rose 1949; Moseley 1972) that faulting of Permian and Triassic rocks in NW England is largely Tertiary in age must be modified. It suggests also that post-early Jurassic rifting in the EISB was insignificant compared with that in the North Sea. Indeed a model (Fig. 6) of slow but continuous post-early Jurassic fault movement (?driven by isostasy and granite buoyancy, Bott 1968) appears as plausible as the more conventional view of pulsed fault movement, e.g. the 'late Cimmerian', end-Cretaceous, Laramide and Miocene/Alpine phases (Ziegler 1990). The rate of post-early Jurassic fault displacement may have declined exponentially with time as part of the later stages of a thermal sag phase.

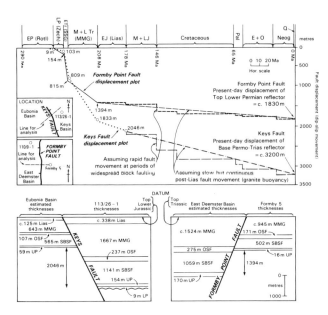

Fig. 6. Analysis of the post-Carboniferous history of fault displacement for the Keys and Formby Point growth faults. Age scale from Harland *et al.* 1990. Simplified cross-sections based on seismic lines and well data, shown for both faults.

Estimates of post-Triassic deposition, erosion and uplift

Northern Ireland is considered to be the closest analogy for the post-early Jurassic history of the EISB, with intermittent and low sedimentation interrupted by periods of erosion and regional uplift (Jackson *et al.* 1987). Map examination shows that formation of the Tynwald Fault Complex must have preceded intrusion of the Fleetwood Dyke Group in the Paleocene (61–65 Ma, Arter and Fagin 1993), with very little subsequent fault displacement (Fig. 2). Nevertheless, regional epeirogenic uplift (as distinct from reversal of fault movement and inversion *per se*, see Cooper *et al.* 1989) probably exceeds 500 m since dyke injection. Francis (1982, fig. 26) suggested that a minimum cover of 0.5 km is required to prevent feeder dykes from reaching the contemporary land surface.

Colter (1978) demonstrated from a study of shale velocities in the Mercia Mudstone Group that the succession in Well 110/3-2 had formerly been buried *c.* 2000 m deeper than at present. Vitrinite reflectance data also confirmed that this thickness of beds had been removed by erosion. Roberts (1989, fig. 3) estimated 1500 m (to 2000 m or more) of post-Jurassic uplift from vitrinite reflectance data on several wells, while Stuart and Cowan (1991) favour '1500 feet' of post-Cretaceous uplift from maturity modelling.

However, it is imperative to note that much if not all of the 1500–2000 m section can be accounted for merely by the thickness of Mercia Mudstone Group and Lias eroded from 110/3-2. An even greater thickness is missing from the Morecambe Field pericline. The preserved thickness of Mercia Mudstone Group and Lias in the Keys Basin is estimated at 3725 m (Fig. 3; Jackson *et al.* 1987), to which could be added perhaps 600 m net of Sinemurian and younger strata removed by erosion from the Keys Basin. The present Mercia Mudstone Group thickness 'near' 110/3-2 amounts to 2138 m (i.e. including the basal 108 m strata cut out by the Crosh Vusta Fault detachment, cf. Jackson *et al.* 1987, fig. 9). Thus, 2187 m (4325 − 2138 m) of Mercia Mudstone Group and Lias have been eroded from the 110/3-2 well site, although detailed unpublished correlations (by DIJ) of the Mercia Mudstone Group do indicate some northwards thickening from the Morecambe Field to the Lias outlier.

Green (1986) has calculated 1–2 or 2–4 km of post-Cretaceous uplift in the central Lake District from apatite fission track analysis. Clearly, amounts of uplift increase from the EISB towards the Lake District. The full thickness of Mercia Mudstone Group (?3.0 km) has been removed from the coastal strip of SW Cumbria and the entire Permo-Triassic from an inner Lake District fringe, though much of this erosion could be pre-Tertiary. Nevertheless, Green's higher value seems rather at variance with the evidence of fault displacement, dyke intrusion and shale velocities quoted above, unless amounts of Tertiary uplift in the Lake District show a fourfold increase over the EISB rate. If Green's lower figure is accepted, and if the fault displacement analysis is valid, it implies that faulting with little uplift took place between the early Jurassic and Paleocene, and uplift with little faulting in post-Paleocene times.

In conclusion, around the Morecambe Field and Well 110/3-2, the authors tentatively suggest 2000–2500 m of post-early Jurassic erosion and uplift (dependent on the present-day level of erosion), of which perhaps 700 m occurred since dyke intrusion. Amounts of inversion (definition of Cooper *et al.* 1989) may be quite low in relation to amounts of uplift.

Timing of trap formation

Fold and fault traps in Carboniferous rocks, if not already in existence by the end of the Westphalian, were formed initially during the Variscan orogeny and modified on several subsequent occasions. Growth faulting with accompanying rollover is of such magnitude (Fig. 6) in the northern domain that fold and fault traps in Permo-Triassic rocks were probably created and sealed during the latest Triassic (and presumably earlier for Permian reservoirs) as a result of footwall uplift and lateral Mercia Mudstone Group seals across growth faults. The initial migration of oil and gas into a palaeo-Morecambe Field structure is known to be pre-mid-Jurassic (Stuart and Cowan 1991). Breaching of the first Morecambe Field trap occurred during the 'late Cimmerian' (Stuart and Cowan 1991) with further modification during the Tertiary. In the southern domain, where growth faulting is less important, trap formation may have commenced slightly later; however, within the same geographically restricted area it is assumed to be broadly coeval with trap formation in the northern domain.

Structural contrasts within the East Irish Sea Basin and adjacent areas

Four broad structural and deformational styles have been identified within the Permian–Lower Jurassic rocks of the EISB and contiguous areas (Figs 7 and 8). The styles are largely independent of the age of the sedimentary fill, since each category has been recognized also in Carboniferous rocks with the exception of the specialized style 3. Style 1, for example, has been noted over an exclusively Carboniferous tract in the Quadrant 109 Syncline (see below). Thus, the styles demonstrate the overriding importance of local tectonic setting and recurrent crustal behaviour during Carboniferous and Permian–early Jurassic subsidence and during Variscan and post-early Jurassic uplift.

The styles are classified according to a limited number of criteria: (a) basin location (within or marginal to the EISB); (b) basin orientation (Caledonoid or otherwise); and (c) the presence or absence of detachments or zones of high ductile strain associated with Permo-Triassic halites (i.e. the nature and facies of the Permo-Triassic evaporite and argillaceous deposits).

The styles are arranged in order of increasing structural complexity and fault density from the simplest (style 1) to the most complex (style 4). Style 1 (folds dominant) and style 2 (faults dominant) comprise solitary or isolated basins located on the fringes of the Clyde Belt and either partially or wholly outside the EISB. Style 3 (where two or more thick Permo-Triassic halites are present) and style 4 (in which halites are thin or absent) consist of linked sub-basins within the EISB. Transitional types are recognized over wide areas; styles 3A and 4B have been differentiated on Fig. 7.

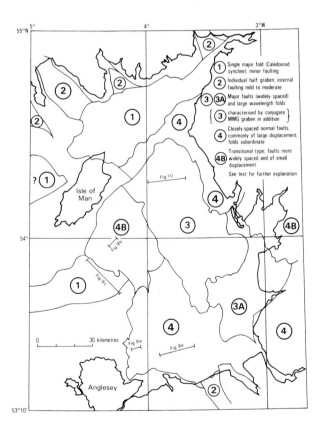

Fig. 7. Offshore distribution of types of structural style (chiefly from seismic examination) within the Permian–Lower Jurassic (and adjacent Carboniferous where appropriate) of the EISB and contiguous areas. Location of seismic lines in Figs 9 and 10 shown in addition.

Style 1

The two prime examples from style 1, the Solway Firth Syncline and the Quadrant 109 Synline, each comprise a single elongate Caledonoid-trending centrocline, some 70 km by

25 km, downwarped between Caledonian massifs (Fig. 2). The syncline can be symmetrical (Solway Firth Syncline) or asymmetrical (Quadrant 109 Syncline with a steeper NW limb, Fig. 9c; Jackson *et al.* 1987, fig. 3d). There is little internal faulting in this style, except locally axial-parallel on the steeper parts of the fold limb and directed basinwards (Fig. 8). NW–SE to NNW–SSE faults of Clyde Belt trend, and growth faults in particular, are scarce by comparison with other structural styles.

Opposite the Criffell Granodiorite and the largely concealed Manx granites (Bott 1978), the Solway Firth and Quadrant 109 synclines form hangingwall synclines to the North Solway Fault and Lambay–Eubonia fault system, respectively (Fig. 2). Although seismic evidence on the basin margins is meagre, the absence of wedge-shaped geometries in the seismo-stratigraphic units suggests that away from the granites the contemporary basin margins were unfaulted with onlap towards the bounding basement highs.

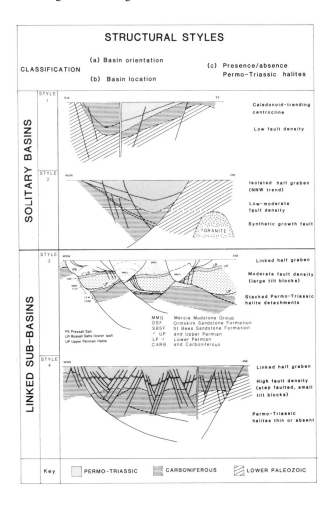

Fig. 8. Summary characteristics of structural styles (cross-sections schematic in part and not to scale). The styles are based upon the three criteria of basin orientation, basin location (peripheral/central to the Clyde Belt) and presence or absence of Permo-Triassic halites. See text for further details.

The Solway Firth Basin (Bott 1964; Bott and Young 1971; Ord *et al.* 1988) is filled with a maximum of 3.3 km Permian–?Lower Jurassic sediment, and is separated from the Carlisle Basin to the northeast by a NW–SE sill crossing the inner Solway Firth. The basin displays a long history of limited axial sagging (margin:axis ratio of 4:5, cf. Beamish and Smythe 1986), rather than fault-controlled subsidence. It overlies a Carboniferous precursor basin, though the post-early Jurassic

synclinal axis is displaced slightly to the NW. During the Carboniferous, the basin may have been a half-graben tilted to the SE towards a growth fault marking the northern boundary of the Ramsey–Whitehaven Ridge. In Permo-Triassic times, the northern margin of the ridge appears to have acted as a hinge line with a thinner succession over the ridge and a continuous fault has not been observed.

During Permian and Triassic deposition, the Solway Firth Basin afforded continuity between the North Channel Basin and EISB across the Rhins of Galloway Uplift and Ramsey–Whitehaven Ridge. Thus the Solway Firth Basin can be viewed as a transfer zone basin, offsetting the Clyde Belt and linking NW–SE internally faulted basins to the north and south (Fig. 1).

The Quadrant 109 Syncline is a Variscan fold lying on or close to the presumed depositional axis of the Craven–Dublin Basin (Ramsbottom *et al.* 1978; BGS 1991). The infill of 7.5 km of Carboniferous sediment (Fig. 9c) shows limited syn-depositional centripetal thickening (margin:axis ratio of 9:10), and on Fig. 9c is internally conformable. Stratigraphic control on reflectors is poor and reliant on ties to 112/30-1, and to four BGS shallow boreholes (Wright *et al.* 1971; Parkin and Crosby 1982). Three boreholes were drilled on strike (70/01, 71/38 and 73/68) and borehole 70/01 yielded non-marine bivalves from the upper *Carbonicola communis* Zone (Late Westphalian A; Wilkinson and Halliwell 1979, p. 52). A prominent and widespread early Namurian reflector is tentatively identified as Near Top Arnsbergian (Fig. 9c). In the fold core an estimated 4000 m Westphalian succession probably includes strata of Westphalian C–D age (and possibly Stephanian); Westphalian C and D deposits are recorded along-strike in Irish well, Well 33/22-1 in the Kish Bank Basin (Jenner 1981).

The successor Permo-Triassic basin, the Eubonia Basin lies along-strike to the northeast. It is a post-Triassic centroclinal fold, though sub-circular in outline in contrast to the elongate Quadrant 109 Syncline (Figs 2, 9b). This suggests recurrent patterns of crustal subsidence.

The Solway Firth and Quadrant 109 synclines are orthogonal to the principal extension direction of the Clyde Belt during the Permian and Triassic. The normally dominant Charnoid and Malvernoid trends and faults of the Clyde Belt have been suppressed in narrow Caledonoid basins, shielded and squeezed by the paired Lower Paleozoic basement ridges. The basinal form has been preserved by a combination of greater deposition during subsidence and reduced erosion during uplift compared to the bounding highs. Post-Variscan and post-early Jurassic uplift has been both regional and differential, and largely along Caledonoid trends.

A related structural style may occur in the anticlinorium of the Caledonoid Riblesdale Fold Belt (Fig. 2; Arthurton 1983, 1984), where Dinantian rocks of basinal facies are bordered by the concealed Lower Paleozoic Rossendale and Bowland blocks (Lee 1988, fig. 8.7; Arthurton *et al.* 1988, fig. 34). Caledonoid folds and periclines are dominant over mapped NW–SE faults (Fig. 1; Aitkenhead *et al.* 1992) in comparison with the Lancaster Fells and Rossendale areas.

The Clitheroe outliers (Earp *et al.* 1961) are the erosional remnant of an exhumed Permo-Triassic wadi or valley (Falcon and Kent 1960, p. 34). The outliers appear to indicate the persistence of a third weak Caledonoid sag or zone of increased sedimentation along the Ribblesdale Fold Belt and through the Leyland Basin. The axis may cross the Clyde Belt transversely to the Berw Basin north of the Llŷn–Rossendale Ridge and Pendle Lineament. The sag is not fault bounded on either Clyde Belt margin and certainly appears to have been operative during early Permian (see earlier discussion) and late Triassic–early Jurassic subsidence and later uplift.

Other, more speculative, examples of weak post-Triassic NE–SW folds posthumously developed above Variscan folds

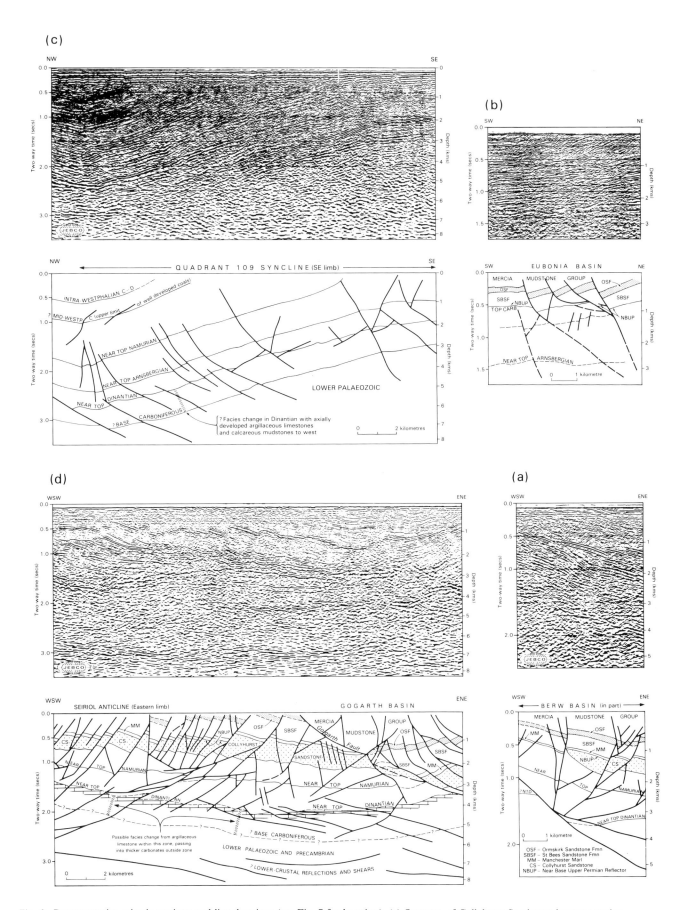

Fig. 9. Representative seismic sections and line drawings (see Fig. 7 for location). (**a**) Overstep of Collyhurst Sandstone by trangressive Manchester Marls, Berw Basin; also illustrates style 4. Courtesy of JEBCO Seismic Ltd. (**b**) Rotation of tilt blocks in Sherwood Sandstone Group above an Upper Permian detachment (Manchester Marls), Eubonia Basin; style 4B. Courtesy of Seismograph Service (England) Ltd. (**c**) To illustrate style 1; dip line across the Quadrant 109 Syncline. Courtesy of JEBCO Seismic Ltd. (**d**) To illustrate style 4; offshore North Wales (Seiriol Anticline and Gogarth Basin). Courtesy of JEBCO Seismic Ltd.

include the Ingleton–Lancaster line across the Lancaster Fells (Fig. 2), the Peel Basin (Fig. 2; Bott 1964; Bott and Young 1971; Hall *et al.* 1984), the 'acute syncline' of the Mauchline Basin near the Kerse Loch Fault (McLean 1978), the St George's Channel Basin (Barr *et al.* 1981) and possibly the poorly known sub-basins of the Ulster Basin.

Style 2

Style 2 areas comprise individual half-graben, trending NW–SE and NNW–SSE on the Clyde Belt margins and dominated by a basement-controlled synthetic growth fault (Figs 7 and 8). They are commonly 50 km by 15 km in size. In the absence of suitable seismic data across the footwall, this style is not illustrated (but see Warren *et al.* 1984, figs 22 and 24; Burgess and Holliday 1979). A local hangingwall depocentre is juxtaposed across the fault with a footwall inlier at a buried or partially buried Lower Paleozoic ridge orthogonal to the Clyde Belt. Examples include the Vale of Clwyd (Wilson 1959; Warren *et al.* 1984), the Stranraer Basin (Mansfield and Kennett 1963; Bott 1964, fig. 7; Bott 1965; Kelling and Welsh 1970) and the Vale of Eden (Bott 1978). In these instances the half-graben is tilted towards the east and the synthetic bounding fault (Fig. 2). However, this arrangement is not universal, e.g. the Dumfries Basin (Brookfield 1978) shows a reverse polarity and is fault bounded on the west against the Criffell Granodiorite.

The synthetic faults appear to have been reactivated during Stephanian–Autunian times (McLean 1978; Ziegler 1990). Lower Permian successions are well developed, and deposition took place both by passive infilling of existing topography (Vale of Eden, Arthurton and Wadge 1981, pp. 71, 107) and in fault-controlled sedimentary wedges (e.g. Luce Bay and Dumfries basins). Later Permo-Triassic deposits are commonly thinner and successions more deeply eroded than in the EISB and Solway Firth Basin reflecting their marginal position and greater post-Triassic uplift and erosion. Invariably they overlie a Carboniferous sequence which is also thinner and of more proximal facies (especially Stranraer Basin and Strangford Lough) compared to the main Craven–Dublin and Solway–Northumberland Trough basins (Ramsbottom *et al.* 1978).

A suite of antithetic faults may occur in addition to the growth fault (e.g. Vale of Clwyd, Collar 1974; Warren *et al.* 1984) though such faults appear to be absent in the Luce Bay sub-basin of the Stranraer Basin (Fig. 2). Fault density is generally low in Permo-Triassic rocks and moderate to high in Carboniferous successions (Warren *et al.* 1984).

Folds, if present, are restricted to two types: either an axial syncline whose axis may be oblique to the growth fault (Vale of Eden Syncline, Arthurton and Wadge 1981); or transverse anticlines such as the St Asaph Ridge in the Vale of Clwyd (Warren *et al.* 1984) and the Stranraer isthmus (Fig. 2).

Assessment of fault heaves and elliptical fault surfaces (Walsh and Watterson 1991) suggests that the synthetic growth fault was initiated at or near the crest of the Lower Paleozoic basement ridge where the present-day fault displacement is at a maximum. Examples include the Vale of Clwyd Fault across the Llŷn–Rossendale Ridge and the North Channel Bounding Fault (east) and Loch Ryan Fault across the Southern Uplands–Longford Down massif. Elsewhere the faults were initiated at the roof crest of the east- and west-facing margins of buried Caledonian granites (Dumfries Basin, and the Pennine Fault adjacent to the Weardale Granite, Bott 1978).

Style 3

Structural style 3 is characterized by regularly-spaced linked half-graben, typically where two intersecting fault trends (NW–SE to NNW–SSE and NE–SW) create a distinctive

rhomboid 'trapdoor' outline in plan view (e.g. the Keys, Tynwald and Lagman basins, Figs 2 and 8). However, parallel rectilinear N–S faults delineate elongate half-graben south of the Quadrant 109 Arch–High Haume Anticline line (e.g. the West Deemster Basin). Unlike all other styles, this type is restricted to Permian–Lower Jurassic sequences. Fault density is lower than in style 4, but the size of individual prospects is correspondingly larger. Exploration targets include both fold- and fault-dominant plays in the Gleaston Dolomite (Smith *et al.* 1974; Jackson *et al.* 1987, fig. 6a) and Sherwood Sandstone Group reservoirs.

Gentle folding and warping occurs throughout (Fig. 10). Fold types in the competent reservoirs include broad hangingwall synclines (the Keys Basin, Jackson *et al.* 1987, fig. 3a), complex periclines modified by axial faulting (e.g. the Morecambe Field, Ebbern 1981; Bushell 1986; Stuart and Cowan 1991), domes (e.g. the 113/26-1 gas discovery) and roll-over anticlines (Fig. 10). Smaller anticlinal folds above the regional datum and which are asymmetric verging towards a growth fault are believed to be associated with compression and minor reversal of fault plane movement during post-early Jurassic inversion (Jackson *et al.* 1987, fig. 3b; cf. Cooper *et al.* 1989, fig. 7).

In the incompetent Mercia Mudstone Group additional fold types include disharmonic folds, for example within the Tynwald Fault Complex, generated during inversion against a footwall buttress (Jackson *et al.* 1987, fig. 3a; cf. Cooper *et al.* 1989, fig. 8). Halokinetically induced folds are best displayed in areas of exceptionally thick Mercia Mudstone Group (estimated to total 3025 m in the Keys Basin depocentre). They include small diapiric structures (Jackson *et al.* 1987, fig. 3c) and salt swells and pillows; in part the latter are responsible for the abnormally thick Rossall Salts (lower leaf) in Well 113/26-1 (Jackson *et al.* 1987, fig. 9).

Fault density is moderate and dominated by widely spaced synthetic growth faults of large post-Carboniferous displacement (e.g. Formby Point Fault maximum 3500 m, and Keys Fault, maximum 3600 m). The synthetic faults penetrate the full Permo-Trias succession and delimit large tilted fault blocks in the Ormskirk Sandstone Formation target (Fig. 10; Jackson *et al.* 1987, figs 3a and 3b). The South Morecambe Field is bisected by the outer faults of the Tynwald Fault Complex into western and eastern lobes (Ebbern 1981, fig. 3). The synthetic faults possess a low-angle broadly listric profile, though modified by small and poorly developed ramps (through sandstones) and flats (in mudstones), governed by the competence of the beds. There is a steeper almost planar upper section especially where the Sherwood Sandstone Group lies near the seabed. Fault plane dips of 30–47 ° have been measured for the Formby Point Fault and 25–41° for the Keys Fault, reflecting the post-Carboniferous extension of *c.* 27% which the area has experienced (Jackson *et al.* 1987, fig. 3). At deeper levels the faults may become listric along a brittle–ductile transition in the crust (Jackson *et al.* 1987, fig. 3a).

Detachments are associated with the lower Rossall Salt (e.g. 110/7-1, Jackson *et al.* 1987, fig. 9) and the St Bees Evaporites (halite in cycles BS2 and BS3, in Wells 110/3-2 and 112/25a-1, Jackson *et al.* 1987, fig. 6a). A third and uppermost family of listric faults and bedding plane glides is focused on the Preesall Salt.

This structural style is believed to derive from the one to three stacked halites, which during Permo-Triassic subsidence and post-early Jurassic uplift acted as detachments and/or bedding-parallel zones of high ductile strain, and decoupled minor fault movement. Genesis of the detachments is aided by a major competence contrast directly beneath the two lower halites (Fig. 3). The Rossall detachment rests on the anhydrite dolomite marker near the Sherwood Sandstone Group/Mercia Mudstone Group contact, and the St Bees Evaporites detach-

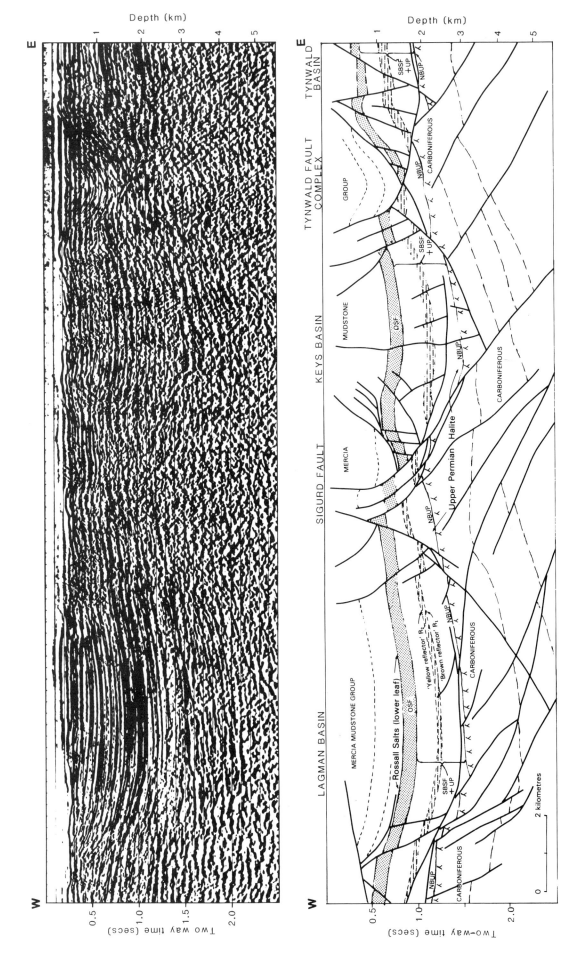

Fig. 10. To illustrate structural style 3; Lagman, Keys and Tynwald basins (see Fig. 7 for location and Fig. 9a for abbreviations). Section courtesy of Western Geophysical.

ment overlies the Gleaston Dolomite near the junction with the Carboniferous.

The widespread development of the Rossall and Upper Permian detachments results in stratified fault systems (and multiple plays). Each detachment-bound unit (the Mercia Mudstone Group, the Sherwood Sandstone Group/Upper Permian, and the Lower Permian/Carboniferous) possesses a slightly different fault pattern. Walker and Cooper (1987) have identified four halite detachments (three Upper Triassic plus Zechstein) at equivalent horizons in the Sole Pit area of the Southern North Sea.

Very large elongate post-depositional conjugate graben with foundered prisms of down-faulted and locally steeply rotated Mercia Mudstone Group are exclusive to this structural style (Fig. 10). Within the graben the Sherwood Sandstone Group is largely or wholly lacking (e.g. Well 110/3-2). Five regionally important examples shown by vertical shading on Fig. 2 are the Tynwald Fault Complex (Jackson et al. 1987, fig. 3a), the Crosh Vusta Fault Complex, the narrow graben separating the South and North Morecambe gas fields (Ebbern 1981; Stuart and Cowan 1991), the Morecambe Field Western Boundary Fault Complex, and the offshore extension of the Kirkby Tear Fault (Jackson et al. 1987). The graben provide an excellent and deep (1000 m) lateral seal for the Sherwood Sandstone Group in the paired footwall blocks. They resemble similar large structures south of the Mid North Sea High (Jenyon 1985) and over the Swarte Bank Hinge (Walker and Cooper 1987, figs 3 and 7) though the Tynwald Fault Complex and Crosh Vusta Fault Complex contain an axial compressional anticline thought to be generated during inversion (Colter 1978, fig. 5; Jackson et al. 1987). Significantly, the graben and foundered prisms in both the Southern North Sea and EISB occur only inside the depositional limits of the St Bees Evaporites/Z2 halites and lower Rossall/Röt halites.

Seismic evidence and the relationship of the Tynwald Fault Complex to the Fleetwood Dyke Group (Fig. 2) demonstrates that formation of the complementary antithetic fault and hence the graben itself, can be dated as post early-Jurassic and pre-Paleocene. The graben may have been created during the Late Cimmerian movements, when, for example, the palaeo-Morecambe Field was breached (cf. Stuart and Cowan 1991, fig. 11).

A small post-depositional Mercia Mudstone Group conjugate graben at the NE end of the Solway Firth Basin (Fig. 2) may indicate local development of an Upper Permian halite.

Elsewhere in the North Irish Sea, style 3 is judged to be present over much of the Kish Bank Basin from the following evidence: (a) rhomboid-shaped half-graben with large tilted fault blocks in the Sherwood Sandstone Group (Jenner 1981, figs 2 and 4); (b) scarce internal faulting (Jenner 1981, fig. 4) near the base of both the Mercia Mudstone Group and Permo-Triassic which suggests the occurrence of detachments (or zones of high ductile strain) and halite; (c) salt swells, and low-angle discontinuities in the upper Mercia Mudstone Group (Jenner 1981, fig. 4).

A related structural style (style 3A, Fig. 7) is recognized towards the SE margin of the EISB over most of the East Deemster Basin, and in the adjacent onshore areas of the Preesall saltfield (Wilson and Evans 1990; Wilson 1990) and Walney Island (Rose and Dunham 1977; Wilson 1990). Although thinner than in style 3 areas, the Preesall, Mythop and upper Rossall halites (Figs 3 and 5) are still considered to act as local detachments or high ductility layers. Similarly, the St Bees Shales (Jackson et al. 1987, figs 6a and 7b) and mudstones in the anhydrite facies of the St Bees Evaporites may absorb fault movement. Upper Permian halites are believed to be absent together with the post-depositional conjugate Mercia Mudstone Group graben.

A complex pattern of widely spaced intersecting faults of variable orientation has been mapped over the poorly exposed outcrop of the Mercia Mudstone Group in the Cheshire Basin (Evans et al. 1968; Earp and Taylor 1986). However, in the saltfield, the density and orientation of faults within the Mercia Mudstone Group is constrained by mining and borehole information (Earp and Taylor 1986, figs 16 and 21). This fault pattern contrasts with the numerous parallel and rectilinear faults (cf. style 4, see below) mapped over the Sherwood Sandstone Group outcrop and on the basin periphery. It is suggested here that the Northwich Halite Formation (and possibly the Mercia Mudstone Group/Sherwood Sandstone Group competence contrast) may decouple fault movement, resulting in structures similar to style 3A of the EISB, particularly where halites are developed lower in the succession (e.g. Congleton, Evans et al. 1968, p. 131).

Style 4

Style 4 areas are also characterized by linked half-graben, but the ubiquitous minor faults (Figs 9a and d) blur the distinction between adjacent half-graben, since Triassic successions are thinner than in style 3 areas and thickness changes across syndepositional faults are less noteworthy. Permo-Triassic halites and their associated regional detachments are absent or poorly developed (except for the younger halites of the Mercia Mudstone Group) and consequently minor faults penetrate the full Permo-Triassic succession.

Tilted fault blocks provide the main exploration target. Discoveries include the Formby Oilfield (Warman et al. 1956, fig. 6; Falcon and Kent 1960) and the newly designated Douglas Oilfield and Hamilton and Hamilton North Gasfields of Block 110/13 (Fig. 2).

The style is best displayed in the EISB southern domain especially northeast of Anglesey, where the fault density is extreme, and the faults are planar and in general trend N–S (Fig. 9d; Jackson et al. 1987, fig. 3b). Fault density and individual throws decrease northwards concomitant with a thickening of Permian and Mercia Mudstone Group halites and mudstones and their increased effectiveness as detachments. Some of the larger growth faults are listric with a well-developed crestal collapse graben (McClay 1989, 1990).

The style reappears in an outer fringe to the EISB (Fig. 7); for example, intense faulting but with a NW–SE to NNW–SSE orientation is seen over the Selker Rocks Anticline, abutting the style 3 area (Jackson et al. 1987, fig. 3a). It is postulated that this eastward increase in fault density reflects the facies change from halite to anhydrite in the Upper Permian of the coastal strip (Jackson et al. 1987, fig. 7b). The anhydrite facies is documented from St Bees Head (Arthurton and Hemingway 1972) and the Haverigg Haws borehole (Rose and Dunham 1977; Jackson et al. 1987, fig. 6a). Much of the Ramsey–Whitehaven Ridge reveals comparable but less severe faulting.

In the southern domain, flights of step-faulted elongate blocks (average width 1.5 km) sequentially tilted to the east or west are arranged in complex arrays separated by polarity switches (Fig. 9d). A keystone graben denotes the switch-over where faults in adjacent arrays are convergent downwards (e.g. the opposing limbs of an anticline elevated above the regional datum, Jackson et al. 1987, fig. 3a, Selker Rocks Anticline). Conversely, where faults from adjacent arrays are divergent downwards, a small centre-of-basin-horst (Fig. 9d; Gibbs 1984) marks the changeover between arrays. In many cases this horst is located near the inflexion point of a large wavelength fold limb believed to be formed during uplift and inversion and implies that in part the fault arrays are late-stage features.

Folds are scarce, though axes are difficult to identify where individual tilt blocks are strongly rotated. In addition to local hangingwall synclines (Fig. 9d) and roll-over anticlines (Fig. 9d on the extreme right), larger wavelength anticlines occur. They appear to be related to local uplift axes and in some cases the previous dip has been reversed (e.g. the asymmetric Seiriol Anticline Fig. 9d). The Selker Rocks Anticline originates from local inversion of the hangingwall depocentre to the Lake District Boundary Fault (Jackson *et al.* 1987, fig. 3a; Roberts 1989).

The 'brittle-type' deformation of style 4 is believed to arise where minor faults during Permo-Triassic subsidence and post-early Jurassic uplift, transmit movement because of the absence of halites. This behaviour is facilitated in the Permo-Triassic rocks of the southern domain where the proportion of competent sandstone (950 m Sherwood Sandstone Group and up to 1150 m Collyhurst Sandstone) in relation to incompetent mudstones and evaporites is much higher than in style 3 areas.

Although seismic resolution of Carboniferous reflectors is imperfect, the dip discordance between Carboniferous and Permo-Triassic reflectors of Fig. 9d highlights two additional aspects of the post-Triassic evolution. First, in many tilt blocks the dips in the Permo-Triassic are commonly steeper than the underlying Carboniferous and second, the net throw of Carboniferous and Permo-Triassic reflectors across the section is low. Post-Triassic reversal of dip and local sub-Permian detachments are suggested as explanations.

The Gogarth Fault (Fig. 9d) is an important listric growth fault in the southern domain with a large symmetrical crestal collapse graben in the hangingwall (McClay 1990). During Permo-Triassic deposition, the regional dip in the footwall block on the west is thought to have been westward (?accompanied by a suite of westerly-throwing minor faults antithetic to a half-graben fault beyond the western end of the section). Local post-Triassic inversion has reversed this earlier dip and imposed the present eastward tilt (the eastern limb of the Seiriol Anticline), such that dips in the Permo-Triassic exceed those in the Carboniferous. Other probable effects of inversion have been to generate small thrusts, new short-cut and contractional faults, increase the displacement on existing faults and lower the fault plane dip by rotation (McClay 1989). Note, however, that the shallow-dipping fault planes of Figs 9a and 9d derive in part from the alignment of the seismic line at 15° to true dip.

Carboniferous analogies onshore suggest three lithologies for decoupling or for accommodating fault movement. First, coal rashings and mudstone seat-earths in Westphalian A and B measures behave as weak horizons and provide local detachments, especially where close to the Base Permo-Triassic unconformity and the competence contrast with thick Collyhurst Sandstone. Second, the thick incompetent Namurian mudstone successions (Sabden Shales and Holywell Shales, Jackson *et al.* 1987) will absorb fault movement. Finally, the Bowland Shales and Brigantian–Pendleian equivalents provide sub-regional detachments at competence contrasts with thickly bedded Dinantian platform carbonates (cf. Lawrence *et al.* 1987). The effectiveness of these local detachments is undoubtedly improved where the faults and the Carboniferous have the same directional dip. This may be an additional factor in determining the local tilt block polarity of the Permo-Triassic during inversion.

The upper faults may also root on low-angle faults both normal and reversed in the Carboniferous and older rocks (Fig. 9d). Thrusting is well known from Anglesey (Greenly 1919) and the Menai Strait (Greenly 1938) both in Lower and Upper Paleozoic rocks. High-amplitude, sub-horizontal reflectors are observed below 2.5 s (*c.* 5.5 km) in the south, for example on the buried eastern flanks of the Holy Island Shelf (Figs 2 and 9d). They are tentatively identified as thrusts and local shears in Precambrian and deeper crustal rocks.

Map examination suggests that structural style 4 also occurs in the highly faulted Carboniferous rocks of the Lancashire Coalfield, the Rossendale Anticline, the (West) Cumberland Coalfield and the northern fringes of the Lake District in the Cockermouth–Caldbeck area (Fig. 2). An early estimate of post-Westphalian A stretching using mine plan information ($\beta = 21\%$, over a horizontal distance of 5.35 km) was provided by Jones *et al.* (1938, fig. 19) across the Wigan Graben of the Lancashire Coalfield. In many of these areas, Lower Paleozoic rocks lie buried at relatively shallow depths (e.g. Holme Chapel, Arthurton *et al.* 1988) compared to the EISB. The increased number of faults as the Carboniferous cover thins above the Lower Paleozoic basement suggests that even minor basement faults have been reactivated and propagated upwards. This explanation would also account for the high fault density where thin Permo-Triassic overlies thin Dinantian carbonates, for example in the former haematite mining areas of Egremont and Furness (Rose and Dunham 1977).

Style 4B is a transitional style with smaller fault displacements than in style 4. It has been differentiated (Fig. 7) in Morecambe Bay and the Eubonia Basin where the fault density is relatively high especially at Near Base Upper Permian level (Fig. 9b; Jackson *et al.* 1987, fig. 3a). The style shares several characteristics with style 3A but tilt blocks are more numerous and Mercia Mudstone Group halites are thinner. The pronounced rotation of tilt blocks in the Sherwood Sandstone Group of Fig. 9b above an Upper Permian detachment is rather unusual for this style.

Conclusions

1. The Clyde Belt is interpreted as part of a short-lived break-up axis linking the Boreal Ocean to Tethys during the suturing of Pangaea and early opening of the North Atlantic.

2. The major extensional boundary faults of the North Irish Sea are located most commonly along the east- and west-facing margins of Caledonian granites, or orthogonal to exposed and partially buried Paleozoic ridges.

3. Variations in fault orientation within the EISB are believed to reflect differences in the pre-Carboniferous subcrop. The pattern of onshore Lower Paleozoic outcrops can be traced offshore with Ordovician–Llandovery rocks postulated in the north and Wenlock–Přídolí rocks in the south.

4. A simple analysis of the fault displacement history of major growth faults in the EISB stresses the importance of growth faulting in relation to observed present-day displacement, and suggests that much of the faulting and folding of Permo-Triassic rocks was accomplished before dyke intrusion in the Paleocene.

5. Four structural styles have been recognized in the EISB and surrounding areas. Style 1 comprises Caledonoid-trending, elongate centroclines with minimal faulting; style 2 comprises individual half-graben outside or on the margins of the EISB; style 3 comprises gentle folds and large tilted fault blocks with stacked halite detachments in the Permo-Triassic; and style 4 comprises step-faulted arrays of small tilt blocks.

6. The influence of Permo-Triassic ductile halite layers and detachments on structural style is shown by a comparison of styles 3 and 4. In style 3, the response during uplift and inversion is by bending, folding and the generation of listric fault fans; whereas in style 4, where halites are ineffective, the response to uplift is brittle-type fracturing and faulting.

We gratefully acknowledge the cooperation of JEBCO Seismic Ltd, Seismograph Service (England) Ltd and Western Geophysical for permission to publish their non-proprietary seismic data. Many thanks to Hugh Ivimey-Cook for the Jurassic ammonite identifications, and to Jed Armstrong, Geoff Warrington, Denys Smith and Nigel Smith for

ideas and discussions over several years. The thought-provoking comments of R. J. Hodgkinson and one other referee improved considerably an earlier version of this script. Our especial thanks to Bob Dunay
for his editorial guidance and unfailing patience. This work was
funded by PED 1, Department of Energy and we acknowledge support
from John Brooks. This paper is published with the permission of the
Director, British Geological Survey (NERC).

References

AITKENHEAD, N., BRIDGE, D. McC., RILEY, N. J. AND KIMBELL, S. F.
1992. *Geology of the Country Around Garstang*. Memoir of the
Geological Survey of Great Britain, Sheet 67 (England and
Wales), HMSO.

ARTER, G. AND FAGIN, S. W. 1993. The Fleetwood Dyke and the
Tynwald fault zone, Block 113/27, East Irish Sea Basin. *In*:
PARKER, J. R. (ed.) *Petroleum Geology of Northwest Europe:
Proceedings of the 4th Conference*. Geological Society, London,
835–843.

ARTHURTON, R. S. 1983. The Skipton Rock Fault—an Hercynian
wrench fault associated with the Skipton Anticline, northwest
England. *Geological Journal*, **18**, 105–114.

—— 1984. The Ribblesdale fold belt, NW England—a Dinantian–
early Namurian dextral shear zone. *In*: HUTTON, D. H. W. AND
SANDERSON, D. J. (eds) *Variscan Tectonics of the North Atlantic
Region*. Geological Society, London, Special Publication, **14**, 131–
138.

—— AND HEMINGWAY, J. E. 1972. The St Bees Evaporites—a carbo-
nate-evaporite formation of Upper Permian age in West Cumberland, England. *Proceedings of the Yorkshire Geological Society*,
38, 565–592.

——, JOHNSON, E. W. AND MUNDY, D. J. C. 1988. *Geology of the
Country Around Settle*. Memoir of the Geological Survey of Great
Britain, Sheet 60 (England and Wales), HMSO.

—— AND WADGE, A. J. 1981. *Geology of the Country Around Penrith*.
Memoir of the Geological Survey of Great Britain, Sheet 24
(England and Wales), HMSO.

BARR, K. W., COLTER, V. S. AND YOUNG, R. 1981. The geology of the
Cardigan Bay–St. George's Channel Basin. *In*: ILLING, L. V. AND
HOBSON, G. D. (eds) *Petroleum Geology of the Continental Shelf of
North West Europe*. Heyden, London, 432–443.

BEAMISH, D. AND SMYTHE, D. K. 1986. Geophysical images of the deep
crust: the Iapetus suture. *Journal of the Geological Society, London*, **143**, 489–497.

BOTT, M. H. P. 1964. Gravity measurements in the north eastern part
of the Irish Sea. *Quarterly Journal of the Geological Society of
London*, **120**, 369–396.

—— 1965. The Deep Structure of the Northern Irish Sea—A Problem
of Crustal Dynamics. *In*: WHITTARD, W. F. AND BRADSHAW, R.
(eds) *Submarine Geology and Geophysics*. Proceedings of the 17th
Symposium of the Colston Research Society, Butterworths, London, 179–204.

—— 1968. The geological structure of the Irish Sea Basin. *In*:
DONOVAN, D. T. (ed.) *Geology of Shelf Seas*. Oliver and Boyd,
Edinburgh, 93–115.

—— 1978. Deep Structure. *In*: MOSELEY, F. (ed.) *The Geology of the
Lake District*. Yorkshire Geological Society, Occasional Publication 3, 25–40.

—— AND YOUNG, D. G. G. 1971. Gravity measurements in the north
Irish Sea. *Quarterly Journal of the Geological Society of London*,
126, 413–434.

BRITISH GEOLOGICAL SURVEY 1990. *Geology of the United Kingdom,
Ireland and the adjacent continental shelf (North Sheet)*. 1:
1 000 000. Compiler, J. A. Chesher.

—— 1991. *Geology of the United Kingdom, Ireland and the adjacent
continental shelf (South Sheet)*. 1:1 000 000. Compiler, J. A.
Chesher.

BROOKFIELD, M. E. 1978. Revison of the Stratigraphy of Permian and
Supposed Permian Rocks of Southern Scotland. *Geologische
Rundschau*, **67**, 110–149.

BURGESS, I. C. AND HOLLIDAY, D. W. 1979. *Geology of the Country
Around Brough-under-Stainmore*. Memoir of the Geological Sur-

vey of Great Britain, Sheet 31 and parts 25 & 30 (England and
Wales), HMSO.

BURLEY, S. D. 1984. Patterns of diagenesis in the Sherwood Sandstone
Group (Triassic), United Kingdom. *Clay Minerals*, **19**, 403–440.

BUSHELL, T. P. 1986. Reservoir Geology of the Morecambe Field. *In*:
BROOKS, J., GOFF, J. C. AND VAN HOORN, B. (eds) *Habitat of
Palaeozoic Gas in Northwest Europe*. Geological Society, London,
Special Publication, **23**, 189–208.

CHADWICK, R. A. AND HOLLIDAY, D. W. 1991. Deep crustal structure
and Carboniferous basin development within the Iapetus convergence zone, northern England. *Journal of the Geological Society,
London*, **148**, 41–53.

COLLAR, F. A. 1974. A geophysical interpretation of the structure of
the Vale of Clwyd, North Wales. *Geological Journal*, **9**, 65–76.

COLTER, V. S. 1978. Exploration for gas in the Irish Sea. *In*: VAN LOON,
A. J. (ed.) Key notes of the MEGS-II (Amsterdam 1978). *Geologie
en Mijnbouw*, **57**, 503–516.

—— AND BARR, K. W. 1975. Recent developments in the Geology of
the Irish Sea and Cheshire Basins. *In*: WOODLAND, A. W. (ed.)
*Petroleum and the Continental Shelf of North-West Europe, Vol. 1:
Geology*. Applied Science, London, 61–75.

COOPER, M. A., WILLIAMS, G. D., DE GRACIANSKY, P. C., MURPHY,
R. W., NEEDHAM, T., DE PAOR, D., STONELEY, R., TODD, S. P.,
TURNER, J. P. AND ZIEGLER, P. A. 1989. Inversion tectonics—a
discussion. *In*: COOPER, M. A. AND WILLIAMS, G. D. (eds) *Inversion Tectonics*. Geological Society, London, Special Publication,
44, 335–347.

COPE, J. C. W., GETTY, T. A., HOWARTH, M. K., MORTON, N. AND
TORRENS, H. S. 1980. *A Correlation of Jurassic Rocks in the British
Isles. Part One: Introduction and Lower Jurassic*. Special Report of
the Geological Society, London, **14**.

DOBSON, M. R. AND WHITTINGTON, R. J. 1987. The geology of Cardigan Bay. *Proceedings of the Geologists' Association*, **98**, 331–353.

DUNHAM, K. C. AND ROSE, W. C. C. 1949. Permo-Triassic geology of
South Cumberland and Furness. *Proceedings of the Geologists'
Association*, **60**, 11–40.

EARP, J. R. AND TAYLOR, B. J. 1986. *Geology of the Country Around
Chester and Winsford*. Memoir of the Geological Survey of Great
Britain, Sheet 109 (England and Wales), HMSO.

——, MAGRAW, D., POOLE, E. G., LAND, D. H. AND WHITEMAN, A. J.
1961. *Geology of the Country Around Clitheroe and Nelson*.
Memoir of the Geological Survey of Great Britain, Sheet 68,
HMSO.

EBBERN, J. 1981. The Geology of the Morecambe Gas Field. *In*:
ILLING, L. V. AND HOBSON, G. D. (eds) *Petroleum Geology of the
Continental Shelf of North-West Europe*. Heyden, London, 485–
493.

EVANS, W. B., WILSON, A. A., TAYLOR, B. J. AND PRICE, D. 1968.
*Geology of the Country Around Macclesfield, Congleton, Crewe
and Middlewich*. Memoir of the Geological Survey of Great
Britain, Sheet 110, HMSO.

FALCON, N. L. AND KENT, P. E. 1960. *Geological Results of Petroleum
Exploration in Britain 1945–1957*. Geological Society, London,
Memoir, **2**.

FISHER, M. J. AND MUDGE, D. C. 1990. Triassic. *In*: GLENNIE, K. W.
(ed.) *Introduction to the Petroleum Geology of the North Sea* (3rd
edn). Blackwell Scientific Publications, Oxford.

FRANCIS, E. H. 1982. Magma and sediment—1. Emplacement mechanism of late Carboniferous tholeiite sills in northern Britain.
Journal of the Geological Society, London, **139**, 1–20.

FRASER, A. J., NASH, D. F., STEELE, R. P. AND EBDON, C. C. 1990. A
regional assessment of the intra-Carboniferous play of northern
England. *In*: BROOKS, J. (ed.) *Classic Petroleum Provinces*. Geological Society, London, Special Publication, **50**, 417–440.

GAWTHORPE, R. L. 1987. Tectono-sedimentary evolution of the
Bowland Basin, N. England, during the Dinantian. *Journal of the
Geological Society, London*, **144**, 59–71.

GIBBONS, W. 1987. The Menai Strait fault system: an early Caledonian
terrane boundary in North Wales. *Geology*, **15**, 744–747.

GIBBS, A. D. 1984. Structural evolution of extensional basin margins.
Journal of the Geological Society, London, **141**, 609–620.

GREEN, P. F. 1986. On the thermo-tectonic evolution of Northern

England: evidence from fission track analysis. *Geological Magazine*, **123**, 493–506.

GREENLY, E. 1919. *The Geology of Anglesey*. Memoir of the Geological Survey of Great Britain, HMSO, London (2 volumes).

—— 1938. The red measures of the Menaian region of Caernarvonshire. *Quarterly Journal of the Geological Society of London*, **94**, 331–345.

HALL, J., BREWER, J. A., MATTHEWS, D. H. AND WARNER, M. R. 1984. Crustal structure across the Caledonides from the 'WINCH' seismic reflector profile: Influences on the evolution of the Midland Valley of Scotland. *Transactions of the Royal Society of Edinburgh: Earth Sciences*, **75**, 97–109.

HARLAND, W. B., ARMSTRONG, R. L., COX, A. V., CRAIG, L. E., SMITH, A. G. AND SMITH, D. G. 1990. *A Geologic Time Scale 1989*. Cambridge University Press, Cambridge.

JACKSON, D. I., MULHOLLAND, P., JONES, S. M. AND WARRINGTON, G. 1987. The geological framework of the East Irish Sea Basin. *In*: BROOKS, J. AND GLENNIE, K. W. (eds) *Petroleum Geology of North West Europe*. Graham & Trotman, London, 191–203.

JENNER, J. K. 1981. The Structure and Stratigraphy of the Kish Bank Basin. *In*: ILLING, L. V. AND HOBSON, G. D. (eds) *Petroleum Geology of the Continental Shelf of North West Europe*. Heyden, London, 426–431.

JENYON, M. K. 1985. Basin-edge diapirism and updip salt flow in the Zechstein of Southern North Sea. *American Association of Petroleum Geologists Bulletin*, **69**, 53–64.

JOHNSON, G. A. L. 1967. Basement control of Carboniferous sedimentation in northern England. *Proceedings of the Yorkshire Geological Society*, **36**, 175–194.

JONES, R. C. B., TONKS, L. H. AND WRIGHT, W. B. 1938. *Geology of the Country Around Wigan*. Memoir of the Geological Survey of Great Britain, HMSO.

KELLING, G. AND WELSH, W. 1970. The Loch Ryan Fault. *Scottish Journal of Geology*, **6**, 266–271.

KENT, P. E. 1948. A Deep Borehole at Formby, Lancashire. *Geological Magazine*, **85**, 253–264.

—— 1949. A structure contour map of the surface of the buried pre-Permian rocks of England and Wales. *Proceedings of the Geologists' Assocation*, **60**, 87–104.

LAWRENCE, S. R., COSTER, P. W. AND IRELAND, R. J. 1987. Structural development and petroleum potential of the northern flanks of the Bowland Basin (Carboniferous), North-west England. *In*: BROOKS, J. AND GLENNIE, K. W. (eds) *Petroleum Geology of North West Europe*. Graham & Trotman, London, 225–233.

LEE, A. G. 1988. Carboniferous basin configuration of central and northern England modelled using gravity data. *In*: BESLY, B. M. AND KELLING, G. (eds) *Sedimentation in a Synorogenic Basin Complex: the Upper Carboniferous of Northwest Europe*. Blackie, Glasgow and London, 69–84.

McCLAY, K. R. 1989. Analogue models of inversion tectonics. *In*: COOPER, M. A. AND WILLIAMS, G. D. (eds) *Inversion Tectonics*. Geological Society, London, Special Publication, **44**, 41–59.

—— 1990. Extensional fault systems in sedimentary basins: a review of analogue model studies. *Marine and Petroleum Geology*, **7**, 206–233.

McLEAN, A. C. 1978. Evolution of fault controlled ensialic basins in north western Britain. *In*: BOWES, D. R. AND LEAKE, B. E. (eds) *Crustal Evolution in northwestern Britain and adjacent regions*. *Geological Journal Special Issue*, **10**, 325–346.

MANSFIELD, J. AND KENNETT, P. 1963. A gravity survey of the Stranraer sedimentary basin. *Proceedings of the Yorkshire Geological Society*, **34**, 139–151.

MORLEY, C. K., NELSON, R. A., PATTON, T. L. AND MUNN, S. G. 1990. Transfer Zones in the East African Rift System and Their Relevance to Hydrocarbon Exploration in Rifts. *American Association of Petroleum Geologists Bulletin*, **74**, 1234–1253.

MOSELEY, F. 1972. A tectonic history of northwest England. *Journal of the Geological Society, London*, **128**, 561–598.

ORD, D. M., CLEMMEY, H. AND LEEDER, M. R. 1988. Interaction between faulting and sedimentation during Dinantian extension of the Solway basin, S. W. Scotland. *Journal of the Geological Society, London*, **145**, 249–259.

PARKIN, M. AND CROSBY, A. 1982. *Geological Results of Boreholes drilled on the Southern United Kingdom Continental Shelf by the Institute of Geological Sciences 1969–1981*. British Geological Survey, Marine Geology Report, **82/2** (2 volumes).

RAMSBOTTOM, W. H. C., CALVER, M. A., EAGAR, R. M. C., HODSON, F., HOLLIDAY, D. W., STUBBLEFIELD, C. J. AND WILSON, R. B. 1978. *A Correlation of Silesian Rocks in the British Isles*. Special Report of the Geological Society, London, **10**.

RHYS, G. H. (Compiler) 1974. *A Proposed Standard Lithostratigraphic Nomenclature for the Southern North Sea and an Outline Structural Nomenclature for the Whole of the (UK) North Sea*. A report of the joint Oil Industry–Institute of Geological Sciences Committee on North Sea Nomenclature. Institute of Geological Sciences Report No. **74/8**, HMSO, London.

ROBERTS, D. G. 1989. Basin inversion in and around the British Isles. *In*: COOPER, M. A. AND WILLIAMS, G. D. (eds) *Inversion Tectonics*. Geological Society, London, Special Publication, **44**, 131–150.

ROSE, W. C. C. AND DUNHAM, K. C. 1977. *Geology and Hematite Deposits of South Cumbria*. Economic Memoir of the Geological Survey of Great Britain, Sheet 58 and part 48 (England and Wales), HMSO.

ROSENDAHL, B. R., REYNOLDS, D. J., LORBER, P. M., BURGESS, C. F., McGILL, J., SCOTT, D., LAMBIASE, J. J. AND DERKSEN, S. J. 1986. Structural expressions of rifting: lessons from Lake Tanganyika, Africa. *In*: FROSTICK, L. E., REID, I., RENAUT, R. W. AND TIERCELIN, J. J. (eds) *Sedimentation in the African Rifts*. Geological Society, London, Special Publication, **25**, 29–43.

SMITH, D. B., BRUNSTROM, R. G. W., MANNING, P. I., SIMPSON, S. AND SHOTTON, F. W. 1974. *A Correlation of Permian Rocks in the British Isles*. Special Report of the Geological Society, London, **5**.

SMITH, N. J. P. (Compiler) 1985. *Pre-Permian Geology of the United Kingdom (South), maps 1 and 2*. British Geological Survey, Keyworth.

SMITH, R. A. 1986. *Permo-Triassic and Dinantian Rocks of the Belfast Harbour Borehole*. Report of the British Geological Survey, vol. **18** (6), 1–13.

SOPER, N. J. AND MOSELEY, F. 1978. Structure. *In*: MOSELEY, F. (ed.) *The Geology of the Lake District*. Yorkshire Geological Society, Occasional Publication 3, 45–67.

——, WEBB, B. C. AND WOODCOCK, N. H. 1987. Late Caledonian (Acadian) transpression in north-west England: timing, geometry and geotectonic significance. *Proceedings of the Yorkshire Geological Society*, **46**, 175–192.

STUART, I. G. AND COWAN, G. 1991. The South Morecambe Field, Blocks 110/2a, 110/3a, 110/8a, UK East Irish Sea. *In*: ABBOTTS, I. L. (ed.) *United Kingdom Oil and Gas Fields 25 Years Commemorative Volume*. Geological Society, London, Memoir, **14**, 527–541.

THOMPSON, D. B. 1970. The stratigraphy of the so-called Keuper Sandstone Formation (Scythian–?Anisian) in the Permo-Triassic Cheshire Basin. *Quarterly Journal of the Geological Society of London*, **126**, 151–181.

TODD, S. P., MURPHY, F. C. AND KENNAN, P. S. 1991. On the trace of the Iapetus suture in Ireland and Britain. *Journal of the Geological Society, London*, **148**, 869–880.

WALKER, I. M. AND COOPER, W. G. 1987. The structural and stratigraphic evolution of the northeast margin of the Sole Pit Basin. *In*: BROOKS, J. AND GLENNIE, K. W. (eds) *Petroleum Geology of North West Europe*. Graham & Trotman, London, 263–275.

WALSH, J. J. AND WATTERSON, J. 1991. Geometric and kinematic coherence and scale effects in normal fault systems. *In*: ROBERTS, A. M., YIELDING, G. AND FREEMAN, B. (eds) *The Geometry of Normal Faults*. Geological Society, London, Special Publication, **56**, 193–203.

WARMAN, H. R., ROBERTS, K. H., BRUNSTROM, R. G. W. AND ADCOCK, C. M. 1956. *Report on Oil and Gas in the United Kingdom*. Symposium sobre Yacimientos de petroleo y gas, XX Congreso Geologico Internacional, Mexico, vol. **v**, Europe, 317–357.

WARREN, P. T., PRICE, D., NUTT, M. J. C. AND SMITH, E. G. 1984. *Geology of the Country Around Rhyl and Denbigh*. Memoir of the British Geological Survey, Sheets 95 and 107, HMSO.

808

WARRINGTON, G., AUDLEY-CHARLES, M. G., ELLIOTT, R. E., EVANS, W. B., IVIMEY-COOK, H. C., KENT, P., ROBINSON, P. L., SHOTTON, F. W. AND TAYLOR, F. M. 1980. *A Correlation of Triassic Rocks in the British Isles*. Special Report of the Geological Society, London, **13**.

WHITTAKER, A., HOLLIDAY, D. W. AND PENN, I. E. 1985. *Geophysical Logs in British Stratigraphy*. Special Report of the Geological Society, London, **18**, 1–74.

WILKINSON, I. P. AND HALLIWELL, G. P. (Compilers) 1979. *Offshore micropalaeontological biostratigraphy of southern and western Britain*. Report of the Institute of Geological Sciences, **79/9**.

WILLS, L. J. 1970. The Triassic succession in the central Midlands in its regional setting. *Quarterly Journal of the Geological Society of London*, **126**, 225–283.

WILSON, A. A. 1990. The Mercia Mudstone Group (Trias) of the East Irish Sea Basin. *Proceedings of the Yorkshire Geological Society*, **48**, 1–22.

—— AND EVANS, W. B. 1990. *Geology of the Country Around Blackpool*. Memoir of the Geological Survey of Great Britain, Sheet 66 (England and Wales), HMSO.

WILSON, C. D. V. 1959. Geophysical investigations in the Vale of Clwyd. *Liverpool and Manchester Geological Journal*, **2**, 253–270.

WRAY, D. A. AND COPE, F. W. 1948. *Geology of Southport and Formby*. Memoir of the Geological Survey of the United Kingdom, HMSO.

WRIGHT, J. E., HULL, J. H., McQUILLIN, R. AND ARNOLD, S. E. 1971. *Irish Sea investigations, 1969–70*. Report of the Institute of Geological Sciences, **71/19**.

ZIEGLER, P. A. 1987. Manx-Furness Basin. *Tectonophysics*, **137**, 335–340.

—— 1990. *Geological Atlas of Western and Central Europe* (2nd edn). Shell Internationale Petroleum Maatschappij B.V., The Hague.

Discussion

Question (D. Naylor, Trinity College, Dublin):

Thicknesses of Liassic strata in the western regions of the Irish Sea may be even greater than recorded in the Mochras borehole, or in the East Irish Sea Basin. Seismic data and other evidence suggest in excess of 2.5 km of Lias is preserved against the bounding faults of the Kish Basin, and similar thicknesses have been encountered in drilling further south in the Irish sector. Do the authors consider that the Clyde trend remained the dominant control on Liassic sedimentation or were other regional factors at play?

Answer (D. I. Jackson):

Four Lias outliers are shown on Fig. 2; seismic thicknesses indicate rapid subsidence at the Keys and Solway Firth basin sites, and more condensed sequences in the Carlisle and Berw occurrences. Thus, early Jurassic palaeogeography was broadly comparable with that in the late Triassic. The persistence of a weak Clyde Belt trend down to the present day is shown by the axial preservation of Lias outliers at Frith Farm and Prees in the Cheshire Basin, in the Knowle Basin and the Lias/Inferior Oolite outlier of Bredon Hill in the Worcester Graben.

Nevertheless, Lias isopachs appear to indicate a regional northerly reduction from the Celtic Sea/Kish Bank Basin (the figures in your question are a welcome addition to our knowledge) and Mochras borehole (Dobson and Whittington 1987, fig. 2) to Northern Ireland. This is the reverse of the regional trend of Permian and early Triassic isopachs. Early Jurassic sedimentation may represent a transition phase in the reversal of palaeoslope from dominance by the Clyde Belt trend during the Triassic (?as a failed break-up axis in the suturing of Pangaea) to dominance by the Celtic Sea trend during the mid-late Jurassic (associated with North Atlantic rifting). This switch to a dominant Celtic Sea sedimentation axis may be manifested as the Near Top Triassic disconformity of the North Irish Sea.

Two additional strands of evidence are:

(a) the depth to the Base Permo-Trias in the St George's Channel Basin comfortably exceeds that of any other basin, onshore or offshore, in the southern UKCS (Smith 1985, map 2; Dobson and Whittington 1987, p. 334). Much of the fill comprises Middle and Upper Jurassic and indicates an acceleration of the Lower Jurassic subsidence patterns (Barr *et al.* 1981);

(b) if the ages are reliable, the thick Triassic halites dated as Carnian and Norian in Wells 103/2-1 and 106/28-1 (Barr *et al.* 1981, fig. 3) and the apparent absence of a thick Preesall Salt/Northwich Halite Formation equivalent imply that accelerated subsidence of the St George's Channel Basin may have started as early as the Carnian.

A detailed study of early Jurassic stage thicknesses is probably required to determine whether the Clyde Belt or Celtic Sea trend was predominant during Liassic sedimentation.

Geochemical modelling of the East Irish Sea Basin: its influence on predicting hydrocarbon type and quality

M. HARDMAN, J. BUCHANAN, P. HERRINGTON and A. CARR

British Gas plc Exploration and Production, 100 Thames Valley Park Drive, Reading, Berkshire RG6 1PT, UK

Abstract: A number of oil discoveries have been made in recent years in the East Irish Sea Basin (EISB). These have renewed industry interest in a basin which many thought to be essentially gas prone. British Gas, over a number of years, has developed and refined a basin model for the area based upon in excess of 23 000 km of seismic data and 40 wells. This model shows that the present-day distribution of oil and gas is controlled by complex interrelationships between the thermal and tectonostratigraphic history of the EISB. An integrated approach has been adopted in creating the model, which includes data generated from many sources including AFTA, detailed diagenetic and geochemical studies and more conventional geological, geophysical, and petrophysical analysis. The paper will address several aspects of the current basin model, including oil–source rock correlation, the origins of hydrogen sulphide within the EISB, and the problems and solutions in validating a basin model when so much of the rock record is missing.

Advances in basin modelling techniques over recent years have enabled quantitative models of the complex geological history of the East Irish Sea Basin (EISB) to be created and tested. These models have been used to study a number of geological problems within the basin. These problems include the relationship between the timing of hydrocarbon generation and the timing and magnitude of inversion, the presence of abundant dead oil staining throughout the basin and, more recently, the relationship between oil and gas charged structures. Integration of a number of techniques, including apatite fission track analysis, detailed chronodiagenetic work and conventional geological observation and geochemical analysis has refined the range of possible basin models.

Structural setting

The EISB is situated in the offshore area to the north of Wales, west of Lancashire and southeast of the Isle of Man (Fig. 1). The basin forms the central part of a series of Permo-Triassic basins in the west of Britain including the North Channel, Solway and Cheshire basins. These may be the preserved remnants of an extensive basin system which previously linked the Celtic Sea basins to the south and the basins off western Scotland. The terminology of Jackson *et al.* (1987) has been used to name the structural elements within the EISB.

The geological history of the EISB is both long and complex. The present-day basin fill sequence of Permo-Triassic age unconformably overlies a previously uplifted, folded and eroded sequence of Carboniferous rocks which contain both oil and gas prone source rocks. The Permo-Triassic was itself inverted during late Cretaceous–early Tertiary times. Most previous publications in the area also refer to a Cimmerian inversion episode which was first proposed by Bushell (1986), the effects of which have been eroded in the later more severe Tertiary inversion. The stratigraphic terminology used herein and shown in Fig. 2 is that proposed by Warrington *et al.* (1980), Jackson *et al.* (1987) and Wilson (1990).

The basin is dominated by faults which trend NW–SE, NE–SW and, in the south of the basin, N–S. Extensional faults trending NW–SE, such as the Keys Fault, and the NE–SW trending faults defining the Ramsey–Whitehaven Ridge were certainly active by early Permian times, and may in fact have formed much earlier as Carboniferous extensional faults. The southeastern part of the basin is dominated by the Crosh Vusta and Formby Point Fault; these trend NW–SE and N–S res-

pectively and dip in a westerly direction (Knipe *et al.* 1993).

Both Permian and Triassic sediments thicken into the hangingwalls of these major faults whilst present-day normal fault displacements of the top Sherwood Sandstone Group across these major faults can reach 4000 ft.

The major extension direction during the Permo-Triassic is thought to be orientated ENE–WSW to NE–SW (Knipe *et al.* 1993), with 10 to 16% extension indicated by offset in the Pre-Permian basement.

Nowhere in the EISB have faults been identified which exhibit a true reverse displacement as a result of inversion. This indicates that inversion related displacement on individual faults is always less than the normal fault displacement. Knipe *et al.* (1993) have documented a series of anticlines which trend NE–SW and parallel adjacent normal faults. These are tentatively identified as inversion anticlines. Sinistral sheer is indicated by these small-scale structures, consistent with that experienced by basins in southwest Britain at this time (Ziegler 1987). Knipe *et al.* (1993) estimate a shortening of 2.5% which if taken up on faults dipping at 45° (as indicated in the seismic data) gives a minimum estimate for the uplift of 1.5 km.

Previous basin models

With the exception of Roberts (1989), previous basin models for the East Irish Sea such as Bushell (1986), and Lewis *et al.* (in press), have all included Cimmerian and Tertiary uplift events. Estimates of the Tertiary inversion have varied considerably from author to author; Coulter and Barr (1975) estimate 4000 to 5000 ft, whilst Lewis *et al.* (in press) suggest 2.7 to 3.2 km (8900 to 10 500 ft).

In an attempt to refine the magnitude of the 'Tertiary Inversion' and its timing, whilst also attempting to quantify if possible the 'Cimmerian Inversion', a multi-disciplinary approach has been taken. A number of different sources of data have been utilized; these include conventional geochemical data, fluid inclusion, paragenetic sequences and radiometric dating of authigenic cements, combined with apatite fission track data.

In order to demonstrate the approach taken and also to illustrate some results and remaining problems, two unreleased wells have been chosen. Both were drilled in recent years and are proprietary to British Gas. The wells termed X and Y were the first wells to encounter a live oil column in the basin.

From *Petroleum Geology of Northwest Europe: Proceedings of the 4th Conference* (edited by J. R. Parker).
© 1993 Petroleum Geology '86 Ltd. Published by The Geological Society, London, pp. 809–821.

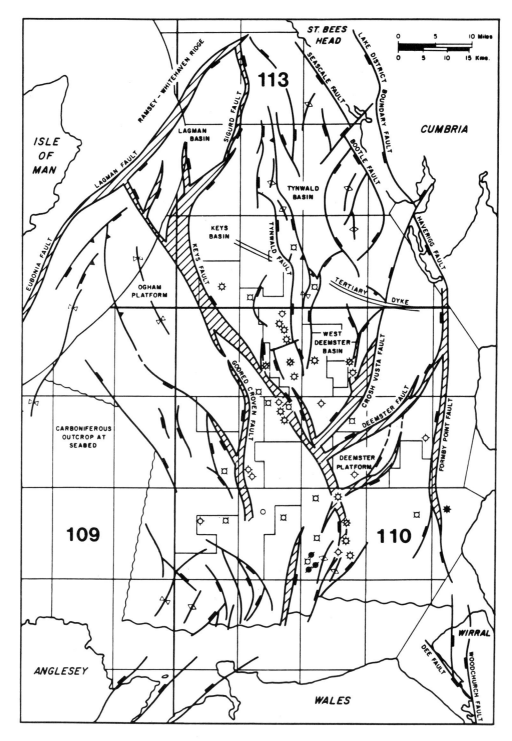

Fig. 1. Structural elements map, East Irish Sea Basin (modified after Meadows and Beach 1993).

Geochemical data

A number of wells have encountered Carboniferous sediments during drilling, and while the importance of these sediments as the source rocks for the hydrocarbons found in the basin is understood, there has been no systematic study of their characteristics. The Carboniferous sediments in the EISB were uplifted and eroded during the Hercynian orogeny, and as a result the upper part of the Carboniferous is reddened (Jackson *et al.* 1987). The age of the Permian subcrop mainly consists of sediments varying in age between Dinantian and Westphalian B, although unpublished seismic evidence indicates that there are structures containing potentially younger Westphalian sediments.

Maturity When plotted against depth, the vitrinite reflectance dataset shows a spread of values reflecting variations in either the amounts of uplift or the geothermal gradients across the basin (Fig. 3). The maturity gradients obtained from wells with significant amounts of Carboniferous sediments are very similar, indicating that the geothermal gradients were very similar across the whole basin (Fig. 3) and thus uplift is believed to be the primary cause of variations in the maturity–depth relationship. Using the method described by Dow (1977), the amount of uplift appears to vary between 1000 and 10 000 ft, the variation reflecting the complex structural history of the basin.

In terms of hydrocarbon generation the reflectance values

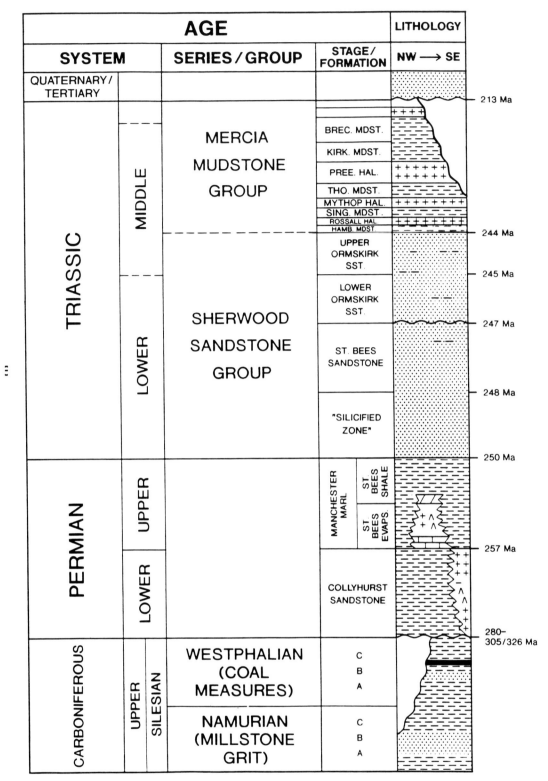

Fig. 2. Lithostratigraphy, East Irish Sea, with ages used in basin models.

show that the maturity varies between the early part of the oil window (0.5% R_0 to 0.7% R_0), and the zone of dry gas generation (2.0% R_0 to 3.0% R_0). The relatively low maturities (<0.70% R_0) of some wells indicate that, regardless of the source potential of the sediments, some parts of the basin do not significantly contribute to overall hydrocarbon budget. The low maturity results from only small amounts of Mesozoic burial, while in the case of wells with maturities in excess of 2.0% R_0, burial during the Mesozoic must have been substantial.

Source potential The generation capability of source rocks in individual wells is extremely variable. The present-day potential for hydrocarbon generation is largely poor. However, there are sections with moderate to good potential. In general the poor present-day potential is largely the result of the high maturity of the sediments, combined with the degree of reddening which took place in late Westphalian–early Permian times. Westphalian sediments (Table 1) generally have high organic carbon contents and are dominated by inertinitic (non-productive) kerogens. There is therefore some doubt as to the volume of gaseous hydrocarbons that will have been generated from the Westphalian, particularly as coals are unusually rare in the section.

Table 1. Source rock data for the East Irish Sea

Age	TOC%	Potential yield $S_2(mg/g)$	HI
Westphalian	54.24	6.42	14
	3.77–77.3	1.10–11.51	12–28
Namurian	1.93	1.61	92
	0.22–8.32	0.30–6.40	2–226

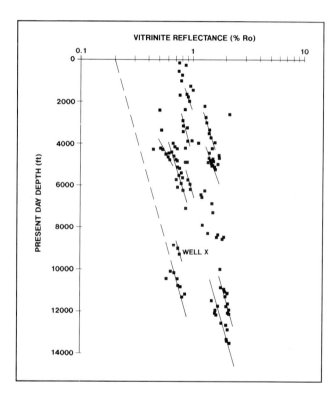

Fig. 3. Vitrinite reflectance versus present day burial depth for Carboniferous sediments in the East Irish Sea Basin.

Namurian sediments, however, appear to have had higher hydrocarbon potential. In particular, the Sabden Shale equivalents deposited during Arnsbergian to Kinderscoutian times (E2-R1) contain small yet significant amounts of sapropelic, i.e. oil-prone, kerogen. At present these have only moderate generation potential, this potential having been realized in reaching the high level of maturity seen in most wells. The presence of oil-prone source rocks in this interval results from the establishment of extensive marine conditions stretching from the Irish Sea into northern Lancashire (Lawrence *et al.* 1987 and Gawthorpe and Fraser 1990).

Hydrocarbons As depicted in Table 2, compositional analyses of East Irish Sea Basin gases show high methane contents. Isotopic values of $\delta^{13}C$ in the range -32.6 to 47.93% and δD values in the range -135 to -168% (Fig. 4) suggest that they were generated from source rocks containing sapropelic kerogen at maturities of at least 0.8% to 1.0% VR_0 (Schoell 1984).

Oils from eight hydrocarbon accumulations provide the basis of this study. GCMS analysis of illustrative 'type' oils are shown in Figs 5 and 7. There appear to be two groups of oils,

the separation being mainly based on their occurrence as either 'live' or 'dead' oils. Bushell (1986) suggested that the dead oils may have formed via biodegradation. Biomarker analyses show that some dead oils contain demethylated triterpanes, compounds that are often generated by biodegradation of oils (Fig. 6). However, in other samples, the clear preservation of low molecular weight n-alkanes does not support the biodegradation hypothesis. The 'live' oils are dominated by aliphatic hydrocarbons and are light, 40 to 45° API, with low sulphur contents.

Table 2. Isotopic values obtained from gases and oils

Well	CH4/C1 to C4	13C	D
Gases			
A	0.94	−43.9	−168
B	0.94	−33.28	−140
C	0.94	−32.6	−135
D	0.93	−42.4	−160
E	0.85	−44.5	−150
Y	–	−47.93	−165

Well	CH4/C1 to C4	13C ARO	13C ALI
Aromatics and aliphatics			
F	–	−29.67	−30.97
G	–	−29.93	−28.86
H	0.90	−28.31	−29.41
I	–	−27.71	−28.81
J	–	−28.51	−28.20
Y	–	−30.57	−30.88

Key

C1 to C4: Normal alkanes
D: Deuterium
13C: Carbon-13
13C ARO: Oil-aromatics
13C ALI: Oil-aliphatics

Oil-source correlation has been attempted using isotopic and biomarker data. Both the isotopic (Table 2 and Fig. 7) and triterpane biomarker data indicate that all the oils appear to have been generated from the same source rock facies (Fig. 8). Comparative data obtained from E_2 to R_1 Namurian source rocks indicate that they were the main source rocks for the oils in this basin. More specifically the biomarker data obtained for the oils indicate that they were generated from source rocks with maturities of approximately 1.0% R_0.

Diagenetic data: paragenesis

Well X Figure 8 illustrates the paragenetic sequence for the Ormskirk Sandstone in Well X. The Upper Ormskirk Sandstone in Well X is of particular note as there is an increase in the abundance of quartz cements from the bottom to top of the waterleg. On entering the oil-bearing section of the reservoir the abundance of quartz overgrowths show a marked decline. However, within the oil leg there is an upwards increase in the abundance of quartz overgrowths. This distribution is suggestive of a diagenetic model in which quartz cementation occurs initially in response to the migration of formation waters immediately below the Mercia Mudstone caprock. This results in a steady increase in the concentration of quartz cement towards the top of the reservoir. The emplacement of oil then occurred and diagenesis ceased in the oil zone. A continued (or

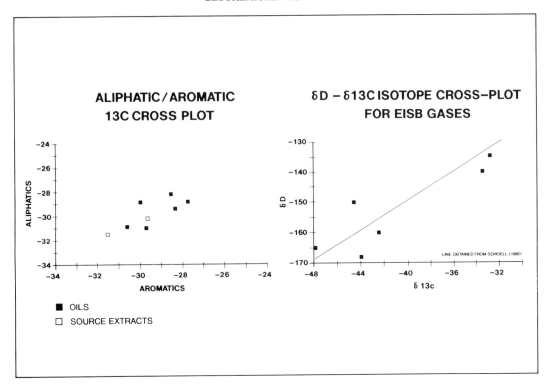

Fig. 4a. δD–δ¹³C Isotope cross-plot for East Irish Sea Basin gases suggesting a common marine source. **(b)** Aliphatic–aromatic ¹³C cross-plot for oils and source rock extracts, East Irish Sea Basin.

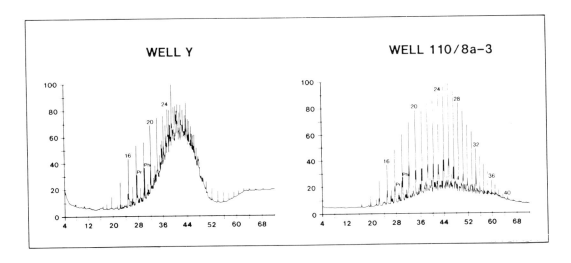

Fig. 5. Gas chromatograms of oils from Well 110/8a-3 and Well Y illustrating severe alteration of the oil in Well Y.

second) phase of quartz cementation below the oil leg results in the observed distribution, i.e. quartz overgrowths both pre- and post-date hydrocarbon emplacement.

Quartz overgrowths seen in the water leg occur in the periphery of oversized pores which often contain remnant calcite cements in their centres. Many of these quartz over-growths are stained by residual hydrocarbon. Since these quartz overgrowths are thought to be 'late' (i.e. to have formed after the emplacement of oil in the reservoir) and are stained by residual oil, either two phases of hydrocarbon migration or one phase of emplacement followed by late stage re-migration/breaching must have occurred.

Similarly, observation of hydrocarbon stained and non-stained carbonate cements and their petrographic relationships suggest that the finely crystalline ferroan dolomite pre-dates

hydrocarbon emplacement, whilst the medium crystalline rhombs of dolomite/ankerite post-date both the ferroan dolo-mite and oil emplacement.

SEM examination suggests that a very minor late stage fibrous illite overgrows all other authigenic phases and there is limited petrographic evidence for a late stage carbonate disso-lution event.

Well Y The Upper Ormskirk Sandstone in Well Y underwent similar diagenesis as shown in Fig. 8. Once again the volu-metric abundance of quartz cement shows a similar relation-ship to that observed in Well X. There is also, however, a marked decrease in the amount of carbonate cement present in the water leg when compared to the oil leg. This is thought to be due to pore fluids upwelling and cooling. As this fluid cools

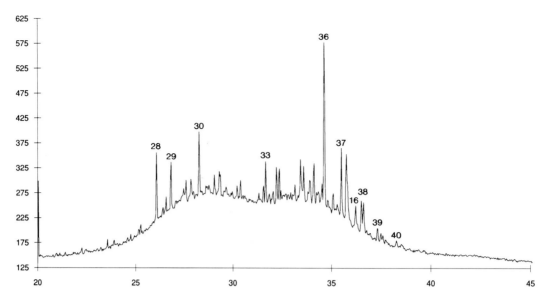

Fig. 6. Demethylated hopane (*m/z* 177) GCMS profile of a dead oil in the East Irish Sea Basin. The presence of demethylated triterpanes is possibly indicative of biodegradation. 36–40 = C_{28} to C_{34} 17 α (H) – 25 norhopanes; 33 = C_{27} demethylated tricyclic terpanes; 30 = C_{24} demethylated triterpanes; 29 = C_{23} demethylated triterpanes; 28 = C_{22} demethylated triterpanes; 16 = 17β (H), 2 α (H) – 30 normorelane.

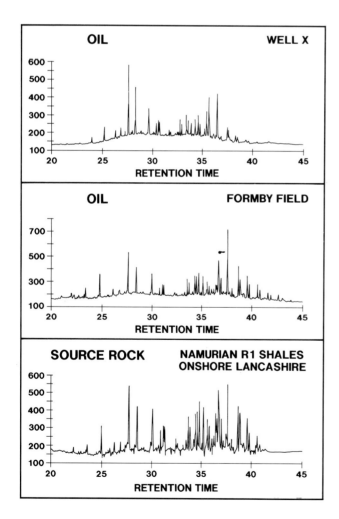

Fig. 7. Panel illustrating the geochemical correlation of the oil from Well X with oil from the onshore Formby field and source rocks from the Namurian E_2–R_1 interval from an onshore well in West Lancashire.

it is likely to become undersaturated in carbonate (Giles and De Boer 1989). Such a situation is likely to lead to dissolution of the authigenic carbonate and produce the observed distribution.

Fluid inclusion analysis

Fluid inclusions have been identified in authigenic phases in both wells and are classified as primary or pseudo secondary in character (Shepherd *et al.* 1985). In this study only inclusions with a size greater than 10 μm and containing both a liquid and vapour phase at room temperature were used. Using a Linkham THM 600 heating/freezing stage, at least three determinations of the homogenization temperature for each fluid inclusion were made. The mean of these temperatures was then calculated. If any individual measurement was found to lie ±2°C outside the inclusion mean, then metastable homogenization was inferred to have taken place and the result discarded. The homogenization temperatures thus measured and calculated are shown in Fig. 9.

In this study the homogenization temperature has been regarded as the minimum trapping temperature of the fluid inclusion. Although a pressure correction is commonly made (e.g. Potter 1977), Hanor (1980) has demonstrated that the application of standard corrections to inclusions which contain a small content of methane are erroneous. Consequently pressure corrections have not been made.

Fluid inclusions have been identified and homogenization temperatures measured in quartz, dolomite, ankerite, calcite and anhydrite cements. In excess of forty inclusions have given statistically acceptable homogenization temperatures in the two wells. In addition, fluid inclusion data is available from other wells in the basin which mirrors the results presented below.

Quartz Homogenization temperatures obtained from two-phase quartz hosted inclusions in Wells X and Y are in the ranges 82.3–126.4°C and 108.9–135.6°C respectively. The frequency distribution of this data is shown in Fig. 10. In addition, a rare but significant occurrence of large, single phase

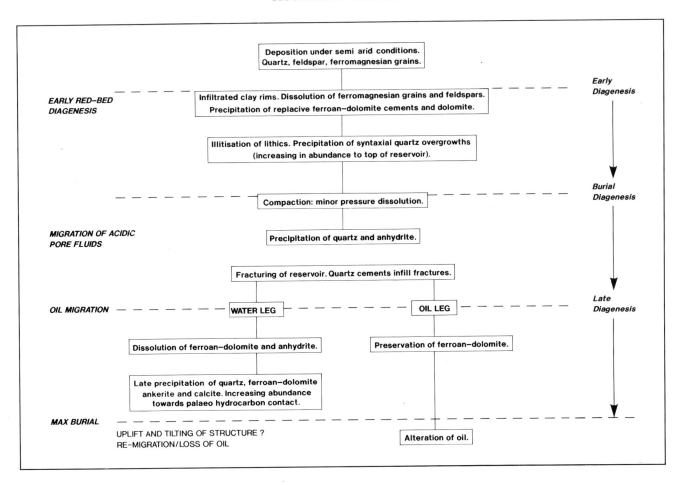

Fig. 8. Paragenetic sequence for the Ormskirk Sandstone in Wells X and Y.

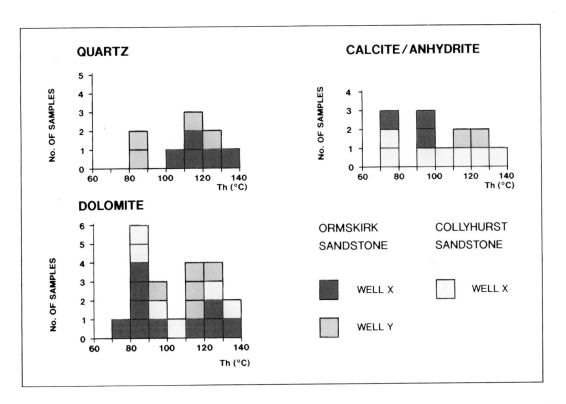

Fig. 9. Frequency histograms showing the distribution of homogenization temperatures measured in quartz, dolomite and calcite/anhydrite authigenic cements in Wells X and Y.

liquid inclusions has been identified. The paragenetic sequence established for the wells includes an early phase of quartz cementation of phreatic origin (the large inclusions), followed by further quartz precipitation at higher temperatures during burial diagenesis (the two phase inclusions). Examination of Fig. 9 indicates two obvious peaks in the T_{hom} distribution. There are at least two possible interpretations of this distribution. The peak at 80–90°C represents precipitation of quartz and entrapment of the inclusion during burial diagenesis. The higher peak at 110–130°C either represents re-equilibration of earlier formed inclusions in response to deeper burial and hence higher temperatures or, alternatively, the higher temperature inclusions may represent the precipitation of quartz at high temperatures as a result of the introduction of migrating hot fluids sourced from deep within the basin. In this case the higher temperatures are not necessarily burial related but are the result of a 'thermal anomaly'.

Dolomite and ankerite   Fluid inclusions within dolomite and ankerite cements are generally hosted by medium to coarsely crystalline rhombs which sometimes coalesce to form a sub-poikilotopic cement. Almost all of the two-phase inclusions occur along cleavage planes, with the largest inclusions delineating the margins of growth zones.

Homogenization temperatures are in the ranges 83.0–131.6°C in Well X and 98.3–125.3°C in Well Y. Fig. 9 shows the frequency distribution of the data. As shown on this diagram the T_{hom} frequency distributions of the Sherwood and Collyhurst Sandstones are similar. Within the distribution the largest peak occurs at 80–90°C, this is interpreted as the temperature of precipitation of the cements. T_{hom} data in the range 100–120°C are thought to have formed from re-equilibration with continued burial. As with quartz-hosted inclusions there are two possible interpretations for the more minor peak at 120–140°C.

The possibility that the range seen in T_{hom} data could somehow be related to the three mineralogies of dolomite–ankerite cements (non-ferroan dolomite, ferroan dolomite and ankerite) has been assessed by careful integration of the petrography and fluid inclusion data. This has shown that the whole range of T_{hom} data is present both in ferroan and non-ferroan cements and suggests that the precipitation of these phases took place over a narrow range of temperatures.

Calcite and anhydrite cements   A far more limited number of inclusions hosted by calcite and anhydrite cements were suitable for T_{hom} determination. Fig. 9 shows the results of these determinations. As with the more numerous quartz and dolomite–ankerite inclusions the data can be interpreted in at least two ways. The higher temperatures are related either to re-equilibration due to depth related temperature increase or via a 'hot flush'.

K–Ar Dating of authigenic illites

Obviously radiometric dating of authigenic illite allows a quantitative constraint to be placed upon the paragenetic sequence which can punctuate the timing of the cement phases. Integration with fluid inclusion temperatures then provides an absolute date and temperature to constrain the burial history model. Authigenic illite is the most widely used phase when this technique is applied. In Well X illite is present in both the Ormskirk and Collyhurst Sandstones, in amounts up to 1%. Ten samples from Well X were analysed and yielded age dates. Unfortunately, however, illite was not sufficiently abundant in Well Y to allow separation.

Following the work published on illite growth by Lee *et al.* (1989) and Hamilton *et al.* (1987), which showed that progressively finer separated fractions of illite gave progressively

younger dates and that the finest fraction effectively gives the date when illite precipitation ceased, the samples were separated and the $<0.5\,\mu m$ illite fraction was dated.

In Well X the illite ages obtained from the Collyhurst Sandstone indicate Jurassic cessation ranging from 208 to 182 Ma and those from the Upper Ormskirk Sandstone occupy a similar though slightly more extensive range from 208 to 140 Ma. This confirms the petrographic observation that illite was formed early in the diagenetic history of the well.

There was no apparent systematic variation of illite age with depth in the well. The ages obtained are similar to the previously published age of 190–175 Ma obtained from the very extensive illites present in the South Morecambe Field (Bushell 1986).

Apatite fission track analysis

Seven samples were collected from the Triassic, Permian and Carboniferous aged sediments in Well X, whilst two Triassic samples were collected in Well Y. Apatite fission track analysis was then carried out using the sample preparation and experimental procedure outlined by Green (1986). The chronological time scale of Harland *et al.* (1989) has been used in conjunction with Warrington *et al.* (1980) and Smith *et al.* (1974) to analyse the results. Apatite yields in all samples were fair (10–15 grains) to excellent (20+ grains) in the sandstones, whilst a poor yield (5–10 grains) was obtained from the one sample from the Mercia Mudstone Group. The quality of individual grains analysed was generally very good, although two samples from the Carboniferous in Well X gave a rather low number of confined track lengths. Nevertheless, the high quality of the data from the other samples gives a high reliability to the thermal data.

All samples in the two wells give apatite fission track ages that are significantly reduced from the stratigraphic ages (Fig. 10). This indicates that they have experienced elevated temperatures since deposition. Since the fission track ages decrease only moderately with depth (Fig. 11), and yet are very significantly reduced from the stratigraphic age, the implication is that most of the samples have been totally annealed at temperatures of 100°C or greater. The consistency of the fission track ages below depths of 1.0 km further supports this.

Corrected ages of the totally annealed samples from Wells X and Y display a consistent value of around 65 ± 10 Ma (Fig. 11), which is considered to represent the time at which cooling from elevated temperatures began, i.e. late Cretaceous–early Tertiary. This is consistent with data presented in Lewis *et al.* (in press) and Green (1989).

Unfortunately data from the samples analysed in Wells X and Y are insufficient to provide a constraint on the palaeogeothermal gradient at the time of maximum palaeotemperatures. They merely identify the late Cretaceous/early Tertiary as the latest possible time that the currently preserved sediments in wells X and Y experienced temperatures greater than or equal to 110°C.

The two shallowest samples in each well appear not to have reached palaeotemperatures quite as high as 110°C based upon track length distributions; the temperatures encountered by these samples appear to have been in the range 100–110°C. This allows an estimate to be made of the amount of cooling that has occurred in each well since the onset of cooling. In Well X the cooling is estimated to be 73°C, whilst in Well Y it is 68°C.

The temperature data available defines a reliable geothermal gradient of 26°C/km for these wells which agrees with Bushell's (1986) determination of 27°C/km.

Lewis *et al.* (in press) contend that palaeogeothermal gradients during the late Cretaceous–early Tertiary were 'normal' and similar to those of the present day. If this proposal is

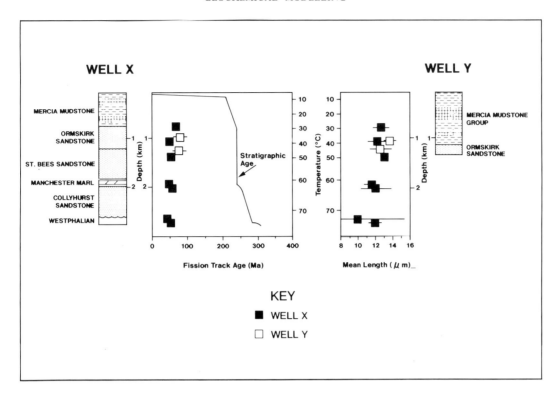

Fig. 10. Apatite fission track analysis parameters plotted against sample depth and present day temperatures for samples from Wells X and Y. Note the major variation fission track age and stratigraphic age of the samples as indicated on the central panel.

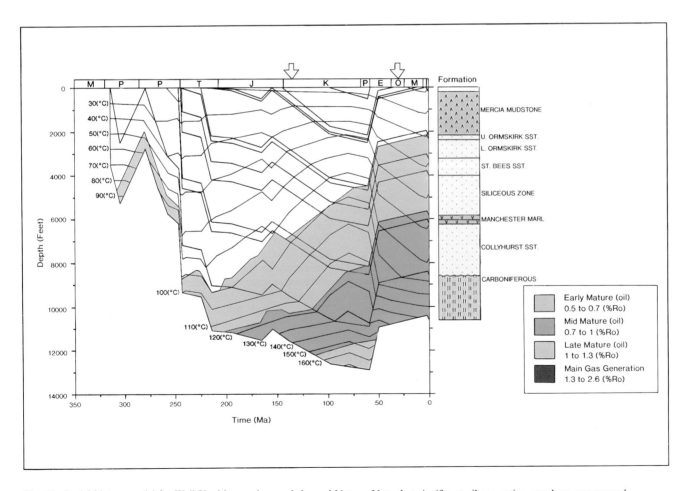

Fig. 11. Burial history model for Well X with superimposed thermal history. Note that significant oil generation may have commenced immediately prior to the Cimmerian inversion; however, the major phase of hydrocarbon generation will have occurred in late Jurassic–late Cretaceous times.

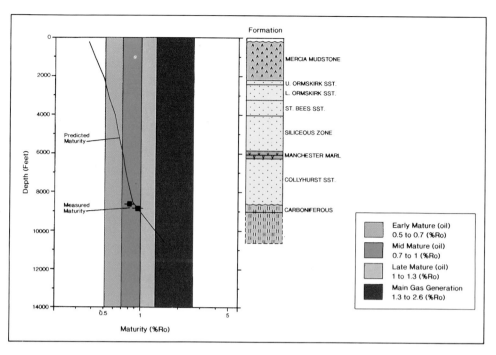

Fig. 12. Computed maturity data from the Well X burial history model presented in Fig. 12, plotted against the measured maturity data from the well.

correct and applied to Wells X and Y, then the maximum palaeotemperatures (110°C assumed) were achieved by slightly in excess of 3 km of post-Triassic burial. Obviously this figure for uplift of 3 km is a 'best guess' since the temperature floor for the apatite fission track technique is 110°C, as defined by annealing. Higher temperatures would result in an even larger inversion if the 'normal gradient' assumption is correct, whereas a higher geothermal gradient would result in significantly less inversion.

Uplift estimates based upon shale velocities

Colter (1978) and Ebbern (1981) suggested that correlation within the Mercia Mudstone Group was difficult as a result of extreme lithological variability within the section. With the relative high density of drilling in the South Morecambe Field, Stuart (1991) has shown that correlation within the Mercia Mudstone Group is possible. More recently this informal correlation has been extended onshore. This has allowed comparison with the stratigraphy proposed for the onshore Fylde area by Wilson (1990), with which it has significant similarities. In summary, a shale has been identified within the Singleton Mudstone Formation which is the best currently available within the basin for estimating uplift based upon interval velocities. The unit is, however, far from ideal for utilizing in such a study, since it is generally thin (100–200 ft) and in areas is both silty and salty.

Utilizing this shale in the manner published by Marie (1975) and Glennie and Boegner (1981), it is suggested that the maximum depth of burial of the Singleton Mudstone in Well X and Well Y is approximately 7400 and 7900 feet respectively. Given the current depths of the Singleton Mudstone in the wells of 1900 and 2800 ft this implies an uplift of 5500 and 5100 ft.

Integration: the burial history model

Data have been generated from a number of techniques which may serve to constrain the burial history models applicable to this area. It is important to stress that no single data source

supplies data of sufficient quality or quantity alone to constrain the range of possible solutions. Furthermore, it must be remembered that basin modelling does not provide a unique solution. Using the constraints discussed previously a range of possible solutions can be proposed.

Using a time scale based upon Harland *et al.* (1989) and modified by Warrington *et al.* (1980) and Smith *et al.* (1974), the stratigraphic succession and other data were modelled using the Basin Mod software package.

It was impossible to obtain any correlation between output from the model and measured parameters/constraints if the 'Cimmerian' phase of inversion was of a greater magnitude than the late Tertiary. Consequently further modelling attempted to define the range of solutions accurately which would satisfy the geological constraints whilst keeping Tertiary inversion dominant over that produced during the 'Cimmerian'. Overall, the models were insensitive to any but extreme variations in the magnitude of the Cimmerian event. As a consequence, the results do not equivocally prove the need for the Cimmerian inversion. The reality of such an event is still based upon the observations made by Bushell (1986) and in subsequent proprietary studies.

Figure 11 shows the preferred burial history model for Well X. In this model, sedimentation from the Permian to Late Triassic is that encountered in the well. Sedimentation is thought to have continued until late Jurassic times, after which a small amount of inversion probably occurred during the Cimmerian event. The amount of uplift produced by this Cimmerian event was 500 feet and is notable for occurring much later than that modelled by Bushell (1986) for the South Morecambe gasfield. Subsidence again recommenced with approximately 200 ft of sediments being deposited during late Jurassic to late Cretaceous times.

The Tertiary was dominated by a major inversion episode commencing at 65 Ma with an initial rapid phase of inversion followed by a second much slower uplift as documented previously by Lewis *et al.* (in press).

The thermal history indicated in Fig. 12 utilizes the transient heat flow option within Basin Mod. The present-day temperatures within the well, when integrated with the thermal

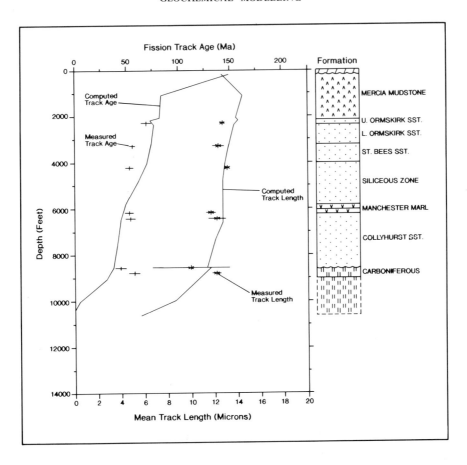

Fig. 13. Computed apatite fission track analysis data (track age and track length) plotted against real data 'measured/computed' from Well X, the burial history model again being that shown in Fig. 12.

properties of the sediments penetrated, indicate a heatflow of 1.35 Hfu with a surface temperature of 6.5°C (Fig. 11). Based upon a consideration of the palaeolatitudes of the EISB through geological time, the surface temperature was modelled as reaching a maximum of 18°C in the Carboniferous. This variation in surface temperature appears to have only a minor effect upon the computed results of the model. Two more contentious thermal events are included within the model. Both of these immediately post-date the inversion phases. The most severe of these occurs following early Tertiary inversion which in the East Irish Sea was accompanied by the emplacement of the Fleetwood dyke swarm. The event is modelled as a short duration pulse during which the heatflow was elevated to 1.5 Hfu. The second thermal event occurs at about 130 Ma and is a second short duration (2 Ma) thermal pulse. One of these two events is required to obtain an accurate match between the measured data and calculated output from the model. In particular the thermal pulses are required to explain the high fluid inclusion temperatures and also the annealed apatite fission tracks. The accuracy of computed versus calculated maturity and apatite fission track data generated by such a model is illustrated in Figs 12 and 13.

In order to generate an accurate match both to the data obtained from the reservoir (AFTA, chronodiagenesis) and the source rock (vitrinite reflectance), the effects of the thermal pulses must be confined to the more porous Sherwood and Collyhurst Sandstones. The mechanism thought likely to result in such a stratigraphically confined heating pulse involves either the release of hot brines from depths greater than those penetrated by the wells, or heating within the aquifers associated with the Tertiary volcanic activity. In the first scenario, the release of hot brines is more likely when the basin experi-

enced major changes in the overall stress regime, i.e. during Cimmerian or Tertiary inversion.

The effect of excluding either of the thermal events is that deeper burial is required to enable an accurate fit to the fission track data. However, in the case of Well X this results in overestimating the maturity. Similarly restricting the burial to reproduce the maturity does not result in a good fit between computed and measured fission track data.

The amount of uplift predicted for the Singleton Mudstone in the thermal pulse model is 2500 ft. Obviously this differs significantly from that predicted by the shale velocities (5500 ft). This discrepancy is due to the difficulties in estimating a 'true' uninverted velocity depth trend. This in turn is a function of lateral lithological variation within the formation. Consideration of Fig. 3, however, indicates approximately 2000 ft of inversion at Well X. Given the errors inherent in using this technique, this is considered a reasonable fit.

While Wells X and Y both contain oil, the South and North Morecambe fields contain sweet gas. Although the Ormskirk and St Bees Sandstones in South Morecambe contain a dead oil stain, there is a clear difference between the hydrocarbons present in the different structures within the basin. Furthermore, sour gas occurs in an increasing number of accumulations. This hydrocarbon type was first encountered in Well 110/7a-3. The reasons for these differences lie in the different burial histories of the various structural elements present in the EISB. In areas with more than 10 000 ft of Mesozoic burial, gas is the main type of hydrocarbon found, while in areas with less Mesozoic burial, oils become increasingly common. It must be stressed, however, that the occurrence of hydrocarbons is not solely a function of Mesozoic burial but, in the case of sour gas, depends also on the effects of the thermal flush event, as outlined below.

H₂S prediction

As stated previously, the East Irish Sea Basin contains significant occurrences of sour gas. The volumes of H_2S within these accumulations is extremely variable. Well 110/7a-3 was the first exploration well to encounter sour gas, with H_2S levels in excess of 4500 ppm being measured during testing. Sulphur isotope data show that the extremely low levels of H_2S in South Morecambe (1–3 ppm) and the highly sour accumulations such as 110/7a-3 have a different origin. This variability is attributed to a combination of the basinwide variation in burial and thermal history, combined with the extent of the area affected by the hot-flush heating event.

Although the presence of sour gas so close to sweet accumulations appears to be enigmatic, the hot flush basin model offers an explanation. Machel (1987) has shown that sulphur reduction to form hydrogen sulphide can occur in two ways in the subsurface: biogenic sulphate reduction (BSR) and thermal sulphate reduction (TSR). Machel suggests that TSR occurs at geologically meaningful rates at temperatures of 100–140°C whilst BSR occurs at temperatures below 92°C.

Application of the hot flush model combined with Machel's H_2S models suggests that any accumulation in the EISB that originally contained a moderate to high sulphur crude component, including wells X and Y, and was subjected to the appropriate temperatures (100–140°C) will have undergone TSR. This will produce a light crude oil or gas with an 'altered' chemistry, bitumen, HCO_3 and hydrogen sulphide. Therefore, in order to predict the type of hydrocarbon that will be found at the present day, it is imperative that the hydrocarbon fill prior to the early Tertiary hot flush is known. Those accumulations containing sweet gas will remain sweet, whereas those containing oil will now contain a mixture of oil–gas and hydrogen sulphide with bitumen.

Conclusions

As proposed, the burial history of the East Irish Sea Basin is complex. Maximum depth burial was achieved in early Tertiary times (65 Ma). Subsequently, during the early Tertiary to Recent, the basin was inverted. The Tertiary inversion appears to have occurred in two stages, an early rapid phase of inversion of 15 Ma duration followed by a slower phase of 50 Ma duration.

The magnitude of Tertiary inversion varies markedly across the basin (1000 to 10 000 ft). Such a wide range in uplift within a relatively small geographical area is suggestive of a mechanism in which individual major tectonic elements separate areas with markedly different inversion histories, rather than a model which involves a 'blanket' or uniform regional inversion.

The existence of the Cimmerian inversion event (Bushell 1986) remains an enigma. Although a number of geological and petrographical relationships strongly imply that such an event occurred, the model presented here is relatively insensitive to its presence or absence, providing that the uplift is modest (less than 1000 ft).

The proposed model implies that much of the inversion will have been taken up along discrete major fault planes; however, at present no reverse fault relationships have been identified along such faults.

The hydrocarbons, both oil and gas, were sourced largely from shales equivalent to the Sabden Shale of Namurian (E_2–R_1) age. Rare Westphalian coals may have provided a poor secondary local gas source in some areas.

The present-day distribution of hydrocarbon types in the basin, i.e. sweet gas in the centre/sour gas and oil on the margins, is thought to be controlled by the maturity and timing of generation in the various source kitchens, combined with the effects of alteration associated with the early Tertiary inversion and heating event.

The Tertiary heating event is an essential component of the model. As discussed earlier, numerically the model can be made to work with heating events after either the Cimmerian or Tertiary inversion episodes. However, general geological considerations favour the early Tertiary. The heating event relates to the passage of hot-brines through the aquifers of the Sherwood and Collyhurst Sandstones. It is responsible for the formation of the 'high' T_{hom} fluid inclusions and annealed fission tracks. Furthermore, it offers a mechanism, if somewhat controversial, for the generation of sour gas and abundant bitumen within the East Irish Sea Basin.

The authors thank the management of British Gas plc Exploration and Production for allowing publication of this paper. We also acknowledge the contributions made by numerous geoscientists within British Gas to the understanding of the East Irish Sea Basin. AFTA work was carried out by Geotrack and chronodiagenetic work by Gaps. Thanks also to Bob Dunay for his help, advice and patience.

References

BUSHELL, T. P. 1986. Reservoir geology of the Morecambe Field. *In*: BROOKS, J., GEOFF, J. C. and VAN HOORN, B. (eds) *Habitat of Palaeozoic Gas in NW Europe*. Geological Society, London, Special Publication, **23**, 189–208.

COLTER, V. S. 1978. Exploration for gas in the Irish Sea. *Geologie en Mijnbouw*, **57**, 503–516.

—— and BARR, K. W. 1975. Recent developments in the geology of the Irish Sea and Cheshire Basins. *In*: WOODLAND, A. W. *Petroleum and the Continental Shelf of North West Europe*, Volume 1, Applied Science, London, 61–75.

DOW, W. G. 1977. Kerogen studies and geologic interpretations. *Journal of Geochemical Exploration*, **7**, 79–99.

EBBERN, J. 1981. The geology of the Morecambe Gas Field. *In*: ILLING, L. V. AND HOBSON, G. D. (eds) *Petroleum Geology of the Continental Shelf of North West Europe*. Heyden, London 485–493.

FRASER, A. J. AND GAWTHORPE, R. L. 1990. *In*: HARDMAN, R. F. P. AND BROOKS, J. *Tectonic Events Responsible for Britain's Oil and Gas Reserves*. Geological Society, London, Special Publication, **55**, 49–86.

GILES, M. R. AND DE BOER, R. B. 1989. Secondary porosity: creation of enhanced porosities in the subsurface from the dissolution of carbonate cements as a result of cooking formation waters. *Marine and Petroleum Geology*, **6**, 261–269.

GLENNIE, K. W. AND BOEGNER, P. L. E. 1981. Sole Pit inversion tectonics. *In*: ILLING, L. V. AND HOBSON, G. D. (eds) *Petroleum Geology of the Continental Shelf of North-West Europe*. Heyden, London, 110–120.

GREEN, P. F. 1986. On the thermo-tectonic evolution of Northern England: evidence from fission track analysis. *Geological Magazine*, **123**, 493–506.

—— 1989. Thermal and tectonic history of the East Midlands shelf (onshore UK) and surrounding regions assessed by apatite fission track analysis, *Journal of the Geological Society*, **146**, 755–773.

HAMILTON, P. J., FALLICK, A. E., MACINTYRE, R. M. AND ELLIOTT, S. 1987. Isotopic tracing of the provenance and diagenesis of Lower Brent Group sands, North Sea. *In*: BROOKS, J. AND GLENNIE, K. (eds) *Petroleum Geology of North West Europe*. Graham & Trotman, London, 939–949.

HANOR, J. S. 1980. Dissolved methane in sedimentary brines; potential effect on the PVT properties of fluid inclusions. *Economic Geology*, **75**, 603–617.

HARLAND, W. B., ARMSTRONG, R. L., CRAIG, L. E., SMITH, A. G. AND SMITH, D. G. 1990. *A Geologic Time Scale*, Cambridge University Press.

JACKSON, D. I., MULLHOLLAND, P., JONES, S. M. AND WARRINGTON, G. 1987. The geological framework of the East Irish Sea Basin. *In*: BROOKS, J. and GLENNIE, K. (eds) *Petroleum Geology of North West Europe*. Graham & Trotman, London, 191–203.

KNIPE, R., COWAN, G. AND BALENDRAN, V. S. 1993. The tectonic history of the East Irish Sea Basin with reference to the More-cambe Fields. *In*: PARKER, J. R. (ed.) *Petroleum Geology of Northwest Europe: Proceedings of the 4th Conference*. Geological Society, London, 857–866.

LAWRENCE, S. R., COSTER, P. W. AND IRELAND, R. J. (1987). Structural development and petroleum potential of the northern flanks of the Bowland Basin (Carboniferous), North-West England. *In*: BROOKS, J. AND GLENNIE, K. (eds) *Petroleum Geology of North West Europe*, Graham & Trotman, London, 225–233.

LEE, M., ARONSON, J. L. AND SAVIN, S. 1989. Timing and conditions of Permian Rotliegende Sandstone diagenesis, southern North Sea: K/Ar and oxygen isotopic data. *American Association of Petroleum Geologists Bulletin*, **73**, 195–215.

LEWIS, C. L. E., GREEN, P. F., CARTER, A. AND HURFORD, A. J. in press. Elevated K/T palaeotemperatures throughout Northern England: three kilometres of Tertiary erosion? *Earth and Planetary Science Letters*.

MACHEL, H. G. 1987. Some aspects of diagenetic sulphate–hydrocarbon redox reactions. *In*: MARSHALL, J. D. (ed.) *Diagenesis of Sedimentary Sequences*. Geological Society, London, Special Publication, **36**, 15–28.

MARIE, J. P. P. 1975. Rotliegendes stratigraphy and diagenesis. *In*: WOODLAND, A. W. (ed) *Petroleum and the Continental Shelf of North-West Europe*. Applied Science, London, 205–211.

MEADOWS, N. S. AND BEACH, A. 1993. Structural and climatic controls on facies distribution in a mixed fluvial and aeolian reservoir: the Triassic Sherwood Sandstone in the Irish Sea. *In*: NORTH, C. AND PROSSER, J. (eds) *Characterization of Fluvial and Aeolian Reservoirs*. Geological Society, London, Special Publication, **73**, in press.

POTTER, R. W. II 1977. Pressure corrections for fluid-inclusions homogenization temperatures based on the volumetric properties of the system NaCl–H_2O. *United States Geological Survey Journal of Research*, **5**, 333–336.

ROBERTS, D. G. 1989. Basin inversion in and around the British Isles. *In*: COOPER, M. A. AND WILLIAMS, G. D. (eds) *Inversion Tectonics*. Geological Society, London, Special Publication **44**, 131–150.

SCHOELL, M. 1980. The hydrogen and Carbon isotopic composition of methane from natural gases of various origins. *Geochimica et Cosmochimica Acta*, **44**, 649–661.

SHEPHERD, T., RANKIN, A. H. and ALDERTON, D. H. M. 1985. *A Practical Guide to Fluid Inclusion Studies*. Blackie, Glasgow.

SMITH, D. B., BRUNSTROM, R. G. W., MANNING, P. I., SIMPSON, S. AND SHOTTON, F. W. 1974. *A Correlation of Permian rocks in the British Isles*. Geological Society, London, Special Report, **5**.

STUART, I. A. and COWAN, G. 1991. Morecambe Gas Field, Blocks 110/2a, 110/3a, 110/8a, UK. *In*: ABBOTTS, I. L. (ed.) *United Kingdom Oil and Gas Fields 25 Years Commemorative Volume*. Geological Society, Memoir, **14**, London, 527–541.

WARRINGTON, G., AUDLEY-CHARLES, M. G., ELLIOT, R. E., EVANS, W. B., IVIMEY-COOK, H. C., KENT, P. E., ROBINSON, P. C., SHOTTON, F. W. AND TAYLOR, F. M. 1980. *A Correlation of Triassic Rocks in the British Isles*. Geological Society, London, Special Report, **13**.

WILSON, A. A. 1990. The Mercia Mudstone Group (Trias) of the East Irish Sea Basin. *Proceedings of the Yorkshire Geological Society*, **48**, Part 1, 1–22.

ZIEGLER, P. A. 1987. Manx-Furness Basin. *Tectonophysics*, **137**, 335–340.

Discussion

Question (K. Dale, Marathon International Petroleum):

I thank the authors and British Gas for sharing this information with the industry. I noted with some interest the views expressed on the origin of hydrogen sulphide gases within the basin. Stuart & Cowan's (1991) paper on the Morecambe Field stated that H_2S concentrations were minimal. As it is an important safety consideration, may I urge the authors to share their current understanding of the magnitude of any H_2S risks to exploratory drilling in the basin with readers of this conference volume.

Answer (M. Hardman):

As Stuart & Cowan (1991) stated, hydrogen sulphide concentrations in the South Morecambe Field are minimal (0–6 ppm). Elsewhere in the basin, concentrations of hydrogen sulphide are much larger—for instance, Well 110/7a-4 encountered 4500 ppm H_2S while testing the Ormskirk Sandstone. H_2S concentrations greater than 4500 ppm have also been encountered.

Currently, these extremely 'sour' accumulations are restricted to the south of the East Irish Sea. This distribution may well reflect the relative density of exploratory drilling within the basin.

Obviously such high concentrations of toxic gas pose a serious safety threat, both for exploratory and any subsequent development/production activities.

Controls on reservoir quality in the Triassic Sherwood Sandstone of the Irish Sea

N. S. MEADOWS[1] AND A. BEACH[2]

[1]Geochem Group Limited, Chester Street, Chester CH4 8RD, UK
[2]Alastair Beach Associates, 11 Royal Exchange Square, Glasgow G1 3AJ, Scotland, UK

Abstract: The Lower Triassic Sherwood Sandstone Group constitutes the principal reservoir target in the East Irish Sea Basin. Interpretation of sedimentary facies allows recognition of major low sinuosity braided fluvial channel, minor ephemeral fluvial channel, aeolian dune and sandsheet, unconfined fluvial sheetflood and playa lake deposits that are discussed in terms of facies associations. Reservoir quality is highly variable, with porosity and permeability data revealing significant differences between fluvial, sheetflood and aeolian facies associations. The latter, in particular, preserve very high reservoir quality reflecting an open grain fabric and paucity of blocky cements. Variations in porosity between the facies associations relate primarily to the distribution of quartz cement which is significantly more abundant in fluvial channel sandstones than other facies. Authigenic illite, while severely limiting permeabilities in some reservoir zones, is not facies specific and does not contribute to the observed variations. It is suggested that the disparity in the abundance of quartz cement between facies associations relates to two distinct grain types that correlate broadly with fluvial and aeolian associations, respectively, and probably derive from different provenances. The evolution of the basin, in controlling the distribution of the major facies, influenced the distribution of the two grain populations and, thus, also the reservoir quality of the sandstones.

The East Irish Sea Basin (EISB) is a post-Variscan structure lying mainly offshore northwest England (Fig. 1), with a sedimentary fill comprising Permian, Triassic and, possibly, Lower Jurassic rocks. The principal reservoir sequence within these sediments is formed by the Lower Triassic Sherwood Sandstone Group (SSG) and in particular the Ormskirk Sandstone Formation (OSF), which is the uppermost formation of the group, and the upper parts of the St Bees Sandstone Formation (SBSF). Directly overlying the Sherwood Sandstone, and forming the seal to the reservoir interval, are the claystones and halites of the Mercia Mudstone Group (Fig. 2).

The Permo-Triassic basins of the Irish Sea area, as defined by Jackson *et al.* (1987) whose structural and stratigraphical terminology is used herein, are part of a series of such basins along the western side of the UK. The formation of these basins has been related to an early phase of Atlantic opening (Ziegler 1982), while their early Triassic (Sherwood Sandstone) fill has been ascribed to the action of a major northward-flowing river system that carried material from the Variscan massifs of southwest England and northern France (Fitch *et al.* 1966; Audley-Charles 1970). The deposits of major fluvial channels are not, however, the only sedimentary facies present within the SSG of the EISB. Within the upper part of the group (upper SBSF and OSF), sediments that can be ascribed to ephemeral fluvial channel, aeolian and sheetflood processes are also recorded, as are sequences representing deposition within standing bodies of water including both channel abandonment and playa lakes. The characteristics and relationships of these facies are described more fully below.

The EISB can be divided into two distinct structural provinces comprising a southern area dominated by N–S-trending faults and a northern area bounded by approximately orthogonal sets of NE–SW- and NNW–SSE-trending extensional faults. These latter include the Keys, Sigurd and Lagman faults to the west and the Deemster, Crosh Vusta, Haverigg and Bootle faults to the east (Fig. 1). Seismic data across this northern province reveal that it comprises a series of rotated half-graben and that the SSG sedimentary sequences can be seen to thicken into the hangingwalls (Meadows and Beach, 1993). In particular, maximum growth is recorded into the Keys, Sigurd and Lagman faults and also into the Formby Point Fault in the southeast.

The northern province includes the North and South Morecambe Fields from which many of the data used in this paper

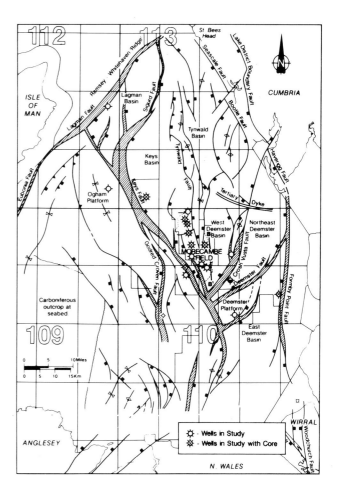

Fig. 1. Generalized structural configuration of the East Irish Sea Basin and surrounding areas at top Sherwood Sandstone level.

have been derived. Specific aspects of the diagenetic history of the reservoir sandstones have been discussed by Macchi (1987), Woodward and Curtis (1987) and Macchi *et al.* (1990). These authors concerned themselves primarily with the morphology, chemistry and distribution of the illite cements that form an important influence on reservoir properties within the Morecambe fields. Although the presence of other cement phases

From *Petroleum Geology of Northwest Europe: Proceedings of the 4th Conference* (edited by J. R. Parker).
© 1993 Petroleum Geology '86 Ltd. Published by The Geological Society, London, pp. 823–833.

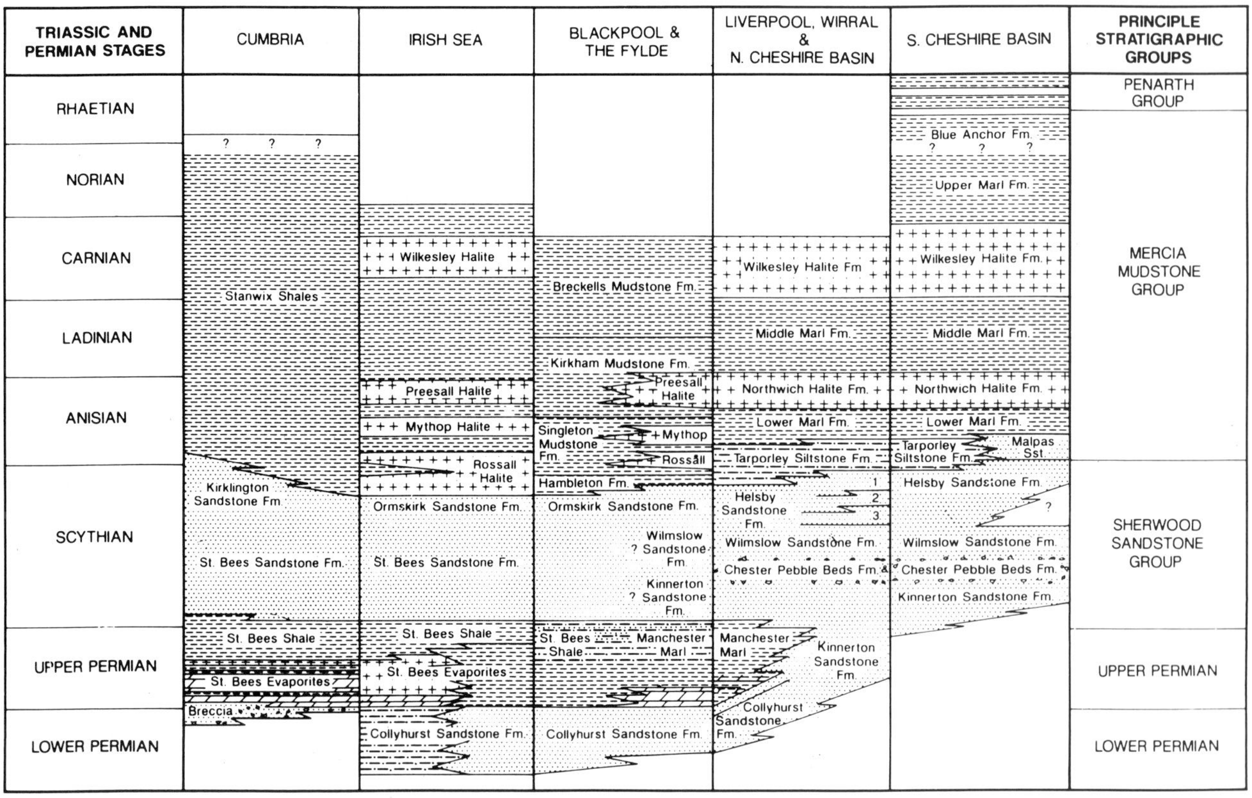

TRIASSIC AND PERMIAN STAGES	CUMBRIA	IRISH SEA	BLACKPOOL & THE FYLDE	LIVERPOOL, WIRRAL & N. CHESHIRE BASIN	S. CHESHIRE BASIN	PRINCIPLE STRATIGRAPHIC GROUPS
RHAETIAN						PENARTH GROUP
NORIAN	? ? ?				Blue Anchor Fm. ? ? ? Upper Marl Fm.	
CARNIAN	Stanwix Shales	+ + Wilkesley Halite + +	Breckells Mudstone Fm.	+ + Wilkesley Halite Fm. +	+ + Wilkesley Halite Fm. + +	MERCIA MUDSTONE GROUP
LADINIAN			Kirkham Mudstone Fm.	Middle Marl Fm.	Middle Marl Fm.	
ANISIAN		+ + Preesall Halite + + + + + Mythop Halite + + +	Preesall Halite Singleton Mudstone Fm. Mythop Rossall	+ Northwich Halite Fm. + + Lower Marl Fm.	+ Northwich Halite Fm. + + Lower Marl Fm.	
SCYTHIAN	Kirklington Sandstone Fm. St. Bees Sandstone Fm.	+ + + Rossall Halite + + Ormskirk Sandstone Fm. St. Bees Sandstone Fm.	Hambleton Fm. Ormskirk Sandstone Fm. Wilmslow ? Sandstone Fm. Kinnerton ? Sandstone Fm.	Tarporley Siltstone Fm. Helsby Sandstone Fm. Wilmslow Sandstone Fm. Chester Pebble Beds Fm.	Tarporley Siltstone Fm. Malpas Sst. Helsby Sandstone Fm. Wilmslow Sandstone Fm. Chester Pebble Beds Fm. Kinnerton Sandstone Fm.	SHERWOOD SANDSTONE GROUP
UPPER PERMIAN	St. Bees Shale St. Bees Evaporites Breccia	St. Bees Shale + + St. Bees Evaporites +	St. Bees Shale Manchester Marl	Manchester Marl Kinnerton Sandstone Fm. Collyhurst Sandstone Fm.		UPPER PERMIAN
LOWER PERMIAN		Collyhurst Sandstone Fm.	Collyhurst Sandstone Fm.			LOWER PERMIAN

Fig. 2. Correlation of Permian and Triassic rocks in northwest England and the Irish Sea. Subdivision of the Helsby Sandstone Formation in N. Cheshire Basin is 1: Frodsham Member; 2: Delamere Member; 3: Thurstaston Member (reproduced with amendments from Smith *et al.* 1974; Warrington *et al.* 1980; Jackson *et al.* 1987).

has been mentioned by previous authors (notably Bushell (1986) and Stuart and Cowan (1991)), the relationship of these phases to sedimentary facies and their effects on reservoir quality, which forms the principal concern of this paper, have not been fully described.

Facies associations

Facies analysis of cores taken from the OSF and the upper part of the SBSF (see Table 1) has revealed the presence of seven facies types that are discussed below in terms of four facies associations representing fluvial channel, sheetflood, aeolian and playa lake deposits. The overall facies assemblage indicates deposition in a semi-arid, continental environment.

Table 1. Facies analysis of cores from the OSF and SBSF

Study wells	Core (m)	Stratigraphic interval
110/2-1	44.8	26.9 m OSF; 17.9 m SBSF
110/2-2	28.5	OSF
110/2-3	110.1	92.0 m OSF; 18.1 m SBSF
110/2-5	44.8	36.2 m OSF; 8.6 m SBSF
110/2-6	253.8	243.4 m OSF; 19.4 m SBSF
110/3-1	15.6	9.0 m OSF; 6.6 m SBSF
110/3-3	262.1	89.9 m OSF; 172.2 m SBSF
110/7-1	12.4	SBSF
110/9-1	4.1	OSF
113/26-1	61.9	OSF

Fluvial channel facies association

Sediments ascribed to deposition within fluvial channels can be subdivided on the basis of internal characteristics and their relationship with other facies, into those representing major, probably perennially-flowing rivers and those representing minor, probably ephemeral channels. It is recognized, however, that these form end-members of a continuum of fluvial channel types and, while the distinction is useful in developing a depositional model, sequences suggestive of intermediate forms also occur in the OSF.

Major fluvial channel sandstones (F1): Deposits characterizing major fluvial channels typically comprise stacked erosively-based and cross-stratified sets of moderately to poorly sorted medium and fine sandstone. These form multi-storey sandstone bodies up to in excess of 20 m in thickness (illustrated by intervals 1038 m to 1059 m and 1182 m to 1204 m in Figs 3a and 3b). The sandstones are dominated by angular to subrounded grains with a moderate to low sphericity (see Figs 4a and 4b) and commonly contain abundant claystone and siltstone intraclasts forming either basal channel lags, or aligned along foreset surfaces. Concomitant with the alignment of intraclasts, a grading of foreset laminae is recorded along with, less commonly, the presence of intrasets comprising small-scale downclimbing sets within the foresets of larger bedforms. Typically, however, individual tabular and trough cross-stratified sets, representing straight and sinuous-crested transverse bar forms respectively, are stacked to form compound channel-filling sandstone bodies. The presence of graded and downclimbing sets suggests the development of more complex, compound, oblique or side-attached bars recording subtle variations in flow competence and the migration of smaller-scale (dune to megaripple) structures across major bedforms, possibly under intermediate- to low-flow stage conditions. Channel-fill sequences commonly, but not exclusively, fine upwards in grain size and exhibit an upward decrease in preserved set thickness that is suggested to represent a decrease

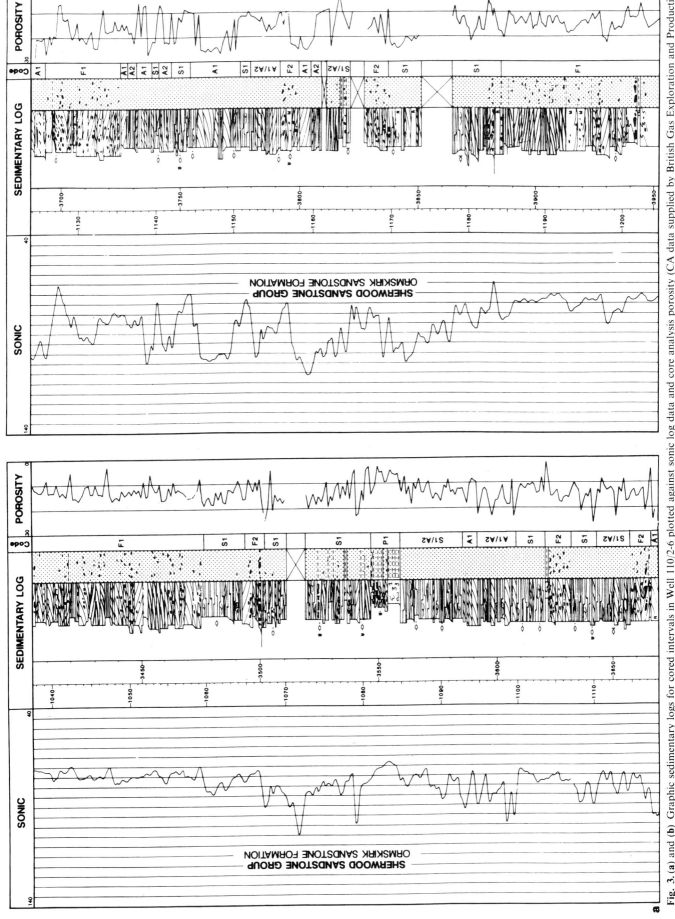

Fig. 3. (a) and (b) Graphic sedimentary logs for cored intervals in Well 110/2-6 plotted against sonic log data and core analysis porosity (CA data supplied by British Gas Exploration and Production).

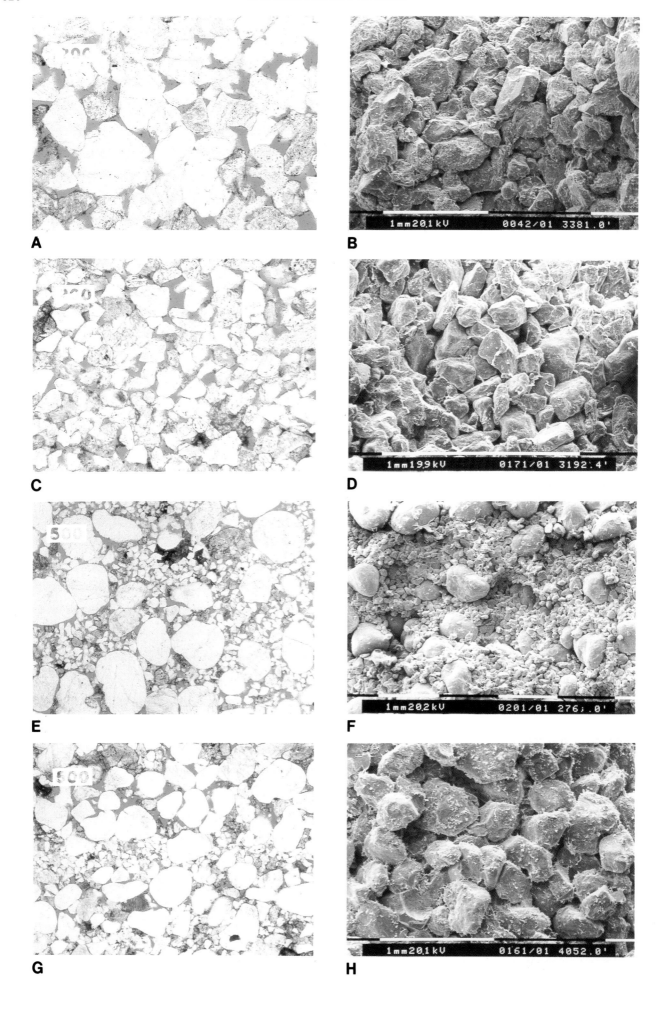

A

B

C

D

E

F

G

H

in current velocities. Locally these sequences pass upwards into abandonment fines comprising finely laminated claystones, siltstones and fine current-rippled sandstones. In such cases the preserved thickness of the channel filling sequence can be used as an approximation for channel depth (Cant 1978; Larue and Martinez 1989). Estimates from thirty such examples indicate a mean channel depth of approximately two metres for the OSF and SBSF rivers.

Interbedded with these sequences there are also intervals of flat laminated or very low angle cross-stratified sandstones preserving primary current lineation. The intimate relationship of these with the more common cross-stratified sandstones would tend to preclude their representing a differing channel type and they are, hence, interpreted as upper-phase plane bed deposits representing high stage bar-top modification or low stage braided channels, where they are subject to flow divergence around the major bedforms. These sequences are, therefore, interpreted as the deposits of perennially-flowing rivers, albeit with channels subject to significant episodic variations in flow regime. Although no perfect modern analogue has been documented for the fluvial system envisaged, several that have been described contain sufficiently similar depositional features to support the interpretation. Notable among these are the South Saskatchewan River in western Canada (Cant 1978) and parts of the Squamish River in British Columbia (Brierley 1989).

Minor fluvial channel sandstones (F2): Sandstones interpreted as having been deposited in relatively minor, possibly ephemeral, fluvial channels exhibit many of the sedimentological characteristics of the major channel sandstones described above. There are, however, significant differences indicative of ephemerality. Principal among these are the preserved thickness of individual and component bar forms (generally less than 0.5 m and only locally more than 1 m), the absence of intrasets and graded foresets, the range of grain types and the interdigitation of these channel deposits with sheetflood, playa and aeolian facies (illustrated by intervals 1064 m to 1067 m and 1156 m to 1158 m in Figs 3a and 3b).

The interdigitation of facies and absence of thick multistorey sandstone bodies in the sequences comprising this facies type suggest, in particular, that these channels were not long-lived morphological features. The grain types recorded comprise a variable mixture of angular to sub-rounded grains similar to those forming the major fluvial channel sandstones, and well-rounded, high sphericity 'aeolian' grains of medium to occasionally coarse sand grade that suggest the fluvial reworking of aeolian sands. These latter grains commonly exhibit frosted surfaces typical of aeolian sands and are, therefore, interpreted as flood-reworked aeolian dune sands.

Aeolian facies association

Sandstones of aeolian origin form a significant component of the SSG and particularly of the OSF. Although volumetrically subordinate to the fluvially deposited sediments their consistently high reservoir quality renders them extremely important in any reservoir evaluation, whether they occur as true aeolian bedforms or as a reworked component within other facies. Aeolian deposits (*sensu stricto*) occur primarily either as dunes or sandsheets.

Aeolian dune sandstones (A1): Sandstones interpreted as representing aeolian dune forms typically exhibit moderate- to high-angle planar to asymptotically-based cross-stratification within sets bounded by planar or undulatory surfaces (e.g. 1149 m in Fig. 3b). The cross-stratification displays a finely developed bimodal grain-size differentiation equivalent to the grainfall lamination of Hunter (1977). Component grains range from fine to coarse sand grade but are most commonly an interlaminated combination of fine to medium grains and medium to coarse grains (see Figs 4g and 4h). The vast majority of these grains are well rounded with high sphericities. Instances of aeolian dune forms comprising significant proportions of more angular low sphericity grains similar to those forming the principal constituent of the major fluvial channel sandstones are, however, locally recorded. Aeolian dune sandstones occur interbedded with other facies, principally aeolian sandsheet and sheetflood deposits, and rarely attain more than a metre in preserved thickness. Where thicker dune intervals do occur (e.g. composite interval from 1145 m to 1151 m in Fig. 3b) they typically comprise a number of sets separated by planar bounding surfaces. These represent either deflation surfaces or the modified stoss slope of the underlying dune where a climbing geometry has developed, although the maximum recorded thickness is less than 3 m.

Aeolian sandsheet sandstones (A2): Very finely flat laminated to very low angle cross-stratified sandstones with laminae-specific bimodal sorting (see Figs 4e and 4f) are interpreted as aeolian sandsheet deposits. They are recognized interbedded with other facies throughout the OSF on a variety of scales. These range from millimetre to centimetre alternations of aeolian laminae and silty sheetflood laminae, to more substantial intervals where aeolian sandsheets are developed up to over a metre in thickness and are generally interbedded with aeolian dune sandstones (e.g. 1144 m to 1155 m in Fig. 3b). These latter sequences commonly exhibit very low angle discordances suggesting the development of low-relief bedforms consistent with analogues from modern aeolian sandsheet environments (e.g. Fryberger *et al.* 1979, 1983). The very fine-scale alternations of wet and dry facies have been suggested to represent subtle variations in the palaeo-water table, while the more substantial aeolian intervals represent significant episodes of lowered water table (Meadows and Beach 1993). The grain types recorded in the aeolian sandsheet intervals are essentially the same as those comprising aeolian dunes.

Sheetflood facies associations

Sediments ascribed to deposition by unconfined fluvial processes form a major component of the OSF although they only rarely form thick single facies sequences and most commonly occur interbedded with other facies (illustrated by the interval 1060 m to 1082 m in Fig. 3a).

Sheetflood deposits (S1): Sheetflood deposits vary from highly porous, generally moderately sorted, flood-reworked aeolian sandstones (e.g. 1172 m to 1173 m in Fig. 3b) to low porosity, poorly sorted finely and irregularly laminated silty and argillaceous fine sandstones (e.g. 1077 m to 1079 m in Fig. 3a). Grain types are also variable, ranging from the typical well-rounded aeolian types to angular low-sphericity types (see Figs 4c and

Fig. 4. Thin section and SEM photomicrographs of sandstones from the Ormskirk Sandstone Formation. (**a**) and (**b**): fluvial channel sandstones comprising angular to sub-rounded grains cemented by quartz overgrowths and preserving moderate porosity but with pore throats commonly constricted; (**c**) and (**d**): sheetflood sandstones comprising mainly sub-angular 'fluvial' grains and occasional well-rounded 'aeolian' grains; (**e**) and (**f**): aeolian sandsheet sandstones, markedly bimodally sorted and comprising well-rounded 'aeolian' grains, preserving moderate to high porosity with minimal quartz cement; (**g**) and (**h**): aeolian dune sandstones, bimodally sorted and comprising well-rounded 'aeolian' grains with minimal cement and an open pore network in coarse-grained laminae; SEM photomicrograph shows aeolian sandstone with moderate illite cement but still retaining pore connectivity. (Thin section photomicrograph scale bars in microns.)

4d). Primary depositional structures are often obscured by soft sediment deformation and bioturbation, but where preserved, include plane bed lamination, current rippling and, locally, wave ripples with form concordant siltstone or micaceous drapes. Subordinate interbeds of cross-stratified sandstone also occur, overlying either planar or slightly scoured erosional surfaces.

The range of structures present suggests deposition from recurrent, rapid bedload sedimentation events that were followed, in some instances, by slack water fall-out of suspension fines. There are, however, also thin desiccated claystone laminae present that provide evidence for the occasional drying out of the sediment surface. A crevasse origin related to flooding of contemporaneously active channels probably accounts for a proportion of the sheetflood deposits although the episodic nature of overbanking processes is difficult to reconcile with the presence of some thick intervals (e.g. >10 m in Well 110/2-6) of these sediments. The irregular bedding that typifies these deposits, resulting mainly from soft sediment deformation, indicates a fairly constant state of sub-surface saturation and is taken to suggest a sediment surface that was mostly coincident with, or near, the palaeowater table. Given these assumptions, it seems likely that, at certain times, areas of the EISB during OSF depositional time formed low relief sandflats subject to flooding and reworking of surficial sediments.

Playa facies associations

Claystone- and siltstone-dominated sequences representing deposition within, or associated with, standing bodies of water occur at several horizons in the OSF. These sediments form a volumetrically small but significant component in that they comprise the only primary non-reservoir units within the sequence.

Playa lake deposits (P1): Blocky claystones and siltstones with thin fine sandstone laminae and thicker beds of cross-stratified sandstone that occur interbedded with sheetflood and aeolian sandsheet facies are interpreted as having been deposited in a playa lake setting (e.g. 1082 m to 1084 m in Fig. 3a). Although similar in character to fluvial abandonment deposits, they are distinguished on the basis of their close association with other interchannel facies. Some of the sandstone beds show wave modification in the form of symmetrical convex-upward ripples and internal, festoon-shaped laminae with both convex- and concave-upward orientations. These internal laminae are commonly form discordant and, apart from thin veneers concordant with the crests, do not usually correspond with the upper ripple surface morphology. These structures indicate an oscillatory mode of formation and are consistent with those of wind-driven wave ripples (de Raaf *et al.* 1977) indicative of very shallow water depths. Additionally, the common occurrence of desiccation features, in the form of sand-filled subvertical fractures up to 10 cm deep, attest to the ephemerality of the lacustrine environment. The presence of evaporite in these sequences has been recorded, mainly as halite pseudomorphs, but is rare and the majority of playas are assumed to have been non-evaporitic.

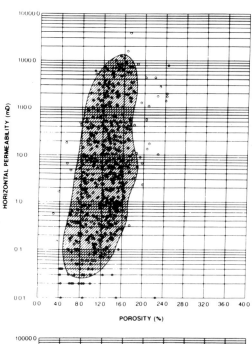

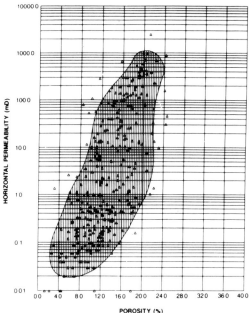

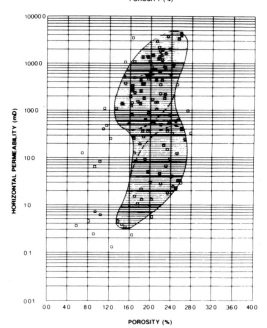

Fig. 5. Porosity and permeability cross-plots derived from conventional core analysis data for Wells 110/2-6 and 110/3-3. Plots relate to the major reservoir facies; fluvial (circles), sheetflood (triangles) and aeolian (squares). Cross-hatched areas represent 95% of each data set; sub-areas in aeolian facies cross-plot represent cross-stratified aeolian dune (higher range) and aeolian sandsheet plus dune toesets (lower range).

Playa margin deposits (P2): Sequences interpreted as representing a playa margin environment are seen essentially as transitional between the deposits of playa lakes and wet, sheetflood-dominated, sandflats. They exhibit the characteristics of both facies with an interstratification of playa deposits, representing lake expansion, and sheetflood, with localized aeolian sandsheet, sediments representing lake contraction and surficial drying-out in the marginal areas.

Reservoir quality

Porosity and permeability

Core analysis data from the two wells in the study with the greatest amount of core (Wells 110/2-6 and 110/3-3) have been plotted for the three major reservoir facies associations (Figs 5a–c) and against sedimentary facies (Figs 3a and 3b). These data reveal a substantial variation in reservoir quality.

The porosity/permeability cross-plot for the fluvial facies (Fig. 5a) reveals a considerable permeability range for a relatively small porosity range. The majority of these data points fall between 6–18% porosity, while permeability values range from less than 0.1 mD to in excess of 1000 mD. Some of the low permeability samples relate to argillaceous and/or micaceous channel abandonment deposits. However, a substantial proportion, especially those with low permeabilities but porosities in excess of 8% can be accounted for by the effects of illite cement. The morphology and distribution of this clay phase has been discussed in detail (Woodward and Curtis 1987; Macchi *et al.* 1990; Stuart and Cowan 1991) and related to diagenetic effects below a palaeogas–water contact in the reservoir. Sandstones above this inferred contact are essentially illite-free, while those below are variably, and in some cases very extensively, affected by illite that bridges pore throats and radically increases the tortuosity of the pore system, while preserving high microporosities. The distribution of the illite is, therefore, independent of facies and influences permeability values throughout the range of facies.

The cross-plot for the sheetflood facies (Fig. 5b) shows a similar range of permeability values to that of the fluvial sandstones, with low permeability data points reflecting the presence of illite and detrital matrix clays in finer-grained intervals. However, the range of porosity values for the sheetflood facies, from approximately 4–22%, reveals an improvement, at the upper end of the range, compared to the fluvial deposits. Similarly, the cross-plot for the aeolian facies (Fig. 5c) reveals a further, and more substantial, increase in porosity values with the majority of samples falling in the range 14–26%. Significantly, in this case, there is a concomitant increase in permeability values with few below 1 mD and an increase in the upper limit to over 3000 mD, although some aeolian sandstones are so friable that core analysis data becomes unreliable. Major variations in the quality of aeolian sandstones are a response to differences in overall sorting characteristics and the finely developed pinstripe lamination of dune toesets and aeolian sandsheets.

The detrimental effect of the illite cement is, thus, much less severe in the aeolian sandstones than it is in other facies, particularly the fluvial sandstones. It is suggested here that this difference is primarily a function of the mean porosity values of the facies in that small decreases in porosity, and particularly in pore throat diameters, increase significantly the chance of illite adopting a pore-bridging habit. Possible reasons for the variation in porosity values include the degree of compaction, extent of blocky cements, grain size and grain shape.

Compaction

The extent of compaction observed in petrographic samples varies from very light to moderate with local evidence of long grain contacts. In order to assess any variations in compaction between the facies associations, the minus cement porosity was calculated for 140 samples taken from all ten cored wells in the study (Fig. 6). These values range from around 12% to nearly 40% in all three facies associations. SEM analysis has revealed no indication of significant grain dissolution in any of the facies and, hence, it must be assumed that sandstones with a high minus cement porosity have undergone minimal compaction, possibly due to very early framework stabilizing cements. The minus cement porosity mode for both fluvial and sheetflood sandstones falls between 24–28%, whereas for aeolian sandstones it occurs in the range 20–24%. This is not, however, considered significant as the variation falls within the standard deviation of all three sets of data. From these calculations it is clear that compaction is not contributing significantly to the differences in porosity between the facies associations.

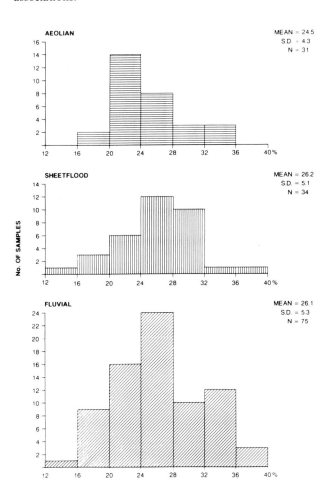

Fig. 6. Minus cement porosity data derived from 140 petrographic samples and plotted according to the major reservoir facies identified from core.

Cement distribution

The major cement phases in the Sherwood Sandstone are clay (mainly illite but also kaolinite and mixed layer illite-smectite), carbonate (ferroan and non-ferroan calcite and dolomite) and quartz. The distribution by facies association for the three groups of cements (clay, carbonate, quartz) is illustrated using normalized data in Fig. 7.

The histogram for total clay cements shows very little difference between the facies except for a slight concentration in sheetflood samples above the 12% clay level. This is probably a response to greater abundances of detrital clays in samples from the finer-grained sheetflood sequences. Similarly, the plot for total carbonate indicates little difference between the facies associations in the lower range of values. It appears, however, that tight carbonate cements (>12%) occur predominantly within sheetflood and fluvial sediments and only very locally in aeolian sandstones. This may relate to the distribution of early diagenetic carbonate which is recorded in some sheetflood sediments, possibly related to incipient calichification, and as reworked clasts within fluvial channels. In both instances this early cement could act as a nucleation site for later carbonate phases.

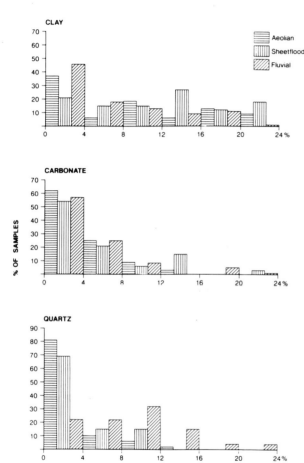

Fig. 7. Distribution of major cement phases according to facies derived from point count data on the same samples illustrated in Fig. 6.

The histogram for quartz cement shows that the majority of both aeolian and sheetflood samples contain less than 4% authigenic silica and that the number of samples with more than this amount tails off very rapidly, especially those from aeolian sandstones. By comparison, the mode for fluvial sandstones lies between 8–12% and a significant proportion of the samples record higher percentages of authigenic quartz. SEM analysis reveals the presence of well developed and commonly interlocking quartz overgrowths in many of the fluvial sandstone samples. In the sheetflood sandstones (and some facies F2 fluvial channel sandstones) it is noticeable that those comprising a high proportion of well-rounded 'aeolian' grains tend to contain the least quartz cement and, vice versa, those dominated by angular grains are generally more quartz cemented. It is of note that the sample mode for quartz cement in the fluvial facies (8–12%) corresponds closely with the difference in core analysis porosity values between aeolian and fluvial sandstones discussed above. It is possible, therefore,

that the distribution of quartz cement is responsible for the differences in reservoir quality between these facies.

Grain types and sorting

There appears to be no significant difference in grain size between the facies, the mean grain size of the sandstones from all facies falling within the range of fine to medium sand. There is a tendency for grains of larger sizes (medium to coarse and coarse sand) to occur more commonly within the aeolian sandstones, generally as discrete laminae within aeolian dune foresets or overlying deflation surfaces in aeolian sandsheets, but this is balanced by the presence of very fine to fine sand laminae in the same sequences. The sorting parameters of the sandstones vary substantially from poor to very good. In general, fluvial sandstones are less well sorted than other facies although the most fine-grained sheetflood sequences commonly contain a scattering of coarse grains and, hence, are also poorly sorted. Aeolian sandstones are well, or locally very well, sorted within individual laminae although overall sorting can be poor or markedly bimodal. Although grain size and sorting vary substantially both within and between facies, there is no consistent variation that could reasonably account for a porosity difference around 10% between fluvial and aeolian sandstones.

The range of grain types encountered, in terms of angularity and sphericity, does however appear to be significant. The characteristic 'aeolian' grains are very well rounded, exhibit a generally high degree of sphericity and commonly evidence aeolian transportation from the microscopic facetting of grain surfaces. These grains occur primarily as the main constituent of aeolian bedforms but also as a component in some sheetflood and minor fluvial channel sandstones. By comparison, the grains that typically dominate the major fluvial channel sandstones are angular to sub-angular, occasionally subrounded, and tend toward low sphericities. It has been suggested from studies of packing parameters that sands composed of high sphericity grains have lower depositional porosities than those with grains of low sphericity (Fraser 1935; Beard and Weyl 1973). In the present case, the greater angularity of the lower sphericity grains probably cancels out this effect and it is likely, as supported by the minus cement porosity data, that depositional porosities were essentially similar for clean sands in all facies.

Discussion

The major sedimentary facies associations in the SSG exhibit differing reservoir properties with the greatest disparity occurring between the deposits of major fluvial channels and those of aeolian dunes. Given the facies non-specific effects of illite, these differences appear to stem primarily from the distribution of quartz cement and, to a lesser extent, carbonate cement and matrix clays. The abundance of authigenic quartz in the form of overgrowths in the fluvial channel sandstones, and some sheetflood sandstones, means that these sediments exhibit a permeability range approximately an order of magnitude lower than comparable aeolian sandstones. In the worst affected instances (those where extensive quartz cement combines with illite) permeabilities have been reduced to the level where these intervals are effectively non-reservoir. By comparison, aeolian sandstones in the worst illite affected zones still retain permeabilities in the range 1–50 mD and are, therefore, still potential reservoir for gas production. The most severe effects of carbonate cement appear to be limited to the cementation of fluvial channel bases and the incipient calichification of some sandflat intervals, while matrix clays severely restrict the reservoir potential of the finer-grained sheetflood sequences.

The cause of the disparity in the distribution of quartz

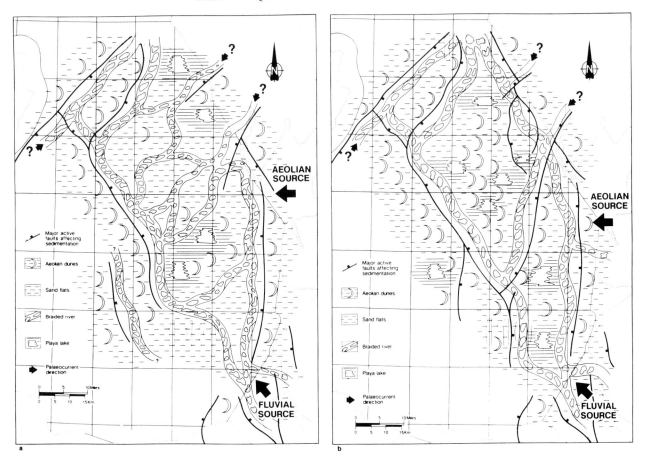

Fig. 8. Schematic palaeogeographies based on well correlations and seismic data for (**a**) the lower and (**b**) the upper Ormskirk Sandstone Formation (from Meadows and Beach 1993), illustrating the sourcing of sediment from the southeast via a fluvial system flowing out of the Cheshire Basin, and from the east by predominantly aeolian transportation.

cement is not immediately clear although it is tempting to suggest that the recognition of two distinct grain populations may be significant. There are several factors that may contribute towards making the precipitation of quartz more likely on the angular to sub-rounded low sphericity grains that characterize the fluvial sandstones and less likely on the well-rounded high-sphericity aeolian grains. The former have a substantially higher total surface area and, hence, a greater abundance of nucleation sites (Stephan 1970). The angularity of the grains implies a greater occurrence of sharp pointed grain contacts and the increased possibility, at least in the early stages of burial, of pressure solution that would make free silica available for reprecipitation (see de Boer 1977). Additionally, the greater abundance of detrital clays in the fluvial and some sheetflood sandstones, as both matrix and clay intraclasts, also increases the local availability of silica during diagenesis in response to smectite to illite transformations (Hower *et al.* 1976; Boles and Franks 1979). Additionally, at least three major phases of quartz precipitation are recorded in the SSG (Burley 1984; Stuart and Cowan 1991) and it seems likely that, once established, quartz would favour existing overgrowths as a nucleation site in preference to initiating new crystals. McBride (1989) indicates that the activation energy required to seed new crystals is much greater than that needed for the continuing development of overgrowths.

Whatever the reason for the disparity in the abundance of quartz cement, the inferred correlation of authigenic quartz with the distribution of the two grain populations implies a reservoir control that is related more to provenance and mode of transport than to depositional process. Where 'aeolian' grains form a substantial component of non-aeolian deposits (mainly sheetflood or minor fluvial channel), then these sedi-

ments assume the reservoir characteristics of true aeolian sandstones. Despite the apparent mineralogical similarity of the two grain populations it is considered unequivocal that they derive from different sources. The presence of well-rounded 'aeolian' grains of the same grade or larger than the more angular 'fluvial' grains would appear to preclude the possibility of the former having been derived by wind reworking of the latter. Similarly, the total volume of rock dominated by 'fluvial' grain types makes it highly unlikely that these were derived by disintegration of the 'aeolian' grains.

The correspondence of the 'fluvial' grain population with the major channel facies suggests that one or more such channels were responsible for transporting these grains into the Irish Sea area. Correlation of core-based facies analysis with sonic logs (illustrated in Figs 3a and 3b) and seismic data (Meadows and Beach 1993) has enabled the establishment of basin-wide palaeogeographies. These suggest the presence of a major fluvial system operating through the EISB during OSF depositional time. Comparison of these palaeogeographies with onshore exposure further suggests that a major element of this system flowed from the south via the Cheshire Basin and ultimately exited from the Irish Sea northwards into the Solway Basin (Fig. 8), although other input routes cannot be excluded. This accords with published accounts of SSG sedimentology in western Britain (Henson 1970; Steel and Thompson 1983; Thompson 1970) and would, therefore, suggest that areas to the south, including the English Midlands, Wales and possibly the Variscan massifs of southwest England (Fitch *et al.* 1966; Audley-Charles 1970), are the source for the 'fluvial' grains.

The provenance of the 'aeolian' grains is more problematical but palaeotransport data for the SSG aeolian sequences in the

Cheshire Basin (Thompson 1969) indicate northeasterly or east-northeasterly winds (relative to modern azimuths). This agrees with trade wind patterns established for other areas of northern Europe in early Triassic times (Clemmenson 1985; 1987). If this were the dominant pattern in the Irish Sea area, then the Carboniferous sequences of the palaeo-Pennine escarpment might be a logical source for the aeolian grains. These sequences include very coarse-grained sandstones and were substantially eroded prior to Triassic deposition, particularly in the northern Pennines.

The palaeogeographical reconstruction for the EISB mentioned above (Meadows and Beach 1993) and for adjacent onshore areas (Macchi and Meadows 1987) suggest a correlation between active Triassic faulting and the course taken by major fluvial channels. Conversely, palaeohighs related to footwall uplift and the formation of ramp anticlines are suggested to have been the loci of aeolian activity (Meadows and Beach 1993). Thus, it is possible that basin evolution, in controlling the distribution of facies, also influenced the distribution of the two differing grain populations upon which reservoir quality has ultimately depended.

The authors wish to express their thanks to British Gas Exploration and Production who made available the core analysis data used in this paper. We would also like to acknowledge the assistance of those others involved in the preparation of the paper, principally Wendy Bryan for typing, Greg Allsop and Claire Harris for draughting and Hywell Jones for photography.

References

AUDLEY-CHARLES, M. G. 1970. Triassic palaeogeography of the British Isles. *Quarterly Journal of the Geological Society, London*, **126**, 49–89.

BEARD, D. C. AND WEYL, P. K. 1973. Influence of texture on porosity and permeability of unconsolidated sand. *Bulletin of the American Association of Petroleum Geologists*, **51**, 349–369.

DE BOER, R. B. 1977. On the thermodynamics of pressure solution—interaction between chemical and mechanical forces. *Geochemica et Cosmochimica Acta*, **41**, 249–256.

BOLES, J. R. AND FRANKS, S. G. 1979. Clay diagenesis of Wilcox sandstones of south-west Texas: implications of smectite diagenesis on sandstone cementation. *Journal of Sedimentary Petrology*, **49**, 55–70.

BRIERLEY, G. J. 1989. River plan form facies models: the sedimentology of braided, wandering and meandering reaches of the Squamish River, British Columbia. *Sedimentary Geology*, **61**, 17–35.

BURLEY, S. D. 1984. Patterns of diagenesis in the Sherwood Sandstone Group (Triassic), United Kingdom. *Clay Minerals*, **19**, 403–440.

BUSHELL, T. P. 1986. Reservoir geology of the Morecambe Field. *In*: BROOKS, J., GEOFF, J. C. AND VAN HOORN, B. (eds) *Habitat of Palaeozoic gas in NW Europe*. Geological Society, London, Special Publication, **23**, 189–208.

CANT, D. J. 1978. Development of a facies model for sandy braided river sedimentation: comparison of the South Saskatchewan River and the Battery Point Formation. *In*: MIALL, A. D. (ed.) *Fluvial Sedimentology*. Canadian Society of Petroleum Geologists Memoir **5**, 627–639.

CLEMMENSEN, L. B. 1985. Desert sand plain and sabkha deposits from the Bunter Sandstone Formation (L. Triassic) at the northern margin of the German Basin. *Geologische Rundschau*, **74**, 519–536.

—— 1987. Complex star dunes and associated aeolian bedforms, Hopeman Sandstone (Permo-Triassic), Moray Firth Basin, Scotland. *In*: FROSTICK, L. E. AND REID, I. (eds) *Desert Sediments: Ancient and Modern*, Geological Society, London, Special Publication, **35**, 213–231.

FITCH, F. J., MILLER, J. A. AND THOMPSON, D. B. 1966. The palaeogeographic significance of isotopic age determinations on detrital micas from the Triassic of the Stockport–Macclesfield district, Cheshire, England. *Palaeogeography, Palaeoclimatology, Palaeoecology*, **2**, 281–312.

FRASER, H. J. 1935. Experimental study of the porosity and permeability of clastic sediments. *Journal of Geology*, **43**, 910–1010.

FRYBERGER, S. G., AHLBRANDT, T. S. AND ANDREWS, S. 1979. Origin, sedimentary features, and significance of low-angle eolian 'sand sheet' deposits, Great Sand Dunes National Monument and vicinity, Colorado. *Journal of Sedimentary Petrology*, **49**, 733–746.

——, AL-SARI, A. M. AND CLISHAM, T. J. 1983. Eolian dune, interdune, sand sheet and siliciclastic sabkha sediments of an offshore, prograding sand sea, Dhahran area, Saudi Arabia. *Bulletin of the American Association of Petroleum Geologists*, **67**, 280–312.

HENSON, M. R. 1970. The Triassic rocks of South Devon. *Proceedings of the Ussher Society*, **2**, 172–177.

HOWER, J., ESLINGER, E. V., HOWER, M. E. AND PERRY, E. A. 1976. Mechanism of burial metamorphism of argillaceous sediment: mineralogical and chemical evidence. *Bulletin of the Geological Society of America*, **87**, 725–737.

HUNTER, R. E. 1977. Basic types of stratification in small eolian dunes. *Sedimentology*, **24**, 361–387.

JACKSON, D. I., MULHOLLAND, P., JONES, S. M. AND WARRINGTON, G. 1987. The geological framework of the East Irish Sea Basin. *In* BROOKS, J. AND GLENNIE, K. W. (eds) *Petroleum Geology of North West Europe*, Vol. 1. Graham & Trotman, London, 191–203.

LARUE, D. K. AND MARTINEZ, P. A. 1989. Use of bed-form climb models to analyse geometry and preservation potential of clastic facies and erosional surfaces. *The American Association of Petroleum Geologists Bulletin*, **73**(1), 40–53.

McBRIDE, E. F. 1989. Quartz cement in sandstones: A review. *Earth-Science Reviews*, **26**, 69–112.

MACCHI, L. 1987. A review of sandstone illite cements and aspects of their significance to hydrocarbon exploration and development. *Geological Journal*, **22**, 333–345.

——, CURTIS, C. D., LEVISON, A., WOODWARD, K. AND HUGHES, C. R. 1990. Chemistry, morphology and distribution of illites from Morecambe Gas Field, Irish Sea, offshore United Kingdom. *The American Association of Petroleum Geologists Bulletin*, **74**(3), 296–308.

—— AND MEADOWS, N. S. 1987. *Field Excursion to the Permo-Triassic of Cumbria and Cheshire*. Excursion Guide 12, Poroperm-Geochem Limited. Chester Street, Chester.

MEADOWS, N. S. AND BEACH, A. 1993. Structural and climatic controls on facies distribution in a mixed fluvial and aeolian reservoir: the Triassic Sherwood Sandstone in the Irish Sea. *In*: NORTH, C. P. AND PROSSER, J. (eds) *Characterization of Fluvial and Aeolian Reservoirs*. Geological Society, London, Special Publication, in press.

DE RAAF, J. F. M., BOERSMA, J. R. AND VAN GELDER, A. 1977. Wave-generated structures and sequences from a shallow marine succession, Lower Carboniferous, County Cork, Ireland. *Sedimentology*, **24**, 451–483.

SMITH, D. B., BRUNSTROM, R. G. W. MANNING, P. I., SIMPSON, S. AND SHOTTON, F. W. 1974. *A correlation of Permian rocks in the British Isles*. Special Report of the Geological Society, London, **5**.

STEEL, R. AND THOMPSON, D. B. 1983. Structures and textures in Triassic (Scythian) braided stream conglomerates ('Bunter' Pebble Beds) in the Sherwood Sandstone Group, North Staffordshire, England. *Sedimentology*, **30**, 341–368.

STEPHAN, H. J. 1970. Diagenesis of the Middle Buntsandstein in South Oldenburg, Lower Saxony, *Meyniana*, **20**, 39–82.

STUART, I. A. AND COWAN, G. 1991. Morecambe Gas Field, Blocks 110/2a, 110/3a, 110/8a, UK. *In*: ABBOTS, I. L. (ed.) *United Kingdom Oil and Gas Fields: 25 years Commemorative Volume*. Geological Society, London, Memoir No. 14, 527–541.

THOMPSON, D. B. 1969. Dome-shaped aeolian dunes in the Frodsham Member of the so-called 'Keuper' Sandstone Formation (Scythian-?Anisian: Triassic) at Frodsham, Cheshire (England). *Sedimentary Geology*, **3**, 263–289.

—— 1970. The stratigraphy of the so-called Keuper Sandstone Formation (Scythian-?Anisian) in the Permo-Triassic Cheshire Basin. *Quarterly Journal of the Geological Society of London*, **126**, 151–181.

WARRINGTON, G., AUDLEY-CHARLES, M. G., ELLIOT, R. E., EVANS, W. B., IVIMEY-COOK, H. C., KENT, P. E., ROBINSON, P. L., SHOTTON, F. W. AND TAYLOR, F. M. 1980. *A correlation of Triassic rocks in the British Isles.* Special Report of the Gological Society, London, **13**.

WOODWARD, K. AND CURTIS, C. D. 1987. Predictive modelling for the distribution of production-constraining illites—Morecambe Gas Field, Irish Sea, offshore UK. *In*: BROOKS, J. AND GLENNIE, K. W. (eds) *Petroleum Geology of North West Europe*, vol. 1. Graham & Trotman, London, 205–215.

ZIEGLER, P. A. 1982. *A geological atlas of Western Europe (2 Volumes)*. Elsevier, Amsterdam.

The Fleetwood Dyke and the Tynwald fault zone, Block 113/27, East Irish Sea Basin

G. ARTER[1] and S. W. FAGIN[2]

[1] *Esso Exploration and Production UK Limited, Mailpoint 25, Esso House, Ermyn Way, Leatherhead, Surrey KT22 8UY, UK*

[2] *Exxon Exploration Company, Houston, Texas, USA*

Abstract: A series of Tertiary dolerite intrusions, the Fleetwood Dyke Group, is intruded into Triassic sediments of the northern East Irish Sea Basin (EISB). Well 113/27-1 penetrated a total of 375 m of dolerite within the Mercia Mudstone Group (MMG). Samples of dolerite have been dated (K–Ar) at between 61 and 65 Ma. Apatite fission track analysis on samples from the same well indicates that approximately 2000 m of uplift occurred in the early Tertiary, at about the same time as dyke intrusion.

Using seismic and marine magnetic data, the Fleetwood Dyke Group has been mapped in the vicinity of Block 113/27 to reveal a complex set of dykes and sills. The presence of these intrusions within the MMG results in velocity pull-ups on the underlying Ormskirk Sandstone Formation (OSF), a potential exploration pitfall.

The geometry of the igneous intrusion has been influenced by the structure of the Tynwald fault zone. This feature is interpreted to have formed in the Triassic as a result of listric faulting detached on Permian evaporites. This interpretation is supported by balanced section reconstruction.

Exploration drilling in the East Irish Sea Basin (hereafter EISB) commenced in 1969. The Morecambe Gas Fields were discovered in 1974 (Ebbern 1981). In 1985, following the Ninth Round of offshore licensing, Esso was awarded operatorship of Block 113/27 located north of the Morecambe Fields. This paper presents interpreted seismic and magnetic data which provide additional information on the origin of the Tertiary igneous intrusions of the Fleetwood Dyke Group and a complex structural feature, the Tynwald fault zone.

Regional setting

Sediments of Early Carboniferous to Late Triassic age have been penetrated by wells drilled in the EISB (Jackson *et al.* 1987). Reconstruction of the post-Triassic burial history of the basin is complicated by the absence of sedimentary rocks younger than Triassic in age. The present-day extent of the basin is defined by the erosional truncation of Permo-Triassic sediments at the base Quaternary (Fig. 1).

The paper is concerned with observations from the vicinity of Block 113/27, in the northern part of EISB. The stratigraphic succession and structural evolution of the area are described with reference to a generalized stratigraphic column for the 113/27 area (Fig. 2). A thorough account of the structure and stratigraphy of the basin is given by Jackson *et al.* (1987) and the same nomenclature is adopted here.

Carboniferous strata ranging from Dinantian to Westphalian have been found in the study area (Fig. 2). Erosional truncation of the Carboniferous at the Hercynian Unconformity is seen on seismic reflection data and is indicated by variable ages of subcropping Carboniferous in wells. The amount of missing section is not known but the maximum preserved thickness of Westphalian in well penetrations in Block 113/27 is 150 m compared with up to 450 m proved, and 1300 m interpreted from seismic data, elsewhere in the basin (Jackson *et al.* 1987).

Regionally, the Upper and Lower Permian have been observed to thicken into major growth faults (Jackson *et al.* 1987). Within the study area seismic and well data indicate a relatively uniform Permian thickness of up to 400 m with slight thickening southwest towards the Keys Fault. Wells drilled in Block 113/27 confirm the Permian palaeogeographic models of Colter and Barr (1975) and Jackson *et al.* (1987). The Lower Permian Collyhurst Sandstone Formation occurs as a claystone with some sandstone and is interpreted as a playa lake

deposit with localized incursions of sand. The Upper Permian comprises two units: the St Bees Evaporites are predominantly halite and are up to 150 m thick and, together with the overlying St Bees Shale, provide a major detachment surface for Triassic listric faulting. The top of the St Bees Evaporites is marked by a decrease in acoustic impedance and is generally a good event on seismic data.

Triassic deposition within the study area was in a graben bounded by major growth faults, the Lagman–Keys faults to the west and the Lake District Boundary Fault to the east. The area is subdivided into two westerly-tilted half-grabens by the Tynwald fault zone. The Lower Triassic Sherwood Sandstone Group (hereafter SSG) is 1200–1500 m thick and comprises fluvial sandstones with subordinate aeolian units. The top of the SSG, the Ormskirk Sandstone Formation, is the principal seismic event for the mapping of Triassic structures.

The youngest sediments are the dolomitic mudstones and halites of the Mercia Mudstone Group of Scythian to ?Carnian age. In an undisturbed section the MMG in Block 113/27 can be correlated with the subdivisions of Jackson *et al.* (1987). The boundaries of the thicker halite units form coherent and correlatable seismic reflectors but the use of these reflectors is limited by frequent zones of halokinesis and minor glide plane faulting, as well as by the presence of igneous intrusions (as detailed below). The MMG is overlain only by a thin (25–40 m) cover of Quaternary sands and clays.

Structural evolution of the Tynwald fault zone

The Tynwald fault zone extends north–south or NNW–SSE, for about 40 km, through the northern part of the EISB (Fig. 1). Seismic interpretation of the faulted area is difficult because the seismic imaging of the steep dips within the fault zone is often poor. A feature of the fault zone in Block 113/27 is the axial anticline which, in appearance, suggests compression related to Tertiary basin inversion. However, high-quality seismic data suggest that the Tynwald fault zone formed by extension on an opposing pair of listric normal faults detached in the St Bees Evaporites (Fig. 3a,b). Balanced cross-section reconstructions and subsequent drilling in Block 113/27 support this interpretation.

Figure 4 is located approximately 10 km north of the lines in Fig. 3. Here the Tynwald fault zone bounds a half-graben. The hangingwall block has moved about 1.0 km to the east by slip on the western boundary fault. This contrasts with the

From *Petroleum Geology of Northwest Europe: Proceedings of the 4th Conference* (edited by J. R. Parker).
© 1993 Petroleum Geology '86 Ltd. Published by The Geological Society, London, pp. 835–843.

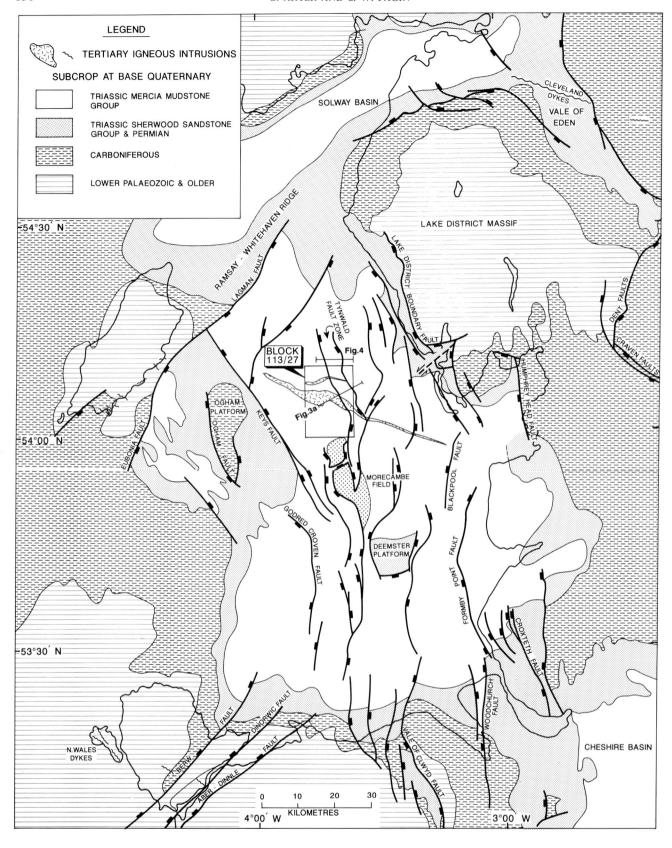

Fig. 1. Simplified structural elements map of the EISB. Compiled from interpreted seismic data and BGS published maps. Detail of the west part of the basin and names of structural features after Jackson *et al.* (1987).

primarily westward direction of extension inferred from balanced section analysis of the seismic profile in Fig. 3a.

Balanced cross-section analysis

The profile in Fig. 3a was chosen for balanced cross-section analysis. The interpreted seismic horizons were depth converted using a layer cake method with velocity information from nearby wells. Because the gross fold form of the hangingwall was the principal balancing constraint, the Top Ormskirk surface was simplified by smoothing through the minor faults within the fault zone.

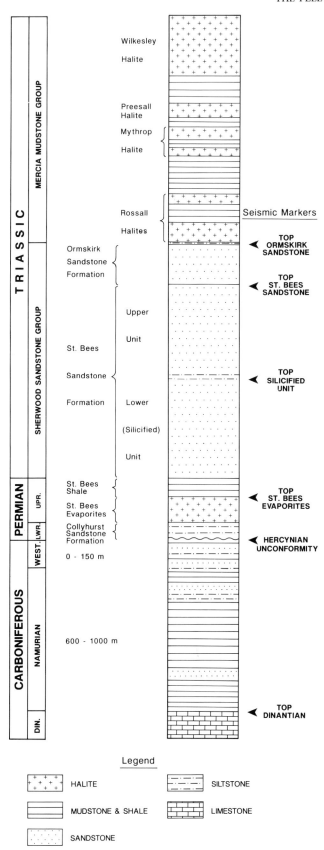

Fig. 2. Generalized stratigraphic column for the 113/27 area.

Figure. 5 illustrates the reconstruction of the depth model from Fig. 3a. Only the SSG above the detachment surface has been modelled. The final restoration shows a satisfactory fit with few gaps or overlaps. Balance to within a few percent of the total mass involved has been demonstrated.

In performing the reconstruction, three decisions had to be made. First, the extent of both eastward and westward transport had to be estimated. Second, the level of the pre-faulted Top Ormskirk regional surface had to be estimated. Third, a strain mechanism to accommodate the deformation must be assumed.

Direction and amount of transport. In Fig. 3a the Tynwald fault zone is located over a large amplitude force fold at the level of the Permian detachment surface. The down-dip extension of the western bounding fault takes a marked downward steepening bend over this force fold. Fault-bend fold theory requires that in response to downward-steepening a hanging-wall fold form must develop which includes a panel that dips away from the fault. Because an eastward-dipping panel is not observed it is not possible to balance the structure with any mechanism which includes extensive (> 500 m) motion to the east. Any attempt to do so will result in a restoration that brings the Top Ormskirk surface well above its likely regional level. The balanced solution presented in Fig. 5 contains only about 400 m of movement to the east but about 2200 m of movement to the west.

The structure of the restored Top Ormskirk. The regional level that was used in the final solution (the bounding surface above the three blocks in Fig. 5a) is a gentle flexure over the crest of the force fold. An alternative model was attempted which included thinning of the SSG over the crest of the fold. This alternative was discarded because it restricted the amount of westward motion of the western block. (In reconstruction, the full section of SSG which exists below where it is first cut by the western fault would be displaced above this regional level after repositioning the western block only 1000–1500 m to the east.)

The strain mechanism. Three modes of strain can be employed to simulate hangingwall deformation: flexural slip, vertical shear and oblique shear. Flexural slip involves slip between bedding planes. It could not be used in this problem because it cannot deform the hangingwall through fault bends greater than 30° in a way that allows the hangingwall to maintain contact with the fault plane. Vertical shear deformation is simple to model but is not a deformation mechanism observable in outcrop or experiment. It was not possible to obtain a satisfactory reconstruction using vertical shear. Oblique shear best simulates the shape of the shear zone developed in clay-cake experiments (e.g. Cloos 1968), and was used with a failure angle of 60° (conforming to the Mohr-Coulomb failure criteria), in modelling the central block deformation. The deformation of eastern and western blocks which was relatively mild was modelled by vertical shear and flexural slip, respectively, for convenience.

Movement history. The development of the graben is shown in six stages (Figs 5a–f). In some of these stages the position of the hangingwall is shown subsequent to fault motion but prior to deformation by oblique shear, to depict the size of the void created. For simplicity the three movements depicted in Figs 5b, 5c and 5e are shown sequentially. They are more likely to have occurred contemporaneously.

Timing of fault movement

On a few seismic lines (Figs 3a, b) it is possible to identify an unconformity surface at 0.1 to 0.2 s two-way time. This surface clearly truncates seismic reflectors within the MMG which have been rotated into the bounding faults of the Tynwald fault zone, and is onlapped by flat-lying reflectors. The age of this unconformity, therefore, constrains the timing of fault

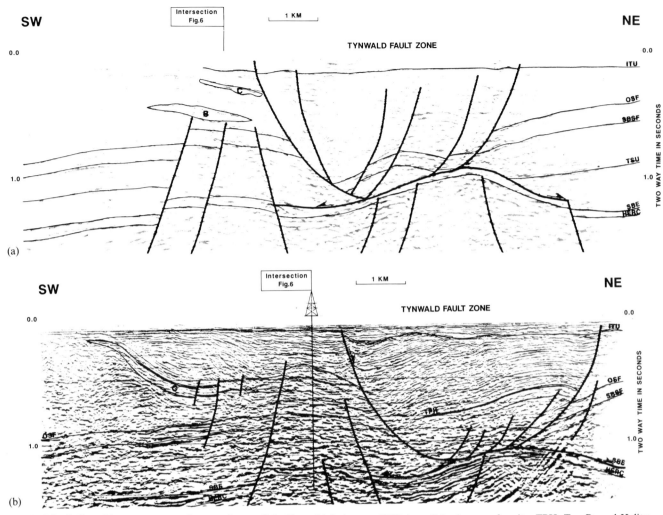

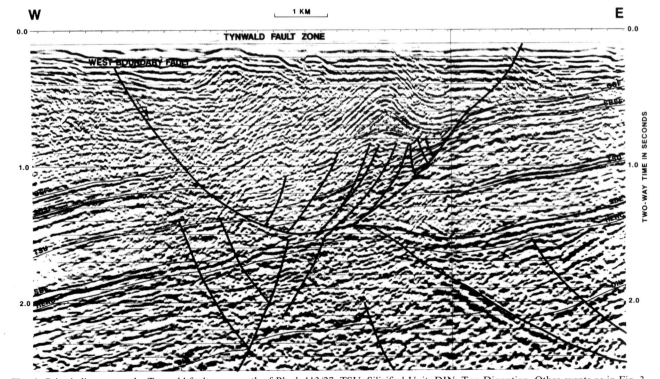

Fig. 3. Seismic lines 1 km apart across the central part of the Tynwald fault zone. ITU: intra-Triassic unconformity; TPH: Top Preesal Halite; OSF: Ormskirk Sandstone Formation; SBSF: St Bees Sandstone Formation; SBE; St Bees Evaporites: HERC: Hercynian Unconformity; B and C: Tertiary intrusions. (**a**) Shows the detachment of the Tynwald fault zone in the SBE; (**b**) shows the onlap of the youngest MMG onto the ITU surface within the fault zone and the well penetration of dyke C.

Fig. 4. Seismic line across the Tynwald fault zone north of Block 113/27. TSU: Silicified Unit; DIN: Top Dinantian. Other events as in Fig. 3. Seismic data courtesy Western Geophysical Co.

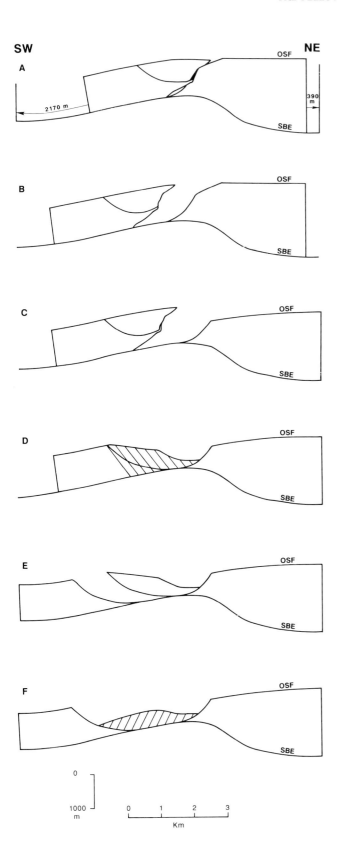

Fig. 5. Balanced section reconstruction of a depth converted section simplified from Fig. 3a. OSF: Ormskirk Sandstone Formation; SBE: St. Bees Evaporties. (**A**) The unfaulted Sherwood Sandstone Group with future fault traces. Gaps and overlaps represent the errors in the reconstruction. (**B**) Movement of the western and central blocks to the west 1040 m. (**C**) Movement of the eastern block 390 m to the east. (**D**) Down-to-the-east oblique shear of the central and part of the western block. (**E**) Movement of the western block 1130 m to the west. (**F**) Down-to-the-west oblique shear of the central block to achieve the present-day depth section.

movement. The unconformity has been tied seismically to borehole control and lies within the MMG, thereby dating the main extension in the Tynwald fault zone as Middle to Late Triassic. The same unconformity surface has also been identified on high resolution seismic data acquired for geotechnical purposes around potential well locations. On these data the unconformity can clearly be separated from the Base Quaternary event and the water bottom multiple. The folding observed within the Tynwald fault zone cannot, therefore, have been caused by basin inversion in the Tertiary.

Confirmation by drilling

A well drilled close to the seismic line in Fig. 3a supports the interpretation of the structure of the Tynwald fault zone. The Permian and Lower Triassic section in the well is markedly different from the type section for the area. The SSG was thinned to about 300 m, with only the Ormskirk Sandstone Formation and possibly the uppermost St Bees Sandstone Formation present. Structural dips of 30–45° and zones of intense shearing were observed in the SSG. The underlying Permian was only 70 m thick with a dolomitic upper unit overlying the Collyhurst Sandstone Formation. The contact between the SSG and the Permian is interpreted as the listric fault.

Tertiary igneous intrusions

The Fleetwood Dyke Group is identified by a pronounced WNW–ESE-trending aeromagnetic anomaly 50 km long, in the northern EISB (Kirton and Donato 1985; Woodland 1977, 1978). Kirton and Donato (1985) modelled profiles across the anomaly as reversely magnetized vertical dykes intruded close to the present-day sea floor and commented that the western part of the anomaly might be due to an inclined dyke. Seismic and well data from Block 113/27 confirm this interpretation.

Two intrusions were penetrated by Well 113/27-1; the main body was 360 m thick and a lower unit was 15 m thick. The intrusions penetrated were a moderately coarsely crystalline dolerite comprising predominantly plagioclase with subordinate clinopyroxene. Two independent K–Ar whole-rock age determinations were made on samples from the well. One set of analyses gave an age of 65.5 ± 1.0 Ma. A second analysis gave an age of 61.5 ± 0.8 Ma (K. Hitchen and J. D. Ritchie, BGS Edinburgh, pers. comm.). These ages suggest that the timing of the Fleetwood Dyke Group intrusion is similar to that of the North Wales Group, given a minimum age of 63.5 Ma by Evans et al. 1973).

Apatite fission track analyses (Gleadow et al. 1983) on dolerite samples from the dyke give a fission track age that is very similar to the K–Ar age determination and indicate that following intrusion, the dyke cooled rapidly to below 50°C and was not subsequently exposed to higher temperatures. Fission track analyses on samples from the Triassic sediments in the same well indicate fission track ages that are much less than the stratigraphic ages of the sediments. These data suggest that the samples reached temperatures in excess of 110°C during Mesozoic burial and were then cooled by uplift in the early Tertiary. The magnitude of the uplift is estimated to be at least 2 km, assuming a geothermal gradient of 30°C km^{-1} in the early Tertiary.

Mapping of the igneous intrusions

A seismic and magnetic survey was recorded over Block 113/27 in 1985. Lines were orientated NW–SE and NE–SW with a grid spacing of 1.0 km; some infill NE–SW lines were recorded at a spacing of 0.5 km.

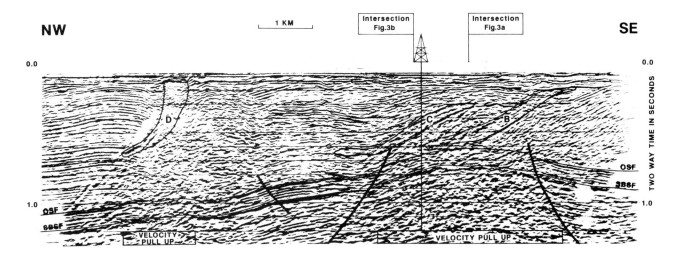

Fig. 6. Seismic expression of Tertiary igneous intrusions. Note the velocity pull-up beneath the thick parts of the intrusions. (Seismic events as in Fig. 3.)

Fig. 6 illustrates the seismic response of the intrusions. The main criteria for identification were the amplitude response of the interface between the dolerite and the MMG and the cross-cutting of reflectors within the MMG. Neither of these characteristics is diagnostic, however, because similar strong amplitude events occur within the MMG due to the impedance contrast between halite and mudstone. Salt diapirism and associated small-scale listric faulting within the MMG create steep rapidly changing dips. A third characteristic for the recognition of intrusions was diffracted energy observed on unmigrated sections which was sometimes due to the lateral termination of an igneous body. Because of the steep dips, shallow depths and laterally discontinuous nature of the reflectors within the MMG, seismic velocity analysis has not proved to be an effective tool for detection of the intrusions. The presence of intrusions within the MMG can cause a velocity pull-up on the underlying OSF surface, a potential exploration pitfall (Fig. 6). Integration of seismic and magnetic data is, therefore, important in areas where intrusions are expected.

Also of concern is the effect of dyke intrusion on surrounding reservoirs. Correlation of halite and mudstone units within the MMG between the wells drilled in Block 113/27, and with the subdivisions of Jackson *et al.* (1987), suggests that the dolerite intrusion penetrated in 113/27-1 has replaced at least one of the main halite units of the MMG. In the well drilled in the Tynwald fault zone the reservoir quality of the SSG was significantly reduced by a pervasive, late-stage halite cement.

Because of the complexity of the intrusions, quantitative magnetic modelling was not attempted within the main survey area. Starting from the well penetration, and using the marine magnetic profiles in combination with the seismic lines, a qualitative interpretation of the distribution of intrusions was produced. This map is superimposed on the marine magnetic anomaly map (Fig. 7). Two sub-parallel zones of intrusions are observed. The southern zone represents the main part of the Fleetwood Dyke Group. To the east of the survey area this is modelled as a simple vertical dyke or dykes (dyke A). Near the west boundary fault of the Tynwald fault zone the intrusion geometry becomes more complex and the seismic imaging is very poor, it appears that the dyke becomes inclined and dips to the northeast within the MMG. West of the Tynwald fault zone the intrusion is well imaged on seismic (Fig. 6) and is interpreted as a pair in inclined dykes intruded within the MMG. The southern edge of dyke C is very close to the Base Quaternary where it causes considerable scattering of

seismic energy (Fig. 3b). No feeder dyke is observed west of the fault zone, the intrusion appears to have spread laterally through the MMG from the area of the west bounding fault.

The northern intrusion is less complex and consists of a west–east-trending dyke which is vertical in the east but becomes inclined in the vicinity of the Tynwald fault zone, where it dips northeast. Locally the dyke is interpreted to be intruded along the fault plane of the west boundary fault. To the west, this dyke feeds some small sills intruded within the MMG.

Magnetic modelling

A profile across dyke A, located just east of the main survey (Figs 7, 8), was selected for 2D magnetic modelling. At this location the position of the dyke is constrained by seismic data and is seen as a vertical zone of diffracted energy on unmigrated data. Previous work in the area (Kirton and Donato 1985) and inspection of the profile in Fig. 8 indicate a significant component of remanent magnetization is required to produce the observed large negative trough with a smaller peak to the north. The values of inclination and declination of $-68°$ and $168°$, respectively, used in the solution are the average values of palaeomagnetic measurements on samples from Anglesey (site T7) and Penrith (site T8) by Dagley (1969). The susceptibility, 1600×10^{-6} cgs, used in the model is the average value of measurements made on samples of dolerite collected from 113/27-1. The direction of the total magnetization vector (sum of induced and remanent) is in the direction of the remanent magnetization, opposite to the present earth magnetic field.

The calculated anomaly in Fig. 8 shows a reasonable agreement with the observed anomaly. Two vertical dykes are required to model the main anomaly. These have widths of 180 and 380 m and are about 500 m apart; the depth to the top of the dykes is about 80 m. Three minor dykes extending upward to 250–350 m below sea-level, are modelled to the south of the main body in an attempt to explain small (10–20 nT) observed anomalies. There appears to be a correlation between the position of these minor dykes and minor faulting within the Tynwald fault zone. Even with the use of marine magnetic profiles such small intrusions might be undetected or interpreted as faults. A similar small anomaly to the north, corresponding to dyke D in Fig. 7, was modelled as a vertical dyke 25 m in width with a top 230 m below sea-level. This dyke is

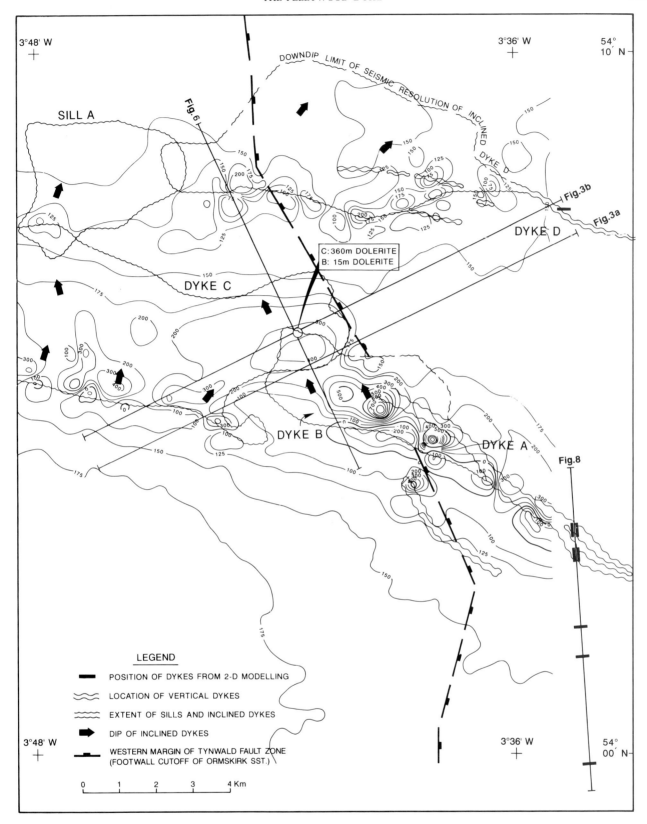

Fig. 7. Marine magnetic survey for Block 113/27, grid interval 1.0 × 1.0 km with local infill of 0.5 km spacing in NE–SW direction. Interpreted distribution of igneous intrusions is from seismic and magnetic data. The mapped area corresponds to the highlighted area in Fig. 1.

not observed on seismic data at the location of the modelled profile.

Controls on dyke intrusion

The trace of the footwall cut-off of the western fault of the

Tynwald fault zone has been plotted on Fig. 7. The segment of the fault zone between the two zones of dyke intrusion has a NNW–SSE trend, compared with the north–south trend of the other parts of the fault zone. The fault geometry suggests a westward transport direction for the western footwall block between the dyke intrusions. This contrasts with an interpreted

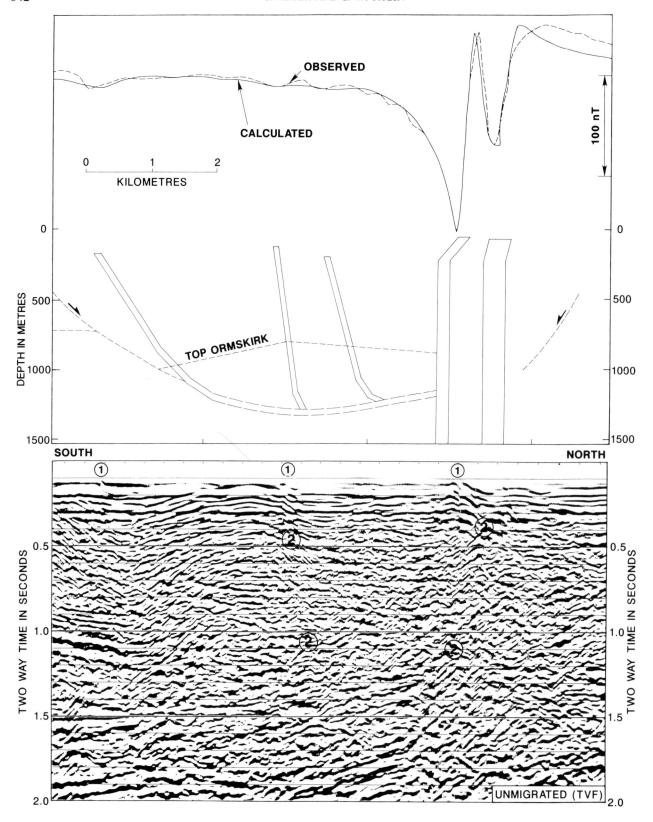

Fig. 8. Observed and calculated marine magnetic anomalies over the Fleetwood Dyke Group, together with unmigrated seismic line. Position of Top Ormskirk reflector and East Fault of the Tynwald fault zone is shown for reference. 1: Shallow disturbance above the dyke; 2: Zone of diffracted energy. The presence of a sill connecting the minor dykes is conjectural and cannot be resolved in the modelling. Seismic and magnetic data courtesy Western Geophysical Co.

eastward transport direction (extension on the west bounding fault) for the northern and southern parts of the fault zone. These observations suggest that the Tertiary dykes have been intruded into transfer fault zones created during Late Triassic extension.

Summary

Intrusions of the Fleetwood Dyke Group previously recognized from aeromagnetic anomaly maps, have now been penetrated by an exploration well drilled in Block 113/27 of the

EISB. The intrusion, dated at between 61 and 65 Ma, is observed on magnetic and seismic reflection data to be a WNW–ESE-trending dyke which, as it crosses the Tynwald fault zone, changes into a series of inclined dykes and sills intruded within the MMG. Observations on the Late Triassic evolution of the Tynwald fault zone, supported by section balancing, suggest that the presence of transfer faults may have locally influenced the location of the intrusions. Dyke intrusion has been responsible for velocity pull-ups which create false structures on time structure maps of the SSG. It is possible that halite mobilized from the MMG by dyke intrusion is responsible for the localized formation of a porosity-reducing halite cement observed in the SSG reservoir.

Richard Lu of Exxon Production Research Company modelled the magnetic profile in Fig. 8. This paper is published with the permission of Block 113/27 co-venturers Esso Exploration and Production UK, Shell UK Exploration and Production and Purbeck Petroleum. The interpretations and conclusions given in this paper are those of the authors and do not necessarily reflect those of the Block 113/27 co-venturers. Patricia Wey is thanked for typing the manuscript.

References

CLOOS, E. 1968. Experimental analysis of Gulf Coast fracture patterns. *American Association of Petroleum Geologists Bulletin*, **52**, 3420–3444.

COLTER, V. S. AND BARR, K. W. 1975. Recent developments in the Geology of the Irish Sea and Cheshire Basins. *In*: WOODLAND A. W. (ed.) *Petroleum and the Continental Shelf of North-West Europe, Vol. 1, Geology*. Applied Science Publishers, London, 61–75.

DAGLEY, P. 1969. Palaeomagnetic results from some British Tertiary dykes. *Earth and Planetary Science Letters*, **6**, 349–354.

EBBERN, J. 1981. Geology of the Morecambe Gas Field. *In*: ILLING, L. V. AND HOBSON, G. D. (eds) *Petroleum Geology of the Continental Shelf of North-West Europe*. Heyden, London, 485–493.

EVANS, A. L., FITCH, F. J. AND MILLER, J. A. 1973. Potassium–argon age determinations on some British Tertiary igneous rocks. *Journal of the Geological Society, London*, **129**, 419–443.

GLEADOW, A. J. W., DUDDY, I. R. AND LOVERING, J. F. 1983. Fission track analysis: A new tool for the evaluation of thermal histories and hydrocarbon potential. *Australian Petroleum Association Journal*, **23**, 93–102.

JACKSON, D. I., MULHOLLAND, P., JONES, S. M. AND WARRINGTON, G. 1987. The geological framework of the East Irish Sea Basin. *In*: BROOKS, J. AND GLENNIE, K. W. (eds) *Petroleum Geology of North West Europe*. Graham & Trotman, London, 191–203.

KIRTON, S. R. AND DONATO, J. A. 1985. Some buried Tertiary dykes of Britain and surrounding waters deduced by magnetic modelling and seismic reflection methods. *Journal of the Geological Society, London*, **142**, 1047–1057.

WOODLAND, A. W. 1977. *Lake District aeromagnetic anomaly map, 1: 250,000 series*. Institute of Geological Sciences, Natural Environment Research Council.

—— 1978. *Liverpool Bay aeromagnetic anomaly map, 1:250,000 series*. Institute of Geological Sciences, Natural Environment Research Council.

The Kish Bank Basin, offshore Ireland

D. NAYLOR, N. HAUGHEY, G. CLAYTON and J. R. GRAHAM

Department of Geology, Trinity College, Dublin 2, Ireland

Abstract: The Kish Bank Basin, a half-graben lying offshore only a few kilometres east and southeast of Dublin in the Irish Sea, is elongated northeast–southwest and measures approximately 48 × 32 km. The basin is bounded to the northwest and north by the Dalkey–Lambay fault system and to the southwest by the Bray Fault. A major dextral strike-slip fault, the northwest-trending Codling Fault, bisects the basin. Three wells have been drilled in the area: Amoco 33/22-1(1977), Shell 33/21-1(1979) and Fina 33/17-1(1986). The Amoco well penetrated a Carboniferous section on the southeastern rim of the basin (Jenner 1981), while the other two holes drilled the Permo-Triassic basin-fill sequence, but without reaching the base of the Permian. The Permo-Triassic succession is similar to that in the East Irish Sea Basin and includes thick Triassic salt sequences. SEM examination of sandstone samples from potential reservoir sections in the 33/21-1 well indicates good porosity in the cleaner sandstones. This is interpreted as largely secondary, following dissolution of early cements. Seismic studies suggest that up to 2.7 km of Jurassic (probably Liassic) strata are present against the western bounding faults.

Maturation studies have been carried out on material from two of the wells. The Silesian section penetrated by 33/22-1 ranges in age from Westphalian B to Westphalian D. It is of 'coal measures' facies and includes numerous seams of medium to high volatile bituminous coal. Much of the Westphalian section is mature with regard to gas generation, with vitrinite reflectance (*Rm*) values ranging from *c.* 0.8% to 1.3%. Maturation levels in the Permo-Triassic section in 33/21-1 range from *c.* 1.3%–1.6% *Rm*, suggesting that any underlying Westphalian rocks in deeper parts of the basin would be mature to marginally post-mature with respect to dry gas generation. The burial history of the basin, and the distribution of source rocks within it, are discussed.

The Kish Bank Basin is one of several basins which lie beneath the Irish Sea between Ireland and Britain. These include the Kish Bank, East Irish Sea, Central Irish Sea, Cardigan Bay and St George's Channel basins (Fig. 1). The Kish Bank Basin is a relatively small (30 × 40 km) isolated sedimentary trough east of Dublin which is separated from the Central Irish Sea Basin to the southeast by an area of outcropping Carboniferous and older strata.

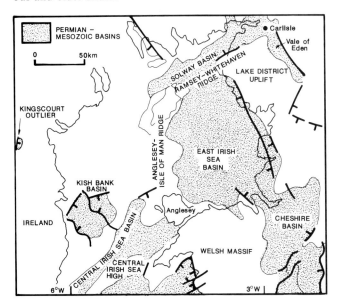

Fig. 1. Permian and Mesozoic basins of the central Irish Sea region.

The presence of the Kish Bank Basin was first suspected from gravity data (Bott 1964). These revealed a rectangular gravity low characterized by steep gravity gradients on the northern and western margins suggestive of fault-bounded basin margins. To the southeast the gravity data showed a gentle gradient onto the Mid-Irish Sea positive element. The picture deduced from the gravity has subsequently been con-

firmed by seismic surveys. A small speculative seismic programme shot by Seiscom Delta was the first to confirm the existence of a deep, if rather small, sedimentary basin. The seismic data clearly demonstrate a half-grabenal form, with fault control of the basin at its northern and western boundaries. The basin is developed in the hangingwalls of two orthogonal faults, the Bray and Dalkey faults (Fig. 2). It is bisected by the NNW-trending Codling Fault, which imposes a noticeable dextral offset. The Codling Fault does not appear to have been active during sedimentation, but is probably a Tertiary structure comparable in trend and age to the Stickle-path and similar faults in southwest England.

On the evidence of research geophysical surveys and seabed sampling, Etu-Efeotor (1976), Dobson (1977) and Dobson and Whittington (1979) suggested that the stratigraphic sequence in the Kish Bank Basin comprised Permo-Triassic, Liassic and Tertiary strata. In 1976, a group operated by Amoco Ireland was awarded Block 33/22, and in 1977 was additionally awarded Blocks 33/16, 33/17 and 33/21. In late 1977, the licence group drilled the first well in the area, 33/22/-1, to a depth of 976.3 m (Jenner 1981).

Irish Shell Petroleum Development Company, as operator of a joint venture group with Agip Ireland Ltd, drilled Well 33/21-1 under a 'farm-in' option agreement with the Amoco Group. The well was drilled in late 1979 to a depth of 2338 m (2303 m subsea). The main objective of the well was to test an anticline with fault-dependent closure at Lower Triassic sandstone level, sealed by Keuper Marl. The test was unsuccessful, and the licence was subsequently relinquished by the Amoco consortium.

In 1986 a third well, 33/17-1, was drilled by a five-company group operated by Charterhouse Petroleum to a TD of 2030 m (1906.5 m subsea). Neither the 33/21-1 nor the 33/17-1 well penetrated the sub-Permian.

Jenner (1981) published details of the Amoco 33/22-1-well, together with an outline map of the basin based on seismic interpretation (Fig. 2A). The Amoco well was drilled on the rim of the basin and encountered, beneath Quaternary sands, an Upper Carboniferous section resting directly on Lower Paleozoic strata. Since that time, summary accounts of the

From *Petroleum Geology of Northwest Europe: Proceedings of the 4th Conference* (edited by J. R. Parker).

basin with speculation regarding stratigraphy and basin development, have been published by Naylor and Shannon (1982) and Shannon (1991). The possible thickness, age and nature of the Jurassic sequence in the basin were considered by Broughan *et al.* (1989). Details of the Shell 33/21-1 (1979) and Fina 33/17-1 (1986) wells are published here for the first time.

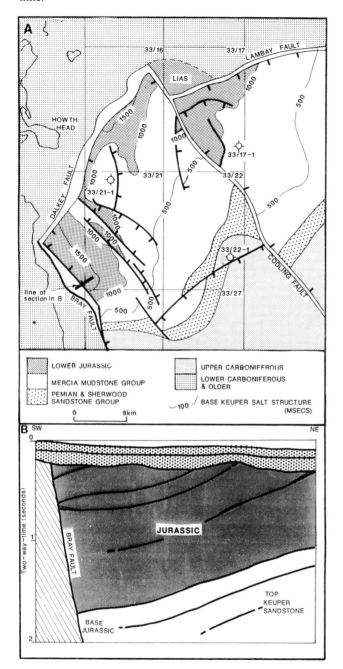

Fig. 2. (A) The geology and structure of the Kish Bank Basin (modified after Jenner 1981); (B) Geoseismic section at the western margin of the basin (modified after Broughan *et al.* 1989).

General stratigraphy

The Amoco 33/22-1 well was drilled outside the limits of the Permo-Triassic basin while the two later wells within the basin penetrated the Permo-Triassic, but did not reach the Carboniferous. In consequence, the lower part of the Permian section has not been seen. The disposition of Carboniferous or older stratigraphic units on the Variscan unconformity surface

beneath the basin is unknown, and subdivisions of the pre-Permian section cannot be mapped seismically with any certainty.

The stratigraphy of the Shell 33/21-1 and Fina 33/17-1 wells is shown in Fig. 3 and Table 1. The correlations shown on the diagram are based on the completion reports prepared by the operators, on the down-hole logs, and on an examination of the cuttings and sidewall cores by the authors. Biostratigraphic data for the Permo-Triassic are derived from the well completion report in the case of 33/21-1, and for Fina 33/17-1 from work carried out by the Applied Geology Unit at Trinity College, Dublin. The quoted depths are below the rotary table (RT) in each case.

The Permo-Triassic section in the Kish Bank Basin correlates reasonably with that of the East Irish Sea and Cheshire basins. The authors have, therefore, decided to use much of the lithostratigraphic nomenclature from those areas at formation level, despite the fact that it can be argued that the Kish Bank is a separate basin. Common informal unit names have been used, rather than burden the literature with local nomenclature. We have also avoided the use of the cumbersome term 'equivalent', since this would not add significantly to the discussion.

Table 1. Formation tops (m) for the Shell 33/21-1 and Fina 33/17-1 wells, Kish Bank Basin

| Lithostratigraphy | Shell 33/21-1 | | Fina 33/17-1 | |
| | Top of unit (m) | | Top of unit (m) | |
	RT	Subsea	RT	Subsea
Quaternary–Tertiary	0	0	0	0
Mercia Mudstone Group				
Upper Keuper Marl	(183)	(148)	385.3	351.8
Upper Saliferous Beds	339	304	782.7	749.2
Middle Keuper Marl	1030	995	?833.3	799.8
Lower Saliferous Beds	1139	1104	?871.4	837.9
Lower Keuper Marls	1282	1247	908.3	872.8
Sherwood Sandstone Group				
Keuper Waterstones Fm.	1305	1270	920.5	887.0
Keuper Sst Fm. Member 3	1326	1291	937.6	904.1
Member 2	1396	1361	?1036.9	1003.4
Member 1	1446	1411	?1060.7	1025.2
St Bees Sandstone Fm.	1570	1535	1135.7	1102.2
Manchester Marl	—	—	1894.0	1860.5
Collyhurst Sandstone Fm.	—	—	1929.7	1896.2

Lower Paleozoic stratigraphy

The Amoco 33/22-1 well (Jenner 1981) is the only well to have penetrated the pre-Permian section of the Kish Bank Basin. An outline stratigraphy of the well is shown in Fig. 5. The well terminated in chloritic slates at 976.3 m, of which 110.6 m were penetrated. This sequence was undated, but was correlated by Jenner on lithological grounds with the Bray Group of the immediate onshore area in County Wicklow. The Bray Group (Gardiner 1970) is of Lower to Middle Cambrian age. However, in the absence of firm dating, it is also possible that the strata encountered in the 33/22-1 well could belong to the Ribband Group (Crimes and Crossley 1968) of Upper Cambrian to Arenig age, which also crops out in County Wicklow and which contains lithological sequences similar to those in the well.

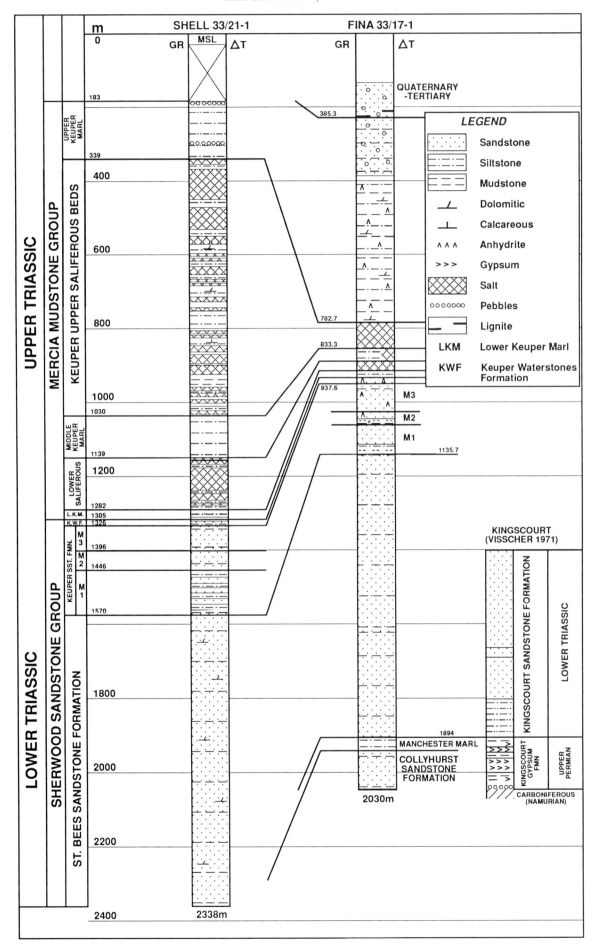

Fig. 3. The stratigraphy of the Shell 33/21-1 and Fina 33/17-1 wells, Kish Bank Basin, and comparison with the onshore section in the Kingscourt Graben (Visscher 1971).

Carboniferous stratigraphy

The biostratigraphic boundaries in the Carboniferous section are based on a palynological re-investigation of the 33/22-1 well by the present authors, and differ somewhat from those published by Jenner (1981) based on the well completion report.

In the 33/22-1 well a Westphalian section (*c.* 740 m thick) rests with unconformity on the Lower Paleozoic sequence. The lowest section, Westphalian B in age, 153 m thick, consists of a basal sandstone, succeeded by interbedded siltstones and sandstones with minor coals. The top of Westphalian B is tentatively placed at the first down-hole record of *Dictyotrilites mediareticulatus* and *Raistrickia fulva* at 734.6 m. The absence of in situ *Radiizonates aligerens* in the Westphalian section suggests that Westphalian A strata were not penetrated.

The Westphalian C section is 388 m thick, and consists of siltstones, sandstones, claystones and coals. The tops of the ranges of the Westphalian C taxa, *Vestipora costata* and *V. pseudoreticulata* are at 347.5 m. Abundant *Cadiospora magna, Raistrickia aculeata* and *Vestipora fenestrata* occur at 384 m, some 12 m below a casing point. These taxa first appear at, or immediately below, the base of Westphalian D, suggesting that the Westphalian C–D boundary in the well is located somewhere between 347.5 m and 384 m.

The first down-hole occurrence of *Cadiospora magna* at 219.5 m marks the top of the section of definite Westphalian D age. The miospore assemblage from 338.3 m contains numerous reworked specimens including *Ahrensisporites guerickei* (Namurian–Westphalian A), *Grandispora spinosa* (late Viséan–early Namurian) and *Radiizonates aligerens* (late Westphalian A).

The uppermost 114 m of the Carboniferous succession are unfossiliferous grey siltstones, red shales and brown sandstones. Although this section was considered by Jenner (1981) to be possibly Stephanian in age, it is equally likely to be a red (or reddened) unit of Westphalian D age comparable to those widely developed in North Wales and West Midlands of England. The age of the interval between 164.6 m and 219.5 m is indeterminate. However, the rare specimens of *Densosporites* spp. and *Lycospora pusilla* recorded indicate an age no younger than Lower Permian.

Permo-Triassic stratigraphy

Collyhurst Sandstone Formation

The correlation in Fig. 3 shows the lowest section of the Shell well to belong to the St Bees Sandstone (Triassic), rather than to the Permian, as suggested as a second alternative in the completion report for the well. There were no diagnostic palynology results at these levels. Correlation with the Fina well makes it likely that the Manchester Marl horizon was not reached in the 33/21-1 well.

The Collyhurst Sandstone was thus penetrated only in the Fina 33/17-1 well, between 1929.7 m and TD. The basal unconformity may, judging from the seismic sections, be only 100 m or less below TD. The interval in the well consisted entirely of fine to coarse sandstone, pale to dark red-brown with rounded to well-rounded grains. The rock is friable, with good to excellent visible porosity.

Manchester Marl

This is a thin interval (96.6 m) in the 33/17-1 well which comprises pink to dark-red or brown claystone, slightly calcareous, with interbeds of red-brown siltstone and fine sandstone. The unit name is used informally, without implying precise correlation with the English sequences.

The samples and sidewall cores which were processed for palynomorphs from these lower two formations proved barren, except for poorly preserved disaccate pollen which were judged to be from cavings.

Sherwood Sandstone Group

St Bees Sandstone Formation The dominant lithology in this formation is red, very fine- to fine-grained sandstone, moderately well sorted. In the 33/17-1 well the sandstones become slightly silty with depth and contain occasional poorly cemented interbeds. White, well-rounded, fine- to medium-grained, sandstone beds occur towards the top of the interval in Shell 33/21-1. Throughout the section there are interbeds of pink to grey firm blocky, anhydritic claystone, occasionally to commonly sandy. The lower part of the interval in 33/17-1 is partially halitic. The bulk of the sandstones are sub-angular, and this together with the red coloration, evaporitic content and lack of fauna suggests water-laid deposition in an arid environment. Zones of less consolidated, well-rounded sands suggest an increasing aeolian influence upwards in the sequence.

Below 2114 m in Shell 33/21-1 and 1496 m in Fina 33/17-1, the sandstones are tightly cemented. This represents the 'silicified zone' of the East Irish Sea Basin (Colter and Barr 1975; Colter 1978). Colter suggested that silicification had been controlled by argillaceous and/or micaceous material in the original sediment. Gamma ray activity is generally higher in this zone and this is true in the two Kish basin wells. Porosities are generally in the 14–18% range in the upper St Bees interval, but fall to 8% or less in the 'silicified zone'.

Keuper Sandstone Formation This interval is predominantly sandy in both wells, with subordinate claystone intercalations. The miospores *Striatoabietites aytugii, Triadispora* spp. and *Lundbladispora* spp. were recorded from 960.1 m in the Fina well; these all range upwards from the upper Scythian. It is possible with some difficulty to distinguish three members within the formation in each well and these have been informally designated Members 1 to 3. Whether these equate with the Thurstaston, Delamere and Frodsham members recognized in the East Irish Sea and Cheshire basins is open to conjecture. From the base upwards these are:

● *Member 1*—variegated white to red-brown, very fine to coarse sandstones, rounded to sub-rounded and friable in the upper part, probably partly fluvial and partly aeolian in origin;
● *Member 2*—dominantly argillaceous fluvial sandstones, light to medium grey, green-grey to red-brown, firm to hard.
● *Member 3*—dominantly red-brown, occasionally grey, fine to medium, rarely coarse, sandstones. These are rounded to sub-angular, moderately well sorted, poorly cemented, slightly anhydritic or dolomitic, or argillaceous. There are subordinate interbeds of red-brown, firm, non-calcareous claystones, silty in part, and white to pink anhydrite. The sandstones in this interval may be predominantly aeolian in origin.

Keuper Waterstones Formation This formation provides a strong seismic marker. In the 33/21-1 well the interval comprises dark red-brown claystone and shale with siltstone and sandstone intercalations. There is an upwards gradation into loosely cemented sandstone which forms the upper part of the unit. The dominant lithology in the Fina 33/17-1 well is white to light-grey or brown, fine-grained sub-rounded to well-rounded and well-sorted, hard sandstone, with white or pink anhydrite or red-brown claystone interbeds. The interval failed to yield diagnostic flora or fauna in either well.

Mercia Mudstone Group

This interval is considerably thinner in Fina 33/17-1 (535.2 m) than in the Shell 33/21-1 well (1122 m). There is considerably less halite in the Fina well and the intervening marl division is less clearly defined. Direct correlation with stratigraphic subdivisions in the East Irish Sea Basin (Jackson *et al.* 1987) is not possible, and informal nomenclature is used, without implying correlation with specific English sequences. The depositional environment throughout was continental and arid to semiarid, with alternations between fluvial, lacustrine, evaporitic and aeolian conditions.

Lower Keuper Marl This unit is dominated in both wells by red-brown and grey, soft to firm claystones which are frequently silty or micaceous, and occasionally halitic or anhydritic. The unit yielded miospores which indicate a Scythian age.

Lower Saliferous Beds This section consists predominantly of halite with minor amounts of red claystone. The halite is clear, white or light brown, crystalline hard and blocky. In the 33/21-1 well, a late Scythian age is indicated for the interval 1139–1171 m and a more general Scythian age for the interval 1171–1282 m on the basis of microflora assemblages recorded from claystone interbeds.

The first down-hole appearance of *Spinotriletes echinoides* in a ditch cuttings sample from 905.3 m in 33/17-1 indicates an age not older than upper Scythian. A similar age is confirmed by the presence of *Striatoabietites aytugii* and *Jugasporites conmilvinus* from a sidewall core at 931.8 m. The occurrence of *Lueckisporites virkkiae* in the same sample was attributed to reworking of Permian material.

Middle Keuper Marl Red-brown and green-grey claystone and siltstone dominate the interval, with subordinate sandstones. There are intercalations and lenses of light-brown to light-grey fine sandstone in the middle part of the section.

On the basis of palynological results, the unit in the Shell well was thought to be late Scythian in age.

Upper Saliferous Beds This unit comprises almost equal amounts of halite and silty red-brown, occasionally green-grey, claystone. There are also minor amounts of intercalated siltstone. The occurrence of *Verrucosisporites thuringiensis* and *Rugulatisporites mesozoicus* from 795.5 m in 33/17-1 indicates an upper Scythian age for this unit. Miospores from the interval 401–663 m in 33/21-1 containing *Alisporites grauvogeli* and *Microcachrydites sittleri* were interpreted as Anisian.

Upper Keuper Marl This unit is entered at 183 m (top returns) in Shell 33/21-1 and at 385.3 m in Fina 33/17-1. Red-brown claystones dominate the interval, locally grading into siltstone, light-brown to green-grey. There are occasional intercalations of gypsum and of light-grey sandstone. In the Shell well, pebble beds containing clasts of green igneous rock, black chert, limestone and sandstone are found at 183 and 296 m. The first down-hole occurrence of *Densoisporites nejburgii* at 768.1 m in 33/17-1 is taken as the top of the Scythian within the Upper Keuper Marl. In the same well, the top of the Anisian Stage is placed at 658.4 m based on the first down-hole appearance of *Lueckisporites triassicus* and *Microcachrydites fastidiosus*. The highest sample which can be confidently assigned to the Ladinian Stage in 33/17-1 is from 457.2 m, based on the first down-hole occurrence of *Microcachrydites sittleri*.

The nearest onshore Upper Permian–Lower Triassic succession (Fig. 1) to the Kish Bank Basin is the small Kingscourt Outlier, 60 km NNW of Dublin. Here, well dated Late Permian (Tatarian) conglomerate, sandstone, mudstone and gypsum of the Kingscourt Gypsum Formation rest unconformably on strata of Westphalian A age (Visscher 1971). The Kingscourt Gypsum Formation is overlain conformably by sandstone and siltstone of the Early Triassic (Scythian) Kingscourt Sandstone Formation (Visscher 1971). Correlation of the Kingscourt and Kish Bank successions is shown in Fig. 3.

Jurassic (Liassic) stratigraphy

Although neither of the wells encountered Jurassic rocks, seismic mapping, seabed sampling and other evidence (Dobson and Whittington 1979; Jenner 1981; Broughan *et al.* 1989) suggest the presence of Jurassic strata along the western and northern margins of the basin (Fig. 2A). Broughan *et al.* (1989) argue that up to 2700 m of Liassic rocks, probably parallel bedded and mudrock dominant, are present against the Bray Fault (Fig. 2B).

Quaternary–Tertiary stratigraphy

In the Fina 33/17-1 well the interval 173.7 m (top returns) to 385.3 m consists predominantly of variable red-brown to grey sandstone with increasing amounts of claystone with depth. Lignite beds typify the upper part of the section but decrease in importance downwards.

On the basis of very poor miospore recoveries, the interval above 237.7 m is thought to be Pleistocene to Recent in age, while the lower section is regarded as being Pliocene or older.

Structure

The Permo-Triassic and Lower Jurassic rocks of the Kish Bank Basin form a half-graben wedge of sediments against the western bounding faults. The underlying Variscan unconformity surface is not a striking seismic feature, and only rarely can lower reflectors be mapped.

The Kish Bank Basin was probably a depositional centre in Permo-Triassic time. Onlap of Permian units against the Variscan floor is seen on some seismic records. There are also some thickness variations in Triassic units associated with intrabasinal faults and it is probable that the western margin faults were also active at this time, although this is difficult to demonstrate. There is little to be seen on the seismic sections by way of wedging of units against the basin margin faults. The degree to which the marginal faults were overstepped, and the basin linked to other depositional centres, is not known.

Earliest Jurassic marine transgression appears to have taken place without a significant period of movement or erosion. Again, only minor variations of thickness within the Liassic sequence are seen on the seismic sections against the marginal faults (Broughan *et al.* 1989), and the Lambey–Dalkey fault system may have been overstepped (see Naylor (1992) for discussion). The structural attitude of the preserved Liassic strata in the basin suggests that tilting or basin inversion occurred after Lower Jurassic deposition.

Middle and Upper Jurassic strata are absent in the Ulster Basin and in the central Irish Sea north of Cardigan Bay. Tate and Dobson (1989) have suggested that a broad E–W axis of uplift developed across Ireland at the end of the Early Jurassic and extended into the area of the central Irish Sea. Naylor (1992) has proposed that this positive element persisted through into early Cretaceous time and was responsible for the absence of pre-Upper Cretaceous strata across the region. There is a possibility, however, that the absence of Upper Cretaceous rocks in the central Irish Sea may be due to erosion rather than non-deposition (see Naylor (1992) for discussion). In terms of the Kish Bank Basin, therefore, tilting and erosion of the Triassic and Liassic sequences may have

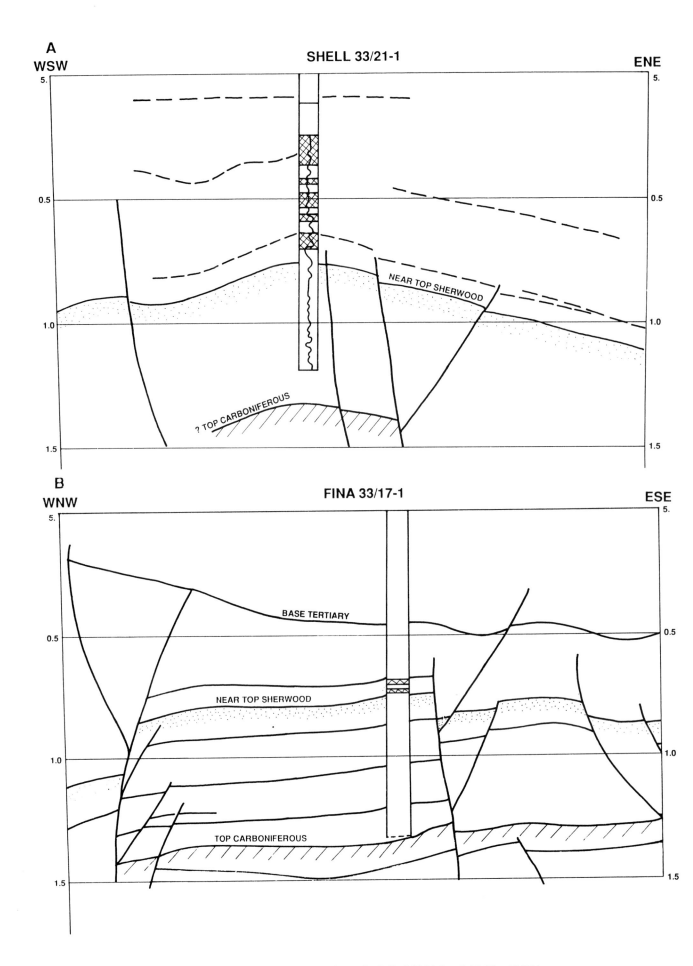

Fig. 4. Line drawings based on seismic sections across the well locations of (**A**) Shell 33/21-1 and (**B**) Fina 33/17-1.

commenced at the end of the Early Jurassic. It is not possible to say whether this early movement phase, or the widespread end Jurassic Late Cimmerian phase affected the area to the greater degree.

Late Cretaceous regional movements probably reactivated existing fault lines in the basin, possibly with some strike-slip movement and the elevation or depression of individual fault blocks. Parts of the basin suffered further subsidence during the Tertiary, with accompanying deposition. An important phase of fault movement occurred in mid-Tertiary time, and this probably included significant dextral offset on the Codling Fault, as well as reactivation of older faults elsewhere in the basin. It is possible that the Codling Fault itself had an earlier progenitor.

The 33/21-1 and 33/17-1 wells were drilled within mapped closures on two different types of structure (Fig. 4). The Shell 33/21-1 well was drilled on a fold structure, although closure at Triassic sandstone level was probably fault dependent. The fold may have been produced by late inversion of the basin. In contrast, the Fina well, east of the Codling Fault, was to test an uplifted fault horst. As might be anticipated, the geometries of the mapped faults which bound the structure are indicative of strike-slip movement. It can be argued that the thinning of the Mercia Mudstone evaporite horizons in 33/17-1, which was noted above, reflects positive movement on the structure during deposition, rather than simply a general eastward thinning of the total section. On the other hand, both structures could to a large extent be a product of Tertiary movements.

Maturation

Maturation levels in the Amoco 33/22-1 well were briefly discussed by Jenner (1981), based on rather few vitrinite reflectance determinations. A more detailed investigation was completed by one of the authors (Haughey 1987), based on coals from 14 ditch cuttings samples, and two sidewall cores (Table 2 and Fig. 5).

Table 2. Vitrinite reflectance (*Rm*) data from the Amoco 33/22-1 and Shell 33/17-1 wells.

Depth (m)	Lithology	Rm (%)	S.D.	Number of grains
Well 33/22-1				
219.5	coal	0.96	0.08	100
347.5	coal	0.83	0.08	100
384	coal	0.83	0.07	100
393.2	coal	0.84	0.08	100
472.4	coal	1.02	0.09	100
576.1	coal	1.02	0.08	100
588.3	coal	1.06	0.07	100
603.5	coal	1.13	0.09	100
637	coal	1.18	0.09	100
649.2	coal	1.15	0.11	100
698	coal	1.22	0.09	100
710.2	coal	1.22	0.10	100
731.5*	siltstone	1.23	0.12	100
734.6	coal	1.21	0.11	100
830.3*	siltstone	1.30	0.13	55
832.1	coal	1.35	0.11	100
Well 33/21-1				
605*	mudrock	1.23	0.13	4
663*	mudrock	1.30	0.16	39
967*	mudrock	1.54	0.13	17
1162.5*	mudrock	1.67	0.23	24
1300*	mudrock	1.51	0.13	5

* sidewall core

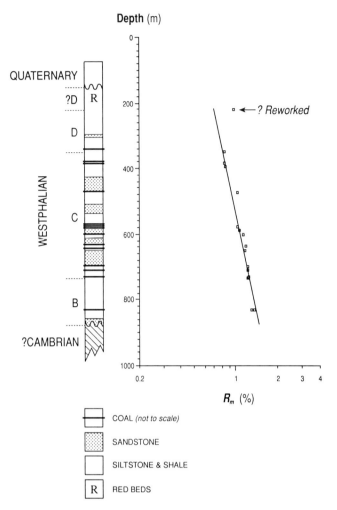

Fig. 5. Vitrinite reflectance (*Rm*) data from Amoco 33/22-1.

The reflectance determinations range from 0.83–1.35% (*Rm*) with a relatively well constrained high gradient (Fig. 5). One anomalously high value obtained from coal fragments associated with red beds at 219.5 m may be due either to reworking, or to oxidation of the vitrinite.

A reworked miospore assemblage from cuttings at 1110 m in the 33/22-1 well includes Westphalian A taxa, suggesting uplift and erosion of earlier Carboniferous rocks during Westphalian time. Reworked miospores ranging in age from Devonian to Westphalian A were also recorded from an onshore Westphalian D section in County Wexford (Clayton et al. 1986), suggesting that the erosional episode was relatively widespread.

Jenner (1981) estimated that 3 km of uplift and erosion were necessary to account for the maturation level and reflectance gradient of the preserved Carboniferous section; the present authors estimate that only c. 1.3 km of cover has been eroded. An uppermost Carboniferous (Westphalian D–Stephanian) section is unlikely to have attained this thickness, so that the observed maturation level is probably due mainly to post-Carboniferous burial. This may have been achieved by Triassic and Liassic cover.

Reservoir diagenesis

A SEM study was made of 20 cuttings samples, 5 sidewall cores and a junk basket sample (2174 m) from Shell 33/21-1. Other sidewall cores from the well were taken in fine-grained lithologies and, therefore, were not relevant to the reservoir study. The stratigraphic position of the samples studied, and a summary of their characteristics, are shown in Table 3.

Fig. 6. Photomicrographs of sandstones from the Shell 32/21-1 well (**a**) 2174 m (junk basket): St Bees Sandstone Formation. Almost total removal of porosity by extensive quartz overgrowths and cements. Small patch of carbonate cement (c) is preserved right-centre. Width of photomicrograph = 1.2 mm. (**b**) 1521 m: Keuper Sandstone Formation, Member 1. Extensive patch of illite from a sandstone which elsewhere has good porosity. (**c**) 1515 m: Keuper Sandstone Formation, Member 1. Sandstone with good porosity and minor development of pore-bridging illite (bottom centre). (**d**) 1362 m: Keuper Sandstone Formation, Member 3. Typical poorly sorted muddy sandstone with negligible porosity.

Table 3. Summary data for samples from Shell 33/21-1. SWS: sidewall sample; md.sst: muddy sandstone; sst: well-sorted sandstone; sst(r): well-sorted sandstone with rounded grains, possibly aeolian; mr: mudrock; x: not present; *: present.

Depth (m)	Lithology	Visual porosity	Detrital clays	Overgrowths	Illite	Other
Keuper Waterstones						
1305	md.sst	fair	common	x	*	
Keuper Sst. Fm. (Mbr. 3)						
1340	sst(r)	good	x	x	*	
1356	md.sst	poor	common	x	*	
1362	md.sst	fair	*	x	* (+ fibrous illite)	carbonate rhombs
Keuper Sst. Fm. (Mbr. 2)						
1395	sst(r)	good	x	x	*	
1398(SWS)	md.sst	poor	common	x	x	
1425	sst	fair	rare	x	*	
1430	md.sst	fair	*	x	*	
Keuper Sst. Fm. (Mbr. 1)						
1450	sst	good	x	quartz minor	*	halite
1486 (SWS)	md.sst	poor	common	x	x	
1494	md.sst	not detected	x	x	x	mica-rich contamination
1500	md.sst	fair	common	x	x	
1509	sst	good	x	x	*	
1515	sst	good	x	x	*	
1518	sst(r)	good	x	x	*	
1521	sst(r)	good	x	x	*	minor quartz
1528	sst(r)	good	x	quartz minor	*	
St Bees Sst. Fm.						
1704 (SWS)	md.sst	poor	common	x	*	
1707	md.sst	fair	*	quartz minor	*	
1716	md.sst	fair	*	x	x	halite
1748(SWS)	mr	poor	common	x	x	
1818	md.sst	fair	common	x	*	
1821 (SWS)	md.sst	poor	common	x	x	
2110	md.sst	fair	*	x	*	halite
2174 (junk basket)	sst	poor	x	quartz extensive	* (+ fibrous illite)	carbonate cement common
2270	md.sst	poor	common	x	*	

All the samples examined were very friable, except for the junk basket sample. This is surprising in view of the maturation values (Table 2). Compactional effects are seen only in some of the muddier sandstones, which appear to have had a different diagenetic history to the relatively mud-free sandstones. It seems probable that the lack of compactional features in the presently porous sandstones is due to the widespread presence of an early pore-filling cement, which has been later dissolved to give secondary porosity. Similar observations and interpretation were made by Burley (1984) in a regional study of the Sherwood Sandstone Group. The rather irregular sizes and distribution of pores observed in several of the samples supports the suggestion of a secondary porosity.

The junk basket sample shows clear evidence for extensive carbonate cementation, as well as the silica cementation which is mainly responsible for the strong induration. Both thin section and SEM study suggest formation of an early carbonate cement and later quartz overgrowths and cementation (Fig. 6a). This level is interpreted as localized preservation of early cement, in this case carbonate, due to almost total loss of permeability by quartz cementation, whereas elsewhere in the section the early cements were dissolved, producing secondary porosity. In some samples (1716 m, 2110 m in 33/21-1) there is evidence for an early phase of evaporite precipitation in the form of halite, which may have fulfilled a similar role to the carbonate cementation elsewhere. This halite generally shows evidence of dissolution.

Modification of secondary porosity by the deposition of diagenetic clays is very variable in occurrence and there is no obvious pattern discernible from the material examined. The clay appears to be exclusively illite, which is precipitated mainly in platy form (Fig. 6b). Fibrous illites are also present but are not common (Table 2). The illite occurs in places as thin coatings of detrital grains, as small plates overgrowing detrital grains, as pore-bridging cement (Fig. 6c), and as patches of small plates within pore spaces. Other mineral phases observed during the SEM study were of minor importance volumetrically.

In general the results of this study are in good agreement with work on cores from the Morecambe Gas Field (Colter and Ebbern 1978, 1979; Ebbern 1981) and elsewhere in the Sherwood Sandstone Group (Burley 1984; Strong and Milodowski 1987). The three sandstone types noted by Colter and Ebbern (1979) can be matched with material from the Kish Bank Basin succession, and are also indicated in the tabulated data of Burley (1984). These are:

(1) sandstones with quartz overgrowths and cement, and also with calcite cementation and generally limited porosity. These were noted as forming prominent horizons within the St Bees Sandstone Formation, as in the Kish Bank Basin (Fig. 6a);

(2) muddy sandstones with little porosity, most clays being detrital (Fig. 6d);

(3) sandstones with high porosities, but variable permeability due to the effects of platy illite (Figs 6b,c). The effect on permeability might be expected to be greater than the effect on porosity as micropores could remain in the illite-cemented horizons (Colter and Ebbern 1979; Woodward and Curtis 1987).

The predominance of replacement and pore-filling illite as the main authigenic phase (type 3 above) matches our observations on the porous sandstones of the Kish Bank Basin. The predominance of illite as the authigenic clay phase may be as much related to depth (temperature) as to the pore water/host grain chemistry. In the Morecambe Fields there is typically an illite-free zone overlying a zone of extensive illite development (Woodward and Curtis 1987) which has been related to a palaeo-gas–water contact. There is no support for a similar situation in the Kish Bank Basin from the very limited data available.

As noted elsewhere (Colter and Ebbern 1978, 1981; Burley 1984; Bushell 1986; Strong and Milodowski 1987), the primary depositional texture was a major control on the diagenetic history. The lack of early pore-filling cements in muddier sandstones has allowed compactional fabrics to dominate and permeabilities are universally low. In the originally more porous, mud-poor sandstones the main controls are interpreted to be early cement formation, reducing the main effects of compaction, and formation of platy illite within the secondary pores.

Hydrocarbon prospectivity

The Amoco 33/22-1 well demonstrated a Carboniferous section with hydrocarbon source rocks on the margin of the Kish Bank Basin. The two wells drilled subsequently to test the basin-fill sequence were apparently drilled on valid structural targets, and came in close to prognosis. Neither encountered significant shows of hydrocarbons despite demonstrating adequate reservoir potential in the Collyhurst Sandstone and in the Triassic Sherwood Sandstone Group, as well as good seal. Possible reasons for the lack of success might be that:

• the Westphalian source sections encountered at the basin rim do not continue beneath the basin and the older Carboniferous rocks lack good source sequences. Neither the 33/21-1 nor 33/17-1 wells penetrated to the pre-Permian, nor do the seismic data at this level allow accurate mapping of the sub-unconformity units. However, the Carboniferous section in Amoco 33/22-1 lacks any section below Westphalian B, suggesting possibly a marginal onlap situation in this part of the Irish Sea. If Lower Carboniferous rocks are present beneath the basin, as suggested by Jenner (1981), they may be in Irish Midlands facies with low source potential, rather than in the more promising facies of the Dublin Basin sequence;
• the structures are a product of Tertiary movements and post-date the main phases of hydrocarbon generation and migration.

On balance, it would seem that the first of these possibilities provides the main explanation for the lack of hydrocarbons in the two Kish Bank Basin tests. Jenner (1981) gives a cumulative thickness of more than 11 m for the coal seams encountered in 33/22-1. If these coal seams extended beneath the Permo-Triassic basin it is probable that some generation of gas would have continued in the Tertiary, even if in insufficient quantities to fill young structures completely. In the case of the 33/17-1 structure there is some evidence on the seismic records to suggest that the structure may have had some expression, and been available for gas entrapment, at an early stage of basin development.

The authors are grateful to Amoco Exploration for permission to publish details of the 33/22-1 and 33/21-1 wells, and for providing sidewall and cuttings material. Our thanks are also due to Arlington Exploration, Clyde Petroleum, Conroy Petroleum, Fina Exploration and Kish Developments for permissions to release results from the 33/17-1 well. We are also grateful to M. Rutherford, N. Ainsworth and M. O'Neill for biostratigraphical work carried out on the 33/17-1 samples in the Applied Geology Unit, Trinity College, Dublin. One of us (DN) would like to thank Fionnuala Broughan and Nigel Anstey for their work, assistance and helpful discussions on the 33/17-1 prospect.

References

BOTT, M. P. H. 1964. Gravity measurements in the north-eastern part of the Irish Sea. *Quarterly Journal of the Geological Society of London*, **120**, 369–396.

BROUGHAN, F. M., NAYLOR, D. AND ANSTEY, N. A. 1989. Jurassic rocks in the Kish Bank Basin. *Irish Journal of Earth Sciences*, **10**, 99–106.

BURLEY, S. D. 1984. Patterns of diagenesis in the Sherwood Sandstone Group (Triassic), United Kingdom. *Clay Minerals*, **19**, 403–440.

BUSHELL, T. P. 1986. Reservoir geology of the Morecambe Field. *In*: BROOKS, J., GOFF, J. C. AND VAN HOORN, B. (eds) *Habitat of Palaeozoic gas in N.W. Europe*. Geological Society, London, Special Publication, **23**, 189–208.

CLAYTON, G., SEVASTOPULO, G. D. AND SLEEMAN, A. G. 1986. Carboniferous (Dinantian and Silesian) and Permo-Triassic rocks in South County Wexford, Ireland. *Geological Journal*, **21**, 366–374.

COLTER, V. S. 1978. Exploration for gas in the Irish Sea. *In*: VAN LOON, A. J. (ed.) Key Notes of the MEGS-II (Amsterdam 1978). *Geologie en Mijnbouw*, **57**, 503–516.

—— AND BARR K. W. 1975. Recent developments in the geology of the Irish Sea and Cheshire basins. *In*: WOODLAND, A. W. (ed.) *Petroleum Geology of the Continental Shelf of North-West Europe. Vol. 1 Geology*. Applied Science Publishers, London, 61–75.

—— AND EBBERN, J. 1978. The petrography and reservoir properties of some Triassic sandstones of the Northern Irish Sea Basin. *Journal of the Geological Society, London*, **135**, 57–62.

—— AND —— 1979. SEM studies of Triassic reservoir sandstones from the Morecambe Field, Irish Sea, UK. *Scanning Electron Microscopy*, **1**, 531–538.

CRIMES, T. P. AND CROSSLEY, J. D. 1968. The stratigraphy, sedimentology, ichnology and structure of the Lower palaeozoic rocks of part of north-eastern Co. Wexford. *Proceedings of the Royal Irish Academy*, **67B**, 185–215.

DOBSON, M. R. 1977. The history of the Irish Sea Basins. *in*: KIDSON, C. AND TOOLEY, M. J. (eds) *The Quaternary History of the Irish Sea*. Geological Journal Special Issue No. **7**. Seel House Press, Liverpool, 93–98.

—— AND WHITTINGTON, R. J. 1979. The geology of the Kish Bank Basin. *Journal of the Geological Society, London*, **136**, 243–249.

EBBERN, J. 1981. Geology of the Morecambe Gas Field. *In*: ILLING, V. AND HOBSON, G. D. (eds) *Petroleum Geology of the Continental Shelf of North-West Europe*. Heyden, London, 485–493.

ETU-EFEOTOR, J. D. 1976. Geology of the Kish Bank Basin. *Journal of the Geological Society, London*, **132**, 708.

GARDINER, P. R. R. 1970. Regional fold structures in the Lower Palaeozoics of South-East Ireland. *Bulletin of the Geological Survey of Ireland*, **1**, 47–51.

HAUGHEY, N. 1987. *Vitrinite reflectance data from the Upper Palaeozoic of Ireland with particular reference to the Silesian coalfields*. PhD thesis, University of Dublin.

JACKSON, D. I., MULLOLLAND, P., JONES, S. M. AND WARRINGTON, G. 1987. The geological framework of the East Irish Sea Basin. *In*: BROOKS, J. AND GLENNIE, K. W. (eds) *Petroleum Geology of North West Europe*. Graham & Trotman, London, 191–200.

JENNER, J. K. 1981. The structure and stratigraphy of the Kish Bank Basin. *In*: ILLING, L. V. AND HOBSON, G. D. (eds) *Petroleum Geology of the Continental Shelf of North-West Europe*. Heyden, London, 426–431.

NAYLOR, D. 1992. The post-Variscan history of Ireland. *In*: PARNELL, J. (ed.) *Basins of the Atlantic Seaboard: Petroleum Geology, Sedimentology and Basin Evolution*. Geological Society, London, Special Publication, **62**, 255–275.

—— AND SHANNON, P. M. 1982. *The Geology of Offshore Ireland and West Britain*. Graham & Trotman, London.

SHANNON, P. M. 1991. The development of the Irish offshore sedimentary basins. *Journal of the Geological Society, London*, **148**, 182–189.

STRONG, G. E. AND MILODOWSKI, A. E. 1987. Aspects of the diagenesis of the Sherwood Sandstones of the Wessex Basin and their influence on reservoir characteristics. *In*: MARSHALL, I. D. (ed.) *Diagenesis of sedimentary sequences*. Geological Society, London, Special Publication, **36**, 325–337.

TATE, M. F. AND DOBSON, M. R. 1989. Late Permian and early Mesozoic rifting and sedimentation offshore NW Ireland. *Marine and Petroleum Geology*, **6**, 49–59.

VISSCHER, H. 1971. The Permian and Triassic of the Kingscourt Outlier, Ireland. A palynological investigation related to regional stratigraphical problems in the Permian and Triassic of Western Europe. *Geological Survey of Ireland Special Paper No. 1*.

WOODWARD, K. AND CURTIS, C. D. 1987. Predictive modelling for the distribution of production-constraining illites—Morecambe Gas Field, Irish Sea, Offshore UK. *In*: BROOKS, J. AND GLENNIE, K. W. (eds) *Petroleum Geology of North West Europe, Vol. 1*. Graham & Trotman, London, 205–215.

The tectonic history of the East Irish Sea Basin with reference to the Morecambe Fields

R. J. KNIPE,[1] G. COWAN[2] and V. S. BALENDRAN[2]

[1] *Department of Earth Sciences, The University, Leeds LS2 9JT, UK*
[2] *British Gas plc Exploration and Production, 100 Thames Valley Park Drive, Reading RG16 1PT, UK*

Abstract: Regional seismic lines from the East Irish Sea area and a 2D radial seismic survey across the Morecambe Fields have been used to assess fault pattern evolution and the history of tilting in the East Irish Sea Basin. Extension and fault-controlled sedimentation began in Permian times and continued throughout the Permo-Triassic. Later inversion and erosion has removed the post-Triassic cover in the area. The main extension direction during basin evolution was ENE–WSW to NE–SW. Under this extension, NE–SW-trending basin-bounding faults in the north and south were dominated by strike-slip and oblique displacements. Offset of the pre-Permian basement suggests the amount of post-Carboniferous net extension varies from 10–16%. The basin can be divided into two discrete structural domains which controlled the geological evolution of the East Irish Sea Basin. Basin evolution in the northern part of the basin was controlled by easterly dipping faults which induced tilting towards the southwest or west. The southeastern part of the basin is dominated by two faults which dip to the west or southwest, inducing tilting towards the east. The complex structural and tilting history of the Morecambe Fields arises from their position at the boundary of these two domains. Detailed analysis of this area illustrates how the propagation of the Keys Fault (from the north) and the Crosh Vusta Fault (from the south) competed to dominate tilting history. The faults which now bound the Morecambe Fields show only minor influence on Permo-Triassic sedimentation patterns, implying that the main displacement on these faults post-dated the deposition of the Sherwood Sandstone. The development of these boundary faults is thought to be related to the accommodation of arching between the Keys Fault and Tynwald fault system. The differing tilting histories of the North and South Morecambe Fields are indicated by the differing orientations of a previously shared palaeo-hydrocarbon–water contact. This is now preserved as the top platy-illite surface and is interpreted as arising from the development of a tilt-transfer zone during the propagation and linking of the northerly and southerly propagating fault arrays. Analysis of the tilting history of the Morecambe area has allowed the top platy-illite surface to be mapped outside well control.

Aims of the study

The East Irish Sea Basin occupies the area east and southeast of the Isle of Man and forms the central part of a linked set of Permo-Triassic basins extending from the Cheshire Basin in the south to the Solway Basin in the north. This paper assesses the regional tectonic setting of the Morecambe Gas Fields and the detailed structural evolution of the fault array within the fields. Details of the geology of the Morecambe Fields can be found in Colter and Barr (1975), Bushell (1986), Stuart and Cowan (1991) and Stuart (1993). This analysis was undertaken to help define structural controls on trap formation and diagenetic layering in the East Irish Sea Basin (EISB), and was carried out during development drilling of the South Morecambe Field, and prior to preparation of the Annex B for the structurally separate North Morecambe accumulation. Both fields contain platy-illite which precipitated below a palaeo-hydrocarbon–water contact, and the mapping of illite distribution is of fundamental importance to field development (Colter and Barr 1975; Bushell 1986; Woodward and Curtis 1987). Good well control in the South Morecambe Field allowed an empirical model to be established which showed that a linear relationship existed between the depth to top platy-illite and the depth to the top of the reservoir, but limited well control over the North Morecambe Field cast uncertainty over the mapping of the top platy-illite surface in the North Morecambe structure. Part of this study was to use the structural evolution of the area to map illite distribution in North Morecambe, the details of which are given in Stuart (1993).

In particular the aims of the study were to assess:

(1) the pattern and orientation of displacements associated with extension and inversion events in the area;
(2) the distribution and location of faulting during different stages of structural evolution;
(3) the tilting history of different parts of the North and South Morecambe Fields;
(4) the structure of the field at the time of platy-illite precipitation;
(5) the present geometry of the palaeo-gas–water contact (top platy-illite surface) in light of the tilting history of North and South Morecambe.

Database

A selection of regional seismic lines from the eastern Irish Sea area together with the results of a 500 km 2D radial seismic survey across the Morecambe Fields (H861-38 to H861-2) have been used to assess the evolution of the fault patterns. Well data from 42 wells and a dipmeter study of 19 wells in the Morecambe Fields (Cowan *et al.* 1993) provided additional structural information. The East Irish Sea Basin occupies the area east and southeast of the Isle of Man and forms the central part of a linked set of Permo-Triassic basins extending from the Cheshire Basin in the south to the Solway Basin in the north (Fig. 1). These basins form the remnants of a more extensive Permo-Triassic and Jurassic basin system which was related to the Celtic Sea trough, the North Sea and the basins off western Scotland. Figure 2 shows the main structural elements of the area, based upon analysis of regional seismic lines (H832-8, SSL/M18, SSL/M-24, H832-28, H832-32, H832-36, H832-50, H832-33, H832-109, H811-377 and H811-205), as well as unpublished in-house seismic interpretations. Structural elements are named according to Jackson *et al.* (1987).

EISB regional structural setting

Extension and fault-controlled sedimentation began in early Permian times and continued throughout the Triassic. Sub-

From *Petroleum Geology of Northwest Europe: Proceedings of the 4th Conference* (edited by J. R. Parker).

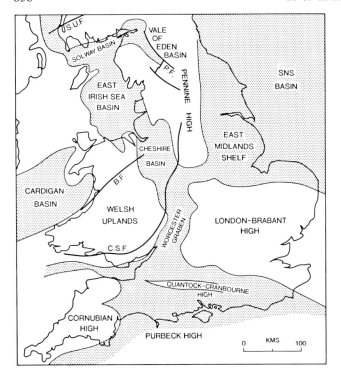

Fig. 1. Permo-Triassic basins of the southern UK. Recent studies (Burley 1987) suggest that the Lake District area may not have been uplifted during the Permo-Triassic. S.U.F.: Southern Uplands Fault; P.F.: Pennine Fault; C.S.F.: Church Stretton Fault; B.F.: Bala Fault; ——: Permo-Triassic basin margin. (Adapted from Burley 1987).

sequent inversion and erosion has removed the post-Triassic cover in the area. The Permo-Triassic of the EISB has a preserved thickness of nearly 3 km in the centre of the basin, but since Triassic sediments of the Mercia Mudstone group outcrop on the sea bed, the pre-inversion basin-fill is likely to have been greater. The geology of the East Irish Sea Basin has been described by Colter and Barr (1975), Colter (1978), Bushell (1986) and Jackson *et al.* (1987).

Stratigraphy

Colter and Barr (1975) proposed that the Sherwood Sandstone Group be divided into three units, the St Bees Sandstone Formation, the 'Keuper' Sandstone Formation and 'Keuper' Waterstones Formation. The 'Keuper' sandstone could, in turn, be divided into three members, the Frodsham, Delamere and Thurstaston Members (Fig. 3). Warrington *et al.* (1980) proposed that the Keuper Waterstones Formation and Keuper Sandstone Formation be termed the 'Ormskirk' Sandstone Formation, and the latter term is used in this paper. Thompson (1970) proposed that the junction between the St Bees Sandstone Formation and the overlying Ormskirk Sandstone Formation represented the equivalent of the Hardegsen unconformity, but analysis of high-resolution dipmeters from the Morecambe Fields show no angular unconformity at this junction although it does appear that a significant depositional hiatus occurred at this time (Stuart and Cowan 1991). However, there are a number of small unconformities of around 1.5–2.0° which can be identified in the wells of the Morecambe Fields, near the lithostratigraphic divisions described by Colter and Barr (1975) and Cowan *et al.* (1993).

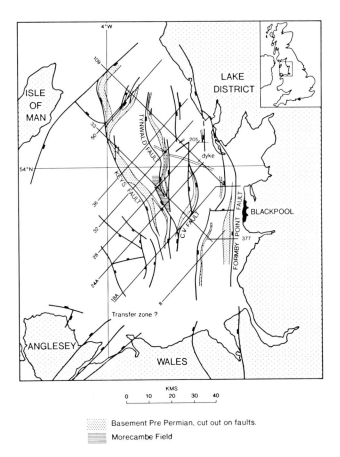

Fig. 2. Regional fault pattern showing seismic lines used (see text for details). C.V.: Crosh Vusta Fault.

SYSTEM/SERIES/GROUP		WARRINGTON *et al* 1980	COLTER & BARR 1975			SOUTH MORECAMBE FIELD RESERVOIR LAYERS
QUATERNARY						
MIDDLE/ UPPER TRIASSIC	MERCIA MUDSTONE GROUP	MERCIA MUDSTONE GROUP	KEUPER MARL AND SALIFEROUS BEDS			NA
LOWER TRIASSIC	SHERWOOD SANDSTONE GROUP	ORMSKIRK SANDSTONE FORMATION	KEUPER WATERSTONES FM.			1
						2
			KEUPER SANDSTONE FORMATION	FRODSHAM		3
				DELAMERE		4-5-6
				THURSTASTON		7-8-9-10
		ST. BEES SANDSTONE FORMATION	ST. BEES SANDSTONE FORMATION			11 | 24
UPPER PERMIAN		ST. BEES EVAPORITES AND EQUIVALENTS	MANCHESTER MARL			NA
LOWER PERMIAN		COLLYHURST SANDSTONE AND EQUIVALENTS	COLLYHURST SANDSTONE AND EQUIVALENTS			
UPPER CARBONIFEROUS		WESTPHALIAN	WESTPHALIAN			
		NAMURIAN	NAMURIAN			

* IN THIS PAPER UPPER ORMSKIRK SANDSTONE FM. CORRESPONDS TO THE
KEUPER WATERSTONES FM. OF COLTER & BARR.
NA: NOT APPLICABLE.

Fig. 3. Stratigraphic terminology of the East Irish Sea Basin.

Structural setting

The area of study is dominated by faults which trend NE–SW, NW–SE and N–S. Each of these orientation elements forms part of the fault array produced at various times during the basin evolution. The northern part of the Irish Sea Basin is dominated by the Keys Fault which trends NW–SE and dips to the NE. The fault links with a NE–SW-trending fault which forms the southeastern border of the Ramsay–Whitehaven Ridge. The other major fault in the northern section of the basin is the Tynwald fault system which also dips towards the east or northeast. Thus, the main structural controls on basin evolution in the northern part of the basin are easterly dipping faults which induce tilting down to the southwest or west. The southeastern part of the basin is dominated by the Crosh Vusta and the Formby Point faults which dip to the west or south-west. Both induce tilting down to the east. The presence of major easterly dipping faults in the north of the basin and westerly dipping faults in the south of the study area has had a profound influence on structural evolution in the basin. The Morecambe Fields are located at the junction of these two fundamental domains which have controlled the geological evolution of the Irish Sea Basin. The complex tilting history experienced by the Morecambe Fields arises directly from their position at the boundary of these two domains. Sherwood Sandstone Group facies distribution maps (Meadows and Beach 1993, figs 8a and b) also indicate the location and persistence of a boundary between these domains.

Both listric and planar faults are present in the basin fault array. The displacement magnitudes on the major faults and the hangingwall geometries indicate that the main extension direction during Triassic–early Jurassic basin evolution was ENE–WSW to NE–SW. Under this regime the basin boundary faults in the north and south which trend SW–NE are likely to have acted as transfer zones where strike-slip and oblique displacements were important. The amount of net extension indicated by the offset of the pre-Permian basement varies from 10–16%. The marked thickening of the Triassic–lower Jurassic (?) sequences into faults implies that extension was most rapid during this period. However, since the stratigraphical framework is entirely lithostratigraphical, estimates of subsidence rates are difficult to make. In-house maturity modelling shows that the top Carboniferous (base Permian) must have been carried to at least 11 000 ft (3000 m) during the Early Jurassic in order to generate hydrocarbons. This is required since platy-illite precipitation beneath a palaeo-hydro-carbon–water contact in the Morecambe reservoir has been dated at 180 Ma using the K/Ar Method (Bushell 1986).

Analysis of the thickness variations between the base Permian and top Sherwood Sandstone Group in terms of two-way travel times (TWTT, isochrons) provides an insight into the evolution of the fault array in the EISB (Fig. 4). In the northern domain, easterly dipping faults induced a thickening of sediments to the west into the faults. Westerly dipping faults controlled sedimentation in the southern domain where thicknesses increase eastwards into the faults.

The complex fault pattern in the central part of the basin, in the vicinity of the Morecambe Fields, reflects the competition between the Keys Fault and the Crosh Vusta Fault to control structural patterns. Two isochron maxima for the interval between Permian to top Sherwood Sandstone reflectors are present along the Keys Fault. A northern and thicker depo-centre occurs where the Keys Fault changes strike to parallel the SE–NW-trending Morecambe Fields boundary fault and a second, less well-developed depocentre exists in the southern central section of the Keys Fault to the west of the Morecambe Fields. This evidence suggests that the displacement on the Keys Fault was concentrated in these two areas, and that the Keys Fault may have developed by the linking of two fault

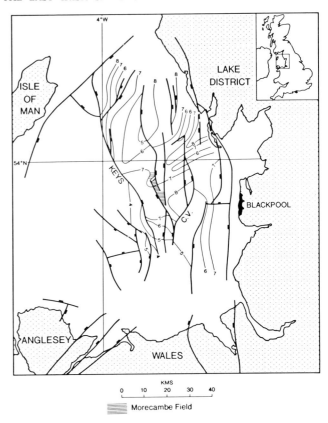

Fig. 4. Permo-Triassic isochrons between top Carboniferous and top Sherwood Sandstone in 1/10 s.

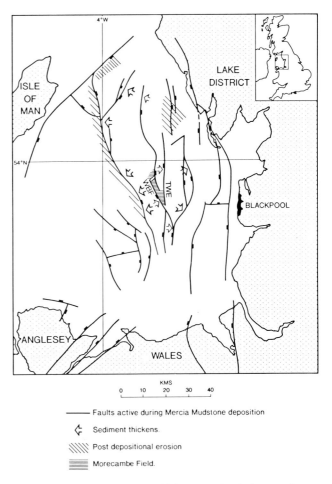

Fig. 5. Map showing the faults which were active during Mercia Mudstone times as indicated by sediment thickening into faults.

LINE H832–28

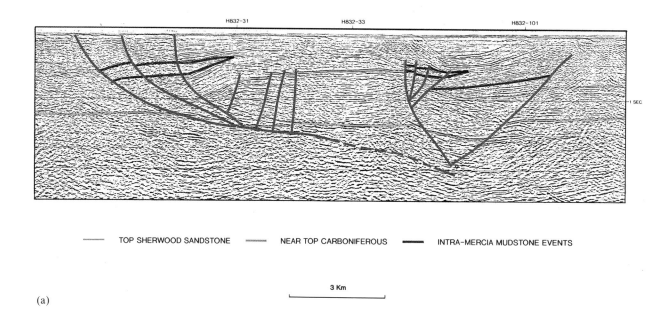

———— TOP SHERWOOD SANDSTONE   ———— NEAR TOP CARBONIFEROUS   ▬▬▬▬ INTRA-MERCIA MUDSTONE EVENTS

3 Km

(a)

GEO–SEISMIC SECTION OF LINE H832–28

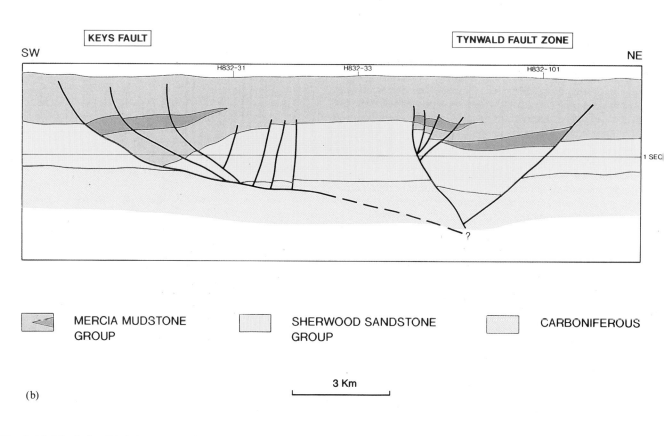

MERCIA MUDSTONE GROUP   SHERWOOD SANDSTONE GROUP   CARBONIFEROUS

3 Km

(b)

Fig. 6. (a) Seismic line H832-28. (b) Geoseismic interpretation of part of line H832-28 running SW–NE through the northern part of the South Morecambe Structure (Fig. 1). The Morecambe Fields are located on a roll-over into the Key Fault to the west. The thickening of the intra-Mercia Mudstone units to the NE and to the SW within the Tynwald fault system suggests that this zone experienced E–W 'rocking' during subsidence and deposition of the Mercia Mudstone Group sediments. Marked thickening of Mercia Mudstone sediment packages towards the Keys Fault can also be seen, but that there is no thickening of the Sherwood Sandstone Group into the Keys Fault. The normal faults which are antithetic to the Keys Fault form the western bounding faults of the Morecambe Field.

segments initially separated by the area now represented by an isochronal ridge. The reduced TWTT in the southern of these depocentres indicates that it developed either slightly later or at a slower rate, than the northern one. This is important for the evolution of the Morecambe Fields since it suggests that the early tilting patterns may be related to the development of a deeper and longer-lived depocentre to the southeast of the field associated with fault activity in the Crosh Vusta Fault.

The isochron map (Fig. 4) and reservoir layer isochore maps from the Morecambe Fields show that the faults which now bound the Morecambe Fields have no discernible influence on sedimentation patterns. This implies that the main displacement on these boundary faults post-dated the deposition of the Sherwood Sandstone. Thus, the boundary faults of the Morecambe Fields will have propagated into a structure already established or initiated by movement on the Keys and Crosh Vusta faults.

Figure 5 presents the results of an analysis of the faults which shows signs of activity during the deposition of the Mercia Mudstone, as indicated by thickening into faults. An important feature indicated from this map is evidence for activity on the fault which forms the eastern boundary of the Tynwald fault zone (fault TWE on Fig. 5). Figure 6a shows part of the seismic line H832-28. It can be seen that the early Morecambe structure formed on a roll-over towards the Keys Fault and that evidence for activity on the Tynwald fault system during this period is also present. Early activity is indicated on the eastern member (A) while a higher stratigraphic section (B) appears to thicken into the western fault indicating a period of slightly later activity on this fault (Fig. 6b). Clearly subsidence was controlled by faulting and E–W rocking of the Tynwald structure. These features together with some slight indication of thickening into the fault which bounds the western segment of the South Morecambe Field suggest that the Morecambe Fields became an 'isolated' NNW–SSE structural entity at this time. The development of these field boundary faults involves the southward propagation of the Tynwald fault system and the northerly propagation of westerly dipping faults (WBF, Fig. 5). This initiation and propagation of the faults which bound the field area may have been in reponse to extension during arching across the area between the active Keys Fault and the Crosh Vusta Fault. It is interesting to note that the increased rate of extension indicated at this time would enhance the chance of fracturing in the dome forming between these faults.

Figure 7 shows a schematic evolution of the eastern Irish Sea area during the Permian to Triassic and summarizes the interpretations made above. The important aspects to note are the following:

(1) The early fault pattern established in the area appears to have produced a broad domal structure in the southern domain (Fig. 7i). The southern part of this domain probably had a simple pattern of dips into the major faults, i.e. the Keys and the Crosh Vusta. The northern part of the southern domain would have had a more complex dip pattern because of the southern tips of the east dipping faults which occupied the central part of the basin. These tip zones were probably associated with NE–SW-trending transfer zones which account for the NE–SW ridges present in Fig. 2.

(2) Increased activity on the Keys and Crosh Vusta faults would have increased the amplitude of the dome developing in the southern domain (Fig. 7ii). It is likely that the area of the Morecambe Fields was developing a reservoir geometry at this stage.

(3) Increased arching in the southern domain between the Keys and Crosh Vusta faults and the propagation into this area of the faults which became the west and east bound-

ary to the Morecambe Fields created the early Morecambe structure (Figs 7ii–7iv).

Morecambe Field structure

The Morecambe Gas Fields occupy a NNE–SSW-trending area approximately 25 km long and some 5–10 km wide located in the central part of the East Irish Sea. The fields are located along a complex transfer zone which marks the junction of the two structural domains which constitute the East Irish Sea Basin. The structural history of the fields reflect the competition and interplay between these domains to control the tilting and burial patterns. The Morecambe Fields are composed of two separate structural closures, the South Morecambe Field and the North Morecambe Field. The North Morecambe Field is bounded by faults to the west, east and south and forms a broad domal structure which is dip-closed to the north. The South Morecambe Field can be divided into two elements which are separated by a north–south-trending graben informally referred to as the 'central graben'. The area to the west of this graben forms the main South Morecambe Horst area which narrows to the south into an area of complex faulting. The eastern boundary fault of this horst block forms the southern tip of the Tynwald Fault (Fig. 2). The second element to the South Morecambe Field is to the east of the Tynwald fault zone or graben and forms the eastern flank, where the Sherwood Sandstone dips east towards the Crosh Vusta Fault. The North and South Morecambe Fields are separated by a deep ENE–WSW-trending graben.

The base Permian to top Sherwood Sandstone depocentre associated with the Keys Fault westwards of the field is not as well developed as the depocentre to the southeast of the field associated with the Crosh Vusta Fault (Fig. 4). The depocentre associated with the Crosh Vusta Fault thus appears to be an older feature which controlled the early tilting patterns across the field. Dipmeter data show a southward structural tilt beneath intra-Ormskirk unconformities in the south of the field (Cowan *et al.* 1993, figs 7 and 14). This is supported by the layer isopach map of reservoir layers 7–10 (the Thurstaston Member of Colter and Barr 1975) which thicken towards the southeast in the south of the field. Layers 4–6 (Delamere Member of Colter and Barr 1975) show northwesterly thickening in the northern part of South Morecambe. Despite this, consistent palaeocurrent vectors suggest flow towards the west, indicating that a palaeoslope towards the Keys Fault was dominant. The isopach pattern shown by layers 1 and 2 (the Keuper Waterstones of Colter and Barr 1975) again show southerly thickening towards the Crosh Vusta Fault. This last thickening pattern is also representative of fault activity during Mercia Mudstone deposition, but there is evidence that the history of the field area becomes more complex during this time because of the propagation of the faults which become the field bounding structures.

Top platy-illite surface

Platy-illite, which is present in the lower parts of the Morecambe reservoir, is thought to have developed below a palaeo-hydrocarbon–water contact. K/Ar dating of the illite shows that it precipitated around 180 Ma, during Early Jurassic times.

Understanding the geometry of the top platy-illite surface is crucial to both the understanding of the structure at the time of hydrocarbon trapping and the subsequent deformation events which disrupted the originally horizontal planar surface. For these reasons this study has used the top platy-illite maps derived from a combination of seismic and well control (Stuart and Cowan 1991) to define the tilting history of the field. Details of the mapping of the top platy-illite surface over the

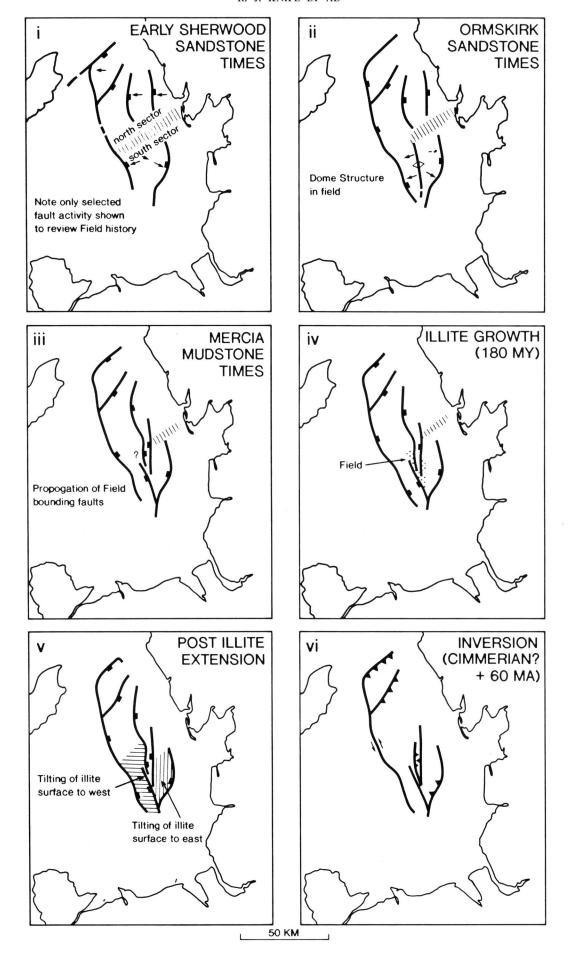

Fig. 7. Regional structural evolution.

undeveloped North Morecambe structure utilizing the results of this structural study are given by Stuart (1993).

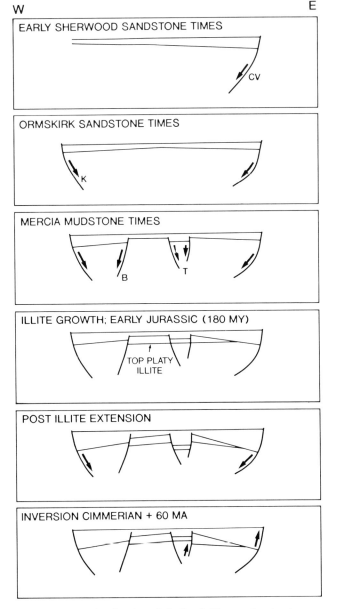

Fig. 8. Cross-sectional review of the South Morecambe structure. CV: Crosh Vusta Fault; B: Morecambe Field Boundary Fault; K: Keys Fault; T: Tynwald fault system.

Synthesis of structural evolution in field area

A cross-sectional model of field development is shown in Fig. 8. The early part of the field development (pre-Ormskirk Sandstone deposition) appears to be caused by a uniform tilting down to the southeast, controlled by the Crosh Vusta Fault (Fig. 7i). This was followed by the development of tilting down to the west or northwest in North Morecambe and the northern part of South Morecambe reflecting the development of a depocentre to the west which was controlled by the Keys Fault during the deposition of reservoir layers 4–6. The southern part of South Morecambe and the eastern flank were still tilting down to the southeast at this time. Later, a broad arch between the Keys and Crosh Vusta faults developed, during the deposition of layers 1–3 of the South Morecambe reservoir (Fig. 7ii). This arch was fractured during the deposition of the

Mercia Mudstone by the propagation southwards of the Tynwald fault system and the northward propagation of the fault which bounds the west of the field. During this phase, rapid extension and subsidence was occurring and the 1500 ft deep SW–NE graben structure separating the North and South Morecambe structures developed, with the relative downthrowing of South Morecambe. The first phase of hydrocarbons probably migrated into a common North and South Morecambe structure immediately prior to the formation of this graben. The structurally high areas at the time of illite growth were in North Morecambe, the NE and SW portion of South Morecambe and an even more pronounced high along the western edge of the eastern flank of the field (Fig. 6, Stuart and Cowan 1991). The very thick preserved section of illite-free reservoir in this area may be a consequence of the fact that illite did not precipitate at all in this area, implying that the southwards extension of the central graben faults formed aquifer seals.

The present dip of the illite surface suggests that after precipitation, the North Morecambe structure continued to tilt down to the west under the influence of the Keys Fault and the east flank of the South Morecambe Field continued to tilt down to the east and southeast under the influence of the Crosh Vusta Fault (Fig. 9). The main horst area of South Morecambe appears to show a somewhat hybrid behaviour after the illite growth, with the northern part tilting down to the west (presumably controlled by the Keys Fault) and the southern part tilting down to the southeast and east similar to the eastern flank. The central portion of the main South Morecambe Horst block appears to have acted as a transfer zone dominated by bending rather than faulting during this development, giving rise to the strike changes in this zone.

The different dips of the top platy-illite surface in North Morecambe and the northern parts of South Morecambe also indicate the operation of a post-illite tilt-transfer zone between these areas (Fig. 9). In this case the transfer involved some displacement on the graben faults which separate these two fields and created preferential downthrow and rotation of the northern side (Fig. 7iv). Uplift and inversion subsequently induced sinistral movement along NW–SE faults as well as reverse displacement and folding along structures trending SW–NE (Fig. 7vi).

Inversion structures

Vitrinite reflectance measurements and spore colour indices from Carboniferous source rocks show much higher maturity values than would be expected from the present depth of burial and it is clear that the basin has been inverted (Colter and Barr 1975). A less reliable estimate of maximum burial has been obtained using the shale velocity analysis technique of Marie (1975), as applied by Colter and Ebbern (1978) and Jackson et al. (1987). These techniques suggest that around 2000 m of inversion has occurred. Recent apatite fission track data published by Green (1986) and Lewis et al. (1991) suggests that over 3 km of uplift has occurred over the basin. They attribute this to regional uplift associated with the Alpine orogeny (see Hardman et al. (1993) for a detailed discussion).

There were probably at least two phases of inversion, one associated with the Cimmerian (Bushell 1986), and one during the Tertiary (Stuart and Cowan 1991). The local complications of this inversion are difficult to assess in detail, but are illustrated by evidence of displacement reduction along faults in the Tynwald fault system. A number of features are present on the seismic lines analysed which may have been caused by inversion of the normal faults. These include tilted and fold sediment wedges. There are no reverse offsets on the faults in the study area indicating that inversion displacements are always less than the normal fault displacements. The most

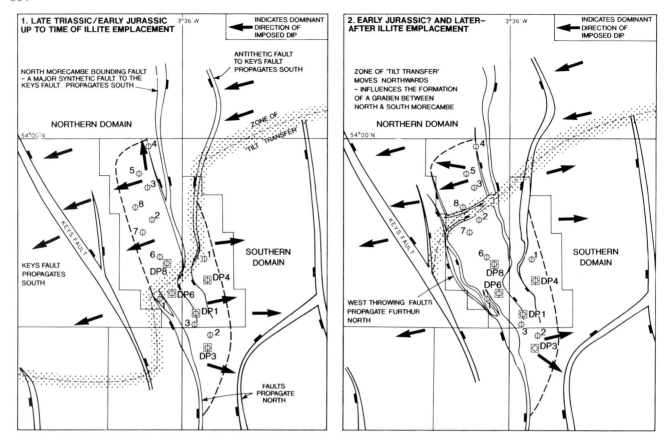

Fig. 9. Diagrammatic illustration of the development of the North Morecambe structure. (1) During the Early Jurassic, illite precipitated in a common structure formed between the southerly propagating Keys Fault and a northerly propagating Tynwald fault system. A zone of tilt transfer, therefore, exists between the westerly dipping structures associated with the Keys Fault and the easterly dipping structures to the south. (2) With continued northwards propagation of the Tynwald fault system the zone of tilt transfer jumps north to form the transfer graben system which structurally separates North and South Morecambe. The top platy-illite surface tilts NW in North Morecambe as opposed to SW in South Morecambe.

important inversion structures, reviewed in Fig. 10, are NE–SW-trending fold structures parallel and adjacent to faults which trend NE–SW (Fig. 11). These folds trend across the Permian to Sherwood Sandstone isochron shown in Fig. 4 and often have dips which are higher than those expected to be formed on the underlying fault geometries by extension or differential compaction. They are thus tentatively interpreted as inversion structures formed by contraction and associated reactivation of normal faults. This interpretation is also supported by their preferential location along SW–NE-trending segments of the fault array.

The amount of uplift associated with these structures is difficult to assess. A minimum shortening estimate of 1.5 km can be made over a distance of 57 km, that is approximately 2.5%. If this displacement occurs on faults with dips of at least 45° as is indicated from the seismic data, then the exhumation associated with the inversion by fault movement is at least 1500 m. Dipmeter analysis suggests that faulting is slightly steeper at top Sherwood Sandstone level: around 50° to 60°. This would suggest around 2000 m of uplift. The pattern of inversion indicates that NW–SE faults acted as transfer zones and the NE–SW faults acted as frontal or oblique inversion. A sinistral sense of shear is indicated by all the local fold structures (Fig. 7iv). This sense of shear is the same as that experienced by southwestern areas of the UK during Eocene to Oligocene inversion (Ziegler 1978) and the timing is confirmed by apatite fission track analysis (Stuart and Cowan 1991;

Fig. 10. Faults showing inversion structures. Synclines associated with faults are the result of fault reactivation associated with uplift due to shortening.

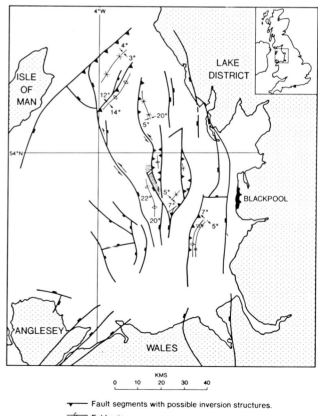

◄━ Fault segments with possible inversion structures.

✕ Fold axis

▤ Morecambe Field

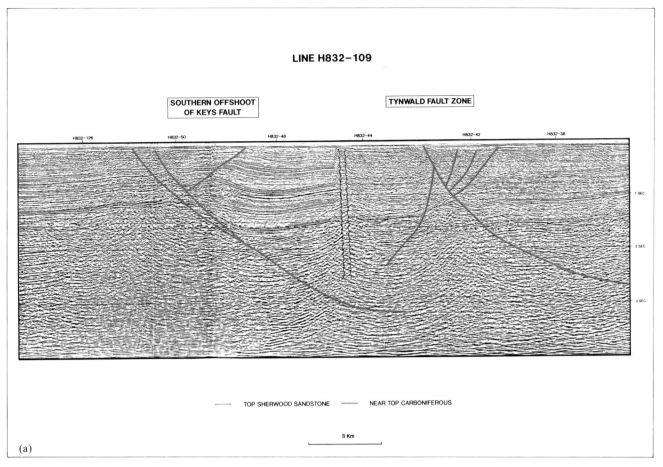

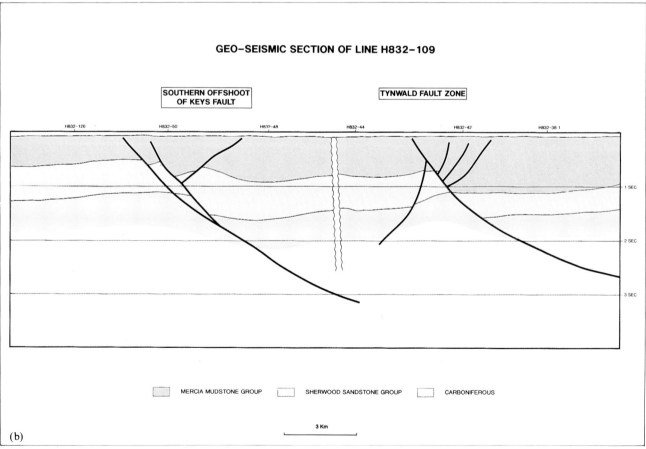

Fig. 11. (a) Seismic line H832-109. (b) Geoseismic section, line H832-109. This line is orientated NW–SE across the northern part of the basin (Fig. 2), between the NE–SW extension of the Keys Fault and the Tynwald fault system. The 'chimney' of noise seen on the seismic section is caused by a Tertiary Igneous dyke. The syncline beneath line H832-48 is an inversion structure formed by contraction reverse reactivation on the Keys Fault (see text).

Hardman *et al.* 1993). Dipmeter analysis suggests that the last phase of tilting imposed a northerly dip, possibly in response to Alpine deformation in the south of England.

Conclusions

(1) This study has produced a new model for the structural evolution of the Morecambe Fields which identifies sub-domains with different tilting histories and relates the growth of platy-illite to the development of the fault array in the region.

(2) The Morecambe Fields are located at the junction of two fundamental domains which controlled the geological evolution of the East Irish Sea Basin. In the northern domain, evolution was controlled by eastward dipping faults, while in the southern domain, basin development was controlled primarily by westerly dipping faults.

(3) The early history of the fields is associated with the development of a dome between the Keys Fault and the Crosh Vusta Fault. Fracturing of this dome resulted in the creation of faults which bound the east and west of the main area of the Morecambe Fields. These events pre-date platy-illite precipitation.

(4) Post-illite evolution of the field involved continued tilting associated with extension and the separation of the North and South Morecambe structures. This was followed by at least two phases of basin inversion, the last uplift occurring during the Tertiary.

We would like to thank the directors of British Gas Exploration and Production for permission to publish this paper. Richard Collier, Richard Hodgkinson and Bob Dunay are thanked for their constructive criticism. Ruth Johns drafted the diagrams.

References

BURLEY, S. D. 1987. *Diagenetic modelling in the Triassic Sherwood Sandstone Group of England and its offshore equivalents in the U.K. Continental Shelf.* Unpublished PhD thesis, Hull University.

BUSHELL, T. P. 1986. Reservoir Geology of the Morecambe Field. *In*: BROOKS, J., GOFF, J. C. AND VAN HOORN B. (eds) *Habitat of Palaeozoic Gas in Northwest Europe.* Geological Society, London, Special Publication, **23**, 189–207.

COLTER, V. S. 1978. Exploration for gas in the Irish Sea. *Geologie en Mijnbouw*, **57**, 503–516.

—— AND BARR, K. W. 1975. Recent developments in the geology of the Irish Sea and Cheshire Basins. *In*: WOODLAND, A. W. (ed.) *Petroleum and the Continental Shelf of North-West Europe, Vol. 1.* Applied Science Publishers, London, 61–75.

—— AND EBBERN, J. 1978. The petrography and reservoir properties of some Triassic sandstones of the Northern Irish Sea Basin. *Journal of the Geological Society, London*, **135**, 57–62.

COWAN, G., OTTESON, C. AND STUART, I. A. 1993. The use of dipmeter logs in the structural interpretation and palaeocurrent analysis of the Morecambe Fields, East Irish Sea Basin. *In*: PARKER, J. R. (ed.) *Petroleum Geology of Northwest Europe: Proceedings of the 4th Conference.* Geological Society, London, 867–882.

GREEN, P. F. 1986. On the thermotectonic evolution of Northern England: evidence from fission track analysis. *Geological Magazine*, **123**, 493–506.

HARDMAN, M., BUCHANAN, J., HERRINGTON, P. AND CARR, A. 1993. Geochemical modelling of the East Irish Sea Basin: its influence on predicting hydrocarbon type and quality. *In*: PARKER, J. R. (ed.) *Petroleum Geology of Northwest Europe: Proceedings of the 4th Conference.* Geological Society, London, 809–821.

JACKSON, D. I., MULHOLLAND, P., JONES, S. M. AND WARRINGTON, G. 1987. The geological framework of the East Irish Sea Basin. *In*: BROOKS, J. AND GLENNIE, K. (eds) *Petroleum Geology of North West Europe.* Graham & Trotman, London, 191–203.

LEWIS, C. L. E., GREEN, P. F., CARTER, A. AND HURFORD, A. J. 1991. Elevated K/T palaeotemperatures throughout Northern England: three kilometres of Tertiary exhumation? *Earth and Planetary Science Letters*, in press.

MARIE, J. P. 1975. Rotliegende stratigraphy and diagenesis. *In*: WOODLAND, A. (ed.) *Petroleum and the Continental Shelf of Europe.* Institute of Petroleum, London, 205–210.

MEADOWS, N. S. AND BEACH, A. 1993. Controls on reservoir quality in the Triassic Sherwood Sandstone of the Irish Sea. *In*: PARKER, J. R. (ed.) *Petroleum Geology of Northwest Europe: Proceedings of the 4th Conference.* Geological Society, London, 823–833.

STUART, I. A. 1993. The geology of the North Morecambe Gas Field, East Irish Sea Basin. *In*: PARKER, J. R. (ed.) *Petroleum Geology of Northwest Europe: Proceedings of the 4th Conference.* Geological Society, London, 883–895.

—— AND COWAN, G. 1991. The South Morecambe Field. *In*: ABBOTTS, I. L. (ed.) *UK Oil and Gas Fields, 25 Years Commemorative Volume.* Geological Society, London, Memoir, **14**, 527–541.

THOMSON, D. B. 1970. The stratigraphy of the so-called Keuper Sandstone Formation (Scythian–?Anisian) in the Permo-Triassic Cheshire Basin. *Quarterly Journal of the Geological Society, London*, **126**, 151–181.

WARRINGTON, G., AUDLEY-CHARLES, M. G., ELLIOTT, R. E., EVANS, W. B., IVIMEY-COOK, H. C., KENT, P. E., ROBINSON, P. L., SHOTTEN, F. W. AND TAYLOR, F. M. 1980. *A correlation of Triassic rocks in the British Isles.* Geological Society, London, Special Report, **13**.

WOODWARD, K. AND CURTIS, C. D. 1987. Predictive modelling of the distribution of production constraining illites—Morecambe Gas Field, Irish Sea, Offshore UK. *In*: BROOKS, J. AND GLENNIE, K. (eds) *Petroleum Geology of North West Europe.* Graham & Trotman, London, 205–215.

ZIEGLER, P. A. 1978. North-western Europe: Tectonics and Basin development. *Geologie en Mijnbouw*, **53**, 43–50.

The use of dipmeter logs in the structural interpretation and palaeocurrent analysis of Morecambe Fields, East Irish Sea Basin

G. COWAN,[1] C. OTTESEN[2] and I. A. STUART[1]

[1] British Gas plc Exploration and Production, 100 Thames Valley Park Drive, Reading, Berkshire RG6 1PT, UK

[2] Z & S Geologi a.s., Sverdrupsgate 23, 4007 Stavanger, Norway

Abstract: Dipmeter logs have been processed in a variety of ways as an aid to understanding the geology of the North and South Morecambe Gas Fields. They have yielded high quality structural information which has contributed to understanding the structural configuration and tectonic history of these fields. Several faults have been identified, cutting well paths, and dipmeter analysis has allowed the correct orientation and direction of throw to be determined. In one well, the deviated well path has intersected each of two listric faults in two separate places, allowing unusually precise definition of these fault planes. Throughout the Ormskirk Sandstone Formation, the predominant palaeocurrent direction has been found to be towards the west or southwest. The most consistent palaeocurrent orientations were measured in intervals of planar cross-stratified channel-fill facies sandstones, but similar results, with greater scatter were observed in ephemeral channel sandstones facies. These consistent palaeocurrent results suggest that the East Irish Sea Basin formed a distinct depo-centre during Ormskirk sandstone times. The implication is that during the deposition of the Ormskirk Sandstone, palaeocurrents were controlled by a regional dip towards the Keys Fault. There are changes in palaeoflow from the base of the Ormskirk Sandstone Formation to the top (a vertical thickness of 800 ft), also changes from the southern to the northern ends of the field. In the underlying St Bees Sandstone Formation, consistent palaeocurrent directions proved more difficult to obtain, since major channel facies sandstones are rarer. No angular unconformity was observed between the St Bees and Ormskirk Sandstone formations. Within the Ormskirk Sandstone interval, however, very consistent, 1° to 2° changes in structural dip occur at the same stratigraphic levels in several wells and indicate small-scale tectonic rotations during the deposition of these sandstones. These are probably related to tectonic reactivation which has been invoked in previous literature to explain periodic rejuvenation of the depositional system during deposition of the reservoir sandstones. The fact that the observation of such small-scale adjustments can be repeated in several wells is an indication of the degree of precision of the results obtained from the dipmeter logs in these wells.

Introduction

This paper presents the results of studies carried out by means of dipmeter logs, both the high-resolution dipmeter tool (HDT) and the stratigraphic high-resolution dipmeter tool (SHDT), that have been run on many of the pre-development and development wells on the South and North Morecambe Gas Fields. These fields lie within the UK Blocks 110/2a, 110/3a, and 110/8a in the East Irish Sea Basin (EISB) (Fig. 1). The term 'Morecambe Field' is used informally to refer to both fields collectively. The geology, exploration history, and general development of South Morecambe has been recently described by Stuart and Cowan (1991), and North Morecambe in Stuart (1993).

An outline of the stratigraphy is shown in Fig. 2. The reservoir sandstones belong to the Triassic Sherwood Sandstone Group. The overlying Mercia Mudstone Group forms the seal; this comprises mudstones and four correlatable salt horizons which were precipitated during periodic drying of the basin. Dipmeters were normally run only through the Sherwood Sandstone Group, so that the bulk of this paper addresses interpretation within the reservoir sequence; however, one dipmeter log was interpreted through the Mercia Mudstone sequence, and this has permitted important conclusions regarding faulting at this level.

Reservoir deposition

The Sherwood Sandstone Group forms a thick fluvial sequence (c. 4800 ft) deposited in a semi-arid environment within the subsiding EISB, and comprises a heterogeneous system of interlayered sandstone facies of variable reservoir quality. Five major facies associations are recognized. Detailed sedimento-

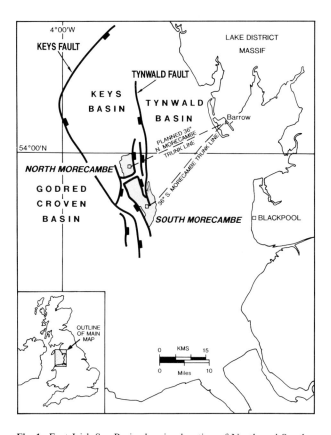

Fig. 1. East Irish Sea Basin showing location of North and South Morecambe Fields.

From *Petroleum Geology of Northwest Europe: Proceedings of the 4th Conference* (edited by J. R. Parker).
© 1993 Petroleum Geology '86 Ltd. Published by The Geological Society, London, pp. 867–882.

logical descriptions have been given in Stuart and Cowan (1991) and Cowan (1993). The major facies associations are as listed below.

(1) Major channel sandstones These consist of stacked erosively-based intervals of planar cross-stratified, medium-grained sandstones. Analysis of dipmeter responses in this facies has provided the clearest indication of palaeocurrent directions in the Sherwood Sandstone Group.

(2) Secondary channel sandstones These comprise thinly bedded, erosively-based intervals, often showing fining-upwards grain-size profiles. These sandstones are often associated with sheetflood sandstones.

(3) Sheetflood sandstones These consist of flat-laminated tabular intervals of fine- to medium-grained sandstones with mudstone/siltstone partings.

(4) 'Non-reservoir fines' This facies grouping includes abandonment facies mudstones, overbank fines and non-evaporitic playa lake mudstones. These are not abundant in the sequence.

(5) Aeolian facies Thin aeolian deposits are common throughout the field. They are composed of very well-rounded, medium- to coarse-grained sandstones exhibiting aeolian-style pinstripe lamination, and were deposited as aeolian dune and sandsheet deposits. Aeolian sandstones are never more than a few feet thick, but their high permeabilities (up to 11 D) render them an important reservoir facies. The aeolian dune facies has high angle cross-lamination and can be readily differentiated from the low angle cross-stratified aeolian sandsheet facies using dipmeter logs.

SYSTEM/SERIES/GROUP		WARRINGTON *et al* 1980	COLTER & BARR 1975	SOUTH MORECAMBE FIELD RESERVOIR LAYERS
QUATERNARY				
MIDDLE/ UPPER TRIASSIC	MERCIA MUDSTONE GROUP	MERCIA MUDSTONE GROUP	KEUPER MARL AND SALIFEROUS BEDS	NA
LOWER TRIASSIC	SHERWOOD SANDSTONE GROUP	ORMSKIRK SANDSTONE FORMATION	KEUPER WATERSTONES FM.	1 2
			KEUPER SANDSTONE FORMATION — FRODSHAM	3
			KEUPER SANDSTONE FORMATION — DELAMERE	4-5-6
			KEUPER SANDSTONE FORMATION — THURSTASTON	7-8-9-10
		ST. BEES SANDSTONE FORMATION	ST. BEES SANDSTONE FORMATION	11 | 24
UPPER PERMIAN		ST. BEES EVAPORITES AND EQUIVALENTS	MANCHESTER MARL	NA
LOWER PERMIAN		COLLYHURST SANDSTONE AND EQUIVALENTS	COLLYHURST SANDSTONE AND EQUIVALENTS	
UPPER CARBONIFEROUS		WESTPHALIAN	WESTPHALIAN	
		NAMURIAN	NAMURIAN	

* IN THIS PAPER UPPER ORMSKIRK SANDSTONE FM. CORRESPONDS TO THE
  KEUPER WATERSTONES FM. OF COLTER & BARR.
  NA: NOT APPLICABLE.

Fig. 2. Stratigraphic nomenclature of the East Irish Sea Basin.

Reservoir layering

The reservoir sequence can be subdivided into correlatable layers. There is an alternation between layers dominated by

major channel sandstone facies, with little else preserved, and layers dominated by sheetflood and secondary channel sandstone. On this basis, the Ormskirk Sandstone Formation has been divided into 10 major layers across South Morecambe. Most of these layers can be correlated across North Morecambe, although a separate layering scheme is used there. The fact that such an alternation exists (as opposed to all facies being randomly intermixed in the manner of classical braided river models) is probably because the sheetflood facies is the distal rather than lateral equivalent of the major channel facies. It has been envisaged that some form of external, tectonic or climatic, basin-wide control was responsible for the observed alternation, whereby a periodic rejuvenation of the system caused a return from sheetflood-dominated to channel-dominated deposition. However, until the present dipmeter studies were performed, no independent confirmation of the tectonic controls had been obtained.

Field structure

The field is situated at the junction of two fundamental structural domains which have controlled the development of the EISB half-graben. The southern domain is controlled by a dense pattern of north–south-trending faults which dip to the west and produce easterly dipping half-grabens. The northern domain is dominated by the Keys Fault (Jackson *et al.* 1987), which lies to the west of the field and has propagated south-wards from the northern margin of the basin. Details of the structural development of the field are given in Stuart and Cowan (1991) and Knipe *et al.* (1993). The significance for the present paper is that the structural history has given rise to a complex pattern of faulting with South Morecambe fault-bounded to the west, North Morecambe fault-bounded to the east. Areas of both fields exhibit complex patterns of internal small-scale faults, some of which have been detected, defined and described by means of dipmeter analysis. Definition of the bounding faults of both fields has been improved by dipmeter analysis.

Reasons for the study

South Morecambe Field came on stream in 1985. Although currently being used as a base load field, it was developed as a seasonal supply facility, which required it to be capable of sustained production at very high rates during a severe winter. Also implicit in this concept was that South Morecambe would be a 'fail-safe' supply; for example, if a major Southern North Sea Basin gas field went off-stream, the South Morecambe development would be required to make up the shortfall. As such, all aspects of the development contained an in-built requirement for very high reliability; reservoir uncertainties, therefore, needed to be reduced to a minimum.

By 1987 the Stage I of South Morecambe development was complete. Many of the Stage I wells exhibited high permeability zones down to the lowest parts of the reservoir, as well as even further down in the water zone. Much of this high permeability rock belonged to the major channel sandstone facies association described above, the remainder being of aeolian origin. There was concern regarding the directional permeability fabric that these channels might impose upon the field. If the channel sandstone bodies dipped into the water zone, there was a perceived risk that they might provide conduits for premature encroachment into producing wells, particularly if these wells were produced at very high rates. This was a particular concern in the case of South Morecambe Reservoir Layer 5, a relatively thin major channel sequence overlain and underlain by lower permeability sheetflood sandstone facies.

There was, therefore, a need to model field performance and

Table 1. The wells comprising the database

Wells	Interval	Tool	Log ft	Core ft	Deviation
110/2-1	4100–2897	HDT	1200	125	<3°
110/2-2	4046–3100	HDT	950	100	<2°
110/2-4*	5098–3605	HDT	1500	—	<3°
110/2-5*	4760–3235	HDT	1530	160	<2°
110/2a-7	4190–2989	SHDT	1200	640	<2°
110/2a-8*	4407–3150	SHDT	1250	40	<2°
110/2a-F1	4400–2754	SHDT	1650	1258	21° WNW
110/2a-F4	6500–4500	SHDT	2000	—	62–50° NW
110/2a-F5	5989–4840	SHDT	1150	573	46–31° W
110/2a-H1	4660–3067	SHDT	1600	1211	35° SW
110/2a-H3	6972–5660	SHDT	1300	701	53–44° NNE
110/3a-A2	5357–4363	SHDT/M	1000	210	34 3° ESE
110/3a-A3	4297–2895	SHDT	1400	1215	12–9° WSW
110/3a-A5	5786–5200	SHDT/M	600	135	34–29° NNW
110/3a-A6	6139–4630	SHDT/M	1500	370	47–39° WNW
110/3a-A7	5792–4458	SHDT	1330	—	35–15° SSW
110/3a-D1	3970–3501	SHDT	470	425	<1°
110/8a-C1	3820–3287	SHDT	530	350	<2°
110/8a-C3	5351–1700	SHDT	3550	55	50–35° ESE
19 wells (4 HDT, 15 SHDT)			25710	7568	(16 wells)

* Indicates wells from the North Morecambe Field.
M: Micro-electrical scanning tool (MEST).

predict fluid movements through the reservoir in order, if necessary, to pre-empt production problems. To assess the probability of the channel sandstones dipping down into the water zone, it was necessary to determine their orientation, by establishing their palaeocurrent directions. There were many uncertainties; it was not known, for example, whether the channel type and orientation was similar throughout the vertical reservoir sequence, or had varied with time. The primary intention of initiating a study of the Morecambe Field dipmeters was therefore to assess palaeocurrent directions. Other issues which it was hoped could be addressed by a study of dipmeter data included fault resolution. Parts of the structure, particularly of South Morecambe, are highly faulted and there was, therefore, a need to extract all possible structural information from the dipmeter data.

Database

The 19 wells listed in Table 1 with corresponding dipmeter raw data and cores were available and incorporated in the study.

Data quality

The detailed results of this study owed much to the excellent data quality as well as the large number of wells studied. In the industry dipmeter logs are often interpreted on a single well basis, the aim being to verify a specific structure, such as a fault or the structural dip of a certain formation. However, when several wells are interpreted together, much more detailed information can be gathered.

Most logs are of excellent quality, even the HDTs logged in 12.25 in. holes in the mid-1970s. The high quality is due to a combination of several factors: in-gauge holes; mostly 8.5 in. hole diameter; the use of conductive drilling muds; the sediments themselves; and perhaps most importantly, the shallow logging depths and consequent low temperatures, pressures and more stable logging speeds.

Method

Data were processed with the INCLINE and RECALL/REVIEW systems (both trademarks of Z & S Consultants Ltd, London). For all data an initial quality check from a field print playback was essential. The data were then pre-processed in various ways, including calliper calibrations, correction for magnetic declination, correction for variable logging speed (if necessary) and depth matched.

Correlation method

Several automatic interval correlations using the least squares method were carried out with differing parameters depending on the interpretational aim. In interval correlation, an interval of fixed length is compared to an interval of the same length on another micro-conductivity curve. The curve pieces are correlated for various displacements, and the displacement causing the highest regression coefficient is recorded. Two correlation results are enough to define a plane. However, as geological surfaces are often non-planar, some or all of the six possible correlations for a 4-electrode dipmeter (HDT) and the 12 crosshole pad-to-pad (PTP) correlations for the 8-electrode SHDT may be required to define the best planar fit to these geological surfaces. Filters can be used to remove spurious dips. In general, two sets of correlation parameters were applied for all wells, though some were correlated with additional, specially chosen parameter sets.

Structural processing

For the structural interpretation, the aim is to correlate major bed boundaries and structural planes. Using the PTP method, correlation results were obtained with the following parameters:

• length of correlation window: 2.6 ft;

- length of step, i.e. correlation increment: 1.7 ft, ensuring only 0.9 ft overlap between successive correlations and less chance of the same feature being correlated twice;
- search angle: 60–75° (i.e. maximum dip of correlated planar features);
- search reference plane: borehole perpendicular plane for sub-vertical wells and horizontal plane for wells deviated more than 10–15°;
- correlation cutoff: 0.3 on individual curve–pair correlation coefficients.

Sedimentological processing

For the sedimentological interpretation, the aim is to correlate minor bed boundaries, drapes, cross-stratification in addition to the more conspicuous structures. The correlation method varied according to tool type, using PTP processing for the HDT data and PTP and side-by-side (SBS) processing for the SHDT data. Since the SHDT pads are only 3 cm apart, small-scale structures can be correlated even if they do not cross the entire borehole.

The parameters chosen were:

- length of correlation window: 4 in. (8 in. for HDT data);
- length of step: 3 in. (5 in. for HDT data);
- search angle and reference plane: 50° relative to structural dip (borehole perpendicular plane for HDT data);
- effective electrode separation (only SHDT): as the SBS method is sensitive to divergence of the electrical field in mudcakes, these results have to be calibrated to the PTP results by increasing the apparent electrode separation. (Electrode separation is 3 cm for SHDT, this is increased to around 3.4 cm.)

Dip removal

In order to compare the dipmeter data and results with Formation MicroScanner (FMS) images, sections of the 2-pad for post-depositional tectonic tilt. This ensures that the individual correlation planes representing sedimentary features are plotted relative to the palaeohorizontal plane. In order to do this it is necessary to identify beds which were deposited horizontally. The flat-laminated sheetflood facies association, which makes up around 35–50% of the sequence, is easily identifiable on dipmeter logs and meets this criterion (Fig. 3).

FMS images

In order to compare the dipmeter data and results with Formation Microscanner (FMS) images, sections of the 2-pad Micro-electrical scanning tool (MEST, a fore-runner of the FMS) logs of 110/3a-A5 and -A6 were loaded, processed and interpreted. The images were normalized both statically and dynamically over 3–6 ft windows (Fig. 4). The MEST offered little improvement over the SHDT in terms of palaeocurrent analysis but hitherto unidentified fractures, both cemented and open, were identified using the FMS tool.

Interpretation methods

Structural interpretation

A dipmeter log is essentially similar to a structural section measured in the field and much the same interpretation methods apply for the two. The dipmeter log is usually more complete than a field section, but the field geologist has actually seen the structures measured, leaving little doubt whether a dip change is brought about by a fault or an unconformity. Structural interpretation is dependent on the

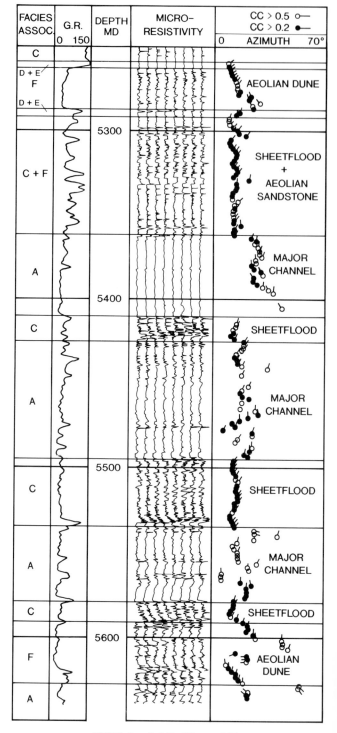

WELL 110/3a–A5

Fig. 3. Example of structural processing (PTP 2.7 ft step). Structural dip has not been removed but the flat-bedded sheetflood facies association can easily be identified by virtue of their low angle and uniform dips as well as highly active resistivity curve motif. These facies define the structural dip.

occurrence of stratification deposited close to the horizontal. In a sandy lithology with an abundance of sedimentary structures, one has to rely on larger boundaries, or units with horizontal lamination. Most of the Morecambe reservoir is thin bedded and as no large-scale depositional slopes (such as delta fronts or alluvial fans) are present, the structural orientation is well defined by the presence of sheetflood sandstones in most units. An exception is stacked cross-bedded channel

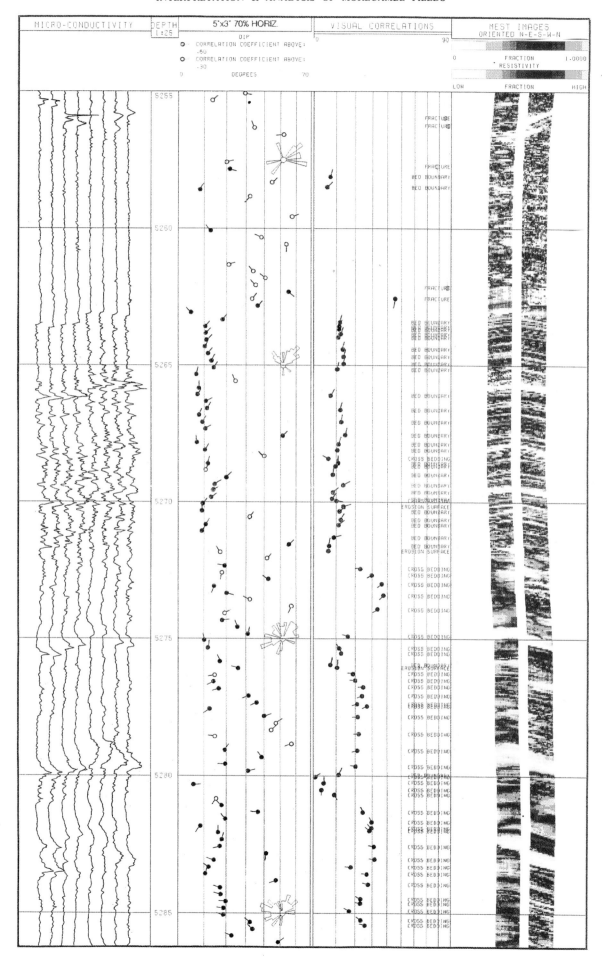

Fig. 4. Micro-electrical scanning tool (MEST, a fore-runner of FMS) plot from Well 110/3a-A5.

sandstone layers within the Ormskirk Sandstone Formation. These layers may be up to 300 ft thick and are dominated by high-angle cross-strata with very few sheetflood intervals.

An important interpretational step was the subdivision into structural blocks of inferred constant dip of palaeohorizontal planes using the depth log plots which display the raw micro-conductivity curves and the correlation results in arrows/tadpoles. The mean orientation was calculated by vector addition within screening cones in space, or by contouring of poles to planes in stereographic plots. At levels or intervals where a change in structural dip was noticed or through possible fault zones, stereograms were plotted and interpreted. This method presents a quick and reliable way for the recognition of both spatial relationships between various structural blocks and the tectonic structures (faults or unconformities) which caused the dip changes.

Fault analysis

This analysis relies heavily on the presence of a fault drag in the hangingwall and/or footwall blocks; the actual fault plane is rarely correlated. Drag is normally best developed in the hangingwall block. Other features which may help the analysis are the presence of parasitic tectonic planes, rotated intermediate fault blocks and the sense of rotation of the hangingwall relative to the footwall block. The slightest drag fold indicates the strike and dip direction of the fault plane if the fault type is known. The dip is only given by a direct fault plane correlation (e.g. main or parasitic plane), otherwise only a minimum dip is indicated by the drag fold. Repeating sections will, in theory, indicate a reverse fault and seismic data will often help define the likely fault type. By combining the fault azimuth with the sense of rotation of the hangingwall block, it is possible to estimate the type of motion on the fault plane.

Analysis of angular unconformities

Throughout the Ormskirk Sandstone Formation, small-scale changes in structural dip were discovered with the aid of vector-azimuth diagrams (example shown in Fig. 5). These depict the azimuth variation with depth, but disregard the actual dip angle. This method is ideal for shallow dipping structures such as those found in the EISB since a spatial change in dip angle by one degree is much more difficult to spot than an accompanying change of dip azimuth by, e.g. 15°. The angular rotations represented by these structural dip changes are small, between 1.5° and 2.0°. Dip changes of this magnitude would be disregarded if observed on a single well. In this case, due to the large data set, it was possible to correlate these angularly unconformable levels and the block rotations represented by them on a field-wide scale (Fig. 6). It was concluded that they represent syn-sedimentary fault block rotations. It can be quite difficult to pinpoint the exact positions of these 'unconformities' since their identification depends upon the presence of continuous flat-lying dip indicators (i.e. sheetflood facies). Consequently, some unconformable levels can only be placed within zones 50–100 ft thick, especially where the boundary is associated with stacked major channel fill facies, in which case the unconformity was assumed to lie at the base of the channel sequence.

Sedimentological interpretation

Sedimentological interpretation was carried out after structural dip removal, using 1:50 scale log plots of micro-conductivity, rotated SBS results (or small-scale PTP results) and other openhole logs. The dipmeter curves and results were calibrated against core data before interpretation of uncored sections. The facies associations derived from sedimentological core

logging were calibrated with typical dipmeter responses (electrofacies). The raw micro-conductivity traces (depth and intensity amplitudes or patterns) are particularly useful here, as well as characteristic dip patterns. Diagnostic features are the average dip magnitude, whether dip is constant, increasing or decreasing upward, uni-, bi- or polymodal directions, thickness of individual coherent dip sets and average correlatability within the pattern (i.e. correlation coefficients). The palaeocurrent interpretation is dependent upon the facies interpretation. Each well was divided into gross facies association intervals, and the dipmeter orientation results for each were then plotted on a Wulff stereogram, after filtering and the removal of spurious dips.

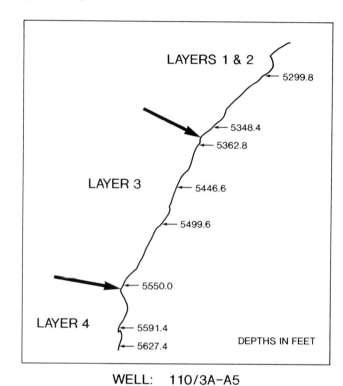

LAYERS 1 & 2 — 5299.8
— 5348.4
— 5362.8

LAYER 3 — 5446.6
— 5499.6

— 5550.0

LAYER 4 — 5591.4
— 5627.4

DEPTHS IN FEET

WELL: 110/3A–A5

Fig. 5. Dip azimuth vector plot of Well 110/3a-A5. This plot shows structural dip azimuth versus depth, disregarding the amount of dip. Non-palaeohorizontal correlation planes are removed by filtering. The changes in mean azimuth (shown as arrows) correspond to the small-scale unconformities which correlate with the sedimentologically defined layer boundaries. This implies that there is a structural control over facies distributions.

Dips and dip patterns were assigned to sedimentological structures on the basis of micro-conductivity curve appearance and the relationship with related correlation planes. This is vital if meaningful palaeocurrent data are to be obtained. For example, a well-defined single plane derived from a marked conductivity change, and followed by a series of less well-defined planes derived from low-profile conductivity oscillations, would be interpreted as an erosion surface followed by a planar cross-stratified bed.

Trough cross-bedding often appears in stereograms as a tri-polar distribution (clover-like) around the palaeohorizontal plane, although on a tadpole plot the plane distribution may look quite scattered and confused. The palaeocurrent in such a tri-polar plot is given by the orientation of the middle sub-population.

Aeolian dune slip facies show very characteristic dip patterns with relatively steep, unimodal dips over 20° lying above an increasing upward dip pattern representing the toe set. The

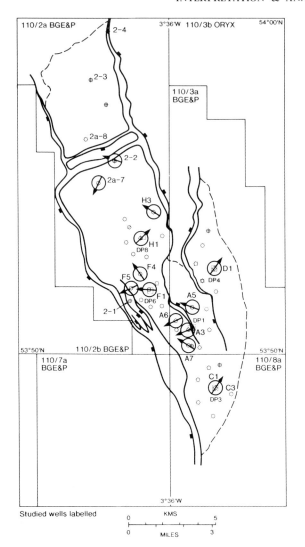

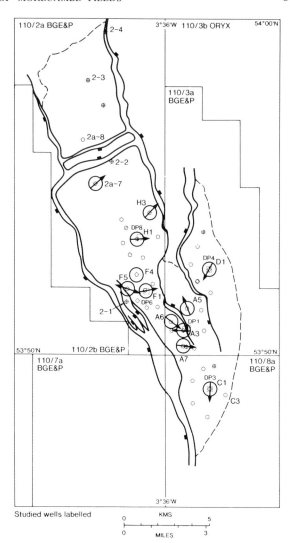

POSSIBLE BLOCK ROTATIONS AT LAYER 2 – 3 EVENT   POSSIBLE BLOCK ROTATIONS AT LAYER 3 – 4 EVENT

Fig. 6. Block rotations at the Layer 2–3 and 3–4 unconformities. Arrows show relative tilt direction of lower structural block compared to upper block.

aeolian sandsheet facies can be distinguished from the fluvial and sheetflood facies as a consequence of their high porosities, which gives a sharp micro-conductivity response and is also reflected in low sonic velocities.

Palaeocurrent orientations from intervals of similar facies association within a stratigraphic unit were plotted in rose diagrams. The average dip azimuths were weighted according to the thickness of the unit, the density of the dip data, consistency of dip within unit and the number of stacked sets within the unit.

Structural interpretation and implications

Faulting

Medium-scale faulting (1–50 ft) is very common, but only two or three wells penetrated large-scale faults (> 100 ft). Examples are given below.

Normal fault with drag

In the St Bees Sandstone Formation of 110/3a-A3 the core showed a well-defined mylonitic, normal fault plane along which the sandstones exhibit drag. Stratigraphic data suggest

the fault throw is around 200 ft. Drag was better developed than normal, reflecting the high porosity in this part of the field. The dipmeter results closely reflected the core data, with bed boundaries being deflected in the direction of the fault plane azimuth over an 80 ft section. The fault plane was correlated as dipping 60° toward ENE (Fig. 7). The structural dip changed across the fault from 7° NNE in the footwall block to 5° N in the hangingwall block. This dip change represents a rotation of 4° toward WSW of the hangingwall block relative to the footwall block. The movement may be described by a NNW-plunging axis of rotation, which is similar to the drag fold axis. The axial coincidence reflects a non-pivotal fault block rotation toward the fault plane (anti-thetic rotation) which is typical of listric faults.

Listric fault

In Well 110/8a-C3 in the southeastern area of South More-cambe an interesting geometric relationship was noted between the easterly deviated well and an easterly dipping system of normal faults. The well deviation decreased from 51° at 2500 ft measured depth (MD) to 33° at 5350 ft MD (total depth), and E to ESE-dipping faults were intersected in the interval 5300–5000 ft MD and 3900–3700 ft MD. Due to the deviation, the correlation search was restricted in direction of the hole azimuth from the normal maximum of 70–80° to 30–45° relative to the horizontal.

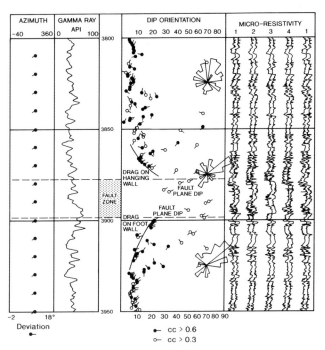

WELL 110/3a–A3

Fig. 7. Dipmeter response to normal fault with a throw of 200 ft. Fault planes exhibit ENE–NE dips of 50–60° and footwall drag and hangingwall drag are clearly seen. The structural dip azimuth changes from NE to NNE across the fault zone.

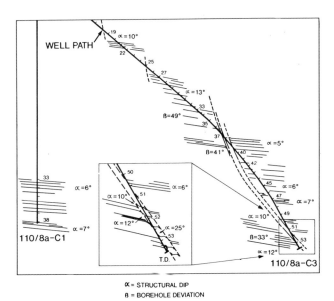

110/8a–C3 LISTRIC FAULT

Fig. 8. Structural section constructed by dipmeter interpretation of the deviated Well 110/8a-C3, which has twice intersected a listric fault zone. The faulting imposes an antithetic westerly rotation of the hangingwall block (dipping 6° ESE) relative to the footwall (11° ESE). The insert depicts the sub-parallel nature between the wellbore and fault which has resulted in an abnormally long section of dip patterns associated with fault drag.

The interpretation is shown in Fig. 8 and essentially features:

(1) the highly deviated well penetrating a steeply E(NE)-inclined fault zone from the 13° ESE-dipping footwall block and into the 5–7° ESE-dipping hangingwall block at 3700 ft MD in the Mercia Mudstone Group. The basis for this interpretation is the change of structural dip associated with dip patterns thought to represent fault drag to the E(NE);

(2) another four E-dipping faults intersected in the next 1500 ft;

(3) the well intersects an E(SE)-dipping fault at 5300 ft in the Ormskirk Sandstone Formation. A 100–150 ft dip pattern showing dips of bed boundaries increasing systematically down-hole from 7° to 35° E is most easily interpreted as a zone of drag on a fault plane oriented sub-parallel to the borehole. The relatively shallow maximum dip of the inferred drag suggests that the fault plane dips approximately 45° rather than 65°. The structural dip apparently changes from 6–7° ESE to 10–12° ESE below the fault (i.e. in the footwall).

Combining the above points, it seems very likely that the well actually penetrates the same fault system twice due to a downward decrease of well deviation and of fault dip, indicating a listric fault system. Supporting evidence is the rotation of the hangingwall (6–7° ESE) relative to the footwall (10–13° ESE). This block rotation involves an antithetic, 5–6° westward rotation, i.e. toward the fault plane. Antithetic faulting was not noticed in the well.

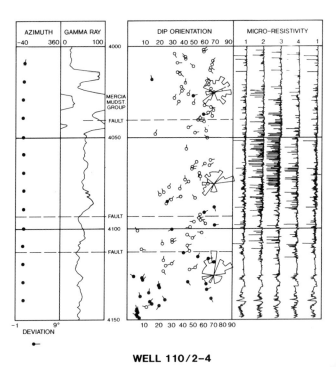

WELL 110/2–4

Fig. 9. Dipmeter log through complex eastern boundary fault zone in North Morecambe Well 110/2-4, processed with a 1 ft step interval. HDT log showing 300 ft of reservoir section displaying confused dip patterns, dominated by N- to E-dipping planes.

Pivotal rotation between forking faults

Well 110/2-4 is situated towards the northeast in North Morecambe. The dipmeter results indicate that a branch of the eastern boundary fault system of the field is penetrated by the well near the top of the Ormskirk Sandstone. Stratigraphic

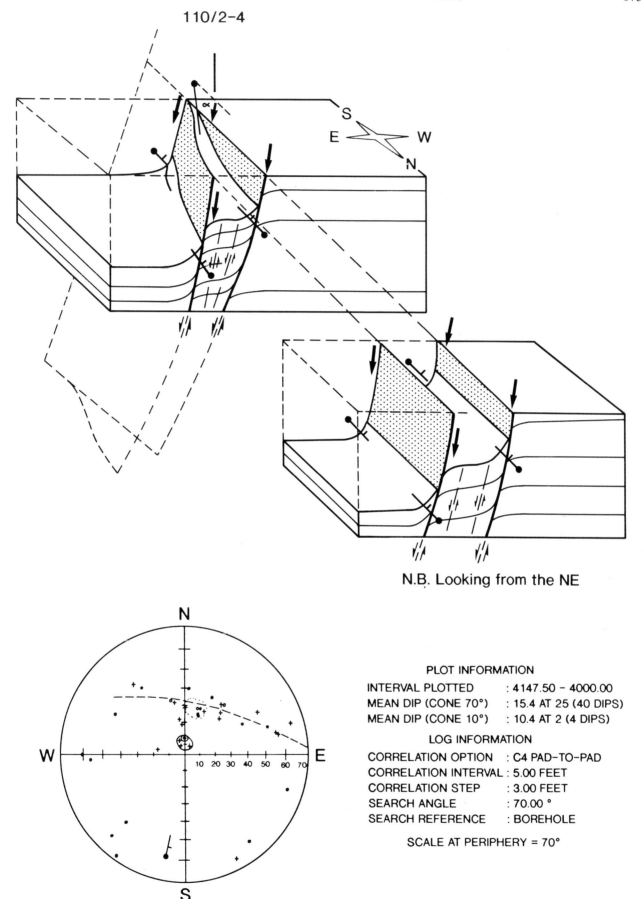

Fig. 10. Block diagram of interpretation of dipmeter response through fault zone shown in Fig. 9. Stereogram displays the poles to planes of dip in the same interval using 3 ft step interval. The fitted great circle represents fault-dragged bed boundaries and shear planes related to the steeply E–ENE-dipping faults, while the northward tilt reflects the structural rotation of the intermediate fault blocks. A dragged-out and sheared intermediate fault block has been tilted to the north due to the forking of the two main E-dipping fault planes. Forking occurs to the south of the well, thus total throw increases to the north.

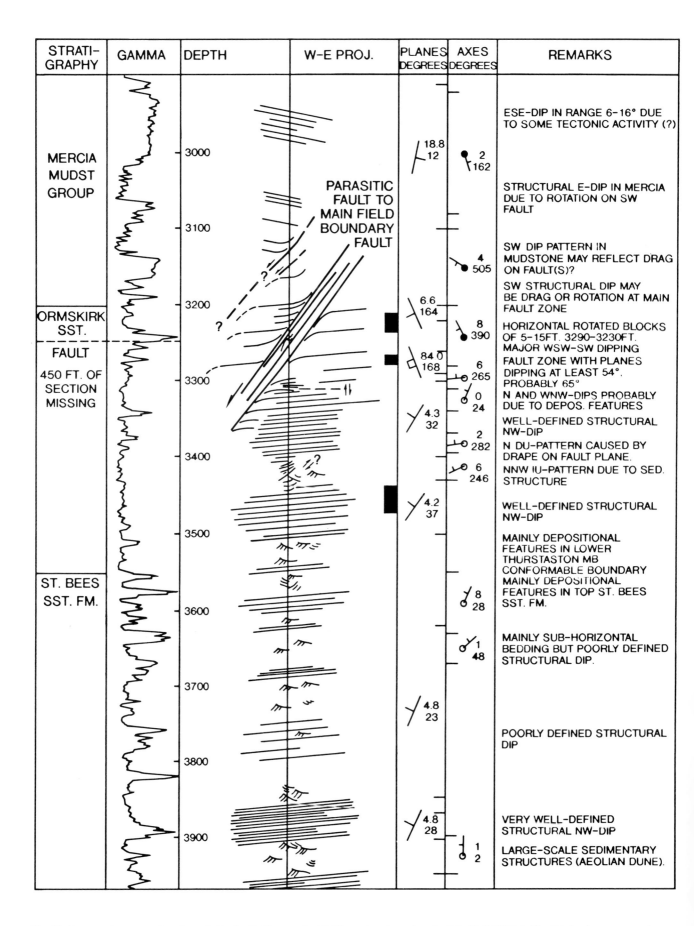

Fig. 11. Structural summary log through the South Morecambe Field's western boundary fault in Well 110/2-1. The third column summarizes mean structural dips, and the fourth column shows symmetry axes extracted from Wulff plots of dip patterns. Filled axes reflect tectonic structures, open are depositional structures. Tick marks on axis sticks indicate deflections of dips away from palaeohorizontal orientation, e.g. fault drag or cross-stratification.

data suggest that 300–350 ft of section has been cut out by the faulting. In the tectonized 100–150 ft section, steep NNE to E dips (25–70°) are dominant, with no shallow dips correlated. Faulting along steeply E–ENE dipping planes at two, possibly three, levels seems probable from the dip patterns. The stereogram of the interval (4150–4000 ft, Fig. 9) reveals two structural axes defined by great circle fits to correlation plane poles (Fig. 10). One describes a 30° northward tilt of most of the involved structures, while the other axis describes the eastward deflection of bedding planes due to extensive fault drag. A 20° WNW mean dip is found 200 ft above the main fault zone. To explain the existence of two perpendicular structural axes in the main fault zone, it is essential to realize their differing natures. The steep northerly plunge of the drag fold axis shows that it has been rotated round the sub-horizontal, E–W-trending, second axis which is thus interpreted as a block rotation axis. It, therefore, appears that the intermediate fault blocks have rotated in a northward fashion while being dragged out along the steep E–ENE-dipping faults, a so-called pivotal fault system. This can also be observed in en échelon fault zones but in this case, seismic data suggest that it is caused by the forking (diverging) of the fault. The merging of the two or three E faults should occur south of the well.

Major fault with shearing and brecciation

Well 110/2-1 penetrated part of the fault zone which forms the western boundary fault of South Morecambe. Dipmeter results define a structural dip of approximately 4° WNW from the underlying St Bees Sandstone Formation into the Lower Ormskirk Sandstone Formation (Layers 4–6). A 60 ft zone making up the boundary between the Upper and Lower Ormskirk Sandstone Formation exhibits steep WSW-dips, the 54° mean of which gives minimum dip of the fault zone. The 65° WSW dips probably represent the fault planes. The zone may be visualized as more or less sheared throughout, although minor, 5–8 ft thick blocks within the zone are signs of brecciation. Drag features in the lower and upper blocks are poorly developed according to the dipmeter results, though the 7° WSW mean dip of a 40 ft section of Upper Ormskirk Sandstone Formation and lowermost Mercia Mudstone Group may suggest a moderate, but penetrative drag in the hangingwall block (Fig. 11).

A second fault zone semi-parallel to the main fault is reflected by poorly defined dip patterns higher in the Mercia Mudstone Group. Structural dip of the hangingwall is approximately 10° E to ESE and represents an overall 12° east-southeastward rotation of this block relative to the 4° WNW structural orientation in the Sherwood Sandstone Group. This would imply that the movement on the WSW-dipping fault zones are composed of an antithetic component as well as a pivotal component of rotation, the latter suggesting that amount of dip slip increases in a southerly direction.

Angular unconformities in the Morecambe area

From a structural point of view, the widespread occurrence of what have been interpreted as subtle field-wide angular unconformities (see above) is a significant characteristic of the Morecambe structure. The present study confirms three levels in the Ormskirk Sandstone Formation as being field-wide angular unconformities, though uncertainty remains in the correlation of the lowermost level, where a fourth semi-field-wide event may be distinguished. Probable correlations are depicted in Fig. 12.

(1) The uppermost angular unconformity clusters around the Upper to Lower Ormskirk Sandstone boundary (Layers 2–3). This marks a major facies boundary between the stacked major channel associations of Layer 3 and the mixed, more argillaceous facies associations of Layers 1 and 2. Layers 1 and 2 represent the final abandonment of the fluvial system prior to the deposition of the overlying salts and mudstones of the Mercia Mudstone Group. It is characterized by a general westerly directed tilt of the older strata, perhaps corresponding to initial movements on the ENE-dipping graben fault west of Morecambe. Three easterly situated wells (-D1, -C1 and -H1) exhibit north-easterly rotations.

(2) The middle level conforms more or less to the Layer 6–7 boundary. This event corresponds to a field-wide change in facies from the sheetflood associations of Layers 4–6 to the stacked fluvial sandstones of Layers 7–10. This event shows less constraint on rotational directions, though an easterly tilt of the older strata is prevalent. The three wells situated in or east of the central graben (-A5, -D1 and -C1) show north or southward rotations, and were thus governed by a different axis of rotation. Whether the east and southernmost areas actually constituted a separate structural block cannot be stated with certainty.

(3) The lowermost level(s) is(are) more problematic. It is possible that two levels are present. It (or they) seems to correspond to a change from mixed fluvial and aeolian associations between Layer 8 and the top of the St Bees Sandstone Formation and the stacked fluvial facies associations of Layer 7.

Two of the minor unconformities correspond to one of the lithostratigraphically defined layers of Colter and Barr (1975).

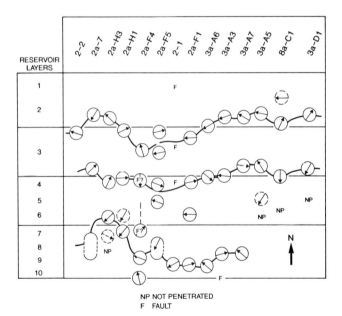

ANGULAR UNCONFORMITIES CUTTING ACROSS LITHOSTRATIGRAPHIC BOUNDARIES

Fig. 12. Stratigraphic levels of inferred angular unconformities marked by circles. The arrows inside the circles reflect the azimuthal direction of tectonic tilt suffered by the older rocks before deposition of the overlying strata. The correlation of the oldest event poses most problems and may be two separate events. Since these events correspond to the major facies changes, the implication is that there is a structural control over facies distributions.

Implications

The fact that general direction of rotation correlates well within the events supports the validity of such correlations and demonstrates that if the dataset is large enough, dipmeter results can be used in very fine-scale structural analyses of regional tectonic activity. It is likely that instead of being an exceptional case study, the Morecambe findings may actually apply to most sedimentary Permo-Triassic basins. The most important implication is that these field-wide correlatable events may be used as semi-chronostratigraphic boundaries. It has been pointed out above that the reservoir sequence appears to consist of an alternation of correlatable layers dominated by major channel sandstone facies; and layers dominated by sheetflood and secondary channel facies associations.

It has been suggested that basin-wide tectonic events were responsible for this alternation. Clearly, these small but consistent tectonic rotations now observed in the dipmeter data are the structural expression, within the field, of these (previously hypothetical) tectonic adjustments which caused the periodic rejuvenation of the depositional system. Identification of these small-scale rotational events has confirmed the basis of the depositional model and has permitted a greatly enhanced confidence in the facies correlations used in reservoir modelling.

The boundary between the St Bees Sandstone Formation and the overlying Keuper Sandstone Formation had been equated in earlier literature with the Hardegsen Unconformity (Colter and Barr 1975; Colter 1978). This boundary lies about 800 ft below the top of the Sherwood Sandstone Group across most of the field. The St Bees Formation shows higher gamma ray and faster sonic log response than the Ormskirk Sandstone Formation, and can thus be correlated basin-wide. Dipmeter logs were carefully investigated across this boundary and it was concluded that there is no evidence of angular unconformity at the top of the St Bees Sandstone (base of the Ormskirk Sandstone) in the field wells.

Sedimentological interpretation and implications

Core was available in 16 wells in which dipmeters were run, allowing calibration of the dipmeter raw curves and dip plots with core, and enabling facies identification to be carried out in uncored sections with a high degree of reliability (Figs 13 and 14). This is vital for palaeocurrent analysis since it is important to identify the likely sedimentary structures causing the dip correlations to provide meaningful palaeocurrent data.

Palaeocurrent data

The palaeocurrent data for the major channel association of the Ormskirk Sandstone Formation in the EISB are summarized in Fig. 15 and Table 2. The consistent westerly flow directions demonstrated here form one of the most important findings of this study. Regional data (summarized in Burley 1987; Fig. 16) suggested that a countrywide north–south, regional palaeoslope controlled deposition during Sherwood Sandstone times. It had been expected that the dominant flow direction would have been to the north, the sediment being fed into the basin through the southern opening from the Cheshire

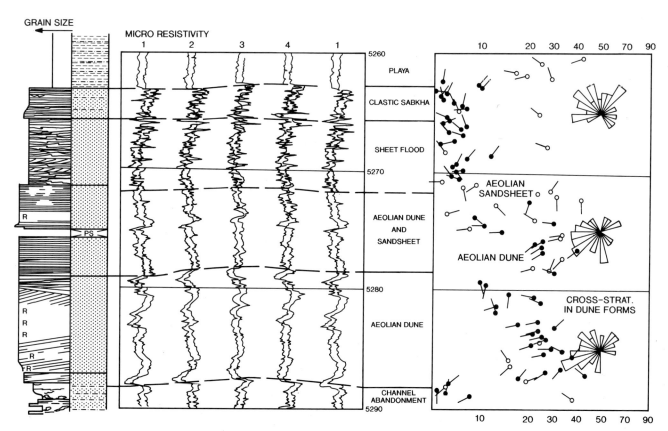

110/3a–A5 AEOLIAN DUNE SANDSTONE WITH LARGE SCALE CROSS–STRATIFICATION OVERLAIN BY SHEETFLOOD SANDSTONES

Fig. 13. Example of dipmeter response calibrated to core. SBS-processed dipmeter with structural dip removed is plotted next to a graphical core log. The upper part of the sequence shows an intraformational conglomerate in a major channel facies sandstone. The individual mudstone clasts each give a strong micro-resistivity response causing the high degree of curve activity shown. These individual peaks cannot be correlated, and thus allow this sub-facies to be easily distinguished from the sheetflood facies sandstones which show similar, but correlatable, high curve activity (see Figs 5 and 17). The lower part of the log shows more typical dipmeter response to channel facies sandstones, with 'flat' curve activity, but many correlatable planes.

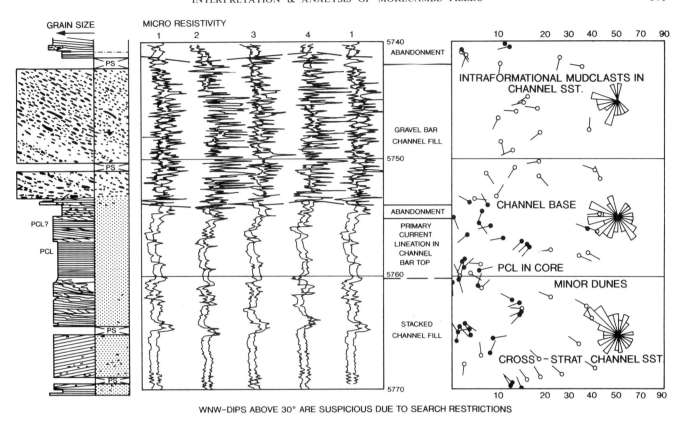

WNW–DIPS ABOVE 30° ARE SUSPICIOUS DUE TO SEARCH RESTRICTIONS

110/3a–A6 MAJOR CHANNEL SANDSTONE WITH AND WITHOUT MUDCLASTS

Fig. 14. Core log and dipmeter response through sheetflood and aeolian sediments. The sheetflood facies shows high curve activity with many correlation planes and low angle dips (compare with Fig. 13). Aeolian dune sediments are characterized by highly resistive layers with many correlatable events. The high-angle dips are characteristic of the dune sediments. The aeolian sandsheet facies shows similar micro-resistivity response but is characterized by low-angle dips. These responses, in conjunction with their gamma ray and sonic character, allow the aeolian sediments to be differentiated from the channel facies.

Table 2. Summary of predominant palaeocurrent directions for major channel sandstone facies

Unit*	2-5	2a-8	2-2	2a-7	D1	A5	2-1	A6	A3	A2	A7	C1	C3
							Wells (N–S) 110/						
1–2	—	—	SW	W	W–WSW	W	—	W	—	W–WSW	W	W–WSW	WNW–W
3	SSW	SSW	WSW	SW	W–WNW	WNW	—	SSW	W	W	W	WNW	WNW
5	—	—	—	SSW	W–WNW	WNW	—	W–S	—	W	W–S	NW	—
7–10	—	SW	W	SW	np	np	WNW	WNW	WNW	W	W	np	np
St Bees	WSW	NW	W	SE	np	np	W	SW	S	np	WNW	np	np

np: not penetrated (well TD above this unit).
—: no consistent palaeocurrent data obtained.
*: South Morecambe reservoir units. Wells 110/2-5 and 110/2a-8 are on North Morecambe, but palaeocurrents listed here are for the correlative South Morecambe unit.

Basin. The results of this study indicate that although the UK regional palaeoslope was towards the north, local depocentres controlled palaeoflow in individual basins.

It can be seen on Fig. 15 that there are systematic variations in space and time through the Ormskirk Sandstone sequence. A feature revealed on Fig. 15 is the rotation of palaeocurrents from the north to the south of the field. There is a strong southwesterly orientation in North Morecambe and the northern limb of South Morecambe, but further south the palaeocurrent vectors swing to the northwest in the southern part of South Morecambe. The net aggregate flow direction of the

major channel units is, therefore, in a westerly direction across the field as a whole.

There is also a tendency for palaeocurrent orientations to rotate anticlockwise upwards through the Ormskirk Sandstone sequence. There are more palaeocurrent vectors in the northwesterly quadrant in the lower part of the Ormskirk Sandstone Formation, but in the upper part of the sequence, more palaeocurrent vectors in the southwesterly sector. Each of the layers shown on Fig. 15 is separated by one of the field-wide unconformities referred to earlier. It is not unexpected that there were adjustments in palaeoflow directions across them,

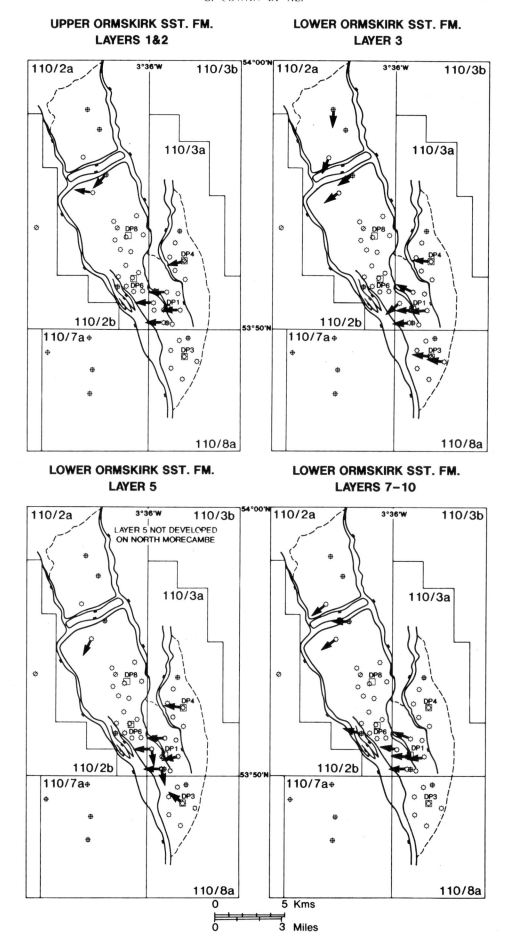

Fig. 15. Aggregate palaeocurrent orientations for major channel facies sandstone in the Ormskirk Sandstone Formation. Arrows represent the average of many measurements.

but the palaeocurrent orientations do not directly parallel the block rotation directions shown in Fig. 6. This is because the principal influence of palaeoflow was the basin-wide palaeo-slope towards the Keys Fault, and the aggregate vectors shown represent many hundreds of individual measurements, some of which do parallel the block rotations.

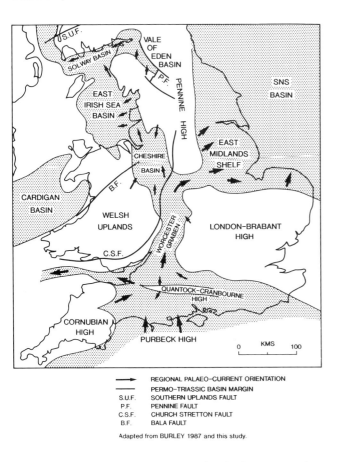

Fig. 16. Permo-Triassic basins showing regional palaeocurrent orientations. Adapted from Burley (1987) and this study.

The palaeocurrent orientations derived from the secondary channel facies sandstones in the Ormskirk Sandstone Formation show westerly flow directions similar to the major channel facies sandstones, but with a much greater spread of palaeocurrent vectors. Sheetflood sandstones are usually flat laminated; where dip patterns reveal small-scale cross-bedding, a very wide spread of palaeocurrent vectors is revealed, although a dominant westerly flow direction is still apparent.

Palaeocurrent data from the St Bees Sandstone Formation is more ambiguous. The thick stacked major channel facies layers encountered within the Ormskirk Sandstone Formation are absent here, and the individual major channel sandstone facies show very variable palaeocurrent orientations. An aggregate westerly or northwesterly palaeoflow is notable, but the well-to-well spread is from northwesterly, through westerly to southerly and even southeasterly, and there are no consistent changes throughout the field as observed in the Ormskirk Sandstone. The thick stacked channel sequences found in the Ormskirk Sandstone probably reflect increased faulting and basin subsidence at the end of St Bees times (Knipe *et al.* 1993).

Palaeocurrent data from aeolian sediments are sparse in comparison to the fluvial sediments, since the aeolian sediments comprise less than 10% of the sequence and it is difficult to obtain reliable palaeocurrent data from the flat-laminated aeolian sandsheet. Figure 17 shows the aeolian dune facies palaeocurrent averages for the Ormskirk Sandstone as a

whole. These show consistent westerly flow directions, probably caused by Triassic southwesterlies, as the UK has rotated by about 10–15° to the NE since the Permian (Glennie 1986). Similar orientations were recorded in barchanoid dune sandstones of the Permo-Triassic Penrith Sandstone in the Vale of Eden Basin (Macchi 1990). These suggest that during the deposition of the Ormskirk Sandstone the UK was still in the northeasterly trade belt.

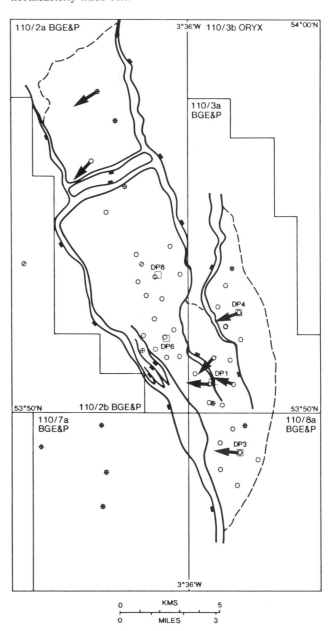

Fig. 17. Aggregate palaeocurrent orientations for aeolian facies sandstones in the Ormskirk Sandstone Formation. Arrows represent the average of many measurements.

The dominantly westerly fluvial palaeocurrent orientation implies that the major channel axes run E–W and that they dip directly into the aquifer along the eastern, dip-closed margin of the South Morecambe Field. The fact that the reservoir is strongly layered, and that a layer rather than a 'pipe' of permeable channel facies dips into the aquifer, reduces the likelihood of water influx. The results of the present study have been input to detailed reservoir modelling to predict water movement during field life. This has shown that although the major channel facies and, more importantly, the aeolian facies, do dip into the aquifer, premature water influx is not predicted to occur.

Conclusions

The logging environment in the EISB provides optimum conditions for the acquisition of dipmeter data. Dipmeter studies of wells from the Morecambe Fields have allowed precise identification of the orientation of fault planes which can be used to extrapolate faults between seismic lines. Many small-scale faults were identified by dipmeter data alone. Dip azimuth vector plots revealed blocks of uniform structural dip separated by low-angle, field-wide unconformities which roughly correspond to the facies controlled reservoir layers. This suggests that the structural evolution of the area controlled the distribution of facies associations within the area. The regionally significant boundary between the St Bees Sandstone and Ormskirk Sandstone Formations is not manifested by an unconformity. Palaeocurrent orientations derived from the major channel facies associations show dominantly westerly palaeocurrent orientations, reflecting a palaeoslope towards the Keys Fault. Aeolian dune facies associations also show westerly palaeowind orientations.

We would like to thank the management of British Gas Exploration and Production Ltd and Z and S Geologi A.S. for permission to publish this paper. The original palaeocurrent interpretations were carried out by Gyrite Brandt and Andor Hjellbakk. Ruth Johns drafted the diagrams.

References

BURLEY, S. D. 1987. *Diagenetic modelling of the Triassic Sherwood Group of England and its offshore equivalents, U.K. Continental Shelf.* PhD thesis, Hull University.

COLTER, V. S. 1978. Exploration for gas in the Irish Sea. *Geologie en Mijnbouw,* **57**, 503–516.

—— AND BARR, K. W. 1975. Recent developments in the geology of the Irish Sea and Cheshire Basins. *In:* WOODLAND, A. W. (ed.) *Petroleum and the Continental Shelf of North-West Europe, Vol. 1.* Applied Science Publishers, London, 61–75.

COWAN, G. (1993). The identification and significance of aeolian deposits in the dominantly fluvial Sherwood Sandstone Group of the East Irish Sea Basin U.K. *In:* NORTH, C. P. AND PROSSER, J. (eds) *The Characterization of Fluvial and Aeolian Reservoirs.* Geological Society, London, Special Publication, in press.

GLENNIE, K. 1986. Early Permian–Rotliegend. *In:* GLENNIE, K. (ed.) *Introduction to the Petroleum Geology of the North Sea.* Blackwell, London, 63–86.

JACKSON, D. I., MULHOLLAND, P., JONES, S. M. AND WARRINGTON, G. 1987. The geological framework of the East Irish Sea Basin. *In:* BROOKS, J. AND GLENNIE, K. (eds) *Petroleum Geology of North West Europe.* Graham & Trotman, London, 191–203.

KNIPE, R., COWAN, G. AND BALENDRAN, V. S. 1993. The tectonic history of the East Irish Sea Basin with reference to the Morecambe Fields. *In:* PARKER, J. R. (ed.) *Petroleum Geology of Northwest Europe: Proceedings of the 4th Conference.* Geological Society, London, 857–866.

MACCHI, L. 1990. *A field guide to the Continental Permo-Triassic rocks of Cumbria.* Liverpool Geological Society, Excursion Guide.

STUART, I. A. 1993. The geology of the North Morecambe Gas Field, East Irish Sea Basin. *In:* PARKER, J. R. (ed.) *Petroleum Geology of Northwest Europe: Proceedings of the 4th Conference.* Geological Society, London. 883–895.

—— AND COWAN, G. 1991. The South Morecambe Field, Blocks 110/2a, 110/3a, 110/8a, UK East Irish Sea. *In:* ABBOTTS, I. L. (ed.) *United Kingdom Oil and Gas Fields, 25 Years Commemorative Volume.* Geological Society, London, Memoir, **14**, 527–541.

WARRINGTON, G., AUDLEY-CHARLES, M. G., ELLIOT, R. E., EVANS, W. B., IVIMEY-COOK, H. C., KENT, P. E., ROBINSON, P. F., SHOTTON, F. W. AND TAYLOR, F. M. 1980. *A correlation of Triassic rocks in the British Isles.* Geological Society, London, Special Report **13**.

The geology of the North Morecambe Gas Field, East Irish Sea Basin

I. A. STUART

British Gas plc Exploration and Production, 100 Thames Valley Park Drive, Reading, Berkshire, RG6 1PT, UK

Abstract: The North Morecambe Gas Field lies in the East Irish Sea Basin close to the producing South Morecambe Field, from which it is separated by a very deep WSW–ENE graben. It is a shallow structure, crest slightly above −3000 ft, with bounding faults to the south and east, but dip-closed to the north and west. The GWC is at −3950 ft. The Triassic Sherwood Sandstone Group reservoir was deposited under semi-arid continental conditions and comprises a complex interplay of channel-fill sandstones, non-channelized sheetflood sandstones and high-permeability aeolian and reworked aeolian sandstones. The vertical alternation of these facies forms the basis for differentiating nine facies-defined reservoir units. Palaeocurrent analysis indicates flow from NE to SW. A complex diagenetic history is recognized, with several phases of dolomite and quartz cementation. Differential compaction is also a significant control on the disposition of reservoir properties. The greatest control on permeability (but not porosity) is platy illite which precipitated beneath a palaeo-gas/water contact at an early stage in the growth of the structure. This gives rise to a diagenetic layering of the reservoir into a high-permeability illite-free layer and a deeper, low-permeability, illite-affected layer. The Mercia Mudstone Group which provides the seal, comprises four mudstone–halite cycles. Thickening of these cycles on the north flank of the structure indicates very early development of the northerly component of dip. The imposition of northerly dip entirely pre-dates the growth of platy illite, and this conclusion has a major impact on the manner in which the surface which separates the illite layers has been mapped on the north flank. The gas, sourced from Westphalian/Namurian sediments, is enriched in CO_2 compared with South Morecambe. This, together with contrasts in vertical cement distribution between the two fields, indicates that the breaching and loss of the first gas charge which is believed to have occurred on South Morecambe did not occur on North Morecambe. Furthermore, although separated from South Morecambe by only a narrow graben, the aquifer pressure is 140 psia less than that of South Morecambe, indicating that the major basin faults are full seals. Proved reserves are 1.08 TCF. The field is currently being developed for first gas in 1994.

The North Morecambe Gas Field lies in the centre of the Permo-Triassic East Irish Sea Basin (EISB), immediately north of the larger South Morecambe Gas Field (Fig. 1). At present only South Morecambe has been developed and previous papers on Morecambe published by British Gas and its subsidiaries have referred primarily or entirely to South Morecambe. This is the first publication to relate specifically to North Morecambe.

Various aspects of the geology of the Morecambe area have been described previously (Colter and Barr 1975; Colter 1978, Colter and Ebbern 1978, 1979; Ebbern 1981; Bushell 1986; Woodward and Curtis 1987; Levison 1988; Stuart and Cowan 1991; Cowan 1993); further papers on specific aspects of Morecambe area geology are included elsewhere in this volume (Knipe *et al.* 1993, Cowan *et al.* 1993).

North Morecambe Field lies wholly within Block 110/2a which also contains part of the South Morecambe. The block was licensed in 1972 solely to Hydrocarbons Great Britain Ltd, but is now held by British Gas Exploration and Production Ltd (formerly known as Gas Council (Exploration) Ltd), all three companies being subsidiaries of British Gas plc. Following the discovery of South Morecambe in 1974, the discovery well for North Morecambe, 110/2-3, was drilled in 1976. The field was delineated by Wells 110/2-4A (dry hole), 110/2-5 and 110/2a-8, and was found to be a separate accumulation from South Morecambe. Subsequent work has shown that, while the geology is similar in broad terms to that of South Morecambe, there are small yet revealing differences between the two fields and these are indicative of the type of geological variation which may be encountered across the basin.

Following development of the larger South Morecambe Field during the 1980s, 'Annex B' approval for North Morecambe was gained in 1992, and first gas is planned for 1994. North Morecambe will be developed by means of a single (normally unmanned) well head platform with minimal pro-

cessing, and a 36-inch pipeline to a new dedicated terminal at Barrow with processing to remove CO_2 and N_2 and treatment of gas to NTS specification. Thus the gas will be produced independently of the South Morecambe facilities (owing to the

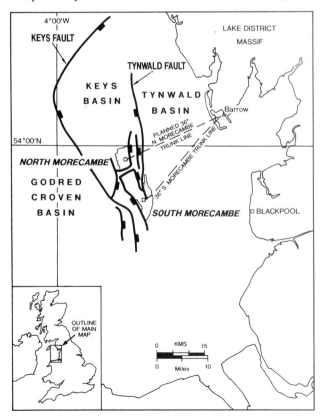

Fig. 1. Location map.

From *Petroleum Geology of Northwest Europe: Proceedings of the 4th Conference* (edited by J. R. Parker).
© 1993 Petroleum Geology '86 Ltd. Published by The Geological Society, London, pp. 883–895.

need to remove inert gases), although the platform will have control and communication links with the South Morecambe complex.

Proved reserves for North Morecambe are 1.08 TCF.

Field stratigraphy

Field stratigraphy is shown on Fig. 2. The nearest penetration of Pre-Triassic is South Morecambe Well 110/8-2, which reached Upper Carboniferous (Namurian or Westphalian). Seismic data indicate a similar Pre-Triassic sequence beneath North Morecambe. The reservoir sandstones belong to the Lower Triassic Sherwood Sandstone Group; the overlying Mercia Mudstone Group (Middle/Upper Triassic) forms the seal. After the Rhaetian transgression an unknown thickness of Jurassic and Cretaceous sediment was deposited but was removed during rapid basin uplift in the early Tertiary. The Mercia Mudstone is now overlain only by a thin veneer of Quaternary sand and gravel.

SYSTEM/SERIES/GROUP		WARRINGTON *et al* 1980	COLTER & BARR 1975
QUATERNARY			
MIDDLE/UPPER TRIASSIC	MERCIA MUDSTONE GROUP	MERCIA MUDSTONE GROUP	KEUPER MARL AND SALIFEROUS BEDS
LOWER TRIASSIC	SHERWOOD SANDSTONE GROUP	ORMSKIRK SANDSTONE FORMATION	KEUPER WATERSTONES FM.
			KEUPER SANDSTONE FORMATION
		ST. BEES SANDSTONE FORMATION	ST. BEES SANDSTONE FORMATION
UPPER PERMIAN		ST. BEES EVAPORITES AND EQUIVALENTS	MANCHESTER MARL
LOWER PERMIAN		COLLYHURST SANDSTONE AND EQUIVALENTS	COLLYHURST SANDSTONE AND EQUIVALENTS
UPPER CARBONIFEROUS		WESTPHALIAN	WESTPHALIAN
		NAMURIAN	NAMURIAN

Fig. 2. Stratigraphy of East Irish Sea Basin with references.

Sherwood Sandstone Group stratigraphy

During early Triassic times, rifting and rapid basin subsidence led to the accumulation of very thick clastic sequences of continental 'red bed' sediments of the Sherwood Sandstone Group. Colter and Barr (1975) described the stratigraphy of the EISB Permo-Triassic using terminology from the Cheshire Basin and divided the Sherwood Sandstone Group into three formations: St Bees Sandstone, Keuper Sandstone and Keuper Waterstones Formations. Colter and Barr (1975) used the term 'Keuper' informally. Warrington *et al.* (1980) and Jackson *et al.* (1987) proposed that the 'Keuper Sandstone' and 'Keuper Waterstones' Formations of the EISB be renamed the 'Ormskirk Sandstone' Formation. The terms proposed by Colter and Barr (1975) are retained as convenient terms for informal stratigraphic units which can be correlated across North and

South Morecambe, but the term 'Ormskirk Sandstone' Formation is used to refer to the full post-St Bees interval.

The lower 4000 ft of Sherwood Sandstone belongs to the St Bees Sandstone Formation. Only the uppermost part (750–800 ft) is classified as Ormskirk Sandstone Formation but this forms the majority of the North Morecambe reservoir, the St Bees being mostly below the GWC. Dipmeters show that the boundary between the St Bees and Ormskirk Sandstone Formations is not an unconformity. However, it corresponds to an abrupt facies change and change in palaeocurrent direction, indicating a hiatus together with tectonic reactivation on the basin margins and/or climatic changes (Cowan *et al.* 1993).

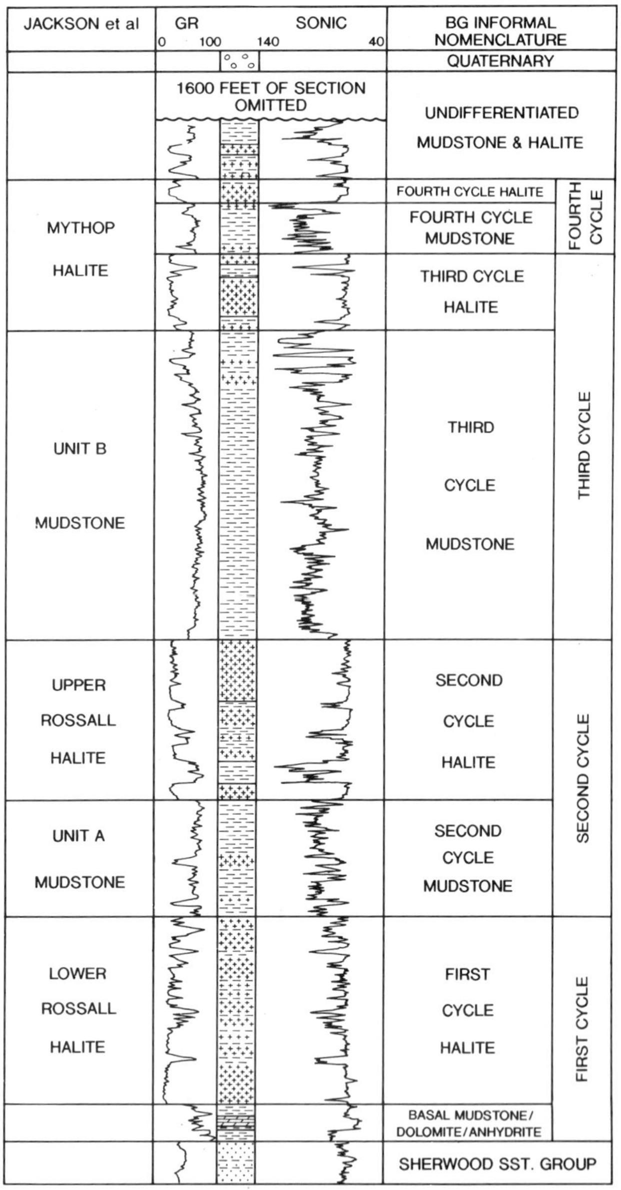

Fig. 3. Mercia Mudstone Group stratigraphy. Based on 110/2-5. Overall thickness first to fourth cycles in this well is 1541 feet. The four cycles can be correlated throughout North and South Morecambe. Above the fourth cycle a further sequence of mudstones and halites occurs but field-wide correlations are not evident. Note some correspondence with the terminology of Jackson *et al.* (1987).

Mercia Mudstone Group stratigraphy

During mid/late Triassic time, repeated phases of basin flooding and drying led ,to the deposition of at least four evaporite cycles of alternating mudstones and halites (Fig. 3).

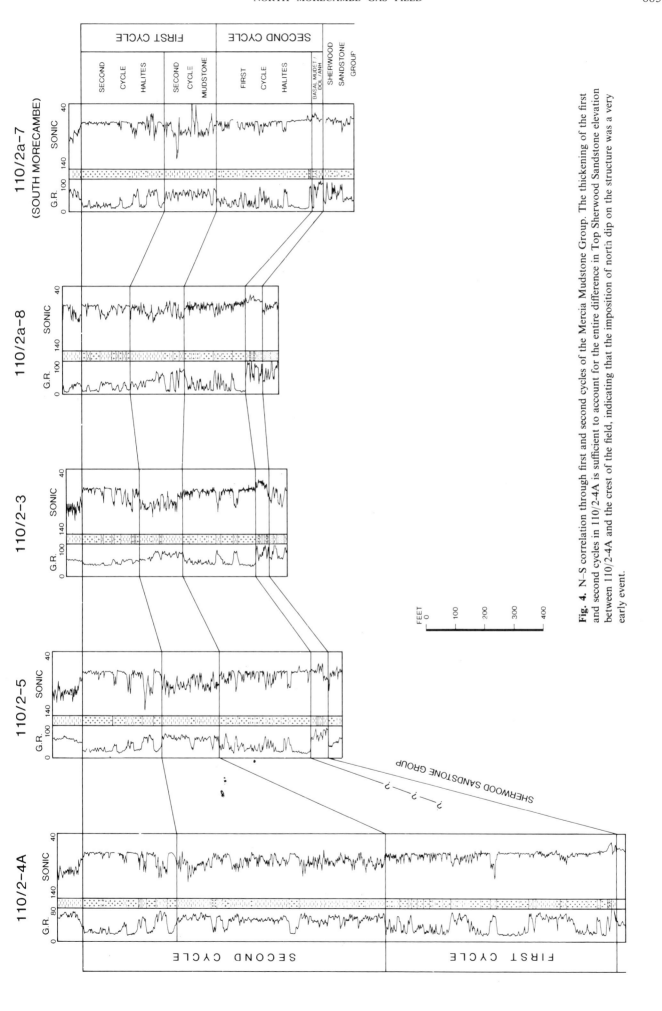

Fig. 4. N–S correlation through first and second cycles of the Mercia Mudstone Group. The thickening of the first and second cycles in 110/2-4A is sufficient to account for the entire difference in Top Sherwood Sandstone elevation between 110/2-4A and the crest of the field, indicating that the imposition of north dip on the structure was a very early event.

These cycles, each several hundred feet thick, are correlatable across the whole of the Morecambe area. No formal nomenclature exists to describe the cycles, but there is some correspondence with units described by Jackson *et al.* (1987, after Evans and Wilson 1975) and Warrington *et al.* (1980). Analysis of this sequence has provided important information about the early development of the North Morecambe structure.

The earliest cycle is the most complex. The basal 100 ft (approximately) comprise mudstone, anhydrite and a minor amount of dolomite. This sequence passes up into the earliest thick halite interval, which differs from the halite portion of later cycles by including numerous mudstone beds, individually thin (10–20 ft) but correlatable over wide areas of the field, and across to South Morecambe. This halite corresponds to the Lower Rossall halite of Jackson *et al.* (1987).

The second, third and fourth cycles are much simpler. Each consists of a thick interval of featureless red mudstone several hundred feet thick, eventually passing upwards (usually with a fairly abrupt transition) to a substantial halite interval. Each halite interval contains subsidiary mudstones which are important as correlation markers. The second cycle corresponds to the 'Unit A Mudstone' and the Upper Rossall Halite of Jackson *et al.* (1987); the mudstone portion of the third cycle corresponds to the 'Unit B Mudstone'; whilst the Mythop Halite incorporates the halite portion of the third cycle and the whole of the fourth cycle.

Above the fourth cycle are further mudstones and halites, and indications in some wells of a fifth cycle which would perhaps correspond to the 'Unit C Mudstone' and the Preesall Halite. However, no consistent correlations have been

achieved at this level across North or South Morecambe Fields.

Across both fields, the thickness of each cycle remains reasonably consistent. Typical thicknesses are:

Fourth cycle: 176 ft (range 90–260 ft)
Third cycle: 643 ft (range 456–765 ft)
Second cycle: 417 ft (range, excluding 110/2-4A, 329–540 ft)
First cycle: 286 ft (range, excluding 110/2-4A, 194–359 ft)

There is no general evidence of thickness variation due to post-depositional halokinesis; where a given cycle becomes relatively thick, the thickening is accounted for in equal measure by the mudstone and halite components of that cycle.

The major exception is in well 110/2-4A (Fig. 4). The third and fourth cycles in this well are of similar thickness to the equivalent cycles throughout the area, but the first and second are substantially thickened in comparison with the norm (second cycle: 690 ft; first cycle: 773 ft; total: 1463 ft). The second cycle is still recognizable as a single cycle, albeit substantially thicker than normal (particularly in the mudstone component), but the interval between the base of the second cycle and the top of the Sherwood Sandstone consists of several thick, alternating, intervals of mudstone and halite, difficult to relate to the 'first cycle' in the rest of the field.

The observations in 110/2-4A have important implications for our understanding of the development of the North Morecambe structure, particularly in relation to the timing of the crucial diagenetic event which caused the formation of platy illite in the reservoir (see *Structural History* section and Fig. 5).

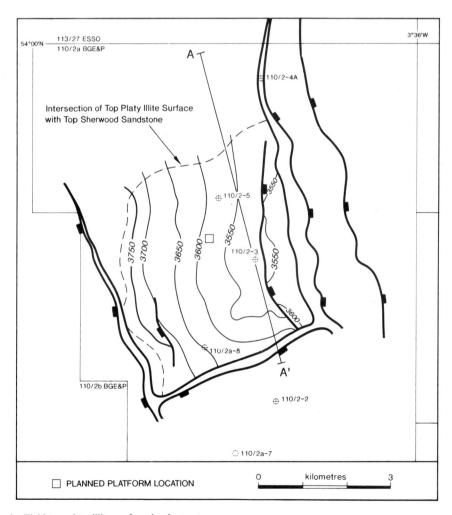

Fig. 5. North Morecambe Field top platy illite surface depth structure map.

Structure

North Morecambe Field lies in the centre of the Permo-Triassic EISB (Fig. 1), a half-graben, the eastern margin of which is formed by major NNE–SSW trending faults which throw Permo-Triassic against older Paleozoic rocks.

Structural description

At Top Sherwood Sandstone Group level (Fig. 6), North Morecambe Field is a structural high with major faults on its eastern and southern boundaries. The crest is slightly above −3000 ft, and the GWC is at −3950 ft. The bounding faults throw the reservoir against the Mercia Mudstone Group. The southern bounding fault limits a very deep WSW–ENE graben which separates the field from South Morecambe. The depth to Top Sherwood Sandstone in this graben is between −4100 ft and −5000 ft, which is deeper than the GWC of either field.

The eastern boundary is a major NNW–SSE normal fault downthrowing to the east by up to 2000 ft. The throw diminishes northwards to about 300 ft at 110/2-4A, in which about 300 ft of reservoir has been cut out due to the well passing through the fault plane. The dipmeter confirms that the well penetrated a pair of fault planes dipping east at about 60° to 70° (Cowan et al. 1992).

West and north of the bounding faults the structure is essentially a uniform surface. Dip is mainly to the west but becomes northwesterly, and eventually northerly, in the northern part of the field (this contrasts significantly with South Morecambe where the main dip is to the east).

Seismic data show numerous small-scale internal faults. The majority are normal dip faults trending N–S and downthrowing to the west in sympathy with the predominant westerly dip. A few antithetic, east-throwing faults have been mapped. In the northern part of the field, where the dip becomes northerly, the small-scale faults remain N–S trending, indicating that faulting post-dates northerly tilting. The only faults not belonging to the N–S array occur in the south of the field where an array of E–W-trending faults, mainly throwing to the south, are subparallel to the field's southern bounding fault.

Structural history

The development of the North Morecambe structure is shown diagrammatically on Fig. 7. It is believed that North and South Morecambe Fields are located at the junction of two fundamental structural domains which control the structural evolution of the EISB (Knipe et al. 1993). To the north the structure is controlled by faults which throw to the east; to the south by faults which throw to the west. The east-throwing Keys Fault, which propagated southwards from the northern margin of the basin, has been a major influence on the development of the North Morecambe structure.

There is no evidence of structural development on North Morecambe during Sherwood Sandstone deposition, although well data indicate that the highest point on the South Morecambe structure was already active at that stage.

Northerly tilting on the north flank occurred very soon after reservoir deposition. The evidence for this is found in the Mercia Mudstone stratigraphy of Well 110/2-4A (see above).

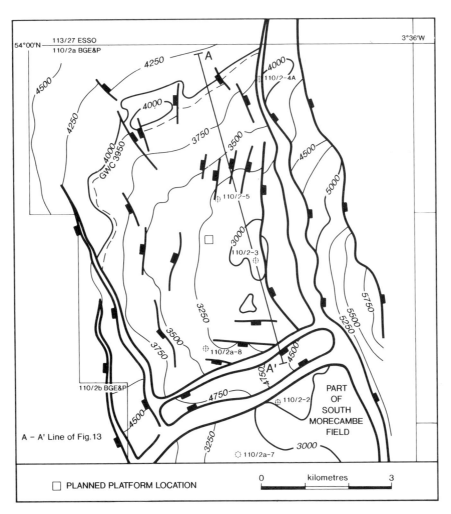

Fig. 6. North Morecambe Field top Sherwood Sandstone depth structure map.

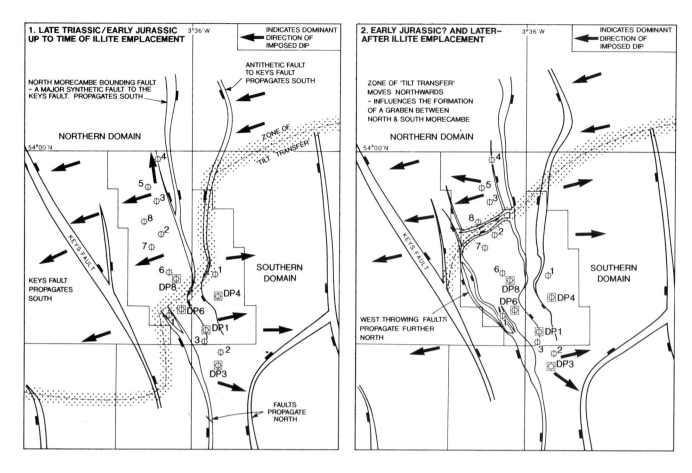

Fig. 7. The development of the North Morecambe structure. (1) During the Early Jurassic illite precipitated in a common structure formed between the southerly propagating Keys Fault and a northerly propagating fault system. A zone of tilt transfer existed between the westerly dipping structures associated with the Keys Fault and the easterly dipping structures to the south. (2) With continued northwards propagation of the southern fault system the zone of tilt transfer jumped north to form the transfer graben system which structurally separates North and South Morecambe.

This well lies down-dip on the north flank of the structure. It is significant that this is the only component of the structure of either North or South Morecambe where there is a strong northerly component of dip. The thickening of the first and second cycles in 110/2-4A is sufficient to account for the entire elevation difference, at Top Sherwood Sandstone level, between this well and the up-dip wells. This indicates that the northerly component of dip must have been imposed during Mercia Mudstone deposition at a very early stage after reservoir deposition.

Thereafter, development of the structure was controlled by growth of NNW–SSE-trending fault systems. North Morecambe and the northern limb of South Morecambe resided within the northern structural domain, characterized by the southward propagation of east-throwing faults. Southward propagation of the ENE-dipping Keys Fault to the west of the field produced tilting to the WSW by late Triassic times. Southward propagation of a major synthetic to the Keys Fault created the eastern boundary fault of North Morecambe and the north limb of South Morecambe.

By the end of the Triassic or early in the Jurassic, a North Morecambe structure had formed (at this point not necessarily separated from the northern limb of South Morecambe), capable of trapping hydrocarbons. The palaeo-GWC corresponding to the present-day top platy illite surface stabilized at this time.

Later structural development was influenced by the northerly propagation of a west-dipping fault array into the north-

ern limb of South Morecambe. The zone of tilt transfer, which had previously separated the northern and southern parts of South Morecambe, now transferred northwards to become the zone separating North and South Morecambe. A deep graben formed along this new zone of tilt transfer, causing complete separation between the two fields. It is significant that the present-day North Morecambe aquifer pressure is 140 PSI lower than that for South Morecambe, even though the bounding faults bring sand against sand within the aquifer.

Continued movement on the Keys Fault array induced minor late internal faulting and, more importantly, tilting of the structure to the WSW. Basin inversion occurred in the early Tertiary, when the structure was elevated to its present depth of 3000 ft. Bushell (1986) and Colter and Barr (1975) estimated post-Triassic inversion to be around 4000–5000 ft, based upon spore coloration and vitrinite reflectance of the underlying Carboniferous (using South Morecambe Well 110/8-2). However, recent fission track analysis suggests that the burial history can be seen in terms either of a substantial amount of burial and uplift (5000 ft to 10 000 ft); or a period of abnormally high temperatures c. 60 Ma, with much less uplift, followed by rapid cooling, and probably caused by the early Atlantic rifting. The onshore expression of this event, which almost fully annealed the fission tracks in apatite grains, has been described by Green (1986).

For further discussion of the structural regime under which the North Morecambe structure developed, and its regional structural context, see Knipe et al. (1993).

Reservoir sedimentology and zonation

Reservoir facies descriptions

The Sherwood Sandstone Group forms a thick sequence (c. 4800 ft) deposited by a fluvial system in a semi-arid environment within the subsiding EISB. It comprises a heterogeneous system of interlayered facies of variable reservoir quality. Five major facies associations are recognized, each of which exhibits textural and reservoir properties directly related to its mode of deposition. Detailed sedimentological descriptions have been given in Stuart and Cowan (1991) and Cowan (1993), and are summarized below.

Facies A: Major channel sandstones. Large-scale intervals, mainly deposited as stacked in-channel transverse bars in high-energy braided rivers. Except where affected by cementation or illite, Facies A displays favourable porosities and permeabilities.

Facies B: Secondary channel sandstones. Small scale intervals deposited by bar migration within ephemeral channels. Usually found in association with sheetflood (Facies C) sandstones. At best, primary reservoir properties are comparable with those of Facies A, but on average much poorer than in Facies A.

Facies C: Sheetflood sandstones. Flat-bedded, laterally extensive tabular intervals, with abundant mudstone or siltstone partings, deposited from extensive non-channelized sheetfloods during high flood stage, distally to the active fluvial system. This association is considered to be a mixed facies comprising fluvial and aeolian adhesion deposits (Cowan 1993). Reservoir properties are generally poor, but where the aeolian influence becomes significant, they are substantially enhanced.

Facies D/E: Non-reservoir fines. This grouping includes abandonment mudstones and overbank fines. These are mainly of limited lateral extent. Certain thicker and more extensive mudstone intervals represent non-evaporitic playa lake deposits.

Facies F: Aeolian sandstones. These were deposited as aeolian dune sands and also tabular intervals of aeolian sandsheet origin. These sandstones are never more than a few feet thick, but are well sorted, well rounded and lack interstitial clays, so exhibit very favourable primary porosities and permeabilities.

Depositional model

As in South Morecambe Field (Stuart and Cowan 1991), the reservoir sequence is organized vertically into thick intervals of stacked major channel sandstone with very little non-channel material, alternating with thick intervals of non-channellized sandstone (mainly Facies C) with very little major channel sand. Both are correlatable across wide areas. At certain times deposition was dominated by major channel systems; large-scale braided rivers flowed in a broad active belt of channel migration which allowed very little opportunity for non-channel facies to be deposited or preserved. At other times deposition was dominated by non-channellized systems, so that thick, laterally extensive intervals of unconfined sheetflood sandstones were developed.

This alternation was controlled by the advance and retreat of a fluvial system, with the sheetflood facies being deposited distally to the channel system. The major channel systems were probably sourced from high ground at the faulted east margin of the EISB and advanced westwards to become established across the Morecambe area. Subsequently, as the channel

systems retreated eastwards towards the basin margin, distal sheetflood systems became established. It seems likely that this fluctuation of the channel/sheetflood system was caused by an external control (boundary fault reactivation or climatic change; dipmeter studies suggest the former, see Cowan *et al.* 1993). This fluctuation occurred several times, resulting in alternating major channel and sheetflood intervals. This is the basis for the system of reservoir units.

Facies correlations and reservoir units

Nine facies-defined reservoir units have been designated across North Morecambe. Each is defined on the following criteria.

(1) Dominated by one facies; consistent primary porosity/ permeability characteristics throughout.
(2) Correlatable with reasonable confidence in all wells (except where faulted out in 110/2-4A).

These units (NMI to NMIX) are not necessarily correlatable beyond North Morecambe, although most can be traced at least to the northern parts of South Morecambe.

In considering the permeability characteristics of reservoir units, the impact of diagenetic platy illite is critical. This formed beneath a palaeo-GWC so that top platy illite cuts across depositional boundaries. However, permeability contrasts between reservoir units are preserved, albeit at much lower absolute permeability values, even where units are illite-affected.

Figure 8 shows the typical reservoir sequence, and the facies dominating each unit. Units NMVIII to NMI constitute the Ormskirk Sandstone Formation (NMI to NMIII are equivalent to the 'Keuper Waterstones Formation of Stuart and Cowan (1991) and earlier authors, NMIV to the 'Frodsham Member', NMV and NMVI to the 'Delamere Member', and NMVII and NMVIII to the 'Thurstaston Member'). The alternation of channellized and non-channellized facies described above is clearly developed throughout this sequence.

Reservoir Unit NMIX. This is the St Bees Sandstone Formation, the vast majority of which lies below the GWC. This unit consists mainly of low-porosity/permeability sandstones of facies B, C and D/E with only minor amounts of facies A. It lies entirely within the illite-affected layer, and permeability is mostly below 0.1 mD, although a few thin, high-porosity aeolian sandstone intervals, correlatable between wells, have higher permeability.

About 350 ft below the top of this unit, a conspicuous mudstone interval occurs which is probably a non-evaporitic playa lake deposit. This is clearly correlatable between 110/2-3 and 110/2-5 and questionably 110/2-4A, and is thus present across much of North Morecambe. The significance of such deposits should not be underestimated, even when, as in this case, they occur beneath the fluid contact, as they can be significant in limiting pressure response and fluid movement in the aquifer.

Reservoir Units NMVII, NMIV, and NMII. These units are dominated by Facies A, and thus form fairly homogeneous field-wide sheets of mutually erosive stacked channel sandstones. Thin intervals of Facies C and D/E form only a minor component and are unlikely to have any field-wide significance since they would have been constantly eroded by the next channel phase. Unit NMIV also contains minor amounts of aeolian sandstone, most abundant in the crestal Well 110/2-3 (with consequently improved reservoir properties).

Thicknesses of these units are: NMVII, 30 to 50 ft; NMIV about 230 ft; NMII 40 to 50 ft. They have homogeneous internal organization and permeability distribution. Where

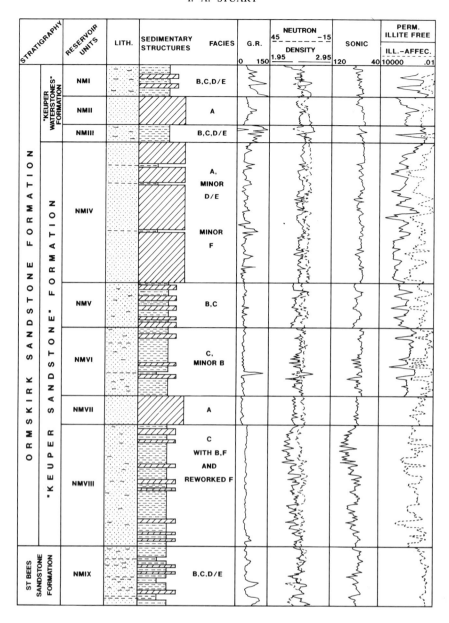

Fig. 8. Summary of reservoir units. Units dominated by stacked major channel facies alternate with units of non-channellized sheetflood with ephemeral secondary channels. Overall thickness of the Ormskirk Sandstone is about 800 ft.

unaffected by platy illite, permeability is good throughout, frequently over 100 mD and up to one Darcy. These units are both overlain and underlain by rock with much poorer reservoir quality.

Reservoir Units NMVIII, NMVI, NMV, NMIII, NMI. These constitute the non-channellized sequences. They comprise Facies C sandstones with associated Facies B.

In Unit, NMVIII, Facies C sandstones are also intimately associated with aeolian sandstones and aeolian reworked detritus. At this level, aeolian influence is evident throughout the Morecambe area. The abundance of aeolian-influenced sediments is a consequence of either a post-St Bees hiatus and abandonment of the fluvial system (which allowed aeolian reworking of fluvial sandstones), or the influx of reworked aeolian detritus following a change in palaeocurrent and source area. At the top of NMVIII the aeolian influence ceases abruptly, indicating a rapid climatic change. The aeolian influence gives rise to favourable reservoir properties in parts of this unit, which forms a 200 ft thick field-wide sheet of sandstone with complex internal organization.

Individual sandbodies are seldom more than a few feet thick, so reservoir properties vary vertically more rapidly than in any other part of the reservoir.

Units NMV and NMVI both form heterogeneous sheets in which secondary channel sandstones are encapsulated as lenses within continuous sheetflood sandstone. The units are separated by the relative influence of Facies B sandstones, which are both more abundant and larger in scale in Unit NMV than in NMVI. As a result, Unit NMV shows the more favourable porosity and permeability. The contrasts are evident in the average core porosity and permeability values from 110/2a-8 (where both units are illite affected), the averages being 12.5% and 2.9 mD for NMV but only 8.6% and 0.55 mD for NMVI.

The basal parts of Units NMIII and NMI each incorporate the abandonment facies of the underlying channel sandstones, and pass upwards into mixed sequences of mudstones, low-permeability Facies C, and very thin poor-quality Facies B sandstones. Permeability is almost entirely below 1 mD even when unaffected by diagenetic illite, apart from a few thin channel (Facies B) intervals of high permeability (over 100 mD) in NMI. Both units form predominantly low-permeability intervals throughout the field, around 50 to 60 ft thick.

Palaeocurrent data

Palaeocurrent directions have been determined from dipmeter analyses, the clearest indications having been obtained from major channel intervals, with a much greater spread of directions in other fluvial facies. Through most of the reservoir the predominant palaeoflow is consistently towards SW to WSW (Cowan *et al.* 1993), indicating the sands entered the field area from the NE/ENE sector. However, in the St Bees Sandstone a very much wider spread of directions was observed, with no predominant field-wide trend. In 110/2-5 indications of a palaeoflow towards the southwest are apparent at some levels but in other wells both northwesterly and southeasterly flow directions are seen, and the overall impression is of a system with a low incidence of stacked major channels and a wide range of individual palaeocurrent directions in the sector NW–SW–SE. Therefore an overall flow towards SW/WSW is not ruled out, nor is it specifically confirmed.

Reservoir diagenesis

Petrography

Quartz forms 70–90% of the detrital mineral assemblage, with feldspars, micas, and rock fragments the other main components. Detrital clays occur in varying amounts, particularly in the sheetflood sand facies. Heavy minerals occur in minor amounts. The most significant diagenetic component is authigenic platy illite, which, although present in relatively small quantities, causes a drastic reduction in permeability. Fibrous illite also occurs in small amounts, both in association with platy illite (sometimes with the fibres growing out of the plates) and locally above the main platy illite zone. Quartz and early dolomite account for 85–95% of total cements; minor cements include feldspar, rare calcite, late stage ankerite, pyrite, gypsum, anhydrite, haematite and kaolinite.

Diagenetic sequence and burial history

The diagenetic sequence is similar in general terms to that of South Morecambe (see Stuart and Cowan 1991). The key diagenetic events at each stage are as follows.

Early diagenesis: Triassic. Precipitation of calcite and nonferroan dolomite micronodules, also minor amounts of quartz and feldspars. Calcite now exists in very minor amounts as remnant masses, but it is probable that part of the intergranular porosity is secondary, formed by calcite dissolution.

Intermediate diagenesis (deep burial): Triassic/Jurassic. Quartz cementation (as syntaxial overgrowths) occurred both before and after precipitation of platy illite. Platy illite precipitation occurred at this stage, dated at 160–180 Ma by K/Ar. Ferroan dolomite/ankerite postdates both quartz and illite. Thermal decarboxylation reactions associated with hydrocarbon generation reduced the pore fluid pH sufficiently to cause dissolution of feldspar and earlier carbonate cements.

Late diagenesis: Cretaceous/Tertiary. Renewed gas generation during the Cretaceous caused the lowering of the GWC into the illite-affected layer, halting the precipitation of cements.

Effects and distribution of platy illite

In terms of impact on reservoir behaviour, the most important diagenetic phase is platy illite. Illite crystals grew radially on detrital grains to form an intricate 'boxwork', blocking pore throats (Fig. 9). Platy illite has no measurable effect on porosity, but reduces permeability by about two orders of magnitude. It formed below a palaeo-GWC after early structural growth and a first phase of gas migration (Bushell 1986; Woodward and Curtis 1987). Owing to renewed gas migration and 'post-illite' steepening of the structure, illite-affected sandstones now occur within the gas leg. The present top platy illite surface (TPIS) is not a horizontal plane because of later structural activity; its present configuration is a function of phases of structural deformation which post-date illite precipitation. Mapping of TPIS is described in a separate section below.

The TPIS divides the reservoir into two non-stratigraphically controlled layers: the interval between TPIS and Top Sherwood Sandstone is the illite-free layer; the interval below TPIS and above the present GWC is the illite-affected layer.

Whereas in parts of South Morecambe there are variations in illite morphology and abundance within the illite-affected layer, no systematic vertical variations in illite morphology occur in North Morecambe. Well developed platy illite is abundant throughout the illite-affected layer. Fibrous illite is rarely developed in the illite-affected layer except for fibres growing from platy illite but this has the same impact on permeability as platy illite alone.

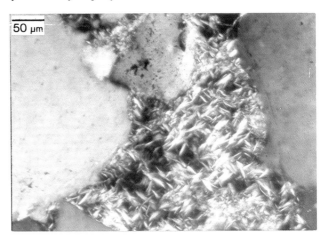

Fig. 9. Thin section photomicrograph Well 110/2a-8. Well developed platy illite bridging a pore.

Effects and distribution of cements

Vertical cement distribution. Figure 10 shows the amounts of quartz, total carbonate cement and visible porosity (determined by point-counting) for the illite layers and total reservoir (a South Morecambe example, 110/2a-7, is also shown). In North Morecambe the illite-free and illite-affected layers contain broadly similar cement quantities in contrast to South Morecambe where cements are more abundant in the illite-free than in the illite-affected layer. The inference (Stuart and Cowan 1991) is that the South Morecambe structure was breached after illite precipitation, whereas, after the initial migration of gas and precipitation of illite, the North Morecambe structure remained sealed and gas-filled down to the palaeo-GWC.

Lateral cement distribution. Table 1 shows well by well average cement contents from point-count data. Well 110/2-3 contains less cement than the other wells. Noting the position of 110/2-3 on the crest of the structure, these data indicate that the crestal region of the structure is relatively less cemented by the late cements, quartz and dolomite, than the flanks of the structure, but reasons for this are as yet unclear.

Facies control on cement distribution. The amounts of carbonate cement within major channel sandstones is highly vari-

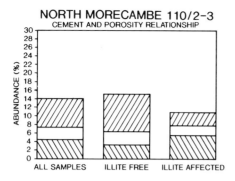

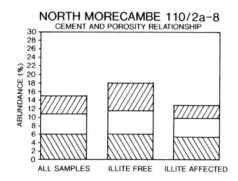

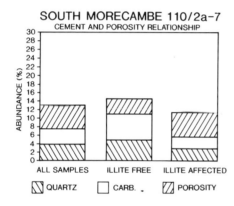

Fig. 10. Histograms showing cements and porosity (based on thin section point-count data). Whereas the illite-affected reservoir of South Morecambe contains noticeably more carbonate and quartz, and correspondingly less porosity, than the illite-free, this relationship is not apparent in North Morecambe where the cement quantities in the illite-free and illite-affected are broadly similar.

Table 1. Average cement contents from point-count data.

Well	Quartz Cement	Total Carbonate	Total Cement
110/2a-8	6.0	4.9	11.3
110/2-5	2.8	8.1	11.5
110/2-3	4.3	3.4	8.2

able. Within individual channel sand bodies early diagenetic ferroan-dolomite nodules tend to be concentrated near the channel base, probably representing aggradation of channel base lags rich in reworked caliche deposits.

Fault control on cement distribution. Severe cementation is likely to affect at least the larger faults. These cements were probably sourced by fluids migrating from depth up fault planes, but cementation is limited to a narrow zone around the fault planes rather than being pervasive over a wide area.

Compaction

There are indications that the reservoir in 110/2-5 and 110/2a-8 has undergone greater compaction than the reservoir in the crestal Well 110/2-3. The cathodoluminescence photomicrographs of Fig. 11 show representative samples from 110/2-3 and 110/2a-8 with differing amounts of compaction. The number of long contacts between detrital grain surfaces is greater in 110/2a-8 than in 110/2-3, implying that 110/2a-8 has suffered greater compaction than 110/2-3.

A possible explanation is that early carbonate cements were better developed in the vicinity of 110/2-3, limiting mechanical compaction. Subsequent dissolution led to the development of significant intergranular porosity. In view of the position of 110/2-3 on the crest of the structure, it seems that the flanks of the structure may have undergone greater compaction than the crest, although the reason for the crest receiving more early carbonate cement than the flanks is not known.

(a)

(b)

Fig. 11. Cathodoluminescence photomicrographs. (a) Well 110/2-3. This sample shows low to moderate compaction as evidenced by the limited number of long and concavo–convex grain contacts present. (b) Well 110/2a-8. Note a slightly higher degree of compaction than in (a). Long and concavo convex grain contacts more common.

Mapping of top platy illite surface (TPIS)

Since TPIS represents a palaeo-GWC the successful mapping of this surface is dependent on our understanding of the structural history of the field. The TPIS map is shown in Fig. 5; the configuration of this surface is also shown on Fig. 12.

On South Morecambe Field, it had been previously found that a plot of well depths to Top Sherwood Sandstone against well depths to TPIS produced a remarkably straight-line relationship (explained by the assumption that the palaeo-structure was a subdued version of the present structure) and this

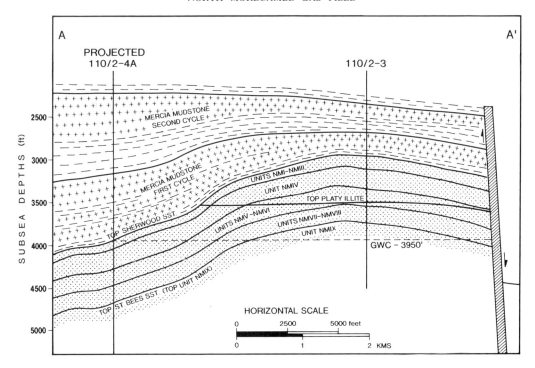

Fig. 12. N–S section through North Morecambe. Line of section shown on Figs 5 and 6.

relationship was used to produce a 'first-pass' map of the TPIS by derivation from the map of Top Sherwood Sandstone, which has proved quite reliable in predicting depth to TPIS in new wells. Such a plot also generates a straight line for the North Morecambe well data, but based on only three wells, 110/2-3, 110/2-5, and 110/2a-8 (110/2-4A being fully illitized all the way up to Top Sherwood Sandstone), see Fig. 13. Based on our understanding of South Morecambe it is reasonable to use this relationship to predict depth to TPIS over most of the North Morecambe structure; however, in view of the observations in the Mercia Mudstone described above, this relationship cannot be used to predict depth to TPIS on the north flank of the structure.

Therefore the North Morecambe TPIS is mapped across the bulk of the field as a WSW-dipping surface, with contours striking NNW–SSE, parallel to the Keys Fault and to the field's main eastern bounding fault. As indicated above, the northerly component of dip on the Top Sherwood Sandstone Surface was imposed during deposition of the earliest part of the Mercia Mudstone Group in the mid/late Triassic and therefore entirely pre-dates the precipitation of platy illite. Therefore the TPIS map does not contain any element of northerly dip. This is a very important contrast with South Morecambe Field, where it has been demonstrated that the structure of TPIS is everywhere a subdued form of the Top Sherwood Sandstone structure, and therefore TPIS always dips in the same direction (with a lesser angle of dip) as Top Sherwood Sandstone at the same point. Had this assumption been applied to North Morecambe, an incorrect configuration of TPIS would have resulted, with a consequent overestimate of the ratio of gas in the illite-free layer to gas in the illite-affected layer.

Some westerly (or WSW) tilting had been imposed on the structure prior to illite precipitation. Therefore the present TPIS dips to the WSW but at a shallower angle than the WSW component of dip on Top Sherwood Sandstone. The internal faulting on the field post-dates, at least in part, the precipitation of illite. Therefore, some displacement of TPIS is mapped along at least the larger-scale internal faults.

It is significant that the TPIS consistently converges with the Top Sherwood Sandstone above the present GWC, so that the

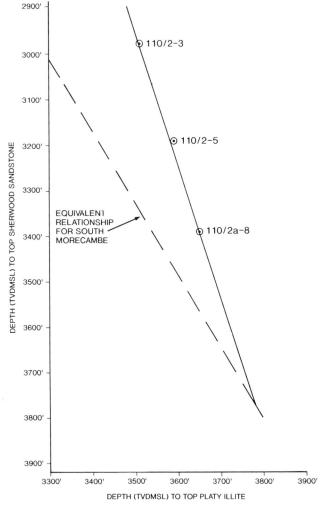

Fig. 13. Crossplot of subsea depth to Top Sherwood Sandstone against subsea depth to TPIS. This is known to produce a remarkably straight line relationship for South Morecambe (broken line), based on a large amount of well control. The more limited North Morecambe well control appears to show a similar relationship.

high-permeability illite-free layer is nowhere directly exposed to the water zone (unlike South Morecambe). This may have important implications for restricting water movement.

Source

Geochemical studies have shown that Upper Carboniferous coals and shales are the major source of gas in the Morecambe structures. The North Morecambe gas contains about 6% CO_2, whereas at South Morecambe the CO_2 content is no more than 0.6%. Given that CO_2 is an early product in the thermal maturation of organic material, and that the two fields share the same source, the inference is that the gas currently residing in the North Morecambe structure is older than the South Morecambe gas. This is consistent with the independent evidence of breaching of the South Morecambe structure but not the North Morecambe structure (see *Vertical cement distribution*).

Conclusions: the separation from South Morecambe Field

Despite its close proximity to South Morecambe (horizontal separation less than 1 km), there are several observable differences in the North Morecambe geology.

- The main dip vector is strongly to the west; on South Morecambe it is mainly to the east, with very gentle west dip on the northern limb.

- There is a strong component of north dip; whereas there is no significant north dip on the South Morecambe structure. Mercia Mudstone Group stratigraphy differs significantly on the northerly-dipping north flank from anywhere else on either field, with substantial thickening of the earliest two mudstone-halite cycles.

- The GWC is 200 ft deeper: −3950 ft as against −3750 ft on South Morecambe.

- The aquifer pressure is 140 psia lower: on South Morecambe it is 1880 psia at −3750 ft, whereas the figure for North Morecambe, extrapolated up to the same depth, is 1740 psia.

- On South Morecambe the illite-free layer consistently contains more cement than the illite-affected layer, indicating a breaching of the structure and loss of the original gas charge after precipitation on platy illite. On North Morecambe the illite layers contain similar amounts of cement.

- The North Morecambe gas contains approximately 6% CO_2, as against 0.6% for South Morecambe.

These contrasts are explained by differences in geological history of the two fields from very soon after deposition of the Sherwood Sandstone reservoir. The northerly component of dip was imposed very early, during the deposition of the lower part of the Mercia Mudstone sequence. Later, after development of a deep graben between the two fields, North Morecambe behaved as a separate structural block, with pronounced westward tilting both before and after platy illite formation, controlled by movement on the Keys Fault. It is believed that the Keys Fault was always the predominant basin structure controlling the development of the North Morecambe structure, whereas the Keys Fault became an influence on South Morecambe structure only at a relatively late stage. Whereas the South Morecambe structure was breached after platy illite precipitation, there is no evidence to suggest that the North Morecambe structure was breached or that the early-generated gas was lost. This explains the major compositional differences between the North and South Morecambe gases. The early-generated gases from the Carboniferous would be expected to be rich in CO_2, which was generated early during the thermal maturation of organic matter. These early, CO_2 rich gases have been lost in South Morecambe, which now contains later gases with much less CO_2; but they have been retained in North Morecambe. With later gas migration, the GWCs in the two fields stabilized at markedly different levels. Depth to Top Sherwood Sandstone in the intervening graben is well below −4000 feet, so that the gas-bearing Sherwood Sandstone is always thrown against Mercia Mudstone. However, the differences in aquifer pressure can be understood only if the fault planes within the aquifer are providing full seals even where the faults juxtapose sand against sand. These aquifer pressure data therefore indicate that the major faults within the EISB do indeed divide the basin into areas with significantly different pressures.

The author would like to thank the management of British Gas Exploration and Production Ltd for permission to publish this contribution. I would also like to acknowledge the contribution made by the many colleagues and co-workers whose work has contributed to our current understanding of the North Morecambe Gas Field.

References

BUSHELL, T. P. 1986. Reservoir Geology of the Morecambe Field. *In*: Brooks, J., GOFF, J. C. AND VAN HOORN, B. (eds) *Habitat of Palaeozoic Gas in Northwest Europe*. Geological Society, London, Special Publication, **23**, 189–207.

COLTER, V. S. 1978. Exploration for gas in the Irish Sea. *Geologie en Mijnbouw*, **57**, 503–516.

—— AND BARR, K. W. 1975. Recent developments in the geology of the Irish Sea and Cheshire Basins. *In*: WOODLAND, A. W. (ed.) *Petroleum and the Continental Shelf of North-West Europe*, Vol. 1. Applied Science, London, 61–75.

—— AND EBBERN, J. 1978. The petrography and reservoir properties of some Triassic sandstones of the Northern Irish Sea Basin. *Journal of the Geological Society, London*, **135**, 57–62.

—— AND —— 1979. SEM studies of Triassic reservoir sandstones from the Morecambe Field, Irish Sea, UK. *Scanning Electron Microscopy*, **1**, 531–538.

COWAN, G. 1993. The identification and significance of aeolian deposits in the dominantly fluvial Sherwood Sandstone Group of the East Irish Sea Basin, UK. *In*: NORTH, C. P. AND PROSSER, J. (eds) *The Characterization of Aeolian and Fluvial Reservoir Sandstones*. Geological Society, London, Special Publication, in press.

——, OTTESEN, C. AND STUART, I. A. 1993. The use of dipmeter logs in the structural interpretation and palaeocurrent analysis of Morecambe Fields, East Irish Sea Basin. *In*: PARKER, J. R. (ed.) *Petroleum Geology of Northwest Europe: Proceedings of the 4th Conference*. Geological Society, London, 867–882

EBBERN, J. 1981. The geology of the Morecambe Gas Field. *In*: ILLING, L. V. AND HOBSON, G. D. (eds) *Petroleum Geology of the Continental Shelf of North-West Europe*, Heyden, London, 485–493.

EVANS, W. B. AND WILSON, A. A. 1975. Outline of Geology on Sheet 66 (Blackpool) of 1:50,000 Series, *Geological Survey of Great Britain*.

GREEN, P. F. 1986. On the thermotectonic evolution of Northern England: evidence from fission track analysis. *Geological Magazine*, **123**, 493–506.

JACKSON, D. I., MULHOLLAND, P., JONES, S. M. AND WARRINGTON, G. 1987. The geological framework of the East Irish Sea Basin. *In*: BROOKS, J. AND GLENNIE, K. (eds) *Petroleum Geology of North West Europe*. Graham & Trotman, 191–203.

KNIPE, R. J., COWAN, G. AND BALENDRAN, V. S. 1993. The tectonic history of the East Irish Sea Basin with reference to the Morecambe Fields. *In*: PARKER, J. R. (ed.) *Petroleum Geology of Northwest Europe: Proceedings of the 4th Conference*. Geological Society, London, 857–866.

LEVISON, A. 1988. The geology of the Morecambe gas field. *Geology Today*, May–June 1988, 95–100.

STUART, I. A. AND COWAN, G. 1991. The South Morecambe Field, Blocks 110/2a, 110/3a, 110/8a, UK East Irish Sea. *In*: ABBOTTS, I. L. (ed.) *United Kingdom Oil and Gas Fields 25 Years Commemorative Volume*, Geological Society, London, Memoir, **14**, 527–541.

WARRINGTON, G., AUDLEY-CHARLES, M. G., ELLIOT, R. E., EVANS, W. B., IVIMEY-COOK, H. C., KENT, P. E., ROBINSON, P. L., SHOTTON, F. W. AND TAYLOR, F. M. 1980. *A Correlation of Triassic Rocks in the British Isles*. Geological Society, London, Special Report, **13**.

WOODWARD, K. AND CURTIS, C. D. 1987. Predictive modelling of the distribution of production constraining illites—Morecambe Gas Field, Irish Sea, Offshore UK. *In*: BROOKS, J. AND GLENNIE, K. (eds) *Petroleum Geology of North West Europe*. Graham & Trotman, London, 205–215.

Atlantic margin exploration: Cretaceous–Tertiary evolution, basin development and petroleum geology

Atlantic margin exploration: Cretaceous–Tertiary evolution, basin development and petroleum geology

Introduction and review

A. M. SPENCER[1] and O. ELDHOLM[2]

[1]Statoil, Postboks 300, 4001 Stavanger, Norway
[2]Department of Geology, University of Oslo, Postboks 1047, 0316 Oslo, Norway

The predominantly rifted continental margin of NW Europe stretches c. 3000 km from Ireland to the Barents Sea and comprises sectors off Ireland (c. 800 km), off the UK and the Faeroe Islands (c. 1000 km) and off Norway (c. 1200 km). The conjugate margin east of Greenland is of similar total length and that southwest of Greenland measures c. 1500 km. Further north, the predominantly sheared margin along the Barents Sea–Svalbard and northeast Greenland sectors extends for c. 1000 km. These Atlantic margins lie in waters as deep as c. 2500 m. In total area the margins cover c. 1.5×10^6 km^2, which is almost five times the area of the prospective part of the North Sea basin. The largest sectors are those off Norway, off northeast Greenland and off Ireland, each of which exceeds the area of the prospective North Sea basin. The margins are extensively underlain by sedimentary basins. The hydrocarbon exploration of these basins has been greatest in the Norwegian sector where a total area of 35 000 km^2 has been licensed, 142 exploration wells drilled and 25 hydrocarbon finds made. The comparable statistics for the other sectors are: UK—35 000 km^2, 97 wells and about seven finds; Ireland—30 000 km^2, 29 wells and about three finds; Southwest Greenland—18 000 km^2, 5 wells, no finds. Off East Greenland there has been neither licensing nor drilling.

The geological events of most importance to the occurrence of the hydrocarbons so far discovered on the Atlantic margins are those which took place in Jurassic to Tertiary times. All of the hydrocarbon finds to date have been generated from Jurassic source rocks, principally oil-prone late Jurassic marine shales which are developed from the Porcupine Basin to the Barents Sea. The most important tectonic event with respect to the known hydrocarbon finds was the late Jurassic to early Cretaceous rifting. This led to restricted marine circulation, allowing source rock accumulation, and created the fault block traps which have proved the most successful exploration targets. Major subsidence occurred in Cretaceous times with sequences 5 km or more in thickness accumulating in the Faeroe–Shetland, Møre, Vøring and Tromsø basins. Late Cretaceous to early Eocene times saw the major events connected with the break-up of the continents: first North America and Greenland separated, followed by the separation of Greenland and Eurasia near the Paleocene–Eocene transition. The rifted margins underwent extension and syn-rift uplift prior to continental break-up and the start of sea floor spreading between Europe and Greenland. The actual break-up and initial phase of sea floor spreading was accompanied by voluminous and extensive subaerial volcanic activity. The last event to have an important effect on the hydrocarbon finds is the late Tertiary, intra-plate uplift of the Scandinavian landmass.

The main proven hydrocarbon plays can be classified with respect to the late Jurassic to early Cretaceous rifting into pre-, syn- and post-rift plays. From the UK sector to north Norway, the pre-rift play is the most successful, with its tilted fault block traps, Jurassic and older reservoirs and adjacent late Jurassic source rocks. One major oil field occurs in syn-rift sandstones off mid-Norway. Significant post-rift gas finds occur in Paleogene reservoirs in the UK sector but off mid-Norway only one exploration well has been targeted at a post-rift, Cretaceous, objective. The discovered, recoverable, hydrocarbon resources, measured in units of 10^6 Sm^3 oil-equivalent, are of the order of 1000 off Norway, 100 off UK and perhaps 10 off Ireland. Current official estimates of the yet-to-find hydrocarbon resources, expressed in the same units, amount to 1500–4000 off Norway and (oil only) 20–1000 off UK.

The theme adopted for this symposium allows the importance of the Cretaceous to Tertiary evolution of the margin to be highlighted. Subsidence then has been responsible for creating and sometimes destroying the maturity of the Jurassic source rocks. Early Tertiary volcanic events have also acted to restrict the areal extent of the prospective hydrocarbon basins and the Tertiary sea floor spreading has placed an absolute, oceanward limit on the extent of the prospective basins. On the other hand, the late Cretaceous–Paleocene regional uplift, erosion and re-deposition, the early Eocene greenhouse, and the thermal pulse leading to continental break-up, present new, hitherto largely unrecognized, challenges for the explorationist. The theme was also chosen to allow a wide participation in the symposium, for these Cretaceous to Tertiary events have been much studied by geoscientists from academic and government institutions.

The thirteen papers published here include an overview of the Mesozoic to Cenozoic plate reconstructions, and reviews containing new information on the Southwest Greenland, Northeast Greenland, Southwest Barents Sea and offshore west Ireland regions. The main group of articles deals with the region to the northwest of the UK, the Faeroe–Shetland Channel, with accounts of the subsidence patterns there, of the Solan Basin, of structuring and transfer zones, of Paleocene sequence stratigraphy, of Paleocene to Miocene compression and of Neogene to Quaternary seismic stratigraphy. A detailed paper describes the seismic structure of the Hatton–Rockall area. A final paper outlines apatite fission track data which suggest kilometre-scale uplift of broad areas of the UK in Tertiary times.

Some of the important implications of the papers included here for future hydrocarbon exploration are these. Although the pre-rift play has been the target of most exploration to date, further finds will be made but will demand the best possible understanding of Mesozoic tectonic history and deposition. As exploration proceeds into still greater water depths only the post-rift plays will normally be within reach of the drill: the requirement is to predict the presence of reservoirs and traps at Cretaceous and Tertiary levels. In one area, off Northeast Greenland, because of the sea ice cover, large areas are still poorly known geologically: Jurassic faults blocks may be present and a Permo-Carboniferous rift basin may contain older hydrocarbon plays. Off Southwest Greenland rift structures of late Cretaceous to Paleogene age have not so far been drilled.

From *Petroleum Geology of Northwest Europe: Proceedings of the 4th Conference* (edited by J. R. Parker).
© 1993 Petroleum Geology '86 Ltd. Published by The Geological Society, London, p. 899.

Evolution of North Atlantic volcanic continental margins

K. HINZ,[1] O. ELDHOLM,[2] M. BLOCK[1] and J. SKOGSEID[2]

[1] Bundesanstalt für Geowissenschaften und Rohstoffe, Postfach 51 01 53, 3000 Hannover 51, Germany
[2] Department of Geology, University of Oslo, Pb. 1047, 0316 Oslo, Norway

Abstract: Structurally, the North Atlantic margins are characterized by Late Cretaceous–Early Tertiary extension, developing predominantly rifted segments south of the Greenland–Senja Fracture Zone, and by oblique extension and wrenching leading to large sheared margin segments further north. The rifted segments document an intense, transient volcanic phase during break-up, emplacing huge extrusive constructions on the outer margin. Commercial exploration has mostly been landward of the shelf edge, but lately there is increased interest in the deep water basins on the outer margins and in the perennially sea-ice-covered shelf off Greenland. Evaluation of these areas requires an understanding of the processes and events prior, during and after the Paleocene–Eocene opening of the North Atlantic. Although the margin is a Cenozoic feature, its segmentation and evolution are governed by the pre-opening history, particularly the Late Jurassic–Early Cretaceous tectonic episode which resulted in large Cretaceous sedimentary basins. Moreover, the first few multi-channel seismic profiles from the NE Greenland inner shelf north of 79°N show a deep Permo-Carboniferous rift basin. The similarity of tectono-magmatic features off NE Greenland and Norway shows the value that conjugate margin studies provide in understanding volcanic margin formation. The events at break-up affected both the subsidence and thermal histories of the marginal basins and were associated with crustal uplift and erosion of the rift region, intrusive activity within the adjacent Mesozoic basins and lava flows spilling onto large areas of the pre-opening crust. The transient magmatism/volcanism had important environmental implications. After break-up, the margin experienced thermal subsidence and little structural deformation, but the extrusive complexes controlled the Paleogene sediment deposition.

To replace the current oil consumption and meet the predicted growth in world demand for energy, mankind needs new reserves for the future. For this reason both industry and governments have again, i.e. after the oil crisis in the early 1970s, become increasingly interested in all available sedimentary basins whose hydrocarbon potential has not yet been assessed. The outer continental margins, in 200–4000 m of water, are considered the most prospective areas for the future. However, these areas have been relatively poorly investigated by commercial exploration due both to high exploration risk and cost and technical difficulties.

To secure a long-term supply of hydrocarbons, intensive geoscientific research is needed on the structure and geological evolution of these frontier areas. A thorough understanding of how the geological processes and events, which have formed the present continental margins, have evolved in space and time is a requirement for both a plausible assessment of the potential hydrocarbon resources, and for a long-term energy policy.

The North Atlantic provides examples of passive margins with large sedimentary basins that may have resource potential. In this paper we present the results of geoscientific research from these margins, in particular the Norwegian–Greenland Sea and adjacent areas (Fig. 1). We first summarize the margin history by referring to overview papers in which a wealth of detailed material is available. Then, recognizing that many margin segments are extensively discussed in this volume and elsewhere, we describe events during crustal break-up which have governed the volcanic margin evolution and have implications for resource evaluation. In particular, we illustrate the break-up events using new seismic data from the previously poorly surveyed NE Greenland margin and we show a tectono-magmatic framework similar to that established on the well-surveyed conjugate Norwegian margin.

North Atlantic tectonic episodes

The continental crust beneath the present continental margins has undergone several episodes of lithospheric extension dur-ing Paleozoic and Mesozoic times until complete continental separation was achieved between Greenland and Eurasia near the Paleocene/Eocene transition. Since the Caledonian collision, the pre-opening history (Hinz *et al.* 1984; Brekke and Riis 1987; Larsen 1987; Ziegler 1988; Doré 1991; Faleide *et al.* 1992) comprises:

- formation of Devonian terrestrial half-grabens and depocentres in Great Britain, Norway and East Greenland;
- Permo-Carboniferous and Middle to Late Triassic rifting and block-faulting in NW Europe and East Greenland;
- Late Jurassic to Early Cretaceous extension resulting in major basin formation during Cretaceous time in the North Sea rift system, SW Barents Sea, and between Norway and Greenland;
- Late Cretaceous rifting that affected the Rockall Trough area, the Labrador–West Greenland continental margin, the Norwegian continental margin, and the western Barents Sea.

The Late Cretaceous tectonic episode led to final separation of Greenland from North America in the Paleocene (chron 27; Chalmers *et al.* 1993), and separation of Greenland from Eurasia during chron 24R. The episode culminated with voluminous igneous activity during break-up emplacing the plateau basalts of the North Atlantic Volcanic Province (NAVP) (Upton 1988) as well as huge extrusive constructions, including wedges of seaward-dipping reflectors prominently displayed in seismic records, along the continent–ocean transition (e.g. Hinz *et al.* 1987; Skogseid and Eldholm 1987; Larsen and Jakobsdottir 1988; Mutter and Zehnder 1988; Eldholm *et al.* 1989; Spence *et al.* 1989). The volcanic margin segments extend along conjugate margin pairs from the SE Greenland–Rockall Plateau in the south to the NE Greenland–Barents Sea region in the north (Fig. 2).

When the excessive igneous activity had abated about 3 Ma after break-up, the young ocean basin widened and deepened forming the present continental margins (Eldholm *et al.* 1989). The plate tectonic evolution is described by Talwani and

From *Petroleum Geology of Northwest Europe: Proceedings of the 4th Conference* (edited by J. R. Parker).
© 1993 Petroleum Geology '86 Ltd. Published by The Geological Society, London, pp. 901–913.

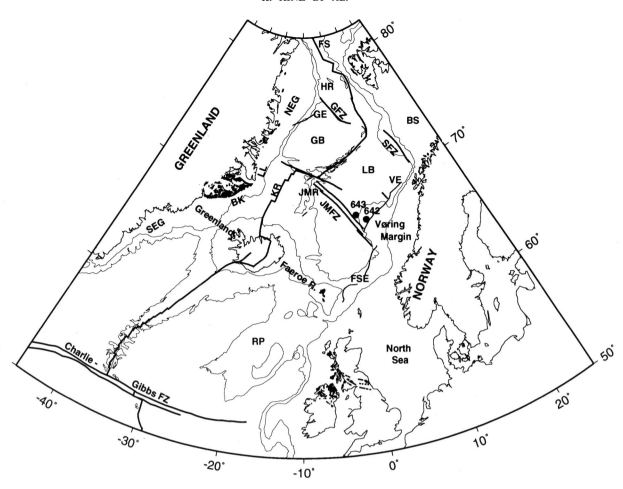

Fig. 1. North Atlantic bathymetry, plate boundaries, main fracture zones, marginal escarpments and ODP sites 642 and 643. KR: Kolbeinsey Ridge; GFZ, SFZ, JMFZ: Greenland, Senja and Jan Mayen fracture zones, respectively; GE, VE, FSE: Greenland, Vøring and Faeroe–Shetland escarpments, respectively; BS: Barents Sea; RP: Rockall Plateau; HR: Hovgaard Ridge; JMR: Jan Mayen Ridge; GB: Greenland Basin; LB: Løfoten Basin; SEG: SE Greenland margin; BK: Blosseville Kyst; LL: Liverpool Land Shelf; NEG: NE Greenland Shelf.

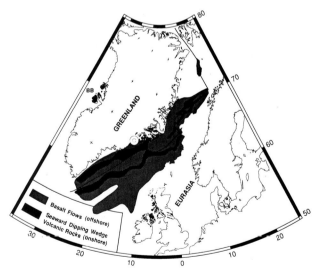

Fig. 2. Early Tertiary extrusives shown in an anomaly 23-time plate reconstruction (from Eldholm 1990). Hot spot location (double circle) according to White and McKenzie (1989).

Eldholm (1977) and Nunns (1983); and summaries of the margin geology have been presented by Roberts *et al.* (1984), Myhre *et al.* (1992) and Larsen (1990) for Hatton Bank, Norway-Svalbard and East Greenland margins, respectively. The post-opening history includes the following.

• A significant reorientation of relative plate motion when sea floor spreading ceased in the Labrador Sea about 35 Ma (chron 13). The Eocene plate motion in the Greenland Sea was achieved by continent–continent translation along a regional transform linking the Atlantic with the Eurasia Basin in the Arctic Ocean. This change in motion initiated separation of NE Greenland and Svalbard (Mhyre and Eldholm 1988) (Fig. 3), although final crustal separation may not have occurred before Miocene times.

• Rifting of the Jan Mayen Ridge apart from the former Liverpool Land margin (Fig. 1) prior to late Oligocene time (Hinz and Schlüter 1980; Larsen 1990). A new spreading centre, associated with volcanism, separated the Jan Mayen Ridge and Greenland about 25 Ma (Gudlaugsson *et al.* 1988).

• Late Neogene regional uplift of the landmasses along the North Atlantic margins during climatic deterioration, leaving substantial Pliocene sediment wedges centred around the present shelf edge.

Most of the post-Caledonian crustal extension and basin formation, including the Early Tertiary opening, took place in areas affected by Caledonian deformation. In fact, an important spatial evolution of the rift zones has been documented on the Norwegian–Greenland Sea margins, where the location of rifts and subsequent basin development have shifted through time but generally stayed within the Caledonian deformation belt (Skogseid *et al.* 1992; Faleide *et al.* 1992). Therefore, the successful continental break-up and formation of the Norwegian–Greenland Sea took place in a region characterized by late Paleozoic and early Mesozoic extension and crustal thinning. The crust was further attenuated by the prominent Late Jurassic–Early Cretaceous rift episode prior to the early Tertiary onset of sea floor spreading.

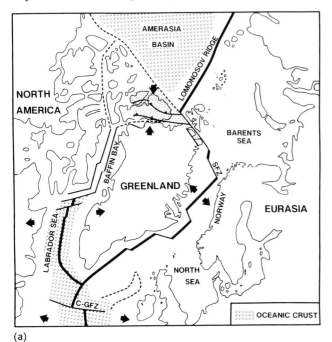

(a)

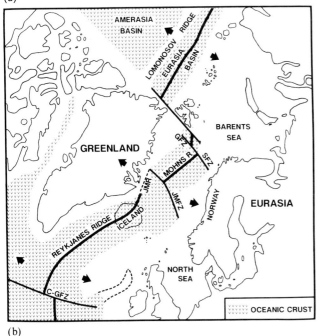

(b)

Fig. 3. North Atlantic–Arctic plate reconstructions (Myhre *et al.* 1992) (**a**) Break-up during chron 24R near the Paleocene–Eocene transition. Note the predominantly rifted plate boundary in the south, and the sheared plate boundary in the north. (**b**) Chron 13, early Oligocene time. GFZ, SFZ, JMFZ and C-GFZ: Greenland, Senja, Jan Mayen and Charlie-Gibbs fracture zones, respectively. JMR: Jan Mayen Ridge.

Another important observation is that the incipient plate boundary between Eurasia and Greenland constitutes two structural megalineaments (Eldholm *et al.* 1989). These are the North Atlantic rift zone, or the Reykjanes–Mohn Line, between the Charlie–Gibbs and Greenland–Senja fracture zones; and a regional shear zone, the De Geer Zone (Harland 1969), between Svalbard and Greenland which continues into

the Arctic along the North Greenland and Canada continental margins. These lineaments intersect near the oldest part of the Senja Fracture Zone (Fig. 3). Thus, the northern North Atlantic margins are structurally composed of predominantly NE-trending rifted segments in the south, and NNW-trending sheared segments in the north.

NE Greenland continental margin

From regional geophysical investigations of varying type, coverage and quality, a tectonic and stratigraphic subdivision into five segments separated by transverse zones has been suggested (cf. Larsen 1990). These segments are: SE Greenland margin, Denmark Strait Ridge (western part of Greenland–Faeroe Ridge), Blosseville Kyst, Liverpool Land Shelf and NE Greenland Shelf (Fig. 1).

The regional geology of the four southern segments is fairly well known (e.g. Hinz and Schlüter 1980; Larsen and Jakobsdottir 1988; Larsen 1990). We consider the petroleum prospectivity to be relatively low, mainly due to limited occurrences of pre-drift Mesozoic sediments. Most of the acoustic basement beneath these segments consists of early Tertiary basaltic rocks formed during break-up and early sea floor spreading. Therefore, we focus on new seismic results north of the Jan Mayen Fracture Zone. On this remote margin north of 72.5°N, intersected by the physiographically distinct Greenland Fracture Zone and the Hovgaard Ridge, we have tentatively identified six structural zones (Fig. 4).

Seismic stratigraphy

Although conditions for multichannel seismic (MCS) surveys are frequently arduous, nearly 10 000 profile km have been collected north of 70°N (Table 1). Due to the absence of drillholes, there is much uncertainty about the age and nature of regional seismic unconformities and markers. Nonetheless, by comparing the MCS data both with seismic stratigraphic interpretations from the Liverpool Land (Hinz and Schlüter 1980) and Blosseville Kyst (Larsen 1984) margins, and with the conjugate Norwegian margin (Skogseid and Eldholm 1989), we have defined five regional unconformities. The seismic stratigraphy (Table 2) is supported by baselap termination of seismic markers onto magnetically datable oceanic crust.

The rift phase unconformity, correlated with the latest Cretaceous to Paleocene rift event, which forms the top of faulted and tilted blocks off Norway, has only been recognized locally north of 75°N where flood basalts are not present (Fig. 4). Older unconformities, i.e. corresponding to the near base Cretaceous rift unconformity off Norway (Table 2) and the mid-Permian peneplain in East Greenland (Surlyk *et al.* 1984), are probably present beneath the outer shelf, but difficult to identify in the sparse dataset.

Unconformity EE, a prominent and extensive horizon, forms the upper boundary of a strongly reflecting sequence that occupies the entire NE Greenland shelf from about 72–76°N (Fig. 5), and the top of wedge-shaped bodies characterized by divergent and seaward-dipping reflectors (Figs 6 and 7). Ocean Drilling Program (ODP) drilling results on the conjugate Vøring margin (Eldholm *et al.* 1987) are consistent with previous interpretations (e.g. Hinz 1981; Hinz *et al.* 1987), that these wedges consist of basaltic flows and volcaniclastic rocks extruded near or above sea-level during break-up. West of the dipping wedges and south of 76°N, EE occupies nearly the entire shelf in close proximity to onshore continental flood basalts between 72°N and Shannon Island (Figs 2 and 4). Therefore, we interpret this smooth and strongly reflective unit to be an offshore equivalent of the East Greenland flood basalts emplaced during chron 24R (Larsen *et al.* 1989).

Unconformity UI, a distinct erosional unconformity, corre-

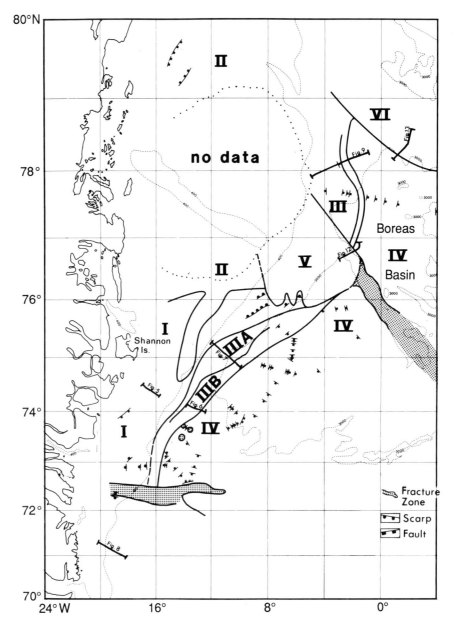

Fig. 4. Tectono-volcanic zonation of the NE Greenland margin. Bathymetry in metres.

lates with the boundary between depositional sequences GR-4 and GR-3 of Hinz and Schlüter (1980) on the Liverpool Land margin. Here, UI terminates on about 9.6 Ma old oceanic crust. UI forms the base of series of depositional units that can be subdivided into two types based on seismic architecture (Fig. 8). Type A is a complex sequence that mostly, but not exclusively, progrades the palaeoshelf by building it outward. Type B is a sequence that mostly, but not exclusively, aggrades the palaeoshelf by building it upward.

By comparison with Antarctic continental margin stratigraphy (Cooper *et al.* 1991), we suggest that type A is mainly caused by sediment deposition from grounded ice sheets in front of glaciers and ice streams during glacial maxima. Thus, it represents episodic, regional glacial advance towards the NE Greenland palaeoshelf edge since at least earliest late Miocene times. Type B is thought predominantly to have been deposited by non-glacial and/or open marine glacial processes. Above UI on the present slope there is a 20–50 km-wide and few hundred metres thick feature with a non-coherent reflection pattern, typical of sediment drifts (Figs 7–9). We suspect that the drift signals the start of strong shallow- and deep-water circulation

induced by a combination of climatic change and the gradual opening of the Fram Strait gateway between Greenland and Svalbard during late middle Miocene through early late Miocene time. This interpretation corresponds with results of Wolf and Thiede (1991) who describe the initiation of the cold East Greenland Current at about 8.3 Ma, increased input of quartz and rock fragments between 9.5 and 7 Ma, and a drastic increase in coarse terrigenous particles at about 4 Ma, suggesting a nearly simultaneous onset of large-scale ice rafting caused by the deterioration of Northern Hemisphere climate.

The prominent and extensive erosional unconformity UII represents a drastic change in the palaeoceanography and in depositional processes. It forms the base of a depositional unit with more frequent occurrence of type A sequences (Fig. 8) suggesting diachronous and episodic outbuilding. UII truncates the older sediment drift and erodes deeply into the pre-late Miocene palaeoslope deposits between 74 and 76°N (Fig. 7). We suggest it represents the onset of a later episode of increased watermass exchange through the Fram Strait which coincides with a distinct lowering of eustatic sea-level 4–5 Ma ago. The corresponding strong erosion and build-up of a

Table 1. MCS surveys on the NE Greenland continental margin north of 70°N

Year	Institution	Vessel	Seismic source		Receiver		Coverage (%)	Sonobuoy station (no.)	Total survey (km)
			volume cubic inches	no. of guns	streamer length (m)	no. of channels			
1974	Federal Institute for Geosciences and Natural Resources (BGR), Germany	M/V LONGVA	1891	16	2400	48	2400	—	730
1975	Federal Institute for Geosciences and Natural Resources (BGR), Germany	M/V LONGVA	1980	16	2400	48	2400	3	1877
1976	Federal Institute for Geosciences and Natural Resources (BGR), Germany	S/V EXPLORA	1430	24	2400	48	2400	4	1374
1980	Geological Survey of Greenland (GGU), Denmark	R/V WESTERN ARCTIC	905 (4500 psi) or Maxipulse	16	3000	60	3000–6000	?	c. 1400
1981	University of Bergen, Norway	R/V HÅKON MOSBY	1380	3	1200	24	1200	c. 10	c. 1100
1982	Geological Survey of Greenland (GGU), Denmark	R/V WESTERN ARCTIC	1600		3000	60	3000	10	c. 850
1983	Federal Institute for Geosciences and Natural Resources (BGR), Germany, and Lamont–Doherty Geological Observatory (L-DGO), USA	S/V PROSPEKTA	2229	24	3000	60	3000	11 ESP	701
		R/V ROBERT D. CONRAD					1000		727
			2933	4	2400	48	700		485 SWAP
1988	Federal Institute for Geosciences and Natural Resources (BGR) in co-operation with the Alfred–Wegener Institute for Polar and Marine Research	P.F.S. POLARSTERN	1562	10	2400	48	2400		1259
1990	Alfred-Wegener Institute for Polar and Marine Research	P.F.S. POLARSTERN	1220	8	800	24	1600–1800		324

$$\sum 9727$$

Table 2. Stratigraphy and igneous units on the Vøring and NE Greenland margins.

NE Greenland margin			Vøring margin	
UIII[3]			UP	Upper Pliocene[1]
UII	Lower Pliocene[3]	*	O'	Base Pliocene[1]
			MM	Middle Miocene[1]
UI	Early lower Miocene[3]	*	A	Lower Miocene[1]
			A'	Middle Oligocene[1]
			TP	Balder Formation tuff[1]
RP	Rift phase unconformity[3]	*	C	Near base Tertiary[2]
			D	Near base Cretaceous[2]
EE	East Greenland flood basalts[3]	*	EE	Early Eocene tholeiitic flood basalts[1]
			IF	Inner flows[1]
			—	Sills, low-angle dikes[1]
			K	Top late Paleocene dacitic flows[1]

*indicates assumed equivalent units; [1]Skogseid and Eldholm (1989); [2]Skogseid et al. (1992); [3]this paper.

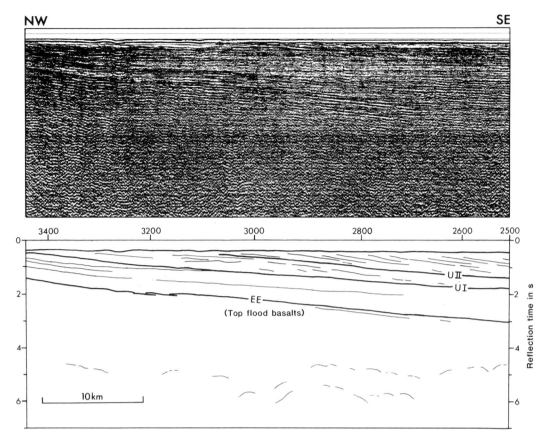

Fig. 5. Reflection seismic record and interpretation from zone I (Line BGR-ARK V/3-14). Location in Fig. 4.

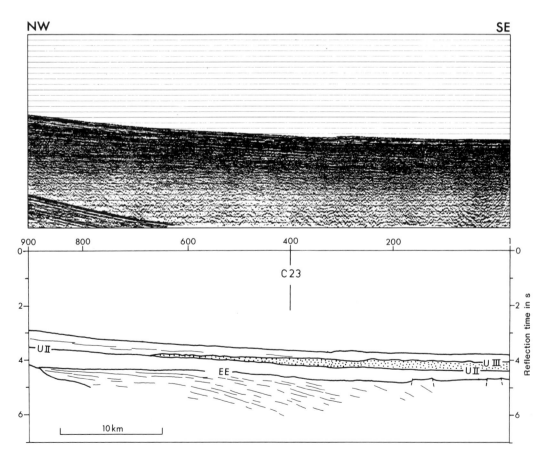

Fig. 6. Reflection seismic record and interpretation (Line BGR-ARK V/3-14) from zone IIIB, and position of magnetic anomaly 23 (C23). Location in Fig. 4.

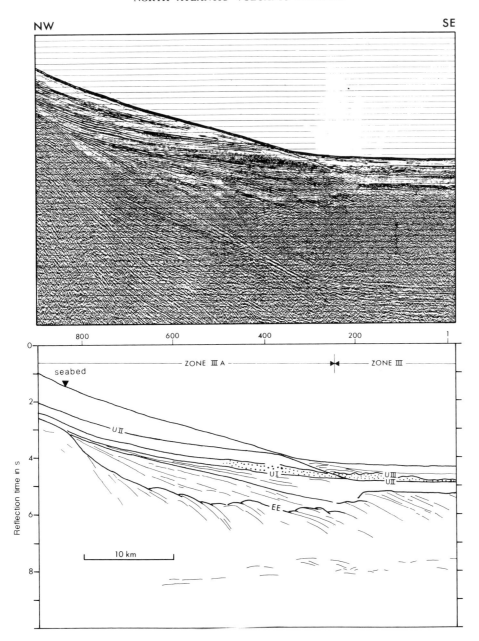

Fig. 7. Reflection seismic record and interpretation (Line NGT-42) across zones IIIA and IIIB. Inferred sedimentary drifts are dotted. Location in Fig. 4.

sedimentary drift and/or a widespread contourite layer above UII, probably relates to increased Northern Hemisphere cooling.

Unconformity UIII (Figs 6, 7 and 9) is thought to represent the onset of the present circulation pattern, i.e. southward transport of polar water by the East Greenland Current and northward transport of warm surface water by the Norwegian Current.

Although the links between depositional processes, ice-volume changes and sea-level fluctuations along the NE Greenland margin are still not well understood, the magnitude of the post-Paleocene sediments is substantial. The maximum thickness is almost 3 s (Fig. 10), or about 5 km, with relatively high seismic velocities ranging from 2–4 km s^{-1}. The main part consists of as much as 2 s (Fig. 11), or 2.5–3.0 km, post-middle Miocene deposits documenting shelf edge progradation of about 80 km.

Main structural units (Fig. 4)

Zone I is part of the 700 km-long NE Greenland Shelf province (Larsen 1990) between the Jan Mayen and Greenland fracture zones. Horizon EE, the East Greenland flood basalts unit, forms the acoustic basement except for a narrow area between 12 and 14°W where there are weak indications of faulted and tilted Cretaceous sediments truncated by unconformity RP. Off Norway a similar horizon, the zone of inner flows, extending from 25–70 km landward of the Vøring Escarpment has been interpreted as basalt flows covering Paleocene and older strata (Talwani *et al.* 1983; Hinz *et al.* 1984; Hinz *et al.* 1987). We suggest that the inner flows and zone I mark the beginning of an excessive volcanic episode resulting in the emplacement of the East Greenland plateau basalts. The volcanism was initiated immediately prior to break-up, within the previously continuous proto-Norway–Greenland rift, and continued during break-up.

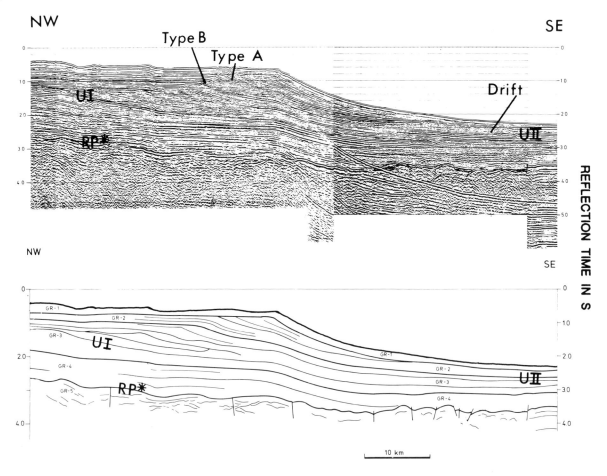

Fig. 8. Reflection seismic record and interpretation (Line BGR 76-11) from the Liverpool Land margin showing type A and B depositional sequences. RP* is the rift unconformity formed during separation of Greenland and Jan Mayen Ridge. Sequences GR-1 and GR-4 from Hinz and Schlüter (1980). Location in Fig. 4.

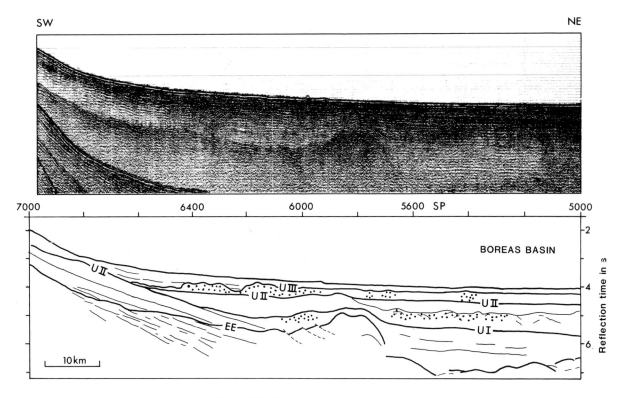

Fig. 9. Reflection seismic record and interpretation (Line BGR 74-31) from Zone III at about 78°N. Inferred sedimentary drifts and contourites are dotted. Location in Fig. 4.

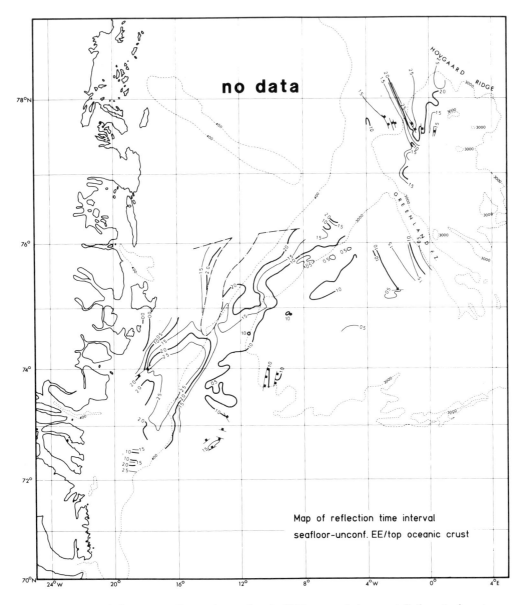

Fig. 10. Sediment thickness, in s(twt), between sea floor and unconformity EE/top oceanic basement. Bathymetry in metres.

Zone II comprises the wide, virtually unexplored, shelf north of 76°N where NE-trending horsts and grabens have been inferred from aeromagnetic data (Larsen 1984, 1990). This interpretation is now supported by the first MCS measurements north of 79°N where Hinz *et al.* (1991) show a very deep rift basin on the inner shelf filled with up to 10 km of high velocity rocks interpreted as Carboniferous and Permian sediments. They also reported structures which in size and seismic characteristics resemble large shelf-edge reefs, or possibly suggest the presence of evaporites.

From the very few seismic profiles north of 76°N and the aeromagnetic data we predict an equivalent of the Cretaceous Vøring Basin just landward of zone III. However, its extent and transition to the late Paleozoic rift basin system on the wide shelf (zone II) is unknown.

Zone IIIA (Figs 6 and 7) is characterized by a suite of seaward-dipping reflectors below a sedimentary cover. Where the oldest magnetic lineation, anomaly 24, is a distinct double-peaked anomaly, zone IIIA lies landward of anomaly 24. Sub-horizontal layering within the feather edge of zone IIIA and escarpments marking its landward edge are observed off Norway and on some lines off NE Greenland. The zone IIIA lavas were emplaced just prior to and during the first stage of sea floor spreading.

Between the Jan Mayen and Greenland fracture zones the 25–50 km-wide zone IIIB, or the outer wedge of dipping reflectors, is characterized by a distinct smooth reflector constituting the top of a basaltic succession beneath a Cenozoic cover. Locally, there are short seaward-dipping horizons, and a pronounced escarpment forms its landward edge. The zone which merges seaward into oceanic crust, overlies the seaward-dipping wedge of zone IIIA (Fig. 7) and is associated with magnetic anomaly 24 north of 75°N.

Whether the volcanic features described above are classified as resulting from 'rift volcanism', 'subaerial sea floor spreading' or 'Icelandic spreading', the seismic and drilling evidence suggest that brief and subsequent periods of voluminous volcanic activity attended the inception and birth of the Norwegian–Greenland Sea as well as other ocean basins. For the conjugate margins off NE Greenland and Norway we document three volcanic zones which successively overlap, suggesting variation and/or episodicity during the volcano-magmatic period. The first volcanic units, inner flows and zone I, were extruded onto continental crust; the later units, IIIA and IIIB, onto crust receiving steadily increased amounts of new igneous material and zone IIIB reflects the last major volcanic event.

Zone IV comprises Cenozoic oceanic crust accreted by sea floor spreading. Zone V is a tectonically complex area along

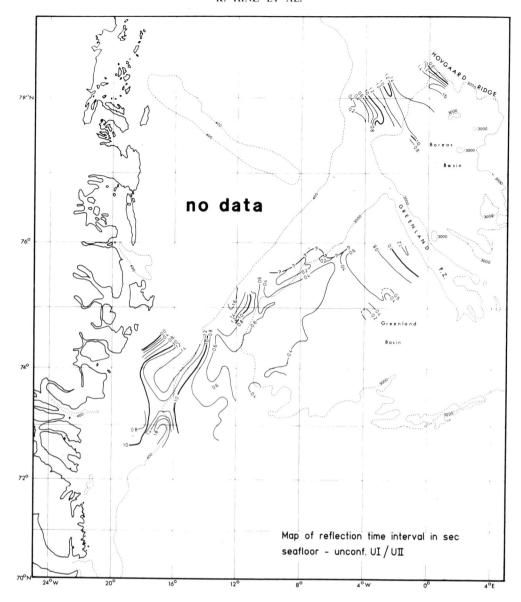

Fig. 11. Sediment thickness, in s(twt), between sea floor and unconformities UI/UII. Bathymetry in metres.

the landward projection of the Greenland Fracture Zone. It appears that Mesozoic rift structures have been inverted, possibly related to wrench movements (Fig. 12). Finally, zone VI is part of the complex Hovgaard Ridge where tilted and faulted blocks support a continental nature for the ridge (Myhre and Eldholm 1988). The inferred Mesozoic deposits are folded along the northeastern flank suggesting transpression (Fig. 13).

Formation of the volcanic margin

To study the effects of the volcanic margin formation on basin development one has to evaluate the entire area that has undergone deformation and magmatic activity prior to and during break-up, as well as the region influenced by post-break-up margin subsidence. This relationship is illustrated in Table 2 and in the composite conjugate margin section in Fig. 14. In particular, ODP Site 642 (Fig. 1) which penetrated the inner part of a dipping reflector wedge has contributed to the understanding of the volcanic margin. The drilling results (Eldholm *et al.* 1989) and the seismic data (Skogseid *et al.* 1992) are consistent with a 15–20 Ma period of Late Cretaceous and Paleocene lithospheric extension prior to break-up during chron 24R. There is also evidence of regional Paleocene

syn-rift uplift, centred along the nascent plate boundary, and the present outer margin stayed high until the transient excess magmatism had abated. Subsequently, oceanic crust was accreted from a spreading axis at bathyal depths.

The transient magmatic episode is documented by a series of igneous features recognized in the seismic data on the conjugate margins:

- high-velocity, 7.2–7.5 km s^{-1}, crustal bodies constituting the lower part of an expanded crust beneath the region covered by basaltic extrusives (Hinz *et al.* 1987; Mutter and Zehnder 1988; Fowler *et al.* 1989; Planke *et al.* 1991). The bodies, which extend from the thinned continental crust well into the oceanic crust, are thought of as melts emplaced during the transient phase:
- local occurrences of acidic to intermediate lavas (e.g. ODP Site 642 lower series) extruded during the late Paleocene rifting;
- sills and low-angle dykes within Cretaceous and Paleocene sediments. These features are restricted to the crust that underwent pre-opening extension and the activity terminated at break-up;
- voluminous tholeiitic flood basalts constituting both the huge extrusive constructions along the continent–ocean

transition and the NAVP onshore plateau basalts. The emplacement was most intense at break-up during chron 24B, diminishing over the next 2 to 3 Ma.

● a regional pyroclastic horizon, the Balder Formation, marking the break-up event in adjacent sediment basins.

By including the volcanic margins, the emplacement of NAVP becomes a major geological event (Fig. 2). The volcanic margins extend for almost 3000 km along the early Eocene plate boundary and quantitative estimates reveal an areal extent of the volcanic province of about 1.3×10^6 km^2 and an extrusive rock volume of about 1.8×10^6 km^3 (Eldholm and Grue pers. comm.). These minimum estimates do not include British Tertiary volcanism and coeval volcanism in the Labrador Sea; however, they rank the North Atlantic among the world's larger igneous provinces.

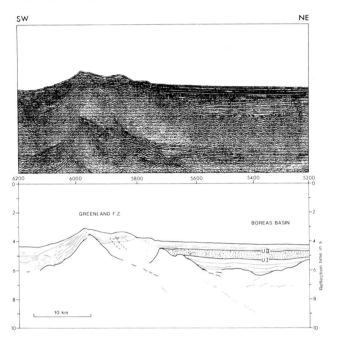

Fig. 12. Reflection seismic record and interpretation of a profile (Line BGR-ARK V/3-10) across the Greenland Fracture Zone. Location in Fig. 4.

A large-scale transient geological event of this magnitude will significantly influence palaeoenvironments by changing oceanographic and atmospheric circulation patterns and compositions. Consequently, the formation of the North Atlantic volcanic margins may have important implications for the hydrocarbon resource potential of the present-day continental margins. First, the event affected the Cretaceous, and older, sedimentary basins on the margin producing regional uplift, erosion, tectonism and emplacement of intrusive and extrusive rock complexes. Second, the effects of the early Tertiary

thermal anomaly on both pre- and post-break-up sediments has to be considered. Third, the post-opening subsidence of the evolving margin will bear an imprint of the volcanic margin formation.

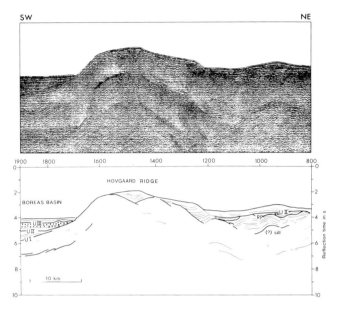

Fig. 13. Reflection seismic record and interpretation of a profile (Line BGR-ARK V/3-10) across the Hovgaard Ridge. Location in Fig. 4.

Eldholm (1990) has shown that the impact of the volcanic margin formation had both near-field and far-field effects. The former deal with the palaeoceanography and depositional conditions in the North Atlantic, while the latter are of global character. In general, sedimentation on evolving continental margins is dependent on watermass circulation and basin sedimentation governed by lateral, or plate tectonic, movements. However, the transient magmatic event introduced an additional component of basin segmentation, first by rift uplift within the Paleocene epicontinental sea, and later by an extrusive barrier between subsiding continental crust and the growing ocean basin. The Paleocene lithospheric extension on the Vøring margin, for example, resulted in faulting and local inversion within the Late Cretaceous grabens, and there is evidence of considerable erosion both during late rifting and immediately after break-up, depositing large volumes of sediment in the Vøring Basin. In fact, most Paleogene sediment has a western source (Fig. 14) (Skogseid et al. 1992).

Evidence for transient igneous activity and associated dynamic uplift along the entire North Atlantic rifted plate boundary suggests that a region of elevated crust between both continental landmasses existed before and after break-up. The elevated region, which probably had a variable along-plate boundary relief, was rapidly denuded changing from subaerial

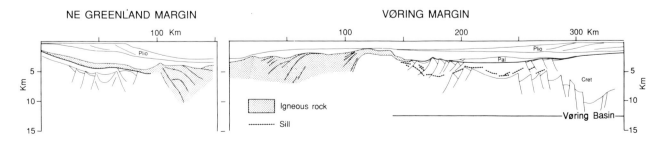

Fig. 14. Conjugate margin transect across the region affected by Late Cretaceous–Early Tertiary extension and magmatism. Constructed from seismic profiles on the Vøring and NE Greenland margins (Skogseid et al. 1992).

to shallow-marine environments. Hence, a major Paleogene sedimentary source region existed first in the central epicontinental sea between Greenland and Eurasia and later on the young continental margins, dominating the sediment supply and deposition at those times. The uplift is also recorded by Paleogene sediments in the Northern North Sea and Faeroe–Shetland basins (Anderton 1993). The basin fragmentation may have led to later changes in sediment composition and depositional environment, and ODP Site 643 (Fig. 1) reveals restricted bottom waters with oxygen deficiency in early Eocene (Kaminski *et al.* 1991). The warm climate (Robert and Chamley 1991) and high biological productivity in the early Cenozoic, the central sediment source, and the possibility of poorly ventilated and restricted sub-basins (Dobson *et al.* 1991) have implications for the resource potential of the Palaeogene margin sequences.

Major unsolved problems

The available drilling results and geophysical data provide abundant new information regarding the architecture of the extrusive and intrusive complexes as well as data on composition, stratigraphy, and timing of the igneous event. However, there are diverging views on the kinds of geodynamic process which furnish the heat for the generation of the excess partial melting. Along some margins, hotspots produced by deep mantle plumes provide a convenient heat source. One view is that 'passive' rifting occurs over a mantle plume (White and McKenzie 1989), while others prefer plume initiation, in some cases accompanied by 'active' rifting (Duncan and Richards 1991; Hill 1991). Nonetheless, there is evidence of volcanic margins apparently unrelated to major hotspots (Hinz 1981; Mutter *et al.* 1988; Coffing and Eldholm 1991).

Another question is the relationship between outer, and younger, seaward-dipping wedges (IIIB) and the main wedges (IIIA). The outer wedge might originate from local untapped melt reservoirs, possibly combined with a shallow extrusion depth. Local volcanic activity after the break-up phase is consistent with tephras in the North Sea (Knox and Morton 1988).

We also refer to a new and intriguing discovery of an area of abnormal upper oceanic crust along the Kolbeinsey Ridge (Fig. 1). This crust is characterized by an internally divergent to planar pattern of reflections having dips toward the spreading centre. The 120 km-wide oceanic crustal structure resembles zone IIIA on the NE Greenland and Norway margins. The crust was apparently formed subsequent to the separation of Jan Mayen Ridge from East Greenland by the mid-oceanic Kolbeinsey Ridge between chrons 7–6 (*c.* 25–19 Ma) and chrons 5A–5 (*c.* 11.5–98 Ma). This conspicuous oceanic crust implies an episode of excess volcanism along the spreading centre from late Oligocene through late middle Miocene time, terminating about the time when glaciation on the North America–Greenland and Eurasian continents started. This observation raises important fundamental questions such as:

● Has an accelerated crustal accretion along the Kolbeinsey spreading ridge, and possibly elsewhere, triggered Northern Hemisphere glaciation because of its impact on atmosphere, ocean and biosphere?
● What causes the excessive volcanic–magmatic episode along the Kolbeinsey spreading centre?

Similar observations exist from other spreading centres suggesting that the global accretion rate of oceanic crust might be of more pulsating nature than previously anticipated. By analogy with volcanic margins, oceanic crust created during periods of high accretion rate, i.e. excess partial melting, might show expanded crustal thickness, elevated basement, upper crustal sub-basement reflectors and smooth basement surface. This phenomenon might also explain volcanic margins away from hotspots if break-up occurs during a period of high accretion rate.

We express our appreciation for the support of the following scientists and institutions for the preparation, execution and interpretation of our marine geoscientific studies in the Norwegian–Greenland Sea: H. Meyer, H. A. Roeser and D. Seidel, Bundesanstalt für Geowissenschaften und Rohstoffe; H. C. Larsen, Greenland Geological Survey; and H. Miller, Alfred-Wegener Institute for Polar and Marine Research. The work has been supported by the Federal Ministry for Research and Technology, Germany and the German–Norwegian VISTA project for basic research.

References

ANDERTON, R. 1993. Basin evolution. *In*: PARKER, J. R. (ed.) *Petroleum Geology of Northwest Europe: Proceedings of the 4th Conference.* Geological Society, London, 31.

BREKKE, H. AND RIIS, F. 1987. Tectonics and basin evolution of the Norwegian shelf between 62°N and 72°N. *Norsk Geologisk Tidskrift*, **67**, 295–321.

CHALMERS, J. A., PULVERTAFT, T. C. R., CHRISTIANSEN, F. G., LARSEN, H. C., LAURSEN, K. H. and OTTESEN, T. G. 1993. The southern West Greenland continental margin: rifting history, basin development, and petroleum potential. *In*: PARKER, J. R. (ed.) *Petroleum Geology of Northwest Europe: Proceedings of the 4th Conference.* Geological Society, London, 915–931.

COFFIN, M. F. AND ELDHOLM, O. (eds) 1991. *Large Igneous Provinces: JOI/USSAC Workshop Report.* The University of Texas at Austin Institute for Geophysics Technical Report No. 114.

COOPER, A. K., BARRETT, P. J., HINZ, K., TRAUBE, V., LEITCHENKOV, G. AND STAGG, H. 1991. Cenozoic-prograding sequences of the Antarctic continental margin: a record of glacio-eustatic and tectonic events. *Marine Geology*, **102**, 175–213.

DOBSON, M. R., HAYNES, J. R., BANNISTER, A. D., LEVENE, D. G., PETRIE, H. S. AND WOODBRIDGE, R. A. 1991. Early Tertiary paleoenvironments and sedimentation in the NE Main Porcupine Basin (well 35/13-1), offshore western Ireland—evidence for global change in the Tertiary. *Basin Research*, **3**, 99–117.

DORÉ, A. G. 1991. The structural foundation and evolution of Mesozoic seaways between Europe and the Arctic. *Palaeogeography, Palaeoclimatology, Palaeoecology*, **87**, 441–492.

DUNCAN, R. A. AND RICHARDS, M. A. 1991. Hotspots, mantle plumes, flood basalts, and true polar wander. *Reviews of Geophysics*, **29**, 31–50.

ELDHOLM, O., 1990. Paleogene North Atlantic magmatic–tectonic events: environmental implications. *Memoire della Societa Geologica Italiana*, **44**, 13–28.

——, THIEDE, J. AND TAYLOR, E. 1989. Evolution of the Vøring volcanic margin. *In*: ELDHOLM, O., THIEDE, J., TAYLOR, E. *et al.* (eds) Proceedings of the Ocean Drilling Program, Scientific Results, **104** College Station, TX (Ocean Drilling Program), 1033–1056.

——, ——, —— *et al.* 1987. *Proceedings of the Ocean Drilling Program*, **104**. College Station, TX (Ocean Drilling Program).

FALEIDE, J. I., VÅGNES, E. AND GUDLAUGSSON, S. T. 1992. Late Mesozoic–Cenozoic evolution of the southwestern Barents Sea in a regional rift-shear tectonic setting. *Marine and Petroleum Geology*, in press.

FOWLER, S. R., WHITE, R. S., WESTBROOK, G. K. AND SPENCE, G. D. 1989. The Hatton Bank continental margin, II. Deep structure from two-ship expanded spread seismic profiles. *Geophysical Journal*, **96**, 295–309.

GUDLAUGSSON, S. T., GUNNARSSON, K., SAND, M. AND SKOGSEID, J. 1988. Tectonic and volcanic events at the Jan Mayen Ridge microcontinent. *In*: MORTON, A. C. AND PARSON, L. M. (eds) *Early Tertiary Volcanism and the opening of the NE Atlantic.* Geological Society, London, Special Publication, **39**, 85–94.

HARLAND, W. B. 1969. Contribution of Spitsbergen to understanding of tectonic evolution of North Atlantic region. *In*: KAY, M. (ed.) *North Atlantic; geology and continental drift.* American Association of Petroleum Geologists Memoir, **12**, 817–851.

HILL, R. I. 1991. Starting plumes and continental break-up. *Earth and Planetary Science Letters*, **104**, 398–416.

HINZ, K. 1981. An hypothesis on terrestrial catastrophes wedges of very thick oceanward dipping layers beneath passive continental margins; their origin and palaeoenvironmental significance. *Geologisches Jahrbuch*, E2, 3–28.

——, DOSTMAN, H. J. AND HANISCH, J. 1984. Structural elements of the Norwegian continental margin. *Geologisches Jahrbuch*, A75, 192–221.

——, MEYER, H. AND MILLER, H. 1991. North-east Greenland Shelf north of 79°N: results of a reflection seismic experiment in sea ice. *Marine and Petroleum Geology*, **8**, 461–467.

——, MUTTER, J. C., ZEHNDER, C. M. AND NGT STUDY GROUP 1987. Symmetric conjugation of continent–ocean boundary structures along the Norwegian and East Greenland margins. *Marine and Petroleum Geology*, **4**, 166–187.

—— AND SCHLÜTER, H-U. 1980. *Continental margin off East Greenland*. Proceedings of the 10th World Petroleum Congress, **2**, 405–418.

KAMINSKI, M. A., GRADSTEIN, F. M., GOLL, R. M. AND GREIG, D. 1990. Biostratigraphy and paleoecology of deep-water agglutinated foraminifera at ODP Site 643, Norwegian–Greenland Sea. *In*: HEMLEBEN, C., KAMINSKI, M. A., KUHNT, W. AND SCOTT, D. B. (eds) *Paleoecology, Biostratigraphy, Paleoceanography and Taxonomy of Agglutinated Foraminifera*. Kluwer Academic Publishers, Dordrecht, 345–386.

KNOX, R. W. O'B. AND MORTON, A. C. 1988. The Record of Early Tertiary N Atlantic Volcanism in Sediments of the North Sea Basin. *In*: MORTON, A. C. AND PARSON, L. M. (eds) *Early Tertiary Volcanism and the opening of the NE Atlantic*. Geological Society, London, Special Publication, **39**, 407–420.

LARSEN, H. C. 1984. Geology of the East Greenland Shelf. *In*: SPENCER, A. M. (ed.) *Petroleum Geology of the North European Margin*. Graham & Trotman, London, 329–339.

—— 1990. The East Greenland Shelf. *In*: GRANTZ, A., JOHNSON, L. AND SWEENEY, J. F. (eds) *The Geology of North America, Vol. L, The Arctic Ocean Region*. Geological Society of America, Boulder, Colorado, 185–210.

—— AND JAKOBSDOTTIR, S. 1988. *Distribution, crustal properties and significance of seaward-dipping sub-basement reflectors off E Greenland*. *In*: MORTON, A. C. AND PARSON, L. M. (eds) *Early Tertiary Volcanism and the Opening of the NE Atlantic*. Geological Society, London, Special Publication, **39**, 95–114.

LARSEN, L. M., WATT, W. S. AND WATT, M. 1989. *Geology and Petrology of the Lower Tertiary Plateau Basalts of the Scoresby Sund Region, East Greenland*. Grønlands Geologiske Undersøkelse Bulletin, **157**.

LARSEN, V. B. 1987. A synthesis of tectonically related stratigraphy in the North Atlantic–Arctic region from Aalenian to Cenomanian time. *Norsk Geologisk Tidskrift*, **67**, 281–293.

MUTTER, J. C., BUCK, W. R. AND ZEHNDER, C. M. 1988. Convective partial melting. 1. A model for the formation of thick basaltic sequences during the initiation of spreading. *Journal of Geophysical Research*, **93**, 1031–1048.

—— AND ZEHNDER, C. M. 1988. Deep crustal structure and magmatic processes: the inception of seafloor spreading in the Norwegian–Greenland Sea. *In*: MORTON, A. C. AND PARSON, L. M. (eds) *Early Tertiary Volcanism and the Opening of the NE Atlantic*. Geological Society, London, Special Publication, **39**, 35–48.

MYHRE, A. M. AND ELDHOLM, O. 1988. The western Svalbard margin (74–80°N). *Marine and Petroleum Geology*, **3**, 134–156.

——, ——, FALEIDE, J. I., SKOGSEID, J., GUDLAUGSSON, S. T., PLANKE, S., STUEVOLD, L. M. AND VÅGNES, E. 1992. Norway–Svalbard Margin: Structural and Stratigraphical styles. *In*: POAG, C. W. AND DE GRACIANSKY, P. C. (eds) *Geologic Evolution of Atlantic Continental Rises*. Van Nostrand Reinhold, New York, 157–185.

NUNNS, A. G. 1983. Plate tectonic evolution of the Greenland–Scotland Ridge and surrounding regions. *In*: BOTT, M. H. P., SAXOV, S., TALWANI, M. AND THIEDE, J. (eds) *Structure and Development of the Greenland–Scotland Ridge: New methods and concepts*. Plenum Press, New York, 11–30.

PLANKE, S., SKOGSEID, J. AND ELDHOLM, O. 1991. Crustal structure off Norway 62° to 70° North. *Tectonophysics*, **189**, 345–365.

ROBERT, C. AND CHAMLEY, H. 1991. Development of early Eocene warm climate, as inferred from clay mineral variations in oceanic sediments. *Palaeogeography, Palaeoclimatology, Palaeoecology*, **89**, 315–331.

ROBERTS, D. G., BACKMAN, J., MORTON, A. C., MURRAY, J. W. AND KEENE, J. B. 1984. Evolution of volcanic rifted margins; Synthesis of Leg 81 results on the west margin of Rockall Plateau. *In*: ROBERTS, D. G., SCHNITKER, D. *et al.* (eds) *Initial Reports of the Deep Sea Drilling Project*, **81**. US Government Printing Office, Washington, DC, 883–911.

SKOGSEID, J. AND ELDHOLM, O. 1987. Early Cenozoic crust at the Norwegian continental margin and the conjugate Jan Mayen Ridge. *Journal of Geophysical Research*, **91**, 11471–11491.

—— AND —— 1989. Vøring Plateau continental margin: seismic interpretation, stratigraphy and vertical movements. *In*: ELDHOLM, O., THIEDE, J., TAYLOR, E. *et al.* (eds) *Proceedings of the Ocean Drilling Program, Scientific Results*, **104**. College Station, TX (Ocean Drilling Program), 993–1030.

——, PEDERSEN, T., ELDHOLM, O. AND LARSEN, B. T. 1992. Tectonism and magmatism during NE Atlantic continental break-up: the Vøring Margin. *In*: STOREY, B. C., ALABASTER, T. C. AND PANKHURST, R. J. (eds) *Magmatism and the Causes of Continental Break-up*. Geological Society, London, Special Publication, **68**, 303–318.

SPENCE, G. D., WHITE, R. S., WESTBROOK, G. K. AND FOWLER, S. R. 1989. The Hatton Bank continental margin, I. Shallow structure from two-ship expanded spread seismic profiles. *Geophysical Journal*, **96**, 273–294.

SURLYK, F., PIASECKI, S., ROLLE, F., STEMMERIK, L., THOMPSON, E. AND WRANG, P. 1984. The Permian Basin of East Greenland. *In*: SPENCER, A. M. *et al.* (eds) *Petroleum Geology of the North European Margin*. Graham & Trotman, London, 303–315.

TALWANI, M. AND ELDHOLM, O. 1977. Evolution of the Norwegian–Greenland Sea: *Geological Society of America Bulletin*, **88**, 969–999.

——, MUTTER, J. C. AND HINZ, K. 1983. Ocean–continent boundary under the Norwegian continental margin. *In*: BOTT, M. H. P., SAXOV, S., TALWANI, M. AND THIEDE, J. (eds) *Structure and Development of the Greenland–Scotland Ridge: New methods and concepts*. Plenum Press, New York, 121–131.

UPTON, B. G. J. 1988. History of Tertiary igneous activity in the N Atlantic borderlands. *In*: MORTON, A. C. AND PARSON, L. M. (eds) *Early Tertiary Volcanism and the Opening of the NE Atlantic*. Geological Society, London, Special Publication, **39**, 429–453.

WHITE, R. S. AND McKENZIE, D. 1989. Magmatism at rift zones: the generation of volcanic margins and flood basalts. *Journal of Geophysical Research*, **94**, 7685–7729.

WOLF, T. C. W. AND THIEDE, J. 1991. History of terrigenous sedimentation during the past 10 m.y. in the North Atlantic (ODP Legs 104 and 105 and DSDP Leg 81). *Marine Geology*, **101**, 83–102.

ZIEGLER, P. A. 1988. *Evolution of the Arctic–North Atlantic and the Western Tethys*. American Association of Petroleum Geologists Memoir, **43**.

The southern West Greenland continental margin: rifting history, basin development, and petroleum potential

J. A. CHALMERS, T. C. R. PULVERTAFT, F. G. CHRISTIANSEN, H. C. LARSEN, K. H. LAURSEN and T. G. OTTESEN[1]

Geological Survey of Greenland, Øster Voldgade 10, DK-1350 Copenhagen K, Denmark
[1] *Present address: DOPAS, Agern Allé 24–26, DK-2970 Hørsholm, Denmark*

Abstract: The development of the continental margin of West Greenland is closely related to the processes that led to the opening of the Labrador Sea. The opening of the Labrador Sea began in the Early Paleocene (anomaly 27N), and not in the Late Cretaceous as previously supposed. Modelling of magnetic data and new interpretation of seismic data indicate that a large area previously regarded as underlain by oceanic crust is in fact underlain by block-faulted continental crust overlain by syn- and post-rift sedimentary sequences. The ocean–continent transition is now placed 100–150 km southwest of the foot of the continental slope instead of at the foot of this slope. Rifting in the Labrador Sea area began, however, in the Early Cretaceous. The earliest sediments are the syn-rift lower and upper members of the Bjarni Formation on the Labrador shelf and their likely equivalents, the pre- to syn-rift Kitsissut and Appat sequences on the Greenland margin. The age of these units is Barremian (or older) to Albian. The units are overlain by widespread mudstone-dominated units, the Markland Formation of the Labrador shelf and the Kangeq Sequence on the Greenland margin. The former is Cenomanian–Danian in age. By analogy the base of the Kangeq Sequence is probably Cenomanian (or Turonian), while the top is known from well ties to be at the Cretaceous–Tertiary boundary. Rifting was subdued during deposition of these mudstone units. Rifting was renewed in the Early Paleocene, and mudstones, siltstones and very fine sandstones were deposited. With the initiation of sea-floor spreading there was considerable igneous activity at the ocean–continent transition, as well as in the onshore area where picrites followed by plagioclase-porphyritic basalts were erupted. After the end of the Paleocene there was little rifting in the region, but compressional structures were formed locally as a response to transpression related to strike-slip movements that transferred plate motion from the Labrador Sea to Baffin Bay. A marked Early Oligocene unconformity separates the syn-drift Paleocene–Eocene succession from the post-drift middle Oligocene–Quaternary sediments. Sediments deposited since the Paleocene are dominated by sands. The main hydrocarbon play types offshore West Greenland are related to tilted fault blocks. Source rocks are anticipated near the base of the Kangeq Sequence, which is also the seal, and reservoirs are sandstones in the Appat and Kitsissut sequences. These two sequences were not reached by any of the exploration wells drilled in the 1970s.

The West Greenland margin is a rifted continental margin, the development of which was related to the processes which led to the opening of the Labrador Sea in late Mesozoic–early Cenozoic time.

The Labrador Sea is about 900 km wide and opens to the southeast into the North Atlantic Ocean. To the north it narrows and shallows, and passes into the Davis Strait, a 300 km wide seaway leading into Baffin Bay (Fig. 1). The Labrador Sea is flanked by typical Arctic continental shelves with banks less than 200 m deep separated by glacially eroded channels.

The Labrador Sea opened by sea floor spreading. The early spreading history is poorly constrained. Until recently it had been generally accepted that spreading began in Late Cretaceous time, as early as anomaly 31 or even 33 (Srivastava 1978; Roest and Srivastava 1989). Initial spreading was in a roughly ENE–WSW direction. At about anomaly 24 time a change of spreading direction took place, probably related to the initiation of spreading in the North Atlantic north of the Charlie-Gibbs Fracture Zone, and spreading took place in a NE–SW direction until starting to slow down at around anomaly 20 time and finally ceasing in the Oligocene.

Recent reprocessing and reinterpretation of seismic data and modelling of magnetic data acquired by Bundesanstalt für Geowissenschaften und Rohstoffe (BGR) in the Labrador Sea in 1977 has led to a new interpretation of the early spreading history (Chalmers 1991). It now appears that (1) sea-floor spreading did not begin in the inner Labrador Sea until the Early Paleocene, the oldest oceanic magnetic anomaly in this part of the sea being 27; and (2) the ocean–continent boundary is not situated at the foot of the bathymetric continental slope as previously supposed (e.g. Tucholke and Fry 1985) but 100–150 km seawards of this.

Knowledge of the geology of the continental shelves is almost entirely due to the results of petroleum exploration in these areas. In the period 1971–83, 28 wells were completed on the Labrador–Baffin Island shelf, leading to six gas discoveries, none of which has been exploited. Only one well recovered any oil. The sediments encountered on the Labrador shelf are predominantly from Early Cretaceous to Recent in age; locally these are underlain by palaeoerosional remnants of Lower Paleozoic sediments. Ordovician sediments underlie the Mesozoic–Tertiary succession over an extensive area off southeast Baffin Island (Balkwill *et al.* 1990).

The West Greenland shelf was the scene of petroleum exploration in the period 1971–78. Following the acquisition of about 40 000 km of reflection seismic data of variable quality, five wells (Fig. 1) were drilled in 1976–77, all of them dry. It had been hoped that Cretaceous reservoir sandstones like those known from onshore outcrops would be encountered, but three of the wells terminated in Paleocene basalts or Precambrian basement after penetrating only Tertiary–Recent sediments (Rolle 1985). One well encountered 26 m of sandstone and mudstone with Campanian palynomorphs before entering basement. The fifth well penetrated the base of the Tertiary 2250 m below seabed and continued 850 m into Campanian–Maastrichtian shale, but had not encountered any Cretaceous sandstones when drilling was terminated. However, none of the wells penetrated the lowest two seismic sequences recognized during recent reinterpretation of seismic data from the 1970s (Chalmers 1989, in press) and also on new seismic lines acquired in 1990. Thus, knowledge of the strati-

From *Petroleum Geology of Northwest Europe: Proceedings of the 4th Conference* (edited by J. R. Parker).

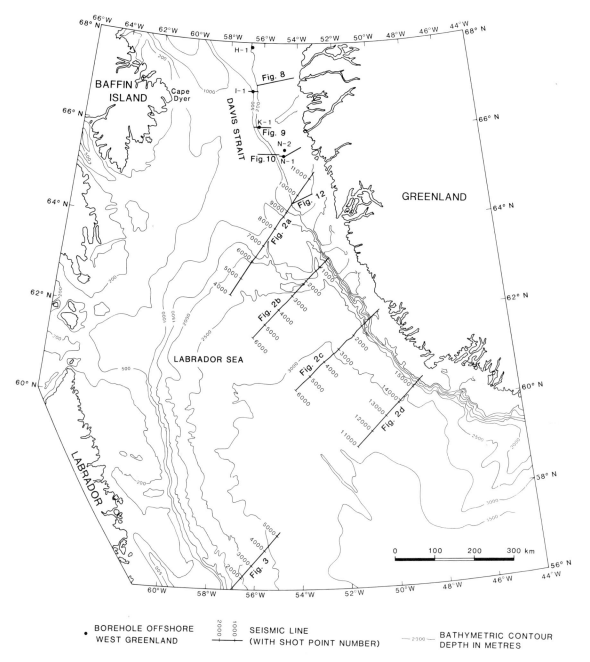

Fig. 1. Bathymetry map of the Labrador Sea showing locations of seismic lines shown on other figures and of wells offshore West Greenland. H-1: Hellefisk-1; I-1: Ikermiut-1; K-1: Kangâmiut-1; N-1: Nukik-1; N-2: Nukik-2.

graphy and petroleum potential of West Greenland is very incomplete, and will remain so until new drilling is carried out.

The land areas on either side of the Labrador Sea and Davis Strait are built up of Precambrian gneisses and other metamorphic rocks. To the north, on the island of Disko and peninsulas of Nûgssuaq and Svartenhuk (69–72°N) in West Greenland, and near Cape Dyer (67°N) in Baffin Island, Cretaceous and Early Tertiary fluvio-deltaic sediments overlain by Early Tertiary flood basalts are extensively exposed.

The revised plate-tectonic history of the Labrador Sea area, together with reinterpretation of a large amount of the industry data acquired offshore West Greenland in the 1970s, has provided the basis for the new ideas and the re-evaluation of petroleum potential that are presented in this paper.

Sea-floor spreading and the ocean–continent transition

In 1977 four seismic and magnetic traverses were made across the Labrador Sea by the BGR (Hinz *et al.* 1979). The seismic

data from one traverse, that ties with the Bjarni H-81 well on the Labrador Shelf, has been reprocessed by GGU in its entirety, while only those parts of the other three lines that lie within Greenland waters have been reprocessed (Figs 1–3).

The most immediately obvious feature of the reprocessed sections is the pattern of tilted fault blocks overlain by syn- and post-rift sediments that can be seen in the deep water parts of the sections off the Greenland shelf (Figs 2 and 4). This is a pattern one associates with rifted continental crust, although it can occur in an oceanic crust environment. The line BGR/77-12 is the most interesting in this context. In earlier interpretations anomalies 31 and 33 had been placed at approximately SP 4450 and 2050 on this line (Roest and Srivastava 1989), the latter in an area where the blocked-faulted pattern is obvious (Figs 2b and 4). However, when modelling of the magnetic data along this line was carried out, the outer part of the line could be modelled as oceanic crust with magnetic zones of alternating polarity, but all attempts to model the area landward of anomaly 27N as alternating stripes of normally and

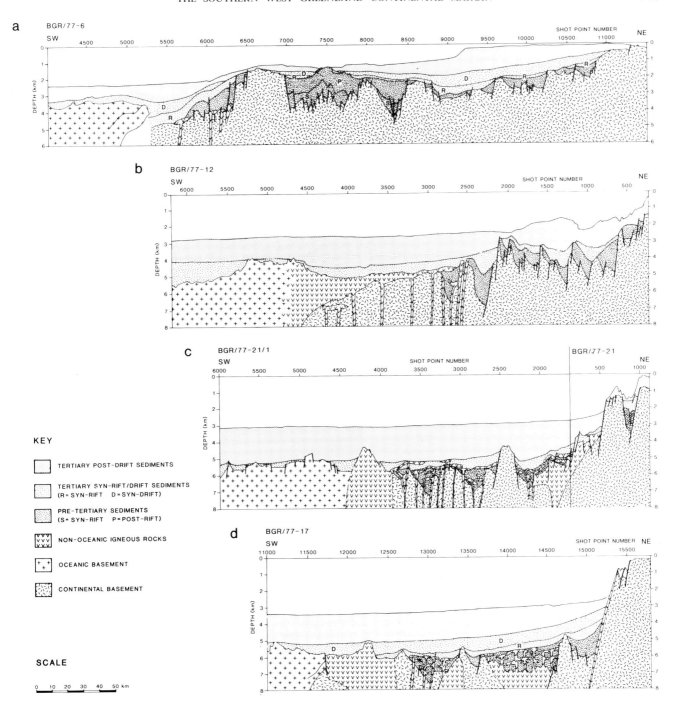

Fig. 2. Interpreted geological cross-sections across the transition from continental to oceanic crust of southern West Greenland, based on BGR seismic lines: (**a**) BGR/77-6; (**b**) -12; (**c**) -21 and -21/1; and (**d**) -17/1. See Fig. 1 for location. The sills within the sedimentary succession on (a), (c) and (d) are visible on the seismic sections and confirmed by magnetic modelling. The interpretation of intrusions within the continental basement is based on magnetic modelling alone, and other interpretations of the high remanent magnetizations may be possible.

reversed magnetized material failed (Chalmers 1991). Between the oceanic crust and the zone with a typical block-faulted structural pattern there is an intermediate zone which has been modelled as continental crust overlain and intruded by reversely magnetized igneous material. This intermediate zone narrows to the northwest so that along the northernmost line (BGR/77-6) the block-faulted terrain disappears abruptly below volcanic rocks with seaward-dipping reflectors and a landward-facing volcanic escarpment (Figs 2a and 5, SP 4800–5250). This is similar to situations known from the Faeroe–Shetland Ridge (Smythe 1983) and the Vøring Plateau (Eldholm *et al.* 1989), and strongly suggests that the ocean–continent transition is located seawards of the escarpment,

and hence seawards of the position suggested by Roest and Srivastava (1989).

Block-faulted continental crust can also be interpreted off the Greenland shelf on the landward parts of the southernmost two lines (Figs 2c and d). Here, however, the crust and overlying sediments have been intensely intruded by dykes and sills and larger igneous bodies (Figs 2c and d, Fig. 6), and the position of the ocean–continent transition is difficult to locate with confidence. On Fig. 2d, seaward-dipping reflectors similar to those known from volcanic continental margins in the North Atlantic (papers in Morton and Parson 1988) are interpreted near SP 11 500.

The southwest part of line BGR/77-17, which ties with the

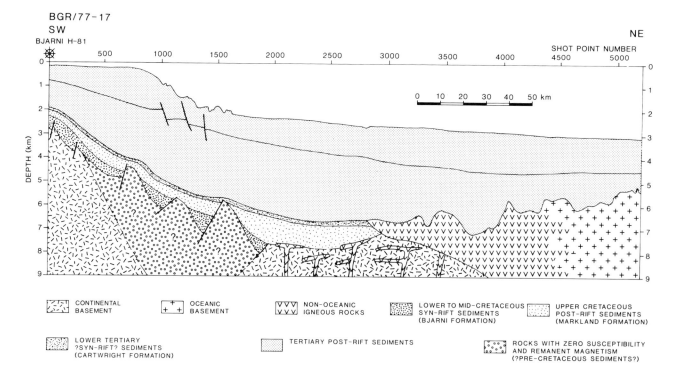

Fig. 3. Interpreted geological cross-section across the transition from continental to oceanic crust off the coast of Labrador. Based on the seismic line BGR/77-17. See Fig. 1 for location.

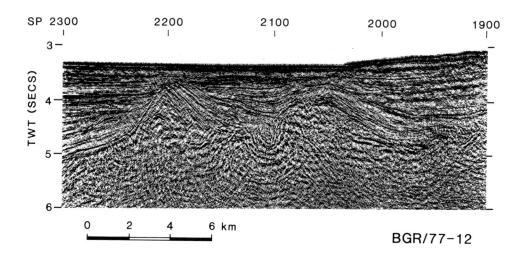

Fig. 4. Part of migrated seismic line BGR/77-12 showing fault blocks on continental crust under deep water. See Figs 1 and 2 for location.

Bjarni H-81 well, has also been reprocessed and interpreted (Fig. 3). The lowest formation in this well, immediately overlying the Precambrian basement, is an Early Cretaceous volcanic unit, the Alexis Formation, that cannot be distinguished on seismic data from the Precambrian basement. This is overlain by the continental–deltaic to marine Bjarni Formation of Barremian (or perhaps older) to Albian age, which in turn is overlain by marine shales of the Markland Formation of Cenomanian–Early Paleocene age (Balkwill 1987; Balkwill *et al.* 1990).

Sediments of the Bjarni and Markland formations can be followed seaward on the seismic data to where basalts onlap, cover and finally conceal them and the rifted continental basement on which they lie (Fig. 3). Roest and Srivastava (1989) place the ocean–continent boundary at SP 1700 on this line, but attempts to model the area on the Labrador side of SP

4500 (anomaly 26R) as alternating remanent magnetizations failed. However, again reflectors that dip towards the ocean can be observed below the top basalt horizon. Whether these are 'dipping reflectors' in the sense of Hinz (1981) is not clear, but it is possible that this is also a volcanic margin.

The extent of oceanic crust according to the model outlined above is shown in Fig. 13. Although linear magnetic anomalies typical of oceanic crust cannot be traced northwest of the Hudson Fracture Zone, attempts to refit the continents prior to sea-floor spreading suggest that there could be an area of oceanic crust northwest of this fracture zone. The absence of visible magnetic stripes here might be a consequence of highly oblique spreading in accordance with the mechanism suggested by Roots and Srivastava (1984).

The consequences of these interpretations are far-reaching as regards the history of the opening of the Labrador Sea and

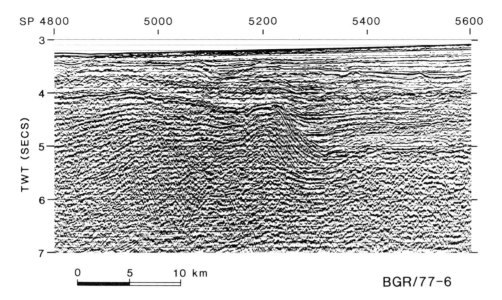

Fig. 5. Part of migrated seismic line BGR/77-6 showing escarpment and dipping reflectors such as are commonly visible in volcanic continental margins in other oceans.

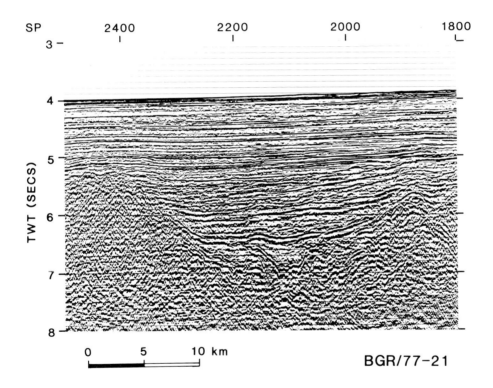

Fig. 6. Part of migrated seismic line BGR/77-21/1. See Figs 1 and 2 for location. Below about 6 s two-way-time (TWT), a structure interpreted as a half-graben can be seen, which is disturbed by irregular high-amplitude reflectors interpreted as sills.

the palaeogeography of the region. It now seems likely that the opening of the Labrador Sea began during chron 27, in the Early Paleocene, about 15 Ma later than previously supposed, although it remains possible that the southernmost Labrador Sea opened earlier. Continental crust is now interpreted to underlie an extensive deep-water area including the area in the inner Labrador Sea where Roest and Srivastava (1989) drew anomalies 31 and 33, and the area of the Hecla Rise which Tucholke and Fry (1985) interpreted as an oceanic rise. The Hecla Rise is now interpreted as an area of shallow continental basement containing grabens and half-grabens (Fig. 2a), some of which have been folded and inverted (Fig. 2a, SP 7500),

probably by transpression associated with strike-slip movements arising from the transfer of sea-floor spreading from the Labrador Sea to Baffin Bay. Southwest of the Hecla Rise is the Gjoa Rise (Fig. 7), which is now interpreted to be a volcanic ocean–continent boundary zone where marginal escarpments and seaward-dipping reflectors can be seen (Fig. 5).

In short, the structure of the West Greenland margin in the Davis Strait area has the character of a transform sheared margin rather than that of a rifted volcanic margin as seen farther south. The situation during the opening of the Labrador Sea may thus have been similar to the present-day Red Sea rift–Dead Sea transform system.

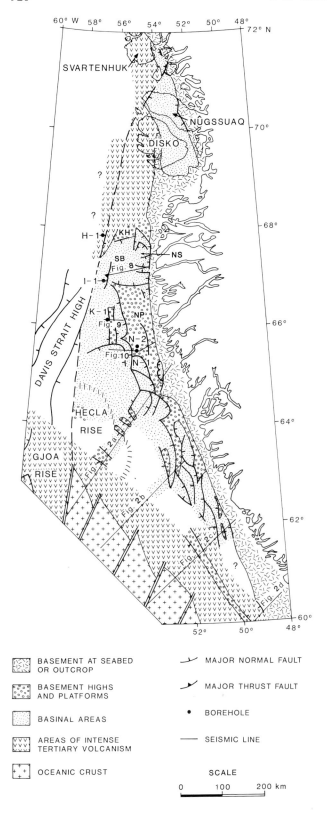

BASEMENT AT SEABED
OR OUTCROP

BASEMENT HIGHS
AND PLATFORMS

BASINAL AREAS

AREAS OF INTENSE
TERTIARY VOLCANISM

OCEANIC CRUST

MAJOR NORMAL FAULT

MAJOR THRUST FAULT

● BOREHOLE

SEISMIC LINE

SCALE

0   100   200 km

Fig. 7. Tectonic map of offshore West Greenland. Areas where crystalline basement is exposed are shown blank as are also areas offshore where the geology is unknown. H-1, I-1, K-1, N-1 and N-2 are sites of wells (cf. Fig. 1). KH: Kangeq High; SB: Sisimiut Basin; NS: Nagssugtôq sub-basin; NP: Nukik Platform. The Kangâmiut Ridge is the north–south ridge with the Kangâmiut-1 well on its west flank.

Structural features of the basement surface off southern West Greenland

Precambrian crystalline basement was reached in two of the

wells drilled in this area, Kangâmiut-1 and Nukik-1. The Nukik-2 well terminated in Early Paleocene basalts which appear from the seismic data to lie almost directly on the basement. The Hellefisk-1 well penetrated almost 700 m of subaerial Early Paleocene basalts before the well was abandoned. By analogy with the onshore exposures, Cretaceous sediments are expected to occur between these basalts and the underlying basement, but the seismic data here are of poor quality and reflections from below the basalts cannot be recognized.

The structural pattern of the basement surface is the result of two major movement systems: ENE–WSW extension related to continental break-up, and NNE–SSW strike-slip movement associated with the transform system in the Davis Strait area (the Ungava Fault complex). Most normal faults trend roughly NNW–SSE, at right angles to the extensional stress that gave rise to continental break-up. Connecting these there are numerous transfer faults at high angles to the general trend (Chalmers, in press). Figure 7 shows a simplification of the complex fault pattern.

The extensional faulting has given rise to rotated fault blocks, grabens and half-grabens, and structural highs. The most important of these are described briefly below and their position indicated on Fig. 7. The largest downthrow recorded on a single fault is over 5000 m (Fig. 2d).

Starting in the north of the area that has been mapped to date, there is a prominent high east of the Hellefisk-1 well—the Kangeq High (see Fig. 7 for location). The southern flank of this plunges into a sediment-filled basin more than 5 km deep—the Sisimiut Basin. The lowest sediments, especially along the eastern side of the basin, are faulted in a complex area of small grabens and fault blocks—the Nagssutôq sub-basin. The well Ikermiut-1 was drilled into the western side of the Sisimiut Basin into structures interpreted by Henderson *et al.* (1981) as shale diapirs but which Chalmers and Pulvertaft (in press) have suggested may have been caused by compression related to strike-slip movements (Fig. 8). Thrusts affecting the basement surface are interpreted in the seismic data which, however, has poor resolution in the deep part of the section.

To the south the Sisimiut Basin is bounded against an area of relatively shallow basement, the Nukik Platform, by an east–west-striking fault zone. The wells Nukik-1 and -2 were drilled into the southern part of this platform. The platform is partially covered by Tertiary basalts into which Nukik-2 was drilled. Nukik-1 terminated in Precambrian basement.

A north–south-striking graben is situated between the Nukik Platform and the Kangâmiut Ridge to the west. The Kangâmiut-1 well was drilled into the west flank of this ridge (Fig. 9). It penetrated 3500 m of Tertiary–Quaternary sediments and 26 m of ?Campanian sandstone and mudstone before terminating in deeply weathered Precambrian basement (Rolle 1985). West of the Kangâmiut Ridge is another basin in which the basement is locally covered by more than 8 km of sediments. The basement gradually shallows to the west towards the Davis Strait High.

The Kangâmiut Ridge ends abruptly to the south at an east–west-striking fault south of which is a basinal area which contains block-faulted structures (Figs 10 and 11). The basin is bounded to the southwest by the Hecla Rise. To the southeast is an area of complicated grabens, half-grabens and basement highs. The continental shelf narrows, basins on the shelf become smaller and relatively starved of Neogene sediments, and the continental slope is a major fault scarp (Fig. 2d). Block-faulted continental crust continues from the foot of the scarp into the deep-water area to the southwest. South of about 62°N, both the basement and the overlying sediments are heavily intruded by igneous material.

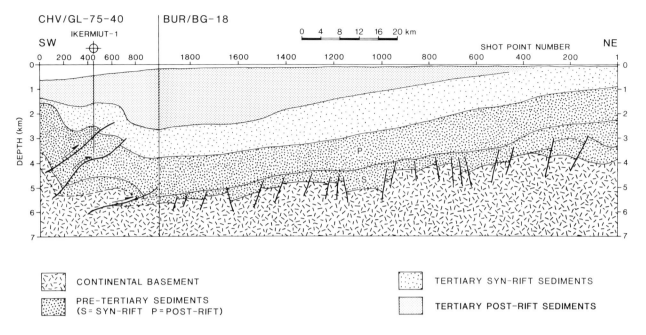

Fig. 8. Interpreted geological cross-section through the Ikermiut-1 well, the Sisimiut Basin and the Nagssugtôq sub-basin. See Figs 1 and 7 for location.

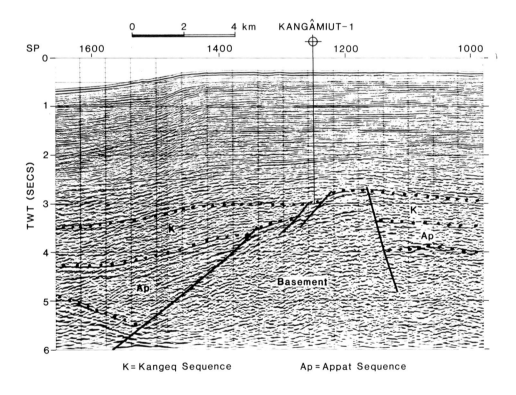

Fig. 9. Seismic section TOT/4-6609 through the Kangâmiut-1 well. See Figs 1 and 7 for location.

Stratigraphy and relation of sedimentation to tectonics

Onshore outcrops

Cretaceous and Early Tertiary sediments and Early Tertiary volcanic rocks related to the development of the West Greenland margin and opening of the Labrador Sea and Davis Strait are exposed on the island of Disko and the peninsulas of Nûgssuaq and Svartenhuk (69–72°N) in West Greenland and near Cape Dyer (67°N) in southeast Baffin Island (Fig. 7).

The Cretaceous sediments in West Greenland were deposited in a delta system; palaeocurrents were towards west and northwest (Pedersen and Pulvertaft 1992). In the southeast and east there are fluvial-deltaic sandstones, mudstones and coal seams of Albian–Santonian age. These pass northwestwards into marine organic-rich mudstones dated by ammonite faunas as Turonian–Maastrichtian (Birkelund 1965; Rosenkrantz and Pulvertaft 1969). The present-day eastern boundary of Cretaceous outcrops is a system of major faults. In the north the occurrence of very coarse conglomerate in Cenomanian

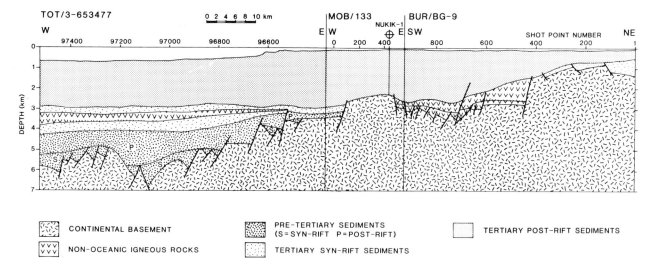

Fig. 10. Interpreted geological cross-section through the Nukik-1 well. Composite section based on seismic lines TOT/3-653477, MOB/133 and BUR/BG-9. See Figs 1 and 7 for location.

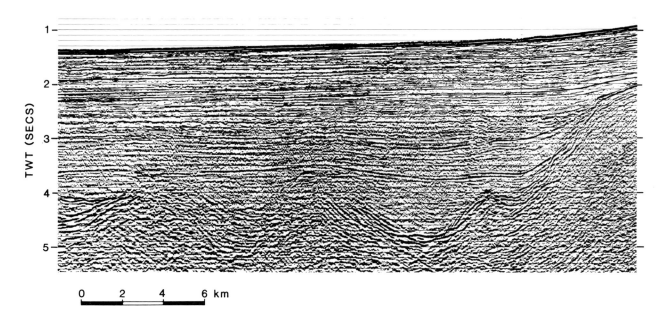

Fig. 11. Migrated seismic line across structures interpreted as rotated fault blocks caused by extensional faulting. This line is confidential and its location is not available.

sediments adjacent to the fault zone indicates active faulting at this time (Rosenkrantz and Pulvertaft 1969), but in the south, Albian–Cenomanian fluvial sands and shales are cut off by the boundary fault without any evidence of the proximity of syn-sedimentary fault scarps, suggesting that fault movement here was post-Cenomanian (Pulvertaft 1979, 1989).

Important faulting took place near the end of the Maastrichtian, resulting in an angular unconformity between Cretaceous and Tertiary sediments in much of the area, deep erosion and the sculpturing of pronounced relief in the surface of the Cretaceous rocks, and the deposition of coarse conglomerates in proximal areas and turbidite fans in more distal settings (Rozenkrantz and Pulvertaft 1969; Christiansen *et al.* 1992). The onset of basalt volcanism in the Early Paleocene (nannoplankton zone NP3, chron 27R; Piasecki *et al.* 1992) took place in a subaqueous environment. The earliest basalts are picritic pillow lavas and breccias which encroached from the west into

a marine embayment, building up a giant 'delta' with foresets up to 500 m high (Clark and Pedersen 1976). Later the pillow breccia 'delta' cut off the embayment and a lake was dammed up in which black mudstones with a very high organic carbon content accumulated (Pedersen 1989). Finally conditions became subaerial, and feldspar–porphyritic flood basalts covered the entire area. Volcanism continued at least until nannoplankton zone NP8, equivalent to chron 25R (Late Paleocene). Further faulting took place after extrusion of the basalts.

Near Cape Dyer there are terrestial sediments of Late Aptian–Early Cenomanian age that appear to have been deposited during active faulting (Burden and Langille 1990). These sediments are overlain unconformably by Early Paleocene mudstone, arkose and conglomerate also deposited during an episode of active faulting. Here also sedimentation ended when picritic basalts were erupted (Clark and Upton 1971).

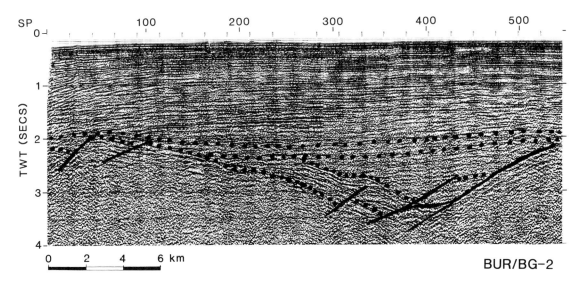

Fig. 12. Seismic section BUR/BG-2, SP 1–500. See Fig. 1 for location.

Offshore

Knowledge of the stratigraphy of the sedimentary succession offshore southern West Greenland is derived mainly from the results of the five exploration wells in this area. However, none of these wells penetrated the lowest sequences interpreted in the seismic sections, and the dating of these is provisional and based on analogies drawn with the Labrador shelf and to a lesser extent with onshore outcrops in West Greenland and eastern Baffin Island.

Of the five wells drilled offshore West Greenland (Fig. 1), three penetrated only Tertiary–Quaternary sediments before being terminated in Early Paleocene basalts (Hellefisk-1 and Nukik-2) or Precambrian basement (Nukik-1). The Kangâmiut-1 well reached the base of the Tertiary and penetrated 26 m of coarse arkosic sand interleaved with mudstone below the Paleocene, before being terminated in deeply weathered Precambrian basement. A Campanian dinoflagellate assemblage has been recovered from these sediments (Rolle 1985). Only one well, Ikermiut-1, drilled a significant section of pre-Tertiary sediments, sampling a 850 m section of Campanian–Maastrichtian mudstone before drilling was stopped (Rolle 1985).

Mesozoic The pre-Tertiary is considered the most important part of the offshore succession both for petroleum exploration and for the understanding of the initial tectonic developments of the West Greenland margin. The seismic stratigraphy of the pre-Tertiary succession has been studied by Chalmers (1989, in press) and Ottesen (1991*b*), and four seismic sequences have been recognized. Resolution of seismic stratigraphy in the deeper sections is, however, poorer than in the Tertiary, and the presence of other sequences is possible. Furthermore, the dating of the pre-Tertiary sequences is problematic because of lack of well data.

The Paleocene mudstones encountered in the Kangâmiut-1 well can be traced to the west on the seismic data and their base is marked by a continuous reflector (Fig. 9). Below this there is another, transparent sequence about 1 km thick, which was not encountered in the Kangâmiut well; this has been named the Kangeq Sequence. Below this is another sequence, named the Appat Sequence. This is over 2 km thick against the fault on the west side of the Kangâmiut Ridge. Both these sequences must be pre-Tertiary.

The Appat and Kangeq sequences can be traced on the seismic grid around the southern end of the Kangâmiut Ridge.

To the east they terminate against the Nukik Platform (Fig. 10). Both sequences continue to the southeast where yet another, older pre-Tertiary sequence can be indentified lying between the Appat Sequence and the basement (Fig. 12); this is called the Kitsissut Sequence and is the oldest sequence yet identified.

To the north the Kangeq Sequence can be traced, though with difficulty because of tectonic complications, to the Ikermiut-1 well where the upper part of the sequence ties with the lowermost 850 m of Campanian–Maastrichtian mudstones encountered in this well. Thus the age of the Kangeq Sequence is established as Campanian or older to Maastrichtian. The Kitsissut and Appat sequences are older than this, but just how old can only be judged by analogies to the Labrador shelf, using relationships to tectonic events as a further guide.

A fourth Mesozoic seismic sequence, as yet unnamed, has been identified by Ottesen (1991*b*) in the Sisimiut Basin. It lies above the Kangeq Sequence, or perhaps is contemporary with its upper part; the relationships are unclear mainly because of the complicated tectonics around the Ikermiut-1 well. This uppermost Mesozoic sequence contains strong progradations towards SSW from the northeast corner of the Sisimiut Basin, and Ottesen has identified the sequence as a delta. None of the wells has penetrated this sequence.

The Kangeq Sequence is a seismically transparent sequence. Its upper part can be traced into mudstones in the Ikermiut-1 well, and because of its transparency the sequence is most likely homogeneous and hence mudstone throughout. The Kangeq Sequence is cut off at faults in many places, but it does not show wedge-shaped geometry or any thickening in the footwall of the faults, hence its deposition does not appear to have been influenced by faulting.

The geometry of the Appat Sequence on the other hand is very obviously related to faults. The sequence shows typical wedge-shapes in cross-section, being thickest in the footwall close to fault planes and thinning away from this towards the crests of the rotated fault blocks (Figs 9 and 12). The reflector characteristics of the Appat Sequence indicate a more heterogeneous lithology than the Kangeq Sequence; the stronger reflections seen in many places could be graben-fill sands.

The Kitsissut Sequence has a restricted distribution. It appears to have been carried down and rotated with the fault blocks. In occasional fault-blocks it is wedge-shaped in cross-section (see Fig. 18). It is much more reflective than the Appat Sequence and tends to show downlap to the west (Fig. 12).

Cenozoic The base of the Tertiary was penetrated in both the Kangâmiut-1 and Ikermiut-1 wells. In the Ikermiut well there is a condensed sequence or hiatus at the Cretaceous–Tertiary boundary covering the interval Late Maastrichtian–Early Paleocene (Rolle 1985). At Kangâmiut-1 there appears to be a hiatus between Campanian and Lower Paleocene, and the Lower Paleocene is thin. However, the sands dated as Campanian could well belong to a syn-tectonic fan developed at the foot of a fault scarp (F. Sommer, pers. comm.; Fig. 9) and the palynomorphs derived from Campanian sediments on the footwall block. The possibility of these sands having an Early Paleocene age must be kept open.

Apart from organic-rich Paleocene mudstones in the Kangâmiut-1 and Ikermiut-1 wells, which have about 2% total organic carbon, the Tertiary sections encountered in the wells are dominated by shallow marine sands; there are no potential source rocks, and very few other potential seal units. A formal seismic stratigraphy has been proposed by Chalmers (in press) and Ottesen (1991*a*) for the entire Tertiary–Quaternary section, using well data (Rolle 1985) to constrain the ages of the many sequences.

The most striking feature revealed during the seismic stratigraphic analysis is a major Early Oligocene unconformity. In the north, around 66°N, the hiatus represented by the unconformity is within the Oligocene. Passing southwards one sequence is overstepped above the unconformity and three sequences are truncated below this surface, so that at 64°30′N the Upper Oligocene lies on Middle Eocene (Chalmers, in press). This unconformity coincides with the major regional R4 unconformity known all over the North Atlantic region (Miller and Tucholke 1983).

Another major super-sequence boundary occurs near the base of the Eocene. This separates the youngest (Paleocene) sequence affected by faulting from the earliest sequence in the overlying post-rift, syn-drift series of sequences which is cut off by the Early Oligocene unconformity. Faulting of the Paleocene sequence can, for example, be seen when the basalts encountered at the bottom of the Nukik-2 well are traced eastwards to where they lie in half-grabens (Fig. 10). Either they were erupted into actively developing half-grabens or the faulting took place after the eruption of the basalts. In either case the conclusion is that there was Paleocene rifting in the area.

Summarizing, the sedimentary-tectonic history is as follows:

(1) pre- to syn-rift deposition of the Kitsissut Sequence by prograding to the west;
(2) faulting and fault-block rotation; syn-rift deposition of the Appat Sequence;
(3) post-rift deposition of the Kangeq Sequence mudstones during an interim phase of thermal subsidence during Campanian or earlier to Maastrichtian time;
(4) hiatus or very reduced sedimentation rate at the Cretaceous–Tertiary boundary;
(5) renewed rifting and sedimentation interrupted by a phase of volcanism in the Paleocene;
(6) post-rift, syn-drift sedimentation in the Eocene and earliest Oligocene. Formation of compressional structures by transpression;
(7) major hiatus accompanied by erosion in the Early Oligocene;
(8) post-drift Late Oligocene to Recent sedimentation.

Age of pre-Tertiary sequences; correlation with conjugate Labrador shelf

The stratigraphy and evolution of the Labrador shelf is well known from the information provided by 28 exploration wells, about 80 000 km of seismic data, and summaries by Balkwill (1987) and Balkwill *et al.* (1990).

The Labrador shelf

The oldest sediments encountered in wells offshore Labrador are Ordovician dolomites and shales. These are probably remnants of a once-widespread platform cover. Although no sediments of this age have been recovered in wells or recognized in seismic sections in the West Greenland shelf, there is evidence that an Ordovician cover extended over West Greenland. Ordovician dolomite and mudstones have been found in a fault breccia on the mainland at 65°45′N (Poulsen 1966; Stouge and Peel 1979), and pebbles of Ordovician chert have been found in Upper Cretaceous conglomerate on the island of Disko (69°25′N) (Pedersen and Peel 1985). This evidence of widespread Ordovician rocks shows that there was at times a seaway between Greenland and Canada long before rifting and the development of the Labrador Sea began.

The lowest unit related to rifting of the Labrador margin is the Alexis Formation which consists of basalts overlying Precambrian basement or Paleozoic sediments.

Syn-rift sedimentation offshore Labrador began with deposition of the Bjarni Formation, which occupies coast-parallel grabens and half-grabens. The formation is divided into two members: the lower Bjarni member consists of non-marine (probably fluvial) sandstones and fluvio-lacustrine shales; the upper Bjarni member consists of sandy, clayey and carbonaceous, predominantly non-marine siltstone and shale, with intercalated beds of sandstone.

The oldest palynomorph assemblages recovered from the Bjarni Formation indicate a Barremian age, but the lowest beds in the deepest basins may well be older than this. The youngest sediments of the Bjarni Formation are of Late Albian or perhaps Cenomanian age. The age of the unconformity between the lower and upper members is probably middle Albian.

The syn-rift Bjarni Formation is overlain by the Markland Formation which is the lowest unit assigned by Balkwill *et al.* (1990) to the drift-phase megasequence on the Labrador shelf. However, along the inner (southwest) side of the Upper Cretaceous basins there are syn-rift, half-graben-confined, wedges of intercalated shale, siltstone and sandstone, that thin seawards and grade into the Markland Formation. The Markland Formation is a very widespread succession of marine shelf shales of Cenomanian–Danian age. It abruptly but conformably overlies the Bjarni Formation in the central parts of structural depressions, but oversteps this formation onto Precambrian basement, Paleozoic strata, or Alexis Formation basalts towards the margins of basins.

Drift-phase sedimentation continued throughout the Late Paleocene and Eocene with deposition of the Cartwright and Kenamu formations, both of which are dominated by shales. Along the southwest flank of the basin areas there are wedges of shallow marine sands that are proximal equivalents of the Cartwright Formation.

The close of drift-phase sedimentation was marked by shelf shallowing, coastal erosion, and seaward progradation of sand. Balkwill *et al.* (1990) noted the coincidence of this stratigraphic level with the postulated Late Eocene (or Early Oligocene) cessation of sea-floor spreading in the Labrador Sea (Srivastava 1978).

Correlation of the West Greenland and Labrador shelves

In the lower part of the Ikermiut-1 well off West Greenland, 1050 m of marine mudstones of Campanian–Middle Eocene age were penetrated before drilling was terminated. These constitute the Ikermiut Formation of Rolle (1985). Rolle correlated the Upper Cretaceous part of the Ikermiut Formation with the Markland Formation, and the Paleocene part with the Cartwright Formation. The Maastrichtian–Lower

Paleocene in the Ikermiut-1 well is a condensed sequence less than 50 m thick, and is the natural place to put a boundary in this well.

In an extension of the new seismic stratigraphic studies of the West Greenland shelf (Chalmers, in press) the upper part of the Kangeq Sequence has been tied with the lower (Upper Cretaceous) part of the Ikermiut Formation in the Ikermiut-1 well. This tie is the basis for correlation of the Kangeq Sequence and discussion of the age of the two lower sequences, the Appat and Kitsissut sequences.

On the basis of the foregoing, the Kangeq Sequence can be correlated as regards both age and lithology with the Markland Formation. If this is so, it is natural to correlate the Appat and Kitsissut sequences with the upper and lower members of the Bjarni Formation, respectively. Although resolution in the deeper parts of the old seismic sections off West Greenland is poor, it can be seen that the geometry and relations to faults are the same for the Appat Sequence and the upper Bjarni member. The way in which the Appat Sequence unconformably overlies the tilted Kitsissut Sequence is similar to the relations observed between the upper and lower Bjarni members in the southern part of the Labrador shelf (compare Fig. 12 with fig. 7.6 of Balkwill *et al.* 1990). However, the geometry of the Kitsissut Sequence suggests that this sequence is partially pre-rift, while the lower Bjarni member is regarded as syn-rift on the basis of conglomerates found in contact with fault surfaces. No information on the lithology of the Kitsissut Sequence is available, but the downlapping reflector configuration sometimes seen in this sequence is not what would be expected from Lower Paleozoic platform cover, the only alternative correlative in this region. Therefore, as a working hypothesis, the Kitsissut Sequence is correlated with the lower Bjarni member on the Labrador shelf. These correlations indicate that the age of the Kangeq Sequence is Cenomanian–Maastrichtian, the Appat Sequence Late Albian, and the Kitsissut Sequence Barremian–Early Albian.

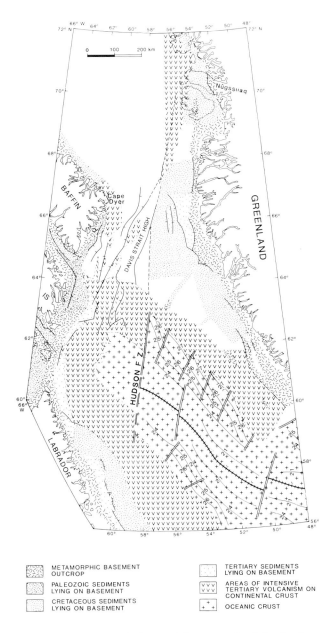

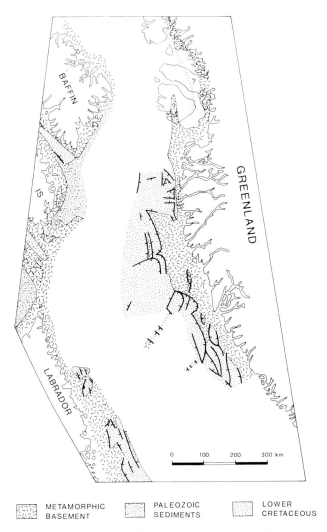

Fig. 13. Summary geological map of the present-day Labrador Sea. Onshore areas where Precambrian crystalline basement is exposed are shown blank, as are also offshore areas where no information is available. The extent of oceanic crust and sea-floor spreading anomalies are based on the interpretation that sea-floor spreading started in the Early Paleocene. Around the oceanic crust is an area which is interpreted as extended and subsided continental crust heavily intruded by, and overlain by, Early Tertiary igneous material. In the continental crust areas, the tones show the age of the *oldest* sediment interpreted to lie on basement. Areas of Tertiary volcanics are shown where they are the deepest events recognizable on seismic data; however, north of 68°N there are expected to be sediments between the basalts and the basement in many places. Other areas of basalts within sediments have been recognized and are shown on Fig. 16.

Fig. 14. Interpreted extent of Early Cretaceous basins in the Labrador Sea area plotted on a pre-spreading reconstruction of the position of Greenland relative to Canada. As on Fig. 13, onshore areas of Precambrian crystalline basement and offshore areas from which no information is available are shown without tone.

Geological development of the Labrador Sea and surrounding basins

The results reviewed above provide the background for a new interpretation of the development of the Labrador Sea and surrounding basins. Unfortunately, data are sparse or lacking in much of the Davis Strait area and most of Baffin Bay, so interpretations in these areas are still very speculative.

Early to Mid-Cretaceous (Fig. 14)

The earliest evidence of rifting is to be found in the Late Jurassic–Early Cretaceous coast-parallel dolerite dyke swarm in southern West Greenland. These dykes are confined to the outer coastal area, and dip seawards at 60° (Watt 1969). Confirmation of the age of these dykes must await the completion of an ongoing dating programme.

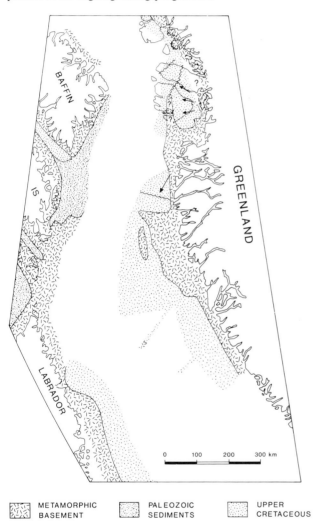

METAMORPHIC BASEMENT   PALEOZOIC SEDIMENTS   UPPER CRETACEOUS

Fig. 15. Interpreted extent of Late Cretaceous basins in the Labrador sea area using the same reconstruction as on Fig. 14.

Assuming that the correlation of the Kitsissut and Appat sequences off West Greenland with the lower and upper members of the Bjarni Formation is correct, fluvio-deltaic (pre- to syn-rift) sedimentation took place along the margins of both West Greenland and Labrador in the Early Cretaceous, starting in the Barremian or perhaps earlier. On the Labrador margin, eruption of basalts accompanied the earliest rift movements, but equivalent lavas have not been recognized on the West Greenland margin. In the onshore areas there is evidence of syn-rift sedimentation at Cape Dyer on Baffin Island in the

Late Aptian–Cenomanian, and farther north in the Bylot Island area deposition of non-marine sediments took place in active grabens in Albian–Cenomanian time (Miall *et al.* 1980). Onshore West Greenland there is no evidence of syn-rift sedimentation until the Cenomanian; on the contrary there is evidence in the south of the onshore area that the present-day boundary fault did not mark the eastern boundary of the area of fluvio-deltaic sedimentation in Albian time (Pulvertaft 1979).

Late Cretaceous (Fig. 15)

Regional marine conditions were established in the Labrador Sea area by the Cenomanian, probably as a result of the global rise in sea-level at this time. A cover of mudstone (Markland Formation, Kangeq Sequence) was deposited over basinal areas. In proximal situations deltas developed leading to deposition of the Freydis member on the inner Labrador margin and a south-southwestward prograding delta in the Sisimiut Basin off West Greenland. Much of the area does not appear to have been affected by rifting during deposition of the Kangeq Sequence and distal parts of the Markland Formation, subsidence taking place by thermal relaxation. Onshore West Greenland there appears to have been a steadily subsiding delta during the Cenomanian–Santonian; the oldest deep marine sediments exposed onshore are of Turonian age.

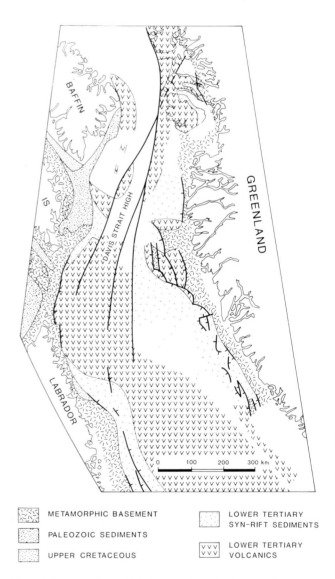

METAMORPHIC BASEMENT   LOWER TERTIARY SYN-RIFT SEDIMENTS

PALEOZOIC SEDIMENTS

UPPER CRETACEOUS   LOWER TERTIARY VOLCANICS

Fig. 16. Interpreted extent of Early Tertiary basins and volcanics in the Labrador Sea area using the same reconstruction as on Fig. 14.

Early Tertiary and start of sea-floor spreading (Fig. 16)

Evidence of renewed rifting along the West Greenland margin is best seen in onshore outcrops in West Greenland where faulting and erosion of the Cenomanian–Santonian deltaic succession probably started in late Maastrichtian time. Near Cape Dyer the deposition of the Early Paleocene Cape Searle Formation is thought to have accompanied faulting in this area (Burden and Langille 1990). Further evidence of regional Late Maastrichtian movement in the region is to be found further north, on Bylot Island, where an unconformity corresponding to Late Maastrichtian and Early Paleocene time has been documented (Miall *et al.* 1980).

Paleocene seismic sequences offshore West Greenland are affected by faulting, and it has been argued that renewed rifting took place throughout this period. The corresponding sediments in the Labrador Shelf are, however, assigned by Balkwill *et al.* (1990) to the drift-phase of sedimentation in this area.

The start of sea-floor spreading and the formation of the earliest oceanic crust in the northern Labrador Sea took place during chron 27N (Early Paleocene). Just before this, during 27R, eruption of voluminous picrite lavas started in West Greenland north of 69°N and near Cape Dyer. Volcanism continued at least until chron 25R (Late Paleocene) in West Greenland. In the offshore area of southern West Greenland Early Paleocene basalts are known from two wells (Hald and Larsen 1987), but seismic-stratigraphic correlation suggests that the volcanic rocks near the Kangâmiut-1 well may be Eocene. Volcanic rocks, probably of Paleocene age, have been interpreted all along the ocean–continent transition zone, both on the Greenland and on the Labrador side of the sea (Figs 2, 3, 7 and 16). The recognition of seaward-dipping reflectors indicates that for a while some of the volcanism was subaerial.

During the Eocene, sea-floor spreading continued, and sedimentation on the opposing shelves was syn-drift until a phase of erosion and relative uplift at the end of the Eocene or in earliest Oligocene time. When sedimentation resumed, spreading had ceased, and the continental shelves grew by simple seawards progradation. Off southern West Greenland the shelf was starved of Neogene sediments, and hence this part of the shelf is narrow today.

The interpretation shown in Figs 14–16 suggests that the Davis Strait area consists of continental crust overlain in places by Lower Tertiary volcanics. This conjecture is supported by reflection seismic data. The movement of Canada away from Greenland in the Early Tertiary was transferred from the Labrador Sea to Baffin Bay by strike-slip faulting— the Ungava fault system (Menzies 1982; Balkwill *et al.* 1990, fig. 7.19; Fig. 16). This strike-slip movement resulted in the formation of complex structures by both transpression and transtension. An example of a compressional structure believed to result from transpression occurs at SP 7500 on Fig. 2a.

Relation of the opening of the Labrador Sea to the opening of the North Atlantic

It is natural to seek a relationship between the opening of the Labrador Sea and the history of the North Atlantic, particularly since recent work on the Labrador Sea and the Greenland margin has suggested that the Labrador Sea opened much closer in time to the opening of the North Atlantic than previously supposed.

Both the opening of the Labrador Sea and the intense volcanism in the region started about 6 Ma earlier than the opening of the North Atlantic and volcanism in East Greenland (Upton 1988; time scale of Berggren *et al.* 1985). Indeed, volcanism in West Greenland had probably ceased by the time the first lavas in East Greenland were erupted (chron 24R,

Early Eocene). However, volcanism in West Greenland started at approximately the same time as that in northwest Scotland and Northern Ireland (Mussett *et al.* 1988).

The North Atlantic margins are volcanic rifted margins, and the Labrador Sea margins also show some of the features that characterize this type of margin. This fact, coupled with the contemporaneity of the initial opening of the respective seas with intense high-temperature (picrite) volcanism, suggests that there may be an underlying relationship between the two regions.

Petroleum prospectivity

During the 1970s, petroleum exploration was carried out on the continental shelf of southern West Greenland. Five wells were drilled, but no hydrocarbon discoveries were made. However, Chalmers and Pulvertaft (in press) have shown that no trapping configuration exists where four of the wells were drilled. Only in Kangâmiut-1 is there a potential reservoir, the ?Campanian sandstones underlying thick Paleocene shales with TOC values over 2%. This reservoir may, however, be breached, as it lies on the down-faulted side of a substantial fault (Fig. 9), along which hydrocarbons could have leaked.

Work carried out since 1987 has shown that the hydrocarbon potential offshore southern West Greenland is far from properly tested. In the first place, it has been shown that an area of more than 50 000 km^2 lying seawards of the continental slope is underlain by continental crust overlain by syn- and post-rift sedimentary sequences. This area was not investigated at all in the 1970s. Much of it lies under water depths between 700 and 950 m, which is within the reach of modern exploratory drilling. Secondly, a pattern of block-faulted and tilted basement and syn-rift sediments overlain by post-rift, syn-drift successions had been mapped over much of the area. The structural style resembles that in many petroleum-producing provinces in rift zones and continental margins elsewhere in the world. Finally, and perhaps most importantly, interpretation of seismic data has revealed seismic sequences older and deeper than any sampled by drilling in the 1970s. It is in these sequences that hopes of hydrocarbon discoveries now lie.

Source rocks

Rolle (1985) showed an average TOC over 2% in the Campanian–Paleocene (Ikermiut Formation) mudstones in the Ikermiut-1 and Kangâmiut-1 wells. It is likely that the Campanian–Maastrichtian mudstones in the Ikermiut-1 well are the equivalent of the upper part of the Kangeq Sequence described earlier in this paper. The character of the organic material in the Ikermiut Formation indicates that much of it originated from terrestial, woody plant debris. As such, it is a potential source rock for gas rather than liquid hydrocarbons.

Ottesen (1991*b*) has mapped the continuation of the Kangeq Sequence to the northeast of the Ikermiut-1 well site, into a south-southwesterly prograding structure interpreted as a major delta (Fig. 15). It is possible, therefore, that the Upper Cretaceous sediments penetrated by the Ikermiut-1 well were deposited in a delta front environment dominated by deposition of terrestrial organic matter. Farther south, off West Greenland, the Kangeq Sequence may have developed as a distal, deep water facies, possibly with a substantially higher quantity of oil-prone amorphous kerogen than farther north.

According to the correlation and dating of the Kangeq Sequence discussed in this paper, the lower part of the sequence is likely to be of Cenomanian–Turonian age, which is the age of a global oceanic anoxic event (Schlanger and Jenkyns 1976; Schlanger *et al.* 1987). Elsewhere in the world, prolific source rocks were deposited at this time (Klemme and Ulmishek 1991). Until the base of the Kangeq Sequence has been drilled, preferably in a distal situation, the source rock potential of offshore West Greenland will not have been tested.

The Appat Sequence probably consists of shales intercalated with sands, as is the case with its suggested correlative on the Labrador shelf. It is possible that some of these shales have source rock properties. The palaeolatitude of West Greenland during the Early Cretaceous was about 45°N (Smith *et al.* 1981). The climate was relatively warm and humid which would have favoured deposition of oil-prone source rocks with type I kerogen.

Direct evidence that generation of liquid hydrocarbons has taken place in the Labrador Sea–Baffin Bay region is provided by the oil seep that occurs off the northeast coast of Baffin Island (MacLean *et al.* 1981).

Source rock maturity

Data on thermal maturity (R_o TAI) and bottom hole temperatures at the five wells are given by Rolle (1985), and additional pyrolysis T_{max} values for two of the wells were provided by Chalmers and Pulvertaft (in press). Geodynamic modelling based on well data was carried out by Issler and Beaumont (1987), and recently optimization using the deterministic model of Welte and Yükler (1981) has been applied to these data.

The temperature and maturity in the wells is strongly influenced by lithology, with a low gradient in sandy units which dominate the upper part of the succession, and a steep gradient in the mudstone-dominated lower parts of the succession which were penetrated only in the Kangâmiut-1 and Ikermiut-1 wells. The wells which penetrated only the shallower sand-dominated successions (Hellefisk-1, Nukik-1 and -2) all terminated in the immature or early mature zone, whereas both the Ikermiut-1 and Kangâmiut-1 wells reach the zone of main oil generation. Typical threshold values for onset of oil generation (e.g. $R_o \geqslant 0.6\%$, $T_{max} \geqslant 435°C$) are reached at about 3700 m below sea-level in the Kangâmiut-1 well and at about 3200 m in the mudstone-dominated Ikermiut-1 well. The optimizing model indicates that the areas surrounding these two wells have a relatively narrow oil window, probably about 400 m, with peak generation conditions occurring within the Upper Cretaceous succession. The anticipated Cenomanian–Turonian source rocks can have generated oil since the Oligocene (Fig. 17).

Reservoirs and seals

The units discussed as potential source rocks appear also to be the only regionally extensive seals. Most shallower Tertiary sediments are more or less coarse clastic, and are also rather poorly structured.

Assuming that the Upper Cretaceous and Lower Tertiary mudstones form the regional seal, any reservoir must lie below them, within the Appat or Kitsissut sequences. The Kitsissut Sequence is expected to be the sandier of these two sequences, but it is only locally preserved. Where reflectors are seen in the Appat Sequence, this is likely to have sandstone intercalations. The equivalent units in the Bjarni Formation on the Labrador shelf were the main targets for petroleum exploration there; three of the gas discoveries and the only oil show are in Bjarni sandstone (Bell and Campbell 1990).

Play types

The most obvious play types identified offshore southern West Greenland are associated with rotated fault blocks produced by extension during either Early Cretaceous or Paleocene time. These can be found over many areas, including deep water areas (Fig. 4). Examples of prospects representative of these play types are shown in Fig. 18.

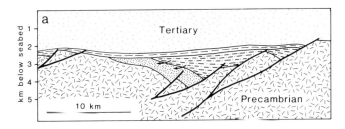

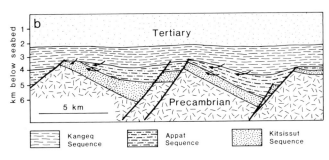

Fig. 18. Two examples of play types offshore southern West Greenland, based on actual examples. (**a**) Mudstones in the Appat Sequence and, if mature, also in the Kangeq Sequence are source, and both these sequences act as seal; sandstones in the Kitsissut Sequence are reservoir. (**b**) Mudstones in the Kangeq Sequence are source and seal; sandstones in the Kitsissut Sequence are reservoir and there may also be reservoir sandstones in the Appat Sequence.

The example shown in Figs 18a and 19 is in an area covered by seismic data from the 1970s, and has been mapped in three dimensions. Source rocks are shales in the Appat Sequence, reservoirs occur in the Kitsissut Sequence, and seal is provided by the Appat and Kangeq sequences. The gross rock volume (GRV) of the Kitsissut Sequence within the area of closure is $3.9 \times 10^9 \text{ m}^3$. Figures for the net to gross ratio and for porosity are unknown, but conservative assumptions of 50% and 20% give a pore space of $390 \times 10^6 \text{ m}^3$, which is equivalent to about 2.5 billion barrels of oil in place. The example shown in Fig. 18b is from an area with only widely spaced seismic lines, so the lateral extent of the structures is not known. In this

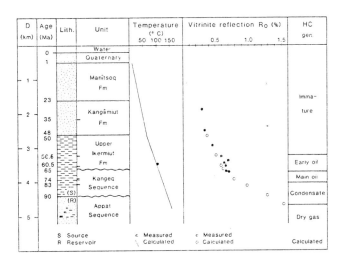

Fig. 17. Thermal modelling of a structural lead west of the Kangâmiut-1 well (cf. Fig. 9) showing depth versus temperature, vitrinite reflectance and hydrocarbon generation. Formation names from Rolle (1985). Based on preliminary modelling by the Danish Modelling Group at the Geological Survey of Denmark.

example sandstones in the Appat and Kitsissut sequences could provide the reservoir, and mudstone at the base of the Kangeq Sequence is both source and seal.

The example shown in Fig. 18b is smaller than many closures in the area, but has been chosen because it represents the structural style best and shows how the Kitsissut Sequence can by syn-rift in westerly fault blocks but pre-rift in easterly fault blocks, suggesting an easterly back-stepping of Cretaceous rifting.

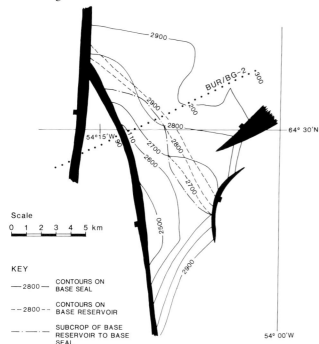

Fig. 19. Structural contour map of top and base reservoir (Kitsissut Sequence) in a structure possibly capable of trapping hydrocarbons. See Figs 12 and 18a for seismic and interpreted cross-sections, respectively.

Other possible play types are related to structures produced by strike-slip faulting. The area to the west of the Ikermiut-1 well could contain trapping structures created by transpression, and elsewhere there are structures believed to be related to transtension that could act as traps, but the current seismic data are too sparse for these to be mapped.

We would like to thank Bodil Sikker Hansen and Jette Halskov for draughting the figures, and Jens Nymose and Jakob Lautrup for photographic work. We would also like to thank Halliburton Geophysical Services, Inc. for permission to reproduce the seismic line shown in Fig. 11. Funding for the reprocessing of the BGR seismic data was provided by the Danish Ministry of Energy Research Programmes, projects 1313/89-5 and 1313/90-0013. We thank K. Hinz of BGR for allowing us access to the field tapes of these seismic data. The paper is published with permission from the Geological Survey of Greenland.

References

BALKWILL, H. R. 1987. Labrador Basin: structural and stratigraphic style. *In*: BEAUMONT, C. and TANKARD, A. J. (eds) *Sedimentary basins and basin-forming mechanisms*. Canadian Society of Petroleum Geologists, Memoir, **12**, 17–43.

——, MCMILLAN, N. J., MACLEAN, B., WILLIAMS, G. L. AND SRIVASTAVA, S. P. 1990. Geology of the Labrador Shelf, Baffin Bay and Davis Strait. *In*: KEEN, M. J. AND WILLIAMS, G. L. (eds) Geology of the continental margins of eastern Canada. *Geology of Canada*, **2**, 293–348.

BELL, J. S. AND CAMPBELL, G. R. 1990. Petroleum resources. *In*: KEEN, M. J. AND WILLIAMS, G. L. (eds) Geology of the continental margins of eastern Canada. *Geology of Canada*, **2**, 677–720.

BERGGREN, W. A., KENT, D. V. AND FLINN, J. J. 1985. Jurassic to Paleogene: Part 2 Paleogene geochronology and chronostratigraphy. *In*: SNELLING, N. J. (ed.) *The chronology of the geologic record*. Geological Society, London, Memoir, **10**, 141–195.

BIRKELUND, T. 1965. *Ammonites from the Upper Cretaceous of West Greenland*. Meddelelser om Grønland, 179, 7 (Bulletin Grønlands Geologiske Undersøgelse, 56).

BURDEN, E. T. AND LANGILLE, A. B. 1990. Stratigraphy and sedimentology of Cretaceous and Paleocene strata in half-grabens on the southeast coast of Baffin Island, Northwest Territories. *Bulletin of Canadian Petroleum Geology*, **38**, 185–195.

CHALMERS, J. A. 1989. A pilot seismo-stratigraphic study on the West Greenland continental shelf. *Rapport Grønlands Geologiske Undersøgelse*, **142**.

—— 1991. New evidence on the structure of the Labrador Sea/Greenland continental margin. *Journal of the Geological Society, London*, **148**, 899–908.

—— in press. Project VEST SOKKEL Phase I. A seismic stratigraphic interpretation of the geology of the continental shelf of southern West Greenland between 64°15′ N and 66°N. *Rapport Grønlands Geologiske Undersøgelse*.

—— AND PULVERTAFT, T. C. R. in press. The southern West Greenland continental shelf—was petroleum exploration abandoned prematurely? *In*: VORREN, T. O. *et al.* (eds) *Arctic geology and petroleum potential*. Special Publication of the Norwegian Petroleum Society, No. 2. Elsevier Amsterdam.

CHRISTIANSEN, F. G. (ed.) 1989. *Petroleum geology of North Greenland*. Bulletin Grønlands Geologiske Undersøgelse, **158**.

——, DAM, G., MCINTYRE, D., NØHR-HANSEN, H., PEDERSEN, G. K. AND SØNDERHOLM, M. 1992. Renewed petroleum geological studies onshore West Greenland. *Rapport Grønlands Geologiske Undersøgelse*, **155**, 31–35.

CLARKE, D. B. AND PEDERSEN, A. K. 1976. Tertiary volcanic province of West Greenland. *In*: ESCHER, A. AND WATT, W. S. (eds) *Geology of Greenland*. Geological Survey of Greenland, 365–385.

—— AND UPTON, B. G. J. 1971. Tertiary basalts of Baffin Island: field relations and tectonic setting. *Canadian Journal of Earth Sciences*, **8**, 248–258.

ELDHOLM, O., THIEDE, J. AND TAYLOR, E. 1989. Evolution of the Vøring volcanic margin. *In*: ELDHOLM, O., THIEDE, J. AND TAYLOR, E. *et al.* Proceedings of the Ocean Drilling Program, **104**, College Station, TX (Ocean Drilling Program), 1033–1065.

HALD, N. AND LARSEN, J. G. 1987. Early Tertiary, low-potassium tholeiites from exploration wells on the West Greenland shelf. *Rapport Grønlands Geologiske Undersøgelse*, **136**.

HENDERSON, G., SCHNIENER, E. J., RISUM, J. B., CROXTON, C. A. AND ANDERSEN, B. B. 1981. The West Greenland Basin. *In*: KERR, J. W. AND FERGUSSON, A. J. (eds) *Geology of the North Atlantic borderlands*. Canadian Society of Petroleum Geologists, Memoir, **7**, 399–428.

HINZ, K. 1981. A hypothesis on terrestrial catastrophies. Wedges of very thick oceanward dipping layers beneath passive continental margins—their origin and paleoenvironmental significance. *Geologisches Jahrbuch*, **E22**.

——, SCHLÜTER, H.-U., GRANT, A. C., SRIVASTAVA, S. P., UMPLEBY, D. AND WOODSIDE, J. 1979. Geophysical transects of the Labrador Sea: Labrador to southwest Greenland. *Tectonophysics*, **59**, 151–183.

ISSLER, D. R. AND BEAUMONT, C. 1987. Thermal and subsidence history of the Labrador and West Greenland continental margins. *In*: BEAUMONT, C. AND TANKARD, A. J. (eds) *Sedimentary basins and basin-forming mechanisms*. Canadian Society of Petroleum Geologists, Memoir, **12**, 45–69.

KLEMME, H. D. AND ULMISHEK, G. F. 1991. Effective petroleum source rocks of the World: stratigraphic distribution and controlling depositional factors. *American Association of Petroleum Geologists Bulletin*, **75**, 1809–1851.

LARSEN, L. M. AND REX, D. C. 1992. A review of the 2500 Ma span of alkaline-ultramafic, potassic and carbonatitic magmatism in West Greenland. *Lithos*, **28**.

MACLEAN, B., FALCONER, R. K. H. AND LEVY, E. M. 1981. Geological, geophysical and chemical evidence for a natural seepage of petroleum off the northeast coast of Baffin Island. *Bulletin of Canadian Petroleum Geology*, **29**, 75–95.

MCALPINE, K. D. 1990. *Mesozoic stratigraphy, sedimentary evolution, and petroleum potential of the Jeanne d'Arc Basin, Grand Banks of Newfoundland*. Geological Survey of Canada, Paper **89-17**.

MENZIES, A. W. 1982. Crustal history and basin development of Baffin Bay. *In*: DAWES, P. R. AND KERR, J. W. (eds) Nares Strait and the drift of Greenland: a conflict in plate tectonics. Meddelelser om Grønland, *Geoscience*, **8**, 295–312.

MIALL, A. D., BALKWILL, H. R. AND HOPKINS, W. S. 1980. *Cretaceous and Tertiary sediments of Eclipse Trough, Bylot Island area, Arctic Canada, and their regional setting*. Geological Survey of Canada, Paper, **79-23**.

MILLER, K. G. AND TUCHOLKE, B. E. 1983. Development of Cenozoic abyssal circulation south of the Greenland–Scotland Ridge. *In*: BOTT, M. H. P., SAXOV, S., TALWANI, M. AND THIEDE, J. (eds) *Structure and development of the Greenland–Scotland Ridge*. Plenum Press, New York, 549–589.

MORTON, A. C. AND PARSON, L. M. (eds) 1988. *Early Tertiary volcanism and the opening of the NE Atlantic*. Geological Society, London, Special Publication, **39**.

MUSSETT, A. E., DAGLEY, P. AND SKELHORN, R. R. 1988. Time and duration of activity in the British Tertiary Province. *In*: MORTON, A. C. AND PARSON, L. M. (eds) *Early Tertiary volcanicm and the opening of the NE Atlantic*. Geological Society, London, Special Publication, **39**, 337–348.

OTTESEN, T. G. 1991*a*. *A preliminary seismic stratigraphic study of the Paleocene–Eocene section offshore southern West Greenland between 66° and 68°N*. Grønlands Geologiske Undersøgelse, Open File Series, **90/1**.

—— 1991*b*. *A preliminary seismic study of the pre-Paleocene section offshore southern West Greenland between 66°N and 68°N*. Grønlands Geologiske Undersøgelse, Open File Series, **91/6**.

PEDERSEN, G. K. 1989. A fluvial-dominated lacustrine delta in a volcanic province, W Greenland. *In*: WHATELEY, M. K. G. AND PICKERING, K. T. (eds) *Delta: sites and traps for fossil fuels*. Geological Society, London, Special Publication, **41**, 139–146.

—— AND PEEL, J. S. 1985. Ordovician(?) gastropods from cherts in Cretaceous sandstones, south-east Disko, central West Greenland. *Rapport Grønlands Geologiske Undersøgelse*, **125**, 30–33.

—— AND PULVERTAFT, T. C. R. 1992. The nonmarine Cretaceous of the West Greenland basin. *Cretaceous Research*, **13**, 263–272.

PIASECKI, S., LARSEN, L. M., PEDERSEN, A. K. AND PEDERSEN, G. K. 1992. Palynostratigraphy of the Lower Tertiary volcanics and marine sediments in the southern part of the West Greenland basin: implications for the timing and duration of the volcanism. *Rapport Grønlands Geologiske Undersøgelse*, **154**, 13–31.

POULSEN, V. 1966. An occurrence of Lower Palaeozoic rocks within the Precambrian terrain near Sukkertoppen. *Rapport Grønlands Geologiske Undersøgelse*, **11**, 26.

PULVERTAFT, T. C. R. 1979. Lower Cretaceous fluvial-deltaic sediments at Kûk, Nûgssuaq, West Greenland. *Bulletin of the Geological Society of Denmark*, **28**, 57–72.

—— 1989. *The geology of Sarqaqdalen, West Greenland, with special reference to the Cretaceous boundary fault system*. Grønlands Geologiske Undersøgelse, Open File Series, **89/5**.

ROEST, W. R. AND SRIVASTAVA, S. P. 1989. Sea-floor spreading in the Labrador Sea: A new reconstruction. *Geology*, **17**, 1000–1003.

ROLLE, F. 1985. Late Cretaceous–Tertiary sediments offshore central West Greenland: lithostratigraphy, sedimentary evolution, and petroleum potential. *Canadian Journal of Earth Sciences*, **22**, 1001–1019.

ROOTS, W. D. AND SRIVASTAVA, S. P. 1984. Origin of the marine magnetic quiet zones in the Labrador and Greenland Seas. *Marine Geophysical Researches*, **6**, 395–408.

ROSENKRANTZ, A. AND PULVERTAFT, T. C. R. 1969. *Cretaceous–Tertiary stratigraphy and tectonics in northern West Greenland*. American Association of Petroleum Geologists, Memoir, **12**, 883–898.

SCHIENER, E. J. 1976. West Greenland coal deposits: distribution and petrography. *Rapport Grønlands Geologiske Undersøgelse*, **77**.

SCHLANGER, S. O., ARTHUR, M. A., JENKYNS, H. C. AND SCHOLLE, P. A. 1987. The Cenomanian–Turonian oceanic anoxic event, 1. Stratigraphy and distribution of organic carbon-rich beds and the marine $\delta^{13}C$ excursion. *In*: BROOKS, J. AND FLEET, A. J. (eds) *Marine petroleum source rocks*. Geological Society, London, Special Publication, **26**, 371–399.

—— AND JENKYNS, H. C. 1976. Cretaceous oceanic anoxic events: causes and consequences. *Geologie en Mijnbouw*, **55**, 179–184.

SMITH, A. G., HURLEY, A. M. AND BRIDEN, J. C. (eds) 1981. *Phanerozoic palaeocontinental world maps*. Cambridge Earth Science Series, Cambridge.

SMYTHE, D. K. 1983. Faeroe–Shetland escarpment and continental margin north of the Faeroes. *In*: BOTT, M. H. P., SAXOV, S., TALWANI, M. AND THIEDE, J. (eds) *Structure and development of the Greenland–Scotland Ridge*. Plenum Press, New York, 109–119.

SRIVASTAVA, S. P. 1978. Evolution of the Labrador Sea and its bearing on the early evolution of the North Atlantic. *Geophysical Journal of the Royal Astronomical Society*, **52**, 313–357.

STOUGE, S. AND PEEL, J. S. 1979. Ordovician conodonts from the Precambrian Shield of southern West Greenland. *Rapport Grønlands Geologiske Undersøgelse*, **91**, 105–109.

TUCHOLKE, B. E. AND FRY, V. A. 1985. Basement structure and sediment distribution in Northwest Atlantic Ocean. *American Association of Petroleum Geologists Bulletin*. **69**, 2077–2097.

UPTON, B. G. J. 1988. History of Tertiary igneous activity in the N Atlantic borderlands. *In*: MORTON, A. C. AND PARSON, L. M. (eds) *Early Tertiary volcanism and the opening of the NE Atlantic*. Geological Society, London, Special Publication, **39**, 429–453.

WATT, W. S. 1969. The coast-parallel dike swarm of Southwest Greenland in relation to the opening of the Labrador Sea. *Canadian Journal of Earth Sciences*, **6**, 1320–1321.

WEAVER, F. H. AND MACKO, S. A. 1987. Source rocks of Western Newfoundland. *Organic Geochemistry*, **13**, 411–421.

WELTE, H. AND YÜKLER, M. A. 1981. Petroleum origin and accumulation in basin evolution—a quantitative model. *American Association of Petroleum Geologists Bulletin*. **65**, 1387–1396.

Discussion

Question (P. Rowe, Total Oil Marine, Aberdeen):

1. Can you give me source rock age, proven (i.e. well) thickness and any geochemical attributes which have been determined in the West Greenland offshore Cretaceous (or references in the literature thereto)?

2. Is there any potential Cretaceous source rock known from the onshore contiguous area?

Answer (F. G. Christiansen):

The question of age, distribution and potential of oil-prone Cretaceous source rocks is very important for the assessment of the prospectivity of southern West Greenland and has been raised by a number of companies that are considering future exploration in the region.

Only two of the wells from the 1970s reached Cretaceous shales, and only the interval Campanian–Maastrichtian was penetrated. Thus models for source rock distribution are at present mainly hypothetical based on general basin development or analogue studies from especially the Labrador Shelf and areas farther south, and from the outcropping onshore sediments in the Disko–Nûgssuaq–Svartenhuk area, West Greenland.

The Campanian–Maastrichtian shales exceed 800 m in the Ikermiut-1 well amd 300 m on the north coast of Nûgssuaq where, however, they are presumed to have a total thickness of at least 600 m. The TOC values are fairly high, typically between 2% and 3% in the well (Rolle 1985; unpublished data) and between 3% and 6% in outcrop (unpublished data). The Hydrogen Index values are low, typically between 40 and 80, pointing towards a gas potential only. However, all of the localities studied are from a setting in front of major deltas with a considerable input of terrestrially-derived organic matter.

Marine shales of Cenomanian–Santonian age have not been penetrated in any of the offshore wells and are only known from scattered outcrops on Svartenhuk. Neither top nor base of this succession on Svartenhuk is well exposed but the thickness is estimated to be a few hundred metres. The few old unpublished data indicate TOC values between 5% and 10%. This succession is one of the main targets for GGU's shallow core (max. 100 m) programme during the 1992 expedition to the area which will be followed by detailed geochemical analyses. Seismic interpretation suggests that marine shales from this age interval may reach a thickness of between 1 and 1.5 km in distal, offshore settings. Based on global analogues and models of major anoxic events (e.g. Schlanger and Jenkyns 1976; Schlanger *et al.* 1987) the Cenomanian–Turonian marine shales are inferred as the most promising oil-prone source interval in West Greenland.

Non-marine oil-prone source rocks of Cretaceous age may be present as lacustrine shales or hydrogen-rich coals within the Appat sequence, which is a possible equivalent to the upper member of the Bjarni Formation of the Labrador Shelf and the Kome, Upernivik Næs, and Atane formations, onshore West Greenland. Most of the coals from Disko and Nûgssuaq are vitrinite-rich with a gas potential only (Schiener 1976) but some hydrogen-rich coals (TOC: *c.* 40%, HI: *c.* 250) have been discovered recently on the north coast of Disko (unpublished data). It should also be noted that resin lumps have been recognized in the fluvial deposits of the Atane Formation.

Pre-Cretaceous source rocks offshore southern West Greenland may also exist, especially within a Lower Paleozoic or Jurassic succession. Lower Paleozoic deposits are well described from the Labrador Shelf and there are a few indications that they also may be present offshore West Greenland. Lower Paleozoic marine shales with a good to excellent source potential such as those reported from North Greenland (Christiansen 1989) or west of Newfoundland (Weaver and Macko 1987) might also be present in the deeper basins.

The globally widespread Upper Jurassic source rocks are known from the Jeanne d'Arc Basin on the Grand Banks of Newfoundland (McAlpine 1990) but their northern extension is not known. Marine Upper Jurassic shales with a source potential still have to be documented on the Labrador Shelf and off southern West Greenland, but the possibility of their existence should not be overlooked.

Late Mesozoic–Cenozoic evolution of the southwestern Barents Sea

J. I. FALEIDE, E. VÅGNES and S. T. GUDLAUGSSON

Department of Geology, University of Oslo, PO Box 1047, Blindern, N-0316 Oslo, Norway

Abstract: Three geological provinces are recognized, separated by major fault zones: the oceanic Lofoten Basin and the Vestbakken volcanic province in the west; the southwestern Barents Sea basin province; and the eastern region which has largely acted as a stable platform since Late Paleozoic times. Since Middle Jurassic times, two structural stages are recognized in the southwestern Barents Sea: Late Mesozoic rifting and basin formation; and Early Tertiary rifting and opening of the Norwegian–Greenland Sea. This evolution reflects the main plate tectonic episodes in the North Atlantic–Arctic break-up of Pangea. Middle–Late Jurassic and Early Cretaceous structuration were characterized by regional extension accompanied by strike-slip adjustments along old structural lineaments, which developed as the Bjørnøya, Tromsø and Harstad basins. Late Cretaceous development was more complex, with extension west of the Senja Ridge and the Veslemøy High, and halokinesis in the Tromsø Basin. Tertiary structuration was related to the two-stage opening of the Norwegian–Greenland Sea and the formation of the predominantly sheared western Barents Sea continental margin. Tectonic activity shifted towards the west in successive phases. The southwestern Barents Sea basin province developed within the De Geer Zone in a region of rift-shear interaction. Initially, oblique extension linked the Arctic and North Atlantic rift systems (Middle Jurassic–Early Cretaceous). Later, a continental megashear developed (Late Cretaceous–Paleocene), and finally a sheared-rifted margin formed during the opening of the Norwegian–Greenland Sea (Eocene–Recent).

The southwestern Barents Sea (Fig. 1) contains some of the deepest sedimentary basins known. They formed in response to phases of regional tectonism in the North Atlantic–Arctic culminating in continental separation of Eurasia and Greenland and the accretion of oceanic crust in early Tertiary times. Several models have been proposed to explain the Late Mesozoic and Cenozoic evolution (Rønnevik *et al.* 1982; Rønnevik and Jacobsen 1984; Berglund *et al.* 1986; Riis *et al.* 1986; Sund *et al.* 1986; Ziegler *et al.* 1986; Brekke and Riis 1987; Gabrielsen and Færseth 1988) but there is little consensus with respect to timing and the importance and direction of strike-slip movements. Regional palaeogeographic and tectonic syntheses (Doré and Gage 1987; Ziegler 1988; Doré 1991) do not account for the structuring of the southwestern Barents Sea in detail. In this study, we aim to establish the structural framework of the southwestern Barents Sea region and to describe the Late Mesozoic–Cenozoic geological evolution of the area. Faleide *et al.* (in press) combine observations in the southwestern Barents Sea with correlation of tectonic and stratigraphic events in the North Atlantic–Arctic to postulate regional models for each major tectonic phase since Middle Jurassic times.

Geological setting

The Barents Sea covers the northwestern corner of the Eurasian continental shelf (Fig. 1). It is bounded by young passive margins to the west and north resulting from the Cenozoic opening of the Norwegian–Greenland Sea and the Eurasia Basin. The western Barents Sea is underlain by large thicknesses of Upper Paleozoic to Cenozoic rocks and comprises three distinct regions.

- The continental margin consists of three segments: a southern sheared margin along the Senja Fracture Zone; a central rifted complex southwest of Bjørnøya associated with volcanism; and a northern sheared and rifted margin along the Hornsund Fault Zone. The continent–ocean transition occurs over a narrow zone along the line of early Tertiary break-up and is covered by a thick Upper Cenozoic sedimentary wedge.
- The Svalbard Platform covered by flat-lying Upper Paleozoic and Mesozoic, mainly Triassic, strata.

- A province between the Svalbard Platform and the Norwegian coast characterized by sub-basins and highs with increasingly accentuated structural relief to the west. Jurassic, Cretaceous and locally Paleocene strata are preserved in the basins.

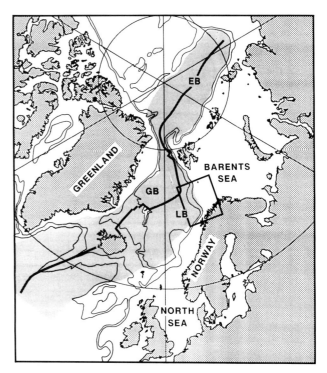

Fig. 1. Location of the southwestern Barents Sea within the North Atlantic–Arctic region. Study area within the box. Present plate boundary shown by thick line and 1000 m and 2000 m water depths by thin lines. Cenozoic oceanic crust is indicated by shading. EB: Eurasia Basin; GB: Greenland Basin; LB: Lofoten Basin.

The regional geology of the western Barents Sea is well known (Rønnevik *et al.* 1982; Rønnevik and Jacobsen 1984; Faleide *et al.* 1984; Riis *et al.* 1986; Gabrielsen *et al.* 1990). Post-Caledonian geological history was dominated by three rift phases: Late Devonian?–Early Carboniferous; Middle Jur-

From *Petroleum Geology of Northwest Europe: Proceedings of the 4th Conference* (edited by J. R. Parker).
© 1993 Petroleum Geology '86 Ltd. Published by The Geological Society, London, pp. 933–950.

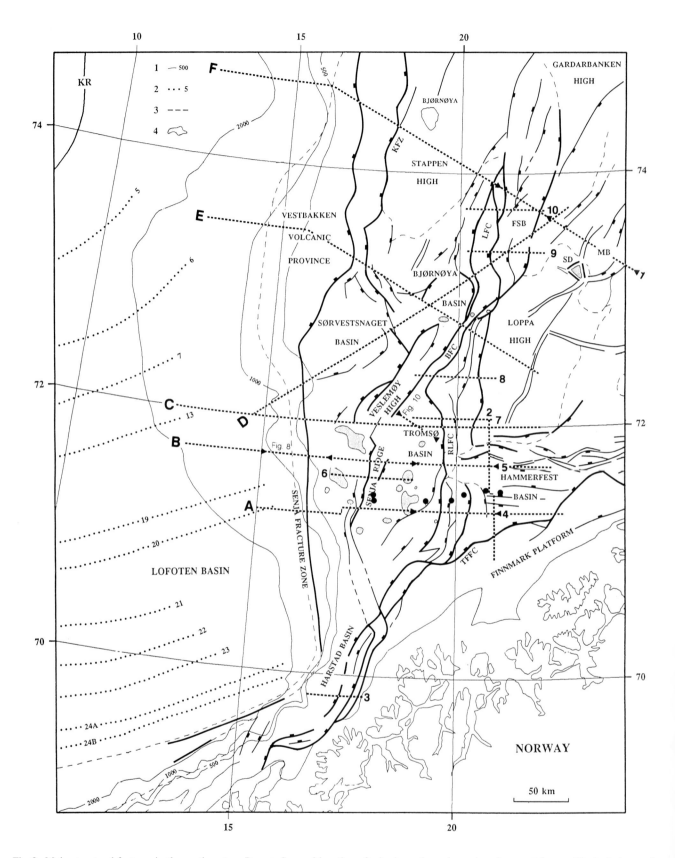

Fig. 2. Main structural features in the southwestern Barents Sea and location of seismic sections shown in subsequent figures. Black circles mark the wells in Fig. 6. 1: bathymetry (m); 2: magnetic lineations; 3: limit of identified oceanic crust in the seismic sections; 4: salt. AFC: Asterias Fault Complex; BFC: Bjørnøyrenna Fault Complex; FSB: Fingerdjupet Sub-basin; KFZ: Knølegga Fault Zone; KR: Knipovich Ridge; LFC: Leirdjupet Fault Complex; MB: Maud Basin; RLFC: Ringvassøy–Loppa Fault Complex; SD: Svalis Dome; TFFC: Troms–Finnmark Fault Complex.

assic–Early Cretaceous; and Tertiary; each comprising several tectonic pulses. During Late Paleozoic times most of the Barents Sea was affected by crustal extension. Later rifting migrated westward, with rifts and pull-apart basins in the southwest and the development of a belt of strike-slip faults in the north. Apart from epeirogenic movements which have produced the present elevation differences, the Svalbard Platform and the eastern part of the basin province have been largely stable since Late Paleozoic times.

Within the study area, three main geological provinces, separated by major fault zones, can be recognized (Figs 2, 3, 4 and 5):

- The oceanic Lofoten Basin formed during the Cenozoic opening of the Norwegian–Greenland Sea and the Vestbakken volcanic province.
- The southwestern Barents Sea basin province of deep Cretaceous and early Tertiary basins (Harstad, Tromsø, Bjørnøya

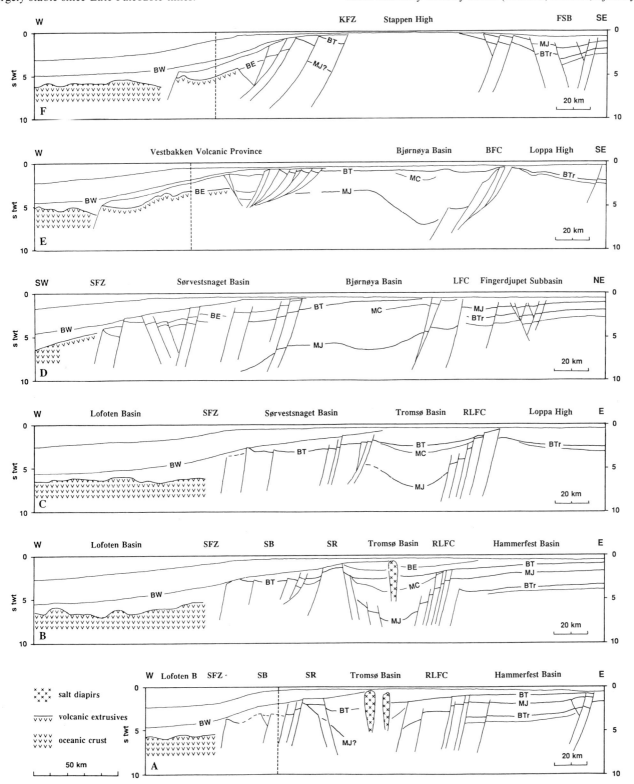

Fig. 3. Regional seismic sections. Location in Fig. 2 and reflector code in Fig. 7. BFC: Bjørnøyrenna Fault Complex; FSB: Fingerdjupet Subbasin; KFZ: Knølegga Fault Zone; LFC: Leirdjupet Fault Complex; RLFC: Ringvassøy–Loppa Fault Complex; SB: Sørvestsnaget Basin; SFZ: Senja Fracture Zone; SR: Senja Ridge; VVP: Vestbakken volcanic province.

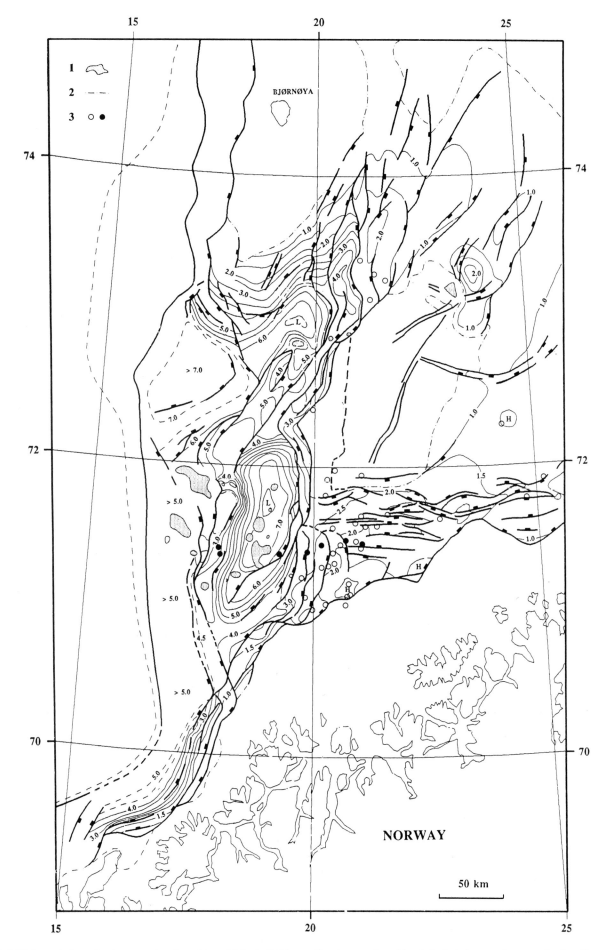

Fig. 4. Mid-Jurassic time-structure map (contour interval 0.5 s twt). The continent–ocean transition is included as a reference only. 1: salt; 2: termination of the Mid-Jurassic reflector; 3: location of wells (black circles mark the wells in Fig. 6).

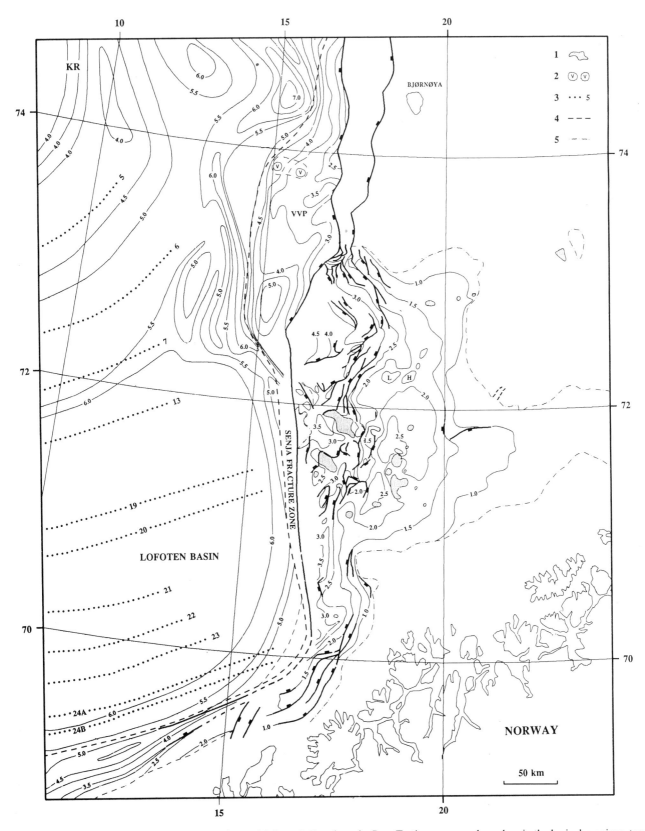

Fig. 5. Base Tertiary time-structure map (contour interval 0.5 s twt). Based on the Base Tertiary sequence boundary in the basinal province, top of Eocene basalts in the Vestbakken volcanic province (VVP) and top of oceanic basement. 1: salt; 2: Mid-Tertiary volcanoes; 3: magnetic lineations; 4: limit of identified oceanic crust in the seismic sections; 5: termination of the Base Tertiary reflector; KR: Knipovich Ridge.

and Sørvestsnaget basins) separated by intra-basinal highs (Senja Ridge, Veslemøy High and Stappen High).
- Mesozoic basins and highs further east between 20–25°E, which have not experienced the pronounced Cretaceous/Tertiary subsidence (Finnmark Platform, Hammerfest Basin, Loppa High, Fingerdjupet Sub-basin).

These provinces are separated by the continental boundary faults along the Senja Fracture Zone and the eastern boundary of the Vestbakken volcanic province, and by the main Jurassic–Cretaceous faults bounding the deep Cretaceous basins. These are the Troms–Finnmark Fault Complex south of 71°N, Ringvassøy–Loppa Fault Complex, Bjørnøyrenna Fault Com-

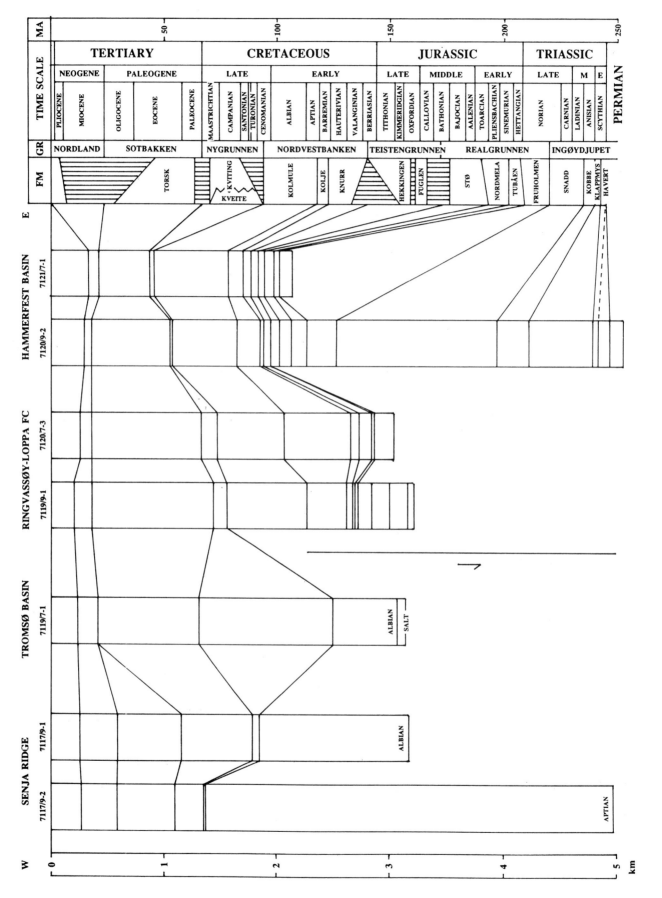

Fig. 6. Well correlation along a west–east transect from the Senja Ridge to the Hammerfest Basin. Location of wells in Fig. 2.

plex and Leirdjupet Fault Complex (Fig. 2). The three provinces are also reflected in the deep crustal configuration. Moho depths in the southwestern Barents Sea basin province typically range between 20–25 km, clearly different from the oceanic crust to the west and the +30 km deep Moho east of the province (Jackson et al. 1990; Faleide et al. 1991).

Data

The present interpretation is based on 40 000 km of multichannel seismic reflection lines, 4000 km of deep seismic profiles, and seismic velocities from 25 expanding spread and 150 sonobouy profiles (Jackson et al. 1990). Forty-seven wells have been drilled in the western Barents Sea basin province, most in the Hammerfest Basin (Fig. 2) where the Mesozoic and Cenozoic successions are well known. Twenty-two of the wells, drilled in 1980–84, have been released by the Norwegian Petroleum Directorate (NPD) as well data summary sheets. Well information is also presented in studies from the Hammerfest Basin (Westre 1984; Grung Olsen and Hansen 1987) and from shallow drilling on the Finnmark Platform, Loppa High, Svalis Dome and Stappen High (Fig. 2) (Mørk et al. 1989).

Seismic interpretation

The Barents Sea sedimentary cover exceeds 15 km in places. Crystalline basement is only rarely recognized on the seismic data while oceanic basement is normally observed west of the continent–ocean boundary. The seismic data demonstrate the existence of thick, relatively undeformed sedimentary sequences of great horizontal extent (Fig. 3) and stratigraphic interpretation and correlation over large distances is often straightforward. In the southwestern Barents Sea, Late Mesozoic–Cenozoic tectonic movements have disturbed the simple pattern and seismic correlation from one structural element to another is locally difficult. In the Tromsø Basin, pre-Tertiary strata are isolated from the surrounding basins and highs by major tectonic boundaries across which correlation is only tentative. Well information is also scarce, but some Upper Cretaceous and Tertiary horizons have been dated from wells in the Tromsø Basin and the Senja Ridge (Fig. 6). Other horizons are less certain. Mid-Jurassic and Early Cretaceous sequence boundaries have been correlated across the Ringvassøy–Loppa Fault Complex to Hammerfest Basin wells. Some of the Tertiary sequence boundaries in the west have been dated by their termination on dated oceanic basement. The seismic stratigraphic framework is shown in Fig. 7.

Structure

Oceanic basin

The Lofoten Basin developed by Cenozoic sea-floor spreading in the Norwegian–Greenland Sea. The oceanic crust is of approximately normal thickness (4.5–6 km) but has an anomalously high velocity adjacent to the continent–ocean boundary, with normal values further seaward (Jackson et al. 1990). The high velocity indicates high-density crust giving rise to large, elongate, positive, free-air gravity anomalies seaward of the Senja Fracture Zone (Eldholm et al. 1987). Linear magnetic sea-floor spreading anomalies terminate against the Senja Fracture Zone (Fig. 2), and show that the oceanic crust becomes younger northwards along the sheared margin. The oceanic basement is overlain by 5–7 km of middle and upper Cenozoic strata (Fig. 3).

Continent–ocean transition

The continent–ocean transition along the Senja Fracture Zone is generally narrow and distinct (Faleide et al. 1991). The oceanic basement in the Lofoten Basin can be followed to about 10 km from a prominent continental boundary fault along the Senja Fracture Zone (Figs 2, 3 and 8). East of the fault a steeply dipping reflector (Fig. 8), continuing to a depth of about 20 km, is correlated with the base of the crust (Faleide et al. 1991).

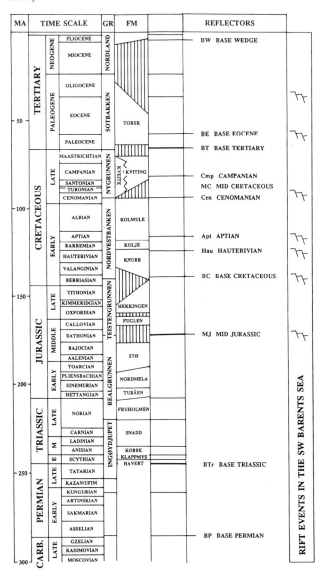

Fig. 7. Stratigraphic summary. Main seismic sequences and reflectors related to the lithostratigraphic framework of Worsley et al. (1988) and rift events in the southwestern Barents Sea basin province.

The Vestbakken volcanic province is an igneous complex which is elevated with respect to the adjacent oceanic crust and masks the continent–ocean boundary at the rifted margin segment (Figs 2 and 3). It is characterized by a smooth acoustic basement surface and some sub-basement reflectors, particularly below its inner part (Faleide et al. 1988; Myhre and Eldholm 1988). By analogy with the Vøring Marginal High (Eldholm et al. 1989), we believe it comprises subaerial flow basalts extruded during break-up in earliest Eocene times. The province includes an outer oceanic part and a stretched continental part, both covered by flows and interbedded sediments. Renewed extensional faulting and volcanism related to the earliest Oligocene change in relative plate motion have partly overprinted the break-up structures (Faleide et al. 1988; 1991).

The landward boundary of the Vestbakken volcanic province is a listric fault complex formed in early Tertiary times (Figs 2 and 3). Locally, flows are recognized within the fault

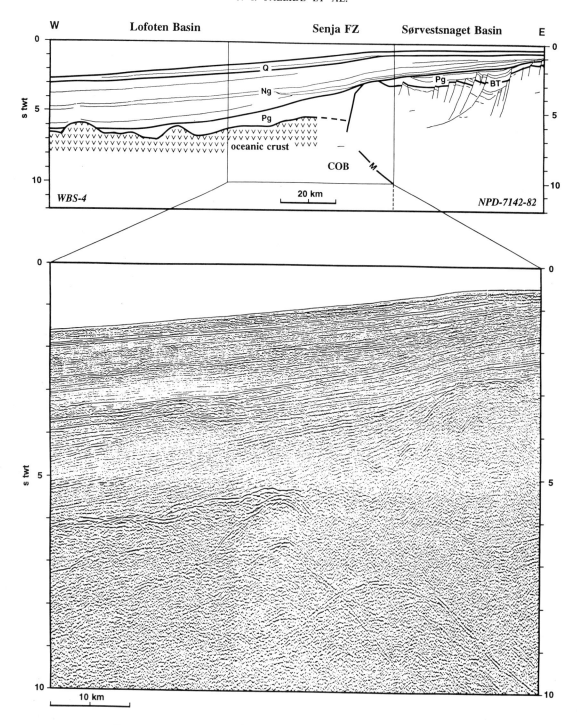

Fig. 8. Deep seismic line across the Senja Fracture Zone. Location in Fig. 2. BT: Base Tertiary; COB: Continent–ocean boundary; M: Moho; Ng: Neogene; Pg: Paleogene; Q: Quaternary.

complex, revealing large-scale vertical movements in Eocene–Oligocene times (Faleide *et al.* 1988). Down-faulting in a pull-apart setting exceeds 2 km along the northern rifted segments west of the Stappen High, diminishing towards the southwest where the volcanic province bounds the Bjørnøya Basin (Fig. 2).

Tertiary marginal basin

The Sørvestsnaget Basin (Fig. 2) was affected by major tectonism during Tertiary break-up. The basin configuration was controlled by older structures underlying the intra-basinal highs (Senja Ridge, Veslemøy High, Stappen High). Seismic profiles reveal a complex history of Cenozoic vertical motions,

sedimentation, and erosion. Structural deformation associated with the early Tertiary opening along the Senja Fracture Zone is evident west of the Senja Ridge, where it caused uplift of the continental crust along the developing transform. The main deformation took place during the initial break-up in earliest Eocene times, but some faults show activity probably as late as the Oligocene. About 1 km of Paleogene sediments were eroded from the elevated outer margin, shedding terrigenous material into the ocean basin. The Paleogene sequence thins and locally pinches out towards an outer high which forms an inner flank to the Senja Fracture Zone (Figs 3 and 8). During Oligocene times the margin became tectonically quiet. Regional subsidence, combined with widespread uplift and erosion of the Barents Sea province in the east, resulted in the

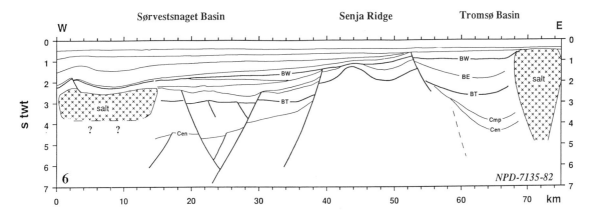

Fig. 9. Interpreted seismic section across the Senja Ridge showing the correlation of the Upper Cretaceous and Tertiary sequences in the Sørvestsnaget and Tromsø basins. Location in Fig. 2 (profile 6) and reflector code in Fig. 7.

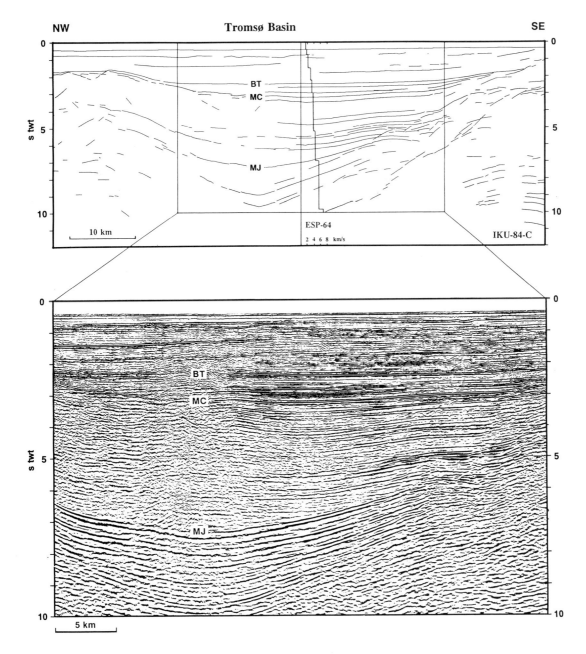

Fig. 10. Deep seismic line in the northern Tromsø Basin. Location in Fig. 2 and reflector code in Fig. 7.

construction of the huge post-Oligocene sedimentary wedge.

The pre-Cenozoic sediments in the Sørvestsnaget Basin correlate with those in the Tromsø Basin. Both Upper and Lower Cretaceous appear to be well developed, although thinner than in the Tromsø Basin. Positive features west of the Senja Ridge have been related to compression, possibly in combination with mobilization of Lower Cretaceous shales which enhanced the structuring (Riis *et al.* 1986; Brekke and Riis 1987). However, we interpret these features as salt structures buried and partly reactivated by the Late Cenozoic wedge: they have negative gravity anomalies and local rim-synclines filled with thick Upper Cretaceous sequences similar to the salt structures in the Tromsø Basin (Fig. 9). The Tertiary strata in the northern Sørvestsnaget Basin are controlled by lineaments that are sub-parallel to the rifted and sheared margin segments (Fig. 5). The most prominent is a NE-trending fault along the Veslemøy High and a NW-trending fault zone separating the basin from the Bjørnøya Basin (Figs 2, 3, 7 and 8). The Tertiary basin west of these faults formed in a pull-apart setting.

Cretaceous basins

The Harstad, Tromsø and Bjørnøya basins underwent large-scale Cretaceous subsidence and sedimentation. A tentative Middle Jurassic reflector has been mapped below 7 s twt (12–14 km) in the deepest part of the Tromsø Basin (Fig. 10) (Gudlaugsson *et al.* 1987: Jackson *et al.* 1990). There is little direct information about the pre-Middle Jurassic basin history, although the deep seismic reflection and refraction profiles indicate a considerable section of pre-Middle Jurassic strata. If a 6.7–7.0 km s^{-1} velocity represents crystalline basement (Jackson *et al.* 1990), then a depth of 18–20 km in the Tromsø Basin corresponds to the deepest continuous reflector (9–9.5 s twt) identified by Gudlaugsson *et al.* (1987) (Fig. 10). Regional considerations (Rønnevik and Jacobsen 1984; Faleide *et al.* 1984) and the presence of salt in the Tromsø and Sørvestsnaget basins (Fig. 9) suggest that the pre-Middle Jurassic sequence probably includes thick Triassic and Jurassic clastic sediments as well as Permo-Carboniferous mixed carbonates, evaporites and clastics. The base of the crust is not identified in these basins, but the gravity field indicates a pronounced crustal thinning beneath the Early Cretaceous rifts (Breivik 1991; Faleide *et al.* 1991).

It is difficult to establish seismic ties to the undrilled Harstad Basin (Fig. 11), but we are confident that a Middle Jurassic rift unconformity can be followed to more than 5 s twt, below thick Cretaceous and Tertiary sequences. The basin-fill is deformed by several phases of predominantly extensional faulting, which probably started in Middle Jurassic and continued during the major subsidence in Early Cretaceous times. Large-scale listric faults, particularly in the southern part of the basin, are associated with hangingwall roll-over anticlines. Renewed normal faulting took place in the Late Cretaceous and Tertiary (Fig. 11).

The NNE-trending Tromsø Basin contains salt diapirs along its axis (Figs 3, 9 and 12). It may have existed as a separate basin during salt deposition in Late Paleozoic times but later merged with the Bjørnøya Basin. The basins developed separately from Late Cretaceous times, when lateral movements took place along the Bjørnøyrenna Fault Complex (Figs 3 and 13) (Gabrielsen *et al.* 1990). The Tromsø Basin evolved mainly in response to Late Jurassic–Early Cretaceous extension. Salt withdrawal in the basin centre locally enhanced the subsidence, particularly during Late Cretaceous times (Figs 9 and 12).

The Bjørnøya Basin is divided by the Leirdjupet Fault Complex into a deep western part and a shallow eastern part, the Fingerdjupet Sub-basin (Figs 3 and 14). Most of the basin-fill is Early Cretaceous in age and the basin appears to be a large half-graben down-faulted along the Bjørnøyrenna Fault Complex (Fig. 3). However, the half-graben geometry may have been accentuated by inversion and tilting when the Stappen High was uplifted in the Tertiary.

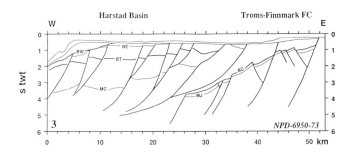

Fig. 11. Interpreted seismic section across the southern part of the Troms–Finnmark Fault Complex and the Harstad Basin. Location in Fig. 2 (profile 3) and reflector code in Fig. 7.

Intra-basinal highs

The intra-basinal highs are among the least understood features in the southwestern Barents Sea. We have grouped them on the basis of similar structural and geophysical signature, but expect that further investigations may reveal substantial differences. The highs were active during several tectonic phases, but became positive features within the Cretaceous basin province, mainly by Late Cretaceous and Early Tertiary faulting and differential subsidence.

The Senja Ridge separates the Tromsø Basin from the Sørvestsnaget Basin (Figs 2 and 3) and is bounded to the west by normal faults active during multiple extensional events. To the east there are fewer and smaller faults. A core of shallow basement just beneath the Tertiary sequence has been suggested to explain the prominent positive gravity anomaly over the ridge (Riis *et al.* 1986; Syrstad *et al.* 1976). However, the depth to basement varies considerably along the ridge and is shallowest near seismic line 7142 (Fig. 3). Both the seismic reflection and refraction data indicate a succession of Mesozoic and possibly Upper Paleozoic sedimentary rocks within parts of the ridge complex. The Senja Ridge has been interpreted in terms of at least two phases of compression, terminating in Late Cretaceous and post-Paleocene times, respectively (Riis *et al.* 1986; Brekke and Riis 1987). Although some transpressional deformation may be involved, we suggest that the Late Cretaceous–Early Tertiary normal faulting and salt mobilization in the Tromsø and Sørvestsnaget basins was responsible for the relief of the ridge with respect to the adjacent basins at Mid-Cretaceous and younger levels (Fig. 9).

The Veslemøy High, earlier considered as the northern part of the Senja Ridge (Gabrielsen *et al.* 1984), is now defined as a separate structural element (Gabrielsen *et al.* 1990). It separates and offsets the Tromsø and Bjørnøya basins (Figs 3 and 4) and is associated with major deep-seated west-facing faults forming a southerly continuation of the Bjørnøyrenna Fault Complex. Relatively thick Lower Cretaceous sediments are present within the high, suggesting some continuity between the Tromsø and Bjørnøya basins prior to Late Cretaceous and Tertiary structuration. However, a prominent positive gravity anomaly over the high indicates shallower basement than in the adjacent basins.

The formation of the Senja Ridge and the Veslemøy High as positive structural elements within the Cretaceous basin province has been related to strike-slip faulting along the Bjørnøyrenna Fault Complex. Riis *et al.* (1986) suggested sinistral shear during the main Early Cretaceous deformation phase to explain compressional faulting and folding in the Senja Ridge. However, Gabrielsen and Færseth (1988) claimed

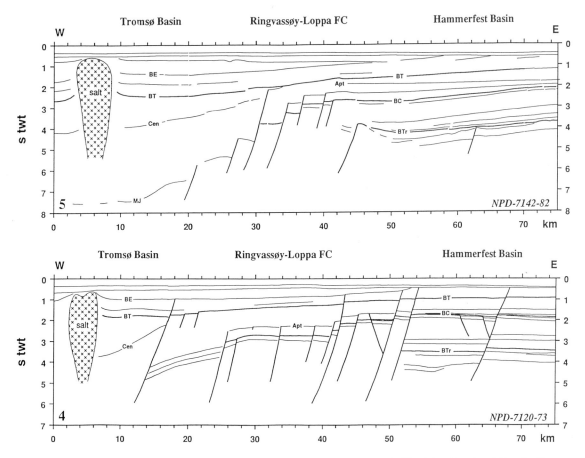

Fig. 12. Interpreted seismic sections across the Ringvassøy–Loppa Fault Complex. Location in Fig. 2 (profiles 4 and 5) and reflector code in Fig. 7.

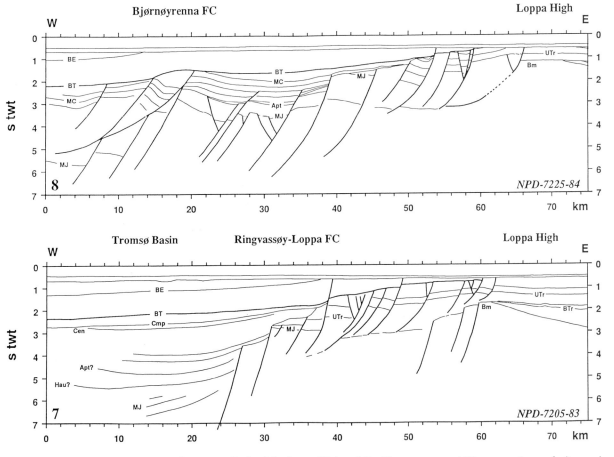

Fig. 13. Interpreted seismic sections across the western flank of the Loppa High and the Bjørnøyrenna and Ringvassøy–Loppa fault complexes. Location in Fig. 2 (profiles 7 and 8) and reflector code in Fig. 7.

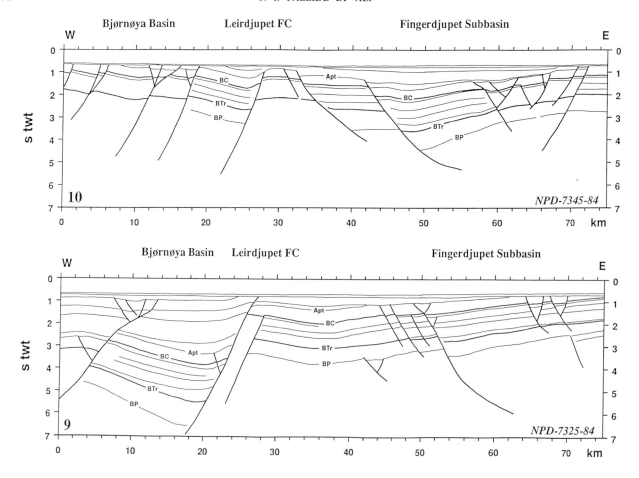

Fig. 14. Interpreted seismic sections across the Leirdjupet Fault Complex separating the main Bjørnøya Basin from the Fingerdjupet Sub-basin. Location in Fig. 2 (profiles 9 and 10) and reflector code in Fig. 7.

that this model fails to explain the compressional structures and suggested dextral transpressional strike-slip movements. Nevertheless, the compressional features are Late Cretaceous in age (Brekke and Riis 1987; Gabrielsen *et al.* 1990). Although the local post-Cenomanian/pre-late Paleocene compressional deformation at the transition between the Senja Ridge and Veslemøy High may indicate a small lateral component, the western boundary faults of both highs are indicative of normal faulting during Late Cretaceous times.

The Stappen High surrounding Bjørnøya (Figs 2 and 3) was part of a north–south-trending elevated area in the western Barents Sea from Late Paleozoic to Jurassic times. Local Cretaceous subsidence and Tertiary uplift were directly associated with activity along the De Geer Zone, and the southern flank of the high was formed by Tertiary inversion of part of the Bjørnøya Basin. Moretti *et al.* (1988) regarded the Stappen High as the eroded crest of a large, rotated fault block which also encompassed the Bjørnøya Basin. However, the down-faulting along the Bjørnøyrenna Fault Complex is mainly Early Cretaceous in age, while most of the uplift and erosion occurred in the Tertiary. The erosion is estimated as 3.5 km at Bjørnøya, decreasing eastwards to 1–2 km (Wood *et al.* 1990). The first phase of Tertiary uplift and tilting is probably related to thermal effects associated with the break-up and early opening of the Norwegian–Greenland Sea. We believe the uplift was initiated in response to early Eocene rifting and volcanism, rather than the earliest Oligocene change in spreading direction as suggested by Wood *et al.* (1990). The Stappen High acted as provenance area for the thick Eocene sequences in the Vestbakken volcanic province and the northern Sørvestsnaget Basin (Fig. 3). Subsequently, the down-faulting of the volcanic province, and renewed volcanism, probably

coincided with the change in relative plate motion. A considerable amount of footwall uplift associated with this event is likely. Finally, the area experienced a phase of post-Oligocene, mainly Pliocene–Pleistocene, uplift which is part of a regional event affecting the entire western Barents Sea.

Cretaceous boundary faults

The eastern extensional boundary faults developed mainly in Early Cretaceous times (Figs 2, 3, 11, 12, 13 and 14). Probable Caledonian basement at shallow depths in the western Loppa High (Gudlaugsson *et al.* 1987) implies a basement relief of as much as 15–16 km between the deep basin province and the stable region to the east. The structural style varies along-strike with listric faulting dominating the southern Troms–Finnmark Fault Complex (Fig. 11) and the Bjørnøyrenna Fault Complex (Fig. 13). The main phase of subsidence started in Middle Jurassic and culminated in Early Cretaceous (Aptian–Albian) times. There is evidence of reactivation in Late Cretaceous times, locally with a slight compressional component, and even Tertiary strata are affected by faulting (Gabrielsen *et al.* 1990). The southern segment, bounding the Harstad Basin, has been considered as part of the Troms–Finnmark Fault Complex (Gabrielsen *et al.* 1990), but in view of the Mesozoic rifting and basin formation, it should rather be considered as the southern continuation of the Ringvassøy–Loppa Fault Complex (Fig. 2). It is characterized by NE- to NNE-trending rotated fault blocks and several listric faults, anticlines and synclines are related to a fault-ramp detachment zone at depth. Steep basement-involved faults are locally observed below the detachment (Fig. 11). The Ringvassøy–Loppa Fault Complex (Figs 2, 3, 12 and 13) comprises rotated fault blocks forming a

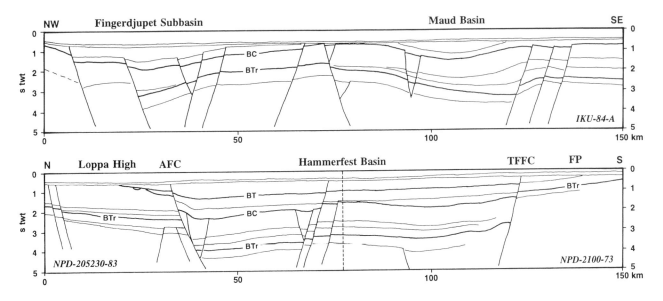

Fig. 15. Interpreted seismic sections crossing basins and highs in the eastern area which have not experienced the pronounced Cretaceous/Tertiary subsidence. Location in Fig. 2 (profiles 1 and 2) and reflector code in Fig. 7. AFC: Asterias Fault Complex; FP: Finnmark Platform; TFFC: Troms Finnmark Fault Complex.

terrace which cross-cuts the Hammerfest Basin (Gabrielsen 1984). Farther north the Bjørnøyrenna Fault Complex (Figs 2, 3 and 13) is recognized by listric faults which appear to flatten into a detachment possibly within Permian rocks. The Leirdjupet Fault Complex (Figs 2, 3 and 13) is a single large fault in the south, associated with flexures and drag phenomena. To the north, it splits into smaller rotated fault blocks.

Eastern platform region

The eastern platform region—the Finnmark Platform, the Hammerfest and eastern Bjørnøya basins and the Loppa High (Fig. 2)—was affected by Jurassic rifting without large-scale post-rift subsidence (Fig. 15). The Hammerfest Basin includes deep, high-angle faults along the basin flanks and listric normal faults detached above or within the Permian sequence in the basin centre. The structuring of the basin was dominated by extension, perhaps with minor strike-slip deformation in Late Jurassic to Early Cretaceous times (Berglund *et al.* 1986; Sund *et al.* 1986; Gabrielsen and Færseth 1988, 1989). A gentle dome parallel to the basin-axis developed during Middle Jurassic–Early Cretaceous rifting (Fig. 15). The doming ended in early Barremian time when the rifting ceased due to tectonic decoupling along the Ringvassøy–Loppa Fault Complex. Onlap of Aptian–Albian reflectors onto the eroded Loppa High indicate that uplift took place along the Asterias Fault Complex (Fig. 15) in Early Cretaceous times.

The Fingerdjupet Sub-basin is characterized by a horst and graben pattern (Figs 2, 3, 14 and 15), generated by Late Jurassic rifting with local reactivation during the Early Cretaceous subsidence.

The Loppa High (Figs 2, 3 and 13) consists of an eastern platform and a western part underlain by shallow Caledonian? metamorphic basement. The western crest has been rejuvenated at least four times since the Devonian, but the present high is a result of post-Middle Jurassic tectonism (Gabrielsen *et al.* 1990). The rotated fault blocks bounding the western crest of the high apparently formed in Late Jurassic to Early Cretaceous times, but have been reactivated later (Fig. 13). Most of the faults appear listric with a detachment below Triassic level. During most of the Cretaceous, the Loppa High

was an island with deep canyons cutting into the Triassic sequence, supplying submarine fans in the basins to the south and west. The erosion events were a response to footwall uplift along the western boundary faults and were possibly amplified by lateral heat transfer from the developing rift basins in the south and west. Seismic subcrop patterns, abnormally high interval velocities and well data indicate that more than 1000 m of Paleogene mudstones were deposited on the Loppa High, and were mostly eroded during later uplift (Wood *et al.* 1990).

Stratigraphy

We have calibrated the seismic stratigraphy with wells in the Hammerfest Basin (Fig. 6) and related the seismic sequences to the main lithostratigraphic units (Worsley *et al.* 1988) (Fig. 7). In the southwestern Barents Sea basin province, Middle Jurassic is the deepest sequence boundary tentatively correlated to the Hammerfest Basin (Figs 3 and 10). However, a pre-Middle Jurassic succession similar to the surrounding areas is probably also present.

Pre-Middle Jurassic

The released wells available at present only penetrate to Permian level. Permo-Carboniferous rocks in the region are expected to be similar to those of Svalbard, Bjørnøya and Northeast Greenland. They comprise three megasequences which are mappable throughout the Barents Sea (Stemmerik and Worsley 1989). The deep seismic reflection and refraction data (Gudlaugsson *et al.* 1987; Jackson *et al.* 1990; Faleide *et al.* 1991) and the presence of Late Carboniferous/earliest Permian salt (Well 7119/7-1; P. Mason, pers. comm.) indicate that the Upper Paleozoic succession extends into the southwestern Barents Sea basin province. Thick Triassic rocks occur throughout the Barents Sea and comprise coarsening-upwards sequences indicating transgressive–regressive depositional cycles (Mørk *et al.* 1989).

The Lower–Middle Jurassic interval is dominated by sandstones which are present throughout the Hammerfest Basin probably thickening into the Tromsø Basin. These sediments probably also covered the Loppa High and Finnmark Platform but were partly eroded during later tectonic activity. Middle Jurassic sandstones form the main reservoir in the Hammerfest Basin (Olaussen *et al.* 1984; Berglund *et al.* 1986; Grung Olsen and Hanssen 1987). The top of the sandstones gives rise to a seismic marker (reflector MJ) which can be followed over large areas of the southern Barents Sea, including the deep Harstad, Tromsø and Bjørnøya basins.

Middle–Upper Jurassic

Major unconformities bound the Teistengrunnen Group, which ranges from late Callovian to late Berriasian in age, although a late Bathonian/early Callovian interval may be present locally (Worsley *et al.* 1988) (Figs 6 and 7). The basal unconformity (reflector MJ) marks the onset of rifting in the southwestern Barents Sea, whereas unconformities within the group reflect interplay between continued Late Jurassic faulting and sea-level changes. The group is so thin that it is generally impossible to resolve these unconformities on the seismic data. It is thinnest on the structural highs in the central Hammerfest Basin and thickens towards the basin boundary faults to the north, south and west. The shales and claystones contain thin interbedded marly dolomitic limestone and rare siltstones or sandstones toward the basin flanks, reflecting relatively deep and quiet marine environments (Worsley *et al.* 1988).

Lower Cretaceous

The Nordvestbanken Group comprises three formations from Valanginian to Cenomanian (Figs 6 and 7). Thickness variations in the Hammerfest Basin are related to basin-parallel structures with the thinnest sequences over the central dome. The group thickens towards the Ringvassøy-Loppa Fault Complex (Fig. 6) and to the north and south before onlapping against the Loppa High and Finnmark Platform, respectively. Shales and claystones dominate, with thin interbeds of siltstone, limestone and dolomite. The marine environments throughout the group are dominated by distal conditions with periodic restricted bottom circulation. Clastic fans built out from the emergent Loppa High, while the Finnmark Platform was a much less pronounced feature (Worsley *et al.* 1988). The lowermost, mainly Hauterivian, Knurr Formation (Figs 6 and 7) probably reflects tectonic events both at its base and top, but it is difficult to resolve the sequence boundaries on the seismic data. The regional Base Cretaceous (BC) reflector corresponds to the top Hauterivian. The overlying Barremian Kolje Formation thickens westwards into the Tromsø Basin (Fig. 6). The top represents a seismic marker in most of the area and has been dated as Top Barremian in the Hammerfest Basin (Westre 1984; Berglund *et al.* 1986; Grung Olsen and Hanssen 1987), but the wells show that a thin Aptian sequence, when present, belongs to the Kolje Formation. Therefore, we have assigned an Aptian age to this sequence boundary (reflector Apt, Fig. 7). The upper, mainly Albian, Kolmule Formation thins westwards in the Hammerfest Basin but is very thick in the Tromsø Basin (Fig. 6).

Upper Cretaceous

The Nygrunnen Group varies in thickness and completeness (Figs 6 and 7). In the Tromsø Basin a 1200 m shale succession has been drilled (Fig. 6) while seismic data indicate the sequence reaches 2000–3000 m in rim synclines in the central basin (Figs 3, 9 and 12). Wells on the Senja Ridge show a thin Upper Cretaceous sequence (Fig. 6), reflecting Late Cretaceous structuring and strong salt-related subsidence in the Tromsø and Sørvestsnaget basins (Fig. 9). Whereas the Tromsø Basin was a depocentre through most of this period, areas further east were either transgressed only during maximum sea-level and/or display only condensed sections (Worsley *et al.* 1988). Wells show that the Campanian interval thins from 250 m to less than 50 m eastward in the Hammerfest Basin (Fig. 6). Thus, the Upper Cretaceous sequence is difficult to resolve on the seismic data east of the Ringvassøy–Loppa Fault Complex and we have only interpreted the Base Tertiary sequence boundary. Claystones with thin limestone stringers in the Tromsø Basin and western part of the Hammerfest Basin change into more calcareous or sandy condensed sequences to the east, with open marine, deep shelf environments in the Tromsø Basin and a shallower starved shelf in the east (Worsley *et al.* 1988).

Paleogene

The Sotbakken Group rests unconformably on the Nygrunnen Group and this depositional break at the Cretaceous–Tertiary transition (Maastrichtian–Danian) occurs throughout the southwestern Barents Sea (Worsley *et al.* 1988). The preserved sequences show a late Paleocene (Thanetian) to early/middle Eocene (Ypresian/Lutetian) age in the Hammerfest Basin but younger, Eocene–Oligocene? sequences may also be present further west (Spencer *et al.* 1984). The sediments are dominated by claystones and interbedded thin siltstones, tuffs and carbonates deposited on an open to deep marine shelf. Seismic data suggest that the lower part is probably present throughout the southwestern Barents Sea with little lithologic variation, and that the younger sequences are preserved only in the Tromsø, Harstad and Sørvestsnaget basins (Figs 3 and 12). In the deepest part of the Tromsø Basin the group attains a thickness of more than 2000 m.

Neogene–Quaternary

The Neogene and Quaternary Nordland Group, resting unconformably on Paleogene and Mesozoic rocks, thickens dramatically in the huge sedimentary wedge at the margin (Figs 3 and 8). The sediments are dated as Late Pliocene to Pleistocene/Holocene in the Hammerfest Basin, where there is little evidence for Miocene and Oligocene sediments (Worsley *et al.* 1988). A more complete succession, extending back to the Miocene in Well 7117/9-1 on the Senja Ridge (Fig. 6), was reported by Spencer *et al.* (1984) who cautioned about dating problems due to reworking. Recently, most of the wedge at the Senja Ridge has been interpreted as being of Late Pliocene/ Pleistocene glacial origin by Eidvin and Riis (1989). This interpretation implies that the glacial sediments, which are typically 100–200 m thick in the Hammerfest Basin, increase to more than 700 m at the Senja Ridge (Fig. 6) and expand to about 4000 m in the Lofoten Basin (Figs 3 and 8).

Geological evolution

The structural evolution of the southwestern Barents Sea since Middle Jurassic time comprises two main stages: Late Mesozoic rifting and basin formation; and Early Tertiary rifting and opening of the Norwegian–Greenland Sea. Several events related to the major plate tectonic movements in the North Atlantic–Arctic region can be recognized:

- Middle–Late Jurassic rifting;
- Early Cretaceous rifting and decoupling of the Harstad, Tromsø and Bjørnøya basins from the basins and highs further east;
- Late Cretaceous–Paleocene faulting in the Harstad and Sørvestsnaget basins;
- Early Eocene opening of the Norwegian–Greenland Sea,

formation of the Vestbakken volcanic province, and extensional faulting and uplift along the Senja Fracture Zone;
- Oligocene structural rejuvenation along parts of the continental margin.

Middle–Late Jurassic

This was a period of regional extension and minor strike-slip adjustments along old lineaments. In the Barents Sea, there is little evidence for Jurassic faulting prior to Middle Jurassic time. A Bathonian–Callovian hiatus separating Middle Jurassic sandstones and Upper Jurassic shales marks the onset of Cimmerian tectonics (reflector MJ, Fig. 7) which initiated the structural differentiation in Fig. 2. The Middle–Late Jurassic phase rifted the Barents Sea through the Hammerfest and Bjørnøya basins along the pre-existing tectonic grain, causing block faulting along east and northeast trends and deposition of Upper Jurassic shales in restricted basins between tilted fault blocks. We believe the subsidence of the Tromsø and Bjørnøya basins was initiated at this time. In the Hammerfest Basin, a thin Teistengrunnen Group (Fig. 6) hinders seismic resolution of Late Jurassic tectonic events, but several closely spaced unconformities in the wells may indicate tectonic movements and sea-level changes. The sequence thins towards the central Hammerfest Basin where doming occurred in response to movements on the basin boundary faults to the north and south (Fig. 15).

Early Cretaceous

At the transition from Late Jurassic to Early Cretaceous times, major rifting combined with a presumably tectonically induced lowstand in relative sea-level affected the entire North Atlantic. The North Atlantic rift basins continued into the southwestern Barents Sea where the Lower Cretaceous section is much expanded in the Harstad, Tromsø and Bjørnøya basins (Figs 2, 3, 11, 12, 13 and 14). At least three tectonic phases affected the area during Early Cretaceous time. The first two phases, Berriasian/Valanginian and Hauterivian/Barremian, also affected the Hammerfest Basin but cannot be resolved on the seismic data. The termination of doming in earliest Barremian time marked the cessation of active rifting in the Hammerfest Basin (Fig. 15). The Tromsø and Bjørnøya basins were probably affected more strongly by these two tectonic phases. The Barremian (Kolje Formation) thickens gradually westwards in the Hammerfest Basin and into the Ringvassøy–Loppa Fault Complex (Figs 3 and 12), reflecting a period of thermal subsidence centred in the Tromsø Basin. Aptian/Albian tectonic activity is evident in the Ringvassøy–Loppa Fault Complex where the Aptian horizon is faulted (Fig. 12). These faults cut the east–west tectonic grain of the Hammerfest Basin and form part of a major hinge zone separating the rapidly subsiding basins in the west from more stable areas in the east. A sequence of high-amplitude reflectors in the northern Tromsø Basin, tentatively correlated with the Aptian, thins and diminishes westwards in the basin (Fig. 13), probably reflecting deep water clastic fans derived from the uplifted Loppa High. The Aptian event is also recognized in the western Hammerfest Basin by uplift and thinning of the Kolmule Formation towards the Ringvassøy–Loppa Fault Complex (Figs 3 and 12). The Leirdjupet Fault Complex to the north experienced footwall uplift and erosion as suggested by the sedimentary unit onlapping a tilted fault block in the Bjørnøya Basin (Fig. 14). This event also affected the adjacent Fingerdjupet Sub-basin. The final Early Cretaceous rift phase was followed by rapid subsidence and infill of the Bjørnøya, Tromsø and Harstad basins by a 5–6 km thick sequence (Kolmule Formation) which covered most of the structural relief by Cenomanian times.

The Early Cretaceous structuring of the southwestern Barents Sea was characterized by extensional faulting with large downthrow to the west, although a minor wrench component was probably present. The Early Cretaceous depocentres in the northern Tromsø Basin and central Bjørnøya Basin, together with the geometry of the Veslemøy High and Senja Ridge (Fig. 4), indicate large-scale extension with an orientation between west and west-northwest, inducing sinistral transtensional strike-slip along the Bjørnøyrenna Fault Complex. The tectonic link between the depocentres is difficult to delineate, but it may run through the saddle separating the Senja Ridge and the Veslemøy High (Fig. 4). Minor dextral strike-slip components were associated with the north-trending fault complexes.

The structuring of the southwestern Barents Sea was coeval with both North Atlantic rifting and the opening of the Amerasia Basin. It appears, however, that the Arctic link was limited to earliest Cretaceous time while the North Atlantic rifting became increasingly important throughout Early Cretaceous time. Moreover, the southern Barents Sea was decoupled from the Greenland Shelf by oblique extension along the De Geer Zone linking the Arctic and North Atlantic rifts (Faleide et al. in press). The thick Lower Cretaceous section in the Harstad, Tromsø and Bjørnøya basins is the result of crustal stretching and thinning within this zone. The rift diminishes to the north where numerous faults are observed in the Fingerdjupet Sub-basin. Further north the extent of the faulting is difficult to establish due to lack of post-Jurassic strata. The extension possibly died out towards the north, but part may have been taken up along a north-northwest-trending fault zone crossing the Svalbard Platform.

Late Cretaceous–Paleocene

This period was dominated by the opening of the Labrador Sea and regional subsidence centred along the North Atlantic rift basins. The trend of these deep, wide, Late Cretaceous basins terminated at the De Geer Zone where dextral oblique-slip produced pull-apart basins in the Wandel Sea Basin and in the southwestern Barents Sea (Faleide et al. in press).

Deep Late Cretaceous basins are observed both in the Sørvestsnaget Basin (Fig. 9) and in the Harstad Basin which is located at the rift-shear intersection (Fig. 11). An important Santonian (86–87 Ma) phase of uplift and faulting in Andøya (Dalland 1981) is probably a further manifestation of activity on the De Geer Zone. The Tromsø and Sørvestsnaget basins, separated by the Senja Rige, continued to subside during the Late Cretaceous while the areas to the east are characterized by a condensed sequence. At the Veslemøy High and in the Bjørnøya Basin, Late Cretaceous extension is manifested by numerous normal faults with throws less than 300 m. An ENE-striking normal fault with a 750–1500 m throw towards the northwest runs along the northwestern border of the Veslemøy High and a similar fault marks the western boundary of the Senja Ridge. Local Late Cretaceous compressional deformation (Gabrielsen et al. 1900) indicates minor wrench components along some of the larger faults, but extension was still dominant and part of the differential subsidence may relate to salt movements in the Tromsø and Sørvestsnaget basins (Fig. 9).

Following a hiatus covering the Cretaceous–Tertiary transition a relatively uniform and widespread sequence was deposited as a sheet covering the entire western Barents Sea in late Paleocene times. Towards the end of the Paleocene the Loppa High became a source area for sediments prograding into the Tromsø Basin and local faulting took place along the western flank of the Senja Ridge.

Eocene

Following the onset of sea-floor spreading in the Norwegian–Greenland Sea and the Eurasia Basin at the Paleocene–Eocene transition, the western Barents Sea margin developed as a shear margin within the De Geer Zone (Faleide *et al.* 1991). Between anomaly 25/24 and anomaly 13 times (earliest Eocene to earliest Oligocene), regional shear between the Norwegian Sea and Eurasia Basin initiated the formation of the continental margins of the Barents Sea and Svalbard and the opening of the southern Greenland Sea. The direction of early opening was at a small angle with the Senja Fracture Zone and the Hornsund Fault Zone (Talwani and Eldholm 1977), causing transform movements with transtensional and transpressional components; the Senja Fracture Zone may have developed as a leaky transform due to transtension (Reksnes and Vågnes 1985; Eldholm *et al.* 1987). At the same time, transpression caused folding and thrusting in western Spitsbergen (Dallmann *et al.* 1988). The central margin segment, however, was subject to rifting and volcanism followed by down-faulting in a pull-apart setting (Faleide *et al.* 1988; 1991).

The Tertiary sediment distribution (Fig. 5) and the main structural elements in the southwestern Barents Sea (Figs. 2 and 3) reflect these events, particularly the Sørvestsnaget Basin and the Stappen High. The narrow southern part of the Sørvestsnaget Basin bounded by the Senja Fracture Zone was uplifted and eroded in early Eocene time, shedding sediments both into the immature ocean basin and east into the Tromsø Basin. However, the main Eocene fill of the Tromsø Basin exhibits a westerly progradational pattern (Spencer *et al.* 1984). In contrast, the wider northern part of the Sørvestsnaget Basin formed in a pull-apart setting related to the releasing bend at the margin (Faleide *et al.* 1991), causing extensional faulting and deposition of a relatively thick Paleogene succession. Clinoforms above the Base Eocene sequence boundary indicate a transport component from the north (Rønnevik and Jacobsen 1984) resulting from initial uplift of the Stappen High during early Eocene time.

The northern Harstad Basin (north of 70°N) comprises a relatively thick sequence of Paleogene sediments deposited adjacent to the oldest oceanic crust along the Senja Fracture Zone, indicating no major uplift during the early opening. On the other hand, the southern part of the Harstad Basin responded to Tertiary continental break-up by uplift and erosion resulting in a thin Paleogene sequence (Fig. 11).

Oligocene–Recent

In earliest Oligocene times (anomaly 13) the relative direction of plate movement changed to west-northwest, causing crustal stretching and sea-floor spreading also in the northern Greenland Sea (Eldholm *et al.* 1987; Myhre and Eldholm 1988). At anomaly 13 time, the Mohns Ridge had not yet migrated the full length of the Senja Fracture Zone. Thus, the fracture zone north of 72°N reflects the change in spreading direction. Faults were reactivated and subsidence continued in the northern Sørvestsnaget Basin where we presently observe the maximum depth of the Base Tertiary horizon (Fig. 5). Furthermore, renewed faulting and volcanism took place locally in the Vestbakken volcanic province. Reksnes and Vågnes (1985) suggested that the Greenland fracture zone was generated by volcanism within a leaky transform zone and that there was not room for the southeastern part of the Greenland fracture zone at anomaly 13 time. This part may correspond to the region on the southwestern Barents Sea margin where we observe renewed faulting and volcanism within the Vestbakken volcanic province and deformation of the oceanic basement adjacent to the northern Senja Fracture Zone.

Since Oligocene times, oceanic crust has been generated along the entire Barents Sea margin, followed by subsidence and burial by Neogene and Quaternary sediments in a clastic wedge sourced by the uplifted Barents Sea area. The widespread and large amount of uplift and erosion is documented

by studies of vitrinite reflectance, apatite fission tracks, shale compaction and diagenesis of clay minerals, as well as by extrapolation of seismic sequence geometries and estimates of rock volumes eroded from the shelf compared with sediment volumes deposited in the sedimentary wedge along the margin (Nyland *et al.* 1992; Riis and Fjeldskaar 1992). Lateral variations in parameters such as porosity, density, seismic velocities and vitrinite reflectance, reveal a general increasing trend in uplift and erosion to the north. By redistributing the sediment volume in the wedge over a palaeo-drainage area similar to the present, average erosion of 1 km in the southern Barents Sea (Nøttvedt *et al.* 1988; Vorren *et al.* 1988, 1991) and 2–3 km in Svalbard (Eiken and Austegard 1987; Vorren *et al.* 1988, 1991) has been estimated. Geochemical and sedimentological studies on samples from wells (Bjørlykke *et al.* 1989; Wood *et al.* 1990; Nyland *et al.* 1992) and shallow cores (Løseth *et al.* 1992) and outcrops on Bjørnøya (Bjorøy *et al.* 1981) and Svalbard (Manum and Throndsen 1978) yield similar estimates.

In the southwestern Barents Sea, erosion estimates range between 0–1000 m in the Tromsø Basin (Riis and Fjeldskaar 1992), 1000–1500 m in the Hammerfest Basin and Loppa High (Berglund *et al.* 1986; Wood *et al.* 1990), 1750–2050 m in the Maud Basin adjacent to the Svalis Dome (Løseth *et al.* 1992) and more than 3000 m at the Stappen High (Wood *et al.* 1990). The methods mostly provide information on maximum burial depth, being imprecise with respect to timing of the uplift and erosion. Most of the wedge appears to be very young and related to several phases of glacial erosion followed by compensating isostatic uplift (Eidvin and Riis 1989; Riis and Fjeldskaar 1992). The late Cenozoic erosion resulted from lowering of the erosional base due to the combined effect of glacio-eustatic lowering of global sea-level and erosion by marine ice sheets (Vågnes *et al.* 1992). Utilizing some simplifying assumptions about eustasy and isostasy, Vågnes *et al.* (1992) estimated the erosion due to glacial driving forces from the present bathymetry of the Barents Sea. When glacial effects are removed, tectonic uplift in the southwestern Barents Sea appears to be small (0–200 m) while it is in the order of 1000 m in the northwestern Barents Sea and Svalbard (Vågnes *et al.* 1992).

Several tectonic and magmatic processes capable of producing the required amounts of uplift may have been active along the western Barents Sea margin. If uplift was related to the early opening of the Norwegian–Greenland Sea, the delay in sedimentation must be explained. Erosion estimates indicate that Eocene and possibly younger sediments were deposited in the area presently occupied by the Bjørnøyrenna. We suggest that these sediments were derived from the uplifted northwestern Barents Sea margin prior to the glaciations, but were removed by submarine glacial erosion and redeposited in front of Bjørnøyrenna. Furthermore, we stress that the large post-break-up, pre-glacial depocentres in the Vestbakken volcanic province and in the Lofoten Basin seaward of the Senja fracture zone strongly suggest that uplifted source areas have been present throughout since the Eocene. The Plio/Pleistocene glaciations increased both the potential amount of material exposed to erosion, and the intensity of erosion in all glaciated regions. This probably explains the very large amount of sediments in glacial submarine deltas (Vågnes *et al.* 1992).

The authors gratefully acknowledge BP Norway, Bundesanstalt für Geowissenschaften und Rohstoffe, IKU Petroleum Research, Elf Aquitaine Norge, Esso Norge, Mobil Exploration Norway, Norwegian Petroleum Directorate, Norsk Hydro, Saga Petroleum and Statoil for excellent cooperation and access to relevant data. We thank O. Eldholm for critical review and A. Breivik for computer assistance. This research project has been funded by BP Norway with additional financial support provided by the Norwegian Petroleum Directorate and the Royal Norwegian Council for Science and the Humanities.

References

BERGLUND, L. T., AUGUSTSON, J., FÆRSETH, R., GJELBERG, J. AND RAMBERG-MOE, H. 1986. The evolution of the Hammerfest Basin. *In*: SPENCER, A. M. (ed.) *Habitat of Hydrocarbons on the Norwegian Continental Shelf.* Graham & Trotman, London, 319–338.

BJORØY, M., MØRK, A. AND VIGRAN, J. O. 1981. Organic geochemical studies of the Devonian to Triassic succession on Bjørnøya and the implications for the Barents Shelf. *In*: BJORØY, M. *et al.* (eds) *Advances in Organic Geochemistry.* John Wiley and Sons, 49–59.

BJØRLYKKE, K., RAMM, M. AND SAIGAL, G. C. 1989. Sandstone diagenesis and porosity modification during basin evolution. *Geologische Rundschau,* **78**, 243–268.

BREIVIK, A. J. 1991. *Årsaker til tyngdeanomaliene på Barentshavmarginen.* Cand. scient. thesis, University of Oslo.

BREKKE, H. AND RIIS, F. 1987. Mesozoic tectonics and basin evolution of the Norwegian shelf between 69°N and 72°N. *Norsk Geologisk Tidsskrift,* **67**, 295–322.

DALLAND, A. 1981. Mesozoic sedimentary succession at Andøya, northern Norway, and relation to structural development of the North Atlantic. *In*: KERR, J. W. AND FERGUSSON, A. J. (eds) *Geology of the North Atlantic Borderlands.* Canadian Society of Petroleum Geologists Memoir, **7**, 563–584.

DALLMANN, W. K., OHTA, Y. AND ANDRESEN, A. (eds) 1988. *Tertiary Tectonics of Svalbard.* Norsk Polarinstitutt Rapport, **46**.

DORÉ, A. G. 1991. The structural foundation and evolution of Mesozoic seaways between Europe and the Arctic Sea. *Paleogeography, Paleoclimatology, Paleoecology,* **87**, 441–492.

—— AND GAGE, M. S. 1987. Crustal alignments and sedimentary domains in the evolution of the North Sea, North-east Atlantic Margin and Barents Shelf. *In*: BROOKS, J. AND GLENNIE, K. W. (eds) *Petroleum Geology of North West Europe.* Graham & Trotman, London, 1131–1148.

EIDVIN, T. AND RIIS, F. 1989. *Nye dateringer av de tre vestligste borehullene i Barentshavet. Resultater og konsekvenser for den tertiære hevingen.* Norwegian Petroleum Directorate Contribution, **27**.

EIKEN, O. AND AUSTEGARD, A. 1987. The Tertiary orogenic belt of West-Spitsbergen: Seismic expressions of the offshore sedimentary basins. *Norsk Geologisk Tidsskrift,* **67**, 383–394.

ELDHOLM, O., FALEIDE, J. I. AND MYHRE, A. M. 1987. Continent–ocean transition at the western Barents Sea/Svalbard continental margin. *Geology,* **15**, 1118–1122.

——, THIEDE, J. AND TAYLOR, E. 1989. Evolution of the Vøring Volcanic Margin. *In*: *Proceedings of the Ocean Drilling Program, Scientific Results* **104**, 1033–1065.

FALEIDE, J. I., GUDLAUGSSON, S. T., ELDHOLM, O., MYHRE, A. M. AND JACKSON, H. R. 1991. Deep seismic transects across the sheared western Barents Sea–Svalbard continental margin. *Tectonophysics.*

——, —— AND JACQUART, G. 1984. Evolution of the western Barents Sea. *Marine and Petroleum Geology,* **1**, 123–150.

——, MYHRE, A. M. AND ELDHOLM, O. 1988. Early Tertiary volcanism at the western Barents Sea margin. *In*: MORTON, A. C. AND PARSON, L. M. (eds) *Early Tertiary Volcanism and the Opening of the NE Atlantic.* Geological Society, London, Special Publication, **39**, 135–146.

——, VÅGNES, E. AND GUDLAUGSSON, S. T. in press. Late Mesozoic–Cenozoic evolution of the southwestern Barents Sea in a regional rift-shear tectonic setting. *Marine and Petroleum Geology.*

GABRIELSEN, R. H. 1984. Long-lived fault zones and their influence on the tectonic development of the southwestern Barents Sea. *Journal of the Geological Society, London,* **141**, 651–662.

—— AND FÆRSETH, R. B. 1988. Cretaceous and Tertiary reactivation of master fault zones of the Barents Sea. *Norsk Polarinstitutt Rapport,* **46**, 93–97.

—— AND —— 1989. The inner shelf of North Cape, Norway and its implications for the Barents Shelf–Finnmark Caledonide boundary: A comment. *Norsk Geologisk Tidsskrift,* **69**, 57–62.

——, ——, HAMAR, G. AND RØNNEVIK, H. C. 1984. Nomenclature of the main structural features on the Norwegian Continental Shelf north of the 62nd parallel. *In*: SPENCER, A. M. (ed.) *Petroleum Geology of the North European Margin.* Graham & Trotman, London, 41–60.

——, ——, JENSEN, L. N., KALHEIM, J. E. AND RIIS, F. 1990. *Structural Elements of the Norwegian Continental Shelf—Part I: The Barents Sea Region.* Norwegian Petroleum Directorate Bulletin, **6**.

GRUNG OLSEN, R. AND HANSSEN, O. K. 1987. Askeladd. *In*: SPENCER, A. M. *et al.* (eds) *Geology of the Norwegian Oil and Gas Fields.* Graham & Trotman, London, 419–428.

GUDLAUGSSON, S. T., FALEIDE, J. I., FANAVOLL, S. AND JOHANSEN, B. 1987. Deep seismic reflection profiles across the western Barents Sea margin. *Geophysical Journal of the Royal Astronomical Society,* **89**, 273–278.

JACKSON, H. R., FALEIDE, J. I. AND ELDHOLM, O. 1990. Crustal structure of the sheared southwestern Barents Sea continental margin. *In*: WEBER, J. R., FORSYTH, D. A., EMBRY, A. F. AND BLASCO, S. M. (eds) Arctic geoscience. *Marine Geology,* **93**, 119–146.

LØSETH, H., LIPPARD, S. J., SÆTTEM, J., FANAVOLL, S., FJERDINGSTAD, V., LEITH, L. T., RITTER, U., SMELROR, M. AND SYLTA, Ø. 1992. Cenozoic uplift and erosion of the Barents Sea—evidence from the Svalis Dome area. *In*: VORREN, T. O. *et al.* (eds) *Arctic Geology and Petroleum Potential.* Norwegian Petroleum Society Special Publication, **2**.

MANUM, S. B. AND THRONDSEN, T. 1978. Rank of coal and dispersed organic matter and its geological bearing in the Spitsbergen Tertiary. *Norsk Polarinstitutt Årbok* **1977**, 159–177.

MORETTI, I., COLLETTA, B. AND VIALLY, R. 1988. Theoretical model of block rotation along linear faults. *Tectonophysics,* **153**, 313–320.

MYHRE, A. M. AND ELDHOLM, O. 1988. The western Svalbard margin (74°–80°N). *Marine and Petroleum Geology,* **5**, 134–156.

MØRK, A., EMBRY, A. F. AND WEITSCHAT, W. 1989. Triassic transgressive–regressive cycles in the Sverdrup Basin, Svalbard and the Barents Shelf. *In*: COLLINSON, J. D. (ed.) *Correlation in Hydrocarbon Exploration.* Graham & Trotman, London, 113–130.

NYLAND, B., JENSEN, L. N., SKAGEN, J. I. SKARPNES, O. AND VORREN, T. 1992. Tertiary uplift and erosion in the Barents Sea; magnitude, timing and consequences. *In*: LARSEN, R. M., BREKKE, H., LARSEN, B. T. AND TALLERAAS, E. (eds) *Structural and Tectonic Modelling and its Application to Petroleum Geology.* Norwegian Petroleum Society Special Publication, **1**, 153–162.

NØTTVEDT, A., BERGLUND, L. T., RASMUSSEN, E. AND STEEL, R. J. 1988. Some aspects of Tertiary tectonics and sedimentation along the western Barents Shelf. *In*: MORTON, A. C. AND PARSON, L. M. (eds) *Early Tertiary Volcanism and the Opening of the NE Atlantic.* Geological Society, London, Special Publication, **39**, 421–425.

OLAUSSEN, S., DALLAND, A., GLOPPEN, T. G. AND JOHANNESSEN, E. 1984. Depositional environment and diagenesis of Jurassic reservoir sandstones in the eastern part of Troms I area. *In*: SPENCER, A. M. (ed.) *Petroleum Geology of the North European Margin.* Graham & Trotman, London, 61–79.

REKSNES, P. A. AND VÅGNES, E. 1985. *Evolution of the Greenland Sea and Eurasia Basin.* Cand. scient. thesis, University of Oslo.

RIIS, F. AND FJELDSKAAR, W. 1992. On the magnitude of the Late Tertiary and Quaternary erosion and its significance for the uplift of Scandinavia and the Barents Sea. *In*: LARSEN, R. M., BREKKE, H., LARSEN, B. T. AND TALLERAAS, E. (eds) *Structural and Tectonic Modelling and its Application to Petroleum Geology.* Norwegian Petroleum Society Special Publication, **1**.

——, VOLLSET, J. AND SAND, M. 1986. Tectonic development of the western margin of the Barents Sea and adjacent areas. *In*: HALBOUTY, M. T. (ed.) *Future Petroleum Provinces of the World.* American Association of Petroleum Geologists Memoir, **40**, 661–676.

RØNNEVIK, H. C., BESKOW, B. AND JACOBSEN, H. P. 1982. Structural and stratigraphic evolution of the Barents Sea. *In*: EMBRY, A. F. AND BALKWILL, H. R. (eds) *Arctic Geology and Geophysics.* Canadian Society of Petroleum Geologists Memoir, **8**, 431–440.

—— AND JACOBSEN, H. P. 1984. Structures and basins in the western Barents Sea. *In*: SPENCER, A. M. (ed.) *Petroleum Geology of the North European Margin.* Graham & Trotman, London, 19–32.

SPENCER, A. M., HOME, P. C. AND BERGLUND, L. T. 1984. Tertiary structural development of the western Barents Shelf. *In*: SPENCER, A. M. (ed.) *Petroleum Geology of the North European Margin.* Graham & Trotman, London, 199–209.

STEMMERIK, L. AND WORSLEY, D. 1989. Late Paleozoic sequence correlations, North Greenland, Svallbard and the Barents Shelf. *In*: COLLINSON, J. D. (ed.) *Correlation in Hydrocarbon Exploration.* Graham & Trotman, London, 99–111.

SUND, T., SKARPNES, O., JENSEN, L. N. AND LARSEN, R. M. 1986. Tectonic development and hudrocarbon potential offshore Troms, Northern Norway. *In*: HALBOUTY, M. T. (ed.) *Future Petroleum Provinces of the World.* American Association of Petroleum Geologists Memoir, **40**, 615–627.

SYRSTAD, E., BERGSETH, S. AND NAVRESTAD, T. 1976. Gravity modelling offshore Troms, Northern Norway. *In*: *Exploration Geology and Geophysics, Offshore North Sea Conference, Article T-1/4.* Norwegian Petroleum Society.

TALWANI, M. AND ELDHOLM, O. 1977. Evolution of the Norwegian–Greenland Sea. *Geological Society of America Bulletin*, **88**, 969–999.

VORREN, T. O., LEBESBYE, E. AND HENRIKSEN, E. 1988. Cenozoisk erosjon og sedimentasjon i det sørlige Barentshav. 18. *Nordiske Geologiske Vintermøte*, January 1988, København (abstract).

——, RICHARDSEN, G. AND KNUTSEN, S. M. 1991. Cenozoic erosion and sedimentation in the western Barents Sea. *Marine and Petroleum Geology*, **8**, 317–340.

VÅGNES, E., FALEIDE, J. I. AND GUDLAUGSSON, S. T. 1992. Glacial and tectonic uplift of the Barents sea. *Norsk Geologisk Tidsskrift*, **72**, 333–338.

WESTRE, S. 1984. The Askeladden gas find—Troms I. *In*: SPENCER, A. M. (ed.) *Petroleum Geology of the North European Margin.* Graham & Trotman, London, 33–39.

WOOD, R. J., EDRICH, S. P. AND HUTCHISON, I. 1990. Influence of North Atlantic tectonics on the large scale uplift of the Stappen High and Loppa High, Western Barents Shelf. *In*: TANKARD, A. J. AND BALKWILL, H. R. (eds) *Extensional Tectonics and Stratigraphy of the North Atlantic Margins.* American Association of Petroleum Geologists Memoir, **46**, 559–566.

WORSLEY, D., JOHANSEN, R. AND KRISTENSEN, S. E. 1988. The Mesozoic and Cenozoic succession of Tromsøflaket. *In*: DALLAND, A., WORSLEY, D. AND OFSTAD, K. (eds) *A lithostratigraphic scheme for the Mesozoic and Cenozoic succession offshore mid- and northern Norway. Norwegian Petroleum Directorate Bulletin*, **4**, 42–65.

ZIEGLER, P. A. 1988. *Evolution of the Arctic–North Atlantic and the Western Tethys.* American Association of Petroleum Geologists Memoir, **43**.

ZIEGLER, W. H., DOERY, R. AND SCOTT, J. 1986. Tectonic habitat of Norwegian oil and gas. *In*: SPENCER, A. M. (ed.) *Habitat of Hydrocarbons on the Norwegian Continental Shelf.* Graham & Trotman, London, 3–19.

The Norwegian continental margin from the Northern North Sea to the Senja Fracture Zone

B. T. LARSEN,[1] J. SKOGSEID,[2] H. BREKKE,[3] P. BLYSTAD,[3] O. RISE[4] and M. LARSEN[4]

[1] *Statoil, PO Box 300, N-4001 Stavanger, Norway*
[2] *Department of Geology, University of Oslo, PO Box 1047, Blindern, N-0316 Oslo, Norway*
[3] *Norwegian Petroleum Directorate, PO Box 600, N-4001 Stavanger, Norway*
[4] *Statoil, PO Box 40, N-9401 Harstad, Norway.*

The Norwegian passive continental margin extends some 1100 km along-strike, from the UK border in the SW to the Senja Fracture Zone in the NE. The margin can be divided into three segments. The Møre segment in the south is a relatively simple, deep and broad Cretaceous basin close to the Norwegian mainland. The middle part is the Vøring/Haltenbanken segment, a broad (400 km) and complex margin with a platform area (Trøndelag Platform), terraces (Halten and Dønna Terrace) and the deep, broad, Cretaceous/Early Tertiary Vøring Basin on the seaward side. The northern part is the Lofoten/Vesterålen segment, where the peninsular Lofoten islands form a high, uplifted and narrow horst between the steep and structured Tertiary margin to the west, and the deep Vestfjord Basin of assumed Cretaceous age to the east.

The development into passive continental margin segments occurred during the opening of the NE Atlantic in the earliest Eocene. The break-up was associated with massive volcanic activity and in particular the Møre and Vøring segments form type examples of volcanic passive margins. The area had a very long history of extensional tectonics and rifting before the final successful opening. Extensional tectonics dominate, both as deep basement-involved faulting, and as more shallow listric faults locally detaching into Triassic evaporite layers; inversion and doming are also seen. In terms of extensional deformation the three segments developed differently, creating a variety of structural features. Across the broadest segment, the Vøring Margin, an apparent lateral migration of rifting events exposes a development from at least Permian time.

From *Petroleum Geology of Northwest Europe: Proceedings of the 4th Conference* (edited by J. R. Parker).
© 1993 Petroleum Geology '86 Ltd. Published by The Geological Society, London, p. 951.

Mesozoic to Cenozoic plate reconstructions of the North Atlantic and hydrocarbon plays of the Atlantic margins

S. D. KNOTT,[1] M. T. BURCHELL, E. J. JOLLEY and A. J. FRASER

BP Exploration, 301 St Vincent Street, Glasgow G2 5DD, UK

[1] *(Present address: Alastair Beach Associates, 11 Royal Exchange Square, Glasgow G1 3AJ, UK)*

Abstract: An integrated approach using plate tectonic analyses and detailed comparative stratigraphy of the North Atlantic has placed new constraints on the Mesozoic to Cenozoic geological history of the Atlantic margin of NW Europe. Key reconstructions from Mesozoic time to the present day have been plotted to show the evolution of the North Atlantic, and in particular the Rockall Trough. The reconstructions show Rockall Plateau attached to Greenland from Late Paleozoic time (380 Ma) to Late Cretaceous time (83 Ma) since when Rockall remained attached to Eurasia. The Rockall Trough probably initiated during end-Carboniferous to Early Permian time and underwent further stretching episodes in the Early Triassic, Early Jurassic, Middle Jurassic, Late Jurassic, Early Cretaceous, mid-Cretaceous and Late Cretaceous to give the present-day Rockall Trough configuration. The Permo-Triassic rift was dominated by oblique opening with a left-lateral component of strike-slip. Jurassic through Early Cretaceous extension was characterized by predominantly left-lateral strike-slip with a minor dip-slip component in the Faeroe basin and north Rockall Trough, and mainly dip-slip extension in central and south Rockall Trough. In Early Cretaceous time (mid-Aptian) the majority of the United Kingdom Continental Shelf (UKCS) Atlantic margin underwent orthogonal opening followed by continued extension in Late Cretaceous to Paleocene time, culminating in the opening of the North Atlantic west Rockall Plateau. The main Late Jurassic and Early Cretaceous rift episodes conveniently divide the stratigraphy into pre-, syn- and post-rift megasequences which form gross play fairways along the North Atlantic margin. Analysis of these fairways permits integration of data from both mature (e.g. North Sea) and immature (e.g. North Atlantic margin) exploration provinces and helps provide a consistent, predictive approach to the assessment of future hydrocarbon potential of the frontier basins lying along the North Atlantic margin.

The North Atlantic region (Fig. 1) had a protracted history of rifting and sea floor spreading. Discrete phases of rifting and spreading had wide reaching effects on the adjoining areas. Rifting initiated in the Early Permian and the onset of sea floor spreading in the central Atlantic was in the Middle Jurassic (Blake Spur Magnetic Anomaly (BSMA)—165 Ma Bathonian, geological time scale after Harland *et al.* 1989). Spreading ridges developed in succession northwards from the central Atlantic (BSMA), between Iberia and Newfoundland (M0—110 Ma, Aptian), then across the Goban Spur and Bay of

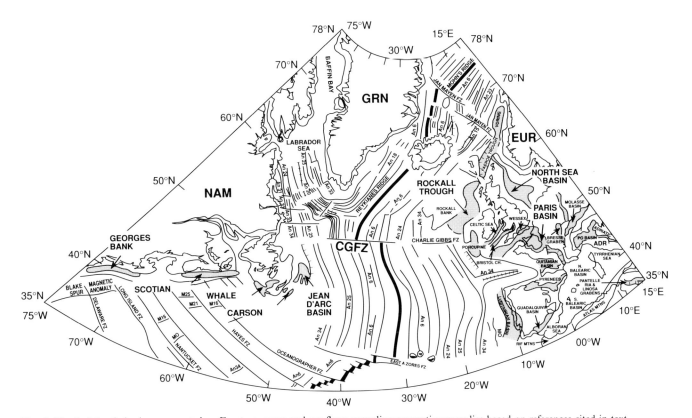

Fig. 1. North Atlantic basins—present day. Fracture zones and sea floor spreading magnetic anomalies based on references cited in text.

From *Petroleum Geology of Northwest Europe: Proceedings of the 4th Conference* (edited by J. R. Parker).

Biscay, and in the south Labrador Sea (An34—83 Ma, Campanian) and finally in the Greenland–Norwegian Sea (An24—
53 Ma, Ypresian) (Srivastava and Tapscott 1986).

Although the spreading ridges formed in a northward propagating succession, relative motion of the plates across the
spreading ridges meant that stress was transmitted over a wide
region and zones of distributed deformation formed along the
plate boundaries to the north of the central Atlantic. One such
zone, the proto-North Atlantic rift (Smythe 1989), is the main
emphasis of this paper. The proto-North Atlantic rift runs
along the western seaboard of the UK, Irish, Faeroese and
Norwegian continental shelves and includes the Rockall
Trough, Faeroe, Møre, Voring, Løfoten, East Greenland and
West Barents Sea basins.

The premise taken here is that plate reorganizations involve
changes in relative motion and/or spreading/rifting rates and
will induce intra-plate stresses simultaneously over a large
region (Cloetingh *et al.* 1990). For example, initiation of rapid
spreading in the central Atlantic induced a rifting phase in the
UKCS during the Late Jurassic. Slip vectors that define the
extension direction across a rift were assumed to approximate
the relative motion of adjacent plates and were obtained from
flow paths describing that motion.

Along the Atlantic margin pre-existing fundamental lineaments played an important role in defining the trend of rift
axes during the Mesozoic and Cenozoic. Most of the proto-
North Atlantic rift along the UKCS western margin lies along
a zone of NE–SW-trending Caledonian terrane boundaries
that stretched, during Early Devonian time, from Newfoundland to Spitsbergen (Fig. 2). The Rockall Trough lay along, or
near to, one of these boundaries. Locally, NW–SE-oriented
transverse lineaments were important too, serving to compartmentalize the rift. Laxfordian and Scourian shear zones in the
Lewisian basement of Scotland trend orthogonally to the
proto-North Atlantic rift; these zones were reactivated as cross
faults during Mesozoic and Cenozoic time controlling both
input of sediment into the rift and deformation in the sedimentary carapace.

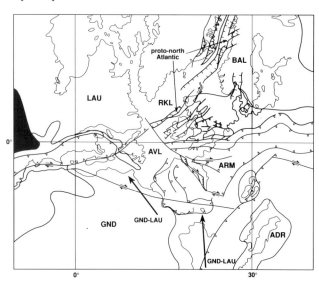

Fig. 2. Pre-rift 'fit' at end Carboniferous time (Stephanian). LAU:
Laurasia; RKL: Rockall; BAL: Baltica; ARM: Armorica; AVL:
Avalonia; GND: Gondwanaland; ADR: Adria; Pale stipple: Continental deposits; dark stipple: shallow marine deposits; thick arrows
indicate amount and direction of relative plate motion. The Rockall
Trough lies along a zone of Caledonian terrane boundaries that
extends to the southwest into Newfoundland.

The integrated approach presented here, using high-resolution plate reconstructions combined with comparative strati-

graphy and basin modelling, provides a powerful tool in
frontier exploration. Hopefully, we have demonstrated how
this approach can be used to allow predictions about the
timing, mode and extent of rifting; the timing of sequence
boundaries; and the distribution of source and reservoir at
both play fairway and prospect scales in the North Sea, UKCS
Western Margin and elsewhere along the North Atlantic
margin.

Plate kinematics

How can plate reconstructions assist in the exploration of a
frontier area such as the UKCS Western Margin? The proto-
North Atlantic rift occupied the diffuse plate boundary
between the Greenland and Eurasian plates for most of its
history and plate reconstructions can provide boundary conditions on the timing, kinematics and amount of relative
motion across this boundary. The main difference between the
reconstructions presented here and previous studies (e.g. Pitman and Talwani 1972; Kristoffersen 1977; Srivastava and
Tapscott 1986; Ziegler 1982, 1990) is the increased resolution
of the database. The magnetic anomaly and fracture zone
databases used resolve plate reorganizations to between 2 and
5 Ma and poles of rotation to within a few tens of kilometres
for post-160 Ma time.

The relative motion of the plates fringing the North Atlantic
can be determined by analysing the oceanic magnetic anomalies and fracture zones, which represent ancient plate boundaries and flow lines, respectively. In this study the most recent
identifications of magnetic anomalies and fracture zones for
the North and central Atlantic (Helman pers. comm; Pindell *et
al.* 1988 and Srivastava *et al.* 1990) and Labrador Sea (Roest
and Srivastava 1989) were used.

Helman (pers. comm.) re-identified all magnetic anomalies
from 1 to 34 in the North and central Atlantic Ocean using the
Lamont-Doherty data bank. He also remapped fracture zones
using bathymetric maps, SEASAT altimetry data and SEA-
SAT-derived gravity images. The parameters for the opening
of the central Atlantic prior to Anomaly 34 were taken from
Pindell *et al.* (1988).

Roest and Srivastava (1989) re-identified sea floor spreading
magnetic lineations, Anomaly 25 (59 Ma) and older in the
Labrador Sea. Roest and Srivastava (1991) remodelled the
motion of Iberia and concluded that it was part of Africa from
Late Cretaceous to mid-Eocene time, a separate plate until
Late Oligocene time and then part of Europe to the present
day. The pre-rift configuration for the North and central
Atlantic (Fig. 2) was modified slightly from Dunbar and
Sawyer (1989) by using a new geological model for the opening
of Rockall Trough (see below).

Reconstruction poles (i.e. poles of rotation that describe the
relative motion of one plate with respect to another for a
specific time interval) for Europe, Africa, Iberia and Greenland
with respect to North America have been used from the
aforementioned studies along with a new geological model for
the opening history of Rockall Trough. The model considers
the Rockall Plateau to have been attached to the Greenland
plate from Late Paleozoic times until 83 Ma, after which
stretching in the Rockall Trough ceased and the Rockall
Plateau became part of the Eurasian plate. In this model,
therefore, the motion of Greenland and Rockall with respect
to Europe had a fundamental influence on the evolution of the
proto-North Atlantic rift.

Tectonostratigraphy

Correlations between tectonic episodes and stratigraphy have
been documented over a number of years (e.g. Pitman 1978;
Ziegler 1982; Dewey 1982; Hubbard 1988; Cloetingh *et al.*

1987). In this contribution, major changes in plate motion (plate reorganizations), major rifting and spreading episodes, and regional unconformities have been documented for the peri-Atlantic basins.

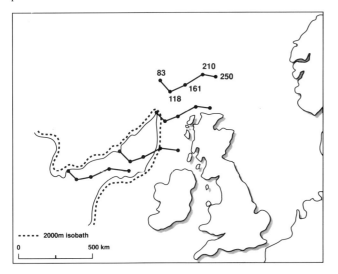

Fig. 3. Flow lines indicating motion of four points on Rockall Plateau with respect to Europe (this study). Note the increase in total stretching (β_T) implied across Rockall Trough moving from northeast ($\beta_T = 4$) to southwest, towards the mouth of the trough ($\beta_T > 6$). The significant stretching episodes (β_t factors for North Rockall Trough) were during the Permo-Triassic ($\beta_t = 1.8$) and the mid-late Cretaceous ($\beta_t = 1.6$). The Jurassic ($\beta_t = 1.1$) to Early Cretaceous ($\beta_t = 1.1$) episodes were predominantly characterized by left-lateral strike-slip with a minor component of extension orthogonal to the rift trend.

The main reorganizations for the motion of Africa with respect to Europe were: early Bathonian; mid-Callovian; early Kimmeridgian; late Berriasian; early Barremian; mid-Aptian and early Campanian. Each of these reorganizations coincided with a major rift event in the peri-Atlantic region, except the Campanian event which marked the Eo-Alpine collision of Africa with Eurasia. For the motion of Greenland and Rockall with respect to Eurasia (Fig. 3) the main reorganizations were: Rhaetian; mid-Callovian; late Oxfordian; late Berriasian; mid-Aptian; Coniacian; and Paleocene. These also coincide with peri-Atlantic rift episodes, except the Paleocene event which coincides with the inception of the Iceland hot spot (White 1992; see also Anderton 1993).

The key changes in magnitude and direction of the relative motions of Africa with respect to Europe, and Greenland and Rockall with respect to Europe are shown in Fig. 4 along with the key tectonic events and the North Sea intra-plate stress curve of Cloetingh et al. (1987) (derived from Vail et al. 1977). The main plate reorganizations broadly coincide with the key tectonic events as recognized by Ziegler (1982).

The synchroneity of some of the changes in Africa–Eurasia and Greenland/Rockall–Eurasia motions is because key events that affected one plate had an effect on adjoining plates and 'the interaction of rigid plates implies that important tectonic phases will be inter-regional and approximately synchronous' (Cloetingh et al. 1990). The mid-Aptian reorganization, seen in both flow paths (Fig. 4), is related to the initiation of spreading in the Bay of Biscay. The Coniacian reorganization, seen in the flow path of Greenland and Rockall with respect to Eurasia (Fig. 3), is related to the initiation of spreading in the Labrador Sea. New results on the timing and cessation of spreading in the Labrador Sea (Chalmers et al. 1993), although placing the initiation of sea floor spreading at An27 time (Danian),

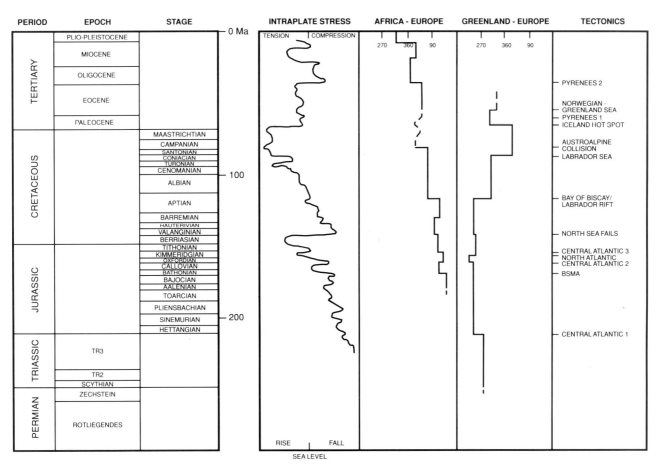

Fig. 4. Tectonostratigraphy of the North Atlantic region showing intra-plate stress curve (Cloetingh et al. 1987), changes in the direction of Africa–Europe and Greenland (Rockall)–Europe motions and correlative tectonic events.

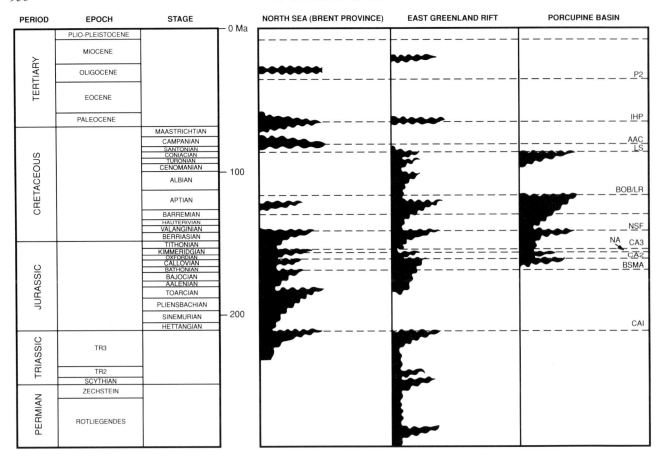

Fig. 5. Chronostratigraphy of three peri-Atlantic basins. Note the coincidence in time of the main erosional unconformities with major plate reorganizations (dashed horizontal lines, see Fig. 4). The abbreviations in the right-hand panel refer to the tectonic events listed in Fig. 4.

younger than previously thought (An33R, Campanian; Roest and Srivastava 1989), do not significantly affect the timing and kinematics of rifting shown in the reconstructions.

Chronostratigraphic charts of the main regional unconformities for the peri-Atlantic basins are shown in Fig. 5. Most regional unconformities (most are type 1 sequence boundaries of Vail et al. 1977) coincide in time with the main plate reorganizations. The reorganizations do not have a time equivalent unconformity in all the peri-Atlantic basins. But basins nearest the central Atlantic (e.g. Lusitania Basin, not shown) have unconformities time-equivalent to changes in the Africa–Eurasia motion. And basins nearest the proto-North Atlantic rift (e.g. East Greenland Rift, Fig. 5) have unconformities time-equivalent to changes in the Greenland/Rockall–Eurasia motion.

A mechanism that can explain the synchroneity of unconformities and plate reorganizations is similar to that proposed by Cloetingh et al. (1987) for the generation of sequence boundaries by fluctuations in intra-plate stress. Tensional stresses across a pre-existing sedimentary basin will induce flexural uplift of the basin centre and subsidence of the margins proportional to the magnitude of the stress. Either a relaxation in tensional stress, or a compressive stress will induce subsidence in the basin and uplift on the basin margins.

The mechanism suggested here is that each rift episode in a sedimentary basin will be preceded by a period of tension and therefore onlap to the basin margin. Fault activity during rifting will relax tensional stresses and the basin will consequently experience effective compression. This will cause flexural uplift on the basin margins and the creation of an erosional unconformity. Therefore, most regional unconformities can be related to the relaxation of tensional stresses during rifting and each rift event related to plate reorganizations

around the North Atlantic. This methodology can be applied in a quantitative way (i.e. estimation of stretching factors and timing of sequence boundaries) to predict the tectonostratigraphic development of less-explored basins.

Some regional unconformities clearly do not coincide with tectonic events (e.g. some Tertiary unconformities in the North Sea). These unconformities are probably related to eustatic falls in relative sea-level controlled either by long-term (periodicity of tens of millions of years) changes in spreading ridge volumes (Pitman 1978) or short-term (periodicity of 100 000 years) changes in the volume of ice sheets (Vail et al. 1977).

Plate reconstructions

A series of plate reconstructions, derived from the reconstruction poles show what is in effect a 'snap shot movie' of the tectonic history of the region from the Mesozoic to the present day. Each reconstruction shows the position of the main plates, key tectonic elements, gross depositional environments and the main hydrocarbon systems at the end of each syn-rift/post-rift couplet and just prior to the next rift episode (e.g. the Zechstein reconstruction shows the scenario after Early Permian rifting and prior to Early Triassic rifting). Stretching factors vary along the proto-North Atlantic, and were estimated by measuring the stretch orthogonal to the rift margin for each rift episode (β_t). The initial width of the rift zone, prior to Mesozoic continental extension, was back-calculated from the present-day width, measured from regional seismic lines and bathymetric maps, and the total stretching factor (β_T), derived from unpublished subsidence analysis. The stretching factors for each rift episode are an average for the entire rift and are not representative of the β factor at a discrete point along that section. The absolute magnitudes of stretching for most rift episodes match reasonably well measurements based on the summation of fault heaves (see below) and basin modelling in the North Sea (Marsden *et al.* 1990).

Paleozoic to Mesozoic framework (Devonian to Late Jurassic)

At end-Caledonian time (Early to Middle Devonian) the Iapetus and mid-European (Rheic) oceans had closed and Laurentia, Baltica, Avalonia and Armorica accreted to form the Old Red Continent assembly (Ziegler 1990). The thickened Caledonian mountain belt underwent extensional collapse giving rise to localized mollase-type basins (e.g. Orcadian basin of Scotland, Hornelen basin of Norway). Devonian reservoir and source facies were deposited in the Orcadian Basin and in other localized basins throughout the relaxed Caledonian belt from Norway, through to Newfoundland. The Late Devonian collision of Gondwanaland with Laurasia induced left-lateral motion along the Great Glen fault system producing small pull-apart basins (e.g. Clair Basin).

Back-arc rifting in Dinantian time, related to passive roll-back of the subducted proto-Tethys slab (Leeder 1988), was followed at end-Dinantian time by a regional inversion as the northward propagating deformation front of the Gondwanaland–Laurasia collision reached northern England (Fraser and Gawthorpe 1990). The culmination of the Hercynian orogeny occurred in Stephanian time as Gondwanaland moved northwards toward Laurasia and inverted Late Devonian and Carboniferous basins in the northern UKCS (Ziegler 1990).

The pre-rift reconstruction (Fig. 2) precludes major opening across the proto-North Atlantic rift at this time (Stephanian). The reconstruction allows for some end-Carboniferous to Early Permian stretching ($\beta_t = 1.5$) but certainly less than the roughly 300 km proposed by Russell (1973, 1976); Russell and Smythe (1978) and Haszeldine (1984).

Rifting continued across the proto-North Atlantic rift and in various localities in the Eurasian foreland into Early Permian time, initiating a series of north–south-trending fault systems (Fig. 6). The most significant extension occurred in East Greenland and further north in the Boreal ocean. Rotliegendes reservoir sandstone was deposited around the margins of the Sole Pit Basin which initiated as part of the Hercynian foredeep. Early Permian sands are likely to have been deposited in the proto-North Atlantic rift as far north as offshore Jameson Land.

Extension gave way to thermal subsidence and marine transgression from the north in Zechstein time. Deposition of marine source rocks in the proto-North Atlantic and throughout the Permian basisns of the Eurasian foreland occurred at the end of the Early Permian. Evaporitic basins developed on the New Red Sandstone continental interior across Central Europe which lay at a latitude of 25°N (Fig. 6).

Significant stretching occurred across the proto-North Atlantic rift ($\beta_t = 1.8$) and in the North Sea ($\beta_t = 1.2$, Marsden *et al.* 1990) in Early Triassic time (Fig. 7). The extension direction, based on the flow path of Greenland and Rockall with respect to Eurasia (Fig. 3), was roughly WNW–ESE, orthogonal to documented Triassic faults in the North Sea (e.g. western bounding fault of the Tern-Eider horst, Yielding *et al.* 1991). Regional thermal subsidence occurred during Late Triassic time in most UKCS rifts, with local halokinetic control of deposystems (Smith *et al.* 1993).

The next rift episode initiated in Rhaetian time with an extension direction substantially different to that experienced in the Early Triassic; indeed, the motion of Greenland and Rockall with respect to Eurasia remained fairly constant in orientation through Early Jurassic to Early Cretaceous time (Fig. 3). Incipient spreading in the central Atlantic region from the Gulf of Mexico to the Azores–Gibraltar fracture zone

induced roughly WSW–ENE-directed stretching in the UKCS. Stretching across the proto-North Atlantic ($\beta_t = 1.1$) was oblique to the rift axis with a large left-slip component (Fig. 8). The expression of this Early Jurassic rift phase in the north Sea may be the broad thermal doming and erosional unconformities (Rhaetian and late Toarcian) in Early Jurassic time. This doming culminated in alkaline volcanic activity in the earliest Middle Jurassic centred at the nexus of the Viking Graben, Central Graben and Witch Ground Graben (Rattray volcanics). Volcanism preceded the next main stretching phase in the North Sea which occurred in mid-Callovian time (Fig. 9). The volcanism may be related to either (1) a deep-seated thermal event located over the junction of the three graben cited above; or (2) mantle thinning and partial melting linked to coeval upper crustal extension in the proto-North Atlantic to the west. Uplift localized at the rift junction led to the deposition of the Brent Group deltaics which prograded northwards down the Viking Graben (Mitchener *et al.* 1992).

Small and localized pull-apart basins are inferred to have developed along the oblique-slip proto-North Atlantic rift during Jurassic time, and led to the deposition of lacustrine to marginal marine source rocks. Shallow marine clastics of Aalenian to Bathonian age were deposited along the western shoreline of the Faeroe Basin. No major Brent-type delta formed in the Faeroe Basin as the uplift due to the North Sea thermal dome was relatively minor this far north.

A major rift episode documented in the North Sea (Rattey and Hayward 1993), in Jameson Land, and inferred from the plate reconstructions in the proto-North Atlantic rift initiated in mid-Callovian time (Fig. 9). The Brent delta prograded northwards into the thermally subsiding Viking Graben during Bathonian time. The overall motif of the Brent Group, however, is one of retrogradation (Mitchener *et al.* 1992). At the onset of the Callovian (*c.* 160 Ma) the west–east-trending facies belts of the Brent Group were dissected by north-trending faults as rifting developed and facies belts consequently followed the new fault trend (Mitchener *et al.* 1992).

Spreading in the central Atlantic continued in Late Jurassic time as oceanic conditions prevailed to the south of the Azores–Gibraltar fracture zone. To the north of the fracture zone restricted marine conditions are inferred as oceanic circulation from the central Atlantic would have been unable to pass the narrow Goban Strait (Fig. 9). This provides an explanation for the anoxic conditions which prevailed in the North Sea rift during the deposition of the Kimmeridge Clay Formation (KCF) and predicts that anoxic conditions would also have been present at this time in the proto-North Atlantic pull-apart basins.

Late Jurassic to earliest Cretaceous (Kimmeridgian to mid-Berriasian)

A major stretching phase initiated in Oxfordian time in the Northern North Sea affecting the central North Sea in Early Kimmeridgian time (Rattey and Hayward 1993). The magnitude of stretching during this phase was greater than the Callovian episode (Callovian $\beta_t = 1.03$; Oxfordian–Kimmeridgian $\beta_t = 1.1$). The extension vector, inferred from the flow path of Greenland and Rockall with respect to Eurasia, was roughly orthogonal to the North Sea (north–south) rift trend, and along the proto-North Atlantic (NE–SW) rift continued to be highly oblique (Fig. 10) with a large left-slip component to extension (Fig. 3).

Footwall uplift and tilting of fault blocks, including those in the Brent Province, occurred throughout the Late Jurassic rift event with re-deposition of footwall-derived shelfal clastics into the evolving basins (e.g. Brae Field, south Viking Graben; Well 206/5-1, Faeroe Basin) (Fig. 11). The anoxic restricted facies of the Kimmeridge Clay Formation continued until at

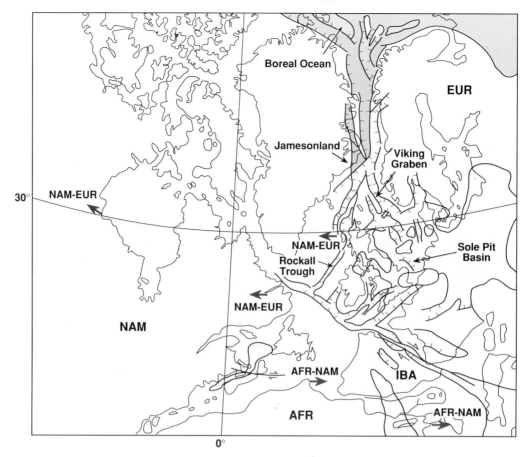

Fig. 6. Late Permian (Zechstein) reconstruction of the North Atlantic region prior to Early Triassic rifting. Mediterranean based in part on Dewey *et al.* (1989). Yellow: continental deposits; green: shallow marine deposits; blue: deep marine deposits; purple: oceanic crust; red: igneous rocks; thick arrows indicate amount and direction of relative plate motion. Small half-arrows indicate sense of shear across rift zones. EUR: Eurasia; NAM: North America; AFR: Africa; IBA: Iberia. See text for elaboration.

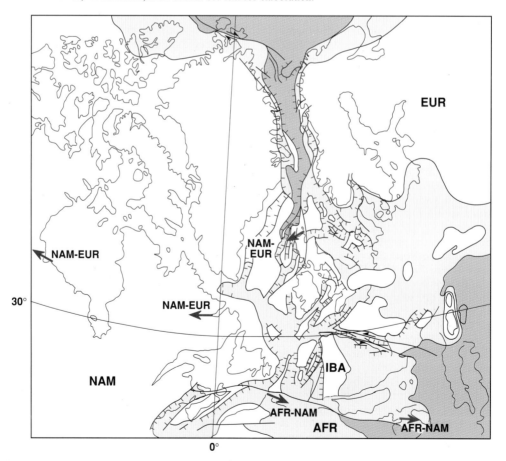

Fig. 7. Late Triassic (Norian) reconstruction prior to Rhaetian rifting. See caption to Fig. 6 for key.

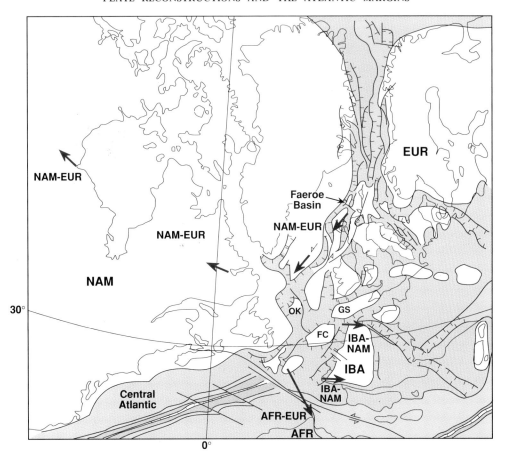

Fig. 8. Middle Jurrassic (Callovian, M25) reconstruction prior to mid-Callovian rifting. See caption to Fig. 6 for key.

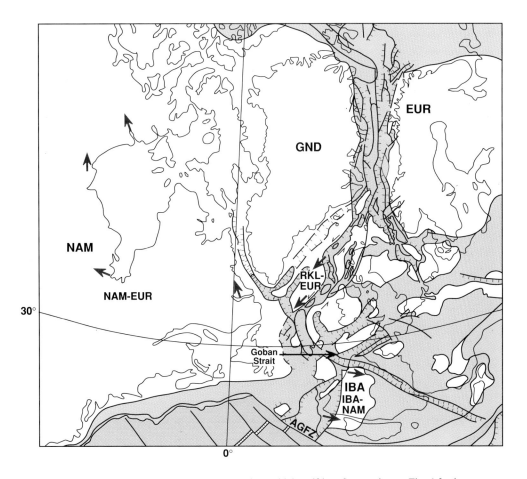

Fig. 9. Late Jurassic (Kimmeridgian, M21) reconstruction prior to Kimmeridgian rifting. See caption to Fig. 6 for key.

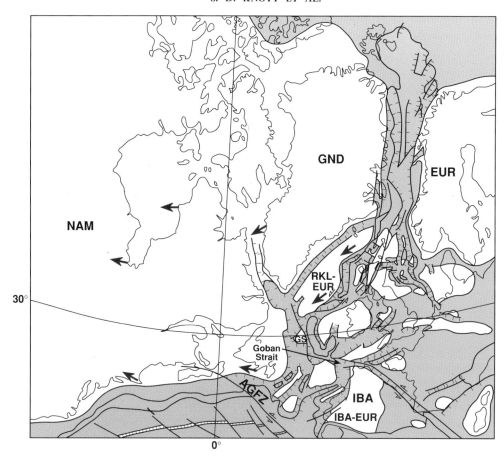

Fig. 10. Early Cretaceous (Berriasian, M16) reconstruction prior to late Berriasian rifting. See caption to Fig. 6 for key.

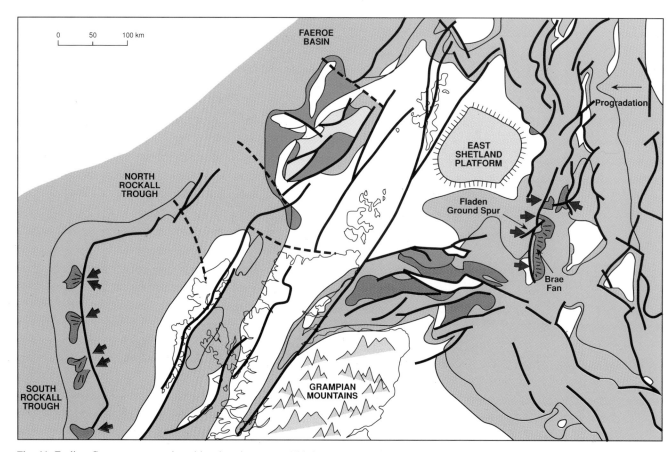

Fig. 11. Earliest Cretaceous gross depositional environments. This is a summary of the main depositional systems over a broad time span rather than a 'snap shot' at one particular time. Yellow: continental deposits; pale green: predominantly retrogradational shelf deposits; dark green: predominantly progradational shelf deposits; blue: deep marine deposits; brown: sand in basinal environment; purple: oceanic crust; red: igneous rocks.

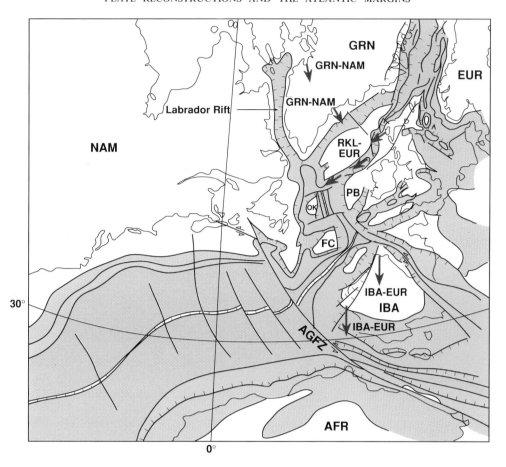

Fig. 12. Early Cretaceous (Aptian, M0) reconstruction prior to mid-Aptian rifting. See caption to Fig. 6 for key.

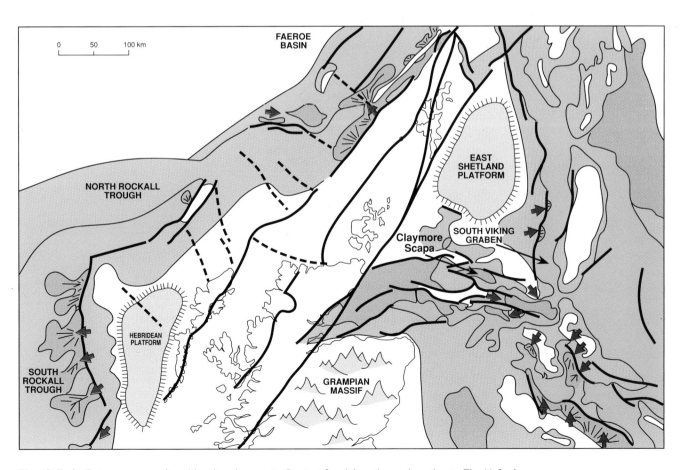

Fig. 13. Early Cretaceous gross depositional environments. See text for elaboration and caption to Fig. 11 for key.

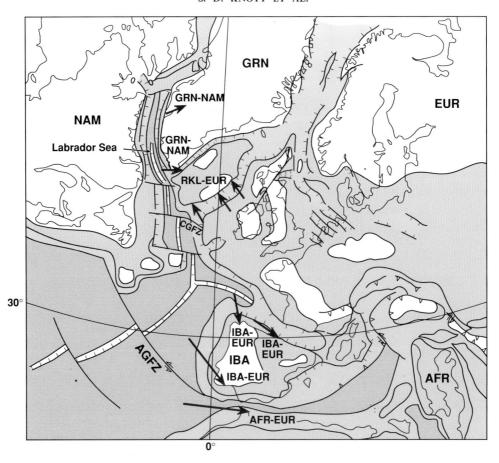

Fig. 14. Late Cretaceous (Santonian, An34) reconstruction after Coniacian rifting. See caption to Fig. 6 for key.

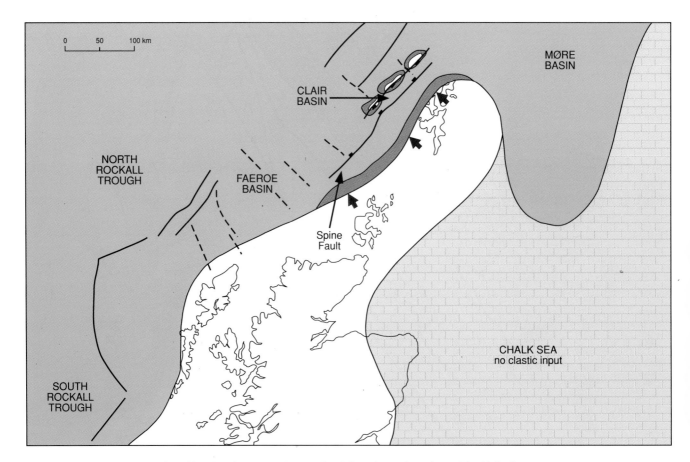

Fig. 15. Late Cretaceous gross depositional environments. See text for elaboration and caption to Fig. 11 for key.

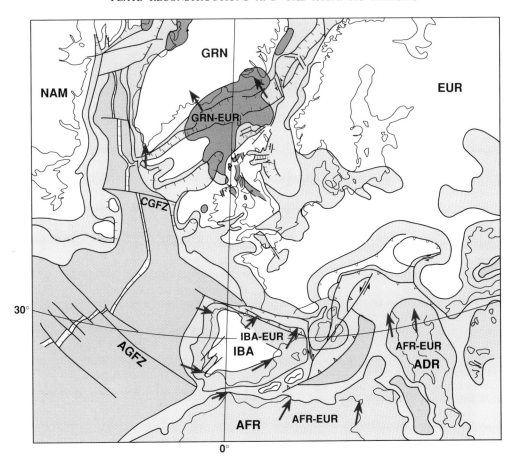

Fig. 16. Paleocene (Thanetian, An26) reconstruction. See caption to Fig. 6 for key.

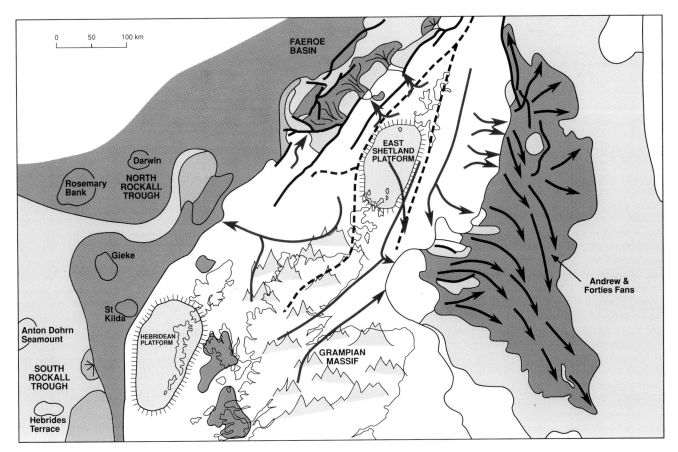

Fig. 17. Late Paleocene gross depositional environments. See text for elaboration and caption to Fig. 11 for key.

least the mid-Tithonian. After this time the central Atlantic ocean currents breached the Goban Strait as it opened in late Tithonian to early Valanginian time restoring oxic conditions to the proto-North Atlantic rift and North Sea and producing the Base Cretaceous Unconformity (see also Rattey and Hayward 1993).

Earliest Cretaceous to mid-Cretaceous (Late Berriasian–mid-Aptian)

Rifting in the North Sea ceased in Tithonian time with only minor movements along the Great Glen fault system and offshore Haltenbanken (Yielding *et al.* 1991), the latter related to compaction and mechanical adjustment during thermal subsidence. To the north, however, the proto-North Atlantic rift (Fig. 12) continued to develop with a major rift episode in mid-late Berriasian time that cut across the north end of the Viking Graben expressed by the ENE-trending fault system, part of which forms the Magnus Field.

The cause of the cessation of rifting in the North Sea in the Tithonian is a matter of speculation. One explanation, offered here, is that cumulative stretching between Orphan Knoll and Porcupine Bank (Fig. 10) at this time ($\beta = 4$, measured from the reconstruction) would have led to major dyke injection and an incipient passive margin would have developed. Lithospheric stretching would have ceased across the UKCS as the relative motion of North America with respect to Eurasia would have been accommodated by the dyke injection across this region. The North Sea rift would have failed, therefore, as the tensional stress relaxed across this portion of the crust. North America was still moving with respect to Eurasia and this motion would have been partly accommodated across the diffuse plate boundary defined by the proto-North Atlantic rift.

The mode of extension in the proto-North Atlantic would have continued to be predominantly left oblique-slip. The large displacement on the ENE-trending fault system north of Magnus, evident on seismic data, is testimony to the catastrophic transfer of displacement from a broad region, including the North Sea and most of the UKCS, out to the narrower proto-North Atlantic rift. Estimates of stretching measured parallel to the slip vector for the mid-late Berriasian rift episode, based on the plate reconstructions (*c.* 30 km across from Magnus to Jameson Land) match closely the total heave across faults known to have been active at this time (32 km) measured on correctly oriented seismic lines and cross-sections over Jameson Land.

The failure of the North Sea rift and incipient spreading (massive dyke injection) between Orphan Knoll and Porcupine Bank would have placed the North Sea and its margins in effective compression. This would have induced uplift on the basin margins; increased subsidence in the basin; and caused contraction of the basin-fill. A manifestation of this compressive stress could be the contractional folds documented along the western flank of the South Viking Graben (Wood and Hall in review; Cherry 1993). Uplift of the margins of the North Sea on the Fladden Ground Spur and Norway is documented by major progradation of clastics of Kimmeridgian to Tithonian age particularly from the Norwegian margin (Fig. 11).

Immediately following rift failure, the North Sea was characterized by thermal subsidence. Initially, the sediment-starved rifts were passively infilled by marine mudstone with local sand-prone submarine fans developed around emergent highs (e.g. Claymore and Scapa fields) (Fig. 13). As mentioned above, fault movement was partly compaction-driven, with some tectonic movement on the Great Glen fault system and environs during the earliest Cretaceous and minor fault displacement that accommodated thermal subsidence.

Mid- to Late Cretaceous (Aptian to Santonian)

The proto-North Atlantic experienced a further rift episode in mid-Aptian time ($\beta_t = 1.1$) related here to the onset of rifting in the Bay of Biscay and the south Labrador Sea (Fig. 14). The extension vector changed markedly from the previous, mainly left-slip motion, to orthogonal opening (Fig. 3). It is the Late Cretaceous opening phase that was the main contributor to the present-day configuration of the attenuated crust in the Rockall Trough (Scrutton 1972; Roberts 1975; Roberts *et al.* 1988; Makris *et al.* 1991). During rifting, significant footwall uplift occurred along the proto-North Atlantic margins. Erosional unconformities and re-deposition of shallow marine clastics and footwall-derived debris flow deposits are predicted (Fig. 15). Aptian age unconformities, transgressed by Late Albian conglomeratic sandstones, have been penetrated along the southern margin of the Faeroe Basin, adjacent to footwall scarps (Fig. 15).

Throughout the Jurassic to Cretaceous history of the Rockall Trough the superimposed rifting episodes have meant that thermal subsidence was of insufficient duration to equilibrate fully the thermal structure of the crust. Each rift pulse overlapped with the preceding one. In Coniacian–Santonian time extension across the Rockall Trough was mainly controlled by spreading (or rifting; see Chalmers *et al.* 1993) in the Labrador Sea. In the proto-North Atlantic the slip vector was oblique to the rift axis with a strong right-slip component (Fig. 3). This was the main stretching phase in the Rockall Trough. Anomaly 34 passes south of the trough (Megson 1987), thus defining the cessation of opening at roughly 83 Ma (end-Santonian). The axis of rifting then switched to the northwest of the Rockall Plateau to the North Atlantic (*sensu stricto*) leaving Rockall Plateau and the failed Rockall Trough attached to Eurasia.

The stretching factor for the Coniacian episode of rifting is estimated to be 1.3 based on the plate reconstruction. During this episode prominent erosional unconformities developed at the rift margins of the proto-North Atlantic and submarine fan aprons developed below active fault scarps. The mud-rich facies of the Cromer Knoll group gave way to chalk deposition in the Southern and Central North Sea and carbonate-rich mud of the Humber and Viking groups in the Northern North Sea and Faeroe Basin (Fig. 15). The UKCS lay at 45° N at this time.

Latest Cretaceous to Paleocene (Campanian–Thanetian)

The Campanian Eo-alpine collision of the leading edge of Adria with the Eurasian continental margin in the eastern Alps transmitted compressive stress into the Eurasian foreland (Fig. 16). The orientation of the slip vector in the collision (NE–SW) is not compatible with the kinematics of structural inversion in the Central Graben and Southern North Sea to the west. Late Cretaceous contraction in the Pyrenees, as Iberia collided with Eurasia, induced a roughly NE–SW-oriented compressive stress in the Eurasian foreland and is thus more likely to have been the cause of the North Sea and Western Margin inversions lying to the north.

Rifting along the length of the North Atlantic (*sensu stricto*), now situated to the northwest of the Rockall Plateau, was superimposed by a major hot spot that initiated in Paleocene time (Joppen and White 1990 and references therein) (Fig. 16). A circular region of uplift, over 2000 km in diameter, was centred over Greenland. The Rockall Trough and Faeroe Basin were uplifted by at least 1.5 km (Joppen and White 1990), which has important implications for source rock maturity and reservoir burial history in this area. Clastic detritus was shed from the area of uplift, which included most of Scotland, into the North Sea and UKCS Western Margin

Tertiary basins (Fig. 17) (Anderton 1993). A base Paleocene unconformity at the margins of the northern UKCS basins is related to this regional uplift (Fig. 5). As western Eurasia, including Britain, moved southeastwards relative to the hot spot, thermal subsidence resumed in the basins.

The presence of the hot spot induced a radial extensional stress over the 2000 km diameter area as a result of dynamic uplift. This tensional stress would have been oriented roughly E–W in the Faeroe Basin. This contrasts with the rift-related NW–SE extension in the basin in Early Paleocene time (Fig. 16). The superimposition of these two stress fields, one related to the hot spot, the other related to rifting, induced reactivation of NW-trending basement structures in the Rockall and Faeroe–Shetland areas which has compartmentalized the basins (Fig. 17).

Prograding shelf systems fringed the west margin of the North Sea and submarine fan deposits ponded in the basins with sandstone reservoirs provided by the erosion of the quartzose hinterland of uplifted Scotland (Fig. 17).

Post-Paleocene framework (Ypresian–present day)

Minor fluctuations in relative sea-level produced erosional unconformities around the North Atlantic basin margin throughout the Tertiary. The initiation of sea floor spreading in the North Atlantic in Early Eocene time (An24) and Greenland–Norwegian Sea (An24R) (Fig. 18) would have placed the UKCS in effective compression between the spreading ridge and the Alpine mountain belt. The associated basin margin uplift may have produced the observed Early Eocene unconformity (see below). In the Early Eocene, oblique-slip deformation occurred along the Iceland–Faeroe and Jan Mayen fracture zones, the Wyville Thomson and West Lewis ridges, and resulted in local uplift in North Rockall, the West Lewis Basin and the Faeroe Basin (Fig. 19). Spreading in the Labrador Sea stopped in Middle Eocene time (An20, Roest and Srivastava 1989—used in this reconstruction) or even later (An13, Chalmers et al. 1993) but spreading continued along the North Atlantic and Norwegian–Greenland Sea to the present day. Volcanism along the northern UKCS terminated in Eocene time (c. 52 Ma, Joppen and White 1989 and references therein) and thermal subsidence continued in the North Sea.

Culmination of the Pyrenean orogeny occurred in Early Oligocene time with thrust emplacement of nappes and transport of piggy-back basins towards the hinterlands of this double-verging orogen (Williams 1985) (Fig. 20). Transmission of the NNE-directed compressive stress, caused by the collision of Iberia with Eurasia, into the European foreland as far north as the UKCS gave rise to roughly orthogonal inversion in southern England (e.g. E–W-trending Purbeck–South Wight monocline) and right-slip transpression in the Sole Pit Trough, Southern North Sea (e.g. NW–SE-trending Dowsing fault zone). The regional, early to mid-Oligocene unconformity probably reflects this phase of tectonic uplift related to Pyrenean compression.

The Burdigalian culmination of the Alpine orogeny in the western Alps (Helvetic nappes of the European carbonate margin) (Fig. 21) was less effective in transmitting compression across Europe, due to the intervening Rhine–Bresse Graben, than the Pyrenean deformation. The minor relative motion between Iberia and Europe during the Early Miocene (Aquitanian to Burdigalian) suggests that thrusting of this age was mainly driven by body forces rather than plate boundary forces.

During the late Neogene the north Atlantic region, including the Central North Sea and west Greenland margin, underwent a phase of rapid subsidence as did large areas of the Mediterranean (Helman pers. comm.). This coincided with major reorga-

nizations of spreading rates and directions along the entire Atlantic spreading system and with late Neogene (Tortonian) collisional tectonics between the African and European plates. Associated with these plate reorganizations was a widespread increase in intra-plate compressive stresses (Cloetingh et al. 1990) which led to increased subsidence rates in the late Neogene (mainly Pliocene) in many peri-Atlantic basins including the North Sea.

In the Rockall Trough, Early Oligocene to Early Miocene shallow marine clastics progressively onlapped the base Oligocene unconformity. A Late Oligocene lowstand wedge is overlain by Late Oligocene to Early Miocene outer shelf sediments in a transgressive systems tract, culminating in the mid-Miocene maximum flooding surface which progressively inundated the Hebridean and Shetland platforms. The overlying mid-Miocene to Pliocene highstand systems tract downlapped and rapidly advanced basinward. The Pleistocene glacio-eustatic sea-level fall led to a downward shift in coastal onlap (but not beyond the shelf slope break), partial incision of the platform and a rapid basinward shift on the shelf system. Subsequently, sea-level rose to the present day and sedimentation resumed across the shelf.

Hydrocarbon habitat

The plate reconstructions and subsequent stratigraphic analysis have shown that the North Atlantic margin has been subject to a long-lived extensional stress regime since the Late Paleozoic. During this time two main phases of rifting appear to have dominated: the Late Jurassic and Early Cretaceous rift events. Together, these have exerted a profound control on the hydrocarbon habitat of the region, both from a trap formation and sediment distribution point of view. These events effectively divide the stratigraphy into three distinct phases of basin evolution which form gross play fairways along the North Atlantic margin (Pegrum and Spencer 1990) (Fig. 22).

Pre-rift (Permian–Mid-Jurassic): the classic tilted fault block play characteristic of all extensional basins. The key element for success in the North Atlantic margin province is the presence of an effective regional reservoir interval (e.g. the Brent delta).

Syn-rift (Late Jurassic–Early Cretaceous): a play dominated by apron fan, basin floor fan and retrogradational shelf sand development. Reservoir presence and effectiveness is essentially controlled by the nature and distribution of the erosion products from active, intra-basinal highs (footwall blocks).

Post-rift (Late Cretaceous–Tertiary): a play characterized by basin floor fan development, fundamentally controlled by extra-basinal processes such as the thermal uplift of Scotland in the Early Tertiary which released vast quantities of coarse siliciclastic sediments into the basinal areas. The regionality of these play types in the North Atlantic margin province is highlighted on a series of sketch cross-sections based on regional seismic data from both the United Kingdom and Norwegian sectors of the margin (Fig. 23).

Along the Løfoten margin (Fig. 23a) the pre-rift is characterized by two phases of tilted fault block generation in the Late Jurassic and Early Cretaceous. The syn-rift is characterized by apron fan development. The post-rift Upper Cretaceous to Tertiary is largely mud prone and is complicated by Late Tertiary uplift extending southwards from the Barents Sea in the north.

The Vøring margin (Fig. 23b) is characterized by the presence of a broad fault terrace not unlike that of the Brent province in the Viking Graben (see below). Both Late Jurassic and Early Cretaceous rifting have generated long wavelength

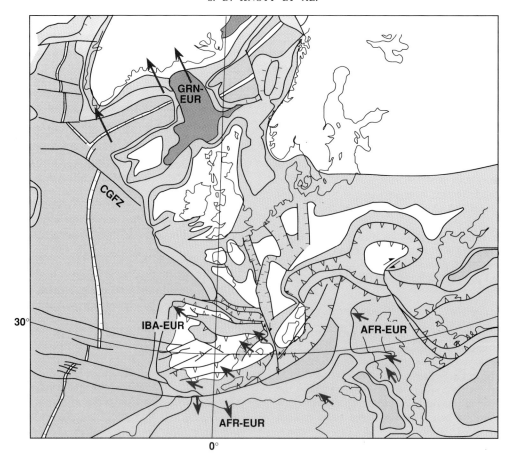

Fig. 18. Eocene (Bartonian, An16) reconstruction. See caption to Fig. 6 for key.

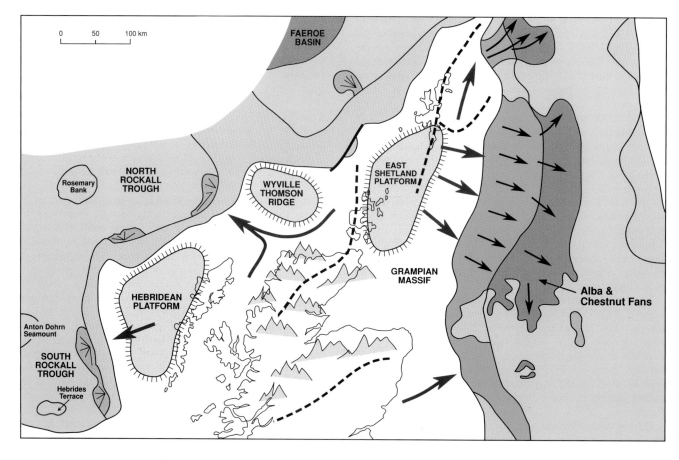

Fig. 19. Mid-Eocene gross depositional environments. See text for elaboration and caption to Fig. 11 for key.

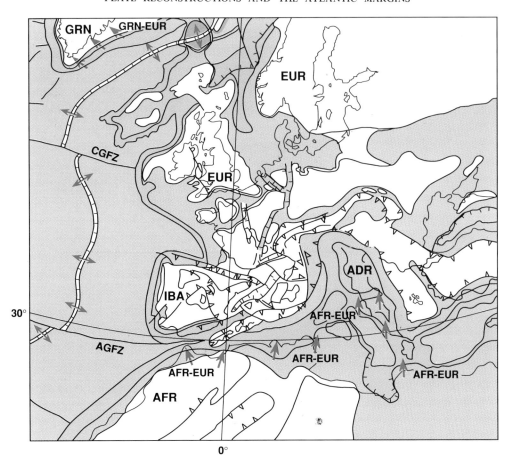

Fig. 20. Oligocene (Chattian, An7) reconstruction. See caption to Fig. 6 for key.

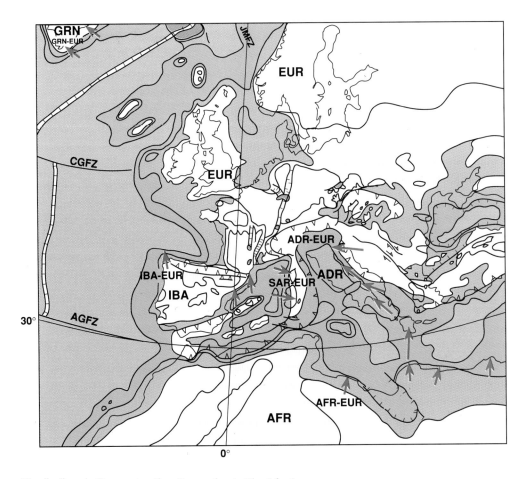

Fig. 21. Miocene (Burdigalian, An6) reconstruction. See caption to Fig. 6 for key.

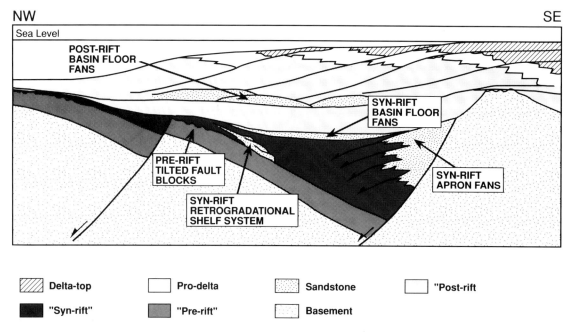

Fig. 22. North Atlantic margin play types.

tilted fault blocks with pre-rift reservoir discoveries on the Halten Terrace in fields such as Heidrun, Smorbukk and Midgard (Fjaeran and Spencer 1991).

The northwest–southeast section across the Møre Basin (Fig. 23c) highlights the large amounts of extension that occurred in the area where the Late Jurassic (North Sea) and Early Cretaceous (Western Margin) rifts intersect. The extension in this part of the margin was dominantly transtensional and of Early Cretaceous age. Subsequent Late Cretaceous dip-slip movements were probably displaced to the northwest. Prospectivity of the pre-rift play is downgraded due to depths of burial exceeding 6000 m attained during the Cretaceous (see Fig. 24). The post-rift is essentially mud prone.

In the North Viking Graben (Fig. 23d) planar normal faults of Late Jurassic age form a series of large tilted fault blocks in the Brent Province. Coupled with the presence of an excellent pre-rift reservoir (the Brent delta) this forms the most successful hydrocarbon province yet encountered in the region. The extent of the Cretaceous–Tertiary thermal subsidence phase is evident and the adjacent thermal uplift of Scotland and the East Shetland platform has ensured a supply of coarse clastics to the basin.

The section across the Faeroe Basin (Fig. 23e) highlights two periods of basin formation; a Devonian–Carboniferous back-basin formed by extensional collapse on an earlier Caledonian thrust; and the main Early Cretaceous rift in the Faeroe Basin. The Jurassic section is thin and fragmented, with attendant implications for problematical reservoir and source rock distribution in this region. The Cretaceous rifting and later subsidence have buried any Permo-Triassic pre-rift reservoirs in the basin centre to depths exceeding 6000 m (Fig. 24).

The section across the Rockall Trough (Fig. 23f) highlights the two main periods of rifting on the Rockall margin: the unsuccessful Triassic–Early Jurassic rift on the platform areas (e.g. West Lewis basin); and the ultimately successful Late Cretaceous rift in the Rockall Trough. The lack of significant Late Jurassic subsidence points to a low potential for the deposition of extensive Upper Jurassic deep marine reservoir and source systems (cf. the North Sea).

Play fairway analysis

The megasequences analysed by play fairway are identified on the summarized chronostratigraphic diagram for the North Atlantic margin (Fig. 25). This diagram rationalizes the interplay between regional stratigraphy and play fairway development and details the range and type of facies developed within each fairway across the province. The play fairways are subdivided into three main play types: pre-rift, syn-rift and post-rift (Fig. 22).

Pre-rift play fairway

Permo-Triassic The key reservoir depositional systems in the Permian (Rotliegendes) and Triassic (Skaggerak Formation) are of aeolian and fluvial origin and are regionally developed. Permo-Triassic reservoirs structured in pre-rift tilted fault blocks will be charged from overlying Middle Jurassic lacustrine and Upper Jurassic marine mudstones either by downward migration or across fault juxtaposition via Jurassic carrier beds.

Lower to Middle Jurassic In the North Sea, the Early Jurassic was marked by a perod of regional thermal uplift, with associated volcanism in the Central North Sea. This uplift is interpreted as a pre-rift thermal bulge, pre-dating the Late Jurassic rifting, which was centred on the intersection of the developing Witch Ground, Central and Viking Graben (Fig. 24). Middle Jurassic sediments were deposited as a major regressive deltaic pulse in the North Viking Graben, shed off the uplifted areas and ponded in the remnant depressions formed by earlier Permian rifting. Further south, in the Central Graben, lakes developed in areas of low relief and reduced sediment supply. A major delta system also developed in the offshore Mid-Norway region forming an additional key area for successful exploration of the pre-rift play. Significantly, no equivalent of the Brent Group or Mid-Norway deltas was developed to the West of Shetland in the Rockall Trough and Faeroe Basin.

The pre-rift Permian to Middle Jurassic fairway contains the largest volume of reserves in the North Atlantic province (32 billion barrels of oil equivalent or 46.5% of the total discovered volume) (Fig. 26a). The field size distribution graph shows a mature fairway with a significant number of billion barrel discoveries (Fig. 26b), reflecting the relative simplicity of the Brent tilted fault block play.

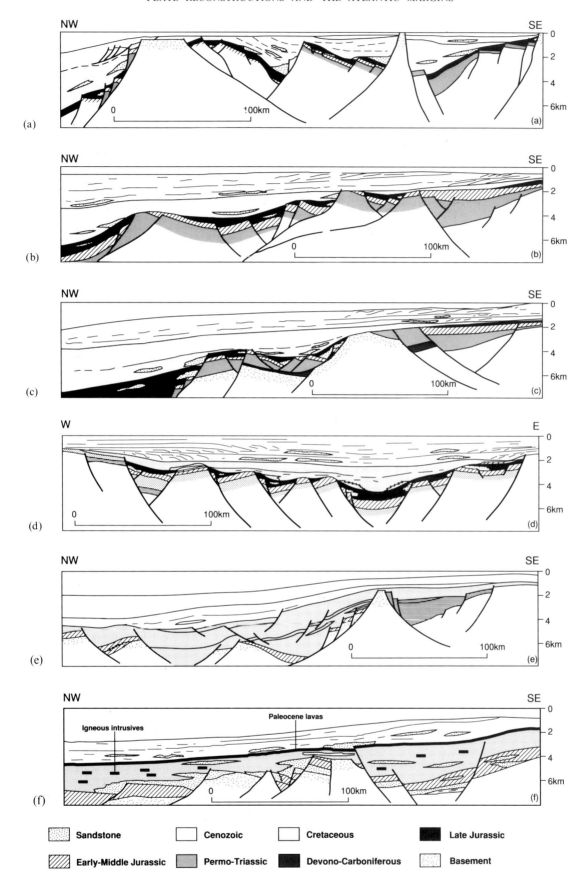

Fig. 23. Geoseismic interpretations across peri-North Atlantic basins. (**a**) Løfoten margin; (**b**) Mid-Norway margin; (**c**) Møre Basin; (**d**) North Viking Graben; (**e**) Faeroe–Shetland Trough; (**f**) Rockall Trough.

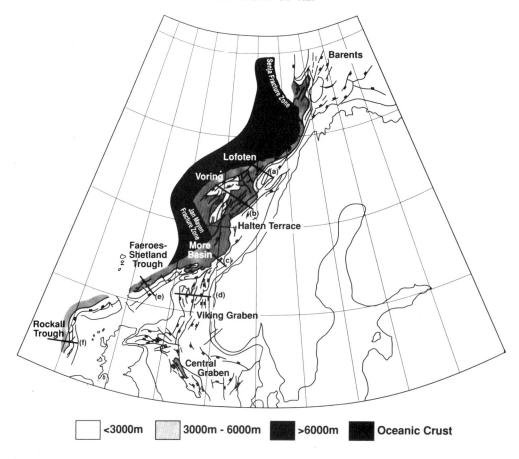

<3000m   **3000m - 6000m**   **>6000m**   **Oceanic Crust**

Fig. 24. North Atlantic continental margin depth to base Cretaceous structure map.

The main control on the regional hydrocarbon potential in the pre-rift play is the presence and effectiveness of pre-rift reservoirs; the Kimmeridge Clay source facies (discussed in the following section) is present and mature over most of the study area. The base Cretaceous depth map (Fig. 24) indicates that large tracts of the pre-rift fairway lie at depths where diagenesis is likely to have rendered reservoirs ineffective. Where the play fairway is at shallower depths, and underlies Upper Jurassic shelfal areas, prospectivity is high risk due to charge problems. Three prospective areas have been identified in this analysis: the Brent Province and Central Graben in the Northern and Central North Sea and offshore Mid-Norway. The Brent Province is a mature play fairway with remaining opportunities confined to small tilted fault block traps clustered around existing fields. The majority of the remaining potential in this play fairway is considered to exist in the Central Graben and Mid-Norway.

Syn-rift play fairway

Late Jurassic During the late Jurassic three main sandstone plays were developed in the syn-rift fairway.

(i) *Shelf sandstones.* These are developed in tilted fault blocks on the margins of the active rift systems; e.g. the Witch Ground Graben (Piper, Claymore etc.) and the eastern flank of the Viking Graben (Troll).

(ii) *Basin-fringing submarine fans.* These occur in combination structural/stratigraphic traps particularly in the South Viking Graben (Brae trend) and South Halibut Trough.

(iii) *Basin floor fans.* These were deposited in the deeper parts

of the sediment-starved rifts, such as the Viking Graben (Miller, Magnus) and South Halibut Trough (Ettrick).

The regional geological model for the syn-rift play fairway indicates that these plays may also be present in parts of the North Atlantic margin; in particular South Rockall and Mid-Norway where fault trends were largely orthogonal to the predicted extension direction.

As discussed earlier, the plate reconstructions indicate the development of a major barrier to open marine circulation in the region of the Goban Straits during the Late Jurassic (Fig. 4d). This strongly affected source rock distribution by restricting ocean water circulation to the north of the constriction, resulting in basinwide anoxia and the deposition of the Kimmeridge Clay Formation source rock facies in basinal settings. The resulting stratigraphic juxtaposition of Upper Jurassic source and deep marine reservoir facies has proven to be a key factor in the success of the early syn-rift play.

Cretaceous Along the North Atlantic Margin, Early Cretaceous sandstones are extensively developed in the hangingwalls of major syn-rift faults, e.g. the Spine fault in the Faeroe Basin and the South Rockall Trough (Fig. 13). The play is limited by the basinward extent and the significant depth of burial of these sandstones (see Fig. 22). In the Løfoten, Mid-Norway, Møre and Rockall basins, easterly derived submarine fan systems sourced from widespread footwall uplift of the basin margins, are inferred to cover large areas of the basin floor.

The plate tectonic reconstructions indicate that the Goban Strait, which had restricted open marine circulation in the region during the Late Jurassic, was breached at the onset of the Early Cretaceous as a result of further stretching between

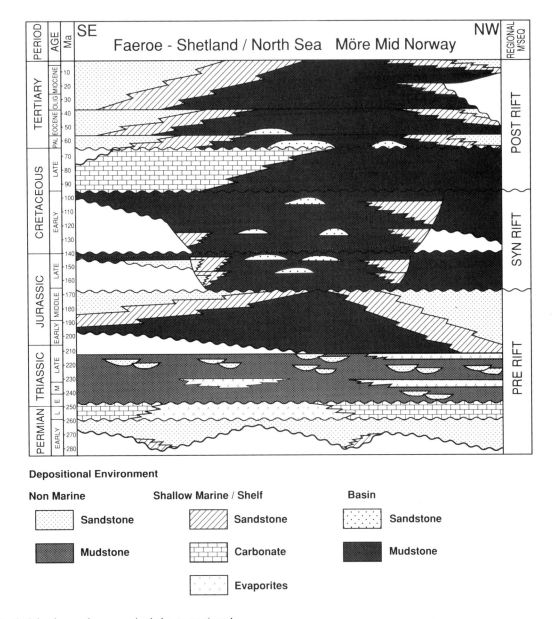

Fig. 25. North Atlantic margin summarized chronostratigraphy.

the Goban Spur and Flemish Cap (Fig. 10). The resulting restoration of oxic conditions to the North Atlantic basins precluded the continued accumulation of the Kimmeridge Clay Formation source facies.

The syn-rift play fairway comprises some 32% of the discovered reserves in the North Atlantic margin province (Fig. 26a). The field size distribution histogram (Fig. 26c) shows that most of the discoveries are concentrated in the small–medium field size category, largely as a result of the stratigraphic complexity inherent in syn-rift plays. The one giant field in this fairway (Troll) dominates the reserves.

The key to success in the syn-rift play is the presence of uplifted intra-basinal provenance areas which act as sources of coarse clastic sediment during rifting. Given the marine setting of the North Atlantic margin during the syn-rift, the deep basinal areas are dominated by basin floor fans. This opens up the possibility for future discoveries in large isolated stratigraphic plays along the margin, in areas where the potential reservoirs are not buried to depths that render them ineffective.

Post-rift play fairway

Two play systems are developed in the post-rift fairway: Upper Cretaceous chalk in the Norwegian sector of the Central North Sea (Ekofisk) and Paleocene and Eocene turbidite sandstones in the remainder of the North Sea and parts of the Western Margin (Forties, Alba). Valid traps are provided for both reservoir systems by structural closures in the fairway centre and in the case of the Tertiary turbidites by stratigraphic pinch-out at the play fairway margins. In particular, the Eocene system offers considerable potential for stratigraphic trapping (e.g. the Alba and Forth fields) although these tend to be on the small size by North Sea standards. The post-rift fairway is charged by upward migration from mature Upper Jurassic source rocks.

Prospectivity in the post-rift is driven primarily by reservoir presence. The key to the success of the post-rift play in the North Sea has been the extra-basinal input of sands from the thermal uplift of Scotland and the East Shetland platform. In the absence of this uplift elsewhere in the North Atlantic region, little prospectivity is envisaged. The post-rift play represents some 22% of the discovered reserves in the province (Fig. 26a); mostly reservoired in the North Sea. The field size histogram (Fig. 26d), in common with that of the syn-rift play, is typified by a poor representation in the larger field classes.

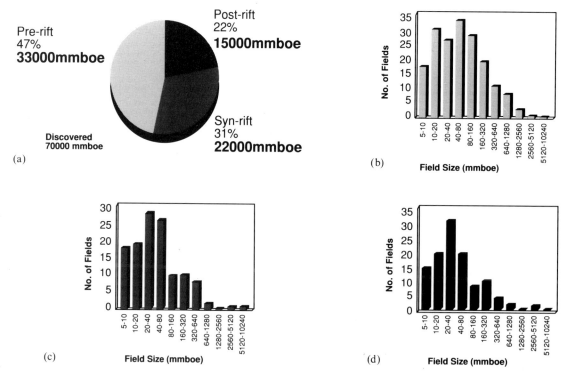

Fig. 26. Play fairway statistics for the UKCS and Norway Atlantic margins. (**a**) Discoveries by fairway; (**b**) field size distribution: pre-rift play; (**c**) field size distribution: syn-rift play; (**d**) field size distribution: post-rift play.

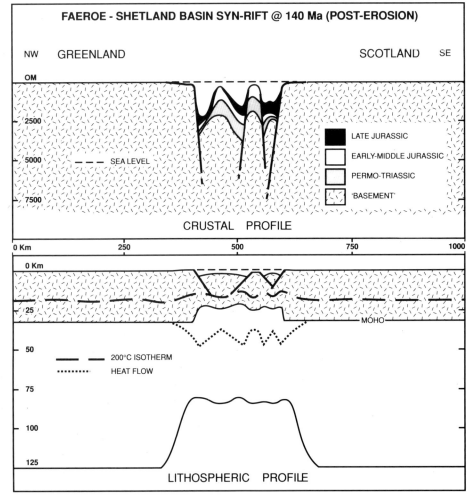

Fig. 27. Cross-section across part of the Faeroe Basin at 140 Ma (Late Berriasian, uppermost Kimmeridge Clay Formation) derived from forward computer model using boundary conditions based on the plate reconstructions and stratigraphic control from well and seismic data. Note vertical scales greatly exaggerated on upper and lower panels. See text for elaboration. The section is located along the same line as Fig. 23e.

Integrated prospect analysis

The integrated approach we have adopted here can be applied to the analysis of individual prospects within the frontier portions of the North Atlantic margin. The cross-section in Fig. 27 shows part of the Faeroe Basin analysed by this approach and is located along the transect of the section shown in Fig. 23e. An iterative computer forward model was used to predict reservoir and source distribution and burial history. Boundary conditions on timing and amount of stretching were derived from the plate reconstructions and stratigraphic control from well and seismic data. A modelled cross-section over this part of the Faeroe Basin (Fig. 27) shows the basin geometry and fill during late Berriasian time. As mentioned above, the Faeroe Basin was an oblique-slip rift at this time (140 Ma), whereas central and south Rockall underwent predominantly dip-slip extension as the rift trend turned to a north–south orientation, parallel to the North Sea. It is predicted that in the central and south Rockall area source rock deposition (Kimmeridge Clay Formation) would have been more laterally continuous compared with the localized nature of pull-apart depocentres inferred for the Faeroe Basin.

The predicted reservoirs for tilted fault block prospects are Permo-Triassic non-marine sandstone and Early to Middle Jurassic marine sandstone (Fig. 27). The location of the fault block in the centre of the basin, and its modelled position below sea-level since Callovian time (160 Ma) indicates that younger reservoir facies are unlikely to be present.

We thank BP Exploration Operating Company Limited for giving permission to publish this work. We are indebted to Marc Helman for providing us with his reconstruction poles for the North and central Atlantic region. Significant contributions to BP's regional overview of the North Sea and Western Margin presented here were made by Roger Anderton, Chris Einchcomb, Steve Flanagan, Paul Hawkes, Richard Hedley, Tony Hayward, Martin Illingworth, David Lynch, Nick Milton, Brian Mitchener, Mark Partington and Peter Rattey.

References

ANDERTON, R. 1993. Sedimentation and basin evolution in the Paleogene of the Northern North Sea and Faeroe–Shetland basins. *In*: PARKER, J. R. (ed.) *Petroleum Geology of Northwest Europe: Proceedings of the 4th Conference*. Geological Society, London, 31.

CHALMERS, J. A., PULVERTAFT, T. C. R., CHRISTIANSEN, F. G., LARSEN, H. C., LAURSEN, K. H. and OTTESEN, T. G. 1993. The southern West Greenland continental margin: rifting history, basin development, and petroleum potential. *In*: PARKER, J. R. (ed.) *Petroleum Geology of Northwest Europe: Proceedings of the 4th Conference*. Geological Society, London, 915–931.

CLOETINGH, S., LAMBECK, K. AND MCQUEEN, H. 1987. Apparent sea-level fluctuations and a palaeostress field for the North Sea region. *In*: BROOKS, J. AND GLENNIE, K. (eds) *Petroleum Geology of North West Europe*. Graham & Trotman, London, 49–57.

——, GRADSTEIN, F. M., KOOI, H., GRANT, A. C. AND KAMINSKI, M. 1990. Plate reorganization: a cause of rapid late Neogene subsidence and sedimentation around the north Atlantic? *Journal of the Geological Society, London*, **147**, 495–506.

DEWEY, J. F. 1982. Plate tectonic evolution of the British Isles. *Journal of the Geological Society, London*, **139**, 371–412.

——, HELMAN, M. L., TURCO, E., HUTTON, D. H. W. AND KNOTT, S. D. 1989. Kinematics of the western Mediterranean. *In*: COWARD, M. P., DIETRICH, D. AND PARK, R. G. (eds) *Alpine Tectonics*. Geological Society, London, Special Publication, **45**, 265–283.

DUNBAR, J. A. AND SAWYER, D. S. 1989. Patterns of continental extension along the conjugate margins of the central and North Atlantic oceans and Labrador Sea. *Tectonics*, **8**, 1059–1077.

FJAERAN, T. AND SPENCER, A. M. 1991. Proven hydrocarbon plays, offshore Norway. *In*: SPENCER, A. M. (ed.) *Generation, accumulation, and production of Europe's hydrocarbons*. Special Publication of the European Association of Petroleum Geoscientists, **1**, Oxford University Press, Oxford, 25–48.

FRASER, A. J. AND GAWTHORPE, R. P. 1990. Tectono-stratigraphic development and hydrocarbon habitat of the Carboniferous in northern England. *In*: HARDMAN, R. F. P. AND BROOKS, J. (eds) *Tectonic events responsible for Britain's oil and gas reserves*. Geological Society, London, Special Publication, **55**, 49–86.

HARLAND, W. B., ARMSTRONG, R. L., COX, A. V., CRAIG, L. E., SMITH, A. G. AND SMITH, D. G. 1989. *A geologic time scale*. Cambridge University Press, Cambridge.

HASZELDINE, R. S. 1984. Carboniferous North Atlantic palaeogeography: stratigraphic evidence for rifting, not megashear or subduction. *Geological Magazine*, **121**, 443–463.

HUBBARD, R. J. 1988. Age and significance of sequence boundaries on Jurassic and Cretaceous rifted continental margins. *American Association of Petroleum Geologists, Bulletin*, **72**, 49–72.

JOPPEN, M. AND WHITE, R. S. 1990. The structure and subsidence of Rockall trough from two-ship seismic experiments. *Journal of Geophysical Research*, **95**, 19821–19837.

KRISTOFFERSEN, Y. 1977. Sea-floor spreading and the early opening of the North Atlantic. *Earth and Planetary Science Letters*, **38**, 273–290.

LEEDER, M. R. 1988. Recent developments in Carboniferous geology: a critical review with implications for the British Isles and N.W. Europe. *Proceedings of the Geologists' Association*, **99**, 73–100.

MAKRIS, J., GINZBURG, A., SHANNON, P. M., JACOB, A. W. B., BEAN, C. J. AND VOGT, U. 1991. A new look at the Rockall region, offshore Ireland. *Marine and Petroleum Geology*, **8**, 410–416.

MARSDEN, G., YIELDING, G., ROBERTS, A. M. AND KUSZNIR, N. J. 1990. Application of a flexural cantilever simple-shear/pure-shear model of continental lithosphere extension to the formation of the northern North Sea Basin. *In*: BLUNDELL, D. J. AND GIBBS, A. D. (eds) *Tectonic evolution of the North Sea Rifts*. Oxford University Press, Oxford, 240–261.

MEGSON, J. B. 1987. The evolution of the Rockall Trough and implications for the Faeroe-Shetland Trough. *In*: BROOKS, J. AND GLENNIE, K. W. (eds) *Petroleum Geology of North West Europe*. Graham & Trotman, London, 653–666.

MITCHENER, B. C., LAWRENCE, D. A., PARTINGTON, M. A., BOWMAN, M. B. J. AND GLUYAS, J. 1992. Brent Group: sequence stratigraphy and regional implications. *In*: MORTON, L. A. C., HASZELDINE, R. S., GILES, M. R. AND BROWN, S. (eds) *Geology of the Brent Group*. Geological Society, London, Special Publication, **61**, 45–80.

PEGRUM, R. M. AND SPENCER, A. M. 1990. Hydrocarbon plays in the northern North Sea. *In*: BROOKS, J. (ed.) *Classic Petroleum Provinces*. Geological Society, London, Special Publication, **50**, 441–470.

PINDELL, J. L., CANDE, S. C., PITMAN, W. C., III, ROWLEY, D. B., DEWEY, J. F., LABREQUE, J. AND HAXBY, W. 1988. A plate-kinematic framework for models of Caribbean evolution. *Tectonophysics*, **155**, 121–138.

PITMAN, W. C., III. 1978. Relationship between eustacy and stratigraphic sequences of passive margins. *Geological Society of America Bulletin* **89**, 1389–1403.

—— AND TALWANI, M. 1972. Sea-floor spreading in the North Atlantic. *Geological Society of America Bulletin*, **83**, 619–654.

RATTEY, R. P. and HAYWARD, A. W. 1993. Sequence stratigraphy of a failed rift system: the Middle Jurassic to Early Cretaceous basin evolution of the Central and Northern North Sea. *In*: PARKER, J. R. (ed.) *Petroleum Geology of Northwest Europe: Proceedings of the 4th Conference*. Geological Society, London, 215–249.

ROBERTS, D. G. 1975. Marine geology of the Rockall Plateau and Trough. *Philosophical Transactions of the Royal Society, London*, **A278**, 447–509.

——, GINZBURG, A., NUNN, K. AND MCQUILLIN, R. 1988. The structure of the Rockall Trough from seismic refraction and wide-angle reflection measurements. *Nature*, **332**, 632–634.

ROEST, W. R. AND SRIVASTAVA, S. P. 1989. Sea-floor spreading in the Labrador Sea: A new reconstruction. *Geology*, **17**, 1000–1003.

—— AND —— 1991. Kinematics of the plate boundaries between Eurasia, Iberia, and Africa in the North Atlantic from the late Cretaceous to the present. *Geology*, **19**, 613–616.

RUSSELL, M. J. 1973. Base-metal mineralization in Ireland and Scotland and the formation of the Rockall Trough. *In:* TARLING, D. H. AND RUNCORN, S. K. (eds) *Implications of Continental Drift to the Earth Sciences*. Academic Press, 581–597.

—— 1976. A possible Lower Permian age for the onset of sea floor spreading in the northern North Atlantic. *Scottish Journal of Geology*, **12**, 315–323.

—— AND SMYTHE, D. J. 1978. Evidence for an early Permian oceanic rift in the northern North Atlantic. *In:* NEUMAN, E.-R. AND RAMBERG, I. B. (eds) *Petrology and geochemistry of continental rifts*. Reidel, Dordrecht, 173–179.

SCRUTTON, R. A. 1972. The crustal structure of Rockall Plateau microcontinent. *Geophysical Journal of the Royal Astronomical Society*, **27**, 259–275.

SMITH, R. I., HODGSON, N. AND FULTON, M. 1993. Salt control on Triassic reservoir distribution, UKCS, Central North Sea. *In:* PARKER, J. R. (ed.) *Petroleum Geology of Northwest Europe: Proceedings of the 4th Conference*. Geological Society, London, 547–557.

SMYTHE, D. K. 1989. Rockall Trough—Cretaceous or Late Palaeozoic? *Scottish Journal of Geology*, **25**, 5–43.

SRIVASTAVA, S. P. AND TAPSCOTT, C. R. 1986. Plate kinematics of the North Atlantic. *In:* VOGT, P. R. AND TUCHOLKE, B. E. (eds) *The Geology of North America, M. The western North Atlantic Region*. DNAG–GSA Series, 379–404.

——, SCHOUTEN, H., ROEST, W. R., KLITGORD, K. D., KOVACS, L. C., VERHOEF, J. AND MACNAB, R. 1990. Iberian plate kinematics: A jumping plate boundary between Eurasia and Africa. *Nature*, **334**, 756–759.

VAIL, P., MITCHUM, R. M., JR. AND THOMPSON, S., III. 1977. *Global cycles of relative changes of sea level*. American Association of Petroleum Geologists Memoir, **26**, 83–97.

WHITE, R. S. 1992. Crustal structure and magmatism of North Atlantic continental margins. *Journal of the Geological Society, London*, **149**, 841–854.

WILLIAMS, G. D. 1985. Thrust tectonics in the south central Pyrenees. *Journal of Structural Geology*, **7**, 11–17.

WOOD, J. L. AND HALL, S. in review. A tectono-stratigraphic model for the development of submarine apron slope fans and basin floor fans during the Late Jurassic–Early Cretaceous of the South Viking Graben. *Basin Research*.

YIELDING, G., BADLEY, M. E. AND ROBERTS, A. M. 1991. Seismic reflections from normal faults in the northern North Sea. *In:* ROBERTS, A. M., YIELDING, G. AND FREEMAN, B. (eds) *The Geometry of Normal Faults*. Geological Society, London, Special Publication. **56**, 79–89.

ZIEGLER, P. A. 1982. *Geological Atlas of Western and Central Europe*. Shell International Petroleum, The Hague.

—— 1990. *Geological Atlas of Western and Central Europe*, 2nd edition. Shell International Petroleum, The Hague.

Subsidence patterns in western margin basins: evidence from the Faeroe–Shetland Basin

J. D. TURNER and R. A. SCRUTTON

Department of Geology & Geophysics, University of Edinburgh, The Grant Institute, West Mains Road, Edinburgh EH9 3JW, UK

Abstract: Anomalously high rates of Paleocene tectonic subsidence have been documented from a number of basins on the NW European continental shelf. This period of accelerated subsidence occurs during what is conventionally regarded as the post-rift phase of basin development. It cannot, therefore, be explained by theoretical models of basin formation in which the post-rift subsidence is driven only by the decay of the thermal anomaly generated during active extension. The Faeroe–Shetland Basin developed as a major Cretaceous depocentre, with Late Cretaceous times in particular characterized by extremely high rates of tectonic subsidence. Subsidence rates on the flanks of the basin slowed dramatically into Paleocene times as a major mid-Paleocene unconformity developed across the eastern side of the basin. Wells from the basin centre, on the other hand, show extremely high and accelerating rates of Late Paleocene subsidence associated with normal faulting. These high rates of observed tectonic subsidence lasted for less than 5 Ma and cannot easily be explained in terms of fluctuations in sediment supply rates or eustatic sea-level. This simultaneous flank uplift and basin centre subsidence acceleration is the same age as both the anomalous Paleocene subsidence reported from other nearby basins, and the oldest magmatism associated with the proto-Icelandic plume. Subsidence behaviour returned to normal, across the entire basin, at the same time as continental break-up was achieved between Greenland and Northwest Europe. This suggests that there is a link between regional tectonics and the anomalous Paleocene subsidence patterns observed in this and other basins on the NW European continental shelf.

Several recent papers have drawn attention to what might be termed 'second-order' features of tectonic subsidence curves calculated from well and seismic data. In particular, periods of anomalously high and accelerating subsidence have been recognized from what is conventionally regarded as the post-rift phase of basin development in extensional sedimentary basins like the North Sea. This is in direct contrast to the subsidence behaviour predicted by theoretical models in which post-rift subsidence is driven only by the decay of the thermal anomaly generated during active extension (e.g. McKenzie 1978).

A dramatic acceleration in tectonic subsidence rates during Paleocene times has been documented by Joy (1993) from the Witch Ground Graben of the North Sea. Similar increases in tectonic subsidence rates have also been reported from the area around the North Sea triple junction and the Porcupine Basin, offshore SW Ireland (White and Latin 1992; White *et al.* 1992). Typically, this accelerated subsidence lasted for between 5 and 10 Ma with subsidence rates up to twice their theoretical value, although there is a large potential error. A second period of anomalously high tectonic subsidence rates, again recognized throughout the North Sea and in many North Atlantic marginal basins, occurred in Quaternary times (Thorne and Watts 1989; Cloetingh *et al.* 1990).

As part of a wider study of the significance of these 'second-order' phenomena, subsidence patterns in sedimentary basins on the western margin of the United Kingdom continental shelf are being investigated. This paper concentrates on a detailed examination of observed subsidence patterns apparent within the Faeroe–Shetland Basin during Late Cretaceous–Early Tertiary times. This period saw a major change in the tectonic setting of the basin, which occurred a little later than the anomalously high Late Paleocene subsidence rates reported from the North Sea and Porcupine basins. A wholly intracontinental rift basin during the Mesozoic, it has been on the eastern margin of the opening North Atlantic ocean since magnetic anomaly C24R, latest Paleocene/earliest Eocene, times (Eldholm *et al.* 1989). The Faeroe–Shetland Basin is ideally situated for a study of the wider extent of anomalous subsidence patterns. It also allows us to explore the impact of continental break-up and, in particular, the adjacent Iceland Plume on subsidence regionally.

Subsidence calculations

For this study, individual wells (Fig. 1) and, where appropriate, synthetic wells from seismic lines have been backstripped assuming Airy compensation according to the method of Steckler and Watts (1978). Backstripping produces a plot of water-loaded depth to basement against time after corrections have been made for the effects of compaction and sediment loading. This provides a picture of how tectonic subsidence, that proportion of subsidence due only to the tectonic force behind basin subsidence, has varied through time. Subsidence curves have been plotted against the timescale of Harland *et al.* (1990) and porosity/depth relationships determined by Sclater and Christie (1980) for sediments from the Central North Sea have been used in decompaction. Theoretical subsidence curves have been calculated assuming uniform, finite-duration extension using the method of Jarvis and McKenzie (1980).

Following the approach of Wood (1982), no eustatic sea-level corrections have been applied to the subsidence curves used here. Although there is a general agreement on the overall trend of eustatic sea-level variations, there are many different estimates for their precise magnitude and frequency. The Late Cretaceous highstand, for instance, has been estimated as being anywhere between 350 m (Hallam 1984) and 100 m (Watts & Steckler 1979) above present day sea-level. We prefer to consider potential eustatic sea-level effects during the interpretation of the subsidence curves.

Late Cretaceous and Tertiary subsidence in the Faeroe–Shetland Basin

The Faeroe–Shetland Basin (Fig. 2a) forms part of a NE/SW-trending basin system dominated by extension and subsidence during Cretaceous times (Earle *et al.* 1989; Duindam and van Hoorn 1987; Mudge and Rashid 1987). It is considered to be underlain by thinned continental lithosphere (Mudge and Rashid 1987) and is flanked by the relatively unthinned continental crust of the Shetland and Faeroe platforms (Bott 1984). The continent/ocean boundary lies further to the west, on the far side of the Faeroe Islands (Bott 1984).

More than 70 wells have been drilled, of which around 50 have been released. Most lie in the shallower, eastern part of

From *Petroleum Geology of Northwest Europe: Proceedings of the 4th Conference* (edited by J. R. Parker).

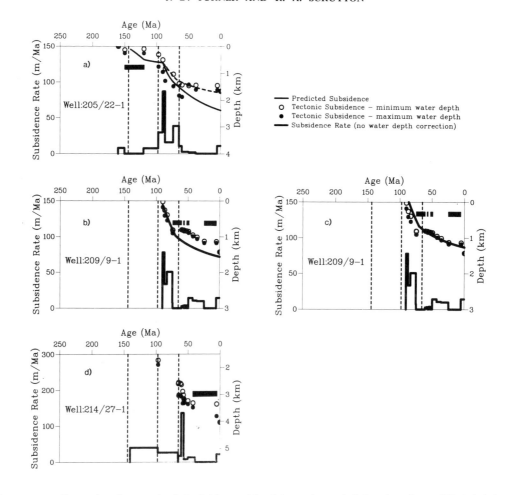

Fig. 1. Subsidence curves. Observed, sediment unloaded subsidence with minimum (open circles) and maximum (filled circles) water depth corrections fitted to theoretical subsidence curves for wells in the Faeroe–Shetland Basin. Sediment unloaded tectonic subsidence rates (no water depth correction applied) are shown along the bottom axis of these plots. Well locations shown on Fig. 2b. Horizontal bars indicate unconformities. Dashed vertical lines indicate Jurassic/Early Cretaceous, Early Cretaceous/Late Cretaceous and Late Cretaceous/Tertiary boundaries. (**a**) BP Well 205/22-1; west of the Rona Ridge, just inside the southern end of the main Faeroe Basin. Fitted to two theoretical subsidence curves, both including Early Cretaceous stretching $\beta = 1.1$, with Late Cretaceous stretching $\beta = 1.2$ (dashed line) and $\beta = 1.5$ (solid line). Note the low and decreasing rates of Early Tertiary subsidence. These rates are much lower than those predicted by theoretical models fitted to the high rates of Late Cretaceous subsidence observed (solid line). If the observed Tertiary subsidence is modelled (dashed line), much lower rates of Late Cretaceous subsidence than those observed are predicted. (**b**) BP Well 209/9-1; northeastern flank of the basin, on the Erlend Platform. Late Cretaceous observed subsidence fitted to Late Cretaceous stretching $\beta = 1.3$. (**c**) BP Well 209/9-1. The observed Tertiary subsidence can be fitted to a theoretical subsidence curve modelled for Late Cretaceous stretching, $\beta = 1.3$, where an episode of Early Tertiary uplift is included in the model. The timing of this apparent uplift corresponds to a mid-Paleocene unconformity developed in this well and across the eastern side of the basin. (**d**) Chevron Well 214/27-1; centre of the main Faeroe Basin. Note the very high rates of Late Paleocene (60.5 to 56.5 Ma) subsidence. N.B. Subsidence rates are plotted against a scale of 0–300 m/Ma in this case, rather than the scale of 0–150 m/Ma used in (a), (b) and (c).

the basin, especially along the Rona Ridge which divides the basin into two (Fig. 2b). To the east, the smaller West Shetland sub-basin is a relatively shallow, asymmetric half graben bounded on its eastern margin by the steeply dipping Shetland Spine Fault. Many wells drilled here have penetrated basement. The oldest rift-related sediments proved are thick Permo-Triassic red beds which are overlain, unconformably, by a Jurassic to Recent cover (see Hitchen and Ritchie 1987 for a more detailed description). West of the Rona Ridge is the main Faeroe Basin. This is estimated to contain up to 10 km of sediments in its deepest parts (Mudge and Rashid 1987). Pre-Upper Cretaceous sediments have been proved in only a few wells here, due to the presence of an extremely thick Tertiary, especially Paleocene, section in the basin centre.

Stretching history

Whilst a general consensus has been reached on the history of tectonic activity within the basin, disagreement remains over

the precise timing, duration and relative importance of individual rifting episodes.

The earliest documented extension dates from Late Permian to Triassic times when the West Shetland sub-basin was initiated as part of a series of NE/SW-trending half-graben along the UK western margin (Earle *et al.* 1989). Active extension waned and eventually ceased during late Triassic times (Earle *et al.* 1989). The extent of Permo-Triassic extension within the main Faeroe Basin remains unproven, but was probably only minor (Duindam and van Hoorn 1987, plate 4). Extension and active fault movements are thought to have resumed during latest Callovian times (Earle *et al.* 1989), although thick Bajocian–Bathonian sands proved in Well 206/5-1 have also been interpreted as products of active faulting (Hitchen and Ritchie 1987). Rifting died out again in latest Jurassic times allowing a drape of anoxic Early Berriasian muds to be deposited over the Jurassic fault block topography (Earle *et al.* 1989).

Early Cretaceous rifting, which established the basin as a

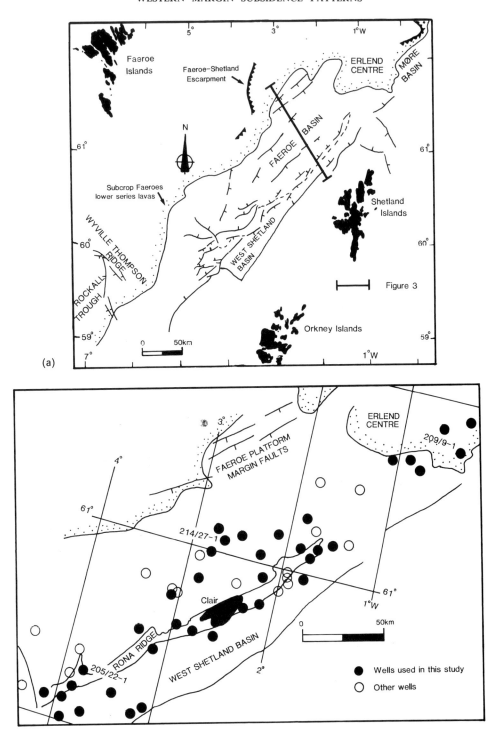

Fig. 2. (a) Tectonic elements of the Faeroe–Shetland Basin area. The Faeroe–Shetland Basin is part of the same Cretaceous basin system as the Rockall Trough, to the south, and the Møre Basin, to the north. After British Geological Survey offshore map sheets. The location of the geoseismic line shown in Fig. 3 is indicated. **(b)** Well location map. Subsidence curves from labelled wells are shown in Fig. 1.

major depocentre, began during latest Berriasian times and was especially intense for the next 10 Ma (Earle *et al.* 1989). After a period of mid-Cretaceous tectonic quiescence, extension resumed during Turonian times (Duindam and van Hoorn 1987). Fault movements continued sporadically into Maastrichtian times with a distinct pulse of Campanian faulting and associated footwall uplift (Hitchen and Ritchie 1987). Paleocene extensional faulting, the youngest reported, controlled a series of small depocentres along the Shetland Spine fault and produced the Flett Basin, a symmetrical sub-basin which developed along the axis of the main Cretaceous basin

(Fig. 3; Hitchen and Ritchie 1987; Mudge and Rashid 1987).

This history of repeated episodes of extension was punctuated by inversion and uplift during Late Maastrichtian to Early Paleocene, and Oligo-Miocene times (Earle *et al.* 1989; Swiecicki *et al.* 1993). Uplift and inversion were especially pronounced in the southwest of the basin where, in places, Pliocene sediments sit unconformably on Paleocene, Cretaceous or even older units (e.g. Duindam and van Hoorn 1987, plate 4, sections D–D′ & E–E′). Around 1.5 km of Oligo-Miocene erosion has been reported from these areas (Swiecicki *et al.* 1993).

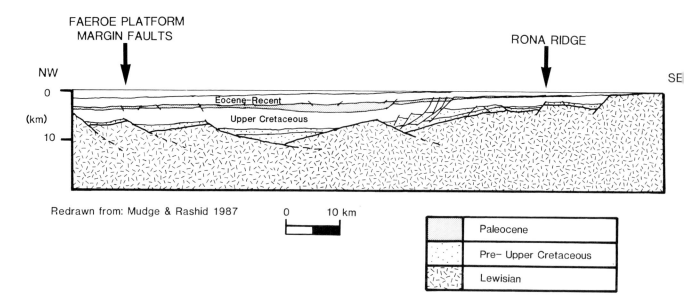

Fig. 3. Geoseismic section across the central part of the Faeroe–Shetland Basin. The line location is shown in Fig. 2. Note the overall westward thickening of the Upper Cretaceous section, overlain by a narrower fault-controlled symmetrical Paleocene sub-basin. Eocene to Recent sediments are preserved as a westward thickening wedge, a geometry also followed by present-day water depths.

Tectonic subsidence

Observed tectonic subsidence curves have been calculated for all released and many unreleased wells in the Faeroe–Shetland Basin (Fig. 2b). Ideally, these curves should be modelled using those periods of extension recognized by other workers, mainly from seismic reflection data, and described above. Three major periods of Mesozoic extension have been considered:

(1) Late Jurassic; Late Callovian–Tithonian (160–146 Ma)
(2) Early Cretaceous; Late Berriasian–Barremian (141–120 Ma)
(3) Late Cretaceous; Turonian–Early Maastrichtian (90–70 Ma)

In practice, the available observed subsidence data are not sufficient to constrain precisely the relative magnitude of these events. Many equally 'best fitting' theoretical subsidence curves can be produced with a range of magnitudes and durations for individual rifting episodes. The approach taken in this paper, therefore, has been to make conservative estimates for the amount of Jurassic and Early Cretaceous extension, so that the bulk of extension is modelled as Late Cretaceous.

Whatever the relative importance of these rifting episodes, a number of general predictions concerning tectonic subsidence rates will hold true for the entire basin. With active extension during Early and Late Cretaceous times, Late Cretaceous and Paleocene subsidence rates should have been high, especially if a period of Paleocene extension is included in the model. Eocene to Recent subsidence rates should then have slowed, exponentially, as post-rift thermal subsidence took over.

The subsidence patterns observed are rather more complicated. For, although Late Cretaceous subsidence rates were uniformly high (Figs 1a, b and d), Paleocene times saw a marked divergence in subsidence behaviour between the flanks and centre of the basin. Subsidence rates in wells from the basin flanks (Figs 1a and b) slowed dramatically, much more rapidly than theoretical models predict, whilst wells in the centre (Fig. 1d) show a sudden acceleration in subsidence rates during Thanetian times (60.5–56.5 Ma). Flank uplift of around 100 m is indicated (Figs 1a, b and c), associated with the development of a mid-Paleocene unconformity across much of the area (Fig. 4b). This unconformity formed at approximately

the same time as the sudden acceleration in tectonic subsidence rates in the basin centre, and in other basins nearby (White *et al.* 1992; White and Latin 1992; Joy 1993). The overall basin geometry reflects this sudden change in subsidence patterns, narrowing so that sedimentation became concentrated along the fault controlled Flett Basin (Fig. 3; Hitchen and Ritchie 1987; Mudge and Rashid 1987).

Tectonic subsidence behaviour returned to normal during Early Eocene times (Figs 1a, b and d). The thickness of Eocene to Recent sediments, and present-day water depths, increase towards the west (Fig. 3), consistent with the effects of thermal subsidence and westward tilting as the basin moved away from the Iceland hotspot during Tertiary times. This pattern of smooth post-rift subsidence was interrupted by the development of a major Oligo-Miocene unconformity in many wells (Figs 1b and d), associated with the uplift and inversion documented from the southwest of the basin (Earle *et al.* 1989; Swiecicki *et al.* 1993). Quaternary times saw another abrupt, acceleration in tectonic subsidence rates, similar to that recognized in other North Atlantic marginal basins (Cloetingh *et al.* 1990) and the North Sea (Thorne and Watts 1989).

Magmatism, unconformities and changes in water depth—relative timing and their relationship to regional tectonics

As well as the peculiar subsidence patterns described above, Late Cretaceous and Early Tertiary times saw major changes in sedimentation patterns and water depths, as well as a considerable amount of igneous activity throughout the Faeroe–Shetland Basin. This all occurred during a major period of tectonic reorganization in the North Atlantic region. The Iceland plume is thought to have been initiated at *c.* 62 Ma (R. White 1989). Plume-related magmatism continued through Danian and Thanetian times, reaching a peak immediately before and during continental break-up at approximately 55 Ma (R. White 1988). Since early Eocene times, magmatism has been restricted to Iceland and the spreading centre itself, the plume's present-day topographic expression covering some $14 \times 10^6 \text{ km}^2$ (Bott 1988). In this section the relative timing of tectonic events, intrabasinal volcanism, water depth and sedimentation pattern changes are described with reference to a series of schematic palaeogeographic reconstructions (Fig. 4). These relationships are summarized in Fig. 5.

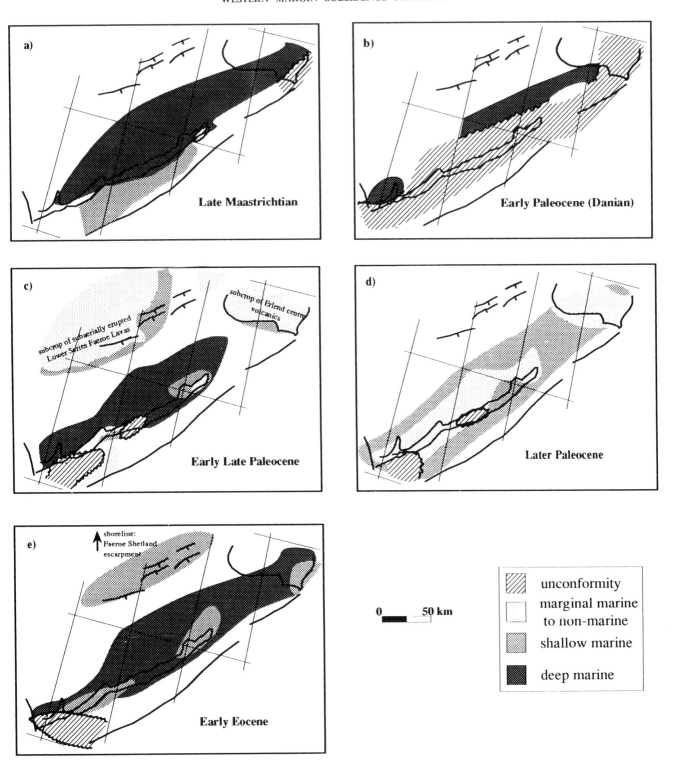

Fig. 4. Schematic palaeogeographic/palaeo-water depth maps for the Faeroe–Shetland Basin from Late Cretaceous to Early Tertiary times: (**a**) Late Maastrichtian, *c.* 66 Ma; (**b**) Danian, *c.* 61 Ma; (**c**) Early Late Paleocene, *c.* 60 Ma; (**d**) Later Paleocene, *c.* 57 Ma; (**e**) Early Eocene, *c.* 53 Ma.

Late Cretaceous tectonic subsidence rates were extremely rapid due to the effects of active Turonian–Early Maastrichtian extension. This rapid subsidence allowed the accumulation of a very thick, > 2000 m in places, sequence of dark grey to black marine muds over the whole basin. These muds were deposited in outer shelf to upper bathyal water depths, which tended to increase during Late Cretaceous times to a maximum in Maastrichtian times (Fig. 4a).

Relatively deep marine sedimentation continued, at least in the centre of the basin, into Danian times after a minor end-Cretaceous hiatus (Fig. 4b). Danian sediments are missing

from many wells in the west of the basin due to the combined effects of this end-Cretaceous hiatus and a more important late Danian/early Thanetian unconformity (Fig. 4b). This late Danian unconformity, accompanied by localized inversion in the southwest of the basin (Swiecicki *et al.* 1993), is an expression of the flank uplift apparent from subsidence analysis (Fig. 1). Although the difficulties involved in comparing radiometric, biostratigraphic and magnetostratigraphic dating schemes make a precise correlation impossible, this period of simultaneous uplift and rapid subsidence does appear to be contemporaneous with the oldest reported plume-related mag-

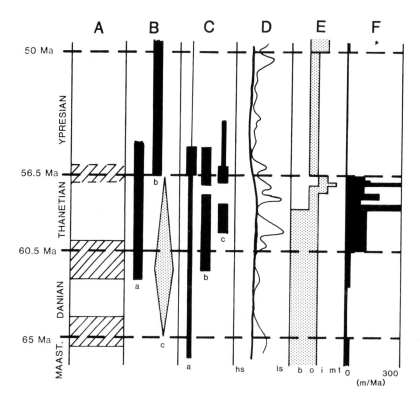

Fig. 5. Latest Cretaceous and Early Tertiary geologic events, regionally, in the Faeroe–Shetland Basin and in Chevron Well 214/27-1. (**A**) Hatched areas represent major intrabasinal unconformities. (**B**) Regional tectonic events: (a) Magmatism in marginal basins associated with the Iceland plume. (b) Ocean floor spreading between Greenland and Northwest Europe. (c) Laramide phase of the Alpine orogeny. (**C**) Intrabasinal igneous activity: (a) Radiometric dates from members of the Faeroe–Shetland sill complex. (b) Ages of lower and upper/middle series Faeroe lavas. (c) Tuffs; phase 1 and phases 2 a–c (Balder) of Knox & Morton (1983). (**D**) Eustatic sea-level fluctuations, 1st and 2nd order, from Haq *et al.* (1987). ls, lowstand, present day sea-level; hs, highstand, 300 metres above present day sea-level. (**E**) Water depths, Well 214/27-1: b, upper bathyal, >175 metres; o, outer shelf, 50–175 metres; i, inner shelf, 10–50 metres, m, marginal marine, 0–10 metres; t, terrestrial, 0 metres. (**F**) Sediment unloaded subsidence rate (m/Ma), Well 214/27-1, equals tectonic subsidence rate with no correction for palaeowater depth. Sources: Eldholm *et al.* (1989); Fitch *et al.* (1988); Hitchen and Ritchie (1987); Knox and Morton (1988); Waagstein (1988); R. White (1988); Ziegler (1990). Plotted against the time scale of Harland *et al.* (1990).

matism in Scotland and West Greenland, between around 62 and 60 Ma (Fig. 5; Mussett *et al.* 1988; Larsen *et al.* 1991). The lower series of Faeroe basalts, which subcrop on the western side of the basin, have also been assigned a Late Danian to Thanetian age, magnetochrons C26R–C25N (Waagstein 1988), as have some members of the Faeroe–Shetland sill complex (Hitchen and Ritchie 1987).

By early Thanetian times, *c.* 60 Ma, marine sedimentation had resumed across almost the entire basin (Fig. 4c). Plume-related igneous activity continued regionally and within the Faeroe–Shetland Basin during Thanetian times. The younger parts of the lower series Faeroe lavas are Thanetian in age (Waagstein 1988). Thanetian radiometric dates have also been obtained from members of the Faeroe–Shetland sill complex (Hitchen and Ritchie 1987) and a minor Late Paleocene tuff, probably equivalent to the phase 1 tuffs of Knox and Morton (1983), is present in some wells (Morton *et al.* 1988). Tectonic subsidence rates on the flanks of the basin (Fig. 1a and b) remained fairly low during Thanetian times but continued to accelerate to about 175 m/Ma in the basin centre (Figs 1d and 5). Water depths decreased across the whole basin, reaching a minimum during late Thanetian times, *c.* 57 Ma, (Figs 4d and 5) when marginal marine to shallow marine conditions prevailed. The sediments deposited at this time may correlate with coals found between the lower and middle/upper series Faeroe lavas on the Faeroe Islands themselves which are thought to have accumulated during a period of igneous quiescence (Lund 1983).

Subsidence rates in the basin centre remained high as water

depths increased and extrusive igneous activity resumed towards the end of Thanetian times (Fig. 5). This phase of igneous activity, between about 57 and 55 Ma was by far the most volumetrically important in the basin's history. The entire thickness of upper/middle series Faeroe lavas, the bulk of the Faeroe–Shetland sill complex and the Balder tuffs were all produced during this period (Waagstein 1988; Gibb and Kanaris-Sotiriou 1988; Morton and Knox 1990). Indeed, the whole region was subjected to enormous amounts of magmatism in latest Paleocene/earliest Eocene times associated with the inception of seafloor spreading between Greenland and Northwest Europe. Tectonic subsidence rates decreased suddenly in the basin centre, returning subsidence behaviour across the entire basin to normal at about the same time as continental break-up was achieved (Fig. 5).

Water depths continued to deepen so that in Early Eocene times (55 Ma onwards) the basin was again dominated by marine sedimentation, with only moderate rates of tectonic subsidence and little or no intrabasinal igneous activity (Figs 4e and 5).

There seems to be a temporal link between these peculiar Paleocene subsidence patterns, igneous activity and regional tectonics (Fig. 5). The acceleration in basin centre subsidence coincides with a period of flank uplift and the initiation of the Iceland plume (mid-Paleocene). Tectonic subsidence patterns returned abruptly to normal across the whole basin at the same time as the inception of sea-floor spreading to the northwest, in earliest Eocene times. The implications of this relationship for the anomalous Paleocene subsidence reported from other nearby basins, together with the models proposed to explain them, are considered in more detail below.

Discussion and implications

The Late Cretaceous to Recent development of the North Sea has traditionally been regarded as the post-rift, thermal sag phase of extensional sedimentary basin evolution. It is only relatively recently that the likely tectonic significance of anomalously high backstripped subsidence rates during Late Paleocene and Quaternary times has been appreciated.

Whilst anomalously high Late Paleocene subsidence rates have been reported from only the Witch Ground Graben and triple junction area of the North Sea (White and Latin 1992; Joy 1993), similar increases in tectonic subsidence rates are apparent in previous studies of the East Shetland Basin and the Viking and Central Grabens (Wood 1982; Zervos 1986; Condon 1988; White, N. J. 1988). Accelerated Late Paleocene subsidence rates have also been reported from the Porcupine Basin (White et al. 1992) and the central parts of the Faeroe–Shetland Basin (this study). The recognition of these peculiar subsidence patterns across so much of the Northwest European continental margin is indicative of a regional, rather than local, control. As subsidence rates began to accelerate in the centre of the Faeroe–Shetland Basin, a major unconformity developed across the basin flanks. Mid-Paleocene unconformities have also been reported from the Western Approaches Trough, Porcupine Basin, Goban Spur, Biscay and Vøring margins (Murray 1979; Snyder et al. 1985; Hillis 1988; Dobson et al. 1991; Skogseid et al. 1992), again indicating a regional control.

An increase in the rate of tectonic subsidence is not necessarily reflected by a commensurate increase in sediment accumulation rate; there may instead be an increase in water depth. Therefore, before any tectonic significance can be attributed to the anomalous subsidence patterns described above, the interdependence of sediment unloaded subsidence rates and water depth, as well as the effects of eustasy, must be considered. In the case of the Late Paleocene acceleration in tectonic subsidence it is possible that it is only apparent, an artefact of the rapid infilling of a deep water Cretaceous basin rather than the result of any additional Paleocene tectonism.

Two short periods of shallowing water depths do indeed correlate with especially high sediment unloaded subsidence rates in Well 214/27–1 (Fig. 5). However, even when extreme palaeo-water depth corrections are made (deepest possible—end Danian; shallowest possible—latest Thanetian), an acceleration in tectonic subsidence rates is still apparent during Late Paleocene times (Fig. 1d). Also, the period of rapid tectonic subsidence continued whilst water depths were actually increasing in latest Paleocene times (Fig. 5). In this case, therefore, it seems that there was a genuine acceleration in tectonic subsidence rates during Thanetian times, its effect amplified, rather than created, by the infilling of a deep water basin. A similar picture emerges from the main North Sea Basin, where a period of anomalously high post-Early Paleocene subsidence rates are indicated, even when subsidence curves are plotted only for those times at which water depths can be precisely constrained, i.e. the present day and periods of coal formation (Bertram and Milton 1989).

During late Danian and Thanetian times the first-order eustatic sea-level curve of Haq et al. (1987) shows a gentle fall followed by a much sharper rise from approximately 58 to 55 Ma (Fig. 5). Sediment unloaded tectonic subsidence rates remained high throughout this period whilst water depths decreased during Thanetian times, reaching a minimum at approximately 57 Ma (Figs 4 and 5). Although differences in biostratigraphic dating schemes make direct comparisons difficult, eustatic sea-level changes appear to be out of phase with

the observed changes in tectonic subsidence rates and water depths, effectively ruling out eustasy as the dominant controlling factor of the observed Late Paleocene subsidence patterns. Furthermore, the magnitude of this eustatic correction, which increases observed tectonic subsidence rates by around 5 m/Ma in early Thanetian times and decreases them by around 15 m/Ma in later Thanetian times, is much smaller than the observed anomaly. The combined effects of eustasy and basin infilling only begin to explain the observed subsidence rates when extreme corrections for palaeowater depths are made. When this is considered, together with the regional evidence for anomalously high rates of Late Paleocene subsidence and contemporaneous flank uplift and erosion, a purely eustatic/sedimentary explanation seems unlikely.

An increase in tectonic subsidence rates of the magnitude described from the Porcupine Basin, North Sea and central parts of the Faeroe–Shetland Basin is consistent with the effects of an additional small pulse of Paleocene extension. Paleocene extension has been suggested for the North Sea and Faeroe–Shetland Basins on the basis of active normal faulting (Mudge and Rashid 1987; Ziegler 1990). Small amounts of Late Paleocene normal faulting have also been reported from the Porcupine Basin (Tate 1990). A period of Paleocene stretching is consistent with that reported from other basins on the margin of the proto-North Atlantic (Skogseid and Eldholm 1989), rifting which culminated in the initiation of seafloor spreading between Greenland and Northwest Europe in earliest Eocene times. However, although active Paleocene extension in the Faeroe–Shetland Basin is generally accepted (Hitchen and Ritchie 1987; Mudge and Rashid 1987), the apparent syn-sedimentary normal faults reported from the Porcupine and North Sea basins are more commonly believed to be due to the effects of differential compaction across preexisting fault scarps (Bertram and Milton 1989; N. White, pers. comm.). Certainly there is no conclusive evidence for active Paleocene lithospheric extension from either the North Sea or Porcupine basins, nor can a simple model of Paleocene extension explain the simultaneous flank uplift and rapid subsidence observed in the Faeroe–Shetland Basin.

The initiation of rapid basinal subsidence coincided with the earliest reported magmatism associated with the Iceland plume (Fig. 5). Subsidence rates returned to more normal levels during the earliest Eocene at about the same time as continental break-up was achieved (Fig. 5). This coincidence led Joy (1992) to suggest that there may be a direct link between the thermal effects of the developing Iceland plume and the high rates of Paleocene subsidence observed. Any model involving heat transfer between the plume and these sedimentary basins has to account for the transfer taking place extremely quickly since the period of rapid subsidence lasted about 4 million years (Fig. 5). Lithosphere heating and cooling effects generally show up over much longer time scales, tens of millions of years. It is therefore more likely that the rapid subsidence rates observed are a mechanical rather than a thermal response.

Cloetingh et al. (1990) suggested that the rapid pulse of Quaternary tectonic subsidence observed in the North Sea (Thorne and Watts 1989), and many other basins around the North Atlantic rim, was due to the response of the lithosphere to a change in intra-plate stress. Their model involves uplift of basin flanks coupled with accelerated subsidence in the basin centre, due to the bending of the lithosphere under a compressive intraplate stress. This model may also have an application to the anomalous Late Paleocene subsidence patterns observed in sedimentary basins on the UKCS. Evidence of flank uplift and simultaneous rapid subsidence, presented here from the Faeroe–Shetland Basin, is consistent with the effects predicted for the development of a compressive intraplate stress field during middle Paleocene times. The combined effects of the relatively slow, but stable, uplift associated with the Iceland plume and the more dynamic effects of the Laramide phase of Alpine deformation could have caused the required compressional stress field at the correct time (Fig. 5). Models using feasible levels of intraplate stresses (100–500 MPa) predict 10–100 m of basin margin uplift (Karner 1986) and increased subsidence in the basin centre of more than 150 m (Cloetingh et al. 1990). These predictions are similar to the observations made from the Faeroe–Shetland Basin. Then as the effects of Laramide compression died away during latest Thanetian times, the region returned to a tensile stress regime associated with renewed igneous activity and the normalization of subsidence behaviour. In some ways this idea is consistent with the view expressed by Ziegler (1990) that Alpine compressional forces impeded the initiation of continental break-up between Greenland and Northwest Europe until latest Paleocene/earliest Eocene times.

The above should only be regarded as a working hypothesis. More work is needed to define the extent and history of development of these anomalous Late Paleocene subsidence patterns, and the accompanying mid-Paleocene uplift, in basins across the whole of the NW European continental margin. Whatever their precise cause, it is becoming clear that these peculiar subsidence patterns are an expression of Paleocene tectonics which affected the whole region. There are therefore a number of important implications for subsidence and heat flow history modelling. Any model appealing to a period of renewed lithospheric extension, or to the thermal effects of the Iceland hotspot, implies a period of heat input which will have altered the maturation history of sediments in these basins. Similarly, a failure to include an estimate for the amount of uplift associated with mid-Paleocene and Oligo-Miocene unconformities in geohistory modelling will result in an underestimate of the degree of maturity of source rocks. Finally, previous studies, in particular of Jurassic extension in the North Sea, which have relied on fitting predicted thermal subsidence curves to a supposed post-rift observed subsidence curve from Cretaceous to Recent in the North Sea (e.g. Latin 1990; Giltner 1987), may have overestimated the amount of stretching from subsidence, ignoring the tectonic implications of these periods of rapid subsidence.

Arco British, Amerada Hess, Amoco UK, BP Exploration, Chevron UK, Elf UK, Esso, the Gas Council, LASMO North Sea, Marathon Oil UK, Mobil North Sea, Phillips Petroleum, Shell UK, Sovereign Oil & Gas, Sun Oil International, Texaco, Total Oil Marine and Western Geophysical kindly provided the data upon which this research is based. Financial support, in the form of a studentship for J. D. T., was provided by the Petroleum Science and Technology Institute. Aidan Joy, Dave Latin, Jakob Skogseid and Nicky White are thanked for providing preprints of their work. J. D. T. also wishes to thank Shell UK for funding his attendance at the Conference itself. The text was much improved by comments from Olav Eldholm and Tony Spencer. This is The Petroleum Science and Technology Institute Contribution Number 7.

References

BERTRAM, G. T. AND MILTON, N. J. 1989. Reconstructing basin evolution from sedimentary thickness; the importance of palaeobathymetric control, with reference to the North Sea. Basin Research, 1, 247–259.
BOOTH, J., SWIECICKI, T. AND WILCOCKSON, P. 1993. The tectono-stratigraphy of the Solan Basin, west of Shetland. In: PARKER, J. R. (ed.) Petroleum Geology of Northwest Europe: Proceedings of the 4th Conference. Geological Society, London, 987–998.
BOTT, M. H. P. 1984. Deep structure and origin of the Faeroe–Shetland Channel. In: SPENCER, A. M. et al. (eds) Petroleum Geology of the North European Margin. Graham & Trotman, London, 341–347.
—— 1988. A new look at the causes and consequences of the Icelandic hot-spot. In: MORTON, A. C. AND PARSON, L. M. (eds) Early

Tertiary Volcanism and the Opening of the NE Atlantic. Geological Society, London, Special Publication, 39, 15–23.
CLOETINGH, S., GRADSTEIN, F. M., KOOI, H. AND KAMINSKI, M. 1990. Plate reorganisation: a cause of rapid late Neogene subsidence and sedimentation around the North Atlantic? Journal of the Geological Society, London, 147, 495–506.
CONDON, P. J. 1988. Seismic Stratigraphy and Distribution of Paleogene Sediments West and East of Shetland. PhD thesis, University of Edinburgh.
DOBSON, M. R., HAYNES, J. R., BANNISTER, A. D., LEVENE, D. G., PETRIE, H. S. AND WOODBRIDGE, R. A. 1991. Early Tertiary palaeoenvironments and sedimentation in the NE Main Porcupine Basin (Well 35/13-1), offshore western Ireland—evidence for global change in the Tertiary. Basin Research, 3, 99–119.
DUINDAM, P. AND VAN HOORN, B. 1987. Structural evolution of the West Shetland continental margin. In: BROOKS, J. AND GLENNIE, K. W. (eds) Petroleum Geology of North West Europe. Graham & Trotman, London, 765–773.
EARLE, M. M., JANKOWSKI, E. J. AND VANN, I. R. 1989. Structural and Stratigraphic Evolution of the Faeroe–Shetland Channel and Northern Rockall Trough. In: TANKARD, A. J. AND BALKWILL, H. R. (eds) Extensional Tectonics and Stratigraphy of the North Atlantic Margins. AAPG Memoir, 46, 461–469.
ELDHOLM, A., THIEDE, J. AND Taylor, E. 1989. Evolution of the Vøring Volcanic Margin. In: ELDHOLM, O., THIEDE, J. et al. (eds) Proceedings of the Ocean Drilling Program, Scientific Results, Ocean Drilling Program, College Station, TX, 104, 1033–1065.
FITCH, F. J., HEARD, G. L. AND MILLER, J. A. 1988. Basaltic magmatism of late Cretaceous and Paleogene age recorded in wells NNE of the Shetlands. In: MORTON, A. C. AND PARSON, L. M. (eds) Early Tertiary Volcanism and the Opening of the NE Atlantic. Geological Society, London, Special Publication, 39, 253–262.
GIBB, F. G. F. AND KANARIS-SOTIRIOU, R. 1988. The geochemistry and origin of the Faeroe–Shetland sill complex. In: MORTON, A. C. AND PARSON, L. M. (eds) Early Tertiary Volcanism and the Opening of the NE Atlantic. Geological Society, London, Special Publication, 39, 241–252.
GILTNER, J. P. 1987. Application of extensional models to the northern Viking Graben. Norsk Geologisk Tidsskrift, 67, 339–352.
HALLAM, A. 1984. Pre-Quaternary Sea-level Changes. Annual Reviews of Earth and Planetary Science, 12, 205–243.
HAQ, B. U., HARDENBOL, J. AND VAIL, P. R. 1987. Chronology of fluctuating sea levels since the Triassic. Science, 235, 1156–1167.
HARLAND, W. B., ARMSTRONG, R. L., COX, A. V., CRAIG, L. E., SMITH, A. G. AND SMITH, D. G. 1990. A Geologic Time Scale 1989. Cambridge University Press.
HILLIS, R. R. 1988. The Geology and Tectonic Evolution of the Western Approaches Trough. PhD thesis, University of Edinburgh.
HITCHEN, K. AND RITCHIE, J. D. 1987. Geological Review of the West Shetland area. In: BROOKS, J. AND GLENNIE, K. W. (eds) Petroleum Geology of North West Europe. Graham & Trotman, London, 737–749.
JARVIS, G. T. AND MCKENZIE, D. P. 1980. Sedimentary basin formation with finite extension rates. Earth and Planetary Science Letters, 48, 42–52.
JOY, A. M. 1992. Right place, wrong time: anomalous post-rift subsidence in sedimentary basins around the North Atlantic Ocean. In: STOREY, B. C., ALABASTER, T. AND PANKHURST, R. J. (eds) Magmatism and the Causes of Continental Break-up. Geological Society, London, Special Publication 68, 387–394.
—— 1993. Comments on the pattern of post-rift subsidence in the central and northern North Sea Basin. In: WILLIAMS, G. D. (ed.) Tectonics and Seismic Sequence Stratigraphy. Geological Society, London, Special Publication, 71, 123–140.
KARNER, G. D. 1986. Effects of lithospheric in-plane stress on sedimentary basin stratigraphy. Tectonics, 5, 573–588.
KNOX, R. W. O'B. AND MORTON, A. C. 1983. Stratigraphic Distribution of Early Palaeogene Pyroclastic Deposits in the North Sea Basin. Proceedings of the Yorkshire Geological Society, 44, 355–363.
—— AND —— 1988. The record of early Tertiary North Atlantic volcanism in sediments of the North Sea Basin. In: MORTON, A. C. AND PARSON, L. M. (eds) Early Tertiary Volcanism and the

Opening of the NE Atlantic. Geological Society, London, Special Publication, **39**, 407–419.

LARSEN, L. M., PEDERSEN, A. K., PEDERSEN, G. K. AND PIASECKI, S. 1992. Timing and duration of the early tertiary volcanism in the North Atlantic: new evidence from West Greenland. *In*: STOREY, B. C., ALABASTER, T. AND PANKHURST, R. J. (eds) *Magmatism and the Causes of Continental Break-up*, Geological Society, London, Special Publication, **68**, 321–334.

LATIN, D. M. 1990. *The Relationship Between Extension and Magmatism in the North Sea Basin*. PhD thesis, University of Edinburgh.

LUND, J. 1983. Biostratigraphic of Interbasaltic Coals from the Faeroe Islands. *In*: BOTT, M. H. P., SAXOV, S., TALWANI, M. AND THIEDE, J. (eds) *Structure and Development of the Greenland–Scotland Ridge*. Plenum, New York, 417–423.

MCKENZIE, D. 1978. Some remarks on the development of Sedimentary Basins. *Earth and Planetary Science Letters*, **40**, 25–32.

MORTON, A. C., EVANS, D., HARLAND, R., KING, C. AND RITCHIE, D. K. 1988. Volcanic ash in a cored borehole W of the Shetland Islands: evidence for Sclandian (late Paleocene) volcanism in the Faeroes region. *In*: MORTON, A. C. AND PARSON, L. M. (eds) *Early Tertiary Volcanism and the Opening of the NE Atlantic*. Geological Society, London, Special Publication, **39**, 263–269.

—— AND KNOX, R. W. O'B. 1990. Geochemistry of Late Paleocene and early Eocene tephras from the North Sea Basin. *Journal of the Geological Society, London*, **147**, 425–437.

MUDGE, D. C. AND RASHID, B. 1987. The Geology of the Faeroe Basin area. *In*: BROOKS, J. AND GLENNIE, K. W. (eds) *Petroleum Geology of North West Europe*. Graham & Trotman, London, 751–763.

MURRAY, J. W. 1979. Cenozoic Biostratigraphy and Paleoecology of Sites 403 to 406 based on the Foraminifers. Initial Reports of the Deep Sea Drilling Project, **48**, 415–430.

MUSSETT, A. E., DAGLEY, P. AND SKELHORN, R. R. 1988. Time and duration of activity in the British Tertiary Igneous Province. *In*: MORTON, A. C. AND PARSON, L. M. (eds). *Early Tertiary Volcanism and the Opening of the NE Atlantic*. Geological Society, London, Special Publication, **39**, 337–348.

SCLATER, J. G. AND CHRISTIE, P. A. F. 1980. Continental Stretching: An Explanation of the Post-Mid-Cretaceous subsidence of the Central North Sea Basin. *Journal of Geophysical Research*, **85**, 3711–3739.

SKOGSEID, J. AND ELDHOLM, O. 1989. Vøring Plateau continental margin: Seismic Interpretation, stratigraphy, and vertical movements. *In*: ELDHOLM, O., THIEDE, J. *et al*. (eds) *Proceedings of the Ocean Drilling Program, Scientific Results*, Ocean Drilling Program, College Station, TX, **104**, 993–1030.

——, PEDERSEN, T., ELDHOLM, O. AND LARSEN, B. T. 1992. Tectonism and magmatism during NE Atlantic continental break-up: the Vøring Margin. *In*: STOREY, B. C., ALABASTER, T. AND PANK-

HURST, R. J. (eds) *Magmatism and the Causes of Continental Break-up*. Geological Society, London, Special Publication, **68**, 305–320.

SNYDER, S. W., MULLER, C., SIGAL, J., TOWNSEND, H. AND POAG, C. W. 1985. Biostratigraphic, paleoenvironmental and paleomagnetic synthesis of the Goban Spur Region. *Initial Reports of the Deep Sea Drilling Project*, **80**, 1169–1186.

STECKLER, M. S. AND WATTS, A. B. 1978. Subsidence of the Atlantic-type Continental margin off New York. *Earth and Planetary Science Letters*, **41**, 1–13.

TATE, M. P. 1990. *Structural Framework and Tectono-Stratigraphic Evolution of the Porcupine Seabight Basin, Offshore Western Ireland*. PhD thesis, University College of Wales, Aberystwyth.

THORNE, J. A. AND WATTS, A. B. 1989. Quantitative Analysis of North Sea Subsidence. *American Association of Petroleum Geologists Bulletin*, **73**, 88–116.

WAAGSTEIN, R. 1988. Structure, composition and age of the Faeroe basalt plateau. *In*: MORTON, A. C. AND PARSON, L. M. (eds) *Early Tertiary Volcanism and the Opening of the NE Atlantic*. Geological Society, London, Special Publication, **39**, 225–238.

WATTS, A. B. AND STECKLER, M. S. 1979. *Subsidence and Eustasy at the Continental Margin of Eastern North America*. American Geophysical Union, Maurice Ewing Series **3**, 218–234.

WHITE, N. J. 1988. *Extension and Subsidence of the Continental Lithosphere*. PhD thesis, University of Cambridge.

WHITE, N. AND LATIN, D. 1992. Subsidence analyses from the North Sea 'Triple Junction'. *Journal of the Geological Society, London*, **150**, 473–488.

——, TATE, M. AND CONROY, J-J. 1992. Lithospheric stretching in the Porcupine Basin, west of Ireland. *In*: PARNELL, J. (ed.) *Basins on the Atlantic Seaboard: Petroleum Geology, Sedimentology and Basin Evolution*. Geological Society, London, Special Publication, **62**, 327–332.

WHITE, R. S. 1988. A hot-spot model for early Tertiary volcanism in the North Atlantic. *In*: MORTON, A. C. AND PARSON, L. M. (eds) *Early Tertiary Volcanism and the Opening of the NE Atlantic*. Geological Society, London, Special Publication, **39**, 3–13.

—— 1989. Initiation of the Iceland Plume and opening of the North Atlantic. *In*: TANKARD, A. J. AND BALKWILL, H. R. (eds) *Extensional Tectonics and Stratigraphy of the North Atlantic Margins*. AAPG Memoir **46**, 149–154.

WOOD, R. J. 1982. *Subsidence in the North Sea*. PhD thesis, University of Cambridge.

ZERVOS, F. A. 1986. *Geophysical Investigation of Sedimentary Basin Development: Viking Graben, North Sea*. PhD thesis, University of Edinburgh.

ZIEGLER, P. A. 1990. *Geological Atlas of Western and Central Europe* (2nd edn). Shell Internationale Petroleum, Maatschappij B. V., The Hague.

Development of the Rockall Trough and the northwest European continental margin

R. S. WHITE

Bullard Laboratories, University of Cambridge, Madingley Road, Cambridge CB3 0EZ, UK

During the Cretaceous and early Tertiary a series of sedimentary basins opened by rifting on the northwest margin of Europe. None of these developed into a fully-fledged ocean until the northern North Atlantic opened at 55 Ma between NW Europe and Greenland. The subsidence, magmatism, and thermal history of these basins and of the continental margin depended critically on the amount and style of stretching and on the temperature of the underlying asthenospheric mantle. Small variations in the mantle temperature can have dramatic effects on both the magmatism produced during rifting and on the subsidence of the basin. An increase of as little as 10% in the mantle temperature can cause uplift above sea-level rather than the more common subsidence when continental crust undergoes large amounts of extension. The initiation of the Iceland mantle plume at about 62 Ma caused abnormally elevated mantle temperatures across a 2000 km diameter region and had a dramatic effect on the subsidence and rifting history of sedimentary basins west of Britain. Their thermal history can be documented from the subsidence and magmatic record in the sedimentary basins and the continental margin.

Using normal incidence and wide-angle seismic data and drilling results from the continental margin and from basins of NW Europe, their subsidence and thermal history can be modelled as a function of the amount of stretching and the temperature of the underlying mantle. The Rockall Trough itself is underlain by 5 km of sediments intruded by Tertiary volcanics. The basement imaged on seismic profiles consists of banded sub-horizontal reflectors, some of which extend over 40 km laterally. These are interpreted as sediments intercalated with submarine lava flows generated during the opening of the Rockall Trough. The stretched non-sedimentary crust beneath the basement is only 6 km thick, indicating extensive rifting ($\beta > 6$). Between 1–3 km of molten rock was generated during rifting as the upwelling asthenosphere decompressed.

The Atlantic continental margin of NW Europe was within the sphere of influence of the newly initiated Iceland mantle plume when it rifted, with consequent massive volcanism and underplating accompanied by uplift above sea-level and an abrupt transition from continent to ocean. As rifting continued, the temperature of the mantle plume decreased, causing a concomitant decrease in the rate of production of igneous rocks and increased rates of subsidence as the formerly inverted regions dropped back below sea-level. In contrast, the Western Approaches and Galicia continental margins to the south were well away from any mantle plumes when they rifted, so generated only relatively minor syn-rift igneous activity, exhibit a normal pattern of subsidence, and developed large tilted fault blocks across a broad transition zone.

From *Petroleum Geology of Northwest Europe: Proceedings of the 4th Conference* (edited by J. R. Parker).
© 1993 Petroleum Geology '86 Ltd. Published by The Geological Society, London, p. 985.

The tectono-stratigraphy of the Solan Basin, west of Shetland

J. BOOTH, T. SWIECICKI and P. WILCOCKSON

Amerada Hess Ltd, 2 Stephen Street, London W1P 1PL, UK

Abstract: The Solan Basin lies in a part of the West Shetland continental margin that has had a complex tectonic history, dominated by extension but punctuated by several episodes of inversion, transpression and extensive erosion. The oldest sedimentary sequence, identified on seismic but not yet penetrated by drilling, may comprise Devono-Carboniferous clastics. In the Permo-Triassic a large system of half-grabens was filled with a thick succession of coarse, continental clastics. These appear to have entered the basin system in two discrete pulses and it is speculated that a third pulse, representing a transition to marginal and fully marine environments, occurred in the Early Jurassic. The area was effectively peneplaned in the late Middle to early Late Jurassic, with the removal of up to 1.5 km of section. Overlying this unconformity is a thin sequence of marginal marine sandstones and organic-rich marine shales, deposited in the latest Jurassic to earliest Cretaceous. Although parts of the region received a major influx of sand derived from the West Shetland Platform in the Early Cretaceous, the equivalent strata throughout most of the Solan Basin are a thin succession of pelagic shales and carbonates. In the early Turonian the basin was inverted. During the Late Cretaceous, extension, related to rifting along the line of the Faeroe–Shetland Trough, resulted in the development of large normal fault systems, providing the space in which a thick sequence of deep marine shales was deposited. In the earliest Paleocene, transpressional reactivation of some faults produced intense, but localized, inversion structures. Generally, however, sedimentation continued uninterrupted through the Paleocene, with the accumulation of deep marine sandstones in the east of the basin. The culmination of Thulean volcanism in the earliest Eocene was marked by the deposition of tuffaceous mudstones, which are overlain by thin coal-bearing, paralic to continental sediments. Regional thermal subsidence began in the early Eocene and continued into the Oligocene with the deposition of a thick sequence of marine clastics. In the Miocene, erosion removed up to 1.2 km of sediment from parts of the Rona Ridge and produced a basin-wide unconformity. This is overlain by Pliocene to Recent glacio-marine sands and gravels.

The Solan Basin is part of the West Shetland continental margin (Fig. 1). To the north lies the Late Cretaceous–Tertiary Faeroe Basin, to the east is the Early Cretaceous West Shetland Basin, while to the south are the Permo-Triassic Sula-Sgeir and Papa basins (Meadows *et al.* 1987). The basin lies at the southwestern end of the Rona Ridge, a feature which separates the Faeroe Basin to the northwest from the shallower West Shetland Basin to the southeast.

Traditionally, the Solan Basin has been regarded as the southwestern extension of the West Shetland Basin (Ridd 1981; Hitchen and Ritchie, 1987). However, although the Solan and West Shetland basins are partly contiguous, their geological histories and internal structures are sufficiently different that they should be regarded as separate basins. Based on its internal structure and differences in the preserved stratigraphy we have divided the Solan Basin into two sub-basins (West and East), separated by the Solan Fault (Fig. 2).

This area has been the subject of hydrocarbon exploration since the early 1970s. Six wildcat wells have been drilled within the Solan Basin proper, with a further twenty in the surrounding area (Fig. 2). Success has generally been limited. However, the discovery in 1990 by Well 205/26a-3 of a major oil accumulation in a pre-Jurassic reservoir has increased interest in this geologically complex area.

The object of this paper is to outline the structural evolution of the Solan Basin, constraining as tightly as possible the age of the tectonic events that have shaped it and integrate this with the stratigraphy of its sedimentary fill.

Basement and structural grain

The basement in this region has been penetrated by several wells and dated as Lewisian (Ritchie and Darbyshire 1984). It consists of acid tonalitic gneisses and pyroxene granulites, with minor amphibolites and garnet-sillimanite gneisses (Hitchen and Ritchie 1987). As elsewhere along this margin, and to the south in the Hebrides (Coward and Enfield 1987; Duindam

and van Hoorn 1987), there is evidence of SE-dipping seismic reflectors deep within the basement that might represent Caledonian thrusts.

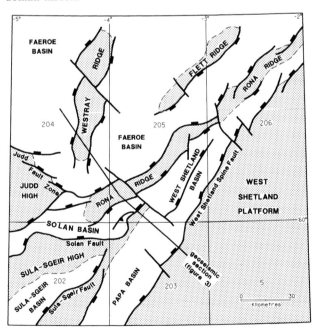

Fig. 1. Tectonic elements of the West Shetland continental margin, Solan Basin area.

The dominant structural grain consists of a series of E- to ENE-trending faults (Fig. 1), most of which display large normal throws at the present day (Fig. 3). These fault systems have existed since at least the Permian and have been repeatedly reactivated, often with changes in their sense of displacement. They are offset by a series of poorly defined NNW-trending transfer zones, of which the best developed is the Judd

From *Petroleum Geology of Northwest Europe: Proceedings of the 4th Conference* (edited by J. R. Parker).
© 1993 Petroleum Geology '86 Ltd. Published by The Geological Society, London, pp. 987–998.

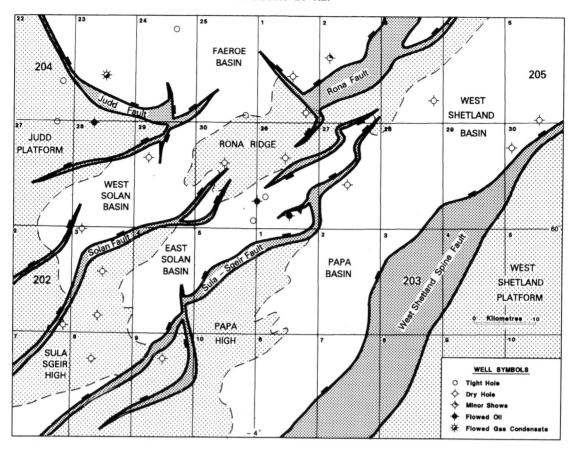

Fig. 2. Tectonic elements and exploration wells in the vicinity of the Solan Basin.

Fault Zone (Kirton and Hitchen 1987; Mudge and Rashid 1987).

Possible Paleozoic sediments

The next oldest rocks drilled in the area are undated, weathered basalts. By analogy with the Midland Valley, southern Scotland and Northern Ireland (Penn *et al.* 1983), these basalts are thought to be Early Permian in age. They are overlain by a thick succession of Permo-Triassic clastics. However, there is seismic evidence (Fig. 4) for the existence of an older, undrilled sedimentary section, patchily preserved in a tilted fault block terrain, which was subsequently truncated by a planar unconformity.

The likely age of this unpenetrated succession is Devono-Carboniferous, by analogy with the Clair oil field, some 100 km to the NE (Ridd 1981; Allen and Mange-Rajetzky 1992). If so, then the unconformity which separates this section from the overlying Permo-Triassic was formed during the Variscan. The older section may be much more competent than that overlying the unconformity, the two intervals having apparently become detached during subsequent basin inversion, with the lower sequence remaining relatively undeformed. This suggests that the Paleozoic sediments are highly indurated, indicating low porosity and limited reservoir potential.

Permo-Triassic to Lower Jurassic

Thick successions of Permo-Triassic clastic sediments are preserved in several half-graben shaped 'basins' in this area: the Papa, Sula-Sgeir and Solan basins (Fig. 1). Each is bounded by large normal faults on their southeastern margins (Figs 2 and 3). These basins appear to be distinctly different from those of similar age found further south in the Minches (Kirton and Hitchen 1987), which were formed by normal reactivation of

Caledonian thrusts that dip to the southeast. The basins discussed here have been modified by later inversion and deep erosion, but were undoubtedly formed in the footwall of normal faults that dipped steeply to the northwest. It is not clear, however, whether the Papa, Sula-Sgeir and Solan basins were separate depocentres or one extensive basin. What are now physically separate wedges of Permo-Triassic sediments, preserved beneath an extensive late Middle Jurassic unconformity, could be erosional remnants of a single large basin.

The Permo-Triassic to Lower Jurassic megasequence (Fig. 5), can be divided into at least three sequences, each related to a pulse of rifting. Where they have been drilled, the two lower sequences consist of fining-upward successions of continental sediments, while the upper one records a transition to fully marine deposition in the Early Jurassic.

Lower Permian to lowermost Triassic

From seismic evidence it appears that the undated weathered lavas which comprise the oldest strata drilled in the area are not the basal unit of this sequence. These lavas are overlain by clastic sediments, consisting of conglomerates and coarse-grained sandstones interbedded with thin siltstone–mudstone layers. These form several large-scale, upward-fining cycles. They are considered to have been deposited as a series of alluvial fans analogous to those of the same age described from the Hebrides by Steel (1974). It is likely that the unpenetrated (sub-basalt) section consists of similar continental clastics.

The alluvial fan facies was succeeded by medium- to coarse-grained, cross-bedded, pebbly sandstones which contain numerous intraformational rip-up clasts. Interbedded anhydritic siltstones and claystones become increasingly common upwards. These sediments were probably deposited within an alluvial braidplain environment and are expected to form the majority of the basin-fill away from active faulted margins.

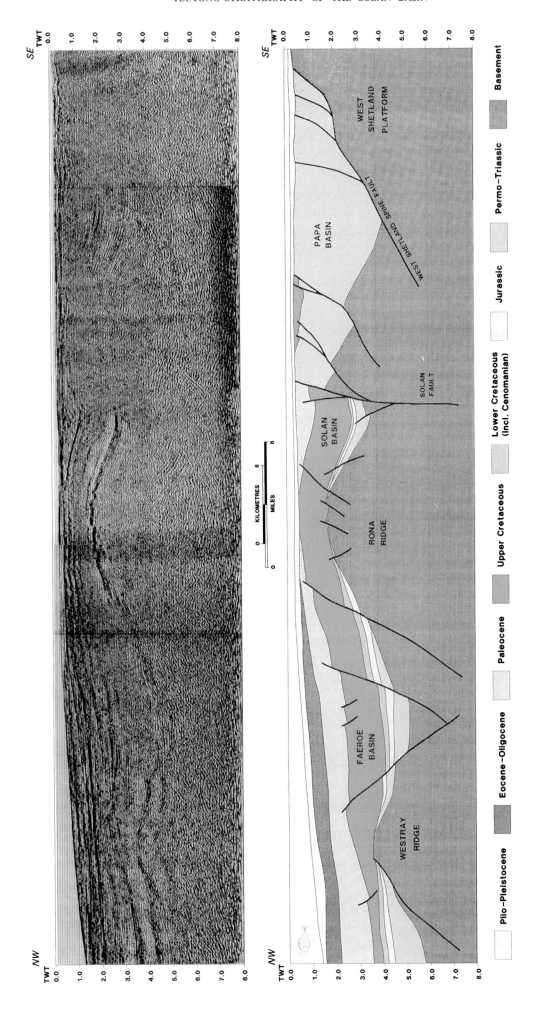

Fig. 3. Regional seismic line and geoseismic interpretation. Location shown in Fig. 1.

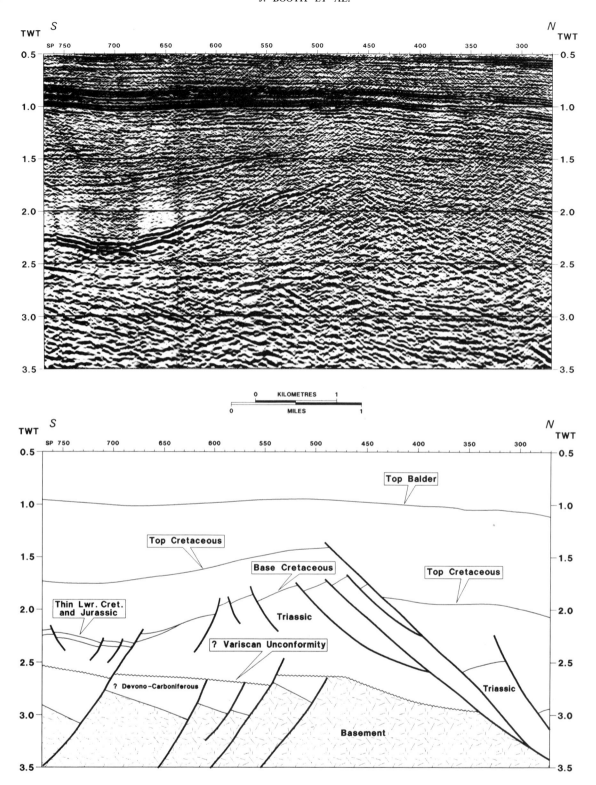

Fig. 4. Seismic line from the West Solan Basin. Note the following features: (1) the tilted fault block terrain truncated by an unconformity; (2) the inverted and truncated Triassic section; (3) the large Late Cretaceous–Paleocene normal faults.

The uppermost portion of this sequence consists of a regionally extensive, 50–100 m thick claystone, which contains a diverse palynoflora, allowing it to be dated confidently as earliest Triassic (Griesbachian). It is interpreted to be a coastal plain deposit.

Lower Triassic to uppermost Triassic

The base of this sequence corresponds to a renewed influx of coarse-grained clastics into the basin. The sequence consists predominantly of fine- to coarse-grained sandstones, often pebbly, cross-bedded and containing intraformational claystone rip-up clasts. These sediments were deposited on an alluvial braidplain with minor ephemeral sheetflooding and interfingering aeolian deposits. Repeated fining-upward cycles are thought to reflect pulses of tectonic activity that repeatedly rejuvenated the hinterland.

Lower Jurassic

Sediments of Early Jurassic age are preserved in the western part of the Solan Basin. These comprise marginal marine to

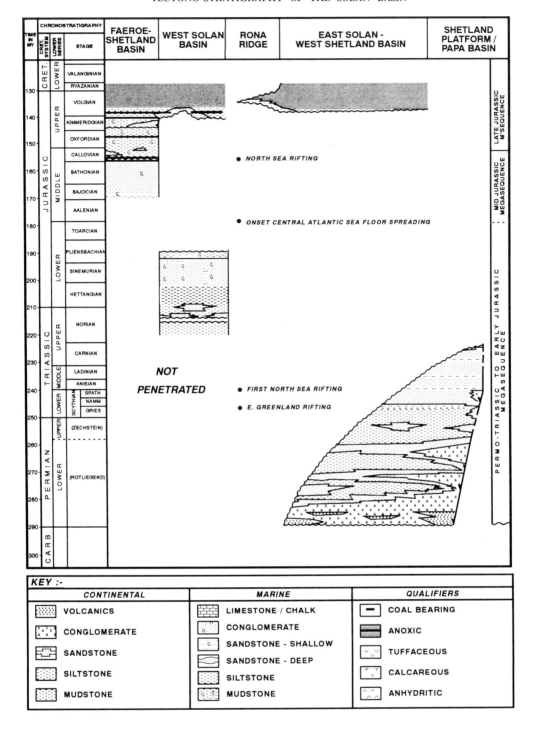

Fig. 5. Permian to Jurassic stratigraphic development.

marine sandstones and interbedded shales grading up into a series of calcareous, occasionally silty, open marine mudstones. They are closely comparable to sediments of this age documented from the Hebrides (Morton 1989).

Middle Jurassic

No Middle Jurassic sediments have been encountered in the Solan Basin. There is, however, evidence for the existence of a proto-Faeroe Basin, of Jurassic to Early Cretaceous age, underlying parts of the Late Cretaceous–Tertiary Faeroe Basin. Coarse-grained clastic sediments, which are equivalent in age to the Bearreraig Formation and Estuarine Group of the Hebrides and the Brent Group of the Northern North Sea, were deposited contemporaneously with faulting in this proto-Faeroe Basin (Haszeldine *et al.* 1987).

It is not known whether a Middle Jurassic section was deposited in the Solan Basin and then removed completely by later erosion, or if the Solan and West Shetland Basins comprised an area of net erosion or sediment bypass at this time. However, reworked Middle Jurassic microfossils have been found within Upper Jurassic sediments in several wells in the area, indicating that Middle Jurassic deposition was much more widespread than the preserved section would suggest.

Late Middle to early Late Jurassic unconformity

A major erosional unconformity truncates the sedimentary sequences described above. In the Solan Basin sediments of Kimmeridgian age lie directly on the Triassic, with abundant reworked Middle Jurassic fossils being recorded in the Upper Jurassic section. Thus, the probable age of the unconformity is

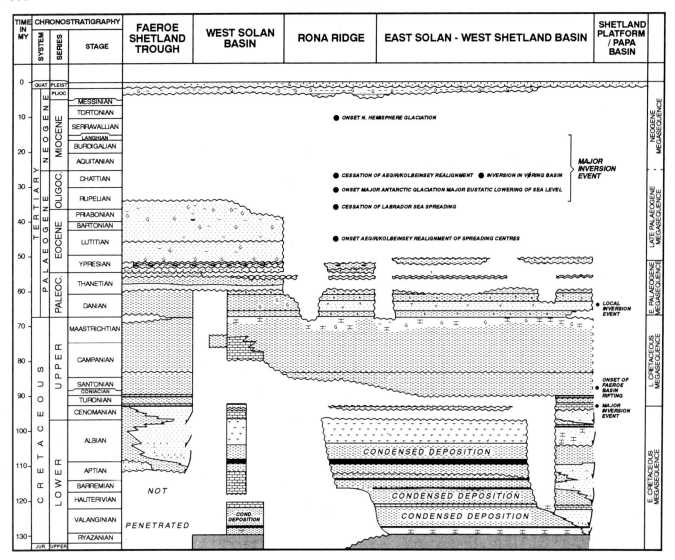

Fig. 6. Cretaceous to Recent stratigraphic development.

constrained to the Middle to early Late Jurassic. Evidence from the Faeroe Basin (Haszeldine *et al.* 1987) and the Hebrides (Morton 1989) indicates the widespread occurrence of an intra-Callovian unconformity. We consider this to have been contemporaneous with the development of the major erosional unconformity in the West Shetland–Solan area.

Apatite fission track analysis from several wells indicates a significant difference in the level of thermal maturity between the Triassic and Upper Jurassic rocks separated by this unconformity. It is estimated that up to 1.5 km of sediment were removed and the area effectively peneplaned at this time. Seismic and dipmeter data indicate an angle of up to 20° between underlying strata and the unconformity surface.

The tectonic origins of this regional unconformity are unknown. It does not appear to have been temporally associated with any significant faulting, nor is there a contemporary unconformity (or other tectonic phenomenon) in the geological record of the Northern North Sea.

Upper Jurassic

The Upper Jurassic section is relatively thin, ranging from only a few metres over the Rona Ridge to 300 m in the Solan Basin depocentre. Tectonic activity appears to have been limited to minor normal faulting. This is in contrast to the proto-Faeroe Basin to the north, where several major fault systems were active, with normal throws of several hundred metres.

The succession (Fig. 5) consists of a thin, patchily distributed marine sandstone, that is occasionally glauconitic with shell coquinas and allochthonous coal fragments. Informally, it is referred to as the Rona Sandstone. There is no evidence of local derivation of this sandstone from active fault scarps in the Solan Basin. This shallow marine sand is transgressed and overstepped by organic-rich shales of the Kimmeridge Clay Formation, which range from Volgian to Ryazanian in age.

The Kimmeridge Clay Formation is a condensed deposit, representing slow accumulation of pelagic muds under oxygen-deficient, sediment-starved conditions. It provides the principal source of hydrocarbons in the region.

Lower Cretaceous

The base of the Lower Cretaceous sequence (Fig. 6) corresponds to a major change in both facies and sediment distribution patterns in the late Ryazanian. This reflects renewed fault activity and rejuvenation of sediment sources in the region; in addition, improved circulation of oxygenated waters within the basin destroyed the restricted, anoxic conditions under which the Kimmeridge Clay Formation had accumulated. Uplift of the margin of the West Shetland Platform and parts of the Rona Ridge led to an input of coarse clastic sediment which, in the West Shetland Basin, reaches a maximum thickness of over 1200 m on the downthrown side of the West Shetland Spine Fault. In the Faeroe Basin more than

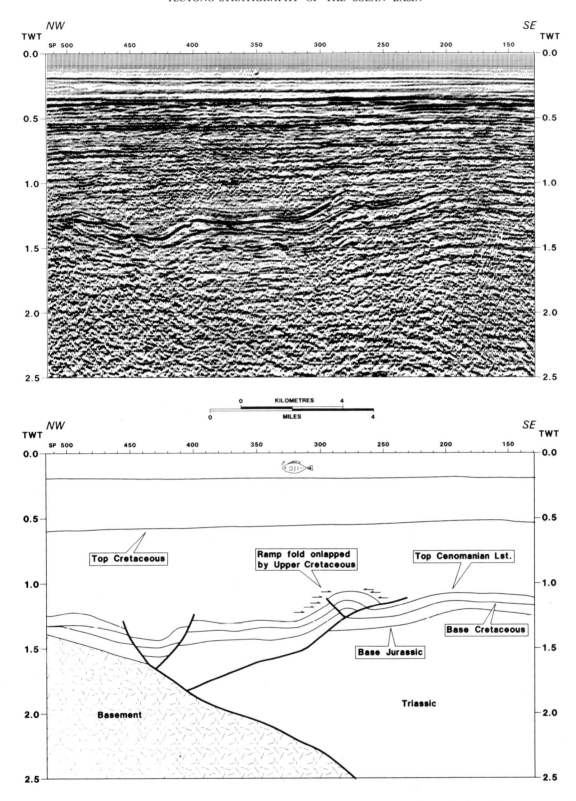

Fig. 7. Seismic line from the East Solan Basin, which provides evidence for dating the major basin inversion event as earliest Turonian. Note the development of a ramp fold in the Late Jurassic–Early Cretaceous section, near the tip of a southeasterly directed backthrust. The prominent peak reflector identified as Top Cenomanian corresponds to a late Cenomanian limestone. The onlapping reflectors are dated as early Turonian–Santonian.

300 m of similar sandstones are present. In contrast, the Lower Cretaceous section in the Solan Basin is 100–200 m thick, and consists predominantly of fine-grained clastics and pelagic carbonates. This suggests that faulting was relatively subdued in the Solan Basin at this time. As a consequence, the reservoir potential of the Lower Cretaceous section in this basin is limited.

The Lower Cretaceous sequence ranges in age from late

Ryazanian to earliest Turonian. It is readily correlatable with the clastic-starved Cromer Knoll Group of the North Sea (Hesjedal and Hamar, 1983).

Mid-Cretaceous basin inversion

The Solan Basin was partially inverted in the mid-Cretaceous as a consequence of basin shortening, resulting from a change in the regional stress field associated with the initiation of the Late Cretaceous Faeroe Basin. This inversion resulted in the extrusion of the Permian–Lower Cretaceous sedimentary fill of the basin towards the northwest, onto the dip-slope of the 'half-graben', with reversal of previously normal faults and the development of thrusts. Along the southern margin of the basin, buttressing against the Sula Sgeir and Solan faults resulted in high-angle reverse faulting and associated folding.

The timing of this event can be closely constrained. The seismic section reproduced in Fig. 7 shows a southeasterly directed backthrust that developed as the basin-fill was extruded and folded. This appears to have reached the palaeo-sea floor, where a ramp fold developed in the Jurassic–Lower Cretaceous section. This must have occurred in relatively deep water as the feature shows no sign of erosion, despite its obvious topographic relief. The prominent reflector generated by the top of the thrusted package corresponds to a limestone of latest Cenomanian age, while the sediments that onlap the ramp-fold, thus post-dating its formation, are of earliest Turonian to Santonian age.

Upper Cretaceous

During the Late Cretaceous the Atlantic rift system attempted to propagate northeastwards along what is now the Faeroe–Shetland Trough (Ziegler 1990). This led to NW–SE-oriented extension and the rapid development of the Faeroe Basin, within which the Upper Cretaceous sedimentary fill may exceed 3000 m in thickness.

With the growth of the Faeroe Basin, large normal faults trending ENE and downthrowing to the north, developed along the northwestern flank of the Rona Ridge. These fault systems are frequently offset by NNW-trending transfer faults (such as the Judd fault zone, Fig. 1). Despite normal throws in excess of 3000 m in places, there is no evidence of contemporary footwall uplift along this margin of the basin.

The Solan and West Shetland Spine faults were reactivated at this time as normal faults downthrowing to the north, with up to 1500 m of Upper Cretaceous sediments being deposited in the Solan and West Shetland basins. There is seismic evidence (Fig. 3) that Upper Cretaceous sediments once extended across the West Shetland Platform, where, perhaps due to deep Tertiary erosion, they are only preserved in small remnants on the downthrown side of large faults.

Within the Solan Basin the Upper Cretaceous sequence (Fig. 6) consists of hemipelagic, deep marine shales, the Shetland Group, with occasionally thin dolomite and limestone bands that become increasingly rare upwards. Chalk Group sediments of Turonian to Coniacian age were deposited on terraces along the West Shetland Spine Fault, while Chalk of Campanian to Maastrichtian age is found on highs in the Sula–Sgeir Basin. It appears that the Late Cretaceous palaeogeography consisted of a relatively shallow West Shetland Platform and a few isolated highs on which pelagic chalks accumulated, giving way northwards, across large normal faults, to areas of deep marine hemipelagic shale deposition.

Earliest Paleocene transpressional event

Although rifting of the Faeroe Basin continued into the early Eocene, it was interrupted during the earliest Paleocene by transpressional reactivation of many of the major ENE-trending (previously normal) fault zones. This did not result in general inversion of the Solan Basin, as had the Mid-Cretaceous event, but rather created several localized, but very 'intense', inversion structures (Figs 8 and 9).

The age of this event can be determined from the seismic section reproduced in Fig. 8. In the centre of this line a ruck fold occurs over two splays of the Solan Fault, which have an

overall normal displacement. The fold was created by reverse motion on these splays (inversion of previously normal faults). The Top Cretaceous reflector is folded, while the Top Balder reflector (which is approximately equivalent to the top of the Paleocene) is unaffected. Furthermore, the first seismic reflector above the Top Cretaceous is folded and onlapped by the overlying sequence, which is undeformed. Given this relationship, it is apparent that the transpressional event occurred in the earliest Paleocene and was not long lived.

The most spectacular structure formed at this time is a 'palm-tree flower' (Fig. 9). This is located where the Solan fault system steps about 4 km to the northwest, which indicates that the transpression which created it was right-lateral. There is evidence, including the absence of much of the Lower Cretaceous succession, for limited relief on the structure as a result of the mid-Cretaceous event; however, its current expression was largely created in the earliest Paleocene.

Paleocene to Lower Eocene

During the early Danian, the West Shetland Basin, the eastern part of the Solan Basin and the flanks of the Rona Ridge were characterized by the deposition of a 125–200 m thick sand unit, informally referred to as the Bryozoan Sand (Hitchen and Ritchie 1987) (Fig. 6). While these sands contain a shallow water macrofauna, the presence of abundant planktonic foraminifera indicate a deep marine setting, as does the abrupt, but conformable, contact with the underlying deep marine Danian shales. It is envisaged that these sands were reworked from a shallow water environment on the West Shetland Platform and transported by mass flow processes into a deep marine setting. This may reflect deposition from a seismically active margin, with individual flows being triggered by earthquakes. Over the rest of the region, deep marine shales were deposited contemporaneously with the Bryozoan Sand Member.

During the later Paleocene an unconformity developed over much of this area. At the same time, submarine fan sands accumulated locally in the western Solan Basin and in the Faeroe Basin (Mudge and Rashid 1987). These were succeeded by Upper Thanetian deep marine shales. The Thulean volcanic event, that was centred on the axis of the Faeroe–Shetland Trough, reached its climax at this time and the unconformity could be a product of thermal doming.

The Thulean volcanic event culminated in the widespread deposition across the region of a thin tuffaceous clastic unit, the Balder Formation, in the earliest Ypresian (Knox and Morton 1988). This was succeeded in the western part of the Solan Basin, on the Judd and Sula–Sgeir highs and on the southwestern end of the Rona Ridge, by a thin unit of continental to paralic coal-bearing strata.

Lower Eocene to Lower Oligocene

This sequence represents the transition from rift to thermal subsidence, with its base marked by a return to marine sedimentation. It is composed predominantly of loosely consolidated glauconitic sands, with minor intercalations of siltstones and mudstones. In the West Shetland Basin and the eastern part of the Solan Basin, these sandstones are underlain by thin deep water shales, reflecting a combination of rapid subsidence and initial sediment starvation.

Oligo-Miocene event

A major mid-Tertiary erosional unconformity extends across the West Shetland Platform and the Solan and West Shetland basins, becoming less pronounced and eventually conformable towards the centre of the Faeroe Basin. In the Solan Basin, Pliocene sediments rest directly on early Oligocene or Eocene

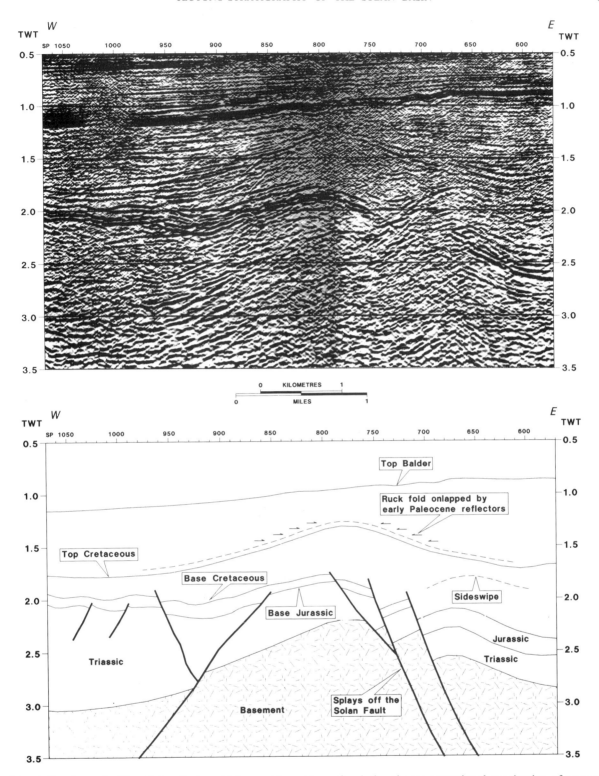

Fig. 8. Seismic line from the West Solan Basin, which provides evidence for dating the transpressional reactivation of some of the major faults as earliest Paleocene. Note the ruck fold developed over two splays off the Solan Fault. The Top Cretaceous reflector is folded but the Top Balder is unaffected. The onlap of this fold by all but the lowermost Paleocene reflectors dates its formation as earliest Paleocene.

strata, thus the age of this unconformity is poorly constrained; it could have been formed at any time during the late Oligocene to Miocene.

The Rona Ridge was uplifted and deeply eroded during the formation of this unconformity (Fig. 3). The amount of sediment removed may have been as much as 1250 m where the erosion was most intense. At the same time, many of the large normal fault systems which had developed in the Late Cretaceous–early Tertiary suffered some minor reverse motion. In deeper parts of the Faeroe Basin (where erosion did not

remove the evidence) large-scale, low-amplitude folds developed over basement horst blocks, suggesting they formed as these blocks were uplifted.

Upper Miocene to Recent

The sediments on top of the Oligo-Miocene unconformity are generally coarse sands and glacio-marine gravels of Pliocene to Quaternary age.

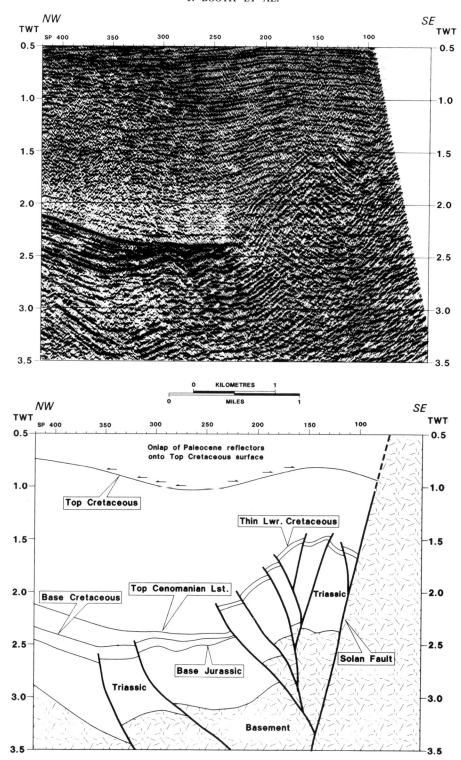

Fig. 9. Seismic line from the East Solan Basin showing a large transpressional flower structure developed where the Solan Fault steps 4 km to the northwest. Note the apparent onlap of the Upper Cretaceous section onto the flower. This is interpreted to be a tectonic feature as the flower was forced upward through this section, akin to salt diapir penetration.

Summary

The Solan Basin is in reality a composite of the remnants of several superimposed sedimentary basins. Its fill can be divided into several discrete packages separated by major, basin-wide unconformities. With a pedantic application of the 'rules' of geology we should regard most of these packages as separate basins. However, this is rather cumbersome and the complication hinders rather than adds to our understanding of this region. The following is a brief summary of the evolution of

this basin (Fig. 10), emphasizing the most significant geological events.

The oldest sedimentary package, which is evident on seismic but has yet to be drilled, is thought to be part of the Devono-Carboniferous Orcadian Basin complex. It is only preserved in small areas of the Solan Basin, within a tilted block terrain that was planed off by a ?Variscan unconformity.

This unconformity is overlain by Permo-Triassic continental clastics and a related sequence of Lower Jurassic marine

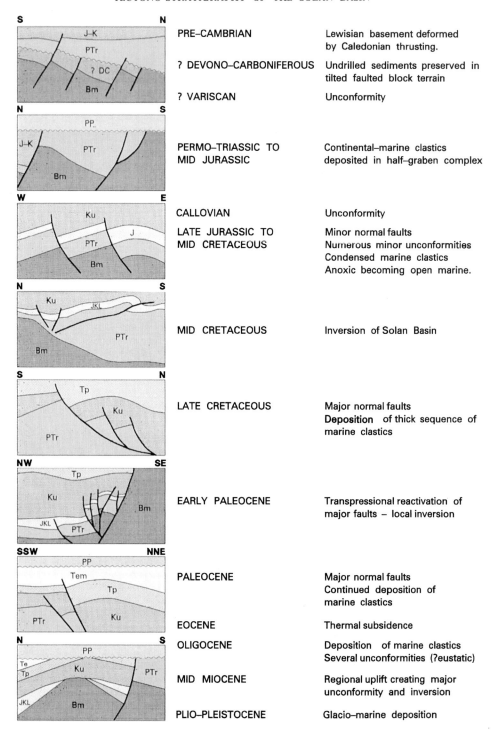

PRE–CAMBRIAN	Lewisian basement deformed by Caledonian thrusting.
? DEVONO–CARBONIFEROUS	Undrilled sediments preserved in tilted faulted block terrain
? VARISCAN	Unconformity
PERMO–TRIASSIC TO MID JURASSIC	Continental–marine clastics deposited in half–graben complex
CALLOVIAN	Unconformity
LATE JURASSIC TO MID CRETACEOUS	Minor normal faults Numerous minor unconformities Condensed marine clastics Anoxic becoming open marine.
MID CRETACEOUS	Inversion of Solan Basin
LATE CRETACEOUS	Major normal faults Deposition of thick sequence of marine clastics
EARLY PALEOCENE	Transpressional reactivation of major faults – local inversion
PALEOCENE	Major normal faults Continued deposition of marine clastics
EOCENE	Thermal subsidence
OLIGOCENE	Deposition of marine clastics Several unconformities (?eustatic)
MID MIOCENE	Regional uplift creating major unconformity and inversion
PLIO–PLEISTOCENE	Glacio–marine deposition

Fig. 10. Summary of the evolution of the Solan Basin. Each of the panels is a sketch of a separate seismic section that illustrates one of the main features in the basin's history. Stratigraphic abbreviations are BM: Basement; DC: Devono-Carboniferous; PTr: Permo-Triassic; J: Jurassic; KL: Lower Cretaceous; KU: Upper Cretaceous; Tp: Paleocene; Te: Eocene; Tm: Miocene; PP: Plio-Pleistocene.

clastics. This package was peneplaned by deep erosion to form an unconformity in the Late Callovian, with the result that it is preserved as a series of wedges that have the superficial appearance of half-grabens. However, it is apparent from seismic and dipmeters that there is an angle of up to 20° between the dip of bedding and the unconformity; the wedge shapes are thus artifacts of the way the area was eroded rather than true half-graben structures. Nevertheless, we suspect that these sediments did indeed accumulate in large half-graben complexes, through their original geometry cannot readily be reconstructed.

After the formation of the Callovian unconformity, fault

activity was limited; however, the whole basin subsided rapidly, but was semi-starved of sediment. A thin sequence of organic-rich shales (Kimmeridge Clay Formation) was deposited across the basin under open marine, anoxic conditions. This overlies a thin lag sand which occurs patchily along the unconformity surface. This pattern of thin sediment accumulation and limited contemporary faulting continued through the Early Cretaceous. Although there are several unconformities in the Upper Jurassic–Lower Cretaceous section they are relatively minor and it is regarded as a single tectonic package.

The Solan Basin was inverted in the Turonian. This resulted in its sedimentary fill being squeezed out northeastwards onto the dip-slope of the basin, with associated minor thrusting, and buttressing along the downthrown side of the Solan Fault causing broad folds.

During the Late Cretaceous a thick sequence of marine sands and shales (up to 3000 m) was deposited over the Solan Basin. This was accommodated by a major normal reactivation of the Solan Fault in response to the regional extension that created the Faeroe Basin, as the North Atlantic tried to propagate northeastwards up the Faeroe–Shetland Trough.

Extension continued until the end of the Paleocene, after which subsidence continued by thermal sag. However, it was briefly interrupted in the earliest Paleocene by a transpressional (? right-lateral) reactivation of some of the ENE-trending fault systems, including the Solan Fault. Where these fault systems are offset by transfer zones and relays this produced local, but very 'intense', inversion structures.

Late in the Tertiary, probably during the mid-Miocene (although, as the wells are not logged over this part of the section and biostratigraphic samples are not usually taken, the actual age is poorly constrained), another major erosional unconformity developed across the entire basin. The cause of this unconformity is unknown but, as it was contemporaneous with uplift of the Rona Ridge (with up to 1.2 km of sediment eroded off this feature), it is probably tectonic. At the same time many of the major normal fault systems around the Solan Basin underwent minor reverse motion and in the Faeroe Basin (where erosion has not removed the evidence) large, broad, low-amplitude folds developed, many of which overlie deep-seated fault blocks (? inversion).

This paper has evolved from review of the exploration potential of the West Shetlands area by Amerada Hess Ltd. We would like to thank our colleagues who contributed much to our understanding of this area, in particular Ian Roche, Steve Boldy, Richard Warren and Mike Golden.

The seismic lines are shown with kind permission of the following partners in Amerada Hess acreage in the Solan Basin, although this work was not conducted on their behalf and does not necessarily represent their technical views of this area: Aran Energy Exploration Ltd; Arco British Ltd; Brabant Oilex Ltd; Deminex UK Oil and Gas Ltd; DSM Energy (UK) Ltd; Fina Exploration Ltd; Hardy Oil & Gas (UK) Ltd; Kerr-McGee Oil (UK) plc; Neste Oy; Texaco Exploration Ltd; The Norwegian Oil Co DNO (UK); Unocal (UK) Ltd.

Review of biostratigraphic data over much of the area has been undertaken by staff of IEDS Ltd and Millenia and their contributions are gratefully acknowledged.

References

ALLEN, P. A. AND MANGE-RAJETZKY, M. A. 1992. Devonian–Carboniferous sedimentary evolution of the Clair area, offshore North-Western UK: impact of changing provenance. *Marine and Petroleum Geology*, **9**, 29–52.

COWARD, M. P. AND ENFIELD, M. 1987. The structure of the West Orkney and adjacent basins. *In*: BROOKS, J. AND GLENNIE, K. W. (eds) *Petroleum Geology of North West Europe*. Graham & Trotman, London, 687–696.

DUINDAM, P. AND VAN HOORN, B. 1987. Structural evolution of the West Shetland continental margin. *In*: BROOKS, J. AND GLENNIE, K. W. (eds) *Petroleum Geology of North West Europe*. Graham & Trotman, London. 765–773.

HASZELDINE, R. S., RITCHIE, J. D. AND HITCHEN, K., 1987. Seismic and well evidence for the early development of the Faeroe–Shetland Basin. *Scottish Journal of Geology*, **23**, 283–300.

HESJEDAL, A. AND HAMAR, G. P. 1983. Lower Cretaceous stratigraphy and tectonics of the SSE Norwegian offshore. *In*: KAASSCHIETER, J. P. H. AND REIJERS, T. J. A. (eds) Petroleum Geology of the SE North Sea and adjacent onshore areas. *Geologie en Mijnbouw*, **62**, 135–144.

HITCHEN, K. AND RITCHIE, J. D. 1987. Geological review of the West Shetland area. *In*: BROOKS, J. AND GLENNIE, K. W. (eds) *Petroleum Geology of North West Europe*. Graham & Trotman, London, 737–749.

KIRTON, S. R. AND HITCHEN, K. 1987. Timing and style of crustal extension north of the Scottish mainland. *In*: COWARD, M. P., DEWEY, J. F. AND HANCOCK, P. L. (eds) *Continental Extensional Tectonics*. Geological Society, London, Special Publication, **28**, 501–510.

KNOX, R. W. O'B. AND MORTON, A. C. 1988. The record of early Tertiary N Atlantic volcanism in sediments of the North Sea Basin. *In*: MORTON, A. C. AND PARSON, L. M. (eds) *Early Tertiary Volcanism and the Opening of the NE Atlantic*. Geological Society, London, Special Publication, **39**, 407–419.

MEADOWS, N. S., MACCHI, L., CUBITT, J. M. AND JOHNSON, B. 1987. Sedimentology and reservoir potential in the West of Shetland, UK, exploration area. *In*: BROOKS, J. AND GLENNIE, K. W. (eds) *Petroleum Geology of North West Europe*. Graham & Trotman, London, 723–736.

MORTON, N. 1989. Jurassic sequence stratigraphy in the Hebrides Basin, NW Scotland. *Marine and Petroleum Geology*, **6**, 243–260.

MUDGE, D. C. AND RASHID, B. 1987. The Geology of the Faeroe Basin area. *In*: BROOKS, J. AND GLENNIE, K. W. (eds) *Petroleum Geology of North West Europe*. Graham & Trotman, London, 751–763.

PENN, I. E., HOLLIDAY, D. W., KIRBY, G. A., SOBEY, R. A., MITCHELL, W. I., HARRISON, R. K. AND BECKINSALE, R. D. 1983. The Larne No. 2 Borehole: discovery of a new Permian volcanic centre. *Scottish Journal of Geology*, **19**, 333–346.

RIDD, M. F. 1981. Petroleum Geology west of the Shetlands. *In*: ILLING, L. V. AND HOBSON, G. D. (eds) *Petroleum Geology of the Continental Shelf of North-West Europe*. Heyden, London, 414–425.

RITCHIE, J. D. AND DARBYSHIRE, D. P. F. 1984. Rb–Sr dates on Precambrian rocks from marine exploration wells in and around the West Shetland Basin. *Scottish Journal of Geology*, **20**, 31–36.

STEEL, R. J. 1974. New Red Sandstone floodplain and piedmont sedimentation in the Hebridean province, Scotland. *Journal of Sedimentary Petrology*, **44**, 336–357.

ZIEGLER, P. A. 1990. *Geological Atlas of Western and Central Europe*, 2nd edn. Shell Internationale Petroleum, Maatschappij B.V., The Hague.

Structuring and transfer zones in the Faeroe Basin in a regional tectonic context

B. RUMPH, C. M. REAVES, V. G. ORANGE and D. L. ROBINSON

Mobil North Sea Ltd, 3 Clements Inn, London WC2A 2EB, UK

Abstract: An integrated structural interpretation of the Faeroe Basin within the regional tectonic setting of the North Atlantic margin was derived from gravity, magnetic and regional seismic data. The tectonic framework of the basin was established during the Caledonian orogeny while its subsequent structural evolution was linked to the development of the Greenland–Scotland Ridge and the Iceland hot-spot. Some major transfer zones had previously been identified in the Faeroe Basin area. The gravity and magnetic data defined a series of transfer zones oriented NW–SE across the basin which were also identified on seismic data. These transfer zones have had major influences on the Mesozoic sedimentation patterns, and define basin segments which most likely had different sedimentary and structural histories, thus introducing a diversity of hydrocarbon plays into the Faeroe Basin.

The Faeroe Basin is located in the Faeroe–Shetland Channel between the Shetland and Faeroe Islands on the eastern Atlantic margin north of Scotland between latitudes 60°N and 63°N, and longitudes 1°E and 6°W (Fig. 1). It is one of a number of basins that developed along the eastern margin of the North Atlantic, and is located between the Møre Basin to the northeast and the Rockall Trough to the southwest.

The Mesozoic structuring of the Faeroe Basin has been interpreted in relation to the plate tectonic history of the North Atlantic and the resultant stress systems that influenced its structural development. The structure of the basin has been interpreted from gravity, magnetic and regional seismic data. The gravity and magnetic data were from marine surveys by the British Geological Survey, speculative surveys by seismic contracting companies, and proprietary data. The seismic data were largely from speculative surveys by Western Geophysical and Simon Horizon, with some additional proprietary data.

Plate tectonic development of the North Atlantic region

The Faeroe Basin has been the subject of many studies in recent years. Crustal, plate-tectonic, volcanic, basinal and stratigraphic details of the basin history have been studied from academic and exploration perspectives. The development of the Rockall Trough, the influence of the Iceland hot-spot, ridge switching in the North Atlantic, especially north of the greater Greenland–Scotland Ridge, and the micro-continental mosaic of the Atlantic margin from Hatton Bank to the Jan

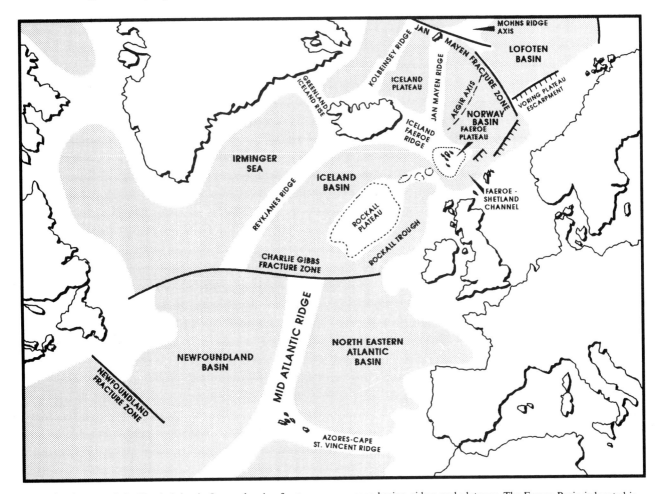

Fig. 1. Major features of the North Atlantic Ocean showing fracture zones, ocean basins, ridges and plateaux. The Faeroe Basin is located in the Faeroe–Shetland Channel.

From *Petroleum Geology of Northwest Europe: Proceedings of the 4th Conference* (edited by J. R. Parker).

Mayen Ridge affected the development and hydrocarbon potential of the basin (Nunns 1983; Scrutton 1986; White and McKenzie 1989; Earle *et al.* 1989; Knott *et al.* 1993).

The intra-plate framework of NW Europe was established during the Caledonian and Variscan orogenies which created Pangea. Readjustments within Pangea up to the end of the Triassic led to the development of a system of rift basins in the North Atlantic region (Duindam and van Hoorn 1987). These basins formed along the margins of the intra-plate shield areas or within the weakened crust of NW Europe.

The Early–Middle Jurassic initiation of seafloor spreading between Africa and North America created two new plate margins (Fig. 2). The net resultant plate motion established a sinistral shear regime oriented parallel to the margin between Africa and Europe. In Late Jurassic time, counterclockwise rotation of Africa relative to Laurasia caused extensional reactivation of NW-trending fractures created in the Variscan and Caledonian orogenies. Fault-driven subsidence occurred in the Faeroe Basin and other basins along the eastern Atlantic margin.

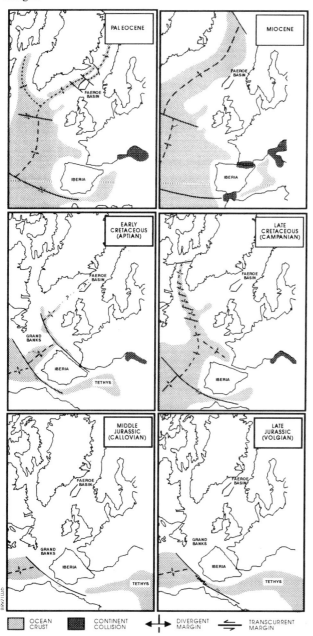

Fig. 2. Plate tectonic reconstructions of the North Atlantic from Middle Jurassic to Miocene showing the position of the Faeroe Basin. The reconstructions were modelled using the ATLAS software package developed by Cambridge Palaeomap Services.

Rapid propagation of seafloor spreading from the northwest corner of Iberia into the Labrador Sea took place during Late Cretaceous to Paleocene times causing compressional reactivation of NE-trending normal faults and the relative uplift of basin margins and intra-basinal highs.

At the beginning of Paleocene times, another seafloor spreading system developed in the North Atlantic and the Norwegian Sea (Fig. 2) and the Faeroe Basin area was now located at the plate margin. Igneous centres associated with the Iceland hot-spot had a strong effect on local subsidence patterns in Paleogene times. After the generation of true oceanic crust to the northwest of the Faeroe Basin in the late Paleocene and Eocene, subsidence was controlled by the cooling and thermal subsidence of the oceanic margin (Duindam and van Hoorn 1987).

Tectonic development of the Faeroe Basin

The tectonic development of the Faeroe Basin area is poorly understood in relation to its crustal structure, the nature of the crust, its relationship to the seafloor-spreading episode in the Rockall Trough and the early sedimentary history. The pre-Tertiary section, especially the Lower Cretaceous and older section, has only been intersected in a limited number of wells west of the Rona Ridge (Fig. 3). Well and published data show a preserved sedimentary section from Devonian–Carboniferous to Recent in the Faeroe Basin. The general geology of the basin has been assessed and reviewed by Haszeldine and Russell (1987), Hitchen and Ritchie (1987), Mudge and Rashid (1987) and Earle *et al.* (1989).

The crust underlying the basin is commonly interpreted to be attenuated continental crust with extensive intrusives, based on the interpretations of velocity layering from crustal refraction data (Bott *et al.* 1976; Zverev *et al.* 1975; Bott and Gunnarsson 1980). A similar interpretation has resulted from newer, more detailed refraction data in the Rockall Trough (Roberts *et al.* 1988). Some authors have suggested the alternative that oceanic deep crust underlies both the Rockall Trough and the Faeroe–Shetland Channel (e.g. Scrutton 1986). However, the structural interpretation of existing crustal seismic data across the Faeroe Basin has indicated a thinning of continental crust of around 20 km compared to the crustal thickness on the Shetland Platform of around 30 km (Nielsen 1983; Zverev *et al.* 1975) and 30–35 km on the Faeroe Platform (Bott 1983).

Tectonic elements

The tectonic elements of the northwest UK continental shelf around the Faeroe Basin (Fig. 3) consist of three major structural trends that are, in order of relative importance, oriented NE–SW, NW–SE and N–S. These trends are related to early Paleozoic regional structural trends established during the Caledonian orogeny.

The first trend is characterized by the Shetland Spine Fault and parallel/sub-parallel structures and faults such as the Rona Ridge, Flett Ridge and Mid-Faeroe Ridge to the northwest; Margareta's Spur and Magnus Embayment to the north; and the Sula Sgeir High and Stack Skerry Horst to the south.

The second set of trends is orthogonal to the first and comprises a series of transfer zones of varying magnitudes, such as the Wyville–Thompson Ridge and Ymir Ridge between the Rockall Trough and the Faeroe Basin (Roberts *et al.* 1983); the Judd Fault at the southern end of the Faeroe Basin; and the Erlend and Magnus transfer zones at the northern end of the Faeroe Basin (Duindam and van Hoorn 1987).

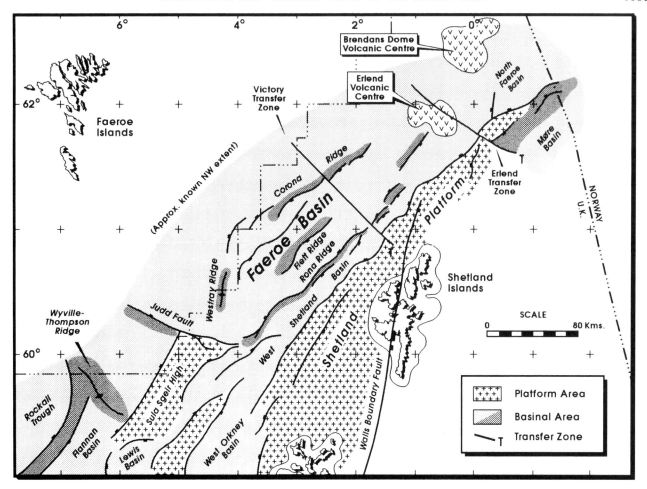

Fig. 3. Tectonic elements of the Faeroe Basin area showing basin and platform areas, major faults and volcanic centres.

The Judd Fault (Fig. 3) defines the southwesterly extent of the Faeroe Basin (the Faeroe Transfer Zone of Duindam and van Hoorn 1987). This transfer zone is a major transcurrent crustal feature that has had continuing effect on the tectonic development of the area. The Erlend Transfer Zone defines the northern limit of the Faeroe Basin (Fig. 3) and is a similar significant crustal feature. It is related to a concentration of intrusive activity. This zone could represent the en échelon continuation of the Tornquist Alignment which may continue further along the northern flank of the Greenland–Iceland–Faeroe Ridge. These transcurrent features may also be genetically related to the larger-scale fracture zones and transverse ridges of the North Atlantic region: the Jan Mayen Fracture Zone; the possible transform fault along the southern margin of the Norway Sea to the north of the Faeroe Islands (Voppel et al. 1979; Nunns 1983) and the complementary possible transform fault to the south of the islands that defines the Iceland–Faeroe Ridge.

The similarity in distance between the transfer zones that define the main part of the Faeroe Basin, and the width of the Iceland–Faeroe Ridge (cf. Figs 1 and 3), suggests a genetic relationship. The transform faults and transfer zones may have had a common origin as en échelon segments and were most likely established during the Caledonian orogeny. Hence, a 'corridor' of crustal weakness may have established the setting required for the development of the mantle plume that, most recently, has resulted in the emergence of Iceland on the mid-ocean ridge (White and McKenzie 1989; Waagstein 1988). The gravity and magnetic data over the Faeroe Basin have defined additional, smaller-scale transfer zones that are discussed below.

The third set of structural trends shows the continuing, though lesser, influence of the predominantly north–south tectonic grain of Caledonian basement of the Shetland Platform. This trend is also reflected in the Caledonian fronts in Norway, Scotland and Greenland, and in the subsequent rifting in the Northern North Sea. The Walls Boundary Fault is the most significant representation of this trend in the Faeroe Basin area. Some fault and structural trends north of the Shetland Islands also show its influence. Similar trends also occur in the western part of the Faeroe Basin, e.g. the Westray Ridge (Fig. 3).

The Faeroe Basin has developed in a complicated setting on the Atlantic margin. The basic structural framework has been interpreted as being established by the Caledonian tectonism. This structural framework is revealed by interpretation of the regional gravity and magnetic data and integrated with regional seismic data.

Gravity and magnetic data

The gravity and magnetic database over the Faeroe Basin area is largely composed of the British Geological Survey's (BGS) data in UK waters, and other speculative and proprietary data. The data are from marine surveys: potential-field and shallow-penetration, high-resolution seismic data (BGS) or speculative exploration seismic programmes (e.g. Western Geophysical) and proprietary data. Ark Geophysics Ltd processed the data into a regional database and produced the gravity and magnetic maps presented in this paper.

The data are irregularly spaced over the basin from less than 2 km to over 10 km between lines, with the widest spacing

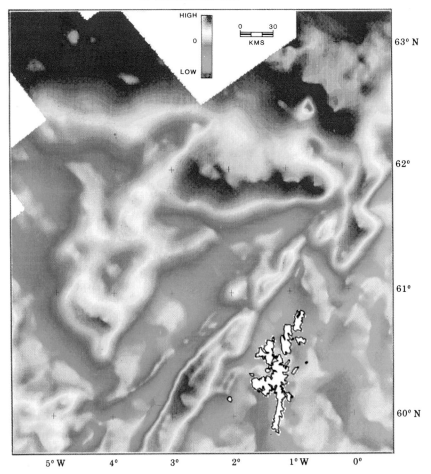

Fig. 4. Bouguer gravity map of Faeroe Basin area, calculated from free air data using Bouguer density of 2.30 g cm^{-3}.

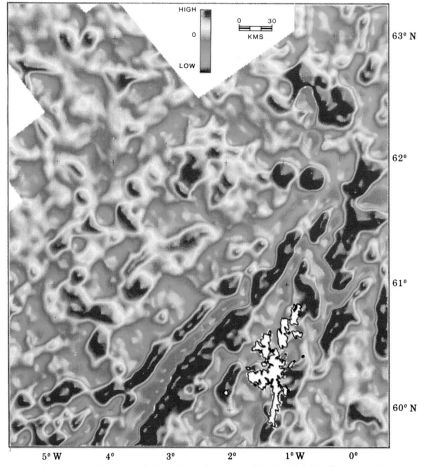

Fig. 5. Residual Bouguer gravity map of Faeroe Basin area, calculated by subtraction of regional field of wavelengths greater than 50 km from Bouguer gravity data.

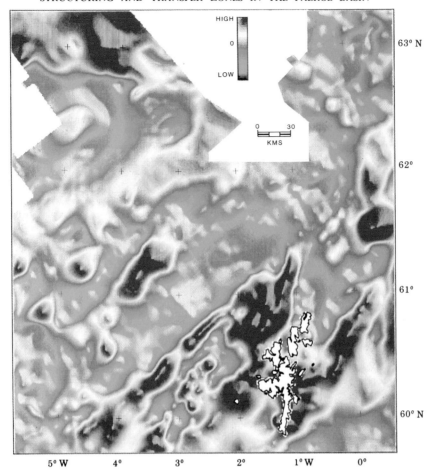

Fig. 6. Magnetic anomaly (reduced-to-pole) map of Faeroe Basin area; reduced-to-pole transformation locates anomalies above source bodies, assuming induced magnetization.

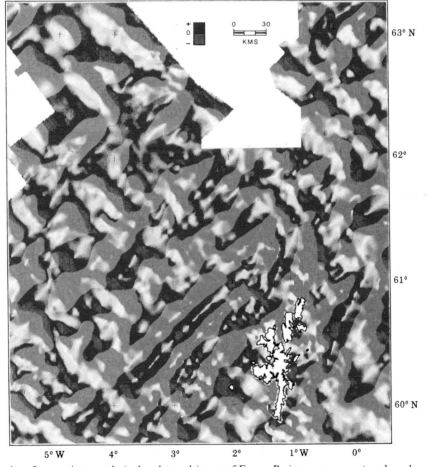

Fig. 7. Second vertical derivative of magnetic anomaly (reduced-to-pole) map of Faeroe Basin area; zero contour boundary between red and blue equivalent to steepest gradients on Fig. 6, and indicates edges of source bodies.

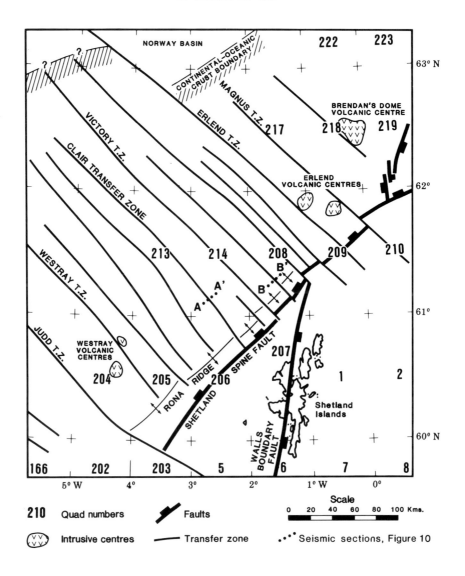

Fig. 8. Tectonic interpretation showing transfer zones and volcanic centres defined by gravity and magnetic data. Seismic sections A–A′, B–B′ shown on Fig. 10.

being in the western part of the basin outside the UK-designated waters. The computer processing of the gridded data (e.g. filtering, etc.) and detailed qualitative or quantitative interpretation is dependent on sufficient data coverage. Unusual anomalies can be produced in the gridding and contouring if the basic field is poorly defined by widely spaced lines, e.g. square magnetic high surrounded by lows at 5°W, 62° 30′N on Fig. 6. Line spacing in this area is around 50–60 km. Similarly, poor definition or inaccuracies in the bathymetry can result in false anomalies in the Bouguer gravity data and any derivative datasets, and hence have to be checked carefully.

Gravity and magnetic data are now commonly displayed and analysed using image-processing techniques. The commonest method of presentation is as colour-contoured, shaded-relief maps. These maps remove the visually-distracting contour lines and easily show the relative amplitudes, orientations and disruptions of anomalies over the area. The Bouguer gravity and the reduced-to-pole magnetic data of the Faeroe Basin area are presented in Figs 4 and 6 as colour-contoured, shaded-relief maps with illumination from the northeast. The

residual gravity data and the second vertical derivative of the magnetic data (reduced-to-pole) are presented on Figs 5 and 7.

Interpretation of gravity data

The most prominent anomaly trend in the Bouguer gravity data is the common NE–SW orientation mentioned above in describing the tectonic elements of the area. The large highs in the north are associated with the volcanic centres at Erlend and Brendan's Dome (Figs 4 and 8), and also to the continent–ocean boundary of the Norway Basin. Indications of orthogonal trends are visible on Fig. 4; for example, the highs parallelling the western limit of the UK-designated waters in Quadrants 204 and 213 terminate in the north of Quadrant 204. The northeastern ends of these anomalies similarly terminate, are off-set or change their character.

Immediately northwest of the Shetland Islands, the parallel highs are caused by the edge of the Shetland Platform and the Rona Ridge. The lower values over the Shetland Platform are probably related to the crustal thickness variation between the continental crust of the Shetland Platform and the thinner,

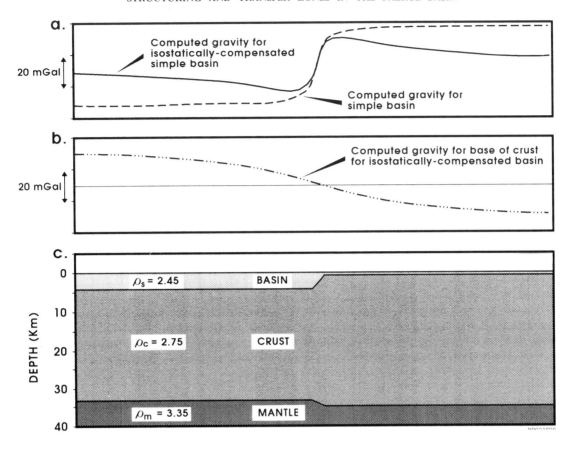

Fig. 9. Gravity modelling of an isostatically-compensated basin: simple basin model with constant crustal thickness of 35 km (Zverev *et al.* 1975) and basin margin fault with 4 km throw; isostatically-compensated basin model with crustal thinning of 2 km across basin margin for assumed simple density model (densities in g cm^{-3}).

heavily intruded (hence denser) crust of the Faeroe–Shetland Channel (e.g. Bott and Gunnarsson 1980). Therefore, the broad regional variation is essentially due to the topography of the crust-mantle boundary resulting from isostatic compensation. The prominent anomaly at the edge of the platform is an isostatic edge-effect anomaly (Fig. 9) and is probably not due to a distinct higher-density block along the edge.

The computed gravity profile is shown (Fig. 9) for a simple basin model and for an isostatically-compensated basin model. The latter shows a prominent marginal anomaly along the platform edge (Fig. 9a). The difference between the two computed profiles is the anomaly caused by the variation in crustal thickness (i.e. the mantle topography) and is shown on Fig. 9b. Note that the isostatically-compensated model has a significantly smaller amplitude at the basin edge. For the isostatically-compensated model, the basin thickness has to be increased by 50% to produce the same anomaly amplitude. Hence, if isostatic compensation is not assumed in areas where it is likely to occur (such as at passive continental margins), then the interpreted basin thicknesses can be grossly inaccurate.

The N–S-oriented anomalies are located in the north of the Shetland Platform, especially associated with the Walls Boundary Fault, and the prominent N–S elongation of the islands is probably due to this structural trend. Regional gravity data to the south show that this fault connects to the Great Glen Fault along a line of anomalies to the east of the Orkney Islands.

The residual gravity data on Fig. 5 again show the prominent NE–SW anomaly orientation, especially along the Shetland Platform and Rona Ridge. Discontinuities in the orthogonal direction can be more easily identified on this map. These

discontinuities are referred to as transfer zones (e.g. Verhoef and Srivastava 1989), in the same context as transform faults in the oceanic basins. An example of the definition of a transfer zone on the residual gravity map is the Clair Transfer Zone (Fig. 8) which offsets the Rona Ridge north of the Clair Field. To the northwest, it is shown by a distinct gradient with a series of small residual gravity anomalies located along its southwestern edge (Fig. 5). Regional data indicate that this zone continues beyond the map area and extends north and west of the Faeroe Islands. The broadly elongate shape of the islands of the Faeroe group is indicative of a structural grain oriented in this direction. NNW–SSE- and NW–SE-trending fissures on the Faeroe Islands were the source of the extensive basalt flows of the area (Noe-Nygaard and Rasmussen 1970).

The transfer zone to the north, the Victory Transfer Zone (Fig. 8), is similarly related to the termination of anomalies in the central part of the basin. The Rona Ridge is offset by this zone, and also sedimentary patterns in the Paleocene are significantly influenced by continuing adjustments along it (Mitchell *et al.* 1993).

The Erlend and Brendan's Dome volcanic centres are identified by local residual gravity highs (cf. Figs 5 and 8) at the centre of the large areas of high gravity values on the Bouguer gravity map (Fig. 4). The Erlend, Magnus and Judd transfer zones have been previously identified as major crustal features (Duindam and van Hoorn 1987) that define the extent of the Faeroe Basin, as mentioned above.

The Westray Transfer Zone shows that the Westray Ridge is composed of two en échelon, offset highs (Fig. 8), rather than a N–S-oriented ridge as previously interpreted (Mudge and Rashid 1987). The gravity and magnetic highs related to

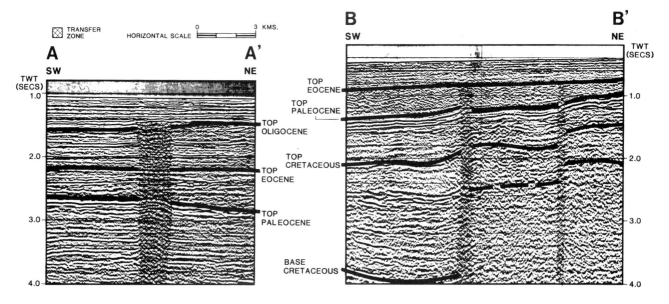

Fig. 10. Seismic expression of transfer zones, central Faeroe Basin. Location of lines shown on Fig. 8 (seismic sections are published with permission of Western Geophysical).

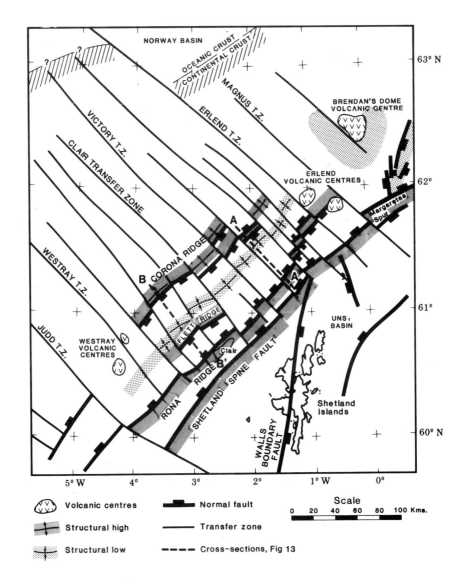

Fig. 11. Structural configuration of the Cretaceous section, Faeroe Basin. Cross-sections A–A', B–B' shown on Fig. 13.

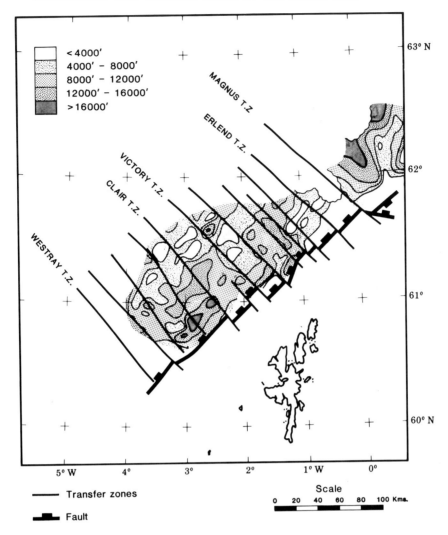

Fig. 12. Thickness of the Cretaceous section showing influence of transfer zones, Faeroe Basin.

the two elements of the Westray Ridge indicate that intrusive centres probably underlie the known sediment-covered structures. The transfer zone may have acted as a conduit for the intrusives.

Interpretation of magnetic data

The complex area of higher frequency, low-amplitude anomalies around the Erlend and Brendan's Dome centres indicate the area with the highest concentration of igneous intrusions. The known edges of flows (e.g. Hitchen and Ritchie 1987, Fig. 9) do not correlate with specific features on the magnetic data. The intrusions occurred in different polarity periods, further complicating the interpretation. The second vertical derivative map (Fig. 7), however, does show the fragmented nature of the area and the dissecting transfer zones that are oriented NW–SE, as similarly shown on the gravity data.

The interpretation map (Fig. 8) shows the major transfer zones that are defined on both the gravity and magnetic data across the Faeroe Basin. The sparser data available in the Faeroese waters, mentioned above, do not define the transfer zones as accurately, but do show that the same general trends continue further to the west.

The magnetic anomaly map shows similar structural trends to the gravity data, especially the NE–SW orientation. Magnetic data are commonly used to identify basic volcanics as these generally produce higher amplitude anomalies than acidic volcanics. The intrusive centres at Erlend and Brendan's

Dome show opposite polarity anomalies. Brendan's Dome has a positive anomaly and the Erlend centres have negative anomalies (compare Figs 4, 6 and 8), suggesting that they were intruded in periods of normal and reversed polarity of the Earth's field. The basalts on the Faeroe Islands have similarly been extruded in periods of different polarity (Waagstein 1988).

Seismic data examples

Transfer zones are shown on NE–SW-oriented seismic lines through the basin (Fig. 10). The transfer zone on the left section shows continuing movement from end Cretaceous times until the early Eocene. The section on the right shows a large displacement at Base Cretaceous level and continuing effect until the Eocene. These examples show that the transfer zones have significantly affected the structural development of the basin.

Basin structuring

Generalized maps of the structure of the central part of the Faeroe Basin at the Base Cretaceous level (Fig. 11) and of the thickness variation of the Cretaceous interval (Fig. 12), show the effect that the transfer zones have had on the Cretaceous section, with large variations of thickness laterally across the zones in some areas. The areas between the transfer zones developed as independent blocks, often with faults in each

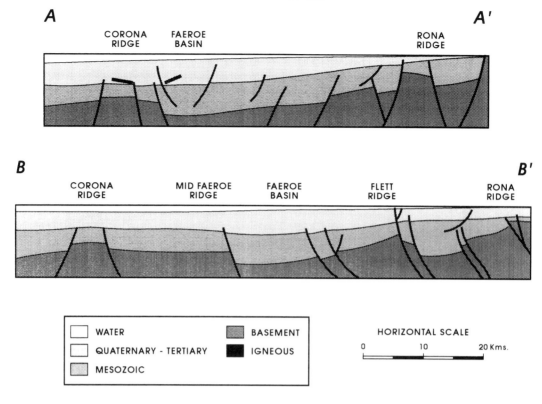

Fig. 13. Diagrammatic cross-sections of the Faeroe Basin. Location of cross-sections shown on Fig. 11.

block terminating within the block or at the transfer zones. This segmented development is comparable to the fracture zones and transform faults of the oceanic basins and the segmented basin development scenario of Lister *et al.* (1986). The blocks between the transfer zones often developed as independent sub-basins with distinct sedimentation patterns within the larger-scale tectonic framework of the basin. The transfer zones were not necessarily continually active throughout the basin's history, but responded to changes in the stress regime resulting from re-adjustments in the plate motions in the North Atlantic area, or from intra-basin mechanisms, such as differential subsidence related to variations in sediment input rate along the basin margin. The zones also acted as conduits for intrusives into the sedimentary section in some areas.

The major structural elements of the central part of the Faeroe Basin (Fig. 11)—the Shetland Spine Fault, Rona Ridge and Flett Ridge—are similar to previous descriptions except for some offsets at transfer zones. The 'Corona Ridge' is used here to identify a NE-trending ridge in the west of the Faeroe Basin (Fig. 11). The Corona Ridge is so termed because it is the structural corollary of the Rona Ridge on the southeastern flank of the basin. The term 'Faeroe Ridge' has been used in relation to various bathymetric and structural features in and near the Faeroe–Shetland Channel and is not appropriate for a distinct structural feature in the central Faeroe Basin. The gravity and magnetic data have identified transfer zones that offset the ridge, and recent better quality seismic data has given better definition of the structure of the ridge.

The diagrammatic cross-sections (Fig. 13) show the general structural configuration of the basin. Some reactivated high-angle reverse faults occur on the flanks of the Rona Ridge with the major faulting being simple normal faulting.

The authors would like to thank Mobil North Sea Ltd and Lasmo (TNS) Ltd for permission to publish this paper. The gravity and magnetic data were obtained from the British Geological Survey, Western Geophysical and proprietary data. Ark Geophysics Ltd produced the shaded-relief colour-contoured gravity and magnetic maps from Mobil's database, and their original interpretation of the data initiated the transfer zone model for the basin. Seismic sections are published with permission from Western Geophysical. We would also like to acknowledge the encouragement and constructive criticism in the preparation of this paper from Fred Alves, Mike Peoples, Bob Dunay and Steve Mitchell.

References

Bott, M. H. P. 1983. The crust beneath the Iceland–Faeroe Ridge. *In*: Bott, M. H. P., Saxov, S., Talwani, M. and Thiede, J. (eds) *Structure and development of the Greenland–Scotland Ridge—new methods and concepts*. Plenum Press, New York, 63–75.

—— and Gunnarsson, K. 1980. Crustal structure of the Iceland–Faeroe Ridge. *Journal of Geophysics*, **47**, 221–227.

——, Nielsen, P. H. and Sunderland, J. 1976. Converted P-waves originating at the continental margin between the Iceland–Faeroe Ridge and the Faeroe block. *Geophysical Journal of Royal Astronomical Society*, **44**, 229–238.

Duindam, P. and van Hoorn, B. 1987. Structural evolution of the West Shetland continental margin. *In*: Brooks, J. and Glennie, K. (eds) *Petroleum Geology of North West Europe*. Graham & Trotman, London, 765–773.

Earle, M. M., Jankowski, E. J. and Vann, I. R. 1989. Structural and stratigraphic evolution of the Faeroe–Shetland Channel and northern Rockall Trough. *In*: Tankard, A. J. and Balkwill, H. R. (eds) *Extensional Tectonics and Stratigraphy of the North Atlantic Margins*. American Association of Petroleum Geologists Memoir, **46**, 461–469.

Haszeldine, R. S. and Russell, M. J. 1987. The Late Carboniferous northern North Atlantic Ocean: implications for hydrocarbon exploration from Britain to the Arctic. *In*: Brooks, J. and Glennie, K. (eds) *Petroleum Geology of North West Europe*. Graham & Trotman, London, 1163–1175.

Hitchen, K. and Ritchie, J. D. 1987. Geological review of the West Shetland area. *In*: Brooks, J. and Glennie, K. (eds) *Petroleum Geology of North West Europe*. Graham & Trotman, London, 737–749.

KNOTT, S. D., BURCHELL, M. T., JOLLEY, E. J. AND FRASER, A. J. 1993. Mesozoic to Cenozoic plate reconstructions of the North Atlantic and the tectonostratigraphic history of the UKCS Western Margin. *In*: PARKER, J. R. (ed.) *Petroleum Geology of Northwest Europe: Proceedings of the 4th Conference*. Geological Society, London, 953–974.

LISTER, G. S., ETHERIDGE, M. A. AND SYMONDS, P. A. 1986. Detachment faulting and the evolution of passive continental margins. *Geology*, **14**, 246–250.

MITCHELL, S. M., BEAMISH, G. W. J., WOOD, M. V., MALECEK, S. J., ARMENTROUT, J. A., DAMUTH, J. E. AND OLSON, H. C. 1993. Paleogene sequence stratigraphic framework of the Faeroe Basin. *In*: PARKER, J. R. (ed.) *Petroleum Geology of Northwest Europe: Proceedings of the 4th Conference*. Geological Society, London, 1011–1023.

MUDGE, D. C. AND RASHID, B. 1987. The geology of the Faeroe Basin area. *In*: BROOKS, J. AND GLENNIE, K. (eds) *Petroleum Geology of North West Europe*. Graham & Trotman, London, 751–763.

NIELSEN, P. H. 1983. Geology and crustal structure of the Faeroe Islands—a review. *In*: BOTT, M. H. P., SAXOV, S., TALWANI, M. AND THIEDE, J. (eds) *Structure and development of the Greenland–Scotland Ridge—new methods and concepts*. Plenum Press, New York, 77–90.

NOE-NYGAARD, A. AND RASMUSSEN, J. 1970. *Geology of the Faeroe Islands (Pre-Quaternary)*. *Geological Survey of Denmark (I Series)*, **25**.

NUNNS, A. G. 1983. Plate tectonic evolution of the Greenland–Scotland Ridge and surrounding regions. *In*: BOTT, M. H. P., SAXOV, S., TALWANI, M. AND THIEDE, J. (eds) *Structure and development of the Greenland–Scotland Ridge—new methods and concepts*. Plenum Press, New York, 11–30.

ROBERTS, D. G., BOTT, M. H. P. AND URUSKI, C. 1983. Structure and origin of the Wyville–Thompson Ridge. *In*: BOTT, M. H. P.,

SAXOV, S., TALWANI, M. AND THIEDE, J. (eds) *Structure and development of the Greenland–Scotland Ridge—new methods and concepts*. Plenum Press, New York, 133–158.

——, GINZBERG, A., NUNN, K. AND McQUILLAN, R. 1988. The structure of the Rockall Trough from seismic refraction and wide-angle reflection measurements. *Nature*, **332**, 632–635.

SCRUTTON, R. A. 1986. The geology, crustal structure and evolution of the Rockall Trough and the Faeroe–Shetland Channel. *Proceedings of the Royal Society of Edinburgh*, **88b**, 7–26.

VERHOEF, J. AND SRIVASTAVA, S. P. 1989. Correlation of sedimentary basins across the North Atlantic as obtained from gravity and magnetic data, and its relation to the early evolution of the North Atlantic. *In*: TANKARD, A. J. AND BALKWILL, H. R. (eds) *Extensional Tectonics and Stratigraphy of the North Atlantic Margins*. American Association of Petroleum Geologists Memoir, **46**, 131–147.

VOPPEL, D., SRIVASTAVA, S. P. AND FLEISCHER, U. 1979. Detailed magnetic measurements south of the Iceland–Faeroe Ridge. *Deutsche Hydrographische Zeitschrift*, **32**, 154–172.

WAAGSTEIN, R. 1988. Structure, composition and age of the Faeroe basalt plateau. *In*: MORTON, A. C. AND PARSONS, L. M. (eds) *Early Tertiary Volcanism and the Opening of the NE Atlantic*. Geological Society, London, Special Publication, **39**, 225–238.

WHITE, R. AND McKENZIE, D. 1989. Magmatism at rift zones: The generation of volcanic continental margins and flood basalts. *Journal of Geophysical Research*, **94**, 7685–7729.

ZVEREV, S. M., KOSMINSKAYA, I. P., KRASILSHCHIKOVA, G. A. AND MIKHOTA, G. G. 1975. [The deep structure of the Iceland and the Iceland–Faeroes–Shetland region in the light of seismic surveys, the North Atlantic Project—72] *Mosk. O-voispt. Prir. Byull., Otd Geol.*, **50**, 99–115 (in Russian).

Paleogene sequence stratigraphic framework of the Faeroe Basin

S. M. MITCHELL,[1] G. W. J. BEAMISH,[1] M. V. WOOD,[1] S. J. MALACEK,[2]
J. A. ARMENTROUT,[3] J. E. DAMUTH[3] and H. C. OLSON[3]

[1] *Mobil North Sea Ltd, 3 Clements Inn, London WC2A 2EB, UK*
[2] *Mobil Exploration and Producing Services Inc., PO Box 900, Dallas, Texas 75221, USA*
[3] *Mobil Research and Development Corporation, Dallas Research Laboratory, Dallas, Texas 75244-4390, USA*

Abstract: During Paleogene times up to 15 000 ft (4570 m) of clastic sediment was deposited in the Faeroe Basin, north of the Shetland Islands. A sequence stratigraphic study has shown that the Paleogene deposition in the Faeroe Basin was cyclic with prominent basinward and landward shifts in sedimentation. The correlation of major unconformity surfaces allowed the section to be subdivided into genetically related packages.

The sequence stratigraphic study utilized approximately 5000 km of seismic data with a line density of approximately 10×20 km. Available well control was integrated into the study by means of synthetic seismograms. Limited palaeontological control utilizing largely dinocysts and radiolaria allowed the identification of 11 Paleocene/Eocene bioevents.

The section was subdivided into nine Paleocene and six Eocene sequences, each separated by Type 1 unconformities. Four of the Paleocene and one of the Eocene packages had evidence of multiple Type 1 unconformities and these are described as sequence sets.

Sequence development has been related to the tectonic subsidence history of the basin. Early in Paleocene times, rapid subsidence resulted in the deposition of thick sequences with distinct shelf, slope and basinal systems. Nine sequences were deposited with a combined maximum thickness of 12 000 ft (3660 m). The following period, late Paleocene to early Eocene, was marked by slower subsidence; thin sequences and ramp systems with seven sequences were deposited with a maximum thickness of 2400 ft (730 m). More rapid subsidence during the late Eocene resulted in five sequences with distinct shelf, slope and basinal systems with a thickness of up to 3500 ft (1070 m). The periods of slower subsidence in the Faeroe Basin may have occurred in response to active rifting in other adjacent basins along the Atlantic margin. The distinct basinal systems which developed during times of more rapid subsidence were more likely to develop sand-prone basin floor deposits.

Regional setting

The Faeroe Basin formed as a result of Late Jurassic/Early Cretaceous rifting on the NW margin of Europe. The resulting basin, located between the Faeroe Islands and the Shetland Islands, was a major depocentre during Paleogene times. Up to 15 000 ft (4570 m) of continentally derived clastic sediments were deposited in a marine to deltaic setting (Fig. 1). The geology of the Faeroe Basin–West Shetland areas has been described by a number of authors including Ridd (1983), Haszeldine *et al.* (1987), Hitchen and Ritchie (1987), Mudge and Rashid (1987) and Earle *et al.* (1989). Little detail regarding the Tertiary stratigraphy has been published, because few wells within the Tertiary depocentre were released by 1987 when most of these papers were published.

An understanding of the basin configuration at the end of the Cretaceous is important because the tectonic elements present then exerted first-order control on Paleogene sedimentation. By the end of the Cretaceous period, subsidence in the older West Shetland Basin had largely ceased (Fig. 1). The Shetland Platform and West Shetland Basin areas were dominantly shelf areas hinged approximately at the Rona Ridge. Basinal clastics were deposited in rapidly deepening water in the Faeroe Basin.

The northeastern and southwestern boundaries of this basin are controlled by major NW-oriented transfer zones which formed to accommodate differential crustal shortening during the Caledonian orogeny (Rumph *et al.* 1993). The Erlend volcanic centre is located along the Erlend Transfer Zone, which marks the northeastern boundary of the basin and was active in the late Paleocene. The southwestern boundary is defined by the Judd Fault, another major transfer zone, separ-

ating the Faeroe Basin from the Sula Sgeir High platform area. Other smaller transfer zones have also been identified by Rumph *et al.* (1993) and were important controls on local sedimentation patterns. The Victory Transfer Zone, extending northwest from the Shetland Islands, strongly influenced sedimentation throughout Paleocene times (Fig. 1), as shown by the difference in Paleocene sediment thickness on either side of the zone (Fig. 2). The Corona Ridge (Rumph *et al.* 1993) exerted a strong influence on sedimentation during the Paleogene and formed a bathymetric high against which sediments often ponded. The thickest Paleocene section in the basin is confined to the region bounded by the Corona and Rona ridges, the Judd Fault and the Victory Transfer Zone. The Eocene isochore map shows a similar pattern, although the thickness variation is less (Fig. 3).

Sequence stratigraphy

The sequence stratigraphic study of the Faeroe Basin utilized 5000 km of seismic lines chosen from proprietary and non-proprietary grids acquired by Western Geophysical, including 3000 km of data reprocessed in 1990. The final grid consisted of 34 dip lines and ten strike lines with a spacing of approximately 10×20 km. The study area for the Paleocene series encompassed the area from the Judd Fault in the southwest to the Erlend volcanic centre in the northeast (Fig. 1). Sequence stratigraphic analysis of the Eocene series encompassed a more limited area between the Judd Fault and the Victory Transfer Zone (Fig. 1). The seismic interpretation was tied to available well control utilizing synthetic seismograms.

The identification and correlation of regionally significant unconformity surfaces allowed the section to be subdivided

From *Petroleum Geology of Northwest Europe: Proceedings of the 4th Conference* (edited by J. R. Parker).

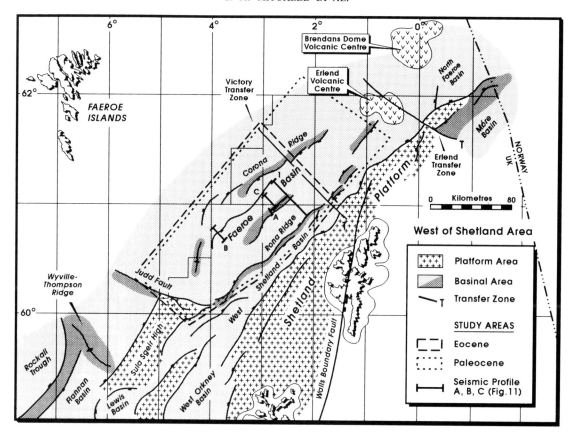

Fig. 1. Tectonic elements map of the Faeroe–Shetland Trough area showing the major tectonic elements present at the end of the Cretaceous period with study areas outlined.

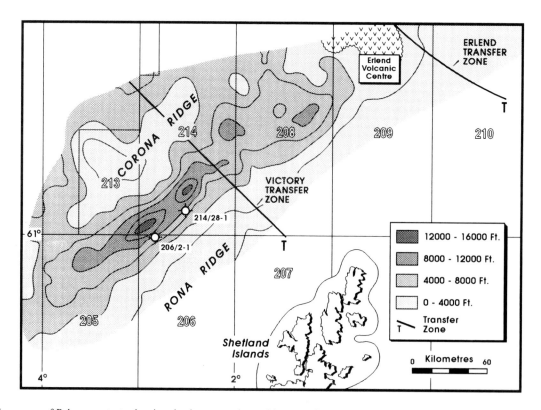

Fig. 2. Isochore map of Paleocene strata showing the depocentre located between the Rona Ridge and Corona Ridge (contour interval 2000 ft).

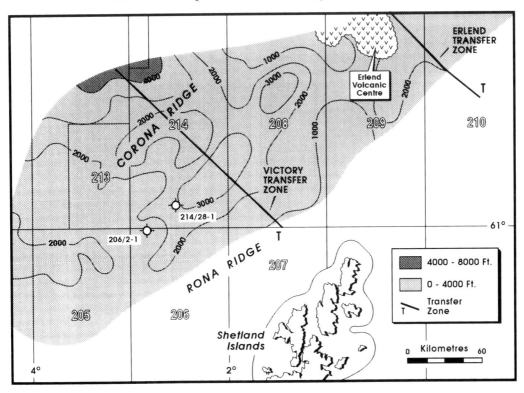

Fig. 3. Isochore map of Eocene strata showing only slight thickening between the Rona Ridge and Corona Ridge (contour interval 1000 ft).

into genetically related packages or 'depositional sequences' (Mitchum 1977). Tertiary deposition in the Faeroe Basin was cyclic with prominent basinward and landward shifts in sedimentation throughout this period of time. The initial interpretation resulted in the subdivision of the section into nine Paleocene and six Eocene packages, each separated by Type 1 unconformities (Fig. 4). Further interpretation demonstrated that four of the Paleocene and one of the Eocene packages comprise multiple sequences bounded by Type 1 unconformities (Fig. 5). These packages are considered 'sequence sets' (P. R. Vail, pers. comm.), each of which includes multiple sequences with similar geometries and depositional styles. Sequences and sequence sets were designated by colour during seismic interpretation, and were assigned numbers on the stacking chart and subsequent maps. The Paleocene interval is composed of sequences 10 to 90, with 20, 30 and 40 being sequence sets. The Eocene interval is composed of sequences 100 to 150, with 150 being a sequence set.

Table 1. Paleocene and Eocene bioevents

Bioevent	Age	Sequence	Event*	Type
11	41.5	120		Dinocyst top
10	45.5	110	P5	Dinocyst abundance
9	49	100	P4	Dinocyst abundance
8	52	100	M8	Dinocyst abundance
7	53	100	M7	Diatom abundance
6	54.5	60	P3	Dinocyst abundance
5	55.5	50	M5	Reappearance of agglutinated forams
4	56.5	40	M4	Dinocyst abundance
3	59	20	P2	Dinocyst abundance
2	61	10	P1	Radiolarian abundance
1	63	10	M1	Foram first occurrence

*Mudge and Copestake 1992

Biostratigraphic tops and abundance data for both dinocysts and radiolaria permitted the identification of 11 Paleocene and Eocene bioevents. These bioevents are designated 1 to 11 from the base (Table 1). Those bioevents characterized by faunal abundance peaks are interpreted to represent condensed intervals (Armentrout *et al.* 1990). The occurrence of these same abundance events in the North Sea demonstrates that they are developed in separate basins and thus may reflect eustatic sea-level events (Mudge and Copestake 1992). The temporal resolution of these bioevents is considered sufficient to constrain the basin framework; however the chronostratigraphy of the area is considered tentative because of the limited number of wells with palaeontology.

Systems tract development and terminology

Highstand and lowstand systems tracts have been recognized on the basis of characteristic lap-out patterns as defined by Van Wagonner *et al.* (1988). Lowstand systems tracts developed in the Faeroe Basin contain discrete, recognizable subdivisions (Fig. 6). The most basinward and normally oldest subdivision is the 'basin-floor thick', a mound-shaped body of sediment on the basin floor that shows downlapping in all directions onto a Type 1 unconformity. This term is used as a description of the seismically defined geometry observed, rather than the term fan, which implies a known lithology and depositional environment. The 'basin-floor thick' may or may not have a morphology similar to that of a submarine fan (Mitchum 1985) and may or may not be sand prone. Lowstand deposition on the depositional slope is termed 'slope-front fill'. When the 'slope-front fill' deposits possess a mound-shaped geometry, they are termed 'slope thicks'. Slope thicks commonly form within or at the base of submarine canyons. They can consist of either accumulated slump deposits, which may be largely mud-prone, or of sand-prone turbidity current and related mass-transport deposits. Local calibration that integrates seismic facies with well control is necessary to distinguish between sand-prone and mud-prone slope and basin-

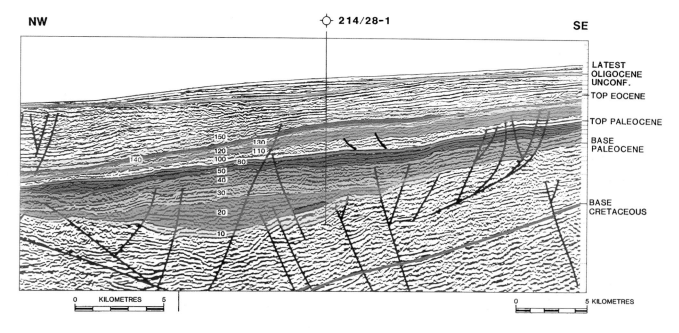

Fig. 4. Interpreted NW–SE-trending seismic profile across the Paleocene depocentre. No single profile will show all of the sequences because the distribution of the units is such that no single area has all fifteen packages present.

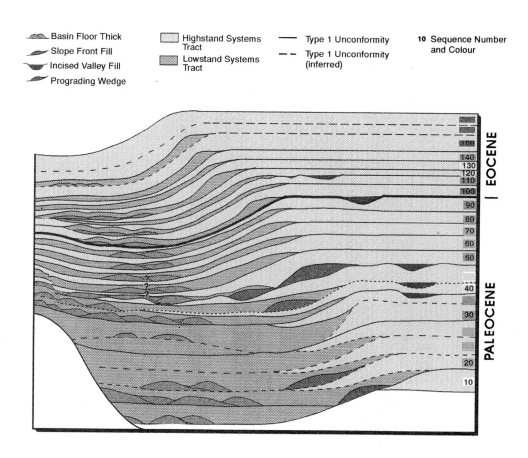

Fig. 5. Schematic profile oriented NW–SE across the Faeroe Basin showing the stacking order of the Paleogene sequences and lowstand subdivisions recogized in each.

bft - Basin Floor Thick   HST - Highstand Systems Tract   MCS - Marine Condensed Section
sff - Slope Front Fill   TST - Transgressive Systems Tract   DLS - Downlap Surface
pw - Prograding Wedge   LST - Lowstand Systems Tract
iv - Incised Valley

NW   SE

Fig. 6. Model illustrating the development of sequences in the Faeroe Basin.

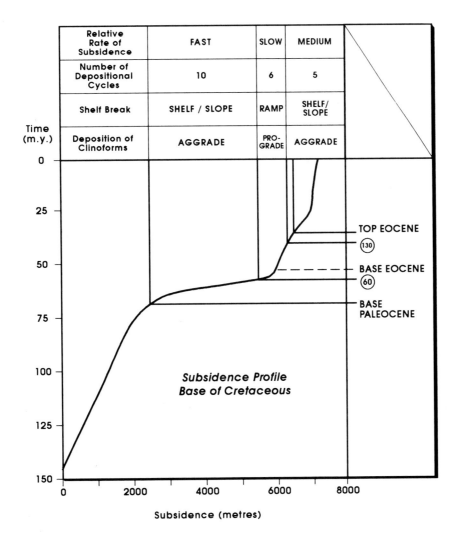

Fig. 7. Subsidence profile from the Faeroe Basin showing the relationship between the relative rate of subsidence and the style of sequence development in the Paleogene.

floor thicks. The lowstand systems tract may include a prograding wedge of fluvial–deltaic deposits. This wedge forms locally within and progrades downslope through an incised river-valley system eroded during a relative sea-level drop. The position and development of prograding wedges in the Faeroe Basin are locally controlled by shelf-margin growth faults (Fig. 6).

Highstand systems tracts may not be preserved or preserved only on the flanks of the basin. In some cases distal portions are interpreted to be preserved well into the basin. Transgressive systems tracts are typically below seismic resolution, although recognizable on electric logs from wells on the palaeoshelf by thin back-stepping parasequences (Vail and Wornardt 1990).

Paleogene sequence variations

Variations in basin subsidence resulted in the deposition of sequences with different styles of systems tract development. Three phases of subsidence and systems-tract development are recognized for the Paleogene section (Fig. 7). These phases provided a means of subdividing the section into sequence sets which are genetically similar.

Phase I: early Paleocene

In the early Paleocene, the Faeroe Basin subsided rapidly, resulting in a well-defined, deep-water basin flanked by a shelf and platform area. This phase of basin development was characterized by: (1) thick sequences; (2) recognizable depositional-shelf breaks; and (3) aggrading clinoform systems.

Sequences 10 to 50 were deposited during this phase, from 66 Ma to 55 Ma, and comprise a total of nine sequences and a combined maximum thickness of 12 000 ft. This set of Paleocene sequences consists of interbedded sandstones and mudstones (Fig. 8). Thick sandstone units have been penetrated in each of the units 10 to 50. The progradation of the units is

from the Shetland Platform and Sula Sgeir High suggesting that these sands were derived from the reworking of older clastics in these areas. Igneous intrusions have been penetrated in sequences 10 to 40 and become more numerous toward the Erlend and Brendan's Dome volcanic complexes (Hitchen and Ritchie 1987).

During this phase, much of the Faeroe Basin stratigraphy was dominated by thick lowstand systems tracts, with highstands, when preserved, located on the flanks of the basin. The geographic extent of the lowstand and highstand systems tract development reflects the contemporaneous basin evolution. The deposition of lowstands alternated in the early sequences between covering the entire basin and being restricted to the area southwest of the Victory Transfer Zone. This variation in sediment distribution reflects continued movement along the transfer zones in the early Paleocene. The absence of deposits over the Corona Ridge is a result of non-deposition rather than erosion. The absence of progradation from the Corona Ridge suggests that it was not subaerially exposed.

Lowstand systems tracts were spatially separated from the subsequent highstand systems tracts during the deposition of sequence 10 (Fig. 9). Lowstand deposition was largely con-

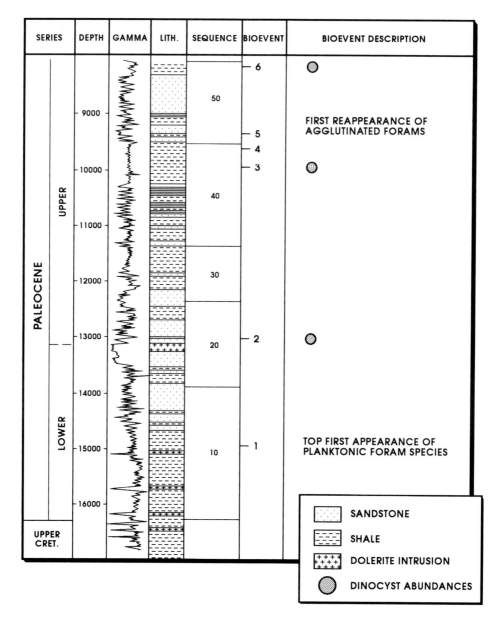

Fig. 8. Type log from Well 214/28-1 with sequences 10 to 50 demonstrating the sand-prone nature of Phase I sedimentation. The sequence boundaries are shown as brought in from the seismic interpretation.

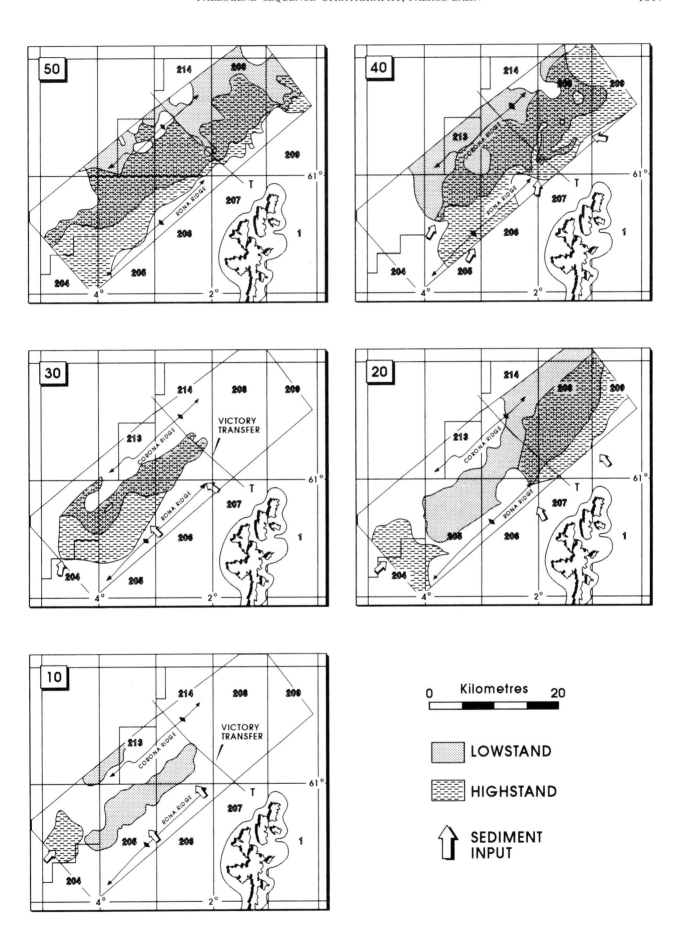

Fig. 9. Distribution of preserved lowstand and highstand systems tracts in the study area for Phase I sequences.

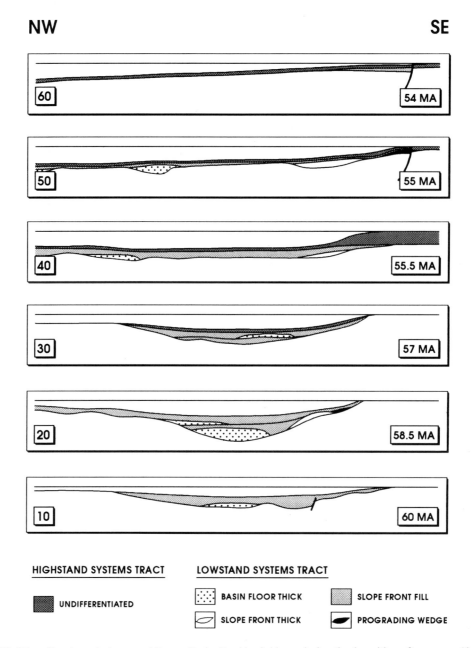

NW   **SE**

HIGHSTAND SYSTEMS TRACT   LOWSTAND SYSTEMS TRACT

UNDIFFERENTIATED   BASIN FLOOR THICK   SLOPE FRONT FILL

SLOPE FRONT THICK   PROGRADING WEDGE

Fig. 10. Schematic NW–SE profiles through the central Faeroe Basin. Rapid subsidence during the deposition of sequences 10 to 50 resulted in thick sequences. The highstand sections for sequences 10 to 30 are only preserved in the southeastern part of the area, near the Sula Sgeir High. Over most of the area, sequence 60 is a thin sequence which exhibits a ramp profile. Sequences 60 to 120 all exhibit this style of development which resulted from slower subsidence.

fined to a restricted area between the Rona and Corona ridges, southwest of the Victory Transfer Zone (Fig. 10). Highstand development is preserved only in the southwestern part of the area, near the Sula Sgeir High.

Sequence set 20 was deposited throughout the basin except for areas such as the Corona Ridge and contains the thickest sequences found in the Paleocene. Again, the only highstand development is largely limited to the southwestern area, near the Sula Sgeir High.

Sequence set 30 represents a period during which lowstand deposition was confined to the area between the Rona and Corona ridges, southwest of the Victory Transfer Zone.

Sequence set 40 deposition took place both northeast and southwest of the Victory Transfer Zone. No section is preserved within this sequence set in the southwest part of the study area. Thick highstand deposition is preserved over a more widespread area than in the earlier sequences.

Subsequently the most erosive unconformity recognized in the Paleogene was cut, followed by the deposition of sequence 50. Lowstand deposits were distributed both northeast and southwest of the Victory Transfer Zone. Within sequence 50, sandstone-prone intervals have been penetrated by wells in both lowstand and highstand systems tracts (Fig. 11). Well 214/28-1 encountered 750 ft of sandstone with a blocky gamma-ray log motif within the lowstand systems tract. Conventional core samples consist of very coarse to fine sandstones exhibiting sedimentary structures, such as graded beds above erosive bases, rip-up clasts, soft-sediment deformation and dewatering features, that are consistent with a mass flow/ gravity deposit origin. The isochron map of the lowstand systems tract at Well 214/28-1 shows a distinct linear trend of thickening perpendicular to the basin margin (Fig. 12). The sandstone body is interpreted as amalgamated mass flow deposits within a well-developed sand-prone slope thick. The

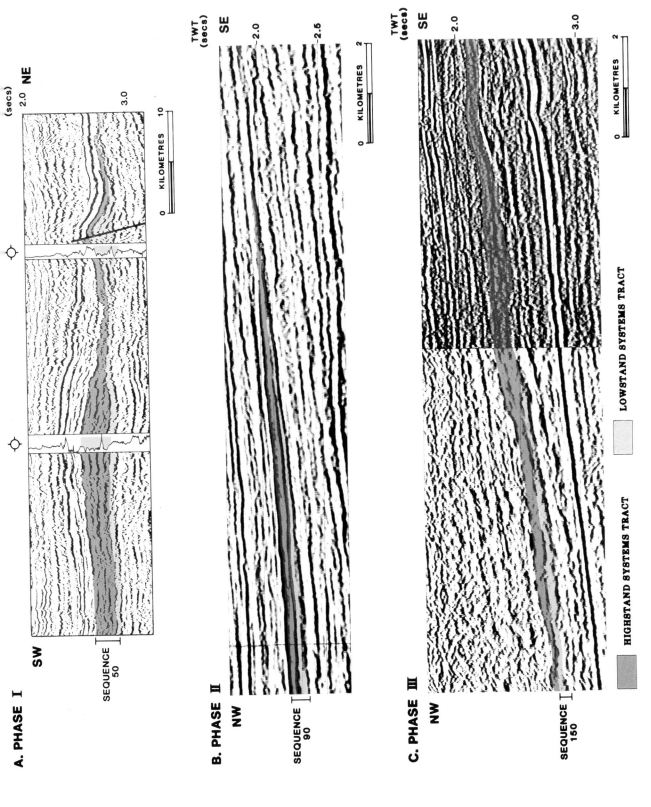

Fig. 11. Seismic profiles illustrating the seismic expression of lowstand and highstand systems tracts in each of the three phases of subsidence. The locations of all 3 profiles are shown in Fig. 1 and that of profile A is also shown in Fig. 12.

isochron thickening is adjacent to the Victory Transfer Zone, suggesting that the location of this deposit was controlled by the transfer zone. Sand in the thick may have been fed down to a submarine canyon which formed in a zone of weakness associated with the transfer zone.

Well 206/2-1 encountered approximately 700 ft of sandstone with at least seven coarsening-upwards cycles within sequence 50. Conventional core samples consist of extensively bioturbated sands interpreted to have been deposited in the lower shoreface environment. These sandstones are intepreted as highstand deposits formed as a deltaic complex that prograded over the lowstand section.

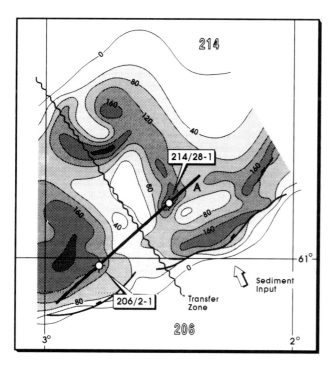

Fig. 12. Isochron map for the lowstand systems tract of sequence 50 (contour interval 40 ms). Seismic profile is shown in Fig. 11.

Phase II: late Paleocene–late Eocene

During the period from 55–39.5 Ma, subsidence slowed (Fig. 7), resulting in thinner cycles of a different depositional style. A ramp profile developed, precluding development of distinct depositional shelf breaks which characterized the early Paleocene. Sequences 60 to 120 are all thin sequences with ramp profiles and a composite thickness of up to 2400 ft. (Fig. 10). Sequence 60 is the thickest within the sequence set and resulted from a subsidence rate that was transitional between Phase I and Phase II.

Sequences 60, 70 and 90 are preserved only in the area southwest of the Victory Transfer Zone (Fig. 13). Sequences 80, 100, 110 and 120 are more extensive although the deposition of lowstand systems tracts may still have been influenced by the transfer zone. Sequence 100 is the only Phase II sequence with the lowstand systems tract seismically resolvable over the Corona Ridge.

The Phase II sequence set is more mudstone prone than sequences 10–50. Sands are normally thin and interpreted to have been deposited in coastal environments (Fig. 14). Sequences 70–100 commonly contain tuffaceous deposits. Bioevent 7 at the base of sequence 100 is characterized by an abundance of diatoms and corresponds to a section of interbedded tuffs, tuffaceous mudstones and mudstones equivalent to the Balder Formation of the Northern North Sea (Deegan

and Scull 1977). The associated seismic event is the most continuous mappable event in the Tertiary section.

Phase II sequences are normally very thin over much of the area, commonly only one or two reflectors thick (Fig. 11). In general, a strong continuous reflector corresponding to the marine condensed section onlaps the sequence boundary. Lowstand system tracts are areally restricted to local areas within the basin (Fig. 13). A thin mounded facies is present locally above the sequence boundary in the basinal setting. This facies is interpreted as gravity deposits forming basin-floor or slope thicks.

Phase III: late Eocene

Subsidence rates increased in the late Eocene from 39.5–36 Ma (Fig. 7). This moderate rate of subsidence re-established a distinct depositional shelf break and allowed the development of predominantly highstand, prograding clinoforms for sequences 130–150. The Type 1 unconformities associated with these sequences are locally highly erosive events. However, there is little evidence of development of substantial basin-floor thick deposits within these sequences in the study area. Well penetrations into these sequences are typically mudstone prone (Fig. 14). Sequence set 150 is a thick accumulation of highstand clinoforms which prograded basinward and apparently filled the basin by the end of the Eocene (Fig. 11). The Base Oligocene Unconformity which marks the top of the Eocene section is also a highly erosive event.

Comparison with global coastal onlap curve

Comparison of the cycle curve developed for the Faeroe Basin with the published coastal onlap curve of Haq *et al.* (1988) shows that there are more sequences recognized in the Faeroe Basin during the Paleocene and fewer in the Eocene (Fig. 15). The study of the Faeroe Basin recognized 13 sequences in the Paleocene, whereas the Haq *et al.* (1988) curve has seven third-order sequences. The Eocene of the Faeroe Basin is divided into seven sequences, whereas the Haq *et al.* curve has 14 third-order sequences. Although the chronostratigraphy of the Faeroe Basin is not as well constrained as that of more mature basins, the base Tertiary and bioevent 7, taken as the top of the Paleocene, are well documented. It is unlikely that there are many more or less cycles present in the Faeroe Basin in the Paleocene or Eocene than those identified in this study.

Sequence development is controlled by the interaction of many factors including tectonic activity, eustatic sea-level changes, climate and sediment supply. The relative rise or fall of sea-level, which results from this interaction, determines the development of unconformities and the style of systems tract deposition. Thus, during a period of eustatic sea-level fall, relative sea-level may rise, still-stand or fall depending on the local rates of accommodation-space formation and sediment accumulation (Jervey 1988).

Unconformities included on the published curve have been identified in sedimentary basins on at least three continents (Vail *et al.* 1984). This does not guarantee that any one basin in the world will coincide with the curve through a significant period of time. As this work in the Faeroe Basin suggests, each individual basin will have its own pattern of sequence development, necessitating the development of local curves applicable to each specific basin studied. Inferred eustatic changes in sea-level are indeed supported by biostratigraphic abundance data although often without the time resolution desired. The 11 bioevents recognized in the Faeroe Basin are not sufficient to allow the calibration of the 20 sequences to other basins with confidence. It is likely that only the largest eustatic changes and long-term trends may be recognizable and correlatable on a global basis. Sequence stratigraphic studies must be

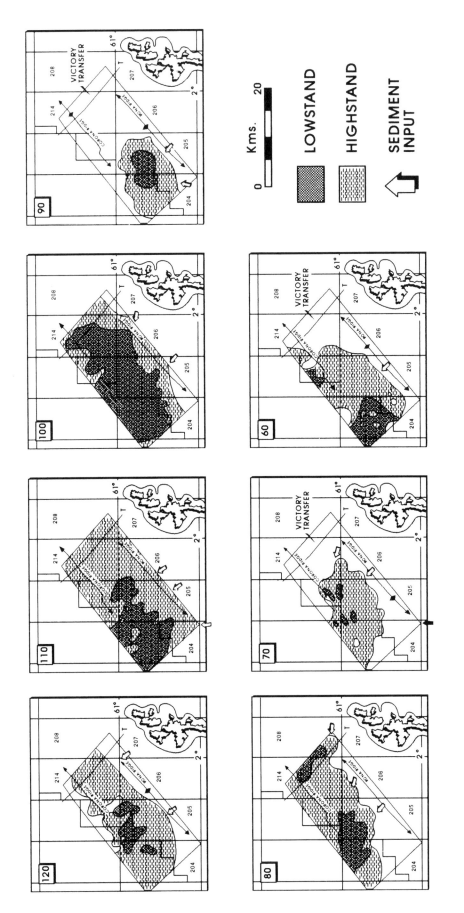

Fig. 13. Distribution of preserved lowstand and highstand systems tracts in the study area for Phase II sequences.

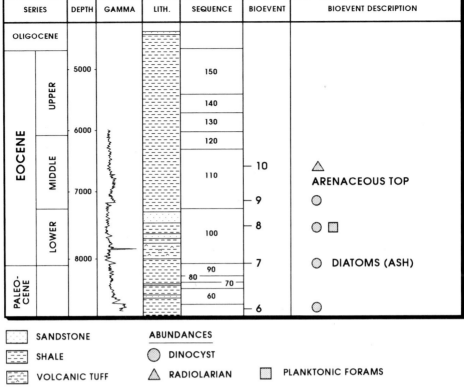

Fig. 14. Type log Well 205/10-2B from sequence 60–150. Sequences 60–120, deposited during Phase II, are characterized by interbedded shales and thin sandstone beds. Sequences 130–150, deposited during Phase III, are typically shale prone.

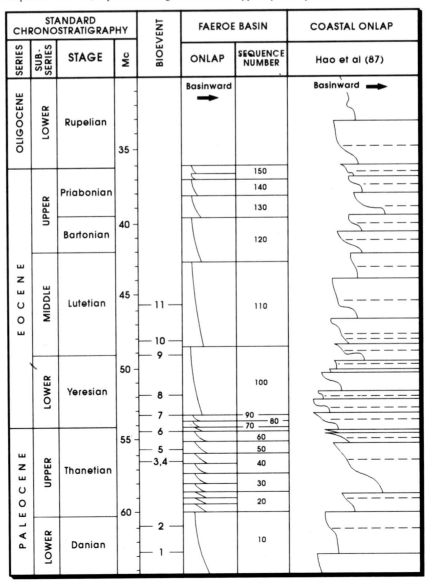

Fig. 15. Comparison of Faeroe Basin sequence onlap chart with the published global coastal onlap chart (Haq *et al.* 1988).

carried out on each individual basin to determine the cycles fundamental to that basin which have determined its unique stratigraphy.

The authors would like to thank Mobil North Sea Ltd, Mobil Research and Development Corp. and Lasmo (TNS) Ltd for permission to publish this paper. We would also like to acknowledge valuable contributions to this study made by L. F. Brown, R. J. Desmarais, R. E. Dunay, L. B. Fearn, S. Haimes, V. G. Orange, C. M. Reaves, B. Rumph and J. Vizgirda. Several of the seismic profiles are published with permission from Western Geophysical.

References

ARMENTROUT, J. M., ECHOLS, R. H. AND LEE, T. D. 1990. Patterns of foraminiferal abundance and diversity: implications for sequence stratigraphic analysis. *In*: ARMENTROUT, J. M. AND PERKINS, B. F. (eds) *Sequence Stratigraphy as an Exploration Tool: concepts and practices in the Gulf Coast.* Gulf Coast Section SEPM Foundation Eleventh Annual Research Conference Program and Abstracts, 53–58.

DEEGAN, C. E. AND SCULL, B. J. 1977. *A Standard Lithostratigraphic Nomenclature for the Central and Northern North Sea.* Institute of Geological Sciences Report No. 77/25.

EARLE, M. M., JANKOWSKI, E. T. AND VANN, I. R. 1989. The structural and stratigraphic evolution of the Faeroe–Shetland Channel and northern Rockall Trough. *In*: TANKARD, A. J. AND BALKWILL, H. R. (eds) *Extensional Tectonics and Stratigraphy of the North Atlantic Margins.* American Association of Petroleum Geologists Memoir, **46**, 461–469.

HAQ, B. U., HARDENBOL, J. AND VAIL, P. R. 1988. Mesozoic and Cenozoic chronostratigraphy and cycles of sea-level change. *In*: WILGUS, C. K., POSAMENTIER, H. W., ROSS, C. A. AND KENDALL, C. G. St. C. (eds) *Sea-level Changes: An Integrated Approach.* SEPM Special Publication **42**, 71–108.

HASZELDINE, R. S., RITCHIE, J. D. AND HITCHEN, K. 1987. Seismic and well evidence for the early development of the Faeroe–Shetland Basin. *Scottish Journal of Geology* **23**, 283–300.

HITCHEN, K. AND RITCHIE, J. D. 1987. Geological review of the west Shetland area. *In*: BROOKS, J. AND GLENNIE, K. W. (eds) *Petroleum Geology of North West Europe.* Graham & Trotman, London, 737–749.

JERVEY, M. T. 1988. Quantitative geological modeling of siliciclastic rock sequences and their seismic expression. *In*: WILGUS, C. K.,

POSAMENTIER, H. W., ROSS, C. A. AND KENDALL, C. G. St. C. (eds) *Sea-level Changes: An Integrated Approach.* SEPM Special Publication **42**, 71–108.

MITCHUM, R. M. 1977. Seismic stratigraphy and global changes in sea level, Part 1: Glossary of terms used in seismic stratigraphy. *In*: PAYTON, C. E. (ed.) *Seismic Stratigraphy: Application to Hydrocarbon Explorations.* American Association of Petroleum Geologists Memoir, **26**, 205–212.

—— 1985. Seismic stratigraphic expression of submarine fans. *In*: BERG, O. R. AND WOOLVERTON, D. G. (eds) *Seismic Stratigraphy II: an Integrated Approach to Hydrocarbon Exploration.* American Association of Petroleum Geologists Memoir, **39**, 117–136.

MUDGE, D. C. AND COPESTAKE, P. 1992. Revised lower Palaeogene lithostratigraphy for the Outer Moray Firth, North Sea. *Marine and Petroleum Geology*, **9**, 53–69.

—— AND RASHID, B. 1987. The geology of the Faeroe Basin area. *In*: BROOKS, J. AND GLENNIE, K. W. (eds) *Petroleum Geology of North West Europe.* Graham & Trotman, London, 751–763.

RIDD, M. F. 1983. Aspects of the Tertiary geology of the Faeroe–Shetland Channel. *In*: BOTT, M. H. P., SAXOV, S., TALWANI, M. AND THIEDE, J. (eds) *Structure and Development of the Greenland–Scotland Ridge.* Plenum Press, New York, 133–158.

RUMPH, B., REAVES, C. M., ORANGE, V. G. AND ROBINSON, D. L. 1993. Structuring and transfer zones in the Faeroe Basin in a regional tectonic context. *In*: PARKER, J. R. (ed.) *Petroleum Geology of Northwest Europe: Proceedings of the 4th Conference.* Geological Society, London, 999–1009.

VAIL, P. R., HARDENBOL, J. AND TODD, R. G. 1984. Jurassic unconformities, chronostratigraphy, and sea-level changes from seismic stratigraphy and biostratigraphy. *In*: SCHLEE, J. S. (ed.) *Interregional Unconformities and Hydrocarbon Accumulations.* American Association of Petroleum Geologists Memoir, **36**, 129–144.

—— AND WORNARDT, W. W. 1990. Well-log seismic sequence stratigraphy: an integrated tool for the '90s. *In*: ARMENTROUT, J. M. AND PERKINS, B. F. (eds) *Sequence Stratigraphy as an Exploration Tool: Concepts and Practices in the Gulf Coast.* Gulf Coast Section SEPM Foundation Eleventh Annual Research Conference, Program and Abstracts, 379–388.

VAN WAGONER, J. C., POSAMENTIER, H. W., MITCHUM, R. M., VAIL, P. R., SARG, J. F., LOUTIT, T. S. AND HARDENBOL, J. 1988. An overview of the fundamentals of sequence stratigraphy and key definitions. *In*: WILGUS, C. K., POSAMENTIER, H. W., ROSS, C. A. AND KENDALL, C. G. St. C. (eds) *Sea-level Changes: An Integrated Approach.* SEPM Special Publication **42**, 39–45.

Late Paleocene to Miocene compression in the Faeroe–Rockall area

L. O. BOLDREEL and M. S. ANDERSEN

Geological Survey of Denmark, Thoravej 8, DK-2400 Copenhagen NV, Denmark

Abstract: Compressional structures are observed at several locations in the Faeroe–Rockall Area. One of these, the Wyville–Thomson Ridge Complex, is part of the Scotland–Greenland Ridge which forms a barrier to the south-flowing deep cold arctic water from the Norwegian Sea. Interpretation of seismic multichannel reflection profiles suggests that the Wyville–Thomson Ridge Complex is the result of compression and that the Ymir Ridge and the Wyville–Thomson Ridge are ramp anticlines connected with a fault plane dipping to the north. A number of small highs offset by reverse faults to the south of the complex are interpreted as foreland thrust folds developed in relation to the tectonic evolution of the ridge complex. At least three Eocene to Miocene compressional phases are recognized. The first took place in late Paleocene–early Eocene and the second in Oligocene times. These compressional phases coincide with pronounced changes in the sea floor spreading geometry in the NE Atlantic. The third phase, in the middle or late Miocene, may possibly be associated with the complex Miocene spreading history of Iceland.

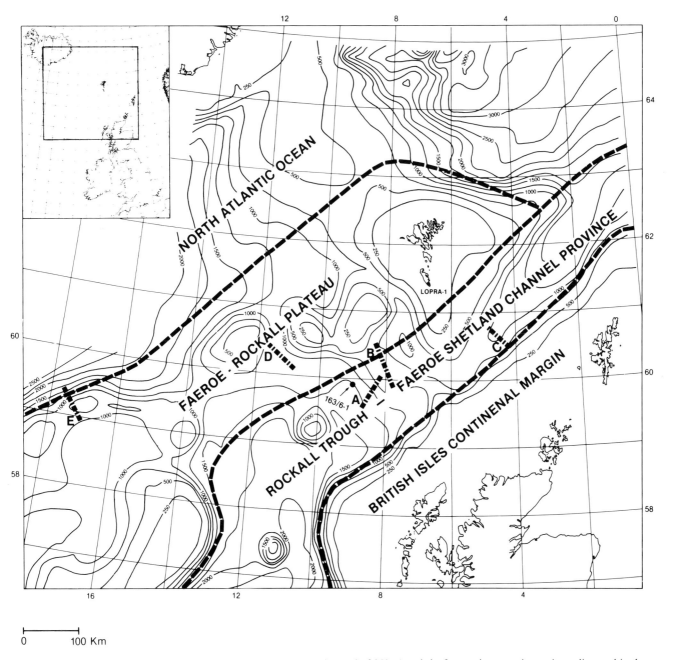

Fig. 1. Bathymetric sketch map of the Faeroe–Rockall area (contour interval of 250 m) and the four major tectonic provinces discussed in the text. Location of the seismic sections A to E in Figs 4–8 is also shown.

From *Petroleum Geology of Northwest Europe: Proceedings of the 4th Conference* (edited by J. R. Parker).

The Faeroe–Rockall Plateau is separated from the NW European continental shelf by the Rockall Trough and the Faeroe Shetland Channel (Fig. 1). The plateau is part of the volcanic passive continental margin formed during the initial opening of the NE Atlantic Ocean. Associated with the rifting episode and the formation of the margin, extensive volcanism occurred and a thick succession of lower Tertiary volcanic rocks is found in most of the area (Roberts, Schnitker et al. 1984; Rasmussen and Noe Nygaard 1970). The surface of the volcanic rocks is recognized as a very strong reflector on seismic reflection profiles and often acts as an acoustic basement. Thus the geology below the surface of the basalt is largely unknown.

Rift basins have been reported to exist immediately below thin basalt cover in parts of the Hatton–Rockall Basin (Andersen et al. 1990) but otherwise direct evidence of crustal rifting preceding the volcanic activity is limited. The rift basins in the Hatton–Rockall Basin predate the volcanics but as no wells have been drilled through the basalt cover the age is not well established at present. It is normally inferred that lithospheric extension takes place prior to the formation of the passive continental margins (e.g. McKenzie 1978). Moreover, it has been suggested that the early Tertiary volcanic activity is associated with the lithospheric extension preceding the early Tertiary continental break-up between Faeroe–Rockall Plateau and Greenland (Roberts et al. 1984a; White et al. 1987; White 1988).

The Faeroe–Rockall Plateau is usually described in terms of extension and the high-amplitude topographic features reflect, to a large extent, the major structures on the surface of the lower Tertiary volcanic rocks (Roberts et al. 1979). However, we have identified a number of compressional structures of post-volcanic age on seismic reflection data (Fig. 2).

In this paper we first review the tectonic setting of the Faeroe–Rockall area. Then, we discuss some of the compression structures, the distribution and the timing of the deformation and the causal relationships in relation to other structural events in the NW European–NE Atlantic region.

Details of the tectonic framework of the Faeroe–Rockall area have been discussed in terms of continental break-up and seafloor spreading in the NE Atlantic Ocean (e.g. Smythe 1983;

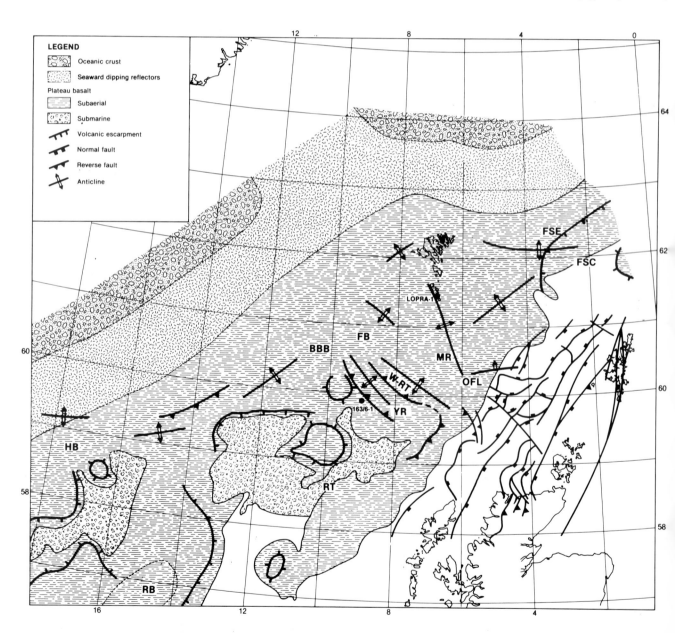

Fig. 2. Distribution of the early Tertiary volcanics, tectonic elements and volcanic escarpments. BBB: Bill Bailey Bank; FB: Faeroe Bank; FSC: Faeroe Shetland Channel; FSE: Faeroe Shetland Escarpment; HB: Hatton Bank; MR: Munkegrunnur Ridge; OFL: Orkney Faeroe Alignment; RB: Rockall Bank; RT: Rockall Trough; W-RT: Wyville–Thomson Ridge; YR: Ymir Ridge. Tectonic elements of the British Isles continental margin according to Earle et al. (1989).

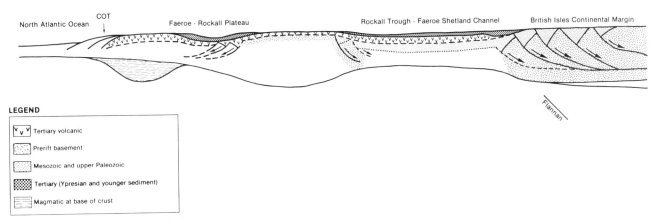

Fig. 3. Generalized geological cross-section of the Faeroe–Rockall area from the British Isles continental margin to the North Atlantic Ocean. Approximate vertical exaggeration: × 8. COT: Continent–Ocean Transition Zone.

Talwani *et al.* 1983; Roberts *et al.* 1984*b*) and in terms of continental rifting in NW Europe (e.g. Earle *et al.* 1989). We propose a division of the study area into four major tectonic provinces: British Isles continental margin, Faeroe Shetland Channel–Rockall Trough Province, Faeroe–Rockall Plateau, and North Atlantic Ocean (Figs 1 and 3), each having a distinct evolutionary history.

British Isles continental margin

Along the continental margin west of the British Isles (Figs 2 and 3), the upper crust is characterized by NE–SW striking upper Paleozoic and Mesozoic sedimentary basins, delimited by faults (Naylor and Shannon 1982). Many of the extensional faults which control the sedimentary basins are listric and appear to be rejuvenations of older compressional faults (e.g. Brewer and Smythe 1986). Tertiary tectonic activity is mainly documented by inversion of Mesozoic basins under the continental slope (Earle *et al.* 1989). Additional evidence of Tertiary tectonic activity is shown by uplift of significant parts of the continental shelf area, shown by the present distribution of subcrops either on the seafloor or under a thin Quaternary cover. The distribution of Tertiary sediments is apparently controlled both by subsidence in the Faeroe Shetland Channel and Rockall Trough province to the west and by sediment supply from uplifted source areas to the east (Ziegler 1990). Therefore the Tertiary sediments are generally restricted to the western part of the province. A number of restricted basins in the western part of the province contain large thicknesses of locally derived clastic sediments of Danian age (Hitchen and Ritchie 1987; Mudge and Rashid 1987). The distribution of these sediments is apparently controlled by slight transverse movements on older (Mesozoic) fault planes (Hitchen and Ritchie 1987).

Faeroe Shetland Channel and Rockall Trough

West of the British Isles continental margin, great thicknesses of Tertiary sediments and volcanic rocks are found in a major Tertiary basin, which is divided into two sub-basins, the Faeroe Shetland Channel and the Rockall Trough, by the Wyville–Thomson Ridge Complex (Figs 1 and 2).

The origin and age of the Faeroe Shetland Channel and the Rockall Trough has been debated, but unequivocal evidence either for the crustal nature or for the age has not been presented. Seafloor spreading as a cause of the origin of the Faeroe Shetland Channel and Rockall Trough has been suggested from plate tectonic reconstructions, from the crustal thickness and the lack of fault blocks dipping away from the margin (Bott 1984; Haszeldine 1984). Others have suggested that the Faeroe Shetland Channel and Rockall Trough were created by rifting of continental crust, based on the velocity structure in the Rockall Trough (Roberts *et al.* 1988; Neish 1990). The suggested age of formation ranges from Carboniferous (Haszeldine 1984) to Late Cretaceous (Ziegler 1990).

In the eastern part of the Faeroe Shetland Channel, Mesozoic sedimentary basins bounded by extensional faults have been identified below the Tertiary sediments (e.g. Earle *et al.* 1989; Hitchen and Ritchie 1987). These basins may be considered as a subsided continuation of the British Isles Continental Margin structural province.

Mesozoic sediments west of the subsided rift basins (Bott 1984; Mudge and Rashid 1987) have not been confirmed by drilling. However, in the bottom of Well 163-6/1 (northern Rockall Trough) approximately 200 m of Paleocene cordierite bearing dacites was found (Morton *et al.* 1988). The presence of this rock type indicates that aluminium-rich rocks are present in a reducing environment below the well site and have been partially incorporated in mafic magmas. In connection ·with the detection of hydrocarbons in the well, this is a strong indication that organic-rich shales are present in the northern part of the Rockall Trough (Morton *et al.* 1988). In the southern part of the Rockall Trough, Danian and possibly older sediments are interpreted from seismic sections (Smythe 1989).

Faeroe–Rockall Plateau

The Faeroe–Rockall Plateau is characterized by widespread lower Tertiary (Thanetian–Ypresian) volcanic rocks (Fig. 2), which are mostly subaerial plateau basalts of tholeiitic character (Rasmussen and Noe-Nygaard 1970; Roberts *et al.* 1983; Smythe 1983; Wood *et al.* 1987). The plateau includes a number of large banks (2000–3000 km^2) (Fig. 1). At the top of the banks the lower Tertiary volcanic rocks outcrop or are covered by a thin layer consisting of Quaternary and Neogene sediments. The banks are separated by deep (500–2000 m) basins or channels, and the bathymetry generally reflects Tertiary subsidence of the basalt surface in these basins, which are partly filled with Paleogene and Neogene sediments. Due to subsidence and limited influx of clastic sediments, water depths increased during Tertiary times. Since the Miocene, deep sea conditions characterized by pelagic sedimentation and deep sea bottom currents (contour currents) have prevailed (Kidd and Hill 1986).

Direct evidence of actual crustal rifting preceding the volcanic activity is limited. Andersen *et al.* (1990) have shown that rift basins exist immediately below the volcanics in the Hatton–Rockall Basin.

North Atlantic Ocean

This structural province (Figs 1 and 2) is characterized by extension preceding the formation of new oceanic crust to the north and west of the Faeroe–Rockall Plateau since the Early Eocene. The continent–ocean transition is characterized by the occurrence of seaward-dipping reflectors, a succession of large wedges of mainly subaerially erupted basalt (Smythe 1983; Roberts *et al.* 1984*b*; Spence *et al.* 1989). The exact location of the continent–ocean transition to the north and the west of the Faeroe Islands has been debated (e.g. Talwani *et al.* 1983). Also, the mechanism responsible for the eruption of large amounts of basaltic lava above sea level along the volcanic continental margins is still debated (White 1988; Mutter and Zehnder 1988; Skogseid and Eldholm *et al.* 1988; Eldholm *et al.* 1989). However, based on conversion of seismic P-waves northwest of the Faeroes, the continent–ocean transition can be located below the inner part of the seaward-dipping reflectors (Bott *et al.* 1975; Smythe 1983).

Stratigraphy of the basalt formation and sedimentary succession from Ypresian to Recent

The oldest widespread volcanics of the Faeroe–Rockall Plateau are the Lower Basalt Series of the Faeroe Islands. It is suggested that the series was extruded during chrons C26R to C25N (Thanetian) (Waagstein 1988). At the 2.2 km deep Lopra-1 well the series had an original thickness of at least 3000 m (Hald and Waagstein 1984; Jørgensen 1984). This series, which was characterized by rhythmic fissure eruption, consists of thick (20–50 m) parallel bedded subaerial basaltic flows with minor intercalations of tuff and volcaniclastic sediments. Refraction seismic data and a vertical seismic profile (VSP) in the Lopra-1 well shows that the Lower Basalt Series lie above a layer which has a lower seismic velocity than the basalts. Furthermore, the Lower Basalt Series is also present on the shelf east of the Faeroe Islands (Kiørboe 1990). It is not proved that the low-velocity layer beneath the Lower Basalt Series consists of sediments but minor gas and oil shows in the lava sequence in the Lopra-1 well indicate the presence, at a deeper level, of marine sediments with an organic content (Jacobsen and Laier 1984). The suggested marine sediments could perhaps continue into the Faeroe Basin where Bott (1984) has observed a similar layer. All the lava flows drilled in the Lopra-1 well are described as subaerial and so have subsided at least 2000 m since the eruption of the basalt. Based on seismic mapping of the Faeroe–Rockall Plateau we estimate that the total subsidence exceeds 3000 m in places. Equivalents of the Lower Basalt Series can possibly be traced as far south as the Hatton–Rockall Basin.

After the formation of the Lower Basalt Series there was a period of quiescence represented by *ca.* 10 m thick lacustrine sediments, characterized by allochthonous coal beds (the coal bearing sequence) of Thanetian age (Lund 1981; oldest part of magnetic chron C24R; Waagstein 1988). The two higher Basalt Series (the Middle and the Upper) above the sediment sequence are suggested to be extruded during C24R (Waagstein 1988). The Middle Basalt Series (*c.* 1.4 km thick) started with explosive volcanism and accumulation of fragmental eruptive products before it turned to continuous volcanism. The volcanism formed shield volcanoes over parts of the old fissures. With the formation of the Upper Basalt Series (>0.9 km), which is only found on the northern islands, the rhythmic volcanism returned. It has been indicated that the Middle and Upper Basalt Series may be confined to a relatively narrow

area along the continental margin (Waagstein 1988) and that the extrusive activity ceased in the Faeroes before the opening of the NE Atlantic during C24R. Thus the volcanic activity continued for 4 to 7 million years on the Faeroes and the major part of the volcanism occurred in the Thanetian. This is in good agreement with age estimates of the basalts from other parts of the Faeroe–Rockall Plateau and eastern margin of the North Atlantic Ocean (Roberts *et al.* 1984*b*; Morton *et al.* 1988; Eldholm *et al.* 1989).

The three-fold division of the lower Tertiary volcanics has not been confirmed outside the Faeroes. However, it is suggested that the Lower Basalt Series is present over most of the Faeroe–Rockall Plateau, whereas the Middle and Upper Basalt Series are restricted to the geographical areas limited by the volcanic escarpments (Fig. 2) found on the surface of the basalt (Smythe 1983). The escarpments formed as volcanic foreset breccias and reveal information concerning the location of the palaeo-coastline.

After the cessation of the volcanism the area subsided and sediments were deposited on top of the Basalt Series in the basins offshore the Faeroe Islands.

Originally, the Tertiary sediments above the basalts were divided into two units separated by the reflector R-4, which was dated as early Oligocene (Roberts 1975; Roberts *et al.* 1983). However, in some areas the R-4 reflector appears to be time-transgressive, and we assume that R-4 may reflect a major environmental change. The available seismic data enable us to divide the post-volcanic sediments into five seismic sequences divided by unconformities (Table 1). As far as possible these unconformities have been correlated to available well data from released commercial wells along the British Isles continental margin and a number of Deep Sea Drilling Project (DSDP) wells outside this province. The preliminary stratigraphy is consistent with the available well data and is also in overall agreement with the seismic stratigraphy for the Rockall Trough proposed by Smythe (1989), which is constrained by a tie to Well 163/6-1 in the northern Rockall Trough and to DSDP wells in the southern part of the Rockall Trough. The preliminary stratigraphy has so far been consistent within the Faeroe–Rockall Plateau, the Faeroe Shetland Channel and the Rockall Trough and it appears to be valid in the Faeroe Bank Basin, although no well is lcoated in the basin.

Table 1. Correlation of seismic unconformities in the Faeroe–Rockall area

		This paper
Smythe (1989)	Generalized reflector name	Approximate age of seismic sequence
		Plio-Pleistocene
R10 latest Miocene	late Miocene	
		late Miocene
R20 late early Miocene – middle Miocene	mid Miocene	
		early Miocene
R30 late Oligocene – early Miocene	late Oligocene	
		Oligocene
R40 earliest Oligocene	Eocene–Oligocene	
		mid- and late Eocene
	Intra Eocene	
		early Eocene
R50 earliest Eocene	Paleocene–Eocene	
		Paleocene and older

The most serious problem arising from the lack of detailed well control probably occurs in the Neogene succession. This succession is represented in most basins by deep water sediments in which unconformities are not correlated in a simple manner to sea-level changes and tectonic activity.

Late Paleocene–Recent tectonic evolution of Rockall Trough and Faeroe–Rockall Plateau

Evidence for differential vertical movements during the volcanic activity are scarce in the Faeroe–Rockall Area. One major exception is the seaward-dipping reflectors on the northern and northwestern flanks of the plateau. These movements, which post-date the basalts of the plateau, are generally attributed to the opening of the NE Atlantic. Within the Faeroe–Rockall Area we often observe nearly parallel bedded reflectors within the lower Tertiary volcanics. We interpret this to mean that the volcanic material erupted subaerially and formed an almost horizontal volcanic plateau of basalt with a very slight dip away from the North Atlantic Province. The plateau terminates abruptly at various escarpments on the east side of the plateau, which marked the shorelines during eruption (Fig. 2; Smythe 1983; Wood *et al.* 1987).

Faulting and folding of the nearly parallel bedded lower Tertiary volcanic basalt have been observed so far in the northern parts of Rockall Trough–Faeroe Shetland Channel and Faeroe–Rockall Plateau tectonic provinces. The locations of the Tertiary compression structures found are shown in Fig.

2. The seismic sections (Figs 4–8) show that the compression affects the basalt and part of the sediments above.

The most impressive compression structure is the Wyville–Thomson Ridge Complex southwest of the Faeroe Islands. The complex consists of two anticlines, the Wyville–Thomson Ridge and the Ymir Ridge, as well as a small intervening Tertiary basin which to the ESE is connected to the Rockall Trough. The complex extends as a whole for approximately 150 km in the NW–SE direction (along-strike) and 50 km in the NE–SW direction and separates the Faeroe–Shetland Channel and the Rockall Trough (Figs 1 and 2).

The southern flank of the Ymir Ridge is expressed for a considerable distance as a high and steep escarpment by the surface of the volcanic rocks. Many reverse faults are seen on the multichannel seismic data (Figs 4 and 5) associated with the southern flanks of the two ridges.

Roberts *et al.* (1983) interpreted the Wyville–Thomson Ridge as a volcanic pile erupted on top of earlier sediments and oceanic crust and resulting from Paleocene volcanic eruptions through a series of fissures aligned with the ridge. The escarpments along the southern part of the ridge were suggested to be due to post-lava faulting. Bott (1984) proposed that the topography of the Wyville Thomson Ridge represented later volcanic loading on a belt of otherwise uniform oceanic crust. Both models are in general accordance with the free air gravity

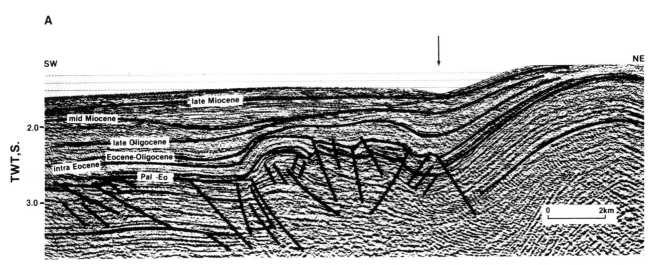

Fig. 4. Migrated seismic section crossing the Ymir Ridge. Location in Fig. 1. Miocene deep water differential sedimentation at the arrow is indicative of contour currents. (Courtesy of Western Geophysical.)

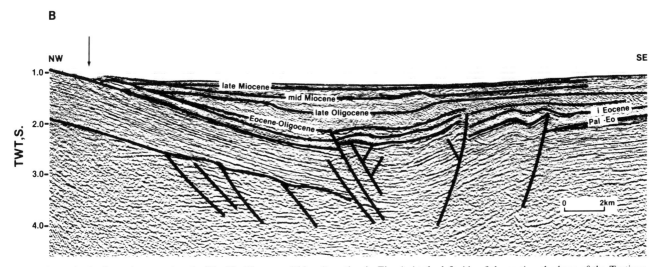

Fig. 5. Migrated seismic section crossing the Wyville–Thomson Ridge. Location in Fig. 4. At the left side of the section the base of the Tertiary volcanics is seen as a distinct erosional unconformity. Miocene deep water differential sedimentation at the arrow is indicative of contour currents. (Courtesy of Western Geophysical.)

and magnetic data but do not account for the observed nearly parallel-bedded reflectors in the volcanics or for the observed compression along the ridge axis.

Wood *et al.* (1987) found that the Ymir Ridge was a post-lava feature with large-scale normal faulting on its south side. They interpreted that the lavas in the basin had subsided up to 2 km more than those on the Rosemary Bank and the Ymir Ridge. They suggested that the subsidence began in the basin in early Tertiary times, and greatly increased during the Oligocene. In order to match most of the features of the data they suggested two models: (1) transpression in which the ridge is relatively uplifted on high-angle reverse faults on its margins, as would be expected to occur above a ramp in the sole detachment; (2) vertically variable extension by a sub-horizontal detachment below the basinal area.

The most recent model (Earle *et al.* 1989) proposes that Tertiary compression in the West Lewis Basin and along the Wyville–Thomson Ridge is closely associated with strike-slip movements along the Orkney–Faeroe alignment (Fig. 2), thus producing the thrust anticline (seen on their Fig. 12), which is inferred to be part of a possible positive flower structure of Oligocene age.

Seismic profiles (e.g. Fig. 4) suggest that the Wyville–Thomson Ridge Complex is an almost purely compressional tectonic feature, which overprints the regional subsidence in the region. The lower Tertiary volcanics form two marked anticlines which constitute the Wyville–Thomson Ridge and the Ymir Ridge. On small highs, which form a foreland thrust belt along the southern flank of the Ymir Ridge, reverse faults offset the volcanics (Fig. 4). These highs are elevated a few hundred metres above the surrounding surface of the basalt to the south of the complex. Similar features – but not as prominent – are seen between the Ymir Ridge and the Wyville–Thomson Ridge (Fig. 5). In general the internal bedding of the volcanics appears to be parallel. However, in places thickness variations are seen in the uppermost part of the volcanic sequence on the major structures, and erosional truncations of the volcanic sequence below Eocene sediments are seen on the two most prominent anticlines (Wyville–Thomson Ridge and Munkegrunnur Ridge, Fig. 2). Erosion of the volcanic sequence is not documented on the remaining minor compression structures shown on Fig. 2. This indicates that updoming of the prominent structures started slightly before the cessation of volcanic activity.

The lower Eocene sequence thins towards the structural culminations and shows distinct changes of thickness across some of the faults in the foreland complexes. However, most of the faults in the foreland complexes cut the lower Eocene sediments but not the middle Eocene or younger sediments above.

The intra-mid-Eocene sequence has only been identified in the vicinity of the major compressional structures. The distribution is apparently controlled by pre-existing structures and possibly also by the supply of sediment derived from the newly formed anticlines. In the structurally most disturbed areas, internal reflectors in this sequence show onlap and downlap towards the substratum, whereas truncation has been observed at the top of the sequence.

The upper Eocene and Oligocene sequence fills pre-existing lows. Locally the upper part of the sequence is truncated, apparently in association with structural uplift.

In the vicinity of the thrust foreland complex south of the Ymir Ridge the reflectors in the lower and intra-mid-Miocene sequence onlap the lower boundary. The sequence decreases in thickness towards the thrust complex and the Ymir Ridge. Above the thrust fault complex, the thinning of this sequence is the result of erosion. Outside the ridge complex the sequence can be divided into two and the upper sequence apparently is missing in the ridge complex.

In the upper mid-Miocene sequence the internal reflectors show onlap towards the lower boundary. This sequence and the younger sequences do not seem to be affected by the tectonic development of the ridges but the sequences are clearly exposed to deep sea bottom currents (Figs 4 and 5).

Summary

We infer that regional subsidence in the Rockall Trough–Faeroe Shetland Channel province and the Faeroe–Rockall Plateau were initiated prior to or during the termination of the volcanic activity, near the Paleocene–Eocene transition.

The initial compression phase (late Paleocene–early Eocene) was apparently concentrated in a relatively narrow zone including the Wyville–Thomson Ridge Complex and the Munkegrunnur Ridge (Fig. 9a). The compression in the Wyville–Thomson Ridge Complex commenced as reverse faulting with the associated formation of ramp anticlines during the latest part of the volcanic activity in this area, presumably in (late) Thanetian. During the last part of this compressional phase, in early Eocene, foreland thrust folds were developed south of the Ymir Ridge and cut by reverse faults, which penetrated to the surface.

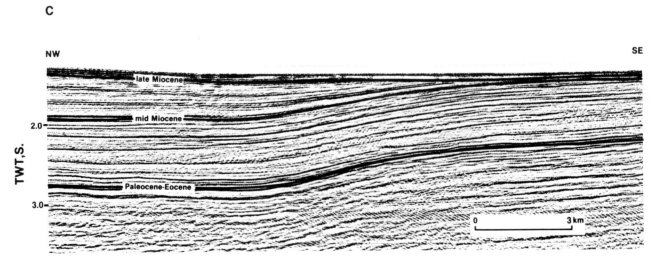

Fig. 6. Migrated seismic section crossing Miocene monoclinal compression structure in the Faeroe–Shetland Channel. Location shown in Fig. 1. (Courtesy of Western Geophysical.)

D

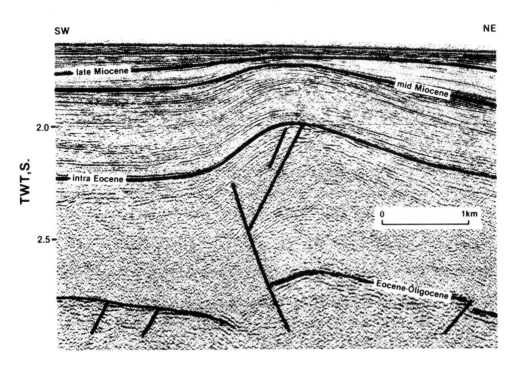

Fig. 7. Migrated seismic section (6 Fold) crossing Miocene anticline southwest of Bill Bailey Bank. Location shown in Fig. 1. The seismic profile was collected using a sleevegun source by the Geological Survey of Denmark and the Department of Geology, University of Aarhus, Denmark.

E

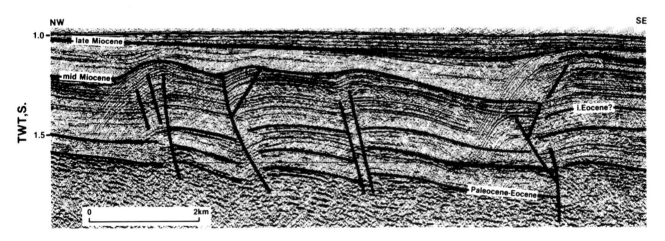

Fig. 8. Migrated seismic section (6 Fold) west of Hatton Bank. The section crosses Miocene high angle reverse faults associated to a major monocline. Location shown in Fig. 1. The seismic profile was collected using a sleevegun source by the Geological Survey of Denmark and the Department of Geology, University of Aarhus, Denmark.

A period of relative structural stability occurred in middle to late Eocene times and presumably also partly in Oligocene times even though there are indications of some structural movements in the middle Eocene.

Renewed compression during the Oligocene seems to have been located within the same geographically narrow zone as the first phase (Fig. 9b). At the Wyville–Thomson Ridge Complex this phase is seen as erosional truncations on the late Oligocene unconformity and further to the east as structural inversion of the West Lewis Basin (Earle *et al.* 1989).

A third phase of widespread compression in middle or late

Miocene times is seen in some compression in the Wyville–Thomson Ridge Complex, and by WSE–ENE to SW–NE oriented compression structures in the Faeroe Shetland Channel (Fig. 6), as well as a number of almost E–W oriented compression structures on the margin of the Faeroe–Rockall Plateau (Figs 7 and 8). The compression is shown by folding and reverse faulting of middle Miocene and older sediments.

Eocene to Miocene aged compressional tectonic fractures, in the form of basin inversion, have been observed in several major basins in NW Europe, and it is generally assumed that they are related to the Alpine deformation (e.g. Ziegler 1990).

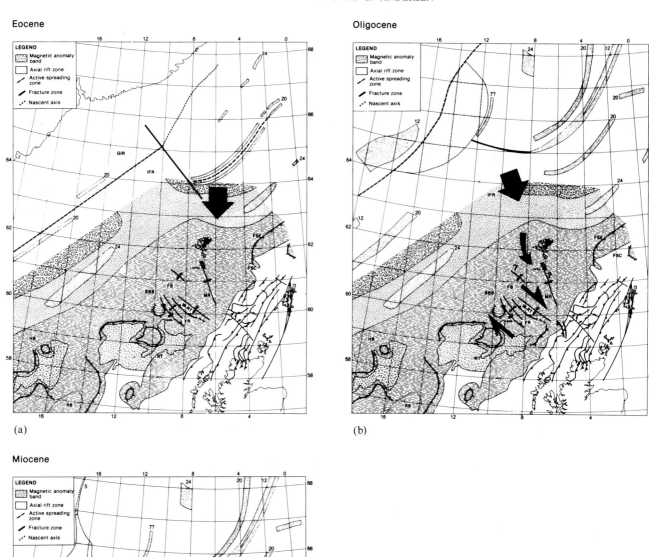

Fig. 9. Inferred structures formed during the three deformation phases. Legend as in Fig. 2. The major structural elements of the North Atlantic according to Nunns (1983). GIR: Greenland Iceland Ridge; IFR: Iceland Faeroe Ridge. Estimated orientation of gravitational ridge push during the deformation phases is shown as heavy arrows. Wrench deformation is also indicated where applicable.

The deformation of the Faeroe–Rockall area does not fit readily into this pattern. Instead, we suggest that seafloor spreading in the North Atlantic influenced the continental margin through a combination of ridge push and rigid plate movements. In Fig. 9 compression structures active during each of the three compression phases are shown, together with

possible 'causal events' related to the North Atlantic spreading.

When seafloor spreading was initiated between Greenland and the Faeroe–Rockall Plateau, during chron 24R, the southern part of the spreading axis (Aegir Ridge) north of the Faeroes was oriented ENE (approximately 30–45°) to the pole of rotation (Nunns 1983). As a consequence of the oblique spreading, approximately N–S gravitational ridge push would have acted on the Faeroe–Rockall Plateau and caused the initial deformation in the Wyville–Thomson Ridge Complex (Fig. 9a).

From the time of magnetic chron C22–20 and until chron C6 (middle Eocene to early Miocene), the continuation of the Reykjanes Ridge (between Jan Mayen and East Greenland) and the Aegir Ridge formed a paired propagating/retreating rift system connected through a pseudo-transform fault (Larsen 1988). Due to the propagation of the western rift segment, the Jan Mayen Block rotated approximately 8° anticlockwise relative to East Greenland and NW Europe. This rotation was also manifested on the Faeroe–Rockall Plateau and caused dextral wrench movements along the Wyville–Thomson Ridge complex and compression in West Lewis Basin (Fig. 9b). Oligocene compression in the Møre and Vøring Basins (Hamar and Hjelle 1984; Ziegler 1988) appears also to be associated with the complex Oligocene spreading geometry.

The third phase of compression is closely related to the continent-ocean transition zone north and west of the Faeroe–Rockall Plateau (Fig. 9c). Therefore it seems natural also to correlate this phase to seafloor spreading in the North Atlantic. It is possible that there is a correlation between this phase of compression on the Faeroe–Rockall Plateau and Miocene rearrangements of the spreading axis on the Faeroe–Iceland–Greenland Ridge (Saemundsson 1979).

We wish to thank Western Geophysical Company for allowing us to use part of their data for illustrations in this paper.

References

ANDERSEN, M. S., BOLDREEL, L. O., GUNNARSSON, K., KJARTANSSON, E., EWING, J., TALWANI, M. AND SAYWER, D. 1990 A seismic investigation of the Rockall Plateau. *Annales Geophysicae*, 63, Special issue.

BOTT, M. H. P. 1984. Deep structure and origin of the Faeroe–Shetland Channel. *In*: SPENCER, A. M. *et al.* (eds) *Petroleum Geology of the North European Margin*. Graham & Trotman, London, 341–347.

——, NIELSEN, P. H. AND SUNDERLAND, J. 1975. Converted P-waves originating at the continental margin between the Iceland–Faeroe Ridge and the Faeroe Block. *Geophysical Journal of the Royal Astronomical Society* 44, 229–238.

BREWER, J. A. AND SMYTHE, D. K. 1986. Deep Structure of the Foreland to the Caledonian Orogen, north-west Scotland: results from the BIRPS WINCH Profile. *Tectonics*, 5, 171–194.

EARLE, M. M., JANKOWSKI, E. J. AND VANN, I. R. 1989. *Structural and stratigraphic evolution of the Faeroe–Shetland Channel and Northern Rockall Trough.* AAPG Memoir, 46, 461–469.

ELDHOLM, O., THIEDE, J. AND TAYLOR E. 1989. Evolution of the Vøring Volcanic margin. *In*: ELDHOLM, O., THIEDE, J. *et al.*, *Proceedings of the Ocean Drilling Program, Scientific Results*. Ocean Drilling Program, College Station, TX, 104.

HALD, N. AND WAAGSTEIN, R. 1984. Lithology and chemistry of a 2-km sequence of Lower Tertiary tholeiitic lavas drilled on Sudøroy, Faeroe Islands (Lopra-1). *In*: BERTHELSEN, O., NOE-NYGAARD, A. AND RASMUSSEN, J. (eds) *The Deep Drilling Project 1980–1981 in the Faeroe Islands*. Foroya Fródskaparfelag, Tórshavn, 15–38.

HAMAR, G. P. AND HJELLE, K. 1984. Tectonic framework of the Møre Basin and the northern North Sea. *In*: SPENCER, A. M. *et al.* (eds) *Petroleum Geology of the North European Margin*. Graham & Trotman, London, 349–358.

HASZELDINE, R. S. 1984. Carboniferous North Atlantic palaeogeography: stratigraphic evidence for rifting, not megashear or subduction. *Geological Magazine*, 121, 443–463.

HITCHEN, K. AND RITCHIE, J. D. 1987. Geological review of the West Shetland Area. *In*: BROOKS, J. AND GLENNIE, K. W. (eds) *Petroleum Geology of North West Europe*. Graham & Trotman, London, 737–750.

JACOBSEN, O. S. AND LAIER, T. 1984. Analysis of gas and water samples from the Vestmanna-1 and Lopra-1 wells, Faeroe Islands. *In*: BERTHELSEN, O., NOE-NYGAARD, A. AND RASMUSSEN, J. (eds) *The Deep Drilling Project 1980–1981 in the Faeroe Islands*. Føroya Fródskaparfelag, Tórshavn, 149–155.

JØRGENSEN, O. 1984. Zeolite zones in the basaltic lavas of the Faeroe Islands. A quantitative description of the secondary minerals in the deep wells of Westmanna-1 and Lopra-1. *In*: BERTHELSEN, O., NOE-NYGAARD, A. AND RASMUSSEN, J. (eds) *The Deep Drilling Project 1980–1981 in the Faeroe Islands*. Føroya Fródskaparfelag, Tórshavn, 71–91.

KIDD, R. B. AND HILL, PH. R., 1986. *Sedimentation on Mid-ocean Sediment Drifts*. Geological Society, London, Special Publication, 21, 87–102.

KIØRBOE, L. 1990. Combined Refraction and VSP Modeling around the Faeroe Islands. *Annales Geophysicae*, 63, Special issue.

LARSEN, H. C. 1988. A multiple and propagating rift model for the NE Atlantic. *In*: PARSON, L. M. AND MORTON, A. C. (eds) *Early Tertiary Volcanism and the Opening of the NE Atlantic*. Geological Society, London. Special Publication, 39, 157–158.

LUND, J. 1981. Eine Ober-Paläozäne Mikroflora von den Färöern, Dänemark. *Couier Forschung Institut Senckenberg*, 50, 41–45.

McKENZIE, D. P. 1978. Some remarks on the development of sedimentary basins. *Earth and Planetary Science Letters*, 40, 25–32.

MORTON, A. C., DIXON, J. E. FITTON, J. G., MACINTYRE, R. M., SMYTHE, D. K. AND TAYLOR, P. N. 1988. Early Tertiary volcanic rocks in Well 163/6-1A, Rockall Trough. *In*: PARSON, L. M. AND MORTON, A. C. (eds.) *Early Tertiary Volcanism and the Opening of the NE Atlantic*. Geological Society, London, Special Publication, 39, 293–308.

MUDGE, D. C. AND RASHID, B. 1987. The Geology of the Faeroe Basin area. *In*: BROOKS, J. AND GLENNIE, K. W. (eds) *Petroleum Geology of North West Europe*. Graham & Trotman, London, 751–764.

MUTTER, J. C. AND ZEHNDER, C. M. 1988. Deep crustal structure and magmatic processes: the inspection of seafloor spreading in the Norwegian–Greenland Sea. *In*: PARSON, L. M. AND MORTON, A. C. (eds) *Early Tertiary Volcanism and the Opening of the NE Atlantic*. Geological Society, London, Special Publication, 39, 35–48.

NAYLOR, D. AND SHANNON, P. H. 1982. *The Geology of Offshore Ireland and Western Britain*. Graham & Trotman, London.

NEISH, J. C. 1990. Rockall Trough; a crustal structure of a complex environment from composite seismic surveys. *Society of Exploration Geophysicists, Sixtieth Annual International Meeting*.

NUNNS, A. G. 1983. Plate tectonic evolution of the Greenland–Scotland ridge and surrounding regions: *In*: BOTT, M. H. P., SAXOV, S., TALWANI, M. AND THIEDE, J. (eds) *Structure and Development of the Greenland–Scotland Ridge*. NATO Conference series, series IV: Marine Sciences, 11–30.

RASMUSSEN, J. AND NOE-NYGAARD, A. 1970. Geology of the Faeroe Islands. Danmarks Geologiske Undersøgelse 1. series, 25.

ROBERTS, D. G. 1975. Marine geology of the Rockall Plateau and Trough. Royal Astronomical Society, *Geophysical Journal*, 278, 447–509.

——, BACKMAN, J., MORTON, A. C., MURRAY, J. W. AND KEENE, J. B. 1984*a*. Evolution of Volcanic Rifted Margins: Synthesis of Leg 81 Results on the West Margin of Rockall Plateau. *In*: ROBERTS, D. G., SCHNITKER, D. *et al.* (eds) *Deep Sea Drilling Program, Initial Reports*, 81, 883–911.

——, BOTT, M. H. P. AND URUSKI, C. 1983. Structure and Origin of the Wyville–Thomson ridge. *In*: BOTT, M. H. P., SAXOV, S., TALWANI, M. AND THIEDE, J. (eds) *Structure and development of the Greenland–Scotland Ridge*. NATO Conference series, series IV: Marine Sciences, 133–158.

——, GINZBERG, A., NUNN, K. AND McQUILLIN, R. 1988. The structure of the Rockall Trough from seismic refraction and wide-angle reflection measurements. *Nature*, 332, 632–635.

——, HUNTER, P. M. & LAUGHTON, A. S. 1979. Bathymetry of the northeast Atlantic: continental margin around the Atlantic: continental margin around the British Isles. *Deep Sea Research*, 26A, 417–428.

——, MORTON, A. C. AND BACKMAN, J. 1984*b*. Late Paleocene-Eocene

Volcanic events in the northern Atlantic Ocean. *In*: ROBERTS, D. G., SCHNITKER, D *et al.* (eds) *Proceedings of the Deep Sea Drilling Program, Initial Reports*, **81**, 913–923.

—— AND SCHNITKER, D. *et al.* 1984. *Deep Sea Drilling Program, Initial Reports*, **81**, 1183.

SAEMUNDSSON, K. 1979. Outline of the geology of Iceland. *Jökull*, **29**, 7–28.

SKOGSEID, J. AND ELDHOLM, O. 1988. Early Cainozoic evolution of the Norwegian volcanic passive margin and the formation of marginal highs. *In*: MORTON, A. C. AND PARSON, L. M. (eds) *Early Tertiary Volcanism and the Opening of the NE Atlantic*. Geological Society, London, Special Publication, **39**, 49–56.

SMYTHE, D. K. 1983. Faeroe–Shetland Escarpment and Continental margin north of the Faeroes. *In*: BOTT, M. H. P., SAXOV, S., TALWANI, M. AND THIEDE, J. (eds) *Structure and Development of the Greenland–Scotland Ridge*. NATO Conference series, series IV: Marine Science, 109–119.

—— 1989. Rockall Trough—Cretaceous or Late Paleozoic? *Scottish Journal of Geology*, **25**, 5–43.

SPENCE, G. D., WHITE, R. S., WESTBROOK, G. K. AND FOWLER, S. R. 1989. The Hatton Bank Continental Margin; I, shallow structure from two-ship expanding spread seismic profiles. *Geophysical Journal of the Royal Astronomical Society*, **96**, 273–294.

TALWANI, M., MUTTER, J. AND HINZ, J. 1983. *Ocean continent boundary under the Norwegian Continental Margin*. NATO Conference series, series IV: Marine Science, 121–131.

WAAGSTEIN, R. 1988. Structure, composition & age of the Faeroe basalt plateau. *In*: PARSON, L. M. AND MORTON, A. C. (eds) *Early Tertiary Volcanism and the Opening of the NE Atlantic*. Geological Society, London, Special Publication, **39**, 225–238.

WHITE, R. S. 1988. A hot-spot model for early Tertiary volcanism in the N Atlantic. *In*: PARSON, L. M. AND MORTON, A. C. (eds) *Early Tertiary Volcanism and the Opening of the NE Atlantic*. Geological Society, London, Special Publication, **39**, 3–13.

——, SPENCE, G. D., FOWLER, S. R., MCKENZIE, D. P., WESTBROOK, G. AND BOWEN, A. N. 1987. Magmatism at rifted continental margins. *Nature*, **330**, 439–444.

WOOD, M. W., HALL, J. AND VAN HOORN, B. 1987. Post-Mesozoic differential subsidence in the north-east Rockall Trough related to volcanicity and sedimentation. *In*: BROOKS, J. AND GLENNIE, K. W. (eds) *Petroleum Geology of North West Europe*. Graham & Trotman, London, 677–687.

ZIEGLER, P. A. 1988. *Evolution of the Arctic–North Atlantic and the Western Tethys*. American Association of Petroleum Geologists Memoir **43**.

——, 1990. *Geological Atlas of Western and Central Europe (2nd edn)* Shell Internationale Petroleum Maatschappij B. V., The Hague.

Preliminary observations of Neogene–Quaternary depositional processes in the Faeroe–Shetland Channel revealed by high-resolution seismic facies analysis

J. E. DAMUTH and H. C. OLSON

Mobil Exploration and Producing Technical Center, PO Box 650232, Dallas, Texas 75265 USA

Abstract: The Neogene and Quaternary sediments of the Faeroe–Shetland Channel and West Shetland slope have been deposited and modified through the interaction of a variety of downslope and parallel-to-slope depositional processes. The upper slope is dominated by mass-transport deposits (debris flows) which progressively diminish downslope. These were apparently deposited during glacial cycles when ice sheets reached the shelf edge and supplied large amounts of terrigenous sediment to the slope. Thin, prograding clinoforms separate packages of debris flows and may represent glacial marine sedimentation during periods of ice retreat from the shelf edge (e.g. interstades). A few submarine canyons occur on the slope and probably provide conduits to the basin for turbidity currents and related mass flows. The middle to lower slope appears to be dominated by glacial marine, hemipelagic and possibly turbidity-current deposits, which have been subjected to reworking by contour currents at many locations. A thick deposit that has the appearance of a deep-sea fan occurs on the middle slope in the northeast part of the area. Although many individual seismic packages within this feature have the appearance of aggradational channel–levee systems, these packages may actually represent contourite deposits. Large migrating sediment waves occur just downslope from this feature and suggest that strong contour currents have interacted with the downslope processes to redistribute sediments on the lower slope. The basin floor has thin conformable sediments that appear to be predominantly glacial marine and hemipelagic with occasional turbidites and debris flows. Thick, extensive debris-flow deposits also occur beneath the basin floor in the northeastern part of the area. The Neogene–Quaternary sediments are separated from the Paleogene section by a major regional unconformity of latest Oligocene or early to middle Miocene age, which forms a major sequence boundary throughout the region.

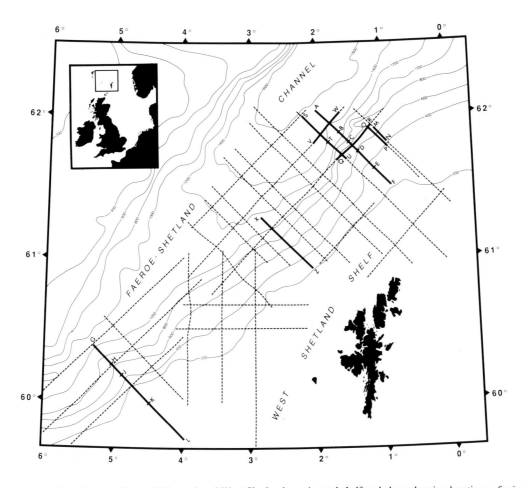

Fig. 1. Bathymetric map of the Faeroe–Shetland Channel and West Shetland continental shelf and slope showing locations of seismic lines (dashed) interpreted. Solid portions of lines show profiles illustrated in Figs 2–7. Bathymetry is redrawn from Roberts *et al.* (1977). Contour interval is 200 m. Inset shows location of area.

From *Petroleum Geology of Northwest Europe: Proceedings of the 4th Conference* (edited by J. R. Parker).
© 1993 Petroleum Geology '86 Ltd. Published by The Geological Society, London, pp. 1035–1045.

Preliminary seismic-sequence and seismic-facies analyses of a grid of high-resolution (50–250 Hz airgun) seismic profiles from the southeastern side of the Faeroe–Shetland Channel (Fig. 1) were undertaken to understand better the Neogene and Quaternary depositional processes in this region. This study was conducted in conjunction with a detailed seismic–strati-graphic analysis using conventional multi-fold seismic of the underlying Paleogene strata of the Faeroe Basin (Mitchell et al. 1993). The Neogene–Quaternary section was studied using high-resolution seismic data because the higher frequencies recorded in these data can potentially provide extremely well imaged examples of the seismic facies, depositional features and stratal geometries and relationships. Thus, depositional features and processes can often be more easily and confidently identified than is possible with standard multi-fold seismic data. These Neogene features and processes can therefore provide important 'modern' analogues that may assist in the interpretation and identification of prospective features in the more deeply buried Paleogene section, which are less well imaged on conventional multi-fold seismic data.

Regional setting and previous work

The Faeroe–Shetland Channel is an elongate basin that trends NE–SW between the West Shetland Shelf and the Faeroe Shelf (Fig. 1). The channel is up to 200 km wide between respective shelf breaks (200 m contour) and deepens to the northeast along its axis from about 1000 m at the southwestern end near the Wyville–Thompson Ridge to more than 1700 m where it enters the Norway Basin. The Faeroe–Shetland Basin beneath the channel is a major depocentre where sediments have been accumulating since the Paleozoic (Duindam and van Hoorn 1987; Haszeldine et al. 1987; Hitchen and Ritchie 1987; Ziegler 1988; Stoker 1990c; Mitchell et al. 1993). A prominent mid-Tertiary unconformity (Fig. 2, latest Oligocene unconformity) forms a major erosional surface throughout the basin and is overlain by a clastic wedge of Miocene (?) to Holocene sediments, which ranges in thickness from less than 60 m beneath the channel floor to more than 400 m beneath the upper continental slope (Mudge and Rashid 1987; Stoker 1990a,b,c; Stoker et al. 1991; Mitchell et al. 1993).

The present watermass circulation in the Faeroe–Shetland Channel consists primarily of the warm (> 9°) Atlantic Surface Water, which flows northeastward into the Norwegian Sea, and the cold (< 0.5°) southwestwardly flowing Norwegian Sea Deep Water (NSDW), which returns cold water to the Atlantic (Crease 1965; Harvey 1965; Wortington 1970; Ellet and Roberts 1973; Meincke 1983; Ellet et al. 1986; Saunders 1990). Current-meter data show that the NSDW extends from about 500 m water depth to the channel floor (Dooley and Meincke 1981) and maintains a very strong SW to WSW current flow with mean velocities of 0.1 to 0.4 m/s and maximum velocities of 0.5 to 0.75 m/s (Akhurst 1991). Miller and Tucholke (1983) inferred that this vigorous southward bottom-water flow from the Arctic through the channel into the North Atlantic was apparently initiated in the late Eocene to early Oligocene as a result of the separation of Greenland and Svalbard, and that bottom-water circulation stabilized to its present pattern in the middle Miocene. However, Eldholm (1990) concluded on the basis of more recent plate-tectonic and palaeoceanographic studies, that Norwegian–Greenland Sea deep waters were isolated throughout the Paleogene. Furthermore, vigorous deep-water exchange from the Arctic Ocean and the Nor-wegian–Greenland Sea to the North Atlantic basins was not initiated until the Neogene, probably during the middle or even late Miocene. A prominent mid-Tertiary erosional uncon-formity (Latest Oligocene Unconformity in Fig. 2) appears to separate the older Paleocene–Eocene succession from the over-lying clastic wedge of late Tertiary age. This is thought to mark the onset of the strong abyssal circulation through the channel (Stoker 1990a,b; Stoker et al. 1991).

Recent studies using various high-resolution seismic records and shallow vibracores (< 10 m penetration) demonstrate that a variety of deep-water depositional processes have interacted to deposit the Pliocene and Quaternary sediments that form the late Tertiary clastic wedge in the Faeroe–Shetland Chan-nel, and in the adjacent portion of the Rockall Trough to the south (Stoker et al. 1989, 1991; Holmes 1990; Stoker 1990a,b,c; Stevenson 1990a,b; 1991a,b; Akhurst 1991). In particular, studies by Akhurst (1991) and Stoker et al. (1991) were conducted at the southwestern end of the Faeroe–Shetland Channel in areas adjacent to, and slightly overlapping, the area

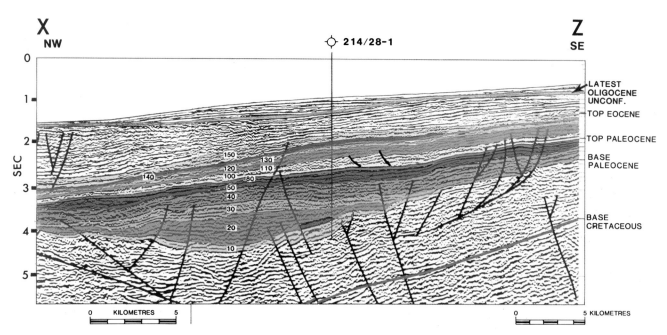

Fig. 2. Multi-fold seismic dip line from Faeroe Basin showing Paleogene seismic sequences interpreted by Mitchell et al. (1993). Location shown in Fig. 1 (X–Z). The latest Oligocene unconformity is marked by the yellow/orange horizon near the top of the section. High-resolution seismic lines in Figs 3–7 image the relatively thin Neogene/Quaternary section above this unconformity (dashed lines labelled LOU). Note the presence of thick Oligocene section beneath the unconformity.

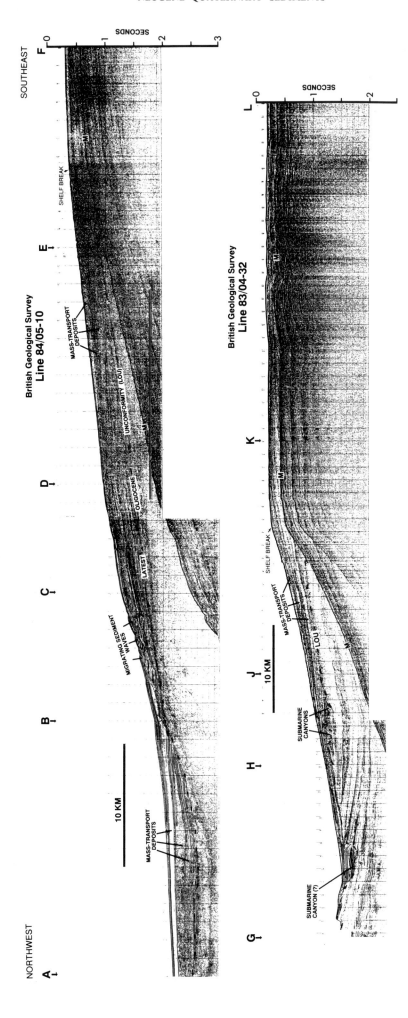

Fig. 3. High-resolution airgun dip seismic lines which extend from the West Shetland shelf to the Faeroe–Shetland Channel floor and illustrate the regional morphology, seismic facies and depositional features described in the text. Profile A–F (top) is from the northeastern part of the study area and Profile G–L (bottom) is from the southwestern end of the study area. The dashed line (LOU) marks the latest Oligocene unconformity and M marks the first water-bottom multiple on these profiles as well as profiles shown in Figs 4–7. Locations of all lines in Figs 3–7 (A–W) are shown in Fig. 1.

of the present study. Stoker *et al.* (1991) showed that mass-transport processes have formed pervasive deposits consisting mainly of debris-flow diamictons composed of redeposited glacigenic depostis. Packages of acoustically transparent debris-flow deposits are separated by acoustically well-stratified sediments that consist primarily of glaciomarine hemipelagites and contourites. These debris flows were generated mainly during glacial cycles when ice-marginal sedimentation was pervasive on the outermost shelf and upper slope. Bottom-current activity (contour currents) was apparently relatively weak during glacial cycles, but became much more vigorous during the transition from glacial to interglacial cycles.

Based on detailed core studies, Akhurst (1991) documented the importance of contour-current activity in the channel during the latest Quaternary, and concluded that contour-current activity has led to decreased rates of sediment accumulation (generally $< 10 \, \text{cm}/10^3 \, \text{yr}$) by entrainment and transport of the finer components of ice-rafted sediments as they were deposited. The strength of these bottom currents apparently fluctuated in cyclical patterns throughout stadial, interstadial, and interglacial conditions with the most intense fluctuations in velocity during both the Holocene and the Last (Eemian) Interglacial. In addition to deposition by ice-rafting and contour-current processes, Akhurst (1991) also documented evidence for mass-transport processes and weak ('low concentration') turbidity-current deposition.

Database

The present study utilized high-resoluton seismic lines selected from grids acquired by the British Geological Survey (Fig. 1). Most of the lines are from the 44-line BGS 85/05 survey in the northeastern half of the study area; the remaining lines are from the BGS 79/14 and 83/04 surveys. Only the lines shown (Fig. 1) were available for the present study. Line spacing generally ranges between 10 to $>25 \, \text{km}$ and averages about 15 km. Thirteen of the lines are NW–SE dip lines that extend from the West Shetland continental shelf (water depth $<200 \, \text{m}$) or upper slope northwestward to the axis of the Faeroe–Shetland Channel in water depths of up to 1700 m. In the northeastern half of the study area, only five strike lines intersect these dip lines and extend southwest–northeast along the basin-floor axis, the lower continental slope and the continental shelf northwest of the Shetland Islands (Fig. 1). Three other strike lines extend through the southwestern half of the study area and continue southwestward beyond the limits of Fig. 1 onto the Wyville–Thompson Ridge and into the Rockall Trough area. Five other lines in the southwestern portion of the study area, two oriented east–west and three north–south, were also utilized.

All seismic lines utilized are unprocessed analogue records shot with small (40 cubic inches) airguns and generally filtered between 50 and 250 Hz. Acoustic penetration is quite variable throughout the grid, as well as along individual lines, but generally ranges between 0.5 and 1.5 s below the sea floor (Fig. 3). Maximum penetration is generally achieved on the lower continental slope and the adjacent basin floor. Upslope, from water depths of about 500 m, and especially landward of the continental shelf (water depth $<200 \, \text{m}$), penetration is severely limited (0.2 s or less beneath the shelf) by two factors: (i) the nature of the shelf deposits, which have been subjected to intense scouring by icebergs during interglacial cycles and glacigenic depositional processes beneath ice sheets during glacial cycles (Belderson *et al.* 1973; Stoker 1990*a,b*; Stoker *et al.* 1991); and (ii) the presence of strong water-bottom multiples that obscure the signal (Fig. 3).

Latest Oligocene unconformity

A very prominent regional unconformity (Fig. 2, latest Oligo-cene unconformity) is observed throughout the region and represents a major sequence boundary that can be consistently traced from beneath the West Shetland Shelf downslope to the axis of the Faeroe–Shetland Channel (dashed horizon labelled LOU in Figs 3–7). This unconformity represents a major erosional surface at most locations, especially in the southwestern end of the channel (Fig. 3, Line 83/04-32, G–L). The section above this unconformity generally shows highly reflective, parallel, acoustically well stratified deposits that thin upslope from a maximum thickness of 0.4–0.5 s beneath the middle of the continental slope. On some lines (e.g. Fig. 3) this unconformity becomes obscured beneath the upper slope (water depth $<400 \, \text{m}$) by the water-bottom multiple. However, on other lines that extend well across the continental shelf, the unconformity is observed to shallow progressively landward to less than 0.02–0.03 s below the sea floor. Here it often becomes obscured by the bubble pulse. In contrast, the section below the unconformity is generally non-reflective to semi-transparent, and acoustic penetration is usually very limited. At many locations, groups of dipping reflections with various orientations appear to represent steeply dipping sets of clinoforms; whereas at other locations the reflections show evidence of extensive faulting and deformation (Fig. 3, Line 85/04-10, A–F). Because of poor acoustic penetration below this unconformity, older, regionally extensive sequence boundaries or other horizons could not be identified.

Stoker (1990*a*) and Stoker *et al.* (1991) suggest that this unconformity may have formed in response to the initiation of intense bottom-water flow from the Arctic to the North Atlantic, which was inferred by Miller and Tucholke (1983) to have occurred after the separation of Greenland and Svalbard during the late Eocene to early Oligocene. We cannot determine the precise age of this unconformity; however, the following lines of evidence suggest that it probably formed somewhat later during the latest Oligocene or early to middle Miocene. A sequence-stratigraphic analysis of the Paleocene section beneath the West Shetland Shelf and Faeroe–Shetland Channel by Mitchell *et al.* (1993) established the approximate top of the Eocene section throughout the present study area using a close-spaced grid of standard multi-fold seismic lines (Fig. 2). We cross-correlated many of our high-resolution seismic lines with this multi-fold data set and determined the depth (in two-way travel time) to the top of the Eocene section at 60 locations throughout the study area from the modern shelf edge downslope to the basin axis. This correlation revealed that the top of the Eocene section generally occurs well below the prominent unconformity at most of these locations (e.g. Fig. 2), and indicates that a thick Oligocene section is present throughout most of the region.

An exception is at the extreme southwestern end of the Faeroe–Shetland Channel where the Eocene and older deposits shallow progressively beneath the unconformity, as well as the sea floor, as they extend up the northeastern flank of the Wyville–Thompson Ridge. In this area the Oligocene has locally been completely eroded and the prominent unconformity cuts into the Eocene section. Eocene strata crop out locally in this region (Stoker 1990*b,c*); however, these Eocene strata rapidly plunge toward the northeast beneath the Faeroe–Shetland Channel floor, resulting in preservation of a thick Oligocene section throughout most of the basin (e.g. Fig. 2).

Another exception is beneath the outer West Shetland Shelf where this Oligocene section thins progressively landward until the Top of Eocene horizon is less than 0.1 s ($<100 \, \text{m}$) below the prominent unconformity. The multi-fold seismic lines (Mitchell *et al.* 1993) show that the thickness of sediment between the top of Eocene horizon and the prominent regional unconformity rapidly increases basinward from the shelf edge, with thicknesses ranging from 1.0 to 2.5 s beneath the base of the modern slope. The Oligocene section becomes even thicker

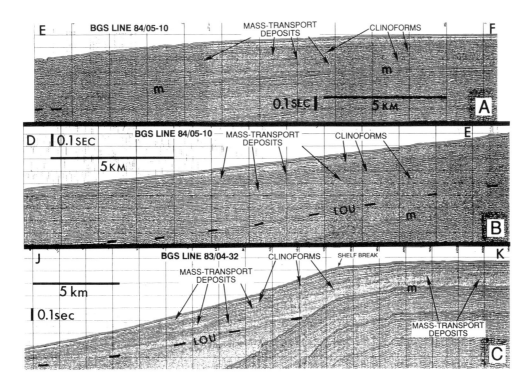

Fig. 4. Examples of seismic facies beneath the upper continental slope. Prograding packages of hummocky facies in lens-shaped or mounded external forms represent mass-transport deposits and are separated by highly reflective clinoforms formed by glacial marine sedimentation (see text).

in the Faeroe Basin beneath the modern channel axis (Fig. 2). This pattern contrasts somewhat with the Neogene–Quaternary clastic wedge above the prominent regional unconformity, which is thickest beneath the present-day upper slope, but rapidly thins basinward beneath the lower slope and Faeroe–Shetland Channel axis (e.g. Stoker *et al.* 1991).

The presence of this thick Oligocene sequence throughout most of the basin suggests that the major unconformity was not formed during the late Eocene or early Oligocene (Stoker 1990*a*; Stoker *et al.* 1991), but is much younger. A younger age for this unconformity is also supported by more recent plate-tectonic and palaeoceanographic data from the Norwegian–Greenland Sea, which suggest that the exchange of deep and intermediate water from the Arctic to the North Atlantic was restricted throughout the Paleogene and that intense bottom-water flow was not initiated until at least the middle Miocene (Eldholm 1990 and references therein). A Miocene age for the initiation of vigorous bottom-water flow through the Faeroe–Shetland Channel and the resultant sea-floor erosion is more consistent with the presence of the thick Oligocene section observed throughout the basin (e.g. Fig. 2).

In addition, subsidence history curves for the Faeroe Basin show a sharp increase in subsidence rate from *c.* 38–25 Ma during the Oligocene (J. M. Vizgirda, pers. comm. 1992). The multi-fold (Mitchell *et al.* 1993) and high-resolution seismic lines show extensive faulting and tilting confined to the Oligocene strata; no faulting or deformation extends upward into the Neogene above the unconformity, nor is extensive faulting observed below the Oligocene. This observation also suggests increased basinal subsidence during the Oligocene. Thus several lines of evidence suggest that the prominent regional unconformity formed near the end of the Oligocene or during the early to middle Miocene in response to increased rates of tectonic subsidence coupled with erosion by vigorous bottom-water flow. These are: (1) the presence of a very thick Oligocene section; (2) the apparent absence of intense bottom-water flow from the Norwegian–Greenland Sea into the North Atlantic until at least middle Miocene; (3) markedly increased

subsidence rate during the Oligocene; and (4) intense faulting and tilting of just the Oligocene section during or shortly after deposition. Because of the uncertainty of the age of this unconformity, we refer to it throughout this paper as the latest Oligocene unconformity (LOU) and suggest that it may mark approximately the boundary between the Paleogene and Neogene sections in the study area.

Neogene-Quaternary seismic facies and depositional processes

Upper slope

The strata above the late Oligocene unconformity/sequence boundary are well imaged throughout the study area, especially seaward of the modern continental shelf edge (e.g. Figs 3 and 4). The northwest–southeast-trending dip lines all show the same seismic facies relationships and depositional patterns from the shelf edge, seaward down the continental slope, to the Faeroe–Shetland Channel floor. The sediments beneath the upper continental slope (shelf edge to 600–800 m water depths) are characterized by extensive zones of hummocky to chaotic (occasionally transparent) seismic facies (Figs 3 and 4). These facies occur in multiple, discrete packages with mound- or lens-shaped external forms. These packages are often separated by thin zones of continuous, parallel, often highly reflective facies (Fig. 4). At many locations these continuous, parallel strata have clearly been eroded or truncated and the hummocky facies lie directly on top of local unconformities. Some of these erosional surfaces appear to represent small submarine canyons or gullies filled with hummocky to chaotic facies (e.g. Fig. 3). Many of these local unconformities and other continuous reflective horizons observed within the upper slope between the hummocky deposits may represent candidate sequence or systems-tract boundaries. Unfortunately, the regional extent for these surfaces could not be determined because of the limited number of seismic lines available (Fig. 1).

Dip lines in the southwestern part of the region show several of these highly reflective, continuous horizons prograding seaward beneath the continental shelf to the modern shelf edge as a series of large clinoforms (Figs 3 and 4). Some of these clinoform surfaces extend down beneath the continental slope to depths of 800 m as thin, continuous parallel reflections, which separate the hummocky, mounded packages described above (Fig. 4C). In addition, these individual clinoforms often downlap onto these hummocky mounded packages. These relationships appear similar to stratal relationships shown in the Exxon model for continental margin systems tracts (Posamentier and Vail 1988) where highstand, prograding slope-clinoform sets downlap onto lowstand basin-floor deposits. However, analysis of additional closely spaced lines in this region will be required to interpret the observed facies confidently in the context of the Exxon sequence-stratigraphic model.

The packages of hummocky to chaotic seismic facies which predominate beneath the upper slope appear to represent a range of large-scale mass-transport deposits (mainly debris flows with some slumps and slides) emplaced by numerous, recurrent episodes of sediment failure. The acoustic characteristics of these deposits include (1) transparent to hummocky, structureless internal reflection character; (2) mounded to lens-shaped external form; (3) erosional surfaces at the base; and (4) external dimensions and thickness. These are all consistent with characteristics that have previously been well documented for debris flows and related mass-wasting deposits (Embley 1976, 1980; Jacobi 1976; Embley and Jacobi 1977; Damuth 1980; Damuth and Embley 1981). Recently, Stoker (1990a) and Stoker et al. (1991) describe deposits with similar acoustic characteristics from the Hebrides Slope at the northeastern end of Rockall Trough and from the southwestern end of the Faeroe–Shetland Channel adjacent to the present study area. They attributed these deposits mainly to debris flows and described shallow vibracores from the youngest deposits, which support their seismic interpretation. These cores show that turbidity-current deposits are also present to a much lesser extent. Stoker (1990a) suggested that these debris-flow and turbidity-current deposits represent redeposited glacial marine sediments that were emplaced mainly during glacial cycles when an ice sheet extended to the shelf edge and could deliver large amounts of terrigenous detritus directly to the upper slope. Based on their acoustic characteristics and their direct analogy to the deposits described by Stoker, we suggest that the deposits of the upper West Shetland Slope are mainly debris flows. Interbedded glacial marine and hemipelagic deposits comprise a relatively minor portion of the total deposits at most locations.

The seismic-facies relationships described by Stoker (1990a) for the Hebrides Slope are quite similar to those we observe beneath the upper West Shetland Slope (Fig. 3). Stoker attributed the continuous, parallel high-amplitude reflections and clinoforms, which separate the hummocky debris-flow packages, to periods of slower glacial marine and hemipelagic sedimentation when the ice sheet retreated (at least temporarily) from the shelf edge. We speculate that ice-sheet retreat may have been coupled with glacio-eustatic sea-level rise during interstadial periods. Sea-level rise coupled with ice-sheet retreat could have temporarily prevented large quantities of terrigenous sediments from being transported across the outer shelf to the slope. If this were the case, then the continuous high-amplitude reflections and clinoforms may represent periods of slower hemipelagic sediment accumulation (i.e. condensed sections), and therefore may be analogous to transgressive and highstand systems-tract deposits in the Exxon sequence-stratigraphic model; whereas the hummocky debris-flow deposits may represent lowstand systems-tract deposits (Figs 3 and 4). However, precise dating and regional mapping

of the clinoforms and other deposits would be required to confirm this interpretation.

Occasional erosional features that have the appearance of submarine canyons are observed in the study area. Submarine canyons form predominantly through mass-wasting and turbidity-current processes. Thus, when active, canyons can provide conduits for transport of sands to the lower slope and basin floor. A major canyon-like feature, now filled and buried, is observed in the northeastern part of the upper West Shetland slope (Figs 5B and C). This feature is up to 200 m deep, more than 3 km wide and filled with deposits that return transparent to chaotic seismic facies, which, based on the above discussion, appear to represent mass-transport deposits (debris flows, etc.). This canyon feature apparently has an interesting, complex history of development. The southeastern wall of the canyon clearly represents a major erosional scarp that has truncated at least 200 m of older strata. These truncated strata appear to be interbedded mass-transport deposits and clinoform surfaces similar to those described above. In contrast, the northwestern wall of the canyon is a constructional feature returning continuous, parallel to subparallel or migrating reflections, and has the appearance of thick levee or overbank sediments deposited on an erosional unconformity that extends upslope to the scarp (Figs 5B and C).

Stevenson (1991b) interpreted this scarp and associated unconformity as a slide scar or zone of sediment removal associated with a large mass-transport deposit informally named the Miller Slide. He interprets the constructional wall of the canyon feature as contourite mounds. Certainly this constructional canyon wall has the appearance of contourite deposits. The occurrence of smaller migrating sediment waves within these deposits (Fig. 5B) suggests that these deposits have been affected by contour-current activity. However, based on the data at hand, we cannot rule out the possibility that this aggradational feature was also at least partially formed by overbanking of downslope flows such as turbidity currents and related mass flows. Further downslope this canyon appears to be an entirely constructional channel feature with both walls consisting of levee-like deposits (Fig. 5D). At this location, the canyon may actually be an aggradational channel/levee system (see below).

Canyon-like features of various sizes are observed along Profile G–L (Fig. 3). Two small buried canyon features occur between H–J. At the end of this profile (beneath G) the southeastern half of an apparent large modern canyon, which is up to 250 m deep and several kilometres wide, is observed, and an older, now buried, canyon-like feature of similar size is present on that same profile just upslope from this modern canyon (Figs 3 and 5A, Profile G–H). The buried feature is filled with highly reflective parallel to subparallel beds. These features clearly have eroded down through large sections of Oligocene strata and their floors and walls form part of the late Oligocene unconformity (LOU). Because of the sparse data coverage, we were unable to map the trends of these and other canyon-like features observed in the study area; thus, their relationships to the continental margin and their true origins remain uncertain. For example, bathymetric contours (Stoker 1990b), as well as GLORIA side-scan sonar images (N. H. Kenyon, pers. comm.), in the area of the modern canyon-like feature at the end of Profile G–H (Figs 3 and 5A) suggest that this erosional feature is actually an oval-shaped depression of uncertain origin, not a linear canyon. Additional mapping of the trends of the other canyon-like features will be required to confirm whether they are submarine canyons.

Middle to lower slope

In contrast to the predominantly hummocky seismic facies of the upper slope, the deposits of the middle to lower slope

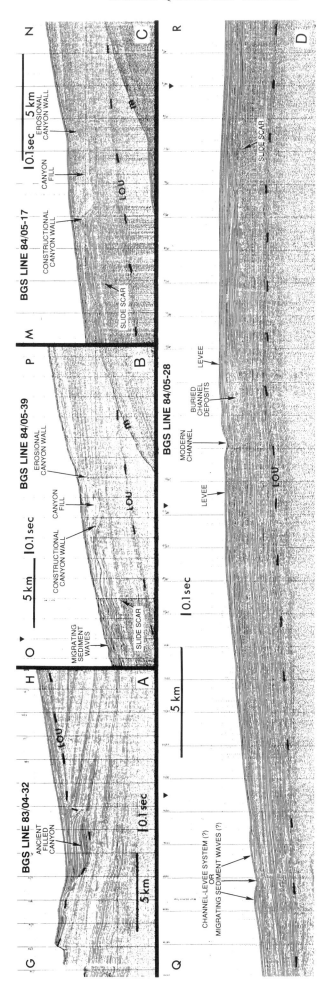

Fig. 5. Examples of submarine canyons and channels. (**A**) Large buried submarine canyon (?). Note eroded, outcropping strata in canyon wall. Erosional scarp near G is apparently not a canyon (see text). (**B**) and (**C**) Buried submarine canyon (?) beneath the upper slope. Hummocky to chaotic deposits filling canyon appear to be mass-transport deposits. Note erosional southeastern wall of canyon as opposed to the constructional, levee-like northwestern canyon wall (see text). (**D**) Deep-sea fan-like feature on the continental slope. Note possible modern channel and levees at apex and buried channel filled with chaotic and hummocky facies just beneath. Feature on the sea-floor at left end of profile may be another channel-levee system, or alternatively, migrating sediment waves (see text). A small triangle at the top of a profile indicates intersection with another illustrated profile.

(water depths of 600–1200 m) are generally characterized by highly reflective, continuous parallel to subparallel seismic facies (Fig. 3). Some of the zones of hummocky facies extend downslope to this region from the upper slope and form interbeds within these parallel facies; however, the number and thickness of hummocky packages progressively decrease downslope. This facies transition indicates that the mass-transport deposits so prevalent on the upper slope become progressively less frequent and widespread across the middle slope until they are only rarely present on the lower slope (Fig. 3). The highly reflective, parallel to subparallel facies of the middle to lower slope deposits appear to be interbedded hemipelagic and glacial marine deposits with occasional turbi-dity-current deposits.

The single strike line along the middle slope that was available for this study shows seismic facies and stratigraphic relationships which suggest that a deep-sea fan-like feature may be present on the northeasternmost part of the West Shetland slope (Fig. 5D). This feature is up to 0.5 s thick and is composed of a series of lens-shaped or mounded seismic packages that contain continuous parallel to subparallel, or occasionally migrating, sub-bottom reflections. These packages and their internal reflections show discordant, onlapping relationships to one another and are similar in appearance to seismic facies patterns associated with overlapping distributary channel-levee systems of modern deep-sea fans (e.g. Bouma *et al.* 1985 and references therein; Kolla and Coumes 1987; Damuth *et al.* 1988; Weimer 1990). Figure 5D shows a feature that has the appearance of a modern fan channel perched atop an associated natural levee system at the present apex of the fan. Channel relief at the sea floor is up to 75 m deep and more than 1 km wide. The associated levee/overbank deposits extend at least 10 km away from this channel in each direction. A broader (>4 km), now buried channel of even greater relief (>200 m) is visible beneath this modern channel and is par-tially filled with chaotic seismic facies which appear to be mass-transport deposits (Fig. 5D). This buried channel may be the downslope continuation of the large submarine-canyon feature observed just upslope in this region (discussed above; Figs 5B and C). Elsewhere, several features that have the appearance of smaller channels and associated levee systems occur both on the present fan surface (e.g Fig. 5D, left end) of a field of migrating sediment waves that occur just down slope (see below). Unfortunately, the bulk of this fan-like feature appears to be located northeastward of the study area beyond the existing seismic coverage. In addition, only one strike line across this feature was available; thus we are unable to confirm whether these apparent channel-levee systems are, in fact, linear features that trend downslope. Therefore, we cannot confirm that this feature truly represents a deep-sea fan. A plausible alternative is that these deposits may represent a series of contourite drift deposits (see below).

At some locations on the lower slope in the northeastern part of the study area (generally between 900–1200 m water depths), the continuous parallel to subparallel seismic facies that characterize the middle slope and the fan-like feature described above abruptly pass into a regular, migrating internal reflection configuration which apparently represents migrating wave- or dune-like bedforms (Fig. 3). These migra-ting waves are best developed along the dip lines where they show amplitudes of up to 50 m and wavelengths of up to 1 km or more (Fig. 6). These migrating waves extend downward from the present sea floor for 0.2–0.3 s (approximately 200–300 m). No strike lines extend through these deposits, and an inadequate number of dip lines were available to map the trends or orientations of the wave crests or troughs. However, Kenyon (1987) reported the presence of five 'slope ridges' on a GLORIA sonograph and seismic lines from this region. These have similar amplitudes and wavelengths to the migrating waves and extend parallel to the West Shetland slope for up to 20 km in about the same location as the waves reported here. Unfortunately, Kenyon's (1987) seismic data do not resolve the internal character of these ridges. He suggested that these features could either be slump folds formed by downslope mass-transport processes, or longitudinal sediment waves formed by contour-current processes.

Based on the regular, migrating sub-bottoms that we ob-serve in these bedforms (Fig. 6), we believe that they represent migrating sediment waves or dunes deposited by (1) contour-current activity, (2) turbidity-current overbanking of canyon or channel walls and levees or (3) a combination of both processes. The presence of a strong southwesterly flow of NSDW through the Faeroe–Shetland Channel floor, the docu-mentation of reworking of sediments by contour-currents based on current measurements, cores and seismic data (Akhurst 1991; Stoker *et al.* 1991) and the apparent orientation and extent of the waves (Kenyon 1987) all argue for a contour-current origin for these waves. In addition, a large field of modern-to-buried migrating waves of similar amplitudes, wavelengths and internal configuration occurs at the north-

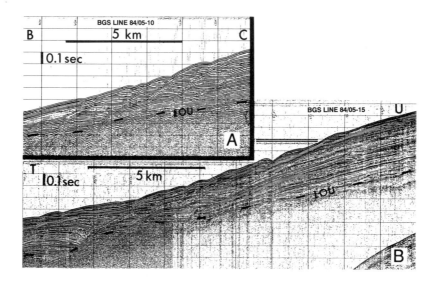

Fig. 6. Examples of migrating sediment waves or drift deposits from the lower continental slope.

eastern end of the adjacent Rockall Trough and has been attributed to contour-current activity related to the overflow of the NSDW from the Faeroe–Shetland Channel (Richards *et al.* 1987; Stoker 1990*a*). Elsewhere in Rockall Trough, the Feni Drift is also a contourite deposit that has comparable sediment waves (Flood *et al.* 1979; Roberts and Kidd 1979). Similar sediment waves have been documented as contourite deposits at other locations around the world (e.g. Damuth 1980; Flood 1988).

However, migrating sediment waves can also be created by overbank flow of turbidity-currents on the backs of submarine-fan channel levees and on submarine canyon walls (Damuth 1979; Normark *et al.* 1980; Carter *et al.* 1990). Recently, Savoye *et al.* (in press) showed examples of large migrating sediment waves developed on a levee of the main channel of the Var Deep-Sea Fan off Nice, France. These sediment waves are similar in size (wavelengths of 1–7 km, amplitudes of 10–50 m) and appearance to the waves on the West Shetland slope, but were created by overbank flow of thick turbidity currents moving down the fan channel. Thus, we cannot rule out the possibility that the sediment waves observed on the West Shetland slope (Fig. 6) may have been, at least in part, formed by overbanking of turbidity currents flowing down the submarine canyon and fan channel (Fig. 5D) just upslope of the wave field. If so, the regular migrating wave-like bedforms observed between 1000 and 1300 m water depths (Fig. 6) just downslope of this fan channel could be a function of the oblique orientation of the dip lines across channel-levee/overbank deposits of the fan, rather than migrating sediment waves of contour-current deposits. In addition, the constructional levee-like northwest wall of the submarine canyon on the upper slope (Fig. 5B) shows zones of smaller, regular migrating waves within the overall levee deposits, which also could have been created by the interaction of overbank flow and contour-current flow.

In summary, although migrating sediment waves can be created by overbanking turbidity flows, given the well documented contour-current flow of the NSDW through this region during the Quaternary, many of the sediment-wave deposits observed in the study area, especially the larger wave fields (e.g. Fig. 6), were probably created primarily by the action of contour-currents. Sediments moving downslope in gravity-controlled flows, as well as glacial marine sediments falling through the water column, were probably entrained within the contour currents and redeposited along the slope as sediment drift deposits. However, more closely spaced seismic lines are needed to determine the true orientation and extent of individual waves and wave sets, and thereby confirm the true origin of these features.

Basin floor

The basin floor of the Faeroe–Shetland Channel (water depths >1300 m) is characterized by highly reflective, continuous parallel facies. Some individual reflections were traced continuously for more than 100 km along the basin axis. These deposits apparently represent predominantly hemipelagic and glacial marine deposits with rare interbedded turbidity-current, and related gravity-controlled flow deposits which have spread out and ponded in the basin axis (Akhurst 1991; Stoker *et al.* 1991). In addition, occasional, thin hummocky to transparent zones of seismic facies are observed between these parallel beds and appear to be minor debris flows or other mass-transport deposits that sometimes flow out onto the channel floor. In contrast to these widely spaced minor debris-flow deposits, three or more very large mass-transport deposits (up to 0.15 s thick) occur beneath the lower slope and channel floor (Fig. 7) just downslope from the deep-sea fan feature and the migrating sediment waves (Fig. 3). Stevenson (1991*b*) interpreted these

deposits as composite mass-flow lobes of the Miller Slide complex.

Large mass-transport deposits such as these are often common on the lower continental rise around the world and have been shown to have moved down slopes of <1° for up to hundreds of kilometres to form deposits extending throughout thousands of square kilometres (Embley 1976, 1980; Jacobi 1976; Embley and Jacobi 1977; Damuth 1980). In addition, these large mass-transport deposits commonly occur in association with large deep-sea fans and often contribute to their construction (e.g. Damuth and Embley 1981).

Conclusions

A prominent regional unconformity, herein referred to as the Latest Oligocene Unconformity (LOU), forms a major sequence boundary that approximately separates the Neogene and Quaternary clastic sediments from the older Paleogene deposits beneath the West Shetland slope and Faeroe–Shetland Channel. Although this unconformity is a major erosional surface, which locally cuts into Eocene strata at the southwestern end of the channel, a thick section of Oligocene sediments is present beneath the unconformity throughout most of the region and indicates that this unconformity did not form until at least the latest Oligocene or the early to middle Miocene. This unconformity apparently formed in response to increased basin subsidence coupled with vigorous erosion of the sea floor related to the initiation of intense bottom-water flow through the channel from the Arctic to the North Atlantic.

The post-Oligocene sediments above this regional unconformity show a variety of seismic facies and morphologic features on high-resolution seismic lines that indicate these sediments have been deposited and modified by a variety of deep-water depositional processes, which include both downslope and parallel-to-slope processes. The upper portion of the West Shetland slope (~200–800 m) is dominated by mass-transport deposits, predominantly debris flows with some slumps and slides, that were apparently deposited during glacial cycles when sea level was low and an ice sheet extended to the shelf edge and supplied large quantities of terrigenous sediment to the upper slope (Stoker 1990*a*). Thin, highly reflective packages of prograding clinoforms often separate mass-flow deposits and may represent glacial marine (ice-rafted) sedimentation during times of ice-sheet retreat from the shelf edge (Stoker 1990*a*), possibly in response to sea-level rise associated with interstades. Submarine canyons and gullies of various sizes are occasionally present, are often filled with mass-transport deposits and may have provided conduits for turbidity currents and related mass flows to reach the basin floor.

In contrast to the upper slope, the middle to lower slope (~800–1200 m) appears to be dominated by glacial marine and hemipelagic sediments along with possible significant deposition by turbidity currents and related gravity-controlled flows. At many locations, especially on the lower slope, these deposits apparently incurred major reworking by contour-current activity. At the northeastern end of the study area, a possible deep-sea fan-like feature is present on the middle slope, which appears to be composed of several overlapping, aggradational channel-levee systems including a 25 km wide modern channel-levee system whose channel is 75 m deep. However, just downslope from this feature, large migrating sediment waves or dunes are observed which appear to represent major contourite deposits. The available data are too sparse to allow delineation and mapping of the shapes and trends of these various features; consequently, we cannot yet confidently determine whether the sediment waves and the levee-like features represent contourites, overbank turbidites, or a combination of both. However, given the strong south-

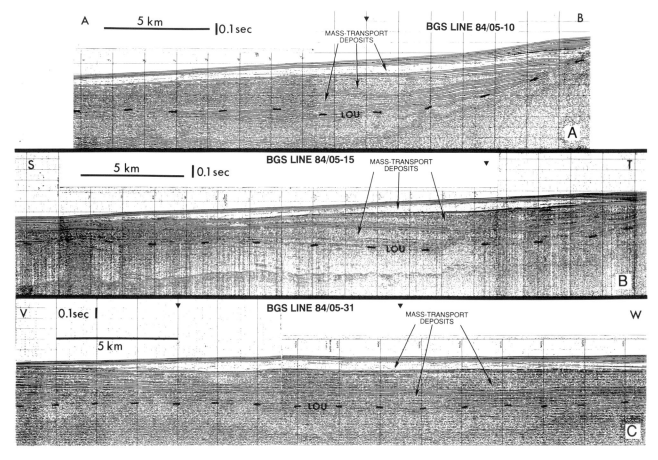

Fig. 7. Examples of large mass-transport deposits from the lower continental slope and channel floor. The hummocky, chaotic and transparent facies with lens- or mound-shaped external forms are interpreted as debris flows and related mass-wasting deposits. Continuous, parallel reflections separating these packages probably represent glacial marine, hemipelagic and turbidity-current deposits. Small triangles show line intersection points.

westerly thermohaline flow of the Norwegian Sea Deep Water through the channel and sedimentary evidence of contourites (Akhurst 1991), we believe that much of the lower slope has at least been shaped by contour-current activity.

The relatively thin sediments of the channel axis are highly reflective and individual beds can sometimes be traced for more than 100 km. These sediments appear to be dominantly glacial marine with some input by turbidity currents and mass-transport deposits. An exception exists at the northeastern end of the study area just downslope from the deep-sea fan feature and the sediment waves where several major mass-transport deposits occur. The data presented in this paper demonstrate the ambiguity or uncertainty of interpreting complex interactions between downslope and parallel-to-slope depositional processes based on inadequate seismic coverage. Analysis of additional infill seismic and core data is currently underway to substantiate the preliminary observations and interpretations put forward here.

We thank Mobil North Sea Ltd., Mobil Research and Development Corp., Lasmo (TNS) Ltd. and the British Geological Survey for permisssion to publish this paper. We thank M. S. Stoker, O. Eldholm, A. M. Spencer, R. B. Bloch, S. M. Mitchell, G. A. Hird, L. K. Vopni and W. P. Alves for reviewing the manuscript critically and offering constructive comments. We also thank N. H. Kenyon, M. C. Akhurst, D. Long, R. Holmes and A. Dobinson for helpful discussions.

References

AKHURST, M. C. 1991. *Aspects of Late Quaternary Sedimentation in the Faeroe–Shetland Channel, Northwest UK Continental Margin.* British Geological Survey Technical Report **WB/91/2.**

BELDERSON, R. H., KENYON, N. H. AND WILSON, J. B. 1973. Iceberg plough marks in the northeast Atlantic. *Palaeogeography, Palaeoclimatology, Palaeoecology*, **13**, 215–224.

BOUMA, A. H., NORMARK, W. R. AND BARNES, N. E. (eds) 1985. *Submarine Fans and Related Turbidite Systems: Frontiers in Sedimentary Geology.* Springer, New York.

CARTER, L., CARTER, R. M., NELSON, C. S., FULTHORPE, C. S. AND NEIL, H. L. 1990. Evolution of Pliocene to Recent abyssal sediment waves on Bounty Channel levees, New Zealand. *Marine Geology*, **95**, 97–109.

CREASE, J. 1965. The flow of Norwegian Sea water through the Faeroe Bank Channel. *Deep-Sea Research*, **12**, 143–150.

DAMUTH, J. E. 1979. Migrating sediment waves created by turbidity currents in the northern South China Basin. *Geology*, 7, 520–523.

—— 1980. Use of high-frequency (3.5–12 kHz) echograms in the study of near-bottom sedimentation processes in the deep-sea: A review. *Marine Geology*, **38**, 51–75.

—— AND EMBLEY, R. W. 1981. Mass-transport processes on Amazon Cone: western equatorial Atlantic. *American Association of Petroleum Geologists Bulletin*, **65**, 629–643.

——, FLOOD, R. D., KOWSMANN, R. O., BELDERSON, R. H. AND GORINI, M. A. 1988. Anatomy and growth pattern of Amazon Deep-Sea Fan as revealed by long-range side-scan sonar (GLORIA) and high-resolution seismic studies. *American Association of Petroleum Geologists Bulletin*, **72**, 885–911.

DOOLEY, H. D. AND MEINCKE, J. 1981. Circulation and water masses in the Faeroes Channels during overflow '73. *Deutsche Hydrographishe Zeitschrift*, **34**, 4–54.

DUINDAM, P. AND VAN HOORN, B. 1987. Structural evolution of the West Shetland continental margin. In: BROOKS, J. AND GLENNIE, K. (eds) *Petroleum Geology of North West Europe*. Graham & Trotman, London, 765–773.

ELDHOLM, O. 1990. Paleogene North Atlantic Magmatic-Tectonic Events: Environmental Implications. *Memorie della Societa Geologica Italiana*, **44**, 13–28.

ELLETT, D. J., EDWARDS, A. AND BOWERS, R. 1986. The hydrography of the Rockall Channel—an overview. *Proceedings of the Royal Society of Edinburgh*, **88B**, 61–81.

—— AND ROBERTS, J. G. 1973. The overflow of Norwegian Sea deep water across the Wyville–Thomson Ridge. *Deep-Sea Research*, **30**, 819.

EMBLEY, R. W. 1976. New evidence for the occurrence of debris flow deposits in the deep sea. *Geology*, **4**, 371–374.

—— 1980. The role of mass transport in the distribution and character of deep ocean sediments with special reference to the North Atlantic. *Marine Geology*, **38**, 23–50.

—— AND JACOBI, R. 1977. Distributions and morphology of large sediment slides and slumps on Atlantic continental margins. *Marine Geotechnology*, **2**, 205–228.

FLOOD, R. G. 1988. A lee wave model for deep-sea mudware activity. *Deep-Sea Research*, **35**, 973–983.

——, HOLLISTER, C. D. AND LONSDALE, P. 1979. Disruption of the Feni sediment drift by debris flows from Rockall Bank. *Marine Geology*, **32**, 311–334.

HARVEY, J. 1965. The topography of the south-western Faeroe Channel. *Deep-Sea Research*, **12**, 121–127.

HASZELDINE, R. S., RITCHIE, J. D. AND HITCHEN, K. 1987. Seismic and well evidence for the early development of the Faeroe–Shetland Basin. *Scottish Journal of Geology*, **23**, 283–300.

HITCHEN, K. AND RITCHIE, J. D. 1987. Geological review of the west Shetland area. *In*: BROOKS, J. AND GLENNIE, K. (eds) *Petroleum Geology of North West Europe*. Graham & Trotman, London, 737–749.

HOLMES, R. 1990. *Foula (Sheet 60°N-4°W) Quaternary Geology.* (1 : 250,000 Offshore Map Series.), British Geological Survey.

JACOBI, R. D. 1976. Sediment slides on the northwestern continental margin of Africa. *Marine Geology*, **22**, 157–173.

KENYON, N. H. 1987. Mass-wasting features on the continental slope of northwest Europe. *Marine Geology*, **74**, 57–77.

KOLLA, V. AND COUMES, F. 1987. Morphology, internal structure, seismic stratigraphy and sedimentation of Indus Fan. *American Association of Petroleum Geologists Bulletin*, **71**, 650–677.

MEINCKE, J. 1983. The modern current regime across the Greenland–Scotland Ridge. *In*: BOTT, M. H. P., SAXOV, S., TALWANI, M. AND THIEDE, J. (eds) *Structure and Development of the Greenland–Scotland Ridge*. Plenum, New York, 637–650.

MILLER, K. G. AND TUCHOLKE, B. E. 1983. Development of Cenozoic abyssal circulation south of the Greenland–Scotland Ridge. *In*: BOTT, M. H. P., TALWANI, M. AND THIEDE, J. (eds) *Structure and Development of the Greenland–Scotland Ridge—New Methods and Concepts*. Plenum, New York, 549–589.

MITCHELL, S. M., BEAMISH, G. W. J., WOOD, M. V., MALACEK, S. J., ARMENTROUT, J. D., DAMUTH, J. E. AND OLSON, H. C. 1993. Paleogene sequence stratigraphic framework of the Faeroe Basin. *In*: PARKER, J. R. (ed.) *Petroleum Geology of Northwest Europe: Proceedings of the 4th Conference*. Geological Society, London, 1011–1023.

MUDGE, D. C. AND RASHID, B. 1987. The geology of the Faeroe Basin area. *In*: BROOKS, J. AND GLENNIE, K. (eds) *Petroleum Geology of North West Europe*. Graham & Trotman, London, 751–763.

NORMARK, W. R., HESS, G. R., STOW, D. A. V. AND BOWEN, A. J. 1980. Sediment waves on the Monterey Fan Levee: a preliminary physical interpretation. *Marine Geology*, **37**, 1–18.

POSAMENTIER, H. W. AND VAIL, P. R. 1988. Eustatic controls on clastic deposition II—sequence and systems tract models. *In*: WILGUS, C. K., HASTINGS, B. S., KENDALL, C. G. St. C., POSAMENTIER, H. W. ROSS, C. A. AND VAN WAGONER, J. C. (eds) *Sea-level Changes: An Integrated Approach*. Society of Economic Paleontologists and Mineralogists, Special Publication, **42**, 125–154.

RICHARDS, P. C., RITCHIE, J. D. AND THOMSON, A. R. 1987. Evolution of deep-water climbing dunes in the Rockall Trough—implications for overflow currents across the Wyville–Thomson Ridge in the (?) Late Miocene. *Marine Geology*, **76**, 177–183.

ROBERTS, D. G. AND KIDD, R. B. 1979. Abyssal sediment wave fields on Feni Ridge, Rockall Trough: long-range sonar studies. *Marine Geology*, **33**, 175–191.

——, HUNTER, P. M. AND LAUGHTON, A. S. 1977. *Bathymetry of the Northeast Atlantic, Sheet 2, Continental Margin Around the British Isles, 1 : 2,400,000*. Institute of Oceanographic Sciences.

SAUNDERS, P. M. 1990. Cold outflow from the Faeroe Bank Channel. *Journal of Physical Oceanography*, **20**, 29–43.

SAVOYE, B., PIPER, D. J. W. AND DROZ, L. in press. Plio-Pleistocene evolution of the Var Fan (Nice, French Riviera). *Marine and Petroleum Geology*.

STEVENSON, A. G. 1990a *Flett (Sheet 61°N-04°W), Sea Bed Sediments, (1 : 250,000 Offshore Map Series.)* British Geological Survey.

—— 1990b. *Miller (Sheet 61°N-02°W), Sea Bed Sediments, (1 : 250,000 Offshore Map Series.)* British Geological Survey.

—— 1991a. *Flett (Sheet 61°N-04°W), Quaternary Geology, (1 : 250,000 Offshore Map Series.)* British Geological Survey.

—— 1991b. *Miller (Sheet 61°N-02°W), Quaternary Geology, (1 : 250,000 Offshore Map Series.)* British Geological Survey.

STOKER, M. S. 1990a. Glacially-influenced sedimentation on the Hebridean slope, northwest United Kingdom continental margin. *In*: DOWDESWELL, J. A. AND SCOURSE, J. D. (eds) *Glacimarine Environments: Processes and Sediments*. Geological Society, London, Special Publication, **53**, 349–362.

—— 1990b. *Judd (Sheet 60°N-6°W), Quaternary Geology, (1 : 250,000 Offshore Map Series.)* British Geological Survey.

—— 1990c. *Judd (Sheet 60°N-6°W), Solid Geology, (1 : 250,000 Offshore Map Series.)* British Geological Survey.

——, HARLAND, R. AND GRAHAM, D. K. 1991. Glacially influenced basin plain sedimentation in the southern Faeroe–Shetland Channel, northwest United Kingdom continental margin. *Marine Geology*, **100**, 185–199.

——, ——, MORTON, A. C. AND GRAHAM, D. K. 1989. Late Quaternary stratigraphy of the northern Rockall Trough and Faeroe–Shetland Channel, northwest Atlantic Ocean. *Journal of Quaternary Science*, **4**, 211–222.

WEIMER, P. 1990. Sequence stratigraphy, facies geometries, and depositional history of the Mississippi Fan, Gulf of Mexico. *American Association of Petroleum Geologists Bulletin*, **74:4**, 425–453.

WORTHINGTON, V. 1970. The Norwegian Sea as a mediterranean basin. *Deep-Sea Research*, **17**, 77–84.

ZIEGLER, P. A. 1988. *Evolution of the Arctic–North Atlantic and the Western Tethys*. American Association of Petroleum Geologists Memoir, **43**.

Seismic structure of the Hatton–Rockall area: an integrated seismic/modelling study from composite datasets

J. KESER NEISH

*Department of Energy (now: Department of Trade and Industry), 1 Palace Street, London SW1E 5HE
(Present address: The Gables, Cowstones, Hatton of Fintray, Aberdeen AB2 0HY, UK)*

Abstract: During 1987–9 the United Kingdom Department of Energy recorded coincident seismic reflection, sonobuoy wide-angle reflection and refraction, and expanding spread refraction data across the Hatton–Rockall Plateau, lying within the continental margin to the west of Scotland. Some 780 km of deep multichannel seismic reflection data were used to define and control the structural model. The presence of Tertiary basalts throughout the area gives rise to severe attenuation of incident energy and substantial velocity inhomogeneities. Information from 53 sonobuoys and 5 reversed expanding spread profiles recorded along the lines allow independent determination of velocity structure. An integrated model is constructed layer by layer from the surface down to Moho. Using information gained from these integrated datasets, the velocity-depth structure of the area can be determined across the entire margin and imaging at depth beneath the basalts substantially improved. Deep, upper mantle reflections are clearly seen beneath Rockall Bank. Evidence of probable Tertiary inversion is seen in Hatton Basin. The main structure of Hatton Bank is fault controlled, and both highly tilted fault blocks and oceanward-dipping reflector sequences are observed. The presence of a two-layer continental crust is revealed, thinned within Rockall Trough and deepening to the east beneath the Hebridean Platform and the west below Hatton–Rockall Plateau.

The Hatton–Rockall area is located within the continental margin to the west of Scotland (Fig. 1). Rockall Trough is a deep bathymetric depression lying between the British mainland and Rockall Plateau, which consists of Rockall Bank, Hatton–Rockall Basin and Hatton Bank.

The Rockall and Hatton Bank passive continental margins have been studied by a number of workers (e.g. Scrutton 1972; Bott *et al.* 1979; Roberts *et al.* 1981, 1988; Joppen and White 1990) but the structure, history, and evolution of the area have remained enigmatic. Seismic data in the area are sparse and

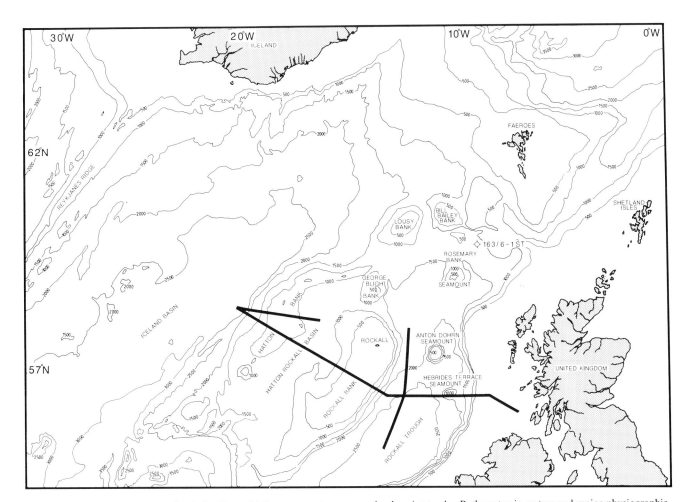

Fig. 1. Continental margin west of Britain. The solid line represents common-depth point tracks. Bathymetry in metres and major physiographic features are shown.

From *Petroleum Geology of Northwest Europe: Proceedings of the 4th Conference* (edited by J. R. Parker).

generally of poor quality, as the pervasive Tertiary volcanics present a substantial imaging barrier. Single well control from the stratigraphic test hole 163/6-1a(st) is some 500 km to the northeast, with shallow core information only available from limited distant Deep Sea Drilling Project sites.

Commencing in 1987, the United Kingdom Department of Energy (now the Department of Trade and Industry) undertook a series of composite seismic surveys designed to study this largely unexplored western margin in a comprehensive, integrated manner.

Data acquisition

During 1987–9 the United Kingdom Department of Energy acquired deep common-depth point seismic data from the UK mainland across the Hebridean Platform, Rockall Trough, and Rockall Plateau (Fig. 1). Fifty-three sonobuoy wide-angle reflection and refraction profiles and five two-ship reversed expanding spread wide-angle reflection and refraction profiles along the common-depth point tracks. Line CDP 87-1 was recorded SE–NW from Colonsay to Barra on the Hebridean Platform, CDP 87-2 E–W across Rockall Trough, CDP 87-3 SE–NW over Rockall Bank, Hatton Basin and Hatton Bank, CDP 88-5 SSW–NNE over Hatton Basin, and CDP 89-1 and CDP 89-2 SW–NE and SSW–NNE, respectively, along Rockall Trough (Fig. 2).

The three differing types of acquisition geometry were required to best confirm and constrain the instabilities of each data type. For example, stacking velocities measured from common-depth point data are a measure of the coherency of the data rather than the solid earth media; sonobuoy and expanding spread profile data can measure these velocities but lack the dip and structural control which can be supplied by the common-depth point method. When recorded coincidentally and integrated, the three differing techniques complement and reinforce one another.

There are three major difficulties in acquiring good quality seismic data within the survey area: deep water, hard sea floor, and Tertiary volcanics. These three factors were addressed in the acquisition stage by the use of large tuned air gun arrays and digital cable technology. This allowed the recording of many channels with an attendant improvement in signal to noise ratio. All parameters were further chosen to enhance the low-frequency output and arrivals. Complete details of all facets of data acquisition and processing may be found in Keser Neish (1990) and Keser Neish and Francis (1990).

The 1987 survey employed a 4860 cubic inch 2000 psi low pressure air gun array towed in a wide geometry to optimize the downgoing wave front. The 1988 survey area, having a primarily hard sea floor (which is an exceptional multiple-generating surface), was surveyed using a 7280 cubic inch 2000 psi pressure air gun array towed in a linear geometry to minimize these effects. The 1989 survey was also shot using a linear configuration but with an air gun capacity of 5692 cubic inches at 2000 psi.

The 1987 and 1988 surveys were recorded into a 4000 m LRS 16A digital cable with 300 raw and 150 array-formed channels to a record length of 20 s at a 4 ms sample rate. The 1989 survey utilized a 4800 m LRS 16A system with 360 raw and 180 array-formed channels and equivalent recording parameters.

Both Ref-Tek and Fairfield expendable sonobuoys were deployed from the common-depth point vessel using the techniques described by Le Pichon et al. (1968). This method allowed simultaneous recording without interruption to common-depth point recording and uses to great advantage the large, stable common-depth point air gun source. Profiles up to 60 km in length were achieved, with 30 km being the average extent.

The expanding spread profile application utilized in this survey was developed by Stoffa and Buhl (1979) according to the theory of Musgrave (1962). The shooting and receiving vessels start from a common mid-point and travel apart on reciprocal bearings, maintaining equivalent shot and receiver distances along the track. ICI open-cast gelignite was used as the explosive source and charge size increased incrementally from 10 kg to 400 kg along each 92 km profile.

Data processing and integration

The aim of all processing was to obtain sufficient primary energy to allow imaging of deep events in the presence of both noise and multiples while retaining amplitude relationships. Information obtained from the processing of each data type was integrated into the processing sequence of other data types. This became very important at the stack and migration phases. Care was taken throughout to preserve amplitude, travel time, and phase information.

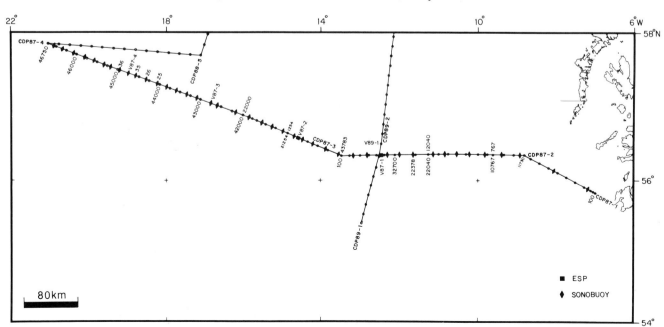

Fig. 2. Shotpoint location map showing detail of common-depth point, expanding spread profiles and sonobuoy sections described in this study.

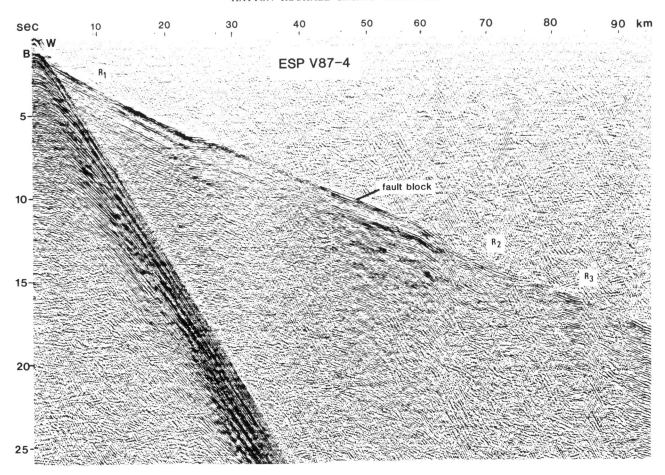

Fig. 3. Expanding spread profile V87-4 located on Hatton Bank (see Figs 2 and 12). W: direct water wave arrival; B: water bottom arrival; R_1, R_2 and R_3: refracted events.

The basic processing flow for all surveys was conventional. An instrument phase compensation filter was applied to remove known signal distortion due to air gun and instrument response. Although pre-stack predictive deconvolution was useful in removing some of the internal multiple energy, a shot gather FK domain multiple attenuation routine was most successful in preserving primary energy. Velocity analyses were at 4 km intervals. Final selection of both stack and migration velocities were controlled by velocity data input from the sonobuoy and expanding spread profile data. The quality of normal move-out calculation was further improved by the use of the long receiver cable. Deconvolution after stack and a radial predictive filter were further used to remove remnant multiple energy and improve signal to noise ratio.

Sonobuoy data processing was quite straightforward. Dispersion, source, instrument phase and hydrophone response corrections were applied prior to a two-window predictive deconvolution. A low-frequency bandpass-filter was chosen to best display the low frequency far offset arrivals. The expanding spread profile data was array formed on acquisition to 75 channels. An FK dip filter was applied to the raw shot records to remove linear noise. The data were sorted into 50 m common offset bins, stacked, and displayed with both low frequency bandpass and radial predictive filters.

All sonobuoy and expanding spread profile velocities were analysed using semblance velocity scans, Le Pichon's Method (Le Pichon *et al.* 1968), the slope-intercept method, *tau*-p inversion and ray trace modelling.

In each instance, individual solutions were independently derived but the final values were those which best satisfied all methods and data types: common-depth point, sonobuoys, and expanding spread profiles. Representative data types are shown for the area of the Hatton Bank common-depth point

seismic section. Figure 3 displays expanding spread profile V87-4, located over the tilted fault blocks which can be clearly seen on the comon-depth point profile. Sonobuoy 36 (Fig. 4) also shows evidence of fault blocks. A representative shot-cable semblance velocity scan from the expanding spread profiles is shown in Fig. 5. These linear velocity analyses are extremely useful in confirming wide-angle reflection and refraction velocities and observing fine structure along the expanding spread profile. Figure 6 is the ESP V87-4 *tau*-p inversion and demonstrates the quality of first-arrival information which can be analysed by this method.

The information from each data type was combined by iterative ray trace modelling using the box method (Cassell 1982). Ray paths, travel times, and amplitudes of normal incident and head waves for dipping layers in inhomogeneous media were calculated using zero-order asymptotic ray theory. The common-depth point profiles were digitized and velocity information added from the techniques discussed above. This method was very useful for resolving complex structures such as those seen in Figs 3 and 12, as complementary data are effectively integrated into a complete solution which does not induce or infer structure.

The 1989 data were further improved by the use of time-variant beam steering to enhance deep energy without damaging the shallower data. A portion of the 1989 data in Rockall Trough was pre-stack depth migrated (Fig. 7). The specifics of this depth migration are discussed in Keser Neish and Francis (1990, 1991). The velocity field generated from the corresponding expanding spread profiles and sonobuoy data analyses and integrated by the method described above was used as the input for this pre-stack depth migration. Depth focusing analysis further confirmed the accuracy of the depth model.

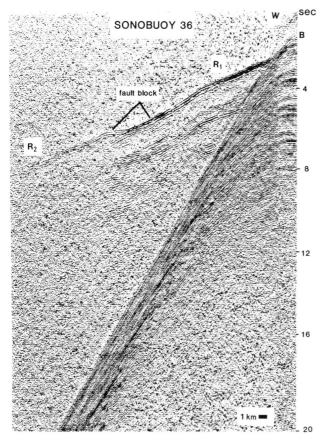

SONOBUOY 36

W sec

B

R₁

fault block

R₂

4

8

12

16

1 km

20

Fig. 4. Sonobuoy 36, Hatton Bank (Figs 2, 8 and 12). W: direct water wave arrival; B: water bottom arrival; R₁ and R₂: refractions breaking from wide-angle reflections tied directly to common-depth point data. Margin fault block structure is clearly visible.

Continental margin seismic data interpretation

The deep structure beneath Britain and the adjacent continental platform to the west of the British Isles has been studied using refraction experiments by a number of authors (Bamford *et al.* 1978; Smith and Bott 1975; Hughes *et al.* 1984) and, more recently, by the common-depth point profiling of the British Institution's Reflection Profiling Syndicate (Smythe *et al.* 1982; Brewer and Smythe 1986; McGeary *et al.* 1987; Cheadle *et al.* 1987).

Line CDP 87-1 traverses the shallow water platform to the west of Britain (Figs 1 and 2). This line served to calibrate the acquisition techniques and analytical methodology over known continental crust. The features described by the workers mentioned above, such as the Flannan thrust, are tied and deep mantle reflectors are observed. The acquisition and interpretation of this line is presented in detail by Keser Neish (1990) and will not be examined further here.

Rockall Trough

The pre-stack depth-migrated section line CDP89-2 (Fig. 7) clearly shows the major features of Rockall Trough. Lithological determination of the shallow structure is based on the work of Roberts (1975), Roberts *et al.* (1981), Masson and Kidd (1986) and Joppen and White (1990). The light yellow and dark orange horizons (Fig. 7) represent probable Quaternary–Pliocene material whose base is a strong regional marker. This unit thickens towards the south and also with increasing water depth. This indicates a relationship with subsidence and

increasing sedimentation. A strong marker again separates the orange sequence from the underlying lavender unit, interpreted as Lower Miocene–Oligocene. This unit deepens towards the south but does not thicken. Disturbance is also observed from shotpoints 100–250. This is interpreted as being related to deep-seated faulting and underlying volcanism. The deeper purple Eocene–Paleocene unit is also affected by this activity but, in general, overlies the Tertiary volcanic section. There is some evidence of intrusion by dykes and sills in the vicinity of shotpoints 650–700 and 850–900, where large-scale faulting is again observed.

The Tertiary volcanics are a major unit observed throughout the area in both intrusive and subaerially extrusive form as described by Roberts (1975), Scrutton (1986), Masson and Kidd (1986), Wood *et al.* (1987), Morton *et al.* (1988) and Joppen and White (1990). This sequence of volcanic rocks presented a strong barrier for seismic energy penetration and deeper imaging and is denoted by shades of blue on the section. The surface of this unit has a high acoustic impedance contrast on all data types and is very strongly diffractive on the common-depth point profiles, but accurate migration has refined the image of this boundary. The unit is thick (1.5–2.0 km) and on CDP 89-2 (Fig. 7) has the apparent form of interleaved basalt flows of varying velocities (Keser Neish and Francis 1990) rather than dykes and sills intruded into sediments. The upper flow surface has the appearance of large (0.5–1.0 km) rounded pillows along most of the line. The identification of the green unit, lying beneath the (violet) Eocene–Paleocene sequence from shotpoints 800–1050, is unclear. Undoubtedly, volcanics are present but most probably in the form of dykes and sills as there is no flow layering apparent and reflectors are short, diffractive, and disrupted. The deep-seated faults have apparently acted as conduits for the transport of volcanic material.

The Tertiary volcanic sequence within Rockall Trough is underlain by a highly reflective, distinctly layered unit (yellow, Fig. 7). This unit has been described by Bentley and Scrutton (1987) as 'deep layered basement' and by Joppen and White as 'acoustically layered basement' and is interpreted as syn-rift basement volcanics interbedded with sediments. On our dataset, the unit exhibits velocities of 5.20–5.60 km s⁻¹ (Fig. 13) and a seismic character similar to that of sedimentary rocks. Such velocity and seismic behaviour is also observed in Permo-Triassic and Devonian sandstones (Trewin 1991). The unit appears to persist throughout the central area of the trough but is interrupted by areas of poorer imaging probably related to fault and magmatic activity. The layered sequence also appears to fault out (shotpoints 850–1050) but this lack of sequence could be due to poor imaging in this area. Given the nature of faulting shown, the layered sequence could have been deposited within fault-controlled basins.

The base of the layered sequence is clearly defined by the cessation of strong layering onto a disruptive surface of discontinuous reflections (beige unit) whose fabric can be seen to follow the fault block trends. This unit is heavily faulted, has a velocity of 5.80–6.20 km s⁻¹, can essentially be followed across the margin, and is interpreted as the upper portion of a two-layered continental crust.

The interpretation of the top lower crust is somewhat problematic on the common-depth point data but indications of top unit form may be seen (dashed line, Fig. 7). Good wide-angle reflection and refraction control is, however, available from expanding spread profiles and sonobuoy data which place the two-way boundary on the form line and confirm unit velocities of 6.45–6.60 km s⁻¹, consistent with an interpretation of lower continental crustal material. The mantle arrival is clearly seen on all expanding spread profile data and provides an effective control on the lower crustal boundary at a depth of 13.7 km with a velocity of 8.0–8.21 km s⁻¹.

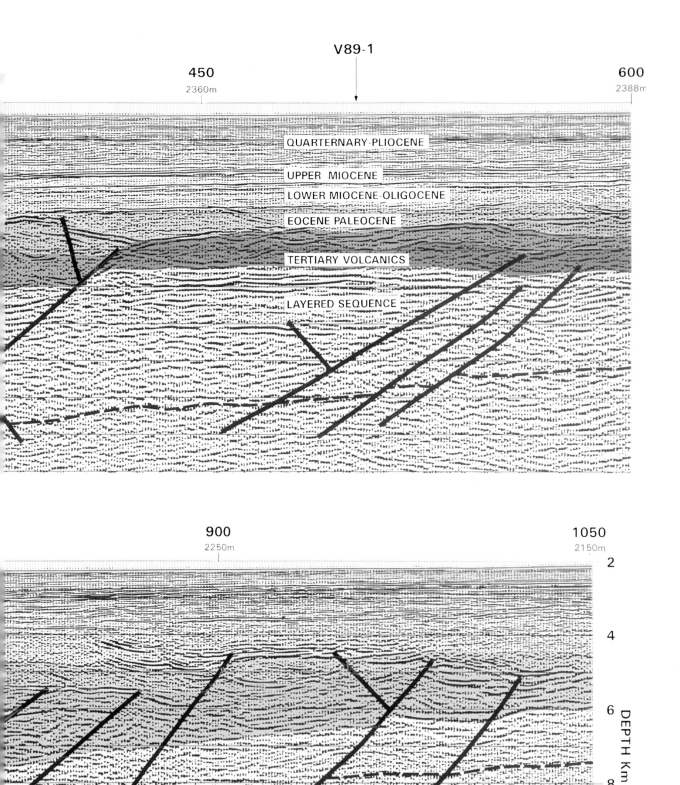

V89-1

450
2360m

600
2388m

QUARTERNARY-PLIOCENE

UPPER MIOCENE

LOWER MIOCENE-OLIGOCENE

EOCENE PALEOCENE

TERTIARY VOLCANICS

LAYERED SEQUENCE

900
2250m

1050
2150m

2

4

6

DEPTH Km

8

10

12

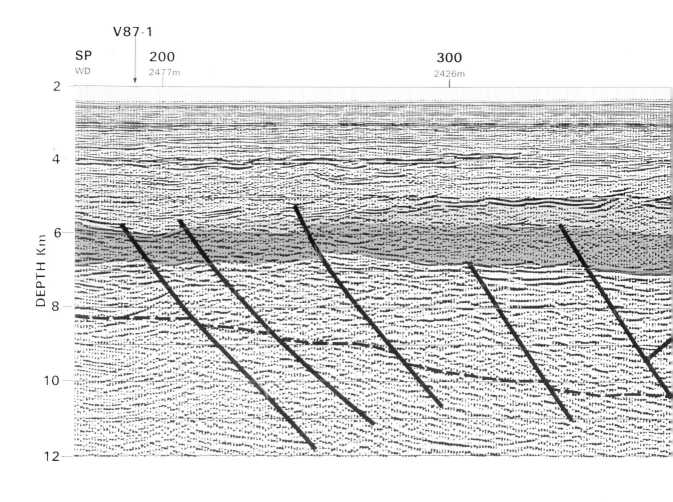

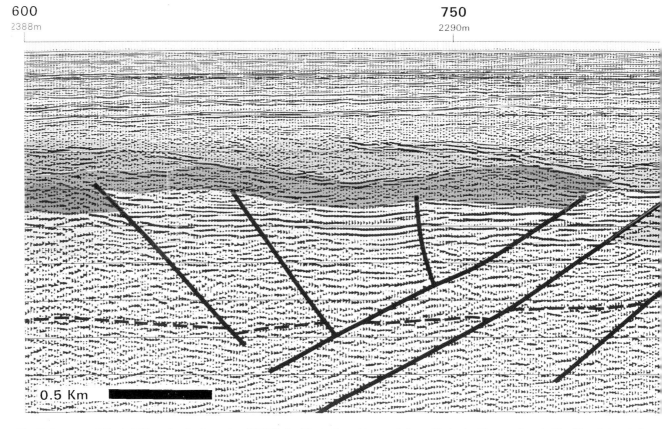

Fig. 7. Interpreted Rockall Trough seismic section CDP 89-2 with stratigraphic correlations. Unit velocities are given in Fig. 13 and descriptions in the text.

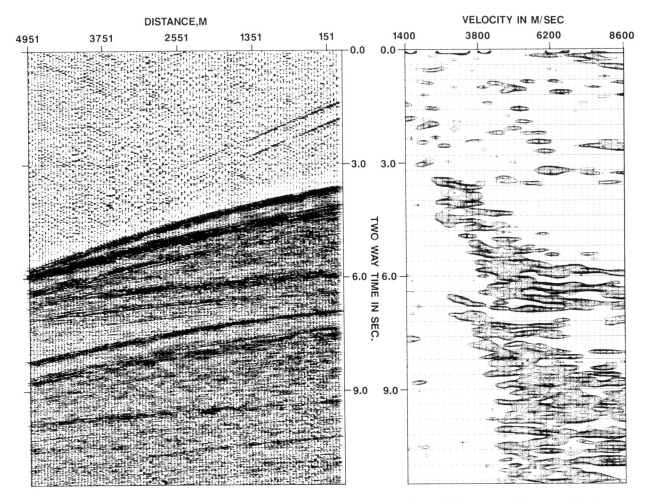

Fig. 5. Shot-cable semblance velocity scan used to study lateral velocity variations and confirm velocity structure. The x–t record is shown on the left and the coherency scan on the right.

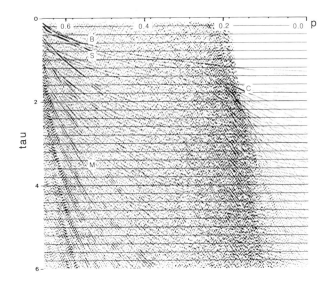

Fig. 6. ESP V87-4 *tau*-p transformation. B: water bottom; s: sedimentary layer arrivals; c: deeper phase arrivals; m: multiples.

Rockall Bank

Line CDP 87-3 (Fig. 2) extends from Rockall Trough, up the eastern margin of Rockall Plateau, and across Rockall Bank. The tilted fault blocks described by Joppen and White (1990) and Makris *et al.* (1991) are also observed here. Roberts

(1975) sampled Lewisian gneiss on Rockall Bank, but Morton and Taylor (1991) postulate an affinity with the Islay terrane.

Figure 8 shows the deep seismic structure of Rockall Bank. A highly reflective lower crust is seen below 4 s. This reflectivity is typical of much of the lower crust surveyed around Britain by the British Institution's Reflection Profiling Syndicate (McGeary *et al.* 1987). Most notable is the deep layered mantle which is observed between 8.0 and 9.2 s twt. This has the characteristics of the deep, upper mantle reflectors described by Warner and McGeary (1987), and Flack *et al.* (1990). ESP V87-2 is located over Rockall Bank and confirms the two-layer crustal structure. The Moho is observed at 7.8 s twt, or a depth of 25.5 km. These results correspond well with the work of Scrutton (1972) and Bunch (1979).

Hatton Basin and Bank

Hatton Basin and Bank lie between Rockall Plateau and the Iceland Basin (Fig. 1). This area has been little studied; however, recent geophysical work (e.g. White *et al.* 1987; Fowler *et al.* 1989; Morgan *et al.* 1989; Spence *et al.* 1989) has shown it to be an area of considerable geological and geophysical interest.

Figures 9a and 9b show the structure across the western portion of Hatton Basin and traverses Hatton Bank (line CDP 87-3, Fig. 2). Most striking is the broadly domal structure of Hatton Bank rising from the deepening edge of Hatton Basin on the east and descending towards the deep ocean basin on the west. The colours represent gross stratigraphic units which can be followed but unfortunately not positively ident-

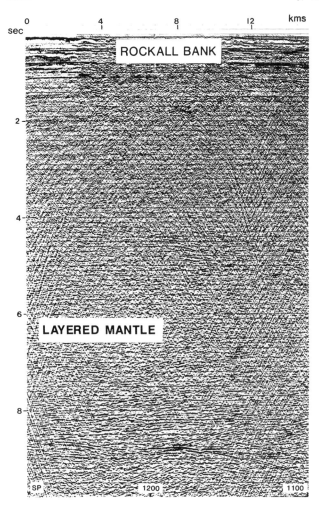

Fig. 8. Detailed Rockall Bank seismic section CDP 87-3, shot points 1100–1300 showing lower crustal reflectivity and layered mantle structure.

ified due to the scarcity of geological information within the area. From the velocity structure and similarities to Rockall units, rough lithologic classifications only can be determined.

Hatton Basin Figure 10 shows the fine structure of the Hatton Basin sedimentary section between shotpoints 43800 and 44100. The more recent units thicken and thin in an irregular manner, hinting at an axial sediment supply. This unit, and the units interpreted in pinks on Fig. 9a, clearly onlap the underlying light-green unit which continues up on to Hatton Bank. This surface appears to have been uplifted and eroded. A rounded, hummocky lenticular body is observed from SP 43800 at 2.5 s twt. This could represent a slump, irregular dumping, or, again, axial sediment supply. Also notable is the gentle westward down-dip of these units, with a corresponding increase in water depth. These factors give rise to strong indications of the presence of Tertiary basin inversion. The basal unit, interpreted as orange in Fig. 9a, displays rapid dip-corrected velocities which would appear to represent metamorphic/upper crustal rocks.

The basin terminates against Hatton Bank, as shown in detail in Fig. 11. Due to over-migration effects the behaviour of the deep basal unit is not clear; however, velocity control indicates that the deep event near shotpoint 44190 is in part a fault-plane image effect and that the unit does not fault to the surface. Interesting v-shaped regions are present at the water bottom, cutting into the sediments (SPs 44390, 44500, 44550,

Figs 9a and 11). These could be glacial gouge marks, channel, or contour current effects.

Hatton Bank. The main structure of Hatton Bank is obviously fault controlled (Fig. 9a) within its gentle dome, with younger veiling sediments lying conformably on the older surfaces. Upon this broad domal surface quite remarkable highly tilted fault blocks are observed (SPs 44820 and 45300, Fig. 9b). These fault blocks occur over a distance of nearly 80 km in an east–west direction and can also be seen on line CDP87-4, some 10 km northwards. Figure 12 shows the detail of these sharp tilted features. Identification is difficult and ray path imaging suffers badly from distortion in this area. Figures 3 and 4 display the expanding spread profiles and sonobuoy data acquired over this region and clearly show the fault block structures, demonstrating the need for dataset integration. The fault block between shotpoints 44900 and 45000 is more distinctly layered than other units seen on the bank. A resolved velocity of 5.1–5.3 km s^{-1} is determined, perhaps indicating older sedimentary rocks (Trewin 1991). The distance and surface between SP 45000 and SP45070 (interpreted as green on Fig. 9b) is overlain by a rather fast, anomalous 'basin' fill material. The fault-bounded block interpreted as orange immediately to the west displays Lewisian-type (Hall and Simmons 1979; Roberts 1975) velocities of 6.1 km s^{-1}. Again, the data at depth are somewhat over-migrated but provide positive evidence of deep-seated faulting. Velocity control on the dark-blue unit is good, serving to reinforce the structure hinted at on the common-depth point data.

Also seen on Fig. 9b is the Hatton Bank oceanward-dipping reflector sequence (SPs 45500–45650), commonly observed on the leading edge of passive margins. The generation of this sequence has been the subject of much discussion (Hinz 1981; Smythe 1983; Roberts *et al.* 1984; White *et al.* 1987) but the boundaries between the oceanward-dipping reflectors and the surrounding rocks remain unclear on this dataset for the formation of oceanward-dipping reflectors over continental crust. The extensive faulting and presence of tilted, rotated fault blocks is also similar to that described by Avedik *et al.* (1982) for the Biscay margin. The degree and depth of faulting seen on line CDP 87-3 across Hatton Bank would appear to suggest reactivation of earlier Caledonide faults and trends.

Discussion

The body of knowledge concerning the evolution of passive continental margins is continually growing. One of the key factors allowing more accurate imaging of the margin west of Britain in this study has been the acquisition of composite seismic datasets using three different geometries, each exploring differing facets of the sub-surface structure. Integration of these data-sets has allowed for more precise modelling methodology, enabling inferred structure to be replaced by measured data.

Analysis and interpretation of these data have revealed a seismic velocity structure extending across the entire continental margin from the UK mainland to the far western edge of Hatton Bank which is consistent with an interpretation of continental crustal rocks (Fig. 13). A distinct two-layer crustal structure is observed, with the upper layer exhibiting velocities of 5.80–6.20 km s^{-1} and the lower unit velocities of 6.45–6.60 km s^{-1}. The upper layer is reasonably consistent in its thickness (approximately 4–5 km) across the margin to Hatton Basin, where it thins to 1 km, and continues thinning across Hatton Bank until it can no longer be positively resolved at the oceanward-dipping reflector sequence. The lower crust thickens beneath the UK mainland (27 km) and Rockall Bank (25.5 km) but thins to a depth of 13.7 km beneath Rockall Trough. The base of the lower crust beneath

87-3

44200 1370m **44000** 1315m **43800** 1258m

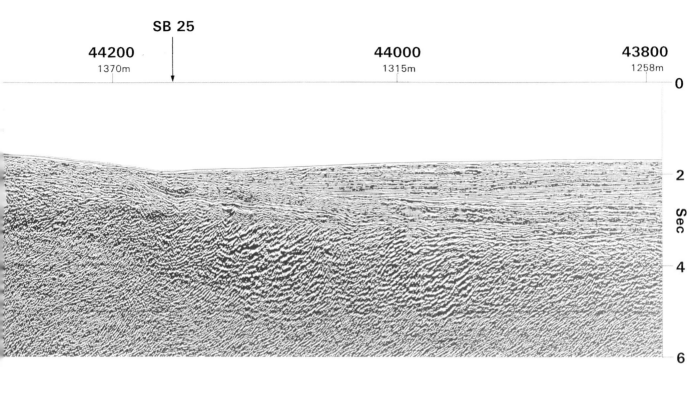

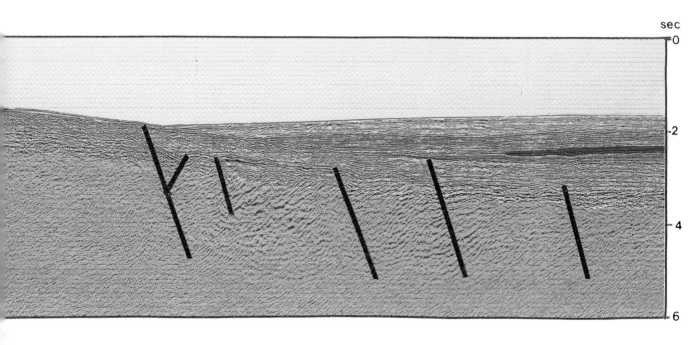

SB 35   SB 26

SP   44600   44400
WD   950m   1080m

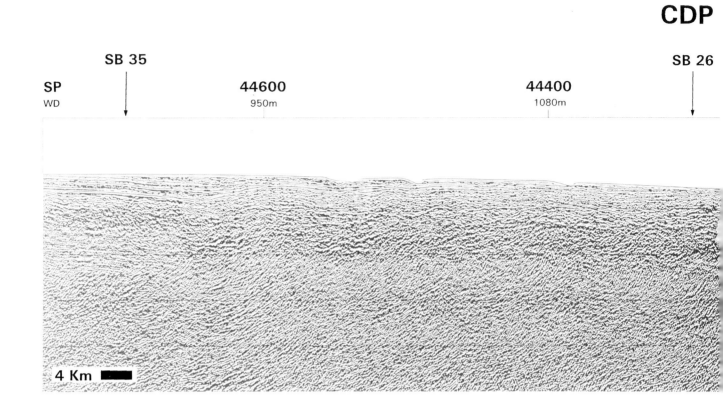

4 Km ▬▬

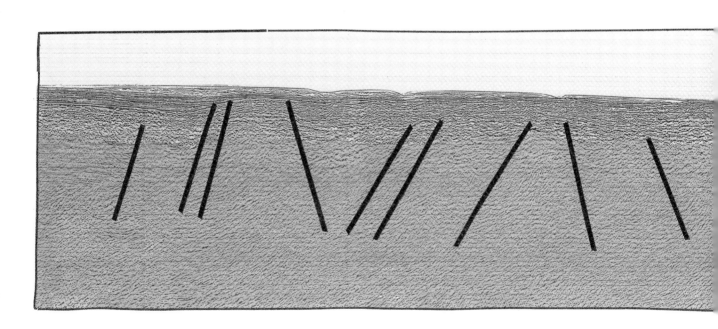

Fig. 9. (a) Uninterpreted and interpreted seismic section showing Hatton Basin and Bank. Location is given in Fig. 2 and structure is described in the text. Yellow: recent (Quaternary–Pliocene) sediments; pinks: old sediments (Miocene, Oligocene, Eocene?); light green: older sedimentary material; dark green: layered? sedimentary material; light blue: disturbed? sedimentary material; orange: upper crust; dark blue: lower crust.

87-3

ESP V87-4

45200
1085m

45000
948m

44800
903m

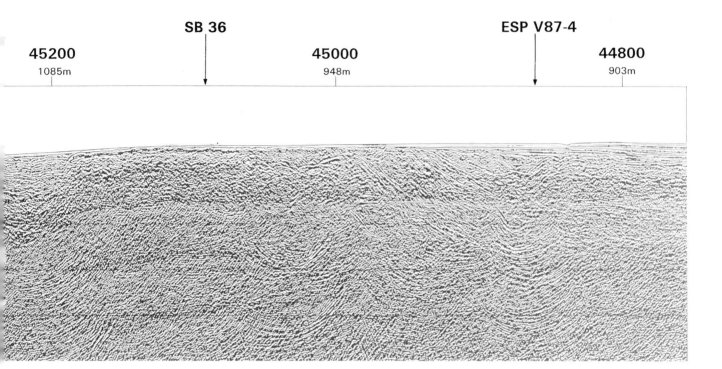

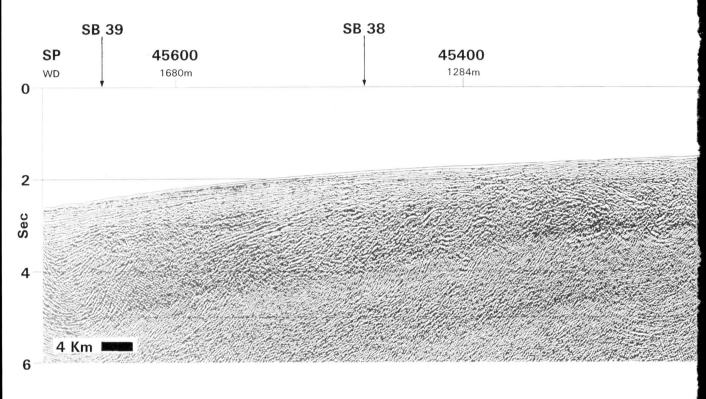

SB 39

SB 38

SP
WD

45600
1680m

45400
1284m

0

2

Sec

4

6

4 Km

sec

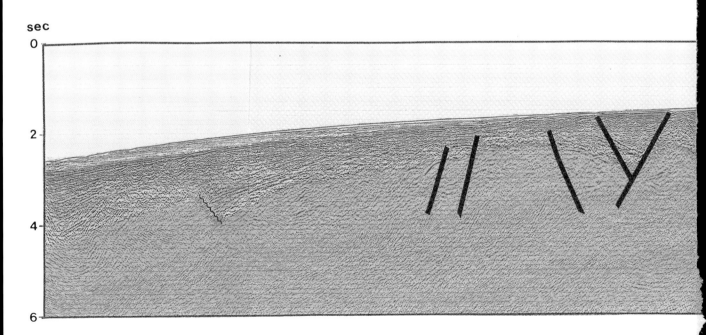

0

2

4

6

Fig. 9. (b) Uninterpreted and interpreted seismic section showing Hatton Bank. Location is given in Fig. 2 and structure is described in the text. Lithological determinations are based on character, correlation and velocity.

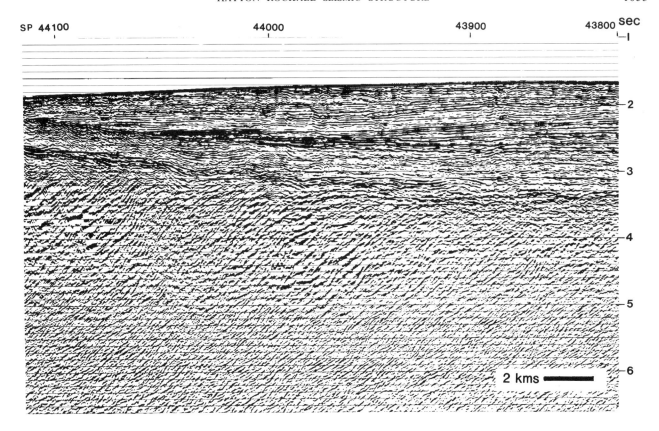

Fig. 10. Detailed Hatton Basin seismic section (CDP 87-3), shot point range 43800–44120, demonstrating basinal stratigraphic relationships.

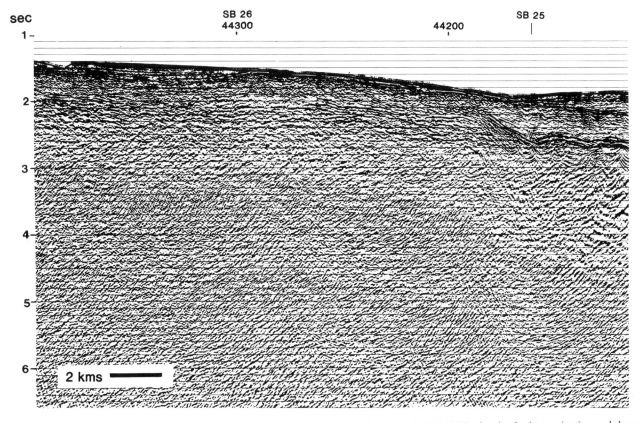

Fig. 11. Detailed Hatton Bank/Hatton Basin seismic sections (CDP 87-3), shot point range 44120–44400, showing basin termination and deeper seismic structure.

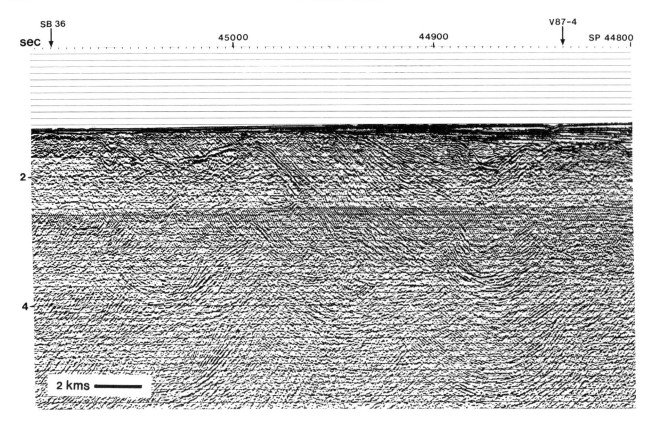

Fig. 12. Detailed Hatton Bank seismic section (CDP 87-3), SPs 44800–45100, clearly displaying highly tilted, rotated fault blocks.

Hatton Bank is observed at a depth of 16 km, whereupon crustal thicknesses decrease in depth towards the oceanic crust of the Iceland Basin (Fig. 1).

Deep crustal and mantle reflectors are observed on the shallow water platform to the west of Britain. Excellent images of lower crustal reflectivity and deep mantle layering are seen persisting for at least 12 km below Rockall Bank. Klemperer and the BIRPS group (1987) and McGeary et al. (1987) have looked in detail at deep crustal layering; considering the volcanicity apparent across the margin, a favoured hypothesis here would be igneous intrusive injection and layering at depth.

The presence of the apparent Tertiary inversion structure in Hatton Basin and on Rockall Plateau has implications concerning the age and history of the formation of the margin, as does the apparent absence of volcanic units in this area. It is evident that the main structure of the broad dome which forms Hatton Bank is fault controlled. Further, the highly tilted rotated fault blocks observed on Hatton Bank indicate extension and reactivation of existing fault trends.

The integrated dataset utilized in this study has served to refine imaging of the deeper volcanic surfaces and layered structure within Rockall Trough. However, the precise history of the evolution of the trough remains elusive. Megson (1987) has presented an excellent compilation of information on the subject. There remain several schools of thought concerning the origin of the crust beneath Rockall Trough: oceanic (e.g. Roberts 1975; Russell and Smythe 1978); continental (Roberts et al. 1988; Makris et al. 1991; Keser Neish 1991) or something in between (Smythe 1989). Results from the Department of Energy surveys allow support for the theory of continental origin, as thinned attenuated continental material is observed beneath Rockall Trough. As advances are made in both geophysical data acquisition and processing technology, more information will become available over this most interesting area as exploration barriers are overcome.

This project was carried out while the author was on the staff of the Exploration Branch of the Petroleum Engineering Division of the United Kingdom Department of Energy. Permission to publish this paper is gratefully acknowledged. Ownership of all data resides with the Secretary of State for Energy.

Many thanks are also given to the marine personnel at Western Geophysical who worked long and hard in adverse conditions: the crews of the M/Vs Western Arctic, Karen Bravo, Sofie Bravo, Western Reliance, and Forth Explorer. All explosive work was exemplarily carried out by 'Pooley' Jones. Kind thanks are also given to Malcolm Francis of Western, for all his assistance in the processing phases and last but not least acknowledgement must be given to Nigel H. Trewin, Aberdeen University, and T. B. D. Greenhill, Dalgaty Exploration Services, for many meaningful discussions and insights into structural relationships.

References

AVEDIK, F., CAMUS, A. L., GINZBERG, A., MONTADERT, L., ROBERTS, D. G. AND WHITMARSH, R. B. 1982. A seismic refraction and reflection study of the continent–ocean transition beneath the North Biscay Margin. *Philosophical Transactions of the Royal Society*, **A305**, 5–25.

BAMFORD, D., NUNN, K., PRODEHL, C. AND JACOB, B. 1978. LISPB-IV-Crustal Structure of Northern Britain. *Geophysical Journal of the Royal Astronomical Society*, **54**, 43–60.

BOTT, M. H. P., ARMOUR, A. R., HIMSWORTH, E. M., MURPHY, T. AND WYLIE, G. 1979. An Explosion Seismology investigation of the Continental Margin west of the Hebrides, Scotland at 58°N. *Tectonophysics*, **59**, 217–231.

BENTLEY, P. A. D. AND SCRUTTON, R. D. 1987. Seismic investigations into the basement structure of Southern Rockall Trough. *In*: BROOKS, J. AND GLENNIE, K. (eds) *Petroleum Geology of North West Europe*. Graham & Trotman, London, 667–675.

BREWER, J. A. AND SMYTHE, D. K. 1986. Deep structure of the foreland to the Caledonian Orogen. North-west Scotland: results of the BIRPS WINCH profile. *Tectonics*, **5**, 171–194.

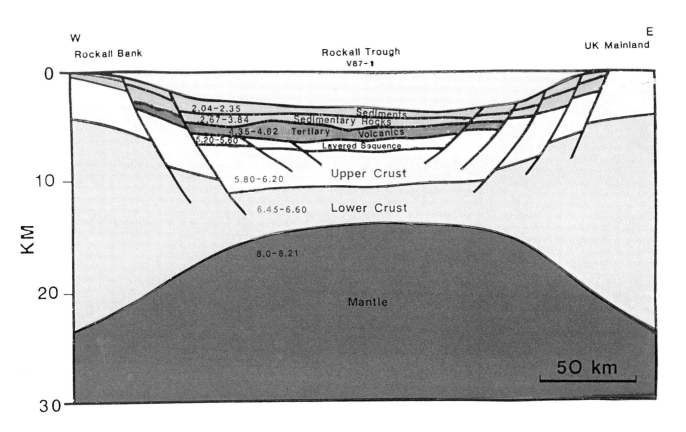

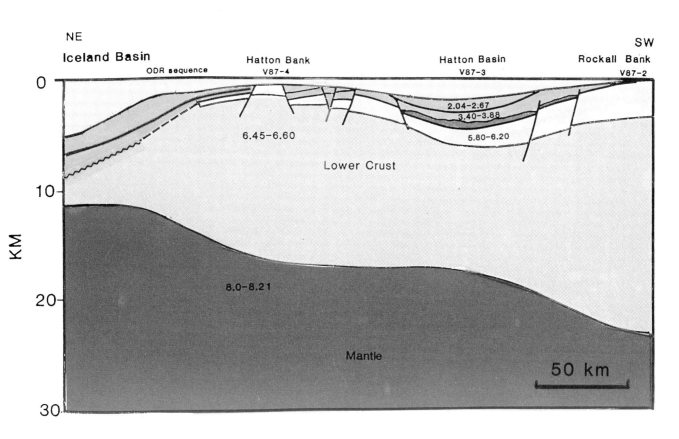

Fig. 13. Rockall–Hatton crustal cross-section along tracks CDP 87-1, CDP 87-2, and CDP 87-3, showing the velocity structure interpretation derived from the data discussed in this study. Both sections have equivalent scale and the vertical exaggeration is 6:1.

BUNCH, A. W. H. 1979. A detailed seismic structure of Rockall Bank (55°N, 15°W)—A synthetic seismogram analysis. *Earth and Planetary Science Letters*, **45**, 453–463.

CASSELL, B. R. 1982. A method for calculating synthetic seismograms in laterally varying media. *Geophysical Journal of the Royal Astronomical Society*, **69**, 339–384.

CHEADLE, M. J., MCGEARY, S., WARNER, M. R. AND MATTHEWS, D. H. 1987. Extensional structures in the western UK continental shelf: a review of evidence from deep seismic profiling. *In*: COWARD, M. P., DEWEY, J. F. AND HANCOCK, P. (eds) *Continental Extensional Tectonics*. Geological Society, London, Special Publication, **28**, 445–465.

FLACK, C. A., KLEMPERER, S. L., MCGEARY, S. G., SNYDER, D. B. AND WARNER, M. R. 1990. Reflections from mantle fault zones around the British Isles. *Geology*, **18**, 528–532.

FOWLER, S. R., WHITE, R. S., SPENCE, G. D. AND WESTBROOK, G. K. 1989. The Hatton Bank continental margin—II. Deep structure from two-ship expanding spread seismic profiles. *Geophysical Journal of the Royal Astronomical Society*, **96**, 295–309.

HALL, J. AND SIMMONS, G. 1979. Seismic velocities of Lewisian metamorphic rocks at pressure to 8 kbar: relationship to crustal layering in N Britain. *Geophysical Journal of the Royal Astronomical Society*, **58**, 337–347.

HINZ, K. 1981. A hypothesis on terrestrial catastrophes. Wedges of very thick oceanward dipping layers beneath passive continental margins—their origin and paleoenvironmental significance. *Geologische Jahrbuche*, **E22**, 3–28.

HUGHES, V. J., WHITE, R. S. AND JONES, E. J. W. 1984. Seismic velocity structure of the northwest Scottish continental margin: some constraints imposed by amplitude studies. *Annales Geophysicae*, **2**, 669–678.

JOPPEN, M. AND WHITE, R. S. 1990. The structure and subsidence of Rockall Trough from two-ship seismic experiments. *Journal of Geophysical Research*, **95**, 19821–19837.

KESER NEISH, J. C. 1990. Rockall Trough: crustal structure of a complex environment from composite seismic surveys. *In*: *Expanded Abstracts of the Technical Program, Sixtieth Annual International Meeting of the Society of Exploration Geophysicists*, San Francisco, **1**, 50–54.

—— AND FRANCIS, M. 1990. A Comparative Study of Deep Seismic Imaging Within the Rockall Trough. *In*: *Expanded Abstracts of the Technical Program, Sixtieth Annual International Meeting of the Society of Exploration Geophysicists*, San Francisco, **1**, 185–188.

—— AND FRANCIS, M. 1991. Rockall Trough: New Images from Deep Seismic Surveys. *In*: *Expanded Abstracts of the Technical Program, Fifty-third Meeting, European Association of Exploration Geophysicists*, Florence, **II**, 104–105.

KLEMPERER, S. L. AND THE BIRPS GROUP. 1987. Reflectivity of the crystalline crust: hypotheses and tests. *Geophysical Journal of the Royal Astronomical Society*, **89**, 217–222.

LE PICHON, X., EWING, J. AND HOUTZ, R. E. 1968. Deep-sea velocity determination made while reflection profiling. *Journal of Geophysical Research*, **73**, 2597–2614.

MAKRIS, J., GINZBURG, A., SHANNON, P. M., JACOB, A. W. B., BEAN, C. J. AND VOGT, U. 1991. A new look at the Rockall Region, offshore Ireland. *Marine and Petroleum Geology*, **8**, 410–416.

MASSON, D. G. AND KIDD, R. B. 1986. Revised Tertiary seismic stratigraphy of the southern Rockall Trough. *Initial Reports of the Deep-Sea Drilling Project*, **94**, 1117–1126.

MCGEARY, S., CHEADLE, J. M., WARNER, M. R. AND BLUNDELL, D. J. 1987. Crustal structure of the continental shelf around Britain derived from BIRPS deep seismic profiling. *In*: BROOKS, J. AND GLENNIE, K. W. (eds) *Petroleum Geology of North West Europe*. Graham & Trotman, London, 33–41.

MEGSON, J. B. 1987. The evolution of the Rockall Trough and implications for the Faeroe–Shetland Trough. *In*: BROOKS, J. AND GLENNIE, K. W. (eds) *Petroleum Geology of Northwest Europe*, Volume 2. Graham & Trotman, London, 653–665.

MORGAN, J. V., BARTON, J. P. AND WHITE, R. S. 1989. The Hatton Bank Continental Margin—III. Structure from wide-angle OBS and multichannel seismic refraction profiles. *Geophysical Journal International*, **98**, 367–384.

MORTON, A. G., DIXON, J. E., FITTON, J. G., MACINTYRE, R. M., SMYTHE, D. K. AND TAYLOR, P. N. 1988. Early Tertiary Volcanic rocks in well 163/6-1, Rockall Trough. *In*: MORTON, A. C. AND PARSON, L. M. (eds) *Early Tertiary Volcanics and the Opening of the NE Atlantic*. Geological Society, London, Special Publication, **39**, 293–308.

——AND TAYLOR, P. N. 1991. Geochemical and isotopic constraints on the nature and age of basement rocks from Rockall Bank, NE Atlantic. *Journal of the Geological Society, London*, **148**, 631–634.

MUSGRAVE, A. W. 1962. Applications of the expanding reflections spread. *Geophysics*, **27**, 981–983.

ROBERTS, D. G. 1975. Marine geology of Rockall Plateau and Trough. *Philosophical Transactions of the Royal Society, London*, **A278**, 447–509.

——, BACKMANN, J., MORTON, A. C., MURRAY, J. W. AND KEENE, J. P. 1984. Evolution of volcanic rifted margins: synthesis of Leg 81 results on the west margin of Rockall Plateau. *Initial Reports of the Deep Sea Drilling Project*, **81**, 883–991.

——, GINZBERG, A., NUNN, K. AND MCQUILLIN, R. 1988. The structure of the Rockall Trough from seismic refraction and wide-angle reflection measurements. *Nature*, **332**, 632–635.

——, MASSON, D. G. AND MILES, P. R. 1981. Age and structure of the southern Rockall Trough—new evidence. *Earth and Planetary Science Letters*, **52**, 115–128.

RUSSELL, M. J. AND SMYTHE, D. K. 1978. Evidence for an Early Permian rift in the northern North Atlantic. *In*: NEUMANN, E. R. AND RAMBERG, I. B. (eds) *Petrology and Geochemistry of Continental Rifts*. D. Reidel, Dordrecht, 173–179.

SCRUTTON, R. A. 1972. The crustal structure of the Rockall Plateau and microcontinent. *Geophysical Journal of the Royal Astronomical Society*, **27**, 259–275.

—— 1986. The geology, crustal structure and evolution of the Rockall Trough and the Faeroe–Shetland channel. *Proceedings of the Royal Society of Edinburgh*, **B88**, 7–26.

SMITH, P. J. AND BOTT, M. H. P. 1975. Structure of the crust beneath the Caledonian Foreland and Caledonian Belt of the North Scottish Shelf Region. *Geophysical Journal of the Royal Astronomical Society*, **40**, 187–205.

SMYTHE, D. K. 1983. Faeroe–Shetland escarpment and continental margin north of the Faeroes. *In*: BOTT, M. H. P., SAXON, S., TALWANI, M. AND THIEDE, J. (eds) *Structure and Development of the Greenland–Scotland Ridge—New Methods and Concepts*. Plenum, New York, 109–120.

—— 1989. Rockall Trough—Cretaceous or late Palaeocene? *Scottish Journal of Geology*, **25**, 5–43.

——, DOBINSON, A., MCQUILLIN, R., BREWER, J. A., MATTHEWS, D. H., BLUNDELL, D. J. AND KEIK, B. 1982. Deep structure of the Scottish Caledonides revealed by the MOIST reflection profile. *Nature*, **299**, 338–340.

SPENCE, G. D., WHITE, R. S., WESTBROOK, G. K. AND FOWLER, S. R. 1989. The Hatton Bank continental margin—I. Shallow structure from two-ship expanding spread seismic profiles. *Geophysical Journal of the Royal Astronomical Society*, **96**, 1273–1294.

STOFFA, P. L. AND BUHL, P. 1979. Two-ship multichannel seismic experiments for deep crustal studies: Expanded spread and constant offset profiles. *Journal of Geophysical Research*, **84**, 7645–7670.

TREWIN, N. H. 1991. *Integration of offshore seismic reflection and refraction data with lithological properties*. Department of Energy Internal Report, London.

WARNER, M. R. AND MCGEARY, S. 1987. Seismic reflection coefficients from mantle fault zones. *Geological Journal of the Royal Astronomical Society*, **89**, 223–230.

WHITE, R. S., WESTBROOK, G. K., FOWLER, S. R., SPENCE, G. D., BARTON, P. J., JOPPEN, M., MORGAN, J., BOWEN, S. N., PRESTCOTT, C. AND BOTT, M. H. P. 1987. Hatton Bank (Northwest UK) Continental margin structure. *Geophysical Journal of the Royal Astronomical Society*, **89**, 265–272.

WOOD, M. V., HALL, J. AND VAN HOORN, B. 1987. Post-Mesozoic differential subsidence in the north-east Rockall Trough related to volcanicity and sedimentation. *In*: BROOKS, J. AND GLENNIE, K. W. (eds) *Petroleum Geology of North West Europe*. Graham & Trotman, London, 667–685.

Cretaceous and Tertiary basin development west of Ireland

P. M. SHANNON,[1] J. G. MOORE,[1] A. W. B. JACOB[2] and J. MAKRIS[3]

[1] *Department of Geology, University College Dublin, Belfield, Dublin 4, Ireland*
[2] *Geophysics Section, Dublin Institute for Advanced Studies, 5 Merrion Square, Dublin 2, Ireland*
[3] *Institut für Geophysik, Universität Hamburg, Bundesstrasse 55, Hamburg, Germany*

Abstract: The Porcupine, Rockall and Hatton–Rockall basins lie on the continental seaboard west of Ireland. Up to 9 km of Cretaceous and Tertiary strata are preserved in the Porcupine Basin, in contrast to 2–3 km in the Rockall and Hatton–Rockall basins. Differences in seismic stratigraphic sequences through the region are attributed to an interplay between the developing North Atlantic Ocean, post-rift thermal subsidence and sea-level changes. Localized Ryazanian fault-controlled alluvial fan clastic rocks in the Porcupine Basin are succeeded by Valanginian to Barremian marine deposits. Sea-level lowstands in Albian and Paleocene–Eocene times, interpreted as rift and ridge-push effects respectively, resulted in delta and submarine fan progradation. Rapid thermal subsidence in Early Oligocene times led to marine deposits and to the onset of geostrophic currents. Miocene slumping triggered shale-dominant turbidite development, followed by tranquil deep marine sediments as subsidence outstripped sedimentation. The Cretaceous sequence in the Rockall Trough is thought to contain extensive igneous bodies overlain by an Upper Cretaceous muddy chalk sequence which is cut by abundant Tertiary sills and dykes. A number of shale-prone Upper Tertiary sequences occur. The Upper Cretaceous to Lower Tertiary succession is thin and frequently absent in the Hatton–Rockall Basin.

Ireland is almost surrounded by Late Paleozoic to Tertiary basins (Fig. 1) which vary in orientation, evolution and petroleum prospectivity (Naylor and Shannon 1982; Shannon 1991a). Basins to the east of Ireland formed in Late Paleozoic times and contain little post-early Mesozoic sediment; those to the south accumulated most sediments during Jurassic and early Cretaceous times; while the basins west of Ireland experienced major periods of subsidence and sedimentation during the Cretaceous and Tertiary.

The largest and least explored offshore basins lie west of Ireland (Fig. 1). The Northwest Offshore Basins, a set of small asymmetric rift basins, extend as an elongate arm from the Porcupine Basin. In view of their small and narrow nature, and the lack of available good quality seismic and well data, they are only fleetingly referred to here (see instead Naylor and Shannon 1982; Shannon 1991a; Trueblood and Morton 1991). The Goban Spur Basin, southwest of Ireland, is also not dealt with here, due to its geographic proximity to the Porcupine Basin and its geological affinity to the Porcupine and Celtic Sea basins (see Cook 1987; Shannon 1991a). This paper concentrates on the Porcupine, Rockall and Hatton–Rockall basins.

The Porcupine Basin trends north–south and widens southwards. It is divided by an east–west basement ridge into a small North Porcupine Basin and a large Main Porcupine Basin (Fig. 1), sometimes referred to as the Porcupine Seabight Basin (Tate and Dobson 1988). Seismic data coverage throughout the Porcupine Basin is extensive but only 25 wells, mostly clustered in the northern region, have been drilled.

The Rockall Trough lies to the west of the Porcupine Basin and has a northeast–southwest Caledonoid trend (Fig. 1). Little petroleum exploration has been carried out. One DSDP well was drilled in the basin (Fig. 1). Rockall Bank forms the western boundary of the basin. Further west still lies the Hatton–Rockall Basin, which is broadly comparable in orientation to the Rockall Trough. Hatton Bank forms the western margin of this basin adjacent to the continent–ocean boundary (Fig. 1). No detailed petroleum exploration has taken place in the Hatton–Rockall Basin, although some regional seismic profiles have been shot and a few shallow boreholes have been drilled (Fig. 1).

This paper concentrates upon the Cretaceous and Tertiary (i.e. broadly post-rift) development of the basins. It is based on several thousand kilometres of very good quality seismic lines from the Porcupine Basin, together with information from approximately 21 wells. Data from Rockall Trough and the Hatton–Rockall Basin consist of 1000 km of recent wide-angle reflection/refraction profiles acquired during the RAPIDS (Rockall and Porcupine Irish Deep Seismic) experiment (Makris *et al.* 1991). These are augmented by seismic reflection lines of various vintages from the Rockall Trough in the vicinity of the RAPIDS profiles, and shallow DSDP borehole information (Naylor and Shannon 1982; Masson and Kidd 1986).

Geological setting

The crust forming the platform regions on the continental shelf of Ireland is approximately 30 km thick. Dredge samples from the area indicate that these (economic basement) rocks consist of Paleozoic and older strata hosting the imprints of many orogenic episodes. The crust under the basins is significantly thinned (Naylor and Shannon 1982; Makris *et al.* 1988; Smythe 1989). While it is generally acknowledged that the thinned crust under the Porcupine Basin is continental in origin (Makris *et al.* 1988), it has been argued (e.g. Roberts *et al.* 1981; Joppen and White 1990) that the crust under the Rockall Trough is oceanic, or of mixed continental/oceanic origin (Megson 1987). However, recent wide-angle reflection/refraction data from the region (Roberts *et al.* 1988; Makris *et al.* 1991) show asymmetrically thinned continental crust 8–10 km thick across the Rockall Trough and 10–22 km thick (Makris *et al.* 1991) beneath the Hatton–Rockall Trough.

The Porcupine Basin has had an extensive pre-Cretaceous syn-rift history (Croker and Shannon 1987; MacDonald *et al.* 1987; Tate and Dobson 1988, 1989a,b; Croker and Klemperer 1989). A pre-Cretaceous origin is also suggested for the Rockall and Hatton–Rockall basins (Bentley and Scrutton 1987; Megson 1987; Shannon 1991a). The main syn-rift phase of basin development ceased at end-Jurassic times. A tilted fault block topography was modified by the Late Cimmerian unconformity, which heralded the onset of thermal subsidence basin development.

Cretaceous development

Up to 5 km of Cretaceous strata are preserved in the Porcupine

From *Petroleum Geology of Northwest Europe: Proceedings of the 4th Conference* (edited by J. R. Parker).
© 1993 Petroleum Geology '86 Ltd. Published by The Geological Society, London, pp. 1057–1066.

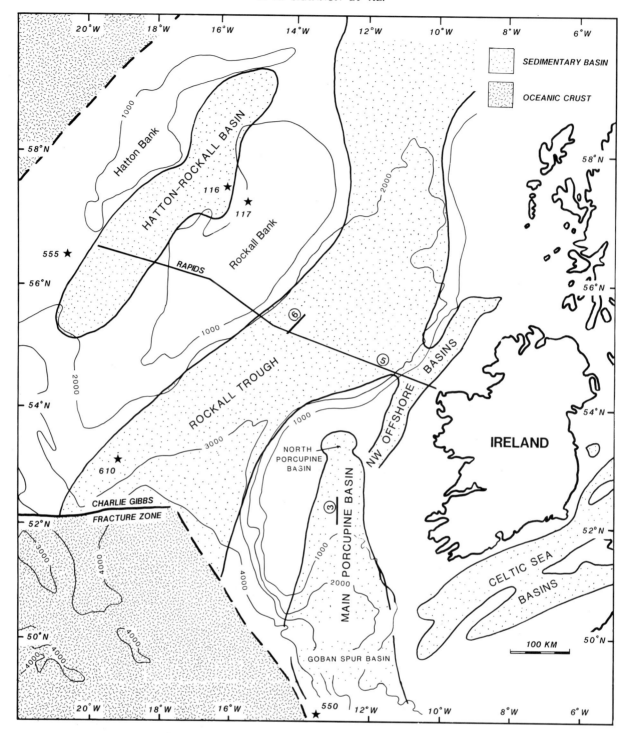

Fig. 1. Location map of the basins west of Ireland. DSDP drill sites are marked with a star. The locations of Figs 3, 5 (the RAPIDS profile) and 6 are indicated by the ringed numbers. Bathymetric contours are in metres.

Basin but only about 1–2 km in the Rockall and Hatton–Rockall basins. A series of seismic sequences has been defined (Fig. 2). The first letter of each refers to the Basin (P for Porcupine, R for Rockall); the second letter (K) refers to the Cretaceous; while the third letter (T for Transition) or number (1–3) refers to the number of the sequence. Facies interpretations of these follow the general principles of seismic stratigraphy (e.g. Sangree and Widmier 1977; Vail *et al.* 1977).

Porcupine Basin

The Cretaceous sequence of the Porcupine Basin rests with angular unconformity upon tilted fault blocks composed of

Jurassic and older strata. It displays a typical steer's head basin profile, extending beyond the limit of the underlying syn-rift deposits (Naylor and Anstey 1987). Four seismic sequences are identified (Fig. 2), distinguished by assemblages of seismic configurations and internal characteristics (Fig. 3).

The oldest sequence (PKT) is only developed locally and displays attributes of both syn-rift and post-rift type. A more complete description is given in Moore (1992). Wedge-shaped downlapping packages with variable seismic amplitudes dominate the edges and bases of the sequence. These give way laterally and vertically to packages of more continuous and increasingly flat-lying low-amplitude reflectors. The sequence is interpreted as fault-controlled alluvial fan deposits, grading

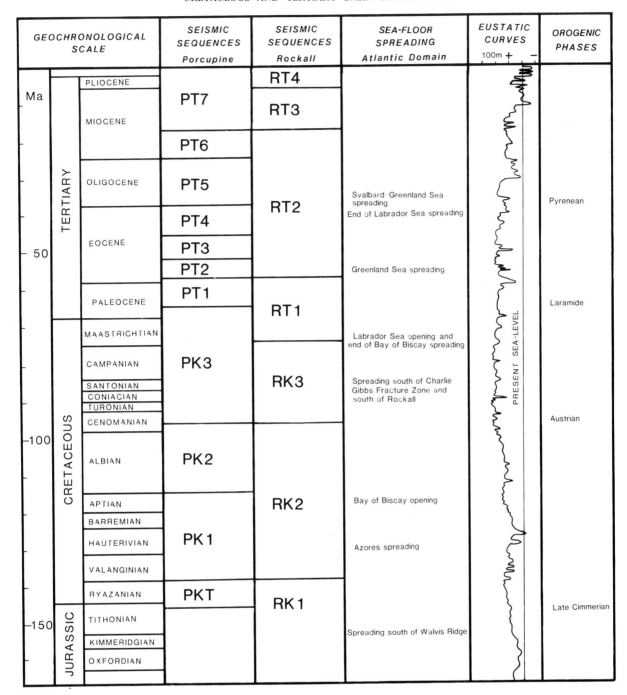

Fig. 2. General sequence stratigraphy of the Cretaceous and Tertiary of Porcupine and Rockall regions. Major sea-floor spreading events for the Atlantic domain, eustatic curves (from Haq *et al.* 1987) and tectonic phases for Central Europe (from Ziegler 1982) are shown.

laterally to mud-prone shallow shelf deposits and succeeded by passive-infill, deeper marine shales that herald the onset of thermal subsidence. It is probably of Ryazanian age.

Throughout most of the basin the second sequence (PK1) consists of a thick package of relatively continuous and flat-lying seismic reflectors. It is up to 1.6 s TWT (2500 m) thick and is of Valanginian to Aptian age. Discontinuous higher amplitude reflectors are sometimes associated with the residual topographic troughs and crests of the underlying syn-rift geometry, while wedge-shaped sub-sequences locally thin away from the basin margins. The variation in seismic character is interpreted as reflecting the underlying topography. In the topographic lows between the tilted fault blocks, sediment was funnelled as turbiditic and sandy submarine fan deposits whilst above the fault blocks the lowest Cretaceous strata are of

shallow marine origin (Croker and Shannon 1987). The wedge-shaped packages towards the basin margins are interpreted as alluvial and deltaic deposits (Fig. 4). Towards the north of the basin the depositional setting appears to have been shallower, with the onset of inner shelf to prodelta depositional environments. Further south, there was probably deeper water (Fig. 4). In the central part of the basin the Median Volcanic Ridge, interpreted by Tate and Dobson (1988) and White *et al.* (1992) as an igneous centre, appears to be partly extrusive and is onlapped by probable Lower Cretaceous marine strata.

Cretaceous sequence PK2 is typically 0.2 s thick (300 m), but reaches thicknesses in excess of 1 s (1500 m) in local depocentres on the eastern margin of the basin. It contains an impressive series of prograding and downlapping packages towards the northern and eastern basin margins. High-ampli-

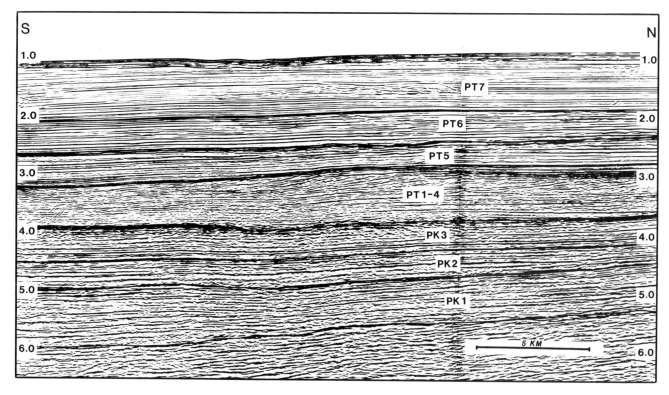

Fig. 3. Seismic section from the Porcupine Basin illustrating the seismic sequences. The deltaic sequences of the Lower Tertiary are seen as a series of southward prograding reflectors. The location of the profile is shown on Fig. 1. The vertical scale is in seconds two-way-travel time (TWT). In this figure, and in Fig. 5, the first letter of each sequence refers to the basin (P for Porcupine, R for Rockall); the second letter (K) refers to the Cretaceous (K) or Tertiary (T); while the third letter (T for Transition) or number (1–3) refers to the number of the sequence.

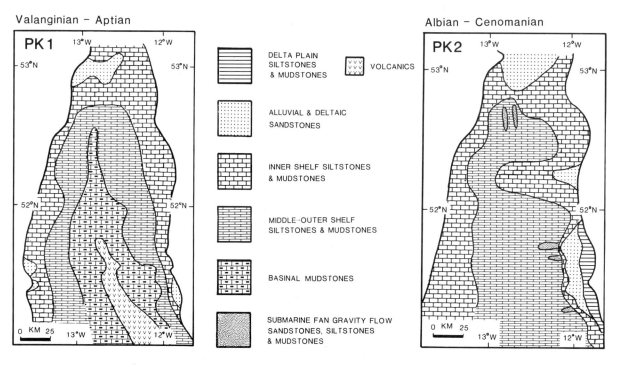

Fig. 4. Palaeogeographic reconstructions for the Lower Cretaceous seismic sequences PK1 and PK2 in the Porcupine Basin.

tude flat-lying reflectors give way southwards to a prograding series of high-amplitude discontinuous reflectors which, in turn, grade into low-amplitude continuous reflectors that extend through most of the centre and south of the basin. Well data confirm a latest Aptian and Albian age and indicate the presence of delta plain, delta front distributary mouth bar and shoreface sand-prone sediments grading southwards to shelf

sandstones, limestones and mudstones (Fig. 4). Deep water conditions are present only in the extreme south (Fig. 4). Occasional, small elongate mounded structures occur basinwards of the deltaic units (Fig. 4). These are interpreted as submarine channellized fan mounds, instigated by mass wastage and basinward slumping from the delta toe. The pattern of sedimentation during sequence PK2 appears to be asymmetric, with deltaic progradation seen only along the north and east margins of the basin.

The latest Aptian and Albian deltaic package is interpreted as reflecting a minor rift episode. It is coeval with the onset of the spreading ridge in the Bay of Biscay (Ziegler 1982). However, in the Celtic Sea basins (Fig. 1), closer to the spreading region, the Albian was a period of tectonic quiescence with relative sea-level rise and devoid of deltaics (Shannon 1991c). Therefore it is unlikely that the deltaics are due to northward-directed ridge-push.

The top Cretaceous seismic sequence (PK3) is characterized by weak continuous reflectors. A strong, basinwide reflector occurs at the top, onlaps the underlying sequences and extends further onto the basement than the underlying strata. The sequence has a uniform seismic character and is typically up to 0.4 s (600 m) thick. Well data from the north of the basin indicate it to be of chalk facies and Cenomanian to Lower Paleocene age. Towards the basin edges it oversteps beyond the pre-Upper Cretaceous basin limits and lies on basement. It marks the resumption of thermal subsidence. Towards the top of the sequence, especially on the eastern margin of the basin, a number of mounded structures occur (Shannon 1991b). These

are interpreted as reefal structures, similar to those recorded in the Aquitaine Basin (Bally 1983).

In the North Porcupine Basin the Chalk is thick and internally homogeneous. It is seismically transparent and displays no internal reflectors. The sequence thins across the dividing boundary ridge and then thickens southwards into the Main Porcupine Basin where it displays internal variations. It is overlain across the region (including the boundary ridge) by relatively uniform Lower Tertiary strata. It is suggested that the dividing ridge was a tectonically active barrier between the two basins in Late Cretaceous to earliest Tertiary times.

Rockall Trough and Hatton–Rockall Basin

The sedimentary succession in the Rockall Trough and Hatton–Rockall Basin is difficult to elucidate due to the poor quality of the seismic data and the paucity of well information. The recent RAPIDS refraction data (Makris et al. 1991) suggest a sedimentary pile up to 5 km thick in both the Rockall Trough and the Hatton–Rockall Basin (Fig. 5). The interpreted sedimentary succession in the Rockall Trough comprises three distinct layers. The oldest sedimentary layer is up to

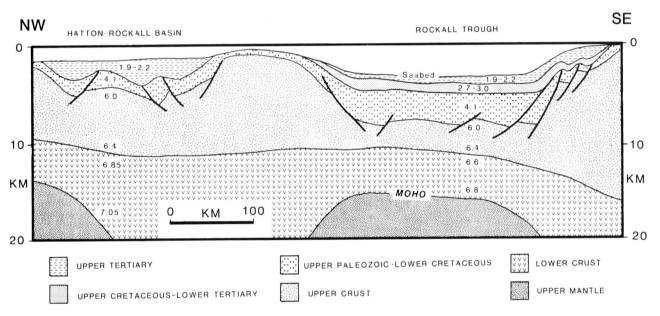

Fig. 5. Cross-section through the Rockall Trough and Hatton–Rockall Basin, based upon an interpretation of the RAPIDS refraction profile by Makris *et al.* (1991). The numbers refer to the refraction velocities (km s^{-1}). The location of the line is shown on Fig. 1.

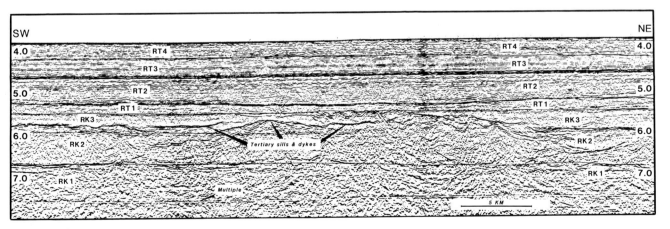

Fig. 6. Seismic section from the Rockall Trough illustrating the sequences described in the text. The character of the pre-Tertiary sequences is frequently masked by the effects of Tertiary sills and dykes. Multiples (indicated) occur in the deeper part of the section. The location of the profile is shown on Fig. 1. The vertical scale is in seconds (TWT).

3.5 km thick and has a velocity of 4.1 km s^{-1}. It extends across the basin and varies in thickness, especially towards the eastern margin. It is thought to range in age from Upper Paleozoic to Lower Cretaceous. The pronounced topography seen in both the Rockall Trough and the Hatton–Rockall Basin at this level is interpreted as a syn-rift, tilted, fault block geometry.

Seismic interpretation by Bentley and Scrutton (1987) suggested that up to 1.2 km of ?Lower Cretaceous to Jurassic strata occur in parts of the Rockall Trough. The region contains Tertiary volcanic sills and dykes (Naylor and Shannon 1982; Joppen and White 1990) which mask the seismic character of the underlying strata. Occasional seismic 'windows', in areas with less abundant igneous bodies, reveal packages of coherent reflectors beneath this section. The available seismic reflection lines from the Rockall Trough allow the definition of three pre-Tertiary seismic sequences (Figs 2 and 6).

The deepest sequence (RK1) is only locally clearly imaged (Fig. 6) and consists of packages of sub-horizontal to shallowly dipping reflectors. Its thickness is unknown. It is thought to be of Lower Cretaceous and older age. It corresponds to the 'layered basement' of Joppen and White (1990) and is thought to comprise an interbedded sequence of sediments and volcanics. It may correspond chronostratigraphically to the Barra Volcanic Zone of Bentley and Scrutton (1987).

The second sequence (RK2) is typically less than 1.2 s (3 km) thick. It is seismically transparent with few coherent reflectors. Its top, frequently masked by interpreted Tertiary sills and dykes (Fig. 6), is only convincingly seen in a few areas. It is interpreted as a Lower Cretaceous succession but the absence of internal character makes it impossible to elucidate its sedimentological make-up.

The third sequence (RK3), typically approximately 0.45 s (500 m) thick, is characterized by low-amplitude, continuous reflectors. Its top corresponds to the 'blue' reflector of Masson and Kidd (1986), thought to be of Upper Cretaceous age. The sequence is frequently cut by irregular, discontinuous high-amplitude seismic reflectors. These sometimes cross-cut the stratigraphy (Fig. 6) and are interpreted as early Tertiary sills and dykes. Sequence RK3 is interpreted as of Upper Cretaceous age, but the slightly differing seismic character, with continuous, albeit low-amplitude, reflectors, suggests a more muddy chalk facies than in the Porcupine Basin.

Tertiary development

The seismic and well data from the Porcupine Basin allows a detailed assessment of Tertiary development. Seismic data from Rockall Trough and the Hatton–Rockall Basin is of reasonable quality at Tertiary level, as it is above the level of the igneous bodies. Some well data, notably from the DSDP sites (Fig. 1, Naylor and Shannon 1982; Masson and Kidd 1986) assist in dating the sequences. Up to 4 km of Tertiary strata are preserved in the Porcupine Basin, and some 2 km in Rockall Trough and the Hatton–Rockall Basin. Our nomenclature for the seismic sequences is similar to that for the Cretaceous, with the basin name (P or R), followed by T (Tertiary) and the sequence number (1–7).

Porcupine Basin

Five seismic sequences (PT1–5) have been mapped within the Lower Tertiary and two (PT6–7) in the Upper Tertiary (Figs 2 and 3). Lower Tertiary sequences are each typically up to 0.5 s (600 m) in thickness, while Upper Tertiary sequences are each up to 0.8 s (300 m) thick.

The earliest Tertiary sequence (PT1) is of Paleocene age. Variable amplitude, continuous, reflectors in the north give way southwards to downlapping prograding foresets. These pass basinwards to continuous, low-amplitude, flat-lying reflectors. The succession is interpreted as a deltaic complex of delta plain mudstones and sandstones approximately 75 km in extent, giving way southwards to a 25 km wide belt of delta front sandstones and then to open marine strata which extend throughout most of the basin (Fig. 7). South of the prograding reflectors, small mounded deposits are interpreted as gravity flow deposits shed southwards from the deltaic complex.

Lower Tertiary sequence (PT2), of lowermost Eocene age, rests with slight unconformity on the underlying sequence, and is similar in its overall geometry. The interpreted deltaic complex extends approximately 20 km further south than the preceding sequence (Fig. 7).

Lower Tertiary sequence (PT3), of middle Eocene age, has a slightly unconformable base and a similar seismic character to the previous two sequences. It represents the most significant progradational phase of the deltaic complex. A broad asymmetric band, up to 75 km wide, of progradational sand-prone delta front sandstones occurs (Fig. 7). The major source area remained to the north, but input from the northwestern basin margin also occurred. In the southwest this sequence is dominated by mounded deposits with intermound canyons and channel fill depoists (Shannon 1992). These are interpreted as submarine fans (Fig. 7).

Lower Tertiary depositional sequence PT4, of uppermost Eocene age, marks a significant retreat in the deltaic complex. Deltaic deposits were restricted to the extreme north of the basin, with medium amplitude, moderately continuous, concordant and relatively flat-lying reflectors extending throughout most of the basin. On the southeastern margin of the basin a series of mounded and incised submarine fan sequences prograded westwards from the shelf edge (Shannon 1992). On the southwestern margin of the basin thinner elongate submarine fan complexes are developed (Fig. 7).

Paleocene–Eocene evolution The Paleocene–Eocene deltaics and coeval submarine fan complexes in the Porcupine Basin represent a set of regressive epidsodes in the basin, which cannot be explained in terms of a simple thermal subsidence basin. Louis and Mermey (1979) suggested that the deltaic deposition took place during a relative sea-level rise, while subsidence curves from a number of Porcupine wells (White *et al.* 1992) indicate a significant phase of rapid subsidence at the base of the Eocene. A range of possible causes could be suggested for these events. While they may have been instigated by Alpine transpressive events similar to those which caused major basin inversion in the North Celtic Sea Basin (Shannon 1991c), this inversion episode is thought to have ended by Eocene times. Alternatively, they may reflect the development of the Iceland hot spot (White and McKenzie 1989). However, the Porcupine Basin lay at the eastern limits of the zone of influence of this feature, and uplift associated with the hot spot is likely to be smaller than that suggested by the magnitude of the deltaics and submarine fans. In both these interpretations it is difficult to explain the observed subsidence pattern. It is therefore suggested that the sediments reflect ridge-push effects associated with spreading ridge adjustments in the North Atlantic. Spreading commenced at this time in the Greenland Sea (Ziegler 1982). The ridge-push is suggested to have produced non-faulted basin margin uplift to provide sediment source areas and rapid basin subsidence.

Sea-level curves (Haq *et al.* 1987) for the Paleocene and Eocene suggest a number of eustatic sea-level falls (Fig. 2). These correspond closely to the Paleocene–Eocene sequence boundaries in the Porcupine Basin. These lowstands may have triggered the instigation of the deltaic and submarine fan pulses. Such eustatic events are likely to have had a more significant influence on the Porcupine Basin, where water depths were shallower and where richer sediment source areas

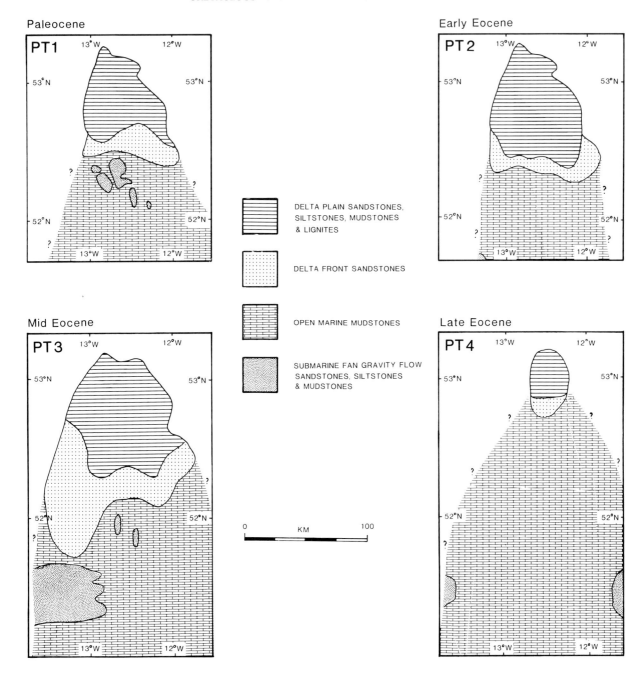

Fig. 7. Palaeogeographic reconstructions for the Lower Tertiary seismic sequences PT1, PT2, PT3 and PT4 in the Porcupine Basin.

were available, than further west in the Rockall and Hatton–Rockall basins.

In Late Eocene–Early Oligocene times the Labrador Sea spreading ridge ceased and the main spreading axis moved eastwards to the Svalbard–Greenland sector (Ziegler 1982). The unconformity at the base of the PT5 sequence coincides with this period. Bentley (1986) mapped a reflector of comparable age in parts of the Rockall Trough. This is confirmed in the present work but the unconformity cannot be mapped regionally with confidence on the basis of the presently available data (see *Rockall Trough*, below). It is therefore unlikely that the unconformity is due entirely to spreading ridge adjustments. The unconformity is directly overlain by deep marine strata and may represent, in part at least, geostrophic effects which developed when the enclosed basin had deepened sufficiently to allow the development of circulating currents.

Oligocene–Pliocene evolution The three overlying seismic sequences (Fig. 2), separated by unconformities, have broadly

similar internal characteristics. All have continuous, moderate amplitude seismic reflectors. They dip gently basinwards and are devoid of significant faults. They contain occasional disturbed zones, channel features and mounded structures. Subsidence outstripped sedimentation in post-Eocene times.

Sequence PT5, of Oligocene age, comprises an outer shelf succession in the north. Further south it is dominated by deep marine basinal facies, cut by deep marine channels and with occasional turbidite mounds (Fig. 8). A large slump complex is developed towards the western margin of the basin (Fig. 8).

The intrusion of sills and dykes occurred during the Oligocene (Seemann 1984; Tate and Dobson 1988). These are recognized on seismic sections as high-amplitude discontinuous reflectors that sometimes transect the stratigraphy.

Upper Tertiary sequence PT6 is dominated by outer shelf to basinal facies. It contains a series of basinward-directed mud-prone slump structures along the western margin (Moore and Shannon 1991) and across the centre of the basin (Fig. 8).

Occasional channels interrupt the shelf to basinal low-

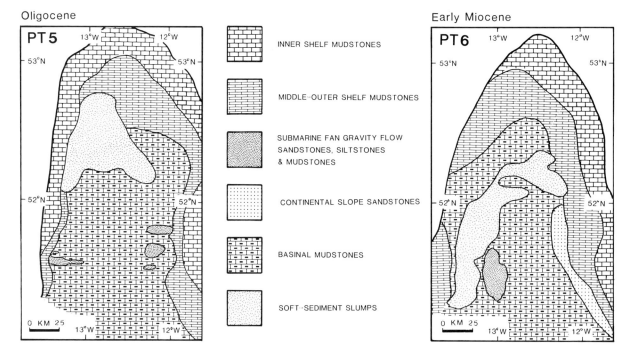

Fig. 8. Palaeogeographic reconstructions for Upper Tertiary seismic sequences PT5 and PT6 in the Porcupine Basin.

energy facies of the uppermost Upper Miocene to Pliocene sequence (PT7). It typically comprises a series of basinal shale-prone deposits, while occasional soft-sediment slump structures occur towards the basin centre.

Rockall Trough

Four seismic sequences (RT1–4) occur within the interpreted Tertiary succession in Rockall Trough (Figs 2 and 5). These appear to correlate to horizons identified in DSDP sites 550 and 610 (Masson and Kidd 1986). However, the long distances involved in such correlation, together with observed changes in reflector character of the lower packages, makes the age assignment somewhat uncertain.

Sequence RT1, approximately 0.3 s (400 m) thick, consists of high-amplitude events of very good continuity. It is interpreted as of Lower Tertiary age, and corresponds, at least in part, to the deltaic progradational sequence PT1 in the Porcupine Basin. However, it has a different character and lacks the typical prograding nature of the coeval Porcupine sequences. We suggest that it consists of a marine shale/limestone facies, probably developed in a deeper water facies than the comparable sequences in the Porcupine Basin.

The second sequence (RT2) is typically of the order of 0.65 s (750 m) thick. Reflector character is variable, with occasional downlapping sequences observed. High-amplitude reflectors are sometimes associated with a lower, occasionally down-lapping, sequence in the north of the basin. Further south, high-amplitude continuous events occur towards the top of the sequence while the downlapping character is replaced by a parallel series of poor reflectors. This sequence is correlated with Eocene–Miocene sequences PT2–6 in the Porcupine Basin. Its base generally corresponds to the 'brown' horizon of Masson and Kidd (1986).

Seismic sequence RT3 is approximately 0.45 s (500 m) thick and has continuous, high-amplitude, flat-lying events. Occasional small mounds occur. The succession is interpreted as of marine deep water origin. It rests with slight unconformity on the underlying package and is correlated with the lower part of

the Miocene sequence PT7 in the Porcupine Basin. Its base corresponds generally with the 'green' reflector of Masson and Kidd (1986).

The uppermost seismic sequence (TR4), typically 0.4 s (350 m) thick, is marked at its base by a local unconformity. The sequence is characterized by lower amplitude reflectors than the underlying package. The reflectors are often continuous and are interpreted as of deep marine origin. The sequence is correlated with the Pliocene part of sequence PT7 in the Porcupine Basin. Its base corresponds generally to the 'yellow' reflector of Masson and Kidd (1986).

On the RAPIDS refraction model (Makris *et al.* 1991) the two upper layers, with velocities of 1.9–2.2 km s^{-1} and 2.7–3.0 km s^{-1} respectively (Fig. 5), are each approximately 1 km thick in the basin centre. The Hatton–Rockall Basin differs from the Rockall Trough in that the 2.7–3.0 km s^{-1} layer is largely absent. The upper low-velocity layer (1.9–2.2 km s^{-1}) is up to 3 km thick and pinches out onto the flanks of the basin.

The two upper lower velocity layers in the Rockall Trough correlate well with seismic reflection profiles in the Rockall Trough. The upper layer corresponds to Oligocene and Upper Tertiary rocks, whereas the parallel lower layer probably represents pre-Oligocene to Upper Cretaceous rocks.

The Upper Cretaceous to Lower Tertiary succession thins westward along the RAPIDS profile. It is approximately 2 km thick in the Porcupine, is interpreted as 1 km thick in the Rockall Trough, and at least locally is absent in the Hatton–Rockall Basin. This is interpreted as the result of intrusion-related thermal uplift associated with the Iceland hot spot (White and McKenzie 1989) and perhaps with the adjacent North Atlantic spreading ridges. Higher refraction velocities (suggestive of igneous intrusions or underplating) in the lower crust beneath the Hatton–Rockall Basin compared to the Rockall Trough (Fig. 5) lend further credence to this interpretation. The resultant thermal buoyancy in the Hatton–Rockall Trough, and to a lesser extent in Rockall Trough, may have lessened the thermal subsidence during Cretaceous and Tertiary times, resulting in the thinner strata in these basins than in the Porcupine Basin further east.

Conclusions

(1) Up to 9 km of Cretaceous and Tertiary marine sandstones, chalks and mudstones are preserved in the Porcupine Basin, with 2–3 km of shale-prone marine strata in the Rockall and Hatton–Rockall basins.

(2) The Cretaceous and Tertiary histories of the basins display many characteristics of thermal subsidence basins which developed as successor basins to pre-Cretaceous rift basins.

(3) Major differences in seismic facies and sedimentary thickness between the various basins are attributed to local rift episodes, to ridge-push effects associated with spreading ridge readjustments in the developing North Atlantic Ocean and to igneous-driven thermal effects of the Iceland hot spot. Eustatic effects played a minor, but discernible, role in the development of facies in the basins.

The authors express their thanks to GECO-PRAKLA Exploration Services for permission to publish Fig. 3 which was part of a speculative seismic survey. Grateful thanks are also due to the Petroleum Affairs Division of the Irish Department of Energy who provided the authors with seismic data from the Rockall region for use in the RAPIDS project. Figure 6 is part of one of these lines.

References

BALLY, A. W. 1983. Seismic Expression of Structural Styles. *American Association of Petroleum Geologists, Studies in Geology* 15, 1.2.3.

BENTLEY, P. A. D. 1986. *Geophysical Studies in the Southern and Central Rockall Trough, Northeast Atlantic.* PhD Thesis, University of Edinburgh

—— AND SCRUTTON, R. A. 1987. Seismic investigations into the basement structure of southern Rockall Trough. *In:* BROOKS, J. AND GLENNIE, K. (eds) *Petroleum Geology of North West Europe.* Graham & Trotman, London, 667–675.

COOK, D. R. 1987. The Goban Spur—exploration in a deep-water frontier basin. *In:* BROOKS, J. AND GLENNIE, K. (eds) *Petroleum Geology of North West Europe.* Graham & Trotman, London, 623–632.

CROKER, P. F. AND KLEMPERER, S. L. 1989. Structure and stratigraphy of the Porcupine Basin: relationships to deep crustal structure and the opening of the North Atlantic. *In:* TANKARD, A. J. AND BALKWILL, H. R. (eds) *Extensional Tectonics and Stratigraphy of the North Atlantic Margin.* American Association of Petroleum Geologists Memoir, 46, 445–460.

—— AND SHANNON, P. M. 1987. The evolution and hydrocarbon prospectivity of the Porcupine Basin, Offshore Ireland. *In:* BROOKS, J. AND GLENNIE, K. (eds) *Petroleum Geology of North West Europe.* Graham & Trotman, London, 633–642.

HAQ, B. U., HARDENBOL, J. AND VAIL, P. R. 1987. Chronology of fluctuating sea-levels since the Triassic. *Science*, 235, 1156–1167.

JOPPEN, M. AND WHITE, R. S. 1990 The structure and subsidence of Rockall Trough from two-ship seismic experiments. *Journal of Geophysical Research*, 95, 19821–19837.

LOUIS, P. R. AND MERMEY, P. 1979. An example of seismic stratigraphy. The Porcupine Basin—Western Ireland. Paper presented at *49th Annual International Meeting of the Society of Exploration Geophysicists, New Orleans.*

MACDONALD, H., ALLAN, P. M. AND LOVELL, J. P. B. 1987. Geology of oil accumulation in Block 26/28, Porcupine Basin, offshore Ireland. *In:* BROOKS, J. AND GLENNIE, K. (eds) *Petroleum Geology of North-West Europe.* Graham & Trotman, London, 643–651.

MAKRIS, J., EGLOFF, R., JACOB, A. W. B., MOHR, P., MURPHY, T. AND RYAN, P. 1988. Continental crust under the southern Porcupine Seabight west of Ireland. *Earth and Planetary Science Letters*, 89, 387–397.

——, GINZBURG, A., SHANNON, P. M., JACOB, A. W. B., BEAN, C. J. AND VOGT, U. 1991. A new look at the Rockall region, offshore Ireland. *Marine and Petroleum Geology*, 8, 410–416.

MASSON, D. G. AND KIDD, R. B. 1986. Revised Tertiary stratigraphy of the southern Rockall Trough. *Initial Reports of the Deep Sea Drilling Project*, 80, 1115–1139.

MEGSON, J. B. 1987. The evolution of the Rockall Trough and implications for the Faeroe-Shetland Trough. *In:* BROOKS, J. AND GLENNIE, K. (eds) *Petroleum Geology of North West Europe.* Graham & Trotman, London, 653–665.

MOORE, J. G. 1992. A syn-rift to post-rift transition sequence in the Main Porcupine Basin, offshore western Ireland. *In:* PARNELL, J. (ed.) *Basins on the Atlantic Seaboard: Petroleum Geology, Sedimentology and Basin Evolution.* Geological Society, London, Special Publication, 62, 333–349.

—— AND SHANNON, P. M. 1991. Slump structures in the Late Tertiary of the Porcupine Basin, offshore Ireland. *Marine and Petroleum Geology*, 8, 184–197.

NAYLOR, D. AND ANSTEY, N. A. 1987. A reflection seismic study of the Porcupine Basin, offshore west Ireland. *Irish Journal of Earth Sciences*, 8, 187–210.

—— AND SHANNON, P. M. 1982. *The Geology of Offshore Ireland and West Britain.* Graham & Trotman, London.

ROBERTS, D. G., MASSON, D. G. AND MILES, P. R. 1981. Age and structure of the southern Rockall Trough—new evidence. *Earth and Planetary Science Letters*, 52, 115–128.

——, GINZBERG, A., NUNN, K. AND McQUILLIN, R. 1988. The structure of the Rockall Trough from seismic refraction and wide-angle reflection measurements. *Nature*, 332, 632–635.

RUSSELL, M. J. AND SMYTHE, D. K. 1978. Evidence for an early Permian oceanic rift in the northern North Atlantic. *In:* NEUMANN, E.-R. AND RAMBERG, I. B. (eds). *Petrology and Geochemistry of Continental Rifts.* Riedel, Dordrecht, 1973–179.

SANGREE, J. B. AND WIDMIER, J. M. 1977. Seismic stratigraphy and global changes of sea-level, Part 9: seismic interpretation of clastic depositional facies. *In:* PAYTON, C. E. (ed.) *Seismic stratigraphy—applications to hydrocarbon exploration.* American Association of Petroleum Geologists Memoir, 26, 165–184.

SEEMANN, U. 1984. Tertiary intrusives on the Atlantic continental margin off Southwest Ireland. *Irish Journal of Earth Sciences*, 6, 229–236.

SHANNON, P. M. 1991a. The development of Irish offshore sedimentary basins. *Journal of the Geological Society, London*, 148, 181–189.

—— 1991b. Irish offshore basins: geological development and petroleum plays. *In:* SPENCER, A. M. (ed.) *Generation, Accumulation and Production of Europe's Hydrocarbons.* Special Publication of the European Association of Petroleum Geoscientists No. 1. Oxford University Press, 99–109.

—— 1991c. Tectonic framework and petroleum potential of the Celtic Sea, Ireland. *First Break*, 9, 107–122.

—— 1992. Early Tertiary submarine fan deposits in the Porcupine Basin, offshore Ireland. *In:* PARNELL, J. (ed.), *Basins on the Atlantic Seaboard: Petroleum Geology, Sedimentology and Basin Evolution.* Geological Society, London, Special Publication, 62, 351–373.

SMYTHE, D. K. 1989. Rockall Trough—Cretaceous or Late Palaeozoic? *Scottish Journal of Geology*, 25, 5–43.

TATE, M. P. AND DOBSON, M. R. 1988. Syn- and post-rift igneous activity in the Porcupine Seabight basin and adjacent continental margin west of Ireland. *In:* MORTON, A. C. AND PARSON, L. M. (eds) *Early Tertiary Volcanism and the Opening of the NE Atlantic.* Geological Society, London, Special Publication, 39, 309–334.

—— AND —— 1989a. Pre-Mesozoic geology of the western and northwestern Irish continental shelf. *Journal of the Geological Society, London*, 146, 229–240.

—— AND —— 1989b. Late Permian to early Mesozoic rifting and sedimentation offshore NW Ireland. *Marine and Petroleum Geology*, 6, 49–59.

TRUEBLOOD, S. AND MORTON, N. 1991. Comparative sequence stratigraphy and structural styles of the Slyne Trough and Hebrides Basin. *Journal of the Geological Society, London*, 148, 197–201.

VAIL, P. R., MITCHUM, R. M. Jr., TODD, R. G., WIDMIER, J. M., THOMPSON, S., SANGREE, J. B., BUBB, J. N. AND HATLELID, W. G. 1977. Seismic stratigraphy and global changes of sea-level. *In:* PAYTON, C. E. (ed.) *Seismic Stratigraphy—Applications to Hydro-*

carbon Exploration. American Association of Petroleum Geologists Memoir, **26**, 49–212.

WHITE, N., TATE, M. P. AND CONROY, J.-J. 1992. Lithospheric stretching in the Porcupine Basin, west of Ireland. *In*: PARNELL, J. (ed.) *Basins on the Atlantic Seaboard: Petroleum Geology, Sedimentology and Basin Evolution*. Geological Society, London, Special Publication, **62**, 327–331.

WHITE, R. AND MCKENZIE, D. 1989. Magmatism at rift zones: the generation of volcanic continental margin and flood basalts. *Geophysical Research*, **94**, 7685–7729.

ZIEGLER, P. A. 1982. *Geological Atlas of Western and Central Europe*. Shell Internationale Petroleum Maatschappij B. V., The Hague.

Elevated palaeotemperatures prior to Early Tertiary cooling throughout the UK region: implications for hydrocarbon generation

P. F. GREEN,[1] I. R. DUDDY,[1] R. J. BRAY[2] and C. L. E. LEWIS[2]

[1] Geotrack International Pty Ltd, PO Box 4120, Melbourne University, Victoria 3052, Australia
[2] Geotrack International, 30 Upper High Street, Thame, Oxfordshire OX9 3EX, UK

Abstract: Elevated palaeotemperatures prior to Early Tertiary cooling, which affected wide areas of the UK region, have been revealed by Apatite Fission Track Analysis (AFTA™). All available evidence suggests that palaeogeothermal gradients were close to present values and that much of the observed heating was due to greater depth of burial, by 1 to 2 km or more of section that was subsequently removed by uplift and erosion. Uplift and erosion were not restricted to local inversion axes. The palaeotemperature data suggest a broad, regional warping, producing kilometre-scale Tertiary uplift and erosion across a wide area, within which recognized inversion axes represent local regions of maximum uplift and erosion. AFTA data show no thermal effects associated with Cimmerian unconformities, and any heating associated with Cimmerian events was of lesser magnitude than Late Cretaceous to Early Tertiary heating. Over much of the UK region, source rocks attained maximum temperatures and reached peak maturity during this later heating episode. The regional extent of heating at this time and its significance to hydrocarbon source rock maturation have not been fully recognized in the past. The timing of the events described here suggests a link to the development of the Atlantic margin, Laramide inversion tectonics and the onset of Alpine tectonism. However, definitive answers to such questions must await further research, particularly involving integration of AFTA and other thermal indicators with structural and geophysical data.

A number of studies have considered the thermal effects of continental rifting from a theoretical point of view (e.g. Weissel and Karner 1989 and references therein) and such effects have been identified in the field by Apatite Fission Track Analysis (AFTA) in a number of settings, e.g. SE Australia (Moore et al. 1986), the Red Sea (Bohannon et al. 1989) and the Gulf of Suez (Omar et al. 1989). These studies suggest that thermal manifestations of rifting, such as increased heat flow during extension or rift margin uplift and erosion, are restricted to a narrow region along the new continental margin. Igneous activity related to rifting can also cause localized heating effects. On the North Atlantic margin, where rifting and separation began in the Early Tertiary (Ziegler 1988, 1990), extensive volcanic activity is not restricted to the margin but extends into the continental interior in the form of the basaltic lavas and minor intrusions of the Tertiary igneous province. Thermal effects associated with rifting could play a significant role in determining the maturation history of suitable hydrocarbon source rocks in this region.

Rifting in the North Atlantic also coincides broadly in time with the Late Cretaceous to Early Tertiary 'Laramide' episode of basin inversion in Northwest Europe (Ziegler 1988, 1990), as well as the early development of the Alpine orogen (Fig. 1). Conventionally, the concept of inversion describes uplift due to reversal of movement along major structures defining discrete axes of inversion. This process is important in hydrocarbon exploration because greater depths of burial along these axes prior to inversion provided sufficient heating for hydrocarbon generation. Inversion can also produce potential trapping structures.

AFTA studies carried out over the last few years have shown that, over much of the UK region, Late Cretaceous to Early Tertiary heating was not restricted to the developing continental margin, to the vicinity of Tertiary igneous activity, or to discrete inversion axes. Elevated palaeotemperatures prior to Early Tertiary cooling, typically around 40 to 100° higher than present temperatures, have now been detected by application of AFTA over a wide region, including northern and eastern England and southern Scotland, the Irish Sea, the Moray Firth, and the Southern North Sea, in locations well away from any obvious source of heat and in rocks of various stratigraphic age from Jurassic to Precambrian.

The purposes of this paper are to review the AFTA evidence for this regional heating episode, to show that considerable support for this episode exists from other types of data and to discuss the implications of this episode to hydrocarbon generation.

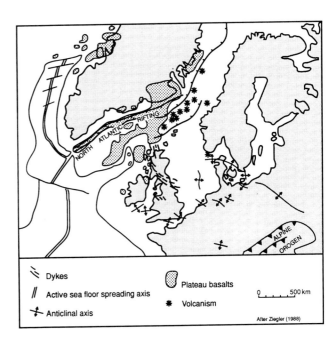

Fig. 1. Simplified Paleocene reconstruction of the UK region, based on Plate 17 of Ziegler (1988), emphasizing the coincidence in time of North Atlantic rifting, igneous activity, Laramide inversion and early development of the Alpine orogen. The onset of regional cooling from elevated palaeotemperatures identified by AFTA also coincides with these events, suggesting some causative link.

From *Petroleum Geology of Northwest Europe: Proceedings of the 4th Conference* (edited by J. R. Parker).
© 1993 Petroleum Geology '86 Ltd. Published by The Geological Society, London, pp. 1067–1074.

Evidence from AFTA for elevated palaeotemperatures prior to Early Tertiary cooling

AFTA is a relatively new method of thermal history analysis, applicable both to sediments and basement rocks, which provides a direct estimate of the time at which a rock began to cool from its maximum palaeotemperature, as well as providing an estimate of maximum palaeotemperature. In many situations, this knowledge is critical to the understanding of the timing of oil generation and migration in relation to trap formation. The kinetic response of the AFTA system is well understood, having been extensively studied and calibrated in laboratory conditions (Green *et al.* 1986, Laslett *et al.* 1987, Duddy *et al.* 1988) and verified by testing against geological constraints (Green *et al.* 1989*a,b*). In general, AFTA can provide estimates of maximum palaeotemperatures with an uncertainty of $\pm 10°C$ (Green *et al.* 1989*a*). The principles involved in the application of the technique have been described for example by Green (1989), Miller and Duddy (1989) and Kamp and Green (1991).

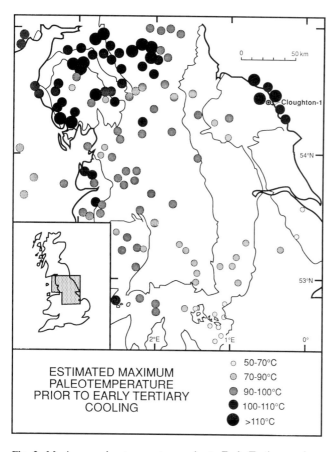

ESTIMATED MAXIMUM PALEOTEMPERATURE PRIOR TO EARLY TERTIARY COOLING

○ 50-70°C
◐ 70-90°C
◕ 90-100°C
● 100-110°C
● >110°C

Fig. 2. Maximum palaeotemperatures prior to Early Tertiary cooling estimated by AFTA in currently outcropping rocks from onshore UK (Hurford 1977; Green 1986, 1989; Lewis *et al.* 1992), and in wells from the Irish Sea (Lewis *et al.* 1992). Data from wells have been corrected to single values appropriate to outcrop level.

By applying AFTA, Green (1986) showed that samples of outcropping Caledonian basement from the Lake District of Northern England reached palaeotemperatures of 70–100°C or higher prior to cooling which began at around 60 Ma. In a subsequent study, Green (1989) applied AFTA to outcrop samples of Triassic to Carboniferous sedimentary rocks from the onshore East Midlands Shelf and the Pennine 'High', to outcropping Paleozoic basement rocks from Central England, and to samples of Jurassic to Carboniferous age from five wells from the East Midlands Shelf. Results again showed evidence

of a period of elevated palaeotemperatures, from which cooling began at around 60 ± 5 Ma. Maximum palaeotemperatures were between 50° and 100°C in outcrop samples, while samples from different wells reached maximum palaeotemperatures 40–70 °C higher than present temperatures.

As discussed by Bray *et al.* (1992), similar effects were subsequently detected in wells from the offshore portion of the East Midlands Shelf. Lewis *et al.* (1992) have found that the effects of elevated palaeotemperatures, prior to cooling beginning in the interval 65 ± 5 Ma, extend over NW England, the Isle of Man and the Irish Sea. (Note that the allowed range in this case allows the possibility of some cooling in the latest Cretaceous.) Early data from the Southern Uplands of Scotland (Hurford 1977) show that this area was also affected by the same event. Unpublished data show similar effects over the whole of NE England and North Wales. The limits of the area affected have yet to be defined. Further work is in progress towards this end.

Figure 2 shows a map of estimated maximum palaeotemperature in outcrop samples from Northern England, based on data from these papers and some unpublished data. Samples from wells have been reduced to a single estimate at outcrop level. This map emphasizes the regional nature of this heating episode, with palaeotemperatures showing little correlation with recognized structures.

Cause of elevated palaeotemperatures prior to Early Tertiary cooling

As discussed in detail by Bray *et al.* (1992), determination of palaeogeothermal gradients at the time of maximum palaeotemperatures can provide insight into the cause of heating and subsequent cooling. For example, if heating was due solely to a period of high heat flow, the geothermal gradient at the time of maximum temperature would be higher than the present-day gradient. Conversely if heating was caused solely by deeper burial followed by uplift and erosion with no change in heat flow, the palaeogeothermal gradient would be the same as the present gradient. Other causes of heating, such as igneous intrusions or passage of hot fluids, might be expected to produce non-linear palaeotemperature profiles.

Determination of maximum palaeotemperatures in a vertical sequence of samples allows direct estimation of the palaeogeothermal gradient. Using this approach, Green (1989) showed that palaeogeothermal gradients in five wells from the onshore East Midlands Shelf prior to Early Tertiary cooling were indistinguishable from present values. Bray *et al.* (1992) analysed these data statistically (using an approach based on maximum likelihood estimation) and came to a similar conclusion. Although data from each well were consistent with a range of values of palaeogeothermal gradient, the best estimates were in each case very close to the present value.

Bray *et al.* (1992) also analysed vitrinite reflectance (VR) data from four of the onshore East Midlands Shelf wells in which AFTA data were presented in Green (1989). Maximum palaeotemperatures determined by VR were generally within 5°C of those determined by AFTA, and the VR data provided estimates of palaeogeothermal gradient which were close to those derived from AFTA; data from an offshore East Midlands Shelf well presented by Bray *et al.* (1992) lead to a similar conclusion.

Therefore, all available evidence suggests that when elevated palaeotemperatures were established prior to Early Tertiary cooling, palaeogeothermal gradients were close to present values throughout the region. This in turn implies that heating was due to increased depth of burial, with subsequent cooling due to uplift and erosion through Tertiary time. Such an interpretation has previously been considered unreasonable for the East Midlands Shelf (Cope 1986), on the grounds that the predicted amounts of uplift (and erosion) were incompatible with the known geological framework. However, any alternative explanation for the observed heating and cooling must account for the observed similarity between present and palaeo-geothermal gradients, and any explanation other than heating due to burial, and cooling due to uplift and erosion, seems unlikely.

Supporting evidence for elevated palaeotemperatures prior to Early Tertiary cooling due to uplift and erosion

We have already mentioned that VR data from onshore and offshore East Midlands Shelf wells suggest similar palaeotemperatures and palaeogeothermal gradients to those indicated by AFTA. Roberts (1989) discussed VR data from a number of Irish Sea wells which he interpreted as indicating elevated palaeotemperatures prior to cooling beginning in the Early Tertiary due to removal of around 2 km of section by uplift and erosion. Fraser *et al.* (1990) and Fraser and Gawthorpe (1990) also discussed the importance of the effects identified by AFTA as discussed above, and Fraser *et al.* (1990) showed VR data from another onshore East Midlands Shelf well (Bardney-1) that show clear evidence of heating effects of similar magnitude to those identified in other East Midlands Shelf wells.

VR values from outcropping Kimmeridge Clay samples from Southern England to the Yorkshire coast, including the area of the onshore East Midlands Shelf covered by the AFTA study of Green (1989), are between 0.44 and 0.59% (Scotchman 1987). These values imply palaeotemperatures of 70° to 100°C for time scales of heating of around 1 Ma using the Burnham and Sweeney (1989) model for evolution of VR as a function of temperature and time. It seems clear from the AFTA results that cooling from these palaeotemperatures began in the Early Tertiary. Similarly, VR values from the Lower Lias of Dorset, Somerset and South Wales presented by Cornford (1986) increase northwards from 0.35 to 0.51%, implying palaeotemperatures of 60–80°C which were interpreted as largely due to burial. Based on geological evidence, Cornford (1986) considered the most likely timing for the onset of cooling from these palaeotemperatures was in the Late Cretaceous to Paleocene.

Cope (1986) reported that VR values from the offshore portion of the East Midlands Shelf were anomalously high, and predicted amounts of uplift (and erosion) which were inconsistent with the known geology of the region. In the light of the evidence discussed to date, together with AFTA and VR data from offshore East Midlands Shelf well 47/29a-1 presented by Bray *et al.* (1992), the 'anomalous' VR levels referred to by Cope (1986) can actually be understood as reflecting the regional burial episode identified by AFTA.

Cope (1986) also discussed the use of shale sonic velocity data to estimate the amounts of section removed by uplift and erosion. A number of studies have reported evidence of this sort pointing to deeper burial (e.g. Marie 1975 for the Sole Pit axis; Colter 1978 for the Irish Sea) but such studies have been largely concerned with identifying amounts of inversion by reference to supposedly stable reference regions. As discussed in detail by Bray *et al.* (1992), use of this technique often leads to an underestimation of the absolute amounts of section removed, as regions such as the East Midlands Shelf have been used as reference regions. Bulat and Stoker (1987) also recognized this potential problem and stressed the regional nature of uplift in the Southern North Sea. A compilation of data from onshore wells by Whittaker *et al.* (1985) also emphasized the regional nature of deeper burial and subsequent uplift and erosion.

In a study of Chalk porosity in wells from the Western Approaches Trough, Hillis (1991) showed that, despite the lack of recognizable inversion strucutres, data from many wells showed evidence of deeper burial. He concluded that 'Tertiary uplift (and erosion) is more widespread than suggested by the distribution of compressional/inversion structures'.

Evidence of deeper burial has also been reported from the Moray Firth area. Roberts *et al.* (1990) suggested that 300–750 m of section was removed during a Paleocene uplift phase, while Pearson and Watkins (1983) speculated on up to 500–700 m of uplift (and erosion). McQuillin *et al.* (1982) suggested tentatively that around 1 km of section had been removed from the Inner Moray Firth by Tertiary uplift and erosion although they preferred a Late Tertiary timing. However, the reasoning of Roberts *et al.* (1990) suggests an earlier timing, and unpublished AFTA studies in this region have confirmed that cooling by uplift and erosion began in the Early Tertiary. The Inner Moray Firth is located adjacent to the Shetland Platform, which has long been inferred to be the site of Paleocene uplift and erosion (e.g. Rochow 1981), and is thought to have acted as the source areas for the Paleocene sands in the North Sea. Hall (1991) has drawn attention to the importance of Early Tertiary (Paleocene to Early Eocene) uplift and erosion in a wider region of the Scottish Highlands. In addition Milton *et al.* (1990) showed, using seismic sequence stratigraphy, that even sites of Paleocene deposition in the North Sea on the flanks of the Shetland Platform were uplifted during the Paleocene.

In summary, many parts of the UK region show evidence of a significant episode of cooling which began in the Early Tertiary, either directly from palaeotemperature indicators or indirectly from inferred episodes of uplift and erosion. The evidence presented above is summarized in Fig. 3 and emphasizes that the episode of regional cooling identified by AFTA is supported by many diverse lines of evidence. It seems likely that the whole of the present onshore UK was affected to some degree. Further AFTA studies in progress support this conclusion.

At present, all available evidence suggests that the elevated palaeotemperatures prior to the onset of cooling were produced by deeper burial, and that cooling was due to uplift and erosion. The flow of hot fluids induced by igneous activity has been suggested as a potential explanation for the elevated palaeotemperatures prior to Early Tertiary cooling. However, the uniformity and regional extent of the thermal effects and the lack of any trends indicating maximum heating close to centres of igneous activity tend to argue against such an origin.

In the following discussion an interpretation in terms of heating due to depth of burial and cooling due to uplift and erosion will be assumed, in the absence of any viable alternative. While other interpretations should always be borne in mind, suitable mechanisms for producing the magnitude and regional extent of the observed heating are not obvious, and any potential explanations must account for the thermal history constraints established by AFTA and VR data.

A revised picture of the Late Cretaceous to Early Tertiary geological development of the UK region?

The conventional picture of the Late Cretaceous evolution of the present onshore UK region, as described for example by Ziegler (1988, 1990), involves a progressive increase in the area of Chalk deposition in deep marine conditions with little terrigenous input. Structural highs such as the Pennines and the Lake District formed islands in the Chalk sea. Minor end-Cretaceous tectonism resulted in the removal of up to around 160 m of Chalk. Clastic sediments preserved above a generally low-angle unconformity record the recommencement of sedimentation in the Paleocene. Volcanic and intrusive activity of the Tertiary igneous province in the mid-Paleocene led to local contact heating and uplift and erosion associated with local doming. During the Paleocene, 'Laramide' inversion affected basinal areas in the Celtic Sea, Bristol Channel and the Hampshire and Channel Basins, as shown in Fig. 3.

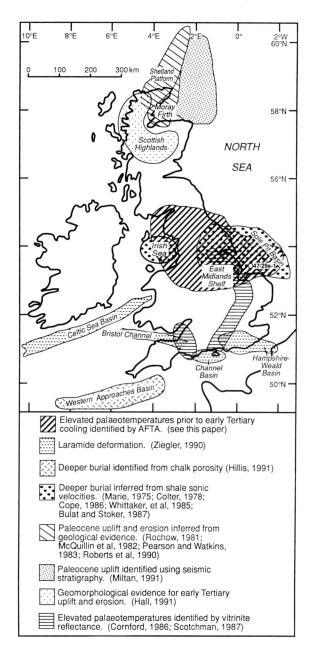

Fig. 3. Compilation of features of the Early Tertiary tectonic evolution of the UK area, as discussed in the text.

However, the picture that emerges from the evidence discussed in this paper is rather different. Palaeotemperatures indicated by both AFTA and VR data in wells such as 47/29a-1 and Cleethorpes-1 require around 1 km of deeper burial before uplift and erosion beginning at c. 60 Ma. As discussed by Bray *et al.* (1992), because thick Chalk sequences are preserved in these wells, the additional section removed during the Tertiary must have been deposited between deposition of the youngest preserved Chalk (c. 74 Ma) and the onset of uplift at c. 60 Ma. Figure 4 shows the post-Carboniferous burial history inferred for Well 47/29a-1 and the accelerated burial between 74 and 60 Ma required by the AFTA and VR data. This should be contrasted to the burial history drawn for this well by Glennie and Boegner (1981), who assumed that the section in this well had undergone no significant uplift and erosion. Over other parts of the UK, particularly where Meso-

zoic sediments are not preserved, it is more difficult to reconstruct the burial history, and greater thicknesses of Jurassic, Triassic and Permian units would be possible. However, based on current evidence, it seems reasonable that the broad style of burial history shown in Fig. 4 would be applicable to most of the present onshore UK.

A significant feature of the information provided by AFTA is that in areas previously thought to have been stable highs, such as the Pennines and the Lake District, Early Tertiary palaeotemperatures in rocks currently at outcrop are among the highest values at around 90–100°C or more. Thus following an interpretation of heating due to burial, these supposed highs must have been local depocentres prior to the onset of uplift and erosion in Paleocene times. Preliminary AFTA data from areas such as North Wales and the Southern Uplands, and possibly the Scottish Highlands, suggest that the same is true there also. Much debate over the years has centred on the extent of the chalk cover over these areas (e.g. George 1974) and it seems that this may have been underestimated in the past. The occurrence of the highest palaeotemperatures in the oldest rocks supports the idea that cooling was due to uplift and erosion, since they indicate deeper exhumation in these regions.

The cooling history for Well 47/29a-1 (Fig. 4) is based on the interpretation of AFTA from five onshore East Midlands Shelf wells presented by Green (1989), and involves equal amounts of cooling in two separate events at c. 60 and c. 30 Ma. In detail, the AFTA data allow only limited constraint on the style of cooling (which is reflected in the distribution of track lengths formed during the cooling phase of the history), and the style illustrated is undoubtedly an oversimplification. However, it is clear that the total cooling from maximum Early Tertiary palaeotemperatures to present-day temperatures did not happen suddenly, and a large proportion occurred in the last 30 Ma. Interpretation of AFTA data from the Irish Sea (Lewis *et al.* 1992) suggests a different style of cooling history with more rapid initial cooling prior to 50 Ma and further slower cooling to the present day. Discrimination between these different styles of cooling in the different regions is at the limits of the capabilities of AFTA techniques, but the data are suggestive of a real difference.

This raises the possibility of a series of discrete cooling events through the Tertiary which has affected different areas to different degrees. Further support for this notion is found in AFTA data from the Cleveland Basin which indicate a different cooling history to those identified in the Irish Sea and East Midlands. This is illustrated in Fig. 5 which shows AFTA data from the Cloughton-1 well compared to data from Irish Sea wells reported by Lewis *et al.* (1992). The thermal histories of comparable samples in the two sequences prior to the onset of cooling are probably roughly equivalent, but the Cleveland Basin data probably reflect an onset of cooling at c. 50 Ma, somewhat later than in the Irish Sea.

In the light of this, and bearing in mind the possibility of slightly earlier cooling in the Irish Sea (65 ± 5 Ma) than in the East Midlands (60 ± 5 Ma), it is possible that cooling took place in a series of discrete episodes rather than in a single event. As a corollary, the protracted nature of the cooling history throughout the region must argue against heating being due to igneous activity and/or associated hot fluid flow.

Since AFTA data only constrain the nature of the net cooling history, it is impossible to assess whether the episodic cooling reflects repeated cycles of uplift and erosion followed by reburial (by progressively lesser amounts of section so that the overall effect is still one of cooling) or simply progressive episodes of uplift and erosion. Bearing in mind the preserved Tertiary stratigraphy in the North Sea, it would seem likely that at least limited re-burial took place during an overall cooling trend. Thus it is possible that some of the sediment

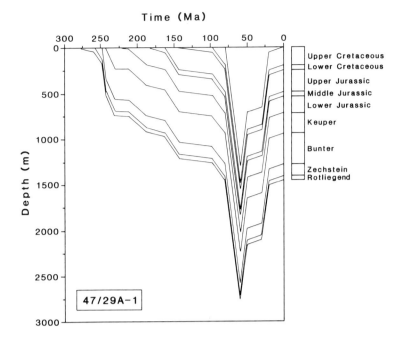

Fig. 4. Reconstructed burial history for the post-Carboniferous section in offshore East Midlands Shelf Well 47/29a-1, based on AFTA and VR data discussed by Bray *et al.* (1992). A similar style of history is inferred for much of the UK onshore region.

eroded from one area could have been deposited in another prior to later erosion in that area. Resolution of such complexity is beyond the scope of present techniques, but the possibilities should be borne in mind.

For the same reason, it is not possible to gain any insight into the thermal history of the area prior to the onset of cooling from maximum palaeotemperatures in the Early Tertiary. Thus it is impossible, for example, to constrain the magnitude of thermal effects associated with the Jurassic–Cretaceous Cimmerian unconformity throughout the region shown in Fig. 2, except to say that palaeotemperatures reached at that time were less than those reached prior to Early Tertiary cooling. In fact, AFTA studies so far in the UK region have shown no area where Cimmerian palaeotemperatures were higher than those reached prior to Early Tertiary cooling, despite the widely held view among explorationists that Cimmerian events were important in terms of hydrocarbon generation.

An interpretation involving kilometre-scale Tertiary uplift and erosion inevitably raises the question of the final destination of the eroded sediment. Since the offshore UK contains copious quantities of Tertiary sediment this does not appear to constitute a problem in qualitative terms. However, attempts to estimate amounts of erosion required to produce preserved sediment volumes seem to require less than the kilometre-scale figures required by the palaeotemperature data (e.g. Rochow 1981). Since the area that appears to have been affected by Tertiary uplift and erosion is so broad, it is possible that eroded sediments were transported large distances and are now in far offshore regions.

Finally it is worth speculating on the relationship between the regional uplift and erosion discussed here and the widely documented episodes of Late Cretaceous to Early Tertiary inversion in the UK and neighbouring regions, as discussed for example by Ziegler (1990). Recognized axes of inversion coincide with local palaeotemperature maxima, for example in the Cleveland Basin and Sole Pit basins, and the inversion movements and regional uplift and erosion seem to be clearly related. It is tempting to visualize the inversion movements as simply local 'ripples' riding on the back of more regional

uplifts, although more detailed work would be required to assess this rigorously.

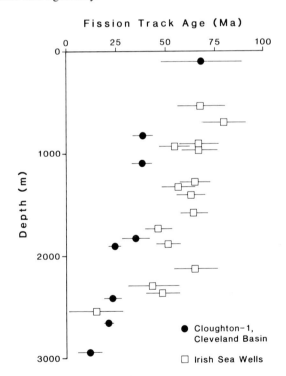

Fig. 5. AFTA data from Cloughton-1 in the onshore Cleveland Basin (unpublished data, courtesy of Clyde Petroleum plc), and a number of Irish Sea wells from Lewis *et al.* (1992). The clear difference in fission track ages from the two basins reflects differences in the time of cooling from maximum palaeotemperatures, as discussed in the text.

Relationship of regional heating and cooling to North Atlantic margin development

Numerous authors (e.g. Roberts 1974; Schwan 1980; Biddle

and Rudolph 1988; Dewey and Windley 1988; Ziegler 1990) have commented on the coincidence in time between Laramide inversion, the onset of Alpine compression, rifting in the North Atlantic between Scotland and Greenland, and activity in the UK Tertiary igneous province, suggesting some causative link between these processes. The onset of regional cooling in the onshore UK region at *c.* 60 Ma (mid-Paleocene) also coincides with these events and suggests that the inferred episode of uplift and erosion may be linked to these other events. Biddle and Rudolph (1988) particularly commented that the widespread distribution of Late Cretaceous to Early Tertiary inversion shows that the causative mechanism(s) operated on a regional scale, and in this regard regional uplift and erosion might be expected to be a logical consequence of the processes driving inversion.

As discussed in the introduction, thermal effects associated with continental rifts are generally thought to be limited to the vicinity of the developing rift flanks. However, the evidence discussed in this paper shows that, at the time of rifting in what was to become the North Atlantic, a region of some 2×10^5 km² at distances of up to 1000 km from the incipient continental margin experienced elevated temperatures, apparently due to greater depth of burial, followed by cooling apparently due to kilometre-scale uplift and erosion. Similar effects have been identified by AFTA in a number of analogous settings, such as the Appalachian Basin, USA (Miller and Duddy 1989), the Bowen Basin, Queensland (Marshallsea 1988), and Southern Africa (Brown *et al.* 1990). These studies all show regional cooling over a vast area more or less coincident with rifting in adjacent regions. Together with the results from the UK discussed here, they highlight what appears to be a new and important class of tectonic processes, which has received little or no attention to date.

One process which might be invoked as a cause of such effects is the development of 'Mantle Plumes', thought to be often associated with rifting and to cause kilometre-scale uplift over a broad region (e.g. White 1988). While such processes have obvious attractions as likely candidates, the offset of the area in which regional thermal effects have been detected in the UK region from the site of the eventual rift seems to pose a practical problem. In addition, while Mantle Plumes could readily bring about uplift (which might reasonably be accompanied by erosion), it is by no means obvious that they could also cause the observed heating with geothermal gradients close to present values required prior to the onset of cooling.

As emphasized by Ziegler (1987) and many papers in the same volume, Late Cretaceous and Early Tertiary compressional structures are widespread throughout northwestern Europe, particularly in the Alpine foreland region but extending up to around 1300 km from the Alpine front. Ziegler (1987, 1990) has speculated on the role of intra-plate compressional stresses transmitted through the continent from the Alpine orogen, and it seems likely that similar processes might be responsible for the effects discussed in this paper. The additional role of North Atlantic rifting might be significant in this regard, as the study area would have been subjected to compressional forces originating from both the northwest and southeast.

At present the nature of processes capable of producing regional heating, such as that observed in the UK and other regions listed above, is not clear. Whatever their origin, such processes are of great importance in hydrocarbon exploration, as explained in the next section.

Implications for hydrocarbon generation

Recognition of a widespread heating episode affecting many regions previously thought to have been relatively stable and cool throughout Mesozoic and Cenozoic times, implies that

hydrocarbon source rocks throughout the UK region reached much higher maturity levels than might be otherwise expected. This enhances the hydrocarbon prospectivity of regions where source rocks might otherwise be supposed to be immature or only marginally mature.

In a number of areas it is clear that source rocks attained maturity and generated oil or gas during heating towards maximum palaeotemperatures prior to Early Tertiary cooling. For example, Fraser *et al.* (1990) and Fraser and Gawthorpe (1990) discussed the generation of oil in the East Midlands of England and showed that oil was generated from Carboniferous source rocks during Mesozoic burial prior to Early Tertiary cooling. In the Irish Sea, the degree of heating was higher, which led to the generation of gas (Stuart and Cowan 1991). In both cases, Early Tertiary cooling effectively terminated the generation of hydrocarbons.

A key factor in the history of the East Midlands Shelf oilfields and the Irish Sea gas fields is that late generation occurred after formation of trapping structures, so that hydrocarbons could accumulate. In other areas of the UK, such as the Northern Pennines, maximum palaeotemperatures were reached in the late Carboniferous (Creaney *et al.* 1980), although late Cretaceous–Early Tertiary heating was also pronounced in this area with peak palaeotemperatures in currently outcropping rocks of around 100°C or more (Lewis *et al.* 1992). For Carboniferous source rocks in particular, the interplay between late Carboniferous and late Cretaceous–Early Tertiary heating is critical in determining the timing of hydrocarbon generation in relation to that of structure formation.

Results to date suggest that over much of Northern, Central and Eastern England, Carboniferous source rocks reached maximum palaeotemperatures, and therefore maximum maturity, prior to Early Tertiary cooling. This includes most areas of present or past exploration activity, including the East Midlands Shelf, Irish Sea, and Cheshire Basin. Previous studies may have seriously underestimated the magnitude of early Tertiary palaeotemperatures in some parts of this region, e.g. Barrett (1988) in the Solway Basin.

Some authors (Bushell 1986; Woodward and Curtis 1987; Stuart and Cowan 1991) have suggested that Cimmerian events may have produced significant maturity in the Irish Sea. Although Carboniferous source rocks in the Irish Sea reached maximum palaeotemperatures prior to Early Tertiary cooling, for reasons discussed earlier we cannot constrain the earlier thermal history (and therefore the history of hydrocarbon generation) directly from AFTA or VR data. Therefore, from the point of view of thermal constraints, the possibility of Cimmerian oil generation remains untestable and information can only be obtained from regional considerations. However, in studies from the UK region to date, no evidence has been found of any thermal manifestations of Cimmerian events.

Although the observed heating has been attributed in this paper to burial, it must be acknowledged that this interpretation is difficult to reconcile with the conventional picture of the geological evolution of the onshore UK region. However, it is also important to realize that evidence of heating is undeniable, in the sense that elevated palaeotemperatures prior to early Tertiary cooling have unequivocally been detected by AFTA, VR and other methods. Thus, whatever alternative interpretation is invoked to explain the heating, the considerations outlined so far in this section in respect of hydrocarbon generation are still valid.

Interpretation in terms of heating due to burial and cooling due to uplift and erosion is more important to questions such as migration pathways, re-migration as a result of discrete uplift periods during the cooling phases, and possible thermal and/or burial depth-related reservoir effects such as diagenetic reactions etc. From these aspects it is vital to ascertain the true origin of the regional heating and cooling discussed in this paper.

Conclusions

A wide region of the present onshore UK experienced an episode of regional cooling beginning in the Early Tertiary at *c.* 60 Ma. Palaeotemperatures in currently outcropping rocks prior to cooling were generally around 70°C or more over much of the region. Palaeogeothermal gradients when maximum palaeotemperatues were established appear to have been close to present values, suggesting that heating was due to burial by up to 1 or 2 km or more of section that was subsequently removed during cooling by uplift and erosion through the Tertiary. The onset of this cooling episode coincides with 'Laramide' inversion and Atlantic rifting. However, thermal effects are not localized. Heating and subsequent cooling affected a wide region, well away from inversion axes, and far inland from the developing continental margin.

Interpretation of this regional heating episode in terms of greater depth of burial requires a profound revision of the geological history of the UK region. In particular, areas previously thought to have been long-term stable highs, such as the Pennines, Lake District and North Wales, must have been local depocentres. No thermal expression of Cimmerian unconformities has been found anywhere in the present onshore UK region.

Recognition of this regional heating episode has major implications for hydrocarbon exploration. Over much of the UK region, source rocks reached maximum palaeotemperatures in this heating phase and in many areas significant oil and/or gas generation occurred. This event exerted profound control on hydrocarbon generation across the UK region and should be considered in appraisal of prospectivity.

The coincidence in time between the onset of regional cooling and events such as the initiation of North Atlantic rifting, Laramide inversion and the onset of Alpine compression suggests a causative link between all these events, although the processes responsible are not clear. Much work remains to be done, particularly involving integration of AFTA and other thermal indicators with structural and geophysical data, in order to understand fully the origin of this episode and its influence on hydrocarbon prospectivity throughout the UK region.

We are grateful to Clyde Petroleum plc, Ledbury, for permission to publish AFTA data from Cloughton-1. We would also like to thank Geoff Laslett, CSIRO Division of Mathematics and Statistics, Clayton, Victoria; and other colleagues in Geotrack and elsewhere, for their valuable input into the ideas presented in this paper.

References

BARRETT, P. A. 1988. Early Carboniferous of the Solway Basin: a tectonostratigraphic model and its bearing on hydrocarbon potential. *Marine and Petroleum Geology*, **5**, 271–281.

BIDDLE, K. T. AND RUDOLPH, K. W. 1988. Early Tertiary structural inversion in the Stord Basin, Norwegian Sea. *Journal of the Geological Society, London*, **145**, 603–611.

BOHANNON, R. G., NAESER, C. W., SCHMIDT, D. L. AND ZIMMERMAN, R. A. 1989. The timing of uplift, volcanism and rifting peripheral to the Red Sea: A case for passive rifting? *Journal of Geophysical Research*, **94**, 1683–1701.

BRAY, R., GREEN, P. F. AND DUDDY, I. R. 1992. Thermal history reconstruction using apatite fission track analysis and vitrinite reflectance: a case study from the UK East Midlands and Southern North Sea. *In*: HARDMAN, R. F. P. (ed.) *Exploration Britain: geological insights for the next decade*. Geological Society, London, Special Publication, **67**, 3–25.

BROWN, R. W., RUST, D. J., SUMMERFIELD, M. A., GLEADOW, A. J. W. AND DEWIT, M. C. J. 1990. An early Cretaceous phase of accelerated erosion on the south western margin of Africa: evidence from apatite fission track analysis and the offshore sedimen-

tary record. *Nuclear Tracks*, **17**, 339–350.

BULAT, J. AND STOKER, S. J. 1987. Uplift determination from interval velocity studies, UK southern North Sea. *In*: BROOKS, J. AND GLENNIE, K. (eds) *Petroleum Geology of North West Europe*. Graham & Trotman, London, 293–305.

BURNHAM, A. K. AND SWEENEY, J. J. 1989. A chemical kinetic model of vitrinite reflectance maturation. *Geochimica et Cosmochimica Acta*, **53**, 2649–2657.

BUSHELL, T. P. 1986. Reservoir geology of the Morecambe Field. *In*: BROOKS, J., GOFF, J. C. AND VAN HOORNE, B. (eds) *Habitat of Palaeozoic Gas in N.W. Europe*. Geological Society, London, Special Publication, **23**, 189–208.

COLTER, V. S. 1978. Exploration for gas in the Irish Sea. *Geologie en Mijnbouw*, **57**, 503–516.

COPE, M. J. 1986. An interpretation of vitrinite reflectance data from the Southern North Sea basin. *In*: BROOKS, J., GOFF, J. C. AND VAN HOORNE, B. (eds) *Habitat of Palaeozoic Gas in N.W. Europe*. Geological Society, London, Special Publication, **23**, 85–98.

CORNFORD, C. 1986. The Bristol Channel Graben: organic geochemical limits on subsidence and speculation on the origin of inversion. *Proceedings of the Ussher Society*, **6**, 360–367.

CREANEY, S. 1980. Petrographic texture and vitrinite reflectance variation on the Alston Block, North-East England. *Proceedings of the Yorkshire Geological Society*, **42**, 553–580.

DEWEY, J. F. AND WINDLEY, B. F. 1988. Paleocene–Oligocene tectonics of NW Europe. *In*: MORTON, A. C. AND PARSON, L. M. (eds) *Early Tertiary Volcanism and the Opening of the NE Atlantic*. Geological Society, London, Special Publication, **39**, 25–31.

DUDDY, I. R., GREEN, P. F. AND LASLETT, G. M. 1988. Thermal annealing of fission tracks in apatite 3. Variable temperature behaviour. *Chemical Geology (Isotope Geoscience Section)*, **73**, 25–28.

FRASER, A. J. AND GAWTHORPE, R. L. 1990. Tectono-stratigraphic development and hydrocarbon habitat of the Carboniferous in Northern England. *In*: HARDMAN, R. F. P. and BROOKS, J. (eds) *Tectonic Events Responsible for Britain's Oil and Gas Reserves*. Geological Society, London, Special Publication, **55**, 49–86.

——, NASH, D. F., STEELE, R. P. AND EBDON, C. C. 1990. A regional assessment of the intra-Carboniferous play of Northern England. *In*: BROOKS, J. (ed.) *Classic Petroleum Provinces*. Geological Society, London, Special Publication, **50**, 417–440.

GEORGE, T. N. 1974. The Cenozoic evolution of Wales. *In*: OWEN, T. R. (ed.) *The Upper Paleozoic and post-Paleozoic rocks of Wales*. University of Wales Press, Cardiff, 341–371.

GLENNIE, K. W. AND BOEGNER, P. L. E. 1981. Sole pit inversion tectonics. *In*: ILLING, L. V. AND HOBSON, G. D. (eds) *Petroleum Geology of the Continental Shelf of North-West Europe*. Heyden, London, 110–120.

GREEN, P. F. 1986. On the thermo-tectonic evolution of Northern England: Evidence from fission track analysis. *Geological Magazine*, **123**, 493–506.

—— 1989. Thermal and tectonic history of the East Midlands shelf (onshore UK) and surrounding regions assessed by apatite fission track analysis. *Journal of the Geological Society, London*, **146**, 755–773.

——, DUDDY, I. R., GLEADOW, A. J. W., TINGATE, P. R. AND LASLETT, G. M. 1986. Thermal annealing of fission tracks in apatite 1. A qualitative description. *Chemical Geology (Isotope Geoscience Section)*, **59**, 237–253.

——, ——, LASLETT, G. M., HEGARTY, K. A., GLEADOW, A. J. W. AND LOVERING, J. F. 1989*a*. Thermal annealing of fission tracks in apatite 4. Quantitative modelling techniques and extension to geological timescales. *Chemical Geology (Isotope Geoscience Section)*, **79**, 155–182.

——, ——, GLEADOW, A. J. W. AND LOVERING, J. F. 1989*b*. Apatite Fission Track Analysis as a paleotemperature indicator for hydrocarbon exploration. *In*: NAESER, N. D. AND MCCULLOH, T. (eds) *Thermal history of sedimentary basins—methods and case histories*. Springer, New York, 181–195.

HALL, A. M. 1991. Pre-Quaternary landscape evolution in the Scottish Highlands. *Transactions of the Royal Society of Edinburgh: Earth Sciences*, **82**, 1–26.

HILLIS, R. 1991. Chalk porosity and Tertiary uplift, Western Approaches Trough, SW UK and NW French continental shelves. *Journal of the Geological Society, London*, **148**, 669–679.

HURFORD, A. J. 1977. Fission track dates from two Galloway granites, Scotland. *Geological Magazine*, **114**, 299–304.

KAMP, P. J. J. AND GREEN, P. F. 1991. Thermal and tectonic history of selected Taranaki Basin (New Zealand) wells assessed by apatite fission track analysis. *AAPG Bulletin*, **74**, 1401–1419.

LASLETT, G. M., GREEN, P. F., DUDDY, I. R. AND GLEADOW, A. J. W. 1987. Thermal annealing of fission tracks in apatite 2. A quantitative analysis. *Chemical Geology (Isotope Geoscience Section)*, **65**, 1–13.

LEWIS, C. L. E., GREEN, P. F., CARTER, A. AND HURFORD, A. J. 1992. Elevated late Cretaceous to Early Tertiary paleotemperatures throughout Northwest England: three kilometres of Tertiary erosion? *Earth and Planetary Science Letters*, **112**, 131–145.

MARIE, J. P. P. 1975. Rotliegendes stratigraphy and diagenesis. *In*: WOODLAND, A. W. (ed.) *Petroleum and the Continental Shelf of North-West Europe*. Applied Science, London, 205–211.

MARSHALLSEA, S. J. 1988. *The thermal history of the Bowen Basin (Queensland): an AFTA study*. PhD Thesis, University of Melbourne.

MCQUILLIN, R., DONATO, J. A. AND TULSTRUP, J. 1982. Development of basins in the Inner Moray Firth and the North Sea by crustal extension and dextral displacement of the Great Glen Fault. *Earth and Planetary Science Letters*, **60**, 127–139.

MILLER, D. S. AND DUDDY, I. R. 1989. Early Cretaceous uplift and erosion of the northern Appalachian Basin, New York, based on apatite fission track analysis. *Earth and Planetary Science Letters*, **93**, 35–49.

MILTON, N. J., BERTRAM, G. T. AND VANN, I. R. 1990. Early Paleogene tectonics and sedimentation in the Central North Sea. *In*: HARDMAN, R. F. P. AND BROOKS, J. (eds) *Tectonic Events Responsible for Britain's Oil and Gas Reserves*. Geological Society, London, Special Publication, **55**, 339–351.

MOORE, M. E., GLEADOW, A. J. W. AND LOVERING, J. F. 1986. Thermal evolution of rifted continental margins: new evidence from fission tracks in basement apatites from southeastern Australia. *Earth and Planetary Science Letters*, **78**, 255–270.

OMAR, G. I., STECKLER, M. S., BUCK, W. R. AND KOHN, B. P. 1989. Fission-track analysis of basement apatites at the western margin of the Gulf of Suez rift: Egypt: evidence for synchroneity of uplift and subsidence. *Earth and Planetary Science Letters*, **94**, 316–328.

PEARSON, M. J. AND WATKINS, D. 1983. Organofacies and early maturation effects in Upper Jurassic sediments from the Inner Moray Firth, North Sea. *In*: BROOKS, J. (ed.) *Petroleum Geochemistry and the Exploration of Europe*. Blackwell, Oxford, 147–160.

ROBERTS, A. M., BADLEY, M. E., PRICE, J. D. AND HUCK, I. W. 1990. The structural history of a transtensional basin: Inner Moray Firth, N.E. Scotland. *Journal of the Geological Society, London*, **147**, 87–103.

ROBERTS, D. G. 1974. Structural development of the British Isles, continental margin and Rockall Plateau. *In*: BURK, C. A. AND DRAKE, C. L. (eds) *The Geology of Continental Margins*. Springer, New York, 343–359.

—— 1989. Basin inversion in and around the British Isles. *In*: COOPER, M. A. AND WILLIAMS, G. D. (eds) *Inversion Tectonics*. Geological Society, London, Special Publication, **44**, 131–150.

ROCHOW, K. A. 1981. Seismic stratigraphy of the North Sea 'Paleocene' deposits. *In*: ILLING, L. V. AND HOBSON, G. D. (eds) *Petroleum Geology of the Continental Shelf of North-West Europe*. Heyden, London, 255–266.

SCHWAN, W. 1980. Geodynamic peaks of Alpinotype orogenies and changes in ocean-floor spreading during Late Jurassic–Late Tertiary time. *American Association of Petroleum Geologists Bulletin*, **64**, 359–373.

SCOTCHMAN, I. C. 1987. Clay diagenesis in the Kimmeridge Clay Formation, onshore UK, and its relation to organic maturation. *Mineralogical Magazine*, **51**, 535–551.

STUART, I. A. AND COWAN, G. 1991. The south Morecambe Field, blocks 110/2a, 110/3a, 110/8a, UK East Irish Sea. *In*: ABBOTTS, I. L. (ed.) *United Kingdom Oil and Gas Fields 25 Years Commemorative Volume*. Geological Society, London, Memoir, **14**, 527–541.

WEISSEL, J. K. AND KARNER, G. D. 1989. Flexural uplift of rift flanks due to mechanical unloading of the lithosphere during extension. *Journal of Geophysical Research*, **94**, 13919–13950.

WHITE, R. S. 1988. A hot-spot model for early Tertiary volcanism in the N Atlantic. *In*: MORTON, A. C. AND PARSON, L. M. (eds) *Early Tertiary Volcanism and the Opening of the NE Atlantic*. Geological Society, London, Special Publication, **39**, 3–13.

WHITTAKER, A., HOLLIDAY, D. W. AND PENN, I. E. 1985. *Geophysical Logs in British Stratigraphy*. Geological Society, London, Special Report, **18**.

WOODWARD, K. AND CURTIS, C. D. 1987. Predictive modelling for the distribution of production-constraining illites—Morecambe Gas field, Irish Sea, offshore UK. *In*: BROOKS, J. AND GLENNIE, K. (eds) *Petroleum Geology of North West Europe*. Graham & Trotman, London, 205–215.

ZIEGLER, P. A. 1987. Late Cretaceous and Cenozoic intra-plate compressional deformations in the Alpine foreland—a geodynamic model. *Tectonophysics*, **137**, 389–420.

—— 1988. *Evolution of the Arctic–North Atlantic and the Western Tethys*. American Association of Petroleum Geologists Memoir **43**.

—— 1990. *Geological Atlas of Western and Central Europe* 1990 (2nd edn). Shell Internationale Petroleum Maatschappij B.V., The Hague.

Crustal extension, subsidence and inversion in NW Europe

A. H. RUFFELL[1] and M. P. COWARD[2]

[1]*Department of Geology, Queen's University, Belfast BT7 1NN, UK*
[2]*Department of Geology, Imperial College, London SW7 2BP, UK*

Examination of a regional grid of deep and commercial seismic data extending from Cardigan Bay (offshore Wales) to the south of the Paris Basin (Aquitaine–Provence–French Alps) has demonstrated that the deep Mesozoic sedimentary basins along this transect are underlain by low-angle crustal detachments. Such features are interpreted to be Variscan thrusts, reactivated as normal faults during Mesozoic crustal extension; they rise to surface in basement rocks surrounding Mesozoic basins, and can be matched to outcrop structures in S. Wales, S. Ireland, Cornwall, Devon and Brittany.

Inversion of the NW European Mesozoic basins occurred in the Cenozoic, but was preceded by minor inversion and uplift of surrounding massifs at various times in the Mesozoic. Such uplift can be demonstrated to have occurred via crustal shortening and re-thrusting along Variscan thrusts at depth. The uplift of basins above such thrusts is concomitant with continued subsidence of basins in the foreland, and can explain the juxtaposition of over-thick sedimentary successions against basins with a predominant unconformity of similar age. In the Celtic Sea area (offshore SE Eire), the North Celtic Sea Basin contains over 2000 m of late Jurassic–early Cretaceous sediments, whilst the South Celtic Sea Basin has virtually no late Jurassic or early Cretaceous preserved: a prominent Aptian–Albian unconformity overlies early–mid-Jurassic sediments. The two basins are separated by the Pembroke Ridge, an ancient 'high' where reactivated Variscan thrusts are thought to crop out (Coward and Trudgill 1989).

The Carboniferous–Permian–Triassic succession of the Plymouth Bay Basin (Western Approaches Trough) is another example of an 'overthickened' sedimentary succession in a basin underlain and bounded by crustal detachments of probable Variscan age. This largely non-marine basin-fill succession contains widely correlatable sedimentary sequences that correspond to successive episodes of infill. The provenance of this clastic material is the Cornubian Massif to the north and Brittany to the south: both uplifted (initially) during Variscan compression.

The Paris Basin is a deep (max. 4000 m) Mesozoic–Cenozoic basin, but unlike the Celtic Seas or Plymouth Bay basins, the timing of infill is less restricted and not associated with a widespread inversion. Models of Mesozoic–Cenozoic inversion in NW Europe involving compression from the Alps are thus difficult to apply: stress has to be applied either along the NW–SE transcurrent faults (e.g. Pays de Bray) or through surrounding massifs. Compression during plate rotation through oceanic separation may be a more plausible method of explaining minor inversion during the Mesozoic in areas north of the Paris Basin.

COWARD, M. P. AND TRUDGILL, B. 1989. Basin development and basement structure of the Celtic Sea Basins (SW Britain). *Bulletin de la Société Géologique de France*, 3, 423–436.

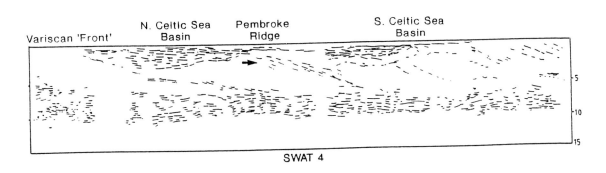

Fig. 1. Deep seismic profile across the Celtic Seas, showing low-angle reflections below Mesozoic basins.

From *Petroleum Geology of Northwest Europe: Proceedings of the 4th Conference* (edited by J. R. Parker).
© 1993 Petroleum Geology '86 Ltd. Published by The Geological Society, London, p. 1075.

Evolution of the English Channel Basin

A. H. RUFFELL,[1] M. HARVEY[2] and G. D. WACH[3]

[1]Department of Geology, Queen's University, Belfast BT7 1NN, UK
[2]Department of Geology, Imperial College, London SW7 2BP, UK
[3]Department of Earth Sciences, University of Oxford, Parks Road, Oxford OX1 3PR, UK

The Mesozoic Channel Basin is structurally linked to the main Wessex Basin (to the north), the Paris Basin (to the south) and Western Approaches Trough to the west. The Mesozoic (and Cenozoic) history of the English Channel area shows the complex interplay between crustal extension–subsidence and later inversion associated with intra-plate tectonics of the European area and the opening of the North Atlantic. Similar structural and sedimentological features are found in the Paris, Wessex, Western Approaches Trough and Channel basins, thus making the latter area the geographic focus for the development of this model of basin evolution.

The development of the Channel Basin is similar to the onshore Paris Basin and Wessex Basin. A syn-rift succession of Permian–Triassic non-marine sediments is succeeded by widespread marine Jurassic sediments of the thermal relaxation phase. Uplift of the basin margins and intra-basinal 'highs' occurred during the earliest Cretaceous, and was facilitated by reactivation of Variscan basement thrusts at depth. Consequently, a common structural/stratigraphic pattern can be observed across the Channel Basin: this comprises a prominent Aptian–Albian unconformity on faulted and folded Jurassic footwall highs, adjacent to thick early Cretaceous successions (Wealden) preserved in the hangingwall. The two successions are separated by Cretaceous–Tertiary inversion axes such as the Isle of Wight–Purbeck Monocline and

northern margin of the Brittany Basin. The final movement on such axes was Oligo-Miocene.

The timing of late Jurassic–early Cretaceous uplift is not absolute: as no sediments of this age are preserved on the footwall of uplifted blocks, the successions on the hangingwall have been studied in detail to assess the nature of unconformity surfaces and relate them to the known tectonic or eustatic changes. Minor unconformities observed in the Wealden succession at outcrop and on seismic data may be related to early uplift. These are subtle features compared to the prominent and widely correlatable early and mid-Aptian and early Albian unconformities associated with the first sediments preserved on footwall blocks since the Jurassic. A tectonic influence on unconformity development is apparent in the mid-Aptian and early Albian, whilst all three unconformities correlate with cycles of global sea-level change on the Exxon Chart.

Early Cretaceous uplift was a gradual-episodic tectonic process, accentuated by patterns of sediment preservation across highs. Minor inversion occurred in the early Albian and was associated with movement along NW–SE transcurrent faults in the basement, and the subsidence of former highs that precipitated the Cretaceous–Tertiary inversion. This early and gradual compression is likely to have been caused by plate rotation and compression associated with the opening of the North Atlantic.

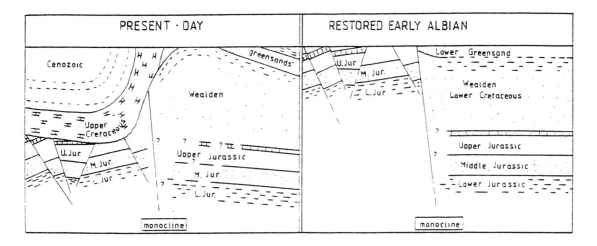

From *Petroleum Geology of Northwest Europe: Proceedings of the 4th Conference* (edited by J. R. Parker).
© 1993 Petroleum Geology '86 Ltd. Published by The Geological Society, London, p. 1077.

Structural styles
and their evolution in
the North Sea area

Structural styles and their evolution in the North Sea area

Introduction and review

I. D. BARTHOLOMEW

Oryx UK Energy Company, Charter Place, Vine Street, Uxbridge, Middlesex UB8 1EZ, UK

The understanding of the structural styles and their evolution in the North Sea area has been greatly advanced in recent years. In the past there were many differing ideas varying between the two extremes of listric dip-slip faulting to strike-slip faulting interpretation styles. Now more certain answers can be given from the marked increase in the use of 3D datasets. This is well demonstrated in the 'Structural styles' section.

The section is arranged by region with papers designed to present contrasting viewpoints and to provoke discussion. The section starts with papers covering the whole North Sea area, continuing with more regional papers moving geographically from north to south.

Regional papers

Following an introductory review talk by **Williams** on the structural models that have been described for the North Sea in the past and the still unanswered questions, **Coward** sets the regional framework of the area by reviewing the Late Caledonian and Variscan tectonic events that formed the underlying structural fabric to the Mesozoic and Cenozoic North Sea Basin. The key point of the paper is to stress that this underlying structural fabric subsequently controlled the faulting styles of the later evolution of the North Sea Basin. This key point is stressed again by **Bartholomew et al.** who make clear that, during the Mesozoic evolution of the North Sea area, the underlying structural grain was oblique to the regional extension direction in most areas. Therefore, the corresponding structures, even though formed in an extensional regime, had to have a component of strike-slip movement. It is demonstrated that even though this strike-slip component of movement was very minor compared to the overall extension, it was enough to cause the formation of pop-up and pull-apart structures in many areas of the North Sea. It is also demonstrated that the histories of these structures were very complex, many structures evolving differently through time depending on which underlying shear zone had been active at any one specific time.

On a different regional theme **Roberts et al.** show that a similar amount of regional extension is obtained using different techniques (flexural backstripping, forward modelling, and fault population statistical analysis). Jurassic extension is estimated to be 15% in the Viking Graben, and 20% in the Central Graben. It is suggested that these relatively low Jurassic extension values indicate that the Triassic extension was equally as important as the Jurassic extension in the stretching history of the North Sea.

East Shetland Basin papers

Three papers are presented from the East Shetland Basin area. **Lee and Hwang** use the East African Rift as an analogue to the East Shetland Basin, paying particular attention to the compartmentalization of the area into zones of similarly hading faults separated by accommodation zones. **Dahl and Solli** highlight the compartmentalization of the Snorre area within the East Shetland Basin by NW–SE and NE–SW cross-fault trends. Compressional features formed during the Triassic and Jurassic extensional episodes are highlighted by both papers as being of local importance. They are interpreted as accommodation features formed as a result of a component of strike-slip movement along either NW–SE or NE–SW fault zones within an overall E–W extensional tectonic regime. The paper by **Demyttenaere et al.** gives an illustration of both NW–SE and NE–SW cross-fault zones using a large 3D dataset over the Cormorant Field. Their importance in delineating and compartmentalizing the field is stressed.

Moray Firth papers

The important observation that the Great Glen Fault did not play a major part in the Mesozoic evolution of the Inner Moray Firth area is made by **Thomson and Underhill**. It was mainly active in the Cenozoic when strike-slip movements along this and other faults has been largely responsible for destroying the hydrocarbon prospectivity in many parts of the basin.

The paper by **Hibbert and Mackertich** describes the complex area where the Inner Moray Firth structural trends intersect with the Central Graben trends. The importance of the understanding of fault timings and trendology in the compartmentalization of relatively small areas is well illustrated from the 3D data over Block 15/21.

Central Graben papers

Seven papers are included on the Central Graben area. Three of these papers concentrate specifically on the Jæren High and East Central Graben and the complexities of the supposedly salt-controlled structures.

Høiland et al. discuss whether thin-skinned extension or salt dissolution was the major controlling factor on the distribution of Late Jurassic sediment traps on the Jæren High. They conclude that a combination of both processes had occurred: the extension being dominant in the Triassic, and then salt dissolution taking over in importance in the Late Jurassic. A similar conclusion is reached by **Penge et al.** for the East Central Graben area as a whole. They also describe the varying geometries of Triassic raft features, all of which initially formed as a result of post-Triassic layer-parallel extension, disconnected from the basement by the underlying Zechstein salt.

Erratt describes various examples of graben margin structures, mainly from the East Central Graben area. He shows that salt withdrawal, commencing in the Triassic, resulted in a complex network of grounded primary withdrawal synclines (Triassic 'pods') and intervening areas of salt preservation that were largely in place prior to the Late Jurassic rifting. A wide range of Upper Jurassic graben margin structures was therefore possible, depending on the location and size of the underlying basement fault relative to the location of the overlying Triassic 'pods' and salt swells.

In all the above papers, the importance, or otherwise, of any wrench component of movement in the evolution of the area is hardly considered. In contrast to this **Sears et al.** describe a

From *Petroleum Geology of Northwest Europe: Proceedings of the 4th Conference* (edited by J. R. Parker).
© 1993 Petroleum Geology '86 Ltd. Published by The Geological Society, London, pp. 1081–1082.

series of structural examples from the Central Graben, all of which are described as Upper Jurassic transtensional and transpressional features related to the oblique movement along deep-seated and much older NW–SE-trending basement faults in an E–W extensional regime. All the examples shown come from 3D seismic surveys. The paper by Stewart (in the Jurassic section) describes, again from 3D seismic survey data, similar such transtensional and transpressional features from the Ula and Gyda areas of the Norwegian Central Graben area. Extensive 3D seismic is also used by **Sundsbø and Megson** to describe similar examples from the Danish Central Graben area. Transtension and transpression is again mooted as the principal mechanism responsible for the generation of the complex structural styles observed.

As well as the oblique nature of the structural styles, a common thread that links these three papers was that the Late Jurassic extensional event reactivated much older fault trends rather than forming completely new faults. The paper by **Platt and Philip** (presented as a poster display) highlights the importance of good quality seismic in order to identify pre-Mesozoic fault trends which were subsequently reactivated. Such a NE–SW trend is envisaged as the main control on the boundary dividing the Forties–Montrose High into a southern area thought to contain a considerable thickness of Carboniferous coals, and a northern area devoid of the Carboniferous coaly sequence.

The complex structural evolution and variation in structural styles of the Norwegian Central Graben are described by **Gowers et al.** who stress the importance of making observations rather than fitting data into a preferred model.

Southern North Sea papers

The two papers presented on the Southern North Sea gave strongly contrasting views on the structural styles and evolution of the area.

An analysis of Mesozoic fault trends from three Southern North Sea blocks by **Arthur** suggest that Mesozoic movements on pre-Zechstein faults were oblique because of their oblique orientation in respect to the principal stress axes, but were purely dip-slip (extensional in the Jurassic and early Cretaceous, and compressional in the late Cretaceous) on post-salt sequences due to the decoupling effect of the salt. The orientations of the pre- and post-salt faults are interpreted as being different.

In the Cleaver Bank area, **Oudmayer and de Jager** clearly demonstrate from 3D seismic survey data that, even though the Zechstein salt does cause decoupling of faults, the general orientations and locations of the faults are coincident in most areas. Tertiary fault patterns observed from attribute displays

are interpreted as Riedel shears indicating that there must have been a strike-slip component of fault movement in the post-salt sequences as well as the pre-salt.

General discussion

The key points of general agreement and contention from the papers given in the 'Structural styles' section are as follows:

Points of agreement

1. The North Sea area contains many Paleozoic and older basement fault trends. These are fundamental in controlling the structural styles formed during later tectonic events.
2. Much use is now being made of sandbox-type models in order to demonstrate the feasibility of a certain structural interpretation. Their use is extremely beneficial in gaining a geometrical understanding of possible structures.
3. The use of 3D datasets and attribute-type analyses has enabled more detailed fault mapping to be attained. This greatly enhances the structural understanding, especially in areas of complex structural geometries such as in the Central North Sea area.

Points of contention

1. Oblique-slip versus extensional faulting—there is much lively debate as to the existence of normal faults versus oblique-slip faults, especially in the Central Graben area where the presence of underlying salt further complicates matters. The key points stressed by the exponents of oblique-slip are that the strike-slip component of movement was only very small compared to the dip-slip component, and that the same structural styles are observed in areas where salt is not present as in areas where salt is present. It is also worth noting that all papers that use 3D seismic data are exponents of oblique-slip movements.
2. The role of salt—most of the Central Graben papers that believe in purely extensional faulting styles concentrate on the decoupling effect of salt relative to the overlying and underlying formations. The Central Graben oblique-slip papers based on 3D datasets, on the other hand, demonstrate a definite link between faults contained within the underlying and overlying formations.

The high quality of papers given in the Structural Styles Session highlights that there is an increasing awareness and understanding of structural geology within the petroleum industry as a whole. Detailed structural interpretation now forms an integral part of the overall exploration process.

Structural models for the evolution of the North Sea area

G. D. WILLIAMS

Department of Geology, University of Keele, Keele, Staffordshire ST5 5BG, UK

Abstract: During the past decade, numerous theories on the geometry and kinematic evolution of fault systems have been expounded. Listric normal faults with shallow detachments generate roll-over anticlines. Alternatively, planar faults bounding rigid blocks, the 'domino' model, may be more realistic in many areas of the North Sea. From onshore analogues, relay ramps connecting en échelon extensional faults are shown to be more common than transfer faults. Recent work on fault displacement gradients has generated realistic models involving planar faults with 'soft linkages' for fault systems. Inversion tectonics and halokinesis are commonly observed in the Southern and Central North Sea, and their effects are sometimes difficult to separate using seismic profiles. Scaled analogue models have provided graphic illustrations of the progressive development of structures. The construction of balanced cross-sections, increasingly used in North Sea interpretations, is a technique that may be used to test the various geometric models. Links between fault system geometries at upper crustal levels and deeper crustal processes have been established using field studies, notably in the Basin and Range Province, coupled with the interpretation of COCORP and BIRPS deep seismic reflection profiles. Geometric, thermal and flexural isostatic factors are used in sophisticated modelling of lithospheric processes in the North Sea. Structural styles vary in different regions of the North Sea due to the underlying basement grain and the presence or absence of Permian and Triassic evaporites.

Significant differences in structural styles characterize the different sub-areas of the North Sea rift system (Fig. 1) (Ziegler 1990; Roberts *et al.* 1990a and b; Underhill 1991; Bartholomew *et al.* 1993). Inherent differences in structural styles are strongly influenced by basement structures. A Caledonian grain has been identified by Frost *et al.* (1981) and Johnson and Dingwall (1981) in the Viking Graben, and this has been elaborated on by Coward (1990). An ENE–WSW, broadly Caledonide trend to the structures of the Inner Moray Firth (Underhill 1991) changes to a dominant EW grain in the Outer Moray Firth with no obvious basement affinities. Using gravity and magnetic data, Frost (1977) and Frost *et al.* (1981) proposed a Caledonide basement control on the NW–SE-trending Central Graben with a component of dextral strike-slip motion on NW–SE-trending faults. The NW–SE trends of the Southern North Sea are compatible with onshore UK Charnoid trends (Lee *et al.* 1990; Walker and Cooper 1986).

A second major controlling factor on structural style is the presence or absence of Permian and Triassic salt deposits. Salts provide detachment horizons and if sufficiently thick tend to 'buffer' deformation so that sub-salt faults do not connect with faults in the cover sequence. Halokinesis is also an important process active from Triassic to Tertiary times in the evolution of the Central and Southern North Sea.

This paper is a review of structural styles and structural theories of the past decade as they have been applied to the North Sea. Structural geologists have tended to use major onshore rifts, notably the Basin and Range of the western USA and the East African Rift System, as analogues for structures interpreted from seismic data. Structural thinking has significantly evolved during the past decade. There has been a proliferation of North Sea-related publications on structural geology. This is to be expected as structure is the dominant control on the vast majority of North Sea oil and gas fields. Because of the wealth of data, therefore, this paper represents only a partial view of the myriad of structural theories.

Structural styles applied to the North Sea

Listric versus planar faults

Wernicke and Burchfiel (1982) produced a classic publication on modes of extensional tectonics which was based upon fieldwork in the Basin and Range Province, literature review

and sub-surface data. Listric faults are considered to be curved in cross-section, soling out into a detachment. In plan, they may also have curvature, concave towards the hangingwall.

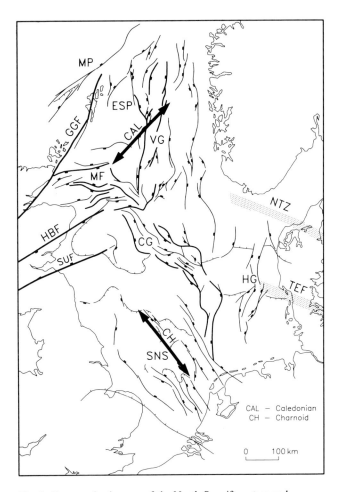

Fig. 1. Structural sub-areas of the North Sea rift system and suggested basement structures. MP: Magnus Province; VG: Viking Graben; MF: Moray Firth; CG: Central Graben; SNS: Southern North Sea; HG: Horn Graben; GGF: Great Glen Fault; HBF: Highland Boundary Fault; SUF: Southern Uplands Fault; NTZ: North Tornquist Fault Zone; TEF: Trans European Fault Zone.

From *Petroleum Geology of Northwest Europe: Proceedings of the 4th Conference* (edited by J. R. Parker).

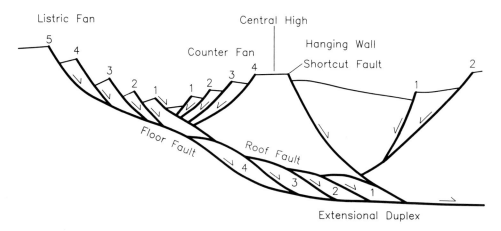

Fig. 2. Nomenclature of linked extensional fault systems (after Gibbs 1983). Fault numbering relates to proposed sequence of development.

Because of this geometry, extensional displacement of the hangingwall generates a fault bend fold called a roll-over anticline. Accommodation space generated in the hangingwall above the fault ramp may be infilled with sediment during faulting to give a syn-rift sequence. Gibbs (1983) took the concept of linked listric faults and generated a nomenclature for various geometric features associated with listiric normal faulting (Fig. 2). Extensional displacement on faults with complex ramp-flat geometries generates predictable basin geometries in the fault hangingwall with a predictable, though complex, stratigraphic development. Based on theories for the sequence of fault development in thrust belts (Boyer and Elliott 1982; Butler 1982), Gibbs (1983) proposed that the general rule in linked extensional fault systems is that new faults are generated within the footwall block giving progressive footwall collapse. Several papers on North Sea structure were generated following the geometric models of Wernicke and Burchfiel (1982) and Gibbs (1983).

Planar faults separating rigid fault blocks are an alternative to listric normal faulting (Le Pichon and Sibuet 1981; Wernicke and Burchfiel 1982; Barr 1987). The so-called 'domino' or 'bookshelf' fault model involves rotation of rigid fault blocks coupled with rotation of intervening planar faults synchronous with slip on the faults (Fig. 3). This is equivalent to a set of adjacent dominoes or books on a shelf being toppled over. This model generates accommodation space below some arbitrary base level, while the shoulders of the tilt blocks may be above base level and may be eroded (Barr 1987). A space problem exists with the base of the model and this may be accommodated by ductile flow of an underlying medium such as salt or, on a crustal scale, the ductile lower crust.

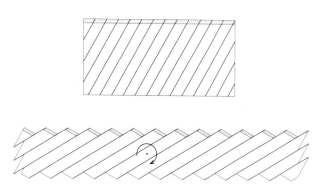

Fig. 3. The domino or bookshelf model for rotational rigid fault blocks (after Wernicke and Burchfiel 1982).

Transfer faults versus relay ramps

Gibbs (1983) included transfer faults as a part of his kinematic model for listric normal faulting. Transfer faults connect en échelon listric normal faults. They are steep faults parallel to (lateral), or oblique to, the dominant fault slip vector. As such, they show strike-slip or oblique-slip displacements (Fig. 4).

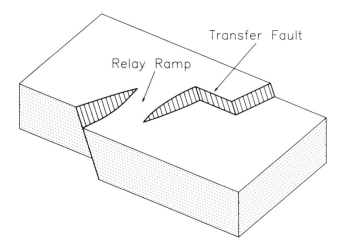

Fig. 4. Three-dimensional model of a transfer fault and a relay ramp connecting en échelon fault segments.

It has become clear from work in the East African Rift (Rosendahl 1987; Morley et al. 1990) that transfer faults are extremely rare and occur only as accommodation structures in 'relay-ramps'. Extensional faults associated with the East African Rift System are largely en échelon in map view and where they overlap, complex inter-relationships occur in strain accommodation zones. Such zones have been termed high relief and low relief accommodation zones by Rosendahl (1987), where the former is a horst block and the latter a graben produced by overlapping, opposed sense faults. Where en échelon overlapping faults have the same sense of throw a relay ramp is the area of increased dip where the footwall of one fault becomes the hangingwall of the other (Fig. 4). A relay ramp is a product of strain accommodation at the tips of overlapping faults. If displacement on each fault is great the relay ramp may fail by secondary faults that are geometrically similar to transfer faults but show dominantly dip-slip displacement (Roberts et al. 1990a). Interactions of overlapping extensional faults generate geometries that may be crucially important for syn-faulting sediment input paths. Relay ramps provide important hydrocarbon migration pathways if major extensional faults are sealing.

Fault displacement gradients

Dahlstrom (1969), Elliott (1983), Hossack (1983) and Williams & Chapman (1983) have proposed that thrust faults show gradients of displacement in map and cross-sectional view, with displacement reducing to zero at the fault tips. A fault may be considered as a surface of slip within a volume; the surface surrounded by a tip line loop. Rippon (1985), using sub-surface British Coalfield data, was first to contour the amount of displacement on individual fault surfaces. A pattern of broadly elliptical contours surrounded the central displacement maximum. Walsh and Watterson (1987, 1988, 1989) and Barnett et al. (1987) have extended these concepts using a vast amount of British Coal and seismic data. Barnett et al. (1987) have proposed an ideal fault geometry for an isotopic medium. Walsh and Watterson (1991), using seismic data, have considered the interaction of several faults in a soft-linked domino model (Fig. 5). In this model, individual fault blocks are allowed to deform and all faults have displacement gradients. An important conclusion is that there is no systematic sequence of faulting and several faults may move synchronously. This contradicts the hard linked fault sequence of Gibbs (1983). Instances when idealized fault models may not be applicable include:

(1) areas where through-going basement faults control the nucleation sites and growth of faults in the cover;
(2) areas where pre-existing faults with a particular slip vector are reactivated by movement along a different slip vector;
(3) areas of strain accommodation in relay ramps where displacement gradients are locally high (Peacock and Sanderson 1991).

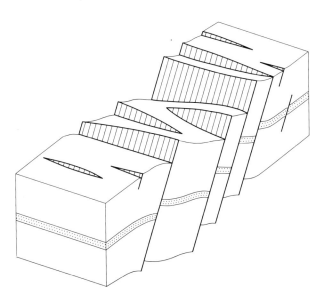

Fig. 5. Soft-linked domino fault model (after Walsh and Watterson 1991).

Inversion tectonics

Several localized areas of the North Sea, notably within the Southern and Central North Sea, have undergone basin inversion: a process where basin depocentres are uplifted and basins are turned inside out (Cooper et al. 1989). The process involves the reactivation by contraction of pre-existing extensional faults although oblique-slip and wrench movements may also generate inverted basins (e.g. Cartwright 1989; Bartholomew et al. 1993). Inversion tectonics are recognized where in the hangingwall of a half-graben the upper part of a fault causes elevation of material above regional, whereas on the lower part of the fault beds are still beneath regional (Fig. 6) (Cooper et al. 1989). A consequence of this geometry is that a syn-rift fill or passive infill must be present to identify inversion positively.

A number of general observations may be made regarding the reactivation of pre-existing faults during inversion tectonics. The process of contractional inversion is selective; only certain faults in fault systems become reactivated, with the others remaining inactive and in net extension. In a similar manner, only certain portions of major faults become reactivated. This may be represented in map view by a contractional fault passing laterally into an extensional fault (Fig. 6).

Halokinesis

The buoyancy model for halokinesis was developed in response to the observation that salt maintains density with burial, while clastic rocks increase density to exceed that of salt at a depth of approximately 3000 ft (Trusheim 1960; Gould and DeMille 1968). Recent work indicates that it is localized anomalies in the overburden above salt which cause salt motion (Jackson and Talbot 1986; Kehle 1988). Seni and Jackson (1983) described the generation of salt structures beneath as little as 300 ft of overburden, where the locus of salt structure growth moves in tandem with the migration of sediment depocentres in Texas and Louisiana.

In the case of the Central and Southern North Sea, a tectonic influence is inferred from the NW–SE alignment of many salt structures matching the orientation of tectonic lineaments beneath the salt. Overburden anomalies would be created in cover sinks above sub-Zechstein grabens. This would cause salt to flow from downthrown areas to upthrown areas, further accentuating initial cover sinks to form local salt withdrawal basins. Flow of salt into the relatively uplifted footwalls of graben structures has been noted by Jenyon (1986, 1988) (Fig. 7).

Scaled analogue models

Scaled analogue models provide graphic representations of geometries of faults and halokinetic structures. They also provide information on the progressive development of fault systems. Listric normal faults (McClay and Ellis 1987; McClay and Scott 1990; Venderville et al. 1987), planar rotational faults (McClay 1990) and inversion structures (Koopman et al. 1987; McClay 1989; Buchanan and McClay, in press) have been modelled using sand box experiments. Oblique extensional and oblique contractional models produced by Richard (1991) provide excellent simulations for structures observed above and below Zechstein salts in the UK Southern North Sea.

The greatest benefit of analogue modelling is their use in understanding the kinematic evolution of fault systems. McClay (1990) suggests that models simulate structures on a variety of scales form entire rift basins to hangingwall deformation above a single fault detachment. Models are, however, limited by the grain size of the modelling materials. Additionally, displacement gradients in basal detachments are not yet possible and it is not possible to model compactional and isostatic effects.

Balanced cross-sections

Increasingly, balanced cross-sections are being used in North Sea structural interpretations. Balanced sections were devel-

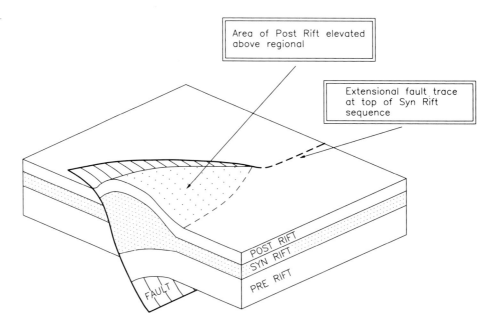

Fig. 6. Three-dimensional model of an extensional fault with a syn-rift sequence partially reactivated by contractional inversion.

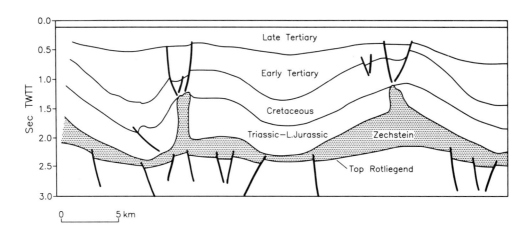

Fig. 7. Halokinetic features in the Zechstein of the UK Southern North Sea interpreted from a seismic section (courtesy of Geco Prakla Ltd).

oped for use in contractional terrains, notably the Rocky Mountains of the USA and Canada (Bally *et al.* 1966; Dahlstrom 1969). Elliott (1983) popularized this technique in the UK. 'A balanced section is not a unique solution. But if it integrates various sorts of data in a quantitative manner and additional work is carried out...a section may be found which is sufficiently restrictive with little room for alternative interpretation' (Elliott 1983).

A balanced cross-section must be retro-deformable and, therefore, should be constructed in a plane containing the mean slip vector for faults. It should also maintain cross-sectional area which in three dimensions means isovolumetric deformation. In extensional tectonics where compacting sediments are a key element of the deformation an appropriate compaction law should be adhered to (e.g. Sclater and Christie 1980) and volume changes from deformed to restored sections should be calculated. Ideally, bed lengths measured from the deformed section should be maintained in the restored version.

Techniques for cross-section restoration in common use are the 'kink fold' method (Suppe 1983) which treats all folds as kink bands. This is particularly useful in contractional terrains. The vertical shear or 'Chevron' construction (Fig. 8a) was

introduced to the UK via a Japec course (Verall 1982). This construction assumes deformation of the hangingwall block by vertical simple shear (analogous to deformation of a vertically stacked deck of cards) by movement over a curved fault above a rigid footwall block. The construction may be used to predict fault geometry at depth if the roll-over shape is known. Alternatively, it may be used to retro-deform an interpreted section where the fault shape and hangingwall geometry are known (see Williams and Vann 1987 for methodology and review). White *et al.* (1986) have adapted the Chevron construction by allowing the hangingwall shear direction to be inclined (the inclined shear construction) (Fig. 8b). Both vertical and inclined shear constructions may be used in extensional and contractional sections and in sections showing inversion tectonics as long as oblique-slip is excluded. These constructions have the property of conserving fault heave and cross-sectional area in the hangingwall block, but bed length is not conserved. Most commercial software packages for balanced section manipulation use the vertical and inclined shear constructions coupled with the ability for depth conversion of seismic and sediment decompaction.

The fundamental drawback of balanced cross-section con-

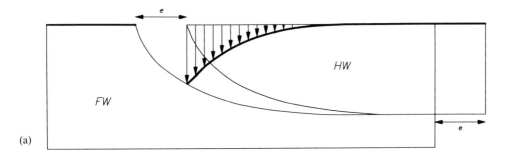

(a)

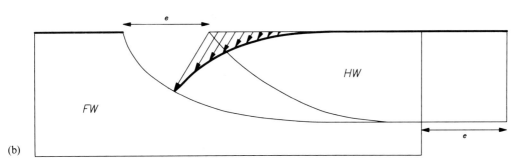

(b)

Fig. 8. (**a**) Vertical shear (Chevron) construction (Verall 1982) and (**b**) inclined shear construction (White *et al.* 1986) to show hangingwall deformation above a listric fault. Both constructions generate area-balanced sections.

struction is that it is presently a two-dimensional technique. Incorrect and misleading interpretations may result from areas that have undergone some component of oblique-slip faulting (Bartholomew *et al.* 1993) or fault reactivations with differing directions of net slip.

Link between upper crustal faulting and deeper lithospheric processes

In the late 1970s it was realized that the metamorphic core complexes of the Basin and Range Province were formed as a result of uplifts linked to regional crustal extension in the Tertiary (Davis 1977; Davis and Coney 1979). Extensional shearing has telescoped rocks and structures that were originally formed at different depths and has caused translations of tens of kilometres of hangingwall blocks (Davis 1977). Uplift coupled with regional extension has exposed lower and mid-crustal mylonites.

COCORP deep seismic reflection data across the Basin and Range (Allmendinger *et al.* 1983, 1987) in association with field observations have enabled the erection of three models for crustal scale processes (Fig. 9).

1. *A sub horizontal, mid-crustal detachment* broadly equates with the brittle–ductile transition in the crust (8–15 km) (Smith 1978).
2. A model involving *extending lenses* or *boudins* (Davis and Coney 1979; Hamilton 1987) is consistent with the COCORP data.
3. A *crustal penetrating shear zone* dipping eastwards was proposed by Wernicke (1981), Wernicke and Burchfiel (1982) and Wernicke (1985). Allmendinger *et al.* (1987) proposed that the COCORP data do not support this model for crustal extension as Moho offset is lacking and there is no evidence for lithosphere-scale, low-angle, normal faults.

The British Institution's Reflection Profiling Syndicate was established by NERC in 1981 to investigate the deep structure

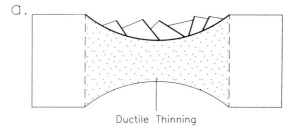

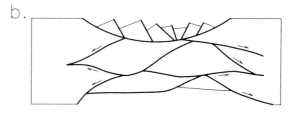

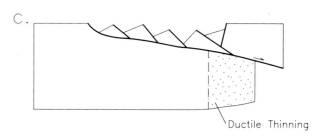

Fig. 9. Models for crustal scale deformation (Allmendinger *et al.* 1987): (**a**) sub-horizontal mid-crustal detachment; (**b**) extending lenses or boudins; (**c**) crustal penetrating shear zone.

of the onshore and offshore UK using deep seismic reflection profiling. By 1990 the syndicate had acquired 14 290 km of high quality deep seismic reflection profile. McGeary (1987) attempted to produce a general description of reflectivity of the continental lithosphere surrounding the UK. Largely non-reflective upper crust is underlain by complex highly reflective lower crust showing predominantly horizontal reflectors. The base of the reflective lower crust marks the position of the Moho and this is used to detect crustal thinning in areas of rifting. There are highly reflective dipping structures in the mantle to depths of *c.* 80 km.

There appears to be little evidence of pre-Mesozoic structures such as inherited Caledonian or Precambrian basement grain in BIRPS data although limitations of the seismic reflection method in imaging steep structures must be taken into account. Recently, Blundell *et al.* (1991) have produced an interpretation of the MOBIL 7 profile that involves differential stretching in the mid- to lower crust in a manner akin to the extending lenses proposed by Hamilton (1984) for the Basin and Range Province.

Lithospheric modelling of the North Sea rift system

Lithospheric modelling of rift systems is heavily reliant on balanced cross-sections as the geometrical evolution of sections is crucial to the loading history of a basin. McKenzie (1978) used an area balance approach to model rifting and subsequent thermal loading of sedimentary basins based on Airy isostasy. More recently, increasingly sophisticated models that incorporate flexural isostasy have involved movement on listric faults coupled to lower crustal plus lithospheric pure shear (Kusznir *et al.* 1991). Using, as a geometric basis, the domino fault model for upper crustal deformation, a flexural cantilever model (Fig. 10) computes the thermal and flexural-isostatic response of the lithosphere (Kusznir *et al.* 1991; Marsden *et al.* 1990). An important consequence of such modelling is the prediction of footwall uplift related to graben-bounding faults and the implications of such uplift for stratigraphic development.

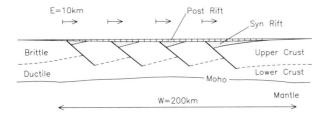

Fig. 10. The flexural cantilever model (Kusznir *et al.* 1991) for geometric, thermal and flexural isostatic modelling of rifting and post-rift basin evolution.

Even the most sophisticated numerical model for lithosphere behaviour is only a two-dimensional representation. As area-balanced cross-sections are the basis for such models, it is essential for geologists to establish extensional vectors so that model input data, the balanced section, is constructed in the correct orientation. Additionally, further work is required in the Central and Northern North Sea on the extent of the Permo-Triassic rifting event. In terms of numerically modelling Jurassic rifting in the North Sea, a precursor rift event would increase the geothermal gradient which would thermally condition the lithosphere, resetting its flexural rigidity.

Structural styles and tectonic events in the North Sea

Basin-related tectonic events in the North Sea span a wide time range and structures show a diachroneity of development. The Viking Graben and Central North Sea underwent a significant Permo-Triassic rifting event (Ziegler 1990; Steel and Ryseth 1991). Rifting and subsidence in the Sole Pit Basin was active

from late Triassic to ?Mid-Jurassic times (Glennie and Boegner 1981). A regional thermal doming was present in the Outer Moray Firth, Mid North Sea High and South Viking Graben areas in mid-Jurassic times. This corresponds with the extrusion of thick sequences of Jurassic volcanics in the area of the Forties Oilfield (Latin *et al.* 1990). Major extension in the Northern North Sea occurred during late Bathonian to Oxfordian times (Ziegler 1990), and in the Central North Sea from late Oxfordian to Kimmeridgian. Large displacements on major graben-bounding faults led to fault-controlled subsidence and at this time the North Sea rifts were deep, underfilled basins with significant palaeobathymetry (Bertam and Milton 1989). Half-grabens were passively infilled by Cretaceous and Tertiary sedimentation. In the Moray Firth, extension persisted to late Volgian times (Underhill 1991). The Sole Pit area of the Southern North Sea underwent a phase of contractional inversion in late Cimmerian times (Glennie and Boegner 1981). During the early Cretaceous, a switch occurred in the locus of rifting as a major extension occurred on the Atlantic margin. This gave rise to large-scale NE–SW-oriented and NW downthrowing faults of the Magnus Province (Ziegler 1990). Tertiary inversion of the Southern and Central North Sea resulted from the transmission of stress from the Alpine Chain (Ziegler 1982) or the Pyrenees (Roberts 1989) via the brittle upper crust.

Structural styles

The Magnus Province shows major planar extensional faults separating widely spaced tilted fault blocks. There has been significant erosion of footwall crests. The half-grabens have limited syn-rift fill and are dominated by Cretaceous and Tertiary sediments showing passive infill (Fig. 11). Faults in the Magnus Province have a distinct Caledonian trend, but are more likely to be related to tensional stress generated by Atlantic opening.

The Viking Graben shows relatively continuous faults that trend approximately N–S and downthrow dominantly to the east. Half-graben-bounding faults appear planar and a rotational 'domino' model is proposed for this region. Beach (1986) proposed a Wernicke-type model based on a deep seismic line (Fig. 12). Beach suggested that a crustal scale shear zone cross-cuts the Moho and dips shallowly eastwards under Norway. This interpretation does not fit gravity modelling of the Moho structure and a flexural cantilever model (Kusznir *et al.* 1991) may be more appropriate. The westerly tilted fault blocks coupled with a mid-crustal detachment may relate to the underlying Caledonian thrust fabric presumed to be relatively flat lying in this region (Coward 1990). The Beryl Embayment is dominated by easterly downthrowing detachment faults which may have been influenced by pre-existing Caledonian thrust structures.

The Inner Moray Firth is a major Jurassic half-graben probably controlled by the Helmsdale Fault and not influenced until late in its history by wrench displacement along the Great Glen Fault (Underhill 1991). Underhill has proposed dip-slip extension on the Helmsdale Fault to generate the Inner Moray Firth Basin (Fig. 13), whereas Roberts *et al.* (1990*a*) proposed an oblique-slip vector with 7 km transtension. The Helmsdale Fault onshore has a Caledonian trend, but as structures are traced to the Outer Moray Firth and Witch Ground Graben an E–W fault trend dominates. In the Outer Moray Firth, steep conjugate faults that are not apparently rotational in nature yield a 'block and basin' type of structure. Johnson and Dingwall (1981) suggested that the orthogonal fault pattern in this area comprise elements of Caledonide and Cimmerian fault trends.

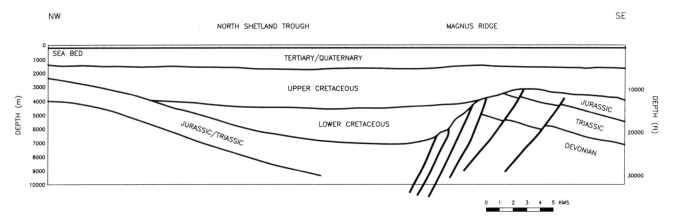

Fig. 11. Sketch to show structural style in the Magnus Province.

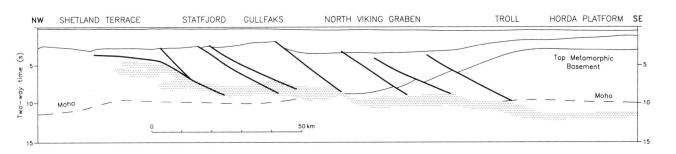

Fig. 12. Proposed structural style in the Viking Graben (adapted from Beach 1986).

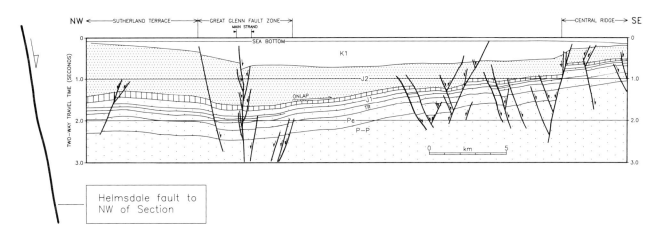

Fig. 13. Structural and stratigraphic interpretation of a NNW–SSE seismic section from the Inner Moray Firth (after Underhill 1991).

The Central Graben transects the Mid North Sea High–Ringkøbing Fyn High in a NW–SE direction. Donato *et al.* (1983) have proposed that buried granitic batholiths beneath the Mid North Sea High are responsible for its continued elevation during North Sea evolution. It is likely that the basement in this area is part of the Variscan foreland. The Tornquist Line and the Trans-European Fault discussed by Coward (1990) and Ziegler (1990) are considered by Roberts *et al.* (1990*b*) to have little or no obvious influence in this area of the North Sea. Similarly the NE–SW-trending Highland Boundary Fault, Southern Uplands Fault and proposed Solway Suture (Blundell *et al.* 1991) appear to have little or no influence on Mesozoic to Tertiary structural development.

In terms of structural style, the basin margin faults are steep in section and en échelon in plan. Several major relay ramps link adjacent overlapping faults (Morley *et al.* 1990; Roberts *et al.* 1990*b*). Generally, steep conjugate faults give rise to non-rotational horst and graben structures. The presence of thick sequences of Zechstein evaporite deposits in certain areas has produced rotational 'domino' style fault blocks of relatively small size (2–4 km wavelength) with a detachment within the Zechstein (Fig. 14) (Sclater and Shorey 1989). Many of these faults do not link directly with faults beneath the Zechstein interval. Large basement faults in the graben areas have triggered halokinesis in the form of salt walls that trend parallel to the basement faults. Pure dip-slip extension has

DEPTH SECTION NO VERTICAL EXAGGERATION

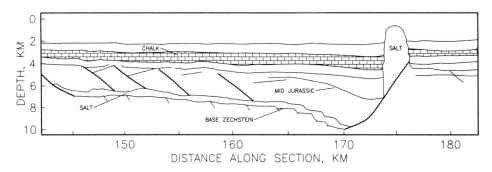

Fig. 14. Structural style in the Central Graben (Sclater and Shorey 1989).

UNRESTORED SIMPLIFIED DEPTH SECTION (PRESENT DAY)

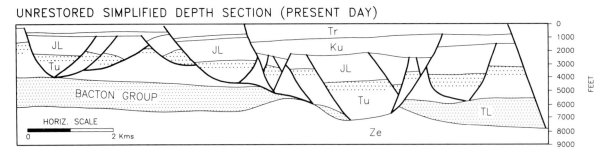

Fig. 15. Structural style in the Southern North Sea (Walker and Cooper 1986).

been proposed by Roberts *et al.* (1990*a* and *b*) although oblique-slip extension reactivating pre-existing basement faults cannot be ruled out (Bartholomew *et al.* 1993). Structural inversion in the Central Graben is localized, and Cartwright (1989) has proposed a net convergence in a NE–SW direction for this event in the Danish Central Graben.

The Southern North Sea is dominated by NW–SE-trending basement structures (Walker and Cooper 1986) that are considered to relate to Precambrian basement with a Charnoid grain. This trend has been detected by gravity and magnetic modelling (Lee *et al.* 1990) and covers a broad area of eastern England. Two major fault zones, the Dowsing fault zone, and the Swarte Bank Hinge, define the margins of the Sole Pit Basin. Charnoid structural trends were utilized during Variscan contractional deformation in this area of the Variscan foreland (Leeder and Hardman 1990). Beneath the Zechstein, faults in the Rotliegend and underlying Carboniferous appear to be dominantly planar, with small extensional separations. Where Zechstein evaporites are sufficiently thick, basement faults do not connect to cover sequences. Over much of the area, thick Triassic evaporites permit listric faulting of the Upper Triassic/Jurassic sequence with detachment above the Triassic Bacton Group. A unique structural pattern exists where steep basement faults are buffered by, and do not cross-cut Zechstein halites. The Triassic Bacton Group shows little faulting, but in certain areas is pulled apart with associated invasion of salt (Walker and Cooper 1986). The post-Bacton Group to Jurassic sequences are extended by listric normal faults detaching at the level of Triassic halite intervals (Fig. 15). This geometry has been partly modelled by Richard (1991) using an oblique-extension and oblique-contraction test rig. Halokinesis is represented by salt walls and salt pillows that appear to be controlled to a large extent by basement trends.

Inversion tectonics is pronounced in certain areas of the Southern North Sea (Badley *et al.* 1989; Ziegler 1990). Where Zechstein salt is thick, contractional fault movements are buffered and are expressed as regional uplift rather than cover faulting (Richard 1991). This can lead to difficulty in separating halokinetic and inversion events. The Sole Pit Basin has been the locus for large-scale inversion in both late Cimmerian and Tertiary times. Total uplift of the Leman Sandstone Formation is of the order of 3 km since the mid-Jurassic.

Conclusions

The North Sea rift system is characterized by differing structural styles in different areas. Dominant controls on structural style are pre-existing basement grain, the presence or absence of thick evaporite deposits and prevailing stress directions during rifting/inversion events. Areas that have remained as highs throughout the rifting history are frequently cored by granitic masses (Donato *et al.* 1983). The structural evolution of the North Sea took place as a series of tectonic events that show a diachroneity of development.

The pre-Mesozoic basement grain has influenced North Sea structuring in several areas. The NW–SE Charnoid grain is sub-parallel to that of the Trans European Fault and the Tornquist fault zone. These structures appear to have influenced Mesozoic rifting in the Central Graben and Southern North Sea. Caledonide trends are detectable in the East Shetland Platform, Viking Graben and possibly in the Inner Moray Firth.

There has been considerable debate on the relative importance of dip-slip and oblique-slip extension in the evolution of the rift system. If steep basement lineaments with a range of orientations are re-used in extension then a mixture of oblique-slip movement vectors will result. Clearly there is a dominance of the pure, dip-slip rifting component, but oblique-slip must have occurred to varying degrees in the North Sea. The establishment of net slip on fault systems is virtually impossible, using even 3D data. Therefore, balanced cross-sections based on vertical 2D profiles will contain varying degrees of error depending on the obliquity between fault slip azimuth and the cross-section. Such errors will be compounded in lithospheric models for the basin based on simple balanced cross-sections constructed orthogonal to rift axes.

Detailed interpretation of 3D data coupled with a better understanding of faulting from onshore analogues has led to increasing sophistication of structural models for the North Sea. Hard linked extension faults developing in a specific sequence are less likely than simultaneous displacements on a number of soft-linked faults. Rotating rigid fault blocks dominate in the East Shetland Platform, Viking Graben and Magnus Province whereas steep, non-rotational faults dominate in the Central Graben, Outer Moray Firth and the sub-Zechstein structure of the Southern North Sea. The presence of evaporites in the Triassic and Permian of the Southern and Central North Sea has led to the development of detachment horizons and listric faults are present in cover sequences.

Inversion tectonics, dominant in the Southern and Central North Sea, lead to basins being turned inside out. The inversion process is selective as only certain faults become reactivated during compression, with other faults remaining in net extension. Inversion takes place by reactivation along individual faults on a local scale or by uplift on a regional scale, for example the Sole Pit area of the Southern North Sea. Basement lineaments are the dominant control on inversion and significant oblique-slip has been involved in the process. The structural history of the North Sea has involved multiple and selective reactivation of pre-existing basement lineaments. The reactivation of a particular fault depends on its orientation relative to the prevailing stress trajectories and its strength. Continued movement on a particular fault under the right conditions will cause strain softening and, therefore, faults that have recently moved are likely to be preferentially reactivated. This explains the observation that extension and inversion frequently occur with a common locus.

Thanks to Eva Palenicek and Terry Doyle for their patience in preparing this manuscript and drafting the figures. Geco Prakla Ltd supplied the author with a comprehensive seismic database for the North Sea.

References

ALLMENDINGER, R. W., HAUGE, T. A., HAUSER, E. C., POTTER, C. J., KLEMPERER, S. L., NELSON, K. D., KNUEPFER, P. AND OLIVER, J. 1987. Overview of the COCORP 40°N Transect, Western United States. The fabric of an orogenic belt. *Geological Society of America Bulletin*, **98**, 308–319.

——, SHARP, S. W., VON TISH, D., SERPA, L., KAUFMAN, S., OLIVER, J. AND SMITH, R. B. 1983. Cenozoic and Mesozoic structure of the eastern Basin and Range Province, Utah from COCORP seismic reflection data. *Geology*, **11**, 532–536.

BADLEY, M. E., PRICE, J. D. AND BACKSHALL, L. C. 1989. Inversion, reactivated faults and related structures: seismic examples from the Southern North Sea. *In*: COOPER, M. A. AND WILLIAMS, G. D. (eds) *Inversion Tectonics*. Geological Society, London, Special Publication, **44**, 201–219.

BALLY, A. W., GORDY, P. L. AND STEWARD, G. A. 1966. Structure, seismic data and orogenic evolution of the southern Canadian Rockies. *Bulletin of Canadian Petroleum Geology*, **14**, 337–381.

BARNETT, J. A. M., MORTIMER, J., RIPPON, J. H., WALSH, J. J. AND WATTERSON, J. 1987. Displacement Geometry in the Volume Containing A Single Normal Fault. *American Association of Petroleum Geologists Bulletin*, **71**, 925–937.

BARR, D. 1987. Structural/stratigraphic models for extensional basins of half-graben type. *Journal of Structural Geology*, **9**, 491–500.

BARTHOLOMEW, I. D., PETERS, J. M. AND POWELL, C. M. 1993. Regional structural evolution of the North Sea: oblique slip and the reactivation of basement lineaments. *In*: PARKER, J. R. (ed.) *Petroleum Geology of Northwest Europe: Proceedings of the 4th Conference*. Geological Society, London, 1109–1132.

BEACH, A. 1986. A deep seismic reflection profile across the northern North Sea. *Nature*, **323**, 53–55.

BERTRAM, G. T. AND MILTON, N. 1989. Reconstructing Basin Evolution from Sedimentary Thickness; the importance of Palaeobathymetric Control, with Reference to the North Sea. *Basin Research*, **1**, 247–257.

BLUNDELL, D. J., HOBBS, R. W., KLEMPERER, S. L., SCOTT-ROBINSON, R., LONG, R. E., WEST, T. E. AND DUIN, E. 1991. Crustal structure of the central and southern North Sea from BIRPS deep seismic reflection profiling. *Journal of the Geological Society, London*, **148**, 445–458.

BOYER, S. E. AND ELLIOTT, D. 1982. Thrust systems. *Bulletin of the American Association of Petroleum Geologists*, **66**, 1196–1230.

BUCHANAN, P. G. AND McCLAY, K. R. (in press). Experiments in basin inversion above reactivated domino faults. *Marine and Petroluem Geology*.

BUTLER, R. W. H. 1982. The terminology of structures in thrust belts. *Journal of Structural Geology*, **4**, 239–245.

CARTWRIGHT, J. A. 1989. The kinematics of inversion in the Danish Central Graben. *In*: COOPER, M. A. AND WILLIAMS, G. D. (eds) *Inversion Tectonics*. Geological Society, London, Special Publication, **44**, 153–176.

COOPER, M. A. *et al*. 1989. Inversion Tectonics—A discussion. *In*: COOPER, M. A. AND WILLIAMS, G. D. (eds) *Inversion Tectonics*. Geological Society, London, Special Publication, **44**, 335–350.

COWARD, M. P. 1990. The Precambrian, Caledonian and Variscan framework to NW Europe. *In*: HARDMAN, R. F. P. AND BOOKS, J. (eds) *Tectonic Events Responsible for Britain's Oil and Gas Reserves*. Geological Society, London, Special Publication, **55**, 1–34.

DAHLSTROM, C. D. A. 1969. Balanced Cross Sections. *Canadian Journal of Earth Sciences*, **6**, 743–757.

DAVIS, G. H. 1977. Characteristics of metamorphic core complexes, S. Arizona. *Abstracts with programs. Geological Society of America*, **9**, 944.

—— AND CONEY, P. J. 1979. Geological development of metamorphic core complexes. *Geology*, **7**, 120–124.

DONATO, J. A., MARTINDALE, W. AND TULLY, M. C. 1983. Buried granites within the Mid North Sea High. *Journal of the Geological Society*, **140**, 825–838.

ELLIOTT, D. 1983. The construction of balanced cross-sections. *Journal of Structural Geology*, **5**, 101–102.

FROST, R. T. C. 1977. Tectonic patterns in the Danish Region. *In*: FROST, R. T. C. AND DIKKERS, A. J. (eds) Fault Tectonics in NW Europe. *Geologie en Mijnbouw*, **56**, 351–362.

——, FITCH, F. J. AND MILLER, J. A. 1981. The age and nature of the crystalline basement of the North Sea Basin. *In*: ILLING, L. V. AND HOBSON, G. D. (eds) *Petroleum Geology of the Continental Shelf of North-West Europe*. Institute of Petroleum and Heyden, London, 43–57.

GIBBS, A. D. 1983. Balanced cross-section construction from seismic sections in areas of extensional tectonics. *Journal of Structural Geology*, **5**, 153–160.

GLENNIE, K. W. AND BOEGNER, P. 1981. Sole Pit inversion tectonics. *In*: ILLING, L. V. AND HOBSON, G. D. (eds) *Petroleum Geology of the Continental Shelf of North-West Europe*. Institute of Petroleum and Heyden, London, 110–120.

GOULD, D. B. AND DeMILLE, G. 1968. Piercement structures in Canadian Arctic Islands. *In*: BRANNSTEIN, J. AND O'BRIEN, G. D. (eds) *Diapirs and Diapirism*. American Association of Petroleum Geologists Memoir, **8**, 183–214.

HAMILTON, W. 1987. Crustal extension in the Basin and Range Provinces, SW USA. *In*: COWARD, M. P., DEWEY, J. F. AND HANCOCK, P. L. (eds) *Continental Extensional Tectonics*. Geological Society, London, Special Publication, **28**, 155–176.

HOSSACK, J. R. 1983. A cross section through the Scandinavian Caledonides constructed with the aid of branch line maps. *Journal of Structural Geology*, **5**, 103–111.

JACKSON, M. P. A. AND TALBOT, C. J. 1986. External shapes, strain rates and dynamics of salt structures. *Geological Society of America Bulletin*, **97**, 305–323.

JENYON, M. K. 1986. Some consequences of faulting in the presence of a salt rock interval. *Journal of Petroleum Geology*, **9**, 29–52.

—— 1988. Fault-salt wall relationships, southern North Sea. *Oil & Gas Journal*, **5**, 76–81.

JOHNSON, R. J. AND DINGWALL, R. G. 1981. The Caledonides: their influence on the stratigraphy of the North-West European Continent Shelf. *In*: ILLING, L. V. AND HOBSON, G. D. (eds) *Petroleum Geology of the Continental Shelf of North West Europe*, Institute of Petroleum and Heyden, London, 85–97.

KEHLE, R. O. 1988. The origin of salt structures. *In*: SCHRIEBER, B. C. (ed.) *Evaporites and hydrocarbons*. Columbia University Press, 345–404.

KOOPMAN, A., SPEKSNIJDER, A. AND HORSFIELD, W. T. 1987. Sandbox studies of inversion tectonics. *Tectonophysics*, **137**, 379–388.

KUSZNIR, N. J., MARSDEN, G. AND EGAN, S. S. 1991. A flexural–cantilever simple-shear/pure-shear model of continental lithosphere extension: applications to the Jeanne d'Arc Basin, Grand Banks and Viking Graben, North Sea. *In*: ROBERTS, A. M., YIELDING, G. AND FREEMAN, B. (eds) *The Geometry of Normal Faults*. Geological Society, London, Special Publication, **56**, 41–60.

LATIN, D. M., DIXON, J. E. AND FITTON, J. G. 1990. Rift-related magmatism in the North Sea Basin. *In*: BLUNDELL, D. J. AND GIBBS, A. D. (eds) *Tectonic evolution of the North Sea Rifts*. Oxford University Press, Oxford, 217–238.

LEE, M. K., PHAROAH, T. C. AND SOPER, N. J. 1990. Structural trends in central Britain from images of gravity and aeromagnetic fields. *Journal of the Geological Society, London*, **147**, 241–258.

LEEDER, M. R. AND HARDMAN, M. 1990. Carboniferous geology of the Southern North Sea Basin and controls on hydrocarbon perspectivity. *In*: HARDMAN, R. F. P. AND BROOKS, J. (eds) *Tectonic Events Responsible for Britain's Oil and Gas Reserves*. Geological Society, London, Special Publication, **55**, 87–105.

LE PICHON, X. AND SIBUET, J. C. 1981. Passive margins: a model of formation. *Journal of Geophysical Research*, **86**, 3708–3720.

MARSDEN, G., YIELDING, G., ROBERTS, A. M. AND KUSZNIR, N. J. 1990. Application of flexural cantilever simple-shear/pure-shear model of continental lithosphere extension to the formation of the northern North Sea Basin. *In*: BLUNDELL, D. J. AND GIBBS, A. D. (eds) *Tectonic evolution of the North Sea Rifts*. Oxford University Press, Oxford, 85–103.

McCLAY, K. R. 1989. Analogue Models of Inversion Tectonics. *In*: COOPER, M. A. AND WILLIAMS, G. D. (eds) *Inversion Tectonics*. Geological Society, London, Special Publication, **44**, 41–62.

—— 1990. Deformation mechanics in analogue models of extensional fault systems. *In*: KNIPE, R. J. AND RUTTER, E. H. (eds) *Deformation Mechanisms, Rheology and Tectonics*. Geological Society, London, Special Publication, **54**, 445–453.

—— AND ELLIS, P. G. 1987. Analogue models of extensional fault geometries. *In*: COWARD, M. P., DEWEY, J. F. AND HANCOCK, P. L. (eds) *Continental Extensional Tectonics*. Geological Society, London, Special Publication, **28**, 109–125.

—— AND SCOTT, A. D. 1990. Hangingwall deformation in ramp-flat listric extensional fault systems. *Tectonophysics*, **188**, 85–96.

McGEARY, S. 1987. Non-typical BIRPS on the margin of the North Sea the SHET survey. *Geophysical Journal of the Royal Astronomical Society*, **89**, 231–238.

McKENZIE, D. 1978. Some remarks on the development of sedimentary basins. *Earth and Planetary Science Letters*, **40**, 25–32.

MORLEY, C. K., NELSON, R. A., PATTON, T. L. AND MUNN, S. G. 1990. Transfer zones in the East Africa Rift System and their relevance to hydrocarbon exploration in rifts. *American Assoication of Petroleum Geologists Bulletin*, **74**, 1234–1253.

PEACOCK, D. C. P. AND SANDERSON, D. J. 1991. Displacements, segment linkage and relay ramps in normal fault zones. *Journal of Structural Geology*, **13**, 721–734.

RICHARD, P. 1991. Experiments on faulting in a two-layer cover sequence overlying a reactivated basement fault with oblique-slip. *Journal of Structural Geology*, **13**, 459–470.

RIPPON, J. H. 1985. Contoured patterns of the throw and hade of normal faults in the Coal Measures (Westphalian) of north-east Derbyshire. *Proceedings of the Yorkshire Geological Society*, **45**, 147–161.

ROBERTS, A. M., YIELDING, G. AND BADLEY, M. E. 1990a. A kinematic model for the orthogonal opening of the Late Jurassic North Sea Rift System. Denmark–Mid Norway. *In*: BLUNDELL, D. J. AND

GIBBS, A. D. (eds) *Tectonic Evolution of the North Sea Rifts*. Oxford University Press, Oxford, 163–189.

——, PRICE, J. D. AND OLSEN, T. S. 1990b. Late Jurassic half-graben control on the siting and structure of hydrocarbon accumulations: UK/Norwegian Central Graben. *In*: HARDMAN, R. F. P. AND BROOKS, J. (eds) *Tectonic Events Responsible for Britain's Oil and Gas Reserves*. Geological Society, London, Special Publication, **55**, 229–257.

ROBERTS, D. G. 1989. Basin inversion in and around the British Isles. *In*: COOPER, M. A. AND WILLIAMS, G. D. (eds) *Inversion Tectonics*. Geological Society, London, Special Publication, **44**, 131–150.

ROSENDAHL, B. R. 1987. Architecture of continental rifts with special reference to East Africa. *Annual Review of Earth and Planetary Sciences*, **15**, 445–503.

SCLATER, J. G. AND CHRISTIE, P. A. F. 1980. Continental Stretching: an explanation of the post mid-Cretaceous subsidence of the Central North Sea Basin. *Journal of Geophysical Research*, **85**, 3711–3739.

—— AND SHOREY, M. D. 1989. Mid-Jurassic through mid-Cretaceous extension in the Central Graben of the North Sea—Part 2: estimates from faulting observed on a seismic reflection line. *Basin Research*, **1**, 201–216.

SENI, S. J. AND JACKSON, M. P. 1983. Evolution of salt structures, East Texas Diapir Province, Part 1: Sedimentary Record of Halokinesis. *American Association of Petroleum Geologists Bulletin*, **68**, 1219–1244.

SMITH, R. B. 1978. *Seismicity, crustal structure and infra plate tectonics of the interior of the western Cordillera*. Memoir of the Geological Society of America, **152**, 11–44.

STEEL, R. AND RYSETH, A. 1991. The Triassic–early Jurassic succession in the northern North Sea: megasequence stratigraphy and intratriassic tectonics. *In*: HARDMAN, R. F. P. AND BROOKS, J. (eds) *Tectonic Events Responsible for Britain's Oil and Gas Reserves*. Geological Society, London, Special Publication **55**, 139–168.

SUPPE, J. 1983. Geometry and kinematics of fault-plane folding. *American Journal of Science*, **283**, 684–721.

TRUSHEIM, K. 1960. Mechanisms of salt migration in N. Germany. *American Association of Petroleum Geologists Bulletin*, **44**, 1519–1540.

UNDERHILL, J. R. 1991. Implications of Mesozoic–Recent basin development in the western Inner Moray Firth, UK. *Marine and Petroleum Geology*, **8**, 359–369.

VENDERVILLE, B., COBBOLD, P. R., DAVEY, P., BROWN, J. P. AND CHOUKROMME, P. 1987. Physical models of extensional tectonics at various scales. *In*: COWARD, M. P., DEWEY, J. F. AND HANCOCK, P. L. (eds) *Continental Extensional Tectonics*. Geological Society, London, Special Publication, **28**, 95–108.

VERALL, P. 1982. *Structural interpretation with applications to Northern Sea problems*. Course Notes No. 3, JAPEC.

WALKER, I. M. AND COOPER, W. G. 1986. The structural and stratigraphic evolution of the northeast margin of the Sole Pit Basin. *In*: BROOKS, J. AND GLENNIE, K. (eds) *Petroleum Geology of North West Europe*. Graham & Trotman, London, 263–275.

WALSH, J. J. AND WATTERSON, J. 1987. Distributions of cumulative displacement and seismic slip on a single normal fault surface. *Journal of Structural Geology*, **9**, 1039–1046.

—— AND —— 1988. Analysis of the relationship between displacements and dimensions of faults. *Journal of Structural Geology*, **10**, 239–247.

—— AND —— 1989. Displacement gradients on fault surfaces. *Journal of Structural Geology*, **11**, 307–316.

—— AND —— 1991. Geometric and kinematic coherence and scale effects in normal fault systems. *In*: ROBERTS, A. M., YIELDING, G. AND FREEMAN, B. (eds) *The Geometry of Normal Faults*. Geological Society, London, Special Publication, **56**, 193–203.

WERNICKE, B. 1981. Low-angle normal faults in the Basin and Range province: nappe tectonics in an extending orogen. *Nature*, **291**, 645–647.

—— 1985. Uniform-sense normal simple shear of the continental lithosphere. *Canadian Journal of Earth Sciences*, **22**, 108–125.

—— AND BURCHFIEL, B. C. 1982. Modes of extensional tectonics. *Journal of Structural Geology*, **4**, 105–115.

WHITE, N., JACKSON, J. A. AND McKENZIE, D. P. 1986. The relationship between the geometry of normal faults and that of the sedimentary layers in their hangingwalls. *Journal of Structural Geology*, **8**, 897–909.

WILLIAMS, G. D. AND CHAPMAN, T. J. 1983. Strains developed in hangingwalls of thrusts due to their slip/propagation rate: a dislocation model. *Journal of Structural Geology*, **5**, 563–571.

—— AND VANN, I. 1987. The geometry of listric normal faults and deformation in their hangingwalls. *Journal of Structural Geology*, **9**, 789–795.

ZIEGLER, P. A. 1982. *Geological Atlas of Western and Central Europe*. Elsevier Science Publishers, Amsterdam.

—— 1990. *Geological Atlas of Western and Central Europe*, 2nd edn. Shell International Petroleum.

The effect of Late Caledonian and Variscan continental escape tectonics on basement structure, Paleozoic basin kinematics and subsequent Mesozoic basin development in NW Europe

M. P. COWARD

Department of Geology, Imperial College, Prince Consort Road, London SW7 2BP, UK

Abstract: Caledonian structures in NW Europe formed by northwestward accretion of continental fragments and magmatic arcs onto the Proterozoic North Atlantic Craton. The main crustal fabrics were formed by Caledonian basin development, subsequent inversion, collision tectonics with thick-skinned and thin-skinned thrusts and large NW–SE strike-slip faults. These fabrics were modified and offset by late Caledonian displacement on NE–SW-trending strike-slip faults.

Late Caledonian (Devonian and Early Carboniferous) tectonics in NW Europe resulted from the lateral expulsion of the English–North Sea–Baltic Block away from an Acadian indentor. The block was approximately triangular in shape bounded by (i) the Ural Ocean to the east, (ii) the left-lateral Great Glen–Midland Valley–North Atlantic shear systems to the NW and (iii) the right-lateral English Channel–South Polish Trough shear systems (in Devonian times) and the South Wales–Southern North Sea–Polish Trough shear zone (in Early Carboniferous times) to the south. During the Early Carboniferous the block expanded NW–SE, as it was released from the indentor, into space created by back-arc extension related to Variscan subduction.

Pull-apart basins which developed along the shear systems include the following. (i) In the Northern North Sea, the West Orkney Basin–East Shetland Platform–Viking Graben formed as a large pull-apart basin in the left-lateral shear systems during Devonian/Early Carboniferous times. (ii) In the Southern North Sea and East Midlands, NW-trending Carboniferous basins developed in a right-lateral shear system, associated with clockwise rotation of Caledonian basement blocks.

The Devonian and Carboniferous basin-fill depended on the position of the block boundary shear systems in the Caledonian and Variscan orogens and their thickened crust. The northern shear boundary affected crust *c.* 50 km thick following Caledonian collision and hence the syn-rift and post-rift sediment fill was dominantly intracontinental. In the south the basin developed in crust of average thickness and hence marine conditions were dominant during the extension. However, Devonian and Carboniferous basins in the Northern North Sea provided channels for sediment fill from the Caledonian mountains into the southern basins.

During the Late Carboniferous, closure of the Ural Ocean and Ural plate collision reversed the sense of shear along the block margins and the wedge-shaped block was pushed back between the Acadian collision zones. This shear reversal caused inversion of the pull-apart and expansion basins in northern Britain. Subsequent reactivation of these basins during the Mesozoic was responsible for the tectonic framework of the North Sea.

Tectonic studies of Mesozoic mountain belts (e.g. Tapponnier *et al.* 1986; Dewey *et al.* 1988; Molnar 1988) show that the continental lithosphere is not rigid but deforms internally during collisional tectonics. In the Himalayan collision zones earthquake activity is not confined to the plate margins but extends several hundred kilometres away from these margins; hence tectonic activity affects much of eastern Asia (Tapponnier *et al.* 1986; Dewey *et al.* 1988). India has indented into Asia resulting in the lateral expulsion of parts of Tibet and western China. Large strike-slip faults cut across China associated with the local development of pull-apart basins (Tapponnier *et al.* 1986). The Asian lithosphere was composed of numerous blocks formed during the accretion of several terranes during Late Paleozoic and Mesozoic times (Coward *et al.* 1988; Dewey *et al.* 1988). These block bounding faults were reactivated during Tertiary collision as thrusts or strike-slip faults and the individual blocks reacted independently to the main Himalayan collision, so that some show shortening while others show lateral displacement or rotation. Similar wide zones of tectonic activity are associated with collision tectonics in the Middle East and Western Europe (Dewey *et al.* 1986) and presumably were associated with older belts such as the Caledonides and Variscides of NW Europe.

This paper analyses aspects of Caledonian and Variscan tectonics in the light of these new ideas of collision tectonics in Asia. In particular it aims to discuss the generation of litho-spheric blocks during initial collision processes and the reactivation of these blocks during subsequent collision, strike-slip and extensional tectonics from Late Paleozoic to Tertiary times.

The pre-Mesozoic basement of NW Europe ranges in age from Archean–Proterozoic gneisses of the Lewisian of NW Scotland and the Svecofennian terrane of Scandinavia, dated at *c.* 2900–1600 Ma (Moorbath *et al.* 1969), to the Devonian and Carboniferous sediments involved in the Variscan fold and thrust belts of central Europe and their Permian molasse. Much of the tectonic framework of NW Europe developed in four major tectonic episodes: Laxfordian & Svecofennian (1800–1600 Ma), Grenville (1100–900 Ma), Caledonian (550–400 Ma) and Variscan (400–300 Ma).

Figure 1 shows the distribution of these tectonic events in four principal domains in Britain. Domain 1 consists of Lewisian rocks of the NW Caledonian foreland, which with their upper Proterozoic cover were originally part of the North American Craton. Their dominant crustal fabric is of Laxfordian age and the fabrics are dominantly sub-horizontal with localized NW–SE-trending steep zones. Domain 2 comprises rocks which were deformed and metamorphosed during the Caledonian tectonic event and can be subdivided into several sub-domains based on the orientation and age of the Caledonian fabric and the presence or absence of Laxfordian and Grenville age basement. Domain 3 is the SE foreland of the

From *Petroleum Geology of Northwest Europe: Proceedings of the 4th Conference* (edited by J. R. Parker).

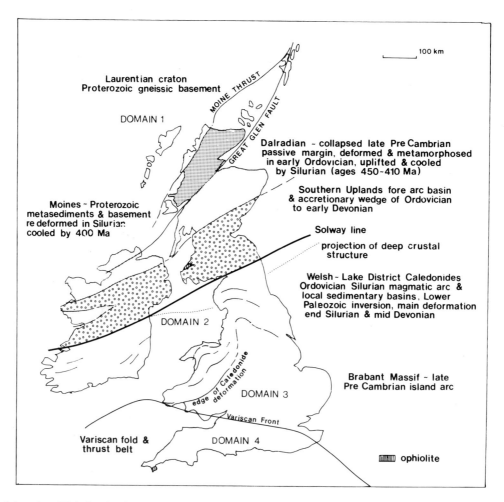

Fig. 1. Structural domains of Britain related to the accretion of various crustal blocks during the Caledonian and Variscan orogenies. From Coward (1990).

Caledonian fold belt and consists of a Late Precambrian magmatic arc, known as the Cadomian magmatic arc or Brabant Massif. Domain 4 consists of rocks deformed and weakly metamorphosed in the Variscan tectonic events during Devonian to Carboniferous times.

Caledonian framework

The Caledonian orogenic belt extends from northern Norway to the Gulf of Mexico and formed as a result of the closure of an early Atlantic (Iapetus) Ocean during the Early Paleozoic by the accretion of magmatic island arcs and old continental fragments onto the North American continental craton. The accretion direction was dominantly towards the NW, perpendicular to the strike of the belt, so that the majority of thrusts and folds verge either NW or SE. However during the orogeny, there were important episodes of left-lateral strike-slip movements.

Crustal fabrics developed during the Caledonian orogeny vary in orientation and define tectonic domains bounded mainly by major strike-slip faults (Fig. 2). NW of the Great Glen Fault the fabrics dip dominantly to the SE as shown on the MOIST, WINCH and DRUM deep seismic surveys (Brewer and Smythe 1984; Hall *et al.* 1984; Cheadle *et al.* 1987; Flack and Warner 1990). In this domain the Caledonides of the NW Highlands of Scotland and their extension offshore consist of Proterozoic metasediments (Moines) metamorphosed during the Grenville event, with minor basic and acid intrusives. The Moines, which are intensely foliated and metamorphosed to upper greenschist and amphibolite facies and

give syn-tectonic metamorphic ages of *c.* 460 Ma (Johnson *et al.* 1985), were thrust to the WNW over a foreland consisting of Lewisian basement, Torridonian arkosic sandstones and Cambro-Ordovician shelf sediments.

Seismic profiles offshore the north coast of Scotland (Brewer and Smythe 1984; Coward and Enfield 1987; Coward 1990) show dipping reflectors in the middle crust which are probably related to the Moine Thrust and to shear zones on its hangingwall. Mineral ages on the Outer Isles Fault (Sibson 1977) suggest that in this region, Laxfordian or later Precambrian fabrics were reactivated during Caledonian times with an overthrust sense. The Outer Isles Fault is characterized by a pronounced SE-dipping crustal fabric (Stein and Blundell 1990) which can be traced in an arcuate pattern north of Lewis into the West Shetland Basin. The fault zone is locally offset by large NW–SE-trending tear faults, which may have been generated as lateral structures during Caledonian, or earlier, deformation.

The most westerly and deepest of the lithospheric reflectors forms the Flannan Zone (Hall *et al.* 1984) which can be traced to a depth of greater than 60 km in the mantle. This zone appears to detach at the base of the crust west of the Hebrides. It has been mapped on several deep seismic sections to the west of Scotland and north of Ireland and has a curved trace which mimics some of the curvature in the Outer Isles Fault.

The Great Glen Fault defines the eastern limit of the Moines. Its age, sense and amount of displacement are disputed. A pre-Middle Devonian left-lateral displacement of *c.* 2000 km was suggested by Morris (1974) and Van der Voo and Scotese (1981) based on palaeomagnetic data. Other authors

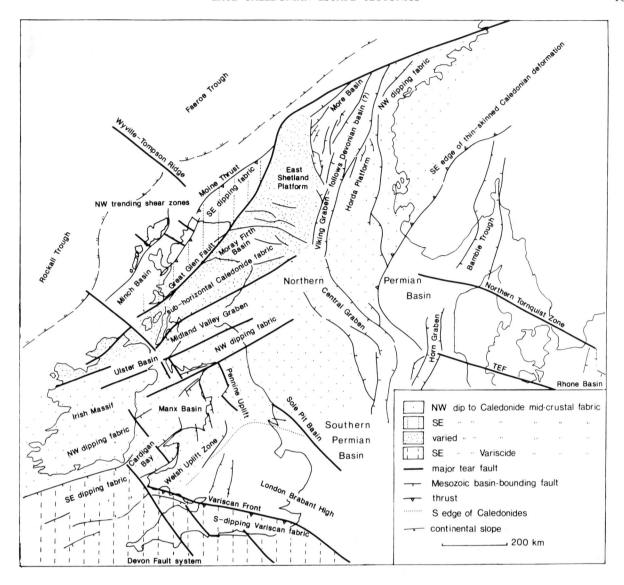

Fig. 2. Map showing the tectonic framework of NW Europe, including the dip of the Caledonian and Variscan mid-crustal fabrics and the trends of the major strike-slip faults. The important Mesozoic basin bounding faults are shown. From Coward (1990).

(Mykura 1976; Smith and Watson 1983) favour a left-lateral displacement of a few hundred kilometres during the late Caledonian, from the offsets of geological structures, followed by later Permo-Carboniferous right-lateral displacement of a few tens of kilometres (e.g. Ziegler 1982, 1989, 1990). Snyder and Flack (1990) and Snyder (in press) argue for a net left-lateral displacement of at least 200 km, based on a restoration of the Flannan reflectors.

SE of the Great Glen Fault the crustal fabrics are more varied in orientation (Fig. 2). Hall (1987) argues for complex interfingering of NW- and SE-dipping thrusts to account for the variably dipping reflectors seen on the WINCH profiles. However, the tectonic history of the region SE of the Great Glen involves a history of extension and subsequent inversion related to the accretion of magmatic arcs and ophiolites onto thinned continental crust. Hence on-shore the structures are extremely complex with no preferred dip direction. SE of the Great Glen in the Grampian Highlands, the rocks consist of late Proterozoic metasediments (Grampian Moines) which pass up, without any major unconformity, into Dalradian sediments. The lower Dalradian consists of syn-rift and post-rift sequences of sandstones, shales and limestones deposited in several deep basins on the NW side of the major rift basin which later became the Iapetus Ocean. As extension acceler-

ated during the middle to upper Dalradian, subsidence increased and the upper Dalradian sediments were deposited in a series of turbidite basins (Anderton 1985; Soper and Anderton 1984). Thinning of the lithosphere was associated with intense igneous activity. The uppermost Dalradian sediments (Latest Precambrian/Early Cambrian) were probably deposited as a distal facies away from the basin margin on the continental slope.

The extensional faults were reactivated in a major compressional episode of Late Cambrian/Ordovician age resulting in large-scale positive structural inversion. The distal sediments of the upper Dalradian were intensely folded into a large fanning fold complex whose axial surface varies in dip across the Highlands. No basement rocks were involved in these large-scale fold structures. The NW boundary of the Dalradian sequence is marked by major shear zones, such as the Port Skerrols Thrust, which most likely represent the original basin bounding faults. The early large scale fold fan is cut by large ductile thrusts, thickening the sedimentary pile and causing high grade Barrovian metamorphism associated with tectonic burial (Wells and Richardson 1979).

SE of the Highlands, subduction continued on the NW margin of the Iapetus Ocean to generate arc-related magmatism and large bodies of gabbro and granite in the Highlands.

Flakes of ophiolitic material and possibly magmatic arcs were obducted over the thickened Dalradian sedimentary pile (Dewey and Shackleton 1984). Ophiolite relics are preserved on Unst in Shetland and along the NW edge of the Midland Valley. South of the Midland Valley a thick Ordovician–Silurian accretionary prism formed which finally docked with a magmatic arc on the southern side of the Iapetus Ocean during Late Silurian/Early Devonian times.

All the structures SE of the Dalradian Highlands are dominated by SE-verging thrusts and folds. Mid-crustal reflectors dominantly dip towards the NW. The SE-verging structures range from the accretionary structures of the Southern Uplands, to the late collisional structures of the Highland Boundary Fault, the Lake District and Wales. The dip of these structures probably reflects the dip of the Iapetus Suture Zone. The Caledonian suture is generally taken to lie along a line from the Solway Firth to the Cheviots (Fig. 1), although prominent deep crustal reflectors lie a few tens of kilometres to the south.

SE of the suture the crust consists of Late Precambrian intrusive and volcanic rocks and volcaniclastic sediments overlain, and intruded by, Lower Paleozoic igneous rocks. In Anglesey Late Precambrian folds and shear zones, with associated high pressure metamorphism (Coward and Siddans 1979), suggest that the basement consists of a collage of accreted Late Precambrian magmatic arcs. Lower Paleozoic sediments vary in thickness, dependent on local basin development, associated with crustal stretching. There were important phases of Lower Paleozoic folding, local uplift and inversion, probably associated with the growth of the magmatic arc. The major deformation in England, Wales and Ireland, south of the suture, occurred during Late Silurian–Early Devonian times and was associated with thickening of the magmatic arc and its associated basins. Large asymmetric SE-facing folds formed above the inverted basins and crustal fabrics dip to the NW. The origin of the magmatic arc is problematical. The magmatic rocks in the Lake District, southern Ireland and North Wales lie on the SE side of the Iapetus Ocean and imply the presence of a SE-dipping subduction zone beneath this magmatic arc, in addition to the NW-dipping subduction zone beneath Scotland. However, the structures and fabrics show that the original normal faults dipped to the NW. This NW dip may follow the earlier Late Precambrian crustal fabric, reworked during Lower Paleozoic extension and inversion.

The Caledonian fabrics change strike across England and Wales; they trend N–S in central Wales, NE–SW in North Wales, E–W in the eastern Lake District and NW–SE beneath northern and eastern England.

The Iapetus Suture Zone and its associated NW-dipping crustal fabrics can be traced on deep seismic data into the Central North Sea (Freeman et al. 1988), where its eastern continuation is not clear. The western boundary of the Scandinavian Craton lies in the Caledonian thrust zones of western Norway, while the southern boundary lies along what is known as the Trans-European Fault Zone (TEF) (EUGENO-S Working Group 1988). In Scandinavia, Caledonian tectonics involved collision between the Scandinavian Craton and the North American Craton. Plate collision began in the Ordovician but continued with several hundred kilometres of crustal shortening and SE-directed overthrusting until the Late Silurian. The TEF lies approximately parallel to the plate convergence vector between the North American and Scandinavian Cratons, deduced from regional overthrust directions. However, close to the TEF, Lower Paleozoic rocks are only observed in well cores and the tectonic transport directions cannot be determined from kinematic indicators. However, the rocks carry a penetrative cleavage and a low-grade metamorphism and do not suggest large amounts of crustal thickening (Ziegler 1982; Coward 1990). The TEF is best interpreted as a zone of strike-slip movement with some transpression, along the plate boundary between the Scandinavian and NW European Blocks rather than a discrete fault.

Coward (1990) suggested that the TEF formed the original boundary between the Scandinavian Plate and the oceanic crust with its associated magmatic arcs to the south. Thus the Scandinavian Caledonides formed a zone of continent–continent collision, while the Caledonides of southern Scotland formed by continent–magmatic arc collision. Thrust tectonics with a shortening rate of c. 1 cm a^{-1} (Hossack and Cooper 1986) occurred in Norway at the same time as the growth of the accretionary prism in the Southern Uplands of Scotland. The boundary between this zone of major overthrusting and simple accretionary thrust tectonics lay at the intersection of the TEF with the active Caledonian thrust zone. This intersection occurs close to the northern end of the Central Graben. Along the TEF there may have been several episodes of transtension and transpression, depending on the relative rates and directions of convergence of the Scandinavian continental plate and the NW European oceanic plate. The TEF appears to be a wide zone of shear and there may be several individual NW–SE-trending shear zones across the central and Southern North Sea.

Strike-slip tectonics, parallel to the Caledonide trend, may have been important during the early stages of collision; some of the palaeomagnetic data of Morris (1974) suggest large strike-parallel movements of Britain relative to North America during the Early Paleozoic. Linear fabrics within the internal and early thrust zones of the Norwegian Caledonides also suggest strike-slip movements. Strike-slip displacement was important during the later stages of Caledonian collision. Bluck (1985), Soper and Hutton (1984) and Hutton (1987) argue for important left-lateral strike-slip movements in the Midland Valley, Southern Uplands and northern England, transposing several different tectonic terranes. Coward (1990) argued that the Moines NW of the Great Glen were transposed to their present position by several hundred kilometres of left-lateral movement during the late stages of Caledonian tectonics.

Variscan framework

The Variscan structures of NW Europe also formed as a result of northwestward accretion of crustal blocks and magmatic arcs onto the Laurentian foreland. The Variscan structures, which are of Devonian–Carboniferous age and hence overlap in time with the late Caledonian (Acadian) deformational events, were associated with the closure of one or more oceans in central and southern Europe, shown by the presence of ophiolites in the internal Variscan belt. Within the Variscan zones of Germany and the Ardennes, small-scale structures indicate dominantly NW-directed overthrusting associated with several hundred kilometres of crustal shortening (Coward 1990). The Trans-European Fault Zone/Tornquist Line was active at this time forming the lateral boundary of the Bohemian Massif (Ziegler 1989, 1990). In southern Britain Variscan tectonics involved a NW-verging thin-skinned fold and thrust belt (Shackleton et al. 1982; BIRPS & ECORS 1986). The mid-crustal reflectors dominantly dip to the south (Fig. 2). However, the Variscan rocks of the Armorican Massif of northern Brittany and Normandy comprise Proterozoic basement (Autran et al. 1980), part of the Icartian Block, which was deformed c. 600 Ma by large-scale shear zones and then overlain by Late Precambrian and Lower Paleozoic sediments; the latter subsequently remained relatively undeformed relative to the Variscan rocks in NE France and Belgium. There must have been a zone of strike-slip deformation along the eastern edge of the Armorican Massif, close to the present trend of the Bray Fault Zone, which separates the zone of thrust tectonics in the German Variscides from the zone of relatively weaker deformation in northern Brittany (Coward and Smallwood 1984). Isotopic data from the German Variscides (Autran *et al.* 1980) show no evidence for Precambrian crystalline basement; the rocks in this region comprise Late Precambrian to Paleozoic sediments resting on the continuation of the Brabant magmatic arc. It is suggested therefore that the Bray Fault Zone lies close to the edge of this old crustal block and acted as a major transform zone during the Variscan tectonics similar to the TEF in Caledonian times.

Devonian basins

Devonian (Old Red Sandstone, ORS) deposits underlie a large area of the Orcadian and East Shetland basins and possibly once formed a continuous basin to link with the Devonian sediments of SW Norway (Fig. 3). Most data on the sedimentary facies and structure of the western part of this basin come from the onshore remnant of the basin in Caithness, Orkney, Shetland and the Moray Firth. The Lower ORS (Siegenian–Emsian) has a restricted distibution and consists of syn-rift sequences which pass up, without any important stratigraphic break, into the Middle ORS (Eifelian–Givetian). The Lower ORS is characterized by coarse lenticular breccias and conglomerates interfingered with finer bedded sandstones, deposited within small intermontaine extensional basins (see Mykura 1976). Local thick fanglomerates suggest deposition in basins bounded by active fault scarps. The Middle ORS, which oversteps the Lower ORS in the west and onlaps the basement, consists of lacustrine flagstones and sandstones, up to 4 km thick in Caithness, formed in a post-rift subsidence phase of sedimentation. Minor unconformities occur at the base of the Middle ORS, possibly related to tilting and differential uplift during extension, rather than to a phase of positive tectonic inversion (Coward *et al.* 1989). There was a gradual regression in later Middle ORS times to dominantly alluvial sedimentation. Palaeoflow was generally towards the NE, parallel to the trend of the basin (Foster 1972) although locally the sediment supply was from the NW as shown by the presence of Torridonian and Cambrian clasts.

There is an important unconformity between the Middle and Upper ORS throughout the Orcadian basin. On the Walls Peninsula of SW Shetland a thick sequence of Middle ORS sandstones was deformed by NE–SW-trending folds before the intrusion of the Sandwich plutonic complex dated at 360 ± 11 Ma (Mykura 1976). On the Orkneys the onset of the Upper ORS sedimentation was marked by an unconformity followed by calc-alkaline lavas dated at *c.* 370 Ma (Halliday *et al.* 1977), followed by high-energy fluvial deposits, locally with aeolian dunes. These Upper ORS deposits are regarded as Frasnian (Rogers *et al.* 1989). Stratigraphic onlaps related to basin inversion have been recognized from seismic data from the West Orkney Basin (Coward and Enfield 1987; Coward *et al.* 1989). The most important effects of ORS inversion are seen close to the Walls Boundary Fault and its offshoots in the Orkney Islands. This inversion is probably associated with strike-slip tectonics associated with the Devonian movements on the proto-Great Glen Fault. From the trends of the folds in the western part of Shetland and the structures within the fault zones on the Orkneys, the shear couple along the proto-Great Glen Fault was left-lateral (Coward and Enfield 1987; Coward *et al.* 1989).

The extension direction varies across the Orcadian Basin. In western Shetland the predominant SW dip of the sediments suggests a NE–SW extension on a moderate–gently NE-dipping normal fault (Serrane 1991). The gently dipping normal fault at the northern edge of this basin is probably a lateral structure. In the East Shetland Basin the extension direction is clearly NE–SW as indicated by the development of small strike-slip faults and small-scale kinematic indicators on faults (Coward *et al.* 1989). However, in the West Orkney Basin the extension direction is clearly NW–SE as denoted by the presence of transfer structures giving a bucket shape to the extensional faults and also by small-scale kinematic indicators on the faults seen onshore. The extensional trajectories curve from NW–SE in the west to NE–SW in the east and the corresponding fault trends also change from west and east (Fig. 3).

In western Norway two sets of Devonian basins occur: (i) the Hitra basins controlled by left-lateral displacement along the Hitra Fault, which is probably a splay off the Great Glen

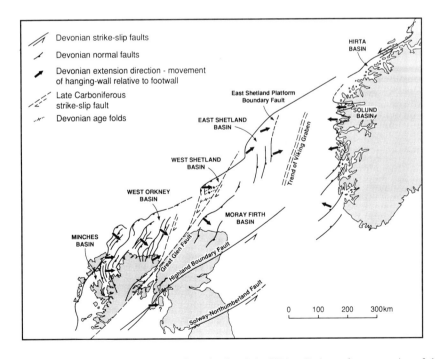

Fig. 3. Map of the Devonian basins and fault systems, north of Scotland and the Viking Graben, after restoration of the Permo-Carboniferous right-lateral strike-slip movements along the Great Glen Fault Zone.

system (Serrane 1991), and (ii) the Solund-Hornelen basins which formed on the hangingwalls of listric normal faults which reactivated Caledonian thrusts (Serrane 1988; Seranne *et al.* 1989). Kinematic indicators in the basal Devonian sediments of the Solund–Hornelen basins indicate the extension direction to be towards the WNW, almost parallel to the lateral walls of the listric faults. In the Hitra basins the extension direction was WSW, parallel to the Hitra fault system.

The SE edge of the Devonian basin system in Scotland is marked by the Midland Valley Basin, infilled with thick fluvial red bed molasse deposits locally up to 9 km thick in the Strathmore region at the northern edge of the Midland Valley (Haughton 1988). Regional fault kinematic studies indicate both dip-slip and strike-slip movements (Haughton 1989; Coward 1990). Bluck (1985) and Haughton (1989) note that many of the clasts in the Lower Devonian sediments cannot be matched with the adjacent Caledonian basement. Haughton (1989) suggests that a basement similar to that of southern Greenland was the source of Devonian sediments in the Midland Valley but that a basement similar to the Grenville of the NE Appalachians was the source of the Lower Paleozoic rocks of the Southern Uplands. Thus there may have been large-scale lateral movements, with displacements of several hundred kilometres. Several of the faults controlling these strike-slip movements may lie within, or south of, the Southern Uplands where large strike-slip fault zones have been recognized. Soper and Hutton (1984) and Hutton (1987) emphasize these strike-slip movements and favour a strike-slip accretion model for the southern Caledonides.

The left-lateral displacements in the Midland Valley and the Southern Uplands are likely to link with either the Trans-European Fault Zone in southern Denmark, which marks the southern edge of the Svecofennian Craton (Soper and Hutton 1984), or with the Hitra Fault Zone in western Norway (Coward 1990). The latter seems most likely during the Devonian as both the Midland Valley and Hitra fault zones bound Devonian basins and show evidence for left-lateral displacement. There is no indication that the Midland Valley and Trans-European fault zones were linked at this time. If they were linked, the fault system would have to bend through *c.* 45°. However, late Devonian sediments occur in the Central North Sea and in the Danish Sector suggesting that there was no compressive or uplift zone of Devonian age where the Midland Valley and Trans-European systems may have joined.

The Midland Valley and Hitra fault zones are offset by several hundred kilometres and the Devonian basins define a large pull-apart basin in this left-lateral system. The shape of this basin is shown in Fig. 3 after restoration of Late Carboniferous right-lateral strike-slip movements on the Great Glen Fault and associated faults in the Orkneys. These right-lateral displacements have been established from offsets of Caledonian structures and geophysical anomalies associated with Caledonian intrusives on Shetland (Flinn 1969; Flinn *et al.* 1979). The restored Devonian basin is *c.* 600 km long and 200 km wide. The associated normal faults trend N–S to NNE–SSW to NNW–SSE on the West Shetland Platform and the extension direction varies from NW–SE at the western and eastern boundaries of the basin to ENE–WSW in the centre of the basin (Fig. 3).

The Great Glen Fault system cuts across the centre of the basin and shows evidence of Devonian left-lateral strike-slip movements in that inversion, post-dating the middle ORS post-rift basin development, is concentrated along the western edge of this fault zone (Fig. 3, see also Coward *et al.* 1989).

The basin is locally deep with >5 s (TWT) of sediments in parts of the West Orkney Basin (Coward *et al.* 1989) in the western part of the pull-apart basin. Gently dipping reflectors, suggesting several kilometres of Lower Paleozoic sediments

(?Devonian), mark the East Shetland Platform. The thick gently dipping reflectors suggest half-grabens trending N–S to NNW–SSW; the reflectors are covered by >1 s (TWT) of sub-horizontal parallel reflectors which may represent the Devonian post-rift fill. The WNW trend of the northern part of the Viking Graben most likely reflects reactivation of Devonian fracture systems.

The stretching factor in the West Orkney Basin is $\beta = 1.4$, as measured from the sums of fault heaves. Under the East Shetland Platform the Devonian sediments are much thicker and the stretching factor may be very large. The Devonian stretching affected crust previously thickened during the Caledonian orogeny. At the peak of Caledonian compression and metamorphism, the crust of the NW Highlands was *c.* 50 km thick (Coward 1983) and hence a considerable amount of crustal stretching was necessary to reduce the topography to sea level. This was only achieved during the early Carboniferous rifting episode. Lower Devonian basins must have been cannibalized during subsequent Devonian stretching so that only the later stages of the stretching history are presvered in the alluvial–lacustrine facies of the Orcadian Basin. Hence it is impossible to estimate crustal stretching from sediment thicknesses.

As the lithosphere had probably not thermally equilibrated following Caledonian thickening, the thermal history of the Orcadian and East Shetland basins must have been very different from that of, for example, the Carboniferous basins of the Southern North Sea or the Mesozoic basins west of Britain, where stretching affected lithosphere of normal thickness or lithosphere thinned by earlier stretching events. The Orcadian–East Shetland basins must have had lower geothermal gradients and hence much cooler basins than Mesozoic basins of the North Sea and west of Britain.

The West Orkney to Viking Graben pull-apart system is only a small part of a much larger strike-slip system affecting the northern Appalachian–Caledonide Belt. An important branch of the northern strike-slip fault may continue west of Norway and link with the Late Caledonian Jan Mayen left-lateral strike-slip system (Harland and Gayer 1972; Ziegler 1989, 1990).

In southern Britain there was important extension of the Lower Paleozoic shelf during the Devonian, producing Lower Devonian growth faults in South Wales (Powell 1989) and North Devon. The normal faults trend approximately E–W. In southern Cornwall, Devonian sediments are characterized by distal turbidites, deposited on thinned crust at the edge of a continental shelf. Holder and Leveridge (1986) estimate Devonian sediments to be >12 km thick in southern Cornwall, suggesting that they were deposited on highly attenuated continental crust or on ocean floor. The Lizard ophiolite complex in south Cornwall is part of this ocean basin and gives ages of 375 ± 34 Ma, suggesting a Middle Devonian age for ocean floor development (Davies 1984). The ocean basin is limited in size; gravity anomalies suggest that the Lizard ophiolite complex is limited to the region south of Devon and Cornwall. Furthermore, the dyke trends in the ophiolite suggest NE–SW extension, at a high angle to the trend of the basin. E–W to ESE–WSW-trending right-lateral strike-slip systems developed along a line from southern Britain to central Germany (Brun and Balé 1989; Strachan *et al.* 1989; Coward 1990); the Lizard ophiolite probably formed in a large pull-apart basin which developed during NE–SW extension along an E–W-trending shear (Barnes and Andrews 1986; Coward 1990). This shear system may pass beneath the northern margin of the Variscan fold and thrust belt to link with the Tornquist Line, Polish Trough and North Dobroagean Zone on the northern edge of the Black Sea (Ziegler 1989, 1990).

Devonian tectonics are therefore characterized by the development of large-scale strike-slip zones and large pull-apart

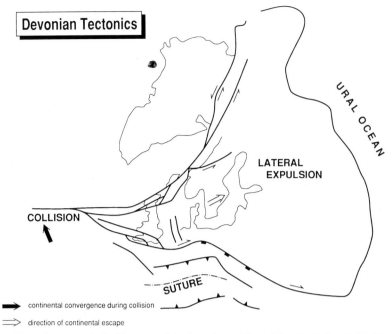

Fig. 4. Simplified map illustrating Devonian tectonics in Europe and the lateral expulsion of the North Sea–Baltic Block, bounded by left-lateral and right-lateral shear systems.

basins. In northern Britain and offshore western Norway the pull-apart basins collapsed the previously thickened Caledonian crust, while in southern Britain they stretched crust of normal thickness to produce oceanic crust. The two systems of strike-slip faults, left-lateral in northern Britain and right-lateral in southern Britain, suggest a driving mechanism which involved lateral continental escape (Fig. 4) similar to that suggested for eastern Tibet by Tapponnier *et al.* (1986) and western Turkey by Sengor *et al.* (1985) and Dewey *et al.* (1986). The Devonian strike-slip motion was coeval with Acadian tectonics in the Central and Northern Appalachians and the lateral expulsion was probably associated with continental collision between the North American and African terrains.

Carboniferous basins

Lower Carboniferous extension continued in the Northern North Sea, Midland Valley of Scotland and Ireland to produce thick Carboniferous deposits beneath parts of the Moray Firth Basin and Midland Valley/Forth Approaches Basin. This extension was probably associated with the reactivation of Devonian fault systems. In southern Britain, south of the Brabant Massif, thick Carboniferous sequences occur in South Wales, the Mendips, SE England and the Ardennes, along the northern edge of the Rhenohercynian Basin. A large turbidite basin developed in north Devon along the northern margin of the Devonian basin.

However, the main change from Devonian to Carboniferous tectonics was the development of extensional basins in central and northern England, the southern North Sea and northern Germany, across the block bounded by the left-lateral and right-lateral Devonian strike-slip faults.

Important basin bounding faults occur along the northern edge of the Solway Firth, the Dent fault system and its continuation along the Lunedale Fault, the Craven/Bowland system and the NE edge of the Goyt Trough. Visean limestones vary in thickness from *c.* 3 km in the Solway and Bowland Basins to 500 m over the Alston fault block (Ord *et al.* 1988; Fraser and Gawthorpe 1990). In the Bowland Basin, basin-wide debris flows demonstrate episodes of extension during the late Chadian and late Asbian (Gawthorpe 1987). In southern Ireland, Dinantian basins occur on the west coast of Clare and in Munster. In North Wales, normal faults formed

parallel to the Caledonian fabric along the Menai Straits and on Anglesey. Within this region the extension direction was dominantly NW–SE (Fig. 5) as shown by the transfer faults along the edge of the Solway and Bowland basins (Gawthorpe 1987). The stretching factor, determined from the cumulative sum of the heaves on large faults, is relatively low, $\beta = 1.1$ or less, although heaves on individual faults may be large.

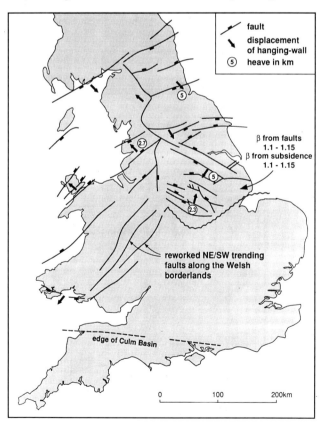

Fig. 5. Map of southern Britain showing the principal normal faults and lineaments active during the early Carboniferous. Fault heaves measured from seismic data published by Fraser and Gawthorpe (1990). Fault displacement directions taken from regional studies of field and seismic data.

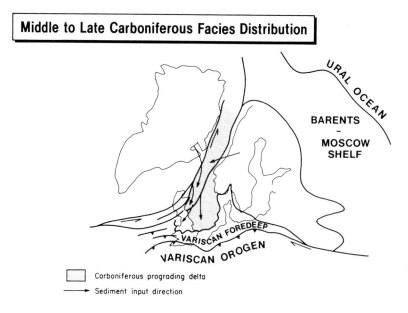

Middle to Late Carboniferous Facies Distribution

☐ Carboniferous prograding delta

⟶ Sediment input direction

Fig. 6. Map showing the distribution of Middle to Upper Carboniferous facies in Europe and illustrating the variation from (i) shelf facies on the Ural passive margin in the E, to (ii) deltaic facies associated with drainage of the Caledonian Mountains into the North Sea and northern Britain and (iii) Variscan foredeep sediments in central Europe.

In the East Midlands oil province and the Southern North Sea, the normal faults trend NW–SE. Curved fault patterns and transfer zones suggest NE–SW extension, perpendicular to that of northern England, Wales and Ireland (Fig. 5). However, the extensional history of the NW- and NE-trending basins is identical (Fraser and Gawthorpe 1990). In the East Midlands the amount of extension measured from fault heaves is $\beta = 1.1$–1.15, slightly higher than that of northern England; the faults are equally large but are more closely spaced. There are insufficient data to estimate Carboniferous extension across the Southern North Sea.

Fraser and Gawthorpe (1990) recognize two main syndepositional systems: (i) a clastic fluvio-deltaic sequence of Late Devonian/Early Chadian age, consisting of ORS facies sediments, probably laterally derived from the basin margins, and (ii) carbonate platforms in which the sediment fill is characterized by the presence of boulder beds and slumps interbedded with carbonate grainstones to deeper water wackestones. The sediments deomonstrate wedge-shaped geometry on seismic profiles and several phases of rifting can be correlated throughout the English basins.

Most of the half-grabens were filled by early Namurian times, when steady subsidence and the influx of fluvio-deltaic sequences of sandstones, coals and marine shales began. Cyclic alternations of marine and terrestrial sediments from the Dinantian through to the Westphalian reflect pulses of thermal subsidence (Leeder 1988) together with eustatic changes of sea level associated with the earliest of the Permo-Carboniferous Gondwana glaciations (Heckel 1990). The basins were infilled by river systems flowing from the NNE, along the line of the Devonian pull-apart system; presumably the strike-slip faults and pull-apart basins NE of Scotland and west of Norway channelled the Carboniferous drainage (Fig. 6).

The late Carboniferous basin in the Southern North Sea and southern Britain is sometimes considered as a foredeep basin to the Variscan thrust structures (e.g. Ziegler 1982, 1990). This theory is rejected here for the following reasons.

(i) The Variscan mountain belt had relatively little topographic relief and Variscan deformation produced no excessive crustal thickening to form a flexural load; there is no obvious driving mechanism for a foredeep in the Southern North Sea at this time;

(ii) There is no indication of depocentre or pinch-out migration as would be expected in an advancing foredeep.

A thermal subsidence model is therefore favoured for the late Carboniferous basin development in the Southern North Sea.

An indication of the amount of thermal subsidence is given by 1 km of Westphalian rocks in the Midland Valley and >2 km in the southern Pennines (Fraser and Gawthorpe 1990). The thickest sequence of post-rift sediments occurs in southern Lancashire, where Fraser and Gawthorpe (1990) record thicknesses of >3.5 km. This thickness of thermal subsidence sediments suggests a stretching factor of >2, far greater than that indicated from the faults.

In the East Midlands, the Middle to Upper Carboniferous post-rift cover is only 1.0–1.5 km thick, suggesting low stretching factors in the order of $\beta = 1.1$–1.15, similar to stretching factors obtained from summing fault heaves. Similarly in the Southern North Sea the Westphalian reaches a thickness of 2 km (Ziegler 1982; Leeder and Hardman 1990), while in northern Germany the combined thickness of Namurian and Westphalian sediments reaches only 2 km in thickness. The thick post-rift sediments in southern Lancashire are therefore anomalous and suggest an extra local control on subsidence other than the stretching observed from Carboniferous growth faults. This may reflect an excess amount of thermal subsidence following the early Carboniferous igneous activity.

The Brabant Massif remained as a tectonic high throughout the early Carboniferous and was onlapped by Namurian and Westphalian sediments, giving a steer's head pattern of onlap similar to that of other thermal subsidence basins. However, no large fault has been recognized along the northern margin of the Brabant Massif in the East Midlands. Along the edge of the Brabant Massif in the Welsh Borderlands, Caledonian faults trend NE–SW (Fig. 5) and show evidence for several episodes of strike-slip reactivation, possibly during the Carboniferous. However, in the East Midlands the Brabant Massif is marked by the lateral termination of NW–SE-trending dip-slip normal faults (Fig. 7). In the Southern North Sea, the edge of the Brabant Massif is marked by the NW–SE-trending Dowsing Fault Zone.

A tectonic model for the Carboniferous extension must take into account the lateral variations in extension direction

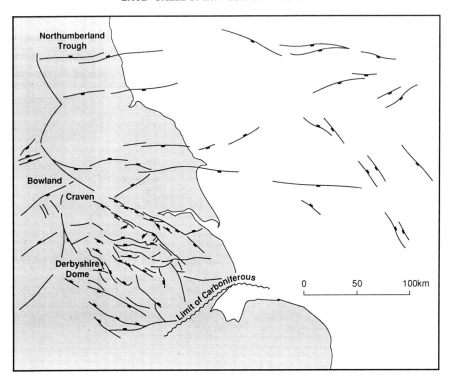

Fig. 7. Map showing the distribution of faults with NW–SE trend in the East Midlands and the Southern North Sea and their relationship to the northern edge of the Brabant Massif. Fault data in the North Sea from Leeder and Hardman (1990).

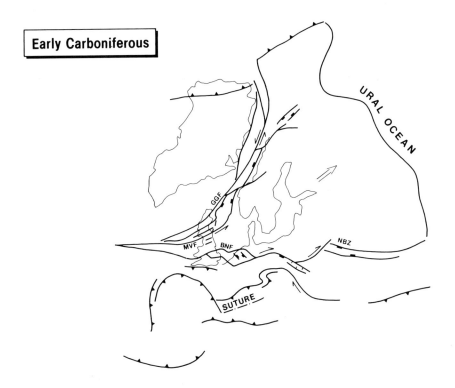

Fig. 8. Simplified map illustrating the early Carboniferous tectonics of Europe which continued the Devonian tectonic process. GGF, Great Glen fault system; MVF, Midland Valley fault system; BNF, Brabant–North Sea fault system; NBZ, North Dobroagean Zone.

throughout central and northern England. Leeder (1987, 1988) suggests that back-arc extension related to northward subduction during the closing of the Variscan Rheic Ocean, caused the Carboniferous stretching north of the Brabant Massif. However, NW–SE-trending back-arc extension would not form the NW–SE-trending faults of the East Midlands, Southern North Sea and northern Germany. Some extra tectonic component is required.

During the Devonian–Carboniferous the Appalachians continued to undergo Acadian–Variscan collision. The lateral expulsion which generated the Devonian basins may also have occurred spasmodically throughout the Carboniferous and may explain the anomalous NW–SE stretching direction. The Lizard pull-apart basin had closed by Middle Devonian times and the ophiolites had been obducted onto the Cornish Devonian flysch by NW-directed thrusts (Holder and Leveridge

1986; Coward 1990). The Western European Block, which had been laterally expelled, therefore required a new strike-slip system along its southern margin (Fig. 8). Caledonian tectonics resulted in a system of fabrics and faults which varied in trend from NE–SW along the Welsh Borderlands to NW–SE in the East Midlands and the Southern North Sea (Coward 1990). These faults were reactivated in a diffuse system of right-lateral shear. The NE–SW-trending fault, such as the Church Stretton Fault, was reactivated as a simple strike-slip fault. However, the shear was more diffuse in the East Midlands and associated with right-lateral rotation of blocks of the Caledonian basements.

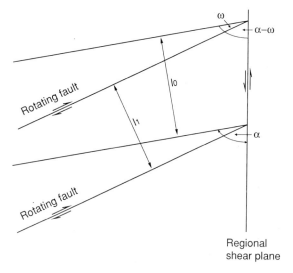

Fig. 9. Diagram to illustrate the shortening $\sqrt{\lambda} = l_1/l_0$, caused by block rotation.

This style of diffuse shear, associated with block rotation, has been used to account for rotations, documented by palaeomagnetic studies, in the western Aegean (McKenzie 1978). Rotation of the blocks causes internal strains. As shown in Fig. 9, if the boundary conditions are fixed so that there is no lateral expansion or contraction and if the blocks rotate so that their bounding faults rotate closer to the orientation of the shear couple, the blocks will need to narrow and lengthen. If the blocks rotate so that their bounding faults rotate away from the shear system, the blocks will widen and shorten. The amount of block narrowing or widening can be calculated from the relationship:

$$l_1/l_0 = \sin(a - \omega)/\sin a$$

where a is the original angle made by the faults, relative to the plane of the regional shear and ω is the angle of rotation. This relationship is summarized in graphical form in Fig. 10.

Thus a diffuse shear system, which rotates older crustal blocks, can cause extension or inversion. It is suggested that during the early Carboniferous a NE-trending right-lateral shear couple affected the Caledonian structures of central Britain, causing rotation from an approximately WNW trend to a NW trend, widening the blocks and allowing stretching in the order of $\beta = 1.1$–1.15. The associated contraction along the length of the blocks was cancelled by a regional NW–SE extension linked to a back-arc basin associated with NW subduction of the Variscan Ocean. Where the shear system was not active, in northern England, Wales and southern Ireland, only NW–SE back-arc extension was active. The right-lateral shear couple continued from the Brabant–North Sea fault systems through the Southern North Sea into northern Germany to join with the Tornquist Line and North Dobroagean Zone into the northern Black Sea. This system of diffuse shear

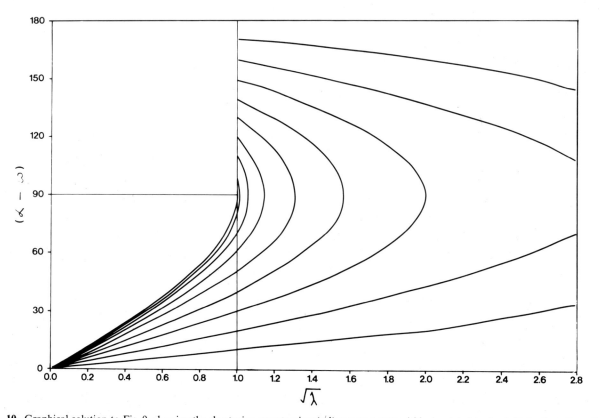

Fig. 10. Graphical solution to Fig. 9, showing the shortening or extension ($\sqrt{\lambda}$) across a rotated block related to the initial orientation of the block relative to simple shear (a) and the amount of rotation (ω). Depending on the orientation of the shear relative to the block, the blocks will widen ($\sqrt{\lambda} > 1$), resulting in further extension, or narrow ($\sqrt{\lambda} < 1$), resulting in inversion. Note that to maintain constant area, a corresponding change in length ($= 1/\sqrt{\lambda}$) should occur along the long axes of the blocks, unless the tectonic situation involves regional extension or contraction, as proposed for the Carboniferous of NW Europe.

and rotation associated with regional NW–SE extension maintains strain compatibility between the different extensional basins. The regional model for Lower Carboniferous tectonics is summarized in Fig. 8.

Late Carboniferous–Permian inversion

End-Carboniferous uplifts and folds characterize the basin-bounding faults and folds of NW Europe (Gawthorpe 1987; Fraser and Gawthorpe 1990). The NE-trending folds are often tight, for example, those that occur in the Pendle–Skipton area associated with inversion of the Skipton–Clitheroe Basin (Gawthorpe 1987). Lower Permian peneplanation of these folds produced a varied subcrop to the base Permian. Fraser and Gawthorpe (1990) estimate > 3000 metres of uplift associated with inversion in the Bowland Basin. Sediments associated with the inversion are characterized by continental red beds and molasse in intermontain troughs, lying on the footwalls to the syn-rift half-grabens.

Basin inversion affected the rift basins of the East Midlands and the Southern North Sea. Leeder and Hardman (1990) describe more than 50 NW–SE-trending folds in the Southern North Sea, some associated with thrusts. Significant erosion and truncation of these inversion anticlines was accompanied by the deposition of the Barren Red Measure facies in typical growth fold basins.

The NE-trending fault zones of Scotland show evidence of right-lateral displacement associated with uplift during the Late Carboniferous and Permian. On the East Shetland Platform, tectonic inversion affects Devonian sediments on the hangingwalls of the N–S-trending faults. The structures show typical inversion geometries, in that the syn-rift sequences show mostly net extension, while the post-rift Middle ORS sequences show net compression. Kinks and small thrusts affect the Middle ORS of the Sumburgh Peninsula and large folds trend NNW across the region from Sumburgh to Lerwick, probably developed on the hangingwalls of reactivated normal faults. From their trend, these folds formed by E–W compression, or more likely were associated with right-lateral strike-slip movements on the Shetland fault system (Coward and Enfield 1987; Coward et al. 1989).

The Late Carboniferous–Permian inversion episode therefore affected all the rift basins, irrespective of their orientation. Some of this deformation may have been associated with NW–SE compression related to Variscan tectonics in NW Europe. However, the N–S-trending folds in the North Sea suggest strike-slip reactivation of the earlier Devonian and Carboniferous strike-slip systems. The inversion tectonics reversed movements on these strike-slip systems, so that the original left-lateral shear along the Great Glen–Midland Valley was replaced by right-lateral shear and the original right-lateral shear couple in central/southern England and the Southern North Sea was replaced by a left-lateral couple. Thus the East Shetland Platform became a pushup zone on a restraining bend related to the right-lateral shear. Similarly the blocks of the East Midlands and the Southern Northern Sea were affected by a diffuse left-lateral shear couple causing anti-clockwise rotation and hence shortening across the blocks.

The overall tectonic pattern is one of the NW European fault wedge being pushed back between the North American and West European blocks, reversing the movement generated during lateral continental escape during the Devonian and Early Carboniferous (Fig. 11). Presumably the region was also affected by some NW–SE compression at this time associated with the late stages of Variscan collision. The driving mechanism for this wedge indentation was most likely to have been plate collision in the Urals in end-Carboniferous and Permian times.

Influence of Paleozoic structures on Mesozoic basin development

Control on subsidence

Crustal stretching during the Early Carboniferous in the Southern North Sea largely controlled the position and intensity of Middle Carboniferous–Permian subsidence. Much of the subsidence in the Southern Permian Basin of the North Sea was probably a result of thermal equilibrium following Carboniferous extension. Irregularities in thickness of the Rotliegendes sediments may result from infilling an irregular topography formed by late Carboniferous inversion, rather than

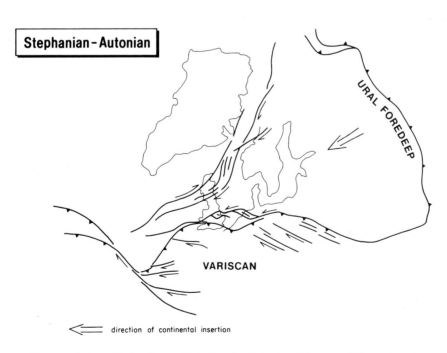

Fig. 11. Simplified map to illustrate the Late Carboniferous–Early Permian tectonics of Europe which involved closure of the Ural Ocean and collision tectonics in the Urals and led to the wedge being driven back between the North American and Variscan plates.

from intense early Permian rifting. In northern England a region of middle to late Carboniferous thermal subsidence overlies the zone of Dinantian stretching, but in southern Lancashire there is a conflict between the amount of stretching and thickness of post-rift fill. The excess subsidence in this region must result from an additional cause, for example, a local thinning of the lithospheric mantle related to high heat flow associated with back-arc basin development.

Generation of anomalously thick crust

Stretching factors estimated from the accumulation of fault heaves for the Jurassic Viking Graben are far less than stretching factors estimated from the variation in thickness of crust across the Viking Graben. β factors from fault heaves are in the order of 1.15–1.25 (Ziegler 1982, 1990), while β factors from crustal thickness are in the order of 1.6–2.0 (Roberts *et al.* 1990). It is unlikely that small-scale faults, undetectable on seismic data, can account for this discrepancy in shortening estimates. Barton and Wood (1984) suggest that the Jurassic extension may have affected crust previously thinned in the Triassic, that is, the Triassic extension may have been in the order of $\beta = 1.5$. If Triassic extension was responsible for the extra thinning of the crust, then Triassic sediments would be thick beneath the Viking Graben. An alternative explanation is that some crustal thinning is of Devonian/Carboniferous age, related to the generation of the proto-Viking Graben.

Modification of lithospheric strength

Stretching initially weakens the lithosphere but following thermal equilibration, the zone of thinned lithosphere will become stronger than the adjacent zones of less thinned lithosphere (Dewey 1982; Gillcrist *et al.* 1987). Thus the zones of Devonian and Carboniferous stretching may become zones of strength until weakened by Permian and Middle Jurassic volcanism. Similarly lithospheric thickening will initially strengthen the lithosphere, but as the lithosphere thermally equilibrates and the lower crust becomes hotter, it will weaken and hence lead to gravitational spreading. Thus the zones of more intense Caledonian thickening in Scotland and Norway became the sites of intense Devonian extension.

Generation of fundamental faults and their subsequent reactivation to generate basin bounding faults

Several late Paleozoic and Mesozoic basins occur on the hangingwalls of earlier thrusts or normal faults. Examples are the Outer Isles and Minches basins on the hangingwall of the Outer Isles Fault (Coward and Enfield 1987), basins on the hangingwalls of Caledonian thrusts and reactivated normal faults along, and to the south of, the Solway Line (Hall *et al.* 1984), and the Celtic Sea and Bristol Channel basins in the hangingwalls of Variscan thick-skinned thrusts (Cheadle *et al.* 1987; Brooks *et al.* 1988). The NW-trending structures in the northern Viking Graben may follow the earlier Devonian fault systems. Devonian and Mesozoic basins have similar trends on the East Shetland Platform.

Large strike-slip fault systems, such as the TEF, the Bray and Great Glen faults appear to have acted as basin-bounding tear faults throughout several episodes of extension and inversion. As argued by Coward (1990), the large tear faults offset crust with different fabrics and strengths and also often offset either the Moho and/or the brittle/ductile transition zone in the middle to lower crust. These tear faults form steeply dipping strength discontinuities in the lithosphere which can control (or 'tram-line') subsequent deformation.

Conclusions

1. The crustal structure of NE Europe formed by the accretion of fragments of continental material plus new magmatic arcs on to a North Atlantic Craton. The emplacement direction was SE–NW. Paleozoic transform systems such as the TEF and Bray fault system, which bound old continental fragments, formed fundamental NW–SE-trending lineaments which were reactivated several times during the Mesozoic and Tertiary.

2. The process of SE–NW accretion was modified by along-strike movements associated with the lateral continental escape of a large triangular shaped fragment of northern Europe. This fragment was bordered by: (i) left-lateral shear systems along its NW margin, forming the Great Glen–Hitra–Jan Mayan fault zones, (ii) right-lateral shear systems along its southern margin, forming the Channel–northern Germany–Polish Trough fault zones and (iii) Ural oceanic crust in the east.

3. Continental escape tectonics were most important during the Devonian and early Carboniferous, during the final stages of Caledonian collision and early stages of Variscan collision.

4. A large Devonian pull-apart basin formed along the NW shear producing the Orcadian–East Shetland basin. The basement structures of the northern part of the Viking graben formed at this time. Along the southern shear zone a Devonian pull-apart basin formed oceanic crust, now preserved as the Lizard Complex. This southern pull-apart basin was partially closed due to Variscan compression during the Middle Devonian.

5. Following closure of the Devonian pull-apart basin, the right-lateral shear system moved to the northern part of the Brabant Massif, to cause the clockwise rotation of Caledonian fabrics. Because of the orientation of the shear couple relative to the basement fabrics, the basement blocks widened during rotation to produce the NW–SE-trending early Carboniferous normal faults of the East Midlands, Southern North Sea and northern Germany.

6. Early Carboniferous continental escape was synchronous with regional back-arc extension with subduction of Variscan oceanic crust. The escaping English–North Sea–Baltic block underwent NW–SE extension synchronous with left-lateral shear along the Great Glen system and right-lateral shear along the diffuse East Midlands–Southern North Sea system.

7. The pull-apart basins affected crust of different thickness. The northern Devonian pull-apart basin stretched crust previously thickened during the Caledonian orogeny, so that even with large stretching values, the basins were filled with terrestrial or lacustrine deposits. The southern pull-apart basins, however, affected crust of normal thickness, generating marine basins and local oceanic crust. The northern pull-apart basin acted as the locus of alluvial transport from the Caledonian mountains to the southern basins; the deltaic facies spread southwards during the early–middle Carboniferous from the southern end of the proto-Viking Graben.

8. Minor inversion events occurred during the Carboniferous, particularly at the beginning of the Westphalian, when Variscan folding affected Britain from South Wales to the Midland Valley. However, the main inversion episode occurred during the late Carboniferous–early Permian, when all the Carboniferous basins were inverted and the shear sense along the major strike-slip faults was reversed. This inversion episode coincided with plate collision affecting the Urals, that is, it probably formed by the escaped block being driven back as an indenting wedge between the North Atlantic craton and Variscan Mountain Belt.

9. Caledonian, Devonian and Carboniferous structures were reactivated during the Mesozoic and Tertiary, for

example, Mesozoic basin trends (Celtic Sea, northern Viking Graben) are reactivated Upper Paleozoic basin trends and much of the Tertiary inversion in the Southern North Sea can be explained by rotational reactivation of the earlier Carboniferous fault blocks.

References

ANDERTON, R. 1985. Sedimentation and tectonics in the Scottish Dalradian. *Scottish Journal of Geology*, 21, 407–436.

AUTRAN, A., BRETON, J.-P., CHANTRAINE, J. AND CHIRON, J.-C. 1980. *Carte tectonique de la France.* 1:1M.

BARNES, R. P. AND ANDREWS, J. R. 1986. Upper Palaeozoic ophiolite generation and obduction in south Cornwall. *Journal of the Geological Society, London*, 143, 117–124.

BARTON, P. AND WOOD, R. J. 1984. Tectonic evolution of the North Sea basin: crustal stretching and subsidence. *Geophysical Journal of the Royal Astronomical Society*, 79, 987–1022.

BIRPS AND ECORS 1986. Deep seismic reflection profiling between England, France and Ireland. *Journal of the Geological Society, London*, 143, 45–52.

BLUCK, B. J. 1985. The Scottish paratectonic Caledonides. *Scottish Journal of Geology*, 21, 437–464.

BREWER, J. AND SMYTHE, D. K. 1984. MOIST and the continuity of crustal reflector geometry along the Caledonide–Appalachian orogeny. *Journal of the Geological Society, London*, 141, 105–120.

BROOKS, M., TRAYNER, P. M. AND TRIMBLE, T. J. 1988. Mesozoic reactivation of Variscan thrusting in the Bristol Channel area, UK. *Journal of the Geological Society, London*, 145, 439–444.

BRUN, J. P. AND BALÉ, P. 1989. Cadomian tectonics in Northern Brittany. *In*: D'LEMOS, R. S., STRACHAN, R. A. AND TOPLEY, C. G. (eds) *The Cadomian Orogeny.* Geological Society, London, Special Publication, 51, 95–114.

CHEADLE, M. J., MCGEARY, S., WARNER, M. R. AND MATTHEWS, D. H. 1987. Extensional structures on the western UK continental shelf: a review of evidence from deep seismic profiling. *In*: COWARD, M. P., DEWEY, J. F. AND HANCOCK, P. (eds) *Continental Extensional Tectonics.* Geological Society, London, Special Publication, 28, 445–465.

COWARD, M. P. 1983. The thrust and shear zones of the Moine thrust zone and the Scottish Caledonides. *Journal of the Geological Society, London*, 140, 795–812.

—— 1990. The Precambrian, Caledonian and Variscan framework to NW Europe. *In*: HARDMAN, R. F. P. AND BROOKS, J. (eds) *Tectonic Events Responsible for Britain's Oil and Gas Reserves.* Geological Society, London, Special Publication, 55, 1–34.

—— AND ENFIELD, M. A. 1987. The structure of the West Orkney and adjacent basins. *In*: BROOKS, J. AND GLENNIE, K. W. (eds) *Petroleum Geology of North West Europe.* Graham & Trotman, London, 687–696.

——, —— AND FISCHER, M. W. 1989. Devonian basins of northern Scotland: extension and inversion related to Caledonian-Variscan tectonics. *In*: COOPER, M. A. AND WILLIAMS, G. D. (eds) *Inversion Tectonics.* Geological Society, London, Special Publication, 44, 275–308.

——, KIDD, W. S. F., PAN YUN, SHACKLETON, R. M. AND ZHANG HU. 1988. The structure of the Tibet Geotraverse, Lhasa to Golmud. *Philosophical Transactions of the Royal Society of London*, A327, 307–336.

—— AND SIDDANS, A. W. B. 1979. The tectonic evolution of the Welsh Caledonides. *In*: HARRIS, A. L., HOLLAND, C. H. AND LEAKE, B. E. (eds) *The Caledonides of the British Isles Reviewed.* Geological Society, London, Special Publication, 9, 187–198.

—— AND SMALLWOOD, S. 1984. An interpretation of the Variscan tectonics of SW Britain. *In*: HUTTON, D. H. AND SANDERSON, D. J. (eds) *Variscan Tectonics of the North Atlantic Region.* Geological Society, London, Special Publication, 14, 89–102.

DAVIES, G. R. 1984. Isotopic evolution of the Lizard Complex. *Journal of the Geological Society, London*, 141, 3–14.

DEWEY, J. F. 1982. Plate tectonics and the evolution of the British Isles. *Journal of the Geological Society, London*, 129, 371–412.

——, HEMPTON, M. R., KIDD, W. S. F., SAROGLU, F. AND SENGOR, A. M. C. 1986. Shortening of continental lithosphere: the tectonics of eastern Anatolia—a young collision zone. *In*: COWARD, M.

P. AND RIES, A. C. (eds) *Collision Tectonics.* Geological Society, London, Special Publication, 19, 3–36.

—— AND SHACKLETON, R. M. 1984. A model for the evolution of the Grampian tract in the early Caledonides and Appalachians. *Nature*, 312, 115–121.

——, ——, CHANG CHENG FA AND SUN YUJIN 1988. The tectonic evolution of the Tibetan Plateau. *Philosophical Transactions of the Royal Society of London*, A327, 379–413.

EUGENO-S Working Group 1988. Crustal structure and tectonic evolution of the transition between the Baltic Shield and the North German Caledonides (the EUGENO-S Project). *Tectonophysics*, 150, 253–348.

FLACK, C. A. AND WARNER, M. R. 1990. Three-dimensional mapping of seismic reflections from the crust and upper mantle, northwest of Scotland. *Tectonophysics*, 173, 469–481.

FLINN, D. 1969. A geological interpretation of aeromagnetic maps of the continental shelf around Orkney and Shetland. *Geological Journal*, 6, 279–292.

——, FRANK, P. L., BROOK, M. AND PRINGLE, I. R. 1979. Basement-cover relations in Shetland. *In*: HARRIS, A. L., HOLLAND, C. H. AND LEAKE, B. E. (eds) *The British Caledonides Reviewed.* Geological Society, London, Special Publication, 8, 109–116.

FOSTER, R. J. 1972. *The Solid Geology of North-East Caithness.* PhD Thesis, University of Newcastle.

FRASER, A. J. AND GAWTHORPE, R. L. 1990. Tectono-stratigraphic development and hydrocarbon habitat of the Carboniferous in northern England. *In*: HARDMAN, R. F. P. AND BROOKS, J. (eds) *Tectonic Events Responsible for Britain's Oil and Gas Reserves.* Geological Society, London, Special Publication, 55, 49–86.

FREEMAN, B., KLEMPERER, S. L. AND HOBBS, R. W. 1988. The deep structure of northern England and the Iapetus Suture Zone from BIRPS deep seismic reflection profiles. *Journal of the Geological Society, London*, 145, 727–740.

GAWTHORPE, R. L. 1987. Tectono-sedimentary evolution of the Bowland Basin, N. England, during the Dinantian. *Journal of the Geological Society, London*, 144, 59–71.

GILLCRIST, R., COWARD, M. P. AND MUGNIER, J.-L. 1987. Structural inversion and its controls: examples from the Alpine foreland and the French Alps. *Geodinamica Acta*, 1, 5–34.

HALL, J. 1987. Geophysical lineaments and deep crustal structure. *Philosophical Transactions of the Royal Society of London*, A317, 33–44.

——, BREWER, J. A., MATTHEWS, D. H. AND WARNER, M. R. 1984. Crustal structure across the Caledonides from the "WINCH" seismic reflection profile: influences on the Midland Valley of Scotland. *Transactions of the Royal Society of Edinburgh, Earth Sciences*, 75, 97–109.

HALLIDAY, A. N., MCALPINE, A. AND MITCHELL, J. G. 1977. The age of the Hoy lavas. *Scottish Journal of Geology*, 141, 609–620.

HARLAND, W. B. AND GAYER, R. A. 1972. The Arctic Caledonides and earlier oceans. *Geological Magazine*, 109, 289–314.

HAUGHTON, P. D. W. 1989. Structure of some Lower Old Red Sandstone conglomerates, Kincardineshire, Scotland: deposition from late-orogenic antecedent streams. *Journal of the Geological Society, London*, 146, 509–525.

HECKEL, P. H. 1990. Evidence for global (glacial-eustatic) control over Upper Carboniferous (Pennsylvanian) cyclothems in midcontinent North America. *In*: HARDMAN, R. F. P. AND BROOKS, J. (eds) *Tectonic Events Responsible for Britain's Oil and Gas Reserves.* Geological Society, London, Special Publication, 55, 35–48.

HOLDER, M. J. AND LEVERIDGE, B. E. 1986. A model for the tectonic evolution of South Cornwall. *Journal of the Geological Society, London*, 143, 125–134.

HOSSACK, J. R. AND COOPER, M. A. 1986. Collision Tectonics in the Scandinavian Caledonides. *In*: COWARD, M. P. AND RIES, A. C. (eds) *Collision Tectonics.* Geological Society, London, Special Publication, 19, 287–303.

JOHNSON, M. R. W., KELLEY, S. P., OLIVER, G. J. H. AND WINTER, D. A. 1985. Thermal effects and timing of thrusting in the Moine thrust zone. *Journal of the Geological Society, London*, 142, 863–873.

LEEDER, M. R. 1987. Tectonic and palaeogeographic models for Lower Carboniferous Europe. *In*: MILLER, J., ADAMS, A. E. AND

WRIGHT, V. P. (eds) *European Dinantian Environments*. Wiley, Chichester, 1–19.

—— 1988. Recent developments in Carboniferous geology: a critical review with implications for the British Isles and NW Europe. *Proceedings of the Geologists' Association*, **99**, 77–100.

—— AND HARDMAN, M. 1990. Carboniferous geology of the southern North Sea Basin and controls on hydrocarbon prospectivity. *In*: HARDMAN, R. F. P. AND BROOKS, J. (eds) *Tectonic Events Responsible for Britain's Oil and Gas Reserves*. Geological Society, London, Special Publication, **55**, 87–106.

MCKENZIE, D. 1978. Some remarks on the development of sedimentary basins. *Earth and Planetary Science Letters*, **40**, 25–32.

MOLNAR, P. 1988. A review of geological constraints on the deep structure of the Tibetan Plateau, the Himalaya and the Karakoram, and their tectonic implications. *Philosophical Transactions of the Royal Society of London*, **A326**, 33–88.

MOORBATH, S., WELKE, H. AND GALE, N. H. 1969. The significance of lead isotope studies in ancient high-grade metamorphic basement complexes as exemplified by the Lewisian rocks of north-west Scotland. *Earth and Planetary Science Letters*, **6**, 245–256.

MORRIS, W. A. 1974. Transcurrent motion determined palaeomagnetically in the northern Appalachians and the Caledonides and the Acadian Orogeny. *Canadian Journal of Earth Sciences*, **13**, 1236–1243.

MYKURA, W. 1976. *Orkney and Shetland*. British Regional Geology, Institute of Geological Sciences, HMSO, Edinburgh.

ORD, D. M., CLEMMEY, H. AND LEEDER, M. R. 1988. Interaction between faulting and sedimentation during Dinantian extension of the Solway basin. *Journal of the Geological Society, London*, **145**, 249–259.

POWELL, C. M. 1989. Structural controls on Palaeozoic basin evolution and inversion in south-west Wales. *Journal of the Geological Society, London*, **146**, 439–446.

ROBERTS, A., PRICE, J. AND OLSEN, T. S. 1990. Late Jurassic half-graben control on the siting and structure of hydrocarbon accumulations: UK/Norwegian Central Graben. *In*: HARDMAN, R. F. P. AND BROOKS, J. (eds) *Tectonic Events Responsible for Britain's Oil and Gas Reserves*. Geological Society, London, Special Publication, **55**, 229–258.

ROGERS, D. A., MARSHALL, J. E. A. AND ASTIN, T. R. 1989. Devonian and later movements on the Great Glen fault system. *Journal of the Geological Society, London*, **146**, 369–372.

SENGOR, A. M. C., GORUR, N. AND SAROGLU, F. 1985. Strike-slip faulting and related basin formation in zones of tectonic escape: Turkey as a case study. *In*: BIDDLE, K. T. AND CHRISTIE-BLICK, N. (eds) Strike-slip deformation, basin formation, and sedimentation. Society of Economic Paleontologists and Mineralogists, Special Publication, **37**, 227–264.

SERRANE, M. 1988. *Tectonique des Bassins Devoniens de Norvege: Mise en Évidence de Bassins Sédimentaire en Extension Formes par Amincissement d'une Croûte Orogénique Épaisse*. Thèse de Doctorat, Université des Sciences et Techniques du Languedoc, Montpellier.

—— 1991. Devonian extensional tectonics versus Carboniferous inversion in the northern Orcadian basin. *Journal of the Geological Society, London*, **149**, 27–37.

——, CHAUVET, A., SEGURET, M. AND BRUNEL, M. 1989. Tectonics of the Devonian collapse basins of western Norway. *Bulletin of the Geological Society of France*, **8**.

SHACKLETON, R. M., RIES, A. C. AND COWARD, M. P. 1982. An interpretation of the Variscan structures in SW England. *Journal of the Geological Society, London*, **139**, 533–541.

SIBSON, R. H. 1977. Fault rocks and fault mechanisms. *Journal of the Geological Society, London*, **133**, 191–213.

SMITH, D. I. AND WATSON, J. V. 1983. Scale and timing on the Great Glen Fault, Scotland. *Geology*, **11**, 523–526.

SNYDER, D. B. in press. Reflections from a relic Moho in Scotland. *Proceedings of the 4th International Symposium on Deep Seismic Reflection Profiling of the Continental Lithosphere, Bayreuth*. American Geophysical Union Geodynamics Series.

—— AND FLACK, C. A. 1990. A Caledonian age for reflectors within the mantle lithosphere north and west of Scotland. *Tectonics*, **9**, 903–922.

SOPER, N. J. AND ANDERTON, R. 1984. Did the Dalradian slides originate as extensional faults? *Nature*, **307**, 357–359.

—— AND HUTTON, D. H. W. 1984. Late Caledonian sinistral displacements in Britain: implications for a three-plate collisional model. *Tectonics*, **3**, 781–794.

STEIN, A. M. AND BLUNDELL, D. J. 1990. Geological inheritance and crustal dynamics of the northwest Scottish Continental Shelf. *Tectonophysics*, **173**, 455–467.

STRACHAN, R. A., TRELOAR, P. J., BROWN, M. AND D'LEMOS, R. S. 1989. Cadomian terrane tectonics and magmatism in the Armorican Massif. *Journal of the Geological Society, London*, **146**, 423–426.

TAPPONNIER, P., PELZER, G. AND ARMIJO, R. 1986. On the mechanics of the collision between India and Asia. *In*: COWARD, M. P. AND RIES, A. C. (eds) *Collision Tectonics*. Geological Society, London, Special Publication, **19**, 115–158.

VAN DER VOO, R. AND SCOTESE, C. 1981. Palaeomagnetic evidence for a large (~2000 km) sinistral offset along the Great Glen Fault during Carboniferous time. *Geology*, **9**, 583–589.

WELLS, P. R. A. AND RICHARDSON, S. W. 1979. Thermal evolution of metamorphic rocks in the Central Highlands of Scotland. *In*: HARRIS, A. L., HOLLAND, C. H. AND LEAKE, B. E. (eds) *The British Caledonides Reviewed*. Geological Society, London, Special Publication, **8**, 339–344.

ZIEGLER, P. A. 1982. *Geological Atlas of Western and Central Europe*. Shell Internationale Petroleum Maatschappij B.V., The Hague.

—— 1988. *Evolution of the Arctic–North Atlantic and the Western Tethys*. AAPG Memoir, **43**.

—— 1989. *Evolution of Laurentia—a study in late Palaeozoic plate tectonics*. Kluwer, Dordrecht.

—— 1990. *Geological Atlas of Western and Central Europe* (2nd edn). Shell Internationale Petroleum, Maatschappij B.V., The Hague.

Regional structural evolution of the North Sea: oblique slip and the reactivation of basement lineaments

I. D. BARTHOLOMEW,[1] J. M. PETERS[2] and C. M. POWELL

Shell UK Exploration and Production, Shell-Mex House, Strand, London WC2R 0DX, UK
(Present addresses: [1] Oryx UK Energy Company, Charter Place, Uxbridge UB8 1EZ, UK
[2] Petroleum Development Oman, PO Box 81, Muscat, Sultanate of Oman)

Abstract: Basement lineaments pre-dating the dominant Mesozoic basin formation of the North Sea are fundamental controlling features within the polyphase tectonic evolution of NW Europe. These inherent zones of weakness were repeatedly reactivated during the evolution of the North Sea rift system. The regional and geometrical constraints imposed by the presence of these pre-existing tectonic grains during the Mesozoic extensional deformation suggest that oblique-slip movement must have occurred along many of the basin-bounding faults in a non-preferential orientation for dip-slip reactivation.

While the North Sea can be considered in broad terms as a series of linked, elongated grabens developed in response to dominantly E–W extensional intra-plate stresses, the orientation, distribution and character of pre-existing shear zones ultimately controlled the geometry of many Mesozoic basins and their faulted margins.

Four types of inherent basement features are recognized as having influenced subsequent basin development: (a) wide, diffuse zones of NW–SE and NE–SW basement shear, e.g. in the Central Graben; (b) narrow shear zones oblique to primary Mesozoic extensional faults, e.g. in the East Shetland Basin; (c) E–W-trending discrete fault systems, e.g. on the Western Platform; and (d) massive palaeohighs, e.g. the Halibut Horst.

The distribution of these distinct structural features throughout the North Sea was controlled by Early Paleozoic plate tectonic evolution. Repeated Mesozoic and Cenozoic reactivation along fundamental zones of weakness, including oblique-slip displacements and basinal inversion, has led to areas of highly complex structuration. Hence, there is the potential for numerous structural and stratigraphic hydrocarbon trapping mechanisms associated with these underlying basement lineaments.

Although the total amount of horizontal crustal movement may be limited, understanding of the role of oblique-slip along pre-existing zones of weakness and the effect of long-lived palaeohighs within the basin is crucial to place Mesozoic and Tertiary hydrocarbon accumulations within a realistic North Sea tectono-stratigraphic framework.

The structural evolution of the North Sea region can be considered as a two-stage process; the initial development of the basin framework during the Early Paleozoic (Coward 1993), and the subsequent repeated reactivation of the existing basement lineaments in response to deformation in the Mesozoic and Cenozoic. The aim of this paper is to describe the structural evolution in terms of one coherent, observation-based model, and to illustrate how the orientation, distribution and character of pre-existing shear zones ultimately controlled the geometry of the Mesozoic basins and their faulted margins.

A series of examples from the North Sea demonstrates the implications of this model for hydrocarbon accumulations in this basin and specifically the relationship between inherent basin architecture, sediment input and hydrocarbon trapping mechanism.

Database

The databases used to compile this regional work include structural interpretation of all the NOPEC regional seismic lines in the series NNST, ESBT, CNST and SNST, structural and seismo-stratigraphic interpretation of a dense 2D seismic grid over the Central Graben and Western Platform areas, examination of structural geometries and hydrocarbon trapping mechanisms from 29 exploration 3D seismic datasets and 4 production 3D seismic datasets targetted specifically at Jurassic and Triassic reservoir levels. The 3D seismic dataset amounts to 8260 km² in total.

The seismic data were correlated using a well data-base of some 600 key wells, markers being identified within these wells using both log correlation and quantitative/qualitative biostratigraphy.

Regional model

The development of an extensive system of elongate rift grabens, and the long-term subsidence history of the northwest European continental shelf, point to an extensional origin for the Meso-Cenozoic North Sea Basin (Barton and Wood 1984; Ziegler 1990). The Mesozoic evolution involved several tectonic pulses but was dominated by a major period of Late Jurassic rapid extension, from the mid-Oxfordian to the Ryazanian. The traditional model of this principal event is a trilete rift system which requires concurrent unique extension directions for the opening of the Moray Firth, Viking Graben and Central Graben (see Fig. 1 for location of basins). Recent work (Roberts *et al.* 1990a) has suggested an alternative model involving the summation of a vector system between these three rift arms of the North Sea Basin. This model requires a considerable wrench component in the Inner Moray Firth to balance regional E–W and WSW–ENE extension in the Viking Graben and Central Graben, respectively.

The model described in the present paper invokes the rifting of the North Sea during the Late Jurassic (Cimmerian) as a response to a simple regional stress regime, having a maximum compressive stress direction oriented approximately N–S to NNW–SSE and a minimum compressive stress direction in an E–W to ESE–WNW direction. This resulted in a regional extension direction of roughly E–W to ESE–WNW over the whole of the North Sea area. A dominant structural style of extensional faulting and block rotation is therefore predicted, which is indeed the case in the main N–S-trending elements of the rift, e.g. in the Viking Graben. However, in many parts of the basin, notably in the Central North Sea, the basin architecture is considerably more complex than would be expected if the region had merely been accommodating crustal extension

From *Petroleum Geology of Northwest Europe: Proceedings of the 4th Conference* (edited by J. R. Parker).

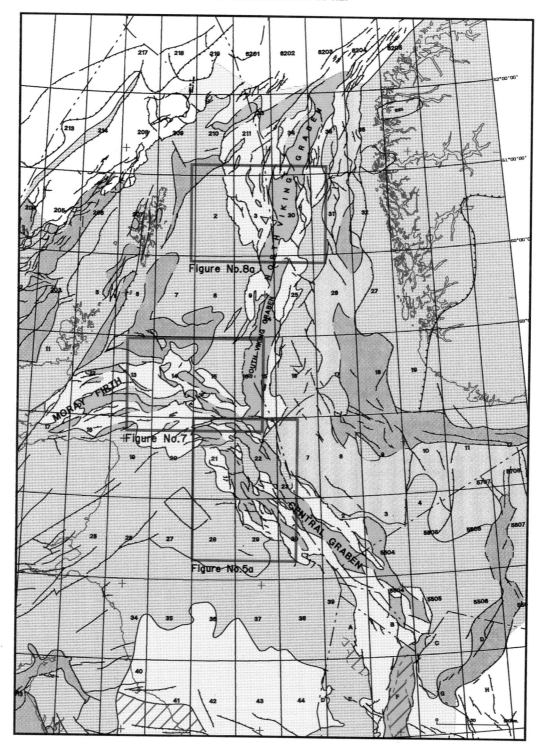

Fig. 1. Late Cimmerian structural framework of the North Sea Basin indicating the basinal depocentres and structurally high platform areas of the trilete rift system (indicated in shades of blue and pink, respectively; purple areas are predominantly Triassic depocentres). The insets indicate the position of Figs 5a, 7 and 8a. Note the difference in orientation between the Viking Graben and Central Graben.

by simple dip-slip deformation. This complex architecture is explained due to the presence of a pre-existing basement framework inherited from Paleozoic deformation which exerted a fundamental control upon the style and geometry of the basins formed in response to the Mesozoic extension.

The formation of two observed suites of basement lineaments can be attributed to the Caledonian and Variscan orogenies, and to the presence of the Trans-European Fault Zone extending NW from Germany into the North Sea Basin

(Fig. 2 and Berthelsen 1988). The basement trends strike NE–SW and NW–SE, respectively. A third suite of E–W-trending basement lineaments originate from pre-Variscan Carboniferous extension, documented along the eastern shore of the UK (Fraser and Gawthorpe 1991). These three suites of faults form the basement framework controlling the gross geometry of the North Sea sub-basins throughout their subsequent evolution. The effect of these basement trends on subsequent Mesozoic geometry was first documented a decade ago by Johnson and

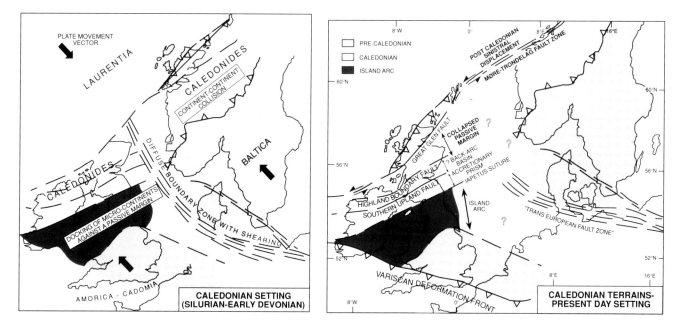

Fig. 2. The Caledonian setting of the North Sea area showing orientation of the Caledonian, Variscan and Trans-European Fault trends which are considered to represent the basement framework controlling Mesozoic geometries within the area.

Dingwall (1981) who recognized the influence of Caledonian trends on Permo-Triassic basin geometry throughout the Northern North Sea.

Prior to the major Late Jurassic phase of deformation, many of the Paleozoic fault zones were reactivated as extensional features during a Late Permian–Triassic rifting event (Marsden *et al.* 1991). Triassic basins along the western edge of the UK, such as the Cheshire Basin and Cardigan Basin, formed along NW–SE- and NE–SW-trending basement shear zones. The fault geometries of these basins conform to a regional N–S maximum compressive stress orientation, with the whole region undergoing E–W extension and the pre-Permian fault zones reactivating obliquely to produce pull-apart basins. This model considers that similar localized pull-apart basins were formed to the east of the UK, with the reactivation of the Trans-European Fault Zone to form the pro-Central Graben during the Triassic. However, clear evidence of oblique movement on basement faults during the Permo-Triassic rift event in the Central North Sea Basin cannot be unequivocally demonstrated due to concurrent and later salt diapirism and subsequent deep burial of the Triassic basinal sequences. East–West extension in the South Viking Graben is documented during this Permo-Triassic deformation (Marsden *et al.* 1991), with both crustal thinning and formation of the basin geometry that subsequently controlled the Mesozoic depocentres.

The presence of pre-existing basement lineaments placed geometric and kinematic constraints upon primary basin formation during Jurassic extensional deformation. Consequently, the Mesozoic basins developed along these inherited structural grains rather than as N–S-trending basins orthogonal to the principal extension direction. As a result of the angular relationship between the basin margins and the principal extension direction, many of the basin-bounding faults were reactivated under non-coaxial stress conditions, and thus are expected to have had oblique-slip components of movement within an overall extensional environment. The nonpreferential orientation of the pre-existing lineaments for reactivation as purely extensional features would also have resulted locally in the formation of transpressional structures. Figure 3 shows the conceptual model of the relationship between the regional Mesozoic E–W extensional direction and the pre-existing structural zones. Basins controlled by pre-existing

NW–SE-striking structures exhibit a dextral sense of oblique shear, with basins bounded by NE–SW-trending lineaments exhibiting a sinistral sense of shear. The type of shear zone varies between diffuse zones and narrower, discrete zones of shear (Fig. 3). However, within this broad framework shown by Fig. 3, the basement structural pattern is seen to vary spatially through the North Sea, the NE–SW Caledonian trend

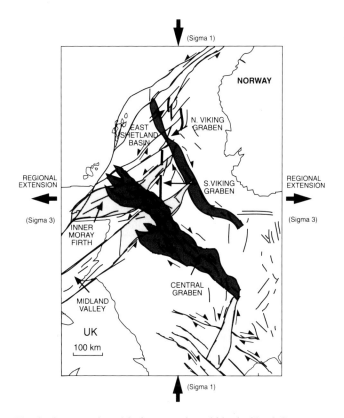

Fig. 3. Conceptual model of transtension within the North Sea Basin during Jurassic regional extension, showing the relationship between pre-existing shear zones and E–W extension. The angular relationship between the basin margins and the principal extension direction causes transtension and transpression within the basins.

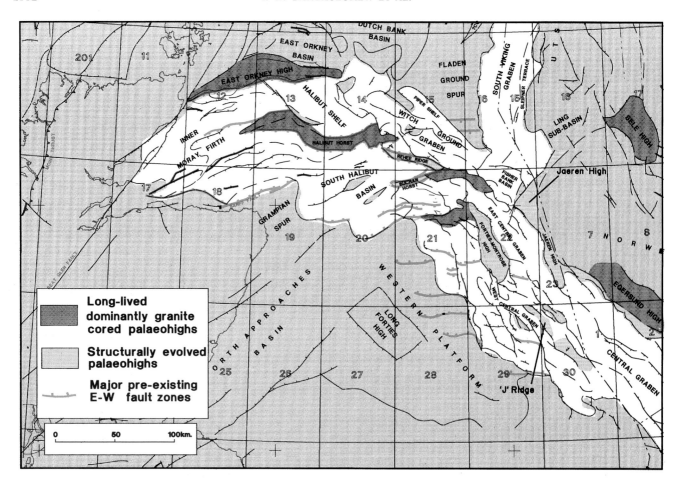

Fig. 4. Pre-existing E–W-trending faults and palaeohighs causing local interference to structures formed within the Late Jurassic regional extension in the North Sea Basin.

being more dominant in the Northern North Sea and Moray Firth Basin whereas the NW–SE trend is more pervasive in the Southern North Sea and Central Graben areas (Glennie and Boegner 1981).

In addition, the principal stress axes may be locally rotated because of the existence of planes of weakness within the rock volume and the presence of granite cored palaeohighs (e.g. the Halibut Horst) such that the principal stress direction is not ubiquitously E–W throughout the entire basin. Figure 4 illustrates the position of palaeohighs modifying the basin geometry of the Central North Sea and the Moray Firth.

Faulting, and rotation of fault blocks associated with active rifting, ceased in the Valanginian, with the re-establishment of the lithospheric thermal equilibrium causing regional post-rift subsidence but with local uplift and development of unconformities. This phase of thermal subsidence from the Valanginian to the present day has been interrupted intermittently by further extension (e.g. on the Halibut Horst bounding faults), and reactivation of the Jurassic extensional faults during sporadic Late Cretaceous–Early Tertiary basin inversion episodes with the formation of unconformities and local erosion. This is especially prevalent in the Southern North Sea (e.g. Van Hoorn 1987; Oudmayer and de Jager 1993).

The phases of inversion are considered to be the result of the interaction of regional stresses induced by the opening of the North Atlantic, progressive closure of the western Tethys and movements along the Trans-European Fault Zone (Cooper and Williams 1989; Ziegler 1990). This interaction resulted in horizontal compression in various sub-basins of the North Sea (Cartwright 1989) overprinting the sedimentary response to passive subsidence during short episodes in the Late Creta-

ceous and Early Tertiary. The continued rejuvenation of basement fault zones characterizes the Alpine inversion phase (Cooper and Williams 1989; Ziegler 1990), and is also seen to affect sedimentation and spatial position of Tertiary fan systems during the Paleocene and Eocene (Hartog-Jager et al. 1993).

Types of basement feature

Four major categories of inherent basement feature are recognized in the North Sea which have fundamentally influenced Mesozoic basin geometry and deformational style:

    (a) diffuse zones of basement shear;
    (b) narrow zones of basement shear;
    (c) E–W-trending basement fault systems;
    (d) palaeohighs.

(a) Diffuse zones of basement shear

Two dominant trends of diffuse shear zone have been identified: trending NW–SE (e.g. in the Central Graben (Sears et al. 1993) and Southern North Sea (Oudmayer and de Jager 1993)); and NE–SW (e.g. the seawards extension of the Midland Valley, the Forth Approaches Basin and the Fisher Bank Basin). The conceptual model of regional extension causing areas of transtension within these diffuse basement shear zones is illustrated by Fig. 3.

The Central Graben can be taken as a prime example of oblique extension along wide basement zones, resulting in an area comprising anastomosing shear zones separated by

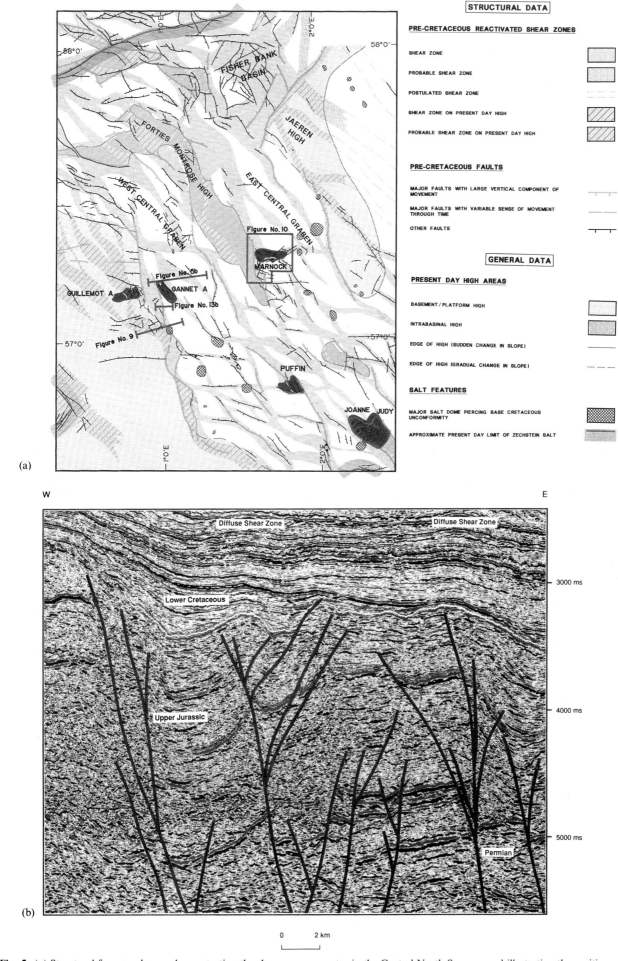

Fig. 5. (**a**) Structural framework map demonstrating the shear zone geometry in the Central North Sea area and illustrating the position and anastomosing nature of the shear zones, the scale of the intershear areas and the present-day highs. (**b**) Seismic profile through a diffuse shear zone indicating changes in deformation mode along faults and the relationship of complex faulting at shallow levels with steep basement faults at depth (example from the 22/21 area of the Central North Sea—see (**a**) for location).

relatively unfaulted lens-shaped areas. These intershear areas are characteristically sigmoidal or rhomboid in shape (Fig. 5a). An individual shear zone can reach 5 km in width at Base Cretaceous level. The relatively unfaulted lens-shaped areas lying between the shear zones are typically 5–10 km wide and 25–30 km in length. The long axes are parallel to the orientation of the main shear zones and thus the basin-bounding faults. In vertical profile, the diffuse zones are seen to be composed of fault splays at shallow levels rooted to steep,

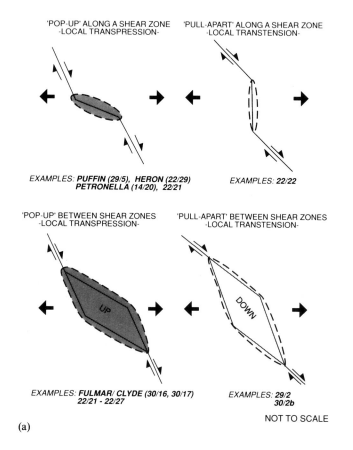

(a)

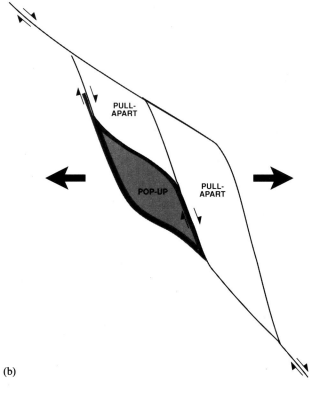

(b)

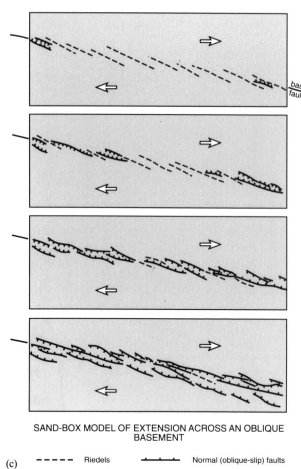

SAND-BOX MODEL OF EXTENSION ACROSS AN OBLIQUE BASEMENT

(c)   - - - - - Riedels   ┣━━━ Normal (oblique-slip) faults

Fig. 6. (**a**) Structural configurations within a transtensional regime under regional E–W extension. Note: the geometries represented are only applicable to areas with many anastomosing shear zones, and are shown as a single fault merely to simplify the figure. (**b**) The formation of both extensional and compressional features simultaneously within a single zone. (**c**) Line drawing of a sandbox experiment illustrating the development of obliquely opening transtensional basins above a single basement lineament within a regional extensional stress regime (from Koopman *et al.* 1987).

planar basement faults (Fig. 5b). The faults commonly have variable apparent offsets in vertical section, often demonstrating extensional offsets at depth and contractional offsets giving apparent reverse throw at shallower levels.

Variation in the dominant trends through time within a diffuse shear zone, with switching of activity between the individual lineaments comprising that shear zone, causes changes in geometry through time. This, combined with the anastomosing nature of the faults within the shear zone, causes a large variation in structural styles within any one particular zone. Figure 6a sketches the structural configurations that are possible. Although the regime is dominantly extensional, pop-up highs and pull-apart basins are frequently found within the shear zones (Figs 6a and 6b). Some inter-shear areas remain as pop-up highs throughout the entire Mesozoic evolution of the basin (Fig. 5a). Figure 6c illustrates the formation of oblique shear above a basement lineament within a regional extensional stress regime during a sandbox experiment (Koopman *et al.* 1987). In contrast to the examples given in Fig. 6a, the experiment illustrates that intra-basin geometries will remain dominantly extensional if only a single basement lineament is controlling the overburden geometry, and if the basement zone has little variation in strike along its length.

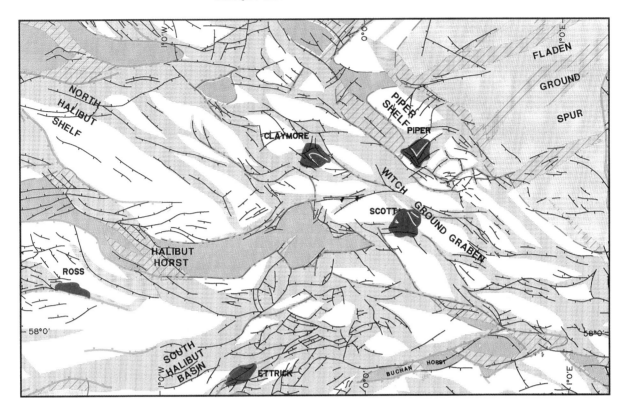

Fig. 7. Structural framework map illustrating the interaction of NW–SE, NE–SW and E–W zones of diffuse shear within the Outer Moray Firth, with the position of hydrocarbon accumulations along reactivated shear zones. (See Fig. 5a for legend.)

In the Outer Moray Firth, the NW–SE and NE–SW structural zones intersect, leading to highly complex interference patterns (Fig. 7). Both trends appear to have been reactivated contemporaneously in the Late Jurassic, with the inter-shear areas forming rhomboid shapes and the fields being concentrated along the reactivated shear zones (e.g. the Piper, Scott and Claymore fields). The area shown in Fig. 7 is north of the Zechstein Salt within the Central North Sea. The complex tectonic style seen within the Outer Moray Firth cannot, therefore, be attributable to halokinesis. The similar geometry of the structures seen in areas containing salt and areas of no salt suggests that it is the reactivation of complex fault systems that has a fundamental effect upon the basin evolution throughout the North Sea Basin. The role of salt tectonics, both temporally and spatially, is linked to activation and geometry of these basement lineaments.

(b) Narrow zones of basement shear

Where pre-existing structural zones were widely spaced and relatively simple in geometry, N–S-trending basins formed orthogonally to the Mesozoic extension direction (e.g. parts of the Viking Graben and the East Shetland Basin areas). The tectonic style is dominated by half-grabens and rotated fault blocks with footwall uplift indicative of dip-slip extensional deformation (Fig. 8b and Badley et al. 1988). Although the N–S-trending basin-bounding faults are the dominant control on basin development, both the nature of the graben margin and the internal geometries of the fault blocks are partially controlled by narrow pre-existing structural trends. Discrete NE–SW and NW–SE fault zones can be traced through the Mesozoic platform areas (Fig. 8a), with 3-D seismic interpretation demonstrating that at the point where these faults intersect the Mesozoic basin, a termination or switching of displacement between individual segments of the N–S-trending fault system occurs.

The position and lateral extent of many of the fault blocks within the basin also appear to be controlled by the underlying basement framework of NW–SE and NE–SW-trending narrow fault zones. A regional seismic line extending from the East Shetland Basin east across the Viking Graben illustrates the discrete nature of the Mesozoic faults, and the dominantly extensional tectonic style, only becoming complex at the intersection of a pre-existing oblique lineament (Fig. 8b). The line also illustrates the role of purely extensional footwall uplift in the immediate footwall of the primary Mesozoic extensional faults in providing localized sediment source areas, as described by Roberts et al. (1990a) for the South Viking Graben.

With the exception of the Brent Field, all the Shell/Esso fields in the North Sea are located over these basement zones oblique to the Mesozoic basin margins (Fig. 8a). The resulting compartmentalization of these fields is complex, with subtle oblique-slip faults forming pressure and migration barriers (e.g. within the Dunlin, Tern, Eider and Cormorant fields). The identification and interpretation of these subtle zones of intersection between the lineaments is, therefore, crucial in establishing the most efficient development plan for a hydrocarbon accumulation in this area.

(c) E–W-trending lineaments

Major E–W lineaments have traditionally been recognized as controlling Mesozoic basin architecture in the Inner and Outer Moray Firth (Fig. 7), and the increasing number of 3D seismic interpretations indicate that these E–W-trending basement features also play a fundamental role in the location of hydrocarbon accumulations along the Western Platform and within the deep Central Graben (Fig. 4).

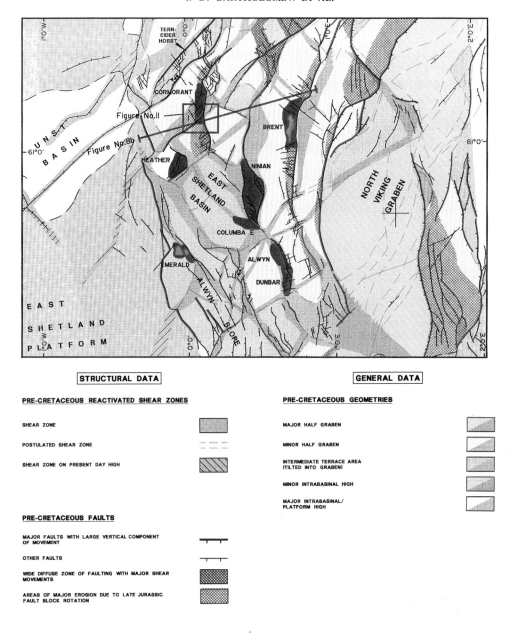

(a)

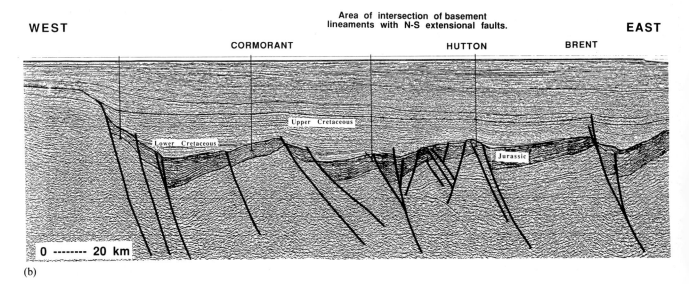

(b)

Fig. 8. (**a**) Pre-Cretaceous basin architecture map demonstrating the relationship between Mesozoic N–S extensional faults and pre-existing narrow fault zones in the East Shetland Basin, Northern North Sea. (**b**) Regional seismic line (courtesy of NOPEC) through the East Shetland Basin demonstrating the dominant extensional geometry of the fault blocks, and interference from pre-existing zones of weakness oblique to the N–S trend. The Upper Jurassic sequence is shaded. For location see (a).

The majority of the faults in the Inner Moray Firth strike E–W or ENE–WSW. It appears that the Inner Moray Firth had the same Paleozoic evolutionary history as the Midland Valley. Both appear to have formed during the Devonian and Carboniferous as NE–SW-trending transtensional pull-apart basins with a minimum compressive stress regime orientated N–S. As a result of this regional stress regime, E–W-trending extensional faults would have formed within both the basins. During the Jurassic, the Midland Valley remained part of the stable platform, whereas in contrast, the Inner Moray Firth was deformed by the Late Jurassic rift phase. There remains considerable debate as to the influence of the Great Glen Fault zone upon the Mesozoic basin geometry with varying estimates as to whether or not the observed geometries are indicative of strike-slip deformation. Regional mapping within the present study suggests a dominant extensional style as described by Underhill (1991), with little evidence of flat-lying strike-slip detachments as interpreted by Bird *et al.* (1987). The reactivation of the E–W and ENE-WSW lineaments as extensional faults implies a Jurassic extension direction of WNW–ESE, the orientation of this extension direction possibly being additionally controlled by the presence of granitic palaeohighs to the northwest, north and south of the basin (the Scottish Landmass, the East Orkney High and the Grampian Spur, respectively) as well as the pre-existing tectonic grain of the basin. However, some of the structural geometries within the Inner Moray Firth do appear to indicate a component of oblique-slip deformation, specifically in the NW part of the basin at the intersection of the Great Glen Fault with the E–W-trending Wick Fault system (Roberts *et al.* 1990*b*). Some degree of oblique-slip movement could be expected within the proposed Mesozoic regional stress regime if imposed upon ENE–WSW-trending pre-existing planes of weakness. It can be invoked without requiring major strike-slip movement on the Great Glen Fault during the Jurassic as the control on basin formation.

In the Central Graben, the Marnock area (Block 22/24), located at the structural transition between the Erskine Ridge and Forties–Montrose High (Figs 4 and 5a), is characterized by complex deformation along a number of E–W-trending fault zones, which control both the gross stratigraphy and lithofacies changes in the Jurassic/Triassic reservoir sequences and consequently the hydrocarbon accumulations (Sears *et al.* 1993). The orientation of E–W faults within a simple E–W regional extensional regime would necessitate them being reactivated as purely strike-slip faults, but detailed mapping suggests that they occur between the en-échelon pull-apart or extensional basins, often with a significant vertical displacement. It appears that few pre-existing lineaments demonstrate pure strike-slip reactivation, and that the observed geometries are more consistent with oblique movements, the component of vertical displacement being the simplest to document. This would suggest that the presence of E–W-trending basement lineaments considerably modifies the proposed regional stress regime of E–W-orientated maximum extension direction over a local area.

(d) Palaeohighs

East–west-trending massive palaeohighs exist within the Central North Sea. Many are probably cored by Caledonian granite and acted as stable buffer zones around which basement shear zones were deflected. The palaeohighs are not generally cut by primary Mesozoic faults but commonly appear to be displaced by older basement lineaments. The Halibut Horst, the East Orkney High and the northern end of the Forties–Montrose High are examples of long-existing palaeohighs affecting both the sediment distribution and basin architecture (Fig. 4). Based on data from wells in the Ross

Field area, the Halibut Horst appears to have controlled sedimentary facies continuously from the Devonian to the Cretaceous, and to have been a major sediment source for both the Inner and Outer Moray Firth during the Jurassic and Cretaceous. In contrast to these granite cored highs, the southern end of the Forties–Montrose High, together with the Erskine Ridge and J-ridge, form a series of NW–SE-trending linked en-échelon palaeohighs which acted as sediment source areas during the Late Jurassic (Fig. 5a). They are not considered to be long-lived highs such as the Halibut Horst, but have evolved during the Mesozoic, being fundamentally linked to pre-existing NW–SE-trending fault zones. The Tern–Eider Horst in the East Shetland Basin is considered to have had a similar tectonic history.

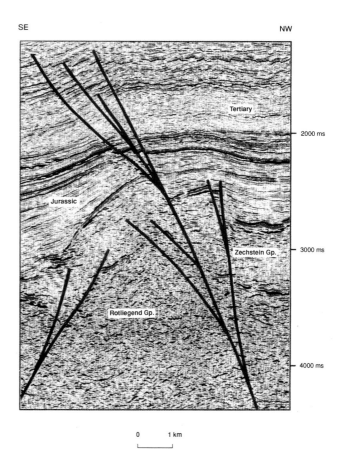

Fig. 9. Fault propagation through Zechstein Salt linking the basement geometry with structure of overlying sequences—example from Quadrant 21, Central Graben margin. For location see Fig. 5a.

The relationship of fault reactivation and salt tectonics

The difficulties in analysing the structural evolution in areas strongly affected by halokinesis of the Zechstein evaporites are considerable. However, regional mapping over the Central North Sea together with salt tectonics studies in the Southern North Sea (Oudmayer and de Jager 1993) have clearly established that salt pillows and diapirs are aligned along the rhomboid structural pattern defined by the dominant NW–SE-trending basement lineaments. There is often a close genetic relationship between diapir formation and the timing of the reactivation of basement fault zones within the Mesozoic rift phase (e.g. Block 21/30 area on the Western Platform). Many of the complex geometries in the Central North Sea traditionally attributed solely to halokinesis can be reinterpreted as the products of salt mobilization triggered by and controlled by movement on steep basement lineaments. If the timing of fault reactivation can be unravelled in an area of intense diapirism, this can have important implications for the position of reservoir sequences in the immediate area of the salt doming. In many areas of the Central North Sea, it appears that faults have propagated through the Zechstein evaporites as distinct fault zones linking the geometry of the formations above the salt directly to those below. In areas covered by 3D seismic, such as Blocks 21/30 and 22/24 in the Central North Sea, the structures observed can best be explained by direct fault propagation through the salt (Fig. 9).

In contrast to these Central North Sea geometries, the Zechstein evaporites in the Southern North Sea basins are seen to act as decollements causing structural disharmony between the basement and overburden, the large displacements on basement lineaments in the Permian section not being reflected by an equivalent displacement in the Jurassic and younger sequences. Nevertheless, a genetically related geometry of some form is virtually always present in the overlying sequences (Van Hoorn 1987).

Examples of basement feature interaction

The recognition of repeated movement on basement features throughout the North Sea Basin is only the first stage in understanding the implications for hydrocarbon exploration. The critical step is unravelling the interaction of the different types of basement feature through time. A number of types of interaction have been recognized throughout different parts of the basin. Each type is detailed below by means of an example.

(a) Interaction of diffuse basement shear zones—the Marnock/Heron area

The 3D amplitude map at Base Cretaceous (Fig. 10) illustrates the triangular geometry of the Marnock Field at the confluence of NW–SE-, NE–SW- and E–W-trending basement fault zones. The structuration by different suites of basement lineaments has implications for the compartmentalization of the hydrocarbon accumulation. The presence of persistent lineaments providing intra-field fault seals allows hangingwall traps to be in a significantly different pressure regime with respect to immediately adjacent footwall traps although theoretically regarded as the same accumulation. In the area south of the Marnock Field, over 2000 psi overpressure is recorded in the immediate hangingwall to a NW–SE-trending basement fault zone, with no connectivity of these pressures through to the footwall which is normally pressured. Further examples within the Central Graben are cited by Gaarenstroom *et al.* (1993).

(b) Interaction of narrow shear zones—the Cormorant Field

The 3D seismic attribute mapping (Fig. 11) undertaken during recent structural analysis of the Cormorant Field (see Fig. 8a

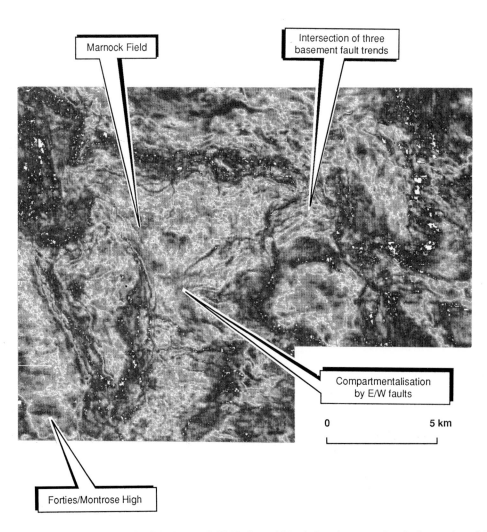

Fig. 10. Amplitude map at base Cretaceous levels of the Marnock Field, Central North Sea, demonstrating the interaction of diffuse shear zones between the Forties–Montrose High and the Erskine Ridge. The hydrocarbon accumulations within the area are compartmentalized by the shear zones acting as pressure boundaries. For location see Fig. 5a.

EN-ECHELON FAULT SUITES
OVERLYING BASEMENT LINEAMENTS

0 _____ 10 km

INTERSECTION OF NW-SE AND NE-SW
BASEMENT LINEAMENTS

Exact location
not given.

Fig. 11. Dip map over the Cormorant Field illustrating the structural compartmentalization of the field by reactivated pre-existing discrete lineaments, and the presence of en-échelon fault suites overlying NW–SE- and NE–SW-trending basement lineaments. The difference between easterly and westerly dipping faults is highlighted by red and blue shading, respectively. For location see Fig. 8a.

for location) reveals the complex interaction of suites of fault systems trending NE–SW and NW–SE throughout the Jurassic reservoir sequences of the field. The majority of the previously interpreted N–S-trending faults considered to be primary Mesozoic extensional faults (Speksnijder 1987) have been resolved into a series of lineaments oriented oblique to the regional extension direction, and considered to be riedel shears with geometries directly controlled by oblique displacements on Caledonian lineaments mapped within the basement under the Cormorant Field. In contrast to the diffuse shear zones characteristic of the Central Graben, the basement shear zones controlling the field geometry are commonly a single, steep discrete lineament. The field is compartmentalized into rhombic fault blocks between intersecting en-échelon fault segments. Detailed mapping within the reservoir section demonstrates some lateral offset on both the NE–SW- and NW–SE-trending faults within the Mesozoic sequence although on any individual seismic profile the fault geometries appear extensional (see Demytennaere *et al.* (1993) for further details).

(c) Change in dominance between diffuse shear zones— the Fulmar/Clyde fields

The structural configuration of the Fulmar/Clyde area at the SE end of the Central Graben is created by a variation in the dominance of trends through time between diffuse shear zones and the switching of activity between the individual lineaments causing depocentres to move through time. Detailed mapping and stratigraphic analysis suggests that the structure acted as a transtensional pull-apart feature in the Late Oxfordian, then switched to a pop-up feature in the Early Volgian. The present model of a switch from overall transtension to overall transpression in this area is illustrated by Fig. 12.

It is important to note that both pull-apart and pop-up structures would be formed contemporaneously in adjacent areas and independent of the necessity for invoking salt tecto-

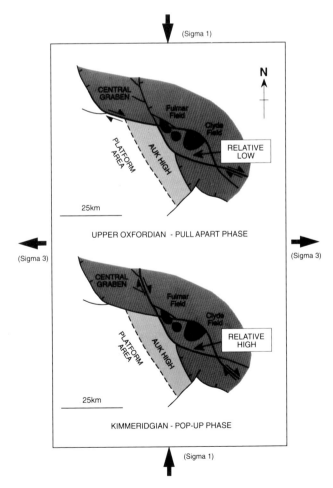

Fig. 12. Model of the Upper Jurassic structural evolution of the Fulmar–Clyde area.

Structural Inversion
Western Platform

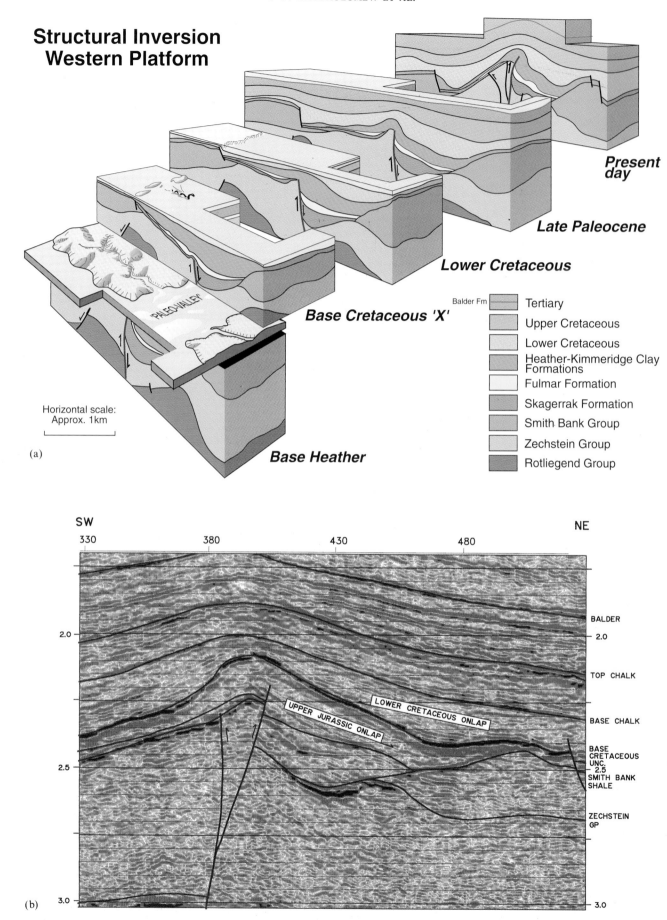

Fig. 13. (a) Repeated reactivation of a basement fault along the margin of the Central Graben in the Guillemot area (Quadrant 21). The model indicates the switch from Upper Jurassic depocentre to Lower Cretaceous palaeohigh. (b) Illustration of a corresponding switch in onlap direction between the Jurassic and Cretaceous on a seismic profile through the Guillemot area, and the continued reutilization of the same fault zone in the Tertiary causing platformation in the Oligocene and Eocene sequences. For location see Fig. 5a.

nics. The exact relation of the pop-up and pull-apart structures is dependent on which shear zone was active at the time (Fig. 6).

(d) Reactivation of basin-controlling faults and basin inversion—the Guillemot area

Along the western edge of the Central Graben, the compartmentalization of the platform area by E–W-trending Permo-Triassic extensional basement lineaments (Fig. 4) also controls the position of oil accumulations. In the Guillemot area dextral oblique-slip on these faults during the regional Jurassic extension resulted in the formation of NW–SE-trending anticlinal axes. This series of en-échelon structural highs form the 21/30-12, 21/30-16, Guillemot A and Guillemot C discoveries. The Guillemot area also demonstrates the effects of continued reactivation of a steep basement lineament through time. Figure 13 illustrates a switch from Upper Jurassic depocentre to Lower Cretaceous palaeohigh, occurring in the hangingwall of the basement lineament seen to have extensional displacement at Rotliegend level. The change in deformational mode along the fault through time is reflected by a corresponding switch in onlap direction within the Jurassic and Cretaceous seismic packages (Fig. 13b). The re-utilization of the same fault during the Tertiary is also demonstrated by Fig. 13b, with fault splays developing and deforming the Oligocene and Eocene sequences.

Applications to hydrocarbon plays

The presence of pre-existing fault trends and their reactivation in response to regional tensional forces played an important role in the supply, deposition and preservation of Jurassic sedimentary packages in the North Sea Basin.

The oblique-slip nature of the movements along many of the fault zones and the intersection of pre-existing structural trends all control the amount of uplift and subsidence occurring both regionally and locally. Sediment entry into the basin is largely controlled by the position of cross-fault trends. Local pop-up highs can act as sediment sources, and then later themselves become depocentres as they become part of an evolving oblique pull-apart basin. It is possible that the whole basin is then inverted by further reactivation along faults, eroding the previously deposited sediment and redistributing it within the new depocentres.

Stratigraphic traps are an integral part of the overall structural model. Truncation and onlapping of sedimentary packages are a commonplace feature and their geometries genetically linked to the history of fault movement (e.g. Figs 13a and 13b).

The timing of trap formation and the remigration of hydrocarbons are also controlled by later fault reactivation in many areas. Traps formed early within the Mesozoic rift phases can be reduced or destroyed by later fault movement and retilting of extensional blocks.

The oblique-slip nature of the fault movements mechanically allows more effective fault sealing. Horizontal components of slip along fault planes may involve cataclasis to a much greater degree than vertical fault planes, especially when the fault zone is reactivated a number of times. This allows sand against sand juxtaposition to be considered as a trapping configuration with an effective fault plane seal. This has been proven in some areas of the Western Central Graben and East Shetland Basin, where very subtle fault zones with almost no apparent vertical displacement form major pressure barriers in the field and have different hydrocarbon–water contacts on either side of the fault zone.

A number of key tools have been used in the present study in order to gain an improved understanding of the structural

evolution of the North Sea and to develop the regional model presented in this paper. These tools are as follows.

1. The use of 3D seismic data provides high-resolution definition of vertical and lateral fault geometries, unconformities, onlapping packages, seismic-stratigraphic facies analysis and the timing of halokinetic movements with respect to fault reactivation. The recognition of oblique-slip structures has often been hampered in the past by aliasing of the rapid lateral changes characteristic of this type of deformation. The increased coverage of 3D data available for seismic interpretation significantly reduces this aliasing risk and enables identification of tectonic structures which, alone or in combination, are characteristic of transtensional and transpressional deformation along the diffuse shear zones. Nevertheless, a mismatch of structural style, lithofacies and palaeoenvironment across what appears to be a minor fault zone may still be difficult to recognize and yet may be crucial in differentiating distinct structural histories between separate intershear areas.

2. Accurate dating incorporating quantitative biostratigraphy to recognize anomalously thick, condensed, or missing sections.

3. Good quality regional data and regional structural/basin architecture maps. This study has highlighted the Central Graben and its margins as areas where the position and asymmetry of present-day highs and lows cannot be used as a simple indicator of Jurassic basin polarity, nor can a simple extensional restoration be used to reconstruct palaeogeographic maps.

4. An understanding of 2D and 3D geometrical style, fault kinematics and the limitations of seismic interpretation in faulted areas with a very complex displacement history.

The combination of these tools allows the explorationist to piece together the complex jigsaw and to predict the presence of favourable trapping geometries with hydrocarbon-bearing sands. Based on the regional evaluation combined with detailed analysis, the model presented indicates that although the total amount of horizontal movement on basement lineaments in the North Sea is probably very limited, the role of oblique-slip tectonics in the Mesozoic structural development of the North Sea is crucial in analysing the hydrocarbon potential of this area.

The authors gratefully acknowledge Shell UK Exploration and Production and Esso Exploration and Production UK for permission to publish the data contained within this paper, NOPEC for permission to publish Fig. 8b, and BP and Agip for permission to publish Fig. 10. Numerous colleagues within Shell are thanked for fruitful discussions throughout the duration of the study, specifically Larry Wakefield, Griff Cordey, Gillian Griffiths and John Parker.

References

BADLEY, M. E., PRICE, J. D., RAMBECH DACHL, C. AND AGDESTINE, T. 1988. The structural evolution of the North Viking Graben and its bearing upon extensional modes of deformation. *Journal of the Geological Society of London*, **145**, 455–472.

BARTON, P. AND WOOD, R. 1984. Tectonic evolution of the North Sea Basin: crustal stretching and subsidence. *Geophysical Journal of the Royal Astronomical Society*, **79**, 987–1022.

BERTHELSEN, A. (CHAIRMAN) AND EUROPEAN GEOTRAVERSE, EUGENO-S WORKING GROUP, 1988. Crustal structure and tectonic evolution of the transition between the Baltic Shield and the North German Caledonides. *Tectonophysics*, **150** (3), 253–348.

BIRD, T. J., BELL, A., GIBBS, A. AND NICHOLSON, J. 1987. Aspects of strike-slip tectonics in the Inner Moray Firth Basin, offshore Scotland. *Norsk Geologisk Tidsskrift*, **67**, 353–370.

CARTWRIGHT, J. 1989. The kinematics of inversion in the Danish Central Graben. *In*: COOPER, M. A. AND WILLIAMS, G. D. (eds) *Inversion Tectonics*. Geological Society, London, Special Publication, **44**, 153–176.

COOPER, M. A. AND WILLIAMS, G. D. 1989. *Inversion Tectonics*. Geological Society, London, Special Publication, **44**.

COWARD, M. P. 1993. The effect of Late Caledonian and Variscan continental escape tectonics on basement structure, Paleozoic basin kinematics and subsequent Mesozoic basin development in NW Europe. *In*: PARKER, J. R. (ed.) *Petroleum Geology of Northwest Europe: Proceedings of the 4th Conference*. Geological Society, London, 1095–1108.

DEMYTTENAERE, R. R. A., SLUIJK, A. H. AND BENTLEY, M. R. 1993. A fundamental reappraisal of the structure of the Cormorant Field and its impact on field development strategy. *In*: PARKER, J. R. (ed.) *Petroleum Geology of Northwest Europe: Proceedings of the 4th Conference*. Geological Society, London, 1151–1157.

FRASER, A. J. AND GAWTHORPE, R. L. 1991. Tectono-stratigraphic development and hydrocarbon habitat of the Carboniferous of northern England. *In*: HARDMAN, R. F. P. AND BROOKS, J. (eds) *Tectonic Events Responsible for Britain's Oil and Gas Reserves*. Geological Society, London, Special Publication, **55**, 49–86.

GAARENSTROOM, L., TROMP, R. A. J., DE JONG, M. C. AND BRANDEN-BURG, A. M. 1993. Overpressures in the Central North Sea: implications for trap integrity and drilling safety. *In*: PARKER, J. R. (ed.) *Petroleum Geology of Northwest Europe: Proceedings of the 4th Conference*. Geological Society, London, 1305–1313.

GLENNIE, K. W. AND BOEGNER, P. L. E. 1981. Sole Pit inversion tectonics. *In*: ILLING, L. V. AND HOBSON, G. D. (eds) *The Petroleum Geology of the Continental Shelf of North-West Europe*. Heyden, London, 110–120.

HARTOG-JAGER, D. DEN, GILES, M. R. AND GRIFFITHS, G. R. 1993. Evolution of Paleogene submarine fans of the North Sea in space and time. *In*: PARKER, J. R. (ed.) *Petroleum Geology of Northwest Europe: Proceedings of the 4th Conference*. Geological Society, London, 59–71.

JOHNSON, R. J. AND DINGWALL, R. G. 1981. The Caledonides: their influence on the stratigraphy of the northwest European Continental Shelf. *In*: ILLING, L. V. AND HOBSON, G. D. (eds) *The Petroleum Geology of the Continental Shelf of North-West Europe*. Heyden, London, 85–97.

KOOPMAN, A., SPEKSNIJDER, A. AND HORSFIELD, W. T. 1987. Sandbox studies of inversion tectonics. *Tectonophysics*, **137**, 379–388.

MARSDEN, G., YIELDING, G., ROBERTS, A. M. AND KUSZNIR, N. J. 1991. Application of a flexural cantilever-shear/pure-shear model of continental lithosphere extension to the formation of the North Sea Basin. *In*: BLUNDELL, D. J. AND GIBBS, A. D. (eds) *Tectonic Evolution of the North Sea Rifts*, Clarendon Press, Oxford, 241–262.

OUDMAYER, B. C. AND DE JAGER, J. 1993. Fault reactivation and oblique-slip in the Southern North Sea. *In*: PARKER, J. R. (ed.) *Petroleum Geology of Northwest Europe: Proceedings of the 4th Conference*. Geological Society, London, 1281–1290.

ROBERTS, A. M., PRICE, J. D. AND OLSEN, T. S. 1990a. Late Jurassic half-graben control on the siting and structure of hydrocarbon accumulations: UK/Norwegian Central Graben. *In*: HARDMAN, R. F. P. AND BROOKS, J. (eds) *Tectonic Events Responsible for Britain's Oil and Gas Reserves*. Geological Society, London, Special Publication, **55**, 229–257.

——, BADLEY, M. E., PRICE, J. D. AND HUCK, I. W. 1990b. The structural history of a transtensional basin: Inner Moray Firth, NE Scotland. *Journal of the Geological Society, London*, **147**, 87–103.

SEARS, R. A., HARBURY, A. R., PROTOY, A. J. G. AND STEWART, D. J. 1993. Structural styles from the southern Central Graben in the UK and Norway. *In*: PARKER, J. R. (ed.) *Petroleum Geology of Northwest Europe: Proceedings of the 4th Conference*. Geological Society, London, 1231–1243.

SPEKSNIJDER, A. 1987. The structural configuration of the Cormorant block IV in context of the northern Viking Graben structural framework. *Geologie en Mijnbouw*, **65**, 357–379.

UNDERHILL, J. A. 1991. Implications of Mesozoic–Recent basin development in the Inner Moray Firth Basin. *Marine and Petroleum Geology*, **8**, 359–370.

VAN HOORN, B. 1987. Structural evolution, timing and tectonic style of the Sole Pit inversion. *Tectonophysics*, **137**, 239–284.

ZIEGLER, P. A. 1990. Tectonic and palaeogeographic development of the North Sea rift system. *In*: BLUNDELL, D. J. AND GIBBS, A. D. (eds) *Tectonic Evolution of the North Sea Rifts*. Clarendon Press, Oxford, 1–36.

Mesozoic extension in the North Sea: constraints from flexural backstripping, forward modelling and fault populations

A. M. ROBERTS,[1] G. YIELDING,[1] N. J. KUSZNIR,[2] I. WALKER[3] and D. DORN-LOPEZ[3]

[1] Badley Earth Sciences Ltd, Winceby House, Winceby, Horncastle LN9 6BP, UK
[2] Department of Earth Sciences, University of Liverpool, Liverpool L69 3BX, UK
[3] Conoco Norway Inc., PO Box 488, N-4001 Stavanger, Norway

Abstract: The magnitude and distribuion of late Jurassic extension in the Northern Viking Graben has been investigated by (i) syn-rift forward modelling using the flexural-cantilever model of continental rifting; (ii) post-rift flexural backstripping of a series of cross-sections; and (iii) the analysis of fault-population statistics.

Application of these three techniques indicates that Jurassic extension on the tilted fault-block terrains of the East Shetland Basin, Tampen Spur and western Horda Platform is on average $c.$ 15% ($\beta = 1.15$). In the graben axis the regional value of β rises to $c.$ 1.3, perhaps locally rising to 1.4. On the eastern Horda Platform Jurassic extension is low, $\beta = c.$ 1.05.

Flexural backstripping of the post-rift part of a cross-section through the Central Graben yields similar estimates of extension. At the flanks of the Jurassic basin β is estimated to be 1.2, rising to a likely maximum of $c.$ 1.3 in the basin centre.

These estimates of extension lie at, or towards, the low end of the previously published range of estimates. The principal reasons for this are (i) the incorporation of the thick sequence of Triassic and Lower Jurassic sediments in the backstripping; and (ii) the use of flexural isostasy (rather than Airy isostasy) in both the backstripping and forward modelling.

The estimates of Jurassic extension obtained in this study do not account for the observed crustal thinning and, therefore, point to there having been a significant pre-Jurassic extensional event in the North Sea, of probable Triassic age. While a residual thermal anomaly from this event may have made a small contribution to post-Jurassic subsidence, compaction of the Triassic–Middle Jurassic sequence has been a significant contribution of the Triassic event to the thickness of the Cretaceous/Tertiary basin.

The analysis of subsidence and extension in the Viking Graben

In recent years a number of papers have been published that attempt to document the magnitude of late Jurassic extension across the Northern Viking Graben. In these papers three principal and independent methods have been used to estimate extension:

1. the summation of fault heaves (e.g. Beach *et al.* 1987; Badley *et al.* 1988; Roberts *et al.* 1990*b*; Ziegler 1990);
2. forward modelling the extensional history (e.g. Barr 1987; Yielding 1990; Marsden *et al.* 1990; Kusznir *et al.* 1991);
3. backstripping of one-dimensional stratigraphic sections (e.g. Beach *et al.* 1987; Giltner 1987; White 1990).

These different techniques have, perhaps not surprisingly, yielded differing estimates of extension; with estimates of stretching factor (β) in the graben axis (where β should be greatest) ranging from 1.15 (Ziegler 1990) to $c.$ 1.5 (White 1990). The discrepancy between these two values equates to a difference in the estimated magnitude of extension in the graben axis of $>300\%$. Similar discrepancies exist for estimates of extension on the graben flanks.

In an attempt to address the apparent discrepancy between the various methods of estimating extension, White (1990) compared his own estimates of β, derived by backstripping, with estimates obtained by quantifying fault-block rotation (a method similar to the summation of heaves). His conclusion was that although his calculations based on fault-block geometry yielded slightly lower β values than did backstripping, there was in fact no significant discrepancy between the two methods used where their errors were considered. With this,

White attempted to lay to rest the long-standing assertion (Ziegler 1983) that subsidence analysis and structural analysis (fault-block geometry) in the North Sea yield irreconcilable results. In his fault-block analysis, however, White assumed that the average Late Jurassic fault-block rotation in the East Shetland Basin, marginal to the Viking Graben, was 10°. Depths to penetrated formation tops at crestal and down-dip locations show that none of the fault blocks in the East Shetland Basin currently dip this steeply, the average dip being $c.$ 6° (e.g. Yielding 1990). Applying White's calculations with a rotation of 6°, rather than 10°, almost halves the extension estimates. Ziegler's (1983) case, as to whether an extension discrepancy exists between the results of subsidence and structural analysis, is therefore reopened.

In this paper a fully integrated study of extension in the Viking Graben is presented, applying backstripping, forward-modelling and fault-summation techniques to the same data-set, in order to investigate whether rigorous application of these techniques yields consistent estimates of β, or whether they point to an 'extension discrepancy'.

The Central Graben, where the geometry of the major fault blocks is commonly obscured by more superficial halokinesis (Roberts *et al.* 1990*a*), is then also briefly considered.

Basin-modelling techniques

While the principal concern of this paper is the quantification of late Jurassic stretching across the Viking and Central Graben, the broader scope of the work from which this derives has been the quantitative modelling of structural and subsidence processes in these areas. The wider results of these basin-modelling studies have been used to make predictions about

From *Petroleum Geology of Northwest Europe: Proceedings of the 4th Conference* (edited by J. R. Parker).
© 1993 Petroleum Geology '86 Ltd. Published by The Geological Society, London, pp. 1123–1136.

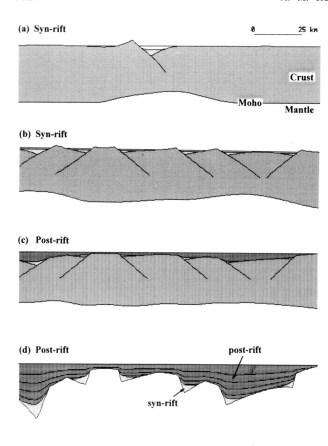

Fig. 1. Examples of the flexural-cantilever basin model. **(a)** Single normal fault, producing hangingwall subsidence and footwall uplift. In the lower crust, ductile extension is distributed across an area wider than the overlying fault (the 'pure-shear width'). **(b)** Multi-faulted model (no erosion), producing domino-style fault blocks within the basin, while the basin margin and fault-block crests are uplifted above sea-level. **(c)** Multi-faulted model, incorporating erosion of the basin margin and fault-block crests, followed by post-rift thermal relaxation. **(d)** Multi-faulted model, as (c), vertically exaggerated, showing syn-rift and post-rift basin-fill.

palaeotopography, palaeobathymetry, sediment source and deposition, and thermal history within the areas under consideration. Two particular basin modelling techiques have been employed in this work:

1. syn-rift modelling using the flexural-cantilever model of continental lithosphere extension;
2. post-rift reverse modelling using flexural backstripping, incorporating reverse modelling of the thermal subsidence.

The flexural-cantilever model

The flexural-cantilever model has previously been described by Kusznir and Egan (1989), Marsden *et al.* (1990), Kusznir *et al.* (1991) and Roberts and Yielding (1991, 1992). A brief description only is therefore given here.

The main elements of the model are summarized in Fig. 1. The model begins with an undeformed profile of the lithosphere and deforms it into a stretched sedimentary basin. The model section is divided into three layers:

1. brittle upper crust, which deforms by faulting;
2. ductile lower crust, which deforms by pure-shear plastic flow;
3. ductile lithospheric mantle, which deforms by pure shear.

The model section is deformed by faults of specified size and position, with extension in the brittle, upper part of the model being balanced by pure-shear extension at depth. Extension is modelled as geologically instantaneous. Both the upper-crustal faulting and the deeper pure shear introduce isostatic loads. These loads are compensated flexurally, i.e. across a finite distance controlled by the strength of the lithosphere, resulting in the characteristic pattern of hangingwall subsidence (half-graben) and footwall uplift (topography) adjacent to each major fault (Figs 1a and b).

In addition to modelling structural processes and compensating for associated loads, the flexural-cantilever model also calculates the geothermal anomaly generated by the lithosphere extension. Loads induced by geothermal perturbation

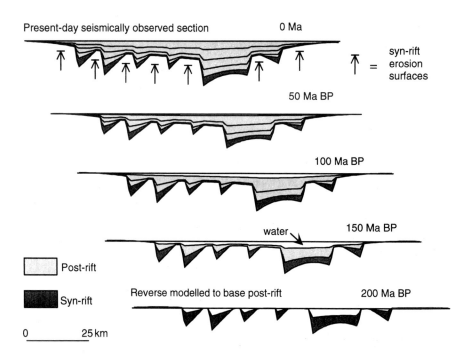

Fig. 2. Schematic illustration of the flexural backstripping method. Starting with a present-day depth section, sediment layers are successively removed to obtain a restoration at the base of the post-rift sequence.

are also isostatically compensated flexurally and can be allowed to relax with time. Thus the model can be taken from the 'active' syn-rift stage into the 'passive', time-dependent, post-rift period. This allows a 'thermal-sag' basin to develop above the older fault-block topography (Figs 1c and d). In this way, the full syn- and post-rift history of a stretched basin can be modelled. Sediment loading and erosional unloading are also flexurally compensated, and sediment compaction effects are included.

Flexural backstripping

The flexural-cantilever model is a particularly powerful method for performing predictive modelling of syn-rift basin geometry. Like all models, however, it involves simplifications which are necessary to keep the model tractable. Most notable amongst the simplifications is the modelling of sedimentary loads and stratigraphy. Because the flexural-cantilever model is a forward model (moving forwards in time) it must first incorporate assumptions about initial conditions in the basin prior to rifting, and must then use further simplified assumptions about the way in which the stratigraphic section develops with time. However, it is difficult to predict sediment distribution quantitatively and it is thus difficult to match precisely the stratigraphic evolution of individual half-graben. During basin-fill the model looks at the stratigraphic evolution of the basin as a whole. Reverse modelling (backwards in time) avoids these problems by starting with present-day stratigraphic data rather than some poorly constrained initial condition.

For this reason, a method has been developed for reverse modelling the thermal-subsidence history of extended basins, using present-day cross-sections to work back in time towards a palinspastic restoration of the syn-rift basin geometry. This procedure is a modified form of flexural backstripping (Watts *et al.* 1982) and takes the mathematics of the time-dependent, thermal-relaxation stage of the flexural-cantilever model (Figs 1c and d) and works it in the opposite time sense. A full description of the model will be published elsewhere, and only the principal geological aspects of the model are described here.

The starting point of the model is a cross-section in depth through a rift basin (see Fig. 2). The stratigraphic units deposited during the period of thermal subsidence are removed one by one, in order to approach a restoration of (or close to) the syn-rift basin geometry. Upon the removal of each unit, three independent calculations are performed in order to derive the next restoration backwards in time. The first step is to 'throw away' the top layer in the section and decompact **all** underlying stratigraphic units in response to the removal of this load. It is important not only to decompact the post-rift sediment pile but also the deeper syn- and pre-rift sediments, should any be present. Ignoring decompaction of stratigraphic section below the post-rift will 'under-expand' the total stratigraphic column and yield an unsatisfactory restoration. Decompaction parameters are lithology dependent and follow the scheme of Sclater and Christie (1980).

Having decompacted the sediment pile in response to the removal of a stratigraphic layer, the isostatic consequences of this are calculated. As in the flexural-cantilever model the isostatic response is calculated flexurally, using finite values for lithosphere strength. Hence, during the removal of a stratigraphic load the isostatic response at any one point is controlled not only by removal of material from above (as in the application of Airy isostasy), but also by removal of material from laterally adjacent areas. This is particularly important for cross-section geometry when removing laterally variable loads. For example, the removal of a large stratigraphic wedge from a graben or half-graben may affect the geometry of an adjacent

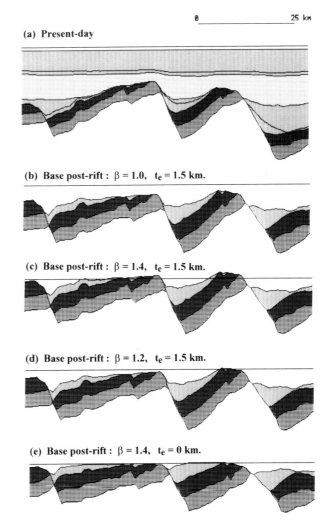

(a) Present-day

0 _____ 25 km

(b) Base post-rift : β = 1.0, t_e = 1.5 km.

(c) Base post-rift : β = 1.4, t_e = 1.5 km.

(d) Base post-rift : β = 1.2, t_e = 1.5 km.

(e) Base post-rift : β = 1.4, t_e = 0 km.

Fig. 3. Illustration of the sensitivity of the backstripping method to β and t_c. (**a**) Present-day depth section across Snorre and Visund fault blocks. The shaded layers are (from top): post-Paleocene, Paleocene, Upper Cretaceous, Lower Cretaceous, Upper Jurassic, Lower–Middle Jurassic, Triassic. Vertical exaggeration c. × 4. (**b**) Section backstripped to 'base Cretaceous' horizon, using β = 1.0, t_c = 1.5 km. (**c**) Backstripped using β = 1.4, t_c = 1.5 km. (**d**) Backstripped using β = 1.2, t_c = 1.5 km. (**e**) Backstripped using β = 1.4, t_c = 0 km.

high. Consider a simple situation such as the single half-graben and footwall high in Fig. 1a. In the forward geological sense a considerable stratigraphic load may accumulate in the half-graben, but little or no deposition will occur on the footwall. If the load in the half-graben is distributed flexurally it will 'weigh down' the adjacent high. Thus, the high will subside further than Airy isostasy would predict. In the reverse sense backstripping will remove the load from the half-graben. With finite lithospheric strength (flexural backstripping) the effects of removing the load will be spread laterally, elevating the adjacent high. Airy backstripping of the high, however, would be incapable of recognizing the flexurally imposed subsidence caused by proximity to a major load. It would, therefore, achieve an unrealistic solution. The authors believe the effects of laterally distributed loading to be important in the detailed subsidence history of sedimentary basins, and thus consider the flexural treatment of loads to be superior to the simpler Airy approach for the analysis of intra-basinal structures.

In addition to flexurally backstripping each layer, the basin's thermal subsidence history is reverse modelled. If the amount of stretching across the section (β) is known, or can be estimated, and the time of rifting is known, then thermal subsidence with time can be calculated (McKenzie 1978).

Values of β may either be estimated from fault heaves and fault-population statistics for the section (see later) or calculated by using the flexural-cantilever model (see later). Thermal relaxation in the forward sense involves cooling and subsidence. Reverse thermal-modelling therefore puts heat back into the model, generating an apparent uplift.

Upon completion of the thermal modelling, the thermal 'uplift' is added to the flexurally-backstripped and decompacted profile. This results in the restored basin model for each time 'snap-shot' (Fig. 2). The basin model, constructed relative to sea-level, illustrates decompacted stratigraphy, intra-basinal structure, predicted palaeobathymetry and predicted areas of emergence.

The sensitivity of the backstripping to different parameters is illustrated by Fig. 3. Figure 3a shows a present-day depth section across the Snorre and Visund fault blocks. Both of these structures show erosion at their crests, indicating that at the end of the rifting period ('base Cretaceous') they were at, or near, sea-level. However, downflank from the crests there is a continuous, marine shale sequence through the Upper Jurassic and Lower Cretaceous. These stratigraphic observations are an important constraint on the backstripping, since they provide information about likely palaeowater-depths. A geologically acceptable restoration is one that honours these observations. The restorations in Figs 3b–d show this section backstripped to 'base Cretaceous' with different values of β (in these examples, constant β was used along the length of the profile). Figure 3b uses $\beta = 1.0$, i.e. the section has been unloaded and decompacted but no heat (syn-rift thermal uplift) has been put back. The eroded fault-block crests restore to well below sea-level, so this restoration is unacceptable. In Figure 3c, $\beta = 1.4$ and large portions of the fault blocks restore above sea-level, including those areas with continuous marine sequences. Hence, this

restoration is also geologically unacceptable. Figure 3d shows the section backstripped with $\beta = 1.2$, and now the fault-block crests are restored near to sea-level but all marine Upper Jurassic sequence remains below sea-level, i.e. this restoration is acceptable. Hence, calibration of the backstripping with the known geology (seismic data and wells) allows an estimate of β to be made. Ideally, predicted water depths in the restoration should be checked against palaeobathymetric estimates from fossil assemblages, but in practice the Upper Jurassic assemblages are sparse and of little use (Barton and Wood 1984; Bertram and Milton 1989). As a general guide, uncertainties in the β estimate are less than ± 0.05.

The restorations in Figs 3b–d used an effective elastic thickness (t_e) of 1.5 km to control the flexural-isostatic response. This small but finite value is consistent with gravity modelling (Barton and Wood 1984) and forward modelling of the structure (Marsden *et al.* 1990). If $t_e = 0$ km (i.e. Airy isostasy) then a significantly different restoration is obtained (see Fig. 3e). A β of 1.4 is now required to restore the eroded fault-block crests near to sea-level. However, strict application of Airy isostasy at this detailed scale has resulted in considerable unrealistic distortion of the structural geometry. Each fault block has been flattened, except at the footwall cut-off of the top of the basement where it is considerably oversteepened. If this restoration were correct, it would imply that a large proportion of the present-day fault-block dip resulted from differential loading in the post-rift phase. This is unlikely since observations of areas of actively extending regions show that domino-style faulting is common, i.e. 'rigid' fault blocks participate in the extension (Jackson and White 1989).

Flexural backstripping can be used to remove the full post-rift sequence. It cannot, however, be used to restore the syn-rift history, for two principal reasons. First, an algorithm for

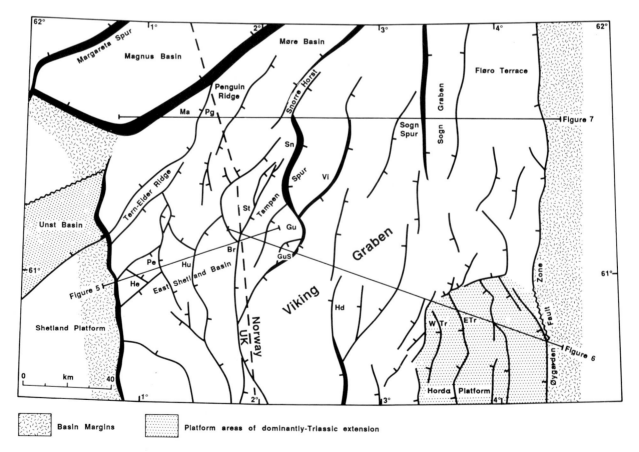

Fig. 4. Structural elements map of the Northern Viking Graben, showing the locations of the three modelled cross-sections (Figs 5, 6 and 7). Oil and gas fields lying on, or close to, the cross-sections are labelled as follows – Br: Brent; Gu: Gullfaks; GuS: Gullfaks Sør; Hd: Huldra; He: Heather; Hu: Hutton; Ma: Magnus; Pe: Pelican; Pg: Penguin; Sn: Snorre; St: Statfjord; ETr: East Troll; WTr: West Troll; Vi: Visund.

realistically restoring the displacement on a fault where both the hangingwall and footwall blocks have participated in the deformation has yet to be devised. All previously published 'section-balancing' schemes assume that only the hangingwall deforms and are thus inapplicable in combination with flexural models. Second, in areas of syn-rift topography, erosion of fault-block crests is the norm (e.g. Yielding 1990; Roberts and Yielding 1991). Reverse modelling such erosion would involve adding an arbitrary load to the section, and would thus deviate from a purely data-derived model. It is, therefore, better to examine syn-rift geometry and erosion in the forward sense, with the flexural-cantilever model. A full basin analysis should thus combine the use of forward and reverse modelling; first using flexural backstripping to restore through the post-rift

history and then using the flexural-cantilever model to match the syn-rift model to the backstripped template.

Basin-modelling of the Viking Graben

In the course of a major study the combined forward- and reverse-modelling techniques described above have been applied to 14 cross-sections across the Norwegian sector of the Viking Graben (60–62°N) and to 8 cross-sections across the UK sector. This paper presents results obtained from three of these lines, located in Fig. 4; two lines are from the Norwegian sector and one is from the UK sector. This modelling study continues from the earlier work by Marsden *et al.* (1990), who applied the flexural-cantilever (forward) model to the full basin

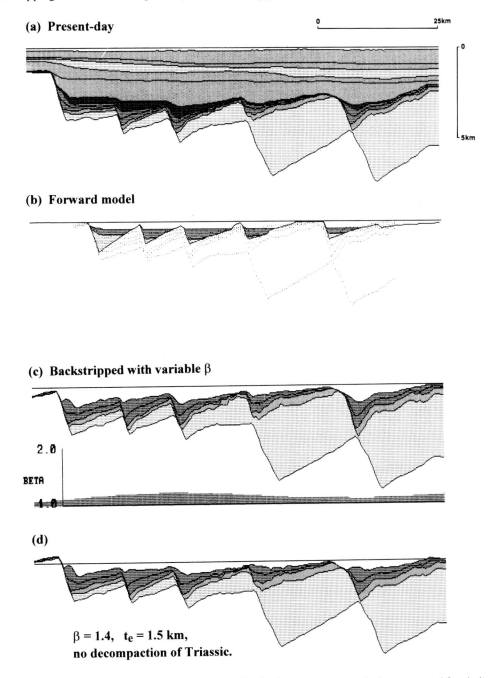

Fig. 5. East Shetland Basin profile (see Fig. 4 for location). (**a**) Present-day depth cross-section, vertically exaggerated for clarity. The section crosses the Heather, Pelican, Hutton, Brent and Gullfaks (part) fault blocks. The uppermost (unshaded) layer is the water layer. The shaded layers are (from top): Tertiary (4 layers), Upper Cretaceous, Lower Cretaceous, Upper Jurassic (2 layers), Middle and Lower Jurassic, and Triassic. (**b**) Forward (flexural-cantilever) model to the base Cretaceous, incorporating extension in the Oxfordian followed by 18 Ma of thermal subsidence. Dotted template shows the observed profile backstripped to base Cretaceous seismic marker with $t_c = 1.5$ km and a constant β of 1.15 (**c**) Backstripped to the base Cretaceous with β-profile (illustrated) derived from the forward model. (**d**) Observed profile backstripped with no decompaction of the Triassic interval. A β of 1.4 is required to restore eroded crest of Brent to sea-level.

history of a section across the Viking Graben. They did not undertake any flexural backstripping. This study concentrates on the Jurassic extensional history of the basin.

East Shetland Basin (UK sector)

The simplest of the three Viking Graben profiles discussed here is that across the East Shetland Basin in the UK sector (see Fig. 4 for location). The present-day cross-section is derived from depth conversion of an ENE–WSW-striking seismic line (Fig. 5a). It extends eastwards from the East Shetland Platform, crossing the west-dipping fault blocks of the Heather, Pelican, Hutton and Brent oil fields, terminating in the east on the Gullfaks structure (Norway).

Modelling of this profile was begun by backstripping the section to the 'base Cretaceous' seismic marker (strictly Late Ryazanian in age) with a constant value of β. After several iterations a value for β of 1.15 (15% extension) was found to give the geologically most-acceptable restoration. To calibrate the restoration, the crest of the eroded/degraded Brent/Statfjord fault block was used. The stratigraphy of the crest of this structure is well known (e.g. Roberts *et al.* 1987; Livera and Gdula 1990; Yielding 1990; Roberts *et al.* 1993) and indicates emergence and erosion in the early late Jurassic followed by complete submergence of the block by the 'base Cretaceous'.

A check on the validity of the restoration is possible by attempting to match its geometry with a flexural-cantilever forward model. Following Roberts *et al.* (1993) the late Jurassic extension has been ascribed to a rapid stretching event in the Oxfordian, allowing a subsequent 18 Ma of post-rift thermal subsidence in the model in order to reach the base Cretaceous. The 'best-fit' forward model, at the 'base Cretaceous' marker, is shown in Fig. 5b. Also shown in Fig. 5b, as a dotted overlay, is the observed section backstripped with a constant β of 1.15. The forward model attempts to match a number of geological features in the backstripped section, specifically: fault-block size, uplift and erosion, hangingwall subsidence and bathymetry. With good seismic data the model will be well constrained by the observable extension on major faults. If, however, fault heave is somewhat unclear on the original seismic data, then some iteration of the model will generally improve the original estimate. In Fig. 5b an attempt has been made to model the structure from the Shetland Platform to the Brent hangingwall. The frontal fault to Gullfaks, beyond the original data (Fig. 5a), has not been included. The model not only incorporates extension and 18 Ma of thermal subsidence but also erosion of any emergent topography. The general fit to the backstripped section is good. The only fault-block crest showing erosion is Brent; Heather, Pelican and Hutton are fully submarine. The dip of the Pelican, Hutton, Brent and Gullfaks structures in the model matches the template well. The Heather structure in the model dips more steeply than the template. This may be because the orientation of the section is somewhat oblique to the Heather bounding fault. In order to match the subsidence in the hangingwall of the Shetland Platform, significant footwall uplift of the platform itself was predicted by the model, prior to erosion (cf. Fig. 1). The predicted erosion of the platform at the basin margin is 1.6 km (see Roberts and Yielding 1991 for further discussion of basin-margin uplift).

The forward model attempts to match the fault-block geometries of model and backstripped section. In so doing, upper-crustal extension is automatically balanced at depth by distributed pure shear. Summing the distributed pure shear for each fault produces a continuous profile of β across the model. This profile of laterally variable β can then be imported into the backstripping routine, as a replacement for the initial value of constant β.

Figure 5c shows the East Shetland Basin section back-

stripped to the base Cretaceous in conjunction with the β-profile derived from Fig. 5b. This restoration is acceptable, taking the crest of Brent close to sea-level, but not above it. Away from the Brent fault block, the restoration shows the crests of Heather, Pelican and Hutton below sea-level at the base Cretaceous, while the larger Gullfaks structure, like Brent, is returned to sea-level. These predictions are in keeping with the known Upper Jurassic stratigraphy on these structures (e.g. Penny 1991; Haig 1991; Eriksen *et al.* 1987). The Shetland Platform is returned close to sea-level, although in this area the platform is probably capped by a Paleocene erosion surface, modifying the original Jurassic basin margin. It is, therefore, not a precise sea-level marker in the base Cretaceous restoration.

The values of β produced by the forward model are shown in Fig. 5c and reach a maximum of 1.20 below the crest of Pelican, decreasing to 1.10 below the crest of Brent. Thus it becomes clear why an averaged β for this section of 1.15 provides a satisfactory initial restoration.

This particular cross-section lies very close to that modelled by White (1990). White achieved estimates for β in the East Shetland Basin ranging between 1.19 and 1.31, with a mean just above 1.25. These estimates imply an extension *c.* 66% greater than that obtained here. It is suggested that White obtained higher values for two reasons. Principally he did not decompact the Middle Jurassic and lower section during subsidence analysis. It is important that these 'pre-rift' sediments, with respect to the Late Jurassic extension, should be decompacted during subsidence analysis. Decompaction will vertically 'expand' the pre-rift sequence to its likely thickness at the time of late Jurassic extension. Failure to incorporate this expansion into the backstripping analysis will require higher values of β to be used in order to return the crests of fault blocks to their correct palaeobathymetric positions. This is illustrated in Fig. 5d, where the Triassic has been treated as non-compactible and a β of 1.4 is required to restore the crest of Brent to sea-level. Thus, failure to decompact the pre-rift section leads to a significant overestimate of β.

In addition to this factor, the likely inaccuracy of White's estimates of β has been compounded by the use of Airy rather than flexural isostasy (though since his analysis was at the centre of each fault block the difference is small).

The lithosphere strength of the models in this paper is controlled by an effective elastic thickness (t_e) of 1.5 km. This is a slightly lower value than previous studies have used (e.g. Marsden *et al.* 1990; Kusznir *et al.* 1991; Roberts and Yielding 1991), but over the 22 lines modelled in this study it has been found to give consistently the most reasonable results when calibrating eroded fault-block crests against sea-level.

Norwegian sector, line 1 (c. 61°N)

The southern of the two illustrated lines across the Norwegian sector of the Viking Graben has been chosen to provide a continuation to the east of the UK line (Fig. 4). The western end of this line (Fig. 6a) terminates on the southernmost part of the Statfjord Field, *c.* 5 km north of the Brent structure in Fig. 5. East of the Statfjord it crosses the southern continuation of Gullfaks and Gullfaks Sør, and then passes over the deep axis of the Viking Graben, where base Triassic can only be estimated. Further to the east the West Troll structure marks the eastern margin of the Horda Platform, a structure dominated by two large Triassic half-graben in the hangingwall of the basin-margin Øygarden Fault. The structure of this line is thus more complex than the East Shetland Basin, comprising fault blocks of opposed polarity on either side of the graben axis.

Initial backstripping of this line was performed with constant β and a best-fit effective elastic thickness of 1.5 km. After

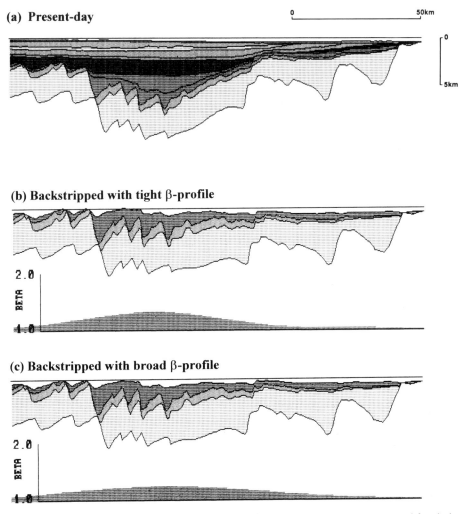

Fig. 6. Norwegian sector, line 1 (see Fig. 4 for location). (**a**) Present-day depth cross-section, vertically exaggerated for clarity. The section crosses the Gullfaks structures, the graben axis and the Horda Platform. The shaded layers are (from top): Tertiary (5 layers); Upper Cretaceous; Lower Cretaceous; Upper Jurassic; Middle and Lower Jurassic; and Triassic. (**b**) Backstripped to the base Cretaceous with a β-profile (illustrated) derived from the forward model. Pure-shear width (Fig. 1a) for each fault in the forward model was 75 km. (**c**) Backstripped to the base Cretaceous with β-profile derived from forward model with pure-shear width increased to 150 km for each fault.

a number of iterations the geologically most-reasonable restoration at the base Cretaceous was once again achieved with a value for β of 1.15. Forward modelling using the flexural-cantilever model was then performed to match the backstripped section, and β-profiles from the forward modelling were used to backstrip the original section again. Two examples of the variable-β backstripping are shown in Fig. 6 (b and c).

In both examples the eroded fault blocks of the Tampen Spur in the west (Statfjord, Gullfaks and Gullfaks Sør) are backstripped close to, but not above, sea-level. Bathymetry in the graben axis is predicted to be in the range 0–800 m. The difference between Figs 6b and 6c is related to the tightness of the β-profile exported from the forward modelling. As shown in Fig. 1, upper-crustal faulting is balanced at depth in the flexural-cantilever model by lower-crustal pure shear. Initially the 'pure-shear width' associated with each fault in the forward model was taken as 75 km (Fig. 1a; see also Kusnir *et al.* 1991). This value can, however, be increased or decreased, giving respectively a broader or a tighter β-profile. In Fig. 6b the β-profile was generated using a pure-shear width of 75 km whereas that in Fig. 6c was generated using a pure-shear width of 150 km.

In both models, β is *c.* 1.10 in the Statfjord area, increases in the graben axis, and decreases to *c.* 1.05 below the Horda Platform. In the graben axis, β reaches a maximum of 1.32

(Fig. 6b), but only 1.24 in Fig. 6c. Total β across the profile is the same in both restorations, constrained by modelling upper-crustal faulting. Backstripping with the combination of two β-profiles from the same forward model thus appears to constrain the likely maximum Jurassic β in the graben axis to *c.* 1.32, while the minimum likely value is *c.* 1.24. Both these values are considerably less than the value of *c.* 1.5 suggested by White (1990).

One final point about Figs 6b and 6c concerns the restoration of the Horda Platform. In neither of these restorations is β sufficiently large to bring the Horda Platform close to sealevel, and yet the shallow-marine Viking Group sediments were deposited in such a setting during the late Jurassic. It is proposed that this discrepancy results from late Jurassic subsidence on the Horda Platform having occurred principally in response to thermal subsidence following significant Triassic extension, rather than to the small amount of Jurassic extension.

Norwegian sector, line 2 (c. 61° 40′ N)

The two lines discussed so far have presented a basin-wide transect at *c.* 61°N. The next line under discussion is *c.* 70 km to the north and comprises another near-complete transect (Fig. 4).

Figure 7a shows the present-day depth conversion of this line. Most of the line lies within the Norwegian sector, but the western 50 km are within the UK sector, giving complete coverage eastwards of the Magnus Basin. The major structures crossed by this line (W to E) are: the Magnus Basin; the Magnus fault block; the Penguin horst; the Snorre horst; the northern extension of Visund; the Sogn Spur; the Sogn Graben; and the Fløro Terrace. The Magnus Basin and Sogn Graben define two separate basinal areas, separated by a complex array of fault blocks.

As with the previous lines, this profile was first backstripped with a constant β (1.15 was again the 'best-fit'), and forward modelling was performed to match the backstripped section. β-profiles from the forward modelling were then used to backstrip the original profile again. Figures 7b and 7c show restorations based on β-profiles generated by forward modelling with a pure-shear width (for each fault in the model) of 75 km and 150 km, respectively.

Both restorations leave the Magnus Basin and Sogn Graben as deep-water basins, predicting in excess of 2 km of bathymetry in the undrilled axis of the Magnus Basin. The intervening fault blocks are all returned to, or close to, sea-level. Of particular note, the severely eroded Snorre horst is returned precisely to sea-level giving much confidence in this particular part of the restoration. The crest of the Magnus fault block is

emergent by c. 400 m. A similar emergence of the crest of Magnus into the early Cretaceous has been predicted by previously published backstripping of the same area (Young 1992).

The forward modelling predicts c.1300 m of crestal erosion on the Magnus structure and nearly 1500 m of erosion on the Snorre horst. This prediction for the Snorre horst agrees well with the seismic interpretation which shows the Jurassic sequence entirely eroded. The erosion of the Jurassic can be readily explained as the consequence of footwall uplift (cf. Yielding 1990; Yielding and Roberts 1992).

The β-profiles shown in Figs 7b and 7c both show peaks beneath the Magnus Basin and the Sogn Graben. With a tighter β-profile (Fig. 7b), β-maxima are 1.23 and 1.29, respectively, whereas the broader β-profile (Fig. 7c) gives maxima of 1.13 and 1.22. These values are considered to bracket the likely range of Jurassic β in the axial parts of this profile.

Basin-modelling conclusions

All three of the Viking Graben profiles modelled (Figs 5–7) can be backstripped to the base Cretaceous with a constant β of 1.15 (and a t_c of 1.5 km). The reason for this becomes apparent with forward modelling. In the areas of the basin away from the graben axes, forward modelling (with pure-shear width of

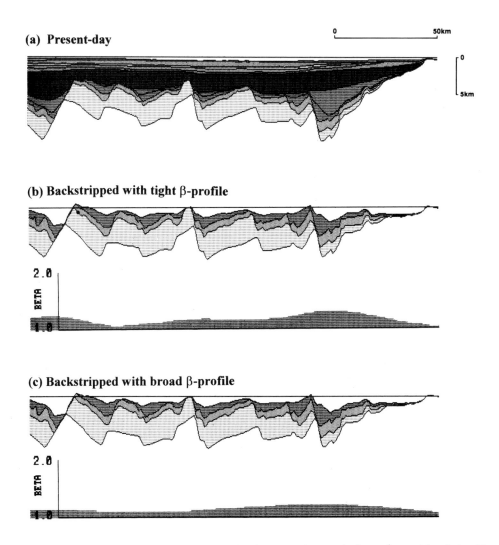

Fig. 7. Norwegian sector, line 2 (see Fig. 4 for location). (**a**) Present-day depth cross-section, vertically exaggerated for clarity. The section crosses the Magnus Basin in the west, the Snorre horst in the centre and the Sogn Graben in the east. Key to layers as in Fig. 6. (**b**) Backstripped to the base Cretaceous with a β-profile (illustrated) derived from the forward model. Pure-shear width for each fault in the forward model was 75 km. (**c**) Backstripped to the base Cretaceous with β-profile derived from forward model with pure-shear width increased to 150 km for each fault. Figures (**b**) and (**c**) are thought to bracket the likely maximum and minimum values for Jurassic β in the graben axes.

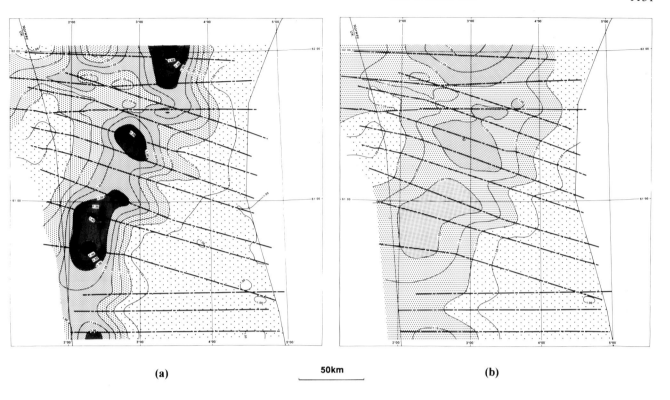

(a)   50km   (b)

Fig. 8. Contoured maps of β (stretching factor) for the late Jurassic extension in the Norwegian sector of the Northern Viking Graben, derived from forward models of the 14 illustrated seismic lines. In (**a**), β-profiles were constructed with pure-shear width (Fig. 1a) for each fault set at 75 km. This has the effect of maximizing β in the axes of the Viking and Sogn Graben, where local maxima in excess of 1.4 are encountered. In (**b**), β-profiles were constructed with pure-shear width for each fault set at 150 km. This has the effect of minimizing β in the graben axes. Note that for both maps the Horda Platform, in the SE, remains an area of consistently low Jurassic extension, less than 5%.

75 km or 150 km) predicts values of β close to 1.15 (in the range 1.10–1.20). In the graben axes, forward modelling suggests β may be higher than this figure, with a maximum permissible value of not much greater than 1.30, and a minimum permissible value not less than c. 1.20. These values are all much lower than those predicted by White (1990). The discrepancy arises principally because compaction of the Triassic–Middle Jurassic sequence has been taken into account and flexural, rather than Airy, isostasy has been used. The resultant values for β are in fact closer to those suggested by Giltner (1987), who also took compaction of the Triassic sequence into account.

Using all 14 lines studied from the Norwegian sector, two maps have been compiled showing the predicted values of Jurassic β (Fig. 8). Figure 8a, constructed with a pure-shear width of 75 km, can be used to constrain maximum permissible β in the graben axis. Locally, where the Viking and Sogn Graben are narrow, the values reach 1.4, but are more commonly c. 1.3. Figure 8b, constructed with a pure-shear width of 150 km, constrains the minimum permissible β in the graben axis. This value is generally slightly above 1.2, except in the approach to the Møre Basin where it reaches 1.3.

Both maps pick out lower values of β of c. 1.15 in the western (Tampen Spur) area. They also highlight the very low values of Jurassic extension associated with the Horda Platform in the SE.

Fault population statistics and estimation of extension

The previous section discussed estimates of β derived from basin-modelling techniques and showed that in the Viking Graben compatible estimates of extension can be derived from forward and reverse modelling. This section addresses the question of whether these results are themselves compatible with simple observations on the original seismic lines.

Roberts *et al.* (1990b) presented an analysis of the five Viking Graben profiles illustrated in Marsden *et al.* (1990), suggesting that the seismically resolvable Jurassic extension on these lines, across the whole graben, was c.15 km. They realized, however, that being based on regional-scale interpretation this was very much a minimum estimate of extension, and that an unquantified extension below the limit of seismic resolution would need to be added to this estimate in order to obtain a more realistic estimate.

Childs *et al.* (1990) presented a methodology for the statistical analysis of fault populations, showing that an observed population distribution, with a lower limit of observational resolution, can be extrapolated to predict the fault population distribution below the observational limit. In practice, the observed fault population on one or more seismic lines can be used to constrain the likely sub-seismic fault population.

Walsh *et al.* (1991) applied this methodology to a range of data from the North Viking Graben, including two of the profiles from Marsden *et al.* (1990). They concluded that c. 40% of the extension on the profiles was likely to be missed by simple fault-heave summation. Thus the minimum estimate of 15 km of extension derived by Roberts *et al.* (1990b) translates to a likely true extension of 25 km. As an average value across the Viking Graben, including the low-extension Horda Platform, 25 km of Jurassic extension yields a β of c. 1.14. This estimate, as an average, is comparable with the results of the basin-modelling described above.

In the course of this study fault-displacement populations have been measured for all of the Norwegian seismic lines. Of all 14 lines, line 2 of this study (Fig. 7) provides the best-constrained and most-reliable fault-population statistics. Figure 9a shows a plot of fault throw (x-axis) against the cumulative number of faults with throw equal to or larger than this value, derived from Norwegian line 2 at the level of the top Statfjord Formation. The data show a near-linear distribution

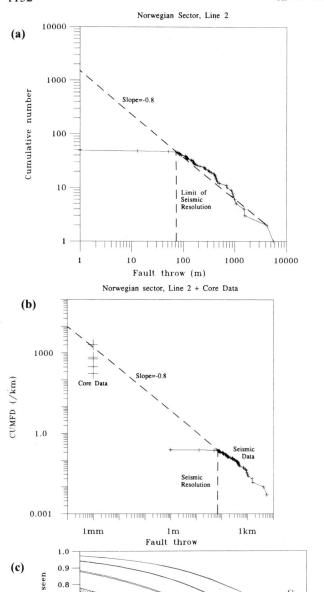

Fig. 9. (a) Fault population plot (log of throw vs. log of cumulative number of faults with equal or greater throw) for Norwegian line 2 (Figs 4 and 7). The majority of the data define an approximately linear segment, with slope = *c.* −0.8. This slope tails off rapidly at fault throw = 70 m, defining the lower limit of seismic resolution for this line. (b) as (a), but expanded to smaller values of throw, in order to incorporate fault-density data from cores in two Northern Viking Graben oil fields. Extrapolation of the slope from the seismic data intersects the upper limit of the core data. (c) Plot of resolved extension as a function of minimum resolved heave, for a range of population slopes (−0.9 to −0.5). For each value of slope the base of the dotted envelope assumes that the fault population scales infinitely downwards, the top of the envelope assumes that the lower fractal limit is at 1 mm heave. Taking a fault dip of 45°, a lower limit of seismic resolution at 70 m and a population slope of −0.8, the prediction is that 65% of the total Jurassic extension is resolved by seismic data, leaving 35% below the limit of seismic resolution.

on this log–log plot, indicating that the throw population is controlled by a power law (i.e. it is fractal or self-similar over a range of scales). The slope of the straight, right-hand segment of the plot, comprising most of the data, is *c.* −0.8. The left termination of this segment corresponds with a fault-throw of about 70 m, defining the effective limit of seismic resolution on line 2. Below this value, fault-throws are not reliably imaged. In Fig. 9b the population slope has been extrapolated down to the scale of fractures in core (mm). The left-hand group of larger crosses illustrates fracture density measured in cores of pre-late Jurassic rocks from two oil fields close to line 2. Extrapolation of the seismically resolvable population for line 2 intersects the upper (more fractured) part of the field for core data.

Using the population slope of *c.* −0.8 and the lower limit of seismic resolution at 70 m, Fig. 9c can be used to predict the proportion of 'missing' extension on line 2, which seismic data do not resolve (cf. Walsh *et al.* 1991; Marrett and Allmendinger 1992). The prediction is that *c.* 65% of extension is resolved by the seismic data, leaving *c.* 35% unseen.

Summation of the Jurassic fault heaves on Norwegian line 2 yields an observable extension of 23 km (from the Magnus fault block eastwards). This occurs within a present-day section length of 180 km, equating with a seismically resolvable *β* of 1.15. Adding the 'missing' 35% extension, predicted to be present below seismic resolution, would give an average total Jurassic *β* across the profile of 1.24. This figure is between the basin-modelling estimates of *β* = 1.15 in the fault-block terrains and *c.* 1.30 in the graben axes.

The observable Jurassic extension on Norwegian line 1 (Fig. 6) (across the western 130 km, where Jurassic extension is significant) is *c.* 16 km. This equates with a seismically resolvable *β* of 1.14. Assuming once more that *c.* 35% of the extension is below seismic resolution gives a total average Jurassic *β* for this line of 1.23. As an average value this is again comparable with the results of basin modelling, which range between 1.10 and 1.32, from flank to axis.

Observable Jurassic extension on the UK line (Fig. 5) is *c.* 10 km, including some erosion of the western basin margin. This is distributed across a present-day section length of 90 km and yields a seismically resolvable *β* of *c.* 1.12. Allowing for *c.* 35% of 'missing' extension gives a total average Jurassic *β* for this line of *c.* 1.20, lower than the sections crossing the graben axis but comparable with the average value of 1.15 (range 1.10–1.20) derived from basin modelling.

The fault-population analysis discussed in this paper and in Walsh *et al.* (1991) and Marrett and Allmendinger (1992) agrees well with the basin-modelling studies. If *β* is derived only from observable fault heaves, then the average value across the basin is lower than that obtained by basin modelling. If, however, an allowance is made for seismically unresolvable extension then estimates of average *β* become comparable with the results of basin modelling.

Basin modelling of the Central Graben

In the Viking Graben the late Jurassic fault-block topography is generally well defined by seismic data, making combined forward and reverse basin modelling and the study of fault-population statistics relatively straightforward. Moving southwards within the North Sea Basin to the Central Graben, the late Jurassic fault-block topography becomes less well defined. The reasons for this have been discussed previously by Roberts *et al.* (1990*a,b*) and relate principally to the presence of Zechstein salt. Throughout the Central Graben, salt movement has had a profound effect on local stratigraphic relationships, masking, to a large extent, stratigraphic relationships attributable to fault-block rotation. In addition, the presence of salt pillows and diapirs commonly degrades the ability to image the sub-salt structure with seismic-reflection data. Basin modelling of basement-involved extension is thus more complex here than in the Viking Graben. It is, therefore, perhaps surprising that the most frequently studied and documented part of the North Sea, in terms of its stretching history, is the Central Graben (Sclater and Christie 1980; Wood and Barton 1983; Ziegler 1983; Barton and Wood 1984; Sclater *et al.* 1986; Hellinger *et al.* 1989; Sclater and Shorey 1989; Latin *et al.* 1990; White and Latin in press). Most of these papers (with the exception of Sclater and Shorey 1989) have concerned themselves with subsidence analysis of individual point locations on traverses across the Central Graben. A diverse range of late Jurassic stretching estimates has been obtained, but these largely divide into two groups, dependent upon the geological model used.

Sclater and Christie (1980), Barton and Wood (1984), Latin *et al.* (1990) and White and Latin (in press) assumed that Jurassic/Cretaceous extension alone can adequately explain the thickness of the late Jurassic, Cretaceous and Tertiary sequences in the Central Graben. Their models assumed that earlier (Triassic or Permo-Triassic) extension had no impact on the subsidence which occurred after the Middle Jurassic. In the axial *c.* 100 km of the Central Graben, where Jurassic extension is concentrated (Roberts *et al.* 1990*a*), these authors all obtained estimates of Jurassic β in the range 1.5–1.6. Sclater *et al.* (1986) and Hellinger *et al.* (1989), however, applied a more complex model to the Central Graben, believing that the compactional and thermal consequences of Triassic extension may have exerted an influence on the thickness of the post-Jurassic stratigraphic sequence. Their estimates of Jurassic β in the axial zone were considerably lower than 1.5–1.6, with Hellinger *et al.* quoting a value as low as 1.17.

In all of these studies Airy isostasy was employed during the backstripping procedure, and in most cases it is not clear how much effort has been made to avoid backstripping locations where halokinesis has had an effect on stratigraphic relationships. It is, therefore, perhaps unclear which estimates are the most reliable, although observations of Triassic half-graben would favour the application of a model which incorporates the stratigraphic effects of Triassic extension.

Flexural backstripping of a Central Graben cross-section

As discussed by Roberts *et al.* (1990*a*) it is impractical to measure Jurassic extension directly in the Central Graben, because of the masking effects of halokinesis. In some areas it is difficult even to identify the basement structure. This analysis of a Central Graben profile is, therefore, restricted to flexural backstripping in the hope that this will circumvent the problems of forward modelling and fault-population analysis in this area.

The present-day section is shown in Fig. 10a. The section strikes ENE and is *c.* 105 km in length, crossing the entire axial zone of the Central Graben, where the Jurassic extension is almost exclusively concentrated. It crosses five significant geological features. The western end of the section lies on the Western Platform, close to the Gannet fields. Moving eastwards the section crosses the symmetric graben of the West Central Graben, the Forties/Montrose High (close to the Montrose Field), the asymmetric half-graben of the East Central Graben, finally crossing the Jurassic basin margin onto the Jæren High.

It is probably impossible to locate a cross-section through the Central Graben which does not cross salt-influenced structures. With this in mind, a cross-section has been chosen in order that, in some areas, the effects of halokinesis can be discounted as an influence on the regional backstripping. On both the Forties/Montrose High and on the Jæren High (along this line of section) no salt is interpreted below the Triassic. This is attributed to salt evacuation from these areas during the Triassic. As a consequence, both the Forties/Montrose High and the Jæren High behaved as the crests of uplifted footwall highs (Roberts and Yielding 1991) during late Jurassic extension. Flexurally backstripping the post-Jurassic sequence above these eroded highs is, therefore, considered to be a valid method of estimating β.

In this part of the East Central Graben the interpreted area of salt is minimal, with significant Triassic evacuation again inferred. In the West Central Graben, however, two significant salt highs are present flanked by Upper Jurassic rim synclines. On the Western Platform a single salt high is present. Clearly

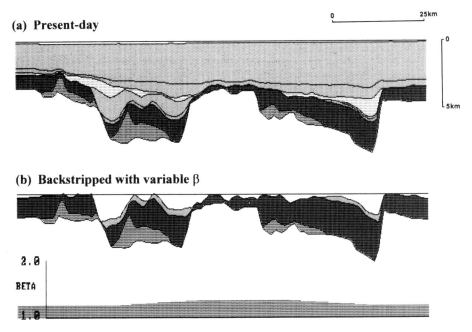

(a) Present-day

0   25km

0

5km

(b) Backstripped with variable β

2.0

BETA

1.0

Fig. 10. Central Graben profile. For location and more discussion of this line see Price *et al.* (1993). (**a**) Present-day depth cross-section, vertically exaggerated for clarity. The section crosses the Western Platform (west), Forties/Montrose High (centre) and Jæren High (east). The uppermost (unshaded) layer is the water layer. The shaded layers are (from top): Tertiary; Upper Cretaceous; Lower Cretaceous; Upper Jurassic; Middle Jurassic; Triassic; and Permian salt. (**b**) Backstripped to the syn-rift event (Oxfordian) at the base of the Upper Jurassic, with $t_c = 1.5$ km and the illustrated β-profile, which varies from 1.2 on the flanks to 1.3 in the centre of the profile.

none of these salt-induced features should be used to peg the backstripping.

It has earlier been suggested that late Jurassic extension in the Viking Graben was principally of Oxfordian age. It has been argued (Roberts *et al.* 1990*a*; Price *et al.* 1993) that the late Oxfordian sands within this part of the Central Graben indicate the likely age of rifting in this area. The Central Graben cross-section has, therefore, been backstripped beyond the base Cretaceous, to the Oxfordian syn-rift stage, on the basis that this should return the eroded flanks of the basin and eroded intra-basinal highs back to sea-level. Initial trial backstripping with constant β across the profile showed that the eroded flanks are returned to sea-level with $\beta = 1.2$, whereas the eroded intra-basinal highs require $\beta = 1.3$. These results suggest that a reasonable approximation to the true β-profile is a bell-shaped profile constrained by these values. Such a β-profile has been constructed and used to backstrip the original section, giving the result shown in Fig. 10b. This restoration honours the known stratigraphy along the profile, returning the eroded highs near to sea-level and keeping Upper Jurassic shales below sea-level. The deeper parts of the West and East Central Graben are likely to have been depocentres for erosion products from the highs.

The results presented in this paper contrast with the recent Airy backstripping of White and Latin (in press), who estimate a maximum Jurassic β in this area of between 1.5 and 1.75. White and Latin largely discounted the influence of Triassic extension in the subsidence of the Central Graben. The estimates of this paper are closer to those of Sclater *et al.* (1986) and Hellinger *et al.* (1989), who treated the subsidence of the Central Graben as a multiple-rift event.

The significance of Triassic extension in the North Sea

The modelling of Jurasic extension in the Viking Graben suggests a likely maximum value for β in the graben axis of *c.*1.4, with a regional maximum of *c.* 1.3 (Figs 6 and 7). This is significantly lower than the total crustal thinning of $\beta = 2$ in the Viking Graben (Klemperer 1988) and, following the previous studies of Giltner (1987), Badley *et al.* (1988), Marsden *et al.* (1990) and Yielding *et al.* (1992), this apparent discrepancy between total thinning and the magnitude of Jurassic extension is most likely to be attributable to appreciable extension in the Triassic.

Likewise, in the Central Graben the maximum estimate of Jurassic β is *c.* 1.3, significantly lower than β obtained from studies of total crustal thinning (Barton and Wood 1984). Once more this points to appreciable extension during the Triassic, as argued by Sclater *et al.* (1986) and Hellinger *et al.* (1989).

The backstripping in this study, both in the Viking and Central Graben, was performed using an Oxfordian rift age to backstrip the post-Jurassic thermal-subsidence history of the basins. No allowance was made for a post-Triassic thermal anomaly at the onset of Jurassic rifting. Decompaction, however, was performed on the full stratigraphic sequence down to the base of the Triassic which, therefore, allowed for the compaction-related stratigraphic effects of Triassic extension. In the Northern Viking Graben, internal consistency has been achieved between the results of this method of backstripping and forward modelling of the late Jurassic rift event. Previous studies (e.g. Badley *et al.* 1988) have attributed the thickness of the post-Jurassic sequence in the Viking Graben to combined relaxation of the post-Triassic and post-Jurassic thermal anomalies. The internal consistency of the forward and reverse models, together with the earlier work of Marsden *et al.* (1990), suggests, however, that the first-order effect of the Triassic stretching event upon the post-Jurassic subsidence history was the generation of a compactible stratigraphic sequence as 'pre-

rift basement' to the Upper Jurassic and younger sequences. This allowed 'overdeepening' of the Cretaceous and Tertiary basin without there necessarily being a major post-Triassic thermal contribution to the Cretaceous/Tertiary subsidence. As argued by Roberts *et al.* (1993) the post-Triassic thermal anomaly may have exerted some small control on stratigraphic relationships within the Upper Jurassic, but in terms of a contribution to regional post-Jurassic subsidence the compaction of the underlying sequence was probably more important.

A realistic stretching model for the North Sea must, therefore, acknowledge significant, and approximately equal, contributions to extension during the Triassic and late Jurassic. Thermal subsidence following Triassic extension generated a thick regionally distributed Triassic–Middle Jurassic stratigraphic sequence, compaction of which made a significant contribution to the observed thickness of the Cretaceous/Tertiary thermal-subsidence sequence.

Conclusions

In the Viking Graben, it is concluded from forward modelling and flexural backstripping that a Jurassic β of 1.15 is a reliable estimate across the tilted fault-block terrains of the East Shetland Basin, Tampen Spur and western Horda Platform. The eastern Horda Platform exhibits a smaller Jurassic β of *c.* 1.05, while in the graben axis a regional value of $\beta = 1.30$ (locally rising to *c.* 1.40) appears to be a reliable estimate.

An analysis of fault-population statistics from the Northern Viking Graben has shown that simple heave summation on Jurassic faults may underestimate the total Jurassic extension by *c.* 35%, as small faults are not imaged on seismic-reflection data. When this 'missing extension' is added to the observable extension, the resulting estimates of total extension are comparable with the more-rigorous forward and reverse basin-modelling techniques.

Application of flexural backstripping to a cross-section through the Central Graben suggests that β here may range from *c.* 1.2 at the flanks of the Jurassic basin to *c.* 1.3 in the basin centre.

All of the estimates of Jurassic extension derived in this study point to there having been a significant pre-Jurassic extensional history in the North Sea. Following previous studies it is concluded that this extension occurred predominantly during the early Triassic.

The estimates of extension in the Viking and Central Graben lie towards or at the lower end of previously published estimates. This derives from two important details of the modelling. Decompaction of the full stratigraphic column is always undertaken, and backstripping is performed with the acknowledgement that small, but finite, values of effective elastic thickness exert a critical control on the geometry of forward- and reverse-modelled cross-sections, and also on the values of β derived therefrom. The combination of forward modelling (with the flexural-cantilever model) and reverse modelling (by flexural backstripping) provides a better method of analysing basin cross-sections than a one-dimensional approach to backstripping using Airy isostasy.

The authors would like to thank Conoco Norway Inc., and Tony Doré in particular, for permission to publish the results of modelling the Norwegian Viking Graben. The depth-converted section from the Central Graben was kindly provided by John Price. Our colleagues Gary Marsden, Mark Newall, Juan Watterson, John Walsh, Mike Badley and Brett Freeman have contributed significantly to our understanding of the modelling techniques employed. We also thank Nicky White for many stimulating discussions on the subsidence history of the North Sea.

References

BADLEY, M. E., PRICE, J. D., RAMBECH DAHL, C. AND AGDESTEIN, T. 1988. The structural evolution of the northern Viking Graben and its bearing upon extensional modes of basin formation. *Journal of the Geological Society, London*, **145**, 455–472.

BARR, D. 1987. Lithospheric stretching, detached normal faulting and footwall uplift. *In*: COWARD, M. P., DEWEY, J. F. AND HANCOCK, P. L. (eds) *Continental Extensional Tectonics*. Geological Society, London, Special Publication, **28**, 75–94.

BARTON, P. AND WOOD, R. 1984. Tectonic evolution of the North Sea basin: crustal stretching and subsidence. *Geophysical Journal of the Royal Astronomical Society*, **79**, 987–1022.

BEACH, A., BIRD, T. AND GIBBS, A. 1987. Extensional tectonics and crustal structure: deep seismic reflection data from the northern North Sea Viking Graben. *In*: COWARD, M. P., DEWEY, J. F. AND HANCOCK, P. L. (eds) *Continental Extensional Tectonics*. Geological Society, London, Special Publication, **28**, 467–476.

BERTRAM, G. T. AND MILTON, N. J. 1989. Reconstructing basin evolution from sedimentary thickness; the importance of palaeo-bathymetric control, with reference to the North Sea. *Basin Research*, **1**, 247–257.

CHILDS, C., WALSH, J. J. AND WATTERSON, J. 1990. A method for estimation of the density of fault displacements below the limits of seismic resolution in reservoir formations. *In: North Sea Oil and Gas Reservoirs II*. Norwegian Institute of Technology, Trondheim. Graham & Trotman, London, 309–318.

ERIKSEN, T., HELLE, M., HENDEN, J. AND ROGNEBAKKE, A. 1987. Gullfaks. *In*: SPENCER, A. M. *et al.* (eds) *Geology of the Norwegian Oil and Gas Fields*. Graham & Trotman, London, 273–286.

GILTNER, J. P. 1987. Application of extensional models to the Northern Viking Graben. *Norsk Geologisk Tidsskrift*, **67**, 339–352.

HAIG, D. B. 1991. The Hutton Field, Blocks 211/28, 211/27, UK North Sea. *In*: ABBOTTS, I. L. (ed.) *United Kingdom Oil and Gas Fields 25 Years Commemorative Volume*. Geological Society, London, Memoir, **14**, 135–144.

HELLINGER, S. J., SCLATER, J. G. AND GILTNER, J. 1989. Mid-Jurassic through mid-Cretaceous extension in the Central Graben of the North Sea—part 1: estimates from subsidence. *Basin Research*, **1**, 191–200.

JACKSON, J. A. AND WHITE, N. J. 1989. Normal faulting in the upper continental crust: observations from regions of active extension. *Journal of Structural Geology*, **11**, 15–36.

KLEMPERER, S. L. 1988. Crustal thinning and nature of extension in the Northern North Sea from deep seisic reflection profiling. *Tectonics*, **7**, 803–822.

KUSZNIR, N. J. AND EGAN, S. S. 1989. Simple-shear and pure-shear models of extensional sedimentary basin formation: application to the Jeanne d'Arc Basin, Grand Banks of Newfoundland. *In*: TANKARD, A. J. AND BALKWILL, H. R. (eds) *Extensional Tectonics and Stratigraphy of the North Atlantic Margins*. American Association of Petroleum Geologists Memoir, **46**, 305–322.

——, MARSDEN, G. AND EGAN, S. S. 1991. A flexural cantilever simple-shear/pure-shear model of continental lithosphere extension: application to the Jeanne d'Arc Basin, Grand Banks and Viking Graben, North Sea. *In*: ROBERTS, A. M., YIELDING, G. AND FREEMAN, B. (eds) *The Geometry of Normal Faults*. Geological Society, London, Special Publication, **56**, 41–60.

LATIN, D. M., DIXON, J. E., FITTON, J. G. AND WHITE, N. J. 1990. Mesozoic magmatic activity in the North Sea Basin: implications for stretching history. *In*: HARDMAN, R. F. P. AND BROOKS, J. (eds) *Tectonic Events Responsible for Britain's Oil and Gas Reserves*. Geological Society, London, Special Publication, **55**, 207–228.

LIVERA, S. E. AND GDULA, J. E. 1990. Brent Oil Field. *In*: BEAUMONT, E. A. AND FOSTER, N. H. (eds) *Structural Traps II, Traps Associated with Tectonic Faulting*. American Association of Petroleum Geologists, Atlas of Oil and Gas Fields, 21–63.

MCKENZIE, D. P. 1978. Some remarks on the development of sedimentary basins. *Earth and Planetary Science Letters*, **40**, 25–32.

MARRETT, R. AND ALLMENDINGER, R. W. 1992. Amount of extension on 'small' faults: An example from the Viking Graben. *Geology*, **20**, 47–50.

MARSDEN, G., YIELDING G., ROBERTS, A. M. AND KUSZNIR, N. J. 1990. Application of a flexural cantilever simple-shear/pure-shear model of continental lithosphere extension to the formation of the northern North Sea Basin. *In*: BLUNDELL, D. J. AND GIBBS, A. D. (eds) *Tectonic Evolution of the North Sea Rifts*. Oxford University Press, Oxford, 241–261.

PENNY, B. 1991. The Heather Field, Block 2/5, UK North Sea. *In*: ABBOTTS, I. L. (ed.) *United Kingdom Oil and Gas Fields 25 Years Commemorative Volume*. Geological Society, London, Memoir, **14**, 127–134.

PRICE, J. D., DYER, R., GOODALL, I., MCKIE, T. WATSON, P. AND WILLIAMS, G. 1993. Effective stratigraphical subdivision of the Humber Group and the Late Jurassic evolution of the Central Graben. *In*: PARKER, J. R. (ed.) *Petroleum Geology of Northwest Europe: Proceedings of the 4th Conference*. Geological Society, London, 443–458.

ROBERTS, A. M., PRICE, J. D. AND OLSEN, T. S. 1990a. Late Jurassic half-graben control on the siting and structure of hydrocarbon accumulations: UK/Norwegian Central Graben. *In*: HARDMAN, R. F. P. AND BROOKS, J. (eds) *Tectonic Events Responsible for Britain's Oil and Gas Reserves*. Geological Society, London, Special Publication, **55**, 229–258.

—— AND YIELDING, G. 1991. Deformation around basin-margin faults in the North Sea/Norwegian rift. *In*: ROBERTS, A. M., YIELDING, G. AND FREEMAN, B. (eds) *The Geometry of Normal Faults*. Geological Society, London, Special Publication, **56**, 61–78.

—— AND —— 1993. Continental Extensional Tectonics. *In*: HANCOCK, P. L. (ed.) *New Concepts in Tectonics*. Pergamon, Oxford.

——, —— AND BADLEY, M. E. 1990b. A kinematic model for the orthogonal opening of the Late Jurassic North Sea Rift System, Denmark–Mid Norway. *In*: BLUNDELL, D. J. AND GIBBS, A. D. (eds) *Tectonic Evolution of the North Sea Rifts*. Oxford University Press, Oxford, 180–199.

——, —— AND —— 1993. Tectonic and bathymetric controls on stratigraphic sequences within evolving half-graben. *In*: WILLIAMS, G. D. (ed.) *Tectonics and Seismic Sequence Stratigraphy*. Geological Society, London, Special Publication, **71**, 87–122.

ROBERTS, J. D., MATTHIESON, A. S. AND HAMPSON, J. M. 1987. Statfjord. *In*: SPENCER A. M. *et al.* (eds) *Geology of the Norwegian Oil and Gas Fields*. Graham & Trotman, London, 319–340.

SCLATER, J. G. AND CHRISTIE, P. A. F. 1980. Continental Stretching: an explanation of the post mid-Cretaceous subsidence of the Central North Sea Basin. *Journal of Geophysical Research*, **85**, 3711–3739.

——, HELLINGER, S. J. AND SHOREY, M. 1986. An analysis of the importance of extension in accounting for the post-Carboniferous subsidence of the North Sea basin. University of Texas Institute for Geophysics, Internal Report.

—— AND SHOREY, M. D. 1989. Mid-Jurassic through mid-Cretaceous extension in the Graben Graben—part 2: estimates from faulting observed on a seismic reflection line. *Basin Research* **1**, 201–215.

WALSH, J., WATTERSON, J. AND YIELDING, G. 1991. The importance of small-scale faulting in regional extension. *Nature*, **351**, 391–393.

WATTS, A. B., KARNER, G. D. AND STECKLER, M. S. 1982. Lithosphere flexural and the evolution of sedimentary basins. *In*: KENT, P., BOTT, M. H. P., MCKENZIE, D. P. AND WILLIAMS, C. A. (eds) The evolution of sedimentary basins. *Philosophical Transactions of the Royal Society, London*, **A305**, 249–281.

WHITE, N. J. 1990. Does the uniform stretching model work in the North Sea? *In*: BLUNDELL, D. J. AND GIBBS, A. D. (eds) *Tectonic Evolution of the North Sea Rifts*. Oxford University Press, Oxford, 217–240.

—— AND LATIN, D. M. (in press). Lithospheric Thinning from Subsidence Analyses in the North Sea 'Triple Junction'. *Journal of Geological Society, London*.

WOOD, R. AND BARTON, P. 1983. Crustal thinning and subsidence in the North Sea. *Nature*, **302**, 134–136.

YIELDING, G. 1990. Footwall uplift associated with Late Jurassic normal faulting in the northern North Sea. *Journal of the Geological Society, London*, **147**, 219–222.

——, BADLEY, M. E. AND ROBERTS, A. M. 1992. The structural evolution of the Brent Province. *In*: MORTON, A. C. HASZELDINE, R. S., GILES, M. R. AND BROWN, S. (eds) *Geology of the Brent*

Group. Geological Society, London, Special Publication, **61**, 27–44.

—— AND ROBERTS, A. M. 1992. Footwall uplift during normal faulting-implications for structural geometries in the North Sea. *In*: LARSEN, R. M. (ed.) *Structural and Tectonic Modelling and its Application to Petroleum Geology.* Norwegian Petroleum Society Special Publication, **1**, 289–304.

YOUNG, R. 1992. Restoration of a regional profile across the Magnus Field in the Northern North Sea. *In*: LARSEN, R. M. (ed.) *Structural and Tectonic Modelling and its Application to Petroleum Geology.* Norwegian Petroleum Society Special Publication, **1**, 221–229.

ZIEGLER, P. A. 1983. Discussion on: Crustal thinning and subsidence in the North Sea. *Nature*, **304**, 561.

—— 1990. Tectonic and palaeogeographic development of the North Sea rift system. *In*: BLUNDELL, D. J. AND GIBBS, A. D. (eds) *Tectonic Evolution of the North Sea Rifts.* Oxford University Press, Oxford, 1–36.

Discussion

Question (N. Morton, Birkbeck College, University of London):

Can the flexural backstripping model you describe, model several phases of extension and the intervening sag phases of basin evolution?

Answer (A. M. Roberts):

As presently formulated, the backstripping procedure inverse-models the *thermal* effects of one extensional event. However, sedimentary layers from any preceding extensional event will be correctly decompacted during the backstripping. The *thermal* effects of earlier extensional events are only significant if the events are closely spaced in time relative to the thermal relaxation time of the lithosphere (*c*. 60 Ma).

Question (A. Fraser, BP Exploration, Glasgow):

In the Magnus Field, turbidites of Kimmeridgian age (contemporaneous with the age of the block faulting) sit on the top of a tilted fault block. Is this situation not better explained by early Cretaceous movement of the Magnus fault block?

Answer (A. M. Roberts):

It is clear that the termination of the main 'Jurassic' faulting episode occurred at slightly different times throughout the North Sea. For example, as on the Magnus Fault, fault activity on the Horda Platform faults continued into the Ryazanian (strictly early Cretaceous). Our restoration across the Magnus structure is at the 'base Cretaceous' seismic marker, i.e. the 'late Ryazanian condensed interval'. It is, therefore, a restoration to the early post-rift phase, not to the time of peak fault activity, and is similar in its construction and assumptions about rift age to that of Young (1992).

Tectonic evolution and structural styles of the East Shetland Basin

M. J. LEE[1] and Y. J. HWANG[2]

[1]Conoco (UK) Ltd, 116 Park Street, London W1Y 4NN, UK

[2]Conoco (UK) Ltd, Rubislaw House, North Anderson Drive, Aberdeen AB2 4AZ, UK

Abstract: Geometric and stratigraphic evidence indicates that the East Shetland Basin evolved by a rift mechanism similar to that defined in the East African Rift. Arcuate half-grabens of probable Devonian extensional origin are the fundamental structural units. The stratigraphic data suggest episodic footwall movements during the mid- to late Jurassic as well as alternating regional subsidence patterns, i.e. 'teeter-totter', against the Horda Platform. The Northern North Sea rift system, which includes the East Shetland Basin, is believed to have opened primarily by orthogonal movement of facing half-grabens rather than by large-scale strike-slip motion between these half-grabens.

Both extensional and compressional (or transpressional) features occur in the East Shetland Basin as syn-rift structures. The extensional structures include the well-known tilted fault blocks and relatively undocumented listric normal faults that occur on the crestal and flanking areas of the larger blocks. The principal compressional stress was associated with oblique shearing at the ends of actively extending half-grabens, which created local reverse faults and anticlinal domes. Some rotated horst blocks, e.g. Murchison Field structure, were also created by the transpressional motion. A broad, four-fold E–W structural zonation has been made for the East Shetland Basin and North Viking Graben based on the degree of half-graben development and structural styles.

The main purpose of this paper is to review the tectonic history of the East Shetland Basin (ESB) using representative stratigraphic data and to present the main structural styles encountered within the basin. Much has been written on these subjects over the last two decades and it is not the intention of this paper to review extensively previous findings. Based on the analysis of key fault geometry and associated thickness data, this paper attempts to re-define the ESB in the context of the East African Rift model as summarized by Rosendahl (1987) and previously applied in the North Sea by Scott and Rosendahl (1989). This paper also presents a structural interpretation of parts of the Statfjord Field that does not necessarily reflect the views of all the participants in the field.

A generalized stratigraphic section for the Northern North Sea is shown in Fig. 1. Fifty known oil fields have been discovered in the ESB. They contain a total of approximately 15 billion barrels of recoverable oil in Jurassic sandstones. Four fields, Statfjord, Brent, Ninian and Gullfaks, each have reserves of more than one billion barrels. Most large fields produce oil from traps closed up-dip by erosional truncation at the Base Cretaceous level. Some fields appear to be laterally sealed by normal faults with oil accumulation on the down-thrown side, e.g. Don Field (Hardman and Booth 1991). This paper reviews both large- and small-scale structural styles encountered in the ESB, which, for the sake of regional discussions, includes parts of the Møre Basin and Magnus Embayment as well as the Tampen Spur and Alwyn Slope.

Regional fault pattern and basin geometry

The main structural features of the ESB and adjacent areas in the Northern North Sea are shown in Fig. 2. The ESB occupies a rhombic-shaped intermediate terrace region between the East Shetland Platform and the deep North Viking Graben. It encompasses an area approximately 180 km long and up to 90 km wide. The Haltenbanken Terrace off mid-Norway can be considered as a non-overlapping, opposing half-graben to the ESB based on its overall geometry. Using Rosendahl's (1987) terminology on half-graben linking modes, the Møre–Trøndelag Fault Zone, the Tampen Spur and the Tern–Eider Ridge structural trend can be considered as a continuous accommodation zone across which half-graben polarity

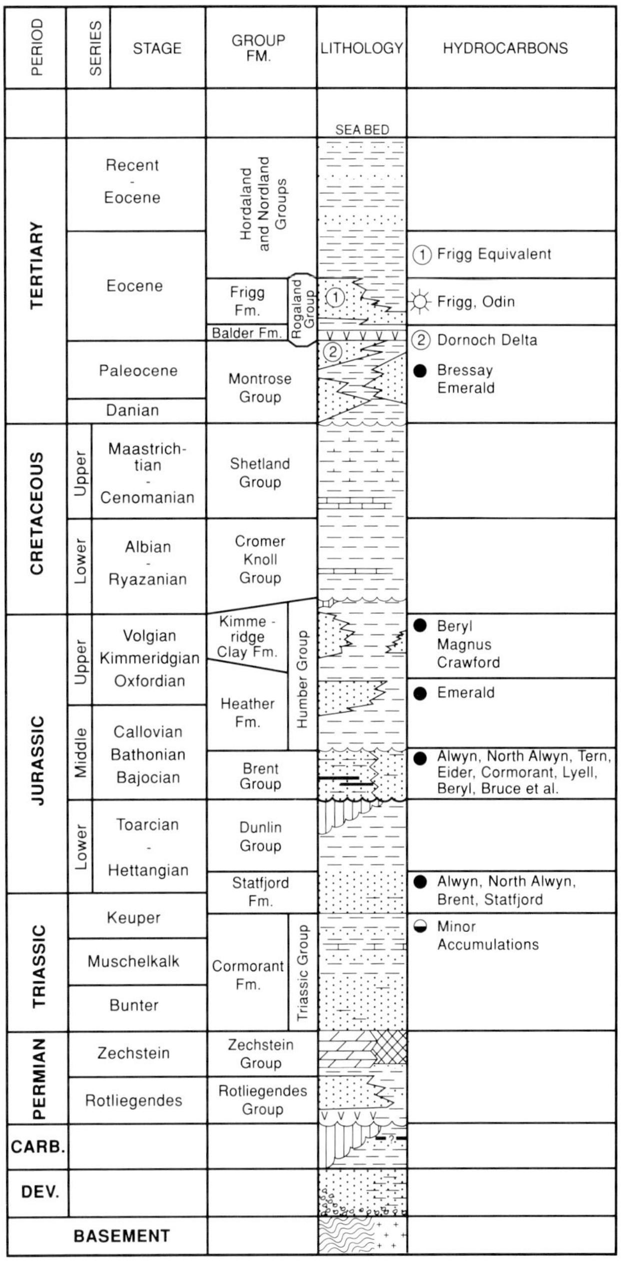

PERIOD	SERIES	STAGE	GROUP FM.	LITHOLOGY	HYDROCARBONS
				SEA BED	
TERTIARY		Recent - Eocene	Hordaland and Nordland Groups		
					① Frigg Equivalent
		Eocene	Frigg Fm.	Rogaland Group ①	☼ Frigg, Odin
			Balder Fm.		
			Montrose Group	②	② Dornoch Delta ● Bressay Emerald
		Paleocene			
		Danian			
CRETACEOUS	Upper	Maastrichtian - Cenomanian	Shetland Group		
	Lower	Albian - Ryazanian	Cromer Knoll Group		
JURASSIC	Upper	Volgian Kimmeridgian Oxfordian	Kimmeridge Clay Fm.	Humber Group	● Beryl Magnus Crawford ● Emerald
			Heather Fm.		
	Middle	Callovian Bathonian Bajocian	Brent Group		● Alwyn, North Alwyn, Tern, Eider, Cormorant, Lyell, Beryl, Bruce et al.
	Lower	Toarcian - Hettangian	Dunlin Group		
			Statfjord Fm.		● Alwyn, North Alwyn, Brent, Statfjord
TRIASSIC		Keuper		Triassic Group	○ Minor Accumulations
		Muschelkalk	Cormorant Fm.		
		Bunter			
PERMIAN		Zechstein	Zechstein Group		
		Rotliegendes	Rotliegendes Group		
CARB.					
DEV.					
		BASEMENT			

Fig. 1. Generalized stratigraphic column of the Northern North Sea (modified from Deegan and Scull 1977).

From *Petroleum Geology of Northwest Europe: Proceedings of the 4th Conference* (edited by J. R. Parker).
© 1993 Petroleum Geology '86 Ltd. Published by The Geological Society, London, pp. 1137–1149.

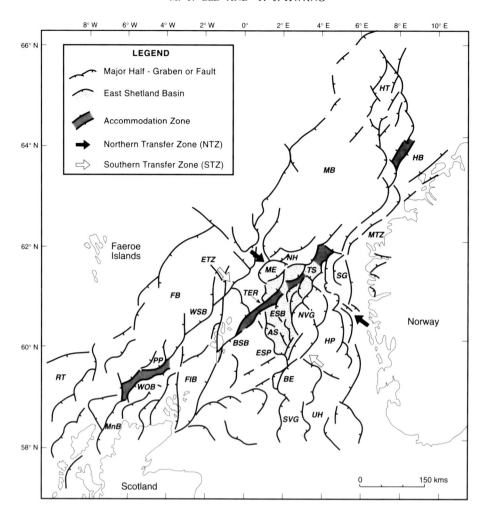

Fig. 2. Structural features map of the Northern North Sea showing the location of the East Shetland Basin in relation to the adjacent basins. Major structural features referred to in text—AS: Alwyn Slope; BE: Beryl Embayment; BSB: Bressay-Sumburgh Basin; ESB: East Shetland Basin; ESP: East Shetland Platform; ETZ: Erlend Transfer Zone; FB: Faeroe Basin; FIB: Fair Isle Basin; HB: Hitra Basin; HP: Horda Platform; HT: Haltenbanken Terrace; MB: Møre Basin; ME: Magnus Embayment: MnB: Minch Basin; MTZ: Møre–Trøndelag Fault Zone; NVG: North Viking Graben; PP: Papa Basin; RT: Rockall Trough; SG: Søgn Graben; SVG: South Viking Graben; TER: Tern–Eider Ridge; TS: Tampen Spur; UH: Utsira High; WOB: West Orkney Basin; WSB: West Shetland Basin.

changes occur. The ESB can then be divided into two parts by this structural trend. The area to the north should in fact be considered as a portion of the NE–SW-oriented Møre and Faeroe Basin rift system rather than the N–S-trending North Viking Graben. The Tampen Spur and the Tern–Eider Ridge alignment can be regarded as remnants of pre-rift rocks that have been partially excluded from syn-rift subsidence and maintained a high relief.

In the main basinal area of the ESB south of the Tern–Eider Ridge most large faults are arcuate and east facing (see Fig. 2). Obliquely offset to the southeast across the North Viking Graben, there is a series of similar arcuate faults in the Horda Platform, which are predominantly west facing. Examples of these faults include the boundary faults of the large half-grabens that contain the Oseberg and Troll fields. The development of this overlapping, opposing half-graben geometry is well documented as a typical rift mechanism in the East African Rift (Rosendahl 1987). In the overlapping area, low-relief accommodation zones, or antiformal welts, often develop to compensate space problems created by semi-contemporaneous subsidence of opposing half-grabens. This type of structural zone can be found within the axial area of the North Viking Graben (Badley *et al.* 1988), away from which the half-graben units subside in opposite directions.

Also shown in Fig. 2 are two NW–SE-oriented transfer zones which are believed to be regionally significant. The northern transfer zone (NTZ) is wider than the southern transfer zone (STZ) and is composed of a number of relatively small faults, some of which are coincident with termination faults of large half-graben systems. The NTZ is partly expressed in a regional gravity map compiled by Hospers *et al.* (1985) in offshore Norway Quadrant 32. The NW-trending boundary or internal faults in the Gullfaks, Statfjord, Murchison and Penguin fields are considered to be the expressions of the NTZ. Collectively the faults in the NTZ control the southwestern terminations of the Magnus Embayment and Møre Basin, as well as the basinal limits of the Margareta's Spur and Nordfjord High.

Across the NTZ, the half-grabens change in polarity in a N–S direction as the major faults become clearly sinusoidal with throws in opposite directions, e.g. the Visund fault block vs. the Oseberg fault block (Fig. 3). The N–S-oriented Søgn Graben trend is deflected to the SW by the NTZ before it joins with the North Viking Graben.

The southern transfer zone (STZ), on the other hand, is somewhat more sharply defined by the alignment of NW–SE-oriented regional faults that include the border faults of the Unst Basin, Alwyn Slope, and Stord Basin. The northern limit of the Utsira High is coincident with the STZ near latitude 60°N. The northwest extension of this transfer zone in the

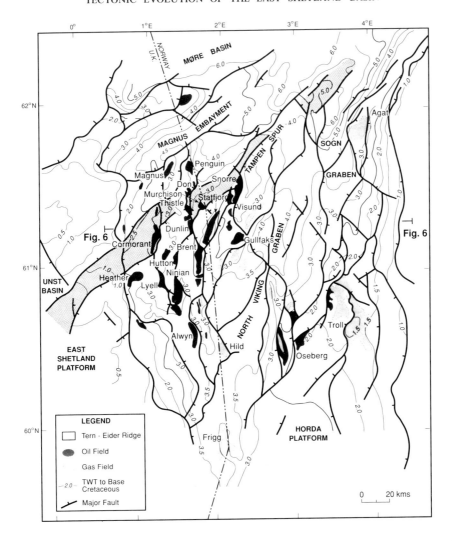

Fig. 3. Structure map of the East Shetland Basin and North Viking Graben. The shaded area represents the Tern–Eider Ridge and its continuations.

Faeroe Basin was previously referred to as the Erlend Transfer Zone by Duindam and van Hoorn (1987).

The position of some key boundary faults and the overall rhombic outline of the basin is partly related to the presence of deeply buried plutons. The southwestern boundary fault of the basin on the Alwyn Slope was probably controlled by the presence of a circular buried granite batholith in the basement (Donato and Tully 1982) that may have prevented continuation of the N–S faulting.

The composite fault pattern shown in Fig. 3 illustrates that, on the whole, the Northern North Sea region, including the ESB, is composed of a series of linked, arcuate half-grabens having varied symmetry and polarity. From a geometric standpoint there are direct analogues between the half-grabens shown in Fig. 3 and those described in the East African Rift (Rosendahl 1987; Scott and Rosendahl 1989). Some of the common features are: (1) curvature of the main bounding faults, which often terminate by splaying or merge with the next set of faults; (2) polarity changes of half-grabens across either low-relief (e.g. North Viking Graben) or high-relief accommodation zones (e.g. Tern–Eider Ridge); and (3) asymmetric distribution of half-grabens across the North Viking Graben and Søgn Graben. The geometric similarity of the ESB as compared to a typical linked half-graben assembly defined in the East African Rift is shown in Fig. 4.

In the ESB a progressive change is noted in the geometry of major half-grabens in an east to west direction. Those half-grabens situated closer to the North Viking Graben, e.g. Brent

area, show well-defined arcuate patterns with clear splay faults. On the other hand, further west near the platform the fault blocks are rather angular, e.g. Heather area, with only limited definition of individual half-grabens (Fig. 3).

According to a deep reflection seismic and gravity interpretation of Holliger and Klemperer (1989) crustal thickness increases progressively from less than 21 km in the North Viking Graben to more than 31 km in the East Shetland Platform. An intermediate thickness range of 21–26 km is indicated in the ESB. Therefore, the extent and degree of half-graben development in the ESB is believed to be a function of crustal stretching and relative distance from the rift axis.

Tectonic evolution

Structural grains and their origin

In the ESB and adjacent areas, four sets of structural trends are recognized. In order of frequency the faults trend N–S, NE–SW, NW–SE and occasionally E–W. A brief description of each of these trends and their possible origin are as follows.

1. N–S trending faults are clearly defined in a number of large tilted fault blocks which contain giant oil and gas fields, e.g. Magnus, Brent, Cormorant, Oseberg, and Troll (see Fig. 3).

    According to Haszeldine (1989) the N–S fracture systems observed onshore UK originated due to the E–W-directed tensional regime which became active in the late Silurian. It

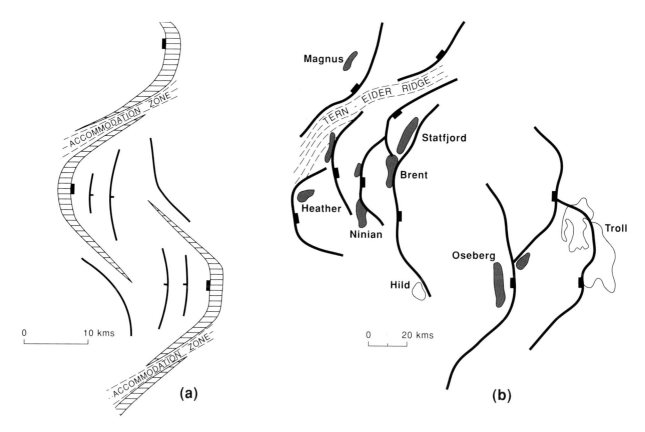

Fig. 4. (**a**) Sketch of a half-graben linking mode observed in the East African Rift (after Rosendahl 1987). Scales vary, but widths of 10 km are typical. (**b**) Simplified fault pattern map of the ESB and North Viking Graben showing the morphology of half-graben units.

is possible that the N–S grains reflect deep seated relict fold axes of the early Caledonian or even older Scourian age metamorphic basement (Park 1973).

2. The NE–SW-oriented fabric is based upon the well-known Caledonian age fold belt as illustrated by Johnson and Dingwall (1981). In many parts of the Northern North Sea this grain is coincident with the strike directions of major Caledonian age thrust and reverse faults.

The nearest known thrust fault considered to be significant for the structural development of the ESB is located in the southern Shetland Isles near Lerwick in the Bressay–Sumburgh Basin (Fig. 2). Here the thrust faults bounding the basin trend N–S and NE–SW forming an arcuate belt within the Dalradian rocks metamorphosed during the Caledonian orogeny. Parts of this thrust fault system appear to have collapsed during the Devonian as a half-graben controlled basin filled with clastics (Coward 1990). The study of movement indicators from the marginal fault zone of this basin by Norton *et al.* (1987) reveals dominant E–W extensional displacement, which suggests reactivation of the older thrust faults during the Devonian.

The NE–SW segment of the above mentioned thrust system can be projected to the northeast and could continue into the ESB. The southern boundary fault of the Tern–Eider Ridge lies on this projection indicating a possible genetic link (Fig. 2). A regional gravity interpretation of the Northern North Sea by Hospers *et al.* (1985) also shows the northeasterly projection at basement level.

3. NW–SE-striking faults are the essential components of the transfer zones previously mentioned. The faults of this orientation parallel the Tornquist Zone (Pegrum 1984; Ziegler 1990) to the south which is a regional NW–SE discontinuity cross-cutting the N–S and NE–SW structural grains. In the vicinity of the ESB, the faults associated with this zone control the southern structural limits of the Sele

High, Utsira High and the Fladen Ground Spur (Pegrum 1984).

4. E–W-oriented faults occur sparsely in the East Shetland Basin as in the Brent Field and the Magnus Embayment areas. The faults of this orientation have not received much attention and their origin remains unclear. Similar E–W-oriented faults do occur in the Central Graben as transfer faults offsetting the main NW–SE structural trends (Cartwright 1987). In the Scottish Midland Valley and adjacent region, the late Carboniferous and Permian age quartz-dolerite dyke swarms show an E–W trend. Major E–W Carboniferous faults are also present. This late Carboniferous and Permian phase of tectonism and volcanism may indicate an early Permian rifting in the northern North Atlantic as pointed out by Russel and Smythe (1983) and would, therefore, be responsible for the generation of the E–W-oriented faults in the ESB.

The above descriptions on the major structural trends render some supporting regional evidence that most, if not all, large fault systems in the ESB originated as pre-Mesozoic features due to anisotropy in the basement. The rhombic-shaped outline of the basin probably owes its origin to a complex interaction of the above four major lineaments throughout their multiple geological history.

Paleozoic evolution

A few wells in the western ESB encountered shallow basement below the Triassic, e.g. Heather Field area. These wells are, however, situated proximal to the East Shetland Platform and do not provide a satisfactory answer to the presence or absence of a Paleozoic precursor in the main part of the ESB.

The Devonian rocks in the ESB and vicinity were deposited as an integral part of the Orcadian Basin which is still ill-

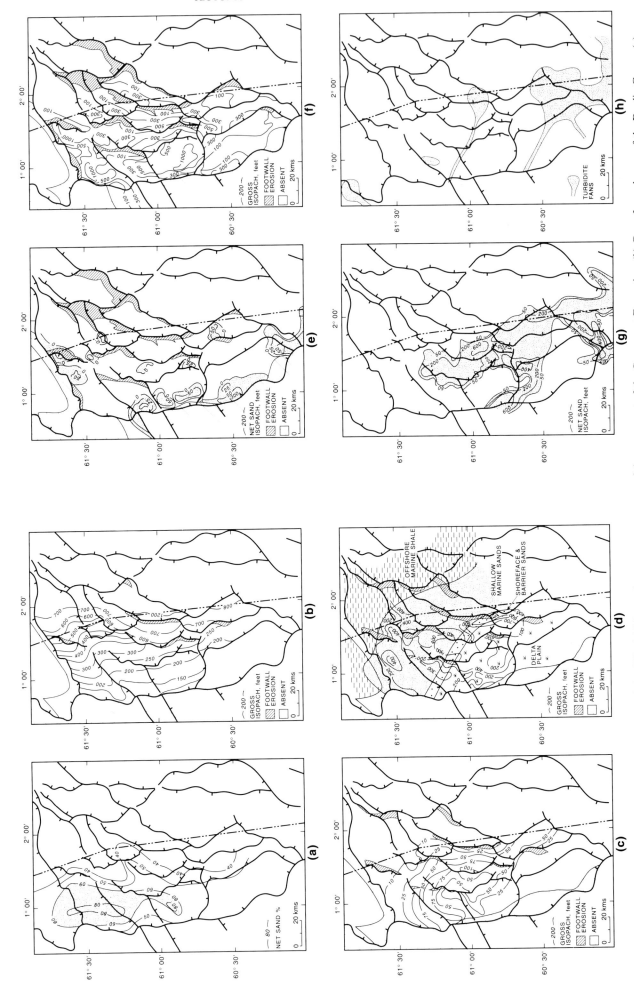

Fig. 5. Regional isopach and sandstone distribution maps of the ESB. (**a**) Net sandstone percentage map of the upper part of the Cormorant Formation. (**b**) Gross Isopach map of the lowermost Brent sandstone. (**c**) Gross isopach map of the Kimmeridge Clay Formation. (**d**) Generalized Bajocian lithofacies map. Note the NW–SE orientation of depositional hinge lines. (**e**) Net sandstone isopach map of the Heather Formation. (**f**) Gross isopach map of the Dunlin Group. (**g**) Distribution of the Paleocene Dornoch delta system. (**h**) Distribution of the lower Eocene turbidite fans.

defined at its northern end. The Orcadian Basin is referred to as a large strike-slip controlled, pull-apart basin bounded to the southeast by the Highland Boundary Fault of Scotland and to the northwest by the Møre–Trøndelag fault system (Ziegler 1990). On the other hand, Norton *et al.* (1987) argue that the Orcadian Basin consists of arcuate half-grabens defined by extensional faults in turn controlled by the Caledonian fabric.

In this paper the extensional model of Devonian basin development is preferred, based on our own mapping in the West Shetland Basin (Fig. 2) together with published information on Devonian basins in northern Scotland (Cheadle *et al.* 1987; Coward 1990). Commercial seismic records show that deposition of the thick wedge-shaped Devonian sediments was largely controlled by dip-slip movements. The bounding faults of the basins are typically arcuate with branch faults joining the border faults of adjacent half-grabens, e.g. the Minch Basin. Some basin complexes, such as those in the West Orkney and West Shetland basins, show sinusoidal alternation of opposing half-grabens, e.g. the Papa Basin vs. the West Orkney Basin (Fig. 2), which is a commonly observed basin-linking mechanism in the East African Rift (Rosendahl 1987).

Carboniferous age sedimentary rocks have not been encountered to date in the Northern North Sea, although Permian age continental red beds have been drilled in several wells in the Magnus Embayment area and nearby Unst Basin. The absence of Carboniferous rocks in the Northern North Sea in general can be explained by either erosion or non-deposition depending upon the relative timing of the Variscan compressional movement. The Permian rocks are probably present in the North Viking Graben and parts of the ESB, as it is well known that the late Permian Zechstein Sea advanced southwards from the Norwegian-Greenland Sea rift (Ziegler 1990).

Mesozoic evolution

It is now well known that rifting in the Northern North Sea commenced during the early Triassic, peaked during the late Jurassic, and terminated by the late Cretaceous (Badley *et al.* 1988; Ziegler 1990). In this paper a series of regional isopach and sandstone percentage maps is presented to demonstrate that rifting and associated block-fault movements occurred intermittently throughout the Triassic and Jurassic and that gross depositional settings were controlled by the underlying structural framework (Fig. 5). The maps are based on well control in both the UK and Norwegian portions of the basin. The stratigraphic nomenclature is taken from Deegan and Scull (1977).

A late Triassic phase of tectonic movement is depicted in Fig. 5a by the sandstone percentage map of the upper part of the Cormorant Formation, from the top of the formation to the uppermost marl section. A general N–S trend of thick sandstone is apparent on the basin margin near the Shetland Platform interrupted by the NE–SW-oriented Tern–Eider Ridge. A second sand-prone trend is indicated parallel to the ridge in the area of the Hutton and Thistle fields. Block faulting during the latest Triassic appears to have been limited in the western margin of the ESB. The Tern–Eider Ridge remained at this time as a topographic high influencing the direction of sediment transport.

The thickness variation of the overlying marine Dunlin Group is a function of regional easterly dip towards the central portion of the North Viking Graben (Fig. 5b). Considerable relief still existed along the Tern–Eider Ridge as indicated by the thickness trends. Significant increases in thickness are also observed across the boundary faults of the Hutton–Ninian–Alwyn and Statfjord–Brent Fields.

The gradual eastward thickening of the Dunlin Group is interpreted as a consequence of thermal subsidence that followed the initial rifting (Badley *et al.* 1988). On the other hand,

in the Beryl Embayment to the south, Richards (1991) recently documented a possible tectonic control on the deposition of the sand-dominated Dunlin Group sediments that indicates the possibility of a certain amount of active rifting and extension during the early Jurassic.

A shift in depositional environment to a deltaic setting followed during the deposition of the overlying Brent Group. On a broad regional scale in the Northern North Sea, Badley *et al.* (1988) reported an overall antithetic thickness relationship between the Brent and Dunlin groups across the North Viking Graben. This apparent 'subsidence teeter-totter' could be considered as an indication of active rifting during Brent sand deposition in the ESB, although an overall thermal subsidence was preferred by Badley *et al.* (1988) based on their interpretation of regional thickness patterns.

In detail, the sub-units of the Brent Group show evidence of subtle fault activities in several large half-graben areas (Fig. 5c). The lowermost Brent sandstone, for example, indicates a wedge-shaped deposition that thickens into the hangingwall areas of the Ninian–Hutton–Murchison fault block, which is coincident with the underlying Dunlin 'hinge-line'. A prominent NE-oriented thick pattern observed north of the Tern–Eider Ridge may be a reflection of NW-facing extensional fault block movement in the northern part of the basin.

A possible tectonic control on the distribution of the Brent delta is shown in Fig. 5d. The composite palaeogeographic map depicted for the Brent suggests that the orientations of the NW–SE-trending delta front and the shallow marine setting beyond it may have been controlled by the transfer zone (NTZ) previously mentioned. This hinge zone also bears a close spatial relationship with the NW–SE Rannoch–Etive thick mapped by Brown *et al.* (1987).

The overlying Humber Group marine sedimentation marks the peak of the Mesozoic rifting and extension period. Initial block faulting and footwall uplift during the deposition of the Heather Formation (Bathonian to Kimmeridgian) resulted in some reworking of the underlying Brent Group sandstones. The sandstones eroded from the emergent footwalls were deposited in two contrasting environments (Fig. 5e). Near the western boundary fault of the basin, N–S-oriented strandline conditions prevailed, whereas farther out in the central part of the basin submarine fault scarps controlled the deposition of relatively thin fan sandstones. A reduced relief is seen on the basin-dividing Tern–Eider Ridge.

The Kimmeridge Clay Formation isopach shown in Fig. 5f best illustrates the syn-rift fault movements within the individual half-grabens. The isopach patterns commonly reveal wedge-shaped sedimentary packages that thicken down the slope toward the hangingwall areas.

During the deposition of the Kimmeridge Clay Formation a shift in extension direction occurred in the North Atlantic that may have resulted from a northward propagation of rifting (Ziegler 1990). In the ESB the effect of this change is reflected in a pronounced rotation of NE-trending fault blocks in the northern portion of the basin. In the half-grabens north of the Tern–Eider Ridge the thickness patterns indicate a NW–SE direction of tilting in contrast to a predominant E–W tilting in the south.

In the ESB, rotation and tilting of half-grabens ceased in the early Cretaceous and a major phase of thermal subsidence followed. The thermal sag was centred along the North Viking Graben with progressive onlap of the marine sediments westward to the East Shetland Platform (Fig. 6).

Cenozoic evolution

Tertiary subsidence in the ESB was largely independent of fault movement. However, the structural transition from platform to basin, which was controlled by the position of large

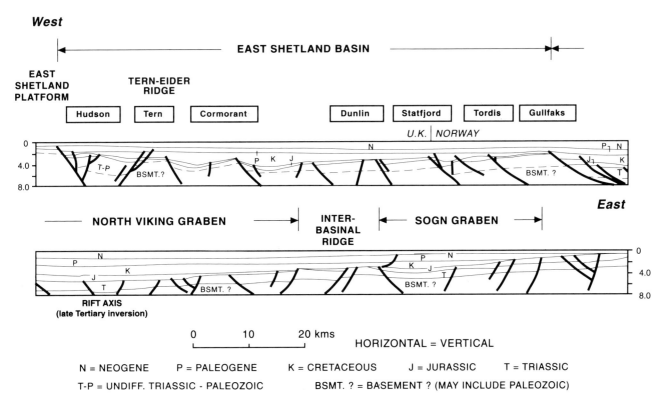

Fig. 6. Regional W–E cross-section (1:1) across the ESB, North Viking Graben, and Søgn Graben. See Fig. 3 for location of the section.

pre-Tertiary half-grabens, continued to exert an influence on sedimentation. Consequently, Paleocene and Eocene sediments, sourced from the west, spilled over the platform into the basin-forming delta and submarine fan deposits (Figs 5g and h).

Deltaic progradations in the ESB took place during the mid- to late Paleocene when the N–S-elongated Dornoch delta was deposited. The delta complex was confined to areas to the south of the Tern–Eider Ridge, which suggests a broad tectonic control on its distribution. The locations of the thickest sand trends also appear to be coincident with the underlying Mesozoic half-grabens, e.g. Brent and Hutton areas, which indicate brief rejuvenation of the major boundary faults (Fig. 5g).

A subtle structural control is also seen in the distribution of the Lower Eocene submarine fan sandstones and the position of the shelf break. Both were strongly influenced by the presence of underlying Mesozoic half-grabens and their bounding faults (Fig. 5h).

Structural styles

Regional considerations

It is proposed that pre-rift anisotropy is the main control on the Mesozoic evolution of the ESB and that the four sets of regional fault trends previously described were simultaneously active during rifting. In this regard, the three-dimensional strain model explained by Reches (1983) can be adapted here to illustrate the near orthorhombic symmetry observed in the Northern North Sea (Fig. 7a). The model depicts a stress field in which the maximum principal stress is vertical, the intermediate principal stress being parallel to the N–S strike direction of the half grabens, and the minimum principal stress lying along the E–W extension direction. The block diagram was modified by Rosendahl (1987) to show that the typical asymmetric subsidence noted in many rift-related half-grabens

is due to an incomplete development of the orthorhombic symmetry or preferential development of part of the system (Fig. 7b). An unequal development of the four sets of faults in the original mosaic results in the asymmetric form of a half-graben.

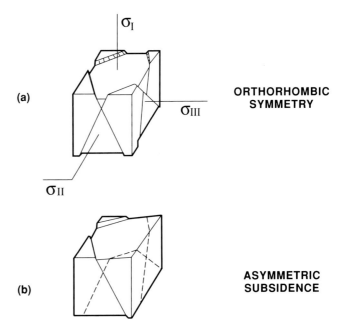

Fig. 7. (a) Model of orthorhombic symmetry showing four sets of normal faults that can accommodate three-dimensional strain (after Reches 1983). (b) Block diagram showing asymmetric subsidence of a half-graben due to an incomplete development of orthorhombic symmetry or development of unequal fault sets (after Rosendahl 1987).

Based on the overall geometry of the major faults shown in Fig. 3, it is postulated that similar basin architecture may have developed in the Norwegian portion of the Northern North Sea. The regional thickness patterns documented by Badley *et al.* (1988) and the lithology of the Triassic and Jurassic rocks described by Vollset and Doré (1984) further suggest that the Horda Platform area in particular was developed as an opposing half-graben system facing the ESB. Several periods of alternating subsidence through time, i.e. 'teeter-totter', both at regional and half-graben scales, can be summarized as follows.

1. In the ESB the Dunlin Group consists mostly of marine shale whereas thick marginal-marine sandy deposits occur on the Horda Platform.
2. The Brent Group delta is thicker and better developed in the ESB: the main Jurassic rifting that began in the early Bathonian did not reach the Horda Platform until the Kimmeridgian.
3. The Humber Group in the ESB is largely composed of shale and siltstones compared to thick sandstones of the age-equivalent Viking Group to the east. Within the ESB, an internal 'teeter-totter' is also denoted by the latest Jurassic faulting and fan sand deposition which is restricted to the Magnus Embayment area (Fig. 5f).

In an attempt to illustrate the concept of the ESB as a rift zone created by linked half-grabens, the main portion of the basin with a simplified fault pattern is shown in Fig. 8. In the ideal half-graben, subsidence is at a maximum in the hangingwall adjacent to the geometric centre of the border fault, where the displacement is mostly dip-slip (Rosendahl 1987). The isopach map of the Kimmeridge Clay Formation, shown in Fig. 5f, confirms a predominant orthogonal dip-slip movement during the late Jurassic rifting. The thickest syn-rift depocentres are preferentially situated within the central axial areas of hangingwall lows suggesting that many half-grabens in the ESB subsided by a near orthogonal mechanism of extension during the late Jurassic rifting.

Towards the arcuate ends of the half-graben, oblique shearing becomes an important local mechanism of faulting at the expense of extension as subsidence continues (see section on compressional structures). The evidence of such oblique motions related to dip-slip movement is also shown in Fig. 8. In a simplified geometric sense the structures trapping the Murchison, Dunlin, Hutton and the north end of the Brent Fields can be considered as segments of rotated horst blocks near the termination points of large half-grabens. In the case of the Hutton Field two merged half-graben units facing the same direction are envisaged. The horst blocks are considered to be antiformal welts created at converging areas of the linked half-grabens to accommodate accumulated horizontal stress. It is believed that, as Hay (1978) pointed out in the Thistle and Murchison areas, a certain amount of rotational movement occurred on the horst blocks resulting in a small offset from the main bounding faults.

Extensional structures

(a) Regional scale syn-rift tilted fault blocks These fault blocks commonly show truncation of the Jurassic and older rocks on the crest of the footwall. The depth section shown in Fig. 6 was derived from the commercially available seismic records. In general the bounding and internal faults of the half-grabens are interpreted to be planar within the limit of seismic resolution. Most large bounding faults appear to have been active until the early Cretaceous. Some large faults near the East Shetland Platform show apparent late tectonic movements through the Paleocene, consistent with regional stratigraphic information.

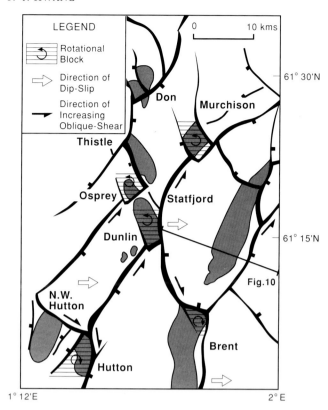

Fig. 8. Map illustrating the origin of rotated horst blocks. The oblique-shearing stress is related to extension within the half-grabens.

Footwall uplift and erosion were greater near the rift axis than in the central part of the ESB as shown by the relative amount of truncation above the top Statfjord horizon. According to Yielding (1990) the major control on the degree of erosion on the crest of the fault blocks is the size of the block, larger fault blocks showing greater degree of uplift and erosion. It is suggested that the size of fault blocks depends on the extent of linkage along arrays of similar polarity half-grabens.

(b) Small-scale listric normal faults The listric normal faults occur on the crestal and flanking areas of several large fault blocks. These faults are relatively small and are believed to be degradation products associated with scarp development and slope failure during active footwall uplift. A schematic model of the geometry of the faults and degradation sheets is depicted in Fig. 9a for the east flank of the Statfjord Field. A seismic line illustrating the same is shown in Fig. 9b. The character of the base of the degradation is a sheet seismic response which does not have a simple event character because different units overlie the main field horizons. Published examples of similar listric faults within the ESB can be found in the Brent Field (Struijk and Green 1991) and Visund Field (Alhilali and Damuth 1987) areas.

The timing and degree of slope failure across the Statfjord and Gullfaks fault blocks is illustrated in the palinspastically restored cross-sections in Fig. 10. These cross-sections were constructed allowing for the effects of compaction and basement unloading assuming an elastic model (Gibson *et al.* 1989). The Base Cretaceous depositional profile shown in Fig. 10b indicates slopes of less than 1° on the western dip-slope of both blocks, whereas slopes varying from 5–12° were present on the crestal area of the Statfjord block. Given this amount of inclination, eastward submarine slides could reasonably be expected to have occurred during the late Jurassic and early Cretaceous time span, coincident with the time of the maximum footwall uplift.

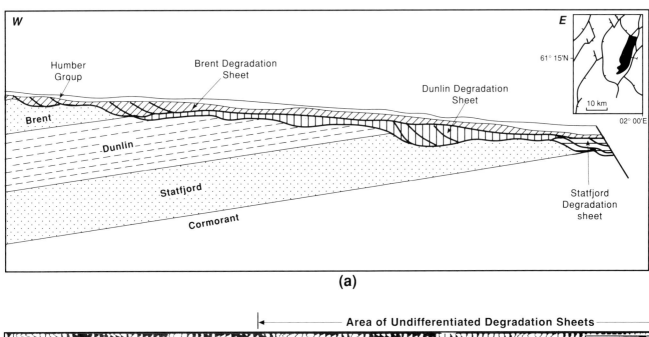

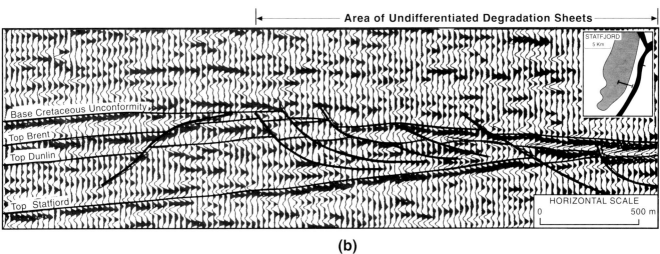

Fig. 9. (**a**) Schematic cross-section showing listric faulting associated with degradation sheets on the east flank of the Statfjord Field. (**b**) Seismic example of listric faults on the east flank of the Statfjord Field. The faults are associated with undifferentiated degradation sheets and slump zones.

The post-Brent erosion across Statfjord appears to be due to both sub-aerial exposure at an early stage and subsequent submarine landsliding (Fig. 10a). The sub-aerial exposure of the Statfjord and the Gullfaks blocks is shown by a reference line depicting a likely base level based on the nature of the sediments. In all the sections, except the present-day picture, the bathymetries shown are relative. In the top Brent section, the interpreted detachment surfaces still show an easterly dip component, implying that Brent/Dunlin fault terraces would have moved downslope by gravity tectonics.

Compressional (or transpressional) structures

(a) Regional compressional features In the ESB, clear indications of compressional structures are generally absent. In other areas of the Northern North Sea, regional compressional structures have been previously reported (see below). These structures are thought to be associated with either the Variscan or Alpine orogenies.

1. Variscan related: Carboniferous age inversion and folding of Devonian strata as observed in the southern East Shetland Platform (Holloway *et al.* 1991) where relatively flat-lying Permo-Triassic and younger rocks rest unconformably on folded Devonian rocks.

2. Alpine related: late Cretaceous to Eocene compressional movements in the 15/9 Gamma Field on the Utsira High (Pegrum and Ljones 1984); in the Stord Basin (Biddle and Rudolph 1988); in the Unst Basin (Johns and Andrews 1985) and on the Mid-Norway Shelf (Caselli 1987).

A broad, low-amplitude late Tertiary age anticlinal flexure has been noted in the axial area of the North Viking Graben (Fig. 6). The anticlinal flexure could indicate a subtle structural inversion to a late Alpine collision or a late Tertiary opening of the Atlantic Ocean.

(b) Local compressional features Occurrences of compressional domes and reverse faults have been reported in Gullfaks Field (Fossen 1989) and in several locations on the Tampen Spur (Hamar and Hjelle 1984). Local 'flower-structures' of probable transpressional origin have also been documented along the margins of the Tern–Eider Ridge (Speksnijder 1987) and possibly on the Hild Field structure as illustrated by Rønning *et al.* (1987).

Local compressional elements occur in many other areas of the ESB. They are an important aspect of the overall structural

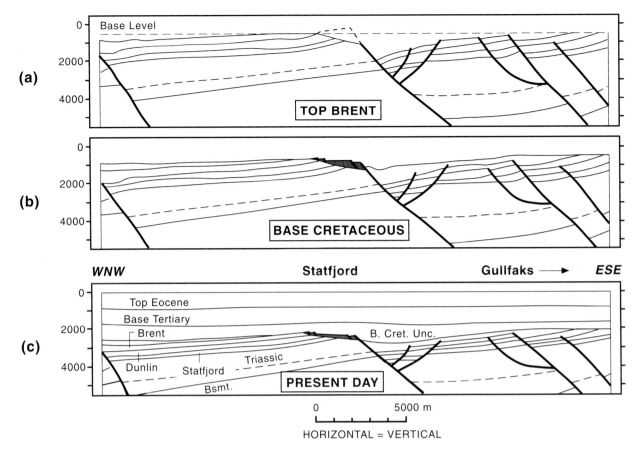

Fig. 10. Palinspastic reconstruction of the Statfjord block showing the development of unstable slope and listric fault on the east flank. See Fig. 8 for location. (**a**) Decompacted and unloaded to top Brent. (**b**) Decompacted and unloaded to near Base Cretaceous. (**c**) Present-day depth section.

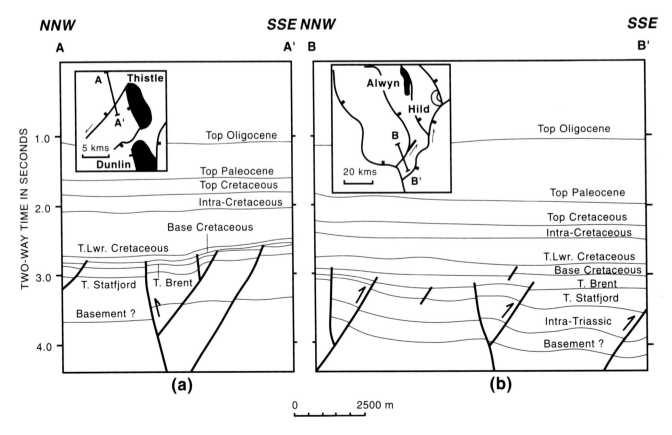

Fig. 11. Geoseismic sections showing compressional (transpressional) structures in: (**a**) northern; (**b**) southern parts of the ESB.

framework. Two examples of local compressional structures are shown in Fig. 11. They are derived from two widely separated areas, one from an area west of the Thistle Field (Fig. 11a), and the other from a fault zone near the southern part of the ESB (Fig. 11b).

The 'flower-structure' shown in Fig. 11a illustrates transpressional motion during the late Jurassic and possibly early Cretaceous. The steep reverse fault is seated in the basement and appears to have been active mainly in the late Jurassic as it is interpreted to be truncated by the Base Cretaceous Unconformity. Late Jurassic folding and related reverse faulting are also shown in Fig. 11b taken from a NE–SW-trending boundary fault of the Alwyn Slope. The folding affected the Statfjord and Brent Group horizons and again appears to have ceased by the early Cretaceous. Although the examples in Figs 11a and 11b are limited to NE–SW-trending faults only, NW–SE faults also show oblique-slip movements (Speksnijder 1987).

The local compressional (or transpressional) structures observed in the ESB are interpreted as syn-rift products of late Jurassic and early Cretaceous age. The model suggests that during active rifting, oblique-shearing occurs at the arcuate ends of each half-graben because of extension and block rotation (Fig. 8). The cumulative stress created by the strike-slip motion should progressively increase towards the axial area of the rift away from the platform area. Thus the half-grabens near the rift axis, such as that containing the Gullfaks Field, could have accumulated substantial horizontal stress along the arcuate bounding faults. The local compressional structures observed in the eastern ESB can then be attributed to syn-rift deformation to accommodate space problems created at the rift margin.

Structural zonation

Based on descriptions of both extensional and compressional features, an E–W structural zonation has been established in the ESB as shown in Fig. 12. From west to east the zones are:

- Zone I   area of rectangular or rhombic fault blocks and incipient half-grabens;
- Zone II   area of similar polarity half-grabens and horst blocks of oblique-shear origin;
- Zone III area of well-defined arcuate half-grabens and syn-rift compressional structures;
- Zone IV   area of low-relief accommodation zones in rift axis, i.e. area of overlapping, opposing half-graben with axial interbasinal ridges, and broad, low-relief late Tertiary anticlines (low-angle listric faults occur primarily in Zones II to IV).

This zonation is an attempt to establish a regional tectonic continuum for the underlying Mesozoic structures within the ESB in accordance with our present understanding of rift propagation. Areas to the north and east of the Magnus Embayment are not included in this scheme due to lack of available data. Obviously this zonation is only a general scheme as local variations in structural style would strongly depend upon the way the half-grabens are linked together to form the larger basin.

Summary and conclusions

1. Geometric and stratigraphic evidence indicates that the ESB evolved by a rift mechanism similar to that defined in the East African Rift. Half-grabens with arcuate geometry are the fundamental structural units.

2. A Devonian extensional origin for the half-grabens is proposed based on the regional fault and subcrop patterns, and by analogy to the Devonian basins of SW Norway and the north coast of Scotland.

3. There are four sets of fault trends: N–S, NE–SW, NW–SE and E–W, each of which can be related to regional basement anisotropy. A three-dimensional strain relationship with a modified orthorhombic symmetry can be applied to explain the development of half-grabens.

4. The ESB can be divided into two contrasting but complementary parts by the NE–SW-trending Tern–Eider Ridge. Polarity changes of opposing half-grabens occur across the ridge. The basin area north of the ridge is structurally related to the NE–SW-trending Møre Basin. An analogue polarity change is documented in the West Orkney Basin area.

5. The main stages of rifting and footwall uplift during the Mesozoic can be demonstrated by the resulting variations in formation thicknesses and sand percentages. Alternating subsidence patterns, i.e. 'teeter-totter' against the Horda Platform, have been observed in the ESB during the Jurassic. The stratigraphic data suggest episodic footwall movements during the mid- to late Jurassic.

6. The distribution of the Brent Group paralic sandstones and the Tertiary submarine fans were strongly influenced by the underlying structural orientations, often resulting in linear sand trends.

7. Both extensional and compressional (or transpressional) features occur in the ESB as syn-rift structures. Strike-slip motion associated with oblique-shearing at the ends of actively extending half-grabens provided the principal compressional stress regime. The extensional structures include the well-known tilted fault blocks and relatively undocumented listric normal faults that occur on the crestal and flanking areas of the large blocks.

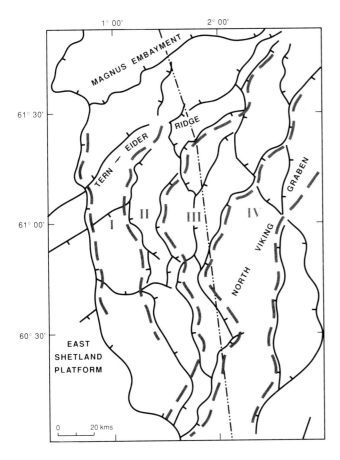

Fig. 12. Simplified structural zonation of the ESB and North Viking Graben. See text for descriptions.

8. Slope failure and degradation along active fault scarps were the main mechanisms of listric faulting.

9. A broad, four-fold E–W structural zonation has been made in the ESB and North Viking Graben based on the degree of half-graben development and structural styles.

10. The currently available stratigraphic and structural data strongly suggest that the Northern North Sea rift opened primarily by orthogonal movement of facing half-grabens rather than by large-scale strike-slip motion between these half-grabens.

Thanks are due to the management of Conoco (UK) for their support and approval to publish this paper. The authors also wish to thank Amerada Hess Norge a.s., Amoco Norway Oil Co., BP Exploration, Chevron (UK) Ltd, Enterprise Oil Norway a.s., Esso Norge a.s., Mobil Development Company Norway a.s., Saga Petroleum a.s., a.s. Norske Shell, and Statoil for permission to release information on the Statfjord Fields. The interpretations and conclusions presented in this report are the responsibility of the authors and do not necessarily reflect the views of Conoco (UK). Special thanks are extended to BP Exploration for providing the results of palinspastic reconstruction on the Statfjord Field.

References

ALHILALI, K. A. AND DAMUTH, J. E. 1987. Slide block(?) of Jurassic sandstone and submarine channels in the basal Upper Cretaceous of the Viking Graben: Norwegian North Sea. *Marine and Petroleum Geology*, 4, 35–48.

BADLEY, M. E., PRICE, J. D., RAMBECH DAHL, C. AND AGDESTEIN, T. 1988. The structural evolution of the northern Viking Graben and its bearing upon extensional modes of basin formation. *Journal of the Geological Society, London*, 145, 455–472.

BIDDLE, K. T. AND RUDOLPH, K. W. 1988. Early Tertiary structural inversion in the Stord Basin, Norwegian North Sea. *Journal of the Geological Society, London*, 145, 603–611.

BROWN, S., RICHARDS, P. C. AND THOMPSON, A. R. 1987. Patterns in the deposition of the Brent Group (Middle Jurassic) UK North Sea. *In*: BROOKS, J. AND GLENNIE, K. W. (eds) *Petroleum Geology of North West Europe*. Graham & Trotman, London, 899–915.

CARTWRIGHT, J. A. 1987. Transverse structural zones in continental rifts—an example from the Danish Sector of the North Sea. *In*: BROOKS, J. AND GLENNIE, K. W. (eds) *Petroleum Geology of North West Europe*, Graham & Trotman, London, 441–452.

CASELLI, F. 1987. Oblique-slip tectonics of Mid-Norway shelf. *In*: BROOKS, J. AND GLENNIE, K. W. (eds) *Petroleum Geology of North West Europe*. Graham & Trotman, London, 1049–1063.

CHEADLE, M. J., MCGEARY, S., WARNER, M. R. AND MATTHEWS, D. H. 1987. Extensional structures on the western U.K. continental shelf: a review of evidence from deep seismic profiling. *In*: COWARD, M. P., DEWEY, J. F. AND HANCOCK, P. L. (eds) *Continental Extensional Tectonics*. Geological Society, London, Special Publication, 28, 445–465.

COWARD, M. P. 1990. The Precambrian, Caledonian and Variscan framework to NW Europe. *In*: HARDMAN, R. F. P. AND BROOKS, J. (eds) *Tectonic Events Responsible for Britain's Oil and Gas Reserves*. Geological Society, London, Special Publication, 55, 1–34.

DEEGAN, C. E. AND SCULL, B. J. 1977. *A standard lithostratigraphic nomenclature for the central and northern North Sea*. Institute of Geological Sciences Report 77/25.

DONATO, J. A. AND TULLY, M. C. 1982. A proposed granite batholith along the western flank of the North Sea Viking Graben. *Geophysical Journal of the Royal Astronomical Society*, 69, 187–195.

DUINDAM, P. AND VAN HOORN, B. 1987. Structural evolution of the West Shetland continental margin. *In*: BROOKS, J. AND GLENNIE, K. W. (eds) *Petroleum Geology of North West Europe*. Graham & Trotman, London, 765–773.

FOSSEN, H. 1989. Indication of transpressional tectonics in the Gullfaks oil-field, northern North Sea. *Marine and Petroleum Geology*, 6, 22–30.

GIBSON, J. R., WALSH, J. J. AND WATTERSON, J. 1989. Modelling of bed contours and cross-sections adjacent to planar normal faults. *Journal of Structural Geology*, 11, 317–328.

HAMAR, G. P. AND HJELLE, K. 1984. Tectonic framework of the Møre Basin and the northern North Sea. *In*: SPENCER, A. M. *et al.* (eds) *Petroleum Geology of the North European Margin*. Graham & Trotman, London, 349–358.

HARDMAN, R. F. P. AND BOOTH, J. E. 1991. The significance of normal faults in the exploration and production of North Sea hydrocarbons. *In*: ROBERTS, A. M., YIELDING, G. AND FREEMAN, B. (eds) *The Geometry of Normal Faults*. Geological Society, London, Special Publication, 56, 1–13.

HASZELDINE, R. S. 1989. Evidence against crustal stretching, north–south tension and Hercynian collision, forming British Carboniferous basins. *In*: ARTHURTON, R. S., GUTTERIDGE, P. AND NOLAN, S. C. (eds) *The role of tectonics in Devonian and Carboniferous sedimentation in the British Isles*. Yorkshire Geological Society, Occasional Publication, 6, 25–33.

HAY, J. T. C. 1978. Structural development in the Northern North Sea. *Journal of Petroleum Geology*, 1, 65–77.

HOLLIGER, K. AND KLEMPERER, S. L. 1989. A comparison of the Moho interpreted from gravity data and from deep seismic reflection data in the northern North Sea. *Geophysical Journal*, 97, 247–258.

HOLLOWAY, S., REAY, D. M., DONATO, J. A. AND BEDDOE-STEPHENS, B. 1991. Distribution of granite and possible Devonian sediments in part of the East Shetland Platform, North Sea. *Journal of the Geological Society, London*, 148, 635–638.

HOSPERS, J., FINNSTRØM, E. G. AND RATHORE, J. S. 1985. A regional gravity study of the northern North Sea (56–62° N). *Geophysical Prospecting*, 33, 543–566.

JOHNS, C. AND ANDREWS, I. J. 1985. The petroleum geology of the Unst Basin, North Sea. *Marine and Petroleum Geology*, 2, 361–372.

JOHNSON, R. J. AND DINGWALL, R. G. 1981. The Caledonides: their influence on the stratigraphy of the North-West European Continental Shelf. *In*: ILLING, L. V. AND HOBSON, G. D. (eds) *Petroleum Geology of the Continental Shelf of North-West Europe*. Heyden & Son, London, 85–98.

NORTON, M. G., MCCLAY, K. R. AND WAY, N. A. 1987. Tectonic evolution of Devonian basins in northern Scotland and southern Norway. *Norsk Geologisk Tidsskrift*, 67, 323–338.

PARK, R. G. 1973. The Laxfordian belts of the Scottish mainland. *In*: PARK, R. G. AND TARNEY, J. (eds) *The early Precambrian of Scotland and related rocks of Greenland*. University of Keele, 65–76.

PEGRUM, R. M. 1984. The extension of the Tornquist zone in the Norwegian North Sea. *Norsk Geologisk Tidsskrift*, 64, 39–64.

—— AND LJONES, T. E. 1984. 15/9 Gamma gas field offshore Norway, new trap type for the North Sea Basin with regional structural implications. *American Association of Petroleum Geologists Bulletin*, 68, 874–902.

RECHES, G. 1983. Faulting of rocks in three-dimensional strain fields II. Theoretical analysis. *Tectonophysics*, 95, 133–156.

RICHARDS, P. C. 1991. Evolution of Lower Jurassic coastal plain and fan delta sediments in the Beryl Embayment, North Sea. *Journal of the Geological Society, London*, 148, 1037–1047.

RØNNING, K., JOHNSTON, C. D., JOHNSTAD, S. E. AND SONGSTAD, P. O. 1987. Hild. *In*: SPENCER, A. M. *et al.* (eds) *Geology of the Norwegian Oil and Gas Fields*. Graham & Trotman, London, 287–294.

ROSENDAHL, B. R. 1987. Architecture of continental rifts with special reference to East Africa. *Annual Review of Earth and Planetary Sciences*, 15, 445–503.

RUSSEL, M. J. AND SMYTHE, D. K. 1983. Evidence for an early Permian oceanic rift in the northern North Atlantic. *In*: NEUMANN, E. R. AND RAMBERG, I. B. (eds) *Petrology and Geochemistry of Continental Rifts*. Riedel, The Netherlands, 173–179.

SCOTT, D. L. AND ROSENDAHL, B. R. 1989. North Viking Graben: an East African Perspective. *American Association of Petroleum Geologists Bulletin*, 73, 155–165.

SPEKSNIJDER, A. 1987. The structural configuration of Cormorant Block IV in context of the northern Viking Graben structural framework. *Geologie en Mijnbouw*, 65, 357–379.

STRUIJK, S. P. AND GREEN, R. T. 1991. The Brent Field, Block 211/29, UK North Sea. *In*: ABBOTTS, I. L. (ed.) *United Kingdom Oil and Gas Fields 25 years Commemorative Volume*. Geological Society, London, Memoir, **14**, 63–72.

VOLLSET, J. AND DORÉ, A. G. 1984. *A revised Triassic and Jurassic lithostratigraphic nomenclature for the Norwegian North Sea. Norwegian Petroleum Directorate, Stavanger, Bulletin* **3**.

YIELDING, G. 1990. Footwall uplift associated with Late Jurassic normal faulting in the northern North Sea. *Journal of the Geological Society, London*, **147**, 219–222.

ZIEGLER, P. A. 1990. *Geological Atlas of Western and Central Europe*. Shell International Petroleum Maatschappij B.V., The Hague, The Netherlands.

A fundamental reappraisal of the structure of the Cormorant Field and its impact on field development strategy

R. R. A. DEMYTTENAERE, A. H. SLUIJK and M. R. BENTLEY

Shell UK Exploration and Production, 1, Altens Farm Road, Nigg, Aberdeen, Scotland, UK

Abstract: Production from the Cormorant Field is hampered by complex in-field faulting. The large number of faults and the presence of fault seals combine to restrict fluid flow within the reservoir and locally compartmentalize the field. A reliable structural description of the field is, therefore, a prerequisite for optimal field management. A revised structural framework has been established for the field, based on new and reprocessed seismic data, and the availability of newly developed seismic attribute mapping tools. Previous structural models have emphasized E–W extension on N–S normal faults. The new model also describes E–W extension, but invokes predominantly dip-slip movement on a network of planar, oblique-slip faults, the geometry of which has been strongly influenced by the reactivation of NE–SW-oriented lineaments in the basements. This has led to a significant revision of both the reservoir sub-compartment geometry, and the connected oil volumes associated with existing producer–injector well pairs.

The Cormorant Field is located in the East Shetland Basin in Blocks 211/21 and 211/26 of the UK Northern North Sea (Fig. 1). The field, discovered in 1972, comprises four undersaturated oil reservoirs which are produced from two surface installations and an underwater manifold centre (Taylor and Dietvorst 1991). The reservoirs consist of sandstones of the Middle Jurassic Brent Group and lie within four structurally discrete, westerly-dipping fault blocks, Blocks 1 to 4 (Figs 1 and 2). The blocks are situated on a N–S-oriented Jurassic fault terrace on the western flank of the Viking Graben.

Production from Cormorant commenced in 1979, since when 368 MMBBL of oil have been produced up to 30/9/91 out of an estimated ultimate recovery of 629 MMBBL of oil (Taylor and Dietvorst 1991). Of the unproduced reserves, some 100 MMBBL are presently still undeveloped as they are not expected to be recovered with the existing wells.

The primary reservoir constraint on the development of remaining oil in Cormorant is the structural complexity of the field. This is a particular problem in Block 4 which contains some 80 MMBBL of undeveloped reserves. In order to optimize the location of further development wells, reservoir simulation models have been constructed for each reservoir block, the framework of which is defined from seismic interpretation and structural modelling.

This paper describes a new structural model for Cormorant, arising from a complete seismic reinterpretation of the field.

Seismic interpretation

The Cormorant area is fully covered by four partly overlapping 3D seismic surveys, shot between 1979 and 1988 (Taylor and Dietvorst 1991). Three key horizons were mapped throughout the field: the Base Cretaceous Unconformity, the top Brent Group and top basement. The Base Cretaceous Unconformity corresponds to the regional Late Cimmerian Unconformity and seismic 'basement' is the eroded top surface of the metamorphic Caledonides. The Base Cretaceous Unconformity and basement reflections are well defined throughout the field, whereas the top Brent Group reflection is of more variable quality and is locally indistinct. Intra- or base-Brent reflections cannot be reliably mapped.

The surveys were interpreted on a Landmark workstation using Shell proprietary software for auto-tracking and attribute studies. Interpretation involved on-screen picking of the three key horizons on seed lines, providing a basis for auto-tracking on all remaining lines. Attributes for dip and azimuth were derived for the key surfaces in order to resolve the pattern of faulting, using the techniques described by Dalley *et al.* (1989). Figures 3 and 4 show azimuth displays for the Base Cretaceous Unconformity and top Brent. These displays high-

light flexures, such as the E–W roll-over of the Base Cretaceous Unconformity across the N–S-oriented fault terrace (Fig. 3) and horizon discontinuities such as faults. The use of auto-tracking increases the degree of consistency in the interpretation and allows subtle surface features to be accurately mapped. The validity of structural features identified by auto-tracking and attribute extraction was checked by back-referencing to the relevant seismic cross-sections.

To enhance surface features, illumination displays were constructed by illuminating the horizon from a distant, imaginary light source. Figure 5 shows the Base Cretaceous Unconformity surface illuminated from the west. Westerly dipping areas are brightly illuminated, whereas easterly dipping zones fall into shadow.

Structural interpretation

The tilted Jurassic fault blocks of the East Shetland Basin are traditionally interpreted as N–S-oriented features resulting from E–W extension associated with the opening of the North Atlantic during the Mesozoic (Ziegler 1982). The major faults seen at Jurassic Level have been interpreted either as listric features soling out above new or reactivated basement faults (e.g. Gibbs 1984) or as more planar, through-going features which directly penetrate the basement (Yielding 1990). Oblique-slip components on faults not oriented N–S have been inferred by Speksnijder (1987), the oblique-slip faults effectively acting as transfer zones within and between the major N–S-oriented fault terraces during E–W extension.

The structural pattern determined from attribute studies on the Cormorant dataset allows significant clarification and refinement of the interpretations described above. Three important general observations are listed below.

1. Although the general N–S orientation of the Cormorant fault terrace is evident in Figs 3–5, the dominant trend of individual faults is NW–SE or NE–SW, rather than N–S. Features previously interpreted as N–S faults, such as the eastern boundary fault to Block 1 can now be resolved into a set of intersecting faults oriented predominantly NE–SW and NW–SE.

2. All major faults, and many of the minor faults, can be traced from the Base Cretaceous Unconformity level into the basement as illustrated by a regional seismic section over Block 1 and the southern tip of Block 4 (Fig. 6). The boundary fault separating the Blocks clearly cuts into the basement. Although the fault becomes initially flatter within the basement, it cannot be seen to sole out. The fault steepens again with depth and merges with the steep, eastern Block 4 boundary fault. Major faults at Base Creta-

From *Petroleum Geology of Northwest Europe: Proceedings of the 4th Conference* (edited by J. R. Parker).
© 1993 Petroleum Geology '86 Ltd. Published by The Geological Society, London, pp. 1151–1157.

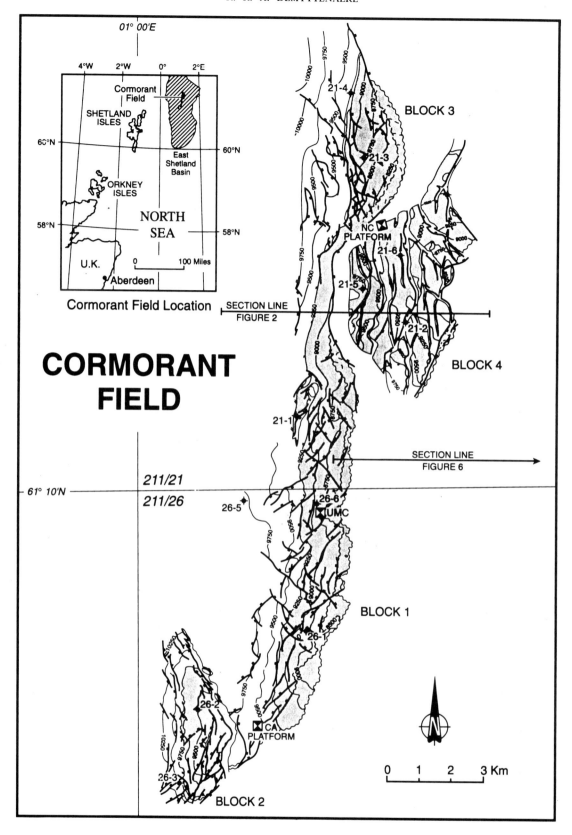

Fig. 1. Depth map of top Brent over the Cormorant area showing location of section lines of Figs 2 and 6.

ceous Unconformity level which cut the basement are illustrated in Fig. 7, which shows an azimuth extraction from
the basement surface. All the major block-bounding faults
are clear and internal block structures are dominated by
NE–SW and NNE–SSW-oriented features, many of which
are congruent with structures at reservoir level. NW–SE-
oriented features are less evident at basement level. This is
illustrated by comparison of the azimuth displays of the

Base Cretaceous Unconformity and basement over Block 3
(Figs 3 and 7). Where NW–SE faults are dominant on the
Base Cretaceous Unconformity display in that area, they
are almost absent at basement level.
3. The geometric relationship between fault traces at the
   basement and Base Cretaceous Unconformity levels can be
   used to identify lateral displacements. The basement azimuth display (Fig. 7) shows a straight 10 km long fault

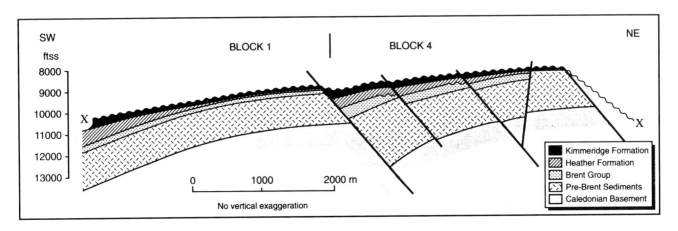

Fig. 2. Schematic geological cross-section over Cormorant Blocks 1 and 4 illustrating the protracted period of faulting activity (see Fig. 1 for location).

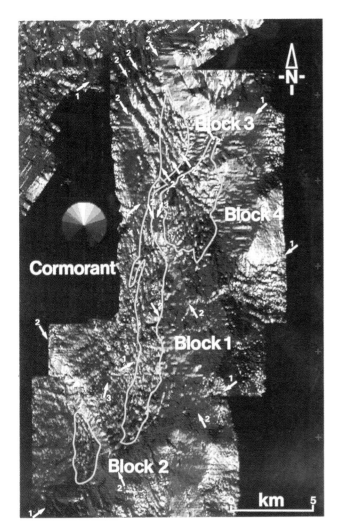

Fig. 3. Azimuth map of the Base Cretaceous Unconformity over the Cormorant area. The colours for the different azimuths are represented in the rose. The Cormorant terrace dips to the west (blue colours on the left) and the eastern boundary faults dip to the east (red-coloured band). The E–W roll-over is indicated by the colour change from blue to red. Three fault orientations are indicated—1: the Caledonian trend NE–SW; 2: a Cimmerian trend NW–SE; 3: a reactivated basement trend NNE–SSW. The eastern edge of the Cormorant terrace is made up of a series of NE–SW and NW–SE faults instead of a single N–S fault (see also Fig. 9). Superimposed on the NNE–SSW lineament in the centre of Block 1 (lineament 3) are a series of approximately N–S-oriented riedel shears (see also Figs 4 and 7).

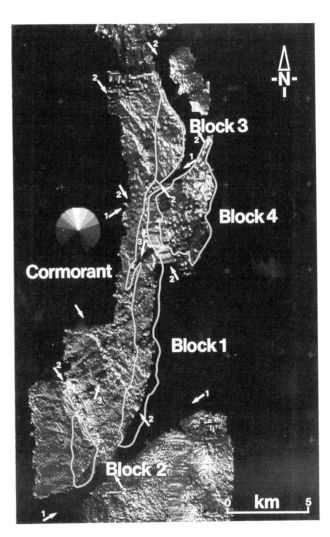

Fig. 4. Azimuth map of the top Brent over the Cormorant area (see Fig. 3 for legend). Compare the diffuse NNE–SSW lineament in the centre of Block 1 with the straight fault on Fig. 7 (lineament 3 in both figures). Internal NW–SE faults dominate the Block 3 fault pattern and are also present in the rest of the field.

striking NNE–SSW and showing a small easterly dip. The same feature at Base Cretaceous Unconformity level (Fig. 3) is still oriented NNE–SSW but is broken down into a series of N–S-oriented, en échelon segments. A schematic view of this is shown in Fig. 8. The en échelon segments are interpreted as riedel shears developed in the Jurassic as a

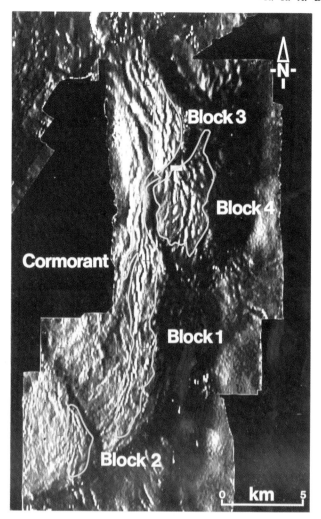

Fig. 5. Illumination map of the Base Cretaceous Unconformity over the Cormorant area. The surface is artificially lit from the west. Rhombic patterns are recognized in the Block 4 area, indicating oblique-slip movements.

result of lateral displacements on the basement lineament. The feature is similar to those produced in analogue sandbox experiments by Naylor *et al.* (1986), and locally produces geometries described by Woodcock and Fisher (1986) in terms of strike-slip duplexes. In this case, the anticlockwise rotation of the fault segments at Base Cretaceous Unconformity level relative to the basement fault suggests a left-lateral offset. Other examples of oblique-slip geometry can be seen in Block 4 on the Base Cretaceous Unconformity illumination display (Fig. 5) where NNE–SSW and N–S faults interlink to produce rhombic fault compartments.

Timing of faulting

Significant structural activity occurred during the deposition of the reservoir sequence in the Cormorant area (Fig. 2). Variations in layer isopachs for sub-layers in the Brent Group (Howe in press) indicate episodic block rotations in various areas of the field throughout Brent Group deposition, with particularly large fault movements taking place prior to the deposition of the uppermost reservoir sand (the Tarbert Formation). Evidence for phases of movement after the deposition of the reservoir sequence occurs in the Humber Group shales, and is indicated by condensation of biostratigraphic zones onto fault block crests. Periods of greatest structural activity are marked by angular discordances. Structural activity continued into the early Cretaceous, as shown by faulting at Base Cretaceous Unconformity level.

The structural development of the Cormorant terrace is, therefore, related to a protracted period of activity lasting from at least the early Jurassic into the early Cretaceous.

Structural model for Cormorant

Inheritance of Caledonian basement structures is believed to have played a direct role in the development of structures at reservoir level. Owing to the horizontal angle between the inherited structural grain in the basement and the sigma 1/sigma 2 plane during E–W extension, an oblique-slip geometry

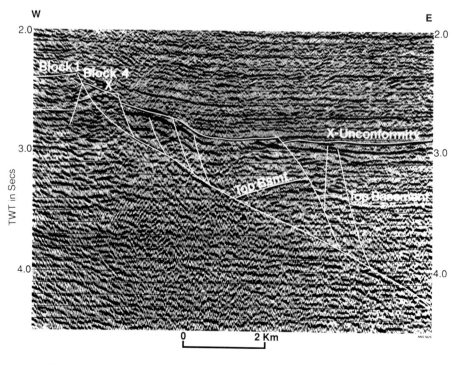

Fig. 6. Regional seismic line over the Cormorant area showing deep-seated block-bounding faults (see Fig. 1 for location).

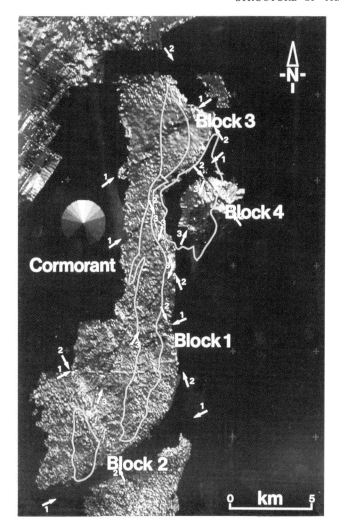

Fig. 7. Azimuth map of the top basement over the Cormorant area (see Fig. 3 for legend). Note the straight 10 km long fault in red striking NNE–SSW in the centre of Block 1, the absence of any important internal NW–SE faults and the internal Caledonian NE–SW faults which disappear at higher levels.

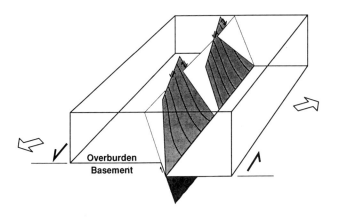

Fig. 8. Schematic model of oblique-slip along a straight basement fault resulting in the formation of riedel shears in the overburden.

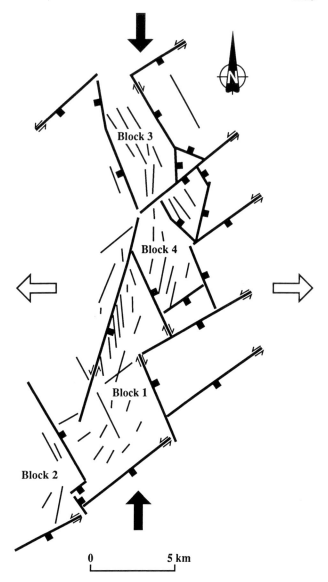

Fig. 9. Kinematic summary of displacements at top Brent Group level.

has been imparted on an otherwise purely extensional setting.

The NE–SW and NNE–SSW lineaments observed at basement level were reactivated during Jurassic E–W extension, in some cases forming the roots to fault systems propagating into the sedimentary cover resulting in oblique-slip movements on the NE–SW and NNE–SSW faults. Although a left-lateral strike-slip component has been identified on these fault sets,

the displacements were predominantly dip-slip and greatest displacements are observed on faults oriented closer to N–S.

Additional extension at reservoir level was taken up on NW–SE faults, conjugate to the predominant NE–SW trend. The large NW–SE-oriented block boundary faults cut through both the basement and cover. However, the absence of small-scale examples of NW–SE faults in the basement, e.g. the Block 3 area on Fig. 7, suggests that this set may be entirely Jurassic in age. Any oblique-slip movements on the NW–SE fault set are likely to have been dextral.

A kinematic summary of displacements at top Brent Group level is indicated in Fig. 9. The model differs from previous interpretations (e.g. Speksnijder 1987) in that all major faults link directly to basement structures, and that no significant N–S structures have been identified.

Cormorant Field development strategy

The structural configuration of the Cormorant fault terrace has had a profound influence on the development strategy for the Cormorant Field, particularly for Block 4. Early development plans for Block 4, based on 2D data and assuming no significant internal structural complexities, proposed a row of crestal oil producers supported by downflank water injectors

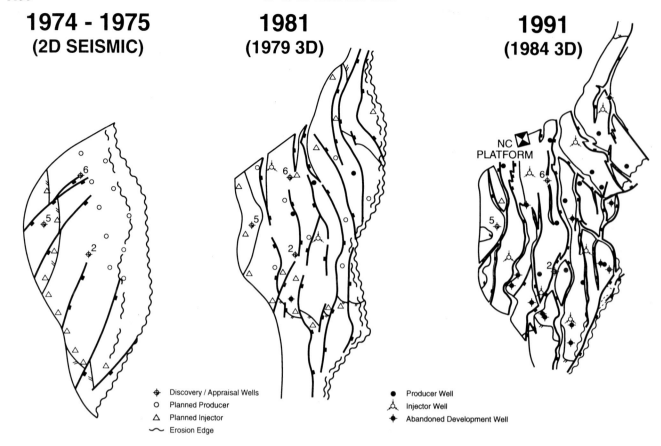

Fig. 10. Three stages in the field development of the structurally complex Block 4.

(Fig. 10). When significant faulting was unexpectedly encountered during development drilling, the development strategy was switched to one aiming to place producer–injector pairs in each major reservoir compartment (Fig. 10 and Ruijtenberg *et al.* 1990).

Crucial to the success of this approach is the ability to map out the major producing compartments. Given the variable quality of the top Brent reflector, even on modern 3D seismic data, interpretation of top Brent structure is strongly guided by the new structural model which draws on observations from the stronger Base Cretaceous Unconformity and basement reflectors. Figure 10 illustrates the 1991 revised sub-compartment geometry in Block 4 based upon the structural re-mapping of the reprocessed 1984 3D seismic. The sub-compartment geometry influences the selection of infill well locations and provides the framework for numerical reservoir simulations used to guide field management decisions.

The new structural map also allows the variations of throw along faults bounding the reservoir compartments to be measured. These variations strongly influence fault sealing capacity determining inter-compartment crossflow, which in turn constrains the description of fluid movements in reservoir modelling. The application of structural data to reservoir simulation in Cormorant Block 4 is described by Bentley and Barry (1991).

Conclusions

The structure of the Cormorant Field has been completely reinterpreted, mostly from new or reprocessed 3D seismic data, using new proprietary techniques for attribute extraction and horizon auto-tracking.

The structure of the field is now seen in terms of Mesozoic E–W extension taken up on a linked system of oblique-slip faults oriented predominantly NE–SW or NW–SE, rather than simple E–W extension on N–S-oriented normal faults. The key features of this interpretation are:

1. the bias towards NE–SW- and conjugated NW–SE-oriented fault sets is caused by reactivation of NE–SW- and NNE–SSW-oriented faults in the underlying Caledonian basement;
2. major block-bounding faults seen at Jurassic levels are identified as relatively steep, through-going structures which penetrate to the basement, rather than features which sole out in the sedimentary cover;
3. although the basin-scale strain model is essentially extensional, the inheritance of lineaments oblique to the extensional direction has resulted in the generation of fault geometries at Jurassic level which are more commonly associated with strike-slip.

The internal structure of the Cormorant fault terrace has had a profound influence on field production, influencing decisions on well locations, as well as underpinning reservoir simulation models used to guide field management. This is particularly marked in Block 4, where an awareness of structural complexities in the reservoir prompted the revision of the original field development plan. The current plan is to drill producer–injector pairs in each sub-block.

The authors would like to thank Shell UK Exploration and Production and Esso Exploration and Production UK for permission to publish this work.

References

BENTLEY, M. R. AND BARRY, J. J. 1991. Representation of fault sealing in a reservoir simulation: Cormorant Block IV, UK North Sea, Society of Petroleum Engineers 22667.

DALLEY, R., GEVERS, E., STAMPFLI, G., DAVIES, D., GASTALDI, C., RUIJTENBERG, P. AND VERMEER, G. 1989. Dip and azimuth displays for 3D seismic interpretation. *First Break*, 7(3), 86–95.

GIBBS, A. 1984. Structural evolution of extensional basin margins. *Journal of the Geological Society, London*, 141, 609–620.

HOWE, B. K. in press. *The Cormorant Oil Field*. American Association of Petroleum Geologists, Memoir.

NAYLOR, M. A., MANDL, G. AND SIJPESTEIJN, C. H. K. 1986. Fault geometries in basement-induced wrench faulting under differential initial stress rates. *Journal of Structural Geology*, 8(7), 737–752.

RUIJTENBERG, P., BUCHANAN, R. AND MARKE, P. 1990. Three-dimensional data improve reservoir mapping. *Journal of Petroleum Technology*, 42(1), 22.

SPEKSNIJDER, A. 1987. The structural configuration of Cormorant Block IV in context of the northern Viking Graben structural framework. *Geologie en Mijnbouw*, 65, 357–379.

TAYLOR, D. J. AND DIETVORST, P. A. 1991. The Cormorant Field, Blocks 211/21a, 211/26a, UK North Sea. *In*: ABBOTTS, I. L. (ed.) *United Kingdom Oil and Gas Fields 25 Years Commemorative Volume*. Geological Society, London, Memoir, 14, 73–81.

WOODCOCK, N. AND FISHER, M. 1986. Strike-slip duplexes. *Journal of Structural Geology*, 8(7), 725–735.

YIELDING, G. 1990. Footwall uplift associated with Late Jurassic normal faulting in the northern North Sea. *Journal of the Geological Society, London*, 147, 219–222.

ZIEGLER, P. 1982. Faulting and graben formation in western and central Europe. *Philosophical Transactions of the Royal Society, London*, A305, 113–143.

Discussion

Question (A. J. Fraser, BP Exploration):

Have you had an opportunity to test your revised model by drilling, and if so with what result?

Answer (R. R. A. Demyttenaere):

This structural interpretation was finalized in March 1991. Since then a few wells have been drilled. The first well was drilled in the Block 1 area. The objective was to drill as high as possible on the crest without encountering a significant eroded Upper Reservoir and to produce unswept oil from an area south of one of the NE–SW-trending fault zones. Water injection support was expected from a well in the SW in the same block. Well results showed that the well had indeed penetrated an undrained area and that erosion was minor as prognosed (NB. previous interpretations did not predict any erosion) and that support came from the injector to the SW. Initial production rates were very high: > 10 000 barrels of oil and no water cut. The top Brent map is currently used as input for a new reservoir simulation model.

A new reservoir simulation model based on the new maps is already available for the Block 2 area. A water injector has been spudded in July 1992. This well, which is currently being drilled in the south of the block, should give support to the crestal producers and sweep additional reserves.

A third well has been drilled in the complex Block 4 area. The target for this water injector was to support at least one, and maybe two, producer wells. No major barriers to waterflood had been recognized on the seismic. The well found an unfaulted section as prognosed and initial injection rate was over 20 000 barrels per day. This, however, quickly dropped to a rate of less than 4000 barrels per day accompanied by a major pressure build-up indicating that there is no communication with the producers. As no significant dip-slip fault has been mapped in between the injector and the producers we infer that another strike-slip fault with no apparent throw at top Brent level is likely to be present.

Prior to the drilling of this isolated injector well it had already been decided to shoot a new 'modern' 3D seismic survey over the area. The progress made in acquisition and processing techniques over recent years makes us confident that we will be able to map the base Brent reflector better on the new survey. This reflector is of critical importance as it is the only reflector which shows the real fault throw at reservoir level. Fault throws at top Brent are not representative of reservoir compartmentalization as a result of syn- and post-depositional erosion.

First results indicate that it will, indeed, be possible to map reliably the base Brent reflector in Block 4 for the first time.

The structural evolution of the Snorre Field and surrounding areas

N. DAHL and T. SOLLI

Saga Petroleum a.s., Kjørbovn 16, PO Box 490, N-1301 Sandvika, Norway

Abstract: The Snorre Field is located on the Tampen Spur, Northern Viking Graben between 61°N and 62°N. Main structural elements are large, westerly tilted fault blocks. The Snorre Structure, one of the largest fault blocks in the East Shetland Basin, extends across Block 34/7 into Block 34/4. The Snorre Fault Block includes the Snorre Field, Tordis Field and the Vigdis Field oil discoveries. Major oil discoveries have so far been made in the Brent Group, the Statfjord Formation and the Lunde Formation. Discoveries have also been made in the Upper Jurassic and Paleocene. Upper Jurassic stratigraphic traps are the main target for further exploration in the area.

The westerly rotated fault blocks on the Tampen Spur are delineated by NNE–SSW- to NE–SW-trending faults. From west to east, these are the Murchison Fault, the Outer Snorre Fault, the Southern Snorre Fault and the Inner Snorre Fault. Other faults recognized are the E–W- and SE–NW-striking cross-faults dividing the elongated fault blocks into minor compartments. The N–S-striking faults are common in the Snorre area. The number of these faults decreases towards the north.

The Snorre Escarpment is the surface expression of the major Inner Snorre Fault representing the eastern border fault of the Snorre Fault Block.

Two major rifting episodes are recognized in the area. In the thermal subsidence phase following the late Permian to early Triassic rifting, large amounts of sediments of the Teist-, Lomvi-, Lunde-, Statfjord Formation, Dunlin- and Lower Brent Group were accumulated. A second major rifting episode resulted in intense fault activity from the early Bathonian to the late Kimmeridgian. Two major pulses of increased tectonic activity are recognized in the second rift episode. In the north, in the Zeta area, there is indication of late Jurassic lateral movements. Late Jurassic faulting, block rotation and footwall uplift, combined with Jurassic to Cretaceous subsidence, make the Snorre Fault Block of the Tampen Spur a very pronounced structural high in the area.

The Snorre Field is situated on the Tampen Spur of the Northern North Sea, between 61°N and 62°N (Fig. 1) (Karlsson 1986; Hollander 1987). The Tampen Spur is bounded by the East Shetland Basin to the west, the Marulk Basin to the northwest, the Marflo Ridge to the east and the Viking Graben to the southeast (Brekke *et al.* 1991). In the south-southeast, the Tampen Spur includes large, westerly rotated fault blocks, one of which is the Snorre Fault Block. Late Jurassic activity of uplifting combined with Jurassic/Cretaceous subsidence make the Snorre Fault Block a very pronounced structural high.

The Snorre Fault Block, which is one of the largest in the Northern North Sea, covers major parts of Blocks 34/7 and 34/4 (Fig. 2). The Tordis and Vigdis fields are additional oil discoveries within the Snorre Fault Block to the south in Block 34/7 (Fig. 1). To the north, a minor oil discovery was made in the Zeta Fault Block (Fig. 3). The Zeta Fault Block is down-faulted and separated from the Snorre Fault Block by the Gamma Fault (Fig. 1). The major oil discoveries have so far been made in the Brent Group deltaic and the Statfjord and Lunde formations fluvial reservoirs. Wells drilled in the Snorre area have also proven hydrocarbons in the Cook Formation of the Dunlin Group and in Upper Jurassic and in Paleocene sands.

The main object of this paper is to summarize the structural style and the structural evolution of the Snorre and surrounding areas and to discuss its geological implications in a tectonic model of the Snorre area.

Structural elements

The main structural elements on the Tampen Spur are NNE–SSW- to NE–SW-striking faults that delineate rotated fault blocks. Major faults related to the Snorre Fault Block are the Murchison Fault, the Outer Snorre Fault, the Southern Snorre Fault and the Inner Snorre Fault (Fig. 1). The Inner Snorre Fault is the eastern boundary fault to the Snorre Fault Block.

The other faults terminate within the fault block (Fig. 1). All major faults show dip-slip movements towards the east.

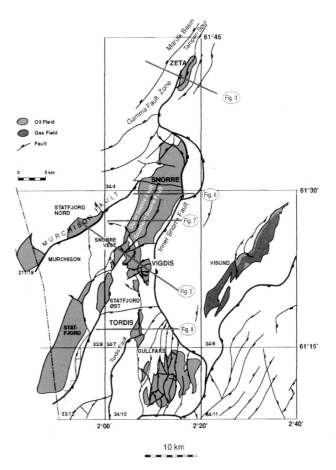

Fig. 1. The Snorre Field and surrounding areas.

From *Petroleum Geology of Northwest Europe: Proceedings of the 4th Conference* (edited by J. R. Parker).
© 1993 Petroleum Geology '86 Ltd. Published by The Geological Society, London, pp. 1159–1166.

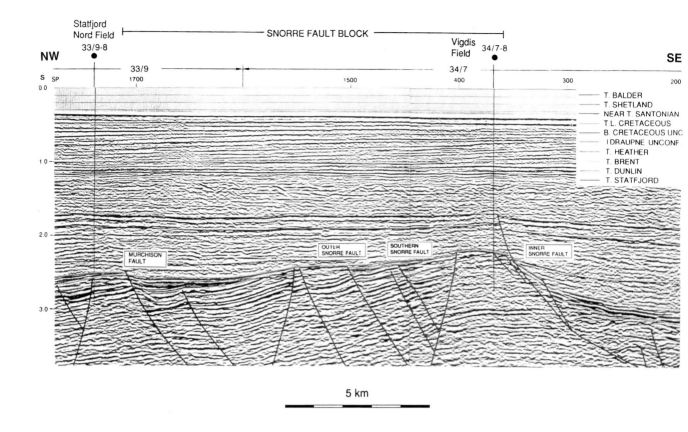

Fig. 2. A seismic section from the Statfjord Nord Field in the west across the Snorre Fault Block to the Vigdis Field in the east.

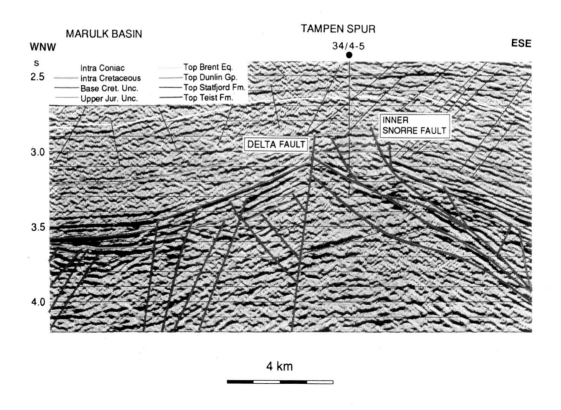

Fig. 3. A seismic section across the Zeta Fault Block.

The Snorre Fault Escarpment is the buried, eroded footwall of the Inner Snorre Fault (Fig. 2). A pronounced relief was established already in the latest Middle to the earliest Late Jurassic. The fault displacement on the pre-rift top Brent horizon increases from c. 1500 m in the Tordis area in the south, to more than 4000 m in the zeta area in the north. The Inner Snorre Fault Escarpment was exposed to erosional forces far into the Cretaceous, modifying the original expression of the Inner Snorre Fault to that of a fault scarp.

Faults striking N–S have a dominant dip to the west (Fig. 1). These faults are frequently cut and displaced by westerly dipping faults trending NNE–SSW to NE–SW.

The E–W- to SE–NW-striking faults are less conspicuous and numerous than the above faults. They act as cross-faults dividing the elongated NNE–SSW fault blocks into minor compartments (Fig. 1).

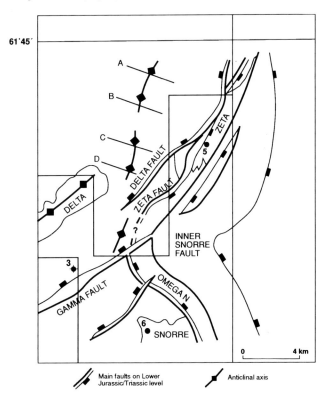

Fig. 4a. Detailed map of the Zeta area.

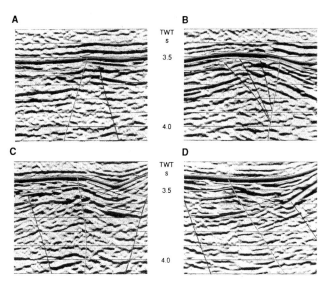

Fig. 4b. Anticline cut by fault of alternating normal and reverse geometry.

In the Zeta area, the major Gamma Fault Zone, including the Delta fault separates the Tampen Spur and the Marulk Basin (Figs 1 and 3). The fault zone consists of a set of en échelon, westerly dipping NE–SW-striking faults which show activity into the Cretaceous. A lateral component of movement in addition to the dominant dip-slip movement in the fault zone is evident in the area. An example of a feature indicating lateral components of movement is shown in Fig. 4a and b where an anticline is intersected by faults with alternating normal and reverse geometries striking oblique to the Gamma Fault Zone.

Structural evolution

Triassic

The Tampen Spur is a late Jurassic to Cretaceous structure and is situated within the continental Permian–Triassic Rift Basin of the Northern North Sea area (Lervik et al. 1989; Badley et al. 1988; Deegan and Scull 1977). Figure 5 summarizes the stratigraphy in the Tampen Spur area.

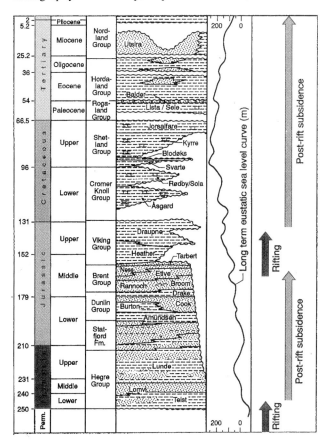

Fig. 5. The stratigraphy in the Tampen Spur area (modified after Nystuen and Fält, in prep.).

Following the rifting, regional subsidence took place during most of the Triassic and the early to Middle Jurassic. In the Tampen Spur area fluvial material was deposited within wide alluvial plains by rivers flowing towards a seaway located to the north (Nystuen et al. 1989; Steel and Ryseth 1990). Large amounts of sediments of the Teist, Lomvi and Lunde formations were accumulated in the rapidly subsiding continental Triassic basin (Fig. 5). The Lomvi Formation has an early to middle Triassic age, while the Lunde Formation is late Triassic, with an age ranging from probably Ladinian to Rhaetian (Nystuen et al. 1989). The Lunde Formation is a reservoir unit in the Snorre Field.

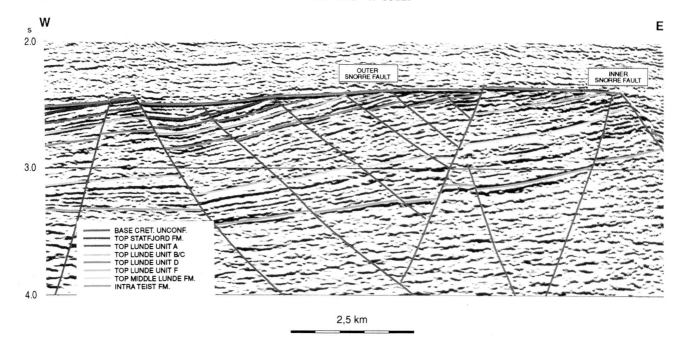

Fig. 6. The Snorre Field. E–W cross-section showing thickening of Triassic strata towards the east.

In the Snorre area, the tectonic influence of the late Permian–eary Triassic rifting is strongly overprinted by later Jurassic tectonic activity. However, elements of the basin configuration are preserved in the sedimentary record and the fault pattern. In the Snorre Field a thickening of both the Lunde Formation and older Triassic strata towards the east is evident (Fig. 6). The eastward thickening diminishes towards the north, as does the thickness of the Triassic sediments. Within the Snorre Fault Block, it is evident that the N–S-striking, west-dipping faults running parallel to the long axis of the Permian–Triassic Rift Basin of the Northern North Sea had most impact on the sedimentation pattern.

Early Jurassic

The subsidence following the late Permian–early Triassic rift episode continued into the Lower Jurassic with a decreasing rate of subsidence. On the Tampen Spur, sediments of the Statfjord Formation accumulated as braided stream deposits on an alluvial plain (Fig. 5) (Nystuen *et al.* 1989). Data from the Snorre Field suggest a Rhaetian to Sinemurian age of the Statfjord Formation. The Statfjord Formation represents the main reservoir in the Snorre Field. It is also a reservoir in parts of the Vigdis Field. In additional to syn-tectonic growth on the major block-bounding faults, growth of the Statfjord Formation towards N–S-striking, west-dipping older Triassic faults is observed in the Snorre area. Some of the faults are cut and displaced by later Jurassic faults (Fig. 7).

During the subsequent marine transgression, fluvial sediments of the Statfjord Formation were directly overlain by the shallow marine siltstone and shales of the Dunlin Group. In the Snorre area oil has been found in the sandy Cook Formation of the Dunlin Group. The eastward thickening as growth towards the N–S-striking, west-dipping fault continued during deposition of the Dunlin Formation (Fig. 8).

Middle Jurassic

Towards the end of Bajocian, the cooling of the crust succeeding the late Permian to early Triassic rifting approached a state of thermal equilibrium (e.g. Badley *et al.* 1988; McKenzie 1978; Sclater and Célérier 1978; Wood and Barton 1983; Barton and Wood 1984; Ziegler 1982; Giltner 1987). In the Snorre area, the Brent delta developed and propagated northwards into the marine basin in the early Middle Jurassic.

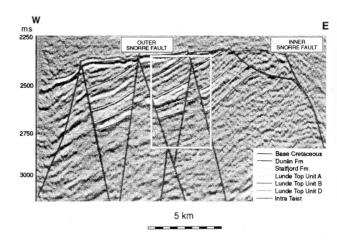

Fig. 7. The Snorre Field. E–W cross-section showing eastward thickening of the Statfjord Formation.

The Brent group is divided into four lithostratigraphic units in the Snorre Fault Block: Rannoch, Etive, Ness and Tarbert Formations. The sandy units of the Brent delta represent the main reservoirs of the Vigdis and the Tordis fields. In the Zeta Fault Block, oil is found in the Brent Formation. However, the shaly Brent Formation in this area is of poor reservoir quality.

The depositional pattern of the Rannoch and Etive formations follows the late Permian to early Triassic rift basin configuration. Towards the end of the Middle Jurassic, renewed fault activity affected the deposition of the Tarbert Formation. Where uneroded, there is no clear fault control on the thickness of the Brent Group. However, a general increase in thickness towards the southeast is observed.

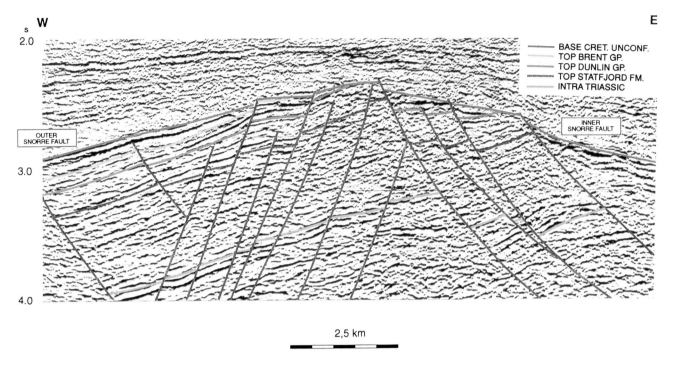

W
s
2.0

OUTER
SNORRE FAULT

3.0

4.0

E

BASE CRET. UNCONF.
TOP BRENT GP.
TOP DUNLIN GP.
TOP STATFJORD FM.
INTRA TRIASSIC

INNER
SNORRE FAULT

2,5 km

Fig. 8. The Tordis Field. E–W cross-section showing eastward thickening of the Dunlin Formation.

Late Jurassic

Towards the end of the Middle Jurassic, in the early Bathonian, a second major rift episode began. Thinning of the crust by rifting was accompanied by syn-rift initial subsidence (Steckler and Watts 1982; McKenzie 1978; Sclater and Célérier 1987; Cochran 1983), resulting in the formation of the Viking Graben. During the subsequent general increase in relative sea-level, the syn-tectonic marine shales of the Heather Formation were deposited within the Northern North Sea marine basin, including the Tampen Spur and the East Shetland Basin (Badley *et al.* 1988; Scott and Rosendahl 1989).

In the Snorre area, the crustal extension was accommodated on mainly easterly dipping faults of NNE–SSW to NE–SW orientation. However, the N–S-striking faults were also reactivated. The Snorre Fault Block was rotated in a southwesterly direction. During extension, the crest of the Snorre Fault Block was uplifted above the erosive base. The Snorre Fault Block represented a topographic high, partly extending above sea-level. In the northern part of the Snorre Field *c.* 1200 m of pre-rift sediments are eroded and Triassic sediments of the Lomvi Formation are directly overlain by Cretaceous sediments.

The pelagic sediments of the Viking Group onlap the uplifted Snorre Fault Block in a direction from SW to NE. The Viking Group is dominated by marine mudstones of the Heather and the Draupne Formation. The Draupne Formation is divided into an upper and lower part by the Intra Draupne Unconformity. The unconformity is a truncational sequence boundary in the crestal parts of the Snorre Fault Block, but it is non-truncational downflank of the Inner Snorre Fault. Several episodes of open marine shelf sand propagating is recorded in wells. The uplift crest of the Snorre Fault Block provides a local source area for the sand, as illustrated in Fig. 9.

In the syn-rift Viking Group, two pulses of crustal extension by rifting are recognized. A first pulse of early Bathonian to late Callovian age affected the deposition of the Heather Formation. As the extension proceeded from the early Oxfordian to the late Kimmeridgian, a lower Draupne Formation

syn-rift wedge developed. The situation is seen in the hangingwall of the Inner Snorre Fault (Fig. 2).

The topography of the Snorre Fault Block, extending above sea-level increased in size and relief as the extension proceeded from the early Bathonian to the late Kimmeridgian. The upper Draupne 'hot shales' of late Kimmeridgian to Ryazanian age onlap and drape the rifted basin topography.

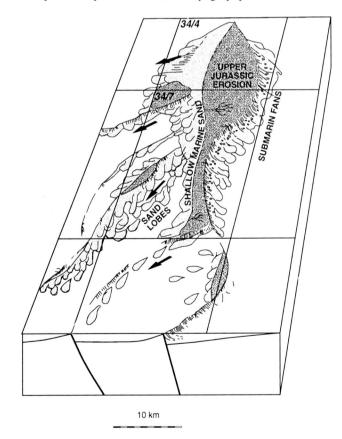

Fig. 9. A schematical presentation of the Upper Jurassic sand developments in the Snorre area.

Early Cretaceous

During uplift and relative sea-level drop in Ryazanian/Valangian time, the newly deposited sediments were exposed to erosion. In the Hauterivian marine transgression the submerged Snorre Fault Block received a cover of carbonate sediments of the Lower Cromer Knoll Group. Following a hiatus, sedimentation of the more shaly Upper Cromer Knoll Group took place probably with minor interruptions until late Aptian–Albian.

Many of the faults active in the late Jurassic ceased to be active before the early Cretaceous. However, the faulting and differential subsidence continued along most of the major faults. The fault movements were not associated with any significant fault-block rotation and may be a consequence of differential compaction in the rapidly subsiding Cretaceous basin.

In the Zeta area, the Gamma Fault Zone showed signs of activity far into the early Cretaceous.

Late Cretaceous

Shales of the Lower Shetland Group were deposited on the submerged Snorre Fault Block during the Turonian–Coniacian subsidence. A hiatus of Santonian age is recorded throughout the area, and deposition of the Upper Shetland Group followed. As opposed to the Viking Group, the Shetland Group onlaps the Snorre Fault Block in a direction from NW to SE. Towards the end of the Cretaceous the pronounced topography of the Snorre Fault Block was filled in with sediments.

Tertiary

The base Tertiary depth map in Fig. 10 shows continued activity on the Inner Snorre Fault in addition to faults of dominantly NW–SE orientation. This indicates a reorientation of the tectonic regime in the late Cretaceous to early Tertiary. Interrupted by a short episode of uplift and erosion, the basin subsidence continued into Tertiary with deposition of the Rogaland Group of Paleocene age and the Hordaland Group of Eocene to Oligocene age. The NW–SE-striking faults observed at the base Tertiary depth map decreased in number and offset from the Paleocene to Eocene. From the Oligocene, seismically observable faults are missing in the area. A major episode of relative uplift and erosion in the Miocene was followed by deposition of sands of the Utsira Formation of the Nordland Group. In the Tertiary, the southern part of the Snorre Fault Block was the structurally highest level.

Tectonic model

The data presented for the Snorre area support a tectonic evolution involving two episodes of rifting since the Permian (Fig. 11). Each rift episode was followed by a long lasting period of regional subsidence related to thermal relaxation. The first rift episode was characterized by westerly dipping faults and a dominantly continental basin-fill thickening towards the east. The second rift episode is related to the formation of the Viking Graben and was accommodated on NNE–SSW- to NE–SW- striking and easterly dipping faults. The absence of faults belonging to the first rift episode in the Zeta area is explained by its marginal position relative to the Permian–Triassic Rift Basin of the Northern North Sea.

The late Jurassic faulting documented in the Snorre Field and surrounding areas supports a domino model for extensions where both faults and fault blocks experienced a westerly rotation (Kusznir et al. 1988). The faults occur in domains of similar fault dip direction. During extension, each fault block uplifts at one side and subsides at the other, giving a net rotation. Barr (1987) and Jackson et al. (1988), who incorporated the domino model of fault-block rotation into the shear model of lithospheric extension (McKenzie 1978), have shown that the fault-block size is the major control on the degree of footwall uplift. Thus, large fault blocks, as the Snorre Fault Block, will typically show the greatest erosion. In the Snorre Field, the present fault dip is 40–45°, while initially c. 50°. This gives a crustal extension of c. 10–20%. The next fault block to the west is located some 25–30 km away. These figures imply, using the concepts of Barr (1987), that the footwall uplift of the Snorre Fault Block should be in the order of c. 1.1–1.5 km. This estimate is broadly in accordance with Yielding (1990) who presented a similar calculation for a profile across the Snorre Fault Block further south, where both extension and the related footwall uplift is less than in the Snorre Field. These numbers fit well with the estimated magnitude of erosion presented in Fig. 12.

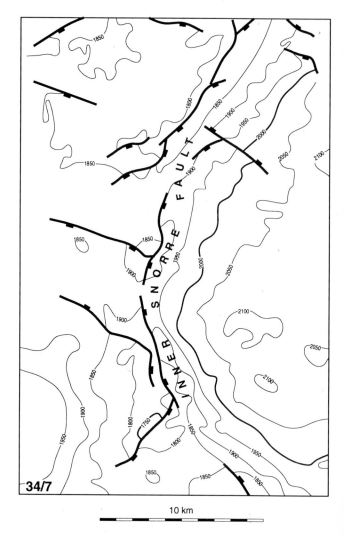

Fig. 10. Base Tertiary depth map.

There is a progressive tilting of the fault blocks during the two tectonic pulses of late Middle to late Upper Jurassic crustal extension. At the end of the late Callovian, the tilt of the pre-rift Brent horizons is c. 2 to 3° less than in the late Kimmeridgian. Based on the above considerations, the foot wall uplift of the Snorre Fault Block is predicted to be 0.6 to 0.7 km in the late Callovian. This implies that more than half of the total uplift of the Snorre Fault Block accompanied the earliest tectonic pulse of early Bathonian to late Callovian age.

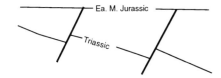

a) Lt. Permian-Ea. Triassic Rifting/Thermal subsidence (first rift episode)

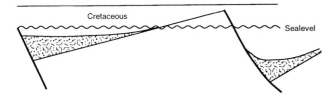

b) Lt. M. Jurassic - Lt. Jurassic Rifting (second rift episode)

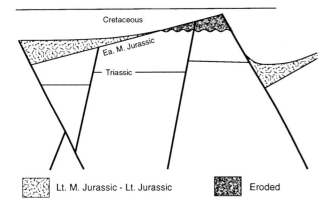

c) First and second episodes of rifting and thermal subsidence

Fig. 11. The structural evolution of the Snorre area during the two episodes of rifting.

SSW to NE–SW orientation. During extension, the crest of the westerly rotating Snorre Fault Block was uplifted. The Snorre Fault Block represented a topographic high in the late Jurassic, partly extending above sea-level. In the Snorre Field up to *c.* 1200 m of pre-rift sediments were eroded.

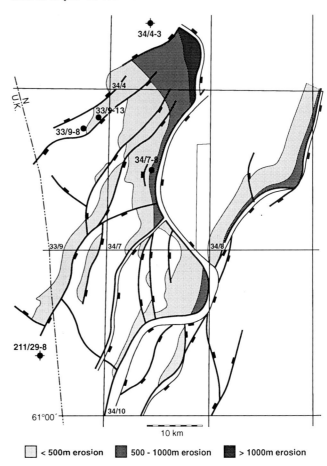

Fig. 12. The Snorre area. Estimated amount of late Jurassic erosion.

Depth profiles across the Snorre Escarpment show the Top Brent hangingwall cut off east of the Snorre Field is vertically about 1.8–2.0 km below the truncation surface in the footwall, taking post-Jurassic fault offset into consideration. A well located down-dip of the Inner Snorre Fault penetrated a 600 m thick Viking Group. Decompacted, the thickness of *c.* 1200 m is significantly less than the fault-block topography. Sedimentation in the late Jurassic was, therefore, unable to keep pace with the rapid fault-related subsidence. With some 1200 m of Viking Group sediments, by the end of the fault activity, the water depth in the East flank may have been about 700–800 m.

Conclusions

In the structural evolution of the Snorre Field and surrounding areas, two rifting episodes are recognized. The late Permian to early Triassic rifting was taken up on faults of dominantly N–S to NW–SE orientation and westerly dip. Large amounts of Triassic to early Middle Jurassic continental sediments were deposited in the following regional subsidence phase. To the north, in the Zeta area, the absence of Triassic to early Middle Jurassic faulting reflects its marginal position in relation to the late Permian to early Triassic Rift Basin of the Northern North Sea.

Towards the end of the Middle Jurassic, in the early Bathonian, renewed tectonic activity announced the beginning of a second major rift episode. The crustal extension was this time accommodated on mainly easterly dipping faults of NNE–

Within the syn-rift pelagic sediments of the Viking Group, two tectonic pulses are identified. The first pulse of early Bathonian to late Callovian age affected the depositional pattern of the Heather Formation by formation of syn-rift sedimentary wedges. As the extension proceeded from the early Oxfordian to the late Kimmeridgian, a syn-rift sedimentary wedge of Lower Draupne sediments was stacked on top of the Heather Formation. During extension, both the faults and the fault blocks experienced a westerly rotation. In a domino model for extension, the predicted uplift of the crestal parts of the Snorre Fault Block in the Snorre Field is estimated to be *c.* 1.1–1.5 km. Half the total uplift is expected to be related to the early Bathonian to late Callovian rifting. Most of the late Jurassic fault activity ceased before the early Cretaceous. Compressional features of early Cretaceous faults are locally observed along E–W- to ENE–WSW-orientated faults. Towards the end of Cretaceous the pronounced relief caused by the late Jurassic rifting was filled in with sediments. Faults of dominantly NW–SE orientation document a reorientation of the tectonic regime from the late Jurassic into the Cretaceous and Tertiary. While the northern part of the Snorre Fault Block experienced most uplift in the late Jurassic, the southern part of the fault block remained in the structurally highest position in the late Cretaceous and Tertiary.

The authors thank J. P. Nystuen, S. Olaussen, B. Tørudbakken and R. Waddams for their helpful comments on the manuscript.

References

BADLEY, M. E., PRICE, J. D., RAMBECH DAHL, C. AND AGDESTEIN, T. 1988. The structural evolution of the North Sea basin: crustal stretching and subsidence. *Journal of the Geological Society, London*, **145**, 455–472.

BARR, D. 1987. Lithospheric stretching, detached normal faulting and footwall uplift. *In*: COWARD, M. P., DEWEY, J. F. AND HANCOCK, P. L. (eds) *Continental Extensional Tectonics*. Geological Society, London, Special Publication, **28**, 75–94.

BARTON, P. AND WOOD, R. 1984. Tectonic evaluation of the North Sea Basin: crustal stretching and subsidence. *Geophysical Journal of the Royal Astronomical Society*, **79**, 987–1022.

BREKKE, H., FÆRSETH, R. B., GABRIELSEN, R. H., GOWERS, H. B. AND PEGRUM, R. G. 1991. Nomenclature of Tectonic Units in the Norwegian North Sea, South of 62° north. *NPD Bulletin*.

COCHRAN, J. R. 1983. Effects of finite rifting times on the development of sedimentary basins. *Earth and Planetary Science Letters*, **66**, 209–302.

DEEGAN, C. E. AND SCULL, B. J. 1977. A standard lithostratigraphic nomenclature for the central and northern North Sea. *NPD Bulletin*, **1**.

GILTNER, J. P. 1987. Application of extensional models to the northern Viking Graben. *Norsk Geologisk Tidsskrift*, **67**, 339–352.

HOLLANDER, N. B. 1987. Snorre. *In*: SPENCER, A. M. *et al.* (eds) *Geology of the Norwegian Oil and Gas Fields*. Graham & Trotman, London, 307–318.

JACKSON, J., WHITE, N. J., GARFUNKEL, Z. AND ANDERSON, H. 1988. Relations between normal fault geometry, tilting and vertical motions in extensional terrains: an example from the southern Gulf of Suez. *Journal of Structural Geology*, **10**, 155–170.

KARLSSON, W. 1986. The Snorre, Statfjord and Gullfaks oilfields and the habitat of hydrocarbons on the Tampen Spur, offshore Norway. *In*: SPENCER, A. M. *ET AL.* (eds) *Habitat of Hydrocarbons on the Norwegian Continental Shelf*. Graham & Trotman, London, 181–197.

KUSZNIR, N. J., MARSDEN, G. AND EGAN, S. 1988. Fault block rotation during continental lithosphere extension: a flexural cantilever model. *Geophysical Journal*, **92**, 546.

LERVIK, K. S., SPENCER, A. M. AND WARRINGTON, G. 1989. Outline of Triassic stratigraphy and structure in the central and northern North Sea. *In*: COLLINSON, J. D. (ed.) *Correlation in Hydrocarbon Exploration*. Norwegian Petroleum Society, Graham & Trotman, London.

MCKENZIE, D. 1978. Some remarks on the development of sedimentary basins. *Earth and Planetary Science Letters*, **40**, 25–32.

NYSTUEN, J. P. AND FÄLT, L-M., in prep. Upper Triassic–Lower Jurassic reservoir rocks in the Tampen Spur area, Norwegian North Sea. Norwegian Petroleum Society.

——, KNARUD, R., JORDE, K. AND STANLEY, K. O. 1989. Correlation of Triassic to Lower Jurassic sequences, Snorre Field and adjacent areas, northern North Sea. *In*: COLLINSON, J. D. (ed.) *Correlation in Hydrocarbon Exploration*. Graham & Trotman, London, 273–289.

SCLATER, J. G. AND CÉLÉRIER, B. 1978. Extensional models for the formation of sedimentary basins and continental margins. *Norsk Geologisk Tidsskrift*, **67**, 253–267.

SCOTT, D. L. AND ROSENDAHL, B. R. 1989. North Viking Graben: An East African perspective. *American Association of Petroleum Geologists Bulletin*, **73**, 155–165.

STECKLER, M. S. AND WATTS, A. B. 1982. Subsidence history and tectonic evolution of Atlantic type continental margin. *In*: SCRUTTON, R. A. (ed.) *Dynamics of Passive Margins*. Geodynamics series 6, AGU, 184–196.

STEEL, R. AND RYSETH, A. 1990. The Triassic–early Jurassic succession in the northern North Sea: Megasequence stratigraphy and intre-Triassic tectonics. *In*: HARDMAN, R. F. P. AND BROOKS, J. (eds) *Tectonic Events Responsible for Britain's Oil and Gas Reserves*. Geological Society, London, Special Publication, **55**, 139–168.

YIELDING, G. 1990. Footwall uplift associated with Late Jurassic normal faulting in the northern North Sea. *Journal of the Geological Society, London*, **147**, 219–222.

WOOD, R. J. AND BARTON, P. 1983. Crustal thinning and subsidence in the North Sea. *Nature* **302**, 134–136.

ZIEGLER, P. A. 1982. *Geological atlas of western and central Europe*. Shell International Petroleum Maatschappij B.V. and Elsevier, Amsterdam.

Controls on the development and evolution of structural styles in the Inner Moray Firth Basin

K. THOMSON and J. R. UNDERHILL

Department of Geology and Geophysics, The University of Edinburgh, Grant Institute, King's Buildings, West Mains Road, Edinburgh EH9 3JW, UK

Abstract: Although it has generally been believed that structural styles in the Inner Moray Firth (IMF) have been largely controlled by strike-slip movements on the Great Glen Fault (GGF), integrated seismic and field studies suggest otherwise. Instead, most structural styles appear to have developed and evolved as a result of dip-slip extension and thermal subsidence consequent upon two phases of rifting during the Permo-Triassic and Late Jurassic and subsequent regional uplift and local inversion during the Tertiary. The integration of demonstrable thickening of Kimmeridgian–Portlandian intervals across the GGF with sedimentological information from onshore outcrops and cored wells suggests that the basin had a half-graben geometry during the Late Jurassic with a depocentre adjacent to the Helmsdale Fault, analogous to half-graben geometries which characterized other Late Jurassic sequences in Greenland and the South Viking Graben. Progressive marine onlap suggests that more gentle regional (thermal) subsidence took place in an underfilled basin, during subsequent Early Cretaceous deposition. However, the seismic data and subcrop information show that the geometries resulting from such classic rift- and thermally-driven phases of extension were modified by Cenozoic regional uplift and inversion in response to intraplate compression resulting from NE Atlantic (Thulean) and Alpine events. These events also appear to have effected minor strike-slip motion on the GGF, with the development of spectacular 'flower structure' and 'helicoidal' geometries, and caused limited oblique-slip reactivation of some extensional structures. The most notable modification of structural styles occurs in areas adjacent to the major basin-bounding faults. Particularly complex structural inversion geometries occur in the northwest corner of the basin adjacent to the Wick Fault while anomalously-trending folds developed in response to space problems in the Sutherland Terrace, between the Helmsdale Fault and the GGF, as a result of opposing senses of slip on these faults.

The Inner Moray Firth (IMF) lies adjacent to the Scottish mainland between the Grampian and East Sutherland coasts (Fig. 1). The continuity of structures and sequences offshore into adjacent coastal outcrops affords the opportunity to integrate field exposures into models of the basin's structural development (e.g. Underhill 1991a). Although well results have been largely disappointing, with only British Petroleum's Beatrice Field proving commercial to date, the data obtained from the IMF give important clues to the nature and timing of deformation in the NW European domain in general and other areas of the North Sea in particular.

Database

The database used in analysing the regional tectonics and local structural styles in the IMF consists of over 8000 km of offshore seismic data supplied by GECO, Horizon, Halliburton, Shelf Exploration, Shell, BP, Arco, Unocal and Esso. Additional onshore seismic was available over the Easter Ross peninsula courtesy of Fina and their partners in onshore licence EXL 208. The seismic data were tied to all released exploration wells in the area and additional confidential data. Use has been made of onshore exposures and electrical well log analyses to add a level of detail below that of seismic data. Additional sonic velocity, vitrinite reflectance and fission track information was integrated into the study in order to quantify some of the effects of relative uplift resulting from specific deformation events.

Structural interpretation

Interpretation of the large seismic database has led to a better understanding of the main controls on Late Paleozoic–Recent basin evolution and the recognition of significant tectonic events in the development of structural styles in the area. These significant events are now discussed in turn.

Late Paleozoic

Seismic data from the IMF highlight the occurrence of a significant unconformity within the Late Paleozoic successions (Fig. 2). It is largely defined by truncation of underlying reflectors and onlap of subsequent units previously identified by Roberts *et al.* (1990). Data from the Beatrice area and onshore analogues from Easter Ross suggest that the dominant structural style beneath this unconformity consists of eroded, thrust-bounded, hangingwall anticlines (Fig. 3; Underhill and Brodie in press). These apparently record a phase of Permo-Carboniferous ('Variscan') contractional deformation prior to partial peneplanation and Permian sedimentation (Fig. 4). As such, these structural styles are analogous to those formed during basin inversion in nearby areas (e.g. Coward *et al.* 1989; Astin 1990; Seranne 1992). Although the most well-defined data for this phase of deformation come from onshore exposures, some supporting evidence exists from recent, good-quality seismic data from the Beatrice area for the overlying succession draping a well-defined topography (Fig. 4), partially created by renewed Permian extension following the inversion episode.

It is likely that the early Devonian deformation in the IMF was controlled by active rifting. Field studies suggest that adjacent areas experienced syn-sedimentary fault activity (Rogers 1987) resulting from regional extensional collapse of the Caledonian orogen (Serrane 1992). Consequently, it is possible that many of the contractional features described above also represent reactivated normal faults. It remains unclear whether contractional deformation is always closely associated with syn-sedimentary faults and the extent to which truncation by the unconformity has occurred within the pre-Permian sequence; thus the exact stratigraphic level expected beneath the unconformity may be highly variable across the IMF and it is probable that significant amounts of Devonian and possibly thick Carboniferous sections will be found locally. Hence, Late Carboniferous successions are likely to be

From *Petroleum Geology of Northwest Europe: Proceedings of the 4th Conference* (edited by J. R. Parker).

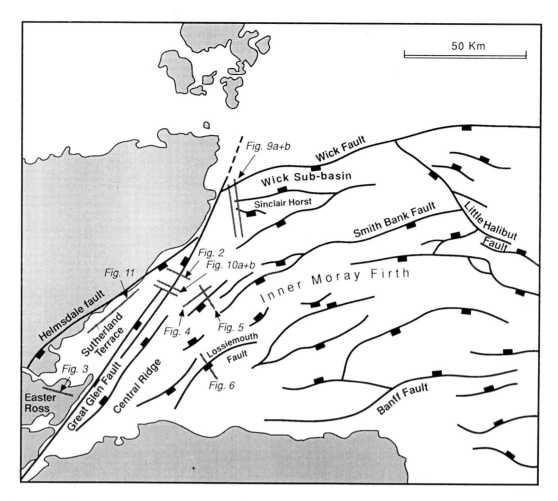

Fig. 1. Location map for the Inner Moray Firth showing the basin's main structural elements. Positions of the various seismic lines used in the article are shown.

preserved in some areas where the unconformity had significant topography or was underlain by pre-existing fault-controlled depocentres.

In contrast with most previous interpretations, there seems little reason to propose that the Great Glen Fault (GGF) played any significant role in the Devonian rift episode or the 'Variscan' or Pre-Permian phase of basin development because of the lack of demonstrable horizontal offset of structures. Indeed, as Rogers *et al.* (1989) have shown, no more that 25–29 km of dextral displacement appears to have affected Devonian sequences.

Permo-Triassic extension

Permian successions within the IMF are hard to separate from other red bed sequences because of the lack of distinctive palaeontological markers and their lack of seismic character. Despite these limitations, the lowest units occurring above the prominent Permo-Carboniferous ('Variscan') unconformity show varying degrees of stratigraphic thickening on the seismic data. Although this can generally be explained by the passive infill of remnant (thrust-related hangingwall) topography, the occurrence of some thickening appears to occur adjacent to reactivated extensional faults (e.g. Fig. 4), indicating a minor component of early, syn-sedimentary tectonic control. Although this would also be consistent with regional extensional events elsewhere in the North Sea, further work is needed to test this hypothesis.

Latest Permian, Triassic and Early Jurassic thermal subsidence

The subdivison of later Permian and Triassic is hampered as Zechstein sequences are also often dominated by reddened sandstones similar to Triassic, Permian, Carboniferous and Devonian strata. Despite these problems, the seismic data show that Late Permian and Triassic sequences are largely dominated by concordant reflectors (layer-parallelism) and broad, basin-wide westerly thickening packages with little or no variation adjacent to faults (Underhill 1991*a*). In our opinion, and in contrast to Frostick *et al.* (1988), the data currently available from wells are too sparse to allow a detailed model to be proposed for Permo-Triassic tectonics and sedimentation. However, the Triassic sequences generally appear to have been laid down in a basin experiencing broad-based subsidence following an earlier extensional episode. As such, they probably represent a phase of thermal subsidence following the minor Permian rifting (contra Frostick *et al.* 1988). The absence of abrupt thickening across and along the GGF again suggests that it played a negligible role in controlling basin development at this time (contra Frostick *et al.* 1988).

The top of the Triassic succession is marked by the most prominent seismic reflector in the area. Its onshore equivalent is represented by the Stotfield Cherty Rock, a significant silcrete horizon marking a notable break in sedimentation (Naylor *et al.* 1989). Overlying Early Jurassic units also consist of relatively thin concordant sequences which show gradual

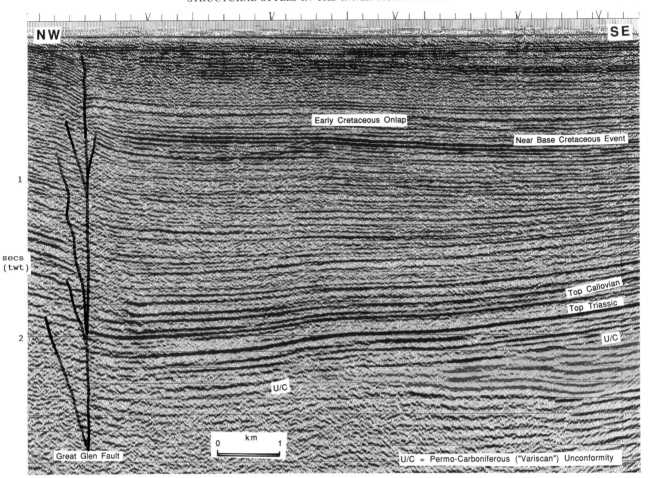

Fig. 2. Seismic line interpretation highlighting the occurrence of an intra-Late Paleozoic unconformity (u/c) defined largely by truncation of underlying strata which separates deformed Devonian sequences (highlighted in green) from overlying Permian red beds. Line shown courtesy of GECO-PRAKLA.

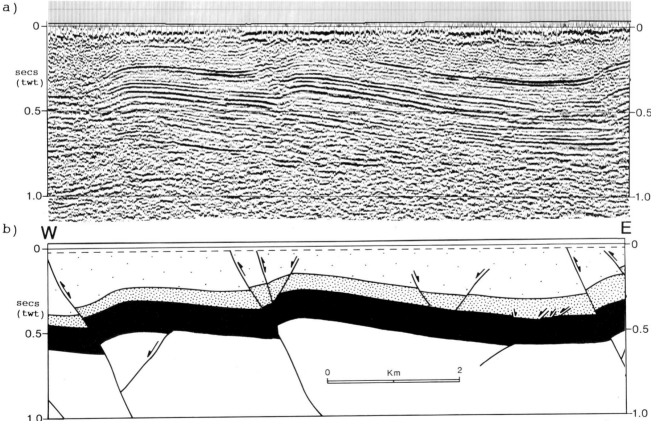

Fig. 3. W–E-trending seismic line and line drawing interpretation showing eroded thrust-bound hangingwall anticlines affecting Devonian sequences in Easter Ross, which are believed to have formed as a result of a phase of Permo-Carboniferous contraction and inversion. Key: open stipple: Late Devonian; close stipple: Top Middle Devonian; black: Middle Devonian fish beds. Line shown courtesy of Fina Exploration.

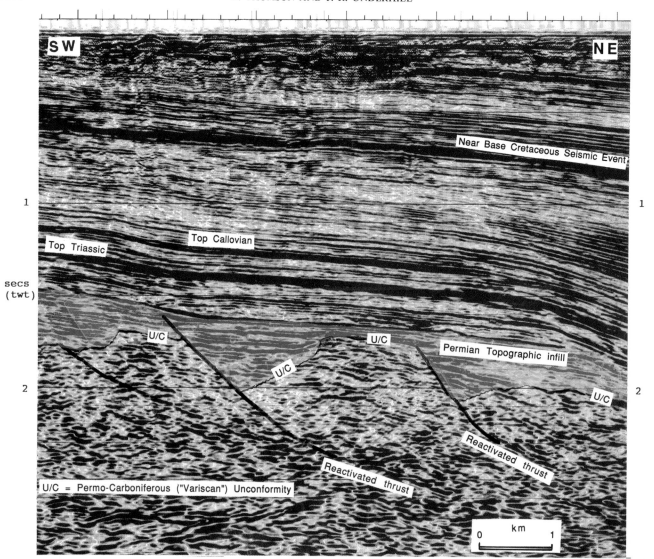

Fig. 4. Seismic line interpretation showing the topographic variation associated with the Permo-Carboniferous erosion surface highlighting the truncation of thrusted and folded Devonian sequences and their subsequent rejuvenation during the Permian as extensional features with a draped infill (highlighted in yellow). Line shown courtesy of BP Exploration.

thickening towards the Scottish mainland with maximum thicknesses occurring in the Sutherland Terrace area, between the GGF and Helmsdale Fault (Andrews and Brown 1987; Andrews *et al.* 1990; Underhill 1991*a,b*). These Early Jurassic sequences are interpreted to represent continued gentle, thermal (post-rift) subsidence following the Permian event.

Early–Middle Jurassic (Toarcian–Aalenian) regional uplift and Middle–Late Jurassic (Aalenian–Oxfordian) thermal subsidence

Evidence for a regional phase of relative uplift affecting the IMF during the Late Toarcian–Early Aalenian due to a regional doming event in the North Sea comes from stratigraphic information beyond the level of seismic resolution. These data show that a series of regional 'mid-Cimmerian' unconformities occur throughout the Middle Jurassic. The earliest and most significant of these unconformities occurs throughout the area, truncates Lower Jurassic successions including the Orrin Formation and shows progressive onlap of Bathonian–Oxfordian sequences towards the east (Underhill and Partington 1993 and in press; Stephen *et al.* 1993).

Although regional doming has not had any significant control on structural styles at a seismic scale, documentation and understanding of its cause and effects appear to have significance for the subsequent development of important

structural styles (e.g. in the Late Jurassic rift episode). Regional mapping of the temporal and spatial variation of this event shows that similar relations exist over a wide area in the North Sea, which has been interpreted to be the result of regional doming ('Central North Sea Dome'; Hallam and Sellwood 1976; Ziegler 1982, 1990*a,b*) above a warm, diffuse and transient plume head (Underhill and Partington 1993 and in press). Evidence that progressive shallowing occurred during the Late Toarcian (i.e. the Orrin Formation in the IMF; Stephen *et al.* 1993) and sequences in other areas of Britain (i.e. Bridport Sands and Dunlin Group), suggests that uplift may have begun in the Lower Jurassic. Stratigraphic evidence suggests that the unconformity reached its maximum areal extent and hence, that the dome reached its climax during the Early Aalenian. Initial subsidence occurred during the Late Aalenian before subsiding gently during the Late Bajocian–Early Kimmeridgian (Underhill and Partington 1993 and in press), perhaps following a renewed phase of (Early Bajocian) uplift which created another relatively widespread erosive unconformity.

Most significantly for the subsequent development of diagnostic structural styles in the IMF, the stratigraphic data suggest that the trilete North Sea rift system already had an expression during the Middle Jurassic prior to the period of volcanism and most significant rifting. In the rift arms, significant deflection of the extent of maximum flooding may be seen along each of the regions subsequently characterized by significant syn-sedimentary extension, suggesting that the IMF and Viking Graben formed topographic depressions prior to the onset of half-graben development (Underhill and Partington 1993 and in press). This suggests that lithospheric thinning and differential subsidence had already occurred along subsequent axes of deformation during uplift of the dome and persisted throughout the phase of subsidence resulting from the decay of this domal uplift.

Late Jurassic (Early Kimmeridgian–Late Portlandian) rifting

A drastic change in structural styles and basin architecture characterizes the Late Jurassic of the IMF. Seismic data demonstrate the development of numerous, classic half-graben across the area interpreted to result from dip-slip extension (Underhill 1991a,b). This phase of extension was the most significant for the development of potential hydrocarbon-bearing footwall structures adjacent to active extensional faults (e.g. Beatrice Field; Fig. 5).

Seismic mapping demonstrates that the dominant structural style of such extensional faults is as linear but discontinuous structures offshore. Onshore mapping gives additional insights into the potential significance of this structural discontinuity. The Helmsdale Fault is similar to the structures mapped on the seismic data in that it, too, is not one continuous fault as generally described, but consists of at least three sub-parallel en-échelon strands linked by overlapping relay ramps rather than transfer faults. Such zones appear to have been important conduits for sediment dispersal during the Late Jurassic (e.g. the early Kimmeridgian, Allt na Cuile Sandstone, Lothbeg Point). Accurate location of such syn-tectonic point sources may help locate additional reservoir sandstones offshore.

The duration of demonstrable syn-sedimentary fault activity may be gauged from the occurrence of divergent seismic reflector geometries and sedimentary evidence from onshore successions. These relations suggest that the area was initially deformed by continuous fault block rotation (Early Kimmeridgian) but was subsequently dominated by a passive sedimentary infill of topography created during discrete rift episodes (characterized by onlapping reflectors) during the Late Kimmeridgian–Late Portlandian (Underhill 1991b) in a similar fashion to that seen elsewhere (e.g. the Brae area in the South Viking Graben (Turner et al. 1987) and East Greenland (Surlyk 1978)).

Although these onlap relationships have previously been used by Exxon workers to derive their original eustatic curve purporting to show global sea-level changes (Vail and Todd 1981; Vail et al. 1984), recent analysis suggests that the onlap relationships are the result of syn-tectonic fault block rotation and limited sedimentary effects in a fully marine domain (seismic marine onlap, Underhill 1991b) and are not the direct result of global sea-level fluctuations in a coastal setting. It is also clear that the surfaces of onlap are not unconformities (per se), but are characterized by complete stratigraphic successions with no breaks. Hence, it seems likely that the surfaces of onlap are only recorded at seismic scale because they represent a considerable period of time. As such they probably represent condensed sections, analogous to those described by Partington et al. (1993).

Further complications to the dominantly extensional structural styles occur in association with Late Jurassic fault activity. Unusual structural geometries characterize several significant faults within the basin as exemplified by the Lossiemouth and Helmsdale faults (Figs 6 and 7). In both cases, hangingwall sequences dip steeply away from the fault before returning to the horizontal through a hangingwall synclinal structure. These are interpreted to form in response to differential compaction as a consequence of varying hangingwall and footwall depositional fills and as a result of a buttressing effect created by the underlying rigid footwall.

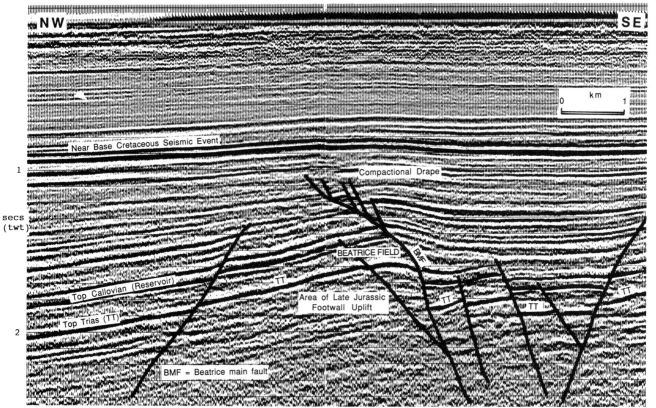

Fig. 5. NW–SE-trending cross-section of Beatrice Field showing the area of Late Jurassic footwall uplift associated with Beatrice Main Fault (BMF) which also shows a compactional drape geometry similar to other examples in the area such as the Lossiemouth (Fig. 6) and Helmsdale (Fig. 7) faults.

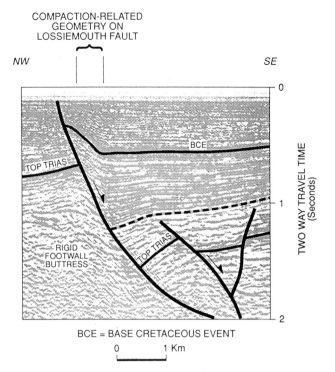

BCE = BASE CRETACEOUS EVENT

0   1 Km

Fig.   6. Seismic line across the Lossiemouth Fault characterizing the unusual increase in depositional dips adjacent to extensional structures in the IMF and interpreted to result from the effects of compactional drape above.

The Late Jurassic sequences are capped by the second most prominent seismic reflector seen in the area: the 'Base Cretaceous Event'. As such, it is convenient to use this regional marker to define a distinct seismic package. However, in reality, this event does not appear to represent an unconformity but rather a condensed section (Rawson and Riley 1982), which lies beyond the level of seismic resolution. Furthermore, it appears to have formed in response to a drastic change in water circulation patterns (Rattey and Hayward 1993), rather than a particularly significant change in tectonic deformation and structural styles.

Cretaceous thermal subsidence

Early Cretaceous sequences are preserved and well-imaged above the 'Base Cretaceous Event'. They are largely characteized by well-defined onlapping reflectors, which show progressive onlap towards the basin's margins. Passive infill of

well-defined Jurassic half-grabens occurred with evidence for significant shifts in depocentre location (compared to the Kimmeridgian) to areas adjacent to the Wick and Little Halibut faults (Andrews *et al.* 1990; Roberts *et al.* 1990). The Upper Cretaceous Chalk sequence can similarly be seen to drape most of the pre-existing topography and although commonly demonstrating progressive onlap onto basement highs, it is only occasionally interrupted by unconformities (Andrews *et al.* 1990).

The lack of evidence for divergent reflectors within the Cretaceous seismic sequences is suggestive of a passive infill of a pre-existing topography. Such an interpretation is supported by palaeontological evidence which suggests that the basin was underfilled at this time (BP proprietary data). Consequently, it seems most plausible that the area was undergoing a phase of thermal subsidence following the Late Jurassic rift episode.

This suggestion contrasts with many previous interpretations which have suggested that the IMF did not undergo Mesozoic thermal subsidence on account of its being interpreted as an isostatically uncompensated strike-slip basin (McQuillin *et al.* 1982; Barr 1985; Bird *et al.* 1987; Roberts *et al.* 1990). This suggestion was based on the regional gravity interpretation of the North Sea by Donato and Tully (1981) which shows a residual negative anomaly over the IMF which in turn has been taken to imply that the basin lies on unthinned lithosphere and formed as a result of upper crustal extension. Although the presence of such an anomaly cannot be disputed, the interpretation is open to question particularly in the light of seismic refraction studies which indicate that the area has experienced significant crustal thinning (Smith and Bott 1975; also see discussion at the end of this article). Indeed, the negative anomaly could be accounted for by the presence of a low density basement of Moine rocks and Caledonian granites beneath the IMF similar to those seen at outcrop on the Scottish Mainland. Although other possibilities also exist such as an heterogeneous stretching model for the IMF with isostatic compensation being offset laterally, the influence of unusual basement density variations may also explain why the negative anomaly can be seen to extend over much of the Scottish Mainland as well as the IMF.

Post-Cretaceous structural history

Regional structural styles

Although no Tertiary sequences occur within the Inner Moray Firth, it is still possible to deduce the basin development of the area during this period using regional tectonic and sedimentary analysis and local seismic evidence. Examination of sea bed

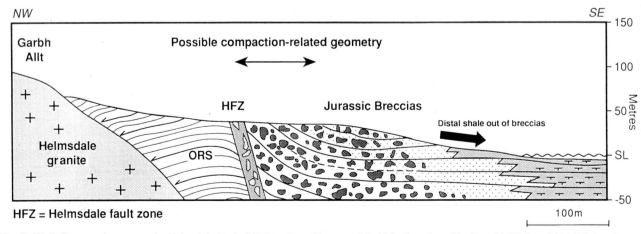

HFZ = Helmsdale fault zone

100m

Fig. 7. W–E Cross-section across the Helmsdale Fault (HFZ) at Dun Glas, near Navidale (based on MacDonald 1985) and highlighting the increase in unusually steep dip to beds immediately adjacent to the fault, which are also believed to be the result of compactional drape adjacent to the Helmsdale Granite footwall buttress (cf. Fig. 6). ORS: Old Red Sandstone.

geological maps and seismic reflection profiles, which preserve rotated half-graben sequences, indicate that the IMF has been uplifted and tilted down to the east (Underhill 1991a). Integration of sonic velocity (Fig. 8), vitrinite reflectance and fission track data further suggest that relative uplift exceeded 1 km in the west of the basin and shows a gradual decrease eastwards to zero near the IMF–Outer Moray Firth transition (e.g. Hillis *et al.* 1992). Once corrected for post-erosional sedimentation, the total erosion values show little variation across the IMF with values of the order of 1 km suggesting that the area experienced significant regional tectonism.

The relative dating of the erosional event can be made using compaction data, evidence of depositional sequences in eastern parts of the Moray Firth and apatite fission track analysis. All imply that relative uplift occurred after the deposition of the Chalk Group (i.e. during the Tertiary). Although Tertiary sediments are only found in the extreme east of the IMF, the areas still further to the east contain large quantities of such sediments. Indeed, the deposition of prograding Paleocene deltaic sequences close to the Inner Moray Firth–Outer Moray Firth transition passing distally eastwards into slope fans and turbidites (Parker 1975; Rochow 1981) is consistent with the uplift and erosion of the Scottish Mainland, East Shetland Platform and the IMF during the Paleocene.

a)

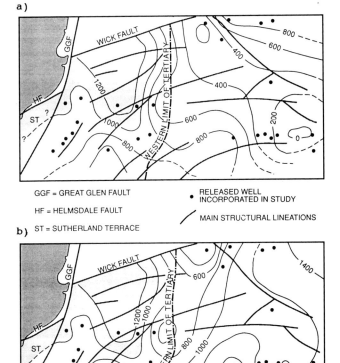

GGF = GREAT GLEN FAULT

HF = HELMSDALE FAULT

b) ST = SUTHERLAND TERRACE

• RELEASED WELL INCORPORATED IN STUDY

╱ MAIN STRUCTURAL LINEATIONS

25 Km

Fig. 8. Maps recording uplift and erosion: (**a**) apparent erosion; and (**b**) total erosion in the Inner Moray Firth, which suggest that the whole area has undergone at least 250 m of uplift after the Cretaceous. The most pronounced uplift appears to characterize the NW corner of the basin where figures exceed 1 km of uplift. The regional uplift is interpreted to result from intra-plate compression. Although only released well locations are shown, additional data from confidential wells have been incorporated into the diagrams.

Local Tertiary structural styles

Several local structural styles appear also to have resulted from Tertiary uplift. Many are highly localized to the NW corner of the basin and have not previously been described. For

example, in the area of UKCS Block 12/16, close to the point where the Wick Fault terminates against the GGF, previously undocumented inversion structures can be found which affect Early Cretaceous sediments. These structures include a hangingwall anticline developed adjacent to the Wick Fault as a result of contractional reactivation (Fig. 9a). A spectacular 'short-cut fault' developed in association with the hangingwall anticline probably in response to increasing compressional stress as a result of buttressing against the Wick Fault.

In the region adjacent to the Wick Fault, numerous normal offsets can be seen on faults displacing Early Cretaceous sediments (e.g. Fig. 9b). However, such movement is not localized to the Wick fault area but can be found to have reactivated older extensional faults in the IMF such as the Smith Bank, Banff, Buckie and Halibut Horst Boundary faults (Roberts *et al.* 1990; Underhill 1991a), suggesting that the whole area experienced tectonic rejuvenation perhaps as a local expression of the regional deformation.

The most significant Tertiary fault activity within the IMF Basin occurs along the narrow Great Glen Fault zone (Fig. 10). Stratigraphic relations on seismic reflection profiles demonstrate that it was active after the Early Cretaceous. It displays typical 'flower structures' (*sensu* Harding 1990) with 'helicoidal' geometries similar to those described from sand-box experiments (Naylor *et al.* 1986) and suggestive of strike-slip movement (Fig. 10). The overall sense of movement can be shown to have a down to the east/southeast component. Although the sense of strike-slip movement cannot be determined directly from seismic profiles, isopach maps of the Mesozoic basin-fill suggest that any subsequent net strike-slip displacement was small, dextral and perhaps less than 10 km (Underhill 1991a).

Independent evidence for there being a period of Tertiary movement affecting the GGF can be found from onshore localities. Numerous dextral offsets of Tertiary dykes can be found adjacent to the Great Glen Fault zone on the island of Mull. This is consistent with the proposed dextral movement suggested by Holgate (1969) during the Tertiary and consistent with the dextral and down to the southeast-Permo-Carboniferous movement proposed by Speight and Mitchell (1979).

Tertiary movement on the GGF may also have contributed to the fold generation evident in the Sutherland Terrace between the Helmsdale and Great Glen faults from both offshore seismic reflection profiles (Fig. 11) and onshore exposures (Fig. 12). Onshore mapping demonstrates that these folds plunge northwest and southeast, at high angles (> 50° and which possibly increase with amount of rotation) adjacent to the Helmsdale Fault. Such fold orientations discount the possibility that they are classical inversion folds due to the hangingwall moving up the Helmsdale Fault and imply that fault-sub-parallel compression occurred in the Sutherland Terrace. It seems likely that dextral movement on the GGF would have created space problems in the Sutherland Terrace as it was compressed into the smaller area to the northeast where the faults merge (Fig. 12). The consequence of such movement was northeast–southwest compression and generation of northwest–southeast-orientated folds. It appears likely that the footwall to the Helmsdale Fault acted as a fixed buttress on which there was a limited amount of sinistral displacement (Fig. 12).

Regional Tertiary events responsible for the structures in the IMF

The proposed Tertiary structures are thought to have formed as a result of the combined effects of rifting of the North Atlantic Ocean and collision of the Alpine system. Together they caused the NW European domain to experience intra-plate deformation. The presence of a mantle 'hot-spot' at the

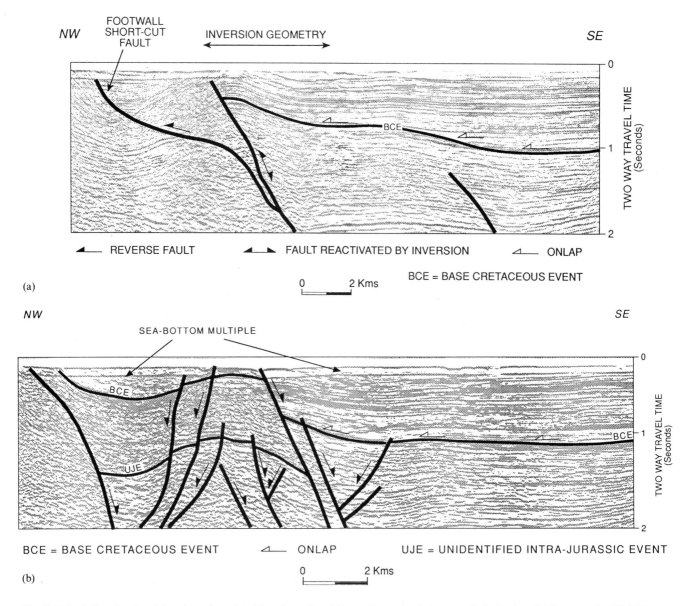

Fig. 9. Seismic line showing: (**a**) an inversion-related hangingwall anticline and associated short-cut fault developed adjacent to the Wick/Great Glen Fault intersection in the NW corner of the basin which is characterized by the greatest amounts of relative uplift and severest late-stage deformation; and (**b**) late horst-block geometries in the area of the Wick Fault probably formed in response to extensional stress related to doming and NE Atlantic opening.

time of rifting (White 1988, 1989) caused a broad bathymetric swell together with gravity and geoidal anomalies of approximate wavelength 2000 km and the production of large quantities of Tertiary volcanics (White 1989).

It is likely that the initial doming would produce extension at the surface, the magnitude of which would decrease radially away from the hot-spot. It is possible that the IMF would have experienced minor amounts of extension by virtue of its position at the margins of the dome (as witnessed by the post-Early Cretaceous extension around the Wick Fault). According to experimental studies (e.g. Dixon 1975; Withjack and Schiener 1982), strike-slip faulting might be expected in association with such domes, and may possibly account for the movements on the GGF and the Helmsdale Fault and hence, initiation of the folding in the Sutherland Terrace.

Upon rifting, the continental margins would not subside but remain relatively elevated due to continued hot-spot activity and the isostatic response of crust thickened by Tertiary volcanics. The consequences would be that the continental margins would experience minor amounts of compression as the margins were tilted back away from the newly formed rift.

This compression would be the probable cause of the inversion seen around the Wick Fault and may have initiated movement of the Great Glen–Helmsdale fault system and its associated folding in the Sutherland Terrace. The evidence of westward-increasing uplift as seen in the IMF could, therefore, be interpreted as a direct consequence of doming and rift shoulder uplift. The highest amounts of relative uplift might, therefore, be expected in Western Scotland or beyond.

However, this model does not explain two major facts which are consequences of the total erosion exceeding the apparent erosion (Fig. 8). Firstly, in the IMF the total erosion shows no marked decrease from west to east. This would require unrealistic amounts of pre-rift doming and post-rift flexure of the rift shoulder. Secondly, decompacted burial curves for wells in the IMF with stratigraphy preserved into the Early Paleocene and in the Late Tertiary require an initial rapid subsidence phase followed by equally rapid uplift in order to satisfy the requirements of both erosion estimates and stratigraphy. Such a rapid burial cannot be easily accounted for by the North Atlantic active rifting model.

In order to explain the Tertiary features of the IMF and be

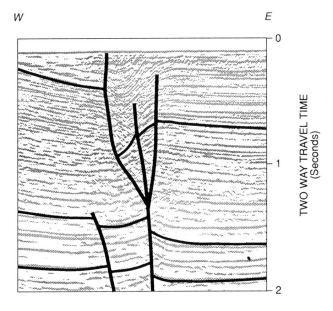

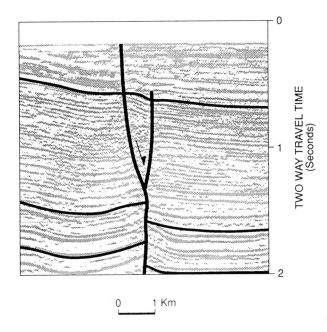

Fig. 10. Seismic lines to highlight the geometry associated with the Great Glen Fault and the nature of the 'negative flower' structure along strike believed to be the result of limited dextral strike-slip motion on the feature during the Tertiary.

consistent with the history of other areas of the northwestern European plate, it may be necessary to have recourse to alternative models which take the presence of the Alpine orogeny at the southern continental margin into account. The opening of the northeastern Atlantic in conjunction with the Alpine orogeny would result in intra-plate compressive stress (Ziegler 1987; Ziegler and Van Hoorn 1989). The orientation of such a compressive stress would be northwest–southeast. This would be consistent with that of England (1988) from the NW–SE and NNW–SSE orientation of Tertiary dyke swarms and Late Cenozoic mesofractures due to lateral escape (extension) in the foreland to the Alpine front (Bevan and Hancock 1986; Hancock and Bevan 1987).

Such a compressive stress regime could possibly account for the inversion seen in the NW corner of the basin (e.g. UKCS Block 12/16) and the dextral strike-slip motion on the Great Glen Fault. In addition, it would allow the explanation of the rapid burial and uplift observed in the IMF during the Tertiary to be explained in terms of differential compression between the crust and lithosphere. As stresses are transmitted with greater efficiency through the lithosphere compared to the crust this will allow heterogeneous compression, with the lithosphere thickening to a greater extent than the crust (Kusznir and Karner 1985; Hillis 1992). If the lithosphere thickens to a greater degree than the crust, rapid subsidence follows with rapid uplift following during the thermal re-equilibration phase. Furthermore, such a lithospheric mechanism for stress transmission through the plate allows the explanation of the differences in timing of Tertiary inversion throughout the plate (Ziegler, 1987). Consequently, it appears possible to integrate the Tertiary history of the IMF with that of other Mesozoic basins in Europe into a compressive intra-plate setting developed in response to North Atlantic rifting over a mantle plume to the northwest and the Alpine collisional events to the south.

Conclusions

(1) Structural styles in the Inner Moray Firth appear to be largely the result of three phases of extensional activity punctuated by periods of contraction, regional uplift and inversion.

(2) Onshore analogues suggest that active rifting which occurred during the Devonian was followed by more broad-based thermal subsidence prior to a phase of Permo-Carboniferous ('Variscan') contraction and inversion.

(3) Renewed extension characterized the Permian with the reactivation of some contractional structures as normal faults. Subsequent Permo-Triassic and Early Jurassic deposition occurred in a broad basin driven by thermal subsidence with little evidence for syn-sedimentary activity.

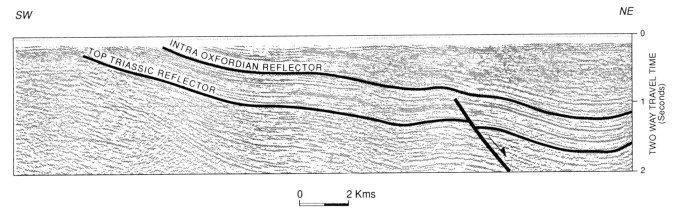

Fig. 11. Offshore SW–NE-trending seismic line from the Sutherland Terrace demonstrating the nature of large-scale folding interpreted to result from opposed senses of strike-slip motion on its bounding Helmsdale and Great Glen faults.

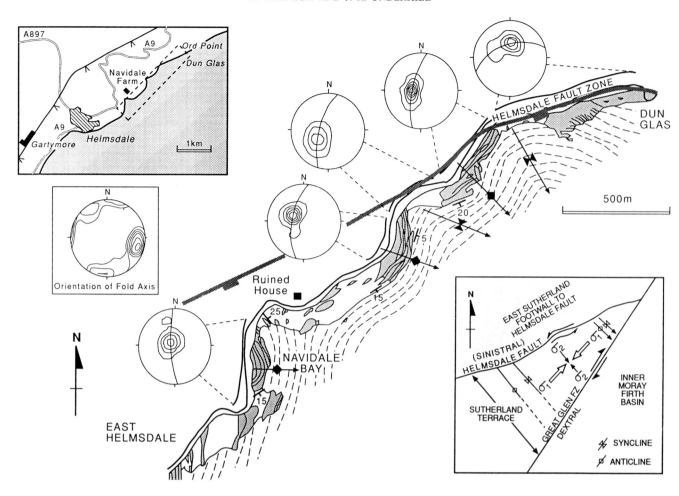

Fig. 12. Onshore map with stereonets showing the nature of individual, gently plunging folds formed at relatively high angles to the Helmsdale Fault. Inserts summarize all available stereonet data from folds along the whole length of the fault and show a cartoon depicting the stress regime interpreted to cause the structural styles in the Sutherland Terrace.

(4) Toarcian–Aalenian regional doming interrupted Early Jurassic gentle subsidence as a result of the rise of a warm and diffuse, transient plume head below the North Sea triple junction. Although its effects lie below seismic resolution, well-based stratigraphic studies show that it led to sub-lithospheric thinning in the Inner Moray Firth, which had major significance for later Jurassic extensional tectonism.

(5) Differential subsidence characterized Aalenian–Late Oxfordian times prior to the onset of significant Early Kimmeridgian–Portlandian extension, which largely followed the course of the earlier thinned crust and created numerous, syn-sedimentary half-graben analogous to those seen in the Viking Graben and Greenland.

(6) Subsequent Early Cretaceous sedimentation appears to have occurred in a gently subsiding, underfilled basin with progressive onlap onto its margin, which is interpreted to be the result of thermal subsidence which probably continued into Late Cretaceous times.

(7) The Inner Moray Firth area was characterized by significant Tertiary tectonism resulting from its intra-plate setting between the North East Atlantic rift and the Alpine fold-and-thrust belt which led to regional- and local-scale complex deformation.

(8) Although little evidence exists for strike-slip control on regional structural styles, limited dextral motions appear to characterize the Great Glen Fault during the (Permo-Carboniferous and more particularly Cenozoic) periods of contraction and inversion and caused the development of spectacular 'flower' geometries.

(9) Unusual structural geometries in the Sutherland Terrace area between the Great Glen Fault and the Helmsdale Fault appear to be the result of local transpression resulting from minor, opposed strike-slip motion on these faults during the Tertiary.

KT and JRU acknowledge the help of Shell, Esso, British Petroleum, Halliburton, Intera, GEOTRACK, GECO, Shelf Exploration, Arco, Horizon, Mobil and Unocal in providing data for the work. We thank Roger Scrutton, John Dixon, Richard Hillis, Sarah Prosser, Jon Turner and Al Fraser for useful discussion and Larry Wakefield, John Parker, Griff Cordey and Ceri Powell (Shell) for logistic support. Ceri Powell and Iain Bartholomew are thanked for providing information and constructive reviews. Mark Cattanach is acknowledged for drafting the diagrams. KT is sponsored by a Shell/Esso Studentship. JRU's relevant research was supported by NERC Grant GR9/692.

References

ANDREWS, I. J. AND BROWN, S. 1987. Stratigraphic evolution of the Jurassic, Moray Firth. *In*: BROOKS, J. AND GLENNIE, K. (eds) *Petroleum Geology of North-West Europe*. Graham & Trotman, London, 785–795.

——, LONG, D., RICHARDS, P. C., THOMSON, A. R., BROWN, S., CHESHER, J. A. AND McCORMAC, M. 1990. *The Geology of the Moray Firth*, British Geological Survey, United Kingdom Offshore Report, HMSO, London.

ASTIN, T. R. 1990. The Devonian lacustrine sediments of Orkney, Scotland; implications for climatic cyclicity, basin structure and maturation history. *Journal of the Geological Society, London*, **147**, 105–120.

BARR, D. 1985. 3-D Palinspastic reconstruction of normal faults in the Inner Moray Firth: implications for extensional basin development. *Earth and Planetary Science Letters*, **75**, 191–203.

BARTHOLOMEW, I. D., PETERS, J. M. AND POWELL, C. M. 1993. Regional structural evolution of the North Sea: oblique-slip and the reactivation of basement lineaments. *In*: PARKER, J. R. (ed.) *Petroleum Geology of Northwest Europe: Proceedings of the 4th Conference*. Geological Society, London, 1109–1122.

BEVAN, T. G. AND HANCOCK, P. L. 1986. A late Cenozoic regional mesofracture system in southern England and northern France. *Journal of the Geological Society, London*, **143**, 355–362.

BIRD, T. J., BELL, A., GIBBS, A. D. AND NICHOLSON, J. 1987. Aspects of strike-slip tectonics in the Inner Moray Firth basin, offshore Scotland. *Norsk Geologisk Tidsskrift*, **67**, 353–369.

COWARD, M. P., ENFIELD, M. A., FISCHER, M. W. 1989. Devonian basins of Northern Scotland: extension and inversion related to Late Caledonian–Variscan tectonics. *In*: COOPER, M. A. AND WILLIAMS, G. D. (eds) *Inversion Tectonics*, Geological Society, London, Special Publication, **44**, 275–308.

DIMITROPOULOS, K. AND DONATO, J. A. 1981. The Inner Moray Firth Central Ridge, a geophysical interpretation. *Scottish Journal of Geology*, **17**, 27–38.

DIXON, J. M. 1975. Finite strain and progressive deformation in models of diapiric structures. *Tectonophysics*, **28**, 89–124.

DONATO, J. A. AND TULLY, M. C. 1981. A regional interpretation of North Sea gravity data. *In*: ILLING, L. V. AND HOBSON, G. D. (eds) *Petroleum Geology of the Continental Shelf of North-West Europe*, Heyden, London, 65–75.

ENGLAND, R. W. 1988. The early Tertiary stress regime in NW Britain: evidence from the patterns of volcanic activity. *In*: MORTON, A. C. AND PARSONS, L. M. (eds). *Early Tertiary Volcanism and the Opening of the NE Atlantic*, Geological Society, London, Special Publication, **39**, 381–389.

FROSTICK, L., REID, I., JARVIS, J. AND EARDLEY, H. 1988. Triassic sediments of the Inner Moray Firth, Scotland: early rift deposits. *Journal of the Geological Society, London*, **145**, 235–248.

HALLAM, A. AND SELLWOOD, B. W. 1976. Middle Mesozoic sedimentation in relation to tectonics in the British area. *Journal of Geology*, **84**, 301–321.

HANCOCK, P. L. AND BEVAN, T. G. 1987. Brittle modes of foreland extension. *In*: COWARD, M. P., DEWEY, J. F. AND HANCOCK, P. L. (eds). *Continental Extensional Tectonics*, Geological Society, London, Special Publication, **28**, 127–137.

HARDING, T. P. 1990. Identification of wrench faults using subsurface structural data: criteria and pitfalls. *American Association of Petroleum Geologists Bulletin*, **74**, 1590–1609.

HILLIS, R. R. 1992. A two-layer lithospheric compressional model for the Tertiary uplift of the southern United Kingdom. *Geophysical Research Letters*, **19**, 573–576.

——, THOMSON, K. AND UNDERHILL, J. R. Quantification of Tertiary erosion in the Inner Moray Firth by sonic velocity data from the Chalk and the Kimmeridge Clay. *Marine and Petroleum Geology*, (in press).

HOLGATE, N. 1969. Palaeozoic and Tertiary transcurrent movement on the Great Glen Fault. *Scottish Journal of Geology*, **5**, 97–139.

KUSZNIR, N. J. AND KARNER, G. 1985. Dependence of the flexural rigidity of the continental lithosphere on rheology and temperature. *Nature*, **316**, 138–142.

MACDONALD, A. C. 1985. *Kimmeridgian and Volgian fault-margin sedimentation in the Northern North Sea area*. PhD thesis, University of Strathclyde.

McQUILLIN, R., DONATO, J. A. AND TULSTRUP, J. 1982. Development of basins in the Inner Moray Firth and the North Sea by crustal extension and dextral displacement of the Great Glen Fault. *Earth and Planetary Science Letters*, **60**, 127–139.

NAYLOR, H., TURNER, P., VAUGHAN, D. J. AND FALLICK, A. E. 1989. The Cherty Rock, Elgin: A petrographic and isotopic study of a Permo-Triassic calcrete. *Geological Journal*, **24**, 205–221.

NAYLOR, M. A., MANDL, G. AND SIJPESTEIJN, C. H. K. 1986. Fault geometries in basement-induced wrench faulting under different initial stress states. *Journal of Structural Geology*, **8**, 737–752.

PARKER, J. R. 1975. Lower Tertiary sand development in the Central North Sea. *In*: WOODLAND, A. W. (ed.) *Petroleum and the Conti-*

nental Shelf of North West Europe. Applied Science Publishers, London, 447–453.

PARTINGTON, M. A., MITCHENER, B. C., MILTON, N. J. AND FRASER, A. J. 1993. Genetic sequence stratigraphy for the North Sea Late Jurassic and Early Cretaceous: distribution and prediction of Kimmeridgian–Late Ryazanian reservoirs in the North Sea and adjacent areas. *In*: PARKER, J. R. (ed.) *Petroleum Geology of Northwest Europe: Proceedings of the 4th Conference*. Geological Society, London, 347–370.

RATTEY, R. P. AND HAYWARD, A. B. 1993. Sequence stratigraphy of a failed rift system: the Middle Jurassic to Early Cretaceous basin evolution of the Central and Northern North Sea. *In*: PARKER, J. R. (ed.) *Petroleum Geology of Northwest Europe: Proceedings of the 4th Conference*. Geological Society, London, 215–249.

RAWSON, P. F. AND RILEY, L. 1982. Latest Jurassic–Early Cretaceous Events and the 'Late Cimmerian Unconformity' in North Sea Area. *American Association of Petroleum Geologists Bulletin*, **66**, 2628–2648.

ROBERTS, A. M., BADLEY, M. E., PRICE, J. D. AND HUCK, I. W. 1990. The structural evolution of a transtensional basin: Inner Moray Firth, NE Scotland. *Journal of the Geological Society, London*, **147**, 87–103.

ROCHOW, K. A. 1981. Seismic stratigraphy of the North Sea 'Paleocene' deposits. *In*: ILLING, L. V. AND HOBSON, G. D. (eds) *Petroleum Geology of the Continental Shelf of NW Europe*. Heyden, London, 255–266.

ROGERS, D. A. 1987. *Devonian correlations, environments and tectonics across the Great Glen Fault*. PhD thesis, University of Cambridge.

——, MARSHALL, J. E. A. AND ASTIN, T. R. (1989). Devonian and later movements on the Great Glen fault system, Scotland. *Journal of the Geological Society, London*, **146**, 369–372.

SERANNE, M. 1992. Devonian extensional tectonics versus Carboniferous inversion in the northern Orcadian basin. *Journal of the Geological Society, London*, **149**, 27–37.

SMITH, P. J. AND BOTT, M. H. P. 1975. Structure of the Crust beneath the Caledonian Foreland and Caledonian Belt of the North Scottish Shelf Region. *Geophysics Journal of the Royal Astronomical Society*, **40**, 187–205.

SPEIGHT, J. M. AND MITCHELL, J. G. 1979. The Permo-Carboniferous dyke swarm of northern Argyll and its bearing on dextral displacement on the Great Glen Fault. *Journal of the Geological Society, London*, **1365**, 3–12.

STEPHEN, K. J., UNDERHILL, J. R., PARTINGTON, M. A. AND HEDLEY, R. J. 1993. The genetic sequence stratigraphy of the Hettangian to Oxfordian succession, Inner Moray Firth. *In*: PARKER, J. R. (ed.) *Petroleum Geology of Northwest Europe: Proceedings of the 4th Conference*. Geological Society, London, 485–505.

SURLYK, F. 1978. Submarine fan sedimentation along fault scarps on tilted fault blocks (Jurassic–Cretaceous boundary, East Greenland). *Gronlands Geologiske undersogelse Bulletin*, **28**, 1–103.

TURNER, C. C., COHEN, J. M., CONNELL, E. R. AND COOPER, D. M. 1987. A depositional model for the South Brae oilfield. *In*: BROOKS, J. AND GLENNIE, K. (eds) *Petroleum Geology of North-West Europe*. Graham & Trotman, London, 853–864.

UNDERHILL, J. R. 1991a. Implications of Mesozoic-Recent basin development in the western Inner Moray Firth, UK. *Marine and Petroleum Geology*, **8**, 359–369.

—— 1991b. Controls on Late Jurassic seismic sequences, Inner Moray Firth, UK North Sea: a critical test of a key segment of Exxon's original global cycle chart. *Basin Research*, **3**, 79–98.

—— AND BRODIE, J. A. A Re-evaluation of the Structural Geology of the Easter Ross Peninsula. *Journal of the Geological Society, London* (in press).

—— AND PARTINGTON, M. A. 1993. Jurassic thermal doming and deflation in the North Sea: implications of the sequence stratigraphic evidence. *In*: PARKER, J. R. (ed.) *Petroleum Geology of Northwest Europe: Proceedings of the 4th Conference*. Geological Society, London, 337–345.

—— AND —— Use of maximum flooding surfaces in determining a regional control on the Intra-Aalenian ('mid-Cimmerian') Sequence Boundary: Implications for North Sea basin development and Exxon's Sea Level Chart. *In*: POSAMENTIER, H. W. AND WIEMER, P. J. (eds) *Recent Advances in Sequence Stratigraphy*.

American Association of Petroleum Geologists Memoir (in press).

VAIL, P. R., HARDENBOL, J. AND TODD, R. G. 1984. Jurassic unconformities, chronostratigraphy, and sea-level changes from seismic stratigraphy and biostratigraphy. In: SCHLEE, J. S. (ed.) International Unconformities and Hydrocarbon Accumulation. American Association of Petroleum Geologists Memoir, 36, 129–144.

—— AND TODD, R. G. 1981. Northern North Sea Jurassic unconformities, chronostratigraphy and sea-level changes from seismic stratigraphy. In: ILLING, L. V. AND HOBSON, G. D. (eds) Petroleum Geology of the Continental Shelf of North-west Europe. Heyden, London, 216–235.

WHITE, R. S. 1988. A hot-spot model for early Tertiary volcanism in the N Atlantic. In: MORTON, A. C. AND PARSONS, L. M. (eds) Early Tertiary Volcanism and the Opening of the N.E. Atlantic. Geological Society, London, Special Publication, 39, 3–13.

—— 1989. Initiation of the Iceland Plume and Opening of the North Atlantic. In: TANKARD, A. J. AND BALKWILL, H. R. (eds) Extensional tectonics and stratigraphy of the North Atlantic margins. American Association of Petroleum Geologists Memoir, 46, 149–154.

WITHJACK, M. O. AND SCHIENER, C. 1982. Fault Patterns Associated with Domes—An Experimental and Analytical Study. American Association of Petroleum Geologists Bulletin, 66, 302–316.

ZIEGLER, P. A. 1982. Geological Atlas of Western and Central Europe. Elsevier, Amsterdam.

—— 1987. Late Cretaceous and Cenozoic intra-plate compressional deformations in the Alpine foreland—a geodynamic model. Tectonophysics, 137, 389–420.

—— 1990a. Tectonic and palaeogeographic development of the North Sea rift system. In: BLUNDELL, D. J. AND GIBBS, A. D. (eds) Tectonic Evolution of the North Sea Rifts. Clarendon Press, Oxford, 1–36.

—— 1990b. Geological Atlas of Western and Central Europe, Second edition. Shell Internationale Petroleum Maapascappij B.V., Netherlands.

—— AND VAN HOORN, B. 1989. Evolution of North Sea Rift system. In: TANKARD, A. J. AND BALKWILL, H. R. (eds) Extensional Tectonics and Stratigraphy of North Atlantic Margins. American Association of Petroleum Geologists Memoir, 46, 471–500.

Discussion

Question (A. Roberts, Badley Ashton, Winceby, Lincs)

In two previous papers I, together with my co-authors, suggested that the opening of the Moray Firth Basin was controlled by strike-slip motion on the Great Glen Fault. In the light of Dr Underhill's recent work I recognize that this suggestion was probably wrong and that extensional movement on the Helmsdale Fault was the dominant control on the architecture of the Inner Moray Firth. Rather than the Helmsdale Fault being a purely dip-slip structure, however, would Dr Underhill be able to reconcile his observations with an overall c. N–S extension direction in the Inner and Outer Moray Firth basins? Such an extension direction would be orthogonal to the large, uplifted Halibut Horst in the centre of the basin and would maintain an extensional (but not necessarily dip-slip) framework throughout the arcuate fault pattern of the Moray Firth. A N–S extension direction would, in addition, maintain kinematic compatibility with dominantly orthogonal opening of the Viking and Central Graben.

Answer (J. R. Underhill)

I would like to thank Alan for his statement and question Although Ken and I cannot rule out a minor component of oblique-slip on the Helmsdale Fault, our field observations do actually imply that the feature was of purely dip-slip origin in the Jurassic. Hence, we would still conclude that the folds which characterize regions adjacent to the fault and complicate interpretations are the result of syn-sedimentary compactional drape and subsequent minor sinistral transpression probably resulting from Tertiary reactivation and inversion.

As a result of our structural interpretations, our best estimates of the overall extensional direction during the Late Jurassic remain NW–SE or NNW–SSE in the local area. However, on the database currently available and analysed for our studies, we cannot rule out the possibility of different local extensional directions affecting other areas such as the Outer Moray Firth and beyond. Given the contrast in directions indicated in this question and by Bartholomew et al. 1993, resolution of the more regional stress field seems likely to stir considerable debate for some time to come.

Question (D. O'Driscoll, Texaco Ltd, Knightsbridge, London)

Can you comment on previously published models which refer to anomalously thinned continental crust under the Outer Moray Firth, Central Graben and South Viking Graben as opposed to thick, competent continental crust under the Inner Moray Firth, and on potential implications for dip-slip/extensional tectonics in the Inner Moray Firth?

Answer (J. R. Underhill and K. Thomson)

We welcome the opportunity to expand upon this important point. It seems that the idea that the Inner Moray Firth is underlain by anomalously thick continental crust is entrenched in the literature and has clouded many workers' thinking about basin development in the area. We agree that the most plausible explanation for the gravity data (originally given in Donato and Tulley 1981) is that the IMF consists of a thick Paleozoic and Mesozoic sedimentary succession, which is underlain by contrasting basement lithologies and local granitic intrusions (Dimitropoulos and Donato 1981) and hence, has highly variable basement densities. Although further interpretation of the gravity data has been taken to imply that no crustal thinning has occurred across the whole area and rather that basin development has occurred under the influence of thick-skinned, strike-slip tectonics, it is worth stressing that the evidence for unthinned crust must be restricted to western parts of the basin. Moreover, valid alternative interpretations of the raw data can be made which are compatible with the seismic data and field evidence for an extensional origin. Indeed, Smith and Bott (1975) have shown (in a rarely-quoted paper) that the results of a crustal refraction study can be used to demonstrate that although 'the [Inner Moray Firth] region lies within the Caledonian belt, it appears that substantial crustal thinning has subsequently occurred, probably contemporaneously with the formation of the sedimentary basin'. Furthermore, they are able to use the data to suggest that the thinning has reduced the crustal thickness to around 23 km beneath the basin (a similar figure to that beneath the Witch Ground Graben; Andrews et al. 1990). Thus, we believe that there is sufficient evidence to show that the Inner Moray Firth demonstrates crustal thinning of similar aspect to that described for the South Viking Graben and Central Graben and hence, was probably also controlled by extensional tectonics. Consequently, although there is a clear difference in detailed gravity signatures shown beneath western parts of the Inner Moray Firth (IMF) and these areas, perhaps implying that the actual amounts of stretching vary across the basin, independent evidence demonstrates that the IMF as a whole experienced crustal thinning. Hence, we think that many previously published interpretations, which have been biased towards the regional gravity data, without taking into account either its non-uniqueness, any local detailed variations or the independent evidence, are likely to be misleading and should be treated with caution.

The structural evolution of the eastern end of the Halibut Horst, Block 15/21, Outer Moray Firth, UK North Sea

M. J. HIBBERT and D. S. MACKERTICH

Amerada Hess Limited, 33 Grosvenor Place, London SW1X 7HY, UK

Abstract: The eastern termination of the Halibut Horst occurs in Block 15/21 of the Outer Moray Firth and shows evidence of a complex structural history. The acquisition of high-resolution 3D seismic datasets across the area of study has permitted mapping of key horizons, better imaging of complex faulting within and around the horst, and enhanced understanding of the structural and stratigraphic evolution.

A marked change in active fault trends from NE–SW during the early/mid-Jurassic to WNW–ESE in the late Jurassic/early Cretaceous is evident. Late Jurassic tectonism resulted in significant erosion of the Halibut Horst Spur and syn-tectonic deposition in basinal areas to the north and south. Continuing reactivation and uplift of the horst during the Cretaceous and Tertiary resulted in a complex inter-relationship between tectonics and sedimentation.

The Halibut Horst is a major sub-surface positive structural feature. It trends broadly east–west and has greatly influenced Mesozoic sedimentation in the Outer Moray Firth. The eastern termination of the horst, the Halibut Horst Spur, occurs in Block 15/21 (Figs 1 and 2).

This paper is intended to build on previous published work

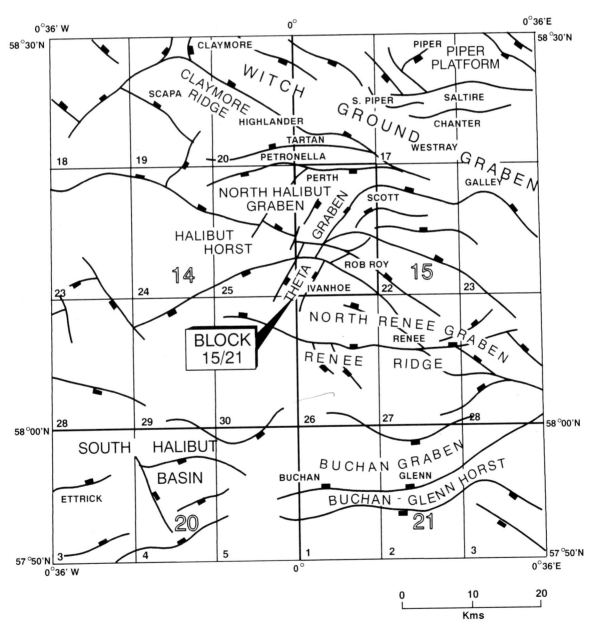

Fig. 1. Outer Moray Firth: regional structural elements.

From *Petroleum Geology of Northwest Europe: Proceedings of the 4th Conference* (edited by J. R. Parker).

which has helped to delineate the complexity of the Outer Moray Firth. It presents some of the results of detailed stratigraphical and structural evaluation undertaken during the exploration and appraisal of Block 15/21 which contains four fields: Ivanhoe, Rob Roy, Hamish and Scott.

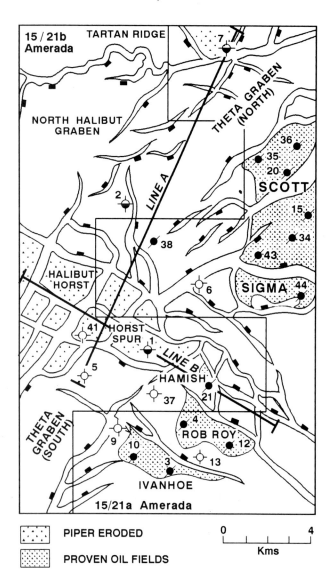

Fig. 2. Structural elements: Block 15/21.

The structural and palaeogeographical evolution of this block has been delineated utilizing an extensive database, comprising two high-resolution 3D seismic surveys, several 2D seismic datasets and over 30 exploration and appraisal wells. A stratigraphic column detailing the nomenclature used in this paper is illustrated in Fig. 3.

Sediment deposition in an extensional setting dominated much of the Devonian to mid-Jurassic history. Periods of uplift and erosion have also occurred in this area, resulting in the absence of Middle–Upper Carboniferous and Lower Jurassic sediments (Fig. 3). Major tectonism during the late Jurassic marked the appearance of a dominant WNW–ESE structural style termed the 'Witch Ground' fault trend (Boldy and Brealey 1990). This trend is clearly visible throughout much of the Outer Moray Firth/Witch Ground Graben (O'Driscoll *et al.* 1990) and has been partly instrumental in forming elongate hydrocarbon-generating basins flanked by rotating tilted fault blocks which form many of the hydrocarbon traps in the

region. This tectonic event is also thought to have initiated the development of the Halibut Horst in its present form.

Late Cretaceous to early Tertiary reactivation of the horst is suggested by the postulated occurrence of sinistral strike-slip fault movement along a NW–SE trend and the initiation of submarine sediment bypass systems during Paleocene times.

Geophysical database and methods

Although more than 30 exploration and appraisal wells have already been drilled within Block 15/21, significant advances in the understanding of the structural evolution of the block have been achieved through the acquisition of 3D seismic surveys. The first survey was acquired in 1984 and covered the Ivanhoe and Rob Roy fields, extending southwards from the Halibut Horst Spur. The remaining part of the block was covered with 3D seismic during 1989–90. A combination of these two datasets has been used as the basis for the seismic mapping discussed in this paper.

Two main attributes have been extracted from the seismic datasets to illustrate the structural evolution and sedimentary history of the block. Firstly, isopach maps have been created for the major geological units from the Jurassic through to the Paleocene. These maps were constructed from the mapped time horizons interpreted from the 3D seismic data. Depth conversion, using simple linear time–depth relationships derived from well data, was then completed before computing the relevant isopach maps. The resultant maps for the Upper Jurassic, Cretaceous and Paleocene intervals are illustrated on Fig. 4. These maps illustrate the main tectonic elements and sedimentary depocentres present during these periods.

The second attribute which has been extracted is the dip of a time horizon. Each seismic horizon was autotracked and the results analysed for the local dip of the horizon (Dalley *et al.* 1989). Areas with no faulting have low dip values, whereas faulted areas give rise to high dip values as the seismic pick passes from the hangingwall of the fault to the footwall. The resultant dip maps on Base Cretaceous and Top Chalk are illustrated on Figs 4d and e. These attribute maps show the dominant fault/lineament trends that transect each seismic horizon. By comparing these maps in sequential time order, and integrating with the isopach maps and geological information from well data, it is possible to interpret the dominant fault trends and their depositional influence through subsequent geological periods.

Structural elements and stratigraphy

The oldest dated sediments encountered in Block 15/21 belong to the Lower Carboniferous Forth Formation (Fig. 3). Sediments of this age lie at depths between 7000–8000 ft on the horst and are estimated to lie at depths in excess of 18 000 ft in the North Halibut Graben. Well data from Triassic and Permian intervals are sparse, but sufficient to enable basic reconstructions of palaeogeography. Jurassic strata form the majority of primary hydrocarbon reservoir targets in this area of the Outer Moray Firth and a significant number of publications and well data are therefore available. Mesozoic sediments are thin (if not absent) over the Halibut Horst and thicken into grabenal areas both to the north and south. Tertiary sediments initially onlapped the horst but from the mid-Paleocene to Recent are recorded across the whole of Block 15/21.

This paper concentrates primarily on the Jurassic and post-Jurassic evolution of the eastern end of the Halibut Horst. Pre-Jurassic structure and sedimentation is discussed as the information is deemed an essential part of the interpretation and understanding of subsequent geological units.

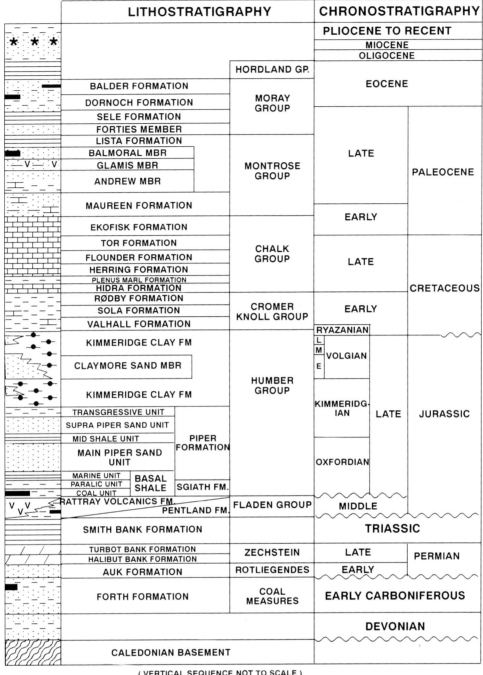

Fig. 3. Stratigraphic column: Block 15/21.

Pre-Jurassic

At the eastern end of the Halibut Horst, Wells 15/21-1 and 15/21b-21 reached total depths in the Lower Carboniferous Forth Formation, which comprised an interbedded sequence of sandstones, shales and coals (Fig. 5). The 15/21-1 well was drilled on the Halibut Horst Spur (Fig. 2), suggesting that during the early Carboniferous this region of the horst was not a positive structural feature, as there is no significant change in sediment type between the 15/21-1 well and the 15/21b-21 well drilled within the graben. Sediment deposition is likely to have occurred in a range of environments from continental to pro-deltaic in an overall extensional tectonic regime (Leeder and Boldy 1990).

Sediments of early Permian age record a period of continental, predominantly clastic, sedimentation. These clastic sediments appear to be only locally developed in the Outer Moray

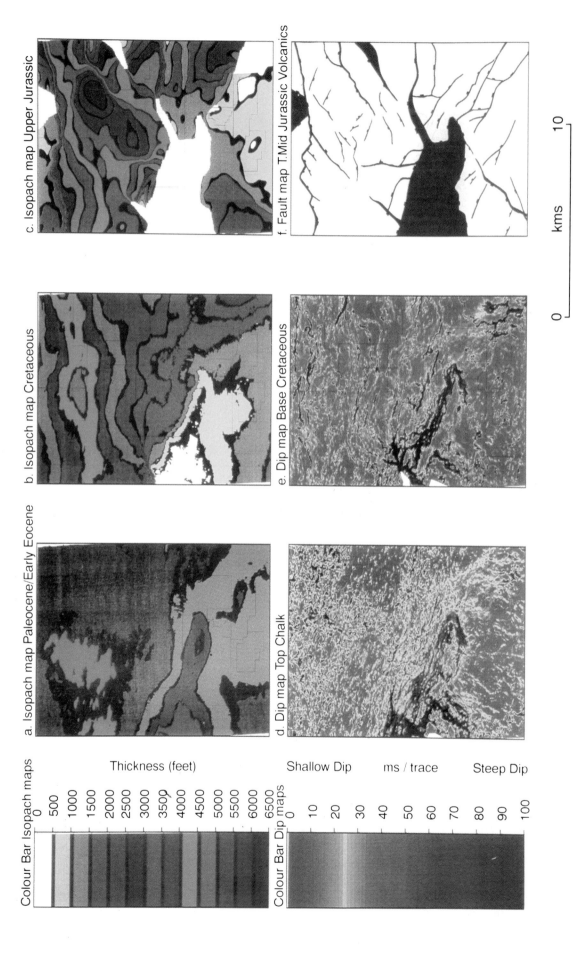

Fig. 4. Seismic isopach and structure dip maps.

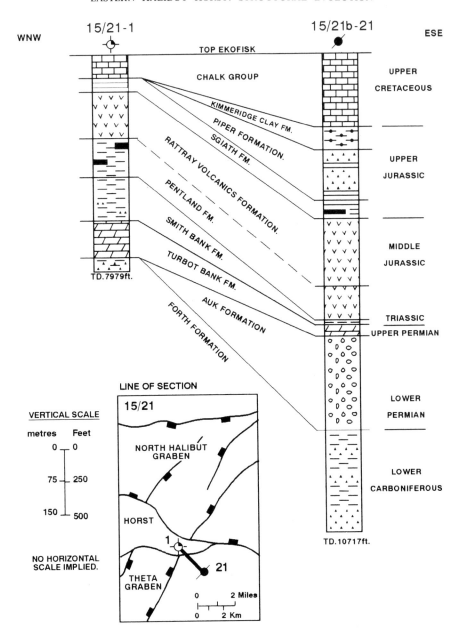

Fig. 5. Well correlation: Halibut Horst Spur.

Firth. Well 15/21b-21 contains over 750 ft of dominantly clastic sediments of the Rotliegendes, Auk Formation (Fig. 5). To the northwest, across the fault which is mapped as bounding the southern part of the Halibut Horst Spur (Fig. 2), Well 15/21-1 did not encounter any Lower Permian sediments.

Triassic strata encountered in this region of the Outer Moray Firth comprise red siltstones, shales and intercalated anhydrites which are assigned to the Smith Bank Formation. The arenaceous sequences of the Skagerrak Formation have not been encountered in this area. Well 15/21-1 penetrated 270 ft of Smith Bank Formation overlying 235 ft of Zechstein dolomites belonging to the Turbot/Halibut Bank formations (Figs 2 and 3). Well 15/21b-21, to the south of the spur and outside of a NE–SW-trending grabenal feature (Fig. 2), penetrated a much thinner sequence of both Triassic and Zechstein strata (Fig. 5).

It is suggested that this graben (termed the Theta Graben by Boldy and Brealey 1990) may have influenced sedimentation during both Triassic and Permian times.

Early to mid-Jurassic

Uplift and erosion which accompanied the early Cimmerian phase of tectonism, resulted in non-deposition (or deposition with subsequent erosion) of Lower Jurassic deposits which are absent in Block 15/21 and across much of the Outer Moray Firth.

The volcanic and interbedded volcaniclastic sequences of the Fladen Group are the oldest preserved Jurassic rocks in this region. Dating has indicated ages no older than Bajocian, with most indicating a Bathonian age (Howitt *et al.* 1975). These volcanics have been intimately associated with extensional rifting, linked with the collapse of the Central North Sea Dome.

The Theta Graben is a NE–SW-trending graben ('Viking' trend of Boldy and Brealey 1990), 2–3 km in width and 10–15 km in length, confined almost wholly to Block 15/21. This graben has a regional significance in as much as it demonstrates a strong tectonic control on Middle and Upper Jurassic sedimentation as well as overprinting on late Cimmerian tecto-

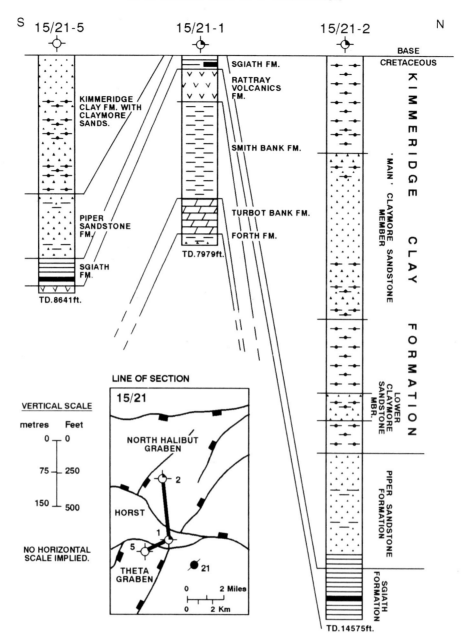

Fig. 6. Well correlation: Halibut Horst/Theta Graben.

nism. Both seismic and well data suggest that the Theta Graben accumulated and localized sediments from at least the late Triassic to the late Jurassic. The structural lineaments which define the margins of this graben are present both to the north and south of the Halibut Horst, as shown in the isopach maps in Fig. 4.

Late Jurassic

After the deposition of the Middle Jurassic Fladen Group sediments, the paralic sequences of the Sgiath Formation were deposited (Fig. 3). In the Rob Roy Field area, it has been noted that deposition of the coal-bearing intervals of the Sgiath Formation was restricted to the structurally lower areas, with the younger Paralic and Marine Units onlapping and overlying adjacent highs (Parker 1991). The preservation of a thick Sgiath Coaly Unit in this region may, therefore, be used as an indication of a structurally low, sediment-receiving area at the time of deposition.

The transition from the Sgiath Formation into the Piper Formation is a gradational one, the marine influence progressively becoming more dominant and culminating in the fully shoreface deposits of the Piper Sandstone Formation. These sands form the principal reservoir rock for the oil fields present in Block 15/21 and indeed for the main Witch Ground Graben fields of Piper, Tartan, Petronella and Highlander. Correlation of the Piper sands is possible on a semi-regional scale throughout the Witch Ground Graben in UK Quadrants 14, 15, 20, 21 and 23 (Harker *et al.* 1987) suggesting the presence of a significant wave-dominated delta system during mid-Oxfordian–early Kimmeridgian times.

Well 15/21-1 (Figs 2 and 5) was drilled on the Halibut Horst Spur and encountered a thin Cretaceous chalk interval overlying the lowermost coal-dominated sediments of the Sgiath Formation. No Jurassic sediments younger than the Sgiath have been found by wells drilled upon the Halibut Horst. The Upper Jurassic isopach map (Fig. 4c), together with the fault trends inferred to have been active during the Upper Jurassic (Fig. 4f), illustrate that both the Halibut Horst and Halibut Horst Spur were sediment-receiving areas (at least until mid-Oxfordian times) as opposed to sediment-sourcing areas. The absence of Jurassic sediments younger than the Sgiath is, therefore, deemed to reflect lack of preservation rather than non-deposition.

SEISMIC LINE 'A'

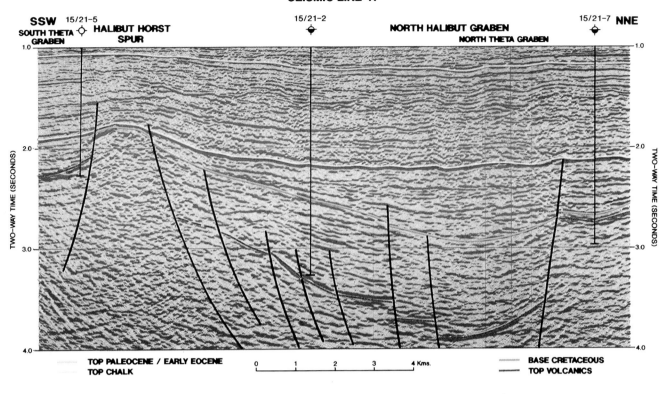

Fig. 7. Seismic line 'A' from the South Theta Graben across the Halibut Horst Spur and across the North Halibut Graben. See Fig. 2 for line location.

To the south of the Halibut Horst, Well 15/21-5 penetrated over 760 ft of combined Sgiath and Piper formations (Fig. 6). Wells in the Ivanhoe/Rob Roy fields contain (when not eroded in crestal areas) an average of over 400 ft of combined Piper and Sgiath formations. To the north of the Halibut Horst, Well 15/21-2 penetrated over 1200 ft of Sgiath and Piper formations and significant thicknesses are encountered in all unfaulted Scott wells (Fig. 6). The existence of this major deltaic succession surrounding the Halibut Horst and Halibut Horst Spur, together with significant sand thicknesses proven by well data, would suggest that Piper Formation sediments were deposited in the region which is now the Halibut Horst Spur.

The Upper Jurassic isopach map (Fig. 4c), which includes the Sgiath, Piper and Kimmeridge Clay formations, is dominated by the NE–SW-trending Theta Graben to the west of the Scott Field. It contains in excess of 6000 ft of Upper Jurassic sediments. The faults controlling sedimentation within this graben were undoubtedly the NE–SW 'Viking' trend faults. It appears most likely that this trend of thick Upper Jurassic sedimentation continued across the Halibut Horst Spur and was subsequently removed during uplift and erosion.

The fault map on the Top Middle Jurassic Volcanics seismic horizon (Fig. 4f) indicates the occurrence of two main fault trends. From well/isopach data it would appear that the NE–SW-trending 'Viking' faults were dominant and influencing sedimentation throughout much of the Upper Jurassic.

The history of fault movement along the Witch Ground trend is likely to have been episodic. These periodic displacements are thought to have been important in controlling the provenance of both the early Volgian Claymore Sands, and the early Cretaceous clastics of the Valhall Formation (such as the Scapa Sandstone Member, McGann et al. 1991).

The Claymore Sandstone Member has been proved in basinal areas to the north and south of the Halibut Horst/Halibut Horst Spur. To the south, Well 15/21-5 encountered over

400 ft of Claymore Sandstone, while to the north, Well 15/21-2 recorded in excess of 1200 ft (Figs 6 and 7). The WNW–ESE-trending Witch Ground trending faults are believed to have been active from 'early' Kimmeridgian (*Mutabilis–Eudoxus* Zone: nomenclature after Boldy and Brealey 1990) when syn-tectonic thickening of the 'latest' Piper Formation units into NW–SE-trending faults is first detected. This trend continues to influence sedimentation and basin development through until at least the end of the early Cretaceous.

The Halibut Horst is interpreted as having undergone erosion during Kimmeridge Clay times (Boldy and Brealey 1990). Much of the sediment within the Claymore Sandstone sequence is thought to have been derived from re-deposition of the Piper Sands. This hypothesis is supported by the structural evolution of the area, mineralogical maturity and heavy mineral compatibility. In areas close to the Tartan Ridge and other intra-basinal fault blocks, Carboniferous and Permian strata may have contributed to the early Volgian sediment flux. In the region of the Halibut Horst/Halibut Horst Spur, source areas for the Claymore Sandstone may lie in areas where Witch Ground faults are transected by older Viking trends. It is suggested that Claymore turbidite influx was localized within the deeper parts of the Kimmeridgian basin. In Block 15/21 this basinal area lies broadly on a trend with the Theta Graben where both Piper and Sgiath Formation sediments are at their thickest (e.g. Wells 15/21-2 and 15/21-5, Fig. 8).

The inference, therefore, is that NE–SW 'Viking' trend faults continued to control sediment accumulation after the development of WNW–ESE 'Witch Ground' trending lineaments.

Post-Jurassic

Cretaceous

Deposition of Kimmeridge Clay facies continued into the

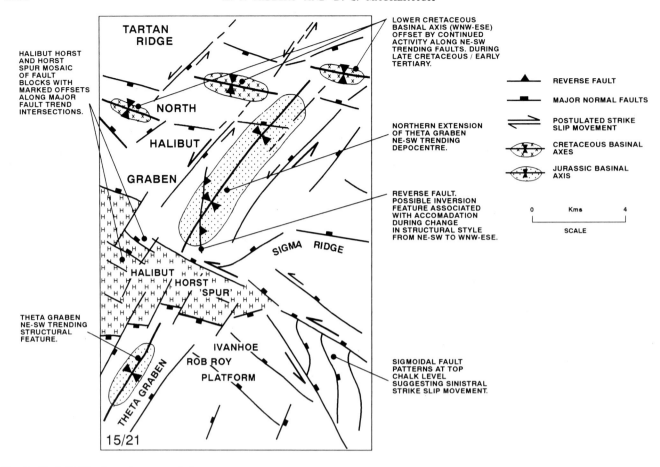

Fig. 8. Block 15/21 schematic structural elements.

Cretaceous (early Ryazanian). The boundary between the Kimmeridge Clay Formation and the Cromer Knoll Group is an unconformity on seismic, especially over the crests of Jurassic tilted fault blocks.

An isopach map for the whole of the Cretaceous interval across Block 15/21 reveals a major E–W-oriented depocentre which is sited to the north of the Halibut Horst (Fig. 4b). This thickening is evident both from well data and on seismic, where reflectors may be seen onlapping the Halibut Horst Spur. The thickening is greatest across the North Halibut Basin, where the maximum thickness is believed to be in excess of 4500 ft. Integration of 3D seismic with regional well isopach data reveals that much of this basin-fill occurred during Lower Cretaceous times, with a relatively uniform overlying thickness of Chalk Group sediments which may be variably reworked on the horst itself.

The location and orientation of the Cretaceous depocentre is noticeably different to that seen for the whole Upper Jurassic (Fig. 4c). Furthermore, there is a marked offset of sub-basinal axes (Figs 4b, e) caused by NE–SW 'Viking' lineaments, which were significant through the Upper Jurassic. This offsetting of 'Witch Ground' trending faults indicates that 'Viking' trend faults were still active (or reactivated) during the Lower Cretaceous (Fig. 8).

The dip map produced on the Base Cretaceous and Top Chalk seismic horizons reveals the structural configuration of the Halibut Horst Spur in some detail (Figs 4d, e). The horst is shown to have a distinctly blocky nature with the main 'Witch Ground Graben' trend faults transected at right angles by the older 'Viking' trend faults. This interaction produces a series of offsetting rectangular fault blocks which contrasts to the much simpler fault pattern previously derived from 2D seismic data, where faults are smoothed out due to fault aliasing.

During the Lower Cretaceous, 'Witch Ground' trending

faults exerted the dominant control on the isopach map (Fig. 4b). This implies that these WNW–ESE lineaments accommodated the major component of basinal extension within the North Halibut Graben, with the 'Viking' trend having a more subtle influence on basin morphology by controlling the location of Lower Cretaceous sub-basins within the North Halibut Graben (Fig. 4e). These sub-basins are significantly smaller than the Upper Jurassic depocentres and it is postulated that strike-slip movement may have been partly instrumental in fragmenting what would otherwise be an elongate Lower Cretaceous fairway (Fig. 8).

The dip map produced at Top Chalk level (Fig. 4d) again shows the structural anisotropy imposed upon the stable massif of the horst block by the two reactivated fault trends. There is, however, a distinct difference in fault trend between the areas north and south of the horst. To the south of the Halibut Horst/Halibut Horst Spur, the Top Chalk seismic reflector displays a relatively uniform dip with a marked NE–SW 'Viking' trend lineament. To the north of the horst the dominant lineament is the 'Witch Ground Graben' trend with a small component of the 'Viking' trend still visible. This minor component is believed to reflect the structural pattern produced by simple thermal sag of the North Halibut Graben between two stable blocks, the Halibut Horst and the Tartan Ridge. The Halibut Horst appears to have acted as a barrier separating zones of varying stress, resulting in local fault patterns which differ from the regional trend.

A significant feature of the Cretaceous seismic event dip maps (Figs 4d, e) is that both exhibit zones of increased dip which display a sigmoidal pattern. The occurrence of these sigmoidal features is limited to an area immediately to the north-northeast of the Rob Roy Field and south of the Scott Field. This zone lies on a trend with the eastern end of the Halibut Horst Spur and also appears to be acting as some

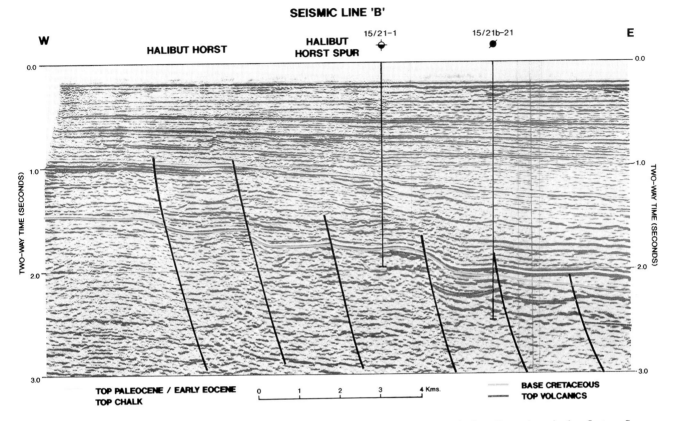

Fig. 9. Seismic line 'B' from the Halibut Horst to the Hamish Field. Note onlap of Maureen and Andrew Formation seismic reflectors. See Fig. 2 for line location.

form of stress boundary. The pattern of these sigmoidal faults suggests that there may have been a degree of sinistral strike-slip accommodation along this zone (Fig. 8).

Tertiary

An isopach map of the Paleocene and early Eocene (Montrose Group and Moray Group) is displayed in Fig. 4a. Paleocene sediments are some 2000–2500 ft thick upon the Halibut Horst, thickening to 2500–3000 ft on the Halibut Horst Spur. In the North Halibut Graben, thicknesses in excess of 4500 ft have been encountered. This thickness contrast may, in part, be as a result of sediment compaction over the Halibut Horst, but is thought to be predominantly due to differential sedimentation on and around the horst.

Isopachs of the Maureen and Andrew formations across Block 15/21, coupled with seismic onlap of sequence bounding reflectors (Fig. 9), suggest that the horst remained a positive submarine feature during advance of the early Tertiary submarine fan systems. Montrose Group sediments helped to infill much of the remaining topographic relief present at the end of the late Cretaceous, but sufficient relief existed to cause the thick Forties submarine channels to be deflected around the Halibut Horst and Halibut Horst Spur.

The submarine fan feeder channels of the Forties Sand Member appear mounded on seismic (Fig. 9) due to differential compaction of Sele Formation claystones over stacked turbidite channel sand sequences. These mounds have been mapped across the area and coincide closely to lines of structural weakness corresponding to the underlying structural boundaries of the Halibut Horst.

Submarine feeder channel systems are present to the north and south of the Halibut Horst, while a complex channel network is present at its eastern termination. The deflection of turbidite channel systems around the horst during Forties Member times is postulated as having been linked with minor

reactivation of the Halibut Horst Spur which is thought to have taken place during mid-late Paleocene times.

It would appear, therefore, that the eastern end of the Halibut Horst was a positive submarine feature, influencing sedimentation until the early Eocene. From the early Eocene to late Oligocene, sediment deposition took place in a deltaic to shallow marine setting with the horst having little effect on the rate or type of sediment deposited.

Conclusions

1. Delineating the occurrence and interaction of both the NE–SW ('Viking' trend) and the WNW–ESE ('Witch Ground' trend) lineaments is fundamental to the interpretation and understanding of the structural evolution of the eastern end of the Halibut Horst.

2. The Halibut Horst is a long-lived structural feature which locally influences sedimentation from the late Jurassic until the early Eocene.

3. The internal structure of the Halibut Horst and Halibut Horst Spur in Block 15/21 is seen to comprise a complex mosaic of fault blocks. The faults which define the structural boundaries of the horst comprise lineaments with trends which are an expression of the interaction of both 'Witch Ground' and 'Viking' trend systems. Embayments are present in the horst which may be a result of strike-slip movement along older 'Viking' lineaments.

4. The Halibut Horst Spur is thought to have been a sediment-receiving area at least until the mid-Oxfordian and probably through until the early Kimmeridgian. Sgiath and Piper sediments were, correspondingly, deposited on the Horst Spur and have subsequently been eroded.

5. 'Witch Ground' (WNW–ESE) lineaments were initiated during the early Kimmeridgian. Erosion of Sgiath and Piper sediments occurred in response to uplift of the Horst Spur with sediments being partly re-sedimented as early Volgian

Claymore Sandstones concurrent with deposition of Kimmeridge Clay shales.

6. Structural style and basin orientation, in Block 15/21, changed from a NE–SW to a WNW–ESE dominance in the period between the early Kimmeridgian and the Ryazanian. This change in structural orientation dominance is likely to have caused fault block rotation which may have resulted in the strike-slip accommodation and local uplift/inversion which is postulated in the North Halibut Graben.

7. Strike-slip reactivation of 'Viking' trend lineaments influenced basin morphology during the early Cretaceous. This reactivation offset the Cretaceous basin axes and resulted in the occurrence of irregular offsets along the faults which flank the Halibut Horst, the Halibut Horst Spur and the Tartan Ridge.

8. Phases of reactivation are evident at the end of the Cretaceous and during the Paleocene when sediment flow regimes are thought to have been deflected around the horst spur.

9. Detailed mapping from 3D seismic data has enabled the production of a suite of depth/isopach/structure maps which, when integrated with multidisciplinary geological studies, enables a comprehensive 'evolutionary picture' to be exposed. Such techniques are helping to delineate the complexity of this region and are assisting in the search for additional reserves in an area which already contains over a billion barrels of proven oil.

The authors would like to thank the management of Amerada Hess Limited and their co-venturers in Block 15/21: Deminex (UK) Oil and Gas Limited, Kerr-McGee Oil (UK) plc and Pict Petroleum plc, for permission to publish this paper. Thanks are also extended to R. Warren and S. A. R. Boldy for constructive criticism of the manuscript. The opinions and interpretations expressed in this paper are acknowledged to be solely those of the authors.

References

BOLDY, S. A. R. AND BREALEY, S. 1990. Timing, nature and sedimentary result of Jurassic tectonism in the Outer Moray Firth. *In*: HARDMAN, R. F. P. AND BROOKS, J. (eds) *Tectonic Events Responsible for Britain's Oil and Gas Reserves*. Geological Society, London, Special Publication, **55**, 259–279.

BROWN, S. 1986. Jurassic. *In*: GLENNIE, K. W. (ed.) *Introduction to the Petroleum Geology of the North Sea*. Blackwell, 133–159.

DALLEY, R. M., GEVERS, E. C. A., STAMPFLI, G. M., DAVIES, D. J., GASTALDI, C. N., RUITENBERG, P. A. AND VERMEER, G. J. O. 1989. Dip and azimuth displays for 3D seismic interpretation. *First Break*, **7**, 86–95.

HARKER, S. D., GUSTAV, S. H. AND RILEY, L. A. 1987. Triassic to Cenomanian stratigraphy of the Witch Ground Graben. *In*: BROOKS, J. AND GLENNIE, K. W. (eds) *Petroleum Geology of North West Europe*. Graham & Trotman, London, 809–818.

HOWITT, F., ASTON, E. AND JACQUE, M. 1975. The occurrence of Jurassic volcanics in the North Sea. *In*: WOODLAND, A. (ed.) *Petroleum and the Continental Shelf of North-West Europe*. Applied Science, Barking, 379–387.

LEEDER, M. R. AND BOLDY, S. A. R. 1990. The Carboniferous of the Outer Moray Firth Basin, Quadrants 14 and 15, Central North Sea. *Marine & Petroleum Geology*, **7**, 29–37.

McGANN, G. J., GREEN, S. C. H., HARKER, S. D. AND ROMANI, R. S. 1991. The Scapa Field, Block 14/19, UK North Sea. *In*: ABBOTTS, I. L. (ed.) *United Kingdom Oil and Gas Fields 25 Years Commemorative Volume*. Geological Society, London, Memoir, **14**, 369–376.

MUDGE, D. C. AND COPESTAKE, P. 1992. Revised Lower Palaeogene lithostratigraphy for the Outer Moray Firth, North Sea. *Marine & Petroleum Geology*, **9**, 53–69.

O'DRISCOLL, D., HINDLE, A. D. AND LONG, D. C. 1990. The structural controls on Upper Jurassic and Lower Cretaceous reservoir sandstones in the Witch Ground Graben, UK North Sea. *In*: HARDMAN, R. F. P. AND BROOKS, J. (eds) *Tectonic Events Responsible for Britain's Oil and Gas Reserves*. Geological Society, London, Special Publication, **55**, 299–323.

PARKER, R. H. 1991. The Ivanhoe and Rob Roy Fields, Blocks 15/21 a-b, UK North Sea. *In*: ABBOTTS, I. L. (ed.) *United Kingdom Oil and Gas Fields 25 Years Commemorative Volume*. Geological Society, London, Memoir, **14**, 331–338.

PRICE, J., DYER, R., GOODALL, I., McKIE, T., WATSON, P. AND WILLIAMS, G. 1993. Effective stratigraphical subdivision of the Humber Group and the Late Jurassic evolution of the UK Central Graben. *In*: PARKER, J. R. (ed.) *Petroleum Geology of Northwest Europe: Proceedings of the 4th Conference*. Geological Society, London, 443–458.

RILEY, L. A., HARKER, S. D. AND GREEN, S. C. H. 1992. Lower Cretaceous palynology and sandstone distribution in the Scapa Field, U.K. North Sea. *Journal of Petroleum Geology*, **15**, 97–110.

TAYLOR, J. C. M. 1986. Late Permian–Zechstein. *In*: GLENNIE, K. W. (ed.) *Introduction to the Petroleum Geology of the North Sea*. Blackwell, 87–111.

Mesozoic evolution of the Jæren High area, Norwegian Central North Sea

O. HØILAND, J. KRISTENSEN and T. MONSEN

Statoil a.s., Postboks 300, 4001 Stavanger, Norway

Abstract: The Jæren High area is located in a structurally complex area where the Central Graben, the Ling Depression, the South Viking Graben and the Witch Ground Graben converge. Wells drilled on the Jæren High indicate rapid changes both in facies and thickness of the Mesozoic sequence. The evolution of the region during Mesozoic times was considerably influenced by the thickness and the distribution of the Zechstein salt. Arches underlain by thick Triassic strata are separated by elongated grabens or troughs where mainly late Jurassic strata rest on ridges of Zechstein salt.

At least three models are possible to explain the distribution of Mesozoic strata in grabens or troughs on the Jæren High. (1) Thin-skinned extension: a relatively thin Zechstein sequence overlain by Triassic strata was faulted, mainly in mid-Jurassic times. The salt started to pierce the footwalls resulting in the development of grabens separating the footwall and hangingwall blocks. These grabens trapped late Jurassic sands and shales, while the Triassic highs suffered erosion and non-deposition. (2) Salt dissolution: this model is based on the dissolution of Zechstein salt near to the surface or in the sub-surface. During early to middle Triassic times salt started to move due to the weight of its overburden. Triassic depocentres and salt ridges developed. Middle and late Jurassic salt dissolution took place due to exposure of the salt at the surface, creating accommodation space for late Jurassic deposition of sands and shales. (3) Combined tectonism and salt dissolution: this model combines both tectonism and salt dissolution. Triassic and Jurassic extension resulted in uplift and rotation of the high. Subsequent erosion resulted in exposure of the salt along elongated ridges and salt dissolution took place due to salt exposure at the surface creating accommodation space for late Jurassic deposition.

The third model, a combination of tectonism and salt dissolution, is the preferred model as it can explain the particular Mesozoic palaeogeography of the Jæren High.

The Jæren High, as defined by Rønnevik *et al.* (1975), is located in the western part of Quadrant 7 on the Norwegian Continental Shelf and the eastern part of Quadrants 22 and 23 in the UK sector (Fig. 1). The Jæren High and its surrounding area has been explored for hydrocarbons since the early 1970s with limited success. Several finds have been made on the western margin of the high, mainly in Tertiary pinch-out plays such as the Everest discovery (UK Blocks 22/10 and 22/14) (Thompson and Butcher 1991). The lack of success in the pre-Cretaceous sequence may be due to the fact that most of the wells have been drilled on structural highs which have been highly eroded in the late Jurassic and hence have Lower Cretaceous rocks resting unconformably on Lower to Middle Triassic rocks.

The structural evolution of the Jæren High is complex and most wells do not reveal a complete Mesozoic evolution due to severe erosion during the late Jurassic. In this paper special attention will be paid to the Mesozoic evolution. A similar geological evolution is thought to have affected the Western Platform west of the Central Graben area (Armstrong and Glennie 1991).

Three different models are proposed to explain the deposition of late Jurassic sediments in troughs between the Triassic highs. These models involve both salt dissolution and tectonism. Salt dissolution has been described in several basins, such as the Delware Basin (Anderson *et al.* 1978; Anderson and Kirkland 1980), the South Oman Salt Basin (Heward 1990) and the North Sea (Smith 1987).

Geological setting

The Jæren High is situated in a structurally complex area where the South Viking Graben, the Witch Ground Graben, the Central Graben and the Ling Depression converge (Fig. 1). The Jæren High itself is a major eastward-tilted fault block at base Zechstein level. The main geological structures observed on the Jæren High today were created during Jurassic and earliest Cretaceous times.

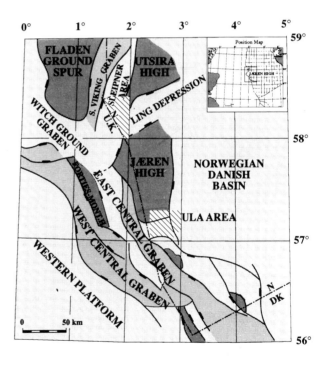

Fig. 1. Structural element map.

Well results and seismic data indicate a relatively thin Zechstein salt sequence compared to the East Central Graben area to the west and the Norwegian Danish Basin to the east (Fig. 2). Some wells have encountered an Upper Permian sequence which consists of carbonates (UK 23/11-1). This suggests that the Jæren High was also a relatively high area with marginal Zechstein facies accumulation during Permian times. An alternative interpretation is that the salt sequence withdrew due to the halokinesis during Triassic times, leaving only a thin carbonate sequence.

From *Petroleum Geology of Northwest Europe: Proceedings of the 4th Conference* (edited by J. R. Parker).

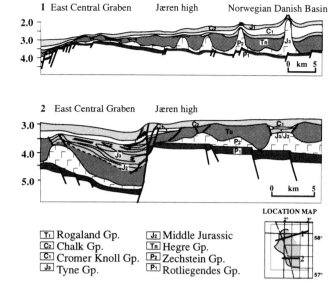

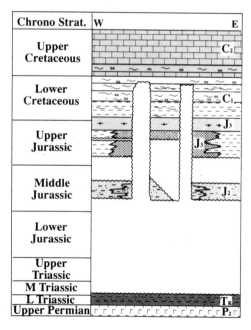

Fig. 2. Geoseismic sections on the Jæren High. Note that Jurassic strata are preserved where Zechstein rocks are thick.

Stratigraphy

A generalized stratigraphy for the region is shown in Fig. 3. About 15 exploration wells have been drilled into the Triassic on the Jæren High and its surrounding area. In at least 10 of these wells, the lower Cretaceous sequence rests unconformably on Lower to Middle Triassic strata. The rest of the wells show Upper Jurassic strata resting unconformably on Lower to Middle Triassic strata.

Fig. 3. Generalized lithostratigraphy of the Jæren High. No scale.

The Upper Cretaceous sequence is dominated by limestones and chalk. The chalk belongs to the upper part of the Tor Formation and the Ekofisk Formation. The thickness of the Upper Cretaceous sequence is approximately 400 m (N 7/1-1) on the high, increasing to 600–700 m (N 7/11-5) downflank from the high.

The Lower Cretaceous sequence is dominated by calcareous shales interbedded with limestone beds. The thickness varies

from less than 50 m (N 7/1-1) up to 200–300 m (N 7/11-5). Seismic data indicate that this sequence can be missing on some of the Triassic highs.

The Jurassic sequence preserved in the salt-related grabens is dominated by Upper Jurassic shales and sandstones which can be up to 150 m thick (N 7/8-3). No Middle Jurassic strata have been drilled in the area, but seismic data suggest that Middle Jurassic strata are preserved in these grabens.

The Triassic sequence is dominated by silty shales and silty sandstones of the Smith Bank Formation. Only a few wells have drilled through the Triassic sequence into the Permian Zechstein Group. In these wells the thickness of the Triassic sequence varies from 200 m (UK 23/11-1) up to 500 m (N 7/8-3).

Seismic mapping

Some 2000 km of seismic data have been interpreted on and adjacent to the Jæren High. A few wells have been tied to the seismic, but these are mainly drilled on Triassic arches and do not give a reliable tie to the Jurassic sequence (Figs 4 and 5). The Jurassic sequence has been defined by a comparison of the seismic character in wells where Jurassic strata have been drilled (Figs 4 and 5).

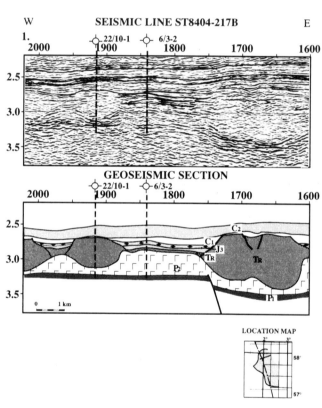

Fig. 4. Seismic line and geoseismic section ST 8404-217 across the northern Jæren High. Note the thickness variation of the Triassic and Zechstein sequences. For legend see Fig. 2.

A striking phenomenon is the relationship between a thick Zechstein salt sequence and the presence of a Jurassic sequence (compare Fig. 6 and Fig. 7). This relationship, where a Jurassic sequence is deposited above a thick Zechstein salt sequence, has been proved by drilling Norwegian wells 7/8-3, 6/3-2 and by wells in the Ula Field. In view of this relationship, the mapping discussed in this paper was concentrated on the interpretation of the Zechstein salt together with the Base Cretaceous seismic event.

The Zechstein salt geometry divides the area into elongated ridges of thick salt and areas where the salt sequence is

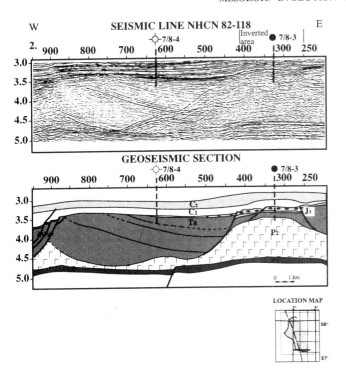

Fig. 5. Seismic line and geoseismic section NHCN 82-118. Note the rotated Triassic fault block (sp.450-850) and the inversion of the Jurassic and Cretaceous strata (sp.250-450). For legend see Fig. 2.

relatively thin (Fig. 7). To the east, towards the Norwegian Danish Basin, the salt is relatively thick with no significant thinning.

The Base Cretaceous map displays arches where early Cretaceous strata lie uncomformably on the Triassic and grabens where late Jurassic sediments have been trapped (Fig. 6). To the east this relationship diminishes where a more complete Jurassic sequence is preserved.

Mesozoic evolution

The late Permian transgression resulted in the deposition of a thick halite sequence in the Northern Permian Basin. Wells drilled on the Jæren High have proved thickness variation and facies changes in this Zechstein sequence. These changes may be primary, but post-depositional salt movements may also explain the variations.

Late Permian regional transtensional movements, perhaps related to lateral movements along the Tornquist Zone (Pegrum 1984), continued into the Triassic, resulting in uplift of the hinterland and rejuvenation of sediment source areas.

Regional and local tectonism, combined with halokinesis, are believed to be the controlling factors in the sedimentation and distribution of the Triassic sequence. Salt movement was probably initiated due to the overburden and to reactivation of faults in the pre-Zechstein basement in response to an early phase of extension, probably in early to mid-Triassic times, across the Central Graben (Armstrong and Glennie 1991).

The sedimentation in the early to mid-Triassic was domi-

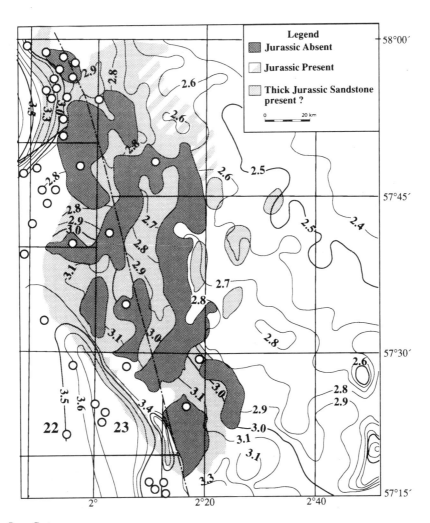

Fig. 6. Structural time map Base Cretaceous.

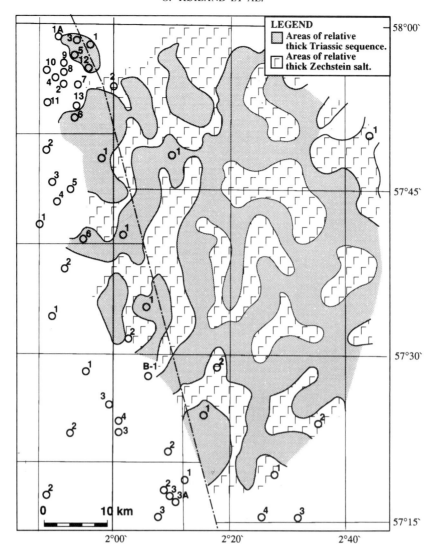

Fig. 7. Relationship between thick Zechstein salt and thick Triassic strata on Jæren High.

nated by thick continental shales and siltstones interbedded with sequences of fluvial sandstones. This sequence is assigned to the Smith Bank Formation and is up to 1000–1200 m thick in some wells. Other wells have penetrated only a few hundred metres.

Mid- to late Triassic sedimentation was dominated by coarse clastic deposits. Thicker fluvial-dominated sandstone sequences ascribed to the Skagerrak Formation are believed to have been deposited all over the area. Later uplift and erosion has often removed most of this sequence, but wells in the Sleipner and Ula area have penetrated sandstones of the Skagerrak Formation (Home 1987; Lervik *et al.* 1989).

It is not clear whether early–mid-Jurassic deposition took place in the Jæren High area, but wells to the southeast have penetrated sediments of this age (N 7/9-1). The Central North Sea area was subjected to mid-Jurassic regional extension, uplift and erosion (Ziegler 1981). This tectonic event was probably an important step in creating the elongated salt ridges and the Triassic highs in the Jæren High area.

A character similar to the seismic response of the Middle Jurassic Bryne Formation is recognized on seismic data at the top of some salt swells on the Jæren High. This suggests the presence of delta/coastal plain Middle Jurassic strata in areas where the salt is high at present. Such strata have been drilled in UK Blocks 23/26 and 23/27 and in the Ula Field (Home 1987).

Late Jurassic extension, rifting and subsidence resulted in a return to marine conditions of deposition so that shallow marine sands and shales were deposited during the late Jurassic. This rifting phase probably led to the final development of the 'mini' grabens lying directly above the salt ridges (Figs 5 and 6). Late Jurassic sands and shales accumulated in these grabens on the Jæren High and its surrounds (N 7/8-3 and 6/3-2).

The Triassic arches were also palaeogeographical highs during the deposition of the early Cretaceous sequence. This early Cretaceous sequence is dominated by calcareous shales and clearly thickens in the areas where late Jurassic sedimentation took place (Figs 2 and 4). Early Cretaceous sandstones have not been encountered so far along these salt 'valleys'.

Late Cretaceous transpressional movements (Pegrum 1984) resulted in inversion in areas where the salt was thick enough to move (Fig. 5). In areas where the salt is absent or too thin to move, no inversion features have been observed at Base Cretaceous level.

Models

Three models are proposed to explain the development of the late Jurassic palaeogeography and sediment traps on the Jæren High. The models are tentative due to the lack of drilling data.

1. Thin-skinned extension model (Fig. 8)

Physical models have been constructed showing the extension of 'brittle' layered sand on viscous, dense, silicon polymer ('salt') due to gliding and spreading along a gently dipping slope (Vendeville and Jackson 1992a,b). Such modelling demonstrates that an initially near-planar fault cutting through the viscous layer is later distorted by a viscous diapir piercing the hangingwall. Such a model can be applied to the Jæren High area, assuming a relatively thin Zechstein salt sequence.

Tectonic extension in early–mid-Triassic times resulted in planar faults cutting through the salt and a slight rotation of the fault blocks to the west occurred (Fig. 8). This extension probably reactivated older fault trends at Rotliegend and older levels. Younger faults may follow these older trends, but they are decoupled from the older: they are listric and die out in the salt sequence.

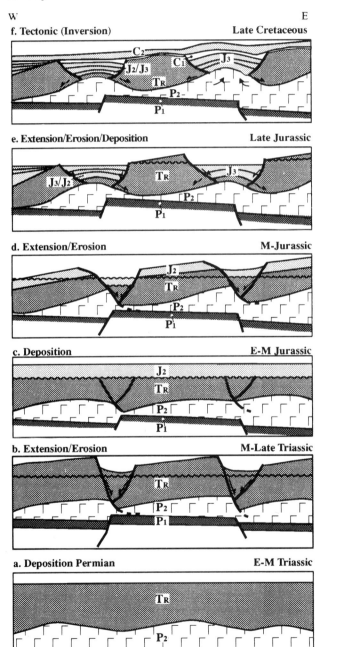

Fig. 8. Model 1: extension. This model attempts to explain the features of the Zechstein to Cretaceous geology of the Jæren High by means of extension in Triassic and Jurassic times. For legend see Fig. 2.

Evidence of such a tectonic event is seen in Well N 7/8-4 (Fig. 5) where a strong seismic reflection representing strata of Scythian age is overlain by younger Triassic strata. Armstrong and Glennie (1991) have suggested that there was an early–mid-Triassic extensional phase on the Western Platform.

Following mid-Triassic–mid-Jurassic deposition (Fig. 8), the regional uplift in mid-Jurassic times resulted in further rotation of the high (Fig. 8). The salt started to pierce the hangingwalls with slight distortion of the fault planes. The graben separating the footwall and hangingwall blocks developed at this time.

Deposition of Middle Jurassic sediments followed by the late Jurassic extension and rifting phase (Fig. 8) created the main elements of the structure of the Jæren High area. During the late Jurassic rifting, further extension and piercing of the footwall resulted in distortion of the fault planes. At the same time, widening of the graben separating the hangingwall and footwall blocks occurred due to the late Jurassic extension, creating accommodation space for late Jurassic sedimentation.

During the early Cretaceous, the grabens continued to subside, while the flanking Triassic arches periodically suffered erosion and non-deposition.

Transpression during the late Cretaceous reactivated the salt sequence in those areas where the original depositional thickness was sufficient to produce mobility, resulting in inversion of some of the late Jurassic grabens (Fig. 8). In areas where the salt was too thin to be mobile or was absent, structural relief on the Base Cretaceous seismic level cannot be observed.

2. Salt dissolution model (Fig. 9)

The salt dissolution model is based on the concept of dissolution of the Zechstein salt initially near to the surface and later in the sub-surface. Such features can also be seen in the Southern Oman Salt Basin where salt withdrawal and salt dissolution has taken place from Cambrian time up to the present day (Heward 1990). There salt removal controlled both the large-scale geometries of sedimentary bodies within the succession overlying the salt and the distribution of sedimentary facies when rates of salt removal balanced or exceeded sedimentation.

On the Jæren High, the Triassic sequences are arranged in similar geometries (Fig. 9). During early–mid-Triassic times, salt started to move due to the overburden load (Fig. 9). Several Triassic depocentres developed while syn-depositional salt withdrawal took place beneath. The salt was concentrated into salt pillows and ridges as Triassic continental clastics infilled the synclines.

During the late Triassic to mid-Jurassic (Fig. 9) the sedimentation was controlled by the existing depocentres. The mid-Jurassic episode of general uplift and erosion removed parts of the Triassic sequence.

During the late Jurassic rifting, uplift and erosion of Triassic strata took place exposing the Zechstein salt at the palaeosurface. This resulted in salt solution sink holes and grabens. These depocentres trapped late Jurassic clastics eroded from the surrounding Triassic highs (Fig. 9). During the early Cretaceous, the salt solution areas continued to trap sediments while the Triassic arches were areas of erosion and non-deposition (Fig. 9).

3. Combined salt dissolution and tectonism model (Fig. 10)

This model explains the Mesozoic evolution of the Jæren High as a result of the combination of tectonism and salt dissolution. In this model, extension is viewed as having had the most significant impact on the evolution of the structures of the Mesozoic rocks on the Jæren High during the Triassic to the

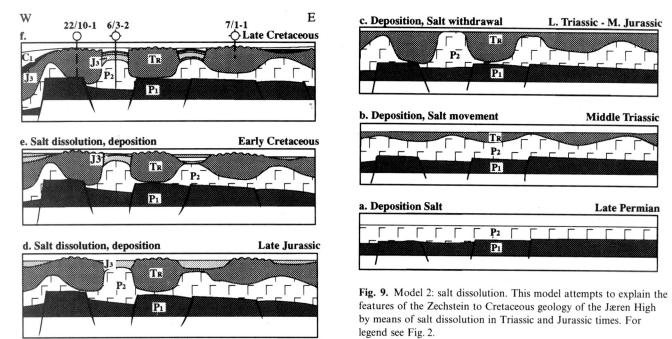

Fig. 9. Model 2: salt dissolution. This model attempts to explain the features of the Zechstein to Cretaceous geology of the Jæren High by means of salt dissolution in Triassic and Jurassic times. For legend see Fig. 2.

late Jurassic. Salt dissolution on the other hand was probably the main controlling factor during the late Jurassic to early Cretaceous.

Early–mid-Triassic extension resulted in fault block rotation and listric faults soled out in the salt (Fig. 10). These faults were decoupled from pre-existing fault trends at deeper levels by the intervening salt layer.

Following the deposition of Middle Triassic–Middle Jurassic sediments, the mid-Jurassic uplift and rotation affected the

Jæren High. The subsequent erosion resulted in exposure of the salt along elongated ridges. Salt dissolution and deposition of Middle Jurassic sediments occurred in elongated grabens where the salt solution controlled the sedimentation (Fig. 10). Late Jurassic rifting led to further exposure and dissolution of the salt. The elongated grabens continued to trap late Jurassic deposits while erosion took place on the Triassic highs. Salt dissolution continued to be the main factor controlling the deposition (Fig. 10).

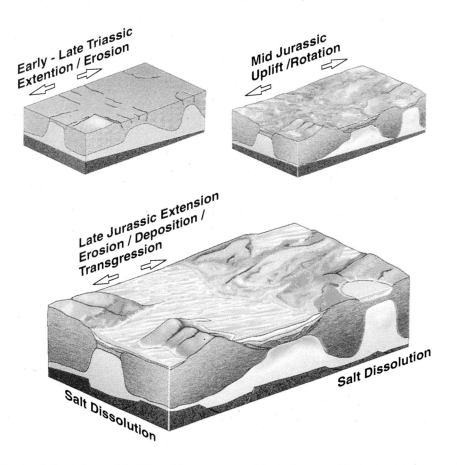

Fig. 10. Model 3: extension + salt dissolution model. This model attempts to explain the features of the Zechstein to Cretaceous geology of the Jæren High by means of extension combined with salt dissolution in Triassic and Jurassic times.

During early Cretaceous times the elongated grabens continued to trap sediment deposition while the Triassic arches suffered erosion and non-deposition.

Stephenson *et al.* (1991) observed in the Sverdrup Basin that salt and anhydrite moved in a tectonic setting which changes between compression and extension in different phases. They concluded that the salt diapirism was most active during compressional phases. The salt was less active during extension but even then it was intruded laterally as tabular bodies.

Such behaviour may be a part of the explanation why, on the Jæren High, salt diapirism was more active in late Cretaceous during transpressional movements (Pegrum 1984) (Fig. 5) than during the late Jurassic extension.

Conclusions

Mid-Jurassic uplift and the late Jurassic rifting phase were fundamental events in the geological evolution of the Central North Sea. These events played important roles in the creation of the Jæren High and its surrounding areas.

Additionally, salt solution is a common phenomenon demonstrated in salt basins such as the South Oman Salt Basin, several basins in North America and the North Sea.

It is difficult to believe that only tectonism alone or salt dissolution alone has created the particular Mesozoic geometries of the Jæren High. A combination of these two mechanisms is, therefore, most likely to have been responsible for the Mesozoic evolution of the high.

The authors would like to thank Statoil which gave us the opportunity to present this paper. Special thanks go to our colleagues A. M. Spencer, B. T. Larsen and L. N. Jensen for their helpful comments.

References

ANDERSON, R. Y., KIETZKE, K. K. AND RHODES, D. J. 1978. Development of dissolution breccias. Northern Delware Basin, New Mexico and Texas. *New Mexico Bureau Mines and Mineral Resources Circular*, **159**, 47–52.

—— AND KIRKLAND, D. W. 1980. Dissolution of salt deposits by brine density flow. *Geology*, **8**, 66–69.

ARMSTRONG, L. A. AND GLENNIE, K. W. 1991. The Kittiwake Field, Block 21/18, UK North Sea. *In*: ABBOTTS, I. L. (ed.) *United Kingdom Oil and Gas Fields 25 Years Commemorative Volume*. Geological Society, London, Memoir, **14**, 339–345.

HEWARD, A. P. 1990. Salt removal and sedimentation in Southern Oman. *In*: ROBERTSON, A. H. F., SEARLE, M. P. AND RIES, A. C. (eds) *The Geology and Tectonics of the Oman Region*. Geological Society, London, Special Publication, **49**, 637–652.

HOME, P. C. 1987. *In*: SPENCER, A. M. *et al.* (eds) *Geology of the Norwegian Oil and Gas Fields*. Graham & Trotman, London, 143–151.

LERVIK, K. S., SPENCER, A. M. AND WARRINGTON, J. 1989. Outline of Triassic Stratigraphy and Structure in the Central and Northern North Sea. *In*: COLLINSON, J. C. (ed.) *Correlation in Hydrocarbon Exploration*. Graham & Trotman, London, 173–189.

PEGRUM, R. M. 1984. Structural development of the south-western margin of the Russian–Fennoscandian Platform. *In*: SPENCER, A. M. *et al.* (eds) *Petroleum Geology of the Northern European Margin*. Graham & Trotman, London, 359–369.

RØNNEVIK, H. C., VAN DEN BOSCH, W. AND BANDLIEN, E. H. 1975. A proposed nomenclature for the main structural features in the Norwegian North Sea. *In*: *Jurassic Northern North Sea Symposium, Norsk Petroleumsforening*, Article 18.

SMITH, R. L. 1987. The structural development of the Clyde Field. *In*: BROOKS, J. AND GLENNIE, K. W. (eds) *Petroleum Geology of North West Europe*. Graham & Trotman, London, 523–531.

STEPHENSON, R. A., VAN BERKEL, J. T. AND CLOETINGH, S. A. P. L. 1991. Relation between Salt Diapirism and In-Plain Stresses in the Sverdrup Basin, Arctic Canada. *Tectonics*. In press.

THOMPSON, P. J. AND BUTCHER, P. D. 1991. The Geology and geophysics of the Everest Complex. *In*: SPENCER, A. M. (ed.) *Generation, Accumulation and Production of Europe's Hydrocarbons*. Special Publication of the European Association of Petroleum Geoscientists, **1**. Oxford University Press, 89–98.

VENDEVILLE, B. C. AND JACKSON, M. P. A. 1992a. The fall of diapirs during thin skinned extension. *Marine and Petroleum Geology*, **9**, 354–371.

—— AND —— 1992b. The rise of diapirs during thin skinned extension. *Marine and Petroleum Geology*, **9**, 331–352.

ZIEGLER, P. A. 1981. Evolution of sedimentary basins in north-west Europe. *In*: ILLING, L. V. AND HOBSON, G. D. (eds) *Petroleum Geology of the Continental Shelf of North-west Europe*. Heyden, London, 3–39.

Extension and salt tectonics in the East Central Graben

J. PENGE, B. TAYLOR, J. A. HUCKERBY and J. W. MUNNS

Amoco (UK) Exploration Company, Amoco House, West Gate, Ealing, London W5 1XL, UK

Abstract: Mesozoic structures in the East Central Graben were formed in an extensional tectonic environment by the complex interaction of deposition, basin extension, halokinesis and subsequent erosion. Detailed analysis of seismic data on the eastern flank of the Forties–Montrose High indicates that the Triassic section exhibits rafting: large fault blocks were translated, rotated and distorted on a ductile Zechstein salt substrate. Rafts can be defined by an empirical structural relationship between the Base Cretaceous Unconformity and the Top Zechstein surfaces. The rafts exerted important controls on the distribution of Triassic reservoir rocks and younger reservoir and source rocks. Halokinetic movement was essentially a passive process related to eastward basin subsidence, with salt upwelling between rafted Triassic blocks. Subaerial exposure following uplift during the Early to Middle Jurassic and contemporaneous erosion and deposition led to the development of significant landscape topography. Onlap during the Middle to Upper Jurassic has preserved features such as cuestas with significant fault-line escarpments capped by Skagerrak sandstones. Adjacent to these escarpments deep erosion resulted in incised valleys, where significant angular unconformities occur. Physical modelling has been used to determine the processes and sequence of events that led to the present form of the East Central Graben. A structural model has been developed which relates the timing and rate of deposition and extension, salt movement and subsequent erosion. The interaction of regional extension, development of Triassic rafts and passive salt movement has not been described before. The model represents an alternative to the conventional interpretation that the observed structural features in the area were primarily the result of salt withdrawal.

The East Central Graben is a Mesozoic half-graben (Fig. 1), which was active from Permian to Paleocene times. It is an important site for hydrocarbon accumulations in Triassic, Jurassic and Tertiary reservoirs. The graben is underlain by Zechstein salt, which has had a profound effect on the distribution of post-Permian sedimentary rocks. The principal objective of this paper is to describe the distribution of Permian and Triassic rocks visible on seismic in the East Central Graben and, using physical modelling and small-scale field analogues, to develop a model to explain this distribution (Fig. 2). Field analogues and physical modelling have been used to evaluate the amount of extension and likely structural forms, and to determine the sequence and consequence of events during extension in the presence of a mobile Zechstein salt substrate.

The extensional model for the formation of East Central Graben structures proposed here will explain the following complex structural and stratigraphic relationships observed on regional seismic lines:

1. variation in the form of Zechstein halokinetic structures with salt walls developed to the west of the graben and intrusive diapirs formed adjacent to the Jaeren High in the east;
2. variation in the form of the Triassic interval, deformed into discrete blocks, called 'rafts' on the western side but passing into more continuous layers on the eastern side;
3. a westward-thinning wedge of Jurassic and Early Cretaceous rocks, which pinches-out towards the western graben edge;
4. a relatively undisturbed blanket of Late Cretaceous and younger strata.

Much of the deformation, which is the focus of this paper, is concentrated in the Zechstein and Triassic intervals. The key concept developed in this paper is the recognition of 'rafts' of Triassic sedimentary rocks, floating on Zechstein salt and separated from each other by 'rift' zones where the Triassic

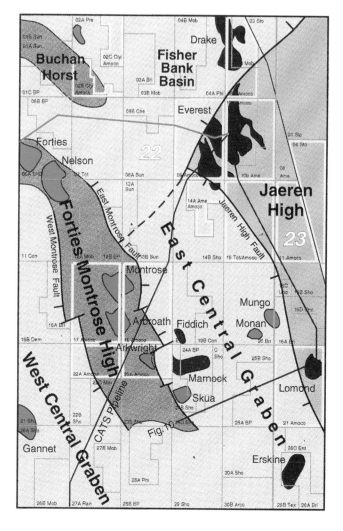

Fig. 1. Location map of the East Central Graben.

From *Petroleum Geology of Northwest Europe: Proceedings of the 4th Conference* (edited by J. R. Parker).
© 1993 Petroleum Geology '86 Ltd. Published by The Geological Society, London, pp. 1197–1209.

interval thins over an elevated salt wall. 'Rafts' are defined as positive discrete blocks of relatively thick, undeformed Triassic sedimentary rocks. They are usually lozenge-shaped in cross-section but can be elongated in plan. The 'rafts' overlie relatively thin salt sections and may 'ground' on pre-Zechstein basement, when salt withdrawal is complete.

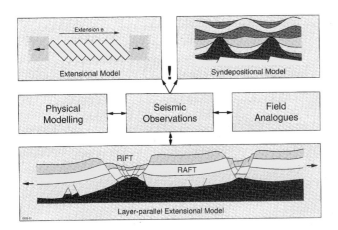

Fig. 2. Interpretative methodology.

'Rafts' are separated by 'rifts', which are elongate zones defined by the axes of elevated and elongated salt walls, above which is a thin zone of Triassic and younger sedimentary rocks. The 'rifts' are zones of intense faulting, the focus of deformation, erosion and end-Triassic to Early Cretaceous sedimentation. 'Rifts' indicate the amount of extension that an originally continuous cover of Triassic section has suffered, while the 'rafts' relate to the original thickness. Thus, careful restoration of 'rafts' by rotation, translation and compensation for the effects of tectonic thinning yields estimates of Mesozoic extension that has occurred in the East Central Graben.

A number of previous models has been proposed for the evolution of structures and controls on sedimentation in the East Central Graben. A dextral transtensional model was proposed by Glennie (1990) and a similar strike-slip model was proposed by Cartwright (1987), in which transverse basement faults influenced the cover rocks. Neither of these models gave much emphasis to the profound influence of salt on the overlying cover rocks during extensional events. Similarly, an orthogonal extension model proposed by Roberts *et al.* (1990*a,b*) did not recognize the influence of salt movement as an important control on the formation and distribution of Mesozoic structures in the East Central Graben.

The importance of salt movements in the East Central Graben has been highlighted by Hodgson *et al.* (1992). This paper emphasized the effects of salt diapirism and salt withdrawal on the deposition of Triassic sediments and the subsequent deformation of the Triassic interval. The model proposed in Hodgson *et al.* (1992) assumes syn-sedimentary deformation of Triassic sediments by differential loading of the Zechstein salt substrate, leading to overdeepening of Triassic depocentres. This model implies near-vertical control on stress fields (i.e. loading by gravity) as the dominant tectonic control, rather than regional basin extension. It also places principal deformation events syn-depositionally within the Triassic period. Literature studies also support initiation of regional salt movement in the Early Triassic period, culminating in initiation of active diapirism by Late Triassic times. However, it is clear from seismic studies summarized in this paper that most salt movement post-dates the Triassic period in the East Central Graben and that the Triassic interval was once con-tinuous and has suffered later (Jurassic–Early Cretaceous) regional extension.

The model proposed here differs because regional extension following Triassic deposition is regarded as the driving mechanism for deformation. Although salt withdrawal is acknowledged as having occurred during Triassic deposition, this withdrawal was limited to local overdeepening of Triassic depocentres. The majority of salt movement is the result of post-Triassic extension with salt movement occurring as a passive rather than an active process, helping to accommodate regional extension in overlying Triassic brittle rocks. This combination of salt movement as an accommodation process to regional extension has not been described before. Intense salt diapirism located above the buried trace of the Jaeren High bounding fault is a post-Jurassic process, effectively unrelated to the regional extension and deformation of Mesozoic strata.

Structural framework of the East Central Graben

The East Central Graben, as defined from regional seismic data, is a Mesozoic half-graben, bounded to the west by the Forties–Montrose High and to the east by the Jaeren High (Fig. 1). There is a shallow eastward dip to the half-graben and there is evidence of a reversal of this half-graben asymmetry as the Triassic depocentre is in the west, while the Jurassic depocentre is on the eastern side. To the north, the half-graben is bounded by the Fisher Bank Basin and has a shallow southward slope, which has been preserved since Triassic times.

The half-graben contains three sedimentary packages (Fig. 3):

1. a basement floor of Devonian–Rotliegendes tilted fault blocks;
2. a complex fill of Zechstein and Triassic sedimentary rocks, overlain by an eastward-thickening wedge of Jurassic–Early Cretaceous section;
3. a Late Cretaceous–Recent blanket of post-graben fill.

Structural styles and tectonic controls

Observations from seismic data

A variety of structural styles is seen in the East Central Graben. In the deeper southeastern part of the half-graben, e.g. Blocks 22/20, 23/26, structural style is dominated by salt-withdrawal features, where rim synclines and diapiric structures have greatly influenced the morphology. However, towards the shallower western margin, e.g. Blocks 22/18 and 22/24, evidence for active diapirism is negligible. In these areas reactive and passive salt walls (*sensu* Vendeville and Jackson 1992*a,b*) are the common form of salt geometry.

The salt walls are characterized by the following features:

1. they occur between Triassic sedimentary rafts; this relationship is not identified in the deeper part of the half-graben;
2. the top surfaces of the walls are flat, triangular or semicircular in shape (Figs 4a–c) with an average wavelength of 4 km;
3. their top surfaces typically exhibit higher reflectivity than the Top Zechstein surfaces elsewhere in the half-graben;
4. their upper bounding surfaces exhibit fault-like displacements. These are predominantly parallel to the steep sides of the more convex salt walls (Figs 4b and d). Conjugate fault pairs may be developed on flat-topped salt features (Fig. 4a).

There is no discernible relationship between the presence of major faulting in the pre-Zechstein section and the location of these salt walls. Although examples of salt swells located directly over basement faults do exist (Fig. 4b), there are many instances where this is not the case, both in the Central Graben and on the Jaeren High (Fig. 4f). In areas where salt is thin to absent, basement structure plays an important role in controlling Triassic structural style.

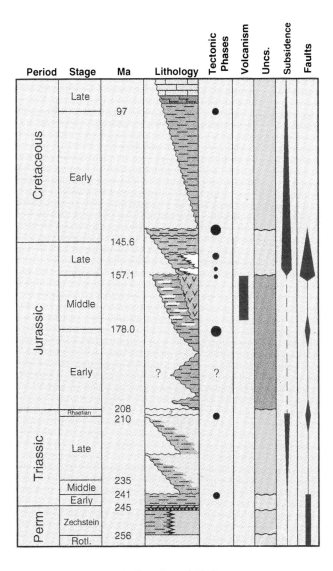

Fig. 3. Stratigraphy of the East Central Graben.

The influence of the Zechstein sequence on the style of the rafting is seen on the Forties–Montrose High (Fig. 4c). At the southern and western margin of the Forties–Montrose High, where thick salt is present, similar structural styles to those seen in the half-graben and Jaeren High are observed (Fig. 4c). To the east and north, where carbonates predominate and salt is thin to absent, movement on Paleozoic fault blocks has determined the preservation of the Triassic section.

Along the shallow margin of the East Central Graben, on parts of the Forties–Montrose High and on the Jaeren High, there is a clear relationship between the Base Cretaceous Unconformity and the Top Zechstein surface. With few exceptions, the crest-lines of Top Zechstein structural highs correspond closely to trough-lines of Base Cretaceous lows. Conversely trough-lines of salt structures correspond to crest-lines of Base Cretaceous structures. This gives the appearance of megascopic pinch-and-swell structures. The few small Base Cretaceous highs which exist above salt highs are associated

with localized zones of inversion caused by the reactivation of the salt walls. The thick section between the Base Cretaceous Unconformity and the Top Zechstein surface is predominantly Triassic in age.

Classification of Triassic rafts

Analysis of Triassic rafts indicates two distinct styles which correspond to the presence or absence of Skagerrak sandstone. In areas where the Skagerrak sandstone has been preserved as seen near the Marnock Field, few signs of 'turtleback' salt-withdrawal features are observed. Typically, the rafts consist of relatively undeformed sequences characterized by parallel internal reflectivity (Figs 4a and b). Internal characterization of these rafts is often complicated by strong multiple contamination. Importantly, the strong parallelism of Triassic reflectors indicates no significant syn-depositional fault rotation of these rafts. With the general absence of Skagerrak sandstone in areas like the Jaeren High and Forties–Montrose High, Triassic morphology is more rounded as ductile deformation predominates over brittle deformation (Fig. 4f).

In areas of thick halite, two types of complex rafts have been recognized. Palinspastic reconstructions demonstrate that they have evolved from larger more simple rafts. The complex rafts have been classified as syn-rotational and contra-rotational:

Syn-rotational rafts (Fig. 4a)

These comprise two fault blocks which dip in the same direction. Significant horizontal separation between each fault block is often not developed, although they are usually vertically offset. The dominant fault separating the paired blocks soles out on the top of a convex salt wall. Their axes are east–west, generally perpendicular to the regional trend of the East Central Graben (NNW–SSE).

Contra-rotational rafts (Fig. 4b)

These are the more common, consisting of rotated, paired blocks, in which each raft has the opposite sense of rotation, with dips toward the axis of symmetry. The halves are separated by a north–south-oriented convex salt wall. Contra-rotational raft pairs are most common along the western margin of the East Central Graben, where they have their long axes parallel to the Forties–Montrose High. Each rafted pair is separated from the high by a roughly north–south-oriented flat-topped salt wall up to 4 km in width. Sets of rafted pairs are separated by roughly east–west-trending salt walls, which are aligned with faulted offsets against the Forties–Montrose High. These salt walls may originally have been related to transtensional basement fault movements. Rafts along this western margin are separated from other Triassic structures deeper into the basin by another series of flat-topped salt walls.

The spatial relationship of the Triassic rafts at the western margin of the East Central graben is most clearly seen in Fig. 5. This diagram consists of two grids which have been superimposed and rotated. The reader views the grids from an azimuth of 340° and at an angle of 67°. The Forties–Montrose High is to the left of the diagram. The grids are colour coded for elevation, red being the shallowest and blue the deepest. The lower grid is the Top Zechstein surface and the upper grid the Base Cretaceous Unconformity. The coloured areas, therefore, represent the Triassic rafts. The black areas between the coloured grids represent the axes of Zechstein salt highs and, therefore, areas of thin Triassic section (rifts).

Each Triassic raft can be roughly re-aligned and juxtaposed by removing the intervening salt walls and correcting for tectonic thinning and rotation, producing the blanket geometry predicted for the Triassic. Once corrections for tectonic

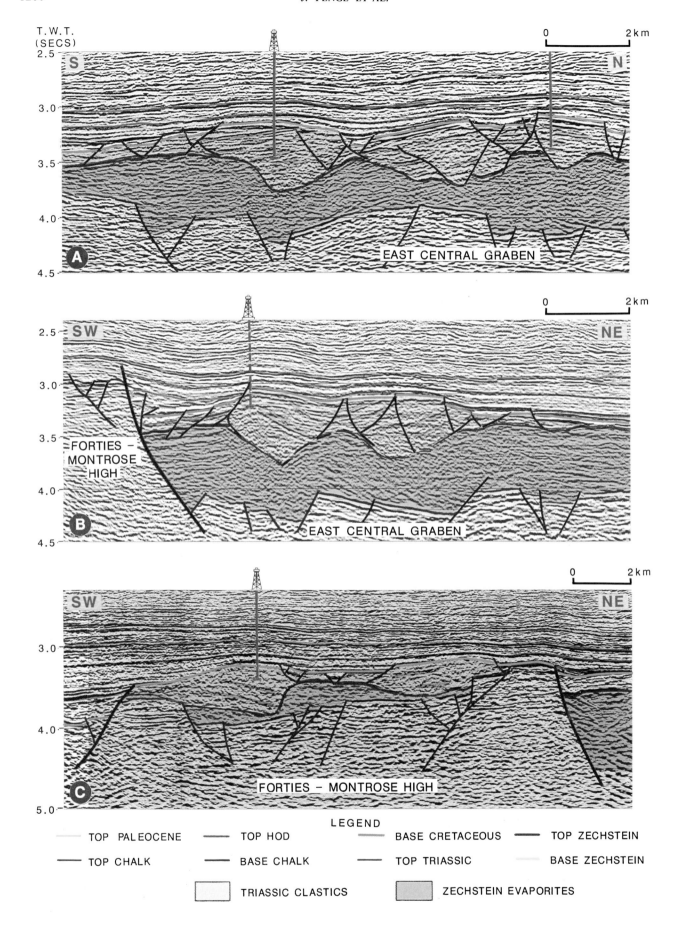

LEGEND

— TOP PALEOCENE   — TOP HOD   — BASE CRETACEOUS   — TOP ZECHSTEIN

— TOP CHALK   — BASE CHALK   — TOP TRIASSIC   — BASE ZECHSTEIN

☐ TRIASSIC CLASTICS   ▨ ZECHSTEIN EVAPORITES

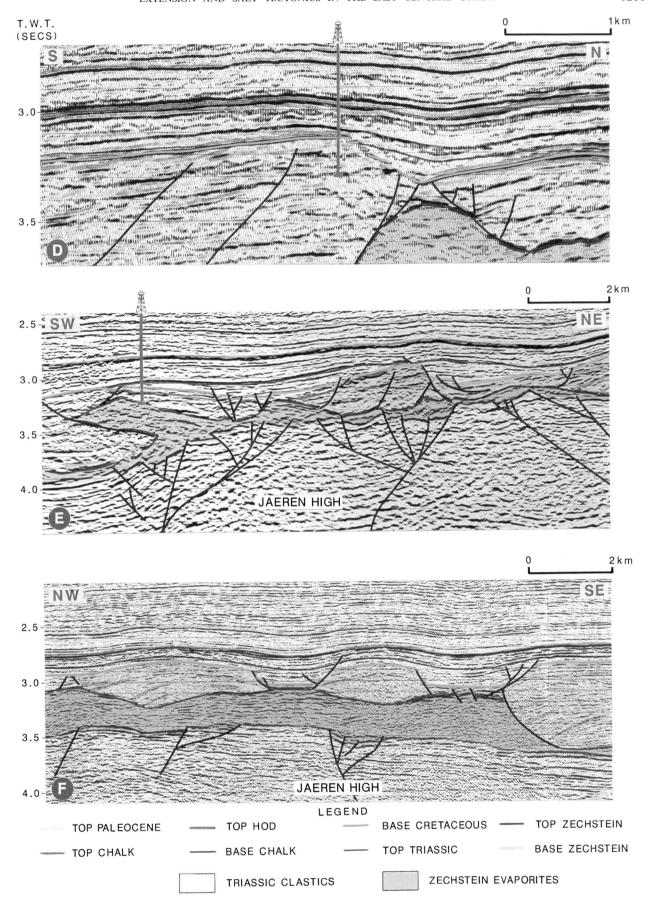

Fig. 4. Morphology of Zechstein and Triassic interval: (a) syn-rotational rafts capped by Skagerrak sandstone, Blocks 22/19 and 22/24, East Central Graben; (b) contra-rotational rafts capped by Skagerrak sandstone, Block 22/18, East Central Graben; (c) contra-rotational rafts of Smith Bank shale, Block 22/23, Forties–Montrose High (courtesy of Simon-Horizon); (d) cuesta capped by Skagerrak sandstone, Block 22/24, East Central Graben (courtesy of BP); (e) cuesta capped by Skagerrak sandstone, Blocks 22/15 and 23/11, Jaeren High; (f) rounded topography of Smith Bank shale rafts, Block 23/11, Jaeren High.

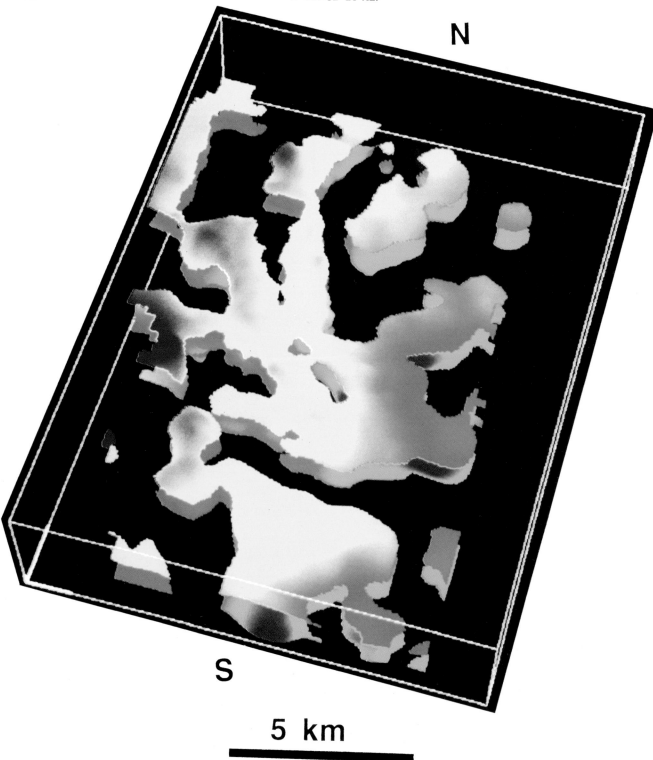

Fig. 5. Three-dimensional perspective of Triassic rafts (coloured) and passive salt walls (black) along the western flanks of the East Central Graben (see text for discussion).

thinning have been made, our estimate of extension for the East Central Graben supports the value of 20% as suggested by Roberts *et al.* (1990*a,b*). To the east of the Montrose Field each rafted block of a contra-rotational pair has dimensions of 1 km thick, 2.6 km (east–west) and 7.2 km (north–south). Further south, rafts, although larger in dimension, exhibit similar ratios.

Areas of thin Mesozoic associated with salt walls have been subjected to a higher degree of deformation than the intervening rafts and display many seismically resolvable faults (Figs 4a–f). None of the observed salt walls pierce the Base Creta-

ceous Unconformity and their crests are at a similar structural elevation. Wells drilled through these Mesozoic thins have encountered both Triassic (Smith Bank or Skagerrak) and Upper Jurassic sediments lying directly above the Zechstein Salt.

Many of the Triassic rafts have been subjected to geomorphological processes, and palaeotopography has been preserved through burial by Upper Jurassic and Cretaceous sediments. Two distinct landforms have been preserved and their relationship to lithology has been demonstrated by well data. In areas capped by Skagerrak sandstone the rafts have the appearance

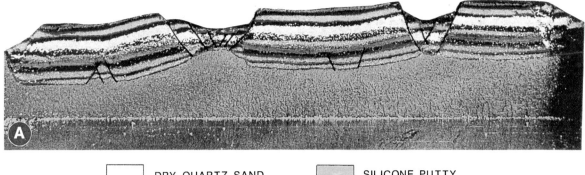

DRY QUARTZ SAND   SILICONE PUTTY

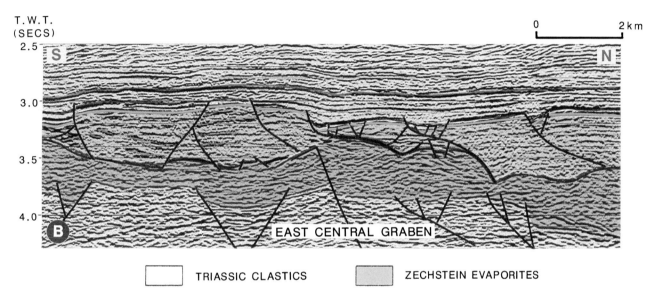

TRIASSIC CLASTICS   ZECHSTEIN EVAPORITES

Fig. 6. Comparison of experimental model to seismic: (**a**) longitudinal section through two-layer model (sand upon silicone) courtesy of B. V. Vendeville (model housing 50 × 50 × 10 cm); (**b**) simple raft, Block 22/23, East Central Graben.

of cuestas, comprising gently dipping conformable Triassic sediments with much steeper fault-line escarpments. These occur along the western margin of the graben (Fig. 4d) and in a few locations on the adjacent Forties–Montrose and Jaeren highs (Fig. 4e). The example in Fig. 4d is a fault-line escarpment some 9 km long with vertical relief of c. 380 m which forms the northern limit of the Marnock Field. Along the margins of this syn-rotational raft, tectonic thinning and reactive salt accommodation occurred, creating a thin and highly fractured Triassic section above the salt walls. Erosional forces during the Early to Middle Jurassic period cut deep incised valleys into the rift zones above the salt. Note that there is evidence of leading-edge erosion due to footwall uplift on the rafts.

On the Jaeren and Forties–Montrose highs the general absence of Skagerrak sandstone from many of the rafts has led to a much greater, more rounded appearance (Fig. 4f). Friable Smith Bank shales, which are more susceptible to erosion, formed a landscape of smooth elongate hills or ridges, often oval in form. The associated valleys separating these features are much broader than those associated with the cuestas within the half-graben.

The presence of rafts within the Triassic section is not a feature of the deeper southeastern part of the East Central Graben. Here, the Triassic is deformed by more listric style faulting and deep rim-synclines have developed adjacent to salt diapirs. No evidence of palaeotopography has been observed in this deeper sector of the basin.

Seismic observations supported by well data demonstrate

that the style of deformation and present-day morphology of the Triassic section is heavily influenced by both lithology and lithological contrasts. A major control is also the presence and thickness of underlying salt.

Observations from physical modelling

Physical models have been used in this study to provide an insight into the mechanisms which were responsible for forming the observed structures and have been used to formulate the structural model presented below. Models constructed by Vendeville et al. (1987) and Vendeville and Jackson (1992a,b) bear close resemblance to structures observed in the East Central graben.

Vendeville et al. (1987) costructed models as two layers with materials of known rheology, the upper unit of dry quartz sand and the lower unit of silicone putty, that could represent the Triassic and Zechstein, respectively (Fig. 6a). The upper sand unit was coloured to form layers to reveal fault throws. To simulate extension, one wall of the housing was removed, allowing the silicone putty to spread and the overlying sand layer to extend. The rate of extension could be controlled by moving the end wall at a constant low speed (c. 1.5 cm per hour).

The results of the experiments were as follows.

1. With instantaneous extension a basal shear stress was applied to sand, causing domino faulting.
2. Under controlled rate extension with stretches in the order

of 20%, no domino faulting was exhibited. Instead a pair of conjugate faults formed with initial dips of 60°, which soled out on the top of the silicone layer.

3. As extension increased minor faults formed parallel to the initial pair. Rift zones thus developed in which severe tectonic thinning took place, and these separated large rafts of undisturbed sand.

A significant feature is the rise of silicone beneath the zones, which is the result of isostatic readjustment due to unloading of the overburden. Small faults at the sand/silicone interface, with throws in the opposite sense to those seen in the rifts, are developed at the margins of these silicone highs. The sand rafts develop a concave upper surface with uplifted rims and a small conjugate fault system at their base. This younger fault system will eventually disect the original raft into two contra-rotational rafts. Further modelling has shown that slow sedimentation rates of the overburden, in addition to slow extension rates, prevent the formation of domino-style faulting and promote the formation of raft and rift-style structures. A thickness to width ratio of 1:2.63 has been measured from this model, which compares with ratios observed from Triassic rafts in the East Central Graben (Fig. 6b).

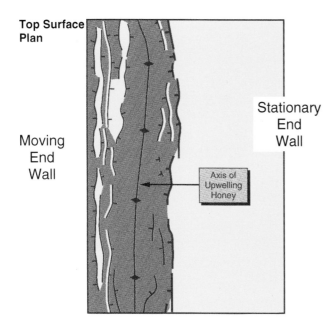

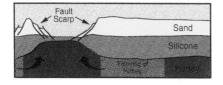

Fig. 7. Uniaxially extending three-layer model (modified after Vendeville *et al.* (1987); model housing 20 × 20 × 20 cm).

A three-layer model by Vendeville *et al.* (1987) provides insight into the behaviour of the Triassic sequence and the Zechstein salt which underlies it (Fig. 7). Here, honey is overlain by a layer of silicon which is, in turn, overlain by a layer of sand. The analogy can be drawn between these layers and the Zechstein Salt, Smith Bank shales and Skagerrak sandstone,

respectively. The overlying sand acts to stiffen the less competent silicone and the honey acts as the decoupling layer. In this case, rafts of sand and silicone are seen to develop.

Finite-element models of up to 30 km in length have been constructed by Amoco's research facility in Tulsa. These numerical models simulate the structures identified in the sandbox models at a scale similar to the seismic lines used in the interpretation.

More recent modelling by Vendeville and Jackson (1992*a,b*) gives an insight into the development of this system with yet further extension. Rotated rafts of sand increase in dip as salt thins with continued extension. Syn-kinematic sedimentation drapes over the leading edge of these rafts. When the blocks ground their sense of rotation begins to reverse.

Tectonic controls on East Central Graben development

The structural features observed from the seismic data are considered to be the product of both halokinesis and extensional tectonics. The resultant features of these mechanisms are as follows.

Halokinesis

Due to its unique physical properties the Zechstein evaporitic sequence has had a major influence on the structural development of the Mesozoic and Tertiary sequences in the East Central Graben.

Zechstein salt is mechanically weak and soluble in water. It has high thermal conductivity and expansivity but low compressibility and is easily mobilized by solid-state flow. Although the salt has a very low density, clastic sediments have lower densities until buried and lithified. Salt will flow at very low temperatures and low differential stresses, and is susceptible to both gravity and tectonic forces.

A variety of mechanisms has been proposed for initiation of salt movement but the most effective method, in the context of the East Central Graben, is differential loading (Jackson and Talbot 1986; Hospers *et al.* 1988). Differential loading is independent of the density contrast between halite and clastic sediments, requiring only that overburden weight is not constant across the upper surface of the salt (due to lateral density variations or topographic relief). This model has been used to support the syn-depositional model. However, it is applicable to the layer parallel model proposed in this paper. The trigger for salt movement in this model is the unloading at the rift zones as the overburden is tectonically thinned. Salt flows passively into this area to maintain isostatic equilibrium, and produces the salt walls observed on the seismic data which are analogous to the silicon highs in the sandbox model. The degree of syn-tectonic deposition during the Jurassic would determine whether the salt wall was accentuated, maintained or reduced in amplitude.

Extensional tectonics

Three principal methods for achieving large horizontal displacements in an extensional basin such as the North Sea have been suggested (Wernicke and Burchfiel 1982). They are:

1. rotation of fault blocks (Fig. 8a);
2. development of curved or listric fault planes (Fig. 8b);
3. development of low-angle normal faults or detachment horizons (Fig. 8c).

Examples of all three mechanisms have been clearly described in other extensional basins. Within the East Central Graben and adjacent highs, a further mechanism is required to form some of the structures observed in the Mesozoic interval above the Zechstein sequence (Fig. 8d). The structures observed are similar in appearance to large-scale boudinage and pinch-and-swell structures. It is suggested that the large-scale rafts in the Triassic can be explained as mega-boudins formed by a process of extension in the plane of layering. A fourth mechanism of achieving large horizontal displacement is therefore proposed in the presence of an underlying ductile layer. This mechanism involves the simultaneous process of extension in the plane of layering and concomitant passive movement of the incompetent material into the resultant accommodation space as follows:

1. extension of a thick clastic sequence along the plane of layering, resulting in necking of the competent sequence;
2. the underlying salt deforms by solid-state flow;
3. necking eventually leads to conjugate rift faulting with separation of the continuous cover sequence into rafts;
4. upwelling of salt into rifted necks occurs due to isostatic compensation;
5. sedimentary rafts float down local or regional gradients on moving salt. These rafts may rotate and tilt until they eventually ground on sub-salt basement.

Intra-raft deformation is limited once rafts are formed but can include conjugate faulting on the raft base and contra-rotation of the raft halves. Footwall uplift of the raft edges close to rift necks may enhance erosion in the rifted areas. Once the rafts ground, rotation can change direction.

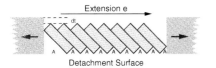

Mode 1
Block rotation above a detachment. Triangular gaps A are left.
(after Gibbs, 1984)

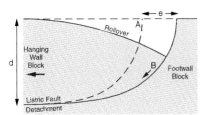

Mode 2
Listric Fault with hanging wall roll-over anticline. Areas A and B are equal. The adjustment in hanging wall shape implies internal strain.
(after Gibbs, 1984)

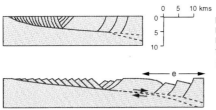

Mode 3
Block rotation above a low-angle extensional fault-shear zone.
(after Wernicke, 1981)

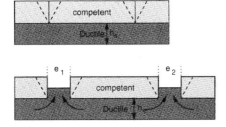

Mode 4
Raft extension of competent layer on a ductile substrate. Substrate flows into rifts in response to differential loading. Rafts more normally bounded by 60° conjugate fault pairs.

Fig. 8. Models illustrating extensional mechanisms.

The two-layer sand/silicone model of Vendeville *et al.* (1987) was constructed to simulate the behaviour of the upper brittle crust above a lower ductile crust. The resultant forms, however, display a remarkable resemblance to models used to

demonstrate the formation of boudinage and pinch-and-swell structures.

Modelling techniques using rocks, rock analogues, photo-elastic rock analogues and finite-element models have confirmed that boudins (either separated by tension, or shear fractures) and pinch-and-swell structures are the result of extension in the plane of layering (Price and Cosgrove 1990). They also re-affirmed Wegmann's (1932) field observations and Ramberg's (1955) experiments that, if the competence contrast was small, the more competent layer formed pinch-and-swell structures before rupturing; conversely, if the competence contrast was large, rupturing occurred after only a small amount of elongation and the more ductile matrix flowed into the neck region. Thickness to width ratios measured from both boudinage and pinch-and-swell structures correspond closely to the observed ratios in the East Central Graben.

Comparison of the sandbox models with Amoco's extensive seismic database has been used to develop schematic models to explain the formation of both syn-rotational and contra-rotational rafts (Fig. 9). The models consist of a competent unit overlying a more ductile unit undergoing continuous extension. The central undeformed section is common to both models illustrated. Model 1 shows the development of structures when initial counter-regional dip occurs, model 2 when the layers remain approximately horizontal. During the early phase of extension both models undergo necking, preferentially located in pre-existing thins. As extension continues, deformation in the rifts becomes more severe and the ductile unit flows into this zone as a result of differential loading. Simple rafts begin to sub-divide into contra-rotational or syn-rotational rafts along conjugate fault zones. The final section for each model illustrates the form of the rafts when the competence contrast is small and ductile deformation predominates. The basement faults, originally located beneath swells in the ductile unit, bear no relationship to the final position of the swells in the model. The effects of erosion have not been taken into account in this sequence.

An extensional model for the formation of structures within the East Central Graben

The model for post-Triassic basin development of the East Central Graben proposed differs from those previously published in the recognition of three significant events:

1. Late Triassic and Late Jurassic layer-parallel extension;
2. rafting of the Triassic section on a mobile Zechstein salt substrate;
3. passive movement of the salt in response to basin extension and modified by differential loading.

These events can be put into chronological context as part of the depositional history and structural evolution of the East Central Graben as summarized below and in Fig. 10. (See Fig. 1 for location.)

Permian (Fig. 10a)

Deposition of platform carbonates on the Forties–Montrose High indicates that it had differentiated early from the East Central Graben, where thick salt (with minor dolomite and anhydrite) was deposited. Thick Zechstein sediments filled in the pre-existing Rotliegendes topography. A thickening wedge of salt occurs on the western and southern side of the Forties–Montrose High but it is absent on the crest, due either to non-deposition or to westward downslope salt evacuation. The western bounding fault of the East Central Graben was active at this time but the Jaeren High did not exist. Thick salt extended eastward from the Forties–Montrose High to the Zechstein depocentre in the Norwegian–Danish Basin.

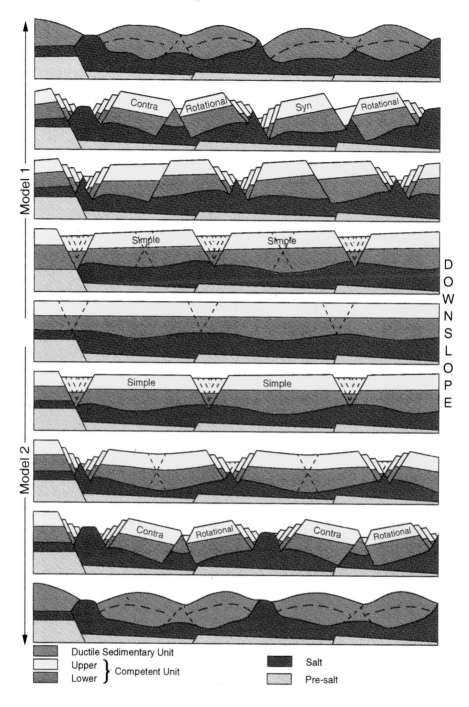

Fig. 9. Two evolutionary models illustrating the formation of complex raft forms.

Early Triassic (Scythian) (Fig. 10b)

The change from evaporitic to clastic deposition was accompanied by regional extension. Thick Smith Bank shales blanketed the area with local depocentres being formed in perturbations on the top salt surface. These depocentres continued to build downwards by differential loading and peripheral salt evacuation. The Rotliegendes interval extended by fault block rotation, while the Zechstein salt was thinned by flowage. There is no clear evidence of salt walls being located above basement faulting at this time (Hospers *et al.* 1988) but this could be due to lateral translation of the rafts and salt walls. Salt walls probably did not reach the surface as there are no halite beds in the Smith Bank shale. Deposition of shale exceeded both salt movement and erosion due to local uplift.

Middle–Late Triassic (Fig. 10c)

There was a gradual transition (locally unconformable) from ubiquitous shale deposition to increasingly proximal fluvial deposition across the whole area. It is likely that Skagerrak sandstones were deposited across the Forties–Montrose High and the Jaeren High area (the latter had still not differentiated at this time). The western bounding fault of the Forties–Montrose High had become inactive, although the high may still have existed since there is evidence of lateral fluvial input from the west. There was a high concentration of fluvial (vs. sheetflood) sandstones on the western side of the East Central Graben, indicating a westerly location of the graben axis. Distal sheetflood and proximal fluvial channel sandstones were derived from the north with minor lateral fluvial input.

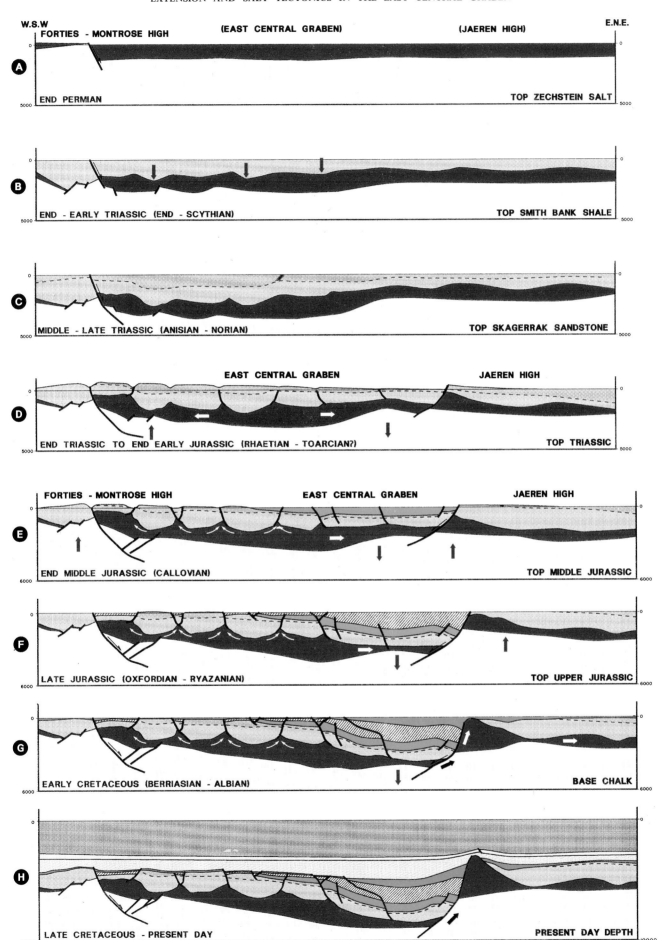

Fig. 10. Evolutionary cross-sections through the East Central Graben.

End Triassic (Fig. 10d)

Deposition ceased at this time and minor extension caused listric movement on the Jaeren High bounding fault. The East Central Graben and Jaeren High finally began to act independently and regional extension caused thinning of ductile halite. On the eastern flank of the Forties–Montrose High, layer-parallel extension of the Triassic clastic sequence and concomitant passive salt flowage into the 'necks' resulted in the formation of simple rafts on a mobile salt substrate as previously described. These rafts were separated by dominantly north–south-oriented rift zones overlying locally rising salt walls which moved passively towards isostatic equilibrium. Analogy with sandbox experiments indicates that the strain rates were low. The regularity in wavelength of rift zones confirms extensional control although possibly enhanced by Triassic depocentres.

During this time the East Central Graben acted as a large fault block with hangingwall subsidence in the east and uplift on its western side. There was net uplift, formation of topography locally in the Triassic and net erosion. Topographic relief probably reached its maximum. Widespread erosion occurred across the half-graben but was concentrated in the rifted necks between rafts. Regional withdrawal of salt from the deeper parts of the basin initiated salt swells adjacent to the Jaeren High bounding fault.

Early–Middle Jurassic (Fig. 10e)

Trends developed at the end of the Triassic continued into the Early to Middle Jurassic, with erosion of the topography to the west of the East Central Graben. Hangingwall subsidence and rising eustatic sea-level led to the deposition of fluvio-deltaic sequences in the deeper eastern and southern parts of the East Central Graben, which onlapped the exposed Triassic landforms.

Salt continued to be evacuated from the deepest part of the half-graben and moved locally into salt walls.

Late Jurassic (Fig. 10f)

Regional extension reached a maximum from Late Oxfordian to Early Kimmeridgian. This extension was relieved by massive growth on the eastern listric bounding fault accompanied by rapid eastward hangingwall subsidence in the half-graben and minor rejuvenation of Rotliegendes basement faults. Progressive deposition of increasingly deep-water Upper Jurassic shales onlapped onto eroded Triassic landforms drowning all but the most prominent Triassic rafts by the end of the Ryazanian. Deposition was concentrated into topographic lows such as rifted necks.

The majority of simple Triassic rafts were further deformed during this extension through internal faulting and block rotation to form contra-rotational and syn-rotational rafts. Regional salt evacuation from the graben centre and localized salt movement accommodated raft rotation and deformation.

Early Cretaceous (Fig. 10g)

Late Jurassic extension declined into the Early Cretaceous with continued listric faulting and associated hangingwall subsidence. The marine incursion which began in the Middle Jurassic continued with progressive overstep of Lower Cretaceous sediments onto previously exposed Triassic rafts. Regional salt evacuation into salt walls occurred to the east of the East Central Graben. Footwall collapse of the Jaeren High contributed to building marginal salt pillows above the Jaeren High bounding fault.

Middle Cretaceous–Present (Fig. 10h)

Rifting ceased in the Early Cretaceous and was replaced by thermal subsidence which has continued to the present day. Triassic rafts were finally drowned by Chalk deposition. The East Central Graben ceased to act as a discrete structural unit by Eocene times when both the Forties–Montrose High and the Jaeren High were drowned. Further burial of Zechstein salt led to massive diapiric upwelling in the deeper southeastern part of the graben. For the first time vertical salt movement became active rather than a passive isostatic response to loading. Diapirs finally breached the sea bed in Paleocene times. Salt in the shallow western part of the half-graben was not buried sufficiently to initiate diapirism and remained in isostatic equilibrium with the overburden. Large-wavelength folds in the Chalk section are related to minor rejuvenation of faults bounding the Triassic rafts. Rejuvenation occurred during Laramide movements. Massive clastic sedimentation has continued to the present day.

Conclusions

Regional seismic data in the East Central Graben demonstrate a complex distribution of Mesozoic and Tertiary sequences. The structures developed in this half-graben can be explained by the interaction of deposition, basin extension, halokinesis and subsequent erosion. Physical modelling and small-scale analogues have been used to clarify the sequence of events and the relative importance of processes that acted on the Mesozoic interval.

Regional extension was the dominant control on the development of structures with the presence of thick halite fundamentally modifying the post-Zechstein section. Post-Triassic layer-parallel extension formed rafts of Triassic rocks floating on salt, which were further extended and deformed in the Late Jurassic period. These rafts have influenced Middle Jurassic to Lower Cretaceous sedimentation. This layer-parallel extension also resulted in the lateral migration of salt walls. Triassic rafts resemble small-scale pinch-and-swell and boudinage structures and have the same extensional origin.

Physical modelling has clarified the processes of raft formation and behaviour, and indicates that end Triassic competence contrasts and strain rates were low but Late Jurassic strain rates were higher. The exact form of the structures produced is governed by, among other factors, the thickness of both the salt sequence and the overburden. Variations of lithology within the rafts modify the morphology of the rafts. Rafts capped by Skagerrak sandstones produce cuestas, while rafts consisting of Smith Bank shale tend to form low-relief rounded hills. The influence of salt thickness may be seen by comparing data from the East Central Graben with the Irish Sea where the salt substrate is thin and layer-parallel extension is clearly the mechanism by which these structures formed. In this area, rafts displaying similar features to those illustrated above have been identified. The major difference is the absence of intervening passive salt walls. The similarities between these two areas supports the hypothesis that layer-parallel extension is the dominant mechanism within the East Central Graben.

The paper is published with permission of Amoco (UK) Exploration Company and the co-authors wish to acknowledge the freedom allowed to discuss and develop the concepts presented in this paper. We would like to thank the following Amoco personnel whose work contributed significantly to the draughting of the paper: Natalie Guainiere, Graham Ball, Chris Raymond and Paul Seaton. Special thanks are due to Bruno Vendeville whose experimental modelling helped clarify many of our observations. We would also like to thank Nigel Higgs (Amoco Production Research) for his support during the writing of this paper.

References

CARTWRIGHT, J. 1987. Transverse structural zones in continental rifts —an example from the Danish sector of the North Sea. In: BROOKS, J. AND GLENNIE, K. W. (eds) Petroleum Geology of North West Europe, Graham & Trotman, London, 441–452.

GLENNIE, K. W. 1990. Outline of North Sea history and structural framework. In: GLENNIE, K. W. (ed.) Introduction to the Petroleum Geology of the North Sea, 3rd edition. Blackwell Scientific Publications, Oxford.

HØILAND, O., KRISTENSEN, J. AND MONSEN, T. 1993. Mesozoic evolution of the Jaeren High area, Norwegian Central North Sea. In: PARKER, J. R. (ed.) Petroleum Geology of Northwest Europe: Proceedings of the 4th Conference. Geological Society, London, 1189–1195.

HODGSON, N. A., FARNSWORTH, J. AND FRASER, A. J. 1992. Salt-related tectonics, sedimentation and hydrocarbon plays in the Central Graben, North Sea, UKCS. In: HARDMAN, R. F. P. (ed.) Exploration Britain: Geological insights for the next decade. Geological Society, London, Special Publication, 67, 31–64.

HOSPERS, J., RATHORE, J. S., FENG, J., FINNSTROM, E. G. AND HOLTHE, J. 1988. Salt tectonics in the Norwegian–Danish basin. Tectonophysics, 149, 35–60.

JACKSON, M. P. A. AND TALBOT, C. J. 1986. External shapes, strain roles and dynamics of salt structures. Geological Society of America Bulletin, 97, 305–323.

PRICE, N. J. AND COSGROVE, J. W. 1990. Analysis of Geological structures. Cambridge University Press, Cambridge, 405–433.

RAMBERG, H. 1955. Natural and experimental boudinage and pinch-and-swell structures. Journal of Geology, 63, 512–526.

ROBERTS, A. M., PRICE, J. D. AND OLSEN, T. S. 1990a. Late Jurassic half-graben control on the siting and structure of hydrocarbon accumulations: UK/Norwegian Central Graben. In: HARDMAN, R. F. P. AND BROOKS, J. (eds) Tectonic Events Responsible for Britain's Oil and Gas Reserves. Geological Society, London, Special Publication, 55, 229–257.

——, YIELDING, G. AND BADLEY, M. 1990b. A Kinematic model for orthogonal opening of the late Jurassic North Sea rift system, Denmark–mid Norway. In: BLUNDELL, D. J. AND GIBBS, A. D. (eds) Tectonic Evolution of the North Sea Rifts. Oxford University Press, Oxford.

VENDEVILLE, B. C., COBBOLD, P. R., DAVY, P., BRUN, J. P. AND CHOUKROUNE, P. 1987. Physical models of extensional tectonics at various scales. In: COWARD, M. P., DEWEY, J. F. AND HANCOCK, P. L. (eds) Continental Extensional Tectonics. Geological Society, London, Special Publication, 28, 95–107.

—— AND JACKSON, M. P. A. 1992a. The fall of diapirs during thin-skinned extension. Marine and Petroleum Geology, 9, 354–371.

—— AND —— 1992b. The rise of diapirs during thin-skinned extension. Marine and Petroleum Geology, 9, 331–353.

WEGMANN, C. E. 1932. Nore sue le boudinage. Bulletin de la Société Géologique de France, 2, 477–491.

WERNICKE, B. AND BURCHFIEL, B. C. 1982. Modes of extensional tectonics. Journal of Structural Geology, 4, 105–115.

Discussion

Question (D. G. Quirk, NAM, Assen, The Netherlands):

My experience of salt tectonics in the Southern North Sea is that there is a much more direct relationship between basement faults, salt walls and post-salt structures than you have shown. How can you mechanically justify 'passive salt movement' in an extensional sense?

Answer (J. Penge et al.):

Figures 4a through 4f illustrate cases where basement faults both underlie salt walls and others where no such relationship exists. Figure 6a is a physical model with a planar base in which passive salt walls are produced in an extensional environment. The mechanism involved is differential loading which occurs due to the tectonic thinning of the overburden. There is no need to invoke basement faulting to produce the structures that have been demonstrated to exist in the East Central Graben.

Question (R. A. James, BP Norway Ltd, Forus, Norway):

Your model depends critically upon the thickness of salt beneath the Triassic 'rafts'. Estimation of this depends in turn upon recognition of the sediment–salt interface at the base of the Triassic. In my experience, picking this interface on seismic is very difficult. Do you have any independent evidence of the depth of top salt, such as from drilling or from gravity data?

Answer (J. Penge et al.):

Well ties to the Top Zechstein exist but are relatively rare in released wells. The Top Zechstein pick is generally well defined beneath the 'rift' zones associated with passive salt walls (Figs 4a–f) and where touchdown occurs beneath Triassic 'rafts'. Our interpretation is further constrained where wells have penetrated a significant Triassic sequence and the Base Zechstein reflector is unambiguous.

However, the model proposed is principally one of layer-parallel extension. In the East Central Graben it takes the form of raft and rifts, not only due to the presence of salt but also to overburden thickness, bulk strain, strain rates and rheological contrast. It was the desire to promote an understanding that several factors have contributed to the formation of this tectonic style which led to the writing of this paper. It is the belief of the authors that neither salt withdrawal nor classic extensional mechanisms alone fully explain the observed features.

Question (N. Hodgson, British Gas Exploration & Production, Reading):

Have the authors observed that the Triassic sediment pod geometries presented on their E–W (dip) seismic lines are also clear on N–S (strike) seismic lines, i.e. the pods are 3-dimensional? If these geometries of N–S orientation are again to be generated by Jurassic faulting, would this not require a significant phase of N–S extension in the Jurassic that has not been previously identified?

Answer (J. Penge et al.):

Figure 5 clearly demonstrated that the rafts ('pods') are 3-dimensional. There is no need to involve a separate phase of north–south extension to explain this observation since the true dip of the East Central Graben is approximately east-southeast and layer-parallel extension has occurred in this direction. Therefore, it is not unreasonable that a north–south component of the extension vector exists. The analogy can be drawn with chocolate-tablet boudinage where extension occurs in more than one direction simultaneously.

Question (A. M. Spencer, Statoil, Stavanger, Norway):

In his concluding remarks, Dr Huckerby stated that 'rifting and rafting [of the Triassic sequence] post-dates the whole of Triassic deposition'. To the east, in the Jaeren High area, intra-Triassic halokinesis accompnaied sediment deposition in basins, as demonstrated by the presence of cross-cutting seismic reflectors within the Triassic sequences (see illustration in Høiland et al. 1993). Have the authors seen such reflections within the Triassic sequences they describe?

Answer (J. Penge et al.):

Not within the study area of the East Central Graben. Internal reflections within the relatively undeformed Triassic rafts are sub-parallel (Figs 4a–f and 6b). In the deeper part of the basin to the southeast where rafting does not occur, the intra-Triassic events are also sub-parallel to the Top Zechstein event.

Relationships between basement faulting, salt withdrawal and Late Jurassic rifting, UK Central North Sea

D. ERRATT

ESSO Exploration and Production UK, Esso House, Ermyn Way, Leatherhead, Surrey KT22 8UY, UK

Abstract: Seismic data are presented to illustrate the variety of structures found at the margins of the Central Graben and intra-graben horsts. The complexity of these structures is due to the interplay of Late Cimmerian basement extension and older salt withdrawal structures within the Zechstein–Jurassic interval. Basement extension occurred in phases from the Permian to Late Jurassic. Salt withdrawal commenced in the Triassic and continued to the Cretaceous, resulting in a network of salt withdrawal synclines and areas of salt preservation, termed 'pods' and 'interpods', respectively. During Late Jurassic extension and rifting, the interaction of basement normal faults and the overlying Zechstein–Jurassic section resulted in structures which vary according to the precise location of the basement fault beneath a pod–interpod complex. Variation in the type and complexity of structuring along basin margins results from the highly variable nature and degree of salt withdrawal prior to Late Jurassic rifting. Complex graben margin structures frequently lend themselves to be interpreted as wrench faults, with different implications for structural and stratigraphic interpretations compared to salt withdrawal interpretations. Careful evaluation of structural style at graben margins therefore has important implications for both prospect mapping and regional structural models

The subject of this paper is the interaction of basement rifting and salt tectonics in the North Sea Central Graben, with respect to the complexities encountered at the edges of graben margin platforms and intra-graben highs.

Earlier papers, most notably by Ziegler (1981), Glennie (1986), Cayley (1987) and Roberts *et al.* (1990*a*), address the relationship between salt and basement tectonics, and the control this has exerted on the stratigraphic development of the basin.

Structural style, particularly at graben margins, has been the subject of papers by Gibbs (1984*a* and *b*, 1987) and, in specific cases (e.g. detachment faults and/or salt withdrawal), these have been the subject of discussion by other authors (e.g. Smith 1987).

Complex graben margin structures frequently display characteristics often associated with wrench faulting, an interpretation which has important implications for the detailed evaluations of the structural and stratigraphic components of graben margin prospects, as well as larger scale models of basin evolution.

The central theme of this paper is a discussion of the importance of the interplay of early tectonic phases, distribution of Zechstein salt and the style of Late Jurassic faulting, with a special emphasis on graben margin structures. Seismic data are discussed in terms of the interaction between salt withdrawal and basement extension.

Regional setting and tectonic history

A detailed discussion of the structural evolution of the Central Graben will not be attempted here, and readers are referred to Ziegler (1990) and references therein for such a discussion. The timing of basement rifting and salt tectonics is, however, pertinent to any discussion of graben margin structures and thus a brief summary of the tectonic history of the Central Graben is given below. A map (Fig. 1), schematic cross-sections (Fig. 2) and stratigraphic column (Fig. 3) accompany the summary.

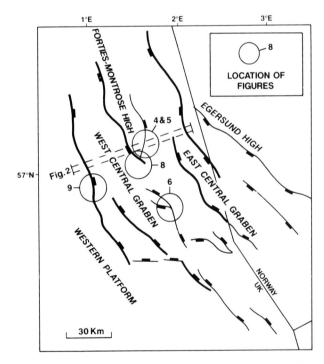

Fig. 1. Structural elements of the Central Graben, showing locations of seismic data referred to in the text.

Permian: initial subsidence

The problems in recognizing Base Zechstein (i.e. basement) faults of pre-Cimmerian age are that they are obscured by the subsequent overprint of halokinesis and Cimmerian extension (Ziegler 1982). Juxtaposition of Zechstein carbonate and evaporite facies on adjacent highs and troughs give some indication of pre-Cimmerian relief (e.g. Cayley 1987; Roberts *et al.* 1990*a*) providing due account is taken of the carbonate facies on the present-day structural highs (e.g. shelfal carbonates on

From *Petroleum Geology of Northwest Europe: Proceedings of the 4th Conference* (edited by J. R. Parker).
© 1993 Petroleum Geology '86 Ltd. Published by The Geological Society, London, pp. 1211–1219.

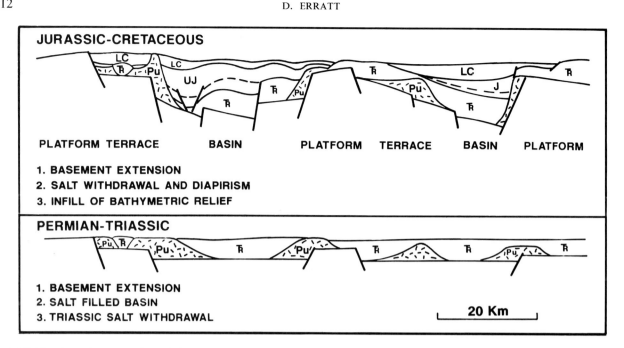

Fig. 2. Highly schematic cross-sections of the Central Graben rift system, illustrating a variety of structural configurations occurring at the margins of platforms and intra-graben highs (vertical exaggeration is approximately 3:1).

the Argyll horst compared to basinal carbonate facies on the neighbouring Auk horst (Bifani 1985).

Triassic–Mid-Jurassic: rifting and salt withdrawal

Roberts *et al* (1990a) refer to well-documented Triassic rift episodes in neighbouring basins in support of inferrred Triassic extension of the Central Graben. Thickening of Triassic strata across the eastern margin of the Forties–Montrose high, with little or no associated thickening of the Upper Jurassic, at least points towards little or no Late Cimmerian overprint. The role of salt withdrawal at such locations remains an important factor in distinguishing between Permian and Triassic rifting.

Throughout this period deposition within the graben system was controlled by salt withdrawal (Cayley 1987; Roberts *et al.* 1990a). The lack of widespread Lower Jurassic strata has been attributed to early Mid-Jurassic updoming of the region (Ziegler 1981).

Late Jurassic: rifting and salt withdrawal

Grounding of primary salt withdrawal synclines both pre-dated and coincided with Late Jurassic extension. The internal stratigraphy of these grounded synclines or 'pods' varies with basin position, but is predominantly composed of Triassic Smith Bank Formation in platform areas. Deposition and preservation of Mid–Late Jurassic sediments occurred in adjacent areas of secondary salt withdrawal ('interpods').

Basement extension resulted in fault-related subsidence of 1000 m or more and fault scarps probably persisted throughout the Early Cretaceous (Bertram and Milton 1989). Thermal subsidence and localized salt diapirism persisted throughout the Cretaceous and Tertiary.

The interaction of Late Jurassic basement rifting and the pre-existing network of pods and interpods in the Permian (Zechstein)–Jurassic 'carapace' (*sensu* Gibbs 1984a) resulted in a wide variety of structural configurations at graben margins, examples of which are presented below.

Graben margin structures: examples

Graben margins (which include the edges of marginal plat-

forms and intra-graben horsts alike) may be broadly subdivided according to the increase in the thickness of Upper Jurassic and/or Lower Cretaceous strata into the adjacent trough, and consequently the likelihood of significant Late Jurassic basement faulting (Fig. 2).

Figures 4 and 5 are from margins with little or no Upper Jurassic/Lower Cretaceous thickening whereas Figs 6–9 are from margins displaying significant thickening in one or both of these intervals.

Eastern margin of the Forties–Montrose High (I)

Figure 4 is from regional seismic data as used by a number of previous authors (Glennie 1986; Roberts *et al.* 1990a; Ziegler 1990). The margin shows the increase in accomodation provided by the Base Zechstein offset taken up entirely by Triassic strata which is unconformably overlain by Upper Cretaceous strata. Total withdrawal of the Zechstein has occurred at the fault, with salt thickness increasing towards the east. The relative structural simplicity of the margin is further illustrated by the alignment of a small offset at Base Cretaceous level with the underlying basement fault.

Eastern margin of the Forties–Montrose High (II)

Figure 5 is a seismic line across the same margin 15 km along-strike from Fig. 4 and illustrates the opposite end member of what might be considered as a spectrum of 'pre-Cimmerian' graben margin configurations. At this location, a continuous cover of non-diapiric Zechstein salt is interpreted to extend across the graben boundary. The thickness change points strongly towards a 'pre-salt' graben margin whereas the absence of diapirism suggests little or no subsequent basement-involved tectonics which might have triggered salt movement. A further contrast with Fig. 4 is provided by the absence of differential Cretaceous subsidence across the margin of the basement high. It is proposed that this supports the interpretations of salt distribution in these two examples. A continuous salt cover across the margin in Fig. 5 has prevented the Cretaceous differential compaction of the carapace where salt is absent as in Fig. 4.

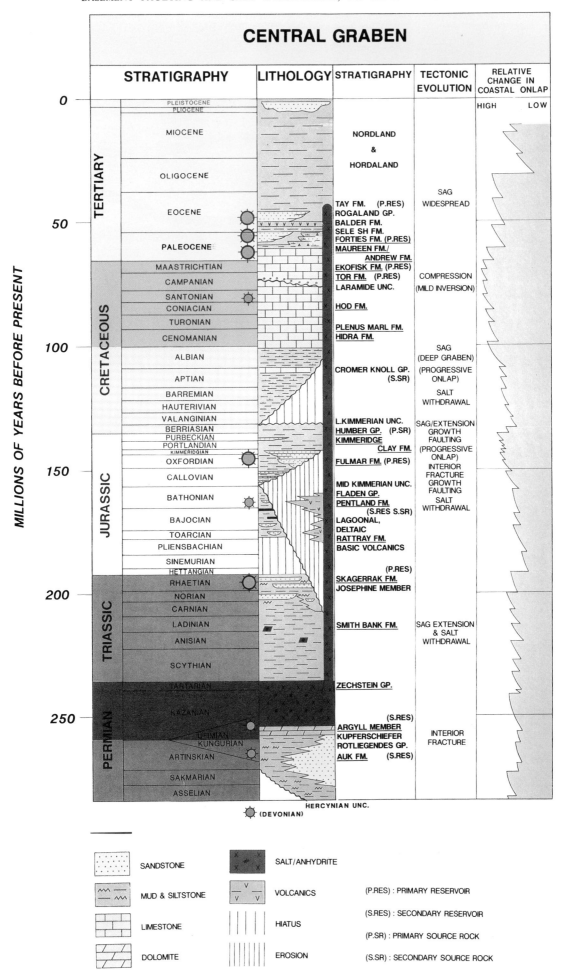

Fig. 3. Central North Sea stratigraphic column (from Cayley (1987), reproduced by permission of the author and Graham & Trotman).

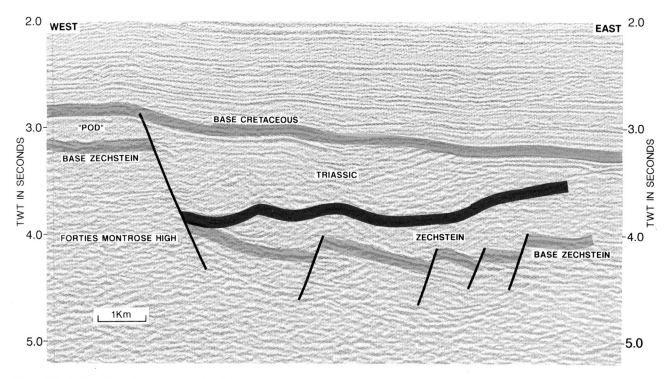

Fig. 4. Seismic data from the eastern margin of the Forties–Montrose High, showing total salt withdrawal at the site of a pre-Jurassic basin margin. (Courtesy of NOPEC, Schlumberger and Geco-Prakla.)

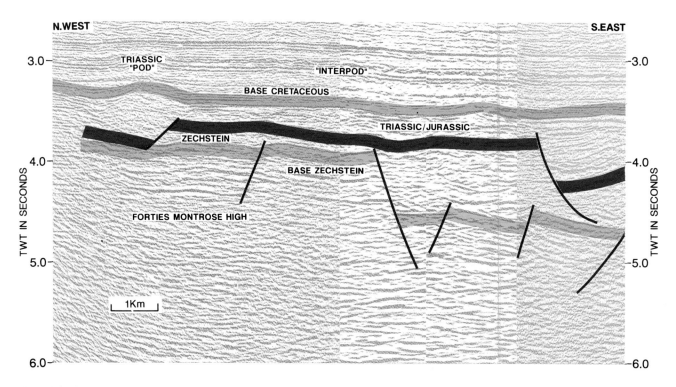

Fig. 5. Seismic data from the eastern margin of the Forties–Montrose High, showing a continuous undeformed salt cover, implying a 'pre-Zechstein' basin margin fault.

Intra-graben horst (I)

The seismic line in Fig. 6 is from a high quality 3D seismic grid and illustrates a basement horst block which shows complex structuring of the carapace by comparison with Figs 4 and 5. Well control indicates that there is no appreciable thickening of the Upper Jurassic section off the flank of the horst. The thickening of the Lower Cretaceous, however, points towards

Late Cimmerian faulting. The most striking feature of the carapace is the control of salt withdrawal on the establishment of individual fault blocks, and the non-alignment of Base Zechstein and Base Cretaceous faults. The non-alignment of basement and carapace fault compartments might be explained by juxtapositions of pods and interpods relative to basement faults, similar to that shown in Fig. 5. Late Cretaceous com-

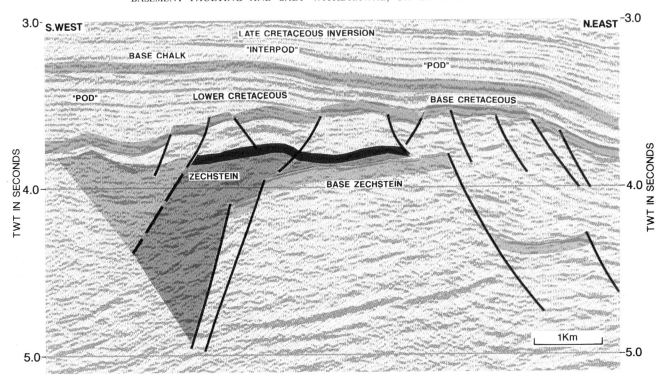

Fig. 6. Seismic data from an intra-graben horst, showing non-alignment of basement (Base Zechstein) and carapace (e.g. Base Cretaceous) fault offsets.

Fig. 7. Seismic line from Fig. 6, showing a reinterpretation of the southwestern flank of the horst as a reverse fault resulting from forced folding of the carapace above a basement normal fault (compare to Fig. 11).

pression has used Zechstein salt as a decollement horizon, displacing the compressional monocline highside of the main basement fault zone. The thickening of the Lower Cretaceous to the southwest of the horst suggests movement and/or relief across the margin. The deformation of the carapace may occur as a series of normal faults stepping down towards the trough, or alternatively as a reverse fault, formed as a result of forced folding of the carapace above a basement normal fault (Fig. 7).

Intra-graben horst (II)

Figure 8 from the Forties–Montrose High, shows a non-alignment of the margin offsets at basement and carapace levels, and as such, is comparable to the structure in Fig. 6, the principal difference being the apparent lack of salt on the horst.

The structure is associated up-dip (towards the platform)

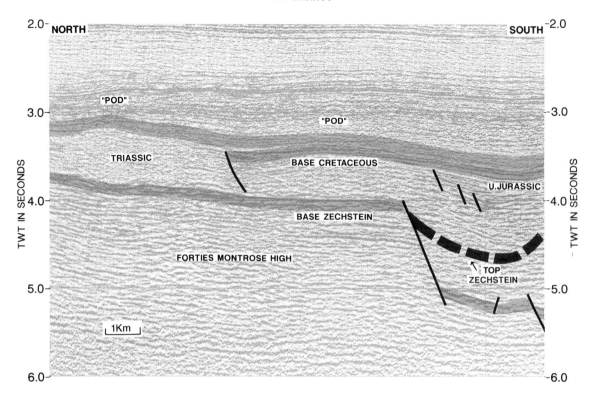

Fig. 8. Seismic data from the Forties–Montrose High, showing non-alignment of basement and carapace faults, and different timing of salt withdrawal on the platform as inferred from the thickness variations of the interpreted Triassic sequence.

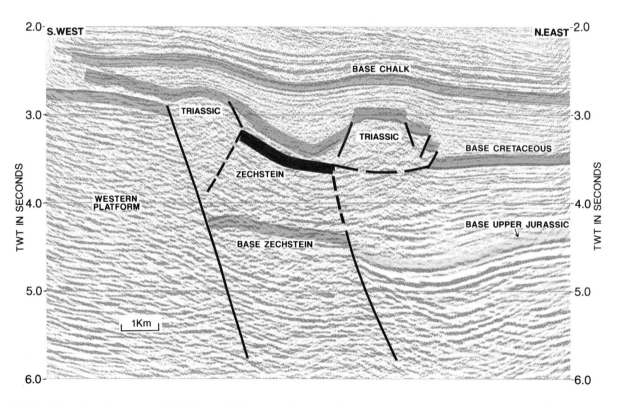

Fig. 9. Seismic data from the margin of the Western Platform, showing a possible allochthonous Triassic sequence overlying the Upper Jurassic of the deep Western Graben.

with a grounded pod of Triassic Smith Bank Shales. The boundary fault between the two carapace fault compartments is interpreted to be a site of salt evacuation from beneath the lower of the two pods, which may have occurred in response to Late Jurassic extension, thus post-dating the Triassic salt withdrawal from beneath the up-dip grounded pod.

Western Platform margin

The final example (Fig. 9) is from the margin of the Western Platform and illustrates a complex graben edge configuration. As in Figs 6 and 8, this example shows a non-alignment of Base Cretaceous and Base Zechstein offsets. However, in this instance the structure is further complicated by the overhang of the Base Cretaceous relative to the Base Upper Jurassic marker of the deep graben (this interpretation has been supported by geophysical modelling studies). The involvement of salt is seen on additional data to the north which clearly show salt diapirism. Of further significance is the transparent seismic character of the 'over-hanging' structure, typical of Smith Bank Pods of horst/platform affinity.

Discussion

Timing of 'Pre-Jurassic' basement rifting episodes

The two seismic lines from the eastern flank of the Forties–Montrose High (Figs 4 and 5) illustrate that there are inherent difficulties in establishing the timing of basement movement, even in areas without any obvious Late Cimmerian overprint. The seismic example in Fig. 5, with continuous non-diapiric Zechstein across the margin, suggests a pre-Zechstein fault. However, it is possible that Triassic rifting triggered salt withdrawal, and subsequent exposure of the diapiric area resulted in the Top Zechstein being 'planed off' to the present-day profile. Figure 4 shows thickening of the Triassic across the margin, with complete salt withdrawal. The change in thickness could be due entirely to pre-rift (i.e. Permian) or syn-rift (i.e. Triassic) salt evacuation. Both Cayley (1987) and Roberts *et al.* (1990a) refer to the possible significance of Zechstein facies associations, and in this example (Fig. 4) Zechstein of entirely anhydrite/dolomite facies penetrated along-strike on the high side of the margin suggests both pre-existing Zechstein topography and an increase in the relief of the margin relative to the example in Fig. 5, thus controling Zechstein facies distributions along the fault zone (i.e. passage from evaporite–carbonate). Establishing whether the carbonates are of shelf or basin facies is a pre-requisite in eliminating the possibility of subsequent erosion of salt, as has occurred on the Auk Horst, while a further consideration would be the possibility of late (i.e. Jurassic–Early Cretaceous) salt withdrawal resulting in a remnant anhydrite–dolomite Zechstein sequence.

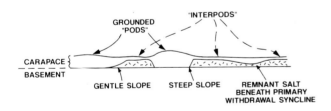

Fig. 10. The variation of structuring associated with salt withdrawal. Horizontal and vertical dimensions of pods vary dramatically across the basin on a scale of hundreds to thousands of metres.

The consequences of salt withdrawal: pod/interpod geometries

Figure 10 illustrates the variability of structuring resulting purely from salt withdrawal which would have been largely in place at the onset of Late Cimmerian rifting. Primary withdrawal synclines may have experienced complete salt withdrawal, resulting in a grounded pod. Alternatively, remnant salt beneath the syncline may result in a 'flat-topped' structure. The transitions between withdrawal synclines and interpods are also liable to vary, from those with steep-sided salt walls to those showing more gentle dips at top salt level.

Interaction of carapace with basement faulting

Extensional forced folds and distributed extension (Fig. 11) Withjack *et al.* (1990) have demonstrated how deformation patterns associated with forced folds are significantly affected by the presence of a ductile detachment layer such as Zechstein salt. In the absence of a detachment horizon, the forced fold develops into a simple through-going normal fault in the carapace, which would provide a relatively simple graben margin geometry (as might occur with rifting beneath a grounded pod). The introduction of a detachment layer produces geometries such as that shown in Fig. 11. This would compare with Figs 6 and 8, and suggests a control on the location of normal faults on the footwall. In the Central Graben this may be a function of the extent of salt on the high as controlled by the distribution of grounded pods. Similarly, as regional extension is transferred from the basement to the carapace via the Zechstein (Roberts *et al.* 1990b), occurrence of grounded pods may act as buffers, with normal faulting transmitted through the Triassic/Jurassic section at their margins, which in turn could act as sites of secondary salt withdrawal.

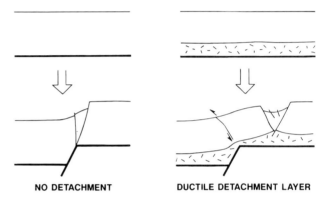

Fig. 11. Experimental models of extensional forced folds (after Withjack *et al.* 1990), showing the formation of a significant reverse fault in the presence of a ductile detachment layer.

Location of basement fault (Fig. 12) Figures 4, 5 and 6 provide evidence that the presence of a 'Permo-Triassic' basement fault will not necessarily control the distribution of preserved Zechstein salt, which may occur either as a continuous cover across the fault or be totally absent due to complete withdrawal. Furthermore the salt may be restricted to the high or low sides of the basement fault. Formation or reactivation of a basement fault during a Jurassic rift phase may be expected to occur randomly with respect to the overlying pod–interpod complexes of the carapace. This is illustrated by the examples of Figs 4 and 5 from the same margin: one showing continuous salt cover; the other total salt withdrawal. A wide variety of structuring should be expected within the carapace, both along the same fault trend, and between faults.

Salt may or may not be involved in faulting at a graben margin and this, coupled with the precise location of the basement fault beneath a pod–interpod system, will be the principal control on the final complexity of the margin (Fig. 12).

Bathymetric relief at graben margins It has been suggested that graben margin structures may have been significantly modified by the bathymetric relief which would have occurred in the scenario described by Bertram and Milton (1989), where Late Cimmerian rifting may have produced water depths of up to 1000 m. This relief, coupled with an unstable lithology such as Zechstein salt, may be responsible for some of the complex structures and stratigraphic relationships at the margins.

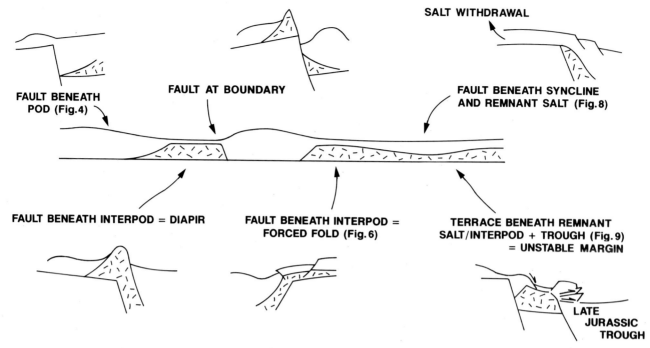

Fig. 12. Interaction of basement tectonics and pod/interpod complexes, with reference to the seismic data presented in the text as examples of the resultant structures at basin margins.

Other models

Wrench/flower structures Many of the structures along the graben margin areas may be interpreted in simple cross-section as flower structures associated with wrench faulting. Harding *et al.* (1985) present the various criteria for distinguishing simple extensional fault styles from those of divergent and convergent wrenches. This paper presents a basement extension–salt dissolution model against which other interpretations might be compared. It will be important to consider the inevitable complications that arise from the interaction of salt and basement normal faults (e.g. the experimental models of Withjack *et al.* 1990) before concluding that wrench faults are the preferred interpretation, with the inevitable conclusions regarding basin history and detailed prospect evaluation that would ensue. The interpretations presented here are consistent with the regional extensional models proposed by Roberts *et al.* (1990c).

Gravity structures Smith (1987) in his discussion of the Clyde Field has already pointed to the difficulties posed by salt withdrawal when evaluating the role of growth fault-related structures, which may look superficially similar to the salt withdrawal 'pods' (e.g. Gibbs 1984b). Gibbs (1987) discussed the potential role of low angle faulting in the development of graben margins. The widespread occurrence of salt on the platforms has now been demonstrated and its importance is primarily manifested through salt withdrawal. At graben margins, the presence of a slip horizon and bathymetric relief might result in the involvement of low angle detachments in the deformation of interpods or primary withdrawal synclines with remnant salt.

Conclusions

The degree of Late Jurassic movement on Base Zechstein faults varies widely across the Central Graben as indicated by the thickness variation of Upper Jurassic and Lower Cretaceous strata between platform areas and adjacent lows.

Salt withdrawal, commencing in the Triassic, resulted in a complex network of grounded primary withdrawal synclines, and intervening areas of salt preservation. This arrangement of pods and interpods was largely in place prior to Late Jurassic rifting.

The interaction of Late Jurassic basement faults with the overlying carapace resulted in a wide range of structures at the graben margins, controlled primarily by the precise location of the underlying basement fault relative to the overlying pod–interpod complex.

The resulting structures in the carapace developed by a number of processes, which included salt withdrawal and diapirism, forced folding, gravity sliding and faulting in response to basement extension displaced by the Zechstein décollement. These features frequently display characteristics which bear a striking similarity to those commonly associated with wrench faults, an alternative interpretation with very different structural and stratigraphic implications for both prospect and field assessment, as well as regional tectonic models. Careful consideration of the alternatives offered by these different models will be important in the assessment of the remaining hydrocarbon potential of the UK Central North Sea.

I thank Dave Curtin and numerous colleagues from Esso UK for helpful suggestions during the preparation of this paper. Phillips Petroleum, Amerada Hess and Geco-Prakla/Nopec are thanked for permission to reproduce seismic data. The paper appears by permission of the management of Esso Expro UK and Shell Exploration and Production UK but the interpretations and conclusions presented are the responsibility of the author.

References

BERTRAM, G. T. AND MILTON, N. J. (1989). Reconstructing basin evolution from sedimentary thickness; the importance of palaeo-bathymetric control with reference to the North Sea. *Basin Research*, **1**, 247–257.

BIFANI, R. 1985. A Zechstein depositional model for the Argyll Field. *In*: TAYLOR, J. C. M. *et al.* (eds) *The role of evaporites in hydrocarbon exploration*. JAPEC Course Notes No. **39**.

CAYLEY, G. T. 1987. Hydrocarbon migration in the Central North Sea. *In*: BROOKS, J. AND GLENNIE, K. W. (eds) *Petroleum Geology of North West Europe*. Graham & Trotman, London, 549–555.

GIBBS, A. D. 1984*a*. Structural evolution of extensional basin margins. *Journal of the Geological Society, London,* **141**, 609–620.

—— 1984*b*. Clyde Field growth fault secondary detachment above basement faults in the North Sea. *American Association of Petroleum Geologists Bulletin,* **68**, 1029–1039.

—— 1987. Development of extensional and mixed mode sedimentary basins. *In*: COWARD, M. P., DEWEY, J. F. AND HANCOCK, P. L. (eds) *Continental Extensional Tectonics*. Geological Society, London, Special Publication, **28**, 19–33.

GLENNIE, K. W. 1986. The structural framework and the pre-Permian history of the North Sea area. *In*: GLENNIE, K. W. (ed.) *Introduction to the petroleum geology of the North Sea*, 2nd edn, Blackwell Scientific Publications, Oxford, 25–62.

HARDING, T. P., VIERBUCHEN, R. C. AND CHRISTIE BLICK, N. 1985. Structural styles, plate tectonic settings and hydrocarbon traps of divergent (transtensional) wrench faults. *In*: BIDDLE, K. T. AND CHRISTIE BLICK, N. (eds) *Strike slip deformation, basin formation and sedimentation*. Society of Economic Paleontologists and Mineralogists, Special Publication, **37**, 51–77.

ROBERTS, A. M., PRICE, J. AND OLSEN, T. S. 1990*a*. Late Jurassic half graben control on the siting and structure of hydrocarbon accumulations: UK/Norwegian Central Graben. *In*: HARDMAN, R.

F. P. AND BROOKS, J. (eds) *Tectonic Events Responsible for Britain's Oil and Gas Reserves*. Geological Society, London, Special Publication, **55**, 229–257.

——, YIELDING, G. AND FREEMAN, B. 1990*b*. Conference report: the geometry of normal faults. *Journal of the Geological Society, London,* **147**, 185–187.

——, —— AND BADLEY, M. E. 1990*c*. A kinematic model for the opening of the Late Jurassic North Sea rift system, Denmark–mid Norway. *In*: BLUNDELL, D. J. AND GIBBS, A. D. (eds) *Tectonic Evolution of the North Sea Rifts*. Clarendon Press, Oxford, 180–199.

SMITH, R. L. 1987. The structural development of the Clyde Field. *In*: BROOKS, J. AND GLENNIE, K. W. (eds) *Petroleum Geology of North West Europe*. Graham & Trotman, London, 523–531.

WITHJACK, M. O., OLSON, J. AND PETERSON, E. 1990. Experimental Models of Extensional Forced Folds. *American Association of Petroleum Geologists Bulletin,* **74**, 1038–1054.

ZIEGLER, P. A. 1981. Evolution of sedimentary basins in North-West Europe. *In*: ILLING, L. V. AND HOBSON, G. D. (eds) *Petroleum Geology of the Continental Shelf of North-West Europe*. Heyden, London, 3–39.

—— 1982. Faulting and graben formation in western and central Europe. *Philosophical Transactions of the Royal Society of London,* **305**, 113–143.

—— 1990. Tectonic and palaeogeographic development of the North Sea rift system. *In*: BLUNDELL, D. J. AND GIBBS, A. D. (eds) *Tectonic Evolution of the North Sea Rifts*. Clarendon Press, Oxford, 1–36.

Comparison of Permo-Triassic and deep structure of the Forties–Montrose and Jaeren highs, Central Graben, UK and Norwegian North Sea

N. H. PLATT and P. PHILIP

GECO-PRAKLA Exploration Services, Boundary Road, Woking, Surrey GU21 5BX, UK

Abstract: Recent acquisition of high resolution, deep penetration seismic in the Central North Sea has permitted imaging of the basement structure. The results indicate strong similarities between the deep structures of the Jaeren and Forties–Montrose highs. The Triassic and Zechstein strata show a distinctive relationship on these graben margin and median highs. Interpretation of the pre-Zechstein structure may indicate the distribution of Devonian and Carboniferous intervals in an area where previous imaging of the deep section has not been successful.

Southern parts of both the Jaeren and Forties–Montrose highs show the presence of a sub-Permian reflective package; this (?possible Carboniferous) sequence is absent in the north, where reflectors may be recognized within the deep section. A prominent horizontal reflector may represent a Middle Devonian limestone marker or the Devonian basal unconformity. This reflector may alternatively mark the top of Upper Paleozoic volcanic rocks. A deeper, westward- or northwestward-dipping reflector may represent the top of the crystalline basement or a deeper fault or shear zone of Caledonide origin.

The potential occurrence of Carboniferous strata at depth over extensive areas of the Central Graben and surrounding highs would have important implications for exploration in the area. Carboniferous coal measures might provide a sub-Zechstein gas-prone hydrocarbon source.

Exploration activity over the last 25 years has led to a good understanding of the post-Variscan structure and Permian–Cenozoic stratigraphy of the North Sea basins (Ziegler and van Hoorn 1989; Glennie 1990). The Central Graben remains a prospective target for exploration, and new data provided by drilling and modern 3D seismic continue to offer new insights into the tectonic evolution of this area (Bartholomew *et al.* 1993; Erratt 1993; Sears *et al.* 1993). Despite this, the deep structure and Paleozoic stratigraphy of the Central Graben and its margins remain poorly documented.

Ziegler (1990*a*) noted that the configuration of the Devonian and Carboniferous basins in the Central and Northern North Sea could not be resolved; palaeogeographic maps (Ziegler 1990*b*) thus offer scant details for the Central Graben area during this period. Deep seismic surveys shot across the Central Graben, notably the GECO NSDP84-85 survey, demonstrated the presence of deep reflectors within the lower crust (Klemperer and White 1989; Klemperer and Hurich 1990; Holliger and Klemperer 1990), but were unable to shed new light on the distribution of Devonian and Carboniferous strata in the area. Recently acquired regional seismic surveys permit assessment of the deep section of axial graben areas, and also provide improved imaging of the sub-Zechstein basement on graben marginal and median highs. This paper uses examples from the CGDT89 (Central Graben Deep Tie) and GNSR91 (GECO North Sea Regional) surveys to illustrate and compare the deep structure of the Jaeren and Forties–Montrose highs.

The Jaeren High forms the eastern margin of the East Central Graben (Fig. 1), lying at the western edge of the Norwegian–Danish Salt Basin. The Forties–Montrose High is a mid-graben feature separating the West and East Central Graben. Both highs are essentially fault bounded. However, comparison of seismic lines across the two basement highs reveals a range of structural styles.

Mesozoic–Tertiary structure of the Jaeren High

Northern areas of the Jaeren High show a strongly faulted western margin, with structural control by two main normal faults of large displacement (Fig. 2). Central parts of the high have a margin comprising a series of smaller extensional faults stepping down westwards towards the East Central Graben (Fig. 3). Southern areas of the Jaeren High show an abruptly faulted western margin locally marked by diapir intrusion along the main graben-bounding fault (Fig. 4).

The Tertiary and Mesozoic sections are both relatively thin over the Jaeren High. An eastward traverse from the East Central Graben onto the Jaeren High shows progressive onlap of the Jurassic, Lower Cretaceous and the lower part of the Chalk successions (Fig. 3). The Paleocene sequence is also reduced in thickness. Over large areas of the Jaeren High, a thin Chalk sequence rests unconformably upon acoustically quiet Triassic strata of the Smith Bank Formation. The top of the underlying Zechstein sequence is difficult to pick, on account of the weak contrast in acoustic impedance between the Smith Bank shales and the Zechstein salt; moreover the contact between these two units appears to demonstrate considerable relief (as shown in Fig. 2).

The structure of the Jaeren High is characterized by lozenge-shaped 'pods' of Triassic strata underlain by thin or absent salt (Figs 2 and 3); small graben occur between the pods and appear to show Jurassic or Lower Cretaceous strata resting on a much thinner Triassic, but thicker salt (Fig. 3). Current thinking suggests that the location of these graben on both the Jaeren High and the Western Platform was controlled by withdrawal of salt into the inter-pod areas associated with the rotation and erosion of the pods and their eventual 'grounding' on the sub-Zechstein succession (Høiland *et al.* 1993; Penge *et al.* 1993; Wakefield *et al.* 1993).

The base Zechstein reflector forms a high amplitude regional marker traceable across the whole study area. Lines cutting the northern part of the Jaeren High show the base Zechstein to be strongly faulted (Fig. 3), while southern areas show only minor faulting (Fig. 4).

Mesozoic–Tertiary structure of the Forties–Montrose High

The Forties–Montrose High also shows variable marginal fault geometry. In the north, where the high runs approximately from east to west, the margins are characterized by stepped faults to both the north and south (Fig. 5). Southern areas of the high show more abruptly faulted margins, although the displacement is still taken up on several extensional faults (Fig. 6).

From *Petroleum Geology of Northwest Europe: Proceedings of the 4th Conference* (edited by J. R. Parker).
© 1993 Petroleum Geology '86 Ltd. Published by The Geological Society, London, pp. 1221–1230.

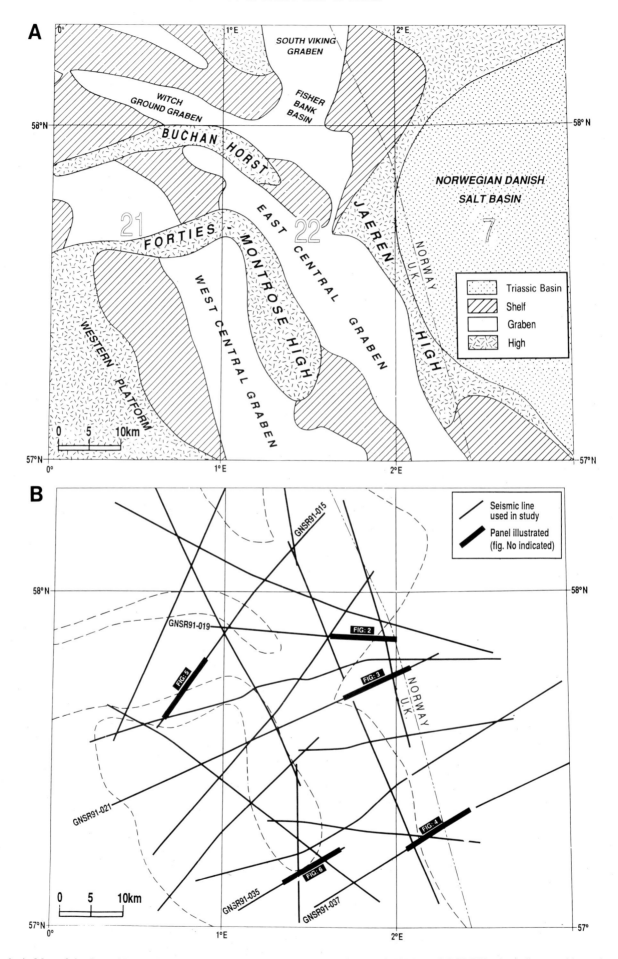

Fig. 1. A. Map of the Central North Sea showing major basement highs and graben. B. GNSR91 and CGDT89 seismic lines used in study, marking panels illustrated in this paper.

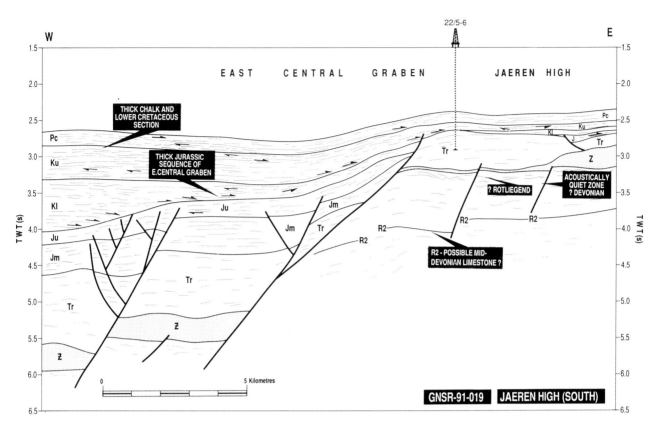

Fig. 2. Drawing of seismic line GNSR91-019 across the Jaeren High (northern area). Pc: Paleocene; Ku: Chalk; Kl: Lower Cretaceous; Ju: Upper Jurassic; Jm: Middle Jurassic; J: Jurassic undifferentiated; Tr: Triassic; Z: Zechstein.

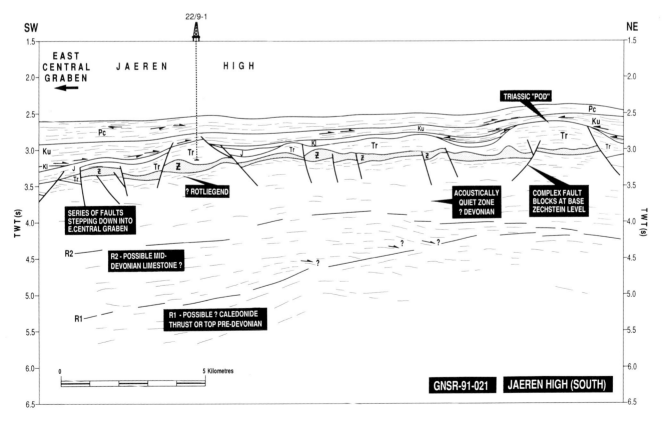

Fig. 3. Drawing of seismic line GNSR91-021 across the Jaeren High (central area). See Fig. 1 for location, Fig. 2 for key.

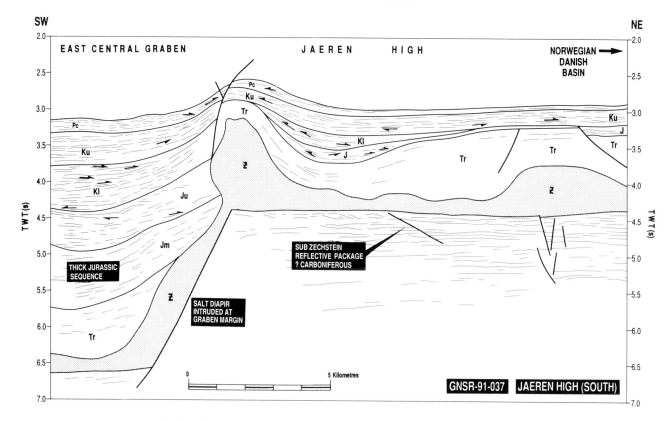

Fig. 4. Drawing of seismic line GNSR91-037 across the Jaeren High (southern area). See Fig. 1 for location, Fig. 2 for key.

Onlap of the Jurassic, Lower Cretaceous and the lower part of the Chalk are evident at the margins of the Forties–Montrose High, as well as a thinning of the Paleocene succession. Lines across the high also reveal the presence of lozenge-shaped 'pods' of Triassic strata similar to those found on the Jaeren High (Fig. 6). However, there is generally less evidence of salt movement and where present the Jurassic–Lower Cretaceous strata appear to drape gentle relief on the top Triassic rather than fill sharply-defined graben as on the Jaeren High.

The northern part of the high shows thick accumulations of highly reflective Middle Jurassic volcanic rocks (Fig. 5). This succession shows a series of high-amplitude sub-parallel reflectors of moderate continuity; the reflectors are erosionally truncated beneath the base Cretaceous unconformity. Their southwestward dip may record progradational growth of a Jurassic volcanic centre located in the Fisher Bank area towards the northeast (Smith and Ritchie 1993). In contrast, volcanic rocks are only locally evident on lines crossing the southern part of the high (Fig. 6).

The base Zechstein seismic marker is strongly disrupted by extensional faulting in the northern and southern areas of the Forties–Montrose High; this contrasts with the situation on southern parts of the Jaeren High where this horizon is largely undeformed (Fig. 4).

Sub-Zechstein stratigraphical development

Comparison of seismic lines across the Jaeren and Forties–Montrose highs provides evidence of regional variations in stratigraphical development. In the northern areas of both highs, a series of high-amplitude sub-parallel reflectors of moderate continuity occurs beneath the base Zechstein marker. This succession, which shows variable time thicknesses of 50–150 ms, has been penetrated by several wells and is composed of red-bed clastic facies assigned to the Lower Permian Rotliegend.

This package appears to lie unconformably upon an acoustically quiet zone characterized by highly discontinuous, low-amplitude, sub-parallel to chaotic reflections 700–1000 ms in time thickness. The nature of this zone is unknown, although the fact that several wells in this area (such as UK 22/10-1 and 22/10-2) encountered Devonian clastic deposits beneath the Rotliegend suggests the presence of an Old Red Sandstone sequence. Facies development may be similar to that in the Midland Valley of Scotland to the west (Fig. 7A).

Several lines across the northern areas of the Jaeren High show the presence of deep reflectors at the base of this acoustically quiet zone. A prominent horizontal reflector (R2) is recognized at two-way-travel times of around 4 s (Figs 2 and 3), while a deeper, gently westward- or northwestward-dipping reflector (R1) is evident on several lines (Fig. 3) between 4 and 5 s two-way-travel time.

The upper reflector R2 shows a positive reflection coefficient (trough-peak configuration on SEG normal polarity) indicating an increase in acoustic impedance. This suggests a transition to rocks of greater density or increased acoustic velocity. One interpretation of reflector R2 is that it marks a base Devonian/top crystalline basement unconformity similar to that proposed for the southern Central Graben by Cartwright (1988) (Fig. 7B); scanty deep well penetrations in the Central Graben area mostly indicate a crystalline basement dominated by phyllites (Frost *et al.* 1981). Alternatively, this reflector may be produced by other acoustically fast lithologies such as carbonate or igneous rocks. The Middle Devonian limestone forms a distinct seismic marker over some areas of the Central Graben and the Mid North Sea High (Pennington 1975) and the top of this limestone appears a good candidate for R2; volcanic rocks and extensive sills are also a characteristic feature of the Midland Valley Devonian, Carboniferous and Permian (Cameron and Stephenson 1985) and could also generate high-amplitude reflections of this type.

The dipping nature of the lower reflector R1 suggests that it may be caused by a low-angle tectonic feature within

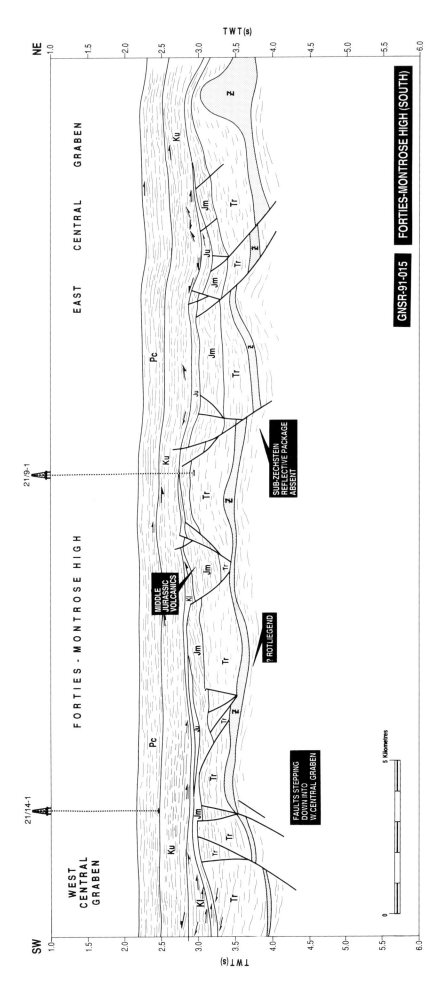

Fig. 5. Drawing of seismic line GNSR91-015 across the Forties-Montrose High (northern area). See Fig. 1 for location, Fig. 2 for key.

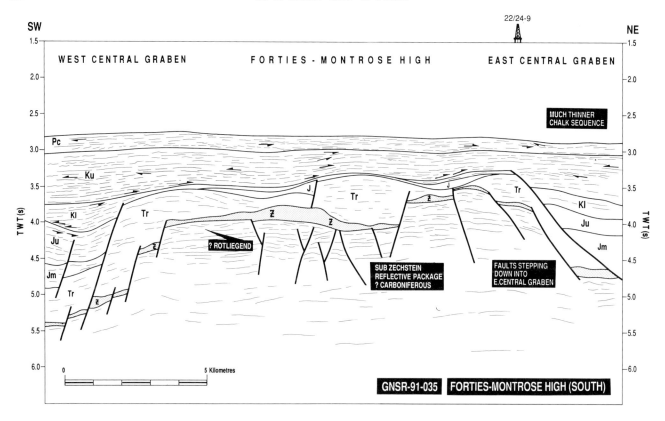

Fig. 6. Drawing of seismic line GNSR91-035 across the Forties–Montrose High (northern area). See Fig. 1 for location, Fig. 2 for key.

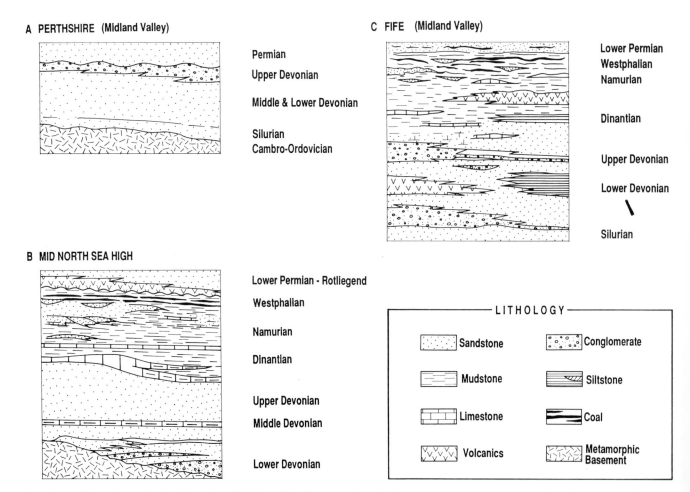

Fig. 7. Possible analogues for the Paleozoic of the Central Graben: (**A**) schematic Devonian succession from the Ochil Hills area, Perthshire, south of the Highland Boundary Fault, Scottish Midland Valley; (**B**) schematic Paleozoic succession from the southern Mid North Sea High/ Danish Central Graben area; (**C**) schematic Carboniferous succession from Fife, Scottish Midland Valley. A and C compiled from published British Geological Survey maps and Cameron and Stephenson (1985), B after Cartwright (1988).

the basement sequence—a Caledonide thrust would be one obvious possibility (Fig. 8). Caledonian structures were recognized in this area by Johnson and Dingwall (1981), who interpreted the en échelon series of northeast–southwest faults in the Viking Graben as reactivated Caledonide features. The faults bounding the north of the Jaeren High have the same trend, although Roberts *et al.* 1990 recently suggested that these faults were entirely Mesozoic structures. Gibbs (1989) and Klemperer and White (1989) interpreted deep reflectors visible in the GECO NSDP84-85 surveys as crustal shear zones of possible Caledonian origin; however, these reflectors occur at two-way-travel times of around 8–10 s and thus appear to be much deeper than the reflectors imaged here.

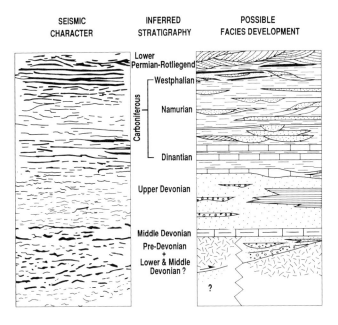

Fig. 8. Seismic character of the pre-Zechstein succession, Jaeren High and Forties–Montrose High, showing possible Paleozoic stratigraphical development, based on analogy with sections shown in Fig. 7. Key as for the Fig. 7.

An alternative interpretation is that the lower reflector R1, representing the top of the crystalline basement, is onlapped by the upper reflector R2, which in this case would record a lithological marker such as the top of the mid-Devonian limestone (Fig. 8) or the top of a volcanic sequence.

Southern areas of both highs show the local development of a package similar to that assigned in the north to the Rotliegend. These areas are marked by the presence underneath this, or directly beneath the Zechstein, of a sub-Permian reflective package averaging 1000 ms in time thickness and showing high-amplitude, parallel reflectors of high continuity (Jaeren High, Fig. 4; Forties–Montrose High, Fig. 6). The nature of this package is not certain, but its seismic character is similar to the Carboniferous of the Southern North Sea, where high-amplitude reflectors result from tuning effects caused by coal horizons. Cartwright (1988) reported a comparable reflective package from the Southern Tail End Graben of the Danish North Sea. Cartwright assigned this a Late Carboniferous to Lower Permian age (Fig. 7B), noting that its thickness (up to 5 km) was probably too great for the sequence to have an entirely Permian age, but that the absence of internal unconformities indicated that it represented a single, genetically related succession. The occurrence of Carboniferous rocks to the west of this in UK Well 39/7-1 led Hedemann (1980) to suggest that they may also occur in other areas of the Central Graben to the north.

Carboniferous strata are also known from the Outer Moray Firth (Andrews *et al.* 1990; Leeder and Boldy 1990), the Forth Approaches Basin, and the Western Platform (Besly 1990). The Carboniferous rocks of these areas are similar to those of the onshore Scottish Midland Valley province (Cameron and Stephenson 1985), which is notable for the presence of thick Carboniferous sequences including coal measures of Namurian–Westphalian age (Fig. 7C). The coal-bearing strata in the Outer Moray Firth appear to be slightly older; these were reported by Andrews *et al.* (1990) to have a Visean to Namurian age.

Internal subdivision of the reflective package may be possible on both highs. Figure 4 shows the sequence near the Jaeren High bounding fault to comprise an upper reflective unit, underlain by an acoustically quieter zone 400–600 ms in time thickness of sub-parallel, moderate- to low-continuity reflectors, before a return to parallel high continuity reflections below. These vertical changes in seismic character may mark subdivisions within a very thick Carboniferous sequence. Figure 6 across the Forties–Montrose High shows that the 500 ms reflective package is divisible into several different units and that it is underlain by a 500 ms acoustically quieter zone, with a gradual loss of seismic definition beneath this.

Interpretation based on analogy with neighbouring areas (Fig. 7) would suggest the presence of a typical Dinantian–Westphalian succession as suggested on Fig. 8. However, it must be stressed that this represents only one possible interpretation and confirmation of a Carboniferous age and more precise dating must both await well penetration and detailed biostratigraphical correlation.

Discussion

Previous seismic obtained in the Central Graben areas has not generally achieved sufficient resolution beneath the salt to permit mapping of the pre-Permian sequence, and well penetrations in the UK sector are limited to a few wells which encountered the Devoninan. Improved seismic imaging of the sub-Zechstein succession thus offers the first opportunity to assess the likely distribution of Upper Paleozoic strata in this area. The results of regional mapping of the deep section are summarized in Fig. 9.

Southern areas of both the Jaeren and Forties–Montrose highs are characterized by the presence of a 2–2.5 km thick pre-Permian reflective package interpreted provisionally as a coal-bearing sequence of Carboniferous age. Local recognition of acoustically quiet intervals beneath this package may indicate an underlying Devonian clastic succession several kilometres in thickness. This interpretation would suggest that the absence of the reflective package in northern parts of the Jaeren and Forties–Montrose highs would indicate the non-deposition or erosion of coal-bearing Upper Carboniferous strata in that area.

In northern parts of the Jaeren High, the occurrence of a sub-Permian acoustically quiet zone overlain by deep reflectors is thought to indicate the presence of a Devonian succession up to 2.5 km in thickness. The sub-Permian basement below northern parts of the Forties–Montrose High remains uncertain, and that beneath the deeper graben areas remains beyond the range of seismic resolution, so that Fig. 9 necessarily remains incomplete.

Nevertheless, the results presented here suggest that structural subdivision of the Jaeren and Forties–Montrose highs may be possible on the basis of their basement structure. North–south variation in stratigraphical development on the Jaeren High is thought to reflect control by major NE–SW-trending faults—similar structures are well known from the northern boundary of the Scottish Midland Valley where syn-sedimentary movement is demonstrable from marked facies

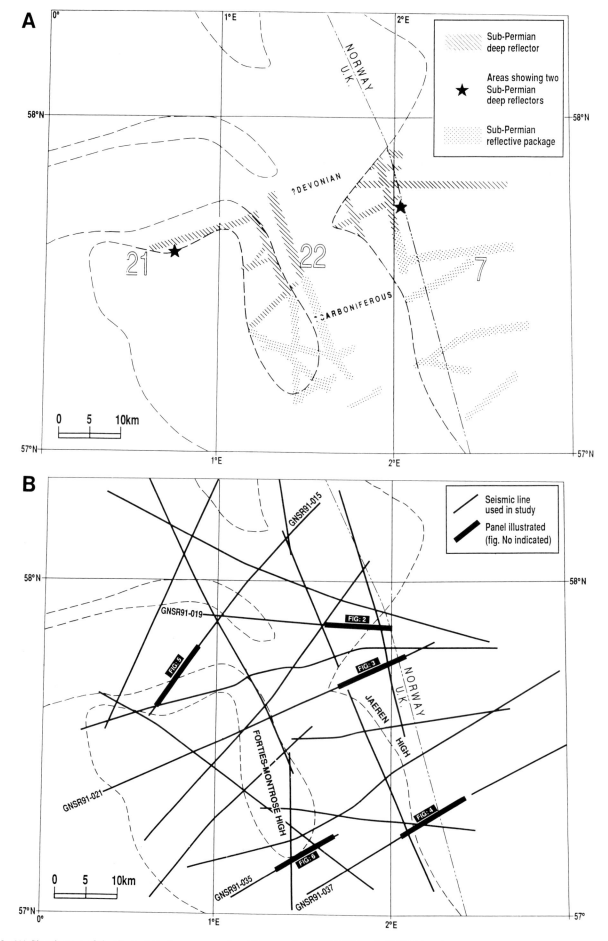

Fig. 9. (A) Sketch map of the Central Graben area produced from study of CGDT89 and GNSR91 regional seismic lines showing the interpreted distribution of Devonian and Carboniferous strata based on regional mapping of the sub-Permian reflective package and deep reflectors. **(B)** GNSR91 and CGDT89 seismic grid for reference.

changes in the Devonian and Carboniferous sequences. Evidence for north–south variations on the graben median high to the west leads us to propose that the Forties and Montrose highs, although linked, may in fact show differing structural evolution and stratigraphical development.

A further implication of these results stems from the potential presence of Carboniferous strata over extensive areas of the Central Graben area and surrounding highs. These may provide an alternative, pre-Zechstein gas-prone source interval in the region. Greater burial of these strata may have resulted in hydrocarbon generation even in areas where Jurassic organic-rich deposits have not reached maturity, potentially providing additional prospectivity in the region.

Conclusions

The Forties–Montrose and Jaeren highs display similar Tertiary and Mesozoic structures. Triassic 'pods' are ubiquitous on highs, with inter-pod areas showing localized accumulation of Jurassic–Lower Cretaceous strata. Middle Jurassic volcanic rocks are evident in northern areas of the Forties–Montrose High, but are rare elsewhere.

Northern areas of the Jaeren High show the presence beneath the Permian of an acoustically quiet zone up to 1000 ms in time thickness, representing a probable true thickness of up to 2.5 km. A horizontal reflector occurs at the base of this zone at two-way-travel times of around 4 s and may record the top of the pre-Devonian crystalline basement or a regional lithological marker such as Middle Devonian carbonate or Upper Paleozoic volcanic rocks. A westward- or northwestward-dipping reflector deeper in the section may represent the basal Devonian unconformity or a deep low-angle fault, possibly of Caledonide origin.

Southern areas of the Jaeren and Forties–Montrose highs show the presence of a regionally extensive sub-Permian reflective package of considerable thickness (2–2.5 km). This essentially conformable sequence is thought to represent a thick Carboniferous coal-bearing succession.

Recognition of north–south structural and stratigraphical variation suggests that, although linked, the Forties and Montrose highs show differing structural evolution and stratigraphical development. The structure and pre-Zechstein stratigraphy of the Jaeren High also varies along strike from north to south, possibly reflecting syn-sedimentary movement on NE–SW-trending faults which were later reactivated during regional Mesozoic extension.

The possible presence of Carboniferous coals would indicate additional gas-prone source rock potential in the Central Graben area, and opens up the possibility of sub-Zechstein gas plays in the region. This would have implications for exploration particularly in areas where Jurassic hydrocarbon source horizons are thermally immature or absent.

We gratefully acknowledge permission to publish from GECO-PRAKLA. Thanks go to Iain Bartholomew for his careful review as well as to cartographers Alan Sutton and Vicky Lock for production of the figures.

References

ANDREWS, I. J., LONG, D., RICHARDS, P. C., THOMSON, A. R., BROWN, S., CHESHER, J. A. AND McCORMAC, M. 1990. *United Kingdom offshore regional report: the geology of the Moray Firth*. HMSO for the British Geological Survey, London.

BARTHOLOMEW, I. D., PETERS, J. M. AND POWELL, C. M. 1993. Regional structural evolution of the North Sea: oblique-slip and the reactivation of basement lineaments. *In*: PARKER, J. R. (ed.) *Petroleum Geology of Northwest Europe: Proceedings of the 4th Conference*. Geological Society, London, 1109–1122.

BESLY, B. M. 1990. Carboniferous. *In*: GLENNIE, K. W. (ed.) *Introduction to the Petroleum Geology of the North Sea* (3rd edition). Blackwell Scientific Publishers, 90–119.

CAMERON, I. B. AND STEPHENSON, D. 1985. *British Regional Geology: The Midland Valley of Scotland* (3rd edition). HMSO for the British Geological Survey, London.

CARTWRIGHT, J. A. 1988. *A Seismic Interpretation of the Danish North Sea*. D. Phil. thesis, University of Oxford.

ERRATT, D. 1993. Relationships between basement faulting, salt withdrawal and Late Jurassic rifting, UK Central North Sea. *In*: PARKER, J. R. (ed.) *Petroleum Geology of Northwest Europe: Proceedings of the 4th Conference*. Geological Society, London, 1211–1219.

FROST, R. T. C., FITCH, F. J. AND MILLER, J. A. 1981. The age and nature of the crystalline basement of the North Sea Basin. *In*: ILLING, L. V. AND HOBSON, G. D. (eds) *Petroleum Geology of Northwest Europe*, Heyden & Son, London, 43–57.

GIBBS, A.D. 1989. A model for linked basin development around the British Isles. *In*: TANKARD, A. J. AND BALKWILL, H. R. (eds) *Extensional Tectonics and Stratigraphy of the North Atlantic Margins*. American Association of Petroleum Geologists Memoir, **46**, 501–509.

GLENNIE, K. W. 1990. Outline of North Sea history and structural framework. *In*: GLENNIE, K. W. (ed.) *Introduction to the Petroleum Geology of the North Sea*. (3rd edition). Blackwell Scientific Publishers, 34–77.

HEDEMANN, H.-A. 1980. Die Bedeutung des Oberkarbons für die Kohlenwasserstoffvorkommen im Nordseebecken. *Erdoel-Kohle-Erdgas*, **33**, 255–266.

HØILAND, O., KRISTENSEN, J. AND MONSEN, T. 1993. Mesozoic evolution of the Jaeren High area, Norwegian Central North Sea. *In*: PARKER, J. R. (ed.) *Petroleum Geology of Northwest Europe: Proceedings of the 4th Conference*. Geological Society, London, 1189–1195.

HOLLIGER, K. AND KLEMPERER, S. L. 1990. Gravity and deep seismic reflection profiles across the North Sea rifts. *In*: BLUNDELL, D. J. AND GIBBS, A. D. (eds) *Tectonic Evolution of the North Sea Rifts*. Clarendon Press, Oxford, 82–101.

JOHNSON, R. J. AND DINGWALL, R. G. 1981. The Caledonides: their influence on the stratigraphy of the North Sea. *In*: ILLING, L. V. AND HOBSON, G. D. (eds) *Petroleum Geology of Northwest Europe*. Heyden & Son, London, 85–97.

KLEMPERER, S. L. AND HURICH, C. A. 1990. Lithospheric structure of the North Sea from deep seismic reflection profiling. *In*: BLUNDELL, D. J. AND GIBBS, A. D. (eds) *Tectonic Evolution of the North Sea Rifts*. Clarendon Press, Oxford, 37–63.

—— AND WHITE, N. 1989. Coaxial stretching or lithospheric simple shear in the North Sea? Evidence from deep seismic profiling and subsidence. *In*: TANKARD, A. J. AND BALKWILL, H. R. (eds) *Extensional Tectonics and Stratigraphy of the North Atlantic Margins*. American Association of Petroleum Geologists Memoir, **46**, 511–522.

LEEDER, M. R. AND BOLDY, S. R. 1990. The Carboniferous of the Outer Moray Firth Basin, Quadrants 14, 15 Central North Sea. *Marine and Petroleum Geology*, **7**, 29–37.

PENGE, J., TAYLOR, B., HUCKERBY, J. A. AND MUNNS, J. W. 1993. Extension and salt tectonics in the East Central Graben. *In*: PARKER, J. R. (ed.) *Petroleum Geology of Northwest Europe: Proceedings of the 4th Conference*. Geological Society, London, 1197–1209.

PENNINGTON, J. J. 1975. The geology of the Argyll field. *In*: WOODLAND, A. W. (ed.) *Petroleum and the Continental Shelf of North West Europe*. Applied Science Publishers, Barking, 285–291.

ROBERTS, A. M., PRICE, H. D. AND OLSEN, T. S. 1990. Late Jurassic half-graben control on the siting and structure of hydrocarbon accumulations: UK/Norwegian Central Graben. *In*: HARDMAN, R. F. P. AND BROOKS, J. (eds) *Tectonic Events Responsible for Britain's Oil and Gas Reserves*. Geological Society, London, Special Publication, **55**, 229–257.

SEARS, R. A., HARBURY, A. R., PROTOY, A. J. G. AND STEWART, D. J. 1993. Structural styles from the Central Graben in the UK and Norway. *In*: PARKER, J. R. (ed.) *Petroleum Geology of Northwest*

Europe: Proceedings of the 4th Conference. Geological Society,
London, 1231–1243.

SMITH K. AND RITCHIE, J. D. 1993. Jurassic volcanic centres in the
Central North Sea. *In*: PARKER, J. R. (ed.) *Petroleum Geology of
Northwest Europe: Proceedings of the 4th Conference.* Geological
Society, London, 519–531.

WAKEFIELD, L. L., DROSTE, H. J., GILES, M. R. AND JANSSEN, R. 1993.
Late Jurassic plays along the western margin of the Central
Graben. *In*: PARKER, J. R. (ed.) *Petroleum Geology of Northwest
Europe: Proceedings of the 4th Conference.* Geological Society,
London, 459–468.

ZIEGLER, P. A. 1990*a*. Tectonic and palaeogeographic development of
the North Sea rift system. *In*: BLUNDELL, D. J. AND GIBBS,
A. D. (eds) *Tectonic Evolution of the North Sea Rifts.* Clarendon
Press, Oxford, 1–36.

—— 1990*b*. *Geological Atlas of Western and Central Europe* (2nd
edition). Shell International Petroleum, The Hague.

—— AND VAN HOORN, B. 1989. Evolution of North Sea rift system. *In*:
TANKARD, A. J. AND BALKWILL, H. R. (eds) *Extensional Tectonics
and Stratigraphy of the North Atlantic Margins.* American Association of Petroleum Geologists Memoir, **46**, 471–500.

Structural styles from the Central Graben in the UK and Norway

R. A. SEARS,[1] A. R. HARBURY,[1] A. J. G. PROTOY[1] and D. J. STEWART[2]

[1] *Shell UK Exploration and Production, Shell-Mex House, Strand, London WC2R 0DX, UK*
[2] *A/S Norske Shell, PO Box 40, Tananger, Norway*

Abstract: A selection of structural styles observed on seismic data from the UK and Norwegian sectors of the Central Graben is presented. The interpretation of these structures is based on a polyphase tectonic model of structural evolution of the North Sea Basin which has been facilitated by the improved quality of pre-Base Cretaceous reflections seen on the increasingly available 3D seismic surveys.

The structural evolution of the Central Graben can be represented by a model invoking two main phases of structural development. Much of the basement framework was created during the Caledonian orogeny which gave rise to NW–SE- and NE–SW-trending diffuse shear zones. Subsequent repeated reactivation along these existing basement lineaments occurred from Late Paleozoic to Cenozoic times. During the late Jurassic to early Cretaceous, the maximum principal stress axis was orientated approximately N–S to NNE–SSW and the minimum principal stress axis orientated E–W to ESE–WNW. Because this stress field was at an angle to the existing framework trends, most of the faults were reactivated with an oblique-slip component which resulted in the formation of transpressional and transtensional features of varying orders of magnitude. Fault movements also locally activated halokinesis which further complicated the structural development.

Seismic examples illustrate how the introduction of an element of oblique-slip can clarify the interpretation of many of the observed structures in the UK and Norwegian Central Graben. Although often subtle in appearance, oblique-slip movements and corresponding zones of transpression and transtension must be identified before a clear picture of localized structural and sedimentary histories can emerge.

The understanding of complex pre-Cretaceous structures in the Central North Sea Graben of the UK and Norway has been greatly improved over the last five years following the interpretation of extensive, good quality 3D seismic surveys. Additionally, unexpected drilling results have led to a revisiting of the traditional structural models. It has become apparent that the Central Graben is not a simple rift basin of solely normal faulting and salt diapirism. Rather, the structural style is the result of the interplay of extensional tectonics, halokinesis and oblique-slip movements during several tectonic phases.

The aim of this paper is to illustrate, with 2D and 3D seismic examples, typical structural features which, alone or in combination, are characteristic of transtensional or transpressional deformation. The introduction of an element of oblique-slip can clarify the seismic interpretation of many of the complex structures found in the UK and Norwegian Central Graben. Bartholomew *et al.* (1993) have described the regional structural model into which the transpressional and transtensional features fit. Reference should be made to this paper for the details of the structural model which is only briefly described here.

Polyphase structural model

The structural evolution of the North Sea can be represented by a model involving two main phases of structural development as described in Bartholomew *et al.* (1993): an Early Paleozoic phase (Johnson and Dingwall, 1981; Coward 1993) during which time the dominant trending lineaments were NW–SE and NE–SW; and a rift-related phase from Late Paleozoic to Cenozoic times, the evolution of which was controlled by repeated reactivation of these pre-existing structural lineaments.

In the Central Graben (Fig. 1) this second phase was initiated during the Late Permian to Triassic, with the principal phase of rift-related E–W extension occurring from the Mid-Oxfordian to Berriasian. The Central Graben, containing elongate NW–SE-trending grabens and intra-basinal highs, has traditionally been interpreted as a dominantly extensional sedimentary basin (Ziegler 1981, 1990a; Cayley 1987). The asymmetrical grabens and highs at Lower Permian basement level (Rotliegend) are bounded by what appear in cross-section to be normal faults (Figs 2 and 3).

Crustal extension across the North sea rift system was, however, not purely orthogonal to the pre-existing fault pattern, but included an oblique-slip component of deformation along many of the fault trends (Ziegler 1990b; Bartholomew *et al.* 1993). Regionally, the maximum principal stress axis was orientated approximately N–S to NNE–SSW, and the minimum principal stress axis oriented E–W to ESE–WNW. This resulted in a regional extension direction roughly WNW–ESE over the whole of the North Sea area. Within the Central Graben, the pre-existing fault pattern caused most of the basins to form along NW–SE trends with a component of dextral strike-slip movement occurring during the extension of these basins. This resulted in oblique-slip faulting styles as described by Bartholomew *et al.* (1993) and Ziegler (1990b).

Regional mapping in the Central North Sea (Bartholomew *et al.* 1993) has established that the basement faults form a fairly open, rhomboidal structural pattern defined by roughly NW–SE- and WNW–ESE-trending lineaments (Figs 4 and 5). During reactivation, shear zones developed along these pre-existing faults. Individual shear zones are often 'dog-leg'-shaped, or consist of a series of smaller en échelon faults which can only be accurately mapped using 3D seismic data. The structures along these shear zones can be interpreted as being the result of oblique-slip movements and the structural geometries observed are very similar to those seen in sandbox experiments (Koopman *et al.* 1987; Tron and Brun 1991). Structures form along and between individual shear zones as a result of local transpression and transtension (Harding, 1983, 1985; Caselli, 1987) and examples of both are illustrated in this paper. Also shown are examples from the Central Graben of synchronous formation of adjacent transtensional and transpressional structures.

In addition to the tectonic structuration, movements of Zechstein salt occurred throughout Mesozoic and Cenozoic times along the graben margins and within the graben along active shear zones (Figs 2–5). This affected sedimentation patterns during the Triassic and Jurassic through the formation of salt withdrawal features and local unconformities

From *Petroleum Geology of Northwest Europe: Proceedings of the 4th Conference* (edited by J. R. Parker).

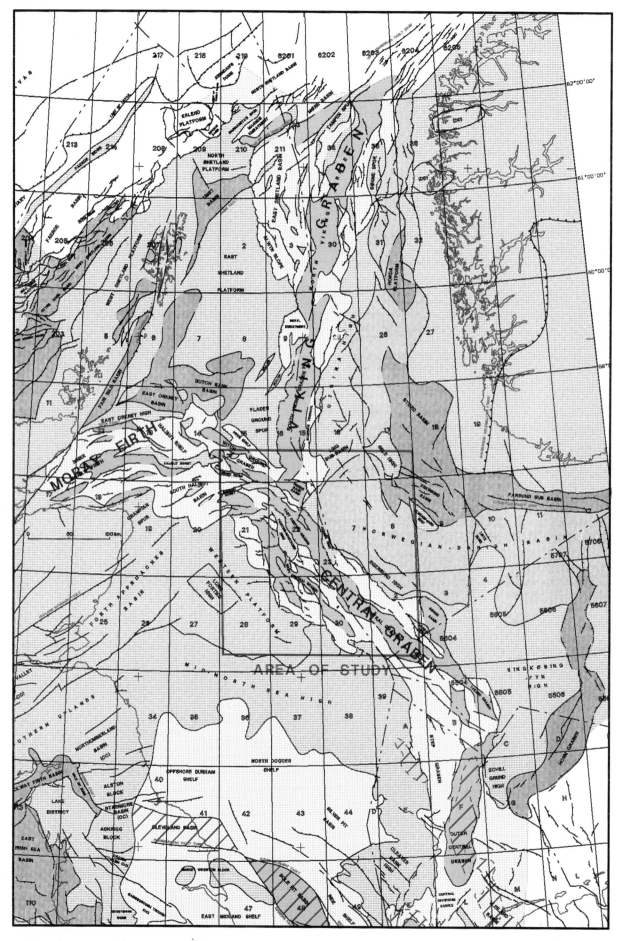

Fig. 1. Cimmerian structural framework map of the North Sea showing the location of the study area within the UK and Norwegian Central North Sea. Shown in dark blue are the Jurassic basins of the Central Graben and Viking Graben, with the Jurassic terraces in lighter blue. In purple are the Triassic basins (dark) and terraces (light). Platforms and intra-basinal highs are in white, Paleozoic faults are indicated in black. Coastal outlines are shown in red.

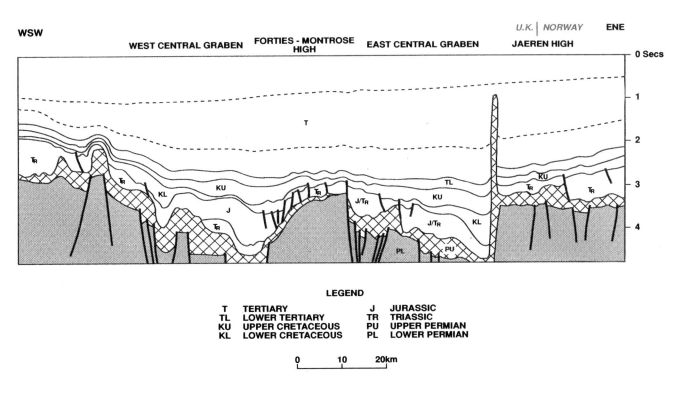

Fig. 2. WSW–ENE regional geological cross-section across the UK Central North Sea based on an interpretation of a regional seismic line. Apparently normal basement faults dominate the section and halokinesis is evident along the graben margins, and within the graben along active shear zones.

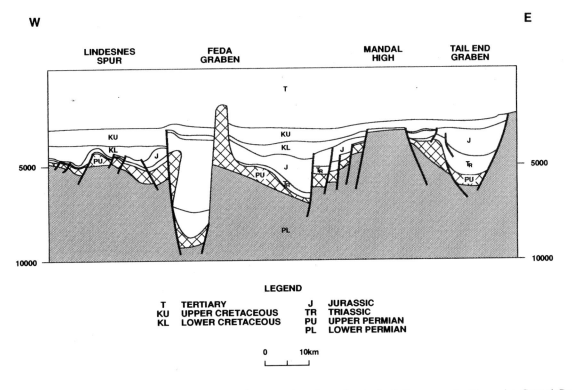

Fig. 3. SW–NE regional geological cross-section based on an interpretation of a regional seismic line across the Norwegian Central Graben. This shows characteristics similar to the UK Central Graben as seen in Fig. 2.

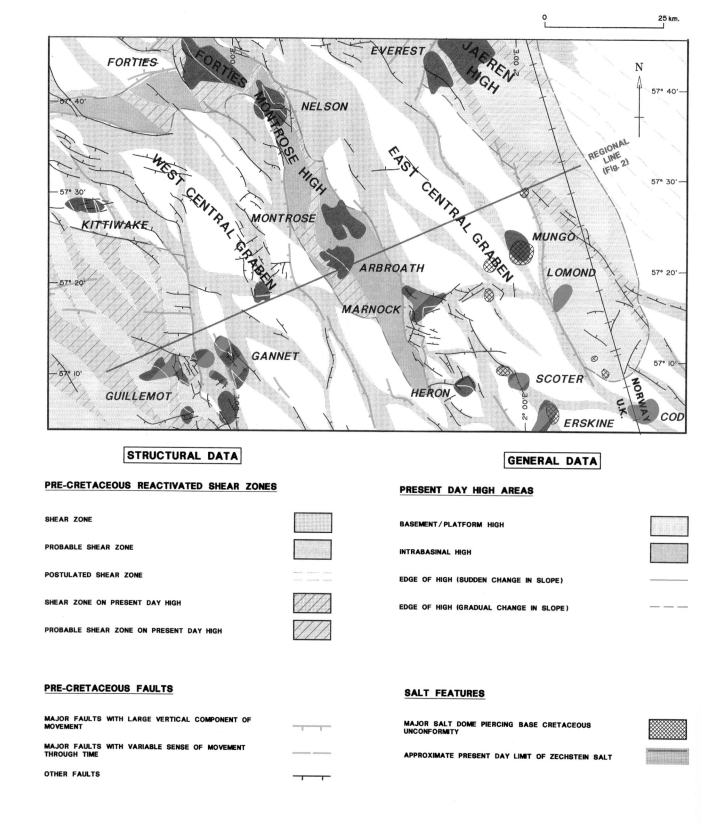

Fig. 4. Northern part of the structural framework map of the Central North Sea with NW–SE-trending faults bounding the main structural features and defining rhomboidal shaped inter-shear zones. The location of the cross-section in Fig. 2 is indicated.

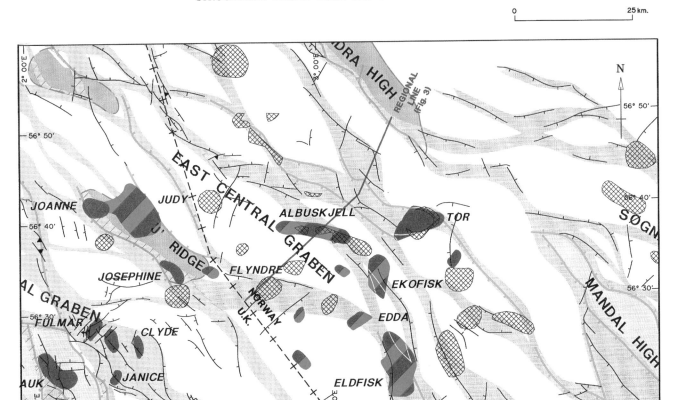

Fig. 5. Southern part of the structural framework map of the Central North Sea covering the southern UK Central Graben and the Norwegian Central Graben. NW–SE-trending faults bound the main structural features and define rhomboidal shaped inter-shear zones. The location of the cross-section in Fig. 3 is indicated. For legend, refer to Fig. 4.

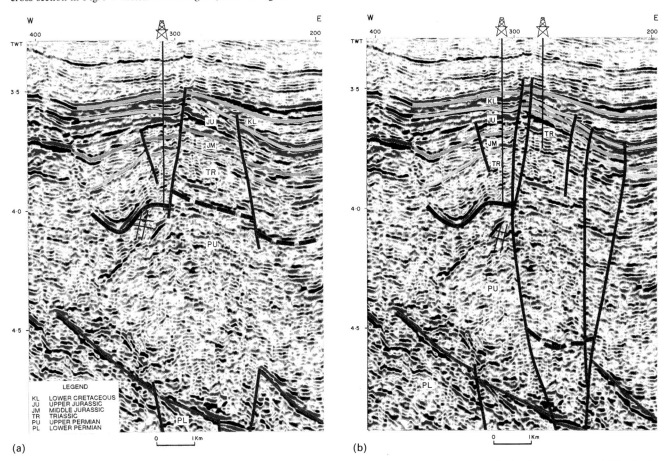

Fig. 6. (**a**) A pure extensional, normal fault interpretation of the pre-Cretaceous of a complex structure, as depicted on an east–west 3D seismic line from southeast of the Forties–Montrose High. Location of this line is shown on Fig. 7. (**b**) An alternative interpretation of the same 3D line introducing an element of transpressional tectonics and reverse faulting.

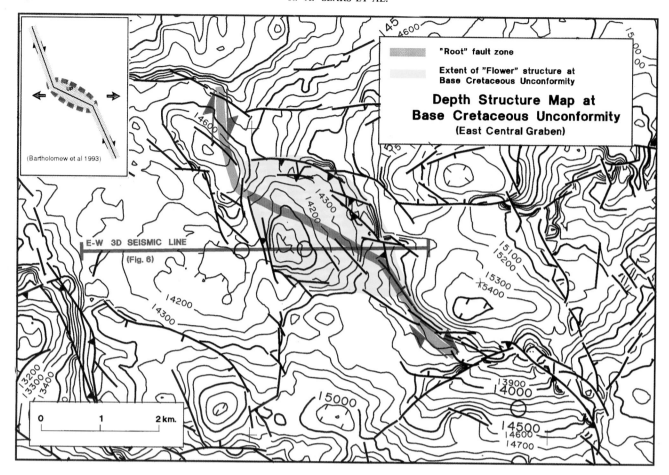

Fig. 7. Depth structure map at Base Cretaceous Unconformity within the Central Graben based on the 3D survey from which Fig. 6 is taken. The yellow area represents the extent of the positive flower structure at Base Cretaceous level as interpreted on Fig. 6b. The splay faults converge at depth into a steep fault zone that displaces basement. The trace of this deep fault zone is shown in blue. The inset shows schematically how local transpression develops along 'dog-leg'-shaped faults subjected to oblique slip.

(Cayley 1987). Halokinesis complicates the interpretation of the shallower structures while the salt can also mask the deeper seismic reflections and the true basement fault movements. As a result, the detailed nature and geometry of these fault zones can often be difficult to interpret.

Seismic identification of oblique-slip movements

The following series of seismic examples illustrates the variety and complexity of the structures that result from reactivation in this multi-phase structural model.

A comparison of extensional and oblique-slip interpretations

In a purely extensional tectonic model, structural configurations are the result of the interaction of extensional basement rifting and salt tectonics.

A seismic line from a modern 3D survey, located southeast of the Forties–Montrose High, is shown in Fig. 6a with an interpretation based on an extensional tectonic model. The small horst block to the east of the indicated well is bounded by normal faults with minor throws. Each stratigraphic interval from Triassic to Lower Cretaceous expands to the west in the hangingwall of the fault where an existing well provides control on the stratigraphy.

What is interpreted as apparent fault throw on a seismic line may, however, not be the true displacement, as fault interpretations on individual lines are often biased towards dip-slip

movements. In contrast, the interpretation of 3D surveys, with the combination of profile criteria such as fault geometry, and map criteria such as fault patterns, lead to the identification of often subtle oblique-slip fault features (Harding 1990).

Based on the interpretation of the full 3D survey, the regional structural interpretation (Fig. 4), comparisons with sandbox models (Koopman *et al.* 1987) and comparisons with sedimentary basins affected by wrench-related tectonics (Biddle and Christie-Blick 1985; Sylvester and Smith 1976), the structures seen here can be better explained by introducing a component of strike-slip movement into the extensional model. This interpretation is shown in Fig. 6b. Here, the fault block on the right is seen as a positive flower structure and the splay faults converge at depth into a steep fault zone that displaces basement. The areal extent of the flower structure and the trace of the 'root' fault zone are shown on the Base Cretaceous Unconformity depth map of Fig. 7. An abrupt stratigraphic mismatch is interpreted across the main NW–SE-trending fault, and a significantly thinner Jurassic sequence and much higher Triassic sequence is interpreted in the flower structure, relative to the neighbouring well. Recent drilling of this structural high, at the location shown on Fig. 6b, has confirmed this interpretation.

Reactivation and uplift along a shear zone

A 3D seismic line situated on a NW–SE-trending ridge feature extending southeast from the Forties–Montrose High to the 'J' Ridge is shown in Fig. 8. The reversal of fault throws at top

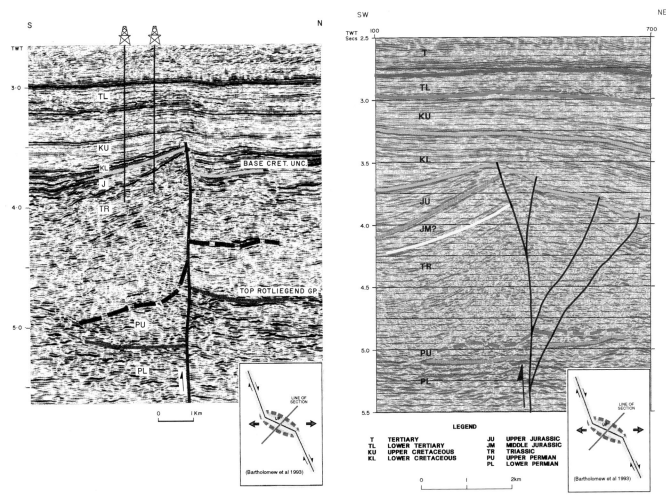

Fig. 8. N–S-oriented 3D seismic line, illustrating reversal of fault throws at top Rotliegend and Base Cretaceous Unconformity levels. For legend, refer to Fig. 2. The inset shows schematically how local transpression develops along 'dog-leg'-shaped faults subjected to oblique slip.

Rotliegend and Base Cretaceous levels indicates that Upper Jurassic/Lower Cretaceous transpressional movements have reactivated older normal faults.

Reactivation along a shear zone is also illustrated by the flower structure on the 3D seismic line from the West Central Graben shown in Fig. 9. The downflank thinning of the Mid-Jurassic wedge suggests that this Jurassic depocentre was uplifted along a NW–SE-trending wrench zone prior to the deposition of the Upper Jurassic. In addition, note the Upper Cretaceous wedge that thickens to the northeast, indicating late Upper Cretaceous reactivation.

Reactivation, prevalent throughout the Central Graben during the Jurassic and Cretaceous, recurred during Tertiary time as illustrated on the 2D seismic example of Fig. 10. The timing of the uplift of the Eocene lobe overlying a reactivated shear zone can be seen from the Mid to Late Eocene onlap on the top surface of the lobe. This Eocene structure has been mapped as an elongate feature, coincident with the trend of the underlying shear zone. There is no strong evidence of thick salt in this area, suggesting this feature is not the result of late salt movement.

Fig. 9. SW–NE-oriented 3D seismic line through an inverted Middle Jurassic basin. Note the onlap of the Upper Jurassic sediments. Reactivation along this fault zone continued well into the Lower Cretaceous. The inset shows schematically how local transpression develops along 'dog-leg'-shaped faults subjected to oblique slip.

Reactivation with subsidence and/or uplift between shear zones

In the southern part of the Central Graben, a Lower Cretaceous 'pull-apart' basin (Fig. 11a) was uplifted during the late Cretaceous (Fig. 11b). The 2D seismic section across this feature (Fig. 12) shows features typical of inverted structures as described by Williams *et al.* 1989, and Koopman *et al.* 1987. This well-developed, inverted Lower Cretaceous basin is bounded by shear zones. Lower Cretaceous transtension resulted in the formation of this 'pull-apart' basin. A switch in the actual faults that were active at either end of the 'pull-apart' basin resulted in inversion of the basin during Upper Cretaceous and Tertiary as indicated by the prominent onlaps within the Cretaceous and Tertiary sequences. This is similar to the Upper Jurassic structural model proposed for the Fulmar Field by Bartholomew *et al.* (1993).

Another example of an inverted basin is shown in Fig. 13, a 2D seismic line from the West Central Graben. This fault-bounded Lower Cretaceous depocentre was inverted during Late Cretaceous time as clearly indicated by the intra-Upper Cretaceous stratal onlap patterns.

Fig. 11. (over the page) (**a**) Structural framework map over the southern UK Central Graben illustrating the rhomboidal Lower Cretaceous pull-apart basin created by east–west extension on a pre-existing NW–SE-dominant fault system. Location of the seismic line (Fig. 12) is indicated. For legend, refer to Fig. 4. The inset shows schematically how local transtension develops between shear zones subjected to oblique slip. (**b**) Structural framework map over the same area as (a), showing the area of Upper Cretaceous inversion created by east–west extension and further reactivation of the pre-existing NNW–SSE-dominant fault system. Location of the seismic line (Fig. 12) is indicated. For legend, refer to Fig. 4. The inset shows schematically how local transpression develops between shear zones subjected to oblique slip.

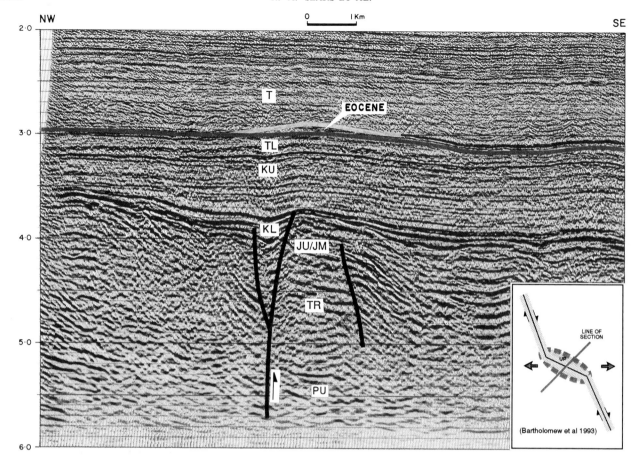

Fig. 10. Seismic line (courtesy of Simon-Horizon) from the Central Graben illustrating reactivation during the Tertiary. For legend, refer to Fig. 9. The inset shows schematically how local transpression develops along 'dog-leg'-shaped faults subjected to oblique slip.

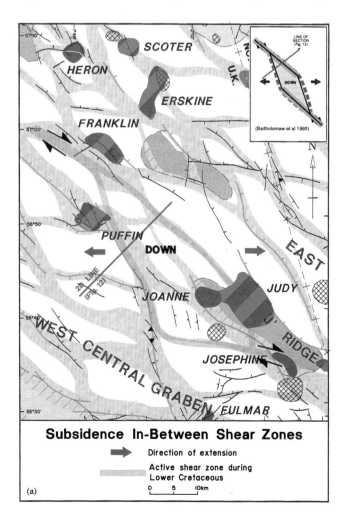

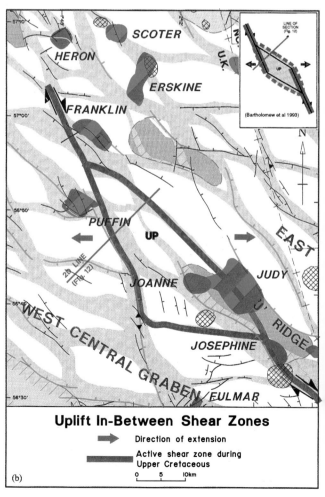

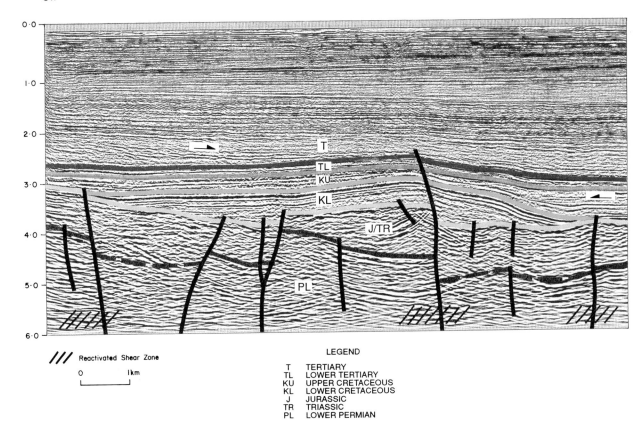

/// Reactivated Shear Zone

0   1km

LEGEND

T   TERTIARY
TL   LOWER TERTIARY
KU   UPPER CRETACEOUS
KL   LOWER CRETACEOUS
J   JURASSIC
TR   TRIASSIC
PL   LOWER PERMIAN

Fig. 12. NE–SW-oriented seismic line illustrating a fault-bounded Lower Cretaceous depocentre. Reactivation along the bounding faults inverted this depocentre, as shown by the Upper Cretaceous and Tertiary onlaps. The location of this line is indicated in Fig. 11.

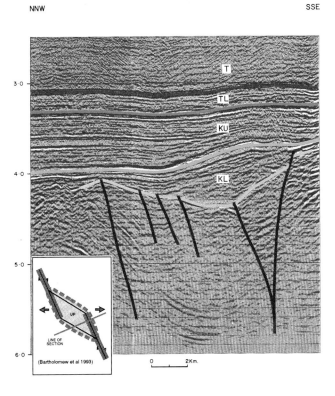

NNW   SSE

(Bartholomew et al 1993)

0   2Km.

Fig. 13. NNW–SSE-oriented 2D seismic line along the western graben margin shows another example of an inverted Lower Cretaceous depocentre. For legend, refer to Fig. 12. The inset shows schematically how local transpression develops between shear zones subjected to oblique slip.

Adjacent transtension and transpression

A 3D seismic example of this complex structural style is taken from the West Central Graben (Fig. 14). On the uninterpreted 3D seismic line (Fig. 14a), the nearly vertical pre-Cretaceous faults which are rooted to the Lower Permian shear fault zones can be clearly seen. Activation of these shear zones during the Lower Cretaceous resulted in the formation of a Jurassic 'pop-up' feature located next to a Lower Cretaceous 'pull-apart' basin (see inset in Fig. 14b).

Another example showing adjacent transpression and transtension is taken from the Albuskjell area of the Norwegian Central Graben where two main fault trends interact. The complex structural configuration resulting from the interplay of halokinesis and several episodes of oblique-slip movement along lineaments rooted to deep-seated basement faults is shown in Fig. 15. Movement along the NNW–SSE-trending fault zone during the late Jurassic and early Cretaceous resulted in the development of a belt of transpressive structures coincident with this fault trend and shown as the shaded area in Fig. 15a. In a line from this 3D survey, this is seen as the flower structure at the position marked 'A' (Fig. 16). The timing of this transpression is clearly seen by the thinning of the Upper Cretaceous section over the crest of this structure.

The WNW–ESE fault trend seen in Fig. 15 was the locus of prolonged transtension, which together with halokinesis, resulted in the development of a fault-bounded Jurassic depocentre immediately adjacent to a Lower Cretaceous graben (the area marked 'B' on Fig. 16). The Upper Jurassic and Lower Cretaceous isochore maps (Figs 17a and b, respectively) show the depositional thicks along this fault trend that resulted from this prolonged transtension.

During the Upper Cretaceous, this WNW–ESE fault trend was reactivated under a changed stress regime, and developed

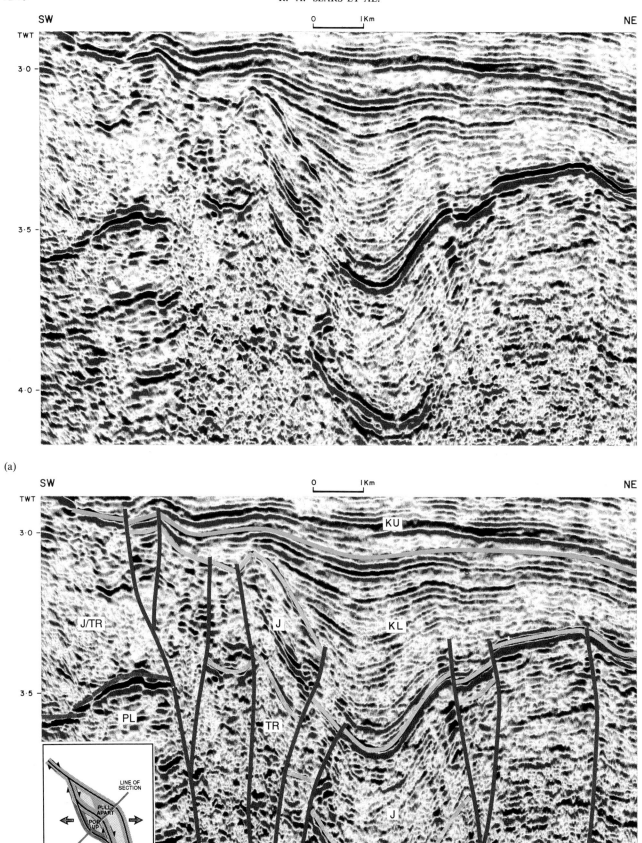

(a)

(b)

Fig. 14. (**a**) An uninterpreted and (**b**) interpreted 3D seismic line from the West Central Graben shows a structurally complex area of adjacent pop-ups and pull-aparts within the same shear zone. Faults were reactivated under a changed stress regime during the Upper Jurassic and Lower Cretaceous. The inset in (b) shows how adjacent transtension and transpression can develop along a shear zone.

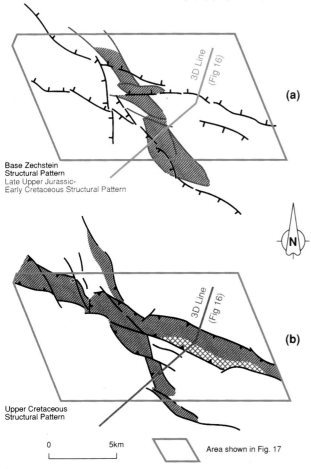

Base Zechstein
Structural Pattern
Late Upper Jurassic-
Early Cretaceous Structural Pattern

(a)

(b)

Upper Cretaceous
Structural Pattern

0   5km

Area shown in Fig. 17

as a sinistral transpression belt, inverting the former graben (Fig. 17c) and creating the prominent flower structure at position 'B' on Fig. 16. The timing of the development of this 'flower' is clearly shown by the marked wedging within the Upper Cretaceous on the crest of the present-day structural high. This interpretation is confirmed by the indicated well, which penetrated the high associated salt wall.

With complex, multi-phase, oblique-slip structuration as recognized here, the timing of transpression and transtension is clearly not the same in all areas. The structures that develop are dependent on the dominant stress regime at the time, and the orientation of the established faults in relation to those stresses (Bartholomew *et al.* 1993).

Conclusions

As more sub-surface data, in particular high quality 3D data-sets, have become available, the observed structure in the UK and Norwegian Central Graben cannot be explained solely by normal faulting and the associated remobilization of Zechstein salts. The detailed analysis of this seismic, correlated with well data, shows that the structural evolution of the Central Graben is clearly polyphase and includes oblique/strike-slip movements and structural inversion.

Fig. 15. Structural map of the Albuskjell area of the Norwegian Central Graben showing the dominant fault patterns at base Zechstein, Upper Jurassic and Cretaceous. The location of the seismic line shown in Fig. 16 is indicated.

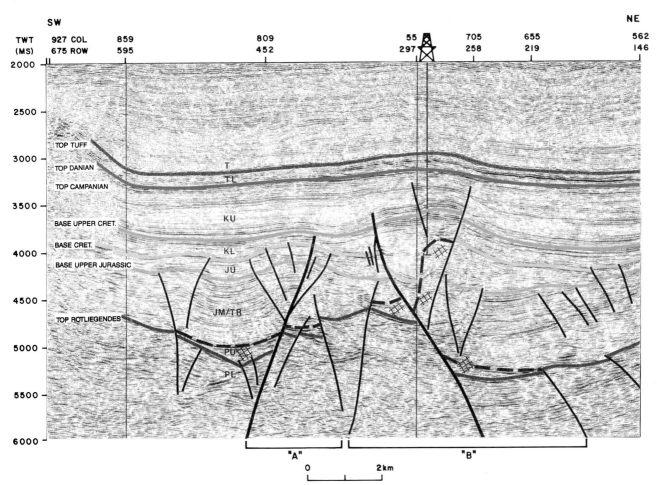

Fig. 16. SW–NE 3D line through the structurally complex area shown in Fig. 15. Repeated reactivation of the dominant fault trends, along with active salt movement, resulted in inversion of the Upper Jurassic and Lower Cretaceous depocentres and emplacement of flower structures along these pre-existing lineaments. For legend, refer to Fig. 9.

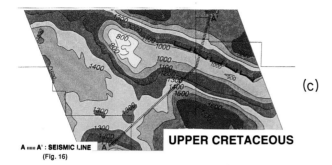

(c)

UPPER CRETACEOUS

A ⟍⟍⟍ A' : SEISMIC LINE
(Fig. 16)

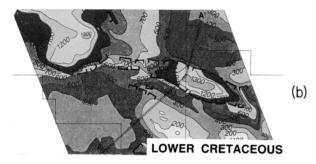

(b)

LOWER CRETACEOUS

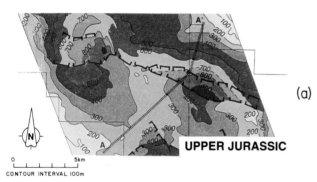

(a)

UPPER JURASSIC

0   5km

CONTOUR INTERVAL 100m

Fig. 17. Isochore maps of the area of this 3D survey illustrating the inversion of the Upper Jurassic and Lower Cretaceous depocentres in the area marked 'B' in Fig. 16. The timing of this reactivation is clearly seen from the pronounced thinning of the Upper Cretaceous over the crest of the present-day high.

This proposed regional tectonic model, when applied to local areas, allows for a better understanding of the past drilling results and for the generation of more accurate pre-drilling prediction such as shown in Fig. 6b. Furthermore, it provides a tectono-stratigraphic framework which can lead to a more accurate prediction of trap definition, reservoir distribution and hydrocarbon habitat. This ultimately leads to the discovery of hydrocarbons in possibly more subtle stratigraphic traps as well as purely structural traps.

We would like to thank I. D. Bartholomew and J. M. Peters for their extensive work on this structural model, and M. Schwander and E. Normann of Norske Shell for their contributions. We also wish to thank Shell UK Exploration and Production, A/S Norske Shell, and Esso Exploration and Producton UK Ltd for permission to publish this paper, and Simon-Horizon for permission to publish Fig. 10.

References

BARTHOLOMEW, I. D., PETERS, J. M. AND POWELL, C. M. 1993. Regional structural evolution of the North Sea: oblique slip and the reactivation of basement lineaments. *In*: PARKER, J. R. (ed.) *Petroleum Geology of Northwest Europe: Proceedings of the 4th Conference.* Geological Society, London, 1109–1122.

BIDDLE, K. T. AND CHRISTIE-BLICK, N. 1985. (eds) *Strike-Slip Deformation, Basin Formation, and Sedimentation.* Society of Economic Paleontologists and Mineralogists, Special Publication, **37**.

CASELLI, F. 1987. Oblique-slip tectonics mid-Norway Shelf. *In*: BROOKS, J. AND GLENNIE, K. W. (eds) *Petroleum Geology of North West Europe.* Graham & Trotman, London, 1049–1063.

CAYLEY, G. T. 1987. Hydrocarbon migration in the Central North Sea. *In*: BROOKS, J. AND GLENNIE, K. W. (eds) *Petroleum Geology of North West Europe.* Graham & Trotman, London, 549–556.

COWARD, M. P. 1993. The effect of Late Caledonian and Variscan continental escape tectonics on basement structure, Palaeozoic basin kinematics and subsequent Mesozoic basin development in NW Europe. *In*: PARKER, J. R. (ed.) *Petroleum Geology of Northwest Europe: Proceedings of the 4th Conference.* Geological Society, London, 1095–1108.

HARDING, T. P. 1983. Graben hydrocarbon plays and structural styles. *In*: KAASSCHIETER, J. P. H. AND REIJERS, T. J. A. (eds) Petroleum Geology of the Southeastern North Sea and the Adjacent Onshore Areas. *Geologie en Mijnbouw*, **62**, 3–24.

—— 1985. Seismic characteristics and identification of negative flower structures, positive flower structures and positive structure inversion. *American Association of Petroleum Geologists Bulletin*, **69**, 582–600.

—— 1990. Identification of wrench faults using subsurface structural data: criteria and pitfalls. *American Association of Petroleum Geologists Bulletin*, **74**, 1590–1609.

JOHNSON, R. J. AND DINGWALL, R. G. 1981. The Caledonides: their influence on the stratigraphy of the northwest European Continental Shelf. *In*: ILLING, L. V. AND HOBSON, G. D. (eds) *Petroleum Geology of the Continental Shelf of North-West Europe.* Heyden, London, 85–97.

KOOPMAN, A., SPEKSNIJDER, A. AND HORSFIELD, W. T. 1987. Sandbox studies of inversion tectonics. *Tectonophysics*, **137**, 379–388.

SYLVESTER, A. G. AND SMITH, R. R. 1976. Tectonic transpression and basement controlled deformation in the San Andreas fault zone, Salton Trough, California. *American Association of Petroleum Geologists Bulletin*, 1625–1640.

TRON, V. AND BRUN, J.-P. 1991. Experiments on oblique rifting in brittle–ductile systems. *Tectonophysics*, **188**, 71–84.

WILLIAMS, G. D., POWELL, C. M. AND COOPER, M. A. 1989. Geometry and kinematics of inversion tectonics. *In*: COOPER, M. A. AND WILLIAMS, G. D. (eds) *Inversion Tectonics.* Geological Society, London, **44**, 3–16.

ZIEGLER, P. A. 1981. Evolution of sedimentary basins in North-West Europe. *In*: ILLING, L. V. AND HOBSON, G. D. (eds) *Petroleum Geology of the Continental Shelf of North-West Europe.* Heyden, London, 3–42.

—— 1990a. *Geological Atlas of Western and Central Europe* (2nd edn). Shell Internationale Petroleum Maatschappij B.V., The Hague.

—— 1990b. Tectonic and palaeogeographic development of the North Sea rift system. *In*: BLUNDELL, D. J. AND GIBBS, A. D. (eds) *Tectonic Evolution of the North Sea Rifts.* Clarendon Press, Oxford, 1–36.

Discussion

Question (D. G. Quirk, NAM, Assen, The Netherlands):

Many of your examples of transtension and transpression do not fit with E–W extension. In such a stress regime, faults that are oriented NNW–SSE, N–S and NNE–SSW will have a component of extensional movement; faults that are oriented WNW–ESE, E–W and ENE–WSW will have a component of compression. Do you not think that your observations can be fitted better to changes in stress directions?

Answer (R. A. Sears):

As pointed out, it is changes in the stress direction that result in transpression or transtension. It is not the absolute stress direction though, but rather the direction of the applied stress relative to the orientation of the active fault.

In the examples presented, it is changes in the orientation of the active fault segments that cause the local transpression or transtension. In particular, compare the inset in Fig. 7 to the mapped fault trace shown. Local transpression only develops in the vicinity of the centre fault segment due to the change in the dominant fault orientation relative to the applied stress. Similar changes in this relative relationship can be seen in the other examples, and it is important to note that in these examples, transpression or transtension develops only locally in an overall extensional basin.

For a more detailed discussion of the occurrence of local transpression and transtension, the reader is referred to Bartholomew *et al.* (1993), Caselli (1983), and Harding (1983, 1985).

The structure of the Norwegian Central Trough (Central Graben area)

M. B. GOWERS,[1] E. HOLTAR[2] and E. SWENSSON[2]

[1] Geoform a.s. Aukehagen 16, 3472 Bødalen, Norway
[2] Norsk Hydro a.s. postboks 200, 1321 Stabekk, Norway

Abstract: The structural geology of the Norwegian Central Trough (also known as the Central Graben) is complex and poorly understood. Previous publicized attempts to explain the structural evolution have failed to account for the variety of observations made from the data by oversimplifying the problem and forcing a particular chosen model. Building from detailed data analysis, seven distinct tectonic phases which affect the area can be defined. Triassic salt tectonics significantly modifies the response to later tectonic phases. Early Jurassic uplift and erosion precede the intense block-fault subsidence of the third phase in the Oxfordian to Volgian. A marked change in the Mid-Volgian heralds the fourth tectonic phase typified by block rotation and related erosion. A distinct early Cretaceous phase of subsidence follows a different pattern from Jurassic subsidence, and is tectonically linked to the sixth phase of inversion. Inversion movements are identified in at least six separate pulses starting in Hauterivian times and continuing into the seventh tectonic phase, that of thermal subsidence initiated in mid-Cretaceous times. These seven phases are largely superimposed on one another, and affect different structural elements within the Central Trough to differing degrees.

The Norwegian Central Trough (Fig. 1) was defined as a structural element by Rønnevik et al. (1975) and further discussed by Gowers and Sæbøe (1985). Although 'Central Trough' is the official nomenclature in Norway, the expression 'Central Graben' has nevertheless gained acceptance as a loosely defined area of extensional faulting.

Although the area has undergone active exploration since 1965, there is remarkably little consensus in the literature as to the structural framework. Publications fall into three distinct categories: descriptions of fields and local structures, stratigraphic and sedimentological papers, and general structural models. Most of these publications include some sort of structural framework map, showing major faults, and a comparison of these maps shows little consensus. Even greater variation is found in the structural models published to explain the structural geology of the area.

Those publications which take a broader view are focused on structural models and mechanisms of basin formation rather than descriptions of the structural geology of the area. The models used appear to be based on regional seismic data and vary considerably. Beach (1985) and Gibbs (1989) postulate controlling detachment faults in opposite directions. Roberts and Yielding (1991) favour a flexural cantilever model and Roberts et al. (1990) interpret an absence of basement faulting at the major Skrubbe Fault, based largely on a footwall uplift model. The fact that the nature of major elements is still in dispute indicates the difficulties in arriving at a reliable structural picture for the area. In general the interpretation of deep seismic data is ambiguous on individual lines, and hence can be interpreted to support many different models if one so chooses.

The spatial frequency of structural features in the Norwegian Central Trough is of the order of 5 km. There is no consistent dip/strike direction, largely due to extensive salt movement and, hence, sideswipe energy on the seismic data is prevalent. In such conditions the use of only a regional seismic grid can be highly misleading. It seems to be possible for individual workers to choose a seismic section which best suits their favoured model and to interpret accordingly. But regional models must be compatible with all the data, and only by working in three dimensions with sufficient data density can a reliable picture be built.

This paper builds from detailed observations made in the area, following on from work presented by Gowers and Sæbøe (1985). Structural models are employed as a useful discussion tool after the data interpretation has been done, in order to avoid 'interpreting the model' into the data. The division of the area into structural units in both time and space is described. This work is based on a vast amount of data, but due to space limitations it is impossible to illustrate or justify all the conclusions by presenting data examples. There is still much work to be done in the Norwegian Central Trough which will lead to the revision and improvement of the structural picture outlined here.

The reader is expected to be reasonably familiar with the stratigraphical framework of the North Sea. The stratigraphic nomenclature used in this paper is shown in Fig. 2. The structural nomenclature shown in Fig. 3 is according to the present official Norwegian guidelines, and the tectonic phases discussed are shown in Fig. 4.

Tectonic phases

One of the major causes of confusion as to the structure of the Central Trough is the common attempt to explain the observations by one mechanism. This study describes at least seven different tectonic phases (Fig. 4) which are significant in the formation of the Central Trough. The structural response to each phase has been significantly steered by structures from preceding tectonic phases. Note that Permian tectonics are often postulated, but this study finds no evidence to suggest that these were significant in this area. Paleozoic trends, however, have probably established zones of weakness which can be recognized in the response to subsequent tectonic phases.

Tectonic phase 1 (TP1): salt tectonics (Triassic)

Salt tectonics has been recognized from an early phase in the exploration of the Central Trough. This is due mainly to the numerous obvious diapirs which penetrate into late Tertiary sediments (see Fig. 1), and are easily seen on the seismic data. A general acceptance that diapirs form when the salt is buried by sufficient overburden has hindered analysis of the real effect of salt tectonics on the development of the basin.

This study indicates that the major movements of the Zechstein salt in the Central Trough area had occurred before the end of the Triassic. These movements created a series of salt highs or ridges with intervening blocks of Triassic sediments resting on thin salt, or directly on the early Permian pre-salt

From *Petroleum Geology of Northwest Europe: Proceedings of the 4th Conference* (edited by J. R. Parker).
© 1993 Petroleum Geology '86 Ltd. Published by The Geological Society, London, pp. 1245–1254.

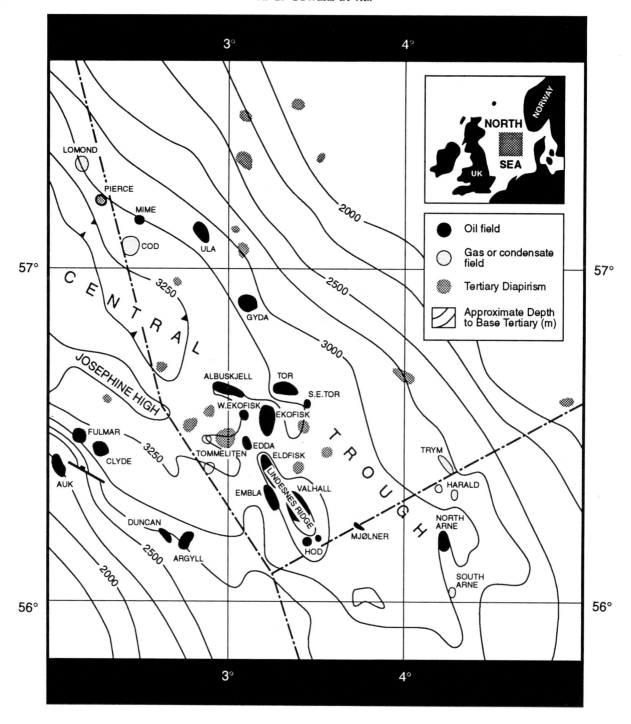

Fig. 1. Approximate depth in metres to base Tertiary corresponding to the form of the thermal subsidence phase TP7. The Central Trough approximates the area deeper than 3000 m.

sediments. The common situation after the late Triassic–early Jurassic uplift and erosion is illustrated in Fig. 5a. Both salt highs and Triassic blocks exhibit a linearity with a NW–SE trend. The pre-Jurassic dating of the salt tectonic which is responsible for this geometry is indicated by the internal geometry of the Triassic and the late Triassic age of the thin sediments found above the salt highs in many instances. Previously, many workers have not recognized these thick Triassic blocks, and have interpreted thick salt over the whole area (Forsberg *et al.* (in press), Spencer *et al.* (1986), Gabrielsen *et al.* (1986)).

Fig. 5b shows a typical geometry where relative subsidence over the salt high area has resulted in greater deposition/

preservation of Upper Jurassic sediments over the salt high than over the Triassic blocks. This type of salt-controlled subsidence is interpreted as subrosion (sub-surface dissolution of salt) in the late Jurassic–early Cretaceous. It has long been recognized on the Jæren High. In the Ula Field subsequent Tertiary local doming of a part of the old salt high has resulted in the obvious 4-way dip closure which was the target of initial drilling (Fig. 5c).

To the south of the Ula Field, the differential subsidence of the salt highs shows a more asymmetric style, as shown in the example in Fig. 5d. Here, the differential subsidence has clearly been active only at one edge of the salt graben, and a fairly uniform thickness of Oxfordian/Kimmeridgian sediments

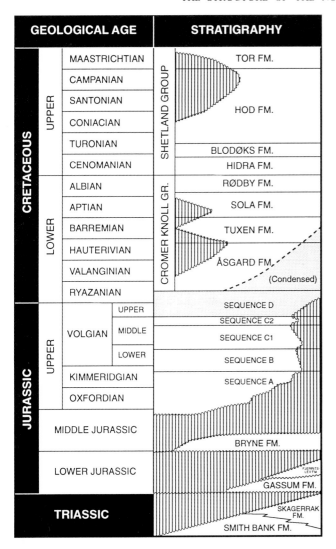

Fig. 2. Stratigraphy in the Central Trough. Late Jurassic after Forsberg *et al.* (in press); other stratigraphy from official Norwegian Petroleum Directorate nomenclature.

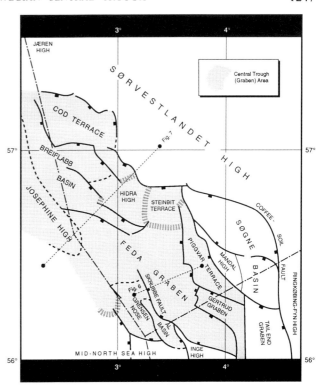

Fig. 3. Structural elements active within the Central Trough in Jurassic and early Cretaceous times. The nomenclature follows the guidelines given by the Norwegian Petroleum Directorate.

cover the Triassic block II. Later tilting and erosion occurred in the Volgian. This development closely follows the style of the two major tectonic phases in the late Jurassic (TP3 and TP4). Here, the boundary between the old salt high and the Triassic block I is the boundary between the stable platform area of the Sørvestlandet High, and the Steinbit Terrace, a boundary probably established already during the deposition of sequence A (Fig. 2).

The boundary between the old salt highs and the Triassic blocks is not always sharp. A gradual transition results in highly asymmetrical Triassic blocks, for example, in the Cod Terrace (Fig. 5e). Clear onlap of Triassic strata to the salt surface indicates salt pillow growth concurrent with Triassic deposition (downbuilding). Rotation of Triassic blocks can be intense, and clearly precedes the deposition of the Jurassic. It is purely salt driven, and thus is not related to the Jurassic rotation seen in Fig. 5d. The rotation is consistent with the 'floating' of the asymmetrical Triassic block in the salt. This explanation demands that, at the time of rotation (before compaction), the Triassic block has an average density less than that of the salt. The Triassic block may also rest directly on the pre-salt Rotliegendes, and rotate due to salt withdrawal and its own weight. This requires that the average density of the Triassic block be greater than that of the salt. Both mechanisms work towards the same result. The 'grounding' of

Triassic blocks due to insufficient salt causes break-up of the blocks in some areas (Fig. 5f) resulting in graben formation along the centre of the Triassic block.

The features described above are illustrative of the main tectonic styles which can be attributed to pre-Jurassic salt movement and its influence on Jurassic tectonics. There are, of course, many variations on these themes, but common to all is the conclusion that the salt tectonics in the Triassic were so intense and widespread that they were an important controlling factor in the structural reponse to subsequent tectonic movements. It follows that a failure to map the real salt distribution leads to erroneous conclusions concerning the structural development of the area.

Tectonic phase 2 (TP2): uplift and erosion

In the deeper basin areas the seismic data are not clear enough to determine the importance of early salt movement. Both the Breiflabb Basin and the Feda Graben show some post-Jurassic diapirism and pre-Jurassic salt swells, indicating that salt was initially deposited over most of the area. However, there appears to be less Triassic remaining in these areas than in surrounding areas, due to non-deposition, or to subsequent erosion. If it can be attributed to the latter then it follows that significantly greater uplift occurred in late Triassic–early Jurassic times in those areas which had the greatest post-Jurassic subsidence rather than in the flanking areas. Such an uplift would be a mirror image of the pattern of Tertiary thermal subsidence (TP7) and would correlate with the early Cimmerian movements elsewhere in the region.

However, no evidence of activity on major faults in Triassic times has been seen. Hence the uplift must have occurred as a gentle basin-wide flexure, rather than faulting. Such an uplift could explain the linearity and the northwest–southeasterly trend of the Triassic salt structuration, and establish a link to the major regional rift-tectonics and subsequent thermal subsi-

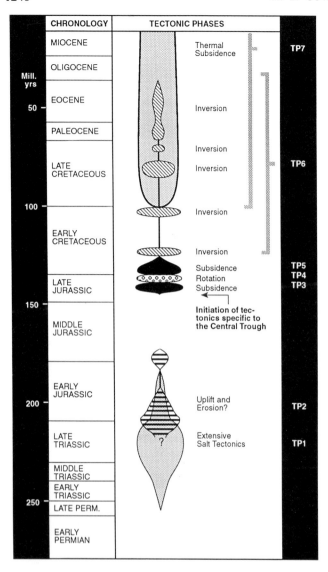

Fig. 4. Summary of the seven main tectonic phases TP1–TP7 observed in the Central Trough. The exact timing of TP1 and TP2 is uncertain, as are the details of the inversion pulses of TP6.

dence. These movements are included as tectonic phase 2 (TP2 in Fig. 4), though they may be coincident with, or even causal to, the salt tectonics of TP1. A second episode of TP2 may well have occurred at the end of the early Jurassic.

Tectonic phase 3 (TP3): symmetrical subsidence

Seismic evidence suggests that the lowermost Upper Jurassic sequences in the area exhibit uniform thicknesses within individual fault blocks or structural elements. In the deep Feda Graben it is often impossible to correlate reliably between different fault blocks due to lack of well control or diagnostic features on the seismic data. However, a consistent pattern of parallel reflectors is observed (Fig. 6), and where drilled, these prove to be Oxfordian to early Volgian, corresponding to sequence A and B of Forsberg et al. (in press) (Fig. 2).

The geometry of these sequences indicates symmetrical subsidence of individual fault blocks with minimal rotation or indeed salt movement. Well control proves thicknesses of over 1100 m were deposited during this phase in the deepest Feda Graben, illustrating that the graben subsidence, though local, was very intense with significant faulting. The sequences penetrated are shale dominated, with only minor sands. This indicates that local erosional products were minimal, and that

footwall uplift was not extensive. This inference is reinforced by the observation of thin condensed sequences of similar age on the flanks of the basin.

This earliest phase of basin formation has not received much attention previously, largely due to the difficulties of interpretation because of the overprint of the more readily recognized rotational tectonics which followed. Söderström et al. (1991) describe this phase in connection with the development of the Mjølner Field, where the associated faulting is demonstrated to follow different trends before and after the end of the early Volgian. In addition they note that initial subsidence was more restricted in area (sequence A) than subsidence in the late Kimmeridgian and early Volgian (sequence B).

Details of these two sub-phases of TP3 are not fully defined over the whole area as the seismic data are often very poor at the great depths involved. A stronger north–south trend to the extensional faults than in subsequent tectonics is observed. In addition a shift of depocentre is evident from sequence A to sequence B from the centre of the basin towards the west and the Skrubbe fault, indicating the start of the dominance of this fault. Note that in the deepest areas there is poor control of the underlying Zechstein and pre-Zechstein structure, and hence the effects of Triassic salt movements on the response to Jurassic tectonics are unclear.

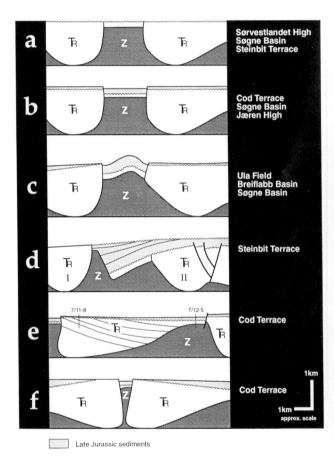

Fig. 5. Sketch sections to show some typical salt structural styles related to Triassic salt movements. (Faulting at base salt is omitted for simplicity.)

Tectonic phase 4 (TP4): rotational faulting

Near the end of the early Volgian a distinct change in tectonic style occurred. The faulting became dominated by a north-west–southeast trend, and individual fault blocks became rotated, with erosion on the footwalls and deposition in the

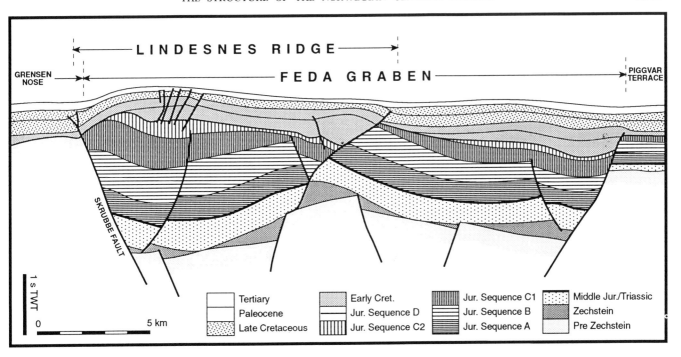

Fig. 6. Simplified sketch section from the southern Feda Graben in which the effects of TP1–TP6 can be seen. (See Fig. 3 for location.)

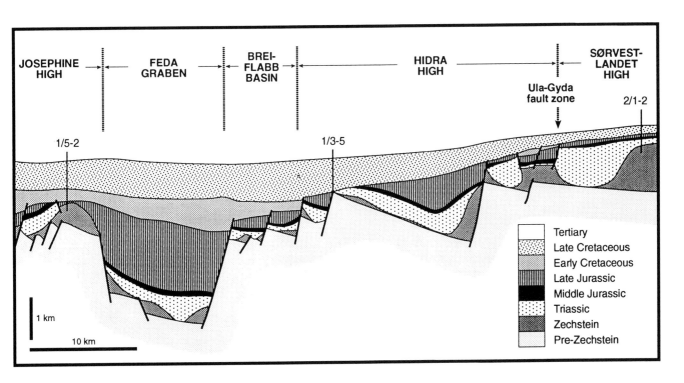

Fig. 7. Sketch section illustrating the variation of structural styles between the Gyda Field and the Josephine High. (See Fig. 3 for location.)

hangingwalls. Fault blocks are downthrown to the axis of the graben, with dips away from the axis. The northeastern flank is very well developed in a series of terrace elements from the Cod Terrace to the Gertrud Graben (Fig. 3), while the south-western flank is narrow and more obscure. The rotational movements are complicated by detachment in salt as discussed previously, but rotation of the pre-salt section shows that the movements are not salt driven.

The rotational movements are most intense and best seen on the Hidra High. Here, the largest rotated block has been eroded such that Upper Cretaceous rests directly on pre-

Zechstein at the apex (1/3-5 well, Fig. 7). Several erosional surfaces combine at this point, while down-dip the Zechstein, Triassic, parallel-bedded late Jurassic, and wedge-shaped late Jurassic (syn-rotational sediments) can all be recognized.

This tectonic phase is one where the net subsidence of the basin was rather small, in contrast to the vast thicknesses accumulated in TP3. The lack of rapid subsidence in the terrace areas allowed shallow marine deposition to take place between the emergent footwalls, and local deposition of reservoir quality sands.

The rotational tectonics ceased abruptly in the late Volgian with a regional subsidence which brought all but the highest areas below wave base. In contrast to the constant ample supply of clastic material experienced throughout the preceding late Jurassic, the late Volgian sediments are condensed in nature, with high organic content. Whereas before, there had been little seabed relief due to constant sediment supply, the final phases of the late Jurassic tectonics probably created considerable local seabed relief, draped by the ubiquitous sequence D (Mandal Formation).

Tectonic phase 5 (TP5): early Cretaceous subsidence

The distribution of the earliest Lower Cretaceous sediments is not well understood, but there is sufficient evidence of renewed basin subsidence at this time, as opposed to the more local structuration of TP4, to warrant the definition of a separate tectonic phase. It is, however, often unclear whether this subsidence is of syn-depositional age, or whether the Lower Cretaceous sediments were deposited in a series of basins created at the end of the Jurassic. The evidence so far indicates both, and the initiation of this tectonic phase may be linked regionally to the renewed clastic input in the late Ryazanian.

The TP5 basins are partially fault bounded, partially flexure bounded, and internally relatively quiescent. The pattern of early Cretaceous subsidence is not directly related to the previous tectonic phases, though the Hidra High, which had been most active in the rotational phase, remained a relatively positive area. Salt subrosion continued to be active in the Cod Terrace resulting in thick early Cretaceous sediments along the Ula and Gyda trend. Thick Lower Cretaceous is also found in the Feda Graben area, and where fault associated, it exhibits a close relationship with the subsequent inversion.

Tectonic phase 6 (TP6): inversion

Inversion has received considerable attention in the literature (see Ziegler 1987) but the precise nature of the inversion in the Central Trough is poorly documented. There are several phases of inversion starting in the early Cretaceous. Indications of late Jurassic inversion are also seen (Fig. 8a), but these can be explained by the TP4 movements and are, therefore, as yet unconfirmed as inversion. Precise timing of these phases is dependent on reliable dating in wells, and reliable interpretation of seismic data in some detail. Better dating is essential to prove that all the various inversion phases are linked in time. It is clear that the different events were highly local and even along the Lindesnes Ridge the effects of the various inversion pulses vary considerably (Fig. 8).

The geometry of the inversion-related structures is dominantly related to previously established normal faulting. Numerous faults show a slight uplift of the early Cretaceous sequences in the hangingwall. In general, the larger the local normal fault movements were in the late Jurassic and earliest Cretaceous, then the larger the subsequent inversion later in the Cretaceous. This relationship is underlined by the intense inversion observed along the Lindesnes Ridge (this area corresponds to a Jurassic sub-element within the Feda Graben), where late Jurassic–early Cretaceous subsidence was greatest (Fig. 8).

The first significant inversion pulse has largely been overlooked, even though it is responsible for the largest reverse fault seen on the Lindesnes Ridge (Fig. 8e), with a reverse throw at Top Jurassic in excess of 1500 m. It appears to be highly focused on a reverse movement on the Skrubbe Fault, and is less prone to flexure than later inversion phases. Exact dating has been difficult. Present analysis indicates movements of pre-Barremian age, probably mostly in the Hauterivian times. It is also possible, but not confirmed, that this early phase started slowly as early as Ryazanian. This inversion

phase has not been reliably indentified in other areas, though this may be due to data quality and resolution problems, rather than to absence of the inversion phase.

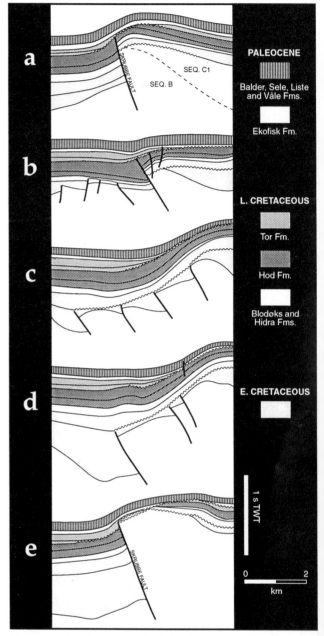

Fig. 8. A series of sequential E–W sketch sections to illustrate the variations in geometry caused by the various inversion phases along the Lindesnes Ridge. (**a**) Northern section showing late Cretaceous fault-related and Tertiary flexural basin inversion; (**c**) central section showing late Cretaceous and Tertiary flexural inversion; (**e**) southern section showing major early Cretaceous fault inversion.

The second inversion pulse is also of early Cretaceous age. It is best seen in the Outer Rough Basin in the Danish sector, where the Well Olaf-1 drilled through a reverse fault generated in this inversion phase. The well, together with the nearby Liva-1 well, shows the early Cretaceous fault subsidence phase starting in the late Ryazanian, followed by a major reversal largely confined to the late Albian. This inversion phase can also be seen in the southern part of the Lindesnes Ridge, but is generally difficult to recognize without precise well control on each side of inverted faults. Vejbæk and Andersen (1987) recognize this phase in minor inversion movements in several areas in the Danish Central Graben. In the north of the

Lindesnes Ridge there is evidence that the effects of this phase, if not a separate phase, continued into the Cenomanian and Turonian as mild basin inversion (Fig. 8).

The third inversion pulse is recognizable on the Lindesnes Ridge. It is clearly separated from the second pulse by a period of inactivity on the Skrubbe Fault during the deposition of a uniform Hidra and Hod Formation from Cenomanian to Santonian times. A thinning of the Tor Formation and severe erosion of the underlying Hod Formation on the Lindesnes Ridge indicate inversion from Santonian to early Maastrichtian times. This inversion was not limited to the Skrubbe Fault where flexure dominated, but affected the whole of the area which became the Lindesnes Ridge. This inversion is far more widespread than the previous phases, with mild inversion being seen in the Søgne Basin, and north into the Breiflabb Basin. The inversion pulse in the mid-Maastrichtian identified by Vejbæk and Andersen (1987) in the Danish sector may be part of this same pulse.

A fourth inversion pulse is recognized in the Paleocene, following the same pattern as the inversion in the previous pulse. Considerable further inversion of the Lindesnes Ridge occurred in the Eocene, giving the Lindesnes Ridge its considerable present relief. In other areas within the Norwegian Central Trough, these later movements, if present, are difficult to separate from differential compaction and salt movements.

Tectonic phase 7 (TP7): thermal subsidence

In the late Cretaceous the whole Central Trough area began to subside and the differences previously seen between the individual structural elements largely disappeared. The exceptions were the inversion movements described above. The subsidence is largely attributed to a thermal cooling post-rift phase inspired by the models of McKenzie (1978). The subsidence is centred in the Breiflabb Basin, and is readily seen in the top Chalk depth map (Fig. 1). The subsidence pattern does not include the Søgne Basin and Tail End Graben, and hence cannot be simply linked to Jurassic graben formation.

Structural elements

The following section reviews the structural elements in space, grouped as basins, highs, and terraces, relating them to the seven tectonic phases outlined above.

The Feda Graben

The Feda Graben was first defined by Gowers and Sæbøe (1985). It is a fault-bounded graben typified by substantial subsidence in TP3. Considerable activity has also occurred in the Feda Graben in the other six tectonic phases such that the resultant structure in this area is particularly complex. The Feda Graben is dominated by the Skrubbe Fault, trending NNW and separating it from the Grensen Nose and the Ål Basin. The Skrubbe Fault is a large normal fault with a throw before inversion of 4000 m or more. The suggestion of Roberts et al. (1990) that the fault does not exist is in conflict with the seismic and well data.

Figure 6 is a sketch section over the southern Feda Graben at a point where it is least complex. The response of the area to TP3 subsidence does not appear to be greatly influenced by the salt. Simple normal faulting dominates, following a variety of trends from N–S to E–W, both new and inherited, with only limited detachment within the salt. The rotational movements of TP4, together with significant erosion, are evident throughout the basin, particularly in the south, though these are much modified by the presence of salt.

The earliest Cretaceous subsidence of TP5 occurred in varying degrees throughout the Feda Graben, apparently in

several local depocentres, particularly at the graben margins. Thick early Cretaceous sediments (over 1500 m) are preserved directly north of the Grensen Nose and are interpreted by some (Brasher, in press) as salt withdrawal effects. However, the poor relationship of the depocentres to the diapirs, and the evidence for a more widespread fault-related subsidence leads us to believe that the role of salt at this time, although important in dictating the response, was not the prime cause of the movements. Anomalous local thicknesses of early Cretaceous sediments are found in down-faulted graben structures along the northern margins of the Feda Graben, well seen under the Albuskjell and Tor fields.

The subsidence was partly linked to the previous movements of TP4, for example the fault in the centre of Fig. 6. Subsequent erosion during inversion masks the true extent of this TP5 subsidence on the Lindesnes Ridge, with folding and faulting resulting in the preservation of substantial thicknesses locally (Well 2/11-1 drilled over 500 m of Lower Cretaceous Valhall Formation). The faulted boundaries of the Feda Graben were affected almost everywhere by the first major inversion episode of TP6. This is most striking along the Skrubbe Fault (Fig. 8), but can be seen on almost all other bounding faults as a gentle flexure of the hangingwall (Figs 6 and 7).

The Skrubbe Fault is exceptional in that it shows such local variety in the inversion movements along its length (Fig. 8), and may indeed have experienced some inversion in the latest Jurassic–earliest Cretaceous. Older E–W Paleozoic tectonic trends experienced mild reactivation, often steering the sites of fault transfer ramps. In addition, the erosion here suggests that the Lindesnes Ridge area was significantly elevated compared to the surrounding areas where erosion was minimal. This same structural element was most active in the several succeeding inversion pulses.

The Gertrud Graben

First identified by Møller (1986) the Gertrud Graben is a pronounced NW–SE-trending asymmetric half-graben in the Danish sector, dipping to the NE where it is separated from the Piggvar Terrace by a major fault linking the Feda Graben and the Tail End Graben. Initially, the area appears to have been part of the platform area, separated from the Feda Graben by a fault, with thin condensed deposits of Callovian to Middle Kimmeridgian age (sequence A). However, during the Kimmeridgian the Gertrud Graben began to subside as a normal fault block together with the Feda Graben to the west as part of TP3. A distinct normal fault trending NNE limits this structural block to the north.

The half-graben nature of the Gertrud Graben was established in the Volgian during the rotational TP4, with significant subsidence along the bounding fault to the northeast. This subsidence continued in the early Cretaceous TP5, and was spectacularly reversed during the inversion of TP6. Vejbæk and Andersen (1987) date the main inversion pulses along this fault as Hauterivian and more particularly Turonian–Santonian in age (pulse 1 and 3 of TP6).

The Ål Basin

The Ål Basin was formed on the footwall of the Skrubbe Fault during TP5. Up until this time the area was an integral part of the Grensen Nose. The basin is strongly asymmetric: the basin floor dips to the east and is abruptly cut off by the Skrubbe Fault at its deepest point. The boundary with the Grensen Nose is faulted in pre-Cretaceous sequences, but in the early Cretaceous it was dominated by flexure, with only minor faulting downthrown to the east, and was probably part of the subsiding Feda Graben. The Ål Basin shows minimal signs of inversion or other tectonic activity after the early Cretaceous.

The identification of deep faulting, and the general absence of thick Jurassic sediments indicates an earlier tectonic history, based on the older north–south and east–west tectonic trends. However, the absence of the pre-Cretaceous Mesozoic sequences, possibly due to footwall erosion related to the Skrubbe Fault, prevents a reliable reconstruction.

The Breiflabb Basin

The Breiflabb Basin is characterized by significant subsidence associated with the TP5 movements. It is bounded to the west by the Josephine High against which the seismic events within the Lower Cretaceous progressively onlap. A major fault at depth appears to separate the northern UK extension of the Breiflabb Basin from the Cod Terrace, but apart from this the boundaries against the Cod Terrace are controlled by subtle faults inherited from the Triassic salt tectonics, detached from the pre-salt basement movements. Evidence of an active phase of salt tectonics in the Triassic is clear from significant diapirs in the north of the basin, especially in the UK sector.

The form of the Breiflabb Basin indicates formation dominated by flexure rather than faulting, with onlap onto the late Jurassic sequence D, a style which is echoed in the Outer Rough Basin in the Danish sector. The one major exception is a large normal fault which forms the boundary between the southern half of the Breiflabb Basin and the Hidra High, and swings across the centre of the basin before dying out. A transitional boundary is found northwards from the Hidra High into the Breiflabb Basin, and in the southwest the structural picture at depth is unclear. The Lower Cretaceous thins southwards up the footwall of a fault forming the boundary between the Breiflabb Basin and the Feda Graben, with typical inversion in the Cretaceous (Fig. 7). The structural details of this area at depth are unclear due to poor data quality.

There are no wells which can reliably calibrate the pre-Cretaceous section in the Breiflabb Basin. Thus, the age of the Upper Jurassic sequences or the related tectonic phases cannot be ascertained. The data available suggest that the Upper Jurassic is not particularly thick in this basin, and that there is a greater thickness of Triassic sediments here than in the Feda Graben to the south.

Mild inversion of late Cretaceous and Tertiary age is also seen, focused on the basin margins and major faults. The Breiflabb Basin was the focus for the TP7 thermal subsidence phase (Fig. 1), which has resulted in the deposition of a very thick Chalk (over 2000 m) and Tertiary sequence. Combined with the thick Lower Cretaceous, these subsidence phases result in a Jurassic section buried at over 7000 m in much of the Breiflabb Basin.

The Søgne Basin

The Søgne Basin was identified as a separate structural element by Gowers and Sæbøe (1985) and stands out as an anomaly compared to the rest of the Norwegian Central Trough as regards tectonic behaviour. The basin is a large rotated easterly dipping fault block limited to the east by the major basement fault, the Coffee Soil Fault. The latter can be followed southwards forming the eastern limit of the Tail End Graben. Northwards the fault diminishes in throw and veers westwards to die out at the northwestern end of the Søgne Basin. The Søgne Basin is divided into a northern and a southern half by an east–west-trending fracture, the southern half showing greater block rotation than the northern half. To the west the fracture is downthrown to the north, and forms the northeastern edge of the Mandal High, while down-dip to the east the polarity reverses and the throw is down to the south, and the

fracture swings round to join the main Coffee Soil Fault. A narrow relay ramp separates the northern and southern parts of the Coffee Soil Fault.

The Søgne Basin shows evidence of extensive Triassic salt tectonics (TP1), with a more or less continuous line of salt highs along its southwestern edge, several diapirs in the centre, and several salt highs and salt swells along the Coffee Soil Fault. There is no evidence for basin formation in the northern half of the basin in the Triassic, since equal thicknesses of Triassic sediments below the late Triassic erosion surface are found on both sides of the fault. However, a wedge of Middle Jurassic sediments indicates that movement on the Coffee Soil Fault and rotation of the Søgne Basin started earlier than activity in the rest of the Central Trough. A thick sequence of parallel-bedded Middle Jurassic in the southern half of the basin indicates a more symmetrical graben formation here.

Continued gentle block rotation in late Jurassic times (concurrent with the symmetrical graben subsidence in the rest of the Central Trough), together with salt withdrawal movements gave a variable, but thin to moderate thickness of Upper Jurassic in the Søgne Basin. Extensive rotational tectonics of Mid-Volgian age (TP4) are not recognized in the Søgne Basin and indeed little active tectonics are seen from this point on, apart from some mild inversion in the late Cretaceous. The Søgne Basin is not significant in the TP7 thermal subsidence pattern (see Fig. 1) and the tectonic development is considered to be more closely related to the Tail End Graben than to the Norwegian Central Trough.

The Mandal High

The Mandal High was introduced by Rønnevik *et al.* (1975) and further defined by Gowers and Sæbøe (1985). Outwardly this high is a stable flat basement block, with Cretaceous sediments resting directly on eroded crystalline basement. In many ways it can be considered as the eroded tip of the Søgne Basin fault block. However, the indications of symmetrical graben formation in the Middle Jurassic to early Volgian in the southern half of the Søgne Basin suggest that the Mandal High may have acted as a distinct stable block already from this time, either as a separate element, or as part of a larger block stretching westward from the area of the present Mandal High.

In common with the Søgne Basin, the Mandal High is interpreted to be little influenced by the various tectonic phases which are prevalent in the rest of the Norwegian Central Trough.

The Grensen Nose

The Grensen Nose was introduced and defined by Gowers and Sæbøe in 1985. It is a structural spur protruding northwards from the Mid North Sea High. Late Cretaceous sediments from TP7 rest on thin to absent early Cretaceous and latest Jurassic sediments. These, in turn, rest on an eroded sequence of variable, but often uncertain age sediments, from Triassic to Carboniferous. Considerable thicknesses of Paleozoic sediments are present. The boundaries of the Grensen Nose are dominated by flexure and, hence, are rather diffuse. The exception is at the Skrubbe Fault north of the Ål Basin.

The Grensen Nose has probably been exposed from at least the early Upper Jurassic with the formation of the Skrubbe Fault, in association with TP3 and TP4. It assumed its present form in the early Cretaceous TP5. The Grensen Nose shows considerable internal faulting, with some old east–west trends being important, together with the NNW–SSE trend. TP6 inversion was also active as local small-scale reverse rejuvenations of the latter fault trend in the Campanian.

The Hidra High

The Hidra High was introduced by Rønnevik *et al.* in 1975. It has been the subject of some confusion in the literature due to its complex nature. The Hidra High is dominated by the rotational tectonics of TP4 (Fig. 7) which are easily seen on the seismic data due to the strong base Zechstein reflector. However, the response within the Mesozoic sediments is complex and obscure due to extensive Triassic salt tectonics, and moderate subsidence in TP3. Intense rotation and erosion in TP4 has removed much of the evidence for earlier history.

The boundary between the Hidra High and the Breiflabb Basin is drawn at the limit of TP5, which affected the latter, but not the former. The boundary with with Sørvestlandet High is drawn at the limit of the effects of both Upper Jurassic tectonic phases, while the boundary with the Cod Terrace is defined by the northern limit of extensive rotation of TP4. The boundary to the southeast with the Steinbit Terrace is transitional and less clear, but also taken as the limit of extensive rotational tectonics. Neither TP5 nor the ensuing TP6 inversions appear to have affected the Hidra High, which ceased activity as a separate structural element after TP4.

The Cod Terrace

The Cod Terrace is dominated by the Triassic salt tectonics of TP1. The boundary between the Cod Terrace and the Sørvestlandet High was present already during TP3, the subsidence being confined to the salt-high areas. Rotation movements in the Volgian were mild, being seen within the Jurassic sediments of the earlier salt highs, and as further movement of Triassic blocks previously rotated due to salt movements.

TP5 is also clearly apparent on the Cod Terrace, again as subsidence over the earlier salt highs, and subsequent inversion movements were concentrated on these same areas of subsidence, intimately related to the salt.

The Steinbit Terrace

Triassic salt tectonics (TP1) were very active in this area, and the boundary to the platform area to the north follows a pronounced east–west-trending salt ridge (Fig. 5d). Salt underlying the whole area is thought to play an important role in the response to Jurassic tectonics. At pre-salt levels the terrace appears as a southwards-dipping zone transitional from the Sørvestlandet High down to the Feda Graben.

To the east, the boundary between the Steinbit Terrace and the Piggvar Terrace is drawn along a significant N–S-trending basement fault which has been active from at least Oxfordian times (TP3). This fault has a large throw in the south (over 2000 m at base Zechstein), but rapidly dies out northwards within 15 km to a minor fault. The southern boundary of the Steinbit Terrace at this point is a flexure down to the deep Feda Graben, the latter experiencing greater subsidence in TP3 than the former. Extensive diapir activity in the south of Block 2/5 obscures the nature of the westerly continuation of this boundary, though faulting is expected, in common with the faulted boundary between the Steinbit Terrace and the Feda Graben in Block 2/4.

Evidence of some TP5 subsidence, and subsequent inversion is found within the Steinbit Terrace, for example in the Tor Field, but the effects of TP4, so intense on the adjacent Hidra High, are minor.

The Piggvar Terrace

In common with most of the other structural elements, the Piggvar Terrace experienced extensive Triassic salt movements. Moderate disturbance from TP3 is also apparent, but the presence of salt seems to have amplified the response of the Mesozoic to TP4. The Jurassic–Triassic sediments are broken into a large number of small rotated fault blocks, detached from the basement, concentrated in a zone in the centre of the Piggvar Terrace, and aligned WNW to coincide with the trend of the northern edge of the Feda Graben.

Minor influence from TP5 and TP6 is seen in the Piggvar Terrace before TP7 subsidence took over. The eastern boundary of the Piggvar Terrace is formed by the major basement fault which limits the Søgne Basin–Mandal High, and hence can be considered as the eastern limit of the area in which the various tectonic phases discussed in this paper are typical.

Discussion

Prior to the establishment of the Central Trough, the area was part of the broad Norwegian–Danish Basin during the deposition of the Zechstein and Triassic deposits. The structural development specific to the Central Trough area can be divided into three major stages: (1) late Triassic to middle Jurassic flexural uplift; (2) late Jurassic to early Cretaceous fragmentation; and (3) late Cretaceous to Tertiary flexural subsidence.

Evidence for stage 1 includes thinning of preserved Triassic and early Jurassic towards the Central Trough from the east; thinning and removal of the otherwise uniform thickness of fluviatile Middle Jurassic deposits following the same pattern; and the alignment in late Triassic times of salt ridges along a NW–SE trend not previously identified as active in the area. The more expected N–S trend dominant in the Norwegian–Danish Basin is seen in the Søgne Basin–Tail End Graben, and this trend does not appear to belong to the Central Trough area as defined in Fig. 3. It is theoretically possible to explain the salt lineament by a progradation of Triassic sediments from the northeast, but this fits poorly with the detailed observations. TP1 and TP2 can thus be interpreted as stage 1 of the tectonics which formed the Central Trough, even though the effects of stage 1 are seen outside the Central Trough area. There is, as yet, no concrete evidence of significant basement faulting related to this stage.

Stage 2 started abruptly with TP3 in the Oxfordian with intense faulting, and continued to the middle of the Cretaceous. The Central Trough area was fragmented during this time, and each of the structural elements defined in Fig. 3 led its own tectonic existence. The tectonic movments were confined to the Central Trough and there was little activity outside the area. Various correlations in time (from one tectonic phase to another) and in space (from one structural element to another) are obvious and must be accounted for when analysing the structural behaviour. Inversion movements of TP6 are closely related to movements of the preceding TP4 and TP5, but in contrast to these the inversion effects continued in several pulses into the Tertiary, long after the onset of the basin-wide thermal subsidence.

Stage 3 consists of TP7, regional subsidence, with minimal faulting, centred on the Breiflabb Basin and Feda Graben, but extending far beyond the area of the Central Trough to link with similar subsidence seen throughout the Northern North Sea Basin. The form of the contours of the subsidence conform with the area of activity in stage 2, and should be linked to the area of hot extended lithosphere. The area excludes the Søgne Basin and indeed the Tail End Graben, illustrating the differences in structural development between the Norwegian and Danish Central Graben areas.

The complex structural evolution outlined above is obviously influenced by a series of different factors through time, and can hardly be expected to be explained by a single extensional model, be it that of McKenzie (1978), the simple shear model of Wernicke (1985) or the flexural cantilever model of Kusznir *et al.* (1991). It is clear that any structural models for the rifting process must take account of both TP3 and TP4. Many attempts to describe the structural evolution of the whole area using models generally fail (i.e. are incompatible with the observations from the data) due to a lack of appreciation of the complexities involved. Local inherited structural configurations have obviously played an important part in the response of the sediments to the applied stress. In particular, the salt movements of TP1 must be fully analysed in three dimensions before structural conclusions can be made from the geometry of the Mesozoic as seen on individual seismic sections.

The purpose of this paper is to introduce a degree of detail to the events which an acceptable structural model should account for. At present no such model has been presented in the literature and this article does not attempt to propose one. The process of unravelling the complex structural history of the Norwegian Central Trough is far from complete.

Conclusions

The Central Trough is a complex area which has been affected by several tectonic phases. These phases may be broadly grouped into three stages: a regional flexural uplift; an intense segmentation due to faulting, with subsidence, rotation and inversion confined to a specific area which is defined as the Central Trough; and finally a regional flexural subsidence. Seven distinct tectonic phases have been recognized. The structural analysis of the area cannot be achieved without detailed analysis of the data in order to separate the effects of the various tectonic phases from one another. Much of the structural analysis previously presented in the literature appears to be model based, and fails to account for the extensive observations which can be made from the data. The term 'central graben' is an inadequate term for the area discussed and the official Norwegian nomenclature form of Central Trough is preferred.

We thank Norsk Hydro for permission to publish this article, and our colleagues in the SNS team who made the analysis presented possible. In addition we thank R. Gabrielsen and I. D. Bartholomew for their comments on the manuscript.

References

BEACH, A. 1985. Some comments on sedimentary basin development in the Northern North Sea. Scottish Journal of Geology, 21(4), 493–512.

BRASHER, J. E. (in press) Local Tectonics and effects on sediment distribution within Eldfisk Field. Presented at Norskpetroleumsforening conference: Petroleum Exploration and Exploitation in Norway, 9–11 Dec. 1991, Stavanger, Norway.

FORSBERG, A. W., GOWERS, M. B. AND HOLTAR, E. (in press) Multidisciplinary stratigraphic analysis in the Upper Jurassic of the Norwegian Central Trough. EAEG conference, Florence, 1991.

GABRIELSEN, R. H., EKERN, O. F. AND EDVARDSEN, A. 1986. Structural development of hydrocarbon traps, Block 2/2, Norway. In: SPENCER, A. M. ET AL. (eds) Habitat of Hydrocarbons on the Norwegian Continental Shelf. Norwegian Petroleum Society, Graham & Trotman, London, 129–141.

GIBBS, A. D. 1989. Structural styles in basin formation. In: TANKARD, A. J. AND BALKWILL, H. R. (eds) Extensional Tectonics of the north Atlantic Margins. AAPG Memoir, 46, 81–93.

GOWERS, M. M. AND SÆBØE, A. 1985. On the structural evolution of the Central Trough in the Norwegian and Danish sectors of the North Sea. Marine and Petroleum Geology, 2, 298–318.

KUSZNIR, N. J., MARSDEN, G. AND EGAN, S. S. 1991. A flexural-cantilever simple shear/pure shear model of continental lithosphere extension: applications to the Jeanne d'Arc Basin, Grand Banks and Viking Graben, North Sea. In: ROBERTS, A. M., YIELDING, G. AND FREEMAN, B. (eds) The Geometry of Normal Faults. Geological Society, London, Special Publication, 56, 41–60.

McKENZIE, D. 1978. Some remarks on the development of sedimentary basins. Earth and Planetary Science Letters, 40, 25–32.

MØLLER, J. J. 1986. Seismic structural mapping of the Middle and Upper Jurassic in the Danish Central Trough. Danmarks Geologiske Undersøgelse, Series A, 13.

ROBERTS, A. M., PRICE, J. D. AND OLSEN, T. S. 1990. Late Jurassic half-graben control on the siting and structure of hydrocarbon accumulations: UK/Norwegian Central Graben. In: HARDMAN, R. F. P. AND BROOKS, J. (eds) Tectonic Events Responsible for Britain's Oil and Gas Reserves. Geological Society, London, Special Publication, 55, 229–257.

—— AND YIELDING, G. 1991. Deformation around basin-margin faults in the North Sea/mid-Norway rift. In: ROBERTS, A. M., YIELDING, G. AND FREEMAN, B. (eds) The Geometry of Normal Faults. Geological Society, London, Special Publication, 56, 61–78.

RØNNEVIK, H. C., VAN DEN BOSCH, W. AND BANDLIEN, E. H. 1975. A proposed nomenclature for the main structural features in the Norwegian North Sea. In: FINSTAD, K. G. AND SELLEY, R. C. (Coord.) Jurassic Northern North Sea Symposium. Norwegian Petroleum Society, Stavanger, JNSS/18, 1–16.

SÖDERSTRÖM, B., FORSBERG, A., HOLTAR, E. AND RASMUSSEN, B. A. 1991. The Mjølner Field: a deep Upper Jurassic oil field in the Central North Sea. First Break, 9(4), 156–171.

SPENCER, A. M., HOME, P. C. AND WIIK, V. 1986. Habitat of hydrocarbons in the Jurassic Ula Trend, Central Graben, Norway. In: SPENCER, A. M. ET AL. (eds) Habitat of Hydrocarbons on the Norwegian Continental Shelf. Norwegian Petroleum Society, Graham & Trotman, London, 111–127.

VEJBÆK, O. V. AND ANDERSEN, C. 1987. Cretaceous–Early Tertiary inversion tectonism in the Danish Central Trough. In: ZIEGLER, P. A. (ed.) Compressional Intra-Plate Deformations in the Alpine Foreland. Tectonophysics, 137, 221–238.

WERNICKE, B. 1985. Uniform-sense normal simple shear of the continental lithosphere. Canadian Journal of Earth Science, 22, 108–125.

ZIEGLER, P. A. (ed.) 1987. Compressional intra-plate deformations in the Alpine Foreland. Tectonophysics, 137, 1–4.

Structural styles in the Danish Central Graben

G. O. SUNDSBØ[1] and J. B. MEGSON

Mærsk Olie og Gas a.s., 50 Esplanaden, DK-1263 Copenhagen K, Denmark
[1] (Present address: Quality Management a.s. Løkketangen 14A, PO Box 102, 1301 Sandvika, Norway)

Abstract: In recent years a substantial part of the Danish Central Graben has been covered by high-resolution 3D and 2D seismic data, and a detailed study of the various tectonic elements, their evolution through time and their importance for hydrocarbon trapping has been made possible. This has revealed that in many instances important structural trends affecting reservoir distribution or production characteristics can only be detected using high-resolution 3D seismic. The evidence is that the Triassic, Middle Jurassic, and basal Upper Jurassic graben trends were predominantly north–south. The area which is now the deepest part of the Tail End Graben was relatively high at these times. The NW trends of the Danish Central Graben are not, on the whole, an accentuation of an older (Triassic/Middle Jurassic) rift along the same direction, but are new trends coming into effect in the Kimmeridgian. These NW–SE trends controlled rifting, with peak activity in the Volgian. In conclusion, the structures of most Danish fields exhibit the superposition of several distinct structural episodes. A detailed understanding of the nature of the individual tectonic events is necessary for unravelling the deformation history, which is important knowledge both in exploration and production phases. The various examples from the Danish Central Graben demonstrate that this requires a degree of seismic resolution that has only recently become achievable.

The Central Graben extends for approximately 450 km from the UK and Norwegian continental shelves, through the Danish and German sectors and southwards towards the Dutch mainland. In the literature, the onset of the development of the Central Graben is commonly ascribed to the Permo-Triassic (Ziegler 1975, 1978, 1988; Skjerven *et al.* 1983) followed by a strong accentuation in the late Jurassic. Locally in the Danish Central Graben more than 4 km of Upper Jurassic sediments were deposited, one of the thickest Upper Jurassic sections in the North Sea. Both the Southern and the Northern Zechstein basins extend into the Danish Central Graben and halokinesis plays an important part in the structural development, particularly in the south where most of the present oil and gas fields are located.

The Cretaceous to early Tertiary inversion movements that strongly affected the Southern North Sea region in the UK and the Dutch sectors (Heybroek 1974, 1975; Clark-Lowes *et al.* 1987), died out northwards and are only mildly expressed in the Norwegian sector (Skjerven *et al.* 1983). These movements were still relatively pronounced throughout the Danish Central Graben and are of importance for the development of several hydrocarbon traps (Megson 1992).

The tectonic framework as it appears today (Figs 1 and 2) is thus largely dominated by the strong late Jurassic rifting as well as by halokinesis and inversion phases. This framework has been described in a number of publications during the last decade (Gowers and Sæbøe 1985; Michelsen *et al.* 1987; Damtoft *et al.* 1987, 1992), together with general outlines of the structural evolution and the dominant structural styles of the region.

The primary sources for structural information for nearly all the previously published studies are the several speculative seismic surveys acquired in 1981 and 1982. Over the last few years, a substantial part of the Danish Central Graben has been covered by high-resolution data (Fig. 3) for exploration, appraisal, and production purposes. This has enabled a more detailed study of the various tectonic elements, their evolution through time, and their importance for hydrocarbon trapping. This study has revealed that in several instances important structural trends affecting reservoir distribution or production characteristics can only be detected using high-resolution 3D seismic.

This paper is not intended to be a comprehensive review of the tectonics of the entire Danish Central Graben, as high-resolution seismic coverage is not yet sufficient to allow for this. Instead, a few selected areas are focused upon, where 3D coverage exists, in order to illustrate the diversity of structural styles found in the Danish Central Graben. In addition, examples will be presented from areas where new information reveals that the previously published work may not accurately describe the structural development.

General tectonic framework

An overview map outlining the major tectonic elements of the Danish Central Graben together with simplified SW–NW cross-sections is included as Figs 1 and 2.

The Coffee Soil Fault

The single most spectacular element of the Danish Central Graben is the Coffee Soil Fault running along the entire eastern margin of the Central Graben in the Danish sector (Figs 1 and 2). This major fault separates the Danish Central Graben from the Ringkøbing–Fyn High. Locally, the fault produces a strong seismic response down to 5–6 s two-way time, making it possible to map the fault plane in some detail below the Mesozoic basin-fill and to get a good estimate of the amount of dip-slip on the fault. The fault plane itself is inclined at a varying but generally low angle, commonly around 30–35°, and appears to be planar when converted to depth. The total vertical movement of this fault, not including that removed by footwall erosion, varies from 2.5 km to more than 4 km, and the horizontal extension (as represented by the fault cut at base Zechstein level) is in places more than 6 km.

As described and discussed in detail by Cartwright (1987, 1989) the Central Graben changes strike orientation within the Danish sector from nearly due N–S in the south to NW–SE further north. This is reflected in the strike direction of the Coffee Soil Fault which is broken into segments of varying strike direction. The changes in strike of the Coffee Soil Fault suggest that the graben itself may be segmented at the same locations. This is corroborated by the changes in graben fill, and in tectonic styles, at the changes in strike of the Coffee Soil Fault.

The timing of the main activity on the Coffee Soil Fault varies along strike; fairly thick Triassic sequences and relatively moderate Upper Jurassic thicknesses are found both in

From *Petroleum Geology of Northwest Europe: Proceedings of the 4th Conference* (edited by J. R. Parker).
© 1993 Petroleum Geology '86 Ltd. Published by The Geological Society, London, pp. 1255–1267.

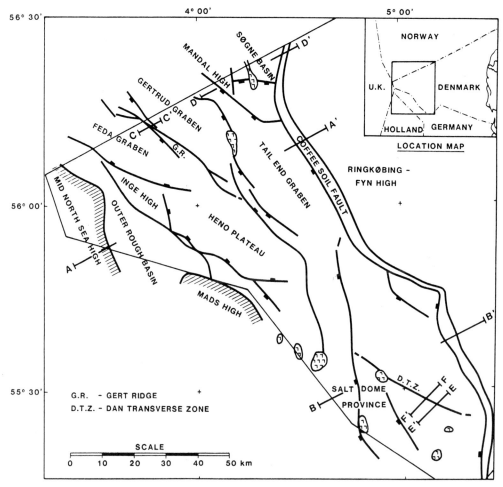

Fig. 1. The main tectonic elements of the Danish Central Graben.

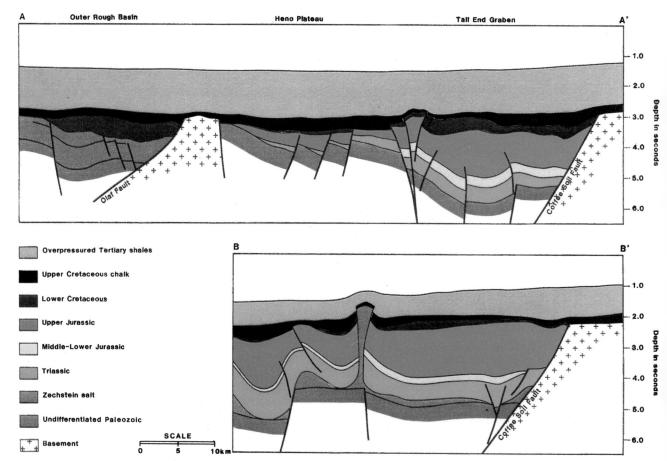

Fig. 2. SW–NE cross-sections through the Danish Central Graben. See Fig. 1 for locations.

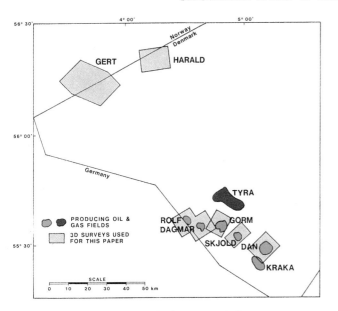

Fig. 3. 3D seismic surveys used in the present study.

units are believed to be related to variations in the slip rates along the Coffee Soil Fault.

Different slip rates for the individual segments of the Coffee Soil Fault would necessarily have caused lateral movements between the related basin compartments. Cartwright (1987) demonstrated how these movements are accommodated at transverse zones cutting across the Tail End Graben (Fig. 5). The present study to a large extent supports this observation and provides more detailed evidence of their existence. The structural examples presented below will elucidate some of the results of the movement along the transverse zones.

The Tail End Graben

The Tail End Graben occupies the eastern part of the Danish Central Graben, immediately westwards of the Coffee Soil Fault. This graben is areally the largest, and also the deepest basin within the Danish sector.

As seen above there is evidence of segmentation of the Central Graben from variations in strike direction of the Coffee Soil Fault bounding the graben towards the east. In particular, the Upper Jurassic Tail End Graben changes in geometry and subsidence rates in the vicinities of these changes in strike direction.

The geometry of the Upper Jurassic sequence varies significantly along the Coffee Soil Fault; a pronounced half-graben with eastward divergent reflectors can be seen along the NW–SE-trending fault segment, while relatively uniform depositional thicknesses occur in the south where the fault strike is closer to a N–S orientation. In the north near the Norwegian sector border the fault again locally resumes a more northerly direction. This northernmost area appears today as a Jurassic

the area of the Southern Zechstein Basin where the graben has a N–S orientation and also in the area of the Northern Zechstein Basin where the graben, at least until well into the late Jurassic, had the same N–S orientation. The central graben segment of NW–SE orientation does not contain a substantial thickness of Triassic deposits, but locally contains up to 4 km of Upper Jurassic shales (Fig. 4). The large variations along strike in stratigraphic thickness of the various

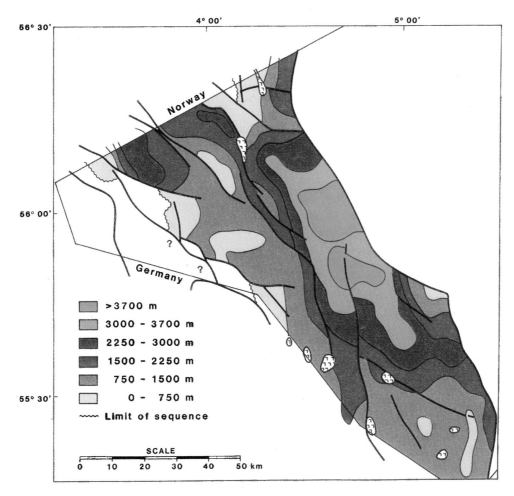

Fig. 4. Jurassic isopach in the Danish Central Graben (modified after Møller 1986).

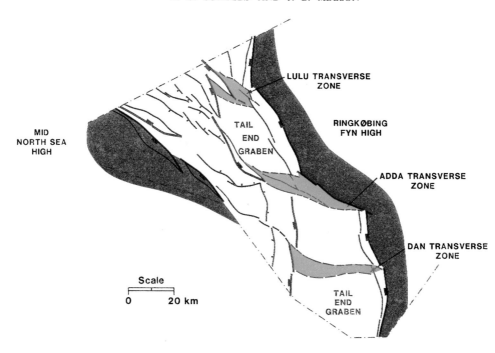

Fig. 5. Transverse zones in the Danish Central Graben (after Cartwright 1987).

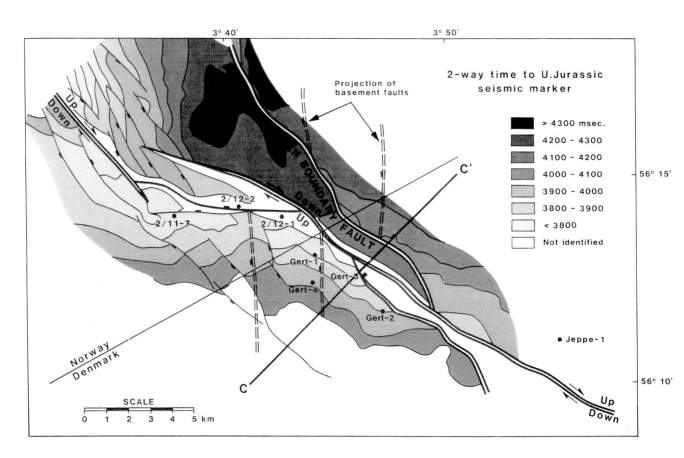

Fig. 6. Structure map of the Gert Ridge at near-base Volgian level. The strike-slip fault offsets both the Feda Graben boundary fault and deeper basement faults in a dextral sense.

half-graben, but this appearance is due solely to strong block rotation towards the end of the late Jurassic and subsequent erosion of sediments originally deposited in a uniformly subsiding basin, as will be discussed later.

Basins west of the Tail End Graben

In the northern part of the Danish Central Graben the Heno Plateau covers an extensive area west of the Tail End Graben. The internal structure of this platform is not well understood, as seismic resolution is generally poor and 3D coverage is limited. Locally, the seismic data indicate the presence of westward dipping NW–SE-trending normal faults forming numerous small late Jurassic half-grabens. This is illustrated in Fig. 2, cross-section A–A′. Further to the west, the Outer Rough Basin is related to normal fault activity on a seismically well defined, southwestward heading low-angle fault, referred to as the Olaf Fault (Cartwright 1987). This basin only contains a thin veneer of Upper Jurassic sediments, while the Lower Cretaceous thickness in places exceeds 1 km. The basin is thus younger than the depocentres lying further to the east. This fault separates a thick Paleozoic sequence underlying the Lower Cretaceous basin in the hangingwall block from crystalline basement rocks on the footwall block (Fig. 2).

Two additional NW–SE-trending late Jurassic grabens are located in the northernmost part of the Danish sector: the Feda Graben and the Gertrud Graben. Both become more prominent features northwards into the Norwegian sector.

In the southern part of the Danish Central Graben the western basin margin falls outside the Danish sector, and only limited information has been available to the authors. This western graben boundary is not entirely fault defined, unlike the eastern boundary, but appears to have been controlled partly by strong flexuring of the graben floor.

Structural examples

The Gert Ridge

Basal Upper Jurassic sands were found oil bearing by the Gert-1 well drilled in 1984, and the extension of this accumulation into the Norwegian sector was subsequently proved by the 2/12-1 well. The Norwegian extension is named the Mjølner Field and has been described by Søderstrøm et al. (1991). Several additional wells have been drilled to delineate the Gert/Mjølner Field.

The Gert Ridge follows a NW–SE trend in the central part of the Central Graben. The ridge crosses the Norwegian–Danish border (Fig. 1) and is expressed at base Cretaceous level as a structural nose plunging from the Norwegian sector southeastwards into the Danish area. At deeper levels within this nose a complex pattern of major faults and high-standing fault slices can be mapped (Fig. 6).

This structure is unique in the area in that, despite the major faulting of Jurassic and early Cretaceous strata along the ridge, there are no indications of late Cretaceous inversion and uplift. All neighbouring fault zones, both to the west and to the east of the ridge, are clearly affected by these late Cretaceous inversion movements.

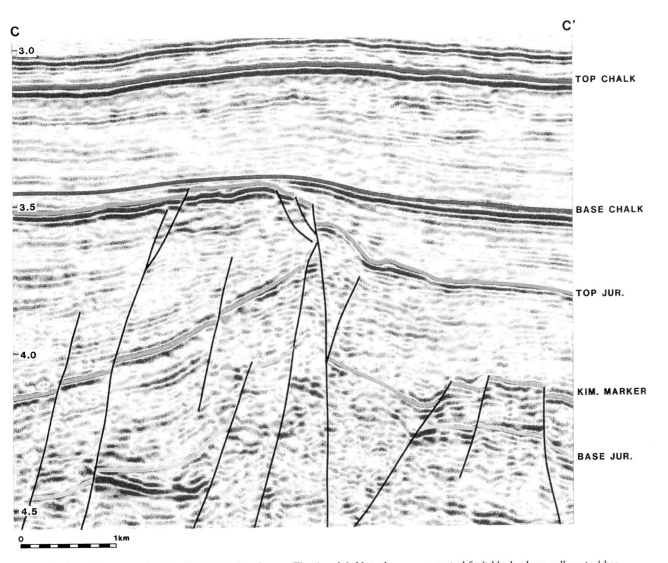

Fig. 7. Seismic section across the Gert Ridge. For location see Figs 1 and 6. Note the narrow, central fault block where well control has demonstrated the presence of basal Upper Jurassic sands overlying Paleozoic rocks in a structurally high position.

In the Danish sector the Gert Ridge separates the late Jurassic Feda Graben from the Gertrud Graben. However, its extension into Norway lies entirely within the Feda Graben. It has been described previously (Søderstrøm *et al.* 1991) as a narrow horst block bounded by normal faults and affected by late Jurassic to early Cretaceous compressional reactivation of one of the major faults. Detailed mapping using 3D seismic reveals that this interpretation cannot satisfactorily explain the observed geometries.

Well information demonstrates the presence of a narrow, high-standing fault slice within the major fault zone. This central fault block, shown on Fig. 7, is bounded towards the SW by a major normal fault which southeastwards forms the margin of the Feda Graben. To the northeast of this same fault slice a narrow graben can be mapped as a southeastward continuation of the Feda Graben, thus producing a repetition of the Feda Graben boundary fault as seen in Fig. 6. The lateral offset of this fault is approximately 3 km in a dextral direction. To explain this geometry it is suggested that the Gert Ridge is formed as a result of latest Jurassic local compression along an oblique-slip fault zone with predominantly strike-slip movement. It is also suggested that this fault did not exist prior to the latest Jurassic. This fault zone strikes in a NW–SE direction and cuts across the eastern margin of the Feda Graben. The strike-slip fault activity appears to be primarily of late Volgian age, but also continued into the early Cretaceous, as demonstrated by local uplift and erosion of Lower Cretaceous sediments along the fault zone. Damtoft *et al.* (1992) relate this structure to wrench tectonics, but assume that the wrench movements occurred along the Feda Graben margin rather than intersecting it.

Evidence for strike-slip movement Documenting strike-slip movement using seismic information is always difficult and requires the combination of a number of criteria, the most important of which are listed and described by Harding (1990). In the Gert area there are several strong indications that this complex structure has a strike-slip origin. None of these will by themselves prove that strike-slip movement was predominant along this fault, but together they provide good support for such an interpretation.

(1) The fault zone is nearly linear and laterally extensive. As shown on Fig. 6 the fault zone can be traced from Block 2/8 in the Norwegian sector, and southwestwards to the western margin of the Tail End Graben.

(2) The sense of the vertical displacement varies along strike. In the area near the Danish–Norwegian border where the Gert Ridge has its maximum relief, faulting is 'down towards the east, while the polarity changes both to the north and to the south of this area.

(3) The major normal fault defining the Feda Graben margin is obliquely intersected and dextrally offset by the strike-slip fault zone, which is therefore later than the Feda Graben as shown in Fig. 8. This has produced the rather peculiar geometry illustrated by Fig. 7, with a narrow fault slice left in a structurally high position.

(4) The fault pattern mapped at top seismic basement is offset across the strike-slip fault. This is illustrated in Fig. 8, which shows a summary interpretation of the tectonic evolution of the area. These basement faults primarily strike N–S, and were reactivated in the initial phases of the late Jurassic rifting. They, therefore, affect the thickness of the lowermost Upper Jurassic sediments, including the Gert Field reservoir sands. It can be demonstrated that the Feda Graben boundary fault post-dates the deposition of the Gert sands.

(5) A pronounced bend on the fault zone is mapped immediately to the north of the Danish–Norwegian border

(Fig. 6). Both compressional and extensional tectonics occur along the fault zone in a pattern that is consistent (Crowell, 1974) with the expected behaviour in a dextral strike-slip zone at this constraining bend.

(6) The thickness of an intra-Upper Jurassic interval that definitely pre-dates the strike-slip movement can be matched across the fault zone by correcting for some 3 km dextral offset.

No definite explanation can be offered for the mechanism causing a 3 km dextral movement along this fault zone. However, the strike-slip fault zone clearly maps as a direct continuation of the major offset in the Coffee Soil Fault corresponding to the Adda Transverse Zone of Cartwright (1987) (Fig. 5).

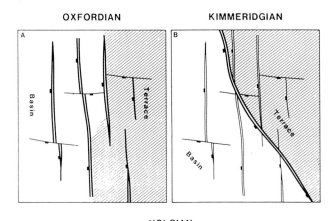

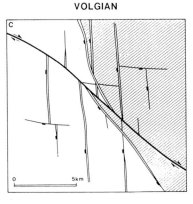

Fig. 8. Schematic representation of the structural evolution of the Gert area during the late Jurassic. (a) Initial late Jurassic rifting reactivated older N–S trending faults. This early fault activity controlled the distribution of the basal Upper Jurassic sand. (b) The Feda Graben developed along a NW–SE trend during the Kimmeridgian–Volgian, bounded towards NE by a major fault. (c) late Volgian strike-slip faulting cut obliquely into the Feda Graben.

It is suggested that this transverse zone marks the southern boundary of the thickest, and youngest, Upper Jurassic depocentre within the Tail End Graben. Therefore, both the Feda Graben and this segment of the Tail End Graben may be seen as late basins formed by extension north of the Adda Transverse Zone. The strike-slip movement could then be a consequence of local adjustment to extensional tectonics. Irrespective of its origin, the recognition of this offset is important for mapping the distribution of the Gert reservoir sand in this area.

An additional outcome of the 3D interpretation is the recognition that reactivation of N–S-trending basement faults in the earliest late Jurassic is the primary control on depositional thickness during the deposition of the basal Upper Jurassic sand. The NW–SE trend of the Feda Graben boundary fault, and the NW–SE-trending strike-slip zone, both clearly post-date the sand deposition.

The Mandal High and Søgne Basin

The southern tip of the Mandal High extends into the Danish sector from the Norwegian sector. Its termination towards the Coffee Soil Fault marks the boundary between the Tail End Graben and the Søgne Basin (Fig. 1). In this area both the Middle Jurassic Bryne Formation and the Upper Cretaceous Chalk have been found hydrocarbon bearing in the Harald Field (formerly Lulu and West Lulu). The occurrence of both these accumulations is related to salt movement along the east side of the Mandal High.

The main fault along the west wide of the Mandal High meets the Coffee Soil Fault where the latter changes strike direction from NW–SE to nearly due N–S. This fault is referred to as one of the main features of the Lulu Transverse Zone by Cartwright (1987) who recognizes the importance of this fault as a hinge zone allowing for differences in the subsidence history of the Tail End Graben to the south, and the Søgne Basin to the north.

From the seismic it is evident that the Mandal High did not emerge as a positive element until Volgian time, and accordingly that the Søgne Basin prior to this was a direct northward continuation of the Tail End Graben. A late Jurassic date for the block rotation is demonstrated by parallelism of the dips of pre-Zechstein and the Upper Jurassic reflectors (Fig. 9) in this part of the Søgne Basin. Therefore, although the Søgne Basin has the appearance of a half-graben related to the rotation of the NW–SE-trending Mandal High, the pre-Volgian sequence was deposited in a N–S trending, normal bounded graben without any significant syn-depositional rotation of the graben floor.

The late graben rotation is interpreted as implying that the N–S orientation of the Søgne Basin became incompatible with the preferred late Jurassic NW–SE rift direction. The subsidence changed to a NW–SE direction along a new fault, which now defines the west side of the Mandal High. Strong footwall uplift occurred along this fault causing severe erosion of the SW part of the Søgne Basin during the formation of the Mandal High.

The main separation into two basins took place along an E–W-trending fault throwing down-to-the-south (Fig. 1), which lies to the north of the fault zone described as the Lulu Transverse Zone by Cartwright (1987). This E–W fault can be demonstrated to have been active in the late Jurassic, and caused a local vertical displacement of up to 1500 m. It is estimated that a corresponding sediment thickness has been eroded off the southern part of the Mandal High.

A late Paleozoic to early Mesozoic connection between the Søgne Basin and Tail End Graben is also suggested by Cartwright (1987) who refers to the Søgne Basin as yet another segment of the Tail End Graben, as does Mogensen et al. (1992).

There is evidence that no hinge line separates the Søgne Basin from the Mandal High as constant dips of pre-Zechstein reflectors are seen throughout this complex. Further to the north in the Norwegian sector, the Mandal High is eroded to a deeper level and no primary reflectors can be recognized. This makes the interpretation of the Søgne Basin and the Mandal High as being originally one rotated graben segment more difficult to substantiate in that area.

The recognition of the latest Jurassic timing of the Mandal High and Søgne Basin rotation and the preceding N–S-dominating fault trend is of importance for understanding the facies distribution of the Middle Jurassic reservoir sand in this area.

The southern salt dome province

The southern salt dome province is that part of the southern Danish Central Graben that extends into the Southern Zechstein Basin (Fig. 1). Salt diapirs, swells, and intrusions are common, and all the producing Chalk fields in this area owe their structure to salt tectonics to a greater or lesser extent (Megson 1992). This part of the Central Graben is located to

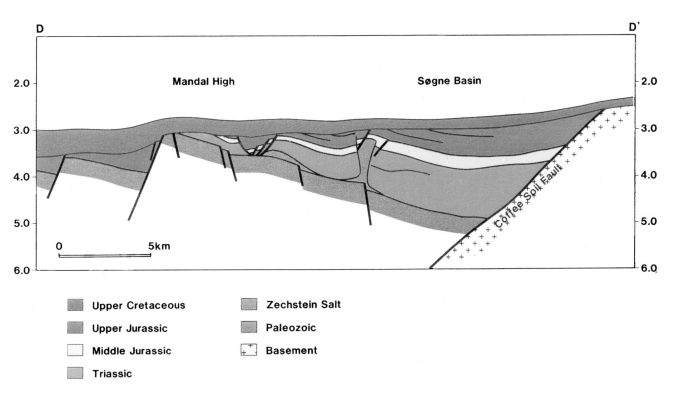

Fig. 9. Cross-section through the Mandal High and the Søgne Basin. For location see Fig. 1.

the south of the major Kimmeridgian/Volgian half-graben development. As a consequence, the pre-Zechstein basin floor is relatively flat, as can be seen in cross-section B–B' (Fig. 2). The primary structural lineaments influencing this basin floor are N–S-trending faults, and the Dan Transverse Zone.

The Dan Transverse Zone This was originally described by Cartwright (1987). Cartwright's assessment of the origin of this feature was as a Tornquist-parallel basement trend which was reactivated through time. The availability of several 3D surveys over fields along this structure (Fig. 3), and a considerable amount of high-resolution 2D seismic data, has made it possible to remap this feature in greater detail. Cartwright (1987) mapped the Dan Transverse Zone with an approximately WNW–ESE direction (Fig. 5) while current mapping reveals the trend to be more NW–SE.

There is a striking alignment of the three fields—Gorm, Skjold and Dan—along the NW–SE trending Dan Transverse Zone (Fig. 10). This area is well covered by 3D surveys, and is hence the best area in which to illustrate the tectonic effects of such a transverse zone.

To the NW of the Gorm Field, a continuation of this NW–SE trend defines the southern boundary of the N–S-trending Jens structure, which is an Upper Cretaceous inversion feature. It is noticeable that to the south of this Dan trend there is another NW–SE trend which links the Rolf and Dagmar salt diapirs. This NW–SE basement trend thus clearly affected the salt movement over much of the area in the salt dome province.

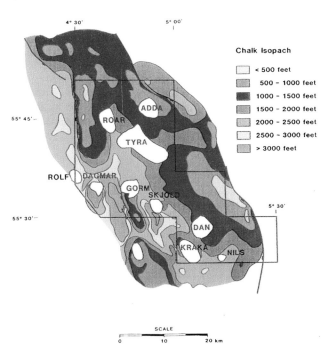

Fig. 10. Chalk isopach in the salt dome province in the southern part of the Danish Central Graben.

Effects of the Dan Transverse Zone at different times In the Zechstein, the Dan Transverse Zone marks the approximate northern limit of mobile Zechstein salt, and is therefore parallel to the edge of the Southern Zechstein Basin in this area. This movement of Zechstein salt at, and to the south of, the Dan Transverse Zone, resulted in short-wavelength thickness variations of all units in the salt dome province. However, an inspection of the long-wavelength components of the isopachs at various times shows how this zone acted as a graben segment boundary through time.

Throughout the Triassic and Lower Jurassic, the Dan Trans-

verse Zone defined the northern boundary of a NW–SE-trending isopach thin approximately 10 km across, with a thicker basin to the north.

In the Upper Jurassic this down-to-the-north movement continued, with the development of the very thick Kimmeridgian–Volgian basin (Fig. 4). Along a traverse across the Dan Transverse Zone the Upper Jurassic thickens rapidly northwards from roughly 1500 m to over 3000 m. At this time there must have been westwards lateral movement to the north due to down-faulting along the Coffee Soil Fault. Consequently sinistral strike-slip motion would be expected at this time along the Dan Transverse Zone, although not necessarily with a large lateral translation.

In the Lower Cretaceous, the basin continued subsiding more strongly to the north of the transverse zone, possibly purely as a consequence of compaction of the thick Upper Jurassic shale section.

During the Upper Cretaceous, the Dan Transverse Zone was clearly a major influence on deposition (Fig. 10), with a complete reversal in basin sense from the Upper Jurassic and Lower Cretaceous. The Chalk isopach shows in general a thicker Chalk section to the south of the Dan Transverse Zone, albeit with local control by salt tectonics. This would suggest syn-depositional movement along this fault trend, as is seen in the Gorm Field (see below).

Salt movements associated with the Dan Transverse Zone There is evidence from seismic data, and from the Dan 3D in particular, that this transverse zone has been a focus for lateral and vertical Zechstein salt movement. This should be expected, as continual differential movement along this fault zone will have resulted in episodic local extension and stress release, and continual opportunities for salt escape.

Figure 11 shows the form of one of these salt features along two parallel seismic lines which cross the Dan Transverse Zone at right angles below the Dan Field. These seismic lines are extracted from the 3D dataset, slightly less than 4 km apart. The variation in basement faulting styles can clearly be seen. The large variations in thicknesses of Triassic and Lower Jurassic units is evidence of complex local movements in which salt was involved. The observed salt bodies have intruded several different stratigraphic levels, and it is possible that one salt complex is composed of individual bodies which may have moved at very different times. It is expected, however, that all mobile salt beneath the Dan Field is of Zechstein age; this is in contradiction to what has been published previously (Jørgensen, in press), where it was postulated that the salt beneath Dan was of Triassic age. This previous view was formed before the 3D seismic data were available.

In summary, it is proposed that there has been repeated episodic lateral movement along this fundamental transverse zone, due to differential vertical movements between areas to the north and south. The Gorm, Skjold, and Dan fields are aligned on this transverse zone. Such a transverse zone is an obvious focus for the formation of local structural highs due to local transtension and transpression, and this would explain the concentration of structures along this zone. However, in addition such a zone should be a focus for vertical movement of hydrocarbons during local stress release and extensional adjustments, by analogy with the salt movements. It could also be predicted that there should be a higher density of faults and tectonic fractures associated with structures at the transfer zone than with similar structures in more quiescent parts of the basin. The Gorm and Skjold fields do in fact show production enhancement due to fracturing. Extending this prediction leads to the proposal that even fields over salt diapirs, such as Skjold, may have a stronger component of faults and fractures in a NW–SE direction than in radial or concentric directions in response to continued reactivation of the transverse zone.

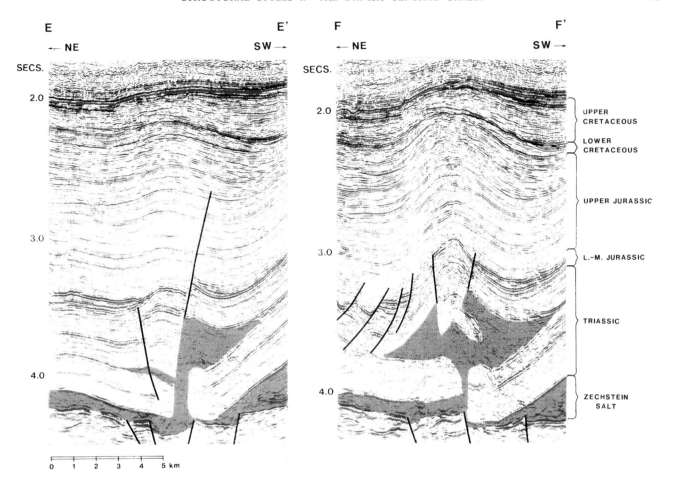

Fig. 11. Seismic sections across the Dan Transverse Zone underneath the Dan Field. The seismic interpretation suggests that Zechstein salt has intruded the Triassic sequence at several levels along the transverse zone. For locations see Fig. 1.

The Gorm Field

The Gorm Field is an oil field in the Chalk reservoir (Megson 1992), which was discovered in 1971. Over 30 wells have been drilled on the structure to date, and the field is now entering a secondary recovery phase.

In 1988, a 3D seismic survey was shot over the Gorm Field, with 25 m sub-surface line spacing. Although a relatively close 2D grid had been available previously, the 3D grid changed the understanding of the field structure considerably.

Comparison of maps from 2D and 3D Figure 12 shows the structure of the Gorm Field as mapped in 1988 from the latest 2D grid. The structure appears domal, with few faults in addition to the main fault. The faults that do exist could be described as being radial or concentric to the dome in trend.

Figure 13 shows the structure as mapped from the 3D dataset. Many more faults are present; in fact a fault statistical analysis suggested that faults as small as 4 ms TWT in throw have been consistently mapped. The most striking feature is the NW–SE fault trend that appears. The clearest example is the one major NW–SE fault that also offsets the main fault trend. In addition there is a clear swathe of small throw NW–SE faults cutting across the field (Fig. 14) and it is proposed that this trend is the trace of the Dan Transverse Zone at top Chalk level.

Importance of NW–SE trend in Gorm: The NW–SE fault trend, although composed of faults of relatively small throw (maximum throw approximately 200 ft), is of considerable import-

ance in the reservoir. In particular this trend appears to control lateral fluid distribution.

After gas injection commenced in the part of the field most strongly affected by NW–SE faults, early gas breakthrough (less than two months) was observed in a producing well to the NW of the injector indicating that these faults and associated fractures provide a pathway for gas. This same trend also locally acts as a partial seal: the gas injectors are to the south of the largest NW–SE-trending fault, and producers drilled to the north of this fault have a much lower gas–oil ratio (GOR) than those to the south. This GOR distribution cannot be explained without the presence of a fault seal. The fault also marks a boundary in primary reservoir characteristics, as the seismic and biostratigraphic data both indicate that this fault was active for some of the time during which the main Maastrichtian reservoir was being deposited. Such early movements on this fault could have enhanced the chance that the fault would seal by maximizing the time over which the fault plane was potentially exposed to circulating fluids.

Where these NW–SE fault trends extend into the eastern upthrown block, there is evidence from 3D reservoir simulation that they are still partial barriers, in spite of vertical throws of considerably less than 50 ft.

Discussion

The example areas presented above illustrate quite different structural styles developed in areas of related tectonic evolution. In spite of the apparent differences between these areas there are several common factors that provide an improved

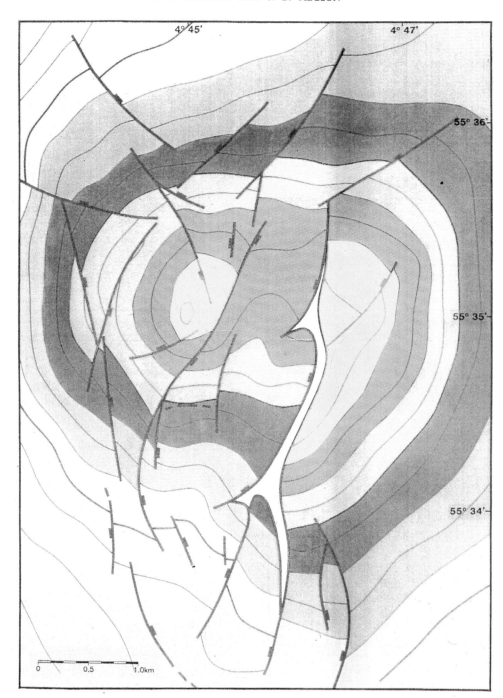

Fig. 12. Top Chalk depth structure map of the Gorm Field, mapped from 2D seismic data in 1988.

understanding of the general development of the Danish Central Graben. The detailed information from the areas covered
with 3D seismic can thus be used for making predictions on
timing of faulting and on fault trends themselves in other parts
of the graben where dense seismic coverage has not yet been
obtained.

Fault control on reservoir distribution and behaviour

The complete dominance of the late Kimmeridgian–Volgian
NW–SE-trending faults in the present-day tectonic picture
obscures the importance of other fault trends, and may easily

cause erroneous assumptions regarding the pre-Kimmeridgian
palaeogeography.

Detailed mapping in different areas within the Danish
Central Graben demonstrates that N–S-trending faults affecting
the pre-Upper Jurassic sequence are also present where they
are not easily recognized from 2D seismic. Dating the main
activity on these faults is difficult in many areas, but in the
Søgne Basin there are significant normal movements on N–S
faults in the Triassic; in the Dan area this fault trend clearly
affects the Lower–Middle Jurassic deposition; and in the Gert
area activity on this trend affected the Oxfordian–early Kimmeridgian. These faults dominated until the Late Cimmerian
rift phase when the NW–SE fault trend became dominant.

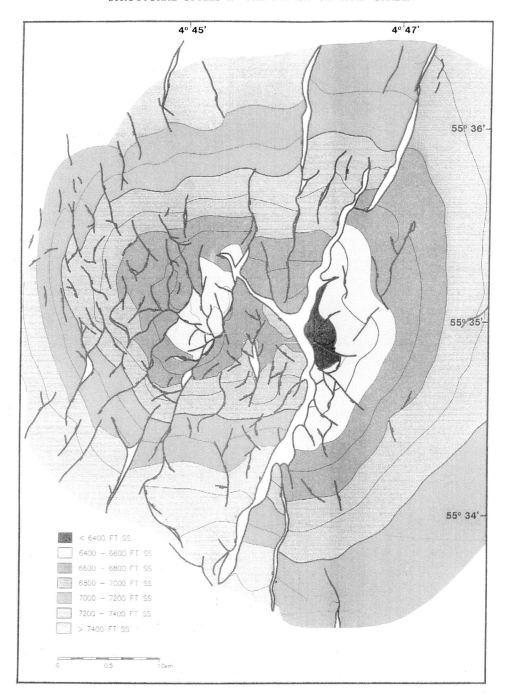

Fig. 13. Top Chalk depth structure map of the Gorm Field, mapped from 3D seismic data in 1991.

Origin of the Danish Central Graben

It is proposed that the Danish Central Graben did not initially form as a NW–SE-trending rift but as a system of predominantly N–S-trending basins and half-grabens of Triassic age. However, the NW–SE trends were present at an early stage as graben segment boundaries, as can be seen from the Dan Transverse Zone. Subsequently, these N–S grabens were linked on a larger scale by late Jurassic NW–SE rifting. One may accordingly question the notion that the formation of the Danish Central Graben originated in the late Permian or early Triassic. Evidently extensive rifting took place prior to the late Jurassic within the area that is now the Danish Central Graben, but this tectonic province may not have existed as a continuous feature until the late Jurassic.

Skjerven *et al.* (1983) in a discussion on the formation of the Central Graben raised the possibility that the Triassic N–S

trend observed in the Norwegian sector extended south into the Danish Central Graben area, and initiated the first transection of the Ringkøbing–Fyn High. They proposed that the NW–SE trend came into existence at a later stage. They were not able to confirm this idea due to lack of data. The observations from the Danish Central Graben described in this paper are additional evidence in favour of this theory.

The observed Triassic thickness distribution in the whole Danish Central Graben demonstrates a thinning of the sequence in the central NW–SE-trending graben segment. Therefore, it appears that the central part of the Ringkøbing Fyn–Mid North Sea High complex remained a relatively high-standing element during the Triassic, and was not completely cut through by the predominantly N–S-trending fault systems. A through-going collapse of the basement high therefore did not occur until the late Jurassic when the NW–SE fault trend became dominant.

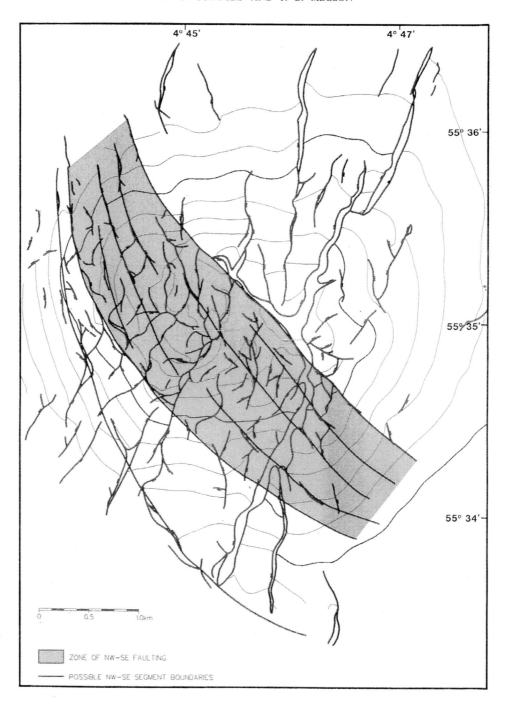

Fig. 14. NW–SE fault trend affecting the Upper Cretaceous–Danian Chalk reservoir of the Gorm Field.

Further support for a late development of the NW-striking portion of the Central Graben is given by the observation that north of the Dan Transverse Zone the deposition of the Middle Jurassic was completely controlled by N–S-trending faults wherever 3D coverage makes detailed mapping possible. The mapping of the Gert area shows that this N–S control continued also for some time into the late Jurassic.

Transverse zones

The existence of transverse zones as an inherent element of rifted basins has been advocated by Gibbs (1984, 1990). This author describes transfer faults as an integral part of his linked tectonic model where basins may be subdivided into crustal domains separated by detachment faults and by strike-slip transfer faults.

In the case of the Tail End Graben the low-angle Coffee Soil Fault locally acts as a detachment *sensu* Gibbs although it is not interpreted as rooting into any horizontal detachment. The differential movement between the various graben segments are accommodated at the transverse zones originally described by Cartwright (1987).

Of the three transverse zones described by Cartwright (Fig. 5) only the Dan Transverse Zone has been covered by extensive 3D seismic. Detailed mapping of this zone clearly demonstrates repeated activity throughout the Mesozoic. The change in polarity through time and the effect this zone has had on the salt movement suggest that this fault zone is not a synthetic normal fault responding to the overall extensional deformation, but is an accommodation feature related to the basin segmentation.

The Adda Transverse Zone as such has not been investigated in the present study as 3D data have not been acquired in this area. However, the Gert Ridge located at a NW extension of this zone provides strong indications for strike-slip movements along this NW–SE trend which may be related to the segmentation of the Danish Central Graben.

Late Jurassic activity on the Lulu Transverse Zone is clearly a response to the differential movements between the Tail End Graben and the Søgne Basin and may accordingly also be considered to be a transverse or transfer fault rather than a synthetic extensional fault.

In addition, the Adda, Lulu, and Dan transverse zones line up with other structural features (Fig. 5), suggesting control of the whole graben complex by several major crustal discontinuities (assigned to the Tornquist Zone by Cartwright (1987)). For instance, the Dan Transverse Zone lines up with the faults defining the western boundaries of the Heno Plateau, Inge High, and Feda Graben. The Adda Transverse Zone lines up with the eastern boundary of the Hano Plateau and the boundary between the Feda and Gertrud grabens.

These crustal discontinuities controlled the location at which tectonic movements took place, although these movements occurred at different times along the same trend.

Conclusions

The NW–SE fault trend dominating the present structural picture of the Danish Central Graben is of a late Jurassic age. Prior to this the predominant rift trend was N–S. The late Jurassic development of the Danish Central Graben is, therefore, not an accentuation and polarization of a Triassic rift system. To a large extent the faults controlling the Triassic to Middle Jurassic deposition were abandoned in the late Jurassic.

Oblique-slip movement appears to be an inherent element in the evolution of the Danish Central Graben. This is consistent with models presented by Shell for the evolution of the UK Central Graben (Bartholomew et al. 1993). In the Danish Central Graben such movements can be linked to changes in the direction of the main graben bounding fault, the Coffee Soil Fault, and may be regarded as accommodation features related to segmentation of the graben floor.

State-of-the-art 3D seismic makes it possible to resolve structural details far beyond those provided by the structural description of the Danish Central Graben published until now. Detailed structural information is important not only in areas of complex tectonics, but also in less disturbed areas since tectonic overprinting easily obscures important trends and lineaments.

Local information obtained from 3D seismic studies give information about the basin-wide tectonic evolution as well as about preferred fracture orientations and can be used as a predictive tool beyond the actual area of coverage.

The authors would like to thank Mærsk Olie og Gas a.s. and partners, Shell Olie og Gasudvinding Danmark, B.V., and Texaco Denmark Inc. for permission to publish this paper.

References

BARTHOLOMEW, I. D., PETERS, J. M. AND POWELL, C. M. 1993. Regional structural evolution of the North Sea: oblique slip and the reactivation of basement lineaments. In: PARKER, J. R. (ed.) Petroleum Geology of Northwest Europe: Proceedings of the 4th Conference. Geological Society, London, 1109–1122.

CARTWRIGHT, J. A. 1987. Transverse structural zones in continental rifts—an example from the Danish sector of the North Sea. In: BROOKS, J. AND GLENNIE, K. W. (eds) Petroleum Geology of NW Europe. Graham & Trotman, London, 441–452.

—— 1989. The kinematics of inversion in the Danish Central Graben. In: COOPER, M. A. AND WILLIAMS, G. D. (eds) Inversion Tectonics. Geological Society, London, Special Publication, 44, 153–175.

CLARK-LOWES, D. D., KUZEMKO, N. C. J. AND SCOTT, D. A. 1987. Structure and petroleum prospectivity of, the Dutch Central Graben and neighbouring platform areas. In: BROOKS, J. AND GLENNIE, K. W. (eds) Petroleum Geology of NW Europe. Graham & Trotman, London, 441–452.

CROWELL, J. C. 1974. Origin of late Cenozoic basins in southern California. In: DICKINSON, W. R. (ed.) Tectonics and sedimentation. SEPM Special Publication, 22, 190–204.

DAMTOFT, K., ANDERSEN, C. AND THOMSEN, E. 1987. Prospectivity and hydrocarbon plays of the Danish Central Trough. In: BROOKS, J. AND GLENNIE, K. W. (eds) Petroleum Geology of NW Europe. Graham & Trotman, London, 403–417.

——, NIELSEN, L. H., JOHANNESSEN, P. N., THOMSEN, E. AND ANDERSEN, P. R. 1992. Hydrocarbon plays of the Danish Central Trough. In: SPENCER, A. M. (ed.) Generation, Accumulation and Production of Europe's Hydrocarbons II. Springer Verlag, 35–58.

GIBBS, A. D. 1984. Structural evolution of extensional basin margins. Journal of the Geological Society, London, 141, 609–620.

—— 1990. Linked fault tectonics of the North Sea. In: BLUNDELL, D. J. AND GIBBS, A. D. (eds) Tectonic Evolution of the North Sea Rifts. Clarendon press, Oxford, 145–157.

GOWERS, M. B. AND SÆBØE, A. 1985. On the structural evolution of the Central Trough in the Norwegian and Danish sectors of the North Sea. Marine and Petroleum Geology, 2, 298–318.

HARDING, T. P. 1990. Identification of Wrench Faults Using Subsurface Structural Data: Criteria and Pitfalls. American Association of Petroleum Geologists Bulletin, 74 (10), 1590–1609.

HEYBROEK, P. 1974. Explanation to tectonic maps of the Netherlands. Geologie en Mijnbouw, 53, 43–50.

—— 1975. On the structure of the Dutch part of the Central North Sea graben. In: WOODLAND, A. W. (ed.) Petroleum and the continental shelf of NW Europe, volume 1: Geology. Applied Science Publishers, Barking, 339–352.

JØRGENSEN, L. N. in press. Dan Field. In: BEAUMONT, E. A. AND FOSTER, N. H (eds) American Association of Petroleum Geologists Treatise of Petroleum Geology, Atlas of Oil and Gas Fields. AAPG, Tulsa.

MEGSON, J. B. 1992. The North Sea Chalk Play: Examples from the Danish Central Graben. In: HARDMAN, R. F. P. (ed.) Exploration Britain: Geological insights for the next decade. Geological Society, London, Special Publication, 67, 247–281.

MICHELSEN, O., FRANDSEN, N., HOLM, L., JENSEN, T. F., MØLLER, J. J. AND VEJBÆK, O. V. 1987. Jurassic–Lower Cretaceous of the Danish Central Trough: depositional environments, tectonism, and reservoirs. Danmarks Geologiske Undersogelse Serie A, 16.

MOGENSEN, T. E., KORSTGÅRD, J. A. AND GEIL, K. 1992. Salt tectonics and faulting in the NE Danish Central Graben. In: SPENCER, A. M. (ed.) Generation, Accumulation and Production of Europe's Hydrocarbons II. Springer Verlag, 163–173.

MØLLER, J. J. 1986. Seismic Structural Mapping of the Middle and Upper Jurassic in the Danish Central Trough. Danmarks Geologiske Undersogelse, Serie A, 13.

SKJERVEN, J., RIJS, F. AND KALHEIM, J. E. 1983. Late Palaeozoic to early Cenozoic structural development of the south-southeastern Norwegian North Sea. Geologie en Mijnbouw, 62, 35–45.

SØDERSTRØM, B., FORSBERG, A., HOLTAR, E. AND RASMUSSEN, B. A. 1991. The Mjølner Field: A deep Feda Graben play in an Upper Jurassic transgressive sand tract. First Break, 9, 156–171.

ZIEGLER, P. A. 1978. North-Western Europe: Tectonics and basin development. Geologie en Mijnbouw, 57, 589–626.

—— 1988. Evolution of the Ascic–North Atlantic rift system. American Association of Petroleum Geologists Memoir, 43.

ZIEGLER, W. H. 1975. Outline of the geological history of the North Sea. In: WOODLAND, A. W. (ed.) Petroleum and the continental shelf of NW Europe. Applied Science Publishers, Barking, 165–187.

Mesozoic structural evolution of the UK Southern North Sea: insights from analysis of fault systems

T. J. ARTHUR

ARCO British Ltd, London Square, Cross Lanes, Guildford, Surrey GU1 1UE, UK

Abstract: An understanding of the Mesosoic fault systems in the UK Southern North Sea is important in modelling the burial history of source and reservoir rocks, in determining the time of formation of structural traps in the Permian and older section and in mapping seismic velocity fields in Mesozoic rocks for accurate time to depth conversion.

The evolution of Mesozoic fault systems in the Southern North Sea can be understood by examination of several faults with different amounts of movement across them, and by analysis of fault patterns. Structural analysis has been undertaken in selected areas where complications due to salt movement are minor. Local fault analysis is illustrated by balanced cross-sections and maps from Blocks 49/28, 48/23 and 48/11. The faults studied are shown to have been active extensional faults during the Jurassic and early Cretaceous, but have in general undergone only limited movement during later Cretaceous and Tertiary phases of subsidence and uplift.

The Mesozoic faults are detached from basal Permian and Carboniferous rocks by decollement in the Zechstein evaporites and are linked by transfer zones to form a basin-wide system of faults. Mesozoic fault systems overlie zones of weakness in pre-Zechstein rocks, but are not directly connected to them, so their orientation is controlled by Mesozoic forces rather than pre-existing structural grain.

Although the Mesozoic and Cenozoic sequences of the UK Southern North Sea are in themselves of relatively minor importance for hydrocarbon exploration, their development has important consequences for the hydrocarbon habitat of the underlying Permian and Carboniferous. The timing of gas generation and the pathways of migrating gas are in part controlled by Mesozoic and later structural evolution of the basin, and an understanding of the relationship between migration timing and trap formation has been shown to be important (e.g. Cleaver Bank area: Alberts and Underhill 1991). Burial and uplift of the Rotliegendes reservoir sandstones during the Mesozoic and Cenozoic has been shown by numerous authors to be an important control on the quality of the reservoir (e.g. Glennie *et al.* 1978; Rossel 1982; Arthur *et al.* 1986). In addition, the Mesozoic structure and stratigraphy of the Southern North Sea control the seismic time structure of the Permian and older seismic reflectors, and correct time to depth conversion of these seismic horizons often depends on correct interpretation of the post-Permian interval. Production of an accurate structural model is critical to assess each of these factors.

Despite having the potential to provide an important key to understanding the structural evolution of the Southern North Sea, the Mesozoic fault systems of the area have not been extensively discussed in the literature. Glennie and Boegner (1981) suggested that strike-slip movement had occurred along Mesozoic faults, but noted that lateral offset of features had not been observed. They recognized that Mesozoic faulting occurred in the same areas as structural, facies and thickness changes in the Permian and Carboniferous, and so were probably associated with long-lived zones of structural weakness. Gibbs (1986) interpreted the major Southern North Sea faults as strike-slip systems and described the complex fault zones as containing flower structures. Although these regional syntheses emphasize a strike-slip element in Southern North Sea tectonics, detailed analyses of individual fault zones by Walker and Cooper (1987) and Badley *et al.* (1989) show little evidence of strike-slip movement along faults.

In this paper detailed analyses of Mesozoic fault systems in Blocks 49/28, 48/23 and 48/11 are presented. These blocks were selected because they have good seismic and well control and because their fault systems show different degrees of development. Furthermore, they are situated away from the main areas of Zechstein halokinesis, thus removing the need to take the effects of salt movement into account. Figure 1 shows the locations of the blocks and the major structural features. The faults are described using maps of fault patterns at Permian Rotliegendes and Triassic Top Bacton Group level together with depth cross-sections. The Top Rotliegendes level reflects sub-salt structure and the Top Bacton demonstrates Mesozoic fault patterns. Balanced restorations illustrate the sense and timing of movement on the faults. The cross-sections are derived from depth-converted seismic time interpretations.

Figure 2 shows the stratigraphic nomenclature for the area and the seismic reflectors picked. Time-to-depth conversion was carried out using a vertical layer cake method, converting time intervals for each layer to thickness and adding them to calculate depth. Time-to-thickness conversion was done using functions relating interval velocity to interval midpoint depth to interpolate between well control points. A vertical depth conversion is not ideal for exact location of faults as it does not compensate for ray bending, and of course the depth calculated for any layer depends not only on the interpreted time structure for that layer but on the time structure interpretation for the overlying layers. Despite these limitations the maximum vertical error due to depth conversion is likely to be no more than 150 ft at the deepest levels, and horizontal error due to assumption of vertical ray path to be of the order of 250 ft.

The depth sections were restored manually, maintaining constant bed length and true stratigraphic thicknesses, with no allowance made for compaction. Wells show that lithology does not change significantly along the sections, and maximum depth of burial changes are gradational, so the error due to lack of compaction correction is not sufficient to invalidate the section balance. Layer-parallel slip has been permitted, but as will be seen from the restorations this is not large scale. No attempt has been made to account for Zechstein salt movement, the top of the salt being assumed as a free surface in the restorations. The structure of the Rotliegendes section has been included on the present-day depth sections to demonstrate the structural relationships and the amount of Zechstein salt present, but to account for salt movement during fault development is a complex three-dimensional problem and was not attempted. It is evident from the restorations and present-day sections, however, that in each case the amount of salt present is sufficient to allow the vertical movement indicated.

From *Petroleum Geology of Northwest Europe: Proceedings of the 4th Conference* (edited by J. R. Parker).

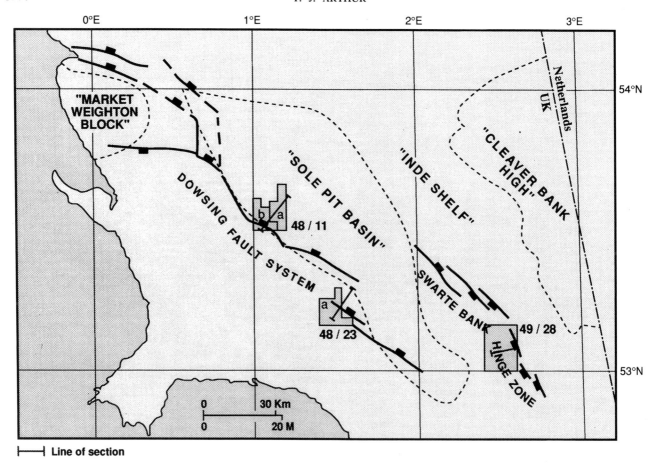

Line of section

Fig. 1. Location map and general structural features.

Chrono-stratigraphy		Litho-stratigraphy	Seismic reflectors
Tertiary / Quaternary		North Sea Group	
Cretaceous	U	Chalk Group	← Top Chalk ← Base Maastrichtian ← Base Chalk
	L	Cromer Knoll Group	
Jurassic	U	Humber Group	← Base Cretaceous ← Top Corallian
	M	West Sole Group	
	L	Lias Group	← Top Triassic
Triassic	U	Haisborough Gp.	
	M		
	L	Bacton Group	← Top Bacton
Permian	U	Zechstein Super Group	← Base Bacton ← Top Rotliegendes
	L	Rotliegendes Group	

Fig. 2. UK Southern North Sea stratigraphy and seismic reflectors.

Block 49/28 structural analysis

The dataset constraining interpretation of Block 49/28 consists of a modern 2D seismic grid with a dip line spacing of 1650 ft and a number of exploration, appraisal and development wells. There are 15 vertical wells and 7 deviated wells in the block.

The seismic data were acquired to image Rotliegendes structure optimally, and lines are oblique to the Mesozoic fault trend. Since the best line of section to analyse a fault is perpendicular to it, the depth section in this block was constructed from maps and is not directly from a seismic section. Figure 3 shows fault patterns within the block and the locations of the depth section and an illustrative seismic section.

The depth section, seismic line and fault map all show that the relationship of Mesozoic to Permian faulting is not simple (Figs 3,4). The fault throw at the Top Bacton seismic horizon is considerably more than the fault throw at Top Rotliegendes level. This means that the main Mesozoic fault (S1) must either be a strike-slip fault or be detached from the older structure by decollement in the Zechstein, or else have a complex history of normal and reverse movement. The map (Fig. 3) and cross-sections (Fig. 4) show that the Mesozoic fault system does not root into a Permian fault, and no inversion-related structure is evident at the fault. Zechstein salt movement to accommodate the faulting is evident, however, so a Zechstein detachment is the preferred interpretation.

The depth section has been restored to end Triassic, Late Cimmerian Unconformity (i.e. near Base Cretaceous) and end early Cretaceous datums (Fig. 4). The end Triassic restoration shows a broadly uniform Triassic interval west of the later main synthetic fault S1, the position of which is shown by a dashed line. The absence of the stratigraphic top of the Triassic over most of the area east of fault S1 due to Cimmerian erosion means that Upper Triassic depositional thickness is not known here. Well data show that the only changes evident in the Haisborough Group in this area are dissolution of the halite intervals where they lie close to the Late Cimmerian Unconformity and regional thickness variations which appear to be unrelated to the faults, so a uniform thickness for this interval has been assumed.

The Late Cimmerian Unconformity restoration, using the Base Cretaceous seismic horizon as datum, shows that almost all movement along faults had been completed by this time. Hangingwall deformation by synthetic faulting took place and a Lower Jurassic interval is preserved west of S1. The Jurassic history of this fault is not known, as only Lias is preserved, but the absence of any marginal facies in the Lias and the parallelism of internal seismic reflectors suggests that this is an erosional remnant and not a local basin and that a significant thickness of Jurassic has been eroded from the area. Diagenetic studies suggest differential uplift and erosion of the order of 1500 ft (Arthur *et al.* 1986).

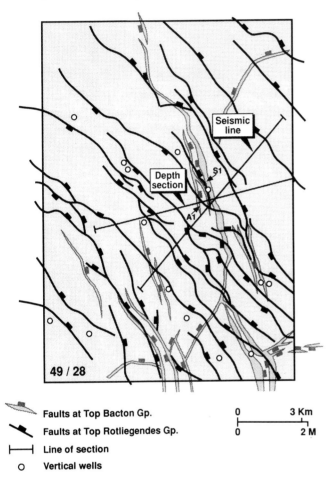

Fig. 3. Block 49/28: fault patterns at Top Rotliegendes Group and Top Bacton Group.

The end early Cretaceous restoration reflects the deposition of a Lower Cretaceous (Hauterivian and Barremian) shale section over the area. The datum is dashed to the west of the hangingwall syncline of the S1 fault because the Base Chalk seismic horizon is here truncated by the Base Maastrichtian Unconformity. The present-day depth section records the post-early Cretaceous events of the area, which as can be seen, did not result in any movement on S1 or the antithetic fault A1. A pre-Maastrichtian Chalk interval is preserved to the east of the antithetic fault A1, but its age has not been determined. Van Hoorn (1987), Walker and Cooper (1987) and Alberts and Underhill (1991) all describe onlapping of Chalk onto the Leman area Cimmerian high, the crest of which lies to the west of the section, so it is probable that only a thin Chalk was originally present and then removed by pre-Maastrichtian erosion. Maastrichtian Chalk then onlapped the Leman high,

and is overlain by a Tertiary section which has not been dated and is seismically unresolvable in this area. From the foregoing discussion it is evident that the Mesozoic fault system in Block 49/28 is relatively straightforward. The fault was initiated in the Jurassic and had ceased moving by the end of the Cimmerian erosional episode, and is clearly a simple normal fault separated from the older section by detachment in the Zechstein salt. Although regional tilting and differential deposition and erosion took place in the Cretaceous and probably in the Tertiary, it appears that the fault system remained inactive. The clear and uncomplicated faulting in Block 49/28 has been used as a model to help in understanding more complex faults in other areas, and as the succeeding analyses will show it has proved very successful.

Block 48/23 structural analysis

The Block 48/23 structural interpretation is not so tightly constrained as that of Block 49/28, but a grid of good seismic data with a 3300 ft dip line spacing and three vertical wells provide a dataset sufficient for a high degree of confidence in the interpretation. The map of fault patterns (Fig. 5) shows that the major pre- and post-Permian faults are generally parallel here, so seismic lines shot for Rotliegendes targets are perpendicular to Mesozoic faults and a depth section suitable for fault analysis can be drawn from a seismic line.

The depth section in this block (Fig. 6) shows a main fault (S2) that is geometrically similar to the 49/28 Mesozoic fault. Again, of the several possible interpretations, downward flattening of the major fault plane into a sub-horizontal detachment in the Zechstein and a simple extensional style is favoured and, as in 49/28, restoration supports this. Upper Cretaceous Chalk is absent from the line of section due, it is believed, to pre-Maastrichtian erosion, so restorations have only been made to the end Triassic and Late Cimmerian Unconformity datums.

The stratigraphic top of the Triassic is present along the whole of the section, so restoration to this datum is straightforward. It is clear that Triassic deposition was uniform over the area. The Top Bacton seismic reflector has not been identified in the fault sliver between S2 and S3 synthetic faults, so the likely position of the horizon has been inferred in this small area.

The second restoration is to the Late Cimmerian Unconformity datum. This shows that fault movement took place during the early and middle Jurassic, but had ceased by the time the Corallian was deposited during the Oxfordian. It has not proved possible to discern whether the S2 or S3 fault moved first, but it is apparent that antithetic faulting involving significant displacement (A2/3) had not developed by the end of the Jurassic. The S1 and S4 synthetic faults had also developed by this time. The Late Cimmerian uplift and erosion was insignificant in this area, and this together with the absence of any intra-Lower Cretaceous markers means that no intermediate restorations between the Late Cimmerian Unconformity and present day have been made.

The present-day depth section shows that early Cretaceous extension and subsidence were accommodated by movement along the S3 fault, with hangingwall deformation accommodated by movement along antithetic fault A2/3, which flattens downwards into a detachment at the Rot Halite at the base of the Haisborough Group. The post-early Cretaceous uplift and erosion of this area can be dated as post-Turonian by reference to Alberts and Underhill (1991). Van Hoorn (1987) dates the uplift as Mid-Tertiary, but there appears to be no stratigraphic evidence in this part of the basin for this conclusion, so in this paper a pre-Maastrichtian age for the uplift is assumed. This is consistent with that seen in the 49/28 area and in the Broad Fourteens Basin in the Netherlands (Van Wijhe 1987).

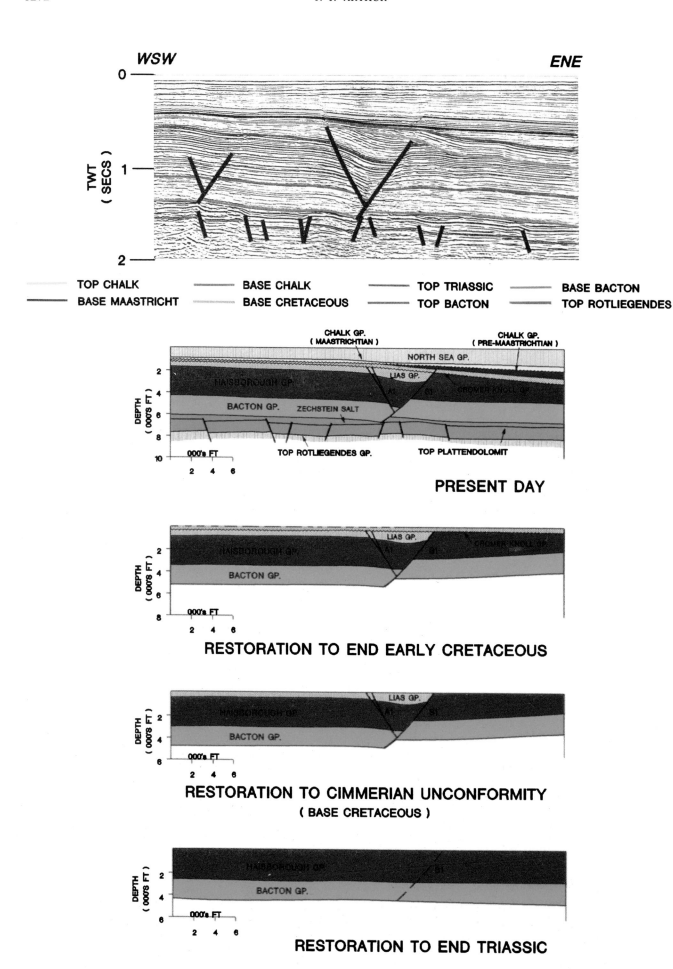

Fig. 4. Block 49/28: seismic line, depth section and restorations.

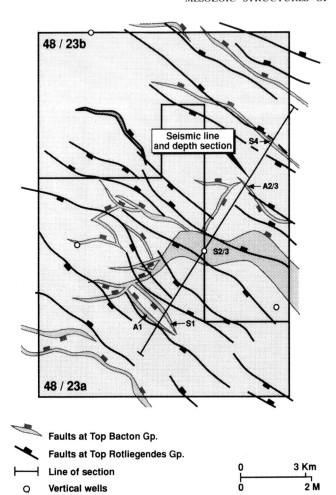

Faults at Top Bacton Gp.

Faults at Top Rotliegendes Gp.

Line of section

Vertical wells

0   3 Km

0   2 M

Fig. 5. Block 48/23: fault patterns at Top Rotliegendes Group and Top Bacton Group.

Block 48/11 structural analysis

Mesozoic faulting in Block 48/11 is more complex than the previous examples, due in part to associated halokinesis, but mainly due to the considerable extension across the system. Indeed, the cross-section analysed records a 23% line-length extension. The interpretation in this block is constrained by a grid of 2D seismic data with a 3300 ft dip line spacing and 11 vertical wells, all but one of which penetrated beneath the Rotliegendes. Seismic lines are perpendicular to Mesozoic trends in this block (Fig. 7), so the depth section was constructed along a seismic line (Fig. 8). Seismic data quality is poor in some parts of the area due to steep dips, intense faulting and sometimes limited impedance contrasts between units, but the well control limits the interpretation ambiguity. The fault system is interpreted as an extensional complex with major faults flattening downwards into detachment levels within the thick salt-dominated Zechstein sequence, and antithetic faulting flattening into a detachment at the Rot Halite immediately above the Bacton Group.

The reconstructions confirm that a purely extensional model is entirely possible for this complex fault system. The section has been restored to datums at the Top Triassic, intra-Liassic and Base Chalk, and a schematic restoration for a time towards the end of the Jurassic has been prepared to illustrate development of the fault system (Fig. 8).

The end Triassic restoration shows that the S1 fault had started to move during the late Triassic, allowing thickening of the Haisborough Group in the footwall, but in general Triassic deposition was uniform. The next restoration, to an arbitrary

Liassic datum, is an illustration of an early stage in the development of the fault system. Movement along the S4 fault allowed thickening of the Lower Jurassic interval on the downthrown side, but no hangingwall deformation had yet occurred.

The next restoration is schematic in that not even an arbitrary datum has been chosen, but a drawing made to illustrate a possible intermediate stage between the Lias restoration and that at the end of the early Cretaceous. During the time period that this drawing illustrates, between the end of the Triassic and the deposition of the Corallian, fault movement was initiated on S4 (as seen on the previous restoration) and later on the antithetic fault A1, which as in 48/23 flattens into a decollement in the Rot Halite immediately above the Bacton Group. Subsequently, movement on S4 stopped, perhaps due to grounding of the hangingwall block on Zechstein carbonates along strike from the line of section following salt withdrawal accommodating the faulting. Extension then proceeded on the A1 fault, which now became the main fault. This extension was associated with growth of a salt swell to the northeast of A1. The salt swell and the A1 fault are intimately related, as the southwest dip imparted by the growth of the swell facilitated slip along the Rot Halite and so allowed the significant slip along the fault.

The next restoration shows the structure at the end of the early Cretaceous. Progressive development of the fault system is readily seen. In the southwest the S1, S2 and S3 faults all moved during the latest Jurassic and fault block crests show minor Cimmerian erosion. Fault S4 shows growth, but it is minor. The A1 fault exhibits the most later Jurassic and early Cretaceous movement, offsetting the S4 fault plane. The relationship between the A1 fault and the salt swell is complex. It appears that regional tilting to the northeast rotated the A1 fault to a dip too shallow to allow movement, and so steeper splays developed, but at the same time growth of the salt swell resulted in net uplift of the A1 footwall crest, causing the depositional thinning of the Middle and Upper Jurassic section evident on seismic data and the subsequent erosion down to the top of the Triassic on the swell crest. An important point to notice from this restoration is that uppermost Jurassic and Lower Cretaceous thicknesses are not controlled by faulting here but by regional tilting and probably salt withdrawal in the northeast.

Comparison of this restoration and the present-day structure shows the Cretaceous evolution of the area. The effects of the Sole Pit inversion, taken in this paper to be pre-Maastrichtian, can be seen in the uplift and erosion of the whole area resulting in exposure of Chalk of Cenomanian to Turonian age at the sea floor (Van Hoorn 1987; Alberts and Underhill 1991). The progressively greater erosion to the northeast is clearly not controlled by movement on the fault but by tilting. The folding seen at Base Chalk level above the salt swell may be tighter than that seen in the underlying section, though seismic data quality is poor and structure at intermediate levels between Top Bacton and Base Chalk is not well imaged. It could be that this folding is due to late reactivation of the salt swell, but as the Chalk structural crest is offset from that at Bacton level the preferred interpretation is that compression has resulted in reverse movement on the A1 fault and folding of the Chalk above the fault tip. However, the poor data quality at Base Cretaceous level means that some contribution from halokinesis, perhaps from out of the line of section, cannot be ruled out.

Discussion

The restoration of cross-sections across three separate Mesozoic fault systems is consistent with the simplest hypothesis for their origin: that movement across them has been purely dip-

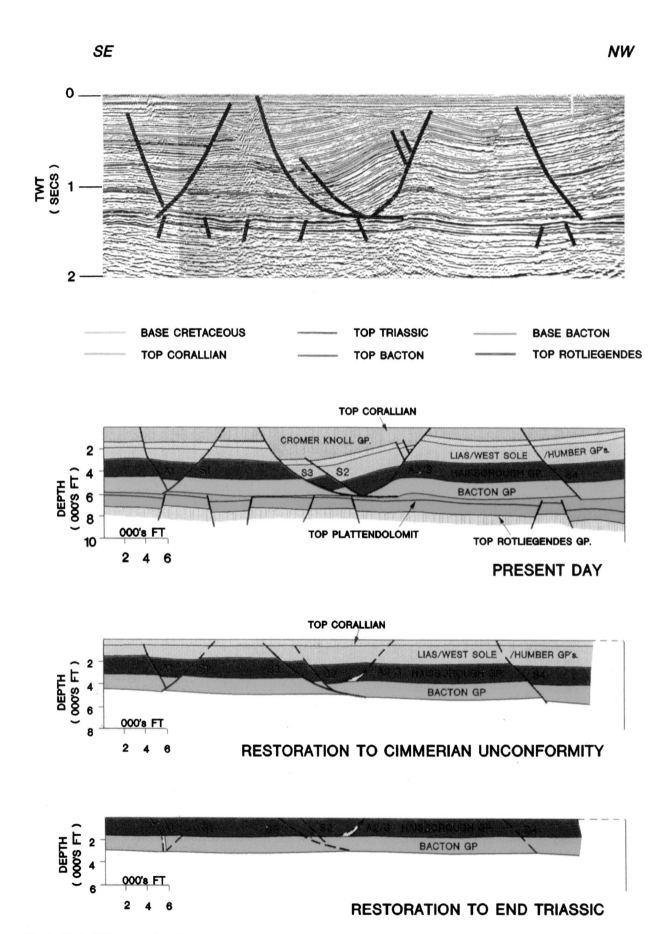

Fig. 6. Block 48/23: seismic line, depth section and restorations.

slip. Examination of mapped fault patterns shows no features characteristic of strike-slip movement, and stratigraphic thickness variations across fault systems are consistent with dip-slip growth of faults. This is in agreement with other detailed studies by Walker and Cooper (1987) on the northeast margin of the Sole Pit area and Badley et al. (1989) in Quadrant 53 on the southern margin of the basin. There remains the possibility that these extensional faults formed as a response to strike-slip movement in the pre-Permian basement. Consideration of the progressive development of strain by simple shear, however, indicates that the development of major extensional faults in cover rocks by strike-slip along a basement fault is unlikely. In a horizontal strain ellipse deforming by simple shear (Fig. 9), which would represent the cover rocks above a basement strike-slip fault, the direction of extension rotates with progressive deformation. This would imply that any extensional fault would become oblique to the direction of extension with time and would therefore develop a strike-slip component to its movement. Furthermore, simple shear deformation would result in horizontal shortening perpendicular to the extension and of the same order of magnitude. There is no evidence of any shortening whatsoever associated with the Jurassic–early Cretaceous extension. Even if the whole area, including the faults, was to be rotated in response to distributed shear, the model of McKenzie and Jackson (1986) indicates that a strike-slip component would be present in the faulting, albeit opposite to the gross sense of shear of the area.

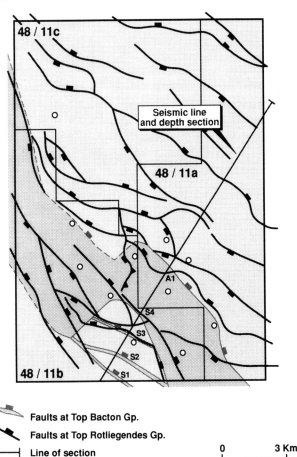

Faults at Top Bacton Gp.

Faults at Top Rotliegendes Gp.

Line of section

Vertical wells

0   3 Km

0   2 M

Fig. 7. Block 48/11: fault patterns at Top Rotliegendes Group and Top Bacton Group.

Supporting evidence that basement shear is not responsible for the Mesozoic extension is provided by published modelling studies. Figure 10 is taken from Richard et al. (1989), in which modelling results for deformation above a basement strike-slip

fault are presented. The model shown consists of layers of sand and glass beads separated from the 'basement' by a layer of silicone, a reasonable analogy for the Southern North Sea where Mesozoic clastics are separated from the Rotliegendes by Zechstein evaporites. The rapid development of strike-slip and not extensional faults is evident from the modelling, and although such models using low cohesion sand tend to produce strike-slip rather than normal faults, the lack of any similarity to Southern North Sea post-salt fault geometries is noteworthy.

Major extensional faults can, of course, develop at releasing bends along strike slip systems. The boundary between a strike-slip system with releasing bends and an extensional system with strike-slip transfer zones is obviously gradational. If we take as a definition of an extensional system that the total length of extensional faults has to exceed the total length of strike-slip faults, then the fault systems described here are obviously extensional.

It is therefore apparent that Mesozoic faults do not record any Triassic, Jurassic or early Cretaceous strike-slip movement in the Southern North Sea, either in the post-Zechstein cover or in the pre-Zechstein basement.

The faults described here do not, in general, show any effects of the Sole Pit inversion, the exception perhaps being the A1 antithetic fault in Block 48/11, which may have moved in a reverse sense in the late Cretaceous or later. Walker and Cooper (1987) noted late Cretaceous fault movement in the area northeast of Sole Pit, and Badley et al. (1989) described reverse movement along the South Hewett Fault. In neither case was strike-slip observed. If late Cretaceous inversion movements were accompanied by, or caused by, compression oblique to the existing faults, or by shearing forces, strike-slip movement would have occurred and be observable on these faults. The absence of evidence of such a component in their movement indicates that compressional forces have acted perpendicular to the existing Mesozoic faults.

This paper is focused on Mesozoic fault systems where they are detached from the Rotliegendes and older rocks by decollement in the Zechstein salt and where this layer is relatively thin. Detachment at Zechstein level means that the response to Mesozoic tectonism is not complicated by pre-existing structure, while halokinetic effects have been minimized by selecting areas of thin and generally immobile salt. However, any structural analysis based on Mesozoic faults must be consistent with the structural configuration of the pre-Zechstein rocks, generally defined in the Southern North Sea by mapping at Top Rotliegendes. The structure at Rotliegendes level is considerably more complicated than the Mesozoic structure (see Fig. 3 for example), reflecting as it does the tectonic events of some 250 million years and probably affected by Variscan structural imprints. Furthermore, seismic reflectors beneath the Top Rotliegendes are often discontinuous or absent, so fault orientation and style are poorly constrained compared to the Mesozoic interval. These factors mean that analysis of Rotliegendes fault systems to derive a unique solution for Mesozoic structural history is not possible, but an examination for consistency with the Mesozoic-based model is appropriate.

Glennie and Boegner (1981) and Van Hoorn (1987) describe fault patterns at Rotliegendes level that reflect strike-slip movement, and this is also the best explanation for the more complex faults in Block 49/28 (Fig. 3). Walker and Cooper (1987), however, observed no indication of strike-slip in Rotliegendes faults northeast of Sole Pit.

The best explanation for this apparent conflict is that faults at Rotliegendes level are often Variscan or Permian in origin, and variable in orientation. Faults oriented perpendicular to Mesozoic extensional forces, such as those northeast of Sole Pit, were reactivated as normal faults in the Jurassic and locally as reverse faults late in the Cretaceous. Faults oblique

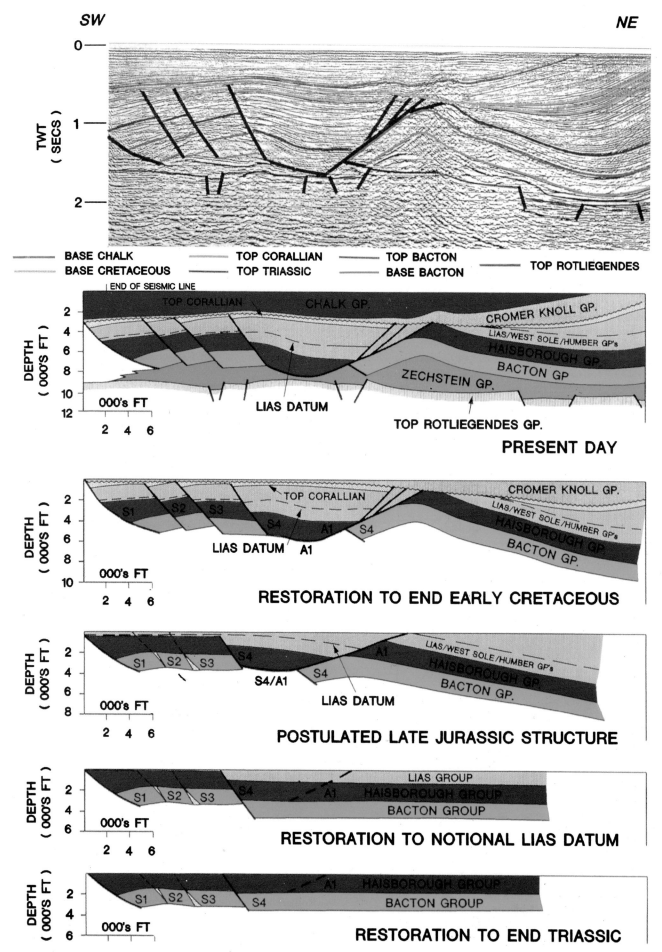

Fig. 8. Block 48/11: seismic line, depth section and restorations.

to these later forces responded by oblique slip, the strike-slip component increasing as the angle between the fault and the direction of principal stress decreases. Strike-slip movement on pre-existing Rotliegendes faults is therefore entirely consistent with pure extension and compression during the Mesozoic.

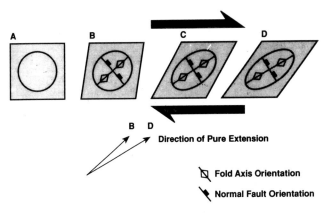

Fig. 9. Theoretical horizontal strain ellipse illustrating deformation by simple shear.

Regional context

The foregoing discussion has focused on individual fault systems and their geometries without reference to their regional significance. The faults described are part of the major fault systems of the UK Southern North Sea, however, and their analysis has importance for the understanding of the basin as a whole.

The fault complex described from Block 49/28 is part of the Swarte Bank Hinge Zone described by Glennie and Boegner (1981), and it links to the northwest with the faults described by Walker and Cooper (1987), northeast of Sole Pit. The 48/11 and 48/23 faults are part of the Dowsing fault system.

Various regional papers (Glennie and Boegner 1981; Van Hoorn 1987; Alberts and Underhill 1991) have noted that these major Mesozoic fault systems coincide with changes in pre-Permian subcrop and Rotliegendes thickness and facies. The location of Mesozoic faults is clearly defined by the location of zones of structural weakness in the underlying Rotliegendes and Carboniferous, but the analyses presented here have shown that the Mesozoic faults are separated from the older sequences by decollement in the Zechstein evaporites. The

faults, therefore, provide a record of structural evolution unaffected by orientation and sense of throw of pre-existing faults. Without entering into an exhaustive review of Southern North Sea structural history, it is appropriate to review how an understanding of the movement on Mesozoic faults can be integrated with the regional geology to elucidate critical phases in the evolution of the area.

Figures 11 and 12 show the fault systems described in this paper in their regional context. The maps are based on regional studies and published work, mainly Van Hoorn (1987), Walker and Cooper (1987) and Alberts and Underhill (1991) as well as the analysis presented in this paper.

Jurassic basin development

Figure 11 illustrates the Jurassic structural framework of the UK Southern North Sea. The Triassic development of the basin as a broad sag was not greatly affected by faulting along the Dowsing and Swarte Bank trends, although some early movement has been documented in this study. During the Jurassic and early Cretaceous, extension of the basin in a northeast–southwest direction is reflected in the development of the linear extensional fault systems detached at Zechstein salt level and sometimes at Triassic halite horizons. Intra-basinal highs such as the Leman and Cleaver Basin uplifts may have been positive features during deposition of the Jurassic sequences, but certainly became uplifted during later Jurassic times, resulting in substantial Cimmerian erosion. The pre-Cretaceous subcrop pattern resulting from this erosion is illustrated in Fig. 11, which shows how the Jurassic growth faults form the boundaries to the Late Cimmerian uplifts.

Cretaceous inversion

Figure 12 illustrates the structural framework of the basin and the effects of inversion during the latest part of the Cretaceous. The absence of strike-slip means that northeast–southwest compression probably caused the uplift and erosion. This must have taken place prior to the Maastrichtian, which oversteps rocks ranging in age from Campanian to Triassic. Some faults striking at a high angle to the compression moved in a reverse sense (Badley *et al.* 1989), but faulting was largely confined to the basin margins and the major Mesozoic faults record minimal late Cretaceous movement. The Chalk was completely eroded where it was depositionally thin, such as the Leman area, or where uplift was considerable, but only partially

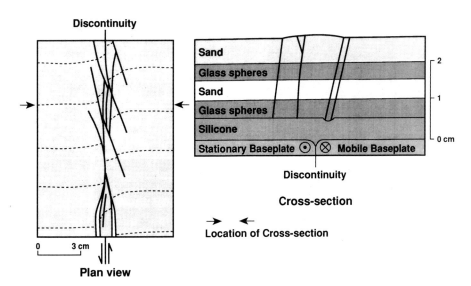

Fig. 10. Analogue modelling of deformation above a basement strike-slip fault using a plastic layer between basement and cover (modified after Richard *et al.* 1989).

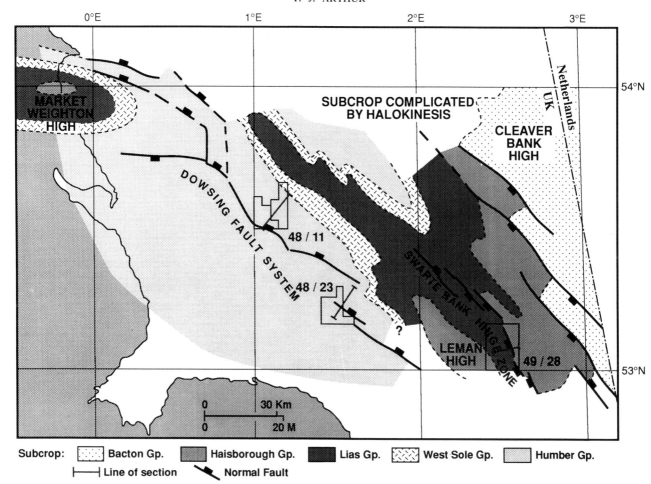

Subcrop: ⠿ Bacton Gp.   ▓ Haisborough Gp.   ■ Lias Gp.   ⧖ West Sole Gp.   ░ Humber Gp.

├──┤ Line of section   ◣ Normal Fault

Fig. 11. Late Cimmerian structural framework and pre-Cretaceous subcrop.

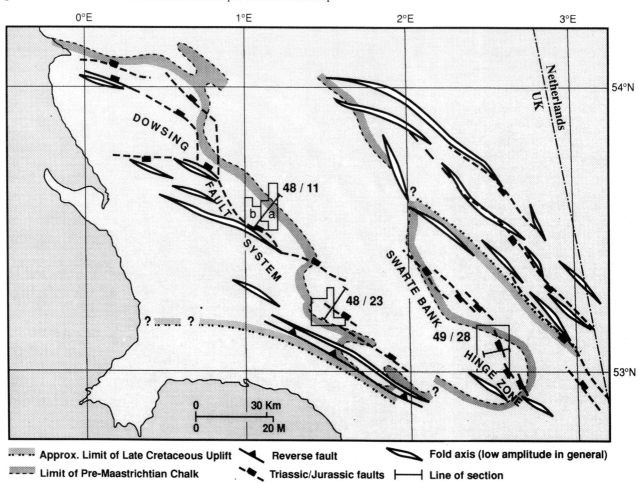

┈┅ **Approx. Limit of Late Cretaceous Uplift**   ◣ **Reverse fault**   ⬭ **Fold axis (low amplitude in general)**

▬ **Limit of Pre-Maastrichtian Chalk**   ◣━ **Triassic/Jurassic faults**   ├──┤ **Line of section**

Fig. 12. Late Cretaceous inversion structural framework.

eroded in other areas. The lateral extent of the uplift is therefore defined not by the presence or absence of Chalk but by the presence or absence of erosion at the pre-Maastrichtian unconformity. This is identifiable with confidence east of 2°E, but not where Cretaceous rocks are absent in the Sole Pit area. West of Sole Pit the lower part of the Chalk is at the sea floor so stratigraphic dating of the uplift is not possible, but the proposal that an extensive uplift occurred before the end of the Maastrichtian (65 Ma) is compatible with dating of uplift in the UK East Midlands at 50–80 Ma from apatite fission track data (Green 1989). This wide uplifted area is elongate parallel to the faults that were active during inversion, reinforcing the case for a compressional origin for the uplift.

Conclusions

Analysis of Mesozoic faults in the UK Southern North Sea provides important information on the geological history of the area. Fault geometries indicate only dip-slip on Mesozoic faults, with oblique-slip occurring on pre-Zechstein faults because they were not perpendicular to the principal stress. Extension in the Jurassic and the early part of the Cretaceous was followed by later Cretaceous compression and uplift. No evidence has been found to indicate a shear component in the the stress fields. This means that line balance of cross-sections can be used to constrain seismic interpretations in Mesozoic fault zones. Burial history during the Jurassic and early Cretaceous was controlled by these faults, so compaction trends and, therefore, seismic interval velocities as well as burial controlled reservoir diagenetic changes during this time can be modelled using discontinuities at these faults. For Cretaceous and Tertiary burial-related effects no such discontinuity would be expected.

This paper derives from the work of many ARCO technical staff, but particularly from seismic interpretations by Andrew Horanic, David Harrison and Ruth Manson, although the interpretations presented and the conclusions drawn are entirely the responsibility of the author. Permission to publish data was given by ARCO British Ltd, Superior Oil (UK) Ltd, Canadian Superior Oil (UK) Ltd, Sun Oil Britain Ltd, Deminex UK Oil & Gas Ltd, Britoil (Development) Ltd, Conoco (UK) Ltd, Goal Petroleum plc, and Pict Petroleum plc.

References

ALBERTS, M. A. AND UNDERHILL, J. R. 1991. The effect of Tertiary structuration on Permian gas prospectivity, Cleaver Bank area, southern North Sea, UK. *In*: SPENCER, A. M. (ed.) *Generation, accumulation, and production of Europe's hydrocarbons*. Special Publication of the European Association of Petroleum Geoscientists, 1. Oxford University Press, Oxford, 161–173.

ARTHUR, T. J., PILLING, D., BUSH, D. AND MACCHI, L. 1986. The Leman Sandstone Formation in UK Block 49/28—Sedimentation, Diagenesis and Burial History. *In*: BROOKS, J., GOFF, J. C. AND VAN HOORN, B. (eds) *Habitat of Palaeozoic gas in N.W. Europe*. Geological Society, London, Special Publication, 23, 251–266.

BADLEY, M. E., PRICE, J. D. AND BACKSHALL, L. C. 1989. Inversion, reactivated faults and related structures: seismic examples from the southern North Sea. *In*: COOPER, M. A. AND WILLIAMS, G. D. (eds) *Inversion Tectonics*. Geological Society, London, Special Publication, **44**, 201–219.

GIBBS, A. D. 1986. Strike-slip Basins and Inversion: a possible model for the Southern North Sea Gas Areas. *In*: BROOKS, J., GOFF, J. C. AND VAN HOORN, B. (eds) *Habitat of Palaeozoic gas in N.W. Europe*. Geological Society, London, Special Publication, 23, 251–266.

GLENNIE, K. W. AND BOEGNER, P. L. E. 1981. Sole Pit Inversion Tectonics. *In*: ILLING, L. V. AND HOBSON, G. D. (eds) *Petroleum Geology of the Continental Shelf of North-West Europe*. Heyden, London, 110–120.

——, MUDD, G. C. AND NAGTEGAAL, P. J. C. 1978. Depositional environment and diagenesis of Permian Rotliegendes sandstones in Leman Bank and Sole Pit areas of the UK Southern North Sea. *Journal of the Geological Society, London*, **135**, 25–34.

GREEN, P. F. 1989. Thermal and tectonic history of the East Midlands shelf (onshore UK) and surrounding regions assessed by apatite fission track analysis. *Journal of the Geological Society, London*, **146**, 755–773.

MCKENZIE, D. AND JACKSON, J. 1986. A block model of distributed deformation by faulting. *Journal of the Geological Society, London*, **143**, 349–353.

RICHARD, P., BALLARD, J. F., COLLETTA, B. AND COBBOLD, P. 1989. Naissance et evolution de failles au-dessus d'un decrochement de socle: modelisation analogique et tomographie. *Compte rendus des Seances de l'Academie des Sciences, Paris*, **309**, Serie II, 2111–2118.

ROSSEL, N. C. 1982. Clay mineral diagenesis in Rotliegend aeolian sandstones of the Southern North Sea. *Clay Minerals*, **17**, 69–77.

VAN HOORN, B. 1987. Structural evolution, timing and tectonic style of the Sole Pit inversion. *Tectonophysics*, **137**, 239–284.

VAN WIJHE, D. H. 1987. Structural evolution of inverted basins in the Dutch offshore. *Tectonophysics*, **137**, 171–219.

WALKER, I. M. AND COOPER, W. G. 1987. The structural and stratigraphic evolution of the northeast margin of the Sole Pit Basin. *In*: BROOKS, J. AND GLENNIE, K. (eds) *Petroleum Geology of North-West Europe*. Graham & Trotman, London, 263–275.

Fault reactivation and oblique-slip in the Southern North Sea

B. C. OUDMAYER[1] and J. de JAGER[2]

[1] Shell UK Exploration and Production, Shell-Mex House, Strand, London WC2R 0DX, UK

[2] Nederlandse Aardolie Maatschappij, Schepersmaat 2, PO Box 28000, 9400 HH Assen, The Netherlands

Abstract: The structural evolution of the Southern North Sea region is controlled by a number of important tectonic phases. The dominant NW–SE structural grain first became evident during dextral wrenching in the Stephanian/Autunian. A second fault direction trends NNE–SSW, and probably also originated in pre-Permian times. Extensional faulting in the Triassic preceded major Middle Jurassic to Early Cretaceous rifting. Subsequent cooling subsidence was interrupted by inversion in the Late Cretaceous, the mid-Paleocene, and again in the mid-Tertiary. During these periods Jurassic basins were uplifted and, locally, subjected to deep erosion.

Despite the multitude of tectonic events, a high degree of fault parallelism can be observed in the Southern North Sea. Fault patterns and directions are strikingly similar, whether viewed on a regional scale or at prospect level. The general structural model is, therefore, one of repeated reactivation of basement faults, which continue to control the structural grain despite changes in tectonic regime. Only rarely and by coincidence do the main structural features have orientations that conform to the regional stress. Consequently the predominant mechanism is considered to be one of oblique-slip.

This phenomenon is particularly well observed on structural highs, such as the Cleaver Bank High, straddling the median line between the Netherlands and the UK. Here, the fault trend at Base Zechstein level runs essentially NW–SE. Repeated fault reactivation has led to anomalously high length-to-throw ratios. Locally short N–S and E–W link-ups accommodate oblique-slip. At higher stratigraphic levels en-échelon faults illustrate the strike-slip nature of the latest movements along the pre-existing NW–SE structural grain.

The presence of salt in the overburden causes faulting at higher levels to be decoupled from basement faulting. Sandbox experiments show that reverse faults and extensional grabens can develop in the post-salt sequence in response to oblique-slip along the pre-salt structural grain. The location of such grabens is invariably offset towards the footwall of the basement fault, the amount of offset dependent on the thickness of the salt.

These sandbox experiments provide valuable analogues for the structural mechanisms that are believed to have operated in the Southern North Sea. Applied to both 2D and 3D seismic interpretation, they allow fault mapping with a greater degree of confidence. Recognition of fault reactivation and oblique-slip will lead to a better understanding of the tectonic history, timing of structuration and trap development.

It has been recognized by various authors that strike-slip faulting by reactivation of old fault trends has played an important role in the structural evolution of the Southern North Sea (e.g. Glennie 1984, 1986; Gibbs 1986; Ziegler 1990). In particular, the predominant NW–SE structural grain of the southern Anglo-Dutch offshore has been identified as being at least as old as the Carboniferous.

A characteristic feature of the Southern North Sea is its inverted basins. The Sole Pit Basin in the UK sector and the Broad Fourteens and West Netherlands Basins in the Dutch sector have been described previously by various authors (e.g. Glennie and Boegner 1981; Van Hoorn 1987; Van Wijhe 1987; Dronkers and Mrozek 1991). The non-inverted Cleaver Bank High and Silver Pit Basin (Fig. 1) have received relatively little attention. The Carboniferous sequence can be imaged reasonably well by seismic, and the relationship between overburden faults in the thick Cretaceous and Tertiary sequences and faults below the Zechstein Salt can be readily studied.

Modern seismic acquisition and interpretation techniques have greatly improved our ability to image the sub-surface, and allow the display of fault patterns at various levels and scales relatively quickly and efficiently. Analyses of these fault patterns at detailed and at regional scale confirm the general structural setting of the area. Enhanced understanding of the structural history has led to a better appreciation of various aspects of the detailed trap development, such as timing of trap formation, fault compartmentalization and fault sealing, which improves the assessment of the geological risks associated with oil and gas prospects.

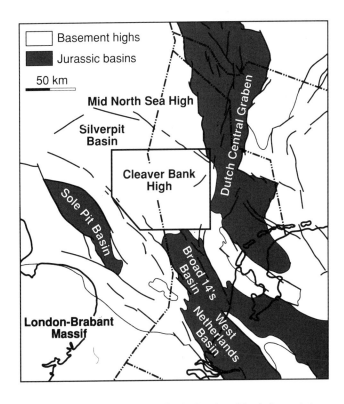

Fig. 1. Main structural elements in the Southern North Sea and the location of the study area.

From *Petroleum Geology of Northwest Europe: Proceedings of the 4th Conference* (edited by J. R. Parker).
© 1993 Petroleum Geology '86 Ltd. Published by The Geological Society, London, pp. 1281–1290.

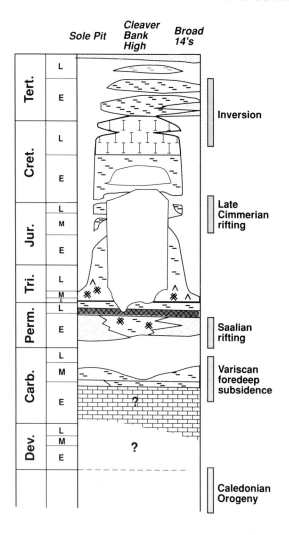

Fig. 2. General stratigraphy and timing of main tectonic phases in the Southern North Sea.

A dramatic increase in heat flow during the Middle Jurassic caused uplift of the Central North Sea Dome (Ziegler 1990). In the Southern North Sea, the main structural elements were now taking shape as fault-bounded Jurassic rift basins (e.g. Sole Pit Basin, Broad Fourteens Basin, West Netherlands Basin and Dutch Central Graben), separated by highs or platforms such as the Cleaver Bank High and the Winterton High.

Thermal subsidence resumed during the Cretaceous. Initially subsidence was strongest in the Jurassic rift basins which remained tectonically active. Gradually, the platform areas were submerged as wells, although initially slowly, as demonstrated by the occurrence of condensed sequences with several hiatuses.

Inversion of the Jurassic basins may have started as early as the Albian (Walker and Cooper 1987), and was certainly active during the Sub-Hercynian tectonic phase from the Turonian to the Campanian. Several intra-Chalk unconformities are related to these early inversion pulses. A second phase of inversion occurred during the Early Paleocene and mainly affected the Broad Fourteens and West Netherlands basins, and to a lesser extent the Dutch Central Graben (Van Wijhe 1987; Dronkers and Mrozek 1991); the Cleaver Bank High may also have experienced some uplift during this phase (Alberts and Underhill 1991). A third pulse of inversion, during the mid-Tertiary, affected mainly the Sole Pit Basin, where practically the entire Chalk and Lower Tertiary sequences are missing, and Upper Tertiary sediments are unconformably overlying sediments locally as old as Triassic (Van Hoorn 1987). This Miocene unconformity can also be recognized over the Cleaver Bank High area and elsewhere in both the UK and Dutch sectors of the Southern North Sea.

From the history of the Southern North Sea it is clear that both the NW–SE and the NNE–SSW trends, that are so obvious on the present-day structural map, already existed in Permo-Carboniferous times, and possibly even earlier. Subsequent structural events resulted in reactivation of pre-existing faults. In general, the maximum regional stress during these later events was not at right angles to the existing structural grain. Consequently, the prevailing tectonic regime during much of the history of the Southern North Sea is one of oblique-slip along reactivated old fault systems.

Structural development of the Cleaver Bank High and Silver Pit areas

Pre-Zechstein fault pattern

A semi-detailed fault map at Rotliegend level confirms the dominant NW–SE fault direction over most of the area under investigation (Fig. 3). The map shows a striking degree of fault parallelism over much of the Cleaver Bank High and over the Silver Pit Basin. This is coupled with an anomalously high length-to-throw ratio. Faults with lengths of up to 25–30 km have a maximum vertical offset of some 300 m only, and mostly the throw is much less.

The long NW–SE faults are concentrated in zones which are occasionally connected by short E–W and N–S link-up faults. Many of these link-up zones are associated with narrow pop-up structures: clear indications of strike-slip along the dominant fault pattern. It is notable that the E–W link-ups usually do not have an expression in the overburden other than mild doming. In contrast, the N–S link-ups are in places related to narrow salt walls and associated overburden faulting.

The pre-Saalian NW–SE structural grain has been reactivated selectively over the Cleaver Bank and Silver Pit areas. Although some faults have not been affected by later tectonic movements, the majority of the faults have experienced post-Permian reactivation, offsetting Rotliegend and Basal Zech-

Geological setting

The Southern North Sea Basin (Fig. 1) has undergone a long and complex geological history, during which it has been affected by orogenic events, foredeep subsidence, rifting, strike-slip faulting, inversion and halokinesis (Fig. 2).

The dominant fault direction over much of the Southern North Sea is NW–SE. Its pre-Saalian origin is clearly reflected in the Carboniferous subcrop pattern over large parts of the Southern North Sea, including the study area (Fig. 3). Although an earlier origin of this structural grain cannot be excluded, first clear evidence for movements along the NW–SE fault system has been assigned to dextral divergent wrenching in Autunian/Stephanian times (Ziegler 1990).

During the Saalian thermal event at the end of the Carboniferous an increase in heat flow caused considerable uplift and erosion, and locally igneous activity. Approximately N–S-trending Early Permian grabens formed in Germany (Gralla 1988). Permian subcrop and Rotliegend isopach maps suggest that the NNE–SSW-trending Dutch Central Graben may also have been initiated at that time.

After this event, the Southern North Sea was subjected to thermal subsidence from Middle Permian times onwards. Generally, only relatively minor normal faulting took place, although isopach maps point towards the beginning of differential subsidence in Late Triassic times (Ziegler 1990).

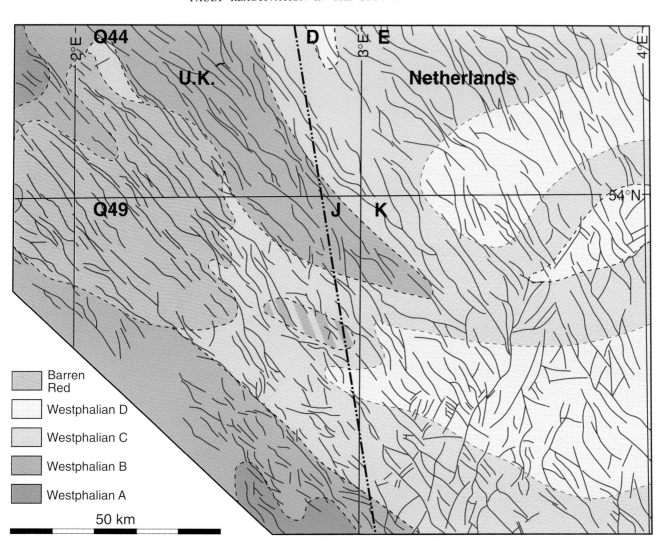

Fig. 3. Semi-detailed Rotliegend fault pattern and Permian subcrop map, illustrating the dominant NW–SE fault trend, parallel to pre-Saalian structuration, over most of the area, and a clear NNE–SSW fault direction in the SE.

stein lithologies in both normal and reverse senses (Fig. 4). Improved resolution from 3D seismic demonstrates a varying degree of intensity of reactivation. Generally, interpretation based only on 2D data allows fault aliasing to occur, resulting in the initial mapping of very long, continuous faults. Remapping based on 3D seismic shows that some of these faults in fact consist of a zone of short, discontinuous faults (Fig. 5). They are arranged in an en-échelon pattern over a deep-seated basement fault, indicating that they resulted from wrenching along a pre-existing lineament. The amount of reactivation that these zones have experienced is probably small compared to that along the reverse faults situated to the SW of this zone. The present reverse component of these faults suggests that at least one phase of reactivation occurred under a different stress regime.

Much of the fault pattern shown in Fig. 3 is still based on 2D seismic data, which were mostly acquired in the dip direction, perpendicular to the prevailing fault trends. The bias in fault interpretation is illustrated in a further example (Fig. 6), where the initial 2D map shows long NW–SE faults only, while on the recent 3D seismic interpretation a NNE–SSW cross-fault direction becomes apparent. Although vertical offset on the NNE–SSW faults is minimal, the faults may have an important effect on compartmentalization of gas fields, inhibiting free communication from one block to the next.

The cross-fault direction recognized on 3D seismic over the

Silver Pit Basin becomes a dominant fault trend in the SE corner of the study area (Fig. 3). These clearly expressed NNE–SSW faults are parallel to the Dutch Central Graben, which manifests itself just NE of the area. Although again an older origin of this general fault direction is suggested by pre-Saalian Carboniferous subcrop patterns, faulting of the Dutch Central Graben is thought to have occurred from Late Triassic times onwards. Analysing the exact timing of fault reactivation at pre-Zechstein level by unravelling the post-Zechstein sequence remains enigmatic due to salt decoupling and the absence of much of the geological record. Where preserved, Lower Triassic Bunter shales are of constant thickness over the area, confirming geological quiescence in the Early Triassic. Much of the Cleaver Bank High and surrounding areas experienced uplift during the Jurassic Late Cimmerian tectonic phase, with erosion cutting progressively deeper into Lower Jurassic and Triassic sediments over the Cleaver Bank High (Van Hoorn 1987). This coincided with a major rifting episode responsible for the formation of the Dutch Central Graben by reactivating the NNE–SSW fault direction concurrent with the NW–SE faulting that created the Sole Pit and Broad Fourteens basins. Age relationships by studying offsets of one fault direction by the other are difficult to establish, even with 3D seismic. A complex pattern emerges, dominated by short faults cutting and abutting against others. This observation is interpreted as a further suggestion that reactivation affected both

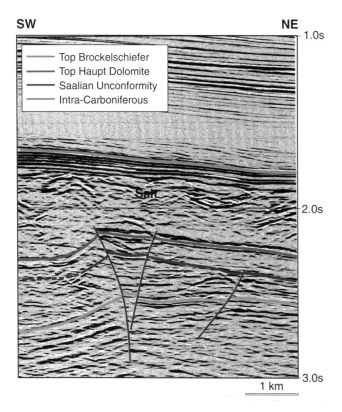

Fig. 4. Seismic section showing non-reactivated Carboniferous fault (right) next to reactivated fault (left) with clear post-Permian reverse component.

fault directions simultaneously, during this and subsequent tectonic events.

Cretaceous structure

Sedimentation over the study area resumed in the Lower Cretaceous with deposition of thin clastics. Locally, the distribution of these deposits was already influenced by halokinesis, but more significant salt tectonics did not occur until Late Cretaceous Chalk deposition. Growth of domes and salt walls is seismically illustrated by many syn-sedimentary unconformities in the Chalk sequence (Fig. 7a) and, therefore, isopach maps for these and later intervals do not represent an unambiguous guide to basement tectonic movements. However, this phase of halokinesis is contemporaneous with the Sub-Hercynian inversion of Jurassic Southern North Sea basins. The observation that the occurrence of halokinesis is related to inversion elsewhere is further illustrated by the Late Campanian to Danian Chalk overstepping both the inversion structures (Van Wijhe 1987) and the salt structures (Fig. 7a), marking a reduced growth of salt domes concurrent with a decrease of tectonic activity and a return to subsidence. This observation is a further indication for linking halokinesis to deep-seated fault activity related to inversion.

The orientation of the salt walls on the Cleaver Bank High suggests that there was reactivation of the NW–SE basement grain, and also of the NNE–SSW direction in the SE of the study area (see Fig. 9). Inversion during this Sub-Hercynian tectonic phase is widely regarded as being associated with early Alpine N–S compression (Ziegler 1990). This resulted in convergent dextral strike-slip movements along the main basin-bounding faults of the NW–SE-trending Sole Pit Basin and the Broad Fourteens Basin (Dronkers and Mrozek 1991).

Locally, steep reverse faults affect the Base Cretaceous and lower parts of the Chalk sequence near salt structures (Fig. 7a),

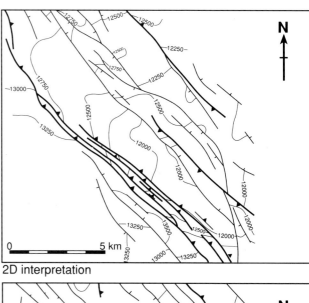

2D interpretation

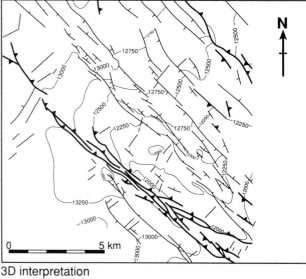

3D interpretation

Fig. 5. Top Carboniferous structure map, showing comparison between maps based on 2D and 3D seismic, illustrating discontinuous nature of a fault zone previously interpreted as being a single, continuous NW–SE fault.

and often bound diapirs at one side. These faults are mostly isolated, and do not propagate upwards to high stratigraphic levels.

Major Late Cretaceous NW–SE reverse faults can also be observed along the northeast margin of the Broad Fourteens Basin. At the northwestern termination of this fault system, a NE–SW-trending Chalk graben can be observed (Fig. 8a). It clearly developed as a small pull-apart graben during Late Cretaceous deposition of the Chalk. This extensional feature is related to a NNE–SSW-trending fault zone at Base Zechstein level. The presence of Zechstein Salt caused decoupling of overburden faults, and locally the position of the Chalk Graben is displaced with respect to the position of the controlling basement faults. The concurrent occurrence of extensional features non-perpendicular to the compressional direction which was responsible for inversion is a clear indication for the occurrence of oblique-slip during the Late Cretaceous.

The structural style of this Chalk graben is very similar to overburden structures produced in sandbox experiments, carried out to model the relation between basement movements on resulting structures in the overburden above a salt layer (Fig. 8b). Sand was used to represent the overburden and silicone putty was used to model the salt. Figure 8b shows the result of a sandbox experiment with pure normal dip-slip movement on a basement fault. The precise results of such experiments depend on the thickness of the original silicone putty layer, overburden thickness and throw on the basement fault. Decoupling of the overburden faults can clearly be seen, and results in an offset of the graben towards the footwall.

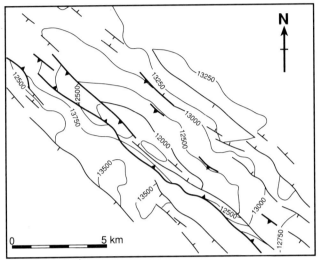

2D interpretation

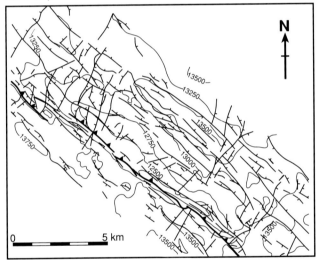

3D interpretation

Fig. 6. Top Carboniferous structure map, showing comparison between maps based on 2D and 3D seismic, illustrating presence of previously unrecognized NNE–SSW subtle cross-fault direction.

Tertiary structure

At the Base Tertiary, faulting occurs in distinct zones (Fig. 10), which are often situated above salt walls and diapirs. These halokinetic structures can be related to a second phase of inversion in the Southern North Sea during the mid-Paleocene and a third phase during the Miocene.

The second phase was predominantly active in the Dutch sector, mainly affecting the Dutch Broad Fourteens and West Netherlands basins, and to a lesser extent the Dutch Central Graben, where a clear unconformity is present between the Chalk and the Tertiary (Van Wijhe 1987). Apatite fission track analysis suggests that parts of the Cleaver Bank High also experienced uplift during this period (Alberts and Underhill 1991), although there is no clear angular unconformity.

The third phase of inversion, in mid-Tertiary times, is mainly active over the Sole Pit Basin and large parts of the UK sector, but also in the West Netherlands Basin, and results in a marked angular unconformity at the Base Miocene. On the Cleaver Bank High the latest Chalk and the Early Tertiary sediments have been affected by further diapir growth during this phase (Fig. 7a).

Large areas of the study area show a remarkable absence of overburden faulting at Base Tertiary level (Fig. 10) and only mild doming. Fault activity appears to have occurred mainly in distinct zones, which are associated with the strongest diapirism. One of these zones coincides with the hinge zone bordering the Sole Pit inversion. Where the SE part of this zone ends, in the southern part of UK Quadrant 49, halokinesis is of markedly shorter wavelength and higher amplitude than in the rest of the study area. The complexity of this structuration is also reflected at Base Zechstein level in a change from predominantly NW–SE faults to an apparently more disorganized pattern, with numerous link-up faults (cf. Fig. 3). This area is situated between the Sole Pit and Broad Fourteens basins and the structural complexity may be related to repeated stress accommodation during various tectonic phases. It also intersects with the NNE–SSW faulting at Base Zechstein level in the Dutch sector, which is also locally marked by steep diapirs and related overburden faulting. Latest movement can be demonstrated to have occurred in the Neogene, when reactivated faults have locally offset the Base Miocene unconformity.

The structure illustrated in Fig. 7a is located in the complex accommodation area between the Sole Pit and Broad Fourteens basins. It again resembles a structure derived from a sandbox experiment, Fig. 7b, which shows the result of reactivation of a basement fault with pure dip-slip, with relatively thin silicone putty. An overburden reverse fault develops above, but detached from, the basement fault. In the hangingwall, at some distance from the main basement fault, a slight thickening of silicone putty develops and a graben forms on top. However, further experiments have shown that with thicker salt present, as on Fig. 7a, oblique-slip reactivation is required for a reverse fault to develop.

Further evidence for oblique-slip comes from fault patterns as imaged on a seismic 3D dataset from the western Dutch offshore in an area where the Zechstein Salt is thin. The fault pattern at Base Zechstein again shows a predominantly NW–SE structural grain, with subordinate N–S and NNE–SSW cross-faults (Fig. 11a). At Base Aptian level, not far above the Zechstein Salt, most deeper faults are expressed as subtle variations in horizon dip only (Fig. 11b). The shallower Base Tertiary dip map shows a much larger number of faults, all extensional and generally parallel to the NW–SE trend. A zone runs diagonally through the centre of Fig. 11c with en échelon faults reminiscent of Riedel fractures. While the faults are small, they are clearly visible on this display. The shear zone occurs directly above the main NW–SE fault trend at Base Zechstein level of Fig. 11a. A seismic line across this feature shows that the Tertiary faults converge towards the fault zone at Base Zechstein level (Fig. 11d), suggesting that the shallow faults are related to deep-seated faulting. The en-échelon pattern indicates dextral divergent strike-slip movements along the master fault.

Conclusions

The predominant NW–SE and also the NNE–SSW fault trends in the Southern North Sea were already established by Permian times. During the Saalian event, Late Cimmerian rifting, and Late Cretaceous and Tertiary inversion, these fault trends have been reactivated selectively. The present-day pre-Zechstein fault pattern is consequently the result of multiple fault reactivation under different stress regimes.

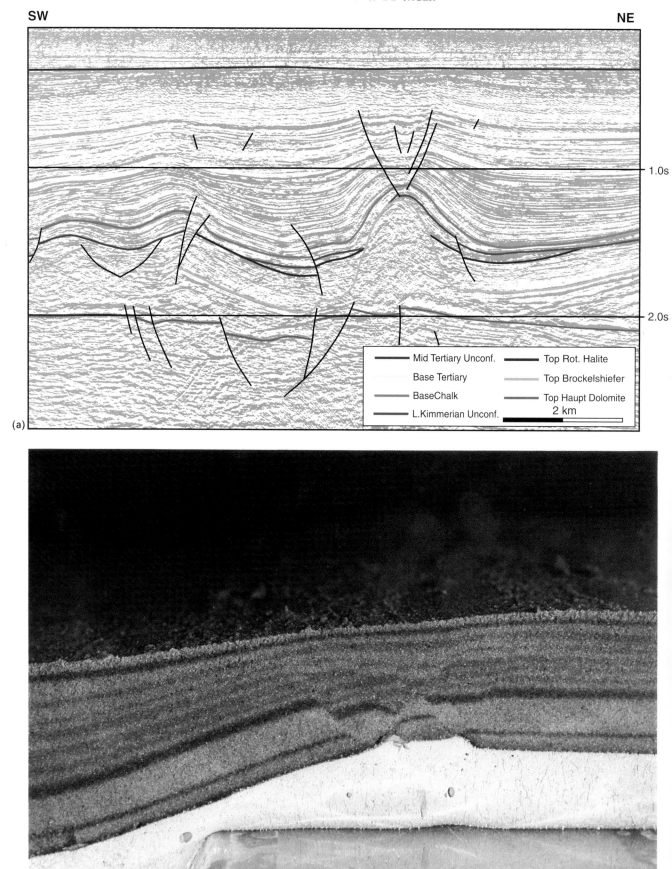

Fig. 7. (**a**) Seismic section from the area with complex overburden deformation. Intra-Chalk unconformities attest to early timing of halokinesis, which ceased in the uppermost Chalk, and resumed during the Early Tertiary. Structural configuration with single, detached reverse fault and overburden graben offset from basement fault resembles experimental results. (**b**) Experimental result of reactivation of a basement fault with pure dip-slip, with relatively thin silicone putty. An overburden reverse fault develops above, but detached from, the basement normal fault. In the hangingwall, at some distance from the main basement fault, a slight thickening of silicone putty develops and a graben forms on top.

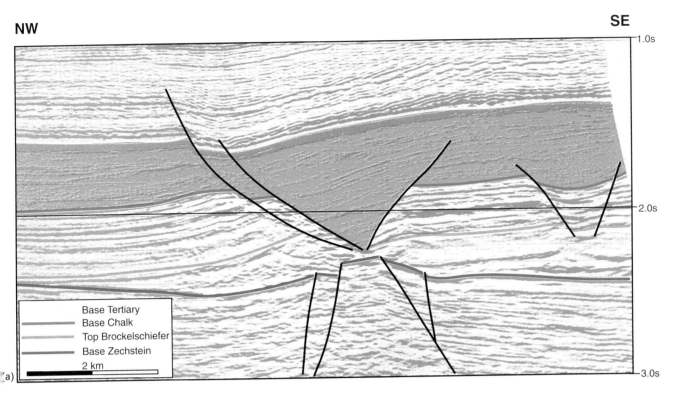

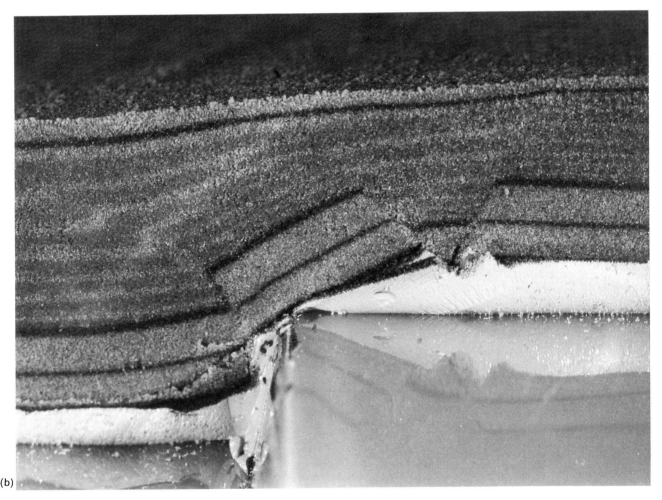

Fig. 8. (**a**) Seismic section illustrating Chalk grabens resulting from extension along NNE–SSW basement faults during Late Cretaceous inversion. Thin Zechstein salt decouples overburden faults from basement faults. (**b**) Experimental result of reactivation of a basement fault with pure dip-slip. Decoupling of the overburden faults can clearly be seen, and results in an offset of the graben towards the footwall. The precise results of these experiments depend on the thickness of the original silicone putty layer, overburden thickness and orientation of offset on the basement fault.

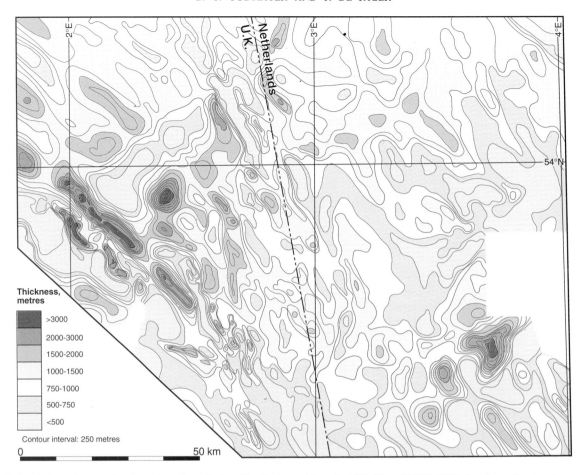

Fig. 9. Zechstein isopach map showing orientation of salt walls, diapirs and lows in NW–SE and NNE–SSW directions. Contour interval 250 m. See text for discussion and Fig. 3 for underlying Rotliegend fault pattern.

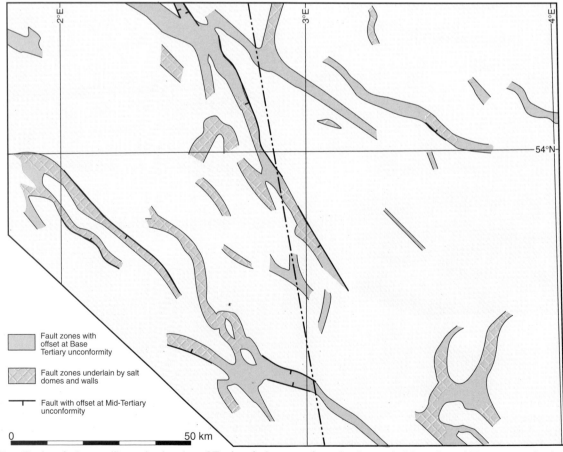

Fig. 10. Base Tertiary fault map, illustrating location of Tertiary fault zones, often related to underlying salt, and Neogene reactivation of some of the faults.

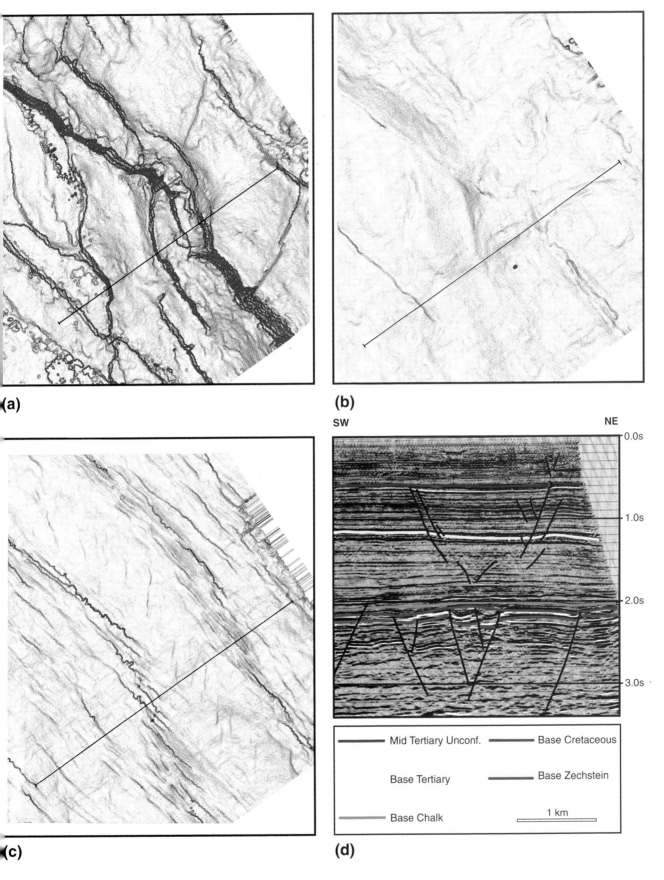

Fig. 11. Horizon dip display maps in ms m⁻¹ from a 3D seismic survey. (**a**) At Base Zechstein level, a NW–SE structural grain is predominant, with minor N–S and NNE–SSW faults. (**b**) At base Aptian level, just above the Zechstein Salt, only a few faults are visible, but the NW–SE grain remains obvious. (**c**) At Base Tertiary level a very dense NW–SE extensional fault pattern is visible. The short en-échelon faults in the centre, that have developed above the main basement fault at Base Zechstein level, are indicative of dextral shear, and suggest that the faulting is strongly controlled by the pre-existing basement grain. (**d**) Seismic section illustrating overburden faulting at Base Tertiary and Base Aptian level, converging towards reactivated pre-Zechstein faults (location marked).

Careful analysis of fault patterns at shallower levels gives indications as to which faults have been reactivated, at what time, and under what stress regime. This information is invaluable for a sound understanding of both regional tectonic setting, and of detailed trap development.

In Late Carboniferous to Early Permian times dextral strike-slip movements took place along NW–SE faults (Ziegler 1960), in an overall extensional setting. During Late Jurassic E–W extension, oblique-slip can be inferred to have taken place along these faults, in a dextral divergent sense. The Late Cretaceous Sub-Hercynian inversion reactivated the NW–SE faults in a compressional regime, while at the same time NNE–SSW fault zones locally express extensional features. The simultaneous occurrence of both compressional and extensional faulting non-perpendicular to each other indicates oblique-slip. During the Tertiary, inversion again affected the Jurassic Southern North Sea basins, both during the mid-Paleocene (Laramide) and the mid-Tertiary tectonic phases. On the Cleaver Bank High, locally dextral divergent strike-slip can again be observed on detailed 3D horizon attribute displays during this period. The latest structuration is expressed as normal faulting in Neogene times.

In areas where salt is present, overburden faults are decoupled from basement faults. However, fault patterns in the overburden are still associated with faulting at depth, and may be indicative of the stresses that have been operating. Sandbox experiments have helped to understand the relationship between these overburden faults and the reactivated deep-seated faults, allowing a better understanding of the structural history of the Southern North Sea.

This paper is published by permission of Shell UK Exploration and Production, Nederlandse Aardolie Mij. BV, Esso Exploration and Production UK Ltd, Energie Beheer Nederland BV, Oranje-Nassau Energie BV, Clyde Petroleum (Netherlands) BV and CLAM Petroleum Co. Special thanks are due to Chris Elders, Daan den Hartog Jager, Peter Ziegler and Bert Dijksman, who assisted in various stages of the preparation of this paper. Jean-Michel Larroque carried out the sandbox experiments in Shell's Research Laboratory in the Netherlands. The artwork was produced by Ian Glenister of Shell Expro's draughting department. Numerous colleagues in Shell UK Exploration and Production and Nederlandse Aardolie Mij. have contributed through their previous work.

References

ALBERTS, M. A. AND UNDERHILL, J. R. 1991. The Effect of Tertiary Structuration on Permian Gas Prospectivity, Cleaver Bank Area, Southern North Sea, UK *In*: SPENCER, A. M. (ed.) *Generation, Accumulation and Production of Europe's Hydrocarbons.* Special Publication EAPG, **1**, 161–173.

DRONKERS, A. J. AND MROZEK, F. J. 1991. Inverted basins of The Netherlands. *First Break*, **9**, 409–425.

GIBBS, A. D. 1986. Strike-Slip Basins and Inversion: a possible model for the Southern North Sea Gas Areas. *In*: BROOKS, J., GOFF, J. C. AND VAN HOORN, B. (eds) *Habitat of Palaeozoic Gas in N.W. Europe.* Geological Society, London, Special Publication, **23**, 23–35.

GLENNIE, K. W. 1984. The Structural Framework and the Pre-Permian History of the North Sea Area. *In*: GLENNIE, K. W. (ed.) *Introduction to the Petroleum Geology of the North Sea.* Blackwells, Oxford, 17–39.

—— 1986. Development of N. W. Europe's Southern Permian Gas Basin. *In*: BROOKS, J. GOFF, J. C. AND VAN HOORN, B. (eds) *Habitat of Palaeozoic Gas in N.W. Europe.* Geological Society, London, Special Publication, **23**, 3–22.

—— AND BOEGNER, P. L. E. 1981. Sole Pit Inversion Tectonics. *In*: ILLING, L. V. AND HOBSON, D. G. (eds) *Petroleum Geology of the Continental Shelf of North-West Europe.* Heyden, London, 110–120.

GRALLA, P. 1988. Das Oberrotliegende in NW Deutschland—Lithostratigraphie und Faziesanalyse. *Geologisches Jahrbuch*, A **106**.

VAN HOORN, B. 1987. Structural Evolution, Timing and Tectonic Style of the Sole Pit Inversion. *Tectonophysics*, **137**, 239–284.

VAN WIJHE, D. H. 1987. Structural Evolution of Inverted Basins in the Dutch Offshore. *Tectonophysics*, **137**, 171–219.

WALKER, I. M. AND COOPER, W. G. 1987. The Structural and Stratigraphic Evolution of the Northeast Margin of the Sole Pit Basin. *In*: BROOKS, J. AND GLENNIE, K. W. (eds) *Petroleum Geology of North West Europe.* Graham & Trotman, London, 263–275.

ZIEGLER, P. A. 1990. *Geological Atlas of Western and Central Europe* (2nd edn). Shell Internationale Petroleum Maatschappij B.V., The Hague.

Fluids:
migration,
overpressure and
diagenesis

Fluids: migration, overpressure and diagenesis

Introduction and review

A. J. FLEET[1] and W. G. CORDEY[2]

[1] BP Exploration, Sunbury Research Centre, Chertsey Road, Sunbury-on-Thames, Middlesex TW16 7LN, UK

[2] Shell UK Exploration and Production, Shell-Mex House, Strand, London WC2R 0DX, UK

Petroleum and water are at the very heart of success or failure for exploration and production but, being fluids, are not easy to observe or predict in the geological environment. The 'Fluids' Sessions, therefore, dealt with many issues which are both at the forefront of technical developments and commercially topical. A consequence of this was that the number of contributions was fewer than had been hoped, particularly on the subject of overpressure. Some studies, because they are very much in their infancy, could only provide a cursory treatment of the topic.

Fluid formation and movement in basins ultimately govern when and where petroleum accumulations occur, reservoir quality in most instances, and the development of overpressure. The composition of petroleum leaving a source rock is very much a function of source-rock type and the subsequent thermal history. The presence of overpressure can dictate the expulsion direction and thus the overall composition of the hydrocarbons entering a migration pathway from one or more source rocks. Phase changes along the migration pathway, or in the reservoir during uplift or burial, determine the phase(s) of the hydrocarbons retained over geological time and must be evaluated together with, or in the context of, tectonism and faulting.

The importance of understanding overpressured regimes cannot be overemphasized, both from the purely exploration aspect and the issues involving safety and the environment. The papers of **Leonard** and **Garenstroom et al.** address the problem. **Leonard** reviews sub-surface pressure in the Norwegian Central Graben and emphasizes the need to build up a three-dimensional understanding of pressure and its relevance to field distribution, generation and migration, reservoir quality and drilling procedures. He argues that all possible causes of overpressure need consideration in evaluating a basin. **Garenstroom et al.** discuss the relationship between overpressure and minimum effective stress and relate the areal distribution of overpressures to the structural framework.

Since the problems associated with predicting petroleum charge to a trap and with interpreting the significance of hydrocarbon seeps in terms of prospects, fairways or dry belts are concepts well entrenched in exploration thinking (e.g. England and Fleet 1991), they receive little explicit attention in this section. **Burrus et al.** and **Høvland** are the exceptions. The former use whole basins, including the Paris Basin, as experimental testbeds to deduce that different expulsion styles occur in different basins and to help in determining whether oil or gas is available from a given source rock. Their work illustrates the strength of using basin-scale 2D fluid flow modelling carefully calibrated against observations. **Høvland** contributes a further dimension to his work on seabed features which indicate evidence for seepage. Not only are these features widespread offshore NW Europe but they take on a broad variety of forms from 'freak' high-amplitude sandwaves to cold-water corals.

Other contributions to the section can be described as innovative or, indeed, in some instances, speculative. The most innovative, at least in bringing new science to bear on petroleum occurrence, is presented by **Ballentine and O'Nions**. They use rare gases (helium, neon and argon) to begin to understand the processes and relative proportions of fluid movement in sedimentary basins. Two of their initial studies, in the Pannonian and Vienna basins, point towards movement of much larger volumes of water than had previously been envisaged and large-scale petroleum gas–water interaction. The fluids discussed by **Warren and Smalley** are less exotic; they turn the spotlight on water and outline its variation in composition. Attention is drawn to the relevance of this variability with respect to reservoir quality, reserves estimates and scale prediction. Their paper is very much a curtain raiser to an atlas of North Sea formation waters due for publication in 1993.

The final three papers of the section focus on the part water plays in diagenesis. **Burley**, in a paper which he confesses to be 'speculative and likely to be highly provocative', suggests that much diagenesis in Jurassic reservoirs in the Northern North Sea can be attributed to hot fluid circulation during the Tertiary. He advances a detailed model which takes into account structural position, fault control on fluid migration and trap. **McAulay et al.** consider in detail the diagenetic development of Brent group reservoir sands in the Hutton and NW Hutton fields. **Gluyas et al.**, in contrast, stand back from the detail and discuss the close inter-relationships through time of reservoir charge and cementation. Both processes can take about 10 Ma and in some reservoirs compete in a 'race for space'. They believe both may be triggered by major changes in basin dynamics such as phases of rapid subsidence and accompanying tectonism.

Overall the papers reflect the wide range of 'Fluid' topics and issues which relate to petroleum geology and more specifically to NW Europe. However, in many instances, they are only able just to scratch the surface or fly inventive kites. The next conference, it is hoped, should see the substantial development of many of the ideas presented here, and hopefully, the use 'in anger' of some of these to make exploration predictions and solve development and production problems.

Reference

ENGLAND, W. A. AND FLEET, A. J. 1991. *Petroleum Migration*. Geological Society, London, Special Publication, **59**.

Distribution of sub-surface pressure in the Norwegian Central Graben and applications for exploration

R. C. LEONARD

Amoco Production Company, PO Box 3092, Houston, TX, USA

Abstract: The Norwegian Central Graben contains a distinctive three-tiered sub-surface pressure system. This pressure system has an important effect on distribution of oil and gas fields, generation and migration of hydrocarbons, reservoir qualities and drilling procedures.

A number of mechanisms have been proposed for the development of overpressure. In this paper, five are briefly discussed: thermal expansion of water, sedimentary compaction, rate of sediment burial, volumetric increases due to generation of hydrocarbons, and column heights of major oil and gas fields. None of the factors are mutually exclusive and all contribute to the development of overpressure in the Norwegian Central Graben.

Utilizing drill-stem tests (DSTs), repeat formation tests (RFTs) and sonic and density logs, the horizontal and vertical boundaries of the three-tiered pressure system of the Norwegian Central Graben have been mapped. A shallow hydropressure system exists down to 1000–2000 m, underlain by a transition zone where pressures abruptly increase to the fracture level where overburden and fluid pressure are in balance. Under the transition zone, a geopressure zone exists to 3000–4000 m, within which pressures increase with depth at a hydrostatic rate. In the centre of the graben, this compartment is underlain by a second transition zone and a deep, high temperature, high pressure zone. The location and depth of the boundaries of the two upper pressure compartments and transition zones are more dependent on temperatures than lithologies or tectonic movements. Within the deep, high pressure zone, variations in pressure gradients indicate the presence of a number of smaller pressure cells, probably separated by faults.

The identification and prediction of overpressured zones can have a significant effect on exploration programmes. From the drilling standpoint, prediction of sharp increases in pressure is a major safety factor, reducing the danger of well blowouts and allowing the operator to be adequately prepared for a number of drilling hazards. In potential reservoirs, overpressure reduces net overburden, thereby reducing porosity loss due to compaction. Sub-surface pressure variations have a significant effect upon hydrocarbon migration as oil and gas will attempt to move from high to low pressure compartments. Defining pressure variations from well to well can give indications of sealing and non-sealing faults.

The pressure configuration has a significant effect upon the distribution of discovered hydrocarbons in the Central Graben. Most recoverable hydrocarbons are found in the first sealed permeable bed in the middle pressure compartment, overlying or adjacent to the transition zone above the deep, high pressure compartment. This includes the Cretaceous to Danian chalk fields in the centre of the graben and the Jurassic sand-reservoired fields on the northeast flank.

Exploration in the deep, high pressure compartment has resulted in only limited success due mainly to poorer potential reservoir conditions, more difficult definition of traps and increased drilling expenses and risk.

Overpressure is a subject that has been of great interest to explorationists and drillers for many years for a number of reasons. Papers by King Hubbert (Hubbert 1940 and 1953) applied groundwater flow concepts and principles of hydrodynamics to calculating sub-surface pressures. However, by the 1960s deeper drilling encountered, in a number of areas, pressures too high to fit the hydrodynamic theory. Initial explanations for these high pressures were that drainage was only slowly occurring from low permeability rocks. However, a rigorous evaluation of a slow-leaking seal (Bradley 1975) demonstrates that even a slow drainage mechanism would eliminate overpressures over geologic time.

Therefore, the overpressures occur due to the formation of pressure seals, described in this paper as 'pressure transition zones'. Pressure distribution in many sedimentary basins around the world have been studied by Amoco Production research teams, headed by Dave Powley and John Bradley. The principles that they have established and their early studies of North Sea pressure compartments have been applied to a more detailed analysis in this paper.

The paper is divided into four sections. First, the origin of overpressure is discussed from the standpoint of five different mechanisms that can be applied in the Central Graben. Next, pressure profiles from several basins around the world are shown to demonstrate certain patterns and concepts. Third, a more detailed look at pressure distributions in the Norwegian Central Graben is shown. Finally, applications to exploration from the study of overpressures are discussed.

Origin of overpressures

Overpressuring is very simply defined as sub-surface pressures in excess of a hydrothermal pressure gradient which is 0.43 to 0.47 psi per foot or 1–1.09 g cm^{-3}, depending on water salinity,

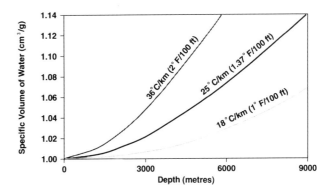

Fig. 1. Effect of thermal gradients on thermal expansion of water. Note that the 36°C km^{-1} gradient is comparable to the gradients in the Norwegian Central Graben (Leonard 1984). Adapted from Magara (1974).

From *Petroleum Geology of Northwest Europe: Proceedings of the 4th Conference* (edited by J. R. Parker).
© 1993 Petroleum Geology '86 Ltd. Published by The Geological Society, London, pp. 1295–1303.

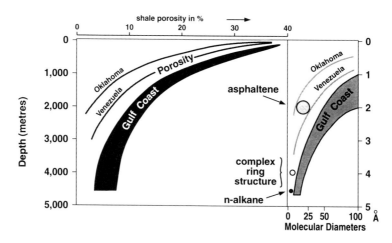

Fig. 2. Inter-relationship of various physical parameters with increasing depth of burial for shale-type sediments. Adapted from Tissot and Welte (1984).

from the surface to any given point. In order for it to develop, fluid flow must be inhibited or prevented and sealed both vertically and laterally.

There have been many mechanisms discussed for causing overpressuring, and in this paper I briefly describe five that I feel are important. It should be noted that none of these five are mutually exclusive. These factors address both thermal expansion of fluids in the sub-surface and reduction in pore size. These factors, in combination, are the main causes of overpressuring.

Aquathermal pressuring or thermal expansion of water is felt to be a major, if not the single, most important mechanism in the development of overpressure. In a closed system, the thermal expansion of water can result in pressure increase of up to 3 psi per foot or approximately 7 times the normal sub-surface pressure gradient in a normally pressured system. As shown on the chart (Fig. 1), this can begin to have a significant effect at depths of less than 3000 m in basins with high gradients such as in the Norwegian Central Graben where the thermal gradient is approximately 36°C km^{-1}. On the other hand, in basins with low geothermal gradients, such as 18°C km^{-1}, as in many Paleozoic basins or on carbonate shelves, this does not begin to have a significant effect until a depth of somewhere in excess of 6000 m. The thermal expansion of water helps to develop overpressure in two ways. First, the tremendous volume of water expansion, even if it is only 5% or 8%, in a large area can cause significant pressure increase. Secondly, the size of the water molecule will increase by a comparable amount. In a comparison of the size of the water molecule vs. the size of the overlying shale pores, the larger the water molecule, the earlier restriction of flow or even sealing takes place.

During sedimentary compaction and porosity reduction, there is, not surprisingly, a reduction in pore diameters. Figure 2 demonstrates porosity reduction from 40% to less than 10% in shales in the Gulf Coast of the United States in conjunction with burial depths of up to 2500 m. Curves from older basins (in the case of Venezuela, a Cretaceous to early Tertiary basin similar in many ways to the Norwegian Central Graben; and in the case of the Arkoma Basin in Oklahoma, a Paleozoic basin) show more rapid increase in shale porosity decrease with depth. Pore diameters decrease to less than 25 Å at a depth of about 3–4 km in the Gulf Coast, at a depth of about 2 km in east Venezuela, and a depth of a little over 1 km in Oklahoma. In the Gulf Coast, shale pores have a molecular diameter of about 10 Å at c. 4.5 km, in Venezuela at about 2.5 km and in Oklahoma c. 2 km. The size of the pores can be contrasted with the size of the oil and water molecules which range from about

5–35 Å for oil, and approximately 3.2–3.5 Å for water. A point is reached where shale pore sizes are too small for an unrestricted flow of water or oil molecules. Clearly this happens to oil at much shallower depths than for water. This paper concentrates on shales or clastic material because the porosity loss and compaction is more of a mechanical process in clastics than in carbonates, where chemical compaction, less predictable with depth, is a factor. In the Central Graben, clastics are the predominant lithology.

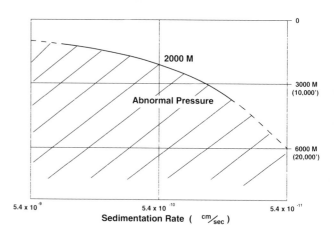

Fig. 3. Relationship between rate of sedimentation and depth to top of overpressured section in the Gulf Coast (USA) from Magara (1978).

Overpressuring will also take place when, due to rapid subsidence, escape of pore waters cannot keep pace with burial. Figure 3 demonstrates a model constructed by Magara (1978) to predict overpressuring in the Northern Gulf of Mexico Tertiary sediments. It demonstrates that overpressuring can be initiated at depths of 1000–6000 m depending upon the depositional rate: a sedimentation rate of 5.4×10^{-10} cm s^{-1} is indicative of a rate of 150 m Ma^{-1}. Due to smaller pore size a more deeply buried section can become overpressured at a much slower rate of deposition. In addition, in especially fine-grained rock, an effective seal will push the curve upwards causing overpressures at a much shallower depth. In the case of the Norwegian Central Graben, the high depositional rate in the Middle Miocene would initiate overpressuring in Mesozoic sediments.

Another factor causing overpressuring is a volumetric

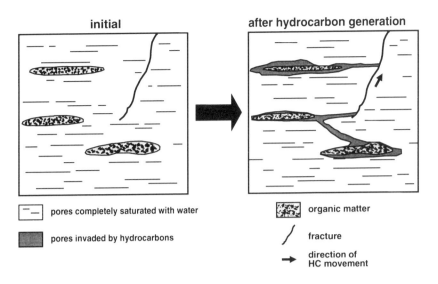

Fig. 4. Volumetric expansion of kerogen followed by fracturing and primary hydrocarbon expulsion. From Tissot and Welte (1984).

increase of organic matter by the generation of gaseous and liquid hydrocarbons from kerogen (Fig. 4). Studies by Ungerer *et al.* (1984) of the French Petroleum Institute have calculated a significant overall volumetric expansion during catagenesis, probably in excess of the volumetric increase of pore waters due to thermal expansion. Additional volumetric expansion can occur during cracking of liquid hydrocarbons to gas. Barker (1990) indicated that a volumetric expansion of up to 80% occurs in the cracking of liquid to gaseous hydrocarbons, at least under reservoir conditions. Gaarenstroom *et al.* (1993) indicated the coincidence of the deep pressure transition zone in the UK Central Graben with the thermal level corresponding to metagenesis, the cracking of oil to condensate and gas. It should be noted that volumetric expansion of source rock will be at a maximum with rich source rocks such as the Kimmeridgian shale and will only have a marginal impact with source rocks containing 1% total organic carbon. Another factor to take into account in looking at the effect of hydrocarbon generation migration, as compared to thermal expansion of water, is that very often the source rock interval is very thin, so while there may be a very significant volumetric expansion in this interval, it is perhaps not as significant as a 5% expansion of water over an interval of 1000–2000 m of sediment.

Another factor that can cause overpressuring is a significant hydrocarbon accumulation. The equations that can be used to calculate the pressure differential due to hydrocarbon accumulation are shown in Fig. 5. This pressure differential is equal to the density of water minus the density of the hydrocarbons times the height of the hydrocarbon column, times a gravitational constant. The result of this is usually only a very small number compared to overpressuring from other factors. However, in the case of a field with very light hydrocarbons such as gas with a very long hydrocarbon column, it can have a significant effect. This effect, however, would only be present over a very limited area. There are also several other factors that have been discussed with regard to overpressuring. One is the transition of montmorillonite to illite which results in volumetric expansion in the order of 2% (Magara 1975). In certain areas major tectonic forces could also have an impact on the sub-surface pressure.

In summary then, there are many ways to cause overpressuring; however none of these are mutually exclusive or absolutely necessary. For example, with a high enough depositional rate, great burial depth or presence of a hydrocarbon-source rock may not be needed. To determine the causes of overpressuring,

rather than taking one simple solution and applying it to all the sedimentary basins in the world, each basin must be looked at separately. The factors that have been discussed previously must be analysed to see first of all from a qualitative standpoint if there will be overpressuring in the basin, and then to try to make a quantitative prediction as to the depths.

Looking specifically at the Norwegian Central Graben, all of the primary factors are present for the development of significant overpressures: (1) a high thermal gradient, resulting in significant thermal expansion of water; (2) sufficient sediment thickness to cause compaction of shale pores; (3) a rapid Tertiary burial rate, particularly in the middle Miocene; (4) major hydrocarbon generation and migration with significant thermal cracking to gas in the deeper portions of the basin; and (5) many accumulations of light hydrocarbons with columns in excess of 200 m also are present.

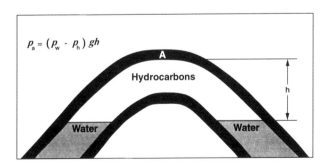

$$P_a = (P_w - P_h)\, gh$$

Fig. 5. Calculation of increased pressure due to hydrocarbon accumulation. P_a: pressure differential; P_w: density of water; P_h: density of the hydrocarbon; g: gravitational constant.

Pressure profiles

Having discussed the causes of overpressuring I would like now to discuss pressure profiles and their distribution in sedimentary basins. Pressures increase at a hydrostatic rate in sedimentary basins. If the right conditions are met, an effective seal to fluid movement is formed and pressure increases rapidly, sometimes even to the fracture point where internal pore pressure is equal to the net overburden plus strength of the overlying rock. This can occur at gradients of up to 2.3 times

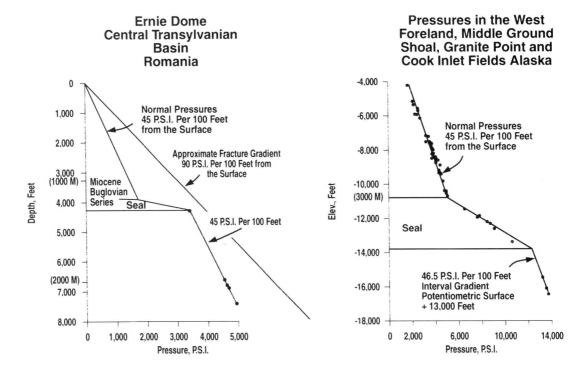

Fig. 6. Pressure/depth gradients, Ernie Dome, Central Transylvania Basin, Romania and Cook Inlet, Alaska from Powley (1984).

the hydrostatic gradient. At this point, pressure again begins to increase with depth at a hydrostatic rate.

Figure 6 shows a pressure profile from the Central Transylvania basin, with a normal pressure gradient down to a depth of something in the order of 1200 m, a very rapid increase over perhaps 100–200 m depth and then resumption of normal pressure gradients. It is interesting to note that this happens at a very shallow depth. This is a basin of a very high rate of deposition with up to several thousand metres of Miocene to Recent sediments and high geothermal gradients (Powley 1984). Another example shown is in the Cook Inlet fields offshore of Alaska where normal pressure gradients are found to be at 3000 m. A seal is formed with a much higher pressure gradient over a depth of *c*. 1000 m (3300 ft) and the normal pressure gradient is resumed below 14 000 ft or about 4000 m.

Norwegian Central Graben

Pressure curves were plotted for wells in the Norwegian Central Graben utilizing drill-stem tests, repeat formation tests, sonic and density logs, and mud weights. In specific instances where wells took a kick and a specific mud weight was needed to stabilize, the mud weight was felt to be an accurate indicator of the sub-surface pressures. When plotted (Fig. 7), the profiles fit into two separate curves: a two-compartment gradient separated by a single seal on the northern flank of the basin and a three-compartment gradient in the central and deepest portion of the study area. The shallowest compartment and the second or middle compartment demonstrate hydrostatic gradients within the compartments and are apparently in pressure communication throughout the basin. The deep pressure readings in the central part of the basin show a number of smaller sealed compartments. This irregular pressure distribution is due to continued burial in each one of these compartments and heating of the water and/or thermal cracking of oil to gas, resulting in a rapid increase in pressure gradients until the pressure reaches the fracture point. At that point in time, fracturing occurs, pressure is bled off and reduced, then it begins to be built up again. What is seen in a scattering of

those points is a number of compartments in these different stages of building up of pressure, fracturing and bleeding off and then building up the pressures again.

The cross-section (Fig. 8) shows that the pressure boundaries do not strictly follow depth or formation. The boundary between the normal pressure zone at the surface and the first overpressure compartment occurs from approximately 1500 m to 2000 m on the northern flank. One of the reasons why the boundary appears to go down to the north is a much lower geothermal gradient; 50% lower in the northern area than it is in the central area (Leonard 1984). This base of the transition zone can be seen on seismic as it forms the upper boundary of the gas clouds that are found over many of the giant fields in the basin. The boundary between the middle pressure compartment and the deep, high pressure compartment also varies with depth dropping down below 4000 m to the north, or 13 000–14 000 ft, and usually occurring at a depth of somewhere in the order of 11 000 ft or 3500 m through most of the basin. Interestingly enough, this boundary does not follow lithology. It is found in the Upper Jurassic shales in some areas, Lower Cretaceous shales in other areas, and the Upper Cretaceous chalk in other areas. One factor that is consistent throughout the basin is that the largest hydrocarbon accumulations are found in the first reservoir overlying or up-dip of the transition zone between the second and third pressure compartments.

Utilizing the well data, a three-dimensional model of the pressure compartments throughout the basin was constructed. The top of the transition zone was mapped, in one case between the normal pressure gradient in the shallow section and the moderately overpressured zone that extends from 1500 m in the centre of the graben to 2500 m on the northern flank. A map of the top of the transition zone overlying the deeper pressure compartment zone shows how the transition zone varies with formation and lithology throughout the basin.

The top of pressure compartment II within the Central Graben ranges from approximately 1500 m or 5000 ft to the centre of the graben down to about 2000 m or 7000 ft on the flank (Fig. 9). The boundary is not always found in the same formation. The seal is found in Miocene-aged sediments in the

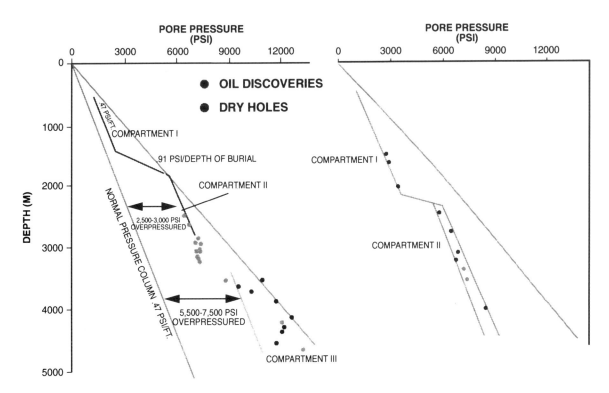

Fig. 7. Pressure/depth gradients, central and north flank of the Norwegian Central Graben.

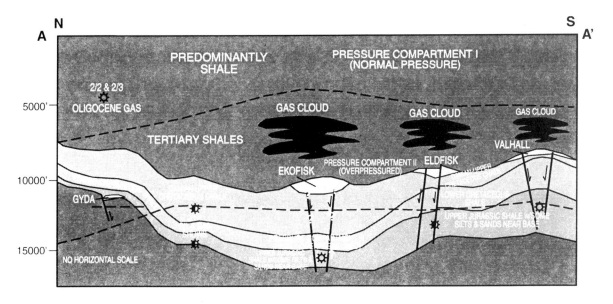

Fig. 8. North/south cross-section, Norwegian Central Graben with boundaries of pressure compartments.

centre of the graben and, to the north, in the base of the Miocene and then within the Oligocene sediments.

A map of the deeper pressure seals is shown in Fig. 10. The transition zone is plotted by lithologic age to show the variations in the area. The pressure seal is found in the Upper Cretaceous Chalk, Lower Cretaceous shale, Upper Jurassic shale and deeper formations. Often, good hydrocarbon shows are found directly beneath the seal (Well N2/1-1). On the northern and eastern sides of the graben, the seal moves beneath the Permian salt and eventually is not encountered. This probably has to do with the lower sub-surface tempera-

tures, slower depositional rate and little or no hydrocarbon generation on the flanks of the graben.

Applications of overpressure studies to exploration

The importance of overpressure in the exploration process can be demonstrated in many ways. From the drilling standpoint, prediction of overpressure is a major safety factor reducing the danger of well blowouts and allowing the operator to prepare adequately for a number of drilling hazards. From a reservoir standpoint, overpressuring affects net overburden, reducing

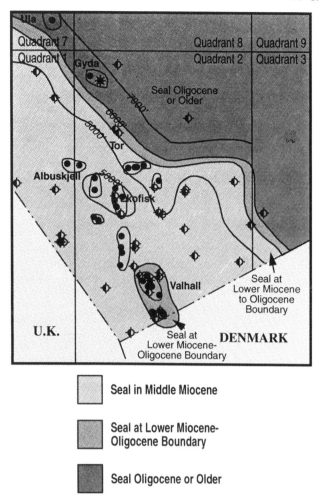

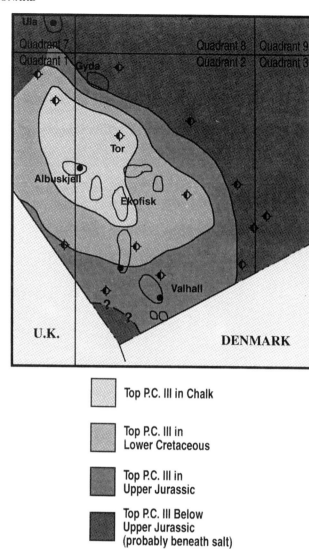

Seal in Middle Miocene

Seal at Lower Miocene-Oligocene Boundary

Seal Oligocene or Older

Fig. 9. Top of transition zone to pressure compartment II, Norwegian Central Graben.

Top P.C. III in Chalk

Top P.C. III in Lower Cretaceous

Top P.C. III in Upper Jurassic

Top P.C. III Below Upper Jurassic (probably beneath salt)

Fig. 10. Top of transition zone to pressure compartment III, Norwegian Central Graben.

porosity due to compaction. Lack of fluid movement can also reduce the supply of free ions associated with cementation and formation of the secondary porosity. The temperature increase experienced in overpressured systems decreases the depth of burial needed for hydrocarbon generation. Transition zone fracturing could be a key to primary and secondary hydrocarbon migration. Defining pressure variations from well to well can give indications of sealing and non-sealing faults.

Some aspects of overpressure compartment mapping are helpful in drilling safety as demonstrated by the pressure plots in the Central Graben. The transition zones (see Fig. 7) are zones of very rapid pressure increase, up to 3000 psi over a depth of perhaps 200 or 300 m. Predicting these zones is clearly a factor that would help in preventing well blowouts. Another factor from the drilling standpoint that these sort of pressure diagrams will help with is that in attempts to drill a deep well, to evaluate prospects in pressure compartment III for example, it is helpful to have an accurate indicator of where these pressure transition zones occur. This allows drilling into the upper pressure transition zone as far as possible, setting casing, increasing the mud weight, drilling through the second compartment into the deeper pressure transition zone, setting casing and then finally drilling into the deeper compartment. Setting casing significantly above the pressure transition zone would result in difficulty in keeping open the hole, given the pressures just above and below the transition zones. So in order to have the maximum hole size for the deep zone, it is very helpful to have a detailed prediction of the pressure distribution throughout the well.

Overpressuring can have an effect on hydrocarbon reservoirs in a number of ways. One example is that overpressuring can reduce the net overburden which results in less mechanical compaction at depth than in normally pressured areas.

An example of the effect of overpressuring on reservoirs is in the high porosity North Sea chalks. As shown on Fig. 11, chalks exhibit a fairly consistent porosity vs. depth relationship world-wide with porosities of perhaps 50–70% in the sea bottom sharply reducing to less than 10% at depths of 1500–2000 m (Scholle 1979). However, North Sea chalks, as shown in the chalks of the Tor Formation of the Valhal Field and Tor Field, have a much higher porosity than would be expected, in excess of 40% at depths of 2500 m and in excess of 20–25% at depths of 3000 m. The reason for this is the overpressuring that supports the grains and prevents compaction. If one removes the effect of overpressuring on the North Sea chalks, the porosities are what you would expect at a given depth of approximately 700–1200 m. The similar high porosities encountered in the chalk in the N2/8-2 well, which did not encounter hydrocarbons, demonstrate that it is overpressuring rather than the presence of hydrocarbons that is the single most important factor in preserving the porosity of the North Sea chalks.

Overpressuring also has major effects on hydrocarbon generation and expulsion. The increase in pressure at the transition zone is usually associated with a sharp increase in temperature gradient. This is not surprising as a seal would provide a thermal blanket for underlying sediments. The other effect is related to fracturing. There is a possible connection between the process of fracturing a seal and injecting fluids into the overlying beds due to overpressuring and microfracturing within the source rocks and expulsion of hydrocarbons.

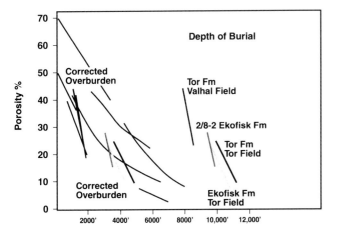

Fig. 11. Porosity/depth relationship of chalks around the world with Norwegian Central Graben examples to demonstrate the effect of overpressure. Adapted from Scholle (1979).

Figure 12 demonstrates an increase of 60% in the thermal gradient under the transition zones from the Block 16 field, offshore Vermillion, Louisiana. There are a number of wells that are used to support this analysis. A normal pressure gradient down to approximately 13 000 ft, or 4000 m is seen. The sharp transition zone occurs over the next approximately 400 m and then the normal pressure gradient is again seen below 4600 m or 15 000 ft. A fairly moderate geothermal gradient of 1.45°F/100 ft down to the transition zone is seen, underlain by a higher gradient of 2.2°F/100 ft. From the standpoint of predicting hydrocarbon generation, understanding the temperature increase with the sharp pressure increase,

in this case, will help predict the depth of hydrocarbon generation and thermal cracking to gas. The other effect, as mentioned before, is related to fracturing. As suggested by Tissot and Welte (1984), hydrocarbon expulsion from source rocks seems to be a pressure-driven discrete hydrocarbon phase movement. Pressure build-up, microfracturing and subsequent pressure release seem to be a major mechanism both in primary migration of hydrocarbons (see Fig. 4) and in forming and defining overpressure compartments. The hypothesis is suggested that the fracture points on the pressure profiles are also keys to hydrocarbon expulsion. Empirical observation that major hydrocarbon reservoirs are often in the first reservoir over the transition zone supports this idea. The same process of expelling hydrocarbons from rocks could be defining the boundaries of pressure compartments. It also seems clear that the most efficient point for expulsion of hydrocarbons is where a source rock generating hydrocarbons is near the border of these major transition zones. Therefore, the natural process in forming the pressure compartments could aid hydrocarbon expulsion.

A study of overpressuring also can have an effect on prediction of hydrocarbon entrapment. The key reason for this is that pressure seals form effective hydrocarbon seals. It can be imagined that if the seals are good enough to entrap water molecules, or at least slow movement for a molecule with a diameter of 3.2–3.5 Å, seals will be even more effective in stopping the movement of hydrocarbons where molecular size is 4–35 Å in diameter.

Figure 13 represents a summary of the distribution of pressures within the Norwegian Central Graben. The boundaries of the pressure seals should also be good hydrocarbon seals. The boundaries between the hydropressure, or the shallow normal pressure compartment, and the geopressure, or the first overpressure compartment zone, are bordered by a pressure seal that is defined by well data and can be seen on the seismic as it is the upper limit of the gas clouds that are found over many of the major oil fields. The deeper pressure compartments are defined by structural mapping of the distribution of the faults and by mapping of the pressures between the compartments. It should be recognized that these pressures seals are seals both from a vertical and from a lateral standpoint and these boundaries between the deep compartments

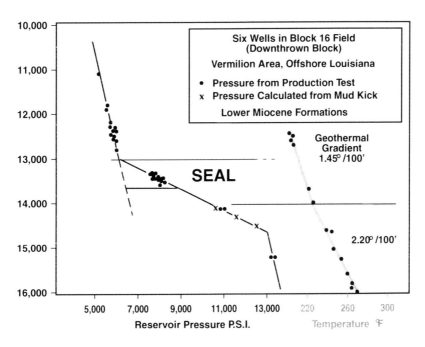

Fig. 12. Comparison of pressure and temperature gradients, Vermillion area, offshore Louisiana (Powley 1980).

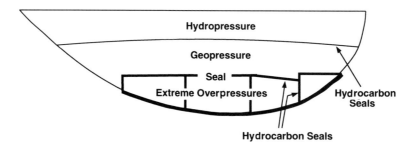

Fig. 13. Summary of pressure distribution in the Norwegian Central Graben.

can form their hydrocarbon traps even where there is not an obvious fault or structural closure to separate them. A weakness in depending upon these pressure seals for hydrocarbon entrapment is timing. By their nature, the pressure seals are probably transitory, as compared to, for example, a sand overlain by an evaporite which has consistently been a seal throughout geological time. Therefore areas of present-day hydrocarbon generation must be considered lower risk when depending upon the pressure seals as hydrocarbon seals than those areas where hydrocarbon generation is calculated to be in an earlier time.

Discussion

Based on the previous discussions, two general hypotheses about the exploration application of pressure systems can be advanced. The first is that the first reservoir above pressure boundaries is usually a major hydrocarbon reservoir. The reasons for this have been demonstrated previously and are related to hydrocarbon migration. In the pressure transition zone, when the fracture zone is reached, fracturing occurs and fluids are bled off into the overlying compartment until the pressure is stabilized. The pressures begin to build up again until the fracture point is reached again and hydrocarbons are bled off again into the shallower horizon. This is an excellent mechanism for movement of hydrocarbons, as well as water, upward within the system. The second hypothesis is that exploration in deep overpressured compartments adds an exploration risk. This is due to several factors. First of all there is difficulty in migrating hydrocarbons into a super-pressure compartment. The second reason would be that the fracturing into the overlying transition zone provides a potential leak. The third possible reason is that often super-high pressure zones are that way because of the lack of a continuous reservoir. So from a source, seal and reservoir standpoint, some added risks are found in the deep overpressured zones.

Reviewing again the cross-section through the Norwegian Central Graben (Fig. 8), these two hypotheses can be applied in this exploration area. The major hydrocarbon fields in the centre of the graben are found in the Upper Cretaceous/ Danian chalk reservoir shown in Ekofisk, Eldfisk and Valhall, the three largest fields. On the north flank of the graben, a series of Jurassic fields (on the left side of the cross-section) such as the Gyda or Ula Field are found. In each case, they are the first hydrocarbon reservoirs above the pressure transition zone. Exploration in the deep pressure compartment has had a much lower rate of success. Hydrocarbon shows are found in virtually every deep well but until very recently, there were no commercial discoveries.

However, exploration throughout the world is taking place in deeper zones and deeper pressure compartments. There has been some success despite the high risk. This includes the Central Graben where several deep discoveries have been made. There are at least three factors to consider when looking at deep pressure systems that can, to some extent, improve

chances of success. Firstly, the further the reservoir is away from the overlying transition zone, the more it reduces the chance of significant vertical leakage when the fracture point is reached and the pressure is temporarily bled off into the overlying zone. Secondly, careful mapping of boundaries of the deep compartments can give an indication of how much hydrocarbons could have been generated within the compartment. In the Central Graben, one problem is that there are fairly small compartments for many of these high pressure zones. If a large contiguous compartment exists, even if hydrocarbons cannot move from outside the compartment into the compartment, perhaps the compartment is big enough to have sufficient hydrocarbons generated to produce a commercial accumulation. Finally, the timing of fracturing and leakage from the compartment is important. While the pressure builds and gradually reaches the maximum pressure, if hydrocarbons are being generated, there is the potential for a commercial accumulation. When fracturing occurs, hydrocarbons are bled off into the overlying compartment. So this risk must be considered something of a wild card when drilling into deeper hydrocarbon compartments.

In Fig. 7 the three compartment systems in the Central Graben have been demonstrated and, while the major reserves are in pressure compartment II, recent exploration efforts have demonstrated several discoveries in the deep, high pressure compartment III. While these are a good deal smaller than in pressure compartment II fields, they are perhaps large enough to be commercial. Figure 14 demonstrates a fault trap indicating some prospects that may be more prospective than others within compartment III. The high standing fault block just below the pressure boundary is at considerable risk because when fracturing occurs, and fluids will bleed off from pressure compartment III to II, this fault block will clearly lose its seal. The closure on the downthrown side of the fault block is much better situated from that standpoint in that it is probably sufficiently removed from the transition zone that fracturing will not deprive it of its seal and also, as a more gentle closure with a large area of catchment for hydrocarbon, it is also perhaps better situated from the hydrocarbon migration standpoint.

Conclusions

1. Overpressures can be caused by a number of factors. None of these can be considered mutually exclusive. When attempting to identify the causes of overpressures in basins, each one of these factors must be separately studied rather than just assuming that the overpressure is caused by just one mechanism. The Norwegian Central Graben contains the correct conditions for every factor (thermal gradient, amount of sediment, sediment burial rate, hydrocarbon generation and thermal cracking to gas and hydrocarbon accumulations), resulting in the highly overpressured, three-tiered system.

Centre of Central Graben

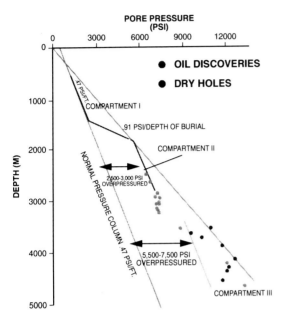

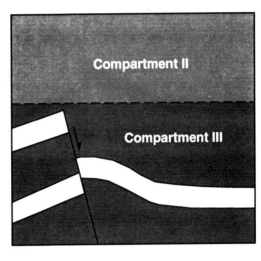

Fig. 14. Evaluation of effect of pressure transition zones and pressure compartments on deep prospects.

2. Sub-surface pressures are characterized by normal pressure gradients and sharp transition zones. This has major effects from the standpoint of drilling safety particularly in that these transition zones can often have an increase in pressure of several thousand psi within a fairly short zone—less than 1000 ft or 300 m.
3. Overpressure analysis, particularly with regard to seals, reservoirs, hydrocarbon generation and migration, should be an integral part of an exploration programme in any sedimentary basin. The distribution of hydrocarbon accumulations, anomalously high porosity of the chalk reservoirs and depth of hydrocarbon generation and thermal cracking are closely related to the pressure compartments of the Norwegian Central Graben.

References

BARKER, C. 1990. Calculated volume and pressure changes during the thermal cracking of oil to gas in reservoirs. *American Association of Petroleum Geologists Bulletin*, **74**, 1254–1261.

BRADLEY, J. S. 1975. Abnormal formation pressure. *American Association of Petroleum Geologists Bulletin*, **59**, 957–973.

GAARENSTROOM, L., TROMP, R. A. J., DE JONG, M. C. AND BRANDENBURG, A. M. 1993. Overpressure in the Central North Sea: implications for trap integrity and drilling safety. *In*: PARKER, J. R. (ed.) *Petroleum Geology of Northwest Europe: Proceedings of the 4th Conference.* Geological Society, London, 1305–1313.

HUBBERT, M. K. 1940. The theory of groundwater motion. *Journal of Geology*, **48**, 785–944.

—— 1953. Entrapment of petroleum under hydrodynamic conditions. *American Association of Petroleum Geologists Bulletin*, **37**, 1954–2026.

LEONARD, R. C. 1984. Generation and migration of hydrocarbons on southern Norwegian Shelf. *American Association of Petroleum Geologists Bulletin*, **68**, 796.

MAGARA, K. 1974. Generation and migration of hydrocarbons on southern Norwegian Shelf. *American Association of Petroleum Geologists Bulletin*, **68**, 796.

—— 1975. Re-evaluation of montmorillonite dehydration as cause of abnormal pressure and hydrocarbon migration. *American Association of Petroleum Geologists Bulletin*, **59**, 292–302.

—— 1978. *Compaction and Fluid Migration.* Elsevier, New York.

POWLEY, D. E. 1980. 'Pressures, normal and abnormal'. American Association of Petroleum Geologists, Advanced Exploration Schools Unpublished Lecture Notes.

—— 1984. 'Pressures, normal and abnormal'. American Association of Petroleum Geologists, Advanced Exploration Schools Unpublished Lecture Notes.

SCHOLLE, P. 1979. Oil from chalk: a modern miracle? *American Association of Petroleum Geologists Bulletin*, **63**.

TISSOT, B. P. AND WELTE, D. H. 1984. *Petroleum Formation and Occurrence* (2nd edn). Springer-Verlag, New York.

UNGERER, P., BESSIS, D., CHENET, P. Y., DURAND, B. NOGARET, E., CHIANELLI, A., OUDIN, J. L. AND PERRIN, J. F. 1984. Geological and geochemical models in oil exploration: principles and practical examples. *In*: DEMAISON, G. AND MURRIS, R. J. (eds) *Petroleum Geochemistry and Basin Evaluation*, American Association of Petroleum Geologists, Memoir **35**, p. 53–77.

Overpressures in the Central North Sea: implications for trap integrity and drilling safety

L. GAARENSTROOM,[1] R. A. J. TROMP,[1] M. C. de JONG[2] and A. M. BRANDENBURG[2]

[1] *Shell UK Exploration and Production, Shell-Mex House, Strand, London WC2R 0DX, UK*
[2] *Shell research (Kon/Shell Exploratie en Produktie laboratorium), Volmerlaan 6, 2288GD Rijswijk ZH, The Netherlands*

Abstract: The Central North Sea Graben is, at pre-Cretaceous levels, a relatively under-explored province of the United Kingdom Continental Shelf. The hydrocarbon potential was recognized in the early 1970s, but both technology and economics prohibited any serious exploration efforts at that time. Exploration for deep targets increased significantly during the second half of the 1980s. This resulted in the discovery and appraisal of several fields, such as the moderately overpressured Marnock and Skua fields and the highly overpressured Erskine, Puffin, Franklin and Heron fields.

From a regional viewpoint, the overpressures observed in the pre-Cretaceous reservoirs in the Central North Sea Graben follow a well-defined trend, increasing with depth. Similarly, the minimum effective stresses derived from leak-off pressures measured in the overlying seals follow a comparable trend related to the present-day depth of burial. This indicates that the present distribution of overpressures is controlled by retention and that the understanding of overpressure generation is of second-order importance.

A more detailed analysis combining the areal distribution of overpressures with the structural framework suggests that the Central North Sea Graben can be subdivided into several pressure cells. The boundaries between the cells are controlled by faults which, in many instances, show signs of repeated reactivation. This reactivation has led in some cases to failure of otherwise well-defined structural traps, whereas in other cases it may have enhanced the trapping integrity. Examples of intermediate cases can also be recognized, which clearly show that the pressure cells form part of a dynamic system, controlled primarily by the structural evolution of the Central North Sea.

A sound understanding of the structural framework and its evolution is therefore of paramount importance in predicting trap integrity and overpressure regimes prior to drilling. The drilling of deep high-pressure/high-temperature wells and the acquisition of modern 3D seismic data permit a continuous improvement in the understanding of overpressures and their distribution.

The geographical definition of the Central Graben is the area of the United Kingdom Continental Shelf between 56° and 58°N and 0° and 3° E. The structure and stratigraphy are summarized in Fig. 1.

The Central North Sea Graben forms part of the North Sea rift system which can be considered as a typical example of linked, elongated grabens. The present-day configuration of the Central Graben is the result of the general polyphase tectonic evolution of NW Europe (Ziegler 1990). The tectonic grain is defined by dominating NW–SE and E–W Hercynian fault trends and subordinate NE–SW Caledonian lineaments.

Hydrocarbon exploration of deep, overpressured, pre-Cretaceous targets has proved the Central North Sea Graben to be a single, large-scale dynamic pressure cell. The top seal is formed by the shales and marls of the Kimmeridge Clay Formation and Cromer Knoll Group; the evaporites and shales from the Zechstein Formation form the seat seal.

In this paper, attention will be focused on the retention aspects of overpressure and the implications for trap integrity in the highly pressured pre-Cretaceous objective sequences. A good understanding of these aspects enables prediction of the (over)pressures in exploration wells. Before addressing retention aspects, potential causes or mechanisms of overpressure generation in the Central North Sea Graben will be briefly reviewed.

Overpressure: causes

Overpressure, or geopressure, is defined as the amount of pore pressure exceeding the hydrostatic pressure. The litho-static pressure or vertical stress is defined as the pressure imposed on the rock by the weight of the overlying lithological column. These and other terms mentioned in this paper are illustrated in Fig. 2.

Several mechanisms for the generation of overpressures have been described in the literature. Although in most situations more than one mechanism is involved, disequilibrium compaction, kerogen transformation and oil cracking are generally considered to be the main processes. In this section the various mechanisms will be briefly described.

Disequilibrium compaction (or undercompaction) Rapid burial of sediments which does not allow pore fluids to dissipate during compaction, is generally accepted as the main overpressure generator in many areas (e.g. Mann and Mackenzie 1990; Waples 1991). Although mostly applicable in deltaic settings such as the Gulf Coast and Nigeria, disequilibrium compaction has also been described as the main pressurizing process in overall normally compacted sediments in the Ventura Basin, offshore Nova Scotia (Mudford and Best 1989).

Kerogen transformation The transformation of kerogen into hydrocarbons, creating an increase in volume, is seen as a principal generator of overpressure in source rocks (Stainforth 1984).

Oil cracking Increased temperatures result in the cracking of oil into gas. This process can take place either in the source rock or in the reservoir, and various authors (Barker 1990; Spencer, 1987; Caillet *et al.* 1991) claim this is one of the principal generators of overpressures. Barker (1990) describes how this process occurs at depths between 12 000 and 17 000 ft.

From *Petroleum Geology of Northwest Europe: Proceedings of the 4th Conference* (edited by J. R. Parker).
© 1993 Petroleum Geology '86 Ltd. Published by The Geological Society, London, pp. 1305–1313.

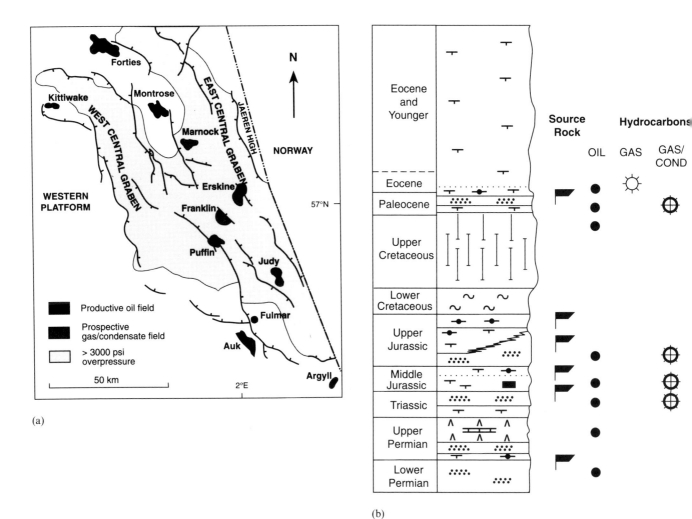

(a)

(b)

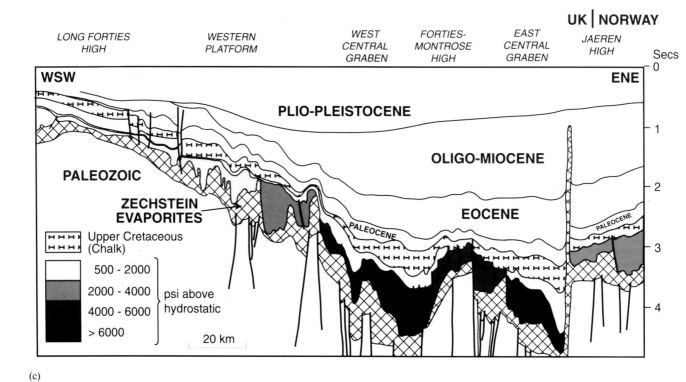

(c)

Fig. 1. (**a**) Location of the area under study and major structural elements; (**b**) generalized stratigraphy of the Central North Sea; (**c**) line drawing of a seismic section across the Central North Sea (from Glennie 1990).

As this process is temperature controlled, it will happen at different depths in different basins. In a sealed system the cracking of only 1% of oil would be sufficient to increase the pore pressure from hydrostatic to above lithostatic. Consequently, in an open system, a large proportion of the gas generated (up to 75%) is likely to escape.

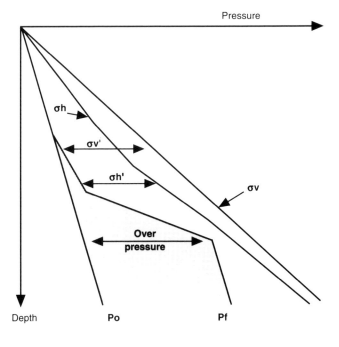

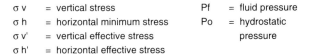

σ v	= vertical stress	Pf	= fluid pressure
σ h	= horizontal minimum stress	Po	= hydrostatic
σ v'	= vertical effective stress		pressure
σ h'	= horizontal effective stress		

Fig. 2. Pressure and stress nomenclature.

Diagenesis The transformation of smectite into illite releases water into the pores of a sediment, thereby increasing the pore pressure. This process takes place at depths between 6000–9000 ft, which in some areas coincides with the onset of overpressures. However, Jansa and Urrea (1990) describe the onset of overpressures some 7000–8000 ft below the smectite–illite transformation in the Ventura Basin, offshore Nova Scotia. Foster and Custard (1980) concluded that the transformation increases the pore pressures due to a decrease in permeability.

A different process of diagenesis was suggested by Hunt (1990) who described the tops of many compartment seals in clastic rocks as appearing to consist of multiple horizontal bands of calcite mineralization forming along a thermocline; the Central North Sea is cited as an example.

Aquathermal processes Pore fluid expansion caused by increased temperature is generally seen as a contributor to overpressure (Barker 1972). Whether this contribution is significant is a matter of debate. Chapman (1980) and Daines (1982) for instance state that high overpressures encountered are very likely to be caused by the very small volume increase resulting from thermal expansion. Several authors argue otherwise. The process is a function of porosity and permeability, which are difficult to measure in shales or claystones.

Inversion tectonics (or overcompaction) This involves the uplift of fault blocks which retain the pore pressure generated

at greater depths. This mechanism may explain the occurrence of higher than expected pore pressures at relatively shallow depths.

The relative importance of these processes is still poorly understood. It is important to realize that conditions exist in the Central Graben for most of these mechanisms. Attempts to scale the processes with computer simulations have so far been severely hampered by the uncertainty and subsequent gross simplifications in boundary conditions. The simple models, however, do suggest that a single process such as transformation from oil to gas will generate sufficient pressure to rupture any seal.

Overpressure: observations

One approach to the study of the effects of overpressure is the analysis of drilling data. The following section lists the data items of importance.

(1) Formation strength estimated from leak-off tests (LOT)

The LOT is performed after drilling out of the casing shoe with the dual objective of establishing the quality of the cement bond and to estimate the strength of the formation in which the casing was set. Procedures differ from operator to operator but the principle is to construct a graph of cumulative volume of mud pumped versus shut-in pressure measured at the stand pipe (Fig. 3). Leak-off is detectable as soon as the relation starts to deviate from the straight line trend, indicating that fractures are being induced in the formation. The pressure at which the formation starts taking fluid (leak-off pressure or LOP) may range from the fracture closure pressure (FCP) to the formation breakdown pressure (FBP).

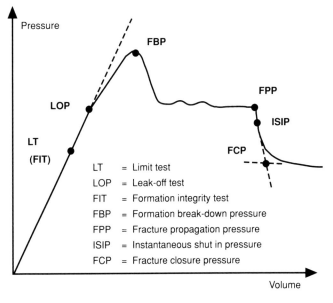

Fig. 3. Nomenclature in casing shoe integrity testing.

A plot of LOP data in the Central North Sea shows a good first-order approximation with depth (Fig. 4). The scatter in the data can be attributed to several causes, such as: inaccuracies of the method, hole geometry, lithology changes, and variations of in situ stresses. Furthermore, a poor cement bond at the casing shoe may result in an underestimation of the formation strength.

The LOP will generally be higher or equal to the minimum in situ stress, in which case the lower envelope of all LOP data can be used to estimate the minimum horizontal stress as a function of depth (Fig. 4). Physically it corresponds to the reopening of previously formed vertical fractures with zero tensile strength. This relationship can be used as a first order approximation of formation strength in the Central Graben.

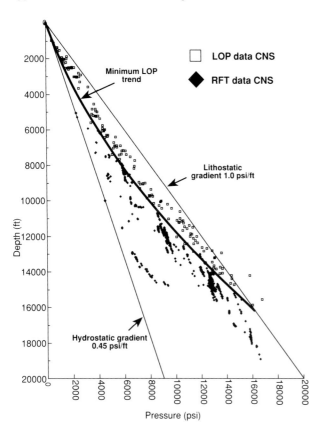

Fig. 4. Plot of fluid pressures (RFT) and leak-off pressures (LOP) as function of depth for the Central North Sea Graben. The trend of minimum horizontal stress shows the clear separation between RFT and LOP data. Approximate hydrostatic and lithostatic trends (0.45 and 1.0 psi ft^{-1} respectively) are shown for reference.

(2) Measurements of fluid pressures

Fluid, or pore pressure is usually measured with wireline tools, such as the repeat formation tester (RFT) or formation micro tester (FMT) tools. Although these tools can provide accurate pressure measurements, their quality can be affected by several factors, such as temperature, seal quality and permeability of the formation.

Reservoir fluid pressure data has been compiled for the area under review. Again, as with the LOT-derived pressures, a depth-related trend can be observed (Fig. 4). However, the scatter in the fluid pressure data (subsequently referred to as the RFT data) is greater than that for the leak-off pressures.

(3) Indirect pressure measurements from drilling parameters

In the absence of LOT or RFT data the only information on formation strength or pressure is empirically derived from a critical analysis of drilling parameters. The key parameter is the combination of mudweight, mudbalance and rate of penetration (ROP). An operator will attempt to optimize the ROP without compromising safety. This means that the margin between mudweight and formation strength/pressure will be kept as low as possible. In many of the high pressure wells drilled in the Central North Sea this is reflected in intermittent

recording of gains and losses while drilling, indicating that formation strength and pore pressure are narrowly balanced (low minimum effective stresses) by the equivalent mudweight used.

Additional information on overpressures may be derived from logging data (e.g. sonic and resistivity logs) or seismic data. This has been successfully demonstrated in the Gulf Coast, but appears less applicable in the Central North Sea Graben where undercompaction is less prevalent.

Seal strength and fluid pressure

The combined plot of fluid pressure and leak-off pressure (Fig. 4) shows that a 'minimum LOP-line' can be drawn, below which nearly all the fluid pressures lie. It is clear that on a prospect-by-prospect basis, the formation strength of the seal has to be greater than the fluid pressure.

The actual efficiency of the seal (the capillary resistance) will be at least partly a function of pressures in the reservoir. The sealing mechanism of any one prospect in the Central North Sea is likely to be a complicated combination of capillary and hydraulic properties (see, for instance, Watts 1987), although the latter is probably the most important in the Central Graben, given the extremely low permeabilities of the Kimmeridge Clay and Cromer Knoll resulting in a very high capillary entry pressure.

The plot of LOP data and RFT data provides empirical evidence that the present-day fluid pressures are limited by the local minimum horizontal stress. Furthermore, the separation between LOP and RFT pressures on a regional scale suggests that differences in the local minimum stress are minor and that minimum stress is mainly controlled by depth.

The relation between LOT and RFT pressures can be examined on a well-by-well basis. In this case the lower envelope of LOP data is used to derive an estimate of the formation strength of a seal at the depth at which a RFT measurement was obtained. Theoretically and in an ideal world, a difference between the two values (the horizontal effective stress, see Fig. 2) of 1 psi would seal a trap, but in practice the risk of breaching increases when the retention capacity is less than 1000 psi (Fig. 5).

Pressure cells and leak points

Observations

The Central North Sea Graben is a single large-scale dynamic pressure cell, where pressure generation locally exceeds the retention capacity. The top seal is formed by the shales and marls of the Kimmeridge Clay Formation and Cromer Knoll Group; the evaporites and shales from the Zechstein Formation form the seat seal. Variable leakage occurs at different points. An example of obvious leakage occurs towards the west and southwest, where the overpressures have been released from the deep graben into Upper Jurassic and younger sands of the Western Platform (e.g. Gannet fields, Cayley 1987).

Relating the distribution of overpressures to the Central North Sea Graben tectonic architecture (Bartholomew *et al.* 1993) shows that in some cases clear pressure cell boundaries can be drawn (Fig. 6). These boundaries are related to the existing fault pattern at basement levels. These basement faults have been reactivated several times.

The exact delineation of individual pressure cells is in most cases much more difficult. For example, wells drilled to pre-Cretaceous objectives in the Blocks 29/7, 29/8 area in the southwest margin of the Central North Sea Graben show a clear stepwise increase of overpressure with depth. Although the controlling faults or pressure 'valves' are difficult to pinpoint individually, Shell/Esso Well 29/7-1 appears to have

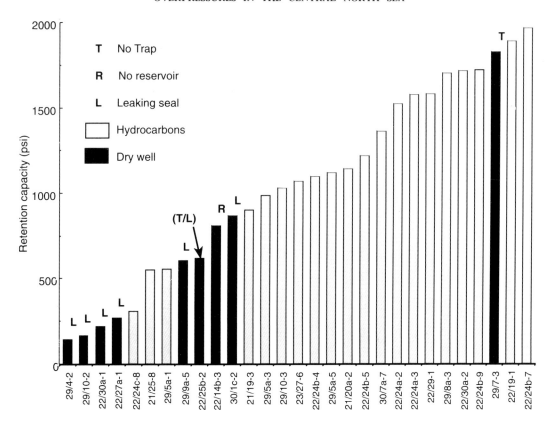

Fig. 5. Retention capacity of several traps drilled in the Central North Sea. The retention capacity is defined as the difference between the RFT pressure and the extrapolated LOT pressure at the shallowest level in the prospect, using the established minimum horizontal stress trend as shown in Fig. 4.

drilled the leak-point of the local pressure cell (Fig. 7). The well was drilled on a valid structure but did not encounter hydrocarbons at pre-Cretaceous levels. The hydrocarbons have leaked to Upper Cretaceous and Paleocene levels due to seal breaching caused by late Tertiary tectonic movements involving the Zechstein salt. Another well within the pressure cell (Shell/Esso Well 29/7-3) shows a stepwise increase in pressures between different pre-Cretaceous reservoir intervals. The pressure jumps observed between Upper Jurassic, Middle Jurassic and Triassic reservoirs are shown in Fig. 7. This is evidence for the argument that the minimum horizontal stress increases with depth and that deeper seals will be able to retain higher pressures.

Pressure cell concept

A schematic summary of possible relationships between adjacent pressure cells is given in Figs 8a–d. In the first case, the fault between the two structures is non-sealing and, consequently, aquifer pressures will fall on the same hydrostatic trend and any leakage will occur at the highest point in the cell. In the second case the fault acts, together with the leaking B structure, as a safety valve for the A structure. The pressure in the latter will be controlled by the dynamic pressure differential over the fault, e.g. the Blocks 29/7, 29/8 area described above. If the dynamic pressure differential between structures A and B is not sufficient, then both structures may breach, as shown in the third example. Assuming that pressure generation is virtually limitless in the Central North Sea Graben, this situation is likely to exist in most adjoining pressure cells. In the last case the fault is completely sealing and the deeper pressure cell A is in communication with a shallower leak point. Examples of this situation can be observed in Fig. 4, where the lower than expected RFT pressures (e.g. at 13 000–

15 000 ft) are found in wells situated close to the Central North Sea Graben margin, e.g. in the Fulmar Field area.

Timing of events

The relative timing of hydrocarbon and overpressure generation and tectonic activity critically affects trap integrity. A detailed understanding of these processes is therefore of crucial importance. Burial graphs for the Central Graben show that rapid burial took place during the Late Tertiary, triggering many pressure generation processes. Onset of overpressure generation will probably have started locally during the Upper Cretaceous but the rate of generation will have peaked during the Late Tertiary (Fig. 9). The cumulative amount of overpressure generated is larger than the retention capacity in the Central North Sea Graben. This will result in periodic releases of overpressures due to repeated hydraulic fracturing. It is therefore important to establish the (palaeo) stress pattern of any potential prospect in order to establish its hydrocarbon trapping integrity.

Several phases of tectonic activity can be observed in the Central North Sea Graben since the Triassic (Ziegler 1990). Some major tectonic events, such as the Mid- and Late Cimmerian, probably pre-date the establishment of the high overpressures. What is critical is the often subtle reactivation of older fault trends during the Early and Late Tertiary. These are likely to control the 'bleeding-off' of the pressure cells. A pressure history graph, such as the conceptual example in Fig. 9, may help to make a qualitative assessment of the retention risks for any prospect in the Central North Sea Graben.

The importance of the relationship between pressure cells and structure is well illustrated by Shell/Esso Well 29/10-2, which drilled a dip-closure at Paleocene and Cretaceous level

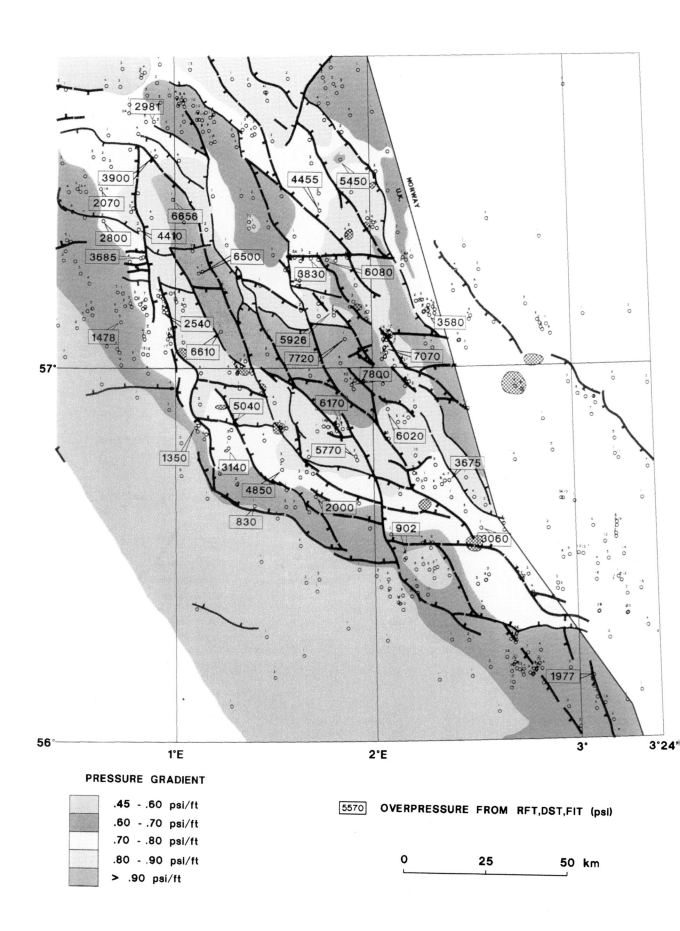

Fig. 6. Tectonic architecture of the Central North Sea and the distribution of overpressures (below Base Cretaceous).

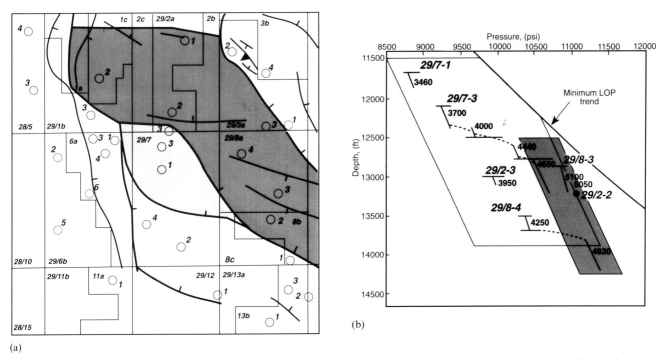

(a)

(b)

Fig. 7. Pressure cells in the Blocks 29/7, 29/8 area, Central North Sea: (**a**) situation map, showing the structural framework of the southwest corner of the Central Graben; (**b**) plot of fluid pressures versus depth, showing stepwise decrease of overpressure between two adjoining pressure cells as well as within the same pressure cell.

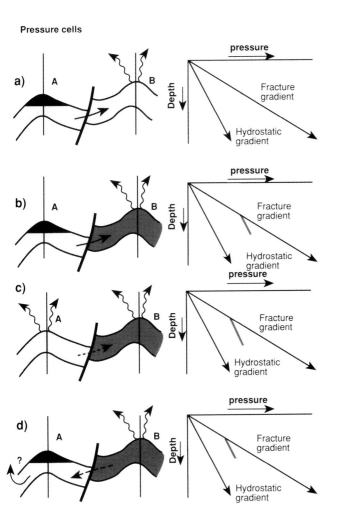

Pressure cells

and a well-defined fault-bounded closure at pre-Cretaceous level. The seismic section (Fig. 10) shows a very clear example of a polyphase inversion structure. The present-day structuration is controlled by the underlying, triangular-shaped Rotliegend basement block. Repeated major and minor activity along the two major fault trends (NNW–SSE and WNW–ENE) has resulted in: (1) a thick accumulation of Fulmar sands in a pull-apart setting during Middle and Upper Jurassic times ('Mid- to Late Cimmerian'); (2) creation of a trap at both pre- and post-Cretaceous levels as the result of the combined effects of Sub-Hercynian and Laramide transpression and uplift; (3) fracturing of the top seal due to continued minor tectonic activity enhanced by halokinesis during the Late Tertiary, causing the prominent pre-Cretaceous trap to leak.

Safety aspects

Several incidents in the North Sea during the latter part of the 1980s prompted a review of practices and procedures in high-

Fig. 8. Pressure cells—possible relationships between adjoining pressure cells. (**a**) No pressure jump over the fault (both cells are in communication); leakage will be controlled by the highest point in cell B where the seal strength is expected to be lowest. (**b**) Dynamic pressure jump over the fault; the pressure in cell A will be controlled by the leak-off in cell B. This prevents a build-up of pressure in cell A to the fracture strength. (**c**) Leakage across the fault is small; the pressure in cell A can build up to the fracture strength of the seal and help maintain the high pressure in cell B. (**d**) Fault is completely sealing; cell A is in pressure communication with a cell with shallower leak-off point. Pressure in cell B is maintained at the fracture gradient.

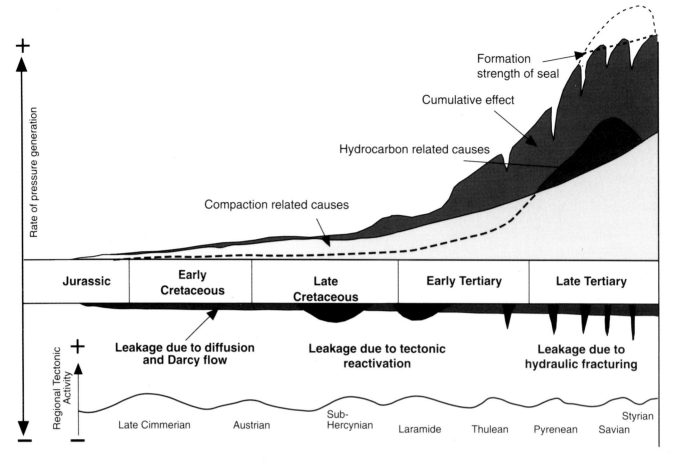

Fig. 9. Rate of overpressure generation. This conceptual sketch shows the relative timing of events that control the overpressure in a hypothetical trap in the Central North Sea Graben.

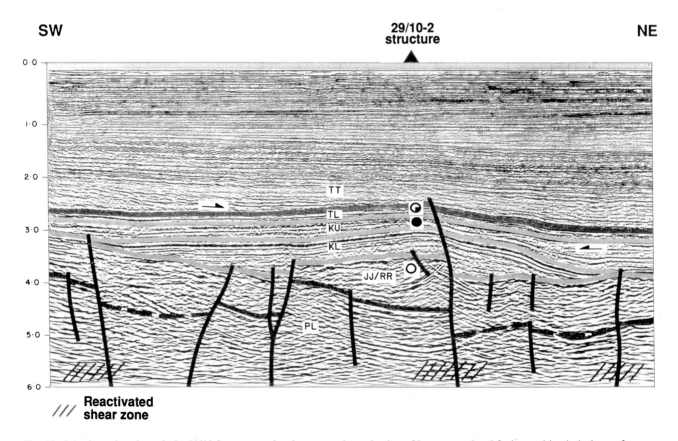

Fig. 10. Seismic section through the 29/10-2 structure, showing repeated reactivation of basement-related faults resulting in leakage of an otherwise well-defined prospect.

pressure drilling. Initially, this resulted in a total risk aversion approach, both in well design as in drilling. As a consequence, wells were 'over-designed' and drilled according to worst-case scenarios. The high mudweights used had a detrimental effect on ROP; this again had an adverse effect on overall drilling performance and cost.

The explorationists were, therefore, urged to develop a well-balanced risk assessment of the likely pressures when describing their prospects. This assessment is still largely based on empirical relations derived from observations and assumptions and must be complemented by a comprehensive sedimentological review to assess the likelihood of vertically stacked pressure cells, often requiring contingent casings to be planned for. The structural interpretation of 3D seismic data is judged to be a prerequisite to the evaluation of prospects where reservoir pressures can approach the formation strength of the overlying seal.

These work practices, coupled with effective feedback of experience obtained drilling these high-pressure wells, can be translated into proper risk management.

Conclusions

A close evaluation of the controls on overpressure in the Central North Sea has revealed the following:

(1) the current distribution of overpressures in the pre-Cretaceous sequences is controlled by differential retention. Potential mechanisms for overpressure generation are plentiful but the contribution of each individual process is difficult to quantify;

(2) the retention capacity is defined by the minimum horizontal stress, which on a regional scale is mainly controlled by depth. The minimum horizontal stress, therefore, controls the maximum pore pressure. This information can be used in pore pressure prediction;

(3) a good understanding of the structural framework is essential in defining likely outlines of individual pressure cells;

(4) in several pressure cells, where formation strength is closely matched by fluid pressure, seal breaching is observed and otherwise well-defined traps have not retained any hydrocarbon column;

(5) trapping integrity will be a function of the local stress pattern and degree and timing of faulting as well as the relative position of the trap with respect to the leak point in the corresponding pressure cell.

The authors would like to acknowledge the invaluable experience base on the high-pressure play created by numerous past and present colleagues in Shell Expro. We have especially benefitted from the work carried out by Mario Nold, Uli Seemann, Jurg Feurer and Maarten Wiemer. Manfred Epting and Gordon Taylor critically reviewed this paper.

The authors wish to thank Shell Research BV., Shell UK Exploration and Production and Esso Exploration and Production UK Limited for permission to publish this paper.

References

BARKER, C. 1972. Aquathermal pressuring-role of temperature in development of abnormal pressure zones. *American Association of Petroleum Geologists Bulletin*, **56**, 2068–2071.

—— 1990. Calculated volume and pressure changes during the thermal cracking of oil to gas in reservoirs. *American Association of Petroleum Geologists Bulletin*, **74**, 1254–1261.

BARTHOLOMEW, I., PETERS, J. M. AND POWELL, C. M. 1993. Regional structural evolution of the North Sea: oblique slip and the reactivation of basement lineaments. *In*: PARKER, J. R. (ed.) *Petroleum Geology of Northwest Europe: Proceedings of the 4th Conference*. Geological Society, London, 1109–1122.

CAILLET, G., SEJOURNE, C., GRAULS, D. AND ARNAUD, J. 1991. Hydrodynamics of the Snorre Field area, offshore Norway. *Terra Nova*, **3**, 180–194.

CAYLEY, G. T. 1987. Hydrocarbon migration in the Central North Sea. *In*: BROOKS, J. AND GLENNIE, K. W. (eds) *Petroleum Geology of North West Europe*. Graham & Trotman, London, 549–555.

CHAPMAN, R. E. 1980. Mechanical versus thermal cause of abnormally high pore pressures in shales. *American Association of Petroleum Geologists Bulletin*, **64**, 2179–2183.

DAINES, S. R. 1982. Aquathermal pressuring and geopressure evaluation. *American Association of Petroleum Geologists Bulletin*, **66**, 931–939.

FOSTER, W. R. AND CUSTARD, H. C. 1980. Smectite–illite transformation—role in generating and maintaining geopressure. *American Association of Petroleum Geologists Bulletin*, **64**, 708.

GLENNIE, K. W. (ed.) 1990. *Introduction to the Petroleum Geology of the North Sea*. Blackwell, Oxford, 34–77.

HUNT, J. M. 1990. Generation and migration of petroleum from abnormally pressured fluid compartments. *American Association of Petroleum Geologists Bulletin*, **74**, 1–12.

JANSA, L. F. AND URREA, V. H. N. 1990. Geology and diagenetic history of overpressured sandstone reservoirs, Ventura Gas Field, Offshore Canada. *American Association of Petroleum Geologists Bulletin*, **74**, 1640–1658.

MANN, D. M. AND MACKENZIE, A. S. 1990. Prediction of pore fluid pressures in sedimentary basins. *Marine and Petroleum Geology*, **7**, 55–65.

MUDFORD, B. S. AND BEST, M. E. 1989. Ventura Gas Field, Offshore Nova Scotia: case study of overpressuring in region of low sedimentation rate. *American Association of Petroleum Geologists Bulletin*, **73**, 1383–1396.

SPENCER, C. W. 1987. Hydrocarbon generation as a mechanism for overpressuring in the Rocky Mountain region. *American Association of Petroleum Geologists Bulletin*, **71**, 368–388.

STAINFORTH, J. G. 1984. Gippsland hydrocarbons—a perspective from the basin edge. *APEA Journal*, **24** (1), 91–99.

WAPLES, D. W. 1991. Generation and migration of petroleum from abnormally pressured fluid compartments. *American Association of Petroleum Geologists Bulletin*, **75**, 326–327.

WATTS, N. L. 1987. Theoretical aspects of cap-rock and fault seals for single- and two-phase hydrocarbon columns. *Marine and Petroleum Geology*, **4**, 274–307.

ZIEGLER, P. A. 1990. *Geological Atlas of Western and Central Europe* (2nd edn). Shell Internationale Petroleum Maatschappij B.V., The Hague.

Deep prospects—imaging, overpressure and associated problems in the UK Central North Sea

J. A. HENDERSON and M. J. N. PEPPER

Ranger Oil (UK) Limited, Ranger House, Walnut Tree Close, Guildford, Surrey GU1 4US, UK

Despite the depth of Mesozoic reservoirs, the extreme temperatures and pressures encountered and the high drilling costs, the High Pressure/High Temperature (HPHT) play in the deepest part of the Central Graben remains one of the most attractive in the North Sea.

Seismic imaging

In the deep Central Graben HPHT play there is a requirement for reliable data on seismic sections at 5 to 6 seconds or greater. It is shown, using a hypothetical single layer model, that at these depths theoretical migration distances can approach 10–15 km where structural dips of 50–60° are present. It is also shown that migration distances predicted by this single layer model closely approximate to those predicted using a realistic Central Graben multilayer model. The single layer model is, therefore, a useful predictive tool for the plannng of seismic surveys providing that salt piercement tectonics are not involved.

Errors in the positioning of seismic events introduced by inaccuracy in the applied migration velocities are potentially large. They can lead to large errors in depth conversion. In addition these errors can be of sufficient magnitude that wells targeted at deep prospects may miss their objective completely.

The structural complexity of the HPHT play means that simple concepts of dip and strike shooting directions are not valid. Large migration misties on 2D data can render the subsequent interpretation meaningless. 3D exploration surveys are the solution, but they can be very expensive and time consuming to acquire and process.

Incorrect interpretation of the Lower Cretaceous may arise because of a sparse well database which has predominantly sampled this interval where it is thin. Limits in the preserved seismic bandwidth may lead to an incorrect identification of the Top Lower Cretaceous. This, combined with a poor understanding of the rock velocities, may lead to significant errors in depth conversion.

Pore pressures

When preparing the exploration well prognosis for a deep HPHT prospect an understanding of compaction and burial history of the sediment pile is critical. As a sediment is buried the rock matrix compacts and porosity is reduced. This process is described for shales in the Central Graben.

The inability of a rock to dewater and reduce its porosity when buried gives rise to overpressuring. In this case part of the overburden pressure is supported by the pore fluids. The rate at which fluids move through a sediment at shallow depths is controlled by permeability, which is logarithmically proportional to porosity. A shale may be described as consolidated when it has permeabilities below 10^{-7} Darcy.

Chemical reactions that occur as a sediment is buried may produce new pore fluids. The restructuring of clay particles will liberate structured water into pore space. Additionally kerogen will restructure, liberating chain and non-planar hydrocarbons. Over the depths at which these reactions occur, permeabilities in the sediment are normally adequate to bleed off the produced pore fluids.

This is not the case at depth when kerogens restructure producing predominantly gas. At these depths shale porosities and permeabilities are so low that produced pore fluids are not dispersed. The shale pore pressure and sediment pore pressure therefore increase to levels that can equal geostatic pressure.

As these pore pressures increase, the sediment porosity increases. This dilation may be seen as a reduction in rock density. When pore pressure is equal to geostatic pressure the sediment may flow like a viscous fluid.

It is important to establish the time at which this reaction occurs. Pore size and pressure can be expected to reduce in time given a constant depth of burial and temperature. Mapping of isopachs and isochrons provides important clues to the evolution of an HPHT prospect. Their use to construct burial history for the sediment also helps to establish whether residual overpressure exists.

From *Petroleum Geology of Northwest Europe: Proceedings of the 4th Conference* (edited by J. R. Parker).
© 1993 Petroleum Geology '86 Ltd. Published by The Geological Society, London, p. 1315.

Source rock permeability and petroleum expulsion efficiency: modelling examples from the Mahakam delta, the Williston Basin and the Paris Basin

J. BURRUS,[1] K. OSADETZ,[2] J. M. GAULIER,[1] E. BROSSE,[1] B. DOLIGEZ,[1]
G. CHOPPIN DE JANVRY,[3] J. BARLIER[4] and K. VISSER[5]

[1] *Institut Français du Petrole BP 311, 92506 Rueil Mamaison, France*
[2] *Institute of Sedimentary Petroleum Geology, Calgary, Canada*
[3] *Total, Paris La Défense, France*
[4] *Elf Aquitaine Production, Pau, France*
[5] *Petrocanada, Calgary, Canada*

Abstract: Petroleum source rock expulsion efficiency classification based solely on geochemical characteristics can be misleading. Source rock permeability characteristics are also important, especially where petroleum expulsion occurs as a separate phase flow. Thermal and maturity history reconstructions and overpressure development, simulated with the TEMISPACK model of the Institut Français du Petrole (IFP) using two-phase (water, petroleum) Darcy flow, illustrates the importance and influence of bulk source rock permeability for expulsion efficiency. Poor dispersed type III Mahakam pelagic shales, rich marine type II Bakken shales and Paris Basin Lower Liassic source rocks, including both rich Lower Toarcian shales and underlying, less rich Hettangian–Lotharingian marls, illustrate how strongly expulsion efficiency is controlled by bulk rock permeability. Simulation of overpressures in Mahakam and Bakken shales requires very low bulk permeabilities, in the nD range. This is too low to allow significant expulsion. These rocks are inferred not to be hydraulically fractured, despite the overpressures, such that expulsion would not be assisted by this mechanism. Lack of overpressures and regional hydrodynamic models suggest that Paris Basin Lower Liassic sources are at least 10 to 100 times more permeable than Mahakam or Bakken sources. Improved permeabilities can be attributed possibly to tectonic fracturing. This facilitates expulsion both into overlying Dogger reservoirs and underlying Keuper reservoirs, consistent with inferred expulsion efficiencies and oil saturation profiles.

Petroleum expulsion from low permeability source rocks is generally viewed as a separate phase transport through the porosity network, largely because alternative mechanisms like molecular diffusion and solution transport are, at best, of minor importance (Tissot and Pelet 1971; Dickey 1975; Momper 1978; Magara 1978; Tissot and Welte 1984). Yet, there remains considerable discussion regarding the relative importance of several poorly constrained factors. Unresolved questions include:

- What quantity of hydrocarbons (HC) remains in the kerogen, trapped by adsorption?
- How do HC reach the closet migration pathway once they leave the kerogen? Is diffusion in the organic network a significant process at this local scale (Stainforth and Reinders 1990)?
- What is the relative importance of source rock geochemical characteristics compared to source rock transport properties in expulsion?

Prevailing ideas pertaining to these questions are as follows. Numerous observations show that HC retention by mature kerogen increases with increasing TOC. This has led to the proposition that expulsion efficiency is dominated by surface interactions between petroleum and kerogen without significant influence from bulk rock permeability properties (McAuliffe 1980; Durand 1988; Pepper 1991).

Microfractures, not pores, with radii commonly inferred to be between 1–10 nm in compacted shales are considered important conduits for primary migration from kerogen to secondary migration pathways. Their importance appears confirmed by bitumen-filled microfractures in mature source rocks, like the La Luna shales (Talukdar *et al.* 1987). Mechanisms opening microfractures include tectonic and hydraulic fracturing. Hydraulic fracturing may result from overpressures due theoretically to either compaction or HC generation (Tis-

sot and Pelet 1971; Meissner 1978; Palciauskas and Domenico 1980; du Rouchet 1981). Few studies have used overpressure history simulations to confirm hydraulic fracturing conditions were reached, or to evaluate the relative importance of tectonic stress as an alternative mechanism for microfracturing. In addition, it is not clear whether microfractures develop vertically, as in Mohr's theory (Meissner 1978), or parallel to bedding, following Griffith's theory (Griffith 1924; Duppenbecker *et al.* 1991), in a given geological setting. In contrast to the 'microfracturation' concept, laboratory experiments have suggested that 'compacted' shales have a measurable permeability to oil through the pore network, estimated to be around 10 nD (Sandvik and Mercer 1990), and that this permeability would be sufficient to allow natural petroleum expulsion to take place, but insufficient for hydraulic fracturing to develop as a consequence of oil generation.

Geochemical mass balance studies suggest that expulsion efficiencies can be as high as 60–90% for oil-prone source rocks with potentials greater than $10 \, kg \, HC \, t^{-1}$ (Cooles *et al.* 1986; Mackenzie *et al.* 1987). Similar calculations suggest very low expulsion efficiencies, approximately 5%, from poorer source rocks with potentials around $5 \, kg \, HC \, t^{-1}$. Such calculations also have high uncertainties due to difficulties attending both analytical method comparisons and source rock porosity and residual petroleum measurements. Still, they are useful 'rules of thumb' in specific geographic regions, but they are neither universal expulsion efficiency models, nor do they show how expulsion efficiency depends on expulsion mechanism.

Following the Darcy approach, the separate phase flow of HC is most dependent on hydraulic heads and overpressures, both of which are generally higher in low permeability sources than in more permeable carrier beds (Hunt 1979; Tissot and Welte 1984). In mature source rocks, overpressures due to compaction can be enhanced by the volumetric expansion associated with HC generation (Tissot and Pelet 1971; Ungerer *et al.* 1983). Capillary pressure (P_c) augments the pressure

From *Petroleum Geology of Northwest Europe: Proceedings of the 4th Conference* (edited by J. R. Parker).

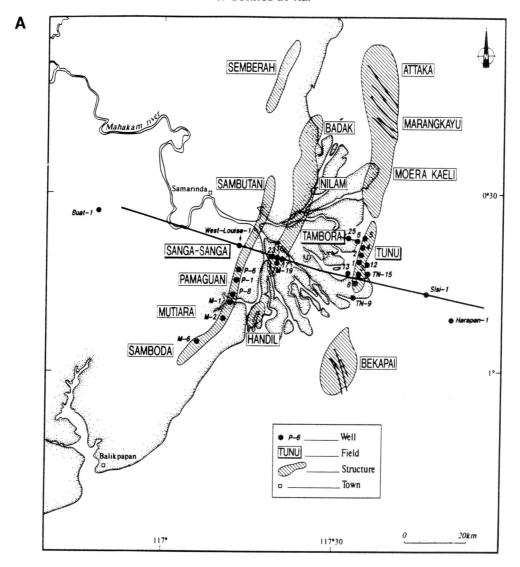

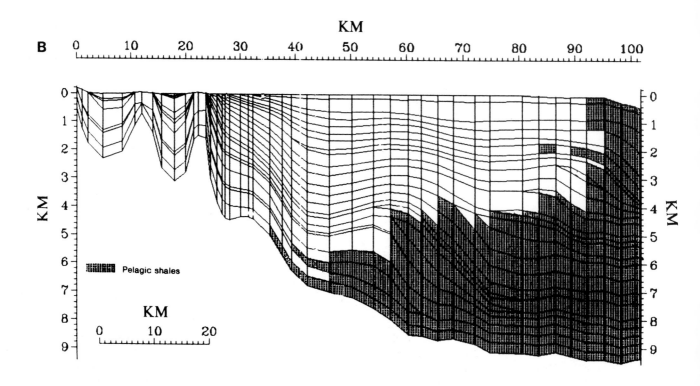

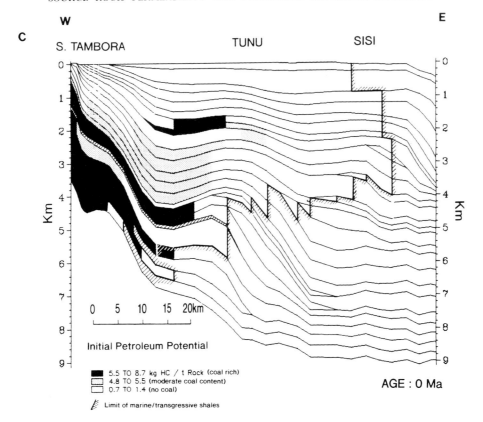

Fig. 1. (**a**) and (**b**) Location map of the Mahakam delta (Indonesia) indicating the principal structural axes, the location of HC accumulations, and the location of the cross-section studied with TEMISPACK. The overpressured shales discussed in the text are shaded. (**c**) Distribution of the initial petroleum potential, reconstructed from the coal content and organic shales content. The highest potential is found in delta plain facies associated with the highest coal content. The reservoir facies are found everywhere above the marine shales, but principally between 2 and 4 km burial ('Tunu main zone').

difference in the HC phase between source rocks, with the smallest pore throats and highest P_c, and carrier beds, with the largest pore throats and the lowest P_c. These mechanisms should all enhance expulsion efficiency, an inference consistent with residual petroleum saturation profiles observed near the margins of thick source rock intervals (Leythaeuser *et al.* 1988). Petroleum phase relationships related to pressure and temperature changes are important to secondary migration (Nogaret 1983; England and Mackenzie 1989) and receive increasing attention during HC expulsion (Burnham and Braun 1990). Overpressured type III source rocks are, for example, likely to expel oil (i.e. C15+) dissolved in a gaseous petroleum at low maturities by a process that significantly fractionates chemical composition (Leythaeuser and Poelchau 1991).

An integrated reconstruction of overpressure, thermal and maturity histories, combined with bulk porosity reduction analysis is required to evaluate the implications of 'separate phase transport' expulsion models. This has led recently to several 'integrated numerical models' that calculate petroleum expulsion and migration by simultaneously solving equations describing compaction, overpressures, thermal history and kerogen cracking. In this study we use a model of this type, IFP's TEMISPACK software (Ungerer *et al.* 1990; Burrus and Audebert 1990). This approach has the principal advantage of simulating expulsion and migration dynamically, through the historical reconstruction of overpressures, temperatures and thermal maturity. A weakness of this approach is illustrated by a previous application of TEMISPACK in the Viking Graben (Burrus *et al.* 1991). The integrated models require additional information, including: fluid properties of petroleum phases, relative permeabilities and capillary pressures, the values of which may not be precisely known. This problem can be

overcome through a sensitivity analysis of poorly specified parameters.

This study examines the relative importance of expulsion mechanisms using a dynamic overpressure history reconstruction. It draws on results from TEMISPACK models in three regions, to be described in detail elsewhere. The three regions include: Indonesian Mahakam delta (Burrus *et al.* 1992*a*); North American Williston Basin (Osadetz *et al.* 1992; Burrus *et al.* 1992*b*); and French Paris Basin (Burrus *et al.* submitted). From these studies, we conclude expulsion efficiency is controlled largely by rock hydraulic properties, principally bulk permeability. In two instances it is possible to constrain precisely stratigraphic unit macro-scale permeabilities by modelling observed overpressures. By replicating observed overpressures we find that hydraulic fracturing thresholds are not attained and may be reached less commonly than is generally inferred. Finally, we conclude that the application of multiphase Darcy law model of hydrocarbon expulsion and migration can result in petroleum potential indications quite different from those inferred using source rock mass-balance calculations or other 'classical' petroleum system evaluation techniques.

Petroleum expulsion modelling using the TEMISPACK model

TEMISPACK is a two-dimensional finite difference model that simulates petroleum evolution along a regional section. It is composed of several modules. The first is a backstripping routine that computes sedimentation rate, layer by layer, along the section using standard compaction curves for each lithofacies. A second module computes overpressure and porosity evolution. This results from the simultaneous solution of three linked equations (Schneider *et al.* 1993), including: the Darcy law equation, the Terzaghi effective stress equation, where effective stress, *s*, is the difference between lithostatic pressure, *S*, and pore pressure, *P*, and a plastic-like rheological law that is a relationship between effective stress and porosity. The model follows classical mechanical analysis (Smith 1971; Nakayama and Lerche 1987; Bethke *et al.* 1987). Permeability is calculated as a function of porosity using the Koseny-Carman law.

Third and fourth modules compute thermal history using transient heat flow equations and kerogen cracking using Arrhenius first-order kinetic models (Tissot *et al.* 1987), respectively. The fourth module allows deduction of the quantity of HC generated. A fifth module, requiring relative permeability curves, solves a generalized Darcy law equation for two fluid phases, petroleum and water. Overpressures are not taken from the second module, but are recalculated to account for HC using a petroleum generation term in source rocks. Capillary pressure effects result in pressure differences between the HC and water phases. In summary, TEMISPACK models petroleum expulsion as a single incompressible phase transport, driven principally by petroleum phase pressure disequilibrium resulting from compaction, HC generation and capillary pressure differences between sources and drains.

Current limitations are due to three considerations. First, petroleum fluid properties are fixed inputs not functions of source rock type, PVT environment or gas–oil ratio. Assigned values must be tested by sensitivity analysis. Porosity is solely affected by mineral grain mechanical compaction without accounting for porosity increases accompanying HC generation. This might result in an overestimation of the contribution HC generation makes to overpressures. Finally, effective stress is entirely supported by mineral grains without consideration of the mechanical role of kerogen, whatever its concentration. These considerations might affect calculations for rocks with very high kerogen contents, like the Bakken Formation.

TEMISPACK models shown below, reconstructed burial and thermal histories as constrained by both current temperatures and kerogen maturity. Subsequently, TEMISPACK calculations of overpressures and permeabilities were adjusted and tested against measured overpressures in shale intervals. Finally, rates of HC expulsion and secondary migration were calculated. Only results pertaining to overpressures and petroleum expulsion are discussed below.

Regional geological settings

Mahakam delta

The Mahakam delta (Fig. 1) fills a Miocene–Pliocene basin with a depositionally regressive succession up to 9 km thick. Below 3.5–4 km depth, shales dominate. Overlying them, sand content increases progressively such that the succession coarsens both upward and from east to west. The deep shales are highly overpressured (Fig. 2). The transition zone from hydrostatically pressured sandy facies underlying overpressured shales is sharp, and characterized by very high overpressure gradients, typically several tens of MPa per 100 m.

Sonic and density logs indicate that undercompaction follows overpressure development (Fig. 2). The Mahakam region is a gas-prone province where complex, stacked sandstone reservoirs contain liquid and gaseous petroleum in varying proportions. The organic matter is a very uniform type III (Durand and Oudin 1979; Oudin and Picard 1982; Huc *et al.* 1988). Organic matter occurs either in coals, interbedded with

sandstones and shales, in the upper hydrostatically pressured zones, or dispersed in the overpressured shales (Vandenbroucke *et al.* 1983). Unlike coals in sandy units, overpressured shales have poor potential, estimated between 1.5 and 2 kg HC t^{-1}.

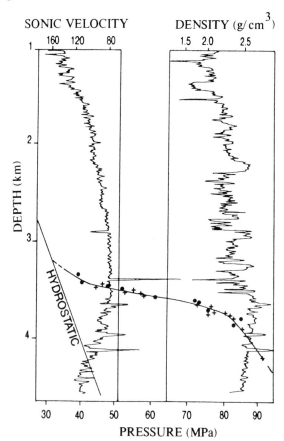

Fig. 2. A typical example of a pressure profile in the Mahakam delta. The upper part of the shales corresponds to a transition zone with high pressure gradient. The composite sonic and density logs represent shales interbedded with sands in the hydrostatic zone, and the overpressured shales below. They clearly indicate that undercompaction and overpressures are linked.

A proposed origin for these petroleums suggests that less mature, exsolved liquid HC, generated above the overpressured zone, are dissolved in more mature, gaseous petroleum migrating vertically across formational boundaries out of the overpressured zone (Durand and Oudin 1979; Oudin and Picard 1982; Vandenbroucke *et al* 1983). Previous workers inferred that very mature petroleum from deeply buried, overpressured shales would be expelled later than petroleum generated in, or above, the overpressure transition zone (Fig. 3). These very mature petroleums would, by themselves, result in gaseous HC accumulations. Liquid HC generated at lower maturities, on the flanks of structures, in or above the overpressure transition zone could be dissolved in gaseous HC migrating out of the overpressured zone. Open hydraulic microfractures caused by overpressures are inferred to facilitate vertical migration through the pelagic shales (Oudin and Picard 1982).

With TEMISPACK we have analysed this setting using observed overpressures to constrain precisely shale permeabilities. Deduction of these permeabilities allows us to compare a petroleum generation model requiring gaseous HC expulsion from overpressured shale sources against a model of HC expulsion from hydrostatically pressured coals either in synclines or on the flanks of productive anticlines.

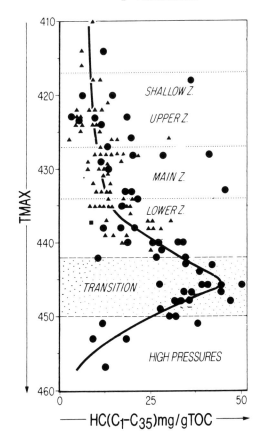

Fig. 3. Residual HC content (g HC per g TOC) as a function of maturity in coals in the Mahakam delta at Handil structure (Durand and Oudin 1979). The decrease of the HC content in the deep overpressured zone has been interpreted as the expulsion of overmature gas. This study suggests that possibility of expulsion of such products is very poor.

Williston Basin

The intracratonic Williston Basin contains a Middle Cambrian to Paleogene sedimentary succession up to approximately 4.8 km thick (Gerhard et al. 1991). Its current sub-circular configuration is primarily due to differential uplift and erosion during Early Devonian, Late Carboniferous and Eocene epeirogenic intervals. Palaeozoic successions are predominantly carbonate rocks while Mesozoic and Cenozoic successions are mostly clastics. This discussion focuses on HC expulsion from a rich uppermost Devonian–earliest Carboniferous source rock, the Bakken Formation.

The Bakken Formation is the only currently overpressured interval in the basin (Meissner 1978). The Bakken typifies rich shale source rocks, with a potential of 70 kg HC t^{-1} (Table 1). Overpressures due to HC generation are inferred to cause hydraulic fracturing allowing vertical migration (Meissner 1978). This explanation appears consistent with currently observed overpressure (around 15 MPa in the Antelope Field at depths around 2.8 km), oil accumulations in the overlying Carboniferous Madison Group carbonates (Fig. 4) and wireline logging data showing that overpressures coincide with mature, oil-wet Bakken shales. However, oil and source rock solvent-extract molecular compositions suggest that oils in the Madison Group were expelled from carbonates within the Madison Group overlying the Bakken shales (Osadetz and Snowdon 1986; Brooks et al. 1987; Snowdon and Osadetz

1988; Osadetz et al. 1992), not from Bakken sources (Meissner 1978). The Bakken does source some oil pools, but these are consistently very mature and restricted to the Bakken Formation (Osadetz et al. 1992).

Table 1. Source rock characteristics

	S2 (kg HC t^{-1} rock)	HI	TOC (%)
Mahakam shale	1.5	100	1.5
Bakken shale	70	690	10
Lower Toarcian Paris B	15–30	550	2.8–5.6
Hettangian–Lotharingian marls	3–7	350	0.8–2

With TEMISPACK we test the assumption that HC generation is responsible for Bakken overpressures and use these overpressures to constrain vertical permeabilities, as in the Mahakam delta example. This allows us to deduce whether overpressures ever reached hydraulic fracturing thresholds, a potentially important condition for expulsion into the overlying Madison Group. From this we can infer if lithological and physical characteristics of the Bakken and Lodgepole petroleum systems corroborate their petroleum geochemistry.

Paris Basin

The intracratonic Paris Basin is another circular depression in which Triassic rocks occur as deep as 3 km (Perrodon and Zabek 1991). The entire succession is normally pressured. Excluding Paleozoic petroleum systems, source rocks occur in Lower Liassic strata, particularly in the Hettangian–Sinemurian and Pliensbachian–Lotharingian intervals, where interbedded marls, shales and carbonates, containing organic matter compositionally intermediate between types II and III, have laterally varying potentials between 3 and 7 kg HC t^{-1} (Espitalié et al. 1987, 1988). Much richer sources occur in a 15 m thick Lower Toarcian shale containing rich type II organic matter with potentials between 15 and 30 kg HC t^{-1}.

The main oil accumulations occur in Upper Triassic Donnemarie and Chaunoy sandstones, underlying Liassic sources, and 300 m above Toarcian sources in Middle Jurassic Dogger carbonates (Poulet and Espitalié 1987). Expulsion is inferred to begin at 2300 m current depth (Fig. 5), at transformation ratios around 30% (Espitalié et al. 1988). Hettangian–Sinemurian and Lower Toarcian source rock expulsion efficiencies of 80% and 10%, respectively, were inferred from Rock–Eval data. Lower Toarcian expulsion efficiency calculations are affected by migration. Immature samples have positive S1 values. Since Rock–Eval pyrolysis does not 'see' heavier and lighter hydrocarbons, these estimates are very approximate, but consistent with concepts that rich oil-prone sources have good expulsion efficiencies (Cooles et al. 1986). Several studies found that the amount of residual oil in source rocks follows TOC content (Fig. 6; Tissot et al. 1971; Espitalié et al. 1988; Ungerer 1990). Lower Toarcian sources (estimated residual oil saturation approximately 10–20% and TOC approximately 5–7%; from Espitalié et al. 1988) and other Liassic sources (estimated residual oil saturation 2–6% and TOC < 3%) follow this trend.

Given published constraints on shale and marl vertical permeabilities in the region, in the absence of overpressures, we employ the 'separate phase' approach used in TEMISPACK to explain expulsion patterns in Paris Basin. These include migration into both overlying and underlying units. Model results are compared to reported expulsion efficiencies and residual oil saturation estimates.

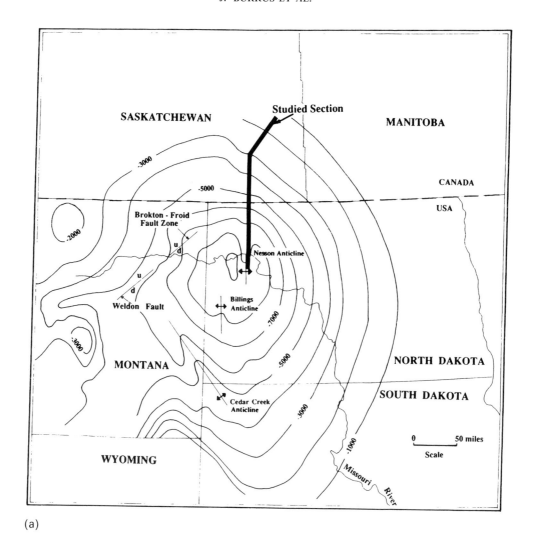

(a)

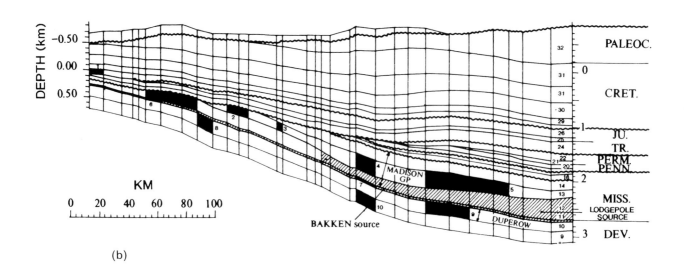

(b)

Fig. 4. (a) Location map of the Williston Basin, and position of the section studied with TEMISPACK. **(b)** The section studied, which reaches the deep part of the basin, in the region of the more mature Nesson Anticline. Layers younger than Devonian are not discussed in this study. The black colour indicates position of oil pools along the section (Osadetz *et al.* 1992). Pools 1 to 6 were traditionally thought to be sourced by the Bakken source. This study confirms an alternative interpretation (Brooks *et al.* 1987; Osadetz *et al.* 1992) that they are mostly fed by the Lodgepole source. Only pool 7, located within the Bakken, appears to be Bakken-sourced.

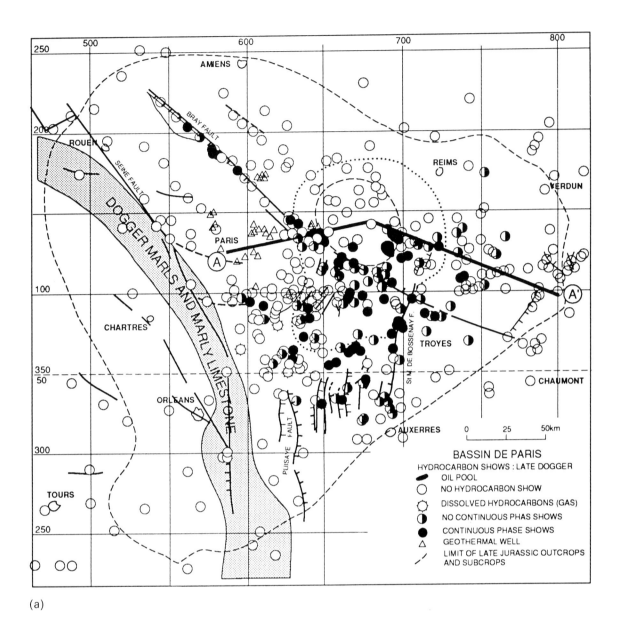

(a)

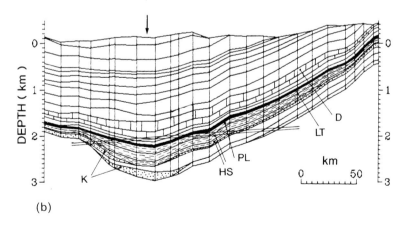

(b)

Fig. 5. (**a**) Location map of the Paris Basin showing the distribution of oil shows in the Dogger reservoir (Poulet and Espitalié 1987). The extent of the oil window (mixed line) and the extent of the effective oil kitchen (with proved expulsion, dotted line) are indicated. The location of the section modelled with TEMISPACK is indicated as AA'. It intersects the centre of the kitchen. (**b**) Simplified cross-section studied. D: Dogger reservoirs (carbonate symbol); LT: Lower Toarcian shales (black colour); PL, HS: Pliensbachian–Lotharingian and Hettangian–Sinemurian sources (shale symbol); K: Keuper reservoirs (sandstone symbol). The approximate top of the 'effective kitchen' taken from Espitalié *et al.* (1988), is indicated by the parallel tilted lines. Results shown in Fig. 14 are obtained in the centre of the basin, indicated by an arrow.

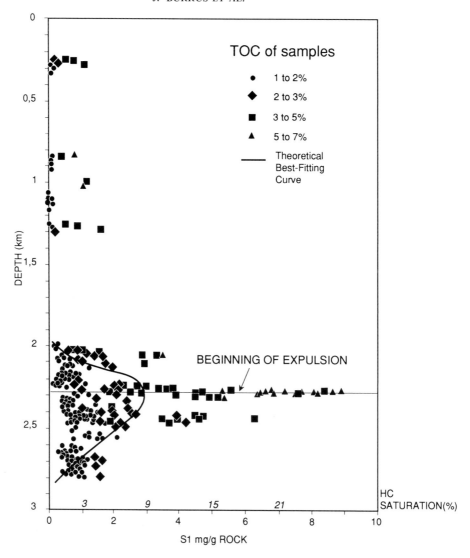

Fig. 6. S1 data for the Liassic shales and marls in the Paris Basin (from Espitalié *et al.* 1988). These data indicate that the onset of expulsion begins around 2.3 km depth. The highest TOC (above 3%) corresponds to the Lower Toarcian. Other data refer to Hettangian–Pliensbachian marls and shales. These data indicate that the Lower Toarcian has much higher residual oil saturations than other source layers. Alternatively, increasing adsorption might be an explanation (see Fig. 14). Espitalié *et al.* (1988) estimated from these data (and maturity data) that the expulsion efficiency would be 80% in the deepest Hettangian–Sinemurian source, and 10% for the Toarcian. S1 data are converted into HC saturations using density and porosity values of Tables 2 and 3. The best fitting theoretical curve is taken from Fig. 14.

Model inputs

Two-dimensional regional sections of each model were simplified into a processable mesh composed, on average, of 35 columns and 35 layers. Note the variations in cell unit scale between models (Figs 1, 4 and 5). Thermal and maturity histories were reconstructed using both proprietary well data and published information.

Idealized Mahakam delta shale mechanical properties (i.e. effective stress/porosity functions) were modelled using detailed analysis of wireline log data, shale mineralogy and RFT tests (Burrus *et al.* 1992*a*, following the procedure of Ungerer *et al.* 1990). The resulting idealized effective stress/porosity relation is assumed to be widely applicable and has been used for the Bakken shale and the Paris Basin Lower Liassic. Shale permeabilities were determined by convergent, iterative calculations of model overpressures resulting from compaction and HC generation, as constrained by observed overpressures.

Using models that best matched observed overpressures, the ratio between pore pressure (P) and lithostatic stress (S) was computed and compared to the commonly acknowledged threshold for hydraulic fracturing, $P/S > 0.85$ (Meissner

1978). Shale permeabilities were neither increased nor modified if the hydraulic fracturing threshold was exceeded. No allowances were made for intermittent or continuous permeability relaxation.

Expulsion–migration parameters employed are as follows. Capillary pressures are consistent with inferred pore radius and the Laplace equation, 0–10 MPa. HC densities and viscosities typical of compressed gases were employed in the Mahakam delta example and those characteristic of light oils were used in the Williston Basin and Paris Basin studies. All three studies use a similar second-order relative permeability function (Fig. 7).

Table 2 shows the 'best fitting' permeability parameters used in each study. Tabulated horizontal and vertical permeabilities are for calculated present-day average porosity in each source. Table 1 lists source rock properties considered in each model. These vary laterally in the Paris Basin model. Table 3 indicates petroleum fluid properties employed. Model oil viscosity is a function of temperature and values are appropriate to present-day source rock temperatures (80°C in Paris Basin and 100°C in Williston Basin, Table 3). Gas viscosity is not a function of temperature.

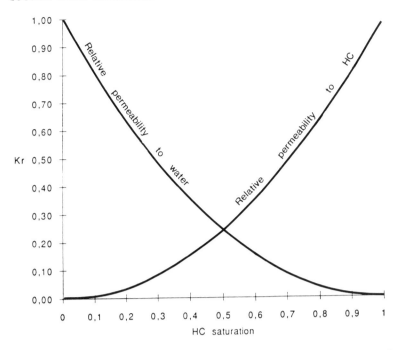

Fig. 7. Relative permeability curves used to compute petroleum and water flow in the 3 examples studied. The oil curve is characterized by a progressive increase at low oil saturation (2–5%). HC are considered immobile below 2% saturation. These curves have little experimental support. They can be justified by the consistent model outputs they provide.

Table 2. Permeability parameters (So: specific surface; K_h, K_v: horizontal and vertical permeabilities)

	So ($m^2 m^{-3}$)	Mean porosity	K_h (Darcy)	K_v (Darcy)
Mahakam shale	100 000 000	0.17	4×10^{-8}	4×10^{-8}
Bakken shale	100 000 000	0.15	4×10^{-6}	4×10^{-9}
Lower Toarcian Paris B.	32 000 000	0.12	3×10^{-6}	3×10^{-7}
Hettangian–Lotharingian marls	32 000 000	0.10	1×10^{-5}	1×10^{-6}

Table 3. Petroleum fluid properties (*: viscosity at present-day source temperature)

	Density ($kg\ m^{-3}$)	Viscosity (Pa s)
Mahakam gas	350	0.0001
Bakken oil	750	0.005*
Paris Basin oil	750	0.003*

Results

Mahakam delta shales

Figure 8 compares observed (open circles) and calculated overpressures, in addition to calculated porosity curves, for three different shale permeabilities at the centre of the section studied (Fig. 1). The solid line (Fig. 8) shows calculations using the 'best' shale permeability model. This model provides the best fit to pressure and porosity variations with depth by accounting for the sharp increase in gradient at the top of the overpressured zone and by qualitatively matching sonic log shape (Fig. 2). The vertical shale permeability for this model is approximately 10 nD (Table 2).

The model shale permeability resulting in the dotted profile is 10 times greater than the 'best' model (Fig. 8). It predicts overpressures that are too small and insufficient undercompaction in the overpressured zone. A third model, the dashed line, has vertical shale permeability one fifth of the 'best' model (Fig. 8). This model results in excessive overpressures and inappropriate porosity variations with depth in the over-

pressured zone. Other models, not shown here, confirm that compaction is the dominant cause of overpressures and that HC generation makes a negligible contribution.

Predicted P/S (pore pressure/geostatic stress) ratios agree well with those estimated from RFT and wireline log data when the 'best' shale permeability model is used (Fig. 9; Burrus et al. 1992a). This model provides particularly good predictions of P/S ratio through the transition zone. Interestingly, the ratio never exceeds 0.85, the hydraulic fracturing threshold, even though permeabilities were not relaxed in the computations. This suggests that hydraulic fracturing, important in previously proposed models, is not necessarily an important process in the overpressured zone.

Figure 10 compares the calculated present-day model HC distributions for two different potential source rock distributions. The first assigns petroleum potential to the overpressured shales and overlying interbedded sands and coals (Fig. 10a). The second assigns petroleum potential only to the interbedded sands and coals above the overpressured zone (Fig. 10b). Although secondary migration analysis is beyond the scope of this discussion, it is obvious that the present HC distribution is similar in both instances and that these accurately reproduce observed distributions (Burrus et al. 1992a). These models suggest three major points. First, that even very light hydrocarbons, including natural HC gases, cannot migrate efficiently through the low permeability overpressured shales. Second, hydraulic fracturing deep in the overpressured zone cannot be demonstrated to occur. Finally, both models suggest the most likely sources of gaseous petroleum pools in this section are the interbedded coals and sandy shales lying above the transition zone.

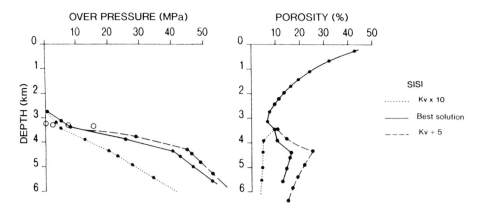

Fig. 8. Computed overpressure profiles at the easternmost structure in Fig. 1 (Sisi). Three permeability assumptions for the marine shales are tested. The best fit with observed overpressures (indicated by open circles, derived from mud density and tests) is the solid line. The higher permeability tested (dotted line) is disproved by the insufficient overpressures. The lower permeability (dashed line) is disproved by the too pronounced undercompaction when compared with Fig. 2.

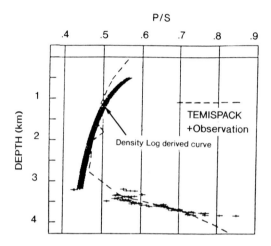

Fig. 9. Computed (dashed line) and observed (logs, RFT) ratio P/S (pore pressure/geostatic stress). The fit is very good. The ratio remains below or equal to 0.85 across and below the transition zone. This suggests that hydraulic fracturing is not a process that is necessarily active at present. Location: as in Fig. 8.

The maturity of sources in sands and coals is lower than that of the underlying overpressured shales. It remains to be determined if sufficient amounts of gas can be generated from type III kerogen at relatively low maturity (TR < 60%) in synclines and on the flanks of culminations above the transition zone. Elsewhere, a five-fold or greater increase in gas yield per unit weight of organic carbon has been observed in type III sources through the reflectance range 0.55 to 0.65% R_0, immediately preceding the onset of liquid hydrocarbon generation (Monnier et al. 1983). Those models invoking a vertical percolation of gas appear incompatible with observed overpressures and inferred shale permeabilities in the transition and overpressured zones. The significance of decreasing residual petroleum values with increasing depth in the overpressured zone (Fig. 3) remains unresolved. Either it truly indicates the expulsion of gases from the overpressured shales by a mechanism not compatible with separate phase flow, like diffusion, or the few data available from these depths (Fig. 3) do not accurately represent gaseous HC contents, possibly due to losses during cuttings sample collection or selection.

Bakken shales

Figure 11 shows theoretical pressure profiles calculated at the south end of the Williston Basin section, essentially in the centre

of the basin. No overpressures are predicted if no oil generation occurs in the Bakken, even when permeabilities are very low. This confirms HC generation is the source of Bakken overpressures. The best fit to the observed overpressure, 15 MPa, is with model permeabilities around 1 nD (Table 2). The fit is sensitive to minor changes in model permeabilities, as small as a factor of ten. Present-day calculated P/S ratio profiles (Fig. 12) are all much below the hydraulic fracturing threshold. P/S ratios peaked at 38 Ma, but never exceeded 0.75. From these computations we infer that Bakken hydraulic fracturing seems incompatible with numerically simulated constraints, especially as our models may overestimate how oil generation contributes to overpressures. This is consistent with observations of Bakken cores, both vertical and horizontal. Cores are not commonly fractured, but some small, oil-coated, bedding-plane fractures are observed (J. A. Le Fever, pers. comm.).

Consequently, our modelling study not only confirms the revised geochemical interpretations of Bakken and Lodgepole source rock effectiveness and oil source-correlations (Fig. 4), but it provides a mechanism, the low permeability of Bakken shales, for expulsion efficiency differences between two rich source rocks of similar thermal history, organic matter type and kinetic parameters (Fig. 13).

Lower Liassic of the Paris Basin

The Paris Basin is not overpressured. The hydraulic heads responsible for regional hydrological circulation in principal aquifers are related to topography and are at least 10 to 100 times smaller than those in preceding examples (Wei et al. 1990). This prevents the deduction of shaly layer vertical permeabilities as performed above. By assigning a vertical permeability of approximately 1 nD to Liassic and other shales and marls, models reach excessive overpressures and prevent oil expulsion. This provides a lower limit on vertical permeabilities. Fortunately, models of present-day hydrodynamic circulation and vertical exchange among aquifers give several aquitard vertical permeability estimates (e.g. Wei et al. 1990) that are confirmed by observed geothermal gradient models (Gaulier et al. in press). These estimates (Table 2) are much higher than those characteristic of the other two models. We infer that this is due to vertical fractures occurring at a spacing of 0.1–1.0 km or less. Lower Toarcian shales, with permeabilities of approximately 0.1 μD, are estimated to be 10 times less permeable than the Hettangian to Lotharingian marls, with permeabilities of approximately 1 μD (Table 2).

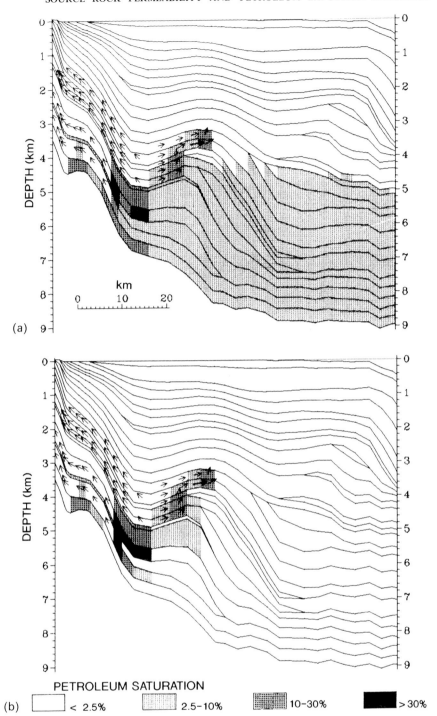

Fig. 10. Calculated distribution of petroleum saturation and flow vectors along the Mahakam section. White: < 2%; light grey: 2–10%; dark grey: 10–30%; black: >30%. Compressed gas assumption (see Tables 1 to 3 for simulation parameters). (**a**) A potential is given to the overpressured shales and to the more sandy facies containing coals. (**b**) No potential is given to the overpressured shales. Since calculated HC distribution is not very different, we conclude that the contribution of gas from the deeper overpressured shales to be negligible, compared with the contribution of the more shallow levels.

Residual oil saturations occur in the model in Liassic sources, and oil accumulations develop in Keuper and Dogger reservoirs. Present-day residual oil saturation profiles (Fig. 14) calculated in the centre of the section (Fig. 5) show significant variations when expulsion is suppressed (Fig. 14, dashed line), if source rock capillary forces are neglected (dotted line), or if they are set to a reasonable value, approximately 1 MPa (solid line). The 'best fit' model is the latter (Fig. 14, solid line). Interestingly, it predicts that oil is expelled down from Hettangian–Sinemurian marls into underlying model elements rep-

resenting sandy Keuper reservoirs, qualitatively matching the Chaunoy and Donnemarie sandstone accumulations (Poulet and Espitalié 1987). More quantitative results would require improved spatial resolution of lithofacies distributions within our model. Comparison of the model that neglects capillary forces (dashed line) to the 'best fit' model (solid line) proves that expulsion into underlying Keuper reservoirs, against buoyancy, is driven by source rock–reservoir capillary pressure contrast, confirming that the source of these oils is the Hettangian–Sinemurian interval, not the Upper Liassic. The driving

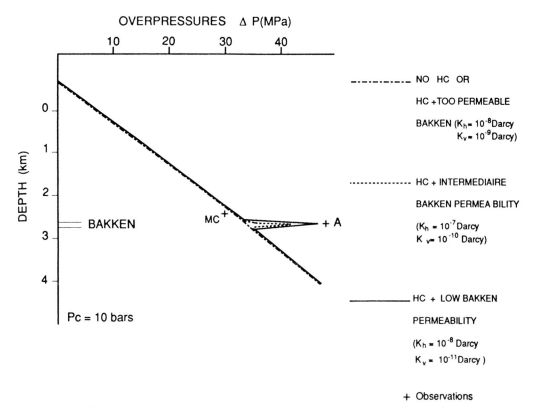

Fig. 11. Computed pressure profiles at the southern extremity of the section in the Williston Basin (Fig. 4). No overpressuring is calculated if oil generation is neglected. Observed overpressures (crosses) are used to adjust the permeability of the Bakken shales. Bakken permeabilities tested are given at 10% porosity.

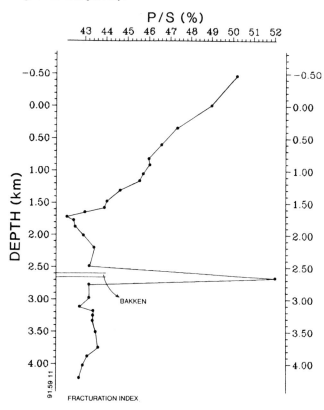

Fig. 12. Computed P/S ratio in the Williston Basin (P: pore pressure; S: geostatic stress) at the same location as the pressure profile of Fig. 11. The ratio decreases with depth, except the narrow peak observed across the Bakken. The ratio remains much below the critical value around 0.85 above which, in the absence of extensional stress, hydraulic fracturing is believed to develop. This result would suggest that hydraulic fracturing is currently not taking place in the Bakken shales.

force for this expulsion disappears if capillary pressures are either set to zero or made equal in both reservoirs and sources. By examining residual oil saturations in Hettangian–Sinemurian model elements, both when expulsion occurs and when it is suppressed, this allows the expulsion efficiency to be estimated. Inferred expulsion efficiencies are 75%, consistent with other estimates discussed above.

Expulsion toward overlying Dogger reservoirs is driven in our model by both buoyancy and capillary pressure differences. Oil saturations in Lower Toarcian model elements are paradoxical (Fig. 14). Oil saturations are higher if expulsion is permitted (solid line) than if expulsion is suppressed (dotted line). This occurs because the Lower Toarcian acts like a reservoir for oils generated in the underlying Lotharingian–Carrixian and Domerian–Pliensbachian sources that have mixed with oils generated indigenously in Lower Toarcian sources. This interpretation is confirmed by S1 data indicating a net import of oil into Lower Toarcian strata. Calculated Lower Toarcian residual oil saturations appear to be at least twice those in the deepest Hettangian–Sinemurian model elements (Fig. 6). This difference seems not to be due to oil adsorption on kerogen, but directly linked to the bulk permeability distribution and results from the balance between hydrocarbon generation in Lower Toarcian strata, the migration of oil from underlying sources and the expulsion of oil toward overlying Dogger reservoirs. Calculated and observed saturations fit well for the Lotharingian–Pliensbachian and Hettangian–Sinemurian intervals (2–6%) but not for the Lower Toarcian (8% and 8–18%, respectively; Fig. 6). Since the Lower Toarcian has much higher TOC than underlying sources, this may indicate preferential adsorption on TOC.

Discussion and conclusions

Petroleum expulsion modelling using separate phase transport, appraises factors not often attributed sufficient importance. Most discussions address source rock geochemical character-

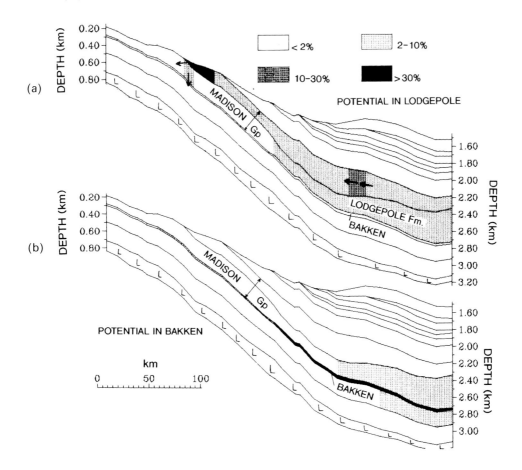

Fig. 13. Calculated distribution of petroleum saturation and petroleum flow vectors in the Williston Basin. White: <2%; light grey: 2–10%; dark grey: 10–30%; black: >30%. Density, viscosity: light oil assumption; capillary pressure in the Bakken: 5 bars (see Tables 1 to 3 for simulation parameters). (a) A potential is given to overlying Lodgepole carbonates only; the Lodgepole-sourced oils explain the accumulations found in the Madison Group, previously attributed to the Bakken source. (b) A potential is given to the Bakken shales only; most oil remains in the Bakken. This simulation confirms the revised oil system proposed by Brooks *et al.* (1987) and Osadetz *et al.* (1992).

istics and give very little consideration to source rock physical properties. This study indicates that expulsion is critically dependent on physical properties, particularly bulk permeability. Current geochemical emphasis is not unexpected, due to the ease with which geochemical parameters are measured, compared to the difficulty of measuring μD–nD permeabilities. This study demonstrates that macroscopic shale permeability estimates can be made by comparing calculated and observed overpressures and undercompaction profiles using integrated computer models. Macroscopic shale and marl permeabilities can vary by a factor of a thousand, indicating the importance of permeability to evaluating expulsion efficiency. Such models also identify whether overpressures are due to compaction disequilibrium or HC generation, and provide a dynamic history of effective stresses that can be used for identifying hydraulic fracturing. We suggest source rock expulsion efficiency classifications using geochemical characteristics alone, without reference to permeability characteristics and history, can be misleading.

Present-day overpressures in the Mahakam delta and in the Bakken shales are attributed to compaction disequilibrium and HC generation, respectively. Duplication of overpressure magnitudes requires very low shale permeabilities in the nD range that result in very low expulsion efficiencies, regardless of initial petroleum potential. Both models seem to discount hydraulic fracturing playing a significant role in expulsion and migration. These results have fundamental implications for oil-source relationships in both the Mahakam delta and the Williston Basin. Model constraints on Bakken expulsion ef-

ficiency and hydraulic fracturing are verified independently by geochemical analyses and core characteristics, corroborating the validity of the Mahakam delta analysis. Despite the simplicity of mechanical models and hydraulic fracturing criteria, model results appear consistent with available observations, suggesting that fracturing due to tectonic stress may be misinterpreted as hydraulic fracturing.

In contrast, the moderate to good source rocks in Lower Liassic marls and shales of the Paris Basin have vertical permeabilities that are at least 100 times higher than Mahakam and Bakken shale permeabilities. As a consequence, the efficiency of expulsion is much improved and pressures are hydrostatic. Expulsion patterns are controlled by buoyancy and capillary pressure gradients between formations. Where capillary pressure dominates, oils can be expelled into underlying Keuper reservoirs. Oils are also expelled into overlying units, towards Dogger reservoirs. Resulting model residual oil saturations match S1 estimates, independently corroborating this analysis.

Present modelling techniques could be improved by considering PVT-dependent petroleum characteristics, adsorption effects, and improved volume balancing during kerogen cracking and porosity evolution. Yet, the ability to confirm revised petroleum system definitions, as in the Bakken example, and the reconstruction of observed expulsion efficiencies, as in the Paris Basin example, seem to confirm the general reliability, applicability and value of integrated modelling. This study emphasizes the need for a 'global' and 'integrated' approach to expulsion analysis. The geochemical characteriza-

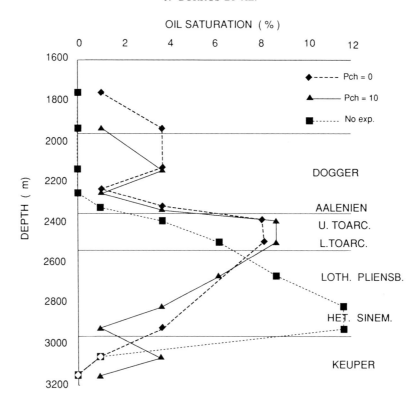

Fig. 14. Calculated oil saturation profiles in the centre of the Paris Basin (see location in Fig. 5). Solid line: 'best' scenario with moderate capillary pressures (10 bars) in the Liassic sources. Dashed line: zero capillary pressure in the Liassic sources. Dotted line: no (or near so) oil expulsion scenario. Simulation parameters: see Tables 1 to 3. These profiles reveal: (a) that downward oil expulsion from the Hettangian–Sinemurian into the Keuper is driven by capillary pressure differences between sources and reservoirs; (b) that Lotharingian–Pliensbachian oil invades the Lower Toarcian, mixes with indigenous Lower Toarcian-sourced oil, then migrates into the Dogger; (c) that the overall expulsion efficiency for the Hettangian–Sinemurian is around 75%, a value consistent with previous estimates; and (d) that the residual oil saturations are much higher (3 times) in the Lower Toarcian than in the underlying Hettangian–Pliensbachian series. This model result is qualitatively consistent with Fig. 6.

tion of kerogen and residual HC is necessary but, in itself, insufficient. By accounting for geological setting, particularly pore pressure history and permeability distribution and evolution, numerical modelling provides valuable criteria by which expulsion and migration can be evaluated.

We thank Elf Aquitaine Production, Petrocanada and Total for their permission to publish this paper.

References

BETHKE, C., HARRISON, W. J., UPSON, C. AND ALTANER, S. P. 1987. Supercomputer analysis of sedimentary basins. *Science*, **239**, 261–267.

BROOKS, P. W., SNOWDON, L. R. AND OSADETZ, K. G. 1987. Families of oils in southeastern Saskatchewan. *In*: CARLSON, C. G. AND CHRISTOPHER, J. E. (eds) *Proceedings of the Fifth International Williston Basin Symposium, Bismark, North Dakota, June 15–17, 1987.* Saskatchewan Geological Society Special Publication **9**, 253–264.

BURNHAM, A. K. AND BRAUN, R. L. 1990. Development of a detailed model of petroleum formation, destruction and expulsion from lacustrine and marine source rocks. *In*: DURAND, B. AND BEHAR, F. (eds) *Advances in Organic Geochemistry: Organic Geochemistry in Petroleum Exploration.* Pergamon Press, 27–39.

BURRUS, J. AND AUDEBERT, F. 1990. Thermal and compaction processes in a rifted basin in the presence of evaporites, the Gulf of Lions case study. *American Association of Petroleum Geologists Bulletin*, **74**, 1420–1440.

——, BROSSE, E., CHOPPIN DE JANVRY, G. AND OUDIN, J. L. 1992a. A regional modelling study of hydrocarbon formation and migra-

tion in the Mahakam delta. *In: Proceedings of the 21st Meeting of the Indonesian Petroleum Association.* IPA 92-11-04, 23–43.

——, DOLIGEZ, B., OSADETZ, K., DEARBORN, D. AND VISSER, K. 1992b. Revised Petroleum Systems in the Williston basin: 2. Evidences from integrated basin modelling. *American Association of Petroleum Geologists Bulletin*, submitted.

——, GAULIER, J. M. AND BARLIER, J. submitted. Oil generation, expulsion and migration in the Paris basin; a regional modeling study. *American Association of Petroleum Geologists Bulletin.*

——, KUHFUSS, A., DOLIGEZ, B. AND UNGERER, P. 1991. Are numerical models useful in reconstructing the migration of hydrocarbons? A discussion based on the Northern Viking Graben. *In*: ENGLAND, W. AND FLEET, A. (eds) *Petroleum Migration.* Geological Society, London, Special Publication, **59**, 89–111.

COOLES, G. P., MACKENZIE, A. S. AND QUIGLEY, T. M. 1986. Calculation of petroleum masses generated and expelled from source rocks. *Advances in Organic Geochemistry.* Pergamon Press, 235–245.

DICKEY, P. A. 1975. Possible primary migration of oil from source rock in oil phase. *American Association of Petroleum Geologists Bulletin*, **59**, 337–345.

DUPPENBECKER, S. J., DOHMEN, L. AND WELTE, D. H. 1991. Numerical modelling of petroleum expulsion in two areas of the Lower Saxony Basin, Northern Germany. *In*: ENGLAND, W. AND FLEET, A. (eds) *Petroleum Migration.* Geological Society, London, Special Publication, **69**, 47–64.

DURAND, B. 1988. Understanding of HC migration in sedimentary basins (present state of knowledge). *In: Advances in Organic Geochemistry: Organic Geochemistry* **13**, Pergamon Press, 445–459.

—— AND OUDIN, J. L. 1979. Exemple de migration des hydrocarbures dans une série deltaïque: le Delta de la Mahakam, Kalimantan, Indonésie. *Proceedings of the 10th World Petroleum Congress*, PD1. Heyden, London, 3–11.

DU ROUCHET, J. 1981. Stress fields, a key to oil migration. *American Association of Petroleum Geologists Bulletin*, **65**, 74–85.

ENGLAND, W. A. AND MACKENZIE, A. S. 1989. Some aspects of the organic geochemistry of petroleum fluids. *In*: POELCHAU, H. S. AND MANN, U. (eds) Geological Modeling. Aspects of Integrated Basin Analysis. *Geologische Rundschau*, **78**, 291–303.

ESPITALIÉ, J., MARQUIS, F. AND SAGE, L. 1987. Organic geochemistry of the Paris Basin. *In*: BROOKS, J. AND GLENNIE, K. (eds) *Petroleum Geology of North West Europe*. Graham & Trotman, London, 71–86.

——, MAXWELL, J. R., CHENET, P. Y. AND MARQUIS, F. 1988. Aspects of hydrocarbon migration in the Paris Basin as deduced from organic geochemical survey. *In: Advances in Organic Geochemistry: Organic Geochemistry*, **13**, Pergamon Press, 467–481.

GAULIER, J. M., BURRUS, J. AND BARLIER, J. in press. Thermal history of the Paris Basin; petroleum implications. *In*: MASCLE, A. (ed.) *HC Exploration in France*. Springer-Verlag, Berlin.

GERHARD, L. C., FISHER, D. W. AND ANDERSON, S. B. 1991. Petroleum Geology of the Williston Basin. *In*: LEIGHTON, M. W., KOLATA, D. R., OITZ, D. F. AND EIDEL, J. J. (eds) *Interior Cratonic Basins*. American Association of Petroleum Geologists Memoir, **51**, 507–560.

GRIFFITH, A. A. 1924. The theory of rupture. *In*: BIEZENO, C. B. AND BURGESS, J. H. (eds) *Proceedings of the First International Congress for Applied Mechanics, Delft*. J. Waltman Publishers, 55–63.

HUC, A. Y., DURAND, B., ROUCACHE, J., VANDENBROUCKE, M. AND PITTION, J. L. 1988. Composition of three series of organic matter of continental origin. *In: Advances in Organic Geochemistry: Organic Geochemistry*, **10**, Pergamon Press, 65–72.

HUNT, J. M. 1979. *Petroleum Geochemistry and Geology*. W. H. Freeman and Co., San Francisco.

LEYTHAEUSER, D. AND POELCHAU, H. S. 1991. Expulsion of petroleum from type III kerogen source rocks in gaseous solution: modelling of solubility fractionation. *In*: ENGLAND, W. AND FLEET, A. (eds) *Petroleum Migration*. Geological Society, London, Special Publication, **59**, 33–46.

——, SCHAEFFER, R. G. AND RADKE, M. 1988. Geochemical effects of primary migration of petroleum in Kimmeridge source rocks from Brae Field area, North Sea: gross composition of C_{15+} soluble organic matter and molecular composition of C_{15+} saturated hydrocarbons. *Geochimica et Cosmochimica Acta*, **52**, 701–713.

MACKENZIE, A. S., PRICE, I., LEYTHAEUSER, D., MULLER, P., RADKE, M. AND SCHAEFER, R. G. 1987. The expulsion of petroleum from Kimmeridge Clay source rocks in the area of the Brae Oilfield, UK continental shelf. *In*: BROOKS, J. AND GLENNIE, K. (eds) *Petroleum Geology of North West Europe*. Graham & Trotman, London, 864–877.

MAGARA, K. 1978. Significance of the expulsion of water in oil-phase primary migration. *Canadian Petroleum Geology Bulletin*, **25**, 195–207.

MCAULIFFE, C. D. 1980. Oil and gas migration: chemical and physical constraints. *American Association of Petroleum Geologists Bulletin*, **63**, 761–781.

MEISSNER, F. F. 1978. Petroleum geology of the Bakken Formation Williston Basin, North Dakota and Montana. Williston Basin Symposium. *The Montana Geological Society, 24th Annual Conference*, 207–227.

MOMPER, J. A. 1978. *Oil migration limitations suggested by geological and geochemical considerations*. American Association of Petroleum Geologists Continuing Education Course Notes, **8**.

MONNIER, F., POWELL, T. G. AND SNOWDON, L. R. 1983. Qualitative and quantitative aspects of gas generation during maturation of sedimentary matter. Examples from Canadian Frontier Basins. *In*: BJOROY, M. (ed.) *Advances in Organic Geochemistry 1981*. John Wiley and Sons, Chichester, 487–495.

NAKAYAMA, K. AND LERCHE, I. 1987. Basin analysis by model simulations: effects of geologic parameters on 1D and 2D fluid flow

systems with application to an oil field. *Gulf Coast Association of Geological Society Transactions*, **37**, 175–184.

NOGARET, E. 1983. Solubilité des hydrocarbures dans le gaz naturel comprimé. Application à la migration du pétrole dans les bassins sédimentaires. *Dissertation, Ecole Nationale Supérieure des Mines Paris*. (Report IFP-31-115).

OSADETZ, K. G., BROOKS, P. AND SNOWDON, L. R. 1992. Oil families and their sources in Canadian Williston Basin (southeastern Saskatchewan and southwestern Manitoba). *Bulletin of Canadian Petroleum Geology*, **40**, 254–273.

—— AND SNOWDON, L. R. 1986. Speculation on the petroleum source rock potential of portions of the Lodgepole Formation (Mississippian), southern Saskatchewan. *In: Current Research, Part B*. Geological Survey of Canada, Paper 86-1B, 647–651.

OUDIN, J. L. AND PICARD, P. F. 1982. Relationship between HC distribution and the overpressured zones. *In: Proceedings of the Eleventh Annual Convention (1982) of the Indonesian Petroleum Association*, 181–202.

PALCIAUSKAS, V. V. AND DOMENICO, P. A. 1980. Micro-fracture development in compacting sediments: relation to hydrocarbon-maturation kinetics. *American Association of Petroleum Geologists Bulletin*, **64**, 927–937.

PEPPER, A. S. 1991. Estimating the petroleum expulsion behaviour of source rocks: a novel quantitative approach. *In*: ENGLAND, W. AND FLEET, A. (eds) *Petroleum Migration*. Geological Society, London, Special Publication, **59**, 9–32.

PERRODON, A. AND ZABEK, J. 1991. Paris Basin. *In*: LEIGHTON, M. W., KOLATA, D. R., OITZ, D. F. AND EIDEL, J. J. (eds) *Interior Cratonic Basins*. American Association of Petroleum Geologists Memoir, **51**, 633–679.

SANDVIK, E. I. AND MERCER, J. N. 1990. Primary migration by bulk hydrocarbon flow. *Organic Geochemistry*, **16**, 1–3, 83–89.

SCHNEIDER, F., BURRUS, J. AND WOLF, S. 1993. Modelling overpressures in low permeability rocks by effective stress porosity relation: physical reality or numerical artifact? *In*: DORÉ, A. G., AUGUSTSON, J. H., HERMANRUD, C., STEWARD, D. J. AND SYLTA, O. (eds) *Basin Modelling: Advances and Applications*. Norwegian Petroleum Society Special Publication 3, Elsevier, Amsterdam, 333–341.

SMITH, J. 1971. The dynamics of shale compaction and evolution of pore fluid pressures. *Mathematical Geology*, **3**, 239–263.

SNOWDON, L. R. AND OSADETZ, K. G. 1988. *Gasoline range (C5 to C8) data and C15+ Saturate Fraction Gas Chromatograms for the crude oils from southeastern Saskatchewan portion of the Williston Basin*. Geological Survey of Canada, Open File Report no. **1785**.

STAINFORTH, J. G. AND REINDERS, J. E. 1990. Primary migration by diffusion through organic matter networks, and its effect on oil and gas generation. *In*: DURAND, B. AND BEHAR, F. (eds) *Advances in Organic Geochemistry: Organic Geochemistry in Petroleum Exploration*. Pergamon Press, 61–74.

TALUKDAR, S., GALLANDO, O., VALLEJOS, C. AND RUGGIERO, A. 1987. Observations on primary migration of oil in the La Luna source rocks of Maracaibo Basin, Venezuela. *In*: DOLIGEZ, B. (ed.) *Migration of Hydrocarbons in Sedimentary Basins*. Editions Technip, Paris, 59–78.

TISSOT, B., CALIFET-DEBYSER, Y., DEROO, G. AND OUDIN, J. L. 1971. Origin and evolution of hydrocarbons in early Toarcian shales, Paris basin, France. *American Association of Petroleum Geologists Bulletin*, **55**, 2177–2193.

—— AND PELET, R. 1971. Nouvelles données sur les mécanismes de genèse et de migration du pétrole, simulation mathématique et application à la prospection. *8th World Petroleum Congress, Moscow*, **4**. Wiley, Chichester, 35–46.

——, —— AND UNGERER, P. 1987. Thermal history of sedimentary basins, maturation indices and kinetics of oil and gas generation. *American Association of Petroleum Geologists Bulletin*, **71**, 1445–1466.

—— AND WELTE, D. H. 1984. *Petroleum Formation and Occurrence* (2nd edn). Springer Verlag, Berlin.

UNGERER, P. 1990. State of the art of research in kinetic modelling of oil formation and expulsion. *Organic Geochemistry*, **16**, 1–25.

——, BEHAR, E. AND DISCAMPS, D. 1983. Tentative calculation of the overall volume expansion of organic matter during hydrocarbon genesis from geochemistry data. Implications for primary migration. *In: Advances in Organic Geochemistry 1981*. John Wiley & Sons, Chichester, 129–135.

——, BURRUS, J., DOLIGEZ, B., CHENET, P. Y. AND BESSIS, F. 1990. Basin evaluation by integrated two-dimensional modelling of heat transfer, fluid flow, hydrocarbon generation and migration.

American Association of Petroleum Geologists Bulletin, **74**, 309–335.

VANDENBROUCKE, M., DURAND, B. AND OUDIN, J. L. 1983. Detecting migration phenomena in a geological series by means of C1-C35 hydrocarbon amounts and distributions. *Advances in Organic Geochemistry 1981*. John Wiley & Sons, Chichester, 147–155.

WEI, H. F., LEDOUX, E. AND MARSILY, G. 1990. Regional modelling of ground water and salt environmental tracers in deep aquifers in the Paris Basin. *Journal of Hydrology*, **120**, 341–358.

Submarine gas seepage in the North Sea and adjacent areas

M. HOVLAND

Statoil, PO Box 300, N-4001 Stavanger, Norway

Abstract: The occurrence of sub-seafloor gas accumulations and the seepage of gas through the seafloor are of interest not only to hydrocarbon prospectors, but also to marine construction engineers and to marine biologists. While direct seep observations (visual and acoustic) are rare, the observations of seep indicators, such as gas-charged sediments, pockmarks, carbonate-cemented sediments, clay diapirs, and dense benthic communities are relatively common.

In the North Sea there are several published direct seep observations and numerous observations of seep indicators. In addition, there are new features which have recently been observed in sandwave fields in the Southern North Sea. These are abnormally high-amplitude ('freak') sandwaves.

In the Skagerrak there are gas-charged sediments, pockmarks, and suspected sub-surface clay diapirs. In the Kattegat there are numerous gas seeps, gas-charged sediments, pockmarks and extensive areas of carbonate-cemented sediments. In the southeastern portion of the Baltic Sea there are gas-charged sediments and pockmarks. Along the Swedish coast (off Stockholm) small pockmarks, gas-charged sediments and anomalous concentrations of algal mats have been studied.

On the shelf off Mid-Norway there are gas-charged sediments, small areas with pockmarks, suspected clay diapirs, and occurrences of cold-water coral reefs thought to be related to seepage.

Biological, geological and geochemical research is currently being undertaken by British, Norwegian, Danish, German, Swedish and Russian scientists in some of the regions mentioned here. The objective of this research is mainly to determine the origin, fate and relevance of both seeping gas and the gas contained in sediments.

The occurrence of gas accumulations in shallow sediments and the seepage of gases through the seafloor are of great interest to marine biologists, marine construction engineers and hydrocarbon prospectors. Submarine seepage was reviewed in depth by Hovland and Judd (1988). This paper provides an up-dated summary of seep locations in the North Sea, the shelf off Mid-Norway, in the Kattegat, the Skagerrak, and the Baltic Sea (Fig. 1).

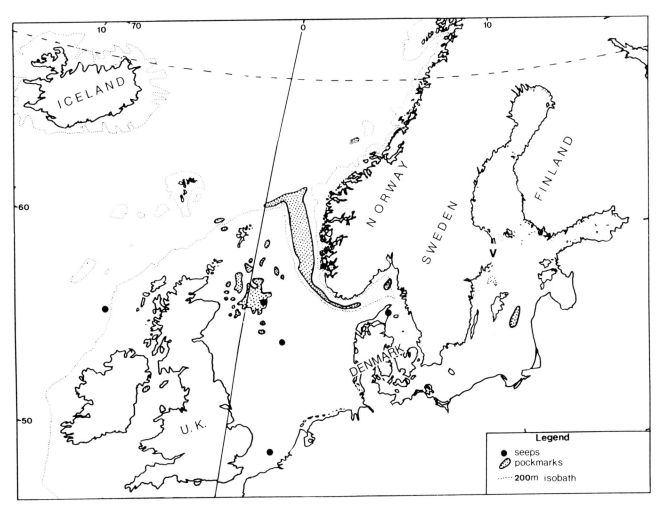

Fig. 1. Map showing location of seeps and pockmarks in the North Sea and adjacent areas. 'V' on the Swedish east coast indicates Vättershaga.

From *Petroleum Geology of Northwest Europe: Proceedings of the 4th Conference* (edited by J. R. Parker).

In addition to the direct observations of seepage, some seep-related features (seep indicators) are also included in this review.

The North Sea

Seepage has been observed directly, and seep indicators recorded across much of the Central North Sea. The southern-most gas-charged sediments in the North Sea occur in the German Bight off the Belgian and Dutch coasts. Direct acoustic observations of seepage plumes have also been made in this area (Hovland and Judd 1988).

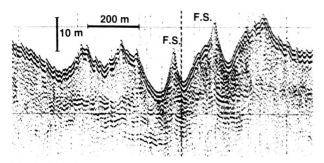

Fig. 2. A shallow reflection seismic (sparker) record across the sand-wave field shown in Figs 3 and 4. Note the two steep-sided and extremely high sandwaves marked F.S. ('freak sandwaves') and the high acoustic reflection strength in some of the sub-surface reflectors indicative of gas-charged sediments.

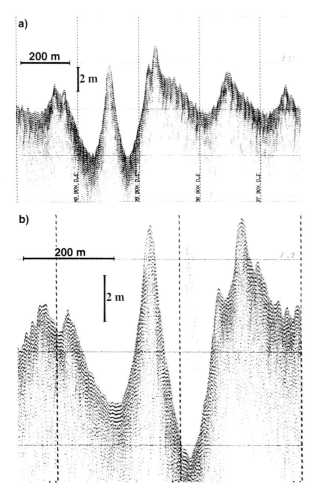

Fig. 3. Two independent high-resolution shallow seismic records across 'freak' sandwaves caused by venting of gas in the Southern North Sea. Note acoustic reflection from the gas plume in the water column. (a) shows a series of three normal sandwaves seen to the right of the 'freak' sandwave.

'Freak' sandwaves

Seabed sediments in parts of the Southern North Sea generally contain a large portion of sand and gravel and are, therefore, normally not prone to pockmark development (pockmarks are caused by gas venting through fine-grained sediments). However, during detailed seafloor mapping for the Zeepipe gas pipeline across a field of large sandwaves off the Belgian and Dutch coasts, gas seeps were found to influence the seabed morphology. Along several kilometres of the surveyed seabed section there were acoustic reflectors with variable reflectivity (Fig. 2). Reflections in the water column adjacent to anomalously high ('freak') sandwaves indicate that gas is seeping out of the seafloor from the underlying gas-charged sediments (Fig. 3b).

The high ('freak') sandwaves build up near gas seep locations. A possible mechanism is that sand grains, which would otherwise fall to rest at locations determined by the sandwave generating tide-water current regime, immediately become resuspended by an additional process—that of seeping fluids. Subsequently, these sand grains are shifted by currents to another ('alternate') nearby location. Over some time, sand will pile up at this 'alternate' location (Fig. 4) and cause the development of 'freak' sandwaves characterized by an abnormal height-to-length ratio and by being less mobile than the surrounding sandwaves.

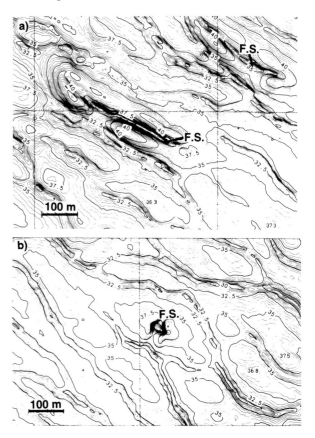

Fig. 4. Two examples of a detailed contour map in the sandwave field shown in Figs 2 and 3: (**a**) two elongated 'freak' sandwaves (F.S.); (**b**) the 'freak' sandwave is isolated and cusped. Note the depressions caused by suspected gas seeps fringing these sandwaves. The contour interval is 0.5 m in both examples.

Tommeliten gas seeps

At the Tommeliten Field in the southern part of the Norwegian sector of the North Sea (Norwegian Block 1/9), continuous methane seeps occur above a piercement salt diapir structure (Fig. 5) (Hovland and Sommerville 1985; Hovland and

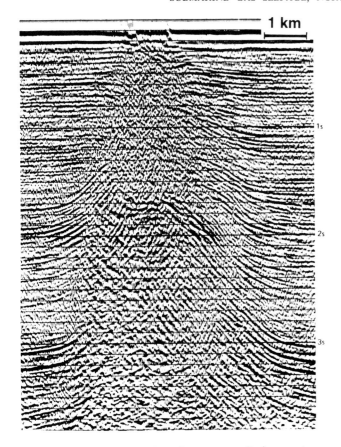

1 km

Fig. 5. A deep reflection seismic section across a salt piercement diapir at the Tommeliten Field in the Central North Sea. Numerous gas seeps occur on the seabed above this structure. The vertical scale is given in seconds two-way-travel time. (From Hovland and Judd 1988.)

Judd 1988). Near the seep locations there are also shallow gas-charged sediments, large patches of benthic communities and small 'eyed' pockmarks, where the 'eye' consists of organisms such as bivalves, echinoderms, starfish, sea anemones and shrimp (Hovland and Thomsen, 1989).

From an estimated total of 120 individual seeps a volume of approximately $24 \, m^3$ per day of gas (at 75 m water depth), mainly methane, emits through the seafloor and enters into the water column (Hovland and Judd 1988). The absolute fate of this gas is unknown, but most of it is expected to survive bubble transport to the sea surface (and thus enter the atmosphere), while some of the volume is dissolved in the seawater, some is oxidized near the seabed, and some is used by CH_4-consuming bacteria in the water column.

Actively seeping pockmarks

In addition to the two largest fields of pockmarks in the North Sea: (1) in the Norwegian Trench (van Weering *et al.* 1973; Hovland and Judd 1988); (2) in the Witch Ground basin (Jansen 1976; Long 1986; Hovland and Judd 1988), there are numerous small basins with soft, fine-grained sediments where pockmarks and other seep-associated features occur.

In UK Block 15/25 an actively seeping pockmark was discovered during a site survey in 1983 (Hovland and Sommerville 1985). The pockmark was first explored by use of ROV (Remotely Operated Vehicle) in 1985. The bubbling gas was determined to be mainly methane with a biogenic isotopic signature. Furthermore, large (up to 1 m × 2 m) concretions of methane-derived carbonate slabs were found inside the 17 m deep pockmark (Hovland and Judd 1988). During recent years, fish biologists and other marine biologists have studied this seep location. It is still producing methane, seemingly at

the same rate as when it was first discovered (Dando *et al.* 1991).

The Skagerrak

Gas-charged sediments and pockmarks were first discovered in the Skagerrak in 1972 (van Weering *et al.* 1973). Detailed reconnaissance pipeline route mapping later provided some more information on these features (Hovland 1991). Large, up to 24 m deep and 500 m wide, pockmarks were thus found above buried positive features suspected to represent clay diapirs. Gas-charged shallow sediments are ubiquitous in the southern and eastern Skagerrak (van Weering *et al.* 1973; Iversen and Jørgensen 1985; Hovland 1991). A continuous shallow gas-charged field may actually exist which extends from these areas to onshore in northern Jutland and into the Kattegat (Fig. 6).

No active (continuous) seepage has been discovered in the Skagerrak so far. However, the presence of gas-charged shallow sediments (10–20 m below seabed) and the existence of numerous pockmarks suggests that intermittent seepage (eruptions) of gas occurs. According to the model of Hovland and Judd (1988), pockmarks represent centres of intermittent gas, porewater, and sediment eruption (Fig. 7). The mechanism triggering the activity is suspected to be build-up of sub-surface gas pressure and pumping of the sediment–porewater–gas system by long-period sea surface storm waves or tidal waves. Although methane gas analysed from sediment samples in the Skagerrak had typical biogenic isotope signatures, heavier hydrocarbons of a thermogenic origin were also found in the sediments (Schmaljohann *et al.* 1990). The shallow gas in the Skagerrak is, therefore, of mixed origin.

The Kattegat

The Jutland–Kattegat shallow gas field contains numerous onshore and offshore active methane seeps (Fenchel 1989; Jørgensen *et al.* 1990). Some of the continuous seeps have recently been studied in shallow water along a beach located south of the city of Fredrikshavn. Other continuous seeps are found in deeper water to the north, northeast and east of Fredrikshavn. All the Kattegat seeps are associated with relatively large surface occurrences of methane-derived carbonates (Jørgensen *et al.* 1990). Some of the carbonate structures in water depths of 40–50 m are described as up to 4 m high pinnacles, mounds and thick crusts of various shapes and sizes. The isotopic signature of the methane seeping offshore is biogenic ($\delta^{13}C < 50\permil$), and is suspected to be generated at sediment depths less than 1000 m (Fenchel 1989).

The Baltic Sea

In the Baltic Sea there are two main regions with occurrence of intermittent gas seepage or gas eruptions: (1) along the southeast coast of Sweden (Söderberg and Floden 1989), and (2) in the Southern Baltic Sea off the northern coast of Poland (Blascisin and Lange 1985).

SE coast of Sweden

Söderberg and Flodén (1989) found several locations with gas escape features during seabed mapping with side scan sonar and a sub-bottom profiler. The features occur along the Swedish coast, south and north of Stockholm, at linear bedrock fault lines buried beneath sediments. The most prominent features were found over intersections between two or more such zones.

At one location (Vättershaga), methane was sampled from bedrock fissures buried below a 3.5 m thick algal mat overlying

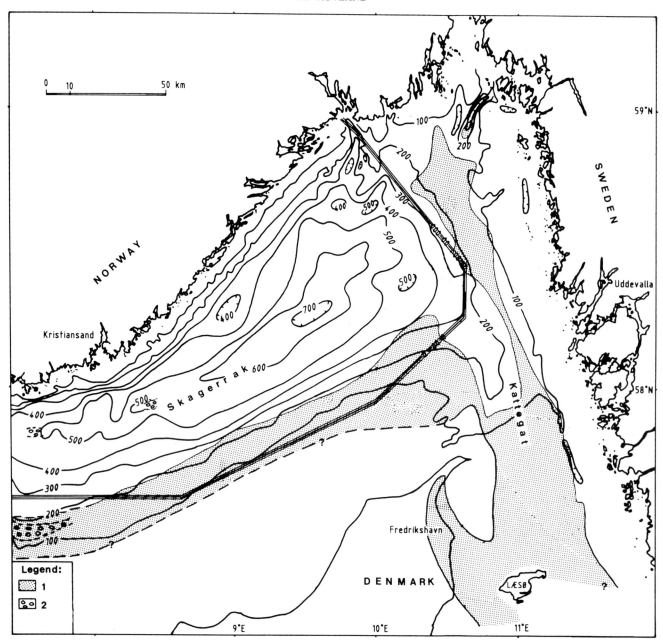

Fig. 6. Map of the Skagerrak and Kattegat showing water depths (m), location of our main survey line, occurrence of shallow gas (1), and pockmarks (2). Shallow gas occurrence is based on van Weering *et al.* (1973), Fenchel (1989), and a survey conducted by Statoil. Pockmark occurrences are based on Maisey *et al.* (1980), van Weering *et al.* (1973) and the Statoil survey.

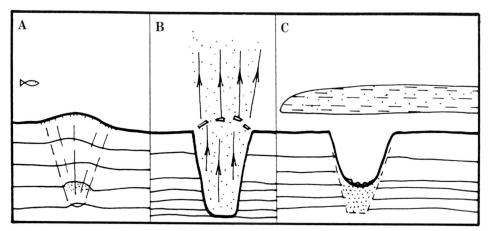

Fig. 7. (**a**) Gas or porewater pressure builds up in a shallow porous layer below an impermeable cohesive sealing layer. (**b**) The excess pressure increases to sealing capacity causing instantaneous eruption and sediment fluidization. Gas, porewater and sediments are ejected into the water column. (**c**) The fine-grained sediments become suspended in the water and are transported away by currents.

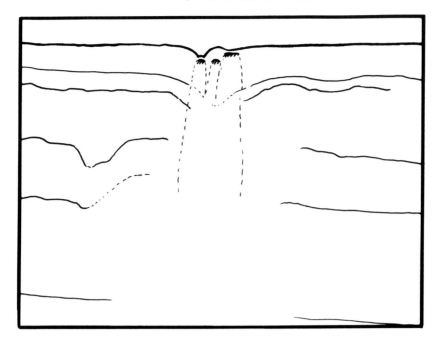

Fig. 8. A sketch of shallow gas accumulation and pockmark craters occurring in the South Gotland Basin of the Baltic Sea. After Blascisin and Lange (1985).

3 m thick sediments. The sampling was performed by driving a 10 m long steel pipe through the algal mat and sediments through a hole in the winter ice (Söderberg and Flodén 1989). The abnormally thick algal mat is believed to have formed partly due to methane seepage (Söderberg and Flodén 1989). Furthermore, the authors suggest that the methane may be formed in deep meta-sedimentary rocks further west and that it migrates and vents through tectonic lineaments in eastern Sweden.

Southern Baltic Sea

Shallow gas and pockmark craters occur in two separate sedimentary basins in the southern Baltic Sea: the Gdansk Basin and the South-Gotland Basin (Fig. 8) (Blascisin and Lange 1985). The craters in the Gdansk Basin range from 10–250 m in width and from 0.5–3 m in depth. Some of them are reported to occur in long chains (Blascisin and Lange 1985).

In the Gotland Basin at a water depth of 110–140 m there are numerous pockmarks up to 6 m deep and 200 m wide. There are also shallow acoustic anomalies (gas-charged sediments) associated with the craters.

The shelf off Mid-Norway

The seabed on the continental shelf off Mid-Norway consists mainly of clay with a rugged upper surface. Most of the ruggedness was caused by ice tectonics and grounded icebergs. However, after discovery of numerous seeps, particularly within one specific area (Hovland 1990a), it now appears that some of the seabed features can be explained by gas-associated mud diapirism. A tentative formation model for some of the ridges and dome-shaped features was presented by Hovland (1990a).

The gas-associated clay diapir model is illustrated with a cartoon in Fig. 9. There are three suspected stages of clay diapir formation and seepage on the Mid-Norway continental shelf:

1. gas permeates from below into a relatively thick unit of plastic clay;

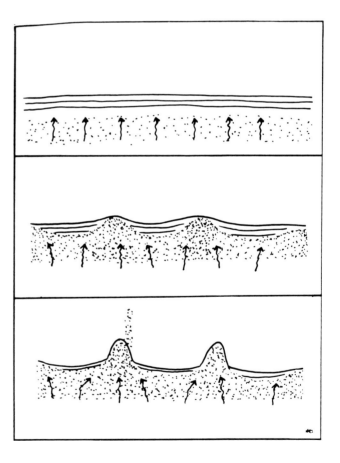

Fig. 9. Cartoon showing three stages in the suspected gas-related clay diapir formation off Mid-Norway: (1) gas migrating from below permeates into the upper plastic clay unit: (2) gas fills more and more of the pore volume in the plastic clay, by substituting water, causing local static instability and surface deformation; (3) after additional buoyancy-induced deformation, tension stress causes vertical weakness zones to develop in the plastic clay, i.e. causing seepage. (After Hovland 1990a.)

2. as gas fills more and more of the clay pore volume, static instability occurs and the clay surface deforms due to differential buoyancy;

3. finally, and after further deformation, vertical weakness zones are established through which gas migrates more or less freely into the water column (Hovland 1990*a*).

When there is no evident gas seepage from the domes or ridges it may be difficult to distinguish between suspected diapirs and morainic features.

From reconnaissance pipeline route surveys for petroleum development off Mid-Norway it has also been found that both isolated and long chains of *Lopheila* reefs exist. The reefs attain heights of 20 m and are often more than 50 m wide. They often occur on ridges of suspected diapiric origin at water depths between 200 m and 300 m (Hovland 1990*b*). Although these cold water coral reefs have been known for many years, a satisfactory explanation for their formation and existence has been lacking. One recent theory is that they subsist on bacteria and microfauna which is based on chemosynthesis resulting from the venting of light hydrocarbons through the seabed (Hovland and Thomsen 1989; Hovland 1990*b*).

Conclusions

This paper has reviewed some of the recent discoveries of seep-related phenomena in north European waters. The results clearly demonstrate the great diversity of seep indicators. It seems that in every region where seabed mapping with side scan sonar and high-resolution reflection seismic is performed, new pockmarks, gas-charged sediments or other seep indicators are found. Considering the relatively small areas of the world's continental shelves that have been mapped, it must be concluded that there are still numerous seep locations, also in northern and arctic waters, remaining to be discovered. Seep detection in the ocean is still in its infancy and the features recorded have mainly been located by chance.

Seep-related phenomena have now gripped the imagination of marine biologists and geochemists and will probably become an area of scientific focus in future years. It is a major challenge to understand these seabed processes and their effects on marine life, and the environment in general.

Statoil is acknowledged for approving publication of data presented here.

References

BLASCISIN, A. M. AND LANGE, D. 1985. *Gasonraojavedenie na dne Baltidickogo i Severnogo More*. The fifth international symposium of the geological history and recent sediment loading process in the Black and Baltic Seas, Varna (proceedings in Russian).

DANDO, P. R., AUSTEN, M., BURKE, R. J. Jr., KENDALL, M. A., KENNICUTT, II, M. C., JUDD, A. G., MOORE, D. C., O'HARA, S. C. M., SCHMALJOHANN, R. AND SOUTHWARD, A. J. 1991. The ecology of a North Sea pockmark with an active methane seep. *Marine Ecology Progress Series*, **70**, 49–63.

FENCHEL, T. 1989. Skorstene i Kattegat. *Hovedområdet, Copenhagen University*, **19** (10), 10–11.

HOVLAND, M. 1990*a*. Suspected gas-associated clay diapirism on the seabed off Mid-Norway. *Marine and Petroleum Geology*, **7**, 267–276.

—— 1990*b*. Do carbonate reefs form due to fluid seepage? *Terra Nova*, **2**, 8–18.

—— 1991. Large pockmarks, gas-charged sediments and possible clay diapirs in the Skagerrak. *Marine and Petroleum Geology*, **8**, 311–316.

—— AND JUDD, A. G. 1988. *Seabed Pockmarks and Seepages*. Graham & Trotman, London.

—— AND SOMMERVILLE, J. H. 1985. Characteristics of two natural gas seepages in the North Sea. *Marine and Petroleum Geology*, **2**, 319–326.

—— AND THOMSEN, E. 1989. Hydrocarbon-based communities in the North Sea? *Sarsia*, **74**, 29–42.

IVERSEN, N. AND JØRGENSEN, B. B. 1985. Anaerobic methane oxidation rates at the sulphate-methane transition in marine sediments from Kattegat and Skagerrak (Denmark). *Limnological Oceanography*, **30**, 944–955.

JANSEN, J. H. F. 1976. Late Pleistocene and Holocene history of the northern North Sea, based on acoustic reflection records. *Netherlands Journal of Sea Research*, **10**, 1–43.

JØRGENSEN, N. O., LAIER, T., BUCHARDT, B. AND CEDERBERG, T. 1990. Shallow hydrocarbon gas in the northern Jutland–Kattegat region, Denmark. *Bulletin of the Geological Society of Denmark*, **38**, 69–76.

LONG, D. 1986. *Seabed sediments, Fladen sheet 58°N–00°E*. British Geological Survey, 1:250 000 map series.

MAISEY, G. H., ROKOENGEN, K. AND RAAEN, K. 1980. *Pockmarks formed by seep of petrogenic gas in the southern part of the Norwegian trench*. Continental Shelf Institute (IKU), Report No. P-258/1/80.

SCHMALJOHANN, R., FABER, E., WHITICAR, M. J. AND DANDO, P. R., 1990. Co-Existence of methane- and sulphur-based endosymbioses between bacteria and invertebrates at a site in the Skagerrak. *Marine Ecological Progress Series*, **61**, 119–124.

SÖDEBERG, P. AND FLODÉN, T. 1989. Geofysiska och geochemiska undersökingar i anslutning till tektoniska lineament i Stockholm Skärgård. Vattenfall FUD rapport, Alvkarleby Vattenfall.

VAN WEERING, T., JANSEN, J. H. F. AND EISMA, D. 1973. Acoustic reflection profiles of the Norwegian Channel between Oslo and Bergen. *Netherlands Journal of Sea Research*, **6**, 214–263.

Discussion

Question (A. H. W. Woodruff, Kerr-McGee, London):

What do you see as the source of the gas observed? Are you observing mainly gas seeps from shallow sources or escaping from deep hydrocarbon reservoirs by faulting and/or microfracture?

Answer (M. Hovland):

The largest pool of gas in sedimentary basins is found in reservoir rocks above the thermogenic hydrocarbon-generating zones, deeper than about 2000 m. Due to buoyancy, this gas will migrate upwards and be stored in shallow porous sediments wherever favourable conditions exist. An often confusing aspect is that bacterial methane is generated in sediments at depths less than 2000 m. This means that the gas stored at shallow depths is often of a mixed origin. Unless there is a deep open fracture/fault system connecting to the surface, it is suspected that gas is stored at shallow depths for long time periods, before it emits through the seabed.

The use of natural He, Ne and Ar isotopes as constraints on hydrocarbon transport

C. J. BALLENTINE[1] and R. K. O'NIONS

University of Cambridge, Department of Earth Sciences, Downing Street, Cambridge CB2 3EQ, UK
[1] (Present address: Paul Scherrer Institut, CH-5232 Villigen PSI, Switzerland)

Abstract: Hydrocarbon accumulations contain rare gases derived from the atmosphere, the crust and, in some cases, the mantle. The distinctive isotopic structure of these different rare gas components allows them to be resolved. The relative abundances of the He, Ne and Ar in the crustal-, mantle- and atmosphere-derived components provides information on the physical processes which have operated in the sub-surface. When combined with mass balance considerations, the rare gases provide powerful constraints on fluid provenance and transport.

The results of case studies from the Pannonian Basin in Hungary, and the Vienna Basin, Austria are reviewed. Natural gas reservoirs in the Pannonian and Vienna basins provide samples from depths between 0.5 and 5.5 km. The atmosphere- and crustal-derived rare gas components show systematic patterns of He/Ar and Ne/Ar fractionation which are in the sense predicted by rare gas partitioning between liquid, such as oil and water, and a gas phase. This could occur either during transport or upon emplacement into the reservoir. The role of diffusive processes in fractionation of rare gases cannot be resolved and is assumed to play only a minor role in hydrocarbon transport in the accumulations investigated.

The rare gases also convey information on the extent of interaction between groundwater and hydrocarbons. The relationship between major species, such as CH_4 and rare gas tracers, provides estimates for the minimum volume of groundwater that has interacted with the hydrocarbon phase. In the case of the Pannonian Basin, the results are consistent with a model in which the natural gas has been transported in a CH_4-saturated water phase, from which the gas has separated upon cooling and/or decompression. The mass balance demonstrates that the interaction between the hydrocarbon gas and groundwater has been on a scale much greater than often envisaged in models of hydrocarbon formation and migration.

Rare gases have been studied in terrestrial materials for many years. Recently, our knowledge of the systematics of rare gases in the Earth's mantle has increased greatly, placing important geochemical constraints on our understanding of the chemical evolution of the Earth (e.g. Allegre *et al.* 1987). Surprisingly little use has been made of rare gases, other than He, in the study of crustal fluid systems. The purpose of this paper is to review recent progress that has been made in the application of rare gases to problems of hydrocarbon migration and transport.

Rare gas isotopic compositions may be used to study the behaviour of crustal fluids, since any fluid within the crust will interact with other fluids of different provenance. For example, hydrocarbons will equilibrate to a greater or lesser extent with the groundwater system containing dissolved atmosphere-derived gases. This will result in the transfer of atmosphere-derived rare gases into the hydrocarbon phase. In many crustal environments mantle-derived volatiles, particularly ^3He, are also present and are usually associated with lithospheric extension and melting (Oxburgh and O'Nions 1987; O'Nions and Oxburgh 1988). Furthermore, decay of the radioelements U, Th and K, either within the fluid reservoir volume or deeper within the crust, will produce radiogenic and nucleogenic rare gas isotopes which may also be present in hydrocarbon reservoirs. With precise isotopic analyses it is now possible in most crustal fluid samples to resolve rare gases from these different sources, quantify their relative contributions, and place constraints on the processes responsible for their input (Fig. 1).

In many instances rare gases are closely associated with major gas species. For example, at mid-ocean ridges ^3He and CO_2 are degassed at a near-constant ratio (Marty and Jambon 1987). Within the continental environment the ratio of $^4He/N_2$ within any one region is often constant, such as in the Hugoton–Panhandle gas fields, Texas (Pierce *et al.* 1964). A methane gas phase which separates from a CH_4-saturated groundwater has a characteristic $CH_4/^{36}Ar$ ratio (Ballentine *et al.* 1991). Where these relationships are preserved, the rare gases provide information about the provenance of the major gas species.

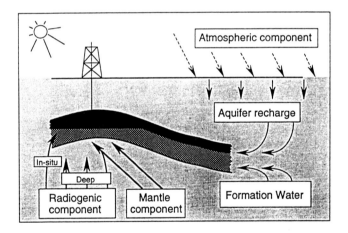

Fig. 1. Schematic diagram of a hydrocarbon reservoir, illustrating the different rare gas components which may occur within hydrocarbon fluids. Atmosphere-derived rare gases are input into the hydrocarbon fluid by interactions between the hydrocarbon phase and the groundwater system which contains dissolved atmosphere-derived rare gases. Radiogenic rare gases produced by the natural decay of the radioelements U, Th and K are also incorporated into hydrocarbon fluids. Within areas of recent continental extension, rare gases derived from the mantle may also be associated with hydrocarbon fluids. The distinct isotopic composition of each of these rare gas components allows the extent of their contribution to the hydrocarbon phase to be resolved.

The relative elemental abundance of rare gases derived from the atmosphere, and those produced radiogenically within the crust, are well defined (e.g. Ozima and Podosek 1983). Since the rare gases are chemically inert, changes in their elemental ratios occur by physical processes and, therefore, must reflect the physical behaviour of the bulk phase in which the rare gases are carried. For example, rare gas fractionation caused by partitioning between two different fluid phases is often

From *Petroleum Geology of Northwest Europe: Proceedings of the 4th Conference* (edited by J. R. Parker).

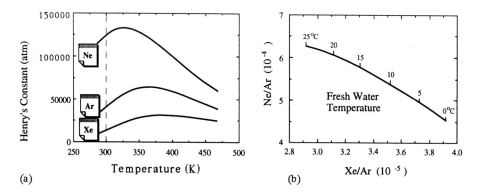

Fig. 2. (a) The Henry's constants for Ne, Ar and Xe illustrate the temperature dependence of their relative solubilities. **(b)** The dissolved atmospheric Ne/Ar and Xe/Ar ratios in freshwater equilibrated with the atmosphere shown as a function of the water temperature. This distinct rare gas pattern enables palaeo-recharge temperatures of groundwaters to be determined to within ±0.5°C.

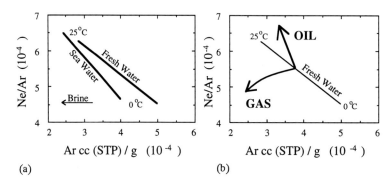

Fig. 3. (a) The Ne/Ar ratio of dissolved atmosphere rare gases in groundwater is plotted against its Ar concentration, for seawater and freshwater. An increase in salinity does not significantly change the relative solubilities of Ne and Ar, but does decrease the concentration of the atmosphere-derived rare gases in solution. This relationship can be used to determine the salinity of groundwater at recharge. **(b)** Interaction between groundwater and another phase will result in partitioning of rare gases between the two phases. The predicted effects of oil–water and gas–water interactions on the dissolved atmospheric Ne/Ar ratio and Ar concentration in the water are shown.

distinct and resolvable from that caused by a diffusive/kinetic fractionation process.

Case studies from hydrocarbon gas fields in the Neogene Pannonian Basin, Hungary and the Vienna Basin, Austria provide the first systematic application of rare gas to the study of a hydrocarbon-rich environment (Ballentine 1991; Ballentine et al. 1991; Ballentine and O'Nions 1992). These studies illustrate how the differently sourced rare gases are resolved using their isotopic composition, how they provide information about the extent of hydrocarbon/groundwater interaction, constrain the mechanism of hydrocarbon transport, and provide an insight into the general behaviour of the fluid regime in sedimentary basins.

Groundwater provenance and rare gase solubility

The introduction of atmosphere-derived rare gases into the sub-surface forms an integral part of these studies. Before discussing the case studies, the solubility of rare gases in groundwater and the information this provides about the provenance of groundwater is briefly reviewed.

The solubility of rare gases in water is strongly temperature dependent. The Henry's constants for Ne, Ar and Xe plotted against temperature are shown in Fig. 2a (Crovetto et al. 1982). The relative proportions of rare gases dissolved in water which has equilibrated with the atmosphere, containing a fixed rare gas composition, is therefore determined by the temperature of the water at equilibration. This is shown for atmosphere-derived Ne/Ar and Xe/Ar ratios in fresh water over the temperature range of 0–25°C (Fig. 2b). Waters which subse-

quently enter the sub-surface, either buried as formation water or as part of aquifer recharge, retain this characteristic rare gas elemental abundance pattern. This property of rare gases has been used successfully in determining the palaeo-recharge temperature of old groundwater systems (e.g. Mazor 1972). Current analytical techniques enable the palaeo-recharge temperature to be determined with a precision of ±0.5°C (e.g. Stute and Deak 1989).

The solubility of rare gases in water also depends on salinity. While an increase in salinity does not greatly affect the solubility of the rare gases, the absolute solubility of the rare gases decreases with an increase in salinity (Smith and Kennedy 1983). This is illustrated in Fig. 3a, where the Ne/Ar ratio of dissolved atmosphere-derived rare gases is plotted against the Ar concentration in freshwater and seawater over a temperature range of 0–25°C. At any given temperature there is a marked decrease in the concentration of dissolved Ar with increasing salinity, but little change in the Ne/Ar ratio. This relationship has been used, for example, to distinguish between formation waters originating as evaporational brines and those originating as seawater in the Palo Duro Basin, Texas (Zaikowski et al. 1987).

Both groundwater palaeotemperature and provenance investigations require that the aqueous rare gas system has remained closed since it last equilibrated with the atmosphere. In some instances it is found that the groundwater contains more atmosphere-derived rare gases than can be accounted for by equilibrium solubility, and this is ascribed to the addition of excess air dissolved in the groundwater system (Heaton and Vogel 1981). On the other hand, if the groundwater equilibrates

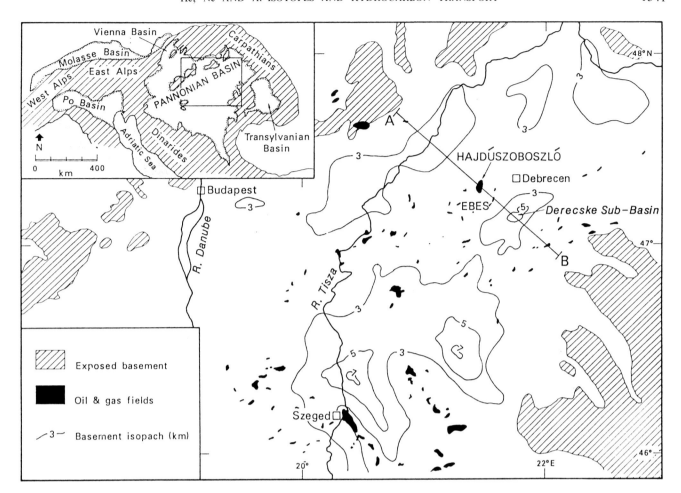

Fig. 4. The locations of the Pannonian Basin and Vienna Basin (inset) shown in relation to the areas of alpine uplift and associated basins. The expanded map shows the location of the Hajdúszoboszló gas field in relation to the basement isopachs, other oil and gas fields and the exposed pre-Pannonian basement (after Ballentine *et al.* 1991). The cross-section A–B is shown schematically in Fig. 8.

with another phase after entering the sub-surface, such as with an oil or gas phase, rare gases dissolved in the groundwater will partition into the oil or gas (e.g. Bosch and Mazor 1988). The rare gases remaining in the water will retain a characteristic rare gas abundance pattern, depending on the phase equilibrated with the water, the extent of interaction between the two phases, and the temperature and salinity of the water during this equilibrium (Fig. 3b). In practice, however, only the hydrocarbon phase, containing the atmosphere-derived rare gases, has been sampled and used to deduce the relationship between the hydrocarbon and any groundwater phase.

The Pannonian and Vienna basins

The two case studies reviewed in this paper are based upon hydrocarbon gas fields from the Pannonian and Vienna basins (Fig. 4). Both basin systems appear to have developed by extension in the early to middle Miocene, creating a series of deep basins separated by shallower basement blocks (Royden 1988; Wessely 1988). Within the Pannonian Basin this was accompanied by surface volcanism. While there is no evidence of this in the Vienna Basin, mantle ^{3}He which is associated with recent melt emplacement is ubiquitous within fluids from both basins (Martel *et al.* 1989; Marty *et al.* 1992). The basins are filled by Neogene–Quaternary sediments which reach maximum depths of 7000 m and 5000 m in the Pannonian and Vienna basins, respectively (Jiricek and Tomek 1981; Berczi 1988). Groundwaters in the Miocene of both basins are highly saline and communicate only locally with shallower systems (Erdelyi 1976; Ottlick *et al.* 1981; Wessely 1983).

The Hajdúszoboszló gas field in the Pannonian Basin (Fig. 4) was used to investigate the detailed systematics of rare gases in a single gas field. It is composed of a series of discrete gas accumulations at different depths within the interbedded clays, marls and sandstones of the Pannonian sedimentary sequence. Samples from six gas fields in the Vienna Basin, located in different stratigraphic and structural units between 0.5 and 5.5 km depth (Fig. 5), provide an impression of rare gas behaviour on the basin scale.

Resolving different rare gas components

The abundance and isotopic composition of He, Ne and Ar have been determined for some 31 samples of natural gas (Ballentine 1991). Both the Pannonian and Vienna Basin samples display similar rare gas isotope systematics. The Vienna Basin samples are used here to illustrate the type of information obtainable from the rare gas isotopes. In the central Vienna Basin, the measured ^{3}He/^{4}Ne ratios (R), normalized to the atmospheric value ($R_a = 1.4 \times 10^{-6}$), vary between $R/R_a = 0.2$ to 0.7 (Fig. 6a). ^{20}Ne within these samples is predominantly atmosphere-derived, as sub-surface radiogenic sources are negligible, and mantle contributions small. The ^{4}He/^{20}Ne ratio in these samples varies between 290 and 32 700, far greater than the atmosphere value of 0.288, which implies that the He in these samples can be considered to be a simple two-component mixture of mantle-derived and radiogenic He, with negligible He contribution from atmospheric sources. Radiogenic He has $R/R_a = 0.02$ and mantle He has

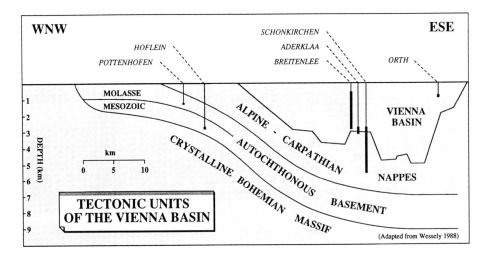

Fig. 5. Schematic cross-section through the Austrian section of the Vienna Basin showing sample locations in relation to the different tectonic units (after Ballentine and O'Nions 1992). Within the central Vienna Basin the temperature gradient is *c.* 31°C km^{-1} and the oil generation window presently lies between 4000–6000 m (Ladwein 1988). The maturity of the Neogene sediments, and the organic content of the calcareous Alpine–Carpathian nappes are both too low for these formations to be significant hydrocarbon sources. It appears that most gas, condensate and oil in the Vienna Basin were probably sourced in the Late Jurassic basinal marl, which matured only during the Miocene subsidence of the Vienna Basin (Ladwein 1988).

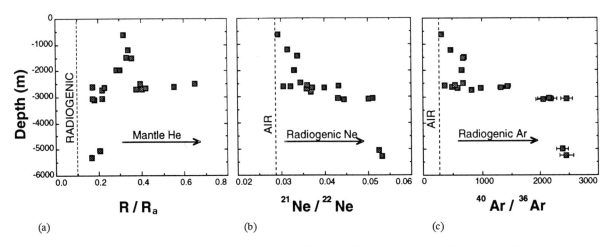

Fig. 6. He, Ne and Ar isotope ratios from the central Vienna Basin (Ballentine 1991), plotted as a function of depth, illustrate how they can be used to resolve rare gases of different provenance. (**a**) R/R_a ratios are the measured $^3He/^4He$ ratio (R) normalized to the atmospheric ratio (R_a). Within the Vienna Basin, atmosphere-derived He contributions are negligibly small and R/R_a ratios reflect a two-component mixture of mantle and crustal-radiogenic He, with end-member values of $R/R_a = 8.0$ and 0.02, respectively. Measured R/R_a values of between 0.2 and 0.7, therefore, represent between a 2 and 8% contribution of mantle ^4He to the total ^4He content. (**b**) $^{21}Ne/^{22}Ne$ ratios are close to the atmospheric value of 0.0290 in the shallowest samples. An increase in the $^{21}Ne/^{22}Ne$ ratio with depth is due to the addition of radiogenic ^{21}Ne. In the deepest sample, with a $^{21}Ne/^{22}Ne$ ratio of 0.05311, 45% of the ^{21}Ne is radiogenic. (**c**) $^{40}Ar/^{36}Ar$ ratios show similar behaviour to the $^{21}Ne/^{22}Ne$ ratios. In the shallowest samples, $^{40}Ar/^{36}Ar$ ratios are close to the atmospheric value of 295.5. An increase in the $^{40}Ar/^{36}Ar$ ratio with depth is due to the addition of radiogenic ^{40}Ar. In the deepest sample, with a $^{40}Ar/^{36}Ar$ ratio of 2460, 88% of the ^{40}Ar is radiogenic.

$R/R_a = 8.0$; therefore these samples contain between 2–8% mantle-derived ^4He (Fig. 6a).

Similarly, the $^{21}Ne/^{22}Ne$ and $^{40}Ar/^{36}Ar$ isotopic ratios may be used to resolve the contribution of Ne and Ar from different sources. Figure 6b and c show the $^{21}Ne/^{22}Ne$ and $^{40}Ar/^{36}Ar$ ratios in the Vienna Basin plotted as a function of depth. In the shallowest samples, both ratios are close to the atmospheric values of $^{21}Ne/^{22}Ne = 0.0290$ and $^{40}Ar/^{36}Ar = 295.5$. There is a general, if not systematic, increase in these ratios with depth to values of 0.05311 and 2460, respectively, in the deepest sample. As with ^{20}Ne, ^{22}Ne and ^{36}Ar are predominantly atmosphere-derived, and because the mantle Ne and Ar contributions are small (Ballentine *et al.* 1991; Ballentine and O'Nions 1992), the increase in the $^{21}Ne/^{22}Ne$ and $^{40}Ar/^{36}Ar$ ratios must be due to additions of radiogenic ^{21}Ne and ^{40}Ar

to an atmosphere-derived component. Within the shallowest samples, the contribution of radiogenic ^{21}Ne and ^{40}Ar is 0% and 6% increasing to 45% and 88%, respectively, in the deepest sample.

Groundwater/hydrocarbon relationships

The ^{20}Ne and ^{36}Ar in gas samples are almost entirely derived from the atmosphere. Other contributions, either from nuclear reactions in the crust or contributions from the mantle, are insignificantly small. They can only reasonably be introduced into the hydrocarbon gas by interaction between the hydrocarbon phase and groundwater which contains the dissolved atmosphere-derived rare gases. The concentration of ^{20}Ne and ^{36}Ar within the hydrocarbon phase, therefore, reflects the

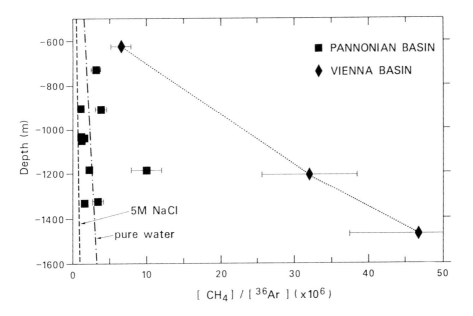

Fig. 7. Variation in measured $CH_4/{}^{36}Ar$ ratios in the Hajdúszoboszló and Vienna Basin gas fields as a function of depth. These values are compared with the $CH_4/{}^{36}Ar$ ratios predicted for CH_4 degassed from air-equilibrated groundwater at the reservoir pressure and temperature, for both pure water and a 5 M NaCl brine (after Ballentine *et al.* 1991). The $CH_4/{}^{36}Ar$ ratios from the Hajdúszoboszló gas field fall close to the theoretical values, suggesting that gases from this field have been transported in solution. Values from the Vienna Basin are much larger and indicate that the extent of interaction between these gases and air-equilibrated groundwater is much less. The Vienna Basin gases are unlikely to have been transported in an aqueous phase.

extent of equilibration between the gas and the groundwater system.

The ratio of $CH_4/{}^{36}Ar$ can be predicted for a gas phase which has degassed from an air-equilibrated water (Ballentine *et al.* 1991). This is determined by the concentration of atmosphere-derived ^{36}Ar dissolved in the water at recharge, and the volume of CH_4 which can be dissolved in the water under reservoir conditions. CH_4 solubility in water depends on pressure and temperature, i.e. depth, and salinity. Assuming a temperature gradient of 35°C km^{-1} and hydrostatic pressure, the predicted $CH_4/{}^{36}Ar$ ratios for pure water and a 5 M NaCl brine saturated with CH_4 at reservoir pressure and temperature may be compared with the measured values in both the Vienna and Pannonian Basin gas samples (Fig. 7).

The $CH_4/{}^{36}Ar$ ratios measured in the Pannonian Basin gas samples are similar to the range of values predicted for a gas phase which has separated from a CH_4-saturated groundwater, which derived its ^{36}Ar through equilibration with the atmosphere. This is consistent with transport of CH_4 in a saturated water phase, from which a gas phase separates upon cooling or decompression. This might occur, for example, at topographic highs such as occupied by the Hajdúszoboszló gas field (Fig. 8). In contrast, samples from the Vienna Basin have $CH_4/{}^{36}Ar$ ratios much higher than the range predicted for a CH_4-saturated water. These values require relatively less interaction between the hydrocarbons and air-equilibrated groundwaters, and suggest that gas transport in the Vienna Basin has not occurred primarily in a CH_4-saturated groundwater.

Constraints on transport

The relative abundance of rare gases in an atmosphere-equilibrated groundwater, and those produced by nuclear reactions in the crust, are well defined. Any exceptional variation in these patterns must reflect the physical processes which have occurred in the sub-surface and, therefore, may be used to identify and quantify these processes. The patterns of fractionation of rare gases in both the Pannonian and Vienna Basin gas samples are very similar. Data from the Vienna Basin are

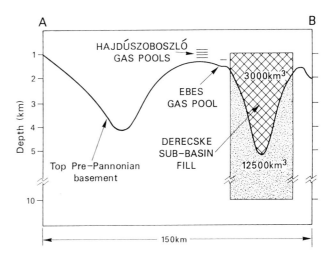

Fig. 8. Schematic cross-section A–B from Fig. 4, showing the location of the Hajdúszoboszló gas field in Pannonian sediments above a basement high (after Ballentine *et al.* 1991). The adjacent Derecske sub-basin reaches a maximum depth of 5.5 km and covers an area of 1250 km². The sedimentary fill of the sub-basin is estimated at 3000 km³ (cross-hatched), compared with a combined sedimentary and basement volume to a depth of 10 km of 12 500 km³ (stippled and cross-hatched). These volumes provide perspective on the estimated volumes of groundwater and rock required to source the atmosphere-derived and radiogenic rare gases, respectively, observed in the Hajdúszoboszló Field. This does not imply that this basin is necessarily the unique source of these rare gases.

used to illustrate the results obtained (Fig. 9). The radiogenic $^4He/{}^{40}Ar$ and $^4He/{}^{21}Ne$ ratios have been normalized to their crustal end-member production ratios in order to obtain a fractionation value F for each sample (a value of $F = 1$ indicates that no fractionation has occurred). These are compared with atmosphere-derived $^{20}Ne/{}^{36}Ar$ ratios in the gas samples, normalized to the expected ratio in groundwater.

Where fractionation is observed in the radiogenic $^4He/{}^{40}Ar$

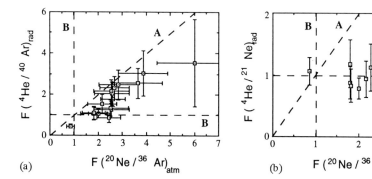

Fig. 9. Comparison of $(^4He/^{40}Ar)_{rad}$ $(^4He/^{21}Ne)_{rad}$ and $(^{20}Ne/^{36}Ar)_{atm}$ normalized to their predicted end-member ratios, and expressed as fractionation values (F). (**a**) Comparison of $F(^4He/^{40}Ar)_{rad}$ and $F(^{20}Ne/^{36}Ar)_{atm}$. All data fall on or close to the line of equal-fractionation, labelled A, suggesting that the radiogenic and atmosphere-derived rare gases must have been intimately mixed *before* rare gas fractionation. (**b**) Comparison of $F(^4He/^{21}Ne)_{rad}$, and $F(^{20}Ne/^{36}Ar)_{atm}$. All data are within error of the line $F(^4He/^{21}Ne)_{rad} = 1$, labelled B, indicating that no fractionation in the radiogenic $^4He/^{21}Ne$ ratio is resolvable, regardless of the extent of fractionation in the atmosphere-derived component. This indicates that diffusive/kinetic fractionation processes have been relatively unimportant. The pattern of rare gas fractionation is typical of that predicted by a gas/liquid phase partitioning model (see text).

ratio, then fractionation of similar magnitude is also observed in the atmosphere-derived $^{20}Ne/^{36}Ar$ ratios (Fig. 9a). Fractionation of the radiogenic $^{21}Ne/^{40}Ar$ ratio (not shown) also occurs coherently with that of the atmosphere-derived $^{20}Ne/^{36}Ar$ ratio. This coherent He/Ar, Ne/Ar fractionation implies that both crustal-radiogenic and atmosphere-derived rare gases have been intimately mixed before the fractionation occurs. In marked contrast, the radiogenic $^4He/^{21}Ne$ ratio does not show any resolvable fractionation from the predicted end-member value, regardless of the extent of fractionation observed in the atmosphere-derived $^{20}Ne/^{36}Ar$ (Fig. 9b). If a kinetic/diffusive process was responsible for the fractionation observed in the He/Ar and Ne/Ar ratios, then it would be expected that fractionation would be observed also in the He/Ne ratio. Diffusive processes cannot, therefore, be resolved in the rare gases and by inference must be assumed to have played only a minor role in the transport of the hydrocarbon gas.

The partitioning of rare gases between liquid and gas phases at equilibrium depends upon the relative solubilities of the rare gases and the gas/liquid volume ratios. For example, if $F_{Ne/Ar}$ = $(Ne/Ar)_{gas}/(Ne/Ar)_{liquid}$, $F_{Ne/Ar} \rightarrow K_{Ne}/K_{Ar}$ as the gas/liquid volume ratio $\rightarrow 0$, where $F_{Ne/Ar}$, K_{Ne} and K_{Ar} are the fractionation value of Ne/Ar in the gas phase and the Henry's solubility constants of Ne and Ar, respectively. The solubility of He and Ne in both water (Crovetto *et al.* 1982; Smith 1985) and oil (Kharaka and Specht 1988) are very similar, whereas the solubility of Ar is quite different. For example in a 5 M NaCl brine at 310K, $K_{He} = 476\,000$ atm, $K_{Ne} = 472\,000$ atm, and $K_{Ar} = 233\,000$ atm. In this case, as the gas/brine volume ratio $\rightarrow 0$, then $F_{He/Ne} \rightarrow 1.01$, $F_{He/Ar} \rightarrow 2.04$ and $F_{Ne/Ar} \rightarrow 2.03$. Rare gas partitioning between an oil or water with a gas phase, therefore, fractionates the He/Ar and Ne/Ar ratios by the same magnitude but would not be resolvable in the He/Ne ratios.

The observed fractionation patterns of rare gases follows that predicted for liquid/gas phase partitioning. They are also consistent with the observation that the Pannonian Basin gases are degassed from a water phase. The Vienna Basin gases, however, do not have such a straightforward relationship with the groundwater system. It is entirely possible that they have been subject to interactions with an oil phase, either within closely associated oil deposits or with the source rocks, from which they are now separated.

Scale of fluid flow

Some limits may be placed on the total volume of groundwater which has interacted with the natural gas accumulation by considering the volume of groundwater required to introduce the atmosphere-derived rare gases present in the gas fields. In addition, a crustal-radiogenic rare gas mass balance may place some limits on the crustal degassing history of the region. Such mass balance calculations have been made from the Hajdúszoboszló Field (Ballentine *et al.* 1991).

The volume of gas in the Hajdúszoboszló gas field is estimated at 33 km³ gas STP. Given the average concentration of atmosphere-derived ^{20}Ne and ^{36}Ar in the gas samples, it contains 7.3×10^3 m³ (STP) ^{20}Ne and 1.3×10^4 m³ (STP) ^{36}Ar. The corresponding volume of water that must degas completely to account for the total reservoir volume of ^{20}Ne and ^{36}Ar is 50 km³ and 17 km³, respectively, given that the concentration of ^{36}Ar and ^{20}Ne in seawater at 25°C is 7.65×10^{-7} cm³ (STP) per g(H₂O) and 1.47×10^{-7} cm³(STP) per g(H₂O), respectively (Ozima and Podosek 1983). The difference between the water volumes calculated from the Ne and Ar concentrations reflects the fact that the $^{20}Ne/^{36}Ar$ ratio in the reservoir gas is not the same as that in atmosphere-equilibrated groundwater. In reality, therefore, the volume of water must be much larger, as the fractionation observed between ^{20}Ne and ^{36}Ar in the reservoir gas requires that the gas/water volume ratio was very small, which in turn results in only partial transfer of the atmosphere-derived rare gases from the water to the gas phase (Ballentine 1991).

The estimated reservoir pore volume is 0.367 km³; these water volume estimates are larger than this by a factor of between 50 and 140. Assuming an average pore space of 5% at a depth of 3.5 km (Dovenyi and Horvath 1988), the *minimum* volume of water to have interacted with the gas phase would occupy a volume of between 350 and 1000 km³ of the sedimentary rock. This can be placed in perspective by considering the Derecske sub-basin to the southeast of the Hajdúszoboszló Field, which has a sedimentary fill of 3000 km³ (Fig. 8).

Rare gases produced radiogenically within the crust also form a significant proportion of the total rare gas inventory in the Hajdúszoboszló gas field. Given the total volume of gas in the Hajdúszoboszló Field and the average concentration of the radiogenic ^4He, then the gas field contains 23×10^6 m³(STP) $^4He_{rad}$. This amount of ^4He is large compared to the volume which could have been produced by α-decay of U and Th within the reservoir itself, and even exceeds the volume expected to have been produced within the sedimentary fill of the Derecske sub-basin. The possibility that deeper regions of the crust have contributed to the ^4He mass balance must be considered. For an average U, Th content of the upper crust of 2.8 ppm and 10.7 ppm (Taylor and McLennan 1985), and an average rock density of $2.7\,cm^3\,g^{-1}$, the volume of rock required to produce the measured $^4He_{rad}$ over 1, 10 and 50 Ma is estimated at 1×10^5, 1×10^4 and $2 \times 10^3\,km^3$, respectively. Over the time of basin formation, the equivalent of the entire crustal rare gas production down to a depth of 10 km over the area of the Derecske sub-basin (Fig. 8) is, therefore, required to produce the volume of $^4He_{rad}$ in the Hajdúszoboszló gas field.

Regional fluid flow overview

The information now available from the rare gases associated with hydrocarbon accumulation in sedimentary basins starts to provide a picture of the fluid behaviour within these extensional basin systems. From the large volume of radiogenic ^4He observed in just one Pannonian Basin gas field at Hajdúszoboszló, some mechanism of storage of these radiogenic rare gases within the deep crust appears to be required. The release of radiogenic rare gases from these regions is presumably related to basin formation. In addition, the radiogenic rare gas ratios of $(^4He/^{21}Ne)_{rad}$, $(^4He/^{40}Ar)_{rad}$ are close to the theoretical production ratios except where the atmosphere-derived $(^{20}Ne/^{36}Ar)_{atm}$ is also shifted from its predicted end-member value. The radiogenic rare gases have not fractionated from the end-member values until *after* mixing with the atmosphere-derived component. This observation requires that the radiogenic rare gases are stored within the crust at close to their production ratios. Furthermore, the release and transport of the radiogenic rare gases also occurs without any relative fractionation. Therefore, the mechanism of transport of the radiogenic rare gases to shallow regions must be principally by advection and most probably as part of a single-phase system.

The volume of water required to introduce the atmosphere-derived rare gases into the Pannonian Basin gas field is significantly larger than the reservoir volume. The radiogenic rare gases, mostly derived from the deeper crust, mix with air-equilibrated groundwater on a regional scale, after which the coherent fractionation of both radiogenic and atmosphere-derived rare gas abundance ratios occurs through liquid–gas phase partitioning. Clear evidence for kinetic fractionation of rare gases has not been found and diffusional transport of hydrocarbons must be assumed to be relatively unimportant in both the Vienna and Pannonian Basin gas fields. In the case of the Pannonian Basin, the concentration of natural gas in the Hajdúszoboszló gas field has occurred, at least in part, by dissolution and transport in groundwater.

This work has been supported by the Royal Society and EEC Contract No. JOUG-006-UK. We are grateful for reviews by C. Clayton and G. Turner. Dept Earth Science. Contribution No. 2558.

References

ALLEGRE, C. J., STAUDACHER, T. AND SARDA, P. 1987. Rare Gas Systematics: Formation of the atmosphere, evolution and structure of the Earth's mantle. *Earth and Planetary Science Letters.* **81**, 127–150.

BALLENTINE, C. J. 1991. *He, Ne and Ar isotopes as tracers in crustal fluids.* Unpublished PhD thesis, University of Cambridge.

—— AND O'NIONS, R. K. 1992. Mantle neon in Vienna Basin hydrocarbon reservoirs. *Earth and Planetary Science Letters* **113**, 553–567.

——, ——, OXBURGH, E. R., HORVATH, F. AND DEAK, J. 1991. Rare gas constraints on hydrocarbon accumulation, crustal degassing and groundwater flow in the Pannonian basin. *Earth and Planetary Science Letters*, **105**, 229–246.

BERCZI, I. 1988. Preliminary sedimentological investigation of a neogene depression in the Great Hungarian Plain. *In*: ROYDEN, L. H. AND HORVÁTH, F. (eds) *The Pannonian Basin: a Study in Basin Evolution.* American Association of Petroleum Geologists Memoir, **45**, 107–116.

BOSCH, A. AND MAZOR, E. 1988. Natural gas association with water and oil as depicted by atmospheric noble gases: case studies from the south eastern Mediterranean coastal plain. *Earth and Planetary Science Letters*, **87**, 338–346.

CROVETTO, R., FERNANDEZ-PRINI, R. AND JAPAS, M. L. 1982. Solubilities of inert gases and methane in H_2O and in D_2O in the temperature range of 300 to 600 K. *Journal of Chem. Phys.*, **76**, 1077–1086.

DOVENYI, P. AND HORVATH, F. 1988. A review of temperature, thermal conductivity and heat flow data for the Pannonian basin. *In*: American Association of Petroleum Geologists Memoir, **5**, 195–233.

ERDELYI, M. 1976. Outlines for the hydrodynamics and hydrochemistry of the Pannonian Basin. *Acta Geologica Academiae Scientiarum Hungaricae*, **20**, 287–309.

HEATON, T. H. E. AND VOGEL, J. C. 1981. 'Excess air' in groundwater, *Journal of Hydrology*, **50**, 201–216.

JIRICEK, R. AND TOMEK. C. 1981. Sedimentary and structural evolution of the Vienna Basin. *Earth Evolution Sciences*, **3–4**, 195–203.

KHARAKA, Y. K. AND SPECHT, D. J. 1988. The solubility of noble gases in crude oil at $25–100°C$. *Applied Geochemistry*, **3**, 137–144.

LADWEIN, H. W. 1988. Organic geochemistry of the Vienna Basin: model for hydrocarbon generation in overthrust belts. American Association of Petroleum Geologists Bulletin, **72**, 586–599.

MARTEL, D. J., DEAK, J., DOVENYI, P., HORVATH, F., O'NIONS, R. K., OXBURGH, E. R., STEGNA, L. AND STUTE, M. 1989. Leakage of helium from the Pannonian Basin. *Nature*, **432**, 908–912.

MARTY, B. AND JAMBON, A. 1987. $C/^3He$ in volatile fluxes from the solid earth; implication for carbon geodynamics. *Earth and Planetary Science Letters.* **83**, 16–26.

——, O'NIONS, R. K., OXBURGH, E. R., MARTEL, D. AND LOMBARDI, S. 1992. Helium isotopes in Alpine regions. *Tectonophysics*, **206**, 71–8.

MAZOR, E. 1972. Paleotemperatures and other hydrological parameters deduced from noble gases in groundwaters; Jordon Rift Valley, Israel, *Geochimica et Cosmochimica Acta*, **36**, 1321–1336.

O'NIONS, R. K. AND OXBURGH, E. R. 1988. Helium, volatile fluxes and the development of continental crust. *Earth and Planetary Science Letters*, **90**, 331–347.

OTTLIK, P., GALFI, J. AND HORVATH, D. 1981. The low enthalpy geothermal resource of the Pannonian Basin, Hungary. *In*: RYBACH, L. AND MUFFLER, L. J. P. (eds) *Geothermal systems: principles and case histories.* 221–245.

OXBURGH, E. R. AND O'NIONS, R. K. 1987. Helium loss, tectonics and terrestrial heat budget. *Science*, **237**, 1583–1588.

OZIMA, M. AND PODOSEK, F. 1983. *Noble Gas Geochemistry.* Cambridge University Press.

PIERCE, A. P., GOTT, G. B. AND MYTTON, J. W. 1964. *Uranium and Helium in the Panhandle gas field, Texas, and adjacent areas.* US Geological Survey Professional Paper No. 454-G.

ROYDEN, L. H. 1988. Late cenogenic tectonics of the Pannonian basin system. *In*: ROYDEN, L. H. AND HORVATH, F. (eds) *The Pannonian Basin: a Study in Basin Evolution.* American Association of Petroleum Geologists Memoir, **45**, 27–48.

SMITH, S. P. 1985. Noble gas solubilities in water at high temperature. *EOS Transactions of the American Geophysical Union*, **66**, 397.

—— AND KENNEDY, B. M. 1983. The solubility of noble gases in water and in NaCl brine. *Geochimica et Cosmochimica Acta* **47**, 503–515.

STUTE, M. AND DEAK, J. 1989. Environmental isotope study (^{14}C, ^{13}C, ^{18}O, D, noble gases) on deep groundwater circulation systems in Hungary with reference to paleoclimate. *Radiocarbon*, **31**, 902–918.

TAYLOR, S. R. AND MCLENNAN, S. M. 1985. *The Continental Crust: in composition and evolution.* Blackwell Scientific, London.

WESSELY, G. 1983. Zur Geologie und hydrodynamik im sudlichen Wiener becken und seiner randzone. *Mitt. Osterr. Geol. Ges.* **76**, 27–68 (in German).

—— 1988. Structure and development of the Vienna Basin in Austria. American Association of Petroleum Geologists, **45**, 333–346.

ZAIKOWSKI, A., KOSANKE, B. J. AND HUBBARD, N. 1987. Noble gas composition of deep brines from the Palo Duro Basin, Texas. *Geochmica et Cosmochimica Acta*, **51**, 73–84.

The chemical composition of North Sea formation waters: a review of their heterogeneity and potential applications

E. A. WARREN and P. C. SMALLEY

BP Research, Sunbury-on-Thames, Middlesex TW16 7LN, UK

Abstract: Significant variations in the salinity, chemical and stable isotopic composition of formation waters can be observed in the existing published data from the North Sea Basin. Variations occur at all scales, from intra-formational and within-field to basin-wide variations in water chemistry within and between formations. Clearly, the present-day North Sea Basin waters are highly heterogeneous. These variations have important implications. Various aspects of formation-water chemistry impact upon reserves estimates (R_w) and scale prediction (e.g. Ba, Sr, SO_4), and also yield valuable information on the processes of water–rock interaction and diagenesis leading to modification of reservoir quality. However, present understanding of the processes causing the compositional heterogeneities is limited by the sparse coverage of the existing published data. Consequently, a new effort is being made to compile formation-water data for all fields in the greatest detail possible. Interim results from this compilation are reported.

Sampling formation waters is a routine operation during exploration, appraisal and production operations in the North Sea. However, unlike the comprehensive geological and petroleum data widely available in the open literature, very few of the water data have been published. This is in stark contrast to the wealth of water data available for other basins, e.g. Gulf Coast (Collins 1975; Carothers and Kharaka 1978; Land and Prezbindowski 1981; Land *et al.* 1988). As these studies have demonstrated, a comprehensive, basin-wide waters dataset can greatly benefit understanding of problems important to the petroleum industry: diagenesis (e.g. Land and Prezbindowski 1981; Land *et al.* 1988); basin plumbing (Bethke and Marshak 1990; Harrison and Summa 1991); reserves estimation (Worthington and Johnson 1987) and scale prediction. This contribution reviews our present understanding of formation-water compositions and variations in the North Sea using the present literature and describes the progress made so far in the compilation of a new atlas of North Sea formation-water data currently in preparation (Warren *et al.* 1993).

There are three main sources of formation-water data in the open literature: stand-alone data for individual fields, the compilation of Norwegian shelf data by Egeberg and Aagaard (1989), and the North Sea R_w catalogue (SPWLA 1989). Inevitably, the quality and amount of information provided in these three sources are variable. However, these data do offer insight into the type and magnitude of variability in formation-water chemistry in the North Sea. The data document four main types of variation: intra-formational variations within fields, variations between different reservoir horizons within fields, regional variations within formations and regional variations between different formations. Each of these is discussed below.

Scales of variation

Fields

Data on individual fields are sparse in the literature, and can be difficult to locate. Many are contained in papers dealing with another related subject (diagenesis, for example) and are not given discussion in their own right. In several cases, formation-water data are reported in journals where one might not naturally begin a search for formation-water chemical data. However, some of these studies contain important information concerning formation-water compositions and variability. Two such are the studies on the Heather Field (Glasmann *et al.* 1989) and the Piper Field (Burley *et al.* 1989).

(i) Intra-formational variations Glasmann *et al.* (1989) documented variations in formation-water resistivity within the Middle Jurassic Brent Group across the Heather Field (Fig. 1). Reported resistivities range from 0.443 in B Block to 0.074 in West Heather, equivalent to a salinity variation of 14 000–100 000 ppm. Although the formation water of the main Heather Field only varies by approximately one-half to one-third seawater salinity (Glasmann *et al.* 1989), the salinity in G Block of the main Heather Field is considerably higher. Glasmann *et al.* (1989) suggested that these present-day variations in formation-water data might be comparable to variations in ancient formation waters indicated by stable isotopic data of authigenic clay minerals. If so, then this implies that variations in water chemistry within a single formation have existed on a field-scale in the geologic past as well.

(ii) Inter-formational variations Formation-water chemistry variations in different stratigraphic intervals were observed within the Piper Field by Burley *et al.* (1989). Chemical and stable isotopic data were reported for the Eocene Sand, Upper Jurassic Piper Formation, Zechstein and Carboniferous sediments in the Piper Field together with Piper Formation water chemical compositions in the upthrown and downthrown blocks of the nearby Tartan Field. Variations in the chemical composition of the formation waters in these different stratigraphic horizons is most obvious in terms of salinity (Fig. 2), which clearly increases with depth. The chemical composition of the formation waters also varies, as revealed by changes in some major ion ratios (Fig. 3). Although Na/Cl ratios are nearly uniform despite an increase in salinity of over an order of magnitude, other ion ratios, such as Ca/Cl and K/Cl, vary significantly but show no correlation with salinity. These suggest that the salinity profile in Fig. 2 is unlikely to result from simple upward vertical diffusion of solutes.

The Heather Field and the Piper Field studies both show that significant chemical variations in formation water compositions occur at a field-scale, both vertically and laterally, in the North Sea.

Regional variations

The magnitude of variations in the chemical compositions of formation waters is illustrated in the comprehensive dataset from the Norwegian shelf published by Egeberg and Aagaard (1989). Major and minor element and stable isotopic compositions are reported for many fields in the Norwegian Continental Shelf (NOCS), although individual well locations are

From *Petroleum Geology of Northwest Europe: Proceedings of the 4th Conference* (edited by J. R. Parker).

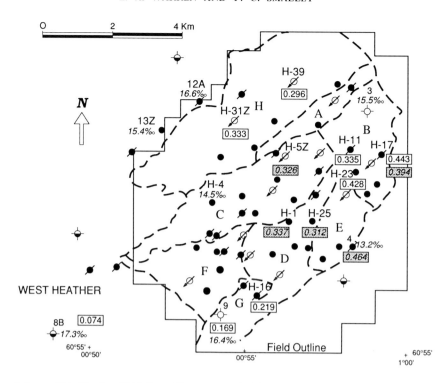

Fig. 1. Water-resistivity variations within Heather Field (from Glasmann *et al.* 1989). Highest values correspond to salinities lower than seawater. G-Block and West Heather have waters with salinities distinctly higher than seawater. Shaded values refer to contaminated samples.

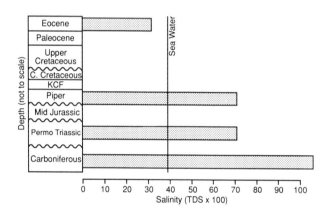

Fig. 2. Salinity variations in different reservoir horizons in Piper Field (data from Burley *et al.* 1989).

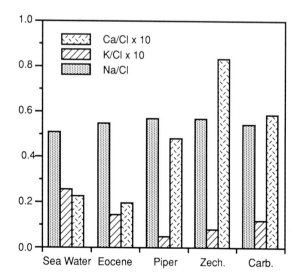

Fig. 3. Formation-water compositional variations in Piper Field. Note the variation in Ca/Cl and K/Cl (data from Burley *et al.* 1989).

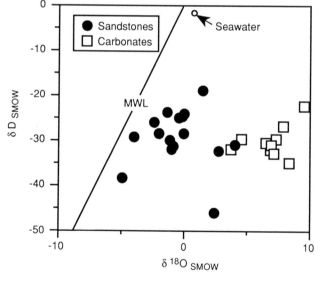

Fig. 4. Stable isotopic compositions of NOCS formation waters (from Egeberg and Aagaard 1989). Present-day meteoric waters plot along the trajectory marked MWL.

not revealed. Such a comprehensive dataset enables detailed examination of variations to be made; that of stable isotopic composition, for example (Fig. 4). None of the formation waters from the NOCS has the stable isotopic compositions of either unmodified meteoric water or seawater (Egeberg and Aagaard 1989), the two obvious sources of water. Furthermore, the stable isotopic composition of the formation waters in carbonate reservoirs appear to be different to those in clastic reservoirs. Aplin *et al.* (in press) plotted similar data for sandstone reservoirs in terms of $\delta^{18}O$ variations with depth, and observed that the deepest samples also had the most evolved oxygen isotopic compositions (Fig. 5). Aplin *et al.* (in press) showed that a similar range and variation in $\delta^{18}O$ occurs in waters from sandstone reservoirs in the United Kingdom

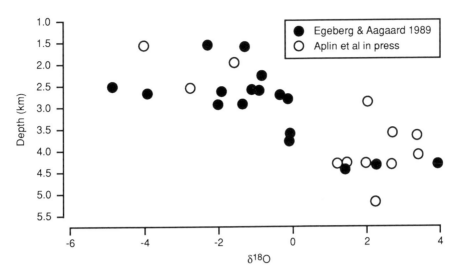

Fig. 5. Variations in oxygen isotope ratios of formation waters from clastic reservoirs; full circles from NOCS, open circles from UKCS (Aplin *et al.* in press).

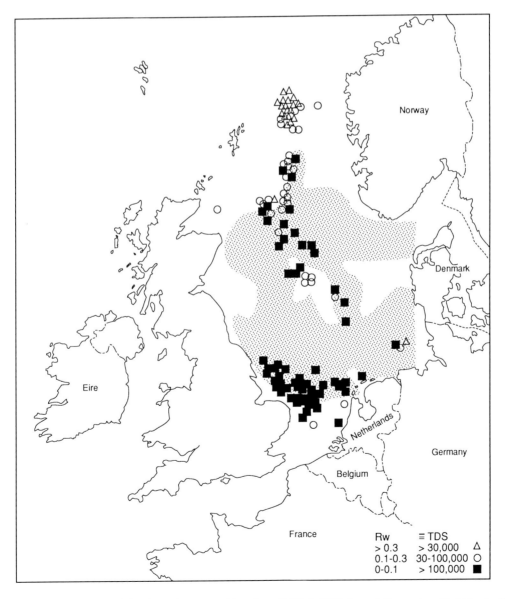

Fig. 6. Water resistivity variations in North Sea formation waters (data from SPWLA 1989). Lowest R_w equates with highest salinity.

Continental Shelf (UKCS). Although the cause of these variations remains a subject of debate (e.g. Aplin *et al.* (in press), clearly the isotopic chemistry of North Sea formation waters is both heterogeneous and complex. Variations in formation-water composition occur on a regional scale and with depth.

Basin-scale variations

A basin-scale image of formation-water compositional variations can be obtained from published salinity data. There are two main sources of data: salinity and density data for individual UKCS fields (Abbotts 1991) and water-resistivity data for wells throughout the North Sea (SPWLA 1989). Converting water resistivity into salinity using a simple conversion (Schlumberger 1989), the resultant salinity data for the North Sea have been plotted for three stratigraphic intervals combined: Paleocene, Middle Jurassic and Rotliegend (Fig. 6). Salinities are highest in the Southern and Central North Sea, and many exceed halite saturation (250 000 ppm). However, salinities are generally lower than seawater throughout the Northern North Sea, although the Heather Field study highlighted heterogeneity (Glasmann *et al.* 1989). Aplin *et al.* (in press) compared the distribution of formation-water salinities

with the distribution of the Zechstein Z3 polyhalite and found that the distribution of salt mimics the highest salinities (Fig. 6). The authors argued that salt dissolution has influenced formation-water chemistry in the underlying Rotliegend sandstone by simple diffusion or density-driven flow, and in overlying strata by salt penetration through diapirism.

Limitations of previous data

The results of all these studies show that there are considerable variations in formation-water compositions both areally and stratigraphically throughout the North Sea Basin. Some simple controls on formation-water composition can be deduced: a major control on salinity appears to be the distribution and diapirism of Zechstein salt. Nevertheless, in general the processes responsible for water variability are not understood (e.g. stable isotopes chemistry, cf. Fig. 4). At least one reason for this is the limited areal and stratigraphic data coverage afforded by the published data. Better data, both in terms of detail and geographical coverage, could enable the processes controlling formation-water composition to be identified and understood, and hence improve the capability for predicting formation-water variations throughout the North Sea. Know-

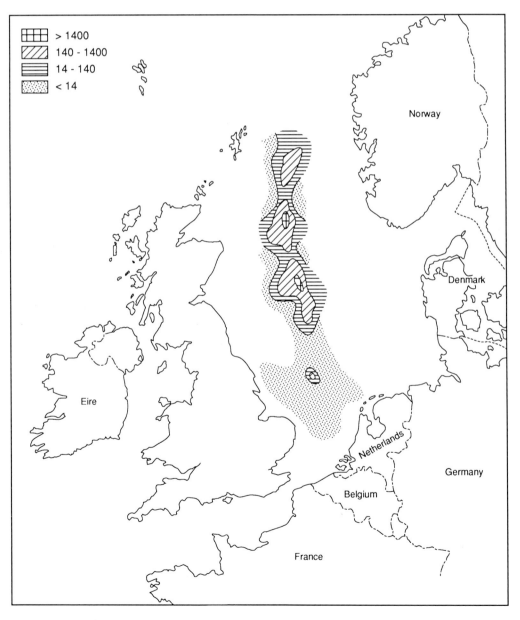

Fig. 7. Barium concentration (mg l^{-1}) in North Sea formation waters (data from Warren *et al.* 1993).

ledge of the lateral and vertical variations in barium concentrations, for example, could greatly improve scale prediction, a costly problem in many Central and Northern North Sea fields. The ideal dataset of North Sea formation waters would combine the comprehensive data of the NOCS dataset (Egeberg and Aagaard 1989) with the basin-wide coverage of the water-resistivity catalogue (SPWLA 1989) and the lateral and vertical detail of the Heather and Piper studies (Glasmann *et al.* 1989; Burley *et al.* 1989). Such a volume is now being prepared (Warren *et al.* 1993) under the auspices of the Petroleum Group of the Geological Society.

The data compiled so far do give insight into the distributions of major solutes in formation waters throughout the North Sea Basin. The distribution of barium is illustrated in Fig. 7 and shows considerable variability. Barium concentrations are highest in the Central North Sea and can be in excess of 1000 ppm. However, barium is negligible in the Southern North Sea. Although the controls on barium concentration have not yet been established for the North Sea, it is clear that salinity (Fig. 6) is not a major control.

Applications

As has been demonstrated in other basins (Collins 1975), a comprehensive North Sea formation-water dataset would have major implications in four areas of study important to the petroleum industry: diagenesis (reservoir quality prediction), fluid-flow modelling (basin plumbing), petrophysics (reserves estimation) and production chemistry (scale prediction).

The reservoir quality of many North Sea oil and gas fields has been greatly modified by the precipitation of mineral cements. Some, such as quartz and calcite, are often sufficiently abundant to reduce porosity severely; others, such as illite, reduce permeability due to the influence of their fibrous morphology on fluid flow. All mineral cements have precipitated from water, an ancient formation water. Although studying present-day formation waters can only be applied to reservoir quality prediction directly in those areas where diagenesis appears to be going on now, the distribution and variation in present-day formation-water chemistry also gives clues concerning the diagenetic processes operating in sedimentary basins.

Similarly, the same data can give direct information on basin plumbing. The field and regional variations described in the published datasets demonstrate that, at present, the formation waters are heterogeneous. This indicates that water–rock interaction processes, e.g. halite dissolution, are acting on formation-water compositions faster than solute transport is homogenizing them. This suggests considerable fluid compartmentalization within the North Sea, not of great surprise considering the complex geology. Despite the abundance of marine sandstones and mudrocks in the post-Triassic sediments, the stable isotopic data suggest present-day waters are predominantly modified meteoric water. This is in contrast to the more marine stable isotopic signatures of Gulf Coast formation waters (Aplin and Warren in press). Displacement of connate seawater by meteoric water is most likely during a lowstand. As yet, the dataset is insufficiently detailed to identify when this might have happened.

The salinity variations observed on both a field and basin scale have great potential impact on calculated water saturations and reserves estimates, especially in fields where salinities are low and a single estimated value of R_w has had to be applied field-wide (Worthington and Johnson 1987). Although a major control on salinity may be the proximity of halite, the processes controlling salinity dispersion are not known in the North Sea. It could be simple diffusion or fluid flow, or a more complex process such as thermohaline convection, as appears to be important in some Gulf Coast areas (Hanor 1987).

Another major application of a suitably detailed dataset is scale prediction. North Sea formation waters are well-known for their scaling potential and the costs of removal have major impact on field economics. Consequently, accurate prediction of scale can impact on equipment design, with subsequent cost savings. The distribution of barium in the North Sea (Fig. 7) gives an indication of where barite scale is most likely to be a problem: in the Northern North Sea and Central North Sea where concentrations are highest. Further data will make prediction of barite scale more accurate and make possible prediction of other scale minerals, e.g. calcite and celestite.

Sample quality and contamination

A major problem which must be considered when interpreting water data is data quality. Water samples are notoriously prone to contamination. This may be due to breakthrough of injected waters (usually seawater in the North Sea) into the sampled formation or contamination by drilling fluids during sample collection. The reliability of water samples often depends on the sample collection procedure: it is well recognized that produced waters are better than a sample collected during a formation test. Other problems with samples may arise due to mineral precipitation between bottom-hole and sample point or in the sampling vessel itself. Poor laboratory analysis may be an additional problem. All these problems can drastically affect the value of the data for understanding and predicting variations in water composition.

One problem is neatly illustrated by recent data from the Piper Formation. (Burley *et al.* 1989). The chemical composition of the formation waters in the Piper Formation in the Piper Field was reported to be very different to those in the Tartan Field. Stable isotopic compositions were also different; the Tartan data appear to fall on a mixing trend between injected seawater and Piper Formation water (Fig. 8). A mixing ratio calculated from Fig 8 was used to estimate a chemical composition of the Tartan water (Table 1). This compared closely with the actual chemical analysis. Consequently, Burley *et al.* (1989) suggested that the differences in formation-water compositions were not indicative of real variation between the two oil fields but were more likely due to seawater breakthrough in the sampled Tartan well.

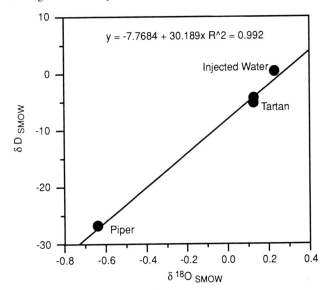

Fig. 8. Stable isotopic composition of Tartan and Piper Field waters (data from Burley *et al.* 1989).

Recent advances in water sampling may overcome some of the problems outlined above, especially contamination. New down-hole Repeat Formation Test (RFT) tools should greatly

Table 1. Comparison of Tartan waters and mixture of Piper formation water and seawater

	Seawater	Piper	Tartan (upthrown)	Tartan (downthrown)	Estimate mixture*
TDS	37700	71106	46920	47160	44380
Na	10770	24600	15600	16200	13540
K	548	207	294	297	480
Ca	548	2080	916	822	804
Mg	1811	528	777	776	1550
Cl	21100	43460	25900	26300	25570
SO_4	2850	23	2200	2300	2280
HCO_3^-	164	415	830	793	214

*mixture of 20% Piper water and 80% seawater

reduce contamination from drilling fluids (Worthington 1992). Better still, direct extraction of formation water from core plugs through centrifugation (cf. Smalley and Oxtoby 1992), particularly when combined with so-called low-invasion coring techniques, not only reduces contamination but enables formation-water variations to be investigated at far smaller scales than at present, i.e. within-well.

Conclusions

Investigation of the limited published formation-water data reveals significant heterogeneity in formation-water composition from the field scale to the basin scale. Variations are observed in salinity, major element chemistry and stable isotopic compositions both within formations and stratigraphically on a regional scale. These variations have important implications on diagenetic processes, basin plumbing, reserves estimation and scale prediction.

Given the sparse coverage of routine water sampling, it is likely that variations in formation-water compositions go unrecognized in many reservoirs. Clearly, an ability to predict the magnitude and distribution of variations in formation-water chemistry has great potential value to the petroleum industry. As yet, the published data either lack sufficient detail or geographical range to be able to do this in many parts of the North Sea.

We would like to thank all the oil companies who have participated in providing data to the North Sea formation waters project. The Mineralogical Society is thanked for granting permission to reproduce Fig. 1.

References

ABBOTTS, I. L. (ed.) 1991. *United Kingdom Oil and Gas Fields 25 Years Commemorative Volume.* Geological Society, London, Memoir, **14**.

APLIN, A. C. AND WARREN, E. A. in press. Mechanisms of silica transport and quartz cementation in deeply-buried sandstones: oxygen isotopic constraints. *Geology.*

——, —— AND GRANT, S. M. in press. Mechanisms of quartz cementation in North Sea reservoir sands: constraints from fluid inclusions. *In*: ROBINSON, A. G. AND HORBURY, A. (eds) *Diagenesis and Basin Development.* American Association of Petroleum Geologists, Memoir.

BETHKE, C. M. AND MARSHAK, S. 1990. Brine migration across North America—the plate tectonics of groundwater. *Annual Reviews of Earth and Planetary Science,* **18**, 287–315.

BURLEY, S. D., MULLIS, J. AND MATTER, A. 1989. Timing diagenesis in the Tartan Reservoir (UK North Sea): constraints from combined cathodoluminescence microscopy and fluid inclusion studies. *Marine and Petroleum Geology,* **6**, 98–120.

CAROTHERS, W. W. AND KHARAKA, Y. K. 1978. Aliphatic acid anions in oil field waters—implications for origin of natural gas. *American Association of Petroleum Geologists Bulletin,* **62**, 2441–2453.

COLLINS, A. G. 1975. *Geochemistry of Oilfield Waters.* Developments in Petroleum Science, **1**, Elsevier, Amsterdam.

EGEBERG, P. K. AND AAGAARD, P. 1989. Origin and evolution of formation waters from oil fields on the Norwegian shelf. *Applied Geochemistry,* **4**, 131–142.

GLASMANN, J. R., LUNDEGARD, P. D., CLARK, R. A., PENNY, B. K. AND COLLINS, I. D. 1989. Geochemical evidence for the history of diagenesis and fluid migration: Brent Sandstone, Heather Field, North Sea. *Clay Minerals,* **24**, 255–284.

HANOR, J. S. 1987. Kilometre-scale thermohaline overturn of pore waters in the Louisiana Gulf Coast. *Nature,* **327**, 501–503.

HARRISON, W. J. AND SUMMA, L. L. 1991. Paleohydrology of the Gulf of Mexico Basin. *American Journal of Science,* **291**, 109–176.

LAND, L. S., MACPHERSON, G. L. AND MACK, L. E. 1988. Geochemistry of saline formation waters, Miocene, offshore La. *Transactions of the Gulf Coast Geological Society,* **38**, 503–511.

—— AND PREZBINDOWSKI, D. R. 1981. Origin and evolution of saline formation water, L. Cretaceous carbonates, south-central Tx. *Journal of Hydrology,* **54**, 51–74.

SCHLUMBERGER, 1989. *Log Interpretation Charts.* Schlumberger Educational Services, Houston.

SMALLEY, P. C. AND OXTOBY, N. H. 1992. Spatial and temporal variations in formation water composition during diagenesis and petroleum charging of a chalk oilfield. *Proceedings of the 7th International Symposium on Water–Rock Interaction, Park City, Utah, USA.* Balkema, Rotterdam, 1201–1204.

SPWLA. 1989. *North Sea R_w Catalogue.* Society of Professional Well Log Analysts, London Chapter.

WARREN, E. A., SMALLEY, P. C., FINDLAY, M. AND TODD, A. C. 1993. *North Sea Formation Waters.* Geological Society, London, Special Publication. (in prep.).

WORTHINGTON, P. F. 1992. Recent advances in formation water sampling: are samples any good? *In*: Chemistry and origins of North sea formation waters: implications for diagenesis and production chemistry (abstract). Meeting held at The Geological Society, London,

—— AND JOHNSON, P. W. 1987. Quantitative evaluation of hydrocarbon saturation in shaly freshwater reservoirs. *Proceedings of Indonesian Petroleum Association,* IPA87-22/06, 1–26.

Models of burial diagenesis for deep exploration plays in Jurassic fault traps of the Central and Northern North Sea

S. D. BURLEY

Department of Geology, University of Manchester, Manchester M13 9PL, UK

Abstract: With increasing burial depth, intergranular porosity declines in a predictable manner according to theoretical and empirically established mathematical relationships. In basins with simple structural and thermal histories, average porosity can often be represented as an exponential function of depth. There are, however, common significant deviations from such a simple model of porosity evolution. Four mechanisms are recognized by which enhanced porosity may be generated or maintained during burial:

- development of overpressures due to restricted pore fluid escape;
- inhibition of mechanical compaction by selective cementation;
- inhibition of cementation as a result of hydrocarbon accumulation; or
- generation of secondary porosity due to either grain or cement dissolution together with simultaneous removal of reaction products.

Secondary porosity may be generated either near the surface by the ingress of dilute meteoric-derived waters or at depth by the generation of 'aggressive' pore fluids. Deep burial diagnesis may encompass both closed and open system porosity generating reactions. In closed systems, increase in temperature concomitant with burial overcomes kinetic barriers enabling dissolution of metastable detrital minerals. By contrast, in more open systems, movement of fluids from either mudstones or evaporites to either overlying or adjacent sandstones introduces chemical disequilibrium that may cause mineral dissolution. Rapid, intermittent movement of fluids along active faults may additionally introduce temperature disequilibrium to further facilitate mineral dissolution in the proximity of faults. Over geological periods of time, however, faults are barriers to fluid flow and provide a mechanism for generating overpressure.

In the Central and Northern North Sea basins, an upper diagenetic regime is separated from a lower diagenetic regime by a regional cover of Cretaceous mudstones and chalks. The Cretaceous sediments provide a regional seal which often coincides with the development of overpressure above deep graben centres reflecting restricted pore water escape. Fluid and solute movement between these two regimes is only possible where faults or gas chimneys penetrate the Cretaceous seal. In the upper regime, porosity loss in Tertiary sediments broadly conforms to the predicted decline of porosity with depth. In the lower regime, Jurassic sandstones on structural highs are characterized by extensive secondary porosity resulting from feldspar grain and carbonate cement dissolution, best preserved where oil accumulation rapidly follows porosity generation. Enhanced porosity can be expected either where sandstones interdigitate with mudstones or in graben centre plays adjacent to faults. This contrasts with graben margin plays where faults are the locus of pronounced cementation and complex diagenetic sequences are developed. In the absence of either oil accumulation or overpressure, secondary porosity either compacts or is occluded by a characteristic deep burial authigenic mineral assemblage. The present heat flow in the North Sea can be interpreted in terms of a residual thermobaric fluid flow from basin centres to basin margins which was probably initiated during the early Tertiary, coincident with the onset of rapid thermal relaxation subsidence.

It is generally accepted that sandstone intergranular porosity declines with increasing depth in a predictable manner according to theoretically and empirically established mathematical relationships (see Fig. 1., Atwater and Miller 1964; Fuchtbauer 1967; Galloway 1974; Nagtegaal 1980). Over very short depth ranges the porosity loss may conform to a straight line relationship of the form

$$\phi = \phi_1 - a.z$$

where ϕ is the porosity at depth z, a is the linear porosity gradient and ϕ_1 is the initial starting porosity of the sediment (e.g. Selley 1978; Magara 1981; Gluyas 1985). However, a linear decline in average porosity with depth implies through extrapolation that zero porosity should be attained at a finite and generally quite shallow depth. Since intergranular porosity is rarely, if ever, reduced to zero before sandstones are buried to the low-grade metamorphic regime, linear depth–porosity models cannot accurately describe the evolution of porosity in a sedimentary basin. This is most apparent in basins with complex burial and thermal histories and in basins with a varied stratigraphic infill.

Average porosity is much better represented over the depth range 0–5 km as an exponential function of depth, described by curves with the form

$$\phi = a.e^{bz}$$

where ϕ is the porosity, z the depth and a and b are constants representing particular sediment properties (Schmoker and Gautier 1988). Such equations have been applied to sedimentary sequences world-wide and do describe average porosity loss as a result of compaction (e.g. Sclater and Christie 1980) and cementation (e.g. Wood 1989) with depth in an approximate way.

There are, however, common significant departures from such a simple model of porosity evolution (see Fig. 1) and this is certainly the case in the North Sea Jurassic, where anomalously high porosities are commonly recorded from deep prospects. It is these departures that contribute greatly to the uncertainty of hydrocarbon exploration, even in so-called mature basins. Four mechanisms are recognized by which enhanced porosity may be generated or maintained:

- development of overpressures due to restricted pore fluid escape;

From *Petroleum Geology of Northwest Europe: Proceedings of the 4th Conference* (edited by J. R. Parker).

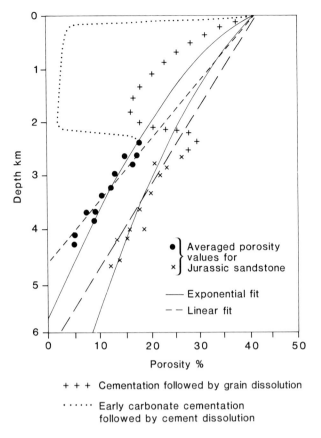

+ + + Cementation followed by grain dissolution

· · · · · · Early carbonate cementation
            followed by cement dissolution

Fig. 1. Linear and exponential porosity loss fit curves for averaged porosity values (measured helium porosity) for two Jurassic sandstone sequences from the Northern North Sea. Over a short depth interval the linear fit matches the data well but predicts zero porosity at depths as shallow as 4.5 km. The dotted and plus lines are speculative trends of porosity evolution assuming the sandstones underwent cement and grain dissolution porosity, respectively. Compare with Fig. 15.

- inhibition of mechanical compaction by selective cementation;
- inhibition of cementation due to emplacement of hydrocarbons in the reservoir; and
- generation of secondary porosity due to either grain or cement dissolution together with simultaneous removal of reaction products.

In a given sedimentary basin one or more of these processes may be operative in generating or maintaining enhanced porosity. An understanding of the structural, sedimentological and burial framework of the basin can provide useful constraints on which processes are most likely to be operative. It is the purpose of this paper to review these processes in the context of the Central and Northern North Sea basins and develop diagenetic concepts which can be applied to reduce risk in deep Jurassic exploration plays. Many of the ideas and concepts presented here are speculative and provocative. The paper is intended not as a definitive statement on burial diagenesis in the North Sea but rather as a forum for stimulating new ideas that may be used in exploration strategies.

Concepts of diagenesis

Diagenetic regimes

Diagenesis is the interplay of chemically, physically and biologically controlled, post-depositional processes by which originally detrital sedimentary assemblages attempt to reach equilibrium with their geochemical environments. Implicit in this view of diagenesis is that the original detrital assemblage, typically comprising grains derived from igneous and metamorphic source areas, is inherently unstable in the aqueous, low temperature geochemical environment that typifies early diagenesis. This is exemplified by the case of detrital oxides, clay minerals and organic matter that accumulate on the sea floor. Geochemically, such an assemblage is highly unstable as soon as sub-oxic conditions are attained in the first few millimetres of burial. Detrital assemblages will thus tend to react very rapidly during diagenesis, although the extent of reaction will depend on the nature of the detrital assemblage, the composition of the depositional pore waters and other external parameters such as sedimentation rate (see Burley *et al.* 1985).

The key point of this view is the inherent instability of sedimentary assemblages in the diagenetic environment. As soon as one of the controlling parameters on the diagenetic system changes, such as pressure, temperature or pore fluid geochemistry, then the sedimentary or earlier-formed diagenetic assemblage will attempt to shift towards a new equilibrium state with its ambient surroundings. Therefore, in the context of a gradually subsiding, evolving sedimentary basin, as a sandstone is buried and pressure and temperature increase, so the mineral assemblage will tend to evolve accordingly (see Fig. 2). Surdam *et al.* (1989) argue that such an evolution is serial and that diagenetic reactions proceed in a predictable sequence (Fig. 3). Similarly, during inversion, as burial temperatures and pressures decline, and dilute, low temperature, surface-derived waters displace basinal fluids, new mineral assemblages will become stable. It may well be that equilibrium is not attained because of kinetic barriers that need to be overcome before reactions proceed. Illite, for example, may be thermodynamically stable at low temperatures, but will not form until temperatures in excess of 70°C are attained because of the presence of a kinetic barrier (Small 1993).

Diagenesis is, therefore, a dynamic system which continually responds to changes in the ambient geochemical environment, either as a result of closed system changes that result from mineral pore water interaction or through changes in external parameters on the system. It is thus essential that any study of diagenesis does not just consider the diagenetic evolution of the particular horizon under investigation as the composition and diagenesis of other lithologies in the basin may influence the diagenesis of that particular formation. Figure 4 summarizes the potential contribution that solute sources (such as mudstones, evaporites and limestones) may make to sandstone aquifers or reservoirs in a lithologically mixed basin fill sequence such as is present in the North Sea basin. There is clearly need to consider all the potential sources of solute that may influence the diagenesis of a given reservoir sequence. The structural style of the play (footwall vs. hangingwall), stratigraphic relationships (pre-rift vs. syn-rift), sedimentological setting (depositional architecture), seal type (degree of footwall erosion and presence of faults) as well as the hydrocarbon migration pathway all contribute to the diagenetic evolution of a particular play. A knowledge of the basin fill, its mineralogy and the various structural styles in the basin are prerequisites to an understanding of which diagenetic reactions are operative in a sedimentary basin and when they take place.

Mechanisms of porosity enhancement

Development of overpressure Mechanical and chemical compaction take place in response to overburden stress when pore fluid pressure is at or remains close to hydrostatic pressure. Providing pore fluids can migrate out of sandstone aquifers during gradual burial, pore pressures will remain close to hydrostatic pressures and sandstones will lose porosity accord

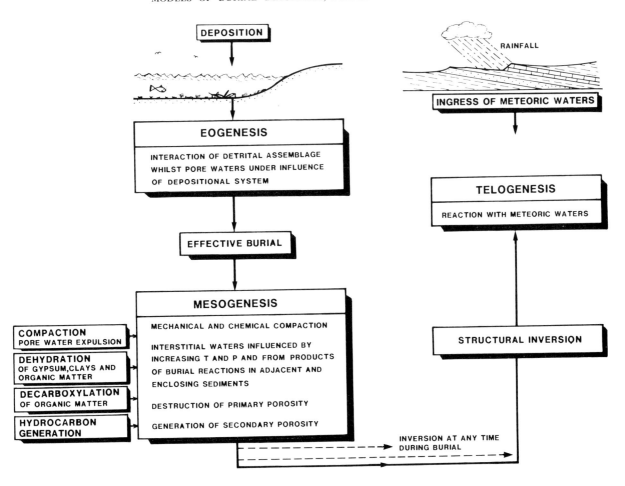

Fig. 2. The regimes of diagenesis (concept based on Schmidt and MacDonald 1979). Diagenesis is a dynamic system that evolves with basin evolution from eogenesis (during which diagenesis is influenced by the depositional environment) through mesogenesis (during which diagenesis is influenced by burial reactions). At any point in the evolution of the basin, structural inversion may take place and enable the ingress of meteoric waters (telogenesis).

ing to a form of the exponential depth–porosity curve. If pore fluid escape is prevented through stratigraphic or fault sealing, then pore pressures greater than hydrostatic pressures, generally termed overpressures, will develop. Overpressure reduces the amount of overburden stress and will therefore reduce porosity loss due to mechanical and chemical compaction. If overpressures exceed lithostatic load then grain inflation fabrics or hydrofractures may develop.

Many areas of the deep Central and Viking Graben in the North Sea are at present highly overpressured (Chiarella and Duffaud 1980; Cayley 1987; Buhrig 1989) and such overpressures certainly contribute to maintenence of enhanced intergranular porosity in some deep plays (e.g. Harris and Fowler 1987). However, for this process to be effective, overpressures must be developed early during burial before mechanical compaction and cementation can reduce porosity. Moreover, the overpressures must be sustained over geological periods of time to prevent compaction resuming and reducing porosity.

Selective cementation The sandstone grain framework may be supported by selective cementation which effectively physically prevents mechanical compaction processes of grain re-orientation and plastic deformation from proceeding. Such cementation may comprise, for example, early fringing calcite rim cements or moderate quartz overgrowth cementation. Early clay mineral cements retard the precipitation of subsequent pore-filling cements, as argued by Hurst and Buller (1984) in Paleocene sandstones for Block 15/9 of the Norwegian North Sea. Ramm (1992) also argues that clay mineral

coats and grain coating microcrystalline silica cements in some Jurassic reservoirs of the Norwegian North Sea may additionally inhibit pressure dissolution because their presence maintains high pore fluid supersaturation with respect to quartz and so removes the chemical potential gradient that drives pressure dissolution.

Selective cementation is also related to the framework grain composition of the sandstone. Arkosic sandstones, for example, contain fewer sites for quartz overgrowth nucleation and will, therefore, retain more intergranular porosity for a given size of quartz overgrowth than for quartz arenites. Bjørlykke *et al.* (1986) observed exactly this relationship in Jurassic sandstone of the Haltenbanken area of the Norwegian North Sea.

However, to be effective in maintaining intergranular porosities, selective cementation must retain permeability in the sandstone and precipitate sufficiently early in the burial history so that high intergranular porosities can be preserved.

Early oil migration It is generally accepted that the displacement of aqueous pore fluids by hydrocarbons will inhibit the progress of diagenetic reactions. Therefore, if a reservoir is charged during early burial, the porosity and permeability characteristics at the time and depth of oil accumulation will be preserved. For such a mechanism to be effective the reservoir must remain sealed since the time of hydrocarbon charging. Should the reservoir seal be breached and the reservoir flushed by later aqueous pore fluids, diagenetic modification will resume. A consequence of early oil accumulation in reservoirs

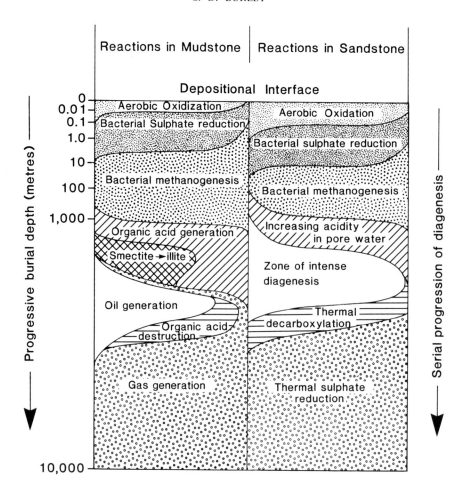

Fig. 3. The theoretical sequence of diagenetic reactions in sandstones and mudstones assuming a serial progression of diagenetic processes driven by temperature increase (modified from Surdam *et al.* 1989).

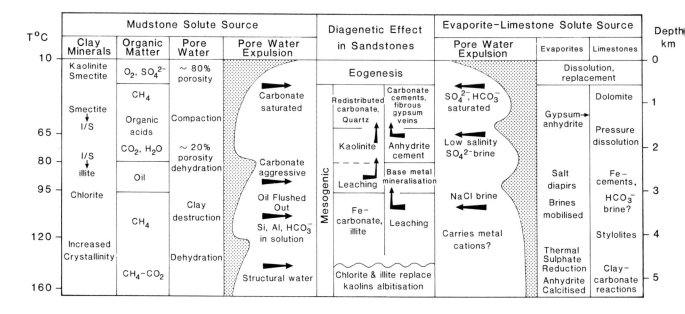

Fig. 4. Summary diagram of potential solute source reactions in mudstones, evaporites and limestones and the possible effects of solute migration into sandstones.

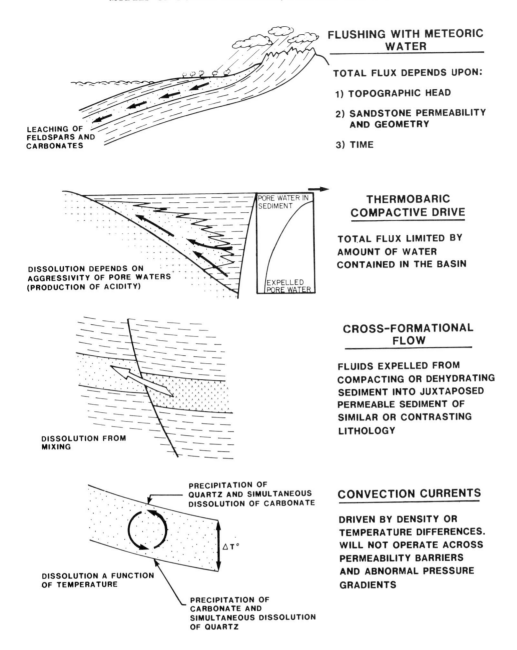

FLUSHING WITH METEORIC WATER

TOTAL FLUX DEPENDS UPON:

1) TOPOGRAPHIC HEAD

2) SANDSTONE PERMEABILITY AND GEOMETRY

3) TIME

LEACHING OF FELDSPARS AND CARBONATES

PORE WATER IN SEDIMENT

EXPELLED PORE WATER

THERMOBARIC COMPACTIVE DRIVE

TOTAL FLUX LIMITED BY AMOUNT OF WATER CONTAINED IN THE BASIN

DISSOLUTION DEPENDS ON AGGRESSIVITY OF PORE WATERS (PRODUCTION OF ACIDITY)

CROSS-FORMATIONAL FLOW

FLUIDS EXPELLED FROM COMPACTING OR DEHYDRATING SEDIMENT INTO JUXTAPOSED PERMEABLE SEDIMENT OF SIMILAR OR CONTRASTING LITHOLOGY

DISSOLUTION FROM MIXING

PRECIPITATION OF QUARTZ AND SIMULTANEOUS DISSOLUTION OF CARBONATE

CONVECTION CURRENTS

DRIVEN BY DENSITY OR TEMPERATURE DIFFERENCES. WILL NOT OPERATE ACROSS PERMEABILITY BARRIERS AND ABNORMAL PRESSURE GRADIENTS

$\Delta T°$

DISSOLUTION A FUNCTION OF TEMPERATURE

PRECIPITATION OF CARBONATE AND SIMULTANEOUS DISSOLUTION OF QUARTZ

Fig. 5. Schematic representation of the end-member mechanisms of generating secondary porosity.

that have remained sealed and have subsequently been buried deeper is that they must be overpressured. Such overpressuring will, of course, help maintain the enhanced porosity resultant of early oil charging. As a corollary of this, if a reservoir is not overpressured, then it is unlikely to have been charged with hydrocarbons early during its burial history unless the initial charge has been lost through leakage.

In some Jurassic plays of the North Sea such a mechanism of maintaining enhanced porosities can be advocated for footwall traps because of the vertical separation between source rock and reservoir (see Fig. 7). For many such tilted fault block plays, when the hangingwall source rock is in the oil generation window at depths of 3 km, the corresponding footwall reservoir may be at depths of only 1 km and, therefore, have high primary intergranular porosities. Providing that there is a vertical migration route between the source rock and the reservoir, then early oil charging can contribute to maintenance of enhanced porosity at depth. This case has been made for the Piper Field (Maher 1980) and a similar mechanism has

been suggested for some of the deep reservoirs of the Central Graben (Cayley 1987).

Some authors have recently argued that quartz cementation may proceed within the oil zone of North Sea Jurassic reservoirs by diffusion of dissolved silica from in situ pressure dissolution through irreducible water films as quartz grains are usually water-wet (Walderhaug 1990; Saigal *et al.* 1992). If such a process is generally operative then the preservation of intergranular porosity through early charging with hydrocarbons will be much less effective.

Generation of secondary porosity The obvious means of developing enhanced porosity is through the dissolution of material from the sandstone. Such enhanced porosity may be generated through the dissolution of either detrital grains or authigenic cements. Clearly, for the dissolution to result in enhanced porosity, the products of the dissolution reaction must be transported away from the site of dissolution and there must, therefore, be some redistribution of the reaction products on a scale greater than that of the reservoir. More-over, once the secondary porosity has been generated, it must be prevented from destruction by both compaction and cementation either through charging with hydrocarbons, maintenance of overpressure or selective cementation. Generation of secondary porosity unaided by one of these additional processes is unlikely to maintain enhanced porosity over geological periods of time necessary for the porosity enhancement to be effective.

Secondary porosity can be generated at any stage of the burial evolution of the sandstone. Jurassic sandstones of the North Sea are generally considered to have potential for secondary porosity generation at three periods during their burial history (Fig. 5). Flushing with meteoric water contemporary with deposition is possible in fluvio-deltaic sandstones or in coastal or shallow marine sandstones overlain by sequences deposited during sea-level fall (Bjørlykke 1984). Such meteoric water flushing is effectively eogenic. Additionally, near-surface flushing with meteoric water is possible in fault blocks that underwent footwall uplift during active extension. The wider fault blocks, with the greatest amount of footwall uplift (Yielding 1990) are expected to have the maximum potential for this telogenic meteoric water flushing. However, both these mechanisms of generating secondary porosity, either through the dissolution of detrital grains or authigenic cements, suffer from the severe drawback that the porosity enhancement takes place near the surface. If secondary porosity generation is to result in effective enhanced porosity in the sub-surface it must be preserved and maintained during burial to depths where hydrocarbon entrapment takes place. It must not be destroyed during burial by the normal processes of compaction and cementation.

Secondary porosity can also be generated by the dissolution of either detrital grains or pore-filling cements in the subsurface as a result of burial diagenetic reactions. In the case of Jurassic reservoirs of the Central and Northern North Sea, detrital feldspars (e.g. Harris 1989; Glasmann 1992) and carbonate cements (Burley 1986; Schmidt and MacDonald 1979) are the minerals mostly considered to have undergone dissolution. Various mechanisms have been proposed for the generation of secondary porosity in the sub-surface, each mechanism with its own protagonists and antagonists (see Bjørlykke 1984). In essence though, these mechanisms fall into four categories: (i) the generation of acidity from decarboxylation of organic matter or dehydration of clay minerals (reverse weathering in effect); (ii) dissolution of unstable detrital minerals as kinetic thresholds are reached as temperature increases; (iii) changes in temperature due to fluid movement; and (iv) mixing of fluids resulting in undersaturation with respect to one or more detrital or authigenic minerals.

Mudstones are frequently envisaged as the source of much of the 'aggressivity' required to dissolve detrital minerals and authigenic pore-filling mineral cements. However, simple mass balance considerations suggest that insufficient organic matter is present in most mudstone sequences to provide the necessary CO_2 to account for all the observed secondary porosity in associated sandstones. Even if there were, other authors argue that the ability of pore waters to cause widespread dissolution cannot be exported from mudstone sources because 'neutralizing' reactions would take place between the pore water and the medium through which they are transported during expulsion (Giles and Marshall 1986).

Alternatively, cooling of ascending hot fluids can cause dissolution of carbonate cements because of their retrograde solubility (Giles and de Boer 1990). Thus, rapid movement of fluids up faults should facilitate dissolution by this mechanism. A variation on this theme of temperature control of mineral solubility takes place in convection cells in which an advecting fluid dissolves feldspars as it descends and warms, but dissolves

carbonates and precipitates quartz cements as it rises and cools (Wood and Hewett 1982). Haszeldine et al. (1984) argued exactly this mechanism for the quartz cements in the Beatrice Field of the Inner Moray Firth.

The main advantage of appealing to mesogenic secondary porosity generation over near-surface processes of porosity generation is that the porosity enhancement takes place at depths of 2–3 km within an already compacted and partially cemented sandstone framework. Such a lithified sandstone framework does not physically collapse as readily as would poorly consolidated sandstone, and providing hydrocarbon migration rapidly follows the porosity generation, enhanced porosity should be maintained to deeper burial depths than would primary porosity. This is also the depth interval over which much of the hydrocarbon generation takes place from the Kimmerigde Clay Formation. The temporal coincidence of hydrocarbon generation with the zone of intense diagenesis (Surdam et al. 1989) supports the contention that the pore fluids associated with migrating hydrocarbons may generate their own enhanced porosity.

Undeniably, there are problems with such an explanation of porosity enhancement. The amount of fluids available in a sedimentary basin is finite and is generally considered insufficient to account for all the 'aggressivity' or solute transport required to account for all the observed or inferred porosity enhancement.

Geological setting of Jurassic fault block plays

The North Sea basin

The North Sea is a failed rift basin that records a long history of crustal extension and subsidence and has a stratigraphic infill that includes evaporites, organic matter-rich mudstones, porous sandstones together with thick deposits of chalk and calcareous mudstones. This varied stratigraphic infill, within the framework of the complex tectonic evolution and highly faulted structural style of the North Sea basin, has fundamental implications for the type of fluids and solutes that can be generated during burial and the fluid migration pathways that may follow.

The basin underwent extension in the end-Jurassic which was followed by a late Mesozoic–Cenozoic period of post-extension thermal relaxation subsidence. Pre-rift sediments in the Permian include evaporite (both gypsum/anhydrite and halite/sylvite) and dolomite deposits (Taylor 1981) while the Triassic includes thick sequences of originally smectite-rich red mudstones (Fisher 1984). Syn-rift sediments in the Jurassic comprise local and regional fluvio-deltaic, shallow marine and deep marine sandstones that now constitute hydrocarbon reservoirs and these are underlain, interbedded with, and overlain by, thick sequences of organic matter-rich, coastal and marine mudstones, the best known of which, the Kimmeridge Clay Formation, constitutes the main source rock. During thermal relaxation subsidence, as the basin as a whole subsided, syn-rift sediments were covered with thick deposits of mudstones, calcareous mudstones and chalks throughout the end-Jurassic and Cretaceous. Continuous sedimentation followed throughout the Tertiary when up to 4 km of marine mudstones and sandstones accumulated.

In the North Sea there is now a growing body of evidence that suggests that the diagenesis of Jurassic and Chalk reservoirs has been strongly influenced by fluids expelled from stratigraphically underlying and structurally lower formations (Burley et al. 1989; Jensenius 1987; Taylor and Lapre 1987). This view is supported by evidence from stable isotopic composition of formation waters in Jurassic reservoirs (Egeberg and Aagaard 1989) which indicates a contribution from a Zechstein source. Figure 6 is a stratigraphic summary diagram

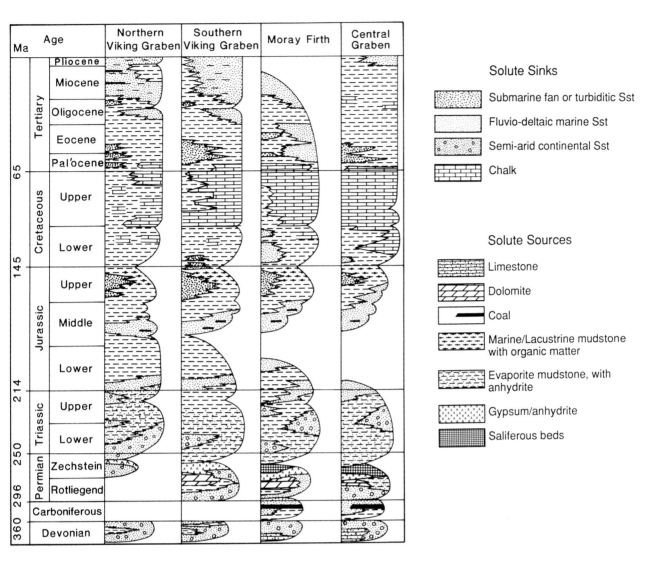

Fig. 6. Stratigraphic summary diagram for the Central and Northern North Sea showing the lateral and vertical distribution of solute sources (Devonian, Carboniferous, Zechstein, Triassic mudstones and Jurassic mudstones) and solute sinks (potential reservoir sandstones and regional Cretaceous mudstone–Chalk sequence). Stratigraphic columns based on Ziegler (1982).

that illustrates the lateral and vertical distribution of solute sources (mudstones, evaporites and limestones) and their relationship to sandstone aquifers or reservoirs in the North Sea basin. From this general compilation it is clear that the potential fluid and solute sources within the basin are varied and numerous when considered in terms of the reactions summarized in Fig. 4. Prediction of solute and fluid sources can only be accomplished through a knowledge of the basin fill, its mineralogy and what diagenetic reactions take place in the basin as a whole.

Jurassic fault block plays of the Central and Northern North Sea

The classic Jurassic play in the Northern North Sea is a fluvial to deep marine sandstone deposited on the footwall or hangingwall of tilted fault blocks. Detailed sedimentological and structural studies of the Jurassic sandstones enable two end-member play types to be identified (Parsley 1984; Johnson and Stewart 1985; Spencer and Larsen 1990):

(i) the pre-rift play, comprising footwall sandstones of essentially Middle Jurassic age;

(ii) the syn-rift play, comprising either footwall or hangingwall Upper Jurassic shallow to deep marine sandstones.

Pre-rift Middle Jurassic plays are almost exclusively footwall sandstones of similar fluvial to paralic depositional facies and reservoir to source rock relationships. However, they vary considerably in aerial size and constitute a gradational series from fault blocks with no footwall erosion to those with major footwall truncation. Upper Jurassic syn-rift plays exhibit greater variety and are of three different types; footwall, shallow marine sandstones (Piper Sandstone type), broadly analogous to the pre-rift plays; hangingwall, shallow to deep marine sandstones banked against a major fault and interdigitating distally with basin centre mudstones (Brae Sandstone type); and shallow to deep marine sandstones in domal traps produced as a result of movement of the underlying Zechstein salt (Fulmar Sandstone type).

Each of these play types is characterized by a particular seal type (Pegrum and Spencer 1991). In the footwall plays the seal depends on the degree of erosion of the footwall crest. Structures with no crestal erosion are sealed by the bounding fault while those with increasing amounts of erosion are progressively sealed by the overlying lithology above the unconformity.

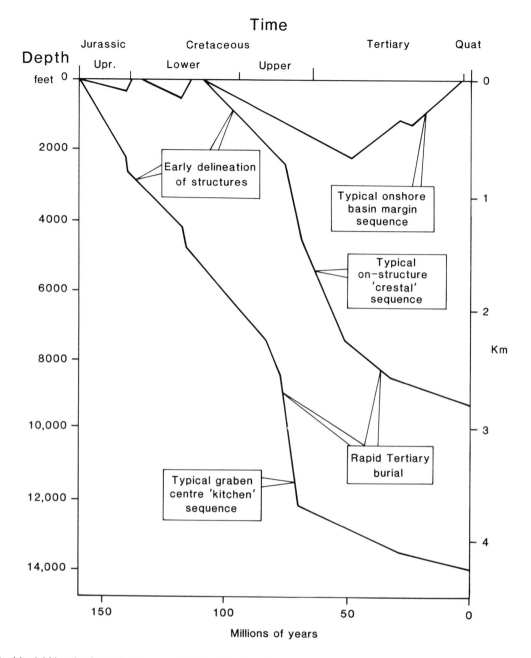

Fig. 7. Idealized burial histories for typical Jurassic tilted fault block basin margin, on-structure footwall crest and graben centre hangingwall sequences. Note that during Upper Cretaceous relaxation subsidence the footwall crest may be at depths of only 1 km while the hangingwall source rock may have entered the hydrocarbon-generation window.

In the hangingwall plays the seal may be provided by either the bounding fault or by conformable cap-rock deposits (or a combination of both, as in the Brae Field). Plays associated with movement of Zechstein salt are typically sealed by unconformable cap rocks although faults may also contribute to the trap seal.

The relationship of the play structure to the seal type essentially dictates the hydrocarbon migration pathway. Cornford *et al.* (1986) recognize three migration pathways in Jurassic tilted fault block structures: (i) upflank migration from either the graben axis or the back half-graben of the tilted fault block structure; (ii) up-fault migration from the hangingwall graben which may also enable the transport of fluids from Jurassic source rocks into the Tertiary (Brooks *et al.* 1983); and (iii) by lateral migration from interbedded and interdigitated mudstones.

Burial histories of Jurassic fault block plays

Early Mesozoic crustal extension and the subsequent relaxation subsidence results in a very distinctive burial history for fault block sequences that is paramount to an understanding of the diagenetic evolution of Jurassic reservoir sandstones. Examination of any number of published burial histories for Middle and Upper Jurassic Central and Northern North Sea sequences reveals the same pattern of burial evolution (Fig. 7; for examples, see Burley 1986; Stewart 1986; Scotchman *et al.* 1989). Basin margin and on-structure crestal sandstones remain within a few hundred metres of the depositional interface throughout the period of extension. This is a direct consequence of fault block rotation on the bounding margin faults (Barr 1987) and the ensuing footwall uplift (Yielding 1990). Depending upon the size of the fault block and the amount of rotation, some footwall crests may even be subaerially exposed. As a result, footwall sequences are characterized by slow subsidence and sedimentation rates, and in the cases of significant uplift, erosion and truncation of footwall crests. This burial history has important implications from deposition through eogenesis to mesogenesis.

Sequences on the footwalls are, therefore, likely to be exposed to the ingress of meteoric water (Burley and Mac-Quaker 1992). Such meteoric water penetration may be contemporary with deposition in fluvial or coastal plain sandstones where slow subsidence rates maintain the depositional system at or above sea-level, in which case the meteoric signature left in the sandstones should be considered eogenic. Alternatively, in the case of shallow and deep marine sequences, meteoric water penetration is only probable in the footwall of larger fault blocks as they rotate during end-Jurassic–early Cretaceous extension (Yielding and Roberts 1991). In this case, meteoric water ingress is established after initial burial and uplift and is telogenic in the sense of Schmidt and MacDonald (1979).

In marked contrast, hangingwall, graben centre sequences are characterized by very rapid burial from the onset of deposition, which continues through the Jurassic. Sandstones and mudstones in such settings are rapidly moved away from the depositional interface and thus reside in the zones of sulphate reduction and bacterial fermentation for relatively short periods of time. These conditions are ideal for source rock accumulation and preservation as the amount of organic matter that accumulates is high (because of the high sedimentation rate), but that which is deposited undergoes minimal alteration during burial and is, therefore, available for hydrocarbon generation (Coleman et al. 1979). Moreover, thick sandstone sequences that are deposited on the hangingwall side of major bounding faults are unlikely to experience subaerial exposure because they are rapidly taken away from the depositional surface.

Later diagenesis is influenced by rate of burial during thermal relaxation subsidence as well as the hydrological fluid flow pathways and aquiclude seals present in the basin. In this respect the relationship of sandstones to mudstone sources, the degree of sandstone connectivity and extent of faulting in the basin are particularly important. Fluids in the thermobaric regime will tend to migrate from high to low pressure environments (Hanor 1987). The net effect of such a pressure drive is inevitably for fluids to migrate from deep graben centres towards basin margins unless trapped in either structural highs or by other barriers to flow (such as stratigraphic pinch-out or faulting). A direct consequence of the separation between hangingwall source and footwall reservoir in tilted fault blocks is a major vertical component to fluid migration. Thus, when a footwall reservoir is at burial depths of only 1 km, the adjacent hangingwall source rock may be at depths of 3 km or more and generating hydrocarbons. The bounding faults that define the tilted blocks have the potential to provide some of the driving energy for fluid migration and may periodically valve fluid movement (Burley et al. 1991). The exact migration pathways of fault-induced fluid flow, however, remain uncertain and not all evidence from Jurassic fault block plays supports the view that the bounding faults are important fluid conduits (see for examples Bjørlykke and Brendsall 1986; Emery et al. 1990).

Diagenetic reactions and hydrocarbon generation are both time/temperature dependent so the burial history of the local basin is critical to the timing of diagenetic reactions and their relationship to hydrocarbon migration. This relationship is well illustrated by a comparison of the burial history of Jurassic sandstones in the Central and Northern North Sea basins. In the former case rapid burial in the Pliocene appears to have been responsible for relatively recent development of overpressures and associated diagenesis in the Central Graben

(Cayley 1987; Saigal et al. 1992). By contrast, the rapid Paleocene burial typical of the Northern North Sea is responsible for initiating reactions that resulted in the intense cementation present in the Outer Moray Firth and South Viking Graben (e.g. Burley et al. 1989).

Diagenesis of Jurassic fault blocks

Early diagenetic assemblages in Jurassic sandstones

A wide diversity of early diagenetic assemblages is recorded from Jurassic sandstones of the North Sea basin (see Burley et al. 1985 and Burley and MacQuaker 1992). These include kaolinite, pyrite, carbonate cements, minor quartz and a variety of iron-bearing clay minerals. The diagenetic assemblage developed is strongly dependent on the geochemistry of the depositional pore waters (marine vs non-marine, oxic vs anoxic; see Burley et al. 1985) but the intensity of early diagenesis is strongly influenced by the time that sediments reside close to depositional interface. This is particularly the case with respect to carbonates (Curtis and Coleman 1986). Carbonate cements form during early diagenesis in response to alkalinity generating reactions (Curtis 1983). A depth-related sequence of carbonate cements corresponds to zones of oxidation, sulphate reduction and bacterial methanogenesis. In marine pore waters, this sequence is Mg-calcite, non-ferroan calcite (associated with disseminated pyrite) and ferroan calcite. The corresponding sequence in non-marine pore waters where sulphate is unavailable for pyrite precipitation is siderite, ferroan dolomite and ferroan calcite. The longer sediment remains in each of these zones the greater the intensity of carbonate cementation. Extensive carbonate cements will, therefore, tend to form at sequence boundaries or during periods of slow sedimentation, sediment bypass or slow subsidence. As a result, sediments on structural highs, at the crests of tilted fault blocks or directly beneath sequence boundaries will tend to be carbonate cemented. By contrast, sediments deposited on the hangingwall of faults where subsidence is rapid, thick sandstone sequences will tend to be less carbonate cemented, unless they record pronounced sequence boundaries.

This relationship between cementation and residence time close to the depositional interface is expressed in the abundance of carbonate nodules in marine mudstone sequences (such as in the Kimmeridge Clay Formation), alternating bands of calcite-cemented sandstones (such as the Bridport Sandstones of the Dorset coast or the Rannoch Sandstones of the Brent Group) and thick carbonate-cemented shallow marine regressive sandstones overlain by flooding surfaces (such as the Bearreraig Sandstone of Skye).

In terms of the burial history curves depicted in Fig. 7, the contrasting burial histories of the two structural settings typical of the tilted fault block dictate that the footwall and hangingwall will undergo very different diagenetic histories. Early diagenesis is very strongly influenced by the length of time the sediment resides close to the depositional interface. The longer the sediment remains within the zones of oxidation, bacterial sulphate reduction and fermentation, the greater the potential for cementation with pyrite, phosphates and carbonates and the more degraded the organic matter will become. Footwall sandstones are likely to be extensively carbonate cemented while the associated or overlying mudstones on structural highs will tend to be poor-quality source rocks. Moreover, footwall crests of large fault blocks are likely to have been subaerially exposed and thus have a high potential for meteoric water recharge.

Burial diagenesis in Jurassic North Sea reservoirs

Despite the inherent potential for complexity, burial-related diagenetic assemblages are characterized by a remarkable degree of uniformity throughout most Jurassic reservoirs. Various studies have investigated depth-related changes from shallow reservoirs (around 2 km crest of structure) to deep reservoir (typically 4.5 km, 5 km in some cases). This is the depth range over which the most important diagenetic reactions are operative and over which oil generation and migration also take place (see Figs 3 and 4).

In the shallower reservoirs with less than 3 km burial, the diagenetic assemblage is dominated by kaolinite, illite–smectite (I/S) clays and minor quartz while downflank or off-structure, calcite and ferroan dolomite–ankerite cements are abundant. Most mechanical compaction is achieved before this depth and the original depositional porosity of the sandstones is reduced by around 50%, although intense pressure dissolution at grain contacts and along microstylolites is not normally established yet. Feldspar dissolution is already initiated (Harris 1989; Glasmann 1992; McAulay et al. 1993), but most sandstones remain arkosic or subarkosic in their bulk composition. Some authors have argued for the widespread development of carbonate cement dissolution porosity at the crest of structures (Burley 1986; Lonoy et al. 1986).

With deeper burial to depths of around 4 km, the kaolinite–I/S clay–calcite diagenetic assemblage is replaced by a dickite–illite–quartz overgrowth–ferroan dolomite/ankerite assemblage (see Burley and MacQuaker 1992; McAulay et al. 1993). Feldspar dissolution is typically intense, leading to the development of diagenetic quartz arenites (Harris 1989; Ehrenberg 1991) and grain dissolution porosity may dominate over intergranular porosity. The intergranular porosity is gradually occluded by later cements that include extensive quartz overgrowths, pervasive authigenic illite and pore-filling and grain-replacive ferroan dolomite/ankerite (Olaussen et al. 1984; Burley 1986; Glasmann et al. 1989). At deeper levels of burial, pressure dissolution at detrital grain contacts and along pronounced microstylolites is widespread (e.g. Olaussen et al. 1984; Bjørlykke and Brendsall 1986).

The most complex diagenetic sequences are developed at the crest of structures and adjacent to major faults (Jourdan et al. 1987; Samways and Marshall 1988; Burley et al. 1989; Burley and MacQuaker 1992). Where fluid flow is directly associated with large bounding faults, diagenetic sequences are cyclical, reflecting the valving of fluids along the faults. This is particularly the case where Jurassic reservoirs are brought into close fault juxtaposition with Zechstein sulphates, as occurs frequently in the Witch Ground and Central grabens. In these areas extensive late ferroan carbonate cements and local barite–base metal sulphide mineralization are developed adjacent to major faults and at the crest of structural highs associated with faults (Baines et al. 1991).

Constraints on the conditions and timing of burial diagenesis in Jurassic reservoirs

An indication of the conditions under which diagenetic reaction took place can be provided by detailed studies of fluid inclusions and oxygen isotopes of authigenic cements in sandstones.

Fluid inclusions in authigenic quartz cements from a variety of Jurassic reservoirs document a range of palaeofluids from very dilute (0.5–4 wt% NaCl equivalents) to very saline (up to 22 wt% NaCl equivalents), and give typical uncorrected homogenization temperatures of 70–140°C (e.g. Haszeldine et al. 1984; Konnerup-Madsen and Dypvik 1988; Burley et al. 1989; Glasmann et al. 1989; Walderhaug 1990; Saigal et al. 1992). In some cases these temperatures are at or very close to the present down-hole temperatures. If a correction needs to be applied to this temperature to take account of the pressure at which the inclusions were trapped (see Potter (1977) for the logic behind this), then precipitation temperatures may actually exceed present down-hole temperatures in many cases.

Oxygen isotopes indicate kaolinite formed from an isotopically lighter fluid than illite (Brint 1989; Glasmann et al. 1989; Hogg 1989; MacAulay et al. 1992). If temperatures of formation are taken from fluid inclusions in co-existing quartz overgrowths (and assuming that the clay minerals and quartz overgrowths are co-precipitates), then kaolinites generally appear to have formed from a pore fluid of $\delta^{18}O$ of less than $-2‰$ SMOW which is indicative of a meteoric water influence (Glasmann 1992). By contrast, the authigenic illite, along with the quartz overgrowths, has precipitated from a palaeopore fluid of $\delta^{18}O$ composition of between 0 and $+4‰$ SMOW, indicative of a much more-evolved pore water, either as a result of mineral–pore water interaction or by mixing of in situ pore waters with pore waters derived from clay mineral or gypsum dehydration.

The timing of diagenetic processes can be constrained either directly by the use of radiogenic isotopes or indirectly via the use of a geothermometer in conjunction with the decompacted burial history of the sample suite by either assuming or modelling the palaeogeothermal gradient. The K–Ar radiogenic decay system is most widely applied to North Sea reservoir sandstones, and several studies report the results of age determinations on authigenic illite clay separates. The majority of illite K–Ar age dates indicate that illite precipitation took place in the mid- to late Tertiary (20–30 Ma) when the reservoir sandstones attained burial depths of 2.5–3.5 km and temperatures exceeding 70°C (e.g. Jourdan et al. 1987; Liewig et al. 1987; Burley and Flisch 1989; Ehrenberg and Nadeau 1989).

If the fluid inclusion homogenization temperatures measured in authigenic quartz overgrowths are used to define the timing of quartz diagenesis via the burial curves, given a palaeogeothermal gradient based on a McKenzie-type model of basin evolution that assumes only conductive heat flow, then quartz cement precipitation also took place during the mid-Tertiary in the Outer Moray Firth (Burley et al. 1989) and the late Tertiary in the Central Graben (Saigal et al. 1992).

Considerable evidence, therefore, suggests that most mesogenic cementation in Jurassic reservoirs took place during the period of rapid late Tertiary burial (in the Paleocene in the Northern North Sea and in the Pliocene in the Central North Sea). This cementation is broadly coincident with hydrocarbon migration (Bissada 1983; England et al. 1989). The fluid inclusion homogenization temperature and palaeosalinity data further suggest that the influence of hot, saline fluid migration was associated with mesogenic cementation. The development of a temperature anomaly during the Tertiary is also recorded in the clay mineralogy (Pearson and Small 1988), vitrinite reflectance (Pearson et al. 1983) and organic geochemistry (Cooper et al. 1975) of Jurassic mudstones from the Central and Northern North Sea basins.

Pressure compartmentalization

Many North Sea Jurassic fault blocks are isolated, seal-bounded fluid compartments that are overpressured and not in hydraulic pressure communication with adjacent fault blocks or the overlying hydrodynamic regime (e.g. Cayley 1987; Buhrig 1989). The nature and origin of the seals which isolate such compartments are poorly known. The fluid compartments must be three-dimensional, and comprise effective horizontal and vertical permeability seals. Lateral seals are most likely to be faults, and top seals are probably stratigraphic, but are suggested by some authors to be diagenetic. Carstens and Dypvik (1981) and Chiarelli and Duffaud (1980) argued that

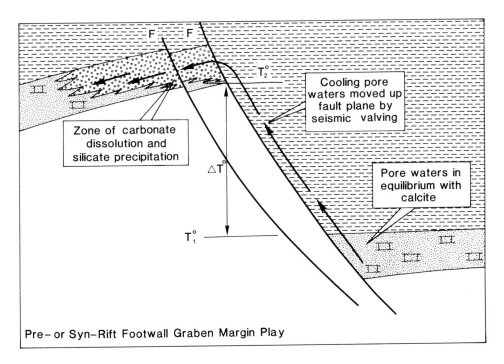

Fig. 8. Conceptual diagenetic model for pre- or syn-rift footwall graben margin play.

early carbonate cementation at sequence boundaries in the Kimmeridge Clay Formation created a widespread impermeable seal. Hunt (1990) envisaged a similar carbonate or siliceous regional seal in the Cretaceous above Central North Sea Chalk reservoirs, but also argued that this seal is essentially horizontal planar and cuts across stratigraphy and structures, and is presumably comparable to the 'carbonate curtain' of Schmidt and MacDonald (1979).

The overpressure associated with such compartments is probably caused by a combination of dehydration reactions, hydrocarbon generation and thermal expansion of fluids (Magara 1981). Sealed compartments effectively maintain a constant volume as they are buried and the extent of overpressure developed within a compartment depends on volume changes associated with dehydration and hydrocarbon-generation reactions. If overpressure development continues, pore pressures in fluid compartments will gradually rise until they approach lithostatic pressure when either the sealing fault will valve fluids (Sibson 1981) or the top seal will fracture (Hunt 1990). In either event, pore fluids escape vertically upwards and the pressure of the compartment is reduced. As a result the fault will re-seal and cementation may reconstitute the top seal, enabling pore pressures to regenerate. Such a mechanism appears to have been operative in the Hild Field of the Norwegian North Sea (Lønøy *et al.* 1986) where a carbonate-cemented gas chimney has developed above the Jurassic reservoir in which extensive cement dissolution porosity has been generated. Release of only a small amount of fluid from a pressure compartment will cause a marked reduction in pore pressure, so the process would have to be repeated many times to transport large amounts of pore fluids or result in a large hydrocarbon accumulation. However, repeated hydraulic fracturing and re-sealing provides a plausible fluid transport process that is directly comparable to the episodic dewatering mechanism of Cathles and Smith (1983) envisaged for large-scale stratabound mineralization around the margins of sedimentary basins.

Conceptual models of burial diagenesis for Jurassic plays

The data reviewed so far now enable concepts of burial

diagenesis to be developed which can be applied to Jurassic fault block plays. This discussion assumes from the foregoing account of diagenetic concepts and the assemblages developed in North Sea Jurassic sandstones that eogenesis is ineffectual in developing enhanced porosity. All porosity enhancement above predicted exponential depth–porosity trends is generated and maintained during burial by a combination of early oil charging, overpressure development and secondary porosity generation.

Five end-member conceptual models of burial diagenesis for Jurassic fault block plays are defined on the basis of relationship to rifting, structural style and seal type (Figs 8–12).

    (i) pre- or syn-rift, footwall graben margin play with no footwall erosion;
    (ii) pre- or syn-rift, footwall graben margin play with footwall truncation;
    (iii) syn-rift, hangingwall graben margin play;
    (iv) pre- or syn-rift, hangingwall graben centre play; and
    (v) syn-rift, hangingwall graben centre domal play influenced by halokinesis.

Each of the conceptual models is associated with a particular fluid migration pathway and solute source. The exact details of these parameters will vary from play to play and an assessment of them can be made for each individual structure.

(i) Pre- or syn-rift, footwall graben margin play with no footwall erosion (Fig. 8) This is the classic footwall tilted fault block play. Delineation of the fault block structure during end-Jurassic extension resulted in separation of the footwall reservoir from the hangingwall solute and hydrocarbon source. The amount of vertical separation may be as great as 2 km. In this setting the basin margin faults that delineate the structure are a probable fluid migration pathway from deep, overpressured fluid compartments where dehydration and hydrocarbon generation reactions are operative. A major vertical component to fluid migration means that structurally higher reservoirs can be charged with hydrocarbons relatively early during the burial history and at shallow burial depths. Additionally, cooling of the migrating pore fluid exclusive of interaction with adjacent

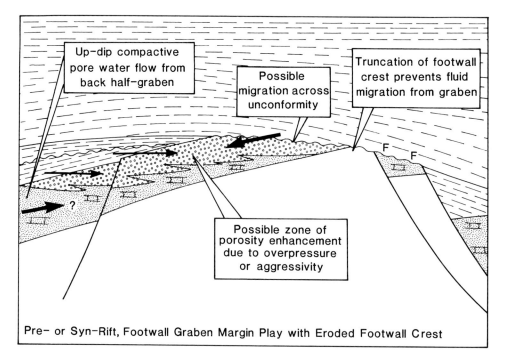

Fig. 9. Conceptual diagenetic model for pre- or syn-rift footwall graben margin play with eroded footwall crest.

lithologies enables propagation of chemical and temperature disequilibria from deep graben sources to structurally high graben margins. The energy for fluid transport may be provided by either seismic pumping (fault displacement drives fluid movement) or by seismic valving (overpressure drives fault dilation). As a consequence of the temperature change associated with vertical fluid migration, carbonates will dissolve on structure as the fluid cools, while silicates (such as quartz cements and clay minerals) should precipitate adjacent to the fault zone. Such silicate cementation is likely to result in reduced reservoir quality next to the fault zone whenever a significant component of vertical transport is involved in the fluid flow. Additionally, if extensive dissolution of carbonate cements takes place on-structure, carbonate cements must locally reprecipitate either downflank of structure, or if the cap-rock seal is locally breached during faulting, above the structure where cementation may contribute to seal integrity. Fault-related fluid movement in this model is likely to result in very complex diagenetic cementation with probable development of diagenetic 'cycles' adjacent to the fault zone. Enhanced feldspar and carbonate cement dissolution may be developed at the crest of the structure where migrating pore fluids are 'ponded'

(ii) Pre- or syn-rift, footwall graben margin play with footwall truncation (Fig. 9) This play-type represents a modification of the above conceptual model in which footwall uplift was associated with erosion of the footwall crest. Such footwalls may have been, and in many cases probably were, subaerially exposed. They may, therefore, have been flushed with meteoric waters during end-Jurassic exposure, but any porosity enhancement that took place near the surface is likely to have been occluded during Cretaceous and Tertiary burial. Eogenic feldspar dissolution voids will have been compacted or cemented. There is, additionally, no fluid migration pathway from the deep graben hangingwall to the truncated crest of the footwall. In this play type, fluid migration must take place either from the back half-graben of the footwall or across the unconformity beneath the cap-rock seal. There is a relatively small component of vertical fluid flow to such a migration

pathway, and so there is no possibility for charging the structure early with hydrocarbons. Equally, there is no potential for rapid temperature change associated with fluid migration, and any migration pathway through the footwall flank is likely to involve significant water–rock interaction.

Enhanced porosity in this play type can, therefore, only be produced either as a result of on-structure overpressure development or through the 'aggressivity' of pore fluids expelled from mudstones above the cap-rock seal. In this latter case, because the overlying organic matter-rich mudstones are generally not oil mature, two different fluid migration pathways must be inferred. The diagenetic sequences that result from this model are likely to be much less complex than those of the footwall play where crestal erosion has not taken place. The lack of a temperature differential between fluid source and solute sink, lack of fault-related fluid flow and lack of a mechanism to pulse fluid flow suggests that diagenetic cycles will not be developed.

(iii) Syn-rift, hangingwall graben margin play (Fig. 10) This is the classic hangingwall play exemplified by the Brae Field. Reservoir sandstones are deposited on the hangingwall in response to footwall uplift. In most cases the hangingwall deposits are submarine and because of active fault displacement and block tilting, are quickly moved from the depositional interface. There is little potential for contemporary flushing with meteoric water. The coarse clastics that constitute the reservoir are submarine fans and related slope deposits that are organized into prograding units.

The bounding faults and an essentially conformable hangingwall mudstone sequence constitute the trap seal. Coarse proximal clastics of the reservoir interdigitate with distal organic matter-rich mudstones that are generally accepted as the main hydrocarbon source. The upper part of the coarse clastic units is commonly cemented with early carbonate cements because of intermittent sedimentation.

Interdigitation of the basinal mudstone source rocks with proximal reservoir lithologies means that fluid migration pathways for sources of 'aggressivity', solute and hydrocarbons are characteristically short and essentially lateral. The greatest

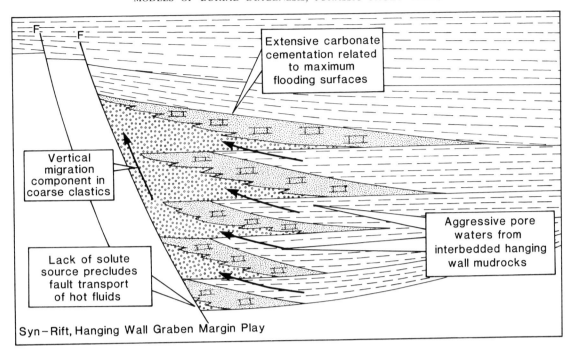

Fig. 10. Conceptual diagenetic model for syn-rift hangingwall graben margin play.

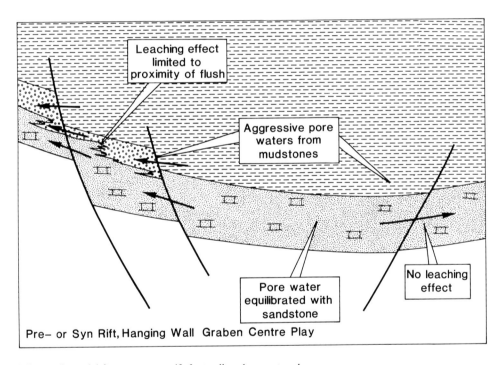

Fig. 11. Conceptual diagenetic model for pre- or syn-rift footwall graben centre play.

potential for feldspar grain dissolution and cement dissolution porosity is located close to the mudstone source. Distally, where mudstones interdigitate with the sandstones, highly channellized enhanced porosities may be developed along the fluid flow fairways in the lower part of the coarse clastic units where there is less carbonate cementation. Proximal to the main bounding faults, the concentration of fluid flow may lead to the establishment of a vertical migration component within the coarse clastic wedge. The absence of a deep solute source beneath the reservoir sequence (unless older Jurassic or pre-Jurassic mudstone, coal or evaporite lithologies are present) precludes a major deep-seated, fault-related, hot fluid flow component.

The short migration pathway from mudstone source to

sandstone reservoir is favourable to the extensive feldspar grain and cement dissolution porosity in the proximal reservoir clastics. Reaction products are likely to be less complex than in footwall plays where there has been no crestal erosion.

(iv) Pre- or syn-rift, hangingwall graben centre play (Fig. 11)
This play is typical of deep graben centres. These have been least explored for in the North Sea because they are often at depths beneath the so-called 'economic porosity basement'. However, thick sandstones are commonly present in the graben centres as hangingwall deposits, and if there are mechanisms operating to develop enhanced porosity at depth, such deposits are a potentially important play.

The seal in this play is provided by a combination of basin

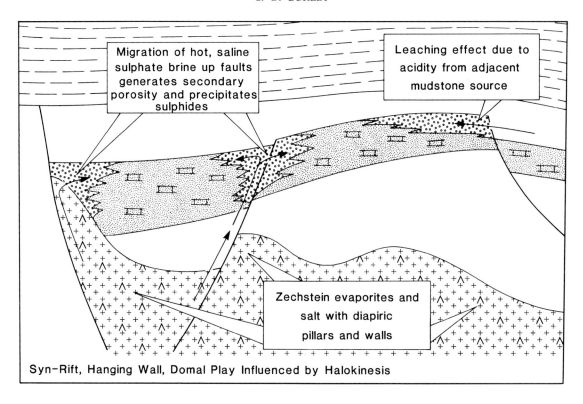

Syn–Rift, Hanging Wall, Domal Play Influenced by Halokinesis

Fig. 12. Conceptual diagenetic model for pre- or syn-rift hangingwall graben centre domal play influenced by halokinesis.

centre or bounding graben margin faults and a conformable mudstone cap rock. The sandstone reservoir is essentially at the same depth as the solute and hydrocarbon source. There is, therefore, no possible vertical component to pore fluid migration in this play. Thus the reservoir cannot be charged with hydrocarbons during early burial, and enhanced porosity cannot be generated as a result of rapid fluid cooling. Faulting in the hangingwall is common and introduces vertical permeability barriers which effectively compartmentalize reservoir sandstones. Cross-formational flow between fault-juxtaposed sandstones is unlikely to generate secondary porosity through fluid mixing because of the probable similarity of fluid types and conditions.

However, the close spatial association of organic matter-rich mudstones where dehydration and hydrocarbon-generation reactions are operative suggests that overpressure development is likely with the extensive development of fault compartmentalization. Enhanced porosity is most likely close to faults where fluids have been directly sourced from graben centre mudstones following fault release of overpressure, and may result directly from either overpressure or from dissolution of detrital grains or pore-filling cements. As there is no temperature differential, grain or cement dissolution can only result from the 'aggressivity' of the expelled pore waters and is likely to be greatest closest to the fault. Carbonate cements should reprecipitate further along the fluid flow path.

The diagenetic evolution of the deep graben centre play is likely to contrast markedly with graben margin plays, with the difference being largely a function of the extent of vertical fluid migration.

(v) Syn-rift, hangingwall domal play influenced by halokinesis (Fig. 12) This play type has many similarities to the graben margin and graben centre plays, being a hangingwall syn-rift play in shallow or deep marine sandstones. It differs, however, in being underlain by a thick sequence of Zechstein evaporites which have undergone varying degrees of diapiric movement that has directly influenced the structural and seal development

of the trap. Many of this type of reservoir are cut by major faults which are often toed in the Zechstein, although fault juxtaposition of the Zechstein with reservoir is rare. The seal is typically an unconformable cap rock, with the unconformity related to the movement of the underlying Zechstein, although a fault component to the seal is common.

Fluid migration can be driven from two directions in this play type. A major compactive, thermobaric drive is forced out of the deep basin centres. This drive may carry hydrocarbons to charge the fault structures, but intergranular flow is unlikely to cause widespread secondary porosity because any 'aggressivity' is likely to be neutralized on route. The only potential for secondary porosity generation from pore fluid aggressivity is from a local mudstone source where acidic fluids can directly charge the structure. This may happen, for example, where Kimmeridge Clay Formation mudstones are juxtaposed against a reservoir sequence on the hangingwall side of a normal fault. A second fluid source is present in the underlying Zechstein evaporite sequence, which may also be in connectivity with Mesozoic fluid sources in deeper parts of the basin. Fluids from this solute source will be hot (due to the greater depth of burial and higher thermal conductivity of salt) and saline (due to dissolved salts, and therefore of relatively high pH). Bicarbonate derived from marine Zechstein carbonates is typically isotopically heavy (Taylor and Lapre 1987), and this should be reflected in any carbonate cements that precipitate from this source. Rapid vertical movement along faults means that a temperature anomaly and geochemical disequilibrium can be transported to the structurally high Jurassic reservoir formations. Cooling of a hot fluid on-structure should result in dissolution of carbonates but precipitation of silicates, while mixing of a saline, sulphate-rich brine with a more dilute, reducing pore fluid derived from the Kimmeridge Clay is likely to precipitate sulphides and ferroan carbonates.

These two fluid sources are thus likely to produce contrasting diagenetic assemblages. The fluid expelled from the Kimmeridge Clay will produce the usual dominant diagenetic assemblage comprising extensive grain dissolution porosity, carbonate dissolution, kaolinite, quartz and isotopically light ferroan carbonate (with a $\delta^{13}C$ value of $> -10‰$ PDB). However, where Zechstein fluids gain access to Jurassic sandstones, they will dissolve in situ carbonate cements and precipitate barite, sulphide and isotopically heavy ferroan carbonate as they cool and mix with the in situ pore fluids. Jurassic sandstones that have not been conduits to either of these fluids or have not ponded such fluids will be dominated by either early carbonate cements (principally sulphate-reduction nonferroan calcites) or will have undergone extensive mechanical compaction contemporary with gradual ferroan calcite and dolomite cementation (principally fermentation and decarboxylation zone carbonates).

Speculations on the diagenetic evolution of the Central and Northern North Sea

Diagenesis of Jurassic reservoir sandstones in the North Sea basin context

One of the major themes emphasized throughout the foregoing discussion is that the diagenesis of Jurassic reservoir sandstones cannot be studied in isolation of that of potential solute source and solute sinks. All lithologies, beneath, adjacent to and in some cases overlying the reservoir sequence, are potential sources of fluids and solutes for diagenetic reactions in the sub-surface.

Despite this pertinent observation, there is remarkably little known of the diagenetic evolution of Carboniferous and Permo-Triassic sediments of the Central and Northern North Sea, and very little factual data available on the pore fluids and solutes they are likely to have generated. They may clearly be sources of CO_2, organic acids, various dissolved cations, H_3O^+, hydrocarbons and isotopically heavy $\delta^{18}O$ pore waters, but information regarding the amounts of these components, their migration pathways and their actual contribution to the diagenesis of Jurassic reservoir sandstones is simply not available. The limited data available indicate that pore water total salinities generally increase with increasing depth and age of formation, and that pore waters present in deeper reservoirs tend to have isotopically heavier $\delta^{18}O$ compositions. Given the potential influence such fluids and solutes may have in Jurassic reservoirs this is an area for considerable further investigation.

Somewhat more information is documented regarding the diagenesis of the overlying Cretaceous and Tertiary sediments that has a direct bearing on the diagenetic evolution of Jurassic reservoir sandstones.

Diagenesis of the Chalk and equivalent Upper Cretaceous marls in the Northern North Sea

Diagenetic modifications to the Chalk are recorded from regional studies of the onshore outcrop and offshore subcrop as well as from several Chalk reservoirs in the Central and Danish North Sea overlying piercement salt diapirs. Here the Chalk is characterized by abundant matrix calcite cement and fracture-filling calcite cements that include compacted calcite veins, uncompacted calcite veins, breccia infillings and stylolite-associated tension gashes, often associated with pyrite (Jensenius 1987; Jensenius and Munksgaard 1989; Taylor and Lapre 1987). During burial, the Chalk is a major sink of bicarbonate according to stable isotope studies which indicate the presence of progressively lighter $\delta^{13}C$ with increasing burial depth. In this sense the Chalk can be regarded as the North Sea version of the 'carbonate curtain' of Schmidt and MacDonald (1979). That is, if carbonate cements are dissolved in crestal Jurassic sandstones, and enhanced porosities are created, the dissolved reaction products must be transported away. The Chalk and associated equivalent calcareous mud-

stones of the Outer Moray Firth and Northern North Sea (the Campanian 'Marls' of Maher (1980) for example) are a major buffer to migrating bicarbonate-charged pore waters and will act as a sink for any bicarbonate transported from underlying sources. The analogy with Schmidt and MacDonald's (1979) carbonate curtain is not exactly comparable because the zone of carbonate cementation in the North Sea chalks and marls does not migrate through stratigraphic units as subsidence proceeds, but is essentially a fixed stratigraphic barrier to vertical bicarbonate transfer. In this sense it can be considered as a stratigraphic-diagenetic equivalent of the regional fluid compartment seal of Hunt (1990). Regionally, therefore, Lower Cretaceous sandstones and calcareous mudstones together with the overlying Upper Cretaceous chalks and marls can be considered as the sink for all upwardly migrating bicarbonate-charged pore waters.

Fault zones, fracture systems and gas chimneys around salt diapirs are the most likely places for carbonate cementation above reservoirs. Evidence to support this interpretation is provided by a combination of trace element, stable isotope, Sr isotope and fluid-inclusion studies. Although much of the matrix calcite is derived from internal pressure dissolution and recrystallization processes, a significant component of the intergranular calcite is derived from a heavy $\delta^{13}C$ source, ascribed to the Kimmeridgian (Taylor and Lapre 1987). There is also a growing body of evidence that suggests the diagenesis of Chalk reservoirs has been strongly influenced by fluids expelled from Zechstein or Triassic sources (Jensenius 1987; Taylor and Lapre 1987). This view is supported by evidence from stable isotopic composition of formation waters in Jurassic reservoirs (Egeberg and Aagaard 1989) which indicates the presence of an isotopically heavy $\delta^{13}O$ fluid source.

These studies of Chalk diagenesis have important implications for the diagenetic evolution of Jurassic reservoir sandstones because they indicate that mass transfer of solute from the Zechstein and Kimmeridgian can take place wherever the regional base Cretaceous carbonate seal is broken by faults, gas chimneys or salt diapirs. Jensenius and Munksgaard (1989) and Rabinowitz et al. (1985) argue that large-scale fluid convection cells are established in Chalk reservoir fracture systems above salt diapirs. These inferred convection cells set up heat flow anomalies that are recorded in the fluid inclusion homogenization temperatures and oxygen isotope signatures of the calcite cements. The hot fluid flow is restricted to the fracture systems and only slightly perturbates the cement characteristics in the matrix of the Chalk.

Diagenesis of Tertiary reservoirs

Diagenetic modification of Tertiary sandstone reservoirs is usually considered as slight, and there are no detailed published studies of the effects of compaction and the cementation in these sandstones. Field descriptions often include brief notes of the diagenetic modification and such short descriptions enable a tentative diagenetic evolution to be proposed. The typical diagenetic assemblage comprises minor amounts of siderite, pyrite, non-ferroan calcite, kaolinite, quartz overgrowths and late ferroan calcites, although calcite cementation in nodules and bands may locally be pervasive. Authigenic chlorite appears to be a relatively common cement (e.g. Hurst and Buller 1984). The shallow Paleocene reservoirs (e.g. Frigg: Héritier et al. 1981) are developed in typically clean sandstones with only very minor amounts of cement and average porosities of 30%. Deeper Paleocene reservoirs (e.g. Everest: Thompson and Butcher 1990; Cyrus: Mound et al. 1991) contain significant quartz and kaolinite cements, and despite the initial development of feldspar dissolution porosity, only have average total porosities of around 20%. As most of the Paleocene sandstones are well sorted, they will have had depositional

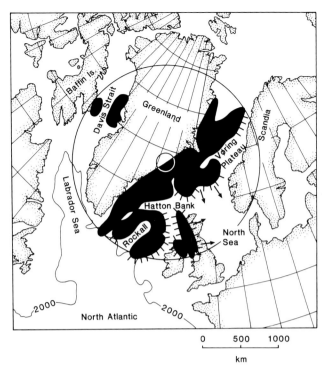

Fig. 13. Reconstruction of the northern North Atlantic region 62 Ma ago showing the inferred position of the mantle plume beneath Greenland, the lateral extent of abnormally hot asthenosphere and areas of uplift. Solid shading indicates the extent of early Tertiary igneous activity around the Atlantic margins. Resulting uplifted areas (radial lines) and possible early Tertiary sediment transport paths (arrows) are indicated (modified from White and McKenzie 1989).

porosities of around 40% bulk sandstone volume (Beard and Weyl 1973). They have, therefore, lost between 10 and 20% of their bulk volume essentially due to mechanical compaction together with minor cementation.

The Tertiary hot fluid flush

Several independent datasets now suggest that during Tertiary times many reservoirs of the North Sea witnessed a distinct event of anomalously 'hot diagenesis'. This event is apparent in

the diagenesis of Jurassic crestal reservoir sandstones, Jurassic mudstones and in the Chalk around faults, gas chimneys and salt diapirs, being present wherever fluids from deep, grabenal parts of the North Sea basin have been able to migrate rapidly vertically along fluid flow conduits to structurally shallow levels. Such a 'hot fluid flush' appears to be intimately associated with hydrocarbon generation and migration. There is a remarkable temporal coincidence of organic maturation, late stage mesogenesis, fracturing and sulphide mineralization and increased burial rate to argue a genetic link is present between some, if not all, of these processes. A considerable body of evidence also now argues that the classic McKenzie model of basin development does not adequately explain the subsidence history of the North Sea Basin and that other crustal or mantle factors may have been operative.

Seafloor spreading in the North Atlantic, south of the Rockall Plateau, commenced some 65 Ma ago. By 62 Ma ago a mantle plume was established beneath the continental lithosphere of East Greenland that created a 1000 km wide mushroom-shaped head of hot mantle (Figs 13 and 14; White and McKenzie 1989). Wherever rifting took place, extensive volcanism resulted from melting of the upwelling asthenosphere, as expressed in the Tertiary igneous province of western Britain and massive extrusion of basalt, underplating and elevation of the rifted margins along the Greenland eastern sea-board, Hatton Bank and Vøring Plateau of the UK–Norwegian continental shelf (Fig. 13; White 1990). It is tempting to suggest a direct link between the increased heat flow associated with the mantle plume and the 'hot fluid flush' recorded from Mesozoic sediments in the North Sea. The temperature increase caused by such a hot spot is, however, very modest, resulting in a temperature anomaly of around 300°C at a depth of 100 km (White 1988). This temperature anomaly would take around 50 Ma to propagate to the surface (White, pers. comm.) so a mantle plume cannot be the direct cause of an early Tertiary hot fluid flush. A much more likely cause is the shallow intrusion of igneous rocks into the sedimentary sequence during the early Tertiary. Unfortunately, other than tuffaceous deposits, there are no recorded examples of igneous rocks in the Tertiary of the North Sea that could account for the 'hot fluid flush'.

An alternative explanation of the 'hot fluid flush' appeals to the effects of regional uplift and tilting contemporary with the igneous activity along the North Atlantic margin. Intrusion

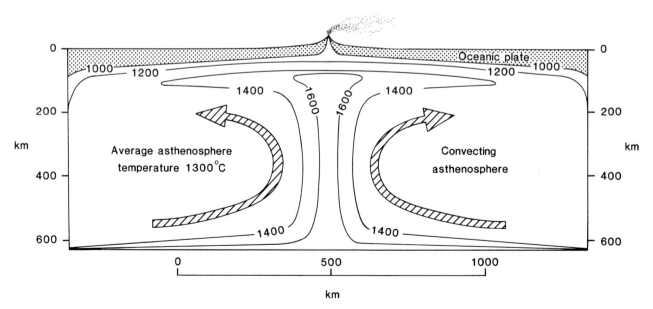

Fig. 14. Diagrammatic cross-section through a mantle plume illustrating the distribution and extent of the temperature anomaly associated with rising hot asthenosphere. Temperature anomaly based on the mantle plume beneath the Cape Verde Rise (modified from White 1988).

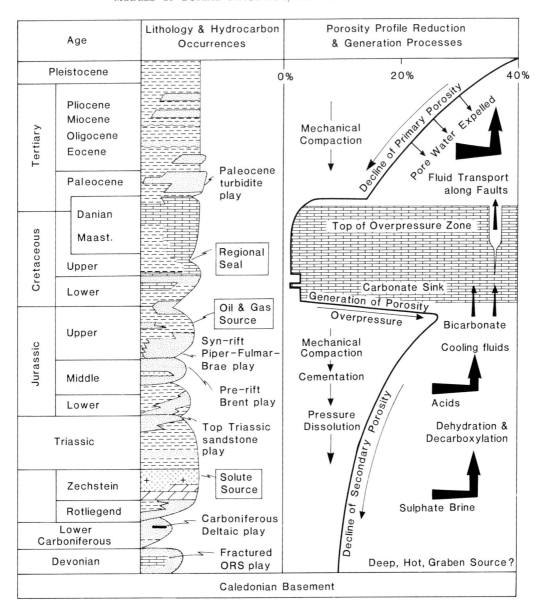

Fig. 15. A speculative model of porosity evolution and diagenetic processes operative in the Central and Northern North Sea. Throughout the Tertiary sequence, primary porosity is lost essentially by mechanical compaction. The Cretaceous section constitutes a regional fluid seal and carbonate buffer that prevents vertical fluid movement from the underlying Jurassic section except where it is cut by faults, gas chimneys or salt diapirs. In the Jurassic, secondary porosity generation beneath the regional Cretaceous seal at the crest of structures is gradually destroyed at depth by compaction and cementation. Note that over short depth intervals the porosity loss curve appears linear or exponential but that over the whole depth interval clearly shows the generation of enhanced porosity in the Jurassic (compare with Fig. 1).

and underplating of up to 2 km of igneous material to the west, northwest and north of the North Sea basin would have certainly influenced sediment provenance and transport pathways for the Tertiary sediments of the Northern and Central North Sea (Fig. 13). This in turn may have been responsible for the contemporary rapid increase in subsidence rates in the Northern North Sea. Joy (1993) similarly argues that the very rapid Tertiary subsidence in the Central and Northern North Sea may be due to the development of a thermal anomaly beneath the North Sea and related to the change from intra-continental basin to one adjacent to a newly developing ocean. Regardless of the exact cause of the rapid burial, such increased subsidence and changes of intra-plate stresses would certainly have resulted in a major readjustment of fluid flow pathways in the Mesozoic sediments of the North Sea (Cloetingh *et al.* 1990) and may have provided the necessary drive for initiating an active thermobaric 'hot fluid flush' from the deeper graben in the North Sea.

A conceptual model for diagenetic evolution in Jurassic reservoirs of the Central and Northern North Sea

The foregoing discussion argues for a distinct organization of diagenetic processes and resulting diagenetic assemblages and fabrics in the North Sea. Figure 15 is a highly speculative attempt to place these diagenetic processes into a regional burial framework that provides a mechanistic interpretation of the dominant processes and spatial distribution in which they operate.

The key feature of this interpretative model is the presence of a widespread Cretaceous and Lower Tertiary blanket cover of chalks, marls and calcareous mudstones that provide a regional seal to upwardly migrating fluids. This regional cover also coincides with the zone of overpressure that is developed across the deep graben centres. The regional seal not only acts as a major barrier to fluid flow but also separates two distinct diagenetic systems:

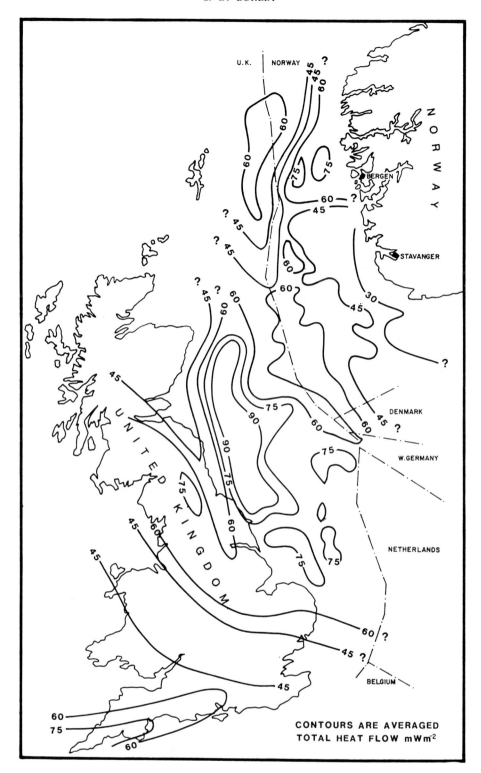

Fig. 16. Total heat flow map for the North Sea (data from Andrews-Speed *et al.* 1985 and Eggen 1984). Note the high heat flow anomalies on either side of the Central and Viking grabens.

1. an upper diagenetic system in the Tertiary that is characterized by the loss of primary porosity which is largely destroyed by mechanical compaction, and is mostly unaffected by introduction of solutes from depth; and

2. a lower diagenetic system in the Mesozoic and Permian sediments that is characterized by the generation of secondary porosity in crestal structural positions below the base Cretaceous and destruction of this sedimentary porosity through a combination of mesogenic cementation, mechanical compaction and chemical compaction. Enhanced por-

osities are only maintained through either the introduction of hydrocarbons shortly after secondary porosity generation or the maintenance of overpressures because of pressure compartmentalization, either due to fault sealing or stratigraphic hydrodynamic traps. Development of pressure compartments may enable convection cells to be established on the scale of tilted fault blocks with flow through large-scale fracture and fault systems.

In this model the Chalk acts as a regional carbonate buffer and prevents significant transfer of bicarbonate from the maturation of the Kimmeridge Clay Formation organic matter-rich mudstones or from cement dissolution in the Jurassic reservoirs through to the Tertiary. The only place where communication between the two diagenetic systems can take place is where large faults cut the Cretaceous seal, or where salt diapirs or gas chimneys penetrate the cover.

Supporting evidence from the present heat flow data

Regional studies of the total heat flow in the North Sea (e.g. Carstens and Finstad 1981; Oxburgh and Andrews-Speed 1981; Eggen 1984; Andrews-Speed et al. 1985) indicate a major positive heat flow anomaly along the shallow margins of the basin (Fig. 16). When these data are resolved into shallow (heat flow in sediments at depths of <1 km) and into deep (heat flow in sediments at depths of >1 km) the pattern of heat flow is rather different (Fig. 17). At shallow depths, the heat flow pattern is directly comparable to the total heat flow while at greater depths, the highest heat flow coincides with the deep graben centres. Andrews-Speed et al. (1985) interpret this distribution of heat flow to reflect a convective downward flow of cool water through the Tertiary sediments to a level close to the base Tertiary. Here the cool water is inferred to intercept the conductive heat flux at the top of the Mesozoic, warm and flow from east to west along the top of the Mesozoic section before ascending in the western part of the North Sea basin, giving higher temperatures at shallow depths and enhancing the shallow heat flow. An alternative, equally plausible interpretation of the heat flow data is that hot waters are being expelled from the deep graben centres, migrate upwards and outwards, but are trapped at the base of the Tertiary (equivalent to the zone of overpressure) and can then only migrate laterally (Fig. 18). Escape of hot fluids to the shallow parts of the basin is only possible at the basin margins where the large basin-bounding faults transect the Tertiary cover and overpressures are released. If this is the case, then significant temperature anomalies should be associated with such faults that cut the Tertiary section and relieve overpressure. Of course, fluid release may not be continuous but pulsed, being controlled by seismic valving processes.

Such large-scale fluid migration may establish convection cells that recirculate hot fluids in the Mesozoic section beneath the base Tertiary. While it is well documented that convection cells are unlikely to be maintained in the intergranular pore networks of individual sandbodies because of permeability heterogeneities (Bjørlykke et al. 1988; although cf. Wood and Hewett 1982) large-scale convection is well documented in hydrothermal systems around fractured granite stocks, recent volcanoes and shallow intrusives. It seems reasonable to argue that such convection systems may be set up in the crest of large fault blocks and circulate through large fracture systems. This may well be the mechanism involved in fluid movement around salt diapirs that has introduced hot fluids into the Chalk reservoirs of the Central North Sea (Jensenius 1987).

The fluid flow convection implied by the present-day heat flow data is probably only a relic of the much more vigorous thermobaric-compactive fluid flow drive that was initiated during the early Tertiary, associated with the onset of rapid thermal relaxation subsidence and the enhanced burial described by Joy (1993) in the Paleocene. Most of this fluid flow may be associated with dehydration, decarboxylation and compaction reactions all initiated as the Mesozoic sections were rapidly buried between the depths of 2–4 km. Such an inferred early Tertiary fluid flow regime is also probably responsible for the temperature anomalies apparent in the reservoir sandstones of many Jurassic reservoirs located on crests of structural blocks and close to the larger bounding faults (Cooper et al. 1975; Carstens and Finstad 1981; Cayley 1987).

Conclusions: implications for deep exploration plays

Much of this account is speculative and likely to be highly provocative. It certainly raises more questions than the answers it provides, and offers an alternative view to the diagenesis of North Sea reservoirs than advocated by Bjørlykke in numerous publications (see Bjørlykke and Aagaard (1991) and references therein).

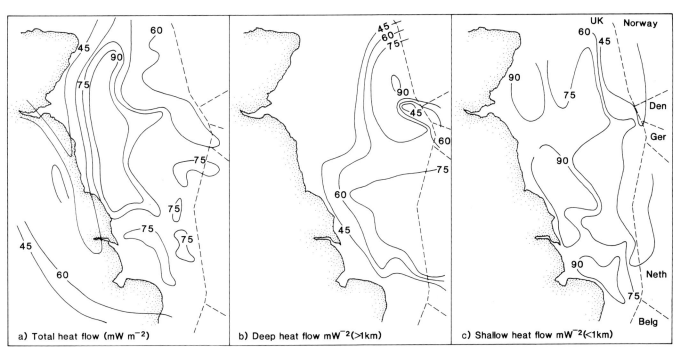

Fig. 17. Resolution of the total heat flow data for the western North Sea into shallow and deep components showing the high heat flow source in the deep graben centres (from Andrews-Speed et al. 1985).

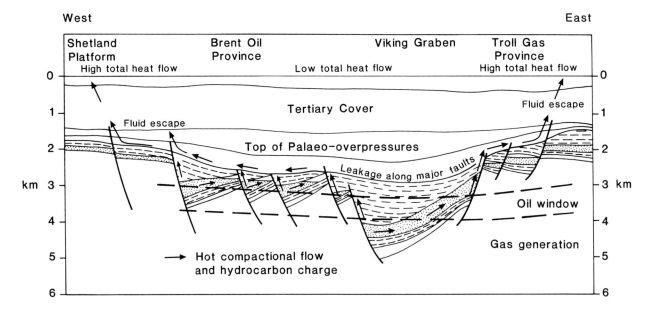

Fig. 18. Schematic cross-section of the Northern North Sea illustrating the interpretative fluid flow responsible for the present heat flow. A residual thermobaric flow is driven from the deep graben towards the basin margins and is responsible for the generation of overpressures beneath the regional Cretaceous seal. Hot fluids can only pass this seal at the basin margins where the bounding faults cut the Cretaceous. The deep hot fluid circulation in the Jurassic section was probably much more vigorous in the early Tertiary when rapid burial took place and fluid generating reactions were initiated. A similar hydrodynamic regime is envisaged for the Central Graben.

If there is any validity to the notion that most of the observed diagenesis in North Sea Jurassic reservoirs is dominated by the effects of hot fluid circulation during Tertiary burial with the associated development of mesogenic secondary porosity, then this raises implications for deep Jurassic play exploration.

The formation of secondary porosity during burial, concomitant with dehydration, decarboxylation and hydrocarbon-generating reactions is likely to be associated with the development of overpressure. Such secondary porosity thus has a high probability of being preserved and its maintenance by overpressure will depress the so-called porosity basement. Undeniably, there will still be progressive loss of porosity with depth, but this porosity loss should be projected from the base Cretaceous. Moreover, the rate of secondary porosity loss is less than for primary porosity because it is developed within an already compacted and cemented sandstone framework. Secondary porosity should, therefore, persist to greater depths than would be expected of primary porosity.

The proposed conceptual diagenetic models have specific implications for the development of enhanced porosity. Feldspar grain dissolution and carbonate cement dissolution should be preferentially developed on structural highs and in hydrodynamic traps. This is simply a function of the fact that compactional, thermobaric fluid flow will tend to migrate upwards and basin marginwards. In this sense the analogy with sediment-hosted mineral deposits is remarkable (Cathles and Smith 1983). Additionally, the potential for secondary porosity preservation and diagenetic fault seal development may be dependent on the component of vertical fluid flow associated with fault dilation. Transfer of temperature disequilibrium in fluids up faults favours precipitation of silicates in secondary porosity generated on structural highs as the hot fluids cool. Such a precipitation mechanism is not favoured if fault-related fluid flow is cross-formational. Fault-bounded deep graben structures may thus be the sites of enhanced porosity preservation. By contrast, structurally high, fault-bounded structures, where a Zechstein source is faulted into close proximity of the reservoir, are likely to be the loci of intense diagenetic cementation.

Stuart Burley is the British Gas Geology Senior Research Fellow at Manchester and expresses his gratitude to BG Exploration and Production (Reading) for continuing to support diagenetic research. This paper draws on the expertise and experience of many colleagues and research students at Manchester, all of whom are gratefully acknowledged. Simon Guscott, Gavin McAulay and Patrick Mckeever read early drafts of the manuscript and constructive reviews by Melvyn Giles (Shell) and John Kantorowicz (Conoco) improved the final version.

References

ANDREWS-SPEED, C. P., OXBURGH, E. R. AND COOPER, C. A. 1985. Temperatures and depth dependent heat flow in the Western North Sea. *American Association of Petroleum Geologists Bulletin*, **68**, 1764–1781.

ATWATER, G. I. AND MILLER, E. E. 1964. The effect of decreasing porosity with depth on future development of oil and gas reserves in South Louisianna. *American Association of Petroleum Geologists Bulletin*, **49**, 334.

BAINES, S., BURLEY, S. D. AND GIZE, A. 1991. Sulphide mineralisation and hydrocarbon migration in North Sea oilfields. *In*: PAGEL, M. AND LEROY, J. (eds) *Proceedings of the 25th Anniversary Mineral Deposita Meeting on Source, Transport and Deposition of Metals, Nancy, 1991*. Balkema, Rotterdam. 87–91.

BARR, K. 1987. Structural/stratigraphic models for extensional basins of half graben type. *Journal of Structural Geology*, **9**, 491–500.

BEARD, D. C. AND WEYL, P. K. 1973. Influence of texture on porosity and permeability of unconsolidated sand. *American Association of Petroleum Geologists Bulletin*, **57**, 349–369.

BISSADA, K. K. 1983. Petroleum generation in Mesozoic sediments of the Moray Firth Basin, British North Sea area. *Advances in Geochemistry 1981*. John Wiley and Sons, London, 7–15.

BJØRLYKKE, K. O. 1984. Formation of secondary porosity: how important is it? *In*: McDONALD, D. A. AND SURDAM, R. C. (eds) *Clastic Diagenesis*. American Association of Petroleum Geologists Memoir, **37**, 277–286.

—— AND AAGAARD, P. 1992. Clay minerals in North Sea Sandstones. *In*: HOUSEKNECHT, D. AND PITMANN, E. D. (eds) *Origin, Diagenesis Petrophysics of Clay Minerals in Sandstones*. Special Publication of the Society of Economic Paleontologists and Mineralogists, **47**, 65–80.

——, ——, DYPVIK, H., HASTINGS, D. S. AND HARPER, A. S. 1986. Diagenesis and reservoir properties of Jurassic sandstones from the Haltenbanken area, offshore mid-Norway. *In*: SPENCER, A. M. (ed.) *Habitat of Hyrdocarbons on the Norwegian Continental Shelf*. Graham & Trotman, London, 275–286.

—— AND BRENDSALL, A. 1986. Diagenesis of the Brent Sandstone in the Statfiord Field, North Sea. *In*: BAUTIER, D. L. (ed.) *Roles of Organic Matter in Sediment Diagenesis*. Special Publication of the Society of Economic Paleontologists and Mineralogists, **38**, 157–167.

——, MO, A. AND PALM, E. 1988. Modelling of thermal convection in sedimentary basins and its relevance to diagenetic reactions. *Marine and Petroleum Geology*, **5**, 338–351.

BRINT, J. F. 1989. *Isotope diagenesis and palaeofluid movement: Middle Jurassic Brent sandstones, North Sea*. PhD thesis, University of Strathclyde, Glasgow, Scotland.

BROOKS, J., CORNFORD, C., NICHOLSON, J. AND GIBBS, A. D. 1983. Geological controls on the occurrence and composition of Tertiary heavy oils, Northern North Sea. (Abstract) *American Association of Petroleum Geologists Bulletin*, **68**, 793.

BUHRIG, C. 1989. Geopressured Jurassic reservoirs in the Viking Graben; modelling and geological significance. *Marine and Petroleum Geology*, **6**, 31–48.

BURLEY, S. D. 1986. The development and destruction of porosity within Upper Jurassic reservoir sandstones of the Piper and Tartan Oilfields, Outer Moray Firth, North Sea. *Clay Minerals*, **21**, 649–694.

—— AND FLISCH, M. 1989. K/Ar geochronology and the timing of detrital I/S clay illitization and authigenic illite precipitation in the Piper and Tartan fields, outer Moray Firth, UK North Sea. *Clay Minerals*, **24**, 285–315.

——, KANTOROWICZ, J. D. AND WAUGH, B. 1985. Clastic Diagenesis. *In*: BRENCHLEY, P. J. AND WILLIAMS, B. P. J. (eds) *Sedimentology: Recent and Applied Aspects*. Geological Society, London, Special Publication, **18**, 189–226.

—— AND MACQUAKER, J. H. S. 1992. Authigenic clays, diagenetic sequences and conceptual diagenetic models in contrasting basin margin and basin centre North Sea Jurassic sandstones and mudstones. *In*: HOUSEKNECHT, D. AND PITMANN, E. D. (eds) *Origin, Diagenesis and Petrophysics of Clay Minerals in Sandstones*. Special Publication of the Society of Economic Paleontologists and Mineralogists, **47**, 81–110.

——, MULLIS, J. AND MATTER, A. 1989. Timing diagenesis in the Tartan reservoir (UK, North Sea); constraints from combined cathodoluminescence microscopy and fluid inclusion studies. *Journal of Marine and Petroleum Geology*, **6**, 98–120.

——, WALSH, J. AND WATTERSON, J. 1991. Active fault participation in the diagenetic modification of sandstone reservoir properties. (abstract) *American Association of Petroleum Geologists Bulletin*, **75**, 1407.

CARSTENS, H. AND DYPVIK, H. 1981. Abnormal formation pressure and shale porosity. *American Association of Petroleum Geologists Bulletin*, **65**, 344–350.

—— AND FINSTAD, K. G. 1981. Geothermal gradients of the Northern North Sea Basin, 59–62°N. *In*: ILLING, L. V. AND HOBSON, G. D. (eds) *Petroleum Geology of the Continental Shelf of North-West Europe*. Heyden, London, 152–161.

CATHLES, L. M. AND SMITH, J. E. 1983. Thermal constraints on the formation of Mississippi Valley-type lead-zinc deposits and their implications for episodic basin dewatering and deposit genesis. *Economic Geology*, **78**, 983–1002.

CAYLEY, G. T. 1987. Hydrocarbon migration in the Central North Sea. *In*: BROOKS, J. AND GLENNIE, K. W. (eds) *Petroleum Geology of North West Europe*. Graham & Trotman, London, 549–555.

CHIARELLA, A. AND DUFFAUD, F. 1980. Pressure origin and distribution in Jurassic of Viking Basin (United Kingdom, Norway). *American Association of Petroleum Geologists Bulletin*, **64**, 1245–1250.

CLOETINGH, S., GRADSTEIN, F. M., KOOI, H., GRANT, A. C. AND KAMINISKI, M. 1990. Plate reorganisation: a cause of rapid late Neogene subsidence and sedimentation around the North Atlantic? *Journal of the Geological Society, London*, **147**, 495–506.

COLEMAN, M. L., CURTIS, C. D. AND IRWIN, H. 1979. Burial rate a kev

to source and reservoir potential. *World Oil*, March 1979, 36–40.

COOPER, B. S., COLEMAN, S. H., BARNARD, P. C. AND BUTTERWORTH, J. S. 1975. Palaeotemperatures in the Northern North Sea Basin. *In*: WOODLAND, A. V. (ed.) *Petroleum and the Continental Shelf of North West Europe*. Applied Science Publishers, Barking, 487–492.

CORNFORD, C., NEEDHAM, C. E. J. AND DE WALQUE, L. 1986. Geochemical habitat of North Sea oils and gases. *In*: SPENCER, A. M. (ed.) *Habitat of Hydrocarbons on the Norwegian Continental Shelf*. Graham & Trotman, London, 39–54.

CURTIS, C. D. 1983. Link between aluminium mobility and destruction of secondary porosity. *American Association of Petroleum Geologists Bulletin*, **67**, 380–384.

—— AND COLEMAN, M. L. 1986. Controls on the precipitation of early diagenetic calcite, dolomite and siderite concretions in complex depositional sequences. *In*: GAUTIER, D. L. (ed.) *Roles of Organic Matter in Sediment Diagenesis*. Special Publication of the Society of Economic Paleontologists and Mineralogists, **38**, 23–33.

EGEBERG, P. K. AND AAGAARD, P. 1989. Origin and evolution of formation waters from oilfields on the Norwegian Shelf. *Applied Geochemistry*, **4**, 131–142.

EGGEN, S. 1984. Modelling of subsidence, hydrocarbon generation and heat transport in the Norwegian North Sea. *In*: DURAND, B. (ed.) *Thermal Phenomena in Sedimentary Basins*. Editions Technip, Paris, 271–283.

EHRENBERG, S. N. 1991. Relationship between diagenesis and reservoir quality in sandstones of the Garn Formation, Haltenbanken, Mid-Norwegian Continental Shelf. *American Association of Petroleum Geologists Bulletin*, **74**, 1538–1558.

—— AND NADEAU, P. N. 1989. Formation of diagenetic illite in sandstones of the Garn Formation, Haltenbanken area, Mid-Norwegian continental shelf. *Clay Minerals*, **24**, 233–253.

EMERY, D., MYERS, K. J. AND YOUNG, R. 1990. Ancient subaerial exposure and freshwater leaching in sandstones. *Geology*, **18**, 1178–1181.

ENGLAND, W. A., MCKENZIE, A. S., MANN, D. M. AND QUIGLEY, T. M. 1989. The movement and entrapment of petroleum fluids in the subsurface. *Journal of the Geological Society, London*, **144**, 327–347.

FISHER, M. J. 1984. Triassic. *In*: GLENNIE, K. W. (ed.) *Introduction to the Petroleum Geology of the North Sea*. Blackwell, Oxford, 85–102.

FUCHTBAUER, H. 1967. Influence of different types of diagenesis on sandstone porosity. *Proceedings of the 7th World Petroleum Congress, Mexico*, vol. 2. Elsevier, London, 353–369.

GALLOWAY, W. E. 1974. Deposition and diagenetic alteration of sandstone in North East Pacific arc-related basins; implications for greywacke genesis. *Geological Society of America Bulletin*, **85**, 379–390.

GILES, M. R. AND DE BOER, R. B. 1990. Secondary porosity: creation of enhanced porosities in the subsurface from the dissolution of carbonate cements as a result of cooling formation waters. *Journal of Marine and Petroleum Geology*, **6**, 261–269.

—— AND MARSHALL, J. D. 1986. Constraints on the development of secondary porosity in the subsurface; re-evaluation of processes. *Journal of Marine and Petroleum Geology*, **3**, 243–255.

GLASMANN, J. R. 1992. The fate of feldspar in Brent Group reservoirs, North Sea: a regional synthesis of diagenesis in shallow, intermediate and deep burial environments. *In*: MORTON, A. C., HASZELDINE, R. S., GILES, M. R. AND BROWN, S. (eds) *Geology of the Brent Group*, Geological Society, London, Special Publication, **61**, 329–350.

——, LUNDEGAARD, P. D., CLARKE, B. K., PENNY, B. K. AND COLLINS, I. D. 1989. Geochemical evidence for the history of diagenesis and fluid migration: Brent Sandstone, Heather Field, North Sea. *Clay Minerals*, **24**, 255–284.

GLUYAS, J. G. 1985. Reduction and prediction of sandstone reservoir potential, Jurassic, North Sea. *Philosophical Transactions of the Royal Society of London*, **135**, 187–202.

HANOR, J. S. 1987. *Origin and Migration of Subsurface Sedimentary Brines*. Society of Economic Paleontologists and Mineralogists Short Course, **21**.

HARRIS, J. AND FOWLER, R. 1987. Enhanced prospectivity of the Mid–Late Jurassic sediments of the South Viking Graben, northern North Sea. *In*: BROOKS, J. AND GLENNIE, K. W. (eds) *Petroleum Geology of North West Europe*. Graham & Trotman, London, 879–889.

HARRIS, N. B. 1989. Diagenetic quartzarenite and destruction of secondary porosity; an example from the Middle Jurassic Brent Sandstone of Northwest Europe. *Geology*, **17**, 361–364.

HASZELDINE, R. S., SAMSON, I. M. AND CORNFORD, C. 1984. Quartz diagenesis and convective fluid movement; Beatrice Oilfield, UK North Sea. *Clay Minerals*, **19**, 391–402.

HÉRITIER, F. E., LOSSEL, P. AND WATHNE, E. 1981. The Frigg Gas Field. *In*: ILLING, L. V. AND HOBSON, G. D. (eds) *Petroleum Geology of the Continental Shelf of North-West Europe*. Heyden, London, 380–394.

HOGG, A. J. C. 1989. *Petrographic and isotopic constraints on the diagenesis and reservoir properties of the Brent Group sandstones, Alwyn South, northern North Sea*. PhD dissertation, University of Aberdeen, Scotland.

HUNT, J. M. 1990. Generation and migration of petroleum from abnormally pressured fluid compartments. *American Association of Petroleum Geologists Bulletin*, **74**, 1–12.

HURST, A. AND BULLER, A. T. 1984. Dish structures in some Palaeocene deep-sea sandstones (Norwegian sector, North Sea): Origin of the dish-forming clays and their effect on reservoir quality. *Journal of Sedimentary Petrography*, **54**, 1206–1211.

JENSENIUS, J. 1987. High-temperature diagenesis in shallow Chalk reservoir, Skjold Field, Danish North Sea: Evidence from their fluid inclusions and oxygen isotopes. *American Association of Petroleum Geologists Bulletin*, **71**, 1378–1386.

—— AND MUNKSGAARD, N. C. 1989. Large scale hot water migration systems around salt diapirs in the Danish Central trough and their impact on diagenesis of Chalk reservoirs. *Geochimica et Cosmochimica Acta*, **53**, 79–88.

JOHNSON, H. D. AND STEWART, D. J. 1985. Role of clastic sedimentology in the exploration and production of oil and gas in the North Sea. *In*: BRENCHLEY, P. J. AND WILLIAMS, B. P. J. (eds) *Sedimentology: Recent Developments and Applied Aspects*. Geological Society, London, Special Publication, **18**. 249–310.

JOURDAN, A., THOMAS, M., BREVART, O., ROBSON, P., SOMMER, F. AND SULLIVAN, M. 1987. Diagenesis as a control of Brent sandstone reservoir properties in the Greater Alwyn Area, East Shetland Basin. *In*: BROOKS, J. AND GLENNIE, K. W. (eds) *Petroleum Geology of North West Europe*. Graham & Trotman, London, 951–961.

JOY, A. M. 1993. Comments on the pattern of post rift subsidence in the Central and Northern North Sea Basin. *In*: WILLIAMS, G. D. AND DOBB, A. (eds) *Tectonics and Seismic Sequence Stratigraphy*. Geological Society, London, Special Publication, **71**, 123–140.

KONNERUP-MADSEN, J. AND DYPVIK, H. 1988. Fluid inclusions and quartz cementation in Jurassic sandstones from Haltenbanken, offshore mid-Norway. *Bulletin of Mineralogy*, **111**, 401–411.

LIEWIG, N., CLAUER, N. AND SOMMER, F. 1987. Rb–Sr and K–Ar dating of clay diagenesis in Jurassic sandstone oil reservoir, North Sea. *American Association of Petroleum Geologists Bulletin*, **71**, 1467–1474.

LØNØY, A., AKSELSEN, J. AND RONNING, K. 1986. Diagenesis of a deeply buried sandstone reservoir: Hild Field, Northern North Sea. *Clay Minerals*, **21**, 497–511.

MACAULAY, C. I., HASZELDINE, R. S. AND FALLICK, A. E. 1992. Diagenetic pore waters stratified for at least 35 million years: Magnus Oilfield, North Sea. *American Association of Petroleum Geologists Bulletin*, **76**, 1625–1634.

MAGARA, K. 1981. Compaction and fluid migration. *Developments in Petroleum Science*, **9**. Elsevier, Amsterdam.

MAHER, C. E. 1980. The Piper Oilfield: *In*: HALBOUTY, M. T. (ed.) *Giant Oil and Gas Fields of the Decade 1968–1978*. American Association of Petroleum Geologists Memoir, **30**, 131–172.

MCAULAY, G., BURLEY, S. D. AND JOHNES, H. 1993. Silicate mineral authigenesis in the Hutton and NW Hutton fields: implications for sub-surface porosity development. *In*: PARKER, J. R. (ed.)

Petroleum Geology of Northwest Europe: Proceedings of the 4th Conference. Geological Society, London, 1377–1393.

MOUND, D. G., ROBERTSON, I. D. AND WALLIS, R. J. 1991. The Cyrus Field. *In*: ABBOTTS, I. L. (ed.) *United Kingdom Oil and Gas Fields 25 Years Commemorative Volume*. Geological Society, London, Memoir, **14**, 295–300.

NAGTEGAAL, P. J. C. 1980. Sandstone framework instability as a function of burial diagenesis. *Journal of the Geological Society, London*, **135**, 101–105.

OLAUSSEN, S., DALLAND, A., GLOPPEN, T. G. AND JOHANNESSEN, E. 1984. Depositional environment and diagenesis reservoir sandstones in the eastern part of Troms 1 area. *In*: SPENCER, A. M. (ed.) *Petroleum Geology of the North European Margin*. Norwegian Petroleum Society, 61–79.

OXBURGH, R. AND ANDREWS-SPEED, C. P. 1981. Temperature, thermal gradients and heat flow in the south-western North Sea. *In*: ILLING, L. V. AND HOBSON, G. D. (eds) *Petroleum Geology of the Continental Shelf of North West Europe*. Heyden, London, 141–151.

PARSLEY, A. J. 1984. North Sea hydrocarbon plays. *In*: GLENNIE, K. W. (ed.) *Introduction to the Petroleum Geology of the North Sea*. Blackwell, Oxford, 205–230.

PEARSON, M. J. AND SMALL, J. S. 1988. Illite/smectite diagenesis and palaeotemperatures in northern North Sea Quaternary to Mesozoic shale sequences. *Clay Minerals*, **23**, 109–132.

——, WATKINS, D., PITTION, J-L., CASTON, D. AND SMALL, J. 1983. Aspects of burial diagenesis, organic maturation, and palaeogeothermal history of an area in the South Viking Graben, North Sea. *In*: BROOKS, J. (ed.) *Petroleum Geochemistry in the exploration of Europe*. Geological Society, London, Special Publication, **12**, 161–173.

PEGRUM, R. M. AND SPENCER, A. M. 1991. Hydrocarbon Plays in the northern North Sea. *In*: BROOKS, J. (ed.) *Classic Petroleum Provinces*. Geological Society, London, Special Publication, **50**, 441–470.

POTTER, R. W. 1977. Pressure corrections for fluid-inclusion homogenization temperatures based on the volumetric properties of the system Na–Cl–H_2O. *Journal of Research of the US Geological Survey*, **5/5**, 603–607.

RABINOWITZ, M., DANOURAND, J-L., JAKUBOWSKI, M., SCHOTT, J. AND CASAN, J. P. 1985. Convection in a North Sea oil reservoir: influences on diagenesis and hydrocarbon migration. *Earth and Planetary Science Letters*, **74**, 387–404.

RAMM, M. 1992. Porosity–depth trends in reservoir sandstones: theoretical models related to Jurassic sandstones, offshore Norway. *Journal of Marine and Petroleum Geology*, **9**, 553–567.

SAIGAL, G. C., BJØRLYKKE, K. AND LARTER, S. 1992. The effect of oil emplacement on diagenetic processes: examples from the Fulmar reservoir sandstones, Central North Sea. *American Association of Petroleum Geologists Bulletin*, **76**, 1024–1033.

SAMWAYS, G. AND MARSHALL, J. 1988. *A seismic pumping model to account for complex diagenesis in faulted sediments*. Society of Economic Paleontologists and Mineralogists Mid-year meeting, Columbus, Ohio, Abstract volume 47.

SCHMIDT, V. AND MCDONALD, D. A. 1979. The role of secondary porosity in the course of sandstone diagenesis. *In*: SCHOLLE, P. A. AND SCHLUGER, P. R. (eds) *Aspects of Diagenesis*. Special Publication of the Society of Economic Paleontologists and Mineralogists, **26**, 175–208.

SCHMOKER, J. W. AND GAUTIER, D. L. 1988. Sandstone porosity as a function of thermal maturity. *Geology*, **16**, 1007–1010.

SCLATER, J. G. AND CHRISTIE, P. A. F. 1980. Continental stretching and explanation of the post-mid Cretaceous subsidence of the Central North Sea Basin. *Journal of Geophysical Research*, **85**, 3711–3739.

SCOTCHMAN, I., JOHNNES, L. H. AND MILLER, R. S. 1989. Clay diagenesis and oil migration in Brent Group sandstones of NW Hutton field, UK North Sea. *Clay Minerals*, **24**, 339–374.

SELLEY, R. C. 1978. Porosity gradients in North Sea oil-bearing sandstones. *Journal of the Geological Society, London*, **135**, 119–132.

SIBSON, R. H. 1981. Fluid flow accompanying faulting: field evidence and models. *In*: SIMPSON, D. W. AND RICHARDS, P. G. (eds)

Earthquake Prediction: An international review. American Geophysical Union, Washington, D.C. Maurice Ewing Series, **4**, 593–603.

SMALL, J. 1993. Experimental determination of the rates of precipitation of authigenic illite and kaolinite in the presence of aqueous oxalate in comparison to the K–Ar ages of illites in reservoir sandstones. *Clays and Clay Minerals*, **41**, in press.

SPENCER, A. M. AND LARSEN, V. B. 1990. Fault traps in the Northern North Sea. *In*: HARDMAN, R. F. P. AND BROOKS, J. (eds) *Tectonic Events Responsible for Britain's Oil and Gas Reserves.* Geological Society, London, Special Publication, **55**, 281–298.

STEWART, D. J. 1986. Diagenesis of the shallow marine Fulmar Formation in the Central North Sea. *Clay Minerals*, **21**, 537–564.

SURDAM, R. C., DUNN, T. L., HEATHER, H. P. AND MacGOWAN, D. B. 1989. Integrated diagenetic modelling; a process orientated approach for clastic systems. *Annual Reviews of Earth and Planetary Science Letters*, **15**, 141–170.

TAYLOR, J. C. M. 1981. Zechstein facies and petroleum prospects in the Central and Northern North Sea. *In*: ILLING, L. V. AND HOBSON, G. D. (eds) *Petroleum Geology of the Continental Shelf of North-West Europe*, Heyden, London, 176–185.

TAYLOR, S. R. AND LAPRE, J. F. 1987. North Sea Chalk diagenesis; its effect on reservoir location and properties. *In*: BROOKS, J. AND GLENNIE, K. W. (eds) *Petroleum Geology of North West Europe.* Graham & Trotman, London, 483–496.

THOMPSON, P. J. AND BUTCHER, P. D. 1991. The Geology and Geophysics of the Everest Complex. *In*: SPENCER, A. M. (ed.) *Generation, Accumulation and Production of Europe's Hydrocarbons.* EAPG Special Publication, **1**, 89–98. Oxford University Press.

WALDERHAUG, O. 1990. A fluid inclusion study of quartz-cemented sandstones from offshore mid-Norway; possible evidence for continued quartz cementation during oil emplacement. *Journal of Sedimentary Petrology*, **60**, 203–210.

WHITE, R. S. 1988. A not-spot model for early Tertiary volcanism in the North Atlantic. *In*: MORTON, A. C. AND PARSON, L. M. (eds) *Early Tertiary Volcanism and the Opening of the NE Atlantic.* Geological Society, London, Special Publication, **39**, 3–13.

—— 1990. Initiation of the Iceland plume and opening of the North Atlantic. *In*: TANKARD, A. J. AND BALKWILL, H. R. (eds) *Extensional Tectonics and Stratigraphy of the North Atlantic Margins.* American Association of Petroleum Geologists Memoir, **46**, 149–154.

—— AND McKENZIE, D. P. 1989. Magmatism at rift zones: the generation of volcanic continental margins and flood basalts. *Journal of Geophysical Research*, **94**, 7685–7729.

WOOD, J. R. 1989. Modeling the effect of compaction and precipitation-dissolution on porosity. *Mineralogical Association of Canada Short Course Handbook in Burial Diagenesis*, **15**, 311–362.

—— AND HEWETT, T. A. 1982. Fluid convection and mass transfer in porous sandstones—a theoretical model. *Geochimica et Cosmochimica Acta*, **46**, 1707–1713.

YIELDING, G. 1990. Footwall uplift associated with Late Jurassic normal faulting in the northern North Sea. *Journal of the Geological Society, London*, **147**, 219–222.

—— AND ROBERTS, A. 1991. Footwall uplift during normal faulting—implications for structural geometries in the North Sea. *In*: LARSEN, R. M. (ed.) *Structural and Tectonic Modelling and its Application to Petroleum Geology.* Norwegian Petroleum Society, Stavanger, 85–98.

ZIEGLER, P. A. 1982. *Geological Atlas of Western and Central Europe.* Shell Reprographics and Elsevier, Amsterdam.

Silicate mineral authigenesis in the Hutton and NW Hutton fields: implications for sub-surface porosity development

G. E. McAULAY,[1] S. D. BURLEY[1] and L. H. JOHNES[2]

[1] Department of Geology, University of Manchester, Manchester M13 9PL, UK

[2] Amoco (UK) Exploration Company Ltd, Amoco House, West Gate, London W5 1XL, UK

Abstract: The Hutton and NW Hutton fields occur in highly faulted, rotated block structures in Block 211/27 of the East Shetland Basin, Northern North Sea. Fourteen wells that core Brent Group reservoir from these fields have been studied in order to characterize the relationship between clay mineral authigenesis and porosity modification in the sub-surface.

Reservoir sands are present in the depth interval 9000–13 000 ft; the shallowest sandstones are subarkoses, contrasting with the deepest sandstones which are mostly feldspar-deficient quartz arenites. The change in composition is a result of diagenetic feldspar dissolution that takes place over the depth interval of study. Authigenic kaolin group minerals and illite are present over the depth range studied, and both exhibit depth-related increases. Three habits of authigenic kaolin group minerals are observed: 'expanded-mica', vermiform kaolinite and blocky dickite. The occurrence of vermiform kaolinite is enhanced in Hutton and the adjacent water zone, and does not occur in abundance below 10 500 ft. This habit formed as a result of syn-depositional meteoric water ingress. Structural models suggest that neither field underwent subsequent meteoric water flushing. At the deeper levels of its occurrence range, vermiform kaolinite shows signs of alteration to blocky dickite. Dickite occurs in both intergranular and grain dissolution pores. A depth-related trend of increasing dickite cementation is strongly linked to increasing K-feldspar dissolution. Enhanced dickite cementation is found adjacent to bed contacts in sandstones within mudstone sequences and in regressive sandbodies close to thick mudstone units. Illite shows a more pronounced depth-related increase across the studied interval. Illite precipitation is also genetically linked with detrital feldspar dissolution and is favoured by diminished pore water flow during deep burial. Authigenic clay distribution is partly controlled by local features such as detrital clay coats, in addition to larger-scale parameters such as primary lithology, facies associations, present burial depth and inferred palaeofluid flow. Intergranular porosity declines with increasing burial depth, while grain dissolution porosity increases with depth from Hutton to the shallow and intermediate depth wells of NW Hutton. Generation of enhanced porosity is not related to the unconformities overlying the Brent Group, but to sub-surface fluid migration pathways. Compaction is not significantly operative over the depth interval 10 000–13 000 ft. By 10 000 ft burial depth, up to 50% of the intergranular volume has been lost by mechanical compaction.

Middle Jurassic Brent Group sandstones of the Northern North Sea and their equivalents are amongst the most intensively studied reservoir lithologies. Authigenic clay minerals and quartz cements severely restrict the production of hydrocarbons in many Brent Group reservoirs, and most previous studies have concentrated on the broad distribution and timing of authigenic clay precipitation in these sequences (see Burley and MacQuaker (1992) for review). Various studies, from the early work of Sommer (1978) and Hancock and Taylor (1978), to recent detailed examinations of Norwegian sector reservoirs (e.g. Ehrenberg 1991a) and the Brent Group review volume (Morton et al. 1992), have described the occurrences of authigenic silicate minerals.

Scotchman et al. (1989) described the nature and timing of diagenesis and oil migration in the NW Hutton Field while Harris (1989, 1990) investigated diagenetic fabrics and porosity trends in the Hutton Field. This present study extends these findings to investigate the distribution of kaolin group polytypes and the relationship between porosity and clay mineral authigenesis over the depth interval 9000–13 000 ft in a traverse from the crest of the Hutton Field to the western downflank areas of the NW Hutton Field. Additionally, quantitative assessments of the extent of feldspar dissolution and quartz cementation have been undertaken in order to determine which reactions have been responsible for the present authigenic silicate mineral assemblage and calculate mass balance estimates for these reactions.

The spatial and temporal distribution of the authigenic silicate cements over the depth interval of study is then used to constrain the hydrodynamic regime under which the silicates

formed with particular reference to the likelihood of telogenetic meteoric water flushing, as advocated by Bjørlykke (1984) and referred to in many subsequent publications.

Geological setting

The NW Hutton Field is located in Block 211/27 along with the Hutton Field, part of which occurs in adjacent Block 211/28 (Fig. 1). The Brent Group reservoir in these fields occurs between 9000 ft and 13 000 ft (Fig. 2). The structural crest to the area is provided by the Hutton Field, details of which are described in Haig (1991). The Hutton Field comprises three main fault blocks, downthrown to the northwest. On the east of the Hutton structure several smaller fault blocks downthrow to the east and southeast and the field is truncated to the east by the Dunlin–Hutton–Ninian fault system. In NW Hutton the overall field structure is SW-dipping, and shows the tilted fault block profile typical of the Brent Group Province (Scotchman et al. 1989; Johnes and Gauer 1991). The NW Hutton Field is highly faulted, with four main production blocks defined by NE–SW-trending sealing faults. To the east, the Hutton Field is separated from the NW Hutton Field by a 1 km wide upthrown fault block that is water-bearing. Hutton is relatively less faulted by comparison with NW Hutton. Details of Brent Group stratigraphy for the Hutton and NW Hutton fields are given in Haig (1991) and Scotchman et al. (1989) respectively. The stratigraphy and facies characteristics for the Brent Group and associated formations are summarized in Table 1.

From *Petroleum Geology of Northwest Europe: Proceedings of the 4th Conference* (edited by J. R. Parker).

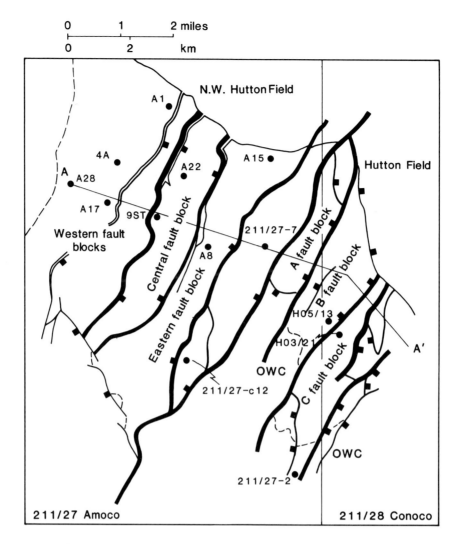

Fig. 1. Simplified structure map of the Hutton and NW Hutton fields (based on Haig 1991 and Johnes and Gauer 1991) illustrating the location of the study wells in relation to structural units. Line A–A′ defines the orientation of the cross-section in Fig. 2.

Table 1. Stratigraphy and facies characteristics of the Brent Group and associated formations

Age	Group	Formation	Facies
Upper Jurassic	Humber	Kimmeridge Clay	Marine muds
Upper Jurassic	Humber	Heather	Marine muds
Middle Jurassic	Brent	Tarbert	Marine muds and sandstones
Middle Jurassic	Brent	Ness	Delta plain muds, channel sands
Middle Jurassic	Brent	Etive	Upper shoreface/barrier bar
Middle Jurassic	Brent	Rannoch	Lower shoreface
Middle Jurassic	Brent	Broom	Sub-littoral sheet sandstone
Lower Jurassic	Dunlin	Drake	Marine muds

Methodology

A total of 14 wells from the Hutton and NW Hutton fields and associated off-structure areas is included in the study. The location of wells sampled is indicated in Fig. 1. Comparable sedimentary facies are present across the 9000 ft to 13 000 ft interval sampled, although complete core coverage of the Brent Group sequence was not always available. Samples were selected from the cores with the aid of detailed sedimentological logs to encompass variation in sedimentary facies and burial depth. Quantitative point-count analysis was performed

on 113 samples from 7 wells, with 300 counts per slide, to determine grain size and sorting parameters, detrital and authigenic components as well as porosity amounts and types. Table 1 details the distribution and number of these samples. Selected samples were further characterized using secondary electron imagery (SEI), back-scatter electron imagery (BSEI) and cathodoluminescence (CL) in a Jeol 6400 scanning electron microscope (SEM) equipped with a quantitative solid-state energy dispersive (ED) X-ray analysis detector. Routine operating conditions were an accelerating voltage of between

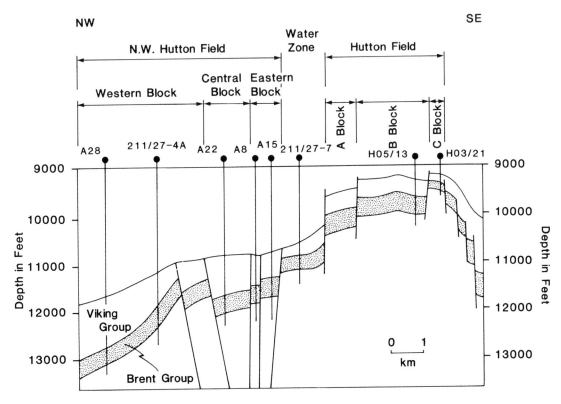

Fig. 2. Simplified structural cross-section (line A–A' in Fig. 1) across the crest of the Hutton Field to the western downflank of the NW Hutton Field.

15 kV and 20 kV, and a beam working distance of 39 mm for ED analysis. For high-resolution imagery this working distance was reduced to 15 mm. Transmission electron microscopy (TEM) techniques were employed for ultra-high resolution imaging purposes.

Identification of clay polytypes required the preparation of a kaolin group clay separate, and XRD analysis of an unoriented powder mount. On the basis of SEM crystal size studies, the 8–5 μm size fraction was optimal for obtaining a pure kaolin group mineral separate; vermiform morphologies are usually coarser than this but contamination with other detrital or authigenic minerals was minimal in this size fraction. To ensure an accurate correspondence between textural identification (SEI, thin section) and polytype assignment, samples dominated by only one or other of the morphologies were selected for separation procedures. XRD cavity mounts were filled with kaolin powder to maximize random orientation of diffracting planes. Traces were collected from a scan interval between $20°$ to $40°$ 2θ using Cu Ka radiation. Significant reflections used to distinguish polytypes were the $20\bar{2}$ and $1\bar{3}1$ at $38.44°$ 2θ Cu Ka and the 131 at $39.31°$ 2θ Cu Ka for kaolinite and the 132 and $20\bar{4}$ at $38.71°$ 2θ Cu Ka for dickite, as listed in Bailey (1980).

Mean grain size in thin section was determined by measuring the maximum diameter of a representative suite of detrital grains while an estimate of sorting was made using the visual comparator of Longiaru (1987). An estimate of the bulk volume intergranular porosity at deposition (the original porosity estimate, OPE) was calculated from the sorting value using the wet sand data of Beard and Weyl (1973). These data were then used to determine the volume of porosity loss due to compaction (CoPL) from the formula:

$$CoPL = OPE - \frac{(100 \times IGV) - (OPE \times IGV)}{(100 - IGV)}$$

(Ehrenberg 1989), where IGV is the present intergranular

volume and is equivalent to the term minus-cement porosity of Rosenfeld (1949). The extent of sandstone volume loss as a result of pressure dissolution was estimated using the grain-overlap method of Houseknecht (1984).

Terminology

Sandstone compositions are classified according to the scheme of McBride (1963). During point counting, distinction between detrital and authigenic clay was made using the criteria of Wilson and Pittman (1977). Clay mineral terminology follows the recommendations of AIPEA. The kaolin group mineral polytypes dickite and kaolinite are distinguished on the basis of powder mount XRD peaks as detailed in Bailey (1980) and are then related to thin-section kaolin populations. Identification of illite is made on the basis of a combination of textural, morphological, optical and XRD criteria. In thin-section and SEM observations, the term illite is not used in a specific compositional sense; where grain-coating and pore-filling, second-order birefringent platy or fibrous clays are observed, the term illitic is used in a general sense to describe such clays, which may include an expanding component.

Two types of porosity are recognized in thin section, comprising intergranular and grain dissolution. Where relics of partially dissolved grains or clay coats around former grains are preserved, intergranular and grain dissolution pores can be readily distinguished. However, when complete dissolution of framework grains has taken place the distinction is more difficult. In such cases, oversized pores (pore of diameter greater than mean grain size of the sample) are assumed to be the result of grain dissolution.

Terminology of diagenetic regimes follows the usage of Schmidt and McDonald (1979). Diagenesis resulting from the interaction of depositional waters with their host sediment is defined as eogenetic but this category also includes contemporary recharge of fresh waters through fluvio-deltaic sandbodies.

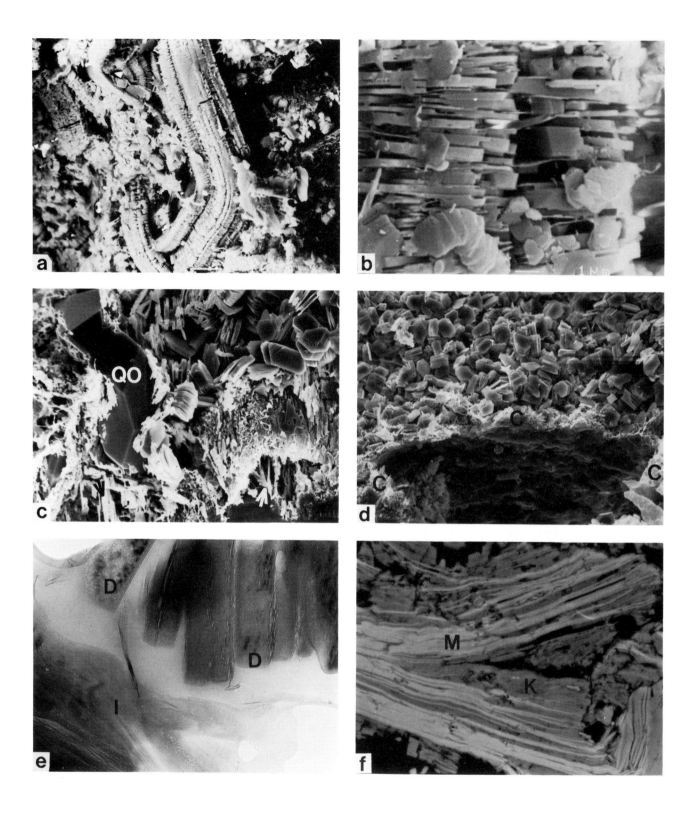

Fig. 3. (**a**) Typical morphology of vermiform kaolinite, with characteristic curving growth habit, and verm length considerably greater than component crystal width. Well 7, 10 545 ft. Width of photomicrograph: 95 μm. (**b**) Detail of vermiform kaolinite, illustrating bimodal crystal size distribution: the widest crystals are also the thinnest, and partition thicker layers comprising several crystals. This bimodality of crystal size is best developed at deeper levels of the vermiform kaolinite range, and may represent a compound kaolinite/dickite morphology. Well 7, 10 561.8 ft. Width of photomicrograph: 13 μm. (**c**) Well-developed crystals of pseudo-hexagonal dickite. The presence of relic feldspar laths (arrowed), and authigenic quartz (QO) suggests that the feldspar dissolution reaction proposed for dickite formation is operable on a very local scale. Well 4A, 12 164 ft. Width of photomicrograph: 72 μm. (**d**) Association of detrital clay coat (C) with authigenic dickite. Dickite formation may be promoted by the presence of detrital clay coats: in addition to providing an energetically favourable nucleation site, clay coats may retard the escape of aluminium and silica liberated by the feldspar dissolution reaction. Well A28, 12 736.2 ft. Width of photomicrograph: 198 μm. (**e**) TEM image of well-developed dickite blocks (D). Irregular wispy laths represent illitic material (I). Well 4A, 12 164 ft. Width of photomicrograph: 1.46 μm. (**f**) 'Expanded-mica' kaolinite grain imaged in backscatter mode: dark grey tones represent kaolinite sheets (K), intergrown along splayed muscovite laths (M). Well A1, 11 166.4 ft. Width of photomicrograph: 88 μm.

Fluids released during compaction, dehydration and decarboxylation reactions are described as mesogenetic. The ingress of meteoric waters subsequent to burial and inversion, such as may occur when rotated fault blocks are subaerially exposed, is defined as telogenetic.

The term 'meteoric water flushing' as used in this paper refers to a dynamic and active aquifer system with a throughput of continental-derived surface waters. Such a flow process may occur either during eogenesis, contemporary with deposition, or during telogenesis in the late Jurassic–early Cretaceous after uplift of rotated fault block crests. Migration of fresh or 'meteoric' waters expelled from fluvio-deltaic sandbodies on compaction is excluded from coverage by this term, as such flow conditions differ crucially from meteoric water flushing in the contexts of flow rate and recharge capacity. Although such fluids may have chemical and isotopic characteristics similar to meteoric waters, they are released during burial and are hydrodynamically mesogenetic.

Results

Kaolin group minerals: recognition, morphology and distribution

Three morphologies of authigenic kaolin group minerals are observed: vermiform, blocky and 'expanded-mica'. On the basis of XRD characterization, the vermiform kaolin morphology (Figs 3a and 3b) was found in all cases to be kaolinite *sensu stricto*, whereas the blocky kaolin morphology (Figs 3c, d and e) was found to be dickite.

Useful criteria in identifying vermiform kaolinite in SEI include the characteristic curving growth habit of sets of crystal aggregates, generally with their length considerably greater than component crystal width (Fig. 3a). This habit is also characterized by the presence of high intercrystalline microporosity. At deeper levels of the occurrence range (Table 2), vermiform kaolinite is frequently composed of bimodal crystal

Table 2. Location of sandstone samples studied together with listing of sandstones quantitatively analysed by modal analysis

Well	Location	Sample number	Depth range sampled	Samples point counted
HO5/13	Crestal Hutton	11	9709–10 087.4 ft	11
2	Water zone, south of Hutton	8	10 545.2–10 574.3 ft	8
A15	East Crest, NW Hutton	27	11 107.5–11 458.7 ft	22
A22	Central NW Hutton	27	11 715.6–12 058.1 ft	13
4AX	West Flank, NW Hutton	24	11 841–12 230 ft	21
A17	West Flank, NW Hutton	23	11 890.7–12 297.6 ft	13
A28	West Flank, NW Hutton	25	12 592.5–13 053 ft	25

elements (Fig. 3b). The widest crystals are also the thinnest, and these partition thicker layers which often comprise several narrower crystals in the same growth plane. Vermiform kaolinite is *not* defined by the largeness of pseudo-hexagonal plates, as observed in SEI, as in some examples individual crystals are only a few microns across. Thus, for recognition in thin section, the curved habit, a lack of micaceous layers, and a coarse, expanded appearance are the most diagnostic criteria for identification.

Qualitatively, the distribution of vermiform kaolinite is both specific and restricted. Enhanced abundances are found in the water zone between and to the south of the two fields, and within the Hutton Field. Vermiform kaolinite is rare in the NW Hutton Field, being restricted to detrital clay-rich overbank splay sandstones and facies cemented by eogenetic ferroan calcite.

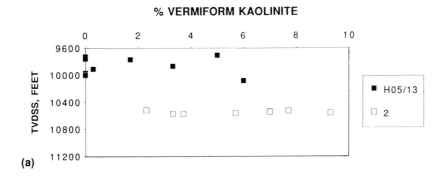

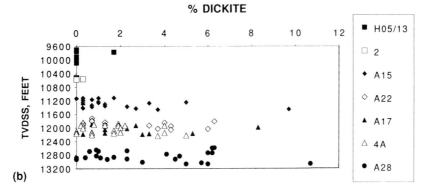

Fig. 4. Abundance of kaolin group minerals with present depth of burial in the studied Brent Group sandstones: (**a**) vermiform kaolin abundance; (**b**) dickite abundance. Mineral abundances determined from petrographic modal analysis.

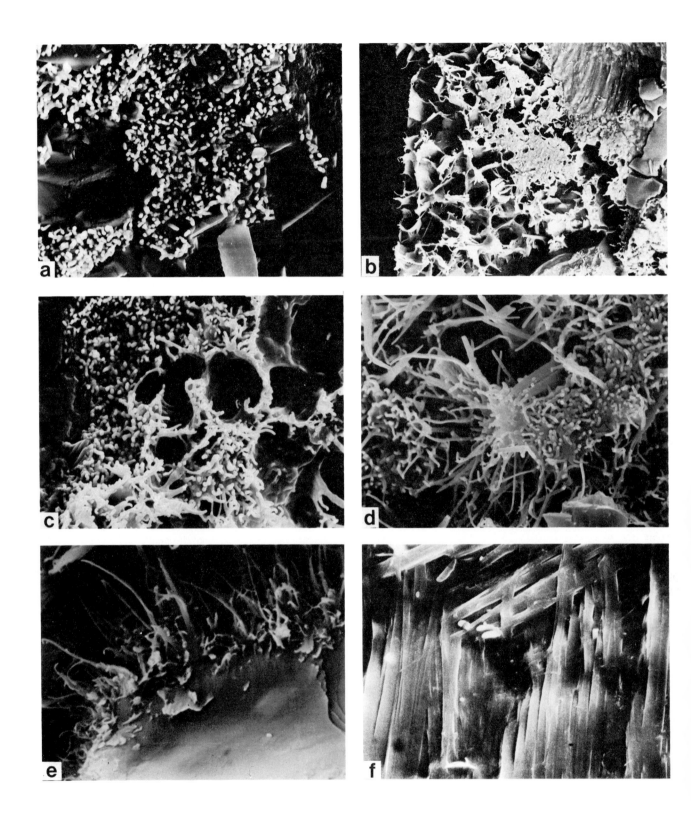

Fig. 5. (**a**) Stubby, nascent illite fibres typical of shallower burial depths. Fibres preferentially nucleate on irregular detrital grain surfaces. Well A8, 11 650 ft. Width of photomicrograph: 64 μm. (**b**) Pore-occluding illitic 'honeycomb' fabric observed at shallower reservoir depths. Well A15, 11 365.3 ft. Width of photomicrograph: 162 μm. (**c**) Detail of illite 'honeycomb' fabric, illustrating that the development of the structure proceeds by the coalescence of individual laths. Well A17, 12 166 ft. Width of photomicrograph: 48 μm. (**d**) Illite habit typical of deeper burial depths, with well-developed individual fibres projecting into pore space. Well A17, 12 166 ft. Width of photomicrograph: 34 μm. (**e**) Development of long, free-standing illite fibres from energetically favourable micaceous substrate. Well A17, 12 219.3 ft. Width of photomicrograph: 18 μm. (**f**) Illite sheet, developed by superposition of parallel 'bandages' of illite fibres. Well 4A, 12 140 ft. Width of photomicrograph: 11 μm.

The most abundant kaolin polytype is dickite. The characteristic morphology of dickite as observed in thin section and in SEI comprises stacks of uniformly sized, pseudo-hexagonal plates, generally with crystal widths of 5–20 μm (Figs 3c, d and e). This constitutes the classic 'blocky' habit of pore-filling kaolin frequently reported from Brent Group sandstones. In SEI the thickness of a block of crystals is rarely observed to exceed crystal width, the opposite relationship to that observed in vermiform kaolinite. Dickite also lacks the finely laminated layer structures developed in vermiform kaolinite. In thin section, dickite may entirely occlude intergranular pores. Orthoclase feldspar dissolution is commonly found adjacent to such areas, although dickite is rarely observed to precipitate within the voids of dissolving feldspar grains. Enhanced dickite cementation occurs close to sandstone–mudstone contacts within mudstone sequences (as shown in Fig. 8), and in regressive marine bars adjacent to mudstone sequences.

The total abundance of authigenic kaolin minerals (kaolinite and dickite) increases from a range of 0–5 volume% in Hutton to a range of 0–10 volume% in the deep western flank of NW Hutton. However, the abundance trends of kaolinite and dickite are very different (compare Figs 3a and 3b). Kaolinite is almost entirely restricted in occurrence to the Hutton Field and the adjacent water zone between the two fields. Even so, mean kaolinite abundance in Hutton is only $c.$ 2% (n = 11) although the water zone Well 2 contains $c.$ 6% mean kaolinite (n = 8). Only small amounts of dickite are present in Hutton (mean value of <1%, n = 11). A trend of increasing dickite abundance is then recorded, from a mean value of $c.$ 2% (n = 25) in the shallow NW Hutton Well A15, rising through $c.$ 2% for the A17 well (n = 13), to >3% (n = 25) in the deep western Well A28.

Within-well variation in the abundance of dickite reveals some consistent trends. Dickite is enhanced in abundance in the Broom Formation and in marine bar facies sandstones. Broom Formation sandstones, overlying thick Dunlin shales, contain a mean of $c.$ 5% dickite (n = 3); other Ness Formation sandbodies (n = 7) in this well—principally crevasse splays and distributary channels—had a mean value of $c.$ 2%. Additionally, localized enhancement of dickite cementation is evident in thin sands adjacent to or within mudstone sequences. An example of this is illustrated in Fig. 9, where point-counting of dickite in closely spaced samples suggests a spatial relationship between the amount of dickite cementation in sandstone and proximity to the contact with mudstone.

There is a depth-related trend in the occurrence of vermiform kaolinite and blocky dickite. Table 3 shows the change from shallow clay assemblages dominated by the kaolinite polytype to deep assemblages dominated by the dickite polytype. However, according to both petrographic and powder XRD studies, dickite and vermiform kaolinite are not mutually exclusive as most of the shallower samples contain both kaolin polytypes.

Table 3. Depth distribution of kaolin polytypes as determined from powder mount XRD

Sample	Depth (ft)	Polytype	Size fraction (μm)
55/13-2392	10087.4	Kaolinite	5–2
2-5415	10541.4	Kaolinite	8–5
2-5619	10561.8	Kaolinite	8–5
7-11027	11027	Kaolinite/dickite	8–5
A15-008	11447.6	Dickite	8–5
c12-6087	11608.6	Dickite	8–5
A8-3929	11667.7	Dickite	8–5
A22-5645	11806.1	Dickite	8–5
A22-8907	12058.1	Dickite	8–5
A28-774	13064	Dickite	8–5

'Expanded-mica' kaolinite (Fig. 3f) is directly associated with detrital micas and is most abundant in the micaceous Rannoch Formation. This morphology is characterized by the growth of sheets of kaolinite parallel to the cleavage surfaces of the host mica grain. Where the mica grain is partially compacted by surrounding detrital grains, mica expansion occurs in a splayed fashion (Fig. 3f). In cases of extreme mica 'expansion', the detrital mica grain may show no trace of the second-order birefringence typical of micas. Such kaolinites are distinguished from vermiform kaolinite by the absence of the curving vermiform habit.

Authigenic illite: recognition, morphology and distribution

Authigenic illite was observed throughout the examined depth range, although it only constitutes a significant authigenic component in the deeper parts of the NW Hutton Field. Where abundant, illitic clay occurs as grain coatings, pore fillings and replacements of partially dissolved feldspar grains. Additionally, regular illite laths are observed to grow from mica cleavage plates (Fig. 5e), and authigenic fibres are recorded growing from detrital clay layers, even at the shallower depths sampled. Sheets of illite are occasionally observed growing on mica surfaces, with discrete layers showing a parallel bandage-like fibre distribution (Fig. 5f).

Pore-filling authigenic illitic clays show a pronounced trend of increasing abundance with depth, as shown in Fig. 6: values progress from a mean of $c.$ 1% (n = 11) in Hutton, through the intermediate depths of NW Hutton (e.g. Well A17, mean $c.$ 4%, n = 13), to a mean value of $c.$ 7% in the deep western flank Well A28 (n = 25).

SEM examination of illitic textures required considerable caution. Preserved core was not generally available, and no samples were prepared using the critical point drying method,

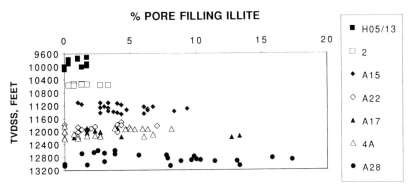

Fig. 6. Abundance of authigenic pore-filling illite with present depth of burial in the studied Brent Group sandstones, as determined by petrographic modal analysis.

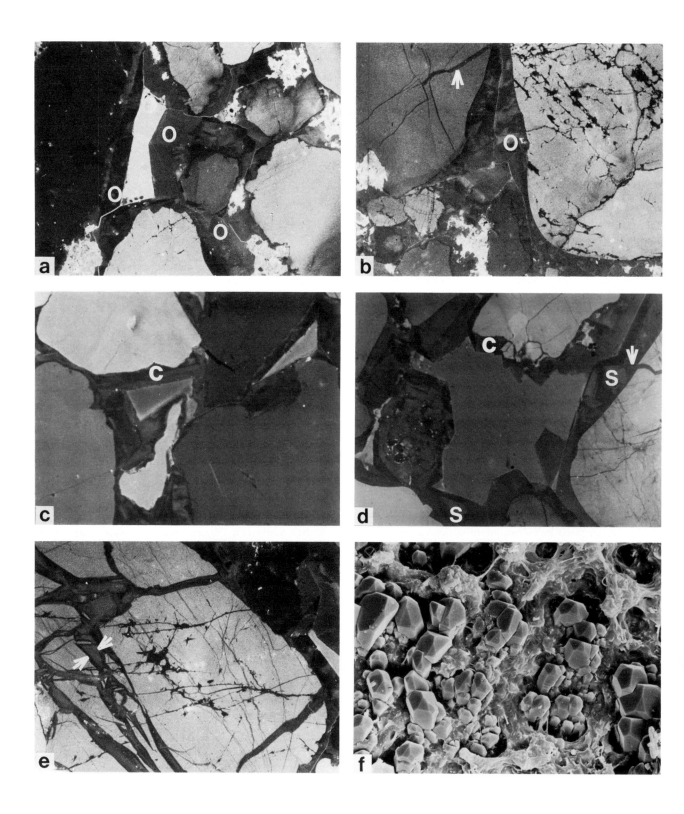

Fig. 7. (**a**) Pervasive development of authigenic quartz overgrowths (O), luminescing darker than their detrital host grains. Overgrowths cement a relatively open-grain framework. Well A22, 11 724.1 ft. Width of CL micrograph: 750 μm. (**b**) Irregularly zoned authigenic quartz cementation (O), occluding porosity on a local scale. Note quartz-healed fracture in detrital grain (arrowed). Well A22, 11 724.1 ft. Width of CL micrograph: 788 μm. (**c**) Concentric zonation with up to three generations of quartz cementation (C). Well A15, 11 428.6 ft. Width of CL micrograph: 457 μm. (**d**) Concentric (C) and sector (S) zonation. Note healed grain fracture cross-cuts first-generation overgrowth only (arrowed), indicating a temporal difference between successive overgrowth precipitation. Well A22, 11 748.1 ft. Width of CL micrograph: 739 μm. (**e**) Healed grain fractures cemented with two authigenic quartz generations (arrowed). Note that the outermost grain overgrowth is unaffected by the grain fracturing. Well A22, 11 724.1 ft. Width of CL micrograph: 724 μm. (**f**) Space-competitive development of numerous euhedral quartz micro-crystallites, preferentially utilizing favourable growth sites. Well A17, 12 014 ft. Width of photomicrograph: 117 μm.

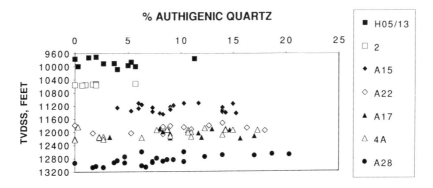

Fig. 8. Abundance of authigenic quartz with present depth of burial in the studied Brent Group sandstones, as determined by petrographic modal analysis.

as recommended by McHardy *et al.* (1982). Consequently, reliable estimates of morphotypes resembling the elongate, truly filamentous 'hairy' illite described from other North Sea Jurassic reservoirs (see McHardy *et al.* 1982; Pallatt *et al.* 1984 and Kantorowicz 1990) cannot be made. However, a general trend of increasing illite fibre length with increasing burial depth appears to be present over the depth interval represented in the Hutton and NW Hutton samples. The typical morphological progression is from stubby, nascent fibres in the shallower wells (Fig. 5a), progressing to abundant, long fibres in the deep A28 and 4A wells (Fig. 5e). The depth-related increase in illite abundance as measured by point-counting methods reflects this, although the lower values are less reliable due to difficulties of observation. At the shallower depths, the development of an illitic 'honeycomb' fabric is produced by the coalescence of individual fibres (Fig. 5c). The coexistence of this fabric with the stubby illite indicates that illite development is heterogeneous on a pore scale.

Authigenic quartz cementation

Authigenic quartz cementation is volumetrically abundant in both the Hutton and NW Hutton fields. Overgrowths are commonly large when examined in CL, attaining widths of up to 120 μm (Fig. 7a), and those developed on adjacent grains may completely occlude porosity on a local scale (Fig. 7b). A complex history of quartz precipitation is indicated by the presence of up to three concentric bright-dark zones as revealed by CL (Fig. 7c). In some cases, overgrowths partially enclose dickite crystals, demonstrating that their formation is synchronous with, or subsequent to, dickite formation.

The distribution of quartz overgrowths shows a depth-related trend of increase over the interval studied (Fig. 8). Authigenic quartz ranges from 0% to 11% in the Hutton Well HO5/13, while the range observed in the deep NW Hutton Well A28 is from 0% to *c.* 20%. From the available core material in both the Hutton and NW Hutton fields the field-wide distribution of major faults does not influence the abundance of authigenic quartz. Several wells were specifically selected for examination because of their proximity to major faults, but petrographic quantification of quartz abundance in such samples did not reveal significantly enhanced quartz cementation. For all wells, apart from Well 2, within-well variability in authigenic silica abundance is considerable (as shown in Fig. 9) and displays no consistent facies-related trends. The shallow cored interval in Well 2, to the south of Hutton, shows reduced modal abundances of authigenic quartz.

Fabric analysis of CL micrographs indicates that the overgrowths cement a relatively open-grain framework (Fig. 7a). Calculations of IGV in well-cemented quartz sandstones (mean quartz cement >10%) suggest that quartz overgrowths

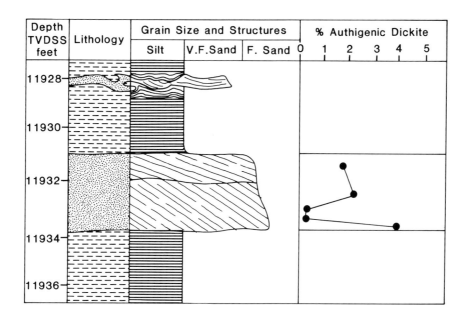

Fig. 9. Enhanced abundance of authigenic dickite at contacts of a thin crevasse splay sandstone with enclosing mudstones.

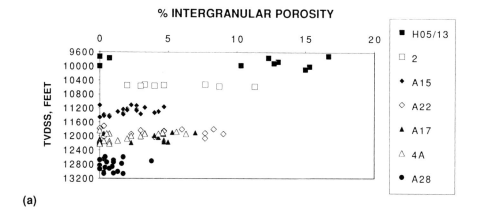

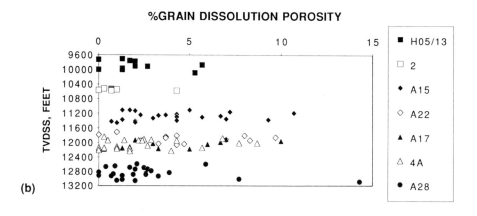

Fig. 10. Relationship between (**a**) intergranular porosity and (**b**) grain dissolution porosity with depth. Visible porosity as determined from thin-section modal anlaysis.

formed in porous sandstones with an average of 26% inter-granular volume (n = 38). This dictates that quartz cementation took place at significant burial depths after appreciable mechanical compaction and the formation of minor volumes of eogenetic cements.

The timing of cementation and compaction is also further constrained by the analysis of CL fabrics. Detrital quartz grains are commonly revealed by CL to have undergone brittle fracturing and subsequent healing with quartz cement. In some instances the brittle fracturing affects only the detrital grains; in others it affects one or more of the overgrowth generations (Fig. 7d), indicating that successive zones of quartz overgrowth development differ in the timing of their precipitation. Several fracture generations affecting detrital grains and not grain overgrowths may also be observed (Fig. 7e).

Not all authigenic quartz develops as entire or substantial grain overgrowths. The growth of numerous euhedral quartz microcrystallites is observed (Fig. 7f), displaying a heterogeneous distribution on the detrital grain surface, which may be controlled by the availability of favourable nucleation sites.

Intergranular porosity

Intergranular porosity declines with increasing structural depth (Fig. 10a). Hutton sandstones have a mean intergranular porosity value of *c*. 10% (n = 11) at depths above 10 000 ft in Well HO5/13 which compares with, for example, a mean value of <4% (n = 13) for Well A17 in NW Hutton at a depth of around 11 500 ft. Mean intergranular porosity for the deep NW Hutton A28 well is reduced to <1% (n = 24). Criteria for the identification of carbonate cement dissolution porosity (e.g. elongate intergranular pores, Schmidt and McDonald

(1979) or regular embayments in detrital grains, Burley and Kantorowicz 1986) were seldom observed, so the intergranular porosity is considered to be mostly relic primary porosity.

Feldspar dissolution

Grain dissolution porosity is chiefly produced by the dissolution of orthoclase feldspar, as identified from skeletal grain remnants in thin section and SEI. Dissolution selectively corrodes sites of high free surface energy (cleavage, twin planes, exsolution lamellae, deformation planes, abraded detrital surfaces, etc.). In the shallow samples from Hutton feldspar, dissolution is minor, as is directly indicated by a low mean value for grain dissolution porosity (see below). At these depths, feldspar grains are rarely completely dissolved. This contrasts with NW Hutton where feldspar dissolution is intense and oversized pores testify to the former presence of significantly more feldspar than is now present.

Plagioclase feldspar is present in only low modal abundances (mean 0.4%, all samples, n = 112), even in the shallow, feldspar-rich sandstones and in those sandstones extensively cemented with early carbonate. Plagioclase is considered not to have contributed significantly to the observed grain dissolution porosity (cf. Harris 1990).

Values of grain dissolution porosity are low in Hutton (mean <3%, n = 11), but they show an increasing trend through shallow to intermediate depths of NW Hutton (e.g. Well A22, mean <5%, n = 13; see Fig. 10b). Thin-section petrography clearly documents that this secondary porosity is not occluded by quartz cements and does not undergo significant compaction. However, at greater burial depths, the feldspar grain dissolution porosity is pervasively occluded by

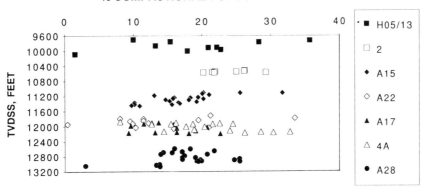

Fig. 11. The relationship between depth of burial and the volume of sandstone lost as a result of compaction. See text for details of compactional porosity loss calculation.

authigenic illite and dickite. Intense feldspar dissolution has undoubtedly taken place in these sandstones and if the amount of illite and dickite occupying oversized pores is added to the total of remaining grain dissolution porosity then the trend of increasing feldspar dissolution with depth is maintained.

Feldspar dissolution has a direct effect on the measured sandstone compositions. At present burial depths in the Hutton Field, Brent sandstones are essentially subarkoses, with a mean detrital quartz (Q), feldspar (F) and rock fragment (R) composition of $Q_{88.2}F_{6.9}R_{4.9}$. The original detrital mineralogy can be further defined by assuming all observed grain dissolution porosity resulted from loss of detrital feldspar. This gives an original mean sandstone composition of $Q_{85.6}F_{9.7}R_{4.7}$. Although the deepest sandstones are predominantly quartz arenites, substitution of feldspar into the depth-related increase in grain dissolution porosity produces similar original sandstone compositions.

Compaction fabric

Most of the sandstones are characterized by a relatively open-grain framework with point to straight grain contacts. Estimates of the amount of porosity loss through compaction lie in the range 0.4–35.7%. Within-well variation in the amount of compaction is greater than 20% for 5 of the 7 wells investigated, and all wells show internal variation of at least 6%. This variation results from the interaction of other parameters that influence compaction on the individual sample scale (such as the presence of detrital micas, detrital clay and early carbonate cements).

CL fabrics indicate that most of the compaction has been accommodated through the effects of grain fabric re-orientation, grain rotation and brittle fracturing. The proportion of ductile grains (micas, mudstone clasts, volcanic rock fragments, etc.) is low (e.g. mean total micas <1%, n = 112) and plastic deformation has not generally contributed to compactive porosity loss.

Estimates of pressure dissolution from studies of the proportion of grain overlap in CL micrographs are very low. In many samples there is no evidence for any pressure dissolution, and even where it is developed the proportion of grain overlap is always less than 5%, which equates with a porosity volume loss of <3% (Mitra and Beard 1980).

The amount of compactive porosity loss displays no systematic variation over the depth interval studied (Fig. 11). Importantly, however, significant compactive porosity loss has been accommodated in even the shallowest samples studied. The shallowest well examined, HO5/13, has a mean compactional porosity loss of *c.* 19% (n = 11). The next deepest well studied,

Well 2, displays mean compactional porosity loss of 24% (n = 8). The deepest well examined, A28, has a mean compaction of *c.* 18% (n = 23). These data indicate that in many cases more than 50% of the original porosity of the sandstone has been destroyed through compaction processes and that most compaction has taken place before a depth of around 10 000 ft has been attained. As the amount of compactive porosity loss does not increase systematically over the depth interval investigated in this study, mechanical compaction is not considered to be active over this depth range.

Discussion

Early silicate cements

The earliest formed silicate cements are vermiform kaolinite and minor amounts of authigenic quartz. The interpretation of vermiform kaolinite as an early diagenetic cement is based on several petrographic criteria. In its restriction to the shallower burial depths examined, it clearly precedes dickite formation, illite formation and feldspar dissolution; these later processes only become significantly operative over the burial depth range studied. Additionally, vermiform kaolinite is occasionally seen enclosed in poikilotopic ferroan calcite in sandstones with a very high minus cement porosity, indicating formation before significant compaction. Minor amounts of authigenic quartz are also enclosed by early calcite cements, and are thus interpreted as being of early diagenetic, pre-burial origin.

Depth-related transition of kaolinite to dickite

Thin-section and SEI petrography show that in addition to being common in the Hutton Field and locally abundant in the adjacent water zones, vermiform kaolinite *sensu stricto* is very rare in the NW Hutton Field. For the samples studied, this corresponds to virtual kaolinite absence below 11 100 ft. SEI examination of the deeper sandstones which contain vermiform kaolinite shows a strongly developed bimodality of plate size, as indicated in Fig. 3b. The combined depth relationships and morphological observations are interpreted to indicate either an in situ alteration of kaolinite to dickite, or a templating effect for dickite growth from the aluminium and silica released by feldspar dissolution (Fig. 12). Such observations are in accordance with the suggestion made by Ehrenberg (1991*b*) of a depth-related transition from kaolinite to dickite in comparable reservoir sandstones.

The known thermodynamics of kaolinite and dickite reinforce the petrographic interpretation of vermiform kaolinite as an early diagenetic precipitate. Anovitz *et al.* (1991) reported

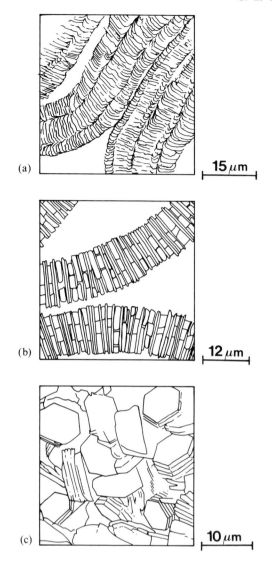

Fig. 12. Schematic diagram of the depth-related changes in kaolin mineralogy and morphology, based on SEM micrographs. (**a**) *Vermiform kaolinite.* Typically forms tripartite stacks of subhedral to euhedral, elongate hexagonal kaolinite plates. Individual crystal size generally 15–30 μm wide but < 1 μm thick. Stacked aggregates commonly exceed 200 μm in length and show tendency to curve. Restricted to burial depths < 11 000 ft. (**b**) *Intergrowth of kaolinite and dickite.* Comprises blocky crystallites of dickite within vermiform aggregate stack. Develops below 10 500 ft. (**c**) *Dickite.* Euhedral crystals, characteristically 5–15 μm in width, and > 1 μm in thickness, displaying 'blocky' habit. Occurs from 10 000 ft but most abundant below 12 000 ft.

that halloysite and dickite are metastable with respect to kaolinite under standard conditions of temperature and pressure. Moreover, Anovitz *et al.* (1991) pointed out that the molar volume of dickite is slightly smaller than that of kaolinite, which would tend to lead to dickite being stabilized relative to kaolinite with increasing pressure. The possibility that active fluid flow is involved in a transformation of kaolinite to dickite is strengthened by the very restricted occurrence of the vermiform morphotype at greater burial depths. Within NW Hutton, the only observed vermiform kaolinite occurs in an overbank splay deposit with a high concentration of detrital clay, or enclosed in early ferroan calcite cements. Although dickite may be thermodynamically favoured at such depths, the inferred low fluid flow rates through such facies or cements would have hindered any dissolution–reprecipitation reactions.

Mass balance of diagenetic reactions taking place with depth

The authigenic silicate mineralogy in the Hutton Field subarkoses at 10 000 ft burial depth is dominated by kaolinite and quartz together with minor amounts of dickite and illite. This authigenic assemblage evolves towards a quartz–dickite–illite assemblage associated with extensive K-feldspar dissolution over the depth range studied, with no observed increase in compactive porosity loss and pressure dissolution. Simple mass balance calculations assuming the subarkose detrital assemblage has evolved to the quartz overgrowth–dickite–illite assemblage can be attempted to determine whether the dissolution of feldspar can account for all the observed reaction products. This can be represented by the reaction

$$4\,KAlSi_3O_8 + 4\,H^+ + 2\,H_2O \rightarrow$$
$$Al_4Si_4O_{10}(OH)_8 + 8\,SiO_2 + 4\,K^+$$

The most valid method of testing how geochemically open this reaction has been during diagenesis (i.e. whether external solute supply is required or whether sufficient reactants are present in the host sandstone) is to examine the ratio of the reaction products, dickite and authigenic quartz. For the above reaction, given molar volumes of 108.9 cm^3 mol^{-1} for K feldspar, 22.7 cm^3 mol^{-1} for quartz, and 99.3 cm^3 mol^{-1} for dickite (Helgeson *et al.* 1978), 435.5 cm^3 of K-feldspar will dissolve to form 99.3 cm^3 dickite and 181.5 cm^3 quartz. Thus, the reaction products dickite and authigenic quartz should be present in the ratio 1 : 1.83 if the reaction took place in a closed system, and a plot of dickite against authigenic quartz should produce a straight line relationship. The extent to which the reaction is geochemically open can be tested on different scales. At one extreme, on the scale of the thin section, petrographic point-count data indicate amounts and solute transport at least on the scale of tens of pore diameters. At the other extreme, on the scale of the reservoir, the plot of all samples (Fig. 13a) shows that for the entire Hutton and NW Hutton fields, closed system diagenesis does not occur. This result is not unexpected, given the petrographic evidence for enhanced dickite cementation in sandstones adjacent to mudstone sequences. A better constrained approach is to consider the potential for mass balance by testing a particular facies on the scale of the reservoir, with the assumption that this minimizes diagenetic variation due to (i) different detrital facies compositions, (ii) different initial pore water compositions and (iii) different facies associations. The Etive Formation samples were considered most appropriate for such a treatment, on account of high channel connectivity (likely to lead to similar pore water chemistry and hydrodynamic evolution) and relatively small thicknesses of enclosing or interbedded mudstones. In Fig. 13b, the scattered distribution of data shows that the reaction products are not present in the proportions indicative of closed system reaction, even within a test facies as appropriate as the Etive Formation. Furthermore, no coherent trend towards incomplete reaction or external aluminium or silica supply may be seen.

Mass balance indicates that insufficient silica is made available by feldspar dissolution to account for the volumes of authigenic quartz recorded in many samples. Another possible source of silica is from pressure dissolution in the sandstones that are quartz cemented. However, maximum grain overlap in even the most intensely compacted sandstones does not exceed 5%. Pressure dissolution of detrital quartz grains in the host sandstones must therefore be discounted as a major source of the observed authigenic quartz. It is thus necessary to invoke an external supply of quartz solute.

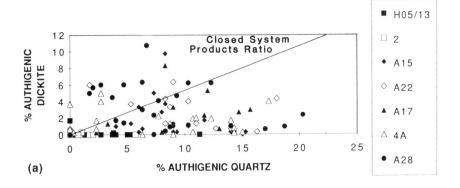

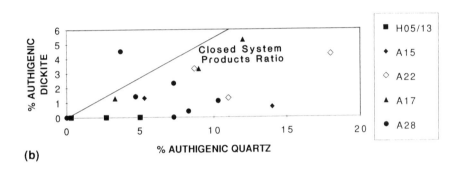

Fig. 13. Mass balance test plot of authigenic dickite against authigenic quartz: (**a**) for all samples; and (**b**) for Etive Formation sandstones.

Relationship of dickite to mudstone intervals

Against the general background of depth-related trends documented above, horizons with enhanced abundances acquire a special significance. The raised levels of dickite cementation found adjacent to bed contacts in sandstones within mudstone sequences, in the Broom Formation and in marine bar facies close to thick mudstone units, indicate that transfer of aluminium in waters expelled from compacting mudstones is likely to be a significant control on dickite growth. This has been postulated theoretically in several works. Curtis (1983) argued that acid waters generated during organic maturation can enhance porosity and deliver aluminium for dickite precipitation as the pH rises. Small (1992) suggested that organic oxalate species have a geologically short half-life under reservoir conditions, but may be active as a dissolution agent in organic-rich mudstones. However, once such fluids are released into sandstones from adjacent mudstones, dissolved Al held in solution by pH buffering or Al-oxalate complexing would rapidly precipitate clays as the oxalate degrades. Ehrenberg (1991a) has documented mudstone-derived contributions to kaolin formation for lithologies similar to the Brent Group in the Garn Formation of the Norwegian Sector.

Pore-scale authigenic clay heterogeneities

Within several sandstone units, localized pore-scale heterogeneities of dickite distribution were observed. Detrital clay coats are associated with areas of dickite growth (Fig. 3d).

Giles and de Boer (1990) have suggested that such areas could provide a favourable nucleation site for the development of kaolin group minerals. Furthermore, it is suggested here that in addition to such a control, clay coats may retard the escape of aluminium and silica in sandstones subjected to a relatively vigorous leaching event. The possibility that these detrital clay/authigenic kaolin associations may also be entirely detrital in origin must be acknowledged, although it is difficult to envisage such structures surviving burial to present depths.

The localized distribution of illite growth, superimposed on a general trend of depth-related increase, is significant. Experimental work by Huang et al. (1986) shows that illite formation is kinetically favoured by virtually stagnant pore fluid conditions. It is suggested that localized illite distribution heterogeneities may be produced by localized areas of pore fluid stagnation. In addition, the presence of even a small amount of detrital clay material or mica (Fig. 5e) may provide an energetically favourable substrate for the nucleation and growth of authigenic illite.

Diagenetic evolution of Brent reservoirs in Hutton and NW Hutton

Footwall uplift and the potential for meteoric water flushing

The potential for kaolin growth as a result of a meteoric water flush, as suggested by Bjørlykke (1984) and Jourdan et al. (1987), can be reassessed in the light of recent structural models of fault block evolution (Marsden *et al.* 1988). In the flexural cantilever model (Marsden *et al.* 1988), fault blocks are assumed to be rigid, and their coupled rotation produces footwall uplift and hangingwall subsidence. A direct geometric consequence of the coupled rotation is that the amount of footwall uplift should be related to the width of the fault block across strike. Yielding (1990) considered predicted uplift values and found them to be in excellent agreement with the observed amount of Late Jurassic erosion on crestal structures for three fields—Snorre, Brent and Heather. Larger fault blocks exhibit greater degrees of crestal erosion—Snorre, for example, has lost up to 1400 m of the pre-Late Jurassic section, and has a flat truncation surface on the seismic depth profile suggesting erosion at wave base. However, the Heather Field, adjacent to the NW Hutton Field, shows little evidence of crestal erosion on regional seismic lines. Well data from the Heather Field indicate that only a third of the Brent Group has been lost by erosion, which compares well with predicted footwall uplift values of around 100 m. Neither the amount of uplift estimated for Heather, nor the measured erosion, indicates subaerial exposure of the fault block crest to be likely. Using theoretical curves relating fault block width and tilt to footwall uplift (Yielding 1990), it is estimated that the Hutton–NW Hutton footwall underwent less than 250 m uplift. Thus, the establishment of vigorous meteoric water recharge through the Brent aquifer as a result of crestal exposure is unlikely in the Heather Field. This conclusion is also applied to NW Hutton on the basis of estimates of eroded Brent Group section comparable to those of the Heather Field, and the low footwall uplift values derived from Yielding's (1990) theoretical curves.

As footwall subaerial exposure is improbable for Hutton and NW Hutton, a telogenetic meteoric water flush during end-Jurassic rifting and fault block rotation recharged from the footwall crest cannot be invoked for kaolin formation and feldspar dissolution. Other models of meteoric water flushing have suggested that meteoric water may flow down faults from the landmass west of the East Shetland Platform and up the Brent Group sandbodies (Glasmann *et al.* 1989). This flow route seems unlikely to provide effective recharge and vigorous aquifer conditions typical of meteoric water flushing and, here, is considered an improbable cause of early kaolin formation.

These considerations do not, however, preclude vermiform kaolinite formation as a result of meteoric water ingress. The topography of the Brent delta, with large distributary channels draining an extensive delta plain to the south (Brown *et al.* 1987; Giles *et al.* 1992), would enable the generation of sufficient hydrodynamic head to flush non-marine and near-shore marine sandstones with continental-derived fresh waters. Such surface or shallow groundwaters of low ionic strength and low to moderate pH are capable of precipitating kaolinite (Livingstone 1963), given sufficient water throughput. Thus, the most plausible interpretation for the formation of vermiform kaolinite in the studied Brent sandstones is within an eogenetic-recharged aquifer system. The finely laminated, skeletal appearance of vermiform kaolinite in this study is certainly compatible with mineral growth under conditions of relatively rapid fluid flux.

Burial-related diagenetic processes

Compaction measurements indicate that an average of 40–50% of intergranular porosity is lost due to mechanical compaction in the samples studied. This compaction was accommodated before intense quartz cementation and, as feldspar grain dissolution pores are not crushed, was also accommodated before feldspar dissolution. The shallowest samples studied at 10 000 ft are as strongly compacted as those at burial depths of 13 000 ft. This observation indicates that over the study depth interval neither mechanical compaction nor pressure dissolution processes have been significantly operative in a systematic way.

By implication, therefore, up to 50% of the intergranular pore volume was destroyed by the combined processes of mechanical compaction over the first 10 000 ft of burial. Consequently, the mean IGV was reduced to around 20–25% of bulk sandstone volume at this depth. Such a result is in accordance with the findings of McBride *et al.* (1991) for Tertiary sandstones of the Texas Gulf Coast and emphasizes the importance of compaction to porosity loss. This amount of compactive porosity loss is close to the tightest grain arrangement that can be accommodated for a rhombohedral packing of perfect spheres (Rittenhouse 1971) but is larger than is normally associated with simple mechanical compaction in well-sorted, quartz-rich sandstones (Wilson and Sibley 1978). The difference is probably due to brittle grain fracturing which further increases the grain packing density. For such compactive porosity loss to have taken place indicates that during the first 10 000 ft of burial palaeopore pressures in the sandstones were maintained at or close to hydrostatic pressure.

K-feldspar dissolution is initiated at depths of at least 10 000 ft and is accompanied by the onset of mesogenetic quartz and dickite cementation. Illitic clays are present over the study interval but only become abundant at depths greater than 12 000 ft. Simple mass balance calculations indicate that all the observed dickite, quartz and illite cements cannot be sourced from the recorded amount of feldspar dissolution. Therefore, on the scale of the studied depth interval, the diagenetic system cannot be considered closed, with both Al and Si having been mobile. Some additional source of these solutes for the observed cements is required, external to the reservoir sandstones investigated. Such a source has not been identified in this study, although pressure dissolution of detrital quartz grains within the sample material is dismissed as a silica source, on the basis of compaction analysis. Potential sources may be from adjacent, interbedded or overlying mudstones as advocated by Boles and Franks (1979), from pressure dissolution at sandstone–mudstone contacts (see Mullis (1992) for the theoretical case) or from structurally deeper Brent Group sandstones that may have been pressure dissolved (as suggested by Olaussen *et al.* (1984) for other Middle Jurassic sandstones in the Norwegian North Sea).

The lack of pressure dissolution fabrics indicates that either quartz cementation provides a stable support for the detrital grain framework or that overpressure was developed during burial and has since been maintained. As there is no geochemical reason why quartz overgrowths should not be affected by pressure dissolution, the latter interpretation is more likely. This is consistent with the widespread development of overpressure within the East Shetland Basin (Buhrig 1989). Non-systematic variations in compaction from well to well may relate to compartmentalization of overpressure. One of the shallowest wells, Well 2, has the highest mean compactional porosity loss, at *c.* 24%. This may in part result from relatively reduced pore pressure conditions in its environs of the water zone. The absence of complex, cyclical diagenetic sequences that are typical of Jurassic reservoir sandstones influenced by faulting (Burley and MacQuaker 1992) implies that the pore fluid flux during burial in the Brent Sandstones of Hutton and NW Hutton has not been controlled by such processes. This conclusion is supported by the general lack of a relationship between faults and cementation throughout the fields.

The relatively simple diagenetic sequence combined with the distribution of authigenic silicate minerals and the absence of evidence for fault-related diagenesis all suggest that the burial diagenesis of Brent Group sandstones in the Hutton and NW Hutton fields records the evolution of a diagenetic system dominated by a slowly evolving thermobaric flow migrating

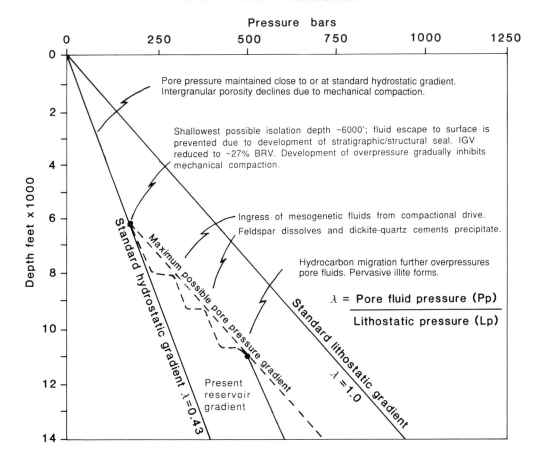

Fig. 14. The inferred sequence of silicate diagenesis in relation to the palaeopore pressure development for the Hutton and NW Hutton fields. The present reservoir gradient is determined from the hydrocarbon column in NW Hutton. Inferred palaeopore pressure is constrained by the maximum possible palaeopore pressure gradient and defines the shallowest possible isolation depth at around 6000 ft.

through the Brent aquifer. Such a flow system was developed subsequent to attainment of the isolation depth and was the result of the invasion of aggressive pore fluids and, subsequently, hydrocarbons that maintained pore pressure (Fig. 14). As zones of extensive feldspar leaching with no associated dickite formation are observed, aluminium and silicon were mobile in this hydrodynamic system. Pore fluid flow rates are inferred to show a significant range of variability, from sufficiently vigorous to remove aluminium and silica from environments of feldspar dissolution, through localized development of dickite on energetically favourable nucleation sites, to abundant dickite formation, and finally to stagnant conditions with extensive illite development. Illite formation is influenced by fluid flow rate (Huang *et al.* 1986) and thermal kinetics, requiring growth threshold temperatures of at least 65°C (Small 1992). Illite distribution is consistent with the development of stagnant pore fluids under conditions of deep burial, long after any kinetic threshold has been passed. If this thermobaric flow model of burial diagenetic evolution is valid, the mechanism by which hydrocarbons migrate up-structure from the deep graben centre (which is their inferred source according to Scotchman *et al.* 1989) must become an area of considerable research interest.

Implications for deep reservoir development

The dominant porosity in the deep NW Hutton Field is secondary and is actively produced during burial through the dissolution of detrital feldspar. It is argued here that the extent and distribution of feldspar dissolution porosity is related to the fluid migration that precedes hydrocarbon migration. Theoretically, therefore, there is not a simple porosity basement as would be predicted by a linear or exponential depth–

primary porosity decay curve. Rather, zones of enhanced porosity should be developed along either sub-surface fluid migration pathways or at their convergence, and in hydrodynamic traps. The generation of grain dissolution porosity in an already compacted and partially cemented sandstone framework dictates that the enhanced porosity has the potential to persist to greater burial depths than would primary intergranular porosity (see also Schmidt and McDonald 1979).

The distribution of deep burial grain dissolution porosity is not related to the unconformity that overlies the Brent Group sequence (cf. Bjørlykke 1984; Emery *et al.* 1990). Any early enhanced porosity that may have been developed near surface would have been due to an eogenetic meteoric water flush and would have been destroyed by either mechanical compaction or cementation during the initial stages of burial. The generation of feldspar dissolution porosity over the interval of study and some evidence of enhanced abundance of dickite close to thick mudstone units suggests that at least some of the pore fluids available during burial were sourced from interbedded and enclosing mudstones. The distribution of porosity in the sub-surface thus in part reflects the ingress of fluids from mudstones and their subsequent migration pathways. Porosity enhancement as a result of feldspar dissolution must then be maintained by the development of pore pressures above hydrostatic. Whether grain dissolution pores remain either uncompacted or collapse is a function of the degree of overpressure developed. This in turn is controlled by the integrity of stratigraphic of structural seals over geological time.

The extent of compaction and silicate mineral authigenesis may vary greatly between adjacent fault blocks because of contrasting pore pressure histories and, if the connectivity of the hydrodynamic system permits, exposure to different migrating pore fluids. Local pore fluid compartmentalization may thus produce distinct porosity–depth relationships in adjacent or juxtaposed fault blocks. The potential application of such concepts of local-scale controls on porosity–depth relationships is illustrated by a consideration of the discussion between Scotchman (1990) and Harris (1990). This discussion debates the contrasting effects of mechanical compaction and cementation in the Lyell and NW Hutton fields, respectively. A possible interpretation of the collapsed grain dissolution pores observed by Harris (1990) is that in the Lyell Field excess pore pressures were not maintained. Conversely, in NW Hutton, where grain dissolution pores are essentially uncompacted but quartz cement is more abundant, pore pressures above hydrostatic pressure were maintained. In any case, the development of stratigraphic or fault pressure compartments is an important concept that should be addressed in exploration strategies for deep fault block plays.

Conclusions

Over the depth interval of study, Brent Group sandstones in the Hutton and NW Hutton fields evolve from an arkosic assemblage to a quartz arenite with a quartz overgrowth–dickite authigenic assemblage associated with extensive K-feldspar dissolution porosity.

Vermiform kaolinite is only abundant in the shallower samples studied, from Hutton and the water zones between the two fields, to a depth of 10 600 ft. Early formation of this kaolinite is indicated by its enclosure within poikilotopic carbonate cements with high minus cement porosities; it pre-dates mesogenetic feldspar dissolution and dickite and illite formation. During eogenesis, vermiform kaolinite is produced by flushing with dilute, low pH meteoric waters shortly after deposition. As a compactional flush is considered unlikely to have the required flow rate, the most plausible hydrodynamic system for the formation of vermiform kaolinite in the studied area is within an eogenetically-recharged aquifer during northward progradation of the Brent delta.

The depth-related trend of increasing dickite abundance shows a good correlation with orthoclase feldspar dissolution, which is considered to contribute some of the aluminium and silica required for its formation. Mudstones are inferred to contribute aluminium and silica for dickite growth on a local scale, as documented by zones of enhanced dickite abundance in regressive marine bars close to mudstone sequences and close to sandstone–mudstone contacts within mudstone sequences. Intergrowths of kaolinite-like and dickite-like plates are observed in the deeper occurrences of vermiform kaolinite. Morphological evidence suggests that a depth-related dissolution–reprecipitation mechanism, in an active fluid flow regime, may account for the rarity of vermiform kaolinite below 11 100 ft. Such dissolution–reprecipitation of vermiform kaolinite below approximately 10 500 ft may thus also increase dickite abundance at depth.

Illite formation shows a general depth-related trend of increasing abundance. A general morphological trend from stubs to fibres to mats is observed. Illite mats are formed from the amalgamation of fibres. Within-sample distribution of illite may be heterogeneous, possibly reflecting localized pore volumes of stagnant fluid. Substrates favourable for nucleation and growth of illite, such as detrital clay layers, also exert significant control on illite development. Illite develops during deep burial after a kinetic temperature barrier has been reached.

Compaction studies indicate that most of the compaction across the depth range has taken place by 10 000 ft burial depth. As pressure dissolution is a minor contributor to authigenic quartz cementation in the studied sandstones, and CL examination reveals an open-grain framework in the quartz-cemented sandstones, most of the quartz cementation is

inferred to have initiated by 10 000 ft burial depth. Locally enhanced abundances of quartz cementation suggest that silica released by K-feldspar dissolution reactions, which increase across the depth range studied, and from deeper pressure dissolution, may be redistributed up the field structure.

An average of 40–50% of IGV has been lost by mechanical compaction which took place before significant quartz cementation and feldspar dissolution. The amount of compactive porosity loss provides a record of the pore pressure during diagenetic evolution. The low IGVs in the Hutton and NW Hutton fields indicate that pore pressure during the early stages of burial remained close to the hydrostatic pressure gradient. However, the subsequent preservation of grain dissolution porosity suggests that pore pressures were maintained above hydrostatic pressure subsequent to the onset of K-feldspar dissolution.

This research was made possible by a grant to Professor C. D. Curtis from Amoco UK Ltd which supported Gavin McAulay's PhD studentship. The continued co-operation and support from Amoco throughout the project is gratefully acknowledged. Stuart Burley is the British Gas Research Fellow at Manchester and wishes to thank BG for supporting diagenetic research.

Professor C. D. Curtis and Iain Scotchman are thanked for their continued enthusiasm and support of the project from its inception. The paper has benefited greatly from discussions with and constructive reading by members of the Clastic Diagenesis Research Group in Manchester, especially Patrick McKeever, Simon Guscott, David Manning, Barbara Ransom and Joe Small. Harry Shaw is thanked for his pertinent and useful reviewer's comments on the manuscript.

References

ANOVITZ, L. M., PERKINS, D. AND ESSENE, E. J. 1991. Metastability in near-surface rocks of minerals in the system Al₂O₃–SiO₂–H₂O Clays. *Clay Minerals*, **39**, 225–233.

BAILEY, S. W. 1980. Structures of Layer Silicates. *In*: BRINDLEY, G. W. AND BROWN, G. (eds) *Clay Minerals and their X-ray Identification*. Mineralogical Society Monograph **5**, 1–124.

BEARD, D. C. AND WEYL, P. K. 1973. Influence of texture on porosity and permeability of unconsolidated sand. *American Association of Petroleum Geologists Bulletin*, **57**, 49–369.

BJØRLYKKE, K. 1984. Formation of Secondary Porosity: How Important Is It? *In*: MCDONALD, D. A. AND SURDAM, R. C. (eds) *Clastic Diagenesis*. American Association of Petroleum Geologists Memoir, **37**, 277–286.

BOLES, J. R. AND FRANKS, S. G. 1979. Clay diagenesis in Wilcox sandstones of southwest Texas: implications of smectite diagenesis on sandstone cementation. *Journal of Sedimentary Petrology*, **49**, 55–70.

BROWN, S., RICHARDS, P. C. AND THOMSON, A. R. 1987. Patterns of deposition in the Brent Group (Middle Jurassic) UK North Sea. *In*: BROOKS, J. AND GLENNIE, K. (eds) *Petroleum Geology of the Continental Shelf of Northwest Europe*. Graham & Trotman, London, 899–913.

BUHRIG, C. 1989. Geopressured Jurassic reservoirs in the Viking Graben: modelling and geological significance. *Marine and Petroleum Geology*, **6**, 31–48.

BURLEY, S. D. AND KANTOROWICZ, J. D. 1986. Thin section and S.E.M. textural criteria for the recognition of cement-dissolution porosity in sandstones. *Sedimentology*, **33**, 587–604.

—— AND MACQUAKER, J. H. S. 1992. Authigenic clays, diagenetic sequences and conceptual diagenetic models in contrasting basin margin and basin centre North Sea Jurassic sandstones and mudstones. *In*: HOUSEKNECHT, D. AND PITMANN, E. D. (eds) *Origin, Diagenesis and Petrophysics of Clay Minerals in Sandstones*. Society of Economic Paleontologists and Mineralogists, Special Publication, **47**, 81–110.

CURTIS, C. D. 1983. Link between aluminium mobility and destruction of secondary porosity. *American Association of Petroleum Geologists Bulletin*, **67**, 380–384.

EHRENBERG, S. N. 1989. Assessing the relative importance of compaction processes and cementation to reduction of porosity in sandstones. A Discussion of: Compaction and porosity evolution of Pliocene Sandstones, Ventura Basin, California. *American Association of Petroleum Geologists Bulletin*, **73**, 1274–1276.

—— 1991a. Kaolinized, potassium-leached zones at the contacts of the Garn Formation, Haltenbanken, mid-Norwegian continental shelf. *Marine and Petroleum Geology*, **8**, 250–269.

—— 1991b. Kaolinite–Dickite Transition in the Garn Formation, Haltenbanken. (Abstract). *In*: STORR, M., HENNING, K.-H. AND ADOLPHI, P. (eds) *Proceedings of the 7th Euroclay Conference, Dresden, 1991*, **1**, 322.

EMERY, D., MYERS, K. J. AND YOUNG, R. 1990. Ancient subaerial exposure and freshwater leaching in sandstones. *Geology*, **18**, 1178–1181.

GILES, M. R. AND DE BOER, R. B. 1990. Secondary porosity: creation of enhanced porosities in the subsurface from dissolution of carbonate cements as a result of cooling formation waters. *Marine and Petroleum Geology*, **6**, 261–269.

——, CANNON, S. J. C., WHITTAKER, M. J., PLEASE, P. M. AND MARTIN, S. V. 1992. Regional review of the Brent Group, UK Sector, North Sea. *In*: MORTON, A. C., HASZELDINE, R. S., GILES, M. R. AND BROWN, S. (eds) *Geology of the Brent Group*. Geological Society, London, Special Publication, **61**.

GLASMANN, J. R., LUNDEGARD, P. D., CLARK, R. A., PENNY, B. K. AND COLLINS, I. D. 1989. Geochemical evidence for the history of diagenesis and fluid migration: Brent Sandstone, Heather Field, North Sea. *Clay Minerals*, **24**, 255–284.

HAIG, D. B. 1991. The Hutton Field, Blocks 211/28, 211/27, UK North Sea. *In*: ABBOTTS, I. L. (ed.) *United Kingdom Oil and Gas Fields 25 Years Commemorative Volume*. Geological Society, London, Memoir, **14**, 135–143.

HANCOCK, N. J. AND TAYLOR, A. M. 1978. Clay mineral diagenesis and oil migration in the Middle Jurassic Brent Sand Formation. *Journal of the Geological Society, London*, **135**, 69–72.

HARRIS, N. B. 1989. Diagenetic quartzarenite and destruction of secondary porosity: an example from the Middle Jurassic Brent sandstone of northwest Europe. *Geology*, **17**, 361–364.

—— 1990. Reply on 'Diagenetic quartzarenite and destruction of secondary porosity: An example from the Middle Jurassic Brent Group of northwest Europe'. *Geology*, **18**, 799–800.

HELGESON, H. C., DELANY, J. M., NESBITT, H. W. AND BIRD, D. K. 1978. Summary and Critique of the Thermodynamic Properties of Rock-Forming Minerals. *American Journal of Science*, **278A**, 1–229.

HOUSEKNECHT, D. W. 1984. Influence of grain size and temperature on intergranular pressure solution, quartz cementation and porosity in a quartzose sandstone. *Journal of Sedimentary Petrology*, **54**, 348–361.

HUANG, W. L., BISHOP, A. M. AND BROWN, R. W. 1986. The effect of fluid/rock ratio of feldspar dissolution and illite formation under reservoir conditions. *Clay Minerals*, **21**, 585–601.

JOHNES, L. H. AND GAUER, M. B. 1991. Northwest Hutton Field, Block 211/27, UK North Sea. *In*: ABBOTTS, I. L. (ed.) *United Kingdom Oil and Gas Fields 25 Years Commemorative Volume*. Geological Society, London, Memoir, **14**, 145–151.

JOURDAN, A., THOMAS, M., BREVART, O., ROBSON, P., SOMMER, F. AND SULLIVAN, M. 1987. Diagenesis as the control of the Brent reservoir properties in the greater Alwyn area (East Shetlands Basin). *In*: BROOKS, J. AND GLENNIE, K. (eds) *Petroleum Geology of North West Europe*. Graham & Trotman, London, 951–961.

KANTOROWICZ, J. D. 1990. The influence of variations in illite morphology on the permeability of Middle Jurassic Brent Group sandstones, Cormorant Field, UK North Sea. *Marine and Petroleum Geology*, **7**, 66–74.

LIVINGSTONE, D. A. 1963. *Chemical composition of rivers and lakes*. US Geological Survey Professional paper 440-G.

LONGIARU, S. 1987. Visual comparators for estimating the degree of sorting from plane and thin sections. *Journal of Sedimentary Petrology*, **57**, 791–794.

MARSDEN, G., YIELDING, G., ROBERTS, A. M. AND KUSZNIR, N. J. 1988. Application of a flexural cantilever simple-shear/pure shear model of continental lithosphere extension to the formation of the northern North Sea Basin. *In*: BLUNDELL, D. J. AND GIBBS, A. D. (eds) *Tectonic Evolution of the North Sea Rifts*. Oxford Scientific Publications, 240–261.

McBRIDE, E. F. 1963. A classification of common sandstones. *Journal of Sedimentary Petrology*, **33**, 664–668.

——, DIGGS, T. N. AND WILSON, J. C. 1991. Compaction of Wilcox and Carrizo Sandstones (Paleocene–Eocene) to 4420 m, Texas Gulf Coast. *Journal of Sedimentary Petrology*, **61**, 73–85.

McHARDY, W. J., WILSON, M. J. AND TAIT, J. M. 1982. Electron microscope and X-ray diffraction studies of filamentous illitic clay from sandstones of the Magnus Field. *Clay Minerals*, **17**, 23–40.

MITRA, S. AND BEARD, W. C. 1980. Theoretical models of porosity reduction by pressure solution for well-sorted sandstones. *Journal of Sedimentary Petrology*, **50**, 1347–1360.

MORTON, A. C., HASZELDINE, R. S., GILES, M. R. AND BROWN, S. (eds) 1992. *Geology of the Brent Group*. Geological Society, London, Special Publication, **61**.

MULLIS, A. M. 1992. A numerical model for porosity modification at a sandstone–mudstone boundary by quartz pressure dissolution and diffusive mass transfer. *Sedimentology*, **39**, 99–107.

OLAUSSEN, S., DALLAND, A., GLOPPEN, T. G. AND JOHANNESSEN, E. 1984. Depositional environment and diagenesis reservoir sandstones in the eastern part of Troms 1 area. *In*: SPENCER, A. M. *et al.* (eds) *Petroleum Geology of the North European Margin*. Norwegian Petroleum Society, 61–79.

PALLATT, N., WILSON, M. J. AND McHARDY, W. J. 1984. The relationship between permeability and the morphology of diagenetic illite in reservoir rocks. *Journal of Petroleum Technology*, **36**, 2225–2227.

RITTENHOUSE, G. 1971. Pore space reduction by solution and cementation. *American Association of Petroleum Geologists Bulletin*, **55**, 80–91.

ROSENFELD, M. A. 1949. Some aspects of porosity and cementation. *Producers Monthly*, 39–42.

SCHMIDT, V. AND McDONALD, D. A. 1979. The role of secondary porosity generation in the course of sandstone diagenesis. *In*: SCHOLLE, P. A. AND SCHLUGER, P. R. (eds) *Aspects of diagenesis*. Society of Economic Paleontologists and Mineralogists, Special Publication, **26**, 175–207.

SCOTCHMAN, I. C. 1990. Discussion on 'Diagenetic quartzarenite and destruction of secondary porosity: An example from the Middle Jurassic Brent Group of northwest Europe'. *Geology*, **18**.

——, JOHNES, L. H. AND MILLER, R. S. 1989. Clay diagenesis and oil migration in Brent Group sandstones of NW Hutton field, UK North Sea. *Clay Minerals*, **24**, 339–374.

SMALL, J. S. 1993. Experimental determination of the rates of precipitation of authigenic illite and kaolinite in the presence of aqueous oxalate in comparison to the K–Ar ages of illites in reservoir sandstones. *Clays and Clay Minerals*, **41**, in press.

SOMMER, F. 1978. Diagenesis of Jurassic sandstones in the Viking Graben. *Journal of the Geological Society, London*, **135**, 65–68.

WILSON, M. D. AND PITTMAN, E. D. 1977. Authigenic clays in sandstones: recognition and influence on reservoir properties and paleoenvironmental analysis. *Journal of Sedimentary Petrology*, **47**, 3–31.

WILSON T. V. AND SIBLEY, D. F. 1978. Pressure solution and porosity reduction in shallow buried quartz arenite. *American Association of Petroleum Geologists Bulletin*, **62**, 2329–2334.

YIELDING, G. 1990. Footwall uplift associated with Late Jurassic normal faulting in the northern North Sea. *Journal of the Geological Society, London*, **147**, 219–222.

The link between petroleum emplacement and sandstone cementation

J. G. GLUYAS,[1] A. G. ROBINSON,[2] D. EMERY,[2] S. M. GRANT[3] and N. H. OXTOBY[4]

[1] BP Norway Ltd UA, Forusbeen 35, 4033 Forus, Norway (Present address: BP Research Centre, Chertsey Rd, Sunbury, Middlesex TW16 7LN, UK)

[2] BP Exploration, 4/5 Long Walk, Stockley Park, Middlesex UB11 1BP, UK

[3] BP Research Centre, Chertsey Rd, Sunbury, Middlesex TW16 7LN, UK (Present address: BP Norway Ltd UA, Forusbeen 35, 4033 Forus, Norway)

[4] BP Research Centre, Chertsey Rd, Sunbury, Middlesex TW16 7LN, UK

Abstract: When a reservoir fills with petroleum, the process commonly takes about 10 Ma. Cementation of a sandstone reservoir can occur in a similarly short time. There is much direct evidence that these two processes can occur at about the same time. In petroleum-bearing sandstone reservoirs it is common to find inclusions of petroleum trapped in diagenetic minerals. The effect on reservoir porosity and permeability caused by the interaction of these two processes is dramatic. Fields with a relatively early petroleum charge, earlier that is than cementation, can be saved from the ravages of reservoir quality destruction. Fields in which the two processes were acting at the same time commonly display porosity–depth gradients which are twice the regional average. If cementation beats petroleum in this 'race for space' there may be no field to produce, as the petroleum may not be recoverable from the low quality reservoir.

We believe that the processes of petroleum generation/migration and cementation may share a common cause. Both processes are commonly associated with major changes within a basin: rapid burial/heating of the sediment pile and/or major faulting which changes the basin 'plumbing'.

The aim of this paper is:

- to investigate the relationship between diagenesis of reservoir sandstones and the emplacement of petroleum;
- to comment on the effect of synchronous reservoir filling (by petroleum) and reservoir cementation on the evolution of reservoir quality;
- to investigate whether the link between the two processes is causal, that is, oil generation/expulsion directly influences cementation; or casual, that is oil generation/expulsion and cementation share a common cause.

Rates of processes

Much evidence exists indicating that the rate of petroleum migration and accumulation can be (geologically) fast. England *et al.* (1987) quote migration rates of $8 \times 10^{-10} \, \text{m s}^{-1}$. Even using conservative estimates of migration route–cross-sectional areas of about $1000 \, \text{m}^2$ and high migration losses (tens of a percent), this figure leads to accumulation rates capable of filling even the largest oil fields in a few million years. Moreover, there are many oil fields developed in young (Miocene, Pliocene) sandstones; direct evidence that accumulation can be completed quickly. Ten million years is, perhaps, then an upper limit to the time it takes to fill most oil fields (W. E. England, pers. comm.).

The rates of diagenetic processes have been investigated in detail by a number of workers over recent years. They conclude that diagenetic processes can occur in geologically short periods. This is particularly true for those processes commonly associated with burial and often termed 'late diagenesis'. Significant quantities (>10%) of quartz, carbonates and clays can be precipitated in a few to a few tens of million years (Lee *et al.* 1989; Glasmann *et al.* 1989; Robinson and Gluyas 1992*a*). The evidence for rapid cementation comes from a number of sources: direct geological evidence, direct geochemical evidence and indirect geochemical evidence. Below are a few examples of these datasets.

Direct geological evidence

The Miocene Sihapas sandstones of the Central Sumatra Basin, Indonesia (Fig. 1) form the main reservoir unit in a number of oil fields. Two such oil fields (Kurau and Melibur) are separated by a reverse fault. Melibur sits in the hangingwall at 275 m (900 ft) and Karau in the footwall at 1500 m (4900 ft). The fault only became active two million years ago. Until that time the Sihapas sandstone was continuous between both fields, yet the diagenesis of the sandstones in the two fields is quite different. The Sihapas sandstone of the Melibur Field is uncemented (average porosity 37%, permeability 100–1000 mD). The same sandstone in the Karau Field is cemented by quartz, clays and small quantities of carbonate (average porosity 20%, permeability 10–100 mD). The bulk cement content of the Sihapas at Karau exceeds 10%. All of this cement must have been precipitated after the reverse fault separated the two localities; that is within 2 Ma. Indeed, cementation may well have been faster since the Karau Field also had to fill with oil.

Direct geochemical evidence

Authigenic illite clay can be dated radiometrically (K/Ar dating). There are now many published data on duration of illite cementation in a variety of sandstones. For example, K/Ar data from the Brent sandstones of the Northern North Sea indicate that illite precipitated in a short period between the Middle Paleocene and the Middle Miocene (Hogg 1989 and references therein; Hogg *et al.* 1992). Similarly, the age of illite cementation in the Rotliegend sandstone of the Southern North Sea has also been measured. Although there is a wide range of dates across the whole of the basin complex, individual fields and groups of fields often show highly restricted ranges of growth time from a few, to a few tens of millions of years (Lee *et al.* 1989; Robinson *et al.* 1992).

From *Petroleum Geology of Northwest Europe: Proceedings of the 4th Conference* (edited by J. R. Parker).

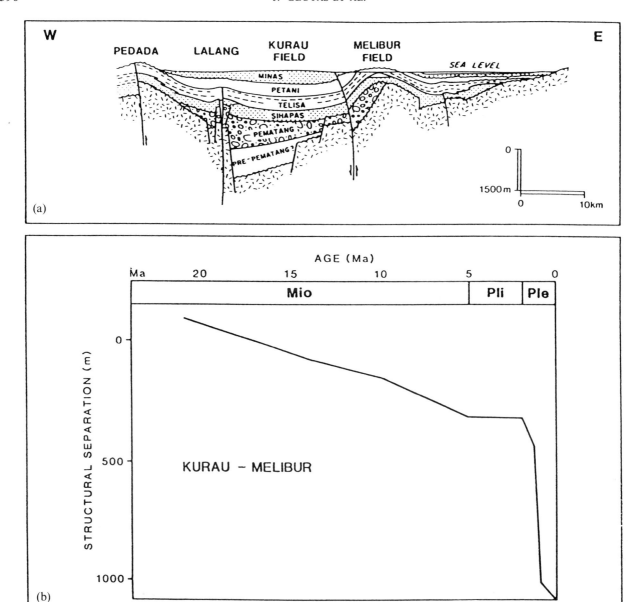

Fig. 1. (**a**) Cross-section of the Central Sumatra, Indonesia basin showing the relative elevation of the Melibur and Karau oil fields; (**b**) crossplot showing the vertical separation between the Sihapas sandstone in the Melibur and Karau fields as a function of time. The sandstone became discontinuous (faulted) between the two fields only two million years ago.

Indirect geochemical evidence

Fluid inclusions are tiny pockets of fluid trapped during cement growth. When viewed at surface conditions they commonly comprise vapour, liquid and possibly solid components. By heating such inclusions the vapour bubble can be made to disappear. The point of disappearance is known as the homogenization temperature and this represents a minimum temperature for cement growth. For inclusions which are methane-saturated, this homogenization temperature is close to the true trapping temperature (Emery and Robinson 1993). Homogenization data are available for many sandstones from many sequences. These data commonly show a restricted range of temperature, and since temperature can be converted to time, using a modelled burial history, a restricted range in time. A selection of these data is presented in Fig. 2.

We conclude from the above that the processes of cementation and petroleum migration/emplacement are both rapid—from a geological viewpoint. Let us now explore the temporal relationship between the two processes.

Petroleum emplacement and diagenesis

There is direct evidence that petroleum emplacement can occur at the same time as cementation of the reservoir sandstones. The evidence is simple. Many primary fluid inclusions contain petroleum (Walderhaug 1990). Thus, mineral precipitation must have occurred in the presence of at least some petroleum.

We have collected detailed information on the distribution of petroleum-filled fluid inclusions in an oil field reservoir sandstone. Figure 3 and Table 1 show these data. The plots show the abundance of fluid inclusions from top to bottom of wells and from crest to flank of the field. The greatest abundance of petroleum-filled fluid inclusions occurs at the crest of the field and in high permeability streaks which lie immediately below low permeability (shaly) intervals. There is also an inverse relationship between cement abundance and petroleum inclusion abundance (Fig. 4). Such relationships are interpreted as products of the filling history of the field. The inference is that the peaks in petroleum inclusion abundance

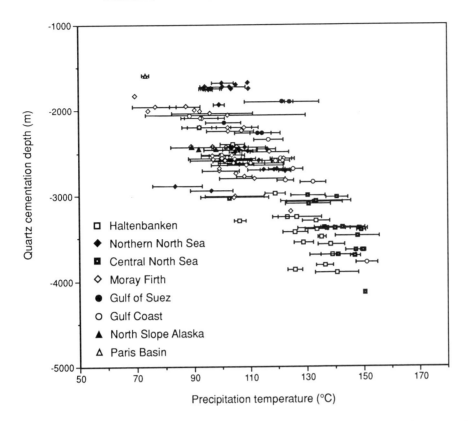

Fig. 2. Mean homogenization temperatures ($\pm 1\sigma$) of fluid inclusions trapped in quartz cement from sandstones ranging in age from Jurassic to Miocene from several areas around the world (from Gluyas *et al.* 1993). These data are taken to indicate that quartz cementation can occur at any temperature but that when cementation does occur it happens over a restricted temperature range, which is equivalent to precipitation occurring in a short time interval.

mark the first-filled portions of the field, including migration routes. These areas were less prone to cementation than the intervals which received a relatively late oil charge and any cement which did precipitate was crowded with petroleum-fluid inclusions. In other words, the presence of petroleum retarded cementation.

The effects on reservoir quality caused by this competition—'the race for space'—between cementing fluids and migrating petroleum are dramatic. Intra-field porosity gradients in fields affected by synchronous cementation and petroleum emplacement are commonly a factor or more greater than regional porosity depth gradients (Fig. 5). For example the regional porosity–depth trend for cemented Jurassic sandstones in the North Sea and Haltenbanken areas is between about 8 and 10% porosity loss for each kilometre of burial (Selley 1978; Gluyas 1985; Grant and Oxtoby 1992). The similarity of such gradients over large areas is likely to be, at least in part, a function of the thermal gradient for the area (Robinson and Gluyas 1992*b*). However, intra-field porosity–depth gradients in North Sea reservoirs affected by synchronous cementation and petroleum emplacement are between about 16 and 20% km^{-1} (Gluyas *et al.* 1990; Emery *et al.* 1993). Moreover, the porosity at the crests of such fields is only marginally less than would be expected from compaction alone, while the porosity in the water-leg defines part of the regional trend.

The crest to flank porosity patterns which develop as a result of synchronous cementation and petroleum emplacement can also be observed on a smaller 'bed' scale. Consider the data in Fig. 6. These data are taken from a field which shows a distinct and dramatic decline in average porosity with depth. However, the best reservoir sandstones are roughly of equal porosity and permeability irrespective of depth. In the case shown the sandstones which form the prime reservoir in each well are either cross-laminated or structureless. Lower quality sand-

stones of similar grain size/sorting differ only in respect of the distribution of small quantities of clay material. The lower quality sandstones are bioturbated.

We interpret the massive difference in present reservoir quality between the cross-laminated and bioturbated sandstones as a 'butterfly effect', in which small initial differences in a property—permeability in this instance—are amplified many times over (Gleike 1988). The cross-laminated sandstones started out marginally more permeable than the adjacent bioturbated sandstones. The cross-laminated sandstones were thus the first at any one locality to receive oil. As cementation had already begun, the cross-laminated sandstones were thus protected, to a large extent, from the cementing fluids. Only later did oil access the bioturbated sandstones by which time the damage had already been done—the bioturbated sandstones were thoroughly cemented.

Petroleum emplacement and diagenesis: is it causal or casual?

In trying to answer this question we are aiming to differentiate a direct causal relationship between oil generation/migration/emplacement and cementation versus one in which cementation and petroleum emplacement share a common cause. We can speculate on how both causal and casual relationships might arise.

There seem to be three common truths in burial diagenesis. We have already examined the short time taken for cementation and the observation that cementation commonly affects large geographical areas (Gluyas *et al.* 1993). The third truth is that cementation of sandstones is largely an open system process which involves importation of matter (Gluyas and Coleman 1992). What is the source of the cement? It might possibly be mudstone (Hower 1981); it might even be maturing

Table 1. Fluid inclusion abundances within wells of a Central North Sea field

Well	Depth (mBRT)	Reservoir zone	Kh (mD)	Oil or water leg	Primary quartz	Secondary quartz	Primary carbonate	Secondary carbonate	Feldspar	Total no. of inclusions
1	3619.70	AA	0.32	water	0	0	0	0	12	12
1	3622.50	AB	20.00	water	0	0	0	0	0	0
1	3623.60	AB	17.00	water	0	0	0	0	0	0
1	3626.20	AB	148.00	water	0	17	0	0	41	58
2	3643.40	AA	36.00	oil	50	23	0	0	109	182
2	3678.50	BA	236.00	?water	161	165	0	0	197	523
2	3683.60	BA	200.00	?water	237	107	0	0	578	922
2	3708.85	BA	92.00	?water	3	4	0	0	0	7
2	3718.80	BA	491.00	?water	0	46	0	0	36	82
3	3407.85	AA	21–44	oil	0	160	0	0	451	611
3	3430.40	BA	1049–1892	oil	2007	1055	0	0	331	3393
3	3434.40	BA	782.00	oil	758	281	0	9	254	1302
3	3444.10	BA	1817.00	oil	822	424	0	0	256	1502
3	3455.30	BA	187.00	oil	452	285	0	0	222	959
3	3467.88	BB	1034–1536	oil	632	185	0	0	314	1131
3	3474.04	BB	1068–1435	oil	505	358	0	0	242	1105
3	3481.00	BB	992.00	oil	330	184	0	0	486	1000
3	3483.50	BB	752.00	oil	198	64	0	0	632	894
3	3489.45	BB	656.00	oil	608	269	0	0	599	1476
3	3492.10	CA	1346.00	oil	484	187	1	0	543	1215
3	3494.55	CA	748.00	oil	368	108	15	0	427	918
3	3502.80	CA	368.00	oil	326	114	0	0	847	1287
4	3802.80	BA	3.44	oil	29	30	0	0	46	105
4	3815.65	BA	11.30	oil	149	84	0	0	151	384
4	3831.00	BB	22.90	water	0	64	0	0	0	64
4	3837.65	BB	5.72	water	9	39	0	0	8	56
4	3841.00	BB	12.70	water	4	6	0	0	32	42
5	3698.10	BA	77.00	oil	203	104	0	0	60	367
5	3703.00	BA	32.00	oil	50	32	0	0	0	82
5	3712.10	BA	20.00	oil	2	3	0	0	63	68
5	3718.10	BA	2.38	oil	0	12	0	0	29	41
6	4347.70	BA	41.10	oil	65	87	0	0	22	174
6	4354.75	BA	54.10	oil	0	20	0	0	15	44
6	4374.10	BB	0.55	?water	0	8	0	0	0	8
6	4382.60	BB	2.38	?water	0	1	14	0	0	15
6	4393.00	CA	0.72	?water	11	12	5	0	46	74
7	4065.65	BB	221.00	water	0	71	0	0	79	150
7	4072.80	BB	13.30	water	213	137	0	0	0	350
7	4077.50	BB	144–292	water	7	3	0	0	21	31
7	4080.10	BB	94.80	water	0	0	0	0	0	0
7	4085.00	CA	37.80	water	0	0	0	0	0	0
7	4090.50	CA	109.00	water	0	0	0	0	0	0
8	5025.40	AA	0.16	water	0	7	0	0	0	7
8	5049.65	AB	0.13	water	0	0	1	0	0	1
8	5054.40	BA	0.79	water	0	0	0	0	0	0
8	5101.45	BB	0.67	water	0	0	0	0	0	0
8	5109.25	BB	11.70	?water	0	8	0	0	90	98
8	5111.60	BB	0.06	water	0	4	0	0	0	4
9	3703.73	AB	19.70	oil	793	609	64	0	316	1782
9	3709.60	BA	211.00	oil	659	489	240	0	165	1553
9	3710.15	BA	402–643	oil	849	935	0	0	41	1825
9	3714.85	BA	758.00	oil	604	431	3	0	365	1403
9	3720.35	BA	485.00	oil	587	332	0	0	434	1353
9	3725.00	BA	3100.00	oil	434	469	0	588	153	1644
9	3731.65	BA	1290.00	oil	348	434	0	0	212	994
9	3737.35	BA	483.00	oil	208	360	199	0	486	1253
9	3743.35	BA	370.00	oil	244	250	0	0	453	947
9	3773.32	BB	673.00	oil	80	270	10	0	384	744
9	3777.50	BB	1090.00	oil	326	270	0	0	234	830
9	3784.52	CA	121–257	oil	289	277	0	0	101	667
9	3794.80	CA	120.00	?oil	154	151	0	0	61	366
9	3798.65	CA	64.50	?oil	84	83	0	0	66	233
9	3804.36	CA	1.45–1.65	?oil	36	74	0	0	8	118

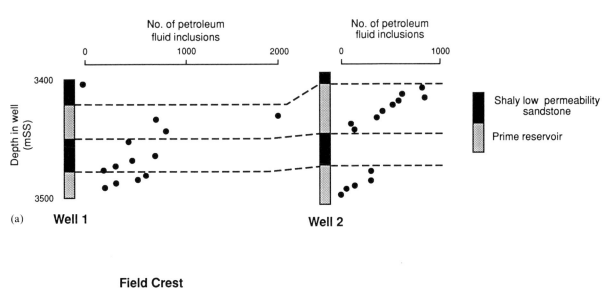

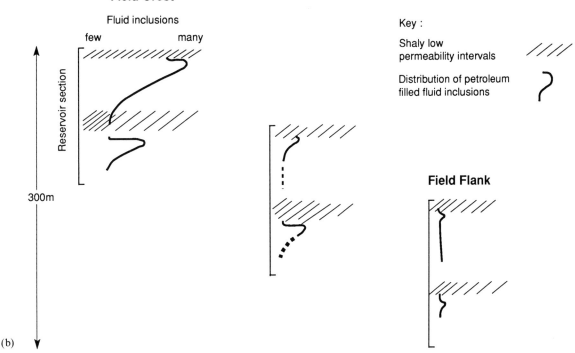

Fig. 3. (a) Distribution of petroleum-filled fluid inclusions within the crestal wells of a Central North Sea oil field; (b) schematic distribution of petroleum-filled fluid inclusions in the oil field. The petroleum-filled fluid inclusions are abundant in the crestal part of the field and in high permeability sandstones on the flanks. The abundance of inclusions was calculated by point counting traverses across appropriate thick (60 μm) wafers of rock illuminated with UV light.

source rock (Curtis 1978). We know for sure that late diagenetic carbonates in sandstones have stable carbon isotope signatures which implicate decarboxylation or organic matter as a source (Gluyas and Coleman 1992). However, no such clear relationship exists for silicate minerals. Much is known of the diagenetic processes which occur in mudrocks; clay transformations which might yield silica and other components (Hower 1981; Foscolos 1990). However, as yet there are no wholly satisfactory explanations of how large volumes of solutes can be transported from mudrock to sandstone (cf. Wood and Hewett 1984; Giles 1987), let alone being able to link in causal fashion oil generation/migration/emplacmeent with reservoir cementation.

It is much more straightforward to examine a casual relationship between diagenesis and petroleum emplacement. If we look deeper than the mere duration of cementation, but rather at the absolute timing of when cementation occurs, a common, simple and repeatable pattern emerges. Cementation is associated with periods of rapid heating and burial (Fig. 7), times when there is a major perturbation to the hydrodynamic regime in a basin (Burley et al. 1989; Gluyas et al. 1993). For example, the Rotliegend sandstones were cemented by illite and other minerals during a phase of rifting which occurred towards the end of the Jurassic (Robinson et al. 1992). Similarly, the Brent sandstones of the Northern North Sea were cemented at the end of a time of rapid burial within the Paleocene/Eocene. Such periods in which water flow is amplified will be periods in which solute transport is promoted. Rapid heating and burial will also promote maturation of potential source rocks. Thus, it seems quite reasonable that the

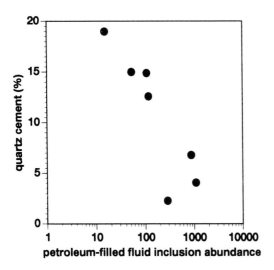

Fig. 4. Antithetic relationship between the petroleum-filled fluid inclusion abundance and quartz cement abundance for the above oil field. Petroleum-filled fluid inclusions are most abundant in the clean, high porosity, high permeability, little-cemented sandstones of the field crest. They are least abundant in well-cemented, low porosity, low permeability sandstones of the field flank. The average data are taken from Table 1.

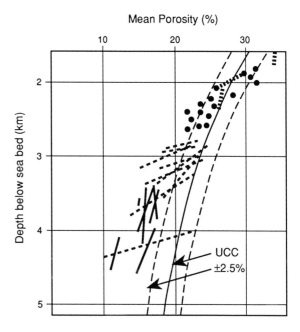

Fig. 5. Regional and intra-field porosity–depth trends. Fields which show no evidence of petroleum being present during cementation have intra-field porosity–depth trends which mimic the regional pattern. Those fields which were cemented during petroleum emplacement have much more rapid porosity declines as a function of depth (from Emery *et al.* 1993). UCC: uncemented sand compaction curve.

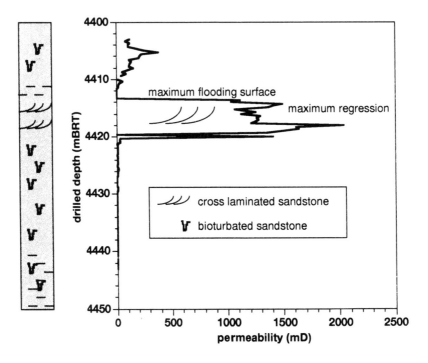

Fig. 6. Permeability–depth plot for a crestal well in a Central North Sea oil field affected by synchronous cementation and oil emplacement. The permeability of the sandstone varies from *c.* 1 D to a few mD between high-quality cross-laminated sandstones and low-quality bioturbated sandstones. The bioturbated sandstones are cemented by quartz. Petroleum-filled fluid inclusion abundance data indicate that minor quantities of quartz cement precipitated in the presence of relatively high petroleum saturations in the cross-laminated sandstone, while in the bioturbated sandstone much quartz precipitated in the presence of relatively low petroleum saturations.

processes of 'burial diagenesis' and petroleum emplacement may share a common cause in the heating, heating rate and burial rate of the sediment pile.

We cannot be sure yet of the source of many of the solutes (largely silica) needed to cement a sandstone. They may originate in mudstones or other sandstones. However, we now recognize that diagenesis is switched on, and perhaps off, in

response to distinct geological events in the basin history and, as such, perhaps the transport mechanism for the solutes is as important as their source.

Conclusions

Cementation of a reservoir sandstone and emplacement of

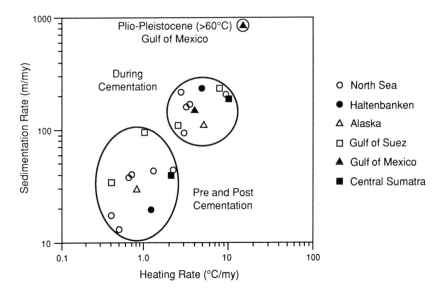

Fig. 7. Estimated rates of sedimentation and heating for sandstones in extensional basins before, during and after periods of quartz cementation. This figure was constructed by converting the temperature of quartz cementation to time using burial and thermal histories and then reading off both the sedimentation and heating rates before, during and after this time (from Gluyas *et al.* 1992).

petroleum in a reservoir can both take place in a geologically short time, around 10 Ma. Often the two processes are synchronous and this can lead to dramatic variations of porosity in a single well or field. These variations in reservoir quality occur because portions of a reservoir which receive the earliest petroleum charge (crest, migration routes) are protected from extensive cementation.

It seems probable that although the processes of cementation and petroleum emplacement are linked in time, one of the processes does not cause the other. Instead the two processes most likely share a common cause: rapid burial and heating promoting both generation of petroleum and flow of solute-bearing water through potential reservoir sandstones.

We thank BP and partners for permission to publish this work.

References

BURLEY, S. D., MULLIS, J. AND MATTER, A. 1989. Timing diagenesis in the Tartan Reservoir (UK North Sea) constraints from combined cathodoluminescence microscopy and fluid inclusion studies. *Marine and Petroleum Geology*, **6**, 98–120.

CURTIS, C. D. 1978. Possible links between sandstone diagenesis and depth-related geochemical reactions occurring in enclosing mudstones. *Journal of the Geological Society, London*, **135**, 107–118.

EMERY, D. AND ROBINSON, A. G. 1993. *Inorganic Geochemistry: Applications to Petroleum Geology*. Blackwell, Oxford, in press.

——, SMALLEY, P. C. AND OXTOBY, N. H. 1993. Synchronous oil migration and cementation: demonstrated by quantitative description of diagenesis. *Philosophical Transactions of the Royal Society*, in press.

ENGLAND, W. E., MACKENZIE, A. S., MANN, D. M. AND QUIGLEY, T. M. 1987. The movement and entrapment of petroleum fluids in the subsurface. *Journal of the Geological Society, London*, **144**, 327–347.

FOSCOLOS, A. E. 1990. Catagenesis of Argillaceous Sedimentary Rocks. *In*: MCILREATH, A. AND MORROW, D. W. (eds) *Diagenesis*. Geoscience Canada, Runge, Ottawa, 177–188.

GILES, M. R. 1987. Mass transfer and problems of secondary porosity creation in deeply buried hydrocarbon reservoirs. *Marine and Petroleum Geology*, **4**, 188–204.

GLASMANN, J. R., CLARK, R. A., LARTER, S., BRIEDIS, N. A. AND LUNDEGARD, P. D. 1989. Diagenesis and hydrocarbon accumulation, Brent Sandstone (Jurassic), Bergen Area, North Sea. *American Association of Petroleum Geologists Bulletin*, **73**, 1341–1360.

GLEIKE, J. 1988. *Chaos*. Sphere Books, London.

GLUYAS, J. G. 1985. Reduction and prediction of sandstone reservoir potential, Jurassic, North Sea. *Philosophical Transactions of the Royal Society*, **A315**, 187–202.

—— & COLEMAN, M. L. 1992. Material flux and porosity changes during sediment diagenesis. *Nature*, **356**, 52–53.

——, LEONARD, A. J. AND OXTOBY, N. H. 1990. Diagenesis and petroleum emplacement: the race for space—Ula Trend North Sea. *13th International Sedimentological Congress*. International Association of Sedimentologists, Utrecht, 193.

——, ROBINSON, A. G. AND GRANT (1993). Geochemical evidence for a temporal control on sandstone cementation. *In*: ROBINSON, A. G. AND HORBURY, A. S. (eds). *Diagenesis and Basin Evolution*. American Association of Petroleum Geologists Special Publication, in press.

GRANT, S. M. AND OXTOBY, N. H. 1992. The timing of quartz cementation in Mesozoic sandstones from Haltenbanken offshore mid-Norway: fluid inclusion evidence. *Journal of the Geological Society, London*, **149**, 479–482.

HOGG, A. J. C. 1989. *Petrographic and isotopic constraints on the diagenesis and reservoir properties of the Brent Group sandstones, Alwyn South, Northern UK North Sea*. PhD thesis, Unversity of Aberdeen.

——, HAMILTON, P. J. AND MACINTYRE, R. M. 1993. Mapping diagenetic fluid flow within a reservoir: K–Ar dating in the Alwyn area (UK North Sea). *Marine and Petroleum Geology*, **10**, 279–294.

HOWER, J. 1981. Shale diagenesis. *In*: LONGSTAFFE, F. J. (ed.) *Clays and the Resource Geologist*. Mineralogical Association of Canada, Short Course Handbook No. 7, 60–80.

LEE, M. C., ARONSON, J. L. AND SAVIN, S. M. 1989. Timing and conditions of Permian Rotliegende sandstone diagenesis, Southern North Sea: K/Ar and oxygen isotope data. *American Association of Petroleum Geologists Bulletin*, **73**, 195–215.

ROBINSON, A. G., COLEMAN, M. L. AND GLUYAS, J. G. 1993. The age and cause of illite cement growth, Village Fields area, Southern North Sea: Evidence from K–Ar ages and $^{18}O/^{16}O$ ratios. *American Association of Petroleum Geologists Bulletin*, **77**, 68–80.

——AND GLUYAS, J. G. 1992a. The duration of quartz cementation in sandstones, North Sea and Haltenbanken basins. *Marine and Petroleum Geology*, **9**, 324–327.

——AND—— 1992b. Model calculations of sandstone porosity loss due to compaction and quartz cementation. *Marine and Petroleum Geology*, **9**, 319–323.

SELLEY, R. C. 1978. Porosity gradients in North Sea oil-bearing sandstones. *Journal of the Geological Society, London*, **135**, 119–132.

WALDERHAUG, O. 1990. A fluid inclusion study of quartz-cemented sandstones from offshore mid-Norway—possible evidence for continued quartz cementation during oil emplacement. *Journal of Sedimentary Petrology*, **60**, 302–310.

WOOD, J. R. AND HEWETT, T. A. 1984. Reservoir diagenesis and convective fluid flow. *In*: McDONALD, D. A. AND SURDAM, R. C. (eds) *Clastic Diagenesis*. American Association of Petroleum Geologists Memoir, **37**, 99–110.

Discussion

Question (C. Garland, BP Exploration, Aberdeen):

It is clear that sandstone cementation occurs mainly in the absence of any hydrocarbon generation. In an oil field, is it possible to distinguish an additional cementation process, or are the same processes continuing in the presence of hydrocarbons?

Answer (J. Gluyas):

It is our opinion that cementation can continue in the presence of petroleum, albeit at a much slower rate than in the absence of petroleum. The occurrence of primary, petroleum-filled fluid inclusions in mineral cements is ample evidence for this process. Quite how petroleum is captured within inclusions is unknown, as are the relative saturations and distributions of oil and water at the time of trapping; although recent work by Tor Nedkvitne and others at Oslo University is beginning to shed some light on the topic (Tor has a paper in press with *Marine and Petroleum Geology*). We do not think that cementation in the presence of petroleum is a different process to cementation in the absence of petroleum—although as we have shown, the presence of petroleum does influence cement distribution and quantity.

Question (M. Osborne, University of Glasgow):

If the link between quartz cementation and oil migration is causal we would expect the crests of fields to be more cemented than the flanks. This is not observed. Surely this means the link is purely co-incidental. The volume of quartz cement present in the Brent Group increases linearly with depth.

Answer (J. Gluyas):

Why should a causal link demand that oil field crests are more cemented? We have speculated that expulsion of petroleum from a source rock may be preceded by expulsion of silica-rich water. Even if this is true it says nothing about which of the fluids might get to a field crest first. The forces acting on the fluids are different both in type and magnitude—petroleum migration is speeded along by its buoyancy; and, there will not be much room for cement if petroleum wins the race!

We disagree with your statement that the link is 'purely co-incidental' because we believe we have shown there to be a common link, via basin processes, between cementation and oil emplacement.

Your observations on the Brent sandstones are what we would expect—indeed what we too have observed. However, we think you need to be very careful in translating this observation directly into a model for process of cementation. We have observed differences in cementation–depth gradient between the regional trend and intra-field trends and between fields in which cementation occurred before, during or after oil emplacement (Fig. 5). Cementation gradients in the absence of petroleum can be explained in terms of cooling silica-saturated fluids (Robinson and Gluyas 1992a).

Field management

Field management

Introduction and review

M. L. B. MILLER[1] and J. H. MARTIN[2]

[1] *Petroleum Science and Technology Institute, Edinburgh, UK (Present address: Curlew Consulting, Whistlebrae, Braco, Dunblane, Perthshire FK15 9RA, UK)*
[2] *Imperial College of Science, Technology and Medicine, Prince Consort Road, London SW7 2BP, UK (Present address: Reservoir Geological Consultant, 150 Croxted Road, London SE21 8NW, UK)*

This 'Field management' section contains 13 papers focusing specifically on two areas of field evaluation:

- appraisal of complex offshore discoveries;
- management and review of developed fields, both offshore and onshore.

In addition, some 13 field case studies are presented within other sections of these volumes, along with many other briefer references to fields within NW Europe, particularly in the North Sea area (Table 1).

A number of themes can be recognized which reflect recent developments within the industry. The first trend which we

Table 1. Cross-references to NW Europe oil and gas fields in *Petroleum Geology of NW Europe: Proceedings of the 4th Conference*

Note: Titles/author listings in plain text indicate those papers which contain detailed accounts of individual fields, or specific aspects of these fields. Those listings in italics refer to specific aspects of fields or groups of fields within thematic or regional reviews.

Field	Title	Author
Adda	*The Lower Cretaceous chalk play in the Danish Central Trough*	*Ineson*
Alba	*Historical overview of Tertiary plays in the UK North Sea*	*Bain*
	The Alba Field: evolution of the depositional model	Newton and Flanagan
Arbroath	*Historical overview of Tertiary plays in the UK North Sea*	*Bain*
Balder	Origin of complex mound geometry of Paleocene submarine-fan sandstone reservoirs, Balder Field, Norway	Jenssen *et al.*
Beryl	*Sedimentation of Upper Triassic reservoir in the Beryl Embayment: Lacustrine sedimentation in a semi-arid environment*	*Dean*
	Beryl Field: geological evolution and reservoir behaviour	Robertson
Brent	*Jurassic exploration history: a look at the past and the future*	*Cordey*
Britannia	*Complex deformation and fluidization structures in Aptian sediment gravity flow deposits of the Outer Moray Firth*	*Downie and Stedman*
Bruce	The Bruce Field	Beckly *et al.*
Buchan	*Genetic sequence stratigraphy for the North Sea Late Jurassic and Early Cretaceous: distribution and prediction of Kimmeridgian–Late Ryazanian reservoirs in the North Sea and adjacent areas*	*Partington et al.*
Caister	*Permo-Carboniferous plays of the Silver Pit Basin*	*Bailey et al.*
	The Caister Fields, Block 44/23a, UK North Sea	Ritchie and Pratsides
Clair	Clair appraisal: the benefits of a co-operative approach	Coney *et al.*
Claymore	*Estimation of recoverable reserves: the geologist's job*	*Corrigan*
Clyde	*Genetic sequence stratigraphy for the North Sea Late Jurassic and Early Cretaceous: distribution and prediction of Kimmeridgian–Late Ryazanian reservoirs in the North Sea and adjacent areas*	*Partington et al.*
	Clyde: reappraisal of a producing field	Turner
Cormorant	A fundamental reappraisal of the structure of the Cormorant Field and its impact on field development strategy	Demyttenaere *et al.*
Crawford	*Low angle faulting in the Triassic of the South Viking Graben: implications for future correlations*	*Morgan and Cutts*
Dan	*Geological aspects of horizontal drilling in chalks from the Danish sector of the North Sea*	*Fine et al.*
Drake	*Paleocene reservoirs of the Everest trend*	*O'Connor and Walker*
Dunlin	*Estimation of recoverable reserves: the geologist's job*	*Corrigan*
Eakring Dukeswood	The Eakring Dukeswood oil field: an unconventional technique to describe a field's geology	Storey and Nash
Embla	The Embla Field	Knight *et al.*
Erskine	*Overpressures in the Central North Sea: implications for trap integrity and drilling safety*	*Gaarenstroom et al.*
Everest	*Historical overview of Tertiary plays in the UK North Sea*	*Bain*
	Paleocene reservoirs of the Everest trend	*O'Connor and Walker*
F/15-A	F/15-A: A Triassic gas field on the eastern limit of the Dutch Central Graben	Fontaine *et al.*
Forth	*Historical overview of Tertiary plays in the UK North Sea*	*Bain*
Forties	*Historical overview of Tertiary plays in the UK North Sea*	*Bain*
	Estimation of recoverable reserves: the geologist's job	*Corrigan*
Franklin	*Overpressures in the Central North Sea: implications for trap integrity and drilling safety*	*Gaarenstroom et al.*

Table 1. (continued)

Field	Title	Author
Frigg	*Historical overview of Tertiary plays in the UK North Sea*	*Bain*
Gannet	*Historical overview of Tertiary plays in the UK North Sea*	*Bain*
Groningen	*The Rotliegend in northwest Germany, from frontier to fairway*	*Burri et al.*
Gryphon	*Historical overview of Tertiary plays in the UK North Sea*	*Bain*
	The geology of the Gryphon Oil Field	Newman *et al.*
Gyda	*Genetic Sequence Stratigraphy for the North Sea Late Jurassic and Early Creta-ceous: distribution and prediction of Kimmeridgian–Late Ryazanian reservoirs in the North Sea and adjacent areas*	*Partington et al.*
	Structural controls on the Late Jurassic age shelf system, Ula Trend, Norwegian North Sea	*Stewart*
Heather	*Estimation of recoverable reserves: the geologist's job*	*Corrigan*
Heron	Overpressures in the Central North Sea: implications for trap integrity and drilling safety	Gaarenstroom et al.
Hutton	Silicate mineral authigenesis in the Hutton and NW Hutton fields: implications for sub-surface porosity development	McAulay *et al.*
	Estimation of recoverable reserves: the geologist's job	*Corrigan*
Hyde	Hyde: a proposed development in the Southern North Sea using horizontal wells	Steele *et al.*
Machar	*Historical overview of Tertiary plays in the UK North Sea*	*Bain*
	The evolution of the fractured chalk reservoir: Machar Oilfield, UK North Sea	Foster and Rattey
Maggie	*Paleocene reservoirs of the Everest trend*	*O'Connor and Walker*
Magnus	*Genetic sequence stratigraphy for the North Sea Late Jurassic and Early Creta-ceous: distribution and prediction of Kimmeridgian–Late Ryazanian reservoirs in the North Sea and adjacent areas*	*Partington et al.*
	Estimation of recoverable reserves: the geologist's job	*Corrigan*
Marnock	*Salt control on Triassic reservoir distribution, UKCS Central North Sea*	*Smith et al.*
	Overpressures in the Central North Sea: implications for trap integrity and drilling safety	*Gaarenstroom et al.*
Maureen	*Historical overview of Tertiary plays in the UK North Sea*	*Bain*
Meillon-Saint Faust	Meillon-Saint Faust gas field, Aquitaine basin: structural re-evaluation aids under-standing of water invasion	Haller and Hamon
Miller	*Genetic Sequence Stratigraphy for the North Sea Late Jurassic and Early Creta-ceous: distribution and prediction of Kimmeridgian–Late Ryazanian reservoirs in the North Sea and adjacent areas*	*Partington et al.*
	Miller Field: reservoir stratigraphy and its impact on development	Garland
Morecambe	The use of dipmeter logs in the structural interpretation and palaeocurrent analy-sis of Morecambe Fields, East Irish Sea Basin	Cowan *et al.*
	The tectonic history of the East Irish Sea Basin with reference to the Morecambe Fields	*Knipe et al.*
Murchison	*Tectonic evolution and structural styles of the East Shetland Basin*	*Lee and Hwang*
	Permo-Carboniferous plays of the Silver Pit Basin	*Bailey et al.*
North Morecambe	*The tectonic history of the East Irish Sea Basin with reference to the Morecambe Fields*	*Knipe et al.*
	The geology of the North Morecambe Gas Field, East Irish Sea Basin	Stuart
NW Hutton	Silicate mineral authigenesis in the Hutton and NW Hutton fields: implications for sub-surface porosity development	McAulay *et al.*
	Estimation of recoverable reserves: the geologist's job	*Corrigan*
Piper	*Estimation of recoverable reserves: the geologist's job*	*Corrigan*
Puffin	*Overpressures in the Central North Sea: implications for trap integrity and drilling safety*	*Gaarenstroom et al.*
Ravenspurn North	Structural and sedimentological controls on diagenesis in the Ravenspurn North gas reservoir, UK Southern North Sea	Turner *et al.*
Saltire	Appraisal geology of the Saltire Field, Witch Ground Graben, North Sea	Casey *et al.*
Salzwedel	*The Rotliegend in northwest Germany, from frontier to fairway*	*Burri et al.*
Skua	*Overpressures in the Central North Sea: implications for trap integrity and drilling safety*	*Gaarenstroom et al.*
Sleipner Ost	*Paleocene reservoirs of the Everest trend*	*O'Connor and Walker*
Smørbukk	Prediction of large-scale communication in the Smørbukk fields from strontium fingerprinting	Stølum *et al.*
Snorre	The structural evolution of the Snorre Field and surrounding areas	Dahl and Solli
Sohlingen	*The Rotliegend in northwest Germany, from frontier to fairway*	*Burri et al.*
Tartan	*Estimation of recoverable reserves: the geologist's job*	*Corrigan*
Thistle	*Estimation of recoverable reserves: the geologist's job*	*Corrigan*
Tiffany et al.	The interaction of structure and sedimentary process controlling deposition of the Upper Jurassic Brae Formation Conglomerate Block 16/17, North Sea	Cherry
Tyra	*Geological aspects of horizontal drilling in chalks from the Danish sector of the North Sea*	*Fine et al.*
Ula	*Genetic Sequence Stratigraphy for the North Sea Late Jurassic and Early Creta-ceous: distribution and prediction of Kimmeridgian–Late Ryazanian reservoirs in the North Sea and adjacent areas*	*Partington et al.*
	Structural controls on the Late Jurassic age shelf system, Ula Trend, Norwegian North Sea	*Stewart*
Valdemar	*Geological aspects of horizontal drilling in chalks from the Danish sector of the North Sea*	*Fine et al.*
	The Lower Cretaceous chalk play in the Danish Central Trough	*Ineson*
Waalwijk	*A review of the Triassic play in the Roer Valley Graben, SE onshore Netherlands*	*Winstanley*
Wytch Farm	Wytch Farm oilfield: deterministic reservoir description of the Triassic Sherwood Sandstone	Bowman *et al.*

note is the industry's ability to contain risk through better understanding of the technical problems, with improved technology able to provide greater flexibility in development planning. Technical disciplines are communicating better and shorter management chains bring the decision-makers closer to the issues.

Several papers describe development prospects whose appraisal history has been prolonged by contradictory delineation well results giving wild fluctuations in anticipated reserves. Other prospects were intially believed to be too small, complex, or poorly productive to warrant development, and two papers consider the consequences of inadequate appraisal. A typical example of fluctuating reserve estimates is described in 'The evolution of a fractured chalk reservoir—Machar Oil Field, UK North Sea' by **Foster and Rattey**. This field was discovered in 1976, but is only now ready for development following a long appraisal history during which the difficulties in understanding the reservoir distribution in a complex chalk system were overcome. High pressure, poor seismic resolution, structural complexities and ill-defined reservoir stratigraphy combine to provide major reserve uncertainties reviewed in the discussion of 'The Embla Field' in the Norwegian sector by **Knight** *et al.* The impact of these uncertainties on the range of development options is clearly shown.

Benefits of new technology and improved communication are a theme of the paper by **Coney** *et al.* 'Clair appraisal: the benefits of a co-operative approach'. Activity in the Clair area has recently been boosted by the technical success of a high angle well which penetrated a productive fracture zone. 'Hyde: a proposed development in the Southern North Sea using horizontal wells' by **Steele** *et al.* illustrates the increasing use being made of horizontal wells to produce from thin Rotliegendes dune sands interbedded with thick low permeability fluvial and lacustrine sediments. This technique will also be used in the development of F/15-A, a Triassic gas field on the eastern limit of the Dutch Central Graben (Fontaine *et al.*, Triassic Session).

Turner in 'Clyde: reappraisal of a producing field' shows how a major field review following some disappointing early development wells went on to form the basis for future development drilling. In our own opinion, external pressures appear to have compromised the original appraisal programme. Caution during field appraisal is perhaps warranted in view of conclusions drawn by **Corrigan** in 'Estimation of recoverable reserves: the geologist's job'. His review of published annual reserve estimates for UK sector fields indicates that structural complexity is the main reason for disappointing field performance particularly in the East Shetland Basin, whereas Paleocene fan fields (Forties Field being the prime example) have outperformed initial expectations.

It appears from many of the papers presented in these volumes as a whole (Table 1) that the lessons are being learnt. Several authors describe the care that is going into the reservoir description, in particular the detailed sedimentological studies required to delineate the Tertiary structural/stratigraphic traps of the Central North Sea, such as Balder (Jenssen *et al.*, Tertiary Session), Alba (Newton and Flanagan, Tertiary Session) and Everest (O'Connor and Walker, Tertiary Session). Others evaluate structural controls on field performance, including Cormorant (Demyttenaere *et al.*, Structural styles Session), Snorre (Dahl and Solli, Structural styles Session) and Morecambe (Cowan *et al.*, Irish Sea basins Session).

Structural complexity is also one of the main features of 'The Bruce Field'. **Beckly** *et al.* cite fault block compartmentalization as the main reason for an extreme range of initial fluid contacts and conclude, realistically, that final elucidation of communication will not be possible until after production has commenced. Reservoir connectivity problems are also addressed by **Stølum** *et al.* who, in their paper 'Prediction of large-scale communication in the Smørbukk fields from strontium fingerprinting', apply new analytical techniques to assess fluid connectivity over geological time. This novel approach appears worthy of additional testing against observed production history in developed fields.

Turning now to the management of producing fields, 'Beryl Field: geological evolution and reservoir behaviour' by **Robertson** illustrates some of the background to an understanding of post-production fluid migration in a structurally complex setting which has enabled the operator to manage an efficient depletion programme through successive drawdown, gas injection and water injection programmes. The future use of horizontal wells has been mentioned above, but of course horizontal drilling has been refined into effective standard development procedure for many fields worldwide, no more so than in Danish offshore chalk fields. **Fine** *et al.* in 'Geological aspects of horizontal drilling in chalks from the Danish sector of the North Sea' discuss the advances in operational production geology that have made drilling of record-breaking horizontal hole segments possible in low permeability chalks. The value of 'real time' wellsite biostratigraphy in 'steering' holes to within a narrow target tolerance is clearly established.

Although discussion of offshore fields dominates these volumes, one should not forget the importance of onshore development in NW Europe. Results of development drilling of the main reservoir of the largest onshore oil field in the region are briefly described in 'Wytch Farm oilfield: deterministic reservoir description of the Triassic Sherwood Sandstone' by **Bowman** *et al.* Additional development well data have enabled the operator to replace an earlier stochastic model with a hybrid model which better represents the overall layering observed in the field. Structural complexity is, of course, not confined to the North Sea. The need for interactive geological and engineering reservoir modelling even relatively late in the lifetime of a field (the 'end game') is also illustrated by 'Meillon-Saint Faust gas field, Aquitaine basin: structural re-evaluation aids understanding of water invasion'. In this paper **Haller and Hamon** indicate how geological appreciation of a fractured reservoir improves its management and may extend its producing life. 'The Eakring Dukeswood oil field: an unconventional technique to describe a field's geology' by **Storey and Nash** describes how production history and water breakthrough trends were used effectively to reconstruct a geological model for a recently abandoned field, developed originally with very little dedicated geological information gathering. Despite this major drawback, the authors not only explain past field history but suggest also that parts of the reservoirs remain poorly swept.

The increasing maturity of the NW Europe petroleum province is illustrated by the number of papers in these volumes which describe regional or sub-regional exploration potential based on evaluations of proven and developed fields. Onshore, the Triassic of the Roer Graben is reviewed in the light of data from the Waalwijk Field by Winstanley (Triassic Session), while Burri *et al.* (Permo-Carboniferous Session) show how the more subtle traps in the Rotliegend of northwest Germany could only be evaluated following close examination of the diagenetic history of the Groningen, Salzwedel and Sohlingen fields. Offshore and further north, the Ula and Gyda fields provide both regional sequence stratigraphic control for the late Jurassic (Partington *et al.* (Jurassic Session)) and structural evidence of the reservoir distribution which has led to the discovery of a number of adjacent hydrocarbon pools (Stewart (Jurassic Session)).

Apart from the technical aspects of appraisal and development, papers included within this 'Field management' section highlight consistently not only the close interaction needed between production geology and geophysics, but also between geoscientist, petroleum engineer and management. A spirit of openness appears to be present. The co-operative approach to appraisal of the Clair discovery, for example, has led to a new appraisal drilling programme in this previously dormant prospect, and development of the Bruce Field is now underway following a similar co-operative approach. Finally, the frank discussion of the risks inherent in plans to bring the Embla Field on stream will also be of interest to all oil company technical and managerial staff contemplating development of marginal or complex fields.

Clair appraisal: the benefits of a co-operative approach

D. CONEY,[1] T. B. FYFE,[2] P. RETAIL[3] and P. J. SMITH[2]

[1] Esso Expro UK, Esso House, Leatherhead, Surrey KT22 8UY, UK
[2] BP Exploration, 301 St Vincent Street, Glasgow G2 5DD, UK
[3] Elf UK, 197 Knightsbridge, London SW7 1RZ, UK

Abstract: The Clair discovery lies 75 km west of the Shetlands, and comprises an elongate ridge of Lewisian basement and an associated roll-over of variable quality Devonian–Carboniferous continental red beds.

Clair was discovered in 1977, when 206/8-1A tested 25° API oil at 1500 BOPD from the red beds at the crest of the roll-over. Well 206/7-1 followed, producing 960 BOPD from fractured basement on the ridge. Ten further wells drilled between 1977 and 1985 indicated an oil in place measurable in billions of barrels, but test results were disappointing. The success of the discovery well was never repeated and commercial test production rates were never achieved.

In 1990, the companies with licence interests over Clair agreed to pool their resources and expertise and collaborate in a joint appraisal programme. A series of reservoir studies quantified the importance of different fracture types to productivity and predicted that the right combination of fractures and matrix could yield much higher flow rates than had previously been achieved. A 3D seismic survey over the central area of the discovery led to a significant improvement in understanding and visualization of the structure.

Two joint appraisal wells were drilled in 1991. The first, a horizontal well in fractured basement, tested at 2100 BOPD after acid wash stimulation.

The second well tested the red beds on the flank of the roll-over and achieved sustained flow rates of over 3000 BOPD from two zones (one of which had been hydraulically fractured) with a significant contribution from previously unidentified open natural fractures. The orientation of these open fractures suggests they are related to a late Cretaceous episode of faulting.

Two further appraisal wells and an extension to the 3D seismic coverage are planned for 1992.

There are still many uncertainties associated with Clair, but the 1991 wells have finally demonstrated that significant and potentially commercial flow rates are achievable. The behavioural change associated with the joint approach has been the catalyst for the success of this new initiative on Clair.

Clair is the largest undeveloped oil discovery on the UK Continental Shelf. It lies some 75 km west of the Shetlands in water depths of up to 150 m (Fig. 1). The Clair structure comprises an elongate NE–SW-trending ridge of Lewisian basement and an associated terrace or roll-over containing a thick sequence of Devonian–Carboniferous continental red beds (Fig. 2).

Clair was discovered in 1977 when BP (on behalf of the BP, Chevron, ICI licence group) drilled 206/8-1A to test 'possible Jurassic and/or Cretaceous sands in a broad anticlinal trap overlying basement'. Instead of the forecast Mesozoic sequence, the well penetrated a 700 m section of Carboniferous to Devonian clastics lying below thick Upper Cretaceous mudstones. The upper 568 m of this sequence was oil bearing

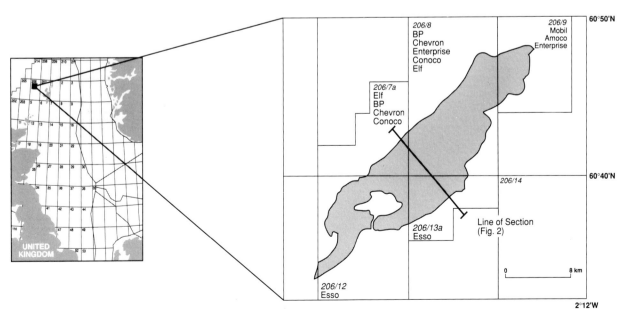

Fig. 1. Location map.

From *Petroleum Geology of Northwest Europe: Proceedings of the 4th Conference* (edited by J. R. Parker).

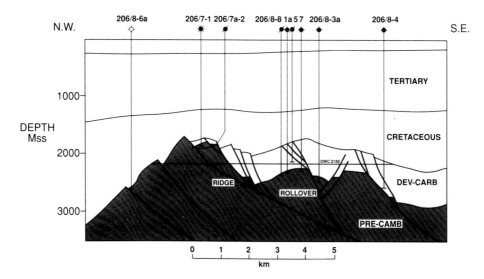

Fig. 2. NW–SE schematic cross-section.

and four drill stem tests were carried out, the best of which flowed at an average rate of 1502 BOPD of 25° API oil. The discovery was greeted with great optimism, which appeared to be borne out by Elf's 206/7-1 well, drilled at the end of 1977 in the adjacent block (for the Elf, Conoco, Gulf, BNOC group). This tested similar relatively heavy oil at 963 BOPD from fractured Lewisian basement underlying a thin sequence of red beds.

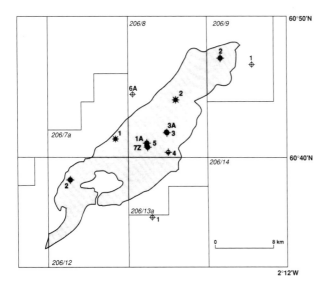

Fig. 3. Drilling activity 1977–85.

Three years of active drilling followed, such that by the end of 1980 a further nine wells had been drilled in and around Clair: two by Esso in Blocks 206/12 and 206/13; two by the Mobil group (Mobil, Amoco and British Gas) in Block 206/9; and five more by the BP group in Block 206/8 (Fig. 3). These wells demonstrated that Clair straddles five UKCS blocks, they confirmed a hydrocarbon column of over 800 m (including two separate gas caps) and indicated an oil in place measurable in billions of barrels. However, as Table 1 shows, the appraisal wells suffered from disappointing test results. The flow rates achieved by the discovery well were never repeated, the wells highlighted the problems of poor and variable reservoir quality and attempts to stimulate by hydraulic fracturing were thwarted by mechanical problems. They also showed that seismic data were of little use below the base of the Cretaceous.

As a result of the disappointing appraisal drilling campaign, the initial optimism that Clair would prove to be comparable to some of the largest North Sea fields waned.

Between 1980 and 1990 only one more well was drilled, 206/8-7 drilled by the BP group in 1985. This was located close to the discovery well on the crest of the Clair roll-over but it was as disappointing as all of the previous appraisal wells. The reservoir section was cut by at least two significant faults with a combined throw of almost 100 m (which were not seen on seismic) and reservoir quality was severely impaired by associated closed fractures. Test flow rates of less than 500 BOPD were recorded and attempts to stimulate the well (in a side-track, 206/8-7Z) were aborted because of mechanical problems.

The joint appraisal strategy and programme

By the end of the 1980s twelve wells had been drilled by four competing licence groups but not one had been able to demonstrate a 'commercial' discovery.

Clair would still be going nowhere had the companies involved not realized that by a joint approach something new and different was possible. Collaboration between the companies was helped by the changes in ownership which had taken place during the 1980s. Enterprise Oil had taken over the interests of British Gas (in 206/9) and of ICI (in 206/8), Chevron had assumed control of Gulf's 206/7 interest and BP had taken over Britoil's (previously BNOC's) interest also in 206/7. These changes reduced the number of companies involved in Clair (from 11 to 8) and led to overlapping interests.

The first major step in the joint programme was the agreement reached in early 1990 between three of the licence groups to shoot a joint 3D seismic survey over the central area of the discovery and to carry out a series of sub-surface and engineering studies. While this activity was going on, two things happened. First, the fourth licence group was brought into the partnership and second, Elf and Conoco increased and spread their involvement by farming in to Block 206/8. By early 1991 all eight companies with interests over Clair were working together under a unique Joint Appraisal Agreement.

The commitment to the joint approach is well illustrated by the management structure (Fig. 4). There is no 'operator' for Clair. All eight companies are represented on a Management Committee (chaired by BP) and on the three work groups which report to the Management Committee: a Reservoir Work Group (chaired by Esso); an Engineering Work Group

Table 1. Clair wells 1977–1985: summary of test flow rates

Well no.	Completion date	Operator	Test flow rate (BOPD)	Interval tested
206/8-1a	July 1977	BP	1502	Devonian–Carboniferous
206/12-2	October 1977	Esso	620	Devonian–Carboniferous
206/9-1	December 1977	Mobil	—	No tests
206/7-1	January 1978	Elf	963	Precambrian basement
206/13-1	January 1978	Esso	—	No tests
206/9-2	March 1978	Mobil	378	Devonian–Carboniferous
206/8-2	August 1978	BP	450	Devonian–Carboniferous
206/8-3a	December 1978	BP	531	Devonian–Carboniferous
206/8-4	February 1979	BP	—	No tests
206/8-5	July 1980	BP	794/1080*	Devonian–Carboniferous
206/8-6a	September 1980	BP	—	No tests
206/8-7	September 1985	BP	445	Devonian–Carboniferous
206/8-7Z	November 1985	BP	108*	Devonian–Carboniferous

* post-fracture stimulation flow rates.

(chaired by Elf); and an Agreements Work Group (chaired by BP). By involving all of the companies in both management and technical work all efforts are focused on solving common problems and the group is able to draw on the resources and the expertise of eight major operating oil companies.

This paper is concerned with the work carried out by the Reservoir Work Group and in particular describes reservoir studies carried out on both the ridge and the roll-over areas prior to drilling, discusses the 1990 3D seismic survey, presents the results and implications of the 1991 wells and outlines the next stage of the joint appraisal programme.

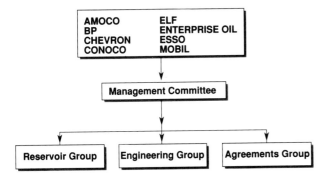

Fig. 4. Joint appraisal management structure.

Reservoir studies on the ridge

Exploration Well 206/7-1 had produced oil from an open-hole test of fractured Lewisian gneissose basement.

Cores from 206/7-1 showed that the fractures were sub-vertical and from the outset it was anticipated that the potential of this production mechanism should be further tested by drilling a horizontal well in basement. This well was intended to drain hydrocarbons from the Devonian–Carboniferous red beds above basement and to evaluate the rate of recharging through the fault and fracture network.

To build confidence in the feasibility of this depletion mechanism a number of reservoir studies were initiated. These included characterization of the fracture pattern (and the relationship between the red beds and basement), structural fracture modelling of the Clair ridge, and predictive simulation of horizontal well behaviour.

Fracture pattern characterization

The relationship between the basement and the red beds was studied in outcrop on the north coast of Scotland and the Orkney Islands, where Devonian Old Red Sandstone 'red beds' rest on Moinian metamorphics. Although of a different age, the fracture style and fracture frequency of the Moinian basement exposed at both locations was considered to be analogous to the Lewisian basement in the Clair area.

Fracture patterns in the outcrops consist of a number of trends of sub-vertical fractures each exhibiting similar ranges of fracture frequency.

In the basement, the average frequency is 1 to 2 fractures per metre. However, some strongly fractured areas (fracture 'corridors' and larger fault zones) have 10–30 fractures per metre. In these fracture corridors and fault zones, the fractures appear to continue up into the overlying Devonian red beds, although the frequency of fractures decreases upwards from the basement into the red beds (from 10–30 to 5–10 fractures per metre).

In the red beds, the background fractures occur at an average frequency of over 1 fracture per metre. The fractures are not usually filled, although apertures are very narrow.

From the field work it was concluded that only the fault zones and corridors are likely to have effective permeability and are therefore potentially important for well drainage, particularly since the fractures in the fault zones and corridors in the red beds align with those in the basement.

Fracture model of the Clair Ridge

Drawing on the conclusions of the field work combined with dipmeter interpretation from Well 206/7-1 and observation from aeromagnetic anomaly mapping, a structural and fracture model for the Clair Ridge area was derived.

In the model, fractures are concentrated within large corridors, classified into three types based on their spacing.

- First order: regional fault zones with 1–1.5 km spacing.
- Second order: intermediate fault zones with 100–200 m spacing.
- Third order: small-scale faults with 30–35 m spacing.

The frequency of fractures within these corridors is relatively independent of the corridor spacing, although it is directly related to corridor width and lithology.

Two main structural trends to the basement faulting pattern are identified from dipmeter interpretations and observations from the aeromagnetic anomaly map (Fig. 5): NNE–SSW oblique faults truncating the ridge, and NE–SW to ENE–WSW normal fault segments. The combination of three scales and two trends generates a regular 30 m diamond-shaped network of fractures.

This structural and fracture model of the Clair Ridge was confirmed and enhanced following the 3D seismic interpretation outlined below.

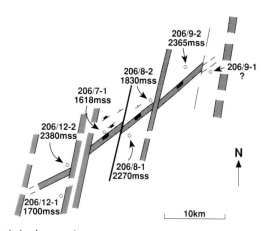

mss - depths to top basement

Fig. 5. Basement fault sketch.

Simulation of horizontal well performance

Predictive numerical simulation of a small area perpendicular to the ridge was undertaken to estimate the possible recovery from a horizontal well in the basement.

For the simulation model (Fig. 6), faulting was simplified to two orders of faults: the first order representing the primary system of regional faults and oriented parallel to the ridge, the second order representing a combination of the smaller-scale faults and oriented both parallel to, and perpendicular to, the ridge.

The model was divided into 10 layers with the model grid comprising alternating 'matrix' cells (of either basement or red beds) and 'fracture' cells representing either first-order faults (10 m wide) or second-order faults (5 m wide). Parallel to the ridge (in the X direction) there were nine cells: five matrix cells each 200 m wide, separated by four, 5 m wide fracture cells. Perpendicular to the ridge (the Y direction) there were 14 matrix cells varying in width from 100 m to 400 m, separated by two, 10 m wide and eleven, 5 m wide fracture cells.

Within the model, basement matrix cells were assumed to be 'dead' cells, and red bed matrix cells were assigned characteristics based on the reservoir studies carried out on the roll-over. A wide range of permeabilities were initially assigned to the fracture cells, from 150 mD for the second-order cells in the red beds to up to 15 000 mD for the first-order cells in the basement. These assumptions on fracture permeability were based on field and well observations and on theoretical values.

The model was constrained to match the basement test results of 206/7-1 (Fig. 7). The fracture permeabilities initially assumed in the model had to be divided by three to successfully reproduce the test permeability thickness product (42 000 mD-feet) which was interpreted to be the contribution from a single fracture corridor. The model match supports a dual porosity interpretation.

The model was then used to predict the potential of a horizontal well drilled into basement and located to intersect a number of fracture zones and to evaluate the sensitivity of such a well to fracture description. Although there was a wide range of predicted profiles (Fig. 8), the simulation model suggested that initial rates of up to 5000 BOPD and recovery of up to 6 MMBBL were possible if both first- and second-order fracture zones were assumed to be contributing.

Therefore, the simulation modelling appeared to support the original assumption that a horizontal well could have significant production and recovery potential.

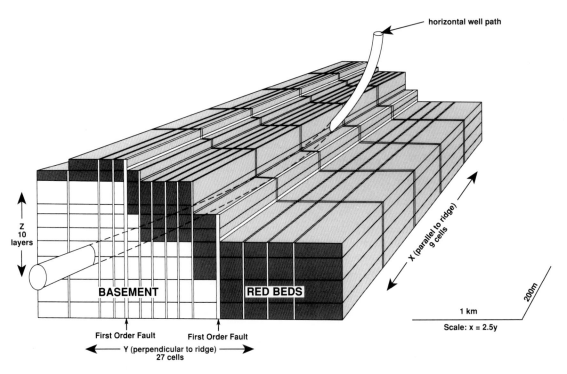

Fig. 6. Ridge simulation model.

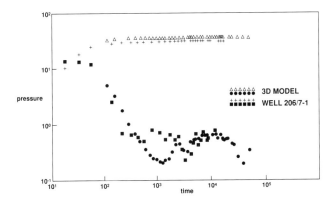

Fig. 7. Match to well test data, 206/7-1.

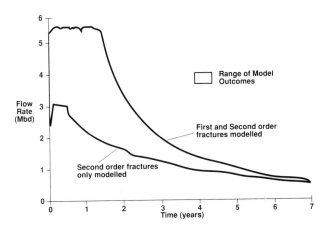

Fig. 8. Horizontal ridge well: predicted potential.

Reservoir studies on the roll-over

The first objective of the roll-over studies was to understand the variable test results in the context of the geology. This understanding was then used to predict whether, and in what circumstances, production rates significantly higher than had been achieved in the past would be possible.

Lithostratigraphy

The lithostratigraphic scheme adopted (Fig. 9) was developed by Allen and Mange-Rajetzky (1992) from over 1700 m of core from eight of the Clair wells and by log correlation of the remainder. Using 206/8-7 as the 'type' well, Allen and Mange-Rajetzky established a zonation of the Clair Group comprising ten lithostratigraphic units, subdivided into the Lower Clair Group (Units I to VI) and the Upper Clair Group (Units VII to X).

Heavy mineral analyses support the differentiation between the Lower and Upper Clair Group. A marked unconformity was recognized between the Lower and the Upper Group and it is tempting to interpret that this unconformity represents the Devonian–Carboniferous boundary.

The sequence is dominated by continental deposits (with fluviatile, aeolian and lacustrine environments recognized) becoming marginal marine at the top. Two depositional cycles are recognized in the Lower Clair Group, interpreted as resulting from two phases of tectonism. Both exhibit characteristics of an areally restricted intermontane basin with limited external drainage.

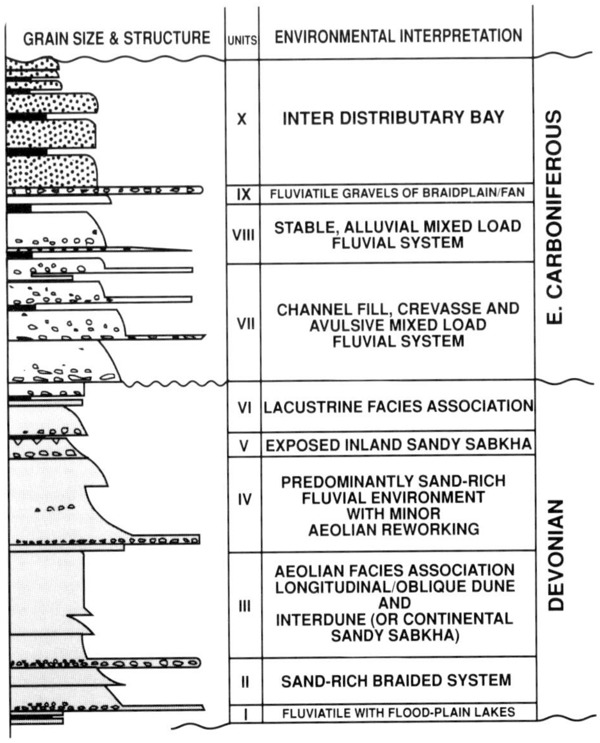

Fig. 9. Lithostratigraphic zonation of Clair Group sequence.

The first cycle starts with a rift lake sequence (Unit I) giving way to more active, high-energy fluvial system (Unit II) then waning to a finer-grained aeolian and interdune sabkha sequence (Unit III). A similar pattern is seen in the second cycle with a high-energy fluvial interval (Unit IV) underlying a sandy sabkha sequence with numerous sheetfloods and evidence of aeolian reworking (Unit V) and finally a fairly low-energy fluvial and lacustrine section (Unit VI). Within Unit VI there is one distinctive marker band of finer-grained lake deposits, referred to as the 'Lacustrine Key Bed', which is seen in a number of the wells and which effectively divides Unit VI into two parts.

The Upper Clair Group contains one major cycle and shows a significant change in depositional pattern. A major fluvial system became established and lasted through Units VII and VIII, evolving into a coarser-grained river system (Unit IX) and finally becoming marginal marine with distributary bay deposits (in Unit X).

There is a major unconformity at the top of the Clair Group, which is overlain by a thick sequence of typical Cretaceous Shetland Group mudstones, forming a regional seal.

Reservoir quality

Average reservoir parameters for Units III to X are shown on Table 2 (Units I + II, where present, only rarely rise above the oil–water contact and can therefore be ignored as potential reservoirs).

There is little variation in average porosity between the units (0.11 to 0.15) and it is the variability of the other two parameters (net reservoir ratio and permeability) which highlight the better quality units. In this context, net reservoir ratio is used as a measure of relative matrix quality and net reservoir is defined as rock with a porosity of greater than 9% and a V_{shale} of less than 35%.

The two units with aeolian influence (III and V) are noticeable for having much higher net reservoir ratios than the other units.

Table 2. Units III to X, mean reservoir parameters

	Net reservoir ratio	Porosity	Matrix permeability (arithmetic mean) (mD)
Unit X	0.20	0.15	<10*
Unit IX	0.27	0.15	<10*
Unit VIII	0.21	0.15	<10*
Unit VII	0.21	0.12	<10*
Unit VI	0.35	0.12	70
Unit V	0.72	0.14	360
Unit IV	0.35	0.12	30*
Unit III	0.84	0.15	80

* Permeability in these units is severely affected by clay minerals.

Permeability varies due to a combination of fracturing and clay mineralogy. The distribution and effect of fractures is discussed below. Clay mineral distribution appears to reflect sedimentological subdivisions and swelling clays primarily affect the units which are dominated by poorly sorted, fluvial sediments. Thus, Unit IV and the majority of the Upper Clair Group contain abundant swelling clays. As a result, these units have very low average permeabilities (of the order of 5–7 mD) and are very sensitive to formation damage.

Therefore, at least over the central area of Clair, the reservoir units of principal interest are Units III, V and VI.

Lithofacies modelling

In order to describe the geometry and characteristics of individual sandbodies (or 'flow units') for reservoir simulation modelling, the lithostratigraphic model was developed into a 3D lithofacies model. Focusing on Units II to VII (i.e. encompassing the three better quality reservoir units) in Wells 206/8-5 and 206/8-7, eight different lithofacies types were recognized and described (Table 3). The relative proportion of each lithofacies type in each of the lithostratigraphic units was calculated from core data. Porosity and permeability distribution for each lithofacies type were derived and applied.

Table 3. Units II to VII, lithofacies types

	Description	Interpretation
1.	Mudstone, siltstone and minor sandstone.	Floodplain including lake/playa.
2.	Fine to very fine sandstone, parallel and ripple cross-laminations.	Peri-lacustrine sheetfloods, bar tops.
3.	Fine to very fine sandstone, interwoven ripple cross-laminations, climbing ripples.	Lacustrine shoreface, beach zones.
4.	Medium to fine sandstone, mud intraclasts, larger cross-sets than lithofacies 2.	Sheet flows, crevasse splays, channel fills.
5.	Medium to very coarse sandstone, granules, thick co-sets.	Channel fills and bars.
6.	Conglomerate, pebbly sandstone.	Channel lags, bar cores.
7.	Fine to medium sandstone, wedge-shaped cross-sets.	Aeolian dunes.
8.	Fine to very fine sandstone with 'wispy' laminae of siltstone and mudstone ('crinkly').	Interdune sabkha, aeolian reworked lacustrine/fluvial deposits.

Finally, using analogues, core and log data from the wells and detailed correlation between closely spaced wells estimates of the three-dimensional geometry of each lithofacies were compiled. These were expressed as ranges of thickness, width and length and where appropriate the rate of thinning and the relative orientation.

Fracture and fault statistics

Cores from the reservoir section of Wells 206/8-7 and 206/8-5 are affected by pervasive fracturing. Two main fracture types are recognized: mineral veins and microfaults (the latter category also including granulation seams).

The distribution of mineral veins (of calcite and pyrite) in both wells appears to be stratigraphically controlled, with veins being particularly prevalent in Unit VI. In 206/8-7 the veins are all cemented and tight, whereas in 206/8-5 they are occasionally vuggy. These vuggy fractures probably explain the observation that the core from 206/8-5 bled oil from fractures when brought to the surface.

In 206/8-7, over a background density of between 2 and 6 microfaults per metre, there are three zones which show a marked increase in intensity, up to 50 per metre (Fig. 10). These three zones are all associated with large-scale faults cutting the wellbore. The background density of microfaults in 206/8-5 is less than in 206/8-7 (between 2 and 4 per metre) and there is one zone of increased intensity which again appears to be related to a fault.

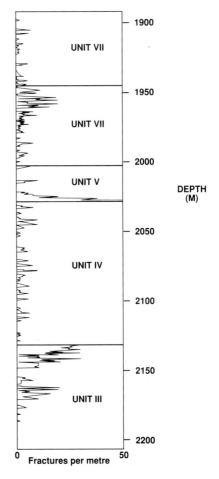

Fig. 10. Well 206/8-7 fracture frequency.

From the two wells studied, there is a clear relationship between the intensity of microfaults and the presence of large-scale faults. There also appears to be a relationship between the number of large-scale faults which cut the wellbore and the background density of microfaults.

In both wells, the microfaults (and the granulation seams) appear to be tight. On average, fractured core plugs have almost exactly half of the permeability of unfractured plugs.

Simulation modelling

Simulation modelling of the red beds sequence focused initially on Wells 206/8-7 and 206/8-5.

The first step was the construction of a lithofacies distribution (derived from the lithofacies modelling described above) on a small-scale cartesian grid using a conditional simulation package developed at Stanford University (Journel and Alabert 1988) and modified by BP. Initial permeabilities were assigned using conventional core plug data, excluding fracture plugs and corrected for overburden.

The effect of background fractures was incorporated by modifying the permeabilities assigned to each lithofacies block using estimates of fracture density (based on a power law relationship between frequencies and length) and fracture permeabilities from core plug data. The model was scaled up to ECLIPSE model blocks and low permeability layers (e.g. the Lacustrine Key Bed) were input. Finally the model for 206/8-7 was 'faulted' with transmissibility multipliers applied to simulate the low permeability fractures close to the faults.

For Well 206/8-7, the best match with the well test data was achieved by invoking both an overall reduction in permeability caused by pervasive fracturing and a further reduction in permeability adjacent to the modelled faults (Fig. 11). Conversely, for Well 206/8-5, modelling based on only the lithofacies distribution and matrix permeabilities (i.e. excluding any effect of fracturing) indicated that a slight enhancement of permeability was required to achieve a match with well test data, implying that some open fractures were present.

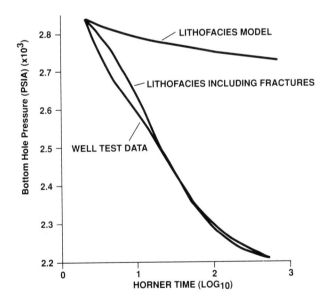

Fig. 11. Match to well test data, 206/8-7.

The simulation model was tested by successfully matching the well test data from 206/8-3A and then used to predict longer-term production potential and to evaluate different well types. The model predicted that an optimally located well would be capable of an initial production rate of 3000 BOPD but with rapid decline indicating that some form of stimulation would be required to achieve sustainable flow rates. Two types of stimulated well were modelled, a hydraulically fractured vertical well and a high-angle (75°) well (Fig. 12).

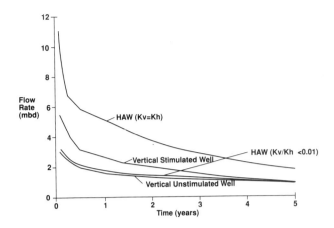

Fig. 12. Roll-over well: predicted potential.

It was predicted that hydraulic fracturing of a vertical well could achieve a two-fold increase in productivity, with initial production rates of greater than 5000 BOPD.

The high-angle well model showed very little improvement over a conventional well because of the very low K_v/K_h ratio (0.0003–0.0004). It was noted, though, that if a high-angle well could be located to encounter an open natural fracture system then its potential would be greatly enhanced. However, since open natural fractures had not been positively identified and since there was, therefore, no information on their distribution and/or orientation a high-angle well was considered a high risk option.

Conclusions of the reservoir studies

The key uncertainties which emerged from the studies were:

- could the modelled production rates be achieved in practice?
- could such rates be sustained?

To address these, the appraisal programme adopted by the Group was to drill two appraisal wells, a vertical stimulated well in the roll-over area to test the Devonian–Carboniferous section and a horizontal well in the ridge to test the efficiency of fractures in the basement as conduits for fluid flow.

3D seismic

The need for 3D seismic was universally recognized, given the difficulty of interpretation below the dominant Base Cretaceous (top reservoir) reflection and the resulting problems with selecting well locations targeted at deeper reservoir zones.

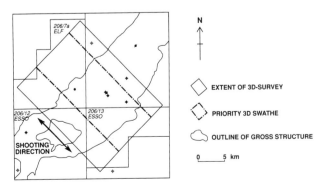

Fig. 13. Area of 1990 3D seismic survey.

The 1990 3D survey was aimed at the central area of the Clair structure (Fig. 13) and covered an area (full-fold stack) of 172.5 km² (11.5 by 15 km). In order to achieve the target of spudding the two planned appraisal wells in May 1991 (i.e. within less than 12 months of acquisition), a 5 km wide swathe of the survey was prioritized for acquisition, processing and interpretation.

Acquisition and processing

Acquisition totalled 2776 sail-line km and took two months to complete (mid-April to mid-June 1990). Acquisition parameters are summarized in Table 4, and of particular note is the use of a single source, triple streamer combination. Shot and group intervals were chosen to combat the severe multiple problems and the relatively high infill was required because of strong currents perpendicular to the shooting direction which led to high feathering angles.

Processing of the central swathe of data was completed by January 1991 (seven months after acquisition).

The processing sequence (Table 4) included further attempts to eliminate multiples with both wave equation multiple attenuation processing and FK multiple attenuation applied prior to dip move-out. Despite all the efforts to eliminate multiples, subsequent interpretation and calibration with well data revealed residual multiple energy still present below the top reservoir reflection, illustrating the intractable nature of the multiple problem in the Clair area.

Table 4. 1990 3D seismic survey

Acquisition summary

Contractor	GECO
Timing	April 18 to June 19, 1990
Total length	2776 sail-line km (including 25% infill)
Average production rate	44 km/day
Source	2920 cubic inch air gun array
Streamer	triple, 2400 m, 192 groups
Shot and group interval	12.5 m
Line spacing	90 m (30 m in-line)
Fold of cover	96

Processing highlights

Contractor	CGG (London)

- FK anti-alias filter on shot records
- Wave Equation Multiple Attenuation (WEMUL)
- FK multiple attenuation
- One pass 3D migration

Interpretation and mapping

Interpretation focused on two horizons, top reservoir and top basement. There is a marked increase in both velocity and density at the top of the reservoir (Base Cretaceous) which generates a high-amplitude reflection interpretable over much of the survey area.

The interface between the Devonian–Carboniferous sequence and Precambrian basement is also associated with a distinct increase in acoustic impedance and provides a clear reflection in many places. Although the character of the top basement reflection is more variable than that at top reservoir, it can still be followed over a significant proportion of the dataset.

Both horizons were interpreted as continuous surfaces using a combination of automatic tracking techniques where appropriate combined with editing and manual picking where necessary (Stewart *et al.* 1991).

Previous 2D sections differ significantly from those derived from the 1990 3D survey. On a typical 3D section (Fig. 14) top basement can be clearly identified and followed rather than 'ghosted' through as on the 2D section (Fig. 15). A very different fault pattern is seen on the 3D section. Different episodes of faults can be distinguished, and in particular it is evident that a number of deep faults which cut basement and the lower part of the reservoir do not extend up to the top of the reservoir. Finally, for the first time, intra-reservoir events can be seen and followed, at least locally, on the 3D data. The combination of intra-reservoir mapping and enhanced fault interpretation helped explain some of the anomalous results of the earlier appraisal wells (e.g. the faults which cut Well 206/8-7 and had such a significant effect on the test results).

As well as conventional contour and fault maps, the interpretation of horizons as continuous surfaces allowed the generation of sophisticated image processed displays combining attributes such as dip, azimuth and illuminated two-way-time (McQuaid *et al.* 1991). These displays achieved a quantum leap in visualization of the Clair structure, enhancing and illustrating the different fault trends and the differences between the structure at top reservoir and at deeper levels.

They also highlighted a NNW–SSE-trending set of faults not previously identified. These are interpreted as late Cretaceous faults and as such represent the latest significant episode of faulting to affect the Clair structure. Their importance in terms of fractures and productivity is discussed below.

1991 well results

Results of 206/7a-2 ridge horizontal well

Well 206/7a-2, drilled by Elf was located within 250 m of 206/7-1 (Fig. 16) and oriented to encounter a number of fracture zones predicted by the studies and identified on the 3D seismic. The well was drilled towards the northwest with a deviated profile crossing through the red beds at 50–70° inclination before being drilled horizontally in the basement (Fig. 17).

The well was drilled 600 m into the basement with the inclination increasing from 70–90° at mid-section. Seven cores taken in the basement section were fractured and bled oil. However, most of the fractures were filled with calcite and few open fractures were present. Logs indicated that the well had encountered several fracture corridors. The well was tested in open-hole at a rate of 670 BOPD of 24° API oil to surface. The permeability-thickness product (KH) was estimated at 10 000 mD-feet, lower than the original basement well. The initial test response does not necessarily support a fracture corridor system.

Acid-stimulation of the well was performed and the flow rate increased to 2350 BOPD before declining to 2040 BOPD. Pressure analysis shows some fracture effect with linear flow in the early producing time. A second acidization resulted in only marginal further improvement with a final flow rate of 2100 BOPD. Post-acidization, production logging indicated that the flow came from several zones located along the horizontal section. KH was estimated at 14 100 mD-feet.

After suspension of the horizontal section of the well, 160 m of the overlying red beds were perforated to test the production in the generally poorly productive Upper Clair Group. Two flow rates of 750 BOPD and 295 BOPD were recorded at different choke settings (44/64" and 24/64", respectively).

The well is considered a technical success despite the lower rates than expected. The feasibility of drilling and coring a 600 m horizontal wellbore in granite-gneiss, encountering fracture corridors and demonstrating the drainage of the red beds through the fracture network were all proven.

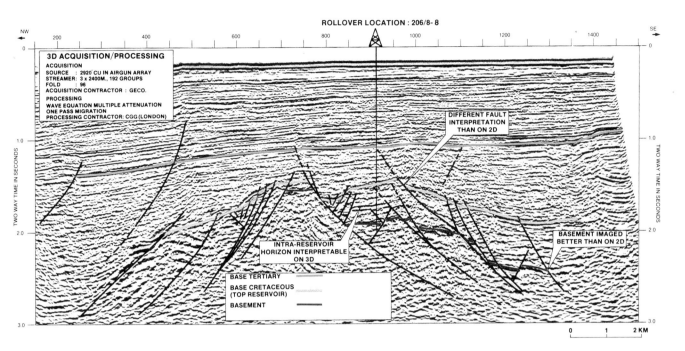

Fig. 14. Inline 263 from 1990 3D seismic survey.

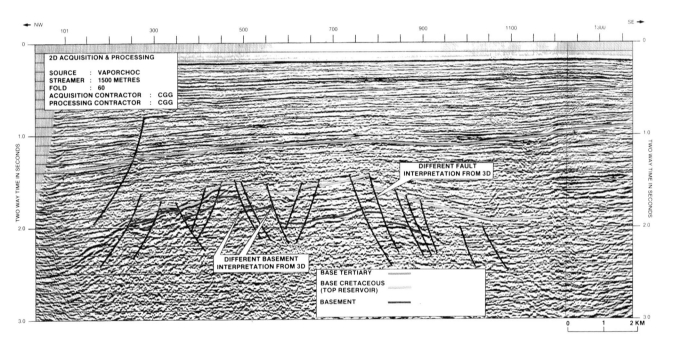

Fig. 15. Typical 2D seismic line.

Results of 206/8-8 roll-over well

Well 206/8-8, drilled by BP, was located on the northwest flank of the roll-over approximately 1 km northeast of the discovery well (206/8-1a) (Fig. 16). This location was chosen because it was in an area which appeared to be free from major faulting, with relatively strong and continuous reflections on the seismic throughout the reservoir sequence (Fig. 14).

The main target zones (Units V + VI) were well developed and oil bearing and, in addition, a significant thickness of Unit III was preserved above the oil–water contact.

Four drill stem tests were carried out (Fig. 18). The first test was on Unit III, but yielded only a disappointing 465 BOPD despite the fact that Unit III appears to have some of the best quality reservoir rock in the well.

The second test across Unit V was much more encouraging. A flow rate of 3100 BOPD (more than twice anything previously recorded on Clair) was achieved without stimulation and was limited only by the equipment on the rig. This test was extended to a six-day flow period and both rate and pressure were maintained throughout. Analysis of the test data indicates that there was a significant contribution from an open

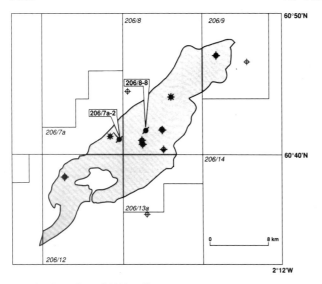

Fig. 16. Location of 1991 wells.

fracture system as well as from the matrix. KH is estimated at 500 000 mD-feet and the interpreted radius of investigation of the test is in excess of 750 m.

DST-3 on the lower part of Unit VI flowed at 1450 BOPD. This test was almost certainly accessing the same reservoir system as DST-2, with the lower rate reflecting the relatively poorer quality of the interval perforated.

The fourth test, of the upper part of Unit VI (above the Lacustrine Key Bed), flowed initially at 1850 BOPD. Following a massive hydraulic fracture the rate increased to 3250 BOPD and as with DST-2 the post-fracture test of DST-4 was extended to six days. Again both rate and pressure were maintained throughout. The interpreted KH (600 000 mD-feet) and radius of investigation (750+ m) are similar to those for DST-2.

There is no unique overall interpretation of the test data from 206/8-8 and it is difficult to estimate the full well potential. It is clearly possible that the massive hydraulic fracture carried out on the DST-4 interval propagated through the Lacustrine Key Bed and that DSTs 2, 3 and 4 (post-frac.) were

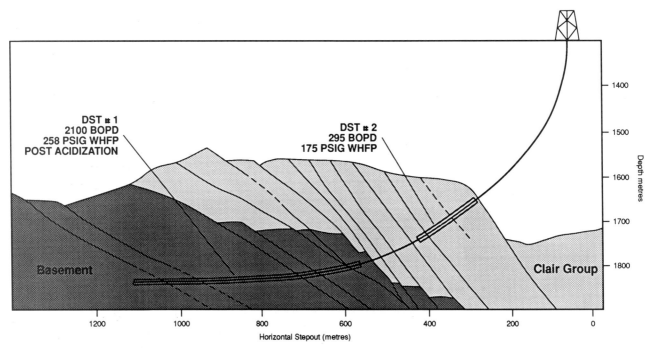

Fig. 17. Horizontal well results (206/7a-2).

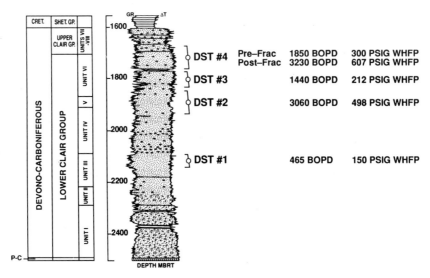

Fig. 18. Vertical well results (206/8-8).

all accessing and testing the same open fracture system, with the different rates reflecting differences in the fracture network or the matrix quality adjacent to the wellbore.

It is fundamentally clear, however, that the greatly improved well performance compared to previous results is due to the presence of a significant open natural fracture system in the Devonian–Carboniferous section which, although hinted at in some of the older wells, had not previously been proven.

Detailed analysis of the fracture system is possible because almost 800 m of orientated core were cut and recovered, on which over 3000 fractures were logged.

Analysis of these leads to three important observations. First, production logging (PLT) data indicate that both better quality matrix and open fractures are required for significant fluid flow. Second, although fractures occur throughout the Clair Group, open fractures are concentrated in the interval from the upper part of Unit IV to the basal part of Unit VII. This helps to explain the results of the test on Unit III, where better quality matrix, but without open fractures, yielded a disappointing flow rate.

The third important observation stems from analysis of the orientations determined for more than 900 of the observed fractures. The dominant trend is NE–SW, parallel to the main structural trend (Fig. 19a). However, the open fractures show a very different pattern with two principal trends, NNW–SSE and WNW–ESE (Fig. 19b). The NNW–SSE set, although less frequent, are the more significant because they are of larger scale (they are seen on the Circumferential Borehole Imaging Log (CBIL) as well as on the cores) and show greater gape. Intervals where this set of open fractures combines with better quality matrix are identified on the PLT log as the main flow intervals in the well. It is also noticeable that, although no major faults cut the wellbore, the critical set of fractures very closely aligns with, and is therefore by implication associated with, the Late Cretaceous faults trending in the same direction recognized for the first time on the 3D seismic. Further work is required to understand fully the inter-relationship of the various fracture sets. However, there does appear to be a direct link between the timing and preservation of the NNW–SSE fracture set and the generation and emplacement of hydrocarbon at the end of the Cretaceous.

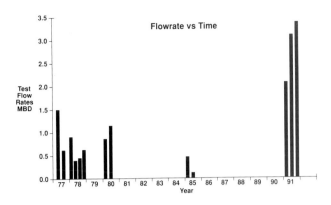

Fig. 20. Comparison between 1991 and historical well test results.

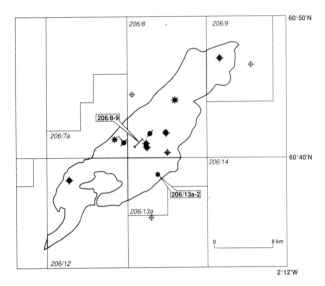

Fig. 21. Location of 1992 wells.

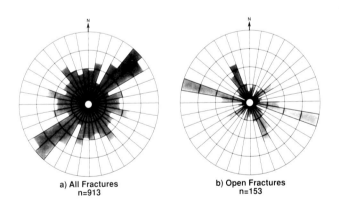

a) All Fractures
n=913

b) Open Fractures
n=153

Fig. 19. Orientation of fractures logged in Well 206/8-8.

Future programme

The test results from the 1991 wells significantly exceeded those from the early appraisal wells (Fig. 20). Well 206/8-8 in particular had shown not only that significant production rates were achievable but also that they could be sustained. Attention has now focused on the question of repeatability. Two further appraisal wells are planned for 1992, the overall objective of both being to demonstrate that the success of 206/8-8 can be repeated (Fig. 21).

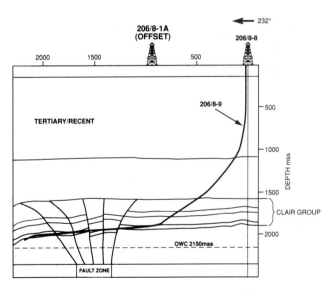

Fig. 22. Sketch section of 1992 core area well (206/8-9).

The reservoir modelling studies had concluded that a high-angle well could be the most productive well type if an open fracture system of known orientation was present in the Devonian–Carboniferous sequence. Well 206/8-8 had proven the presence of such a system and had given clear evidence of its orientation. The first of the 1992 wells is, therefore, planned as a high-angle well in the 'core' area of the discovery (i.e. the area tested by 206/8-8 and by 206/8-1a) targeted at Unit V of the Clair Group and oriented normal to the NNW–SSE-trending open fracture system identified in 206/8-8 (Fig. 22). A 1300 m high-angle (85°) section should allow the wellbore to penetrate at least three fault compartments and will test both the efficiency of a high-angle well completion and the model of open fractures associated with NNW–SSE-trending faults.

The second of the 1992 wells is aimed at demonstrating that the success of 206/8-8 can be repeated outwith the core area. It is to be targeted at a horst feature on the southeastern flank of the structure in Block 206/13a where the seismic character most closely resembles that in the 206/8-8 area. Because this horst feature has not been drilled before, and therefore the intra-reservoir stratigraphy is defined only by seismic correlation, this well is planned as a vertical well with provision for hydraulic fracture stimulation if required.

In addition to the two wells in 1992, an extension to the 3D seismic is planned covering the whole of the northeastern part of the discovery, from the edge of the 1990 survey to the northeastern end of the overall structure as currently mapped.

Conclusions

Fourteen years after the initial discovery, the 1991 wells drilled as part of the joint appraisal programme have finally demonstrated that significant and potentially commercial test production rates are achievable.

This technical success has stemmed from the application and the integration of 3D seismic, reservoir studies, innovative simulation modelling and modern drilling technology. These have all been described in this paper.

More fundamental than any of these technical achievements is the behavioural change that has taken place over the last three years. The spirit of openness, commitment and co-operation which now typifies the Clair partnership has been the catalyst for the technical success which we have enjoyed.

While it is very tempting to bask in the warmth of technical success, this does need to be put in context. One successful well does not make Clair a commercial oil field, and there are no guarantees that Clair ever will be a commercial oil field. What

has been achieved is to create a possibility where none existed. The challenge now facing the Clair Group is to realize the possibility, and the 1992 programme is the next step towards this.

Our thanks go to the management of all eight companies involved in Clair: Amoco, BP, Chevron, Conoco, Elf, Enterprise, Esso and Mobil for their support and permission to publish. The vast majority of the work is documented in unpublished reports compiled by individuals working within, and for, the Clair Group of companies. Notable among those individuals are Tim Bevan, Kes Heffer, Phil Hirst, Adrian Pearce and Richard Fox (of BP) for their work on the roll-over studies and the red beds fracture analysis, and Dominic Monfrin, Nick Birnie, Pierre Doucet and Gerard Massonat (of Elf) who carried out the structural analysis and the simulation modelling of the basement. Finally, and most importantly, we acknowledge the contributions made by the many individuals who have represented their companies on the Clair Reservoir Work Group and who have helped create an unusual and stimulating forum for technical discussion.

References

ALLEN, P. A. AND MANGE-RAJETZKY, M. A. 1992. Devonian–Carboniferous Sedimentary Evolution of the Clair Area, Offshore North-Western UK: Impact of Changing Provenance. *Marine and Petroleum Geology*, **9**, 29–52.

JOURNEL, A. G. AND ALABERT, F. G. 1988. Focusing on Spatial Connectivity of Extreme Valued Attributes: Stochastic Indicator Models of Reservoir Heterogeneities. SPE 18324. *In: Proceedings of the 63rd Technical Conference and Exhibition of the Society of Petroleum Engineers*. Houston, Texas, 621–632.

McQUAID, A. S. J., BROADLY, J. AND NOVAK, M. 1991. Surely that's not Seismic! The Benefits of Image Processing. *In: 61st Annual International Meeting of the Society of Exploration Geophysicists, Expanded Abstracts*. Society of Exploration Geophysicists, 319–321.

STEWART, R. C. S., HUYGHUES-DESPOINTES, T. AND HENDRY, W. J. 1991. The Impact of Modern Interpretation Techniques on the Evaluation of the Clair Discovery, West of Shetlands, UKCS. *In: 61st Annual International Meeting of the Society of Exploration Geophysicists, Expanded Abstracts*. Society of Exploration Geophysicists, 268–270.

Prediction of large-scale communication in the Smørbukk fields from strontium fingerprinting

H-H. STØLUM,[1] P. C. SMALLEY[2] and N-M. HANKEN[3]

[1] *Department of Earth Sciences, University of Cambridge, Downing Street, Cambridge CB2 3EQ, UK*
[2] *Institute for Energy Technology (IFE), PO Box 40, N-2007 Kjeller, Norway (Present address: BP Research Centre, Chertsey Road, Sunbury-on-Thames, Middlesex TW16 7LN, UK)*
[3] *Institute of Biology and Geology, University of Tromsø, Dramsveien 201, N-9037 Tromsø, Norway*

Abstract: This paper discusses a new method which uses intra-field variability of formation water chemistry as a measure of degree of reservoir compartmentalization. If two reservoir units are in good flow communication, it is more likely that any water compositional variations that might have arisen would have homogenized by mixing processes (diffusion, dispersion, flow). On the other hand, a lack of flow communication is likely to inhibit water mixing and thus preserve variations in water compositions. Thus, whatever the *cause* of intra-field variations in water composition, two reservoir units that have a similar water chemistry are more likely to be in good flow communication than two units with different water compositions.

In this study the isotopic composition of strontium ($^{87}Sr/^{86}Sr$) dissolved in the formation water was used to monitor formation water compositional variability. This natural isotopic tracer is particularly useful as it can be measured simply from conventional core samples by extracting residual salts which have precipitated in the pore spaces as a result of formation water evaporation during storage.

The Smørbukk fields (Smørbukk North and Smørbukk South, Haltenbanken area, Norwegian Continental Shelf) comprise three Jurassic reservoir intervals, the Garn, Ile and Tilje formations. In Smørbukk North the main problems are vertical compartmentalization by areally extensive shales and lateral compartmentalization by extensive quartz cementation. In one well, residual salt analysis (RSA) shows that water $^{87}Sr/^{86}Sr$ compositions are extremely homogeneous in each of the Tilje 1 and 2 reservoir zones, indicating good internal vertical communication. However, the Tilje 1 and 2 waters are significantly different from each other, indicating that the shale that separates Tilje 1 and 2 is an effective barrier to vertical communication. A similar relation is seen in a second well, suggesting that the shale barrier is indeed laterally extensive. In the second well, Tilje 1 RSA data are the same as the results for that zone in the first well, indicating excellent lateral communication between the two wells. In Tilje 2, however, there is a distinct between-well difference in formation water composition, indicating that lateral communication is poor in this zone.

In Smørbukk South, lateral connectivity is further complicated by faulting. Using RSA data from three wells it was possible to predict that:

- the Garn has good internal vertical communication, but is laterally divided into two isolated areas, probably the result of a sealing fault;
- the Ile forms three laterally isolated compartments;
- the Tilje is subdivided vertically into three flow units that are interpreted to have poor communication with each other in two of the wells. Laterally, Tilje 1 and 2 form two flow compartments, probably due to faulting.

One of the most important unknowns in the prediction of reservoir behaviour at the appraisal stage is reservoir compartmentalization. This includes vertical compartmentalization (e.g. restriction of vertical fluid flow by sub-horizontal barriers such as shales) and lateral compartmentalization (e.g. due to the presence of sealing faults). Some information on reservoir connectivity can be gained, for example, from pressure testing, but all too often 'hard' data on connectivity only come to light when problems arise during production. Furthermore, reservoir simulation models often rely more on 'soft' data, for example stochastic models of shale distribution. This paper deals with a new way of estimating reservoir connectivity, which can begin to provide data as soon as core samples are available.

The new method of connectivity estimation is applied here to the two Smørbukk fields. They are both in an appraisal stage, with relatively few wells drilled. To what extent are reservoir zones continuous between wells, and what will be the effect of shales seen in wells on vertical fluid flow during production? The present work outlines how a quantitative method based on reservoir fluid analysis can predict flow architecture.

The Smørbukk fields

The two Smørbukk fields (North and South) are situated in the Haltenbanken region, offshore mid-Norway (Fig. 1). The Smørbukk South Field is highly complex and compartmentalized by faulting and both fields are affected by extensive diagenesis due to deep burial (Bjørlykke *et al.* 1986; Ehrenberg 1990). They contain mainly gas and condensate.

The Smørbukk North Field is an elongate structural trap caused by block faulting. It is deeply buried (*c.* 3800–4700 m), and strongly affected by quartz cementation. Three potentially productive formations consist of middle to upper Jurassic sandstones. The upper formation (Garn) has the best reservoir properties, while the lowest formation (Tilje) is most severely affected by cementation diagenesis, and is also the most sedimentologically complex. The formations have different fluid contacts and pressure regimes. There are currently only three wells drilled on the structure and no 3D seismic. Thus very little is known about the degree of faulting in the field.

The strontium isotope data are from the two cored wells, 6506/12-1 and 6506/12-6 (henceforth referred to as wells 1 and 6). The spacing between them is 1.2 km. A cross-section of the field showing the two wells is given in Fig. 2.

From *Petroleum Geology of Northwest Europe: Proceedings of the 4th Conference* (edited by J. R. Parker).

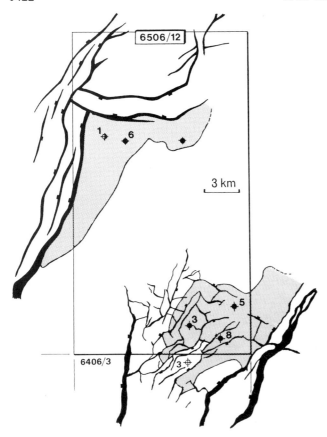

Fig. 1. Map of the two Smørbukk fields. Seismic interpretation from 2D data in the area around Smørbukk North, and from 3D data over Smørbukk South. Outlines of the fields correspond to the Garn Formation contact in Smørbukk South, and deepest contact in Smørbukk North. See Aasheim *et al.* (1986) for regional setting.

The Smørbukk South Field is a domal structure bounded by a major fault to the east. A 3D seismic survey has revealed a highly complex fault pattern over the structure (Figs 3 and 4). Four wells have been drilled, one of which is dry, proving that one of the largest faults on the structure is sealing. The complex fault pattern, and the fact that at least one fault is sealing, make it reasonable to attribute lack of lateral communication primarily to faults, and secondarily to cementation and possible shale-outs. Burial depth is similar to that of Smørbukk North. Quartz cementation is extensive. Stratigraphically and lithologically the two fields are similar, except that the Ror sand is not found in Smørbukk South, where the Ror is a shale sequence.

The strontium isotope data come from three wells, 6506/12-3, -5, and -8 (referred to as wells 3, 5 and 8). The spacing between 3 and 8, and also between 5 and 8, is about 3 km, whereas wells 3 and 5 are separated by about 5 km (see Fig. 1). The areal extent of each of the potential reservoir units is outlined in Figs 3 and 4.

Characterization of formation waters using Sr isotopes

Strontium is composed of four natural isotopes, with atomic masses 88, 87, 86 and 84. Of these, ^{87}Sr is radiogenic, the daughter product of natural decay of the radioactive isotope ^{87}Rb. The abundance of ^{87}Sr in a substance is conventionally expressed as a ratio to the stable non-radiogenic isotope ^{86}Sr. The $^{87}Sr/^{86}Sr$ ratio of strontium dissolved in formation waters in marine sediments is determined by the interplay of two main sources of Sr: seawater (and marine aragonite) and silicate minerals. The Sr isotope ratio of the world's oceans has varied

with time; in the late Jurassic (the age of the studied formations in the Smørbukk fields), it was 0.707 ± 0.001 (Burke *et al.* 1982; Smalley *et al.* 1988). Diagenetic dissolution of carbonate shells in Smørbukk would thus yield strontium with this isotopic signature. However, dissolution of silicates such as mica and feldspar grains is likely to increase the Sr isotope ratio in the water. In general, the $^{87}Sr/^{86}Sr$ ratio of these minerals is high (>0.730; e.g. Burtner 1987; Chaudhuri *et al.* 1987). This is because detrital feldspars and micas usually have high Rb/Sr ratios and, because they are usually old compared to the age of the sediment, they have had time to evolve radiogenic (^{87}Sr-rich) isotopic compositions. Another possible contribution to isotopic heterogeneity of pore water is formation of authigenic clay minerals, which could selectively incorporate heavier or lighter strontium than that of the water. Burtner (1987) and Chaudhuri *et al.* (1987) found Sr ratios varying between 0.707 and 0.730 for a number of oil fields. However, the wide range of possible $^{87}Sr/^{86}Sr$ ratios of minerals in reservoir sandstones can lead to isotopic heterogeneity of waters *within* a reservoir due to local water-rock interaction.

Consider a reservoir compartmentalized by faults or lithological variations into isolated sandbodies or flow units (Fig. 5). Between two separate bodies of formation water, there will often be subtle differences in host sediment mineral composition and diagenetic history, and so they will evolve different $^{87}Sr/^{86}Sr$ signatures. Within each isolated body of formation water, the $^{87}Sr/^{86}Sr$ ratio of dissolved Sr will be homogenized by tracer diffusion, and possibly slow dispersion or fluid flow (Bjørlykke *et al.* 1988). However, homogenization between the two water bodies will be inhibited if the intervening rocks are efficiently tight. In other words, similar $^{87}Sr/^{86}Sr$ values in formation water between wells could be an indication of lateral connectivity of the flow units. Uniform $^{87}Sr/^{86}Sr$ values in different formations within a well would similarly indicate vertical communication between flow units. Conversely, contrasting values would indicate that there are barriers between the wells or between the flow units. Thus the strontium isotopic signature can be used to create a model of flow properties within and between wells. The empirical basis for this method is the fact that strontium values exhibit a strong natural tendency to clustering (Fig. 6). This in turn gives rise to multimodal distributions of large datasets (Fig. 7).

Indications that formation waters have homogenized over geological time do not necessarily imply that over the more limited production lifetime of a field, the formation will behave as one unit. Also efficient barriers over geological time are barriers to diffusion and flow under low pressure gradients. Some barriers of this type may break down due to the larger pressure differences arising during production. Only a systematic study of predictions based on exploration and appraisal wells compared to actual production data from the same fields can give the necessary empirical basis for a general method which will overcome this problem.

One of the main reasons for using Sr isotopes as a tracer for formation water composition is that water $^{87}Sr/^{86}Sr$ can be measured easily from core samples, obviating the need to use expensive (and thus rarely available) DST samples (pressure test data). When a core is drilled, some of the original formation water will be captured inside the core in the sediment pores. As the cores dries up, the water evaporates, while the salts that were dissolved in the pore water precipitate on grain surfaces. These salts contain strontium with the same $^{87}Sr/^{86}Sr$ as the original formation water. This Sr can be retrieved by selecting a few cm^3 of the core, crushing it and performing a gentle extraction with ultrapure water. The $^{87}Sr/^{86}Sr$ ratio can be measured by modern mass spectrometric methods to a standard deviation of $\pm c. 0.00001$. We have termed this method residual salt analysis (RSA). A detailed description of the RSA technique has been given by Smalley *et al.* 1992.

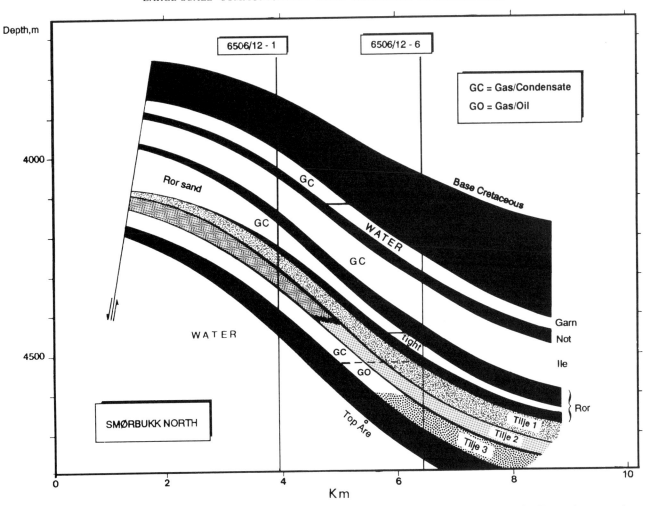

Fig. 2. Cross-section of the Smørbukk North Field, showing the two wells studied. Black zones are non-reservoir units. Reservoir communication in the Tilje Formation is based on $^{87}Sr/^{86}Sr$ data.

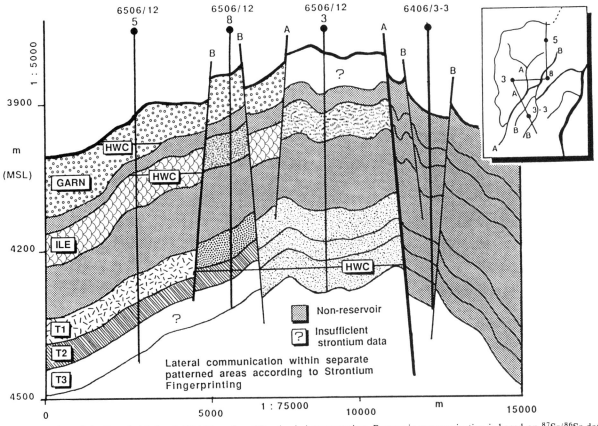

Fig. 3. Cross-section of the Smørbukk South Field based on 3D seismic interpretation. Reservoir communication is based on $^{87}Sr/^{86}Sr$ data.

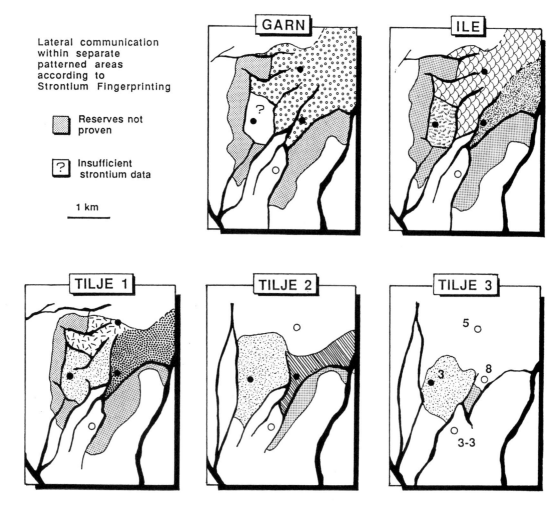

Fig. 4. The main reservoir zones of the Smørbukk South Field in areal view, showing reservoir communication indicated by $^{87}Sr/^{86}Sr$ data. Fault patterns are based on interpretation of 3D seismic.

Potential sources of sampling error

The main potential source of sampling error is contamination by drilling fluids. This danger can be reduced by:

1. avoiding sampling from highly permeable sediments, in which flushing with drilling fluids would have penetrated most deeply into the core. Samples should be taken from well-cemented very fine- to medium-grained sand;
2. sampling only from the central part of the core. Empirical evidence (Johannesen 1989) suggests that Sr isotopic contamination from drilling fluids is often limited to the outer 2–3 cm of the core, and so the problem is minimized by sampling only from the core centre (Fig. 8). Johannesen's data, from the nearby Heidrun Field, are from virtually unconsolidated sands that are more porous and 1–2 orders of magnitude more permeable than the Smørbukk rocks due to a much shallower burial depth. As seen from Fig. 8, even in this extreme case, there is still an uncontaminated zone in the centre of the core. In Smørbukk, the depth of invasion should be much less than that seen in Fig. 8, and consequently is not likely to be a problem. This conclusion is corroborated by other datasets analysed in a similar way. Further, core slabbing and sectioning is carried out using kerosene (low Sr content) as a lubricant which minimizes strontium contamination at this stage. The analysed samples have been obtained by dry sampling;
3. utilizing cores that have been drilled by new low-invasion coring techniques.

Another potential source of error is contamination during RSA. However, the leaching of residual pore salts is carried out using the smallest possible volume of ultrapure water, approximating to the pore volume of the rock sample being analysed. Further, the leaching is carried out quickly (<5 min.). These precautions are designed to minimize any contribution of Sr from mineral grains in the rock itself. This does not appear to be a problem. For example, samples of chalk leached with different amounts of water for various durations showed no contribution of Sr from dissolving carbonate (Smalley *et al.* 1992).

A final potential source of uncertainty, which hampers most studies involving Sr ratios, is uncertainty in data interpretation when there are insufficient data points to distinguish between intra- and inter-sandbody variation. This problem is general to all statistical sampling, and is avoided by using a properly stratified sampling technique and a suitable number of samples.

Potential sources of prediction error

There are two potential sources of prediction error.

1. False negatives, i.e. where single fluid flow units give heterogeneous water compositions. This can occur either (a) where diagenetic water–rock interaction has occurred so recently that the fluid heterogeneities have not yet had a chance to homogenize; or (b) where distances between samples are so great that the waters have not yet had a

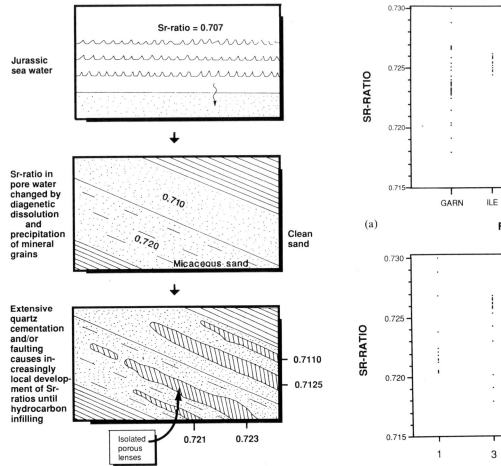

Fig. 5. Schematic sequence showing the origin of $^{87}Sr/^{86}Sr$ clusters in formation waters.

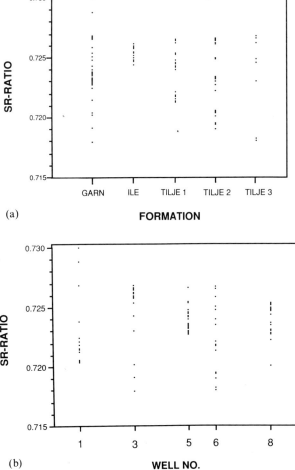

Fig. 6. Natural tendency to clustering of the $^{87}Sr/^{86}Sr$ data as seen in plots against (**a**) formations and (**b**) wells.

chance to homogenize over such a distance. The latter problem is particularly relevant to examination of between-well variations of water composition. These effects can be recognized by the presence of a relationship between water composition and distance from a reference point. Interpretation of such data can be clarified by integration with other types of data (e.g. pressure data).

A more serious source of prediction error is the fact that some barriers to diffusion and slow flow may break up during production. This paper describes an early version of the method which is not able to discriminate between weak barriers and strong barriers likely to be unaffected by production. This inaccuracy is due to lack of calibration of the statistics by a learning dataset. It is therefore likely to give a biased estimate, seeing the reservoir as more compartmentalized than it actually is. In later versions of the method, this bias is corrected by a calibration of the statistical procedure to empirical knowledge about weak and strong barriers from interference tests and production data in several North Sea fields. This information is not released for publication, however, and so we are not able to describe the current version of the method in this paper.

2. False positives, i.e. different flow units coincidentally giving the same $^{87}Sr/^{86}Sr$ value. This could occur where different sandbodies have very similar mineralogies and/or diagenetic histories. Again, other supporting data are needed to facilitate interpretation. In general, water compositional differences are more easily interpreted than similarities.

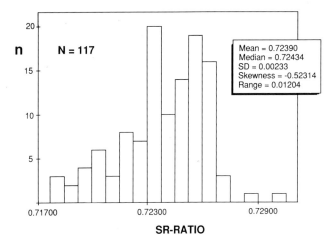

Fig. 7. Histogram of the $^{87}Sr/^{86}Sr$ dataset.

Cluster analysis of strontium isotopic data

Ideally, formation water $^{87}Sr/^{86}Sr$ data should be interpreted with a full knowledge of the mechanisms giving rise to spatial compositional heterogeneities and the processes causing subsequent homogenization. Unfortunately, little is yet known quantitatively about these processes. One aspect in particular that is not yet clear is to what degree water compositions may continue to evolve after petroleum emplacement. In order to

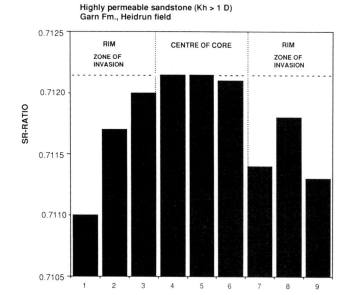

Fig. 8. $^{87}Sr/^{86}Sr$ values from nine samples taken across a core. Notice the plateau of values in the centre of the core, indicating absence of contamination by drilling fluids (from Johannesen 1989). Horizontal axis scaled in centimetres.

avoid subjective and arbitrary interpretations which could change as the understanding of these processes becomes more complete, this paper attempts to use an objective statistical (cluster analysis) approach that is process-independent and standardizes the analysis of the data.

In cluster analysis, the data are treated as sampling a set of discrete populations. The data are placed into a matrix, with the following axes:

- the $^{87}Sr/^{86}Sr$ scale (continuous);
- the wells (discrete);
- the reservoir zones, formations, or potential flow units spanned by the data (discrete).

The $^{87}Sr/^{86}Sr$ values from a given dataset are plotted in terms of well and reservoir unit along an axis of Sr ratios (Figs 9a and b). In comparing several populations, defined by well and reservoir unit, each is displayed along a separate, parallel axis, but with all of the axes spanning the same Sr-ratio interval. The choice of reservoir units depends on the specific hypotheses to be tested.

If the Sr-ratio population represents a single body of formation water, the values will be very similar, and form one tight cluster. If the population comprises more than one isolated flow unit, the data will be more scattered, and will then be structured in clusters along the $^{87}Sr/^{86}Sr$ axis, each cluster

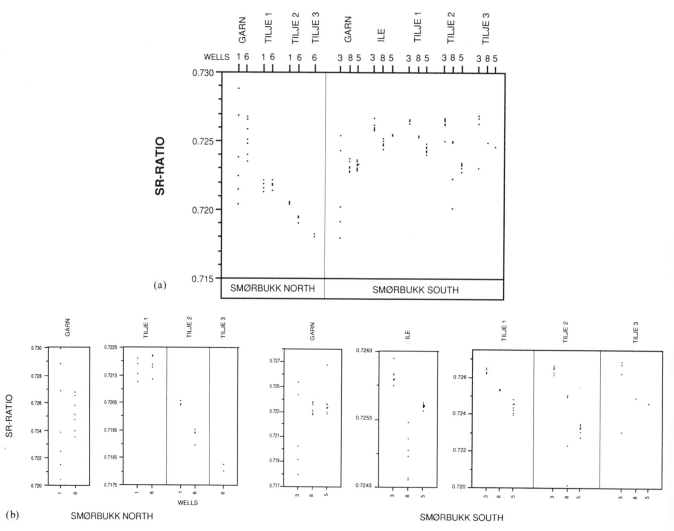

Fig. 9. The $^{87}Sr/^{86}Sr$ dataset for wells from Smørbukk North and South.

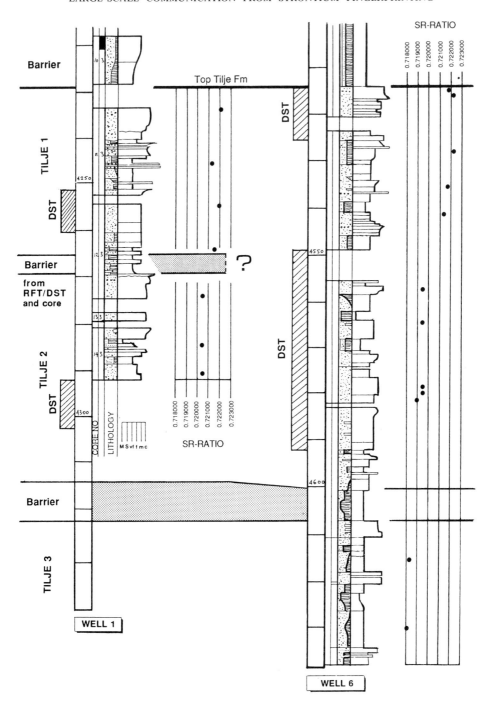

Fig. 10. Core description of the Tilje Formation in Well 6506/12-1 and -6, Smørbukk North Field, showing the three reservoir zones Tilje 1, 2, and 3. Data obtained from well tests (DST and RFT), core and log data, and $^{87}Sr/^{86}Sr$ values. The figure gives depth (m) after core shifts.

representing a separate potential flow unit. In comparing such populations from different reservoir units or wells, there might be some overlap, but provided that the samples have been taken at fairly equal intervals along the core, the medians of the constituent clusters should differ if the populations represent separate geological units. In order to obtain a consistent and standardized evaluation of within-cluster variation and between-cluster separation, a statistical method was used.

Laboratory measurements of $^{87}Sr/^{86}Sr$ ratios were performed by IFE (Institute for Energy Technology), Norway. The data from the two fields consist of 117 measurements. From Smørbukk North, 20 samples were selected from the Tilje Formation and 14 from the Garn Formation. From Smørbukk South there are 39 samples from the Tilje Forma-

tion, 22 from the Ile Formation and 22 from the Garn Formation. As seen from the histogram (Fig. 7), they do not follow a simple distribution. When plotted against wells or formations, however, the data show a marked tendency to clustering (Fig. 6). The objective of the statistical analysis is to quantify this multimodal distribution, by identifying clusters in a standardized way. We tested for clustering by applying a cluster strength statistic, Sc. The identified clusters were then compared using two cluster separation statistics. The method is non-parametric and we developed it for optimal efficiency in analysing this type of data.

Statistics were designed to respond strongly to actual flow barriers, and to calibrate the cut-off values of the parameters according to empirical knowledge. We are unfortunately not

Table 1. Mann-Whitney tests of pairs of clusters

Cluster 1	Cluster 2	T1	T2		
8-Ile	5-Ile	28	84	T < 30	Significant at 0.1% level
3-Ile	8-Ile	77	28	T < 29	Significant at 0.1% level
3-Ile	5-Ile	84	28	T < 30	Significant at 0.1% level
8-Tilje 1	8-Ile	27	6	T = 6	Significant at 1% level
3-Tilje 1	3-Ile	50	15	T < 18	Significant at 1% level
8-Tilje 1	5-Tilje 1	27	6	T = 6	Significant at 1% level
8-Tilje 1	3-Tilje 1	6	21	T = 6	Significant at 2.5% level
1-Tilje 2	6-Tilje 2	6	24	T < 7	Significant at 2.5% level
1-Tilje 2	1-Tilje 2	6	24	T = 6	Significant at 5% level
6-Tilje 1	1-Tilje 1	17	27	T > 13	Null hypothesis accepted at 10% level
6-Tilje 3	6-Tilje 2	3	21		Cluster 1 too small
8-Garn	5-Garn	62	82	T > 54	Null hypothesis accepted at 10% level
3-Tilje 1	3-Tilje 2	28	32	T > 20	Null hypothesis accepted at 10% level
3-Tilje 3	3-Tilje 2	21	23	T > 13	Null hypothesis accepted at 10% level
3-Tilje 3	3-Tilje 1	21	19	T > 13	Null hypothesis accepted at 10% level

Significant: Populations sampled from significantly different populations
Null hypothesis: Populations sampled from identical populations, i.e. most probably the same population

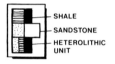
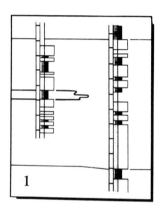
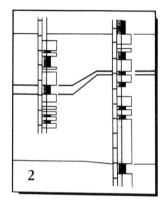
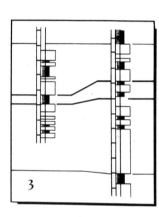
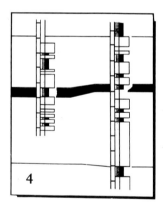
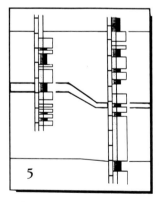
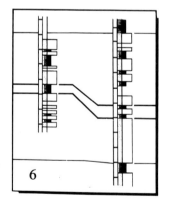
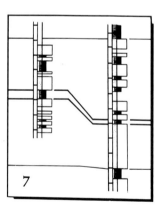

Fig. 11. Possible correlations of the barrier between Tilje 1 and 2 in Wells 6506/12-1 and -6. Core diagrams are schematic versions of Fig. 10.

yet able to publish this empirical information. Therefore, instead of describing this quantitative method in detail, we will address the results in qualitative terms only.

Another relevant question concerns appropriate sample size. How many samples are necessary for practical application of the strontium fingerprinting method? We analysed the data by the Mann-Whitney test, which is a standard non-parametric test designed to give confidence levels for comparisons of small samples (e.g. Conover 1971). This test is far too crude to be useful in practical applications of the method as a predictive tool, but does show that cluster sizes of less than ten values tend to be too small to yield statistically significant compari-

sons of clusters with overlapping data (Table 1). This suggests a two-stage sampling scheme.

Reservoir communication in the Tilje Formation

Description

The Tilje Formation is a lithologically complex sequence of sandy and heterolithic shaly units (Fig. 11). The sandy units, representing a wide range of grain sizes, sorting and sedimentary structures, have porosities generally between 10 and 15%.

There is significant quartz cementation. Permeabilities average around 0.1 mD with the exception of occasional high permeability streaks (approx. 100 mD).

Smørbukk North

The flow unit architecture is the outcome of the interaction between the facies architecture, extensive diagenesis and faulting. Based on DST data, RFT data, core descriptions and logs, the Tilje Formation is inferred to consist of three major (composite) flow units, corresponding to reservoir zones. Unfortunately, details of the test results and interpretations have not been made available for publication, but an overview has been published by Ehrenberg et al. (1992). In Tilje, due to the low average permeability, there is a nested hierarchy of flow units, as large segments within the zones are effectively tight. The highest level is the formation itself, then each of the reservoir zones, then the high-permeability streaks within the zones.

Three reservoir zones were demonstrated in well 1: Tilje 1, 2 and 3. They are separated by two potential flow barriers, corresponding to shaly and cemented intervals. One barrier was also demonstrated in well 6 and correlated to the lower barrier in well 1 on the basis of gamma and neutron-density log signature. In well 6, tests were run over a very large interval, and the test data are ambiguous. The observations are summarized in Fig. 10.

The main questions that need to be answered are thus:

1. Is the upper barrier (between Tilje 1 and 2 in well 1) continuous between well 1 and 6, and if so, which of the shaly intervals in well 6 does it correlate with?
2. Do the reservoir zones form continuous flow units between the wells, or have they been compartmentalized laterally by small sealing faults, or intervening lithological variations (shale-outs, diagenetic effects)?

The strontium dataset is displayed in Figs 9 and 10. The ^{87}Sr/^{86}Sr data for Tilje 1 and 2 in well 1 form two distinct clusters with a large gap between them, thus supporting the test data in indicating that these flow units are separated by the intervening shaly barrier. Well 1 was not cored across Tilje 3. However,

strontium data from well 6 clearly show that the samples below the lower shale in that well form a separate cluster, significantly distant from the range of values given by the samples above the shale. It can thus be predicted that the lower shale is a barrier to vertical fluid flow, and accordingly that in well 6, Tilje 2 and 3 form separate flow units. The ^{87}Sr/^{86}Sr clusters thus successfully predict the geometry established from independent evidence.

In the case of Tilje 1 and 2 in well 6, pressure data are ambiguous. It is not clear whether these zones represent separate flow units, and if so whether they are in flow communication with Tilje 1 and 2 in well 1. There are seven possible geometries for the upper shale (Fig. 11). The ^{87}Sr/^{86}Sr data from the Tilje 1 and 2 interval in well 6 fall in two very distinct clusters. This reveals that there is an intervening barrier to vertical fluid communication, i.e. that the shale is a barrier. The narrow ^{87}Sr/^{86}Sr ranges in the two reservoir zones indicate that they each have good internal vertical communication. This suggests that case 4 is the most likely geometry (Fig. 11).

In the Tilje 1 the ^{87}Sr/^{86}Sr data from both wells have the same range of values, indicating that the zone is continuous between the wells. In Tilje 2 the values from each well form separate clusters, indicating that the formation waters are not in complete lateral communication and suggesting that the Tilje 2 zone may thus not be laterally continuous between the wells. Tilje 2 has poorer reservoir quality than Tilje 1, to the extent that even faults below seismic resolution could form barriers in Tilje 2, whereas this is much less likely in Tilje 1.

A test of the method with pressure data

Statoil (the operator) has published an extensive description of the Smørbukk North Field, summarizing the DST results, and including an interpretation of the pressure communication from DST/RFT data (Ehrenberg et al. 1992). Their interpretation of the pressure data was carried out independently of this study, and we were not aware of this interpretation until after the strontium fingerprinting analysis had been completed. This interpretation is reproduced in Fig. 12, where it is compared to the result of the strontium fingerprinting analysis. The figure

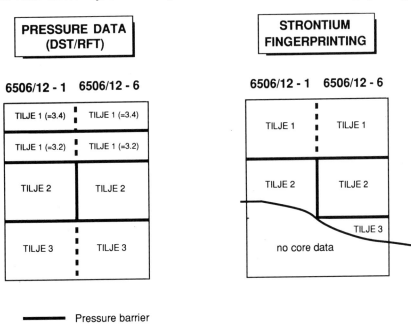

Fig. 12. Comparison of the strontium fingerprinting result with Statoil's interpretation of pressure communication in Tilje Formation, Well 6506/12-1 and 6506/12-6 (Ehrenberg et al. 1992).

shows that the strontium analysis from the cored intervals matches the pressure data interpretation perfectly, except for the subdivision of Tilje (Statoil's Tilje 3.2 and 3.4). This discrepancy is due to the distribution of strontium data, which does not allow conclusions regarding the internal structure of Tilje 1. When the analysis was done, the possibility that a barrier existed within Tilje 1 was not included among the hypotheses to be tested by the fingerprinting method. Thus, Tilje 1 was sampled only as a whole and there is not a sufficient number of samples above and below the potential barrier to form significant clusters (Tilje 3.2 is only represented by two samples in each well, and 3.4 by two and three samples, respectively). If an extensive sampling programme had been carried out in this field, the possibility of a barrier at this level would have been tested explicitly, since a shaly zone is apparent in the cores.

Smørbukk South

The three Tilje zones are found in all the wells. The following interpretation is summarized in Figs 3 and 4. Tilje 1 data from the three wells form distinct clusters separated by gaps, indicating that there is little lateral communication at the scale of the well spacing.

Tilje 2 data from wells 3 and 5 are tightly clustered and separated by a large gap, suggesting no communication. The data from well 8, however, do not form a single cluster. Four of the data points form a subcluster, clearly separated from the clusters of the two other wells, with two outliers, one above and one below this subcluster. The interpretation of this is uncertain, but certainly the values are far lower than those from well 3, indicating no communication between wells 3 and 8. Between wells 3 and 5 there may be restricted lateral communication, but no values from well 8 overlap with those from well 5. More data are needed from well 8 in order to clarify the situation, but wells 8 and 5 most probably are not in lateral communication.

Tilje 3 was only sampled in well 3, since the reservoir does not extend to the two other wells.

Vertically, the large overlap of the clusters in all three zones in well 3 indicates that there are no large-scale barriers to vertical flow in this well. In the other wells, separation of clusters indicates the presence of flow barriers between the zones.

Reservoir communication in the Ile Formation

Description

The Ile Formation is a clean to micaceous, well sorted, fine-grained sandy sequence with highly varying porosities and permeabilities (Fig. 13). It contains 6–7 zones, each a few metres thick, with permeabilities from 10–100 mD. Above and below are transitional intervals with 1–10 mD, while about one third of the formation has permeabilities below 1 mD. The more permeable zones comprise prograding sandbodies with minimal cementation, and could be expected to be laterally extensive.

Smørbukk South

In Smørbukk South, the Ile data from each well form well-defined clusters, with the cluster in well 5 being the strongest of the whole dataset. Thus even if the gaps between the clusters are small, they are still significant. The data suggest that there is probably no lateral communication in the Ile Formation between any of the wells (Figs 3 and 4), and that there are no definite barriers to vertical flow within the formation, at least on a geological time scale. The Sr data suggest that none of

the zones with good permeability is in fluid communication between wells. This indicates that lack of communication is probably due to the presence of sealing faults, rather than diagenetic or sedimentological factors. Due to the highly variable permeability profile within the Ile over vertical distance of a few metres, it is clear that even faults below seismic resolution are potentially sealing barriers in this zone. Compartments could be smaller than the inter-well distances, and this zone may be much less laterally continuous than the strontium data from only three wells are able to reveal.

There are presently no data from the Ile Formation in the Smørbukk North Field.

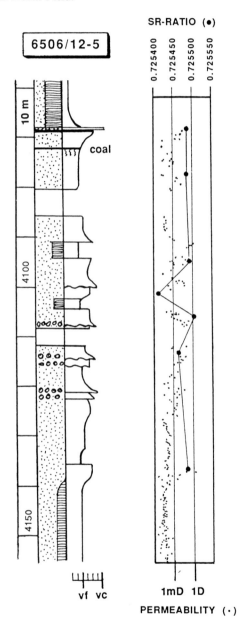

Fig. 13. Core description, permeabilities and ^{87}Sr/^{86}Sr values of the Ile Formation in Well 6506/12-5, Smørbukk South Field. Depth after core shifts.

Reservoir communication in the Garn Formation

Description

The Garn Formation is a clean, well-sorted sandy sequence with 10–15% porosity. Quartz cementation is pervasive (Ehrenberg 1990), as is authigenic illite (Ehrenberg and

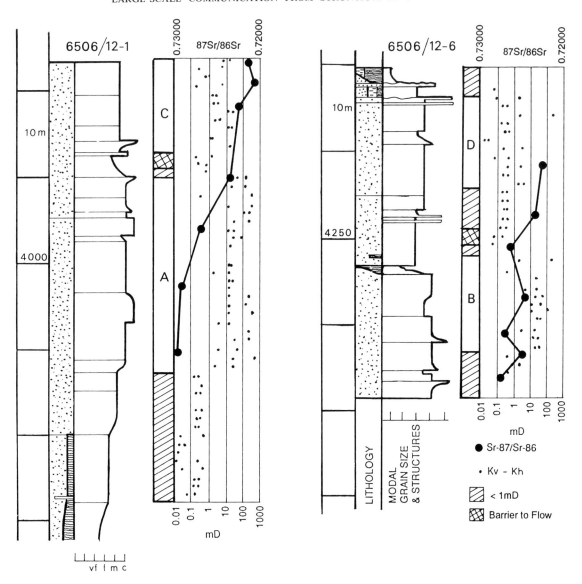

Fig. 14. Core descriptions of the Garn Formation in Wells 6506/12-1 and -6. Units A, B, C and D are four potential flow units observed in the cores. The figure gives depth after core shifts.

Nadeau 1989), and more localized, authigenic kaolinite (Ehrenberg 1991). Permeability reaches about 1 D in some thin zones, but on average it is of the order of 10 mD (Fig. 14).

In each well, the Garn Formation may be subdivided into three units. The upper and lower zones have good reservoir properties whereas the middle unit has low permeability. Thus, if laterally extensive, it might act as a barrier to fluid flow.

Smørbukk North

The lower zone (A and B in Fig. 14), has excellent reservoir properties, and is the main reservoir zone of the field. The upper zone (C and D in Fig. 14) consistently has permeabilities one order of magnitude below those of the lower unit, but is still productive. This raises the following questions:

1. Are the lower (A, B) and/or upper (C, D) zones continuous between wells?
2. Is the middle zone a barrier in any of the wells?
3. If the middle zone is a barrier in both wells, is it continuous between wells?

Unfortunately, the Sr-isotope ratios from well 1 do not form a distinct cluster, and so it is not possible to quantitatively compare the data from the two wells. However, the data from well 6 are within the range of the data from well 1, which indicates that there may be lateral communication within the formation, and so at least the lower zone may be interpreted as being continuous between the wells. In both wells, the data from the lower zone consistently have higher ^{87}Sr/^{86}Sr values than the data from the upper zone, indicating that the middle unit may form a barrier. However, the data also seem to indicate a gradual trend of decreasing Sr-ratio upwards in Garn. The middle unit is not associated with a significant discontinuity in the Sr-ratio profiles. The small discontinuity present in the Sr-ratio profiles, combined with lowered permeability at this level, give reason to think that the middle unit has formed a partial barrier to flow over geological time. In view of the similar overall trend of the data in each well, lower and upper Garn are still most likely to act as a single flow unit during progradation. This conclusion is in agreement with the pressure data interpretation carried out independently by Ehrenberg *et al.* (1992).

The third question cannot be answered on the basis of the present data alone. However, in both wells, the low permeable middle unit is associated with strongly bioturbated facies.

There is no compelling reason why this facies should be continuously distributed between the two wells, which are more than 1 km apart.

More data are needed in order to give firm answers to the three questions above. A two-stage statistical sampling programme in this case would most likely have given unambiguous results.

Smørbukk South

The data from wells 8 and 5 form well-defined clusters, which overlap completely, thus strongly suggesting good communication between these two wells. The data from well 3 do not cluster, and so cannot be evaluated using this method. A larger dataset is needed to evaluate the spread of values in this well. The present data could indicate that well 3 is not in lateral communication with the other wells, or that communication is restricted to the most permeable part of Garn (Figs 3 and 4).

Conclusions

1. Separate reservoir flow units may have discrete formation water compositions, reflected in discrete $^{87}Sr/^{86}Sr$ clusters. Statistical clusters of formation water $^{87}Sr/^{86}Sr$ ratios measured from pore salts can thus be used as an empirical tool to identify intra- and inter-well reservoir flow units and barriers.

2. In the present dataset of 117 $^{87}Sr/^{86}Sr$ ratios there was a natural tendency to clustering. In order to make use of this phenomenon in an objective way, clustering was quantified by means of a cluster strength statistic. Once geologically meaningful clusters were identified, the hypothesis that pairs of clusters were sampled from populations with similar properties (which would most likely mean the same population) was tested by the non-parametric Mann-Whitney test.

3. The method is a potentially useful supplement to more conventional DST, RFT, and interference test data. In the present study the $^{87}Sr/^{86}Sr$ clusters alone would have successfully predicted the vertical flow architecture suggested by test data, and gave additional information in areas where test data were ambiguous.

4. The method was used to test hypotheses about the reservoir geometry and potential flow units. In the case of the Tilje Formation in Smørbukk North the method was used in two steps: (a) to correlate a barrier from one well to another; and (b) to predict whether the potential flow units (Tilje 1 and 2) above and below that barrier would be in lateral communication between the two wells. The method suggests that the zone with the best reservoir properties (Tilje 1) is continuous between the wells, while Tilje 2 appears to be discontinuous.

5. In the Smørbukk South Field, the method was used similarly to generate a three-dimensional prediction of the flow units over the central part of the field based on three wells.

6. For two wells in the Garn Formation, the data were not clustered, and so the method failed to yield useful information about the sampled intervals. This could mean that there is no clustering tendency in those intervals, but it is more likely that lithological heterogeneity within the zones has divided them into small subclusters that would only show up with denser sampling. A two-step sampling procedure would avoid this problem: first, select samples from all cored reservoir zones in all wells, then select more samples from zones which failed to yield clusters in the first pass.

7. The $^{87}Sr/^{86}Sr$ cluster method as described here is likely to lead to biased predictions of compartmentalization due to inability to discriminate between weak and strong barriers. It is therefore not an accurate tool for evaluating reservoir architecture. This problem has been solved in a new, quantitative version of the method, based on empirical tests in developed fields.

We thank the Norwegian Petroleum Directorate for giving permission to publish data from their studies during 1989-90. We also thank Liv Johannesen and Conoco Norway Inc. for permission to publish Fig. 8. Per-Erik Øverli, Elisabeth Tuen, Sten Boye Flood and Erling Skagseth all provided valuable contributions in early phases of the study. The paper has benefited significantly from the reviews by John Martin, James Boles and an anonymous reviewer.

References

AASHEIM, S. M., DALLAND, A., NETLAND, A. AND THON, A. 1986. The Smørbukk gas/condensate discovery, Haltenbanken. In: SPENCER, A. M. (ed.) Habitat of Hydrocarbons on the Norwegian Continental Shelf. Graham & Trotman, London, 299–306.

BJØRLYKKE, K., AAGAARD, P., DYPVIK, H., HASTINGS, D. S. AND HARPER, A. S. 1986. Diagenesis and reservoir properties of Jurassic sandstones from the Haltenbanken area, offshore mid Norway. In: SPENCER, A. M. (ed.) Petroleum Geology of the Northern European Margin. Graham & Trotman, London.

——, MOE, A. AND PALM, E. 1988. Modelling of thermal convection in sedimentary basins and its relevance to diagenetic reactions. Marine and Petroleum Geology, 5, 338–350.

BURKE, W. H., DENISON, R. E., HETHERINGTON, E. A., KOEPNICK, R. B., NELSON, H. F. AND OTTO, J. B. 1982. Variation of seawater $^{87}Sr/^{86}Sr$ throughout Phanerozoic time. Geology, 10, 516–519.

BURTNER, R. L. 1987. Origin and evolution of Weber and Tensleep formation waters in the Greater Green River and Uinta–Piceance Basins, Northern Rocky Mountain area, USA. Chemical Geology, 65, 255–282.

CHAUDHURI, S., BROEDAL, V. AND CLAUTER, N. 1987. Strontium isotopic evolution of oil-field waters from carbonate reservoir rocks in Bindley field, central Kansas, USA. Geochimica et Cosmochimica Acta, 51, 45–53.

CONOVER, W. J. 1971. Practical Nonparametric Statistics. Wiley, New York.

EHRENBERG, S. N. 1990. Relationship between diagenesis and reservoir quality in sandstones of the Garn formation, Haltenbanken, mid-Norwegian continental shelf. American Association of Petroleum Geologists Bulletin, 74, 1538–1558.

—— 1991. Kaolinized, potassium-leached zones at the contacts of the Garn Formation, Haltenbanken, mid-Norwegian continental shelf. Marine and Petroleum Geology, 8, 250–269.

——, GJERSTAD, H. M. AND HADLER-JACOBSEN, F. 1992. Smørbukk field. In: HALBOUTY, M. T. (ed.) Giant Oil and Gas Fields of the Decade 1978–1988. American Association of Petroleum Geologists Memoir, 54, 323–348.

—— AND NADEAU, P. H. 1989. Formation of diagenetic illite in sandstone of the Garn Formation, Haltenbanken, mid-Norwegian continental shelf. Clay Minerals, 24, 233–253.

GIBBONS, K. A. 1991. Use of variations in strontium isotope ratios for mapping barriers: An example from the Troll field, Norwegian continental shelf. American Association of Petroleum Geologists Bulletin, 75, 579–580 (Abstract).

JOHANNESEN, L. 1989. A Strontium Isotopic Study of Formation Waters from the Heidrun Field. PhD thesis, Norwegian Institute of Technology, Trondheim.

SMALLEY, P. C., LØNØY, A. AND RÅHEIM, A. 1992. Spatial $^{87}Sr/^{86}Sr$ variation in formation water and calcite from the Ekofisk chalk oil field: Implications for reservoir connectivity and fluid composition. Part 1. Applied Geochemistry, 7, 341–350.

——, RÅHEIM, A., DICKSON, J. A. D. AND EMERY, D. 1988. $^{87}Sr/^{86}Sr$ in waters from the Lincolnshire Limestone aquifer, England, and the potential of natural strontium isotopes as a tracer for a secondary recovery seawater injection process in oilfields. Applied Geochemistry, 3, 591–600.

The Embla Field

I. A. KNIGHT,[1] L. R. ALLEN,[1] J. COIPEL,[2] L. JACOBS[2] and M. J. SCANLAN[1]

[1] Phillips Petroleum Company Norway, PO Box 220, N4056 Tananger, Norway
[2] Fina Exploration Norway, PO Box 4055, N4004 Stavanger, Norway

Abstract: The Embla Field is the first development of a pre-Jurassic reservoir in the Norwegian sector of the Central Graben. The field is located 5 km south of the Eldfisk Alpha platform, near the Greater Ekofisk production complex, which provides an infrastructure where development of marginal fields becomes commercially viable.

Although the field was initially discovered in 1974 with the 2/7-09 well, inconclusive drill stem test (DST) and geological data left the field potential uncertain. In 1988 a second well, 2/7-20, successfully tested oil and gas from a thick sandstone of indeterminate age. A third well, 2/7-21S, drilled in 1989, gave even better test results from two separate sandstones. Core, log and DST data from these two wells indicate that the reservoir is a faulted and highly fractured sequence of conglomeratic and finer-grained sandstones deposited in a continental environment of probable Paleozoic age.

A 3D seismic programme in 1989 also demonstrated the structural complexity of the field. Situated as it is on the Grensen Nose and bounded on its eastern flank by the Skrubbe Fault/Lindesnes Ridge inversion, it was evident the reservoir had experienced a complex burial and tectonic history. Poor seismic definition of the pre-Cretaceous stratigraphy further dictated the need for a flexible approach to field development.

A phased development programme was initiated, justified by the reserves attributable to the 2/7-20 and 2/7-21S wells. Expendable delineation wells were avoided. Instead, deviated step-out drilling from a central template was favoured, with the intent to subsequently tie the wells back to a remotely controlled surface production facility. First production is scheduled for early 1993.

The Embla Field is wholly located within Block 2/7 of the Norwegian Sector (Fig. 1). The block was awarded as part of Production Licence 018 in the First Round of Norwegian licensing in 1965. Present participants are:

Phillips Petroleum Company Norway a.s.	36.960% (Operator)
Fina Exploration Norway Inc.	30.000%
Norsk Agip a.s.	13.040%
Elf Aquitaine Norge a.s.	7.594%
Norsk Hydro a.s.	6.700%
TOTAL Norge a.s.	3.547%
Den Norske stats oljeselskap a.s.	1.000%
Elf Rep Norge a.s.	0.456%
Elf Rex Norge a.s.	0.399%
Norminol	0.304%

The major geological elements of the Embla Field area and the location of the wells at the base of the Cretaceous (near top reservoir) are shown in Fig. 2.

Exploration and appraisal history

In 1974, the Embla Field discovery well, 2/7-09, was drilled. The initial reservoir objectives were Lower Paleocene/Upper Cretaceous chalk and Jurassic sandstones previously encountered in nearby Feda Graben wells such as 2/7-03. The anticipated chalk reservoir was tested but proved non-productive. A deeper, hydrocarbon-bearing sandstone, which was interpreted at the time to be Upper Jurassic, was encountered at a depth of 4276 m. The well penetrated over 100 m of net sandstone before reaching a total depth of 4418 m. Test results were disappointing with a flow rate of only 36 m³ of oil per day.

In 1986, a review of 2/7-09 suggested that the low flow rate may have resulted from formation damage possibly caused by the use of a water-based drilling fluid. It was estimated that the well had the potential to flow as much as 318 m³ of oil per day. In 1988, an exploration well, 2/7-20, was drilled to determine if an economic hydrocarbon accumulation existed. The reservoir was encountered in an upthrown fault block, over 120 m high

to 2/7-09. Additionally, the well established the presence of a lower sandstone, separated from the upper hydrocarbon-bearing sandstone by a mudstone/siltstone section. The reservoir interval was barren of microfossils, but lithological characteristics suggested it was older than Late Jurassic and it was assigned a questionable Early Permian age. Drill stem test rates were encouraging, as the upper sandstone flowed 566 m³ of oil per day. Following testing, the well was suspended in a manner suitable for a later re-entry and tie-back to a surface production facility.

The 2/7-21S well was drilled in 1989 from a three-slot temporary template installed over the 2/7-20 well location. The objectives of the well were to test the southeasterly continuation of the hydrocarbon-bearing sandstone and to confirm sufficient in-place reserves to justify preliminary design for field development. The well was deviated from the template to a target location at the Base Cretaceous level 1200 m southeast of the wellhead. After drilling both the upper and lower sandstones, the well penetrated an Upper Devonian mudstone. Subsequently, the well reached total depth in altered volcanic rocks of indeterminate age at a measured depth of 5045 m (4689 m TVDSS). A DST performed over the lower sandstone interval flowed up to 785 m³ of oil per day. Upon completion of this test, the well was killed and the upper sandstone was perforated. The two sandstones were then tested co-mingled at a flow rate of 1236 m³ of oil per day. The well was temporarily abandoned, once again suitable for later re-entry and tying back to production facilities.

At this time, a 3D seismic dataset acquired in the winter of 1988–89 (discussed in the following section) became available for interpretation. To confirm the northern continuation of the reservoir, the 2/7-23S well was deviated to a target location 1200 m north of the template. The well encountered a reservoir section almost totally composed of the lower sandstone. The reservoir section was cored throughout, but because of technical problems the well was plugged and abandoned without testing.

During the winter of 1990–91, the 2/7-25S well was drilled from a two-slot extension to the original three-slot template.

From *Petroleum Geology of Northwest Europe: Proceedings of the 4th Conference* (edited by J. R. Parker).
© 1993 Petroleum Geology '86 Ltd. Published by The Geological Society, London, pp. 1433–1444.

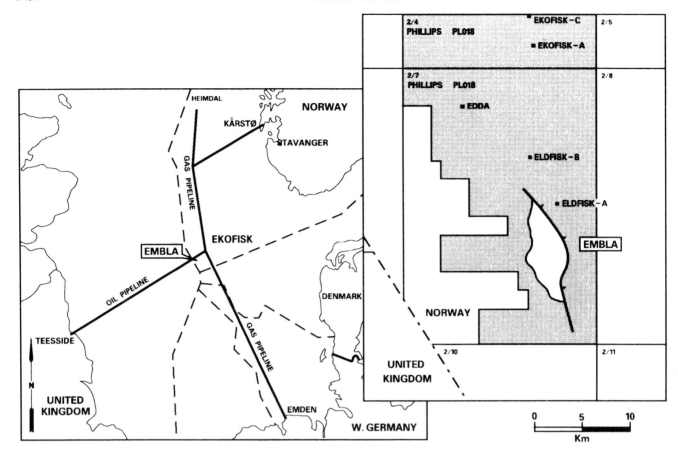

Fig. 1. Field location and infrastructure.

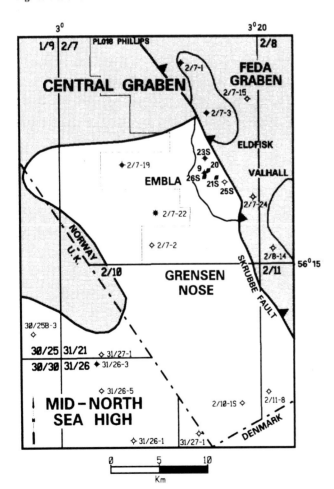

Fig. 2. Major geological elements and location of pre-Cretaceous wells.

This deviated well was designed to step out beyond the 2/7-21S well, approximately 2 km southeast of the template location. After penetrating the Base Cretaceous, 391 m of Upper Jurassic mudstones, siltstones and volcanic rocks were unexpectedly encountered. Below the Upper Jurassic, 33 m of indeterminate age (Paleozoic?) rock was drilled before reaching a total measured depth of 5189 m (4539 m TVDSS). The expected reservoir was absent and the well was abandoned without testing.

The 2/7-26S well was drilled in 1991 to test the reservoir in the western fault block of the Embla structure, and was deviated to a target location approximately 350 m south of 2/7-09. The well demonstrated a reservoir stratigraphy similar to the 2/7-20 well, and both the upper and lower sandstones were extensively cored. The two sandstones were tested separately. Both were hydrocarbon-bearing, but with different oil compositions. This well also had a lower permeability reservoir than previous wells, and tested at modest flow rates of up to 233 m^3 oil per day before being temporarily abandoned for later tie-back.

Field database

Seismic

Acquisition: The pre-1974 2D seismic data which were used to locate the 2/7-09 well showed a structure at the base Cretaceous level but gave poor definition of the pre-Cretaceous stratigraphy. By 1987 better 2D seismic data were available, but the results of the 2/7-20 confirmation well and the implied structural complexity encouraged the Phillips Group to acquire a 3D seismic dataset. It was hoped that the 3D seismic would provide a better understanding of the structure and stratigraphy of the field and would help rationalize the development scheme. In the winter of 1988–89 a 180 km² 3D survey was acquired perpendicular to the Skrubbe Fault trend. After cross-line interpolation, the survey provided a sub-surface sampling interval of 12.5 m by 18.75 m.

Processing: During the initial interpretation of the 3D seismic survey it was recognized that the following three problems were complicating an already difficult structural and stratigraphic interpretation.

1. Multiple energy was masking or interfering with the reservoir and deeper reflections.
2. Rapid lateral velocity variations across the Skrubbe Fault had not been appreciated and compensated for during the original stacking velocity analysis.
3. An extreme velocity gradient across the Skrubbe Fault exceeded the imaging capabilities of conventional time migration.

For the above reasons, the 3D volume was reprocessed during the winter and spring of 1990–91. This effort was based on close attention to the previously mentioned limitations in the original processing, along with a novel parameter verification step. The verification step involved stacking and 3D migration of a subset (cube) of the full 3D volume. Although costly and time intensive, the procedure was validated when the conventionally chosen 2D parameters were found to be less than optimal in 3D. Subsequently, refinements to the multiple suppression and velocity analysis strategies were made and a significantly better quality data volume resulted.

Problems associated with the extreme velocity gradient across the Skrubbe Fault are being addressed through an iterative model-based depth migration scheme. It is planned to refine the velocity model progressively for migration through a combination of new well control and the improved pre-Cretaceous structural imaging provided by depth migration.

Interpretation: The seismic image quality down to Base Cretaceous level is generally excellent, and accurate synthetic ties allow confident mapping of the Top Paleocene (Balder Formation), the Top Shetland Group, the Blodøks Formation (Plenus Marl Equivalent) and the Base Shetland Group. However, poor seismic definition of the reservoir and pre-Cretaceous structure has complicated field appraisal and necessitated reliance on well control. The poor seismic definition is related to the lack of a top reservoir reflection and the absence of any coherent and laterally continuous events within the reservoir.

The absence of a true top reservoir reflector has been partially overcome through interpretation of the Base Cretaceous Unconformity event. This reflector is relatively continuous and unambiguous in the downflank areas. In the crestal area, however, where it is seismically coincident with the top reservoir, a high degree of faulting and several inferred erosional episodes make the event difficult to map with confidence.

Comparison of seismic data to dipmeter logs, vertical seismic profiles and synthetic seismograms suggests that the lack of reservoir definition is a problem associated with complex stratigraphic and structural relationships within the reservoir rather than an issue of seismic quality. Therefore, attempts have been made to work around the poor reservoir and structural imaging problems through use of: (a) dip and azimuth mapping of the Upper Cretaceous Blodøks Formation (Dalley *et al.* 1989); (b) horizon flattening and isochron/ isopach mapping for structural reconstruction; and (c) use of complex seismic attributes for identification of faults and intra-reservoir variation.

Logs and cores

Embla well wireline logging programmes have been partially constrained by the nature of the logging environment. Three particular problems exist. First, the reservoir temperature of 160°C is beyond the operating limit of some wireline tools and survey instruments. Second, the oil-based drilling fluids used to minimize formation damage impede the use of high-resolution dipmeter tools; and third, the high solids content of the heavy drilling fluids restricts the use of sonic scanning devices.

Although limited by these conditions, conventional logging suites of induction resistivity, gamma ray, sonic and nuclear porosity logs have been run successfully. The oil-based dipmeter tool has also been used, primarily in an effort to understand the reservoir structure. Difficulties have been experienced in reconciling the log-derived mineralogy to that actually measured in cores, and discrimination of lithofacies types from the logs has also proved difficult, creating added uncertainty in well correlations and reservoir geometry.

In light of these restrictions and particularly those of the seismic data, it was recognized early in the drilling programme that extensive conventional coring would be essential to provide an accurate characterization of the reservoir. To date, a total of 732 m of both oriented and non-oriented cores have been recovered, and both the upper and lower sandstones are now totally covered.

Field geology

Stratigraphy

The Embla Field is located on the Grensen Nose, a northwest–southeast-trending structural high considered as a northeast extension of the Mid North Sea High. Regional thickness variations in the Upper Permian and Mesozoic stratigraphic sections indicate that the Grensen Nose was part of the Mid North Sea High until subsidence in the central Graben occurred during the Late Cretaceous (Gowers and Sæbøe 1985).

The pre-Cretaceous stratigraphy varies significantly between wells in the Embla Field. To date, the 2/7-26S well has encountered the most complete sequence (Fig. 3). The oldest unit penetrated on the Grensen Nose is a red to greenish-grey, heavily brecciated and strongly altered rhyolitic rock identified in Wells 2/7-21S and 2/7-26S. Virtually all of the original felsic and mafic minerals have been replaced by non-ferroan and ferroan dolomite, silica, anhydrite, kaolinite and locally pyrite. Based upon outcrop analogy with the Midland Valley of Scotland and the overlying stratigraphic section described below, the rhyolite is possibly of Early Devonian age.

The rhyolitic unit is overlain by brown to red, very fine- to fine-grained, moderately to well-sorted micaceous sandstones and silty mudstones. Palynological dating indicates a Late Devonian (Frasnian) age. Miospores are predominant and are associated with marine microplankton (Acritarchs and Chitinozoa) and rare freshwater algae. These Devonian sediments are interpreted to have been deposited in a flood plain/lacustrine environment. No Middle Devonian limestones, such as those observed in the UK sector (e.g. Argyll and Auk fields) have been identified. The section shows similarities to the Midland Valley of Scotland where a volcanic/igneous-rich Lower Devonian is unconformably overlain by Upper Devonian.

Carboniferous or Permian stratigraphic sections have not been positively identified in Embla wells but have been encountered in Grensen Nose wells to the south. A Lower Carboniferous Tournaisian section consisting of interbedded mudstone, sandstone and volcanic rocks was penetrated in the 2/10-01S well. The depositional environment has been interpreted as alluvial plain to marginal marine with interbedded extrusives and tuffs. An interpreted Lower Permian Rotliegendes Group overlies the Carboniferous and consists of continental sandstone and mudstone deposits with interbedded volcanic rocks near the base of the unit. Lower Permian volcanic rocks and overlying Rotliegendes (?) sandstone have also been identified in the 2/7-02 well. The Upper Permian Zechstein Group has not been encountered in any of the Embla wells, but occurs as a thin marine evaporitic sequence of dolomite and anhydrite on the southern Grensen Nose and the Mid North Sea High, grading to a thicker basinal salt sequence in the Feda Graben.

The Triassic stratigraphic section is generally thin to absent (< 100 m) on the Grensen Nose/Mid North Sea High and has not been positively identified in the Embla wells.

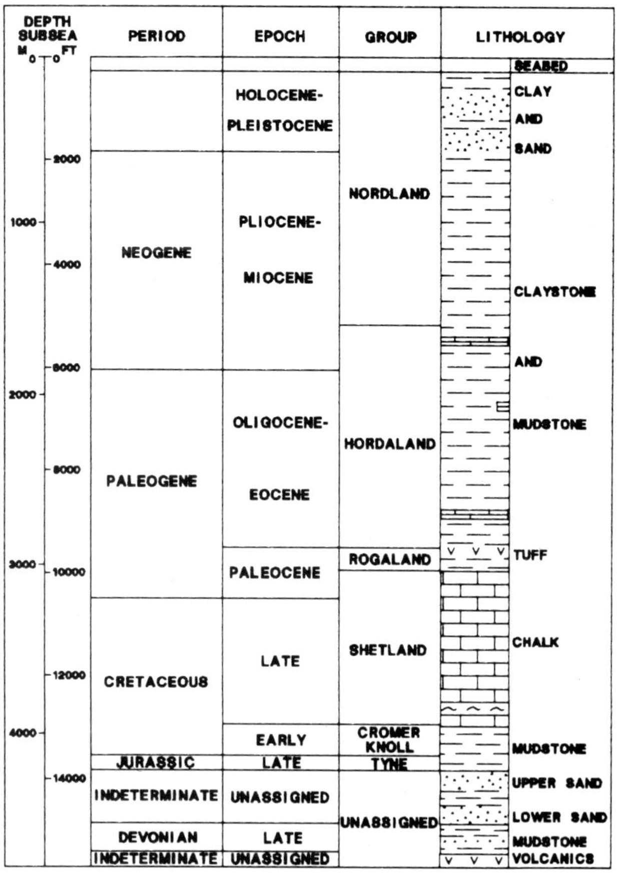

Fig. 3. Generalized 2/7-26S stratigraphy.

The reservoir section of the Embla Field rests directly upon the Upper Devonian mudstone. The reservoir interval is composed of two major sandstone sections of braided fluvial and alluvial fan origin, separated by an intervening mudstone/siltstone sequence deposited in a flood plain/lacustrine setting. The sediments consist of grey–green to red, fine- to medium-grained, often conglomeratic sandstones with interbedded grey–green mudstones and siltstones. Extrusive and intrusive igneous rocks, principally rhyolite, hawaiite and alkalic olivine dolerite are also identified within the reservoir.

The age of the reservoir remains unknown as the sequence is barren of microfossils. The current interpretation favours Early Permian, Carboniferous and/or Late Devonian age. The basis for the interpretation is the previously noted Late Devonian age below the reservoir and the occurrence of a Lower Permian and Carboniferous stratigraphic section in southern Grensen Nose wells. The reservoir sedimentology is also partially analogous to the Argyll Field, although aeolian sediments have not been identified. A Triassic age remains possible, but is not considered as likely.

The Upper Jurassic Tyne Group unconformably overlies the Embla reservoir except near the crest of the subcrop where the

Tyne Group is absent. Only a very thin Mandal Formation mudstone is present high on the structure in the 2/7-20 and 2/7-23S wells, and is absent in the 2/7-21S well. In flank wells, the Tyne Group is predominantly represented by the organic-rich Mandal and Farsund formations, which are over 200 m thick in the 2/7-25S well. Significant Upper Jurassic thickness variations are evident between the Embla wells on the Grensen Nose and wells in the Feda Graben to the east, where Upper Jurassic thickness exceeds 1825 m.

The 2/7-25S well also penetrated a highly altered Upper Jurassic volcanic section interbedded with the Farsund Formation. Petrographic evidence suggests the rock to be an in situ lava flow of hawaiitic composition. The texture is dominated by microlites with scattered micro-phenocrysts of feldspar set in a groundmass of haematite and/or goethite. Some albite feldspar has been preserved, forming the only original mineral species found in the rock suite. Although mineralogically different, the nearest Jurassic volcanic rocks are encountered in UK Quadrants 29 and 30. Petrologically, the greatest similarity is to volcanic rocks in 2/7-02 well, which are currently dated as Permian.

Overlying the Upper Jurassic are mudstones and argillaceous limestones of the Lower Cretaceous Cromer Knoll Group. This interval thins across the structure, varying from 213 m in the 2/7-09 well to 91 m in the 2/7-21S. At the top of the structure where the Upper Jurassic is missing, the Cromer Knoll Group forms the top seal for the underlying reservoir sandstone.

The Upper Cretaceous/Lower Paleocene Shetland Group is represented by approximately 900 m of chalk. The Shetland Group is overlain by over 3000 m of Hordaland Group mudstone and Nordland Group mudstone, clay and sand deposited during Cenozoic subsidence of the Central Graben.

Radiometric age dating and geochemical fingerprinting The unsuccessful palaeontological dating of the reservoir has prompted the use of other age-dating techniques. Potassium–argon (K–Ar) radiometric datings of illite/chlorite fractions in claystones and sandstones from wells 2/7-20 and 2/7-21S indicated Late Triassic (Norian–Rhaetian, 214 ± 6 Ma) to Early Carboniferous (331 ± 3 Ma) ages. These are minimum ages due to an important component of re-crystallized and/or authigenic illite/chlorite.

K–Ar dating of altered basaltic rocks in the reservoir section of Well 2/7-21S indicates an Early Tertiary (Paleocene) age. This apparent age can be interpreted as unreliable data or could reflect alteration caused by later hydrothermal events. As these igneous rocks are interpreted to be extrusive, their position in the reservoir sequence indicates a possible Late Devonian to Permian age, suggesting volcanic activity during the deposition of the reservoir interval. Such volcanic activity is also known from the Oslo Graben and along the northern margin of the Ringkøbing–Fyn High.

Geochemical fingerprinting using X-ray fluorescence spectrometry has also been attempted to constrain estimates of reservoir age. Comparison of samples from the 2/7-21S well to a database of other red bed samples from the North Sea has highlighted the unusual nature of the Embla reservoir. The results are again open to interpretation, but suggest a Permian age.

Structure

The Embla Field has been interpreted as a westward-dipping Paleozoic horst. The structure is elongated northwest–southeast and bounded on the east by the Skrubbe Fault, a Late Jurassic normal fault which was reactivated in a reverse motion during Late Cretaceous inversion (Fig. 4).

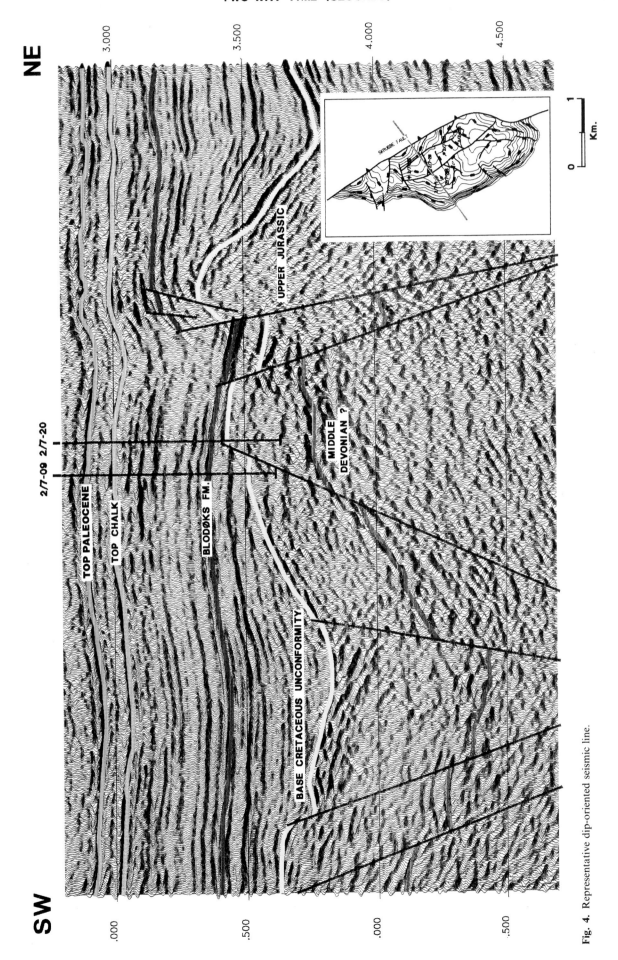

TWO-WAY TIME (SECONDS)

NE

SW

3.000
3.500
4.000
4.500

Km.

0

1

2/7-09 2/7-20

UPPER JURASSIC

MIDDLE DEVONIAN ?

TOP PALEOCENE

TOP CHALK

BLODØKS FM.

BASE CRETACEOUS UNCONFORMITY

SKRUBBE FAULT

.000
.500
.000
.500

Fig. 4. Representative dip-oriented seismic line.

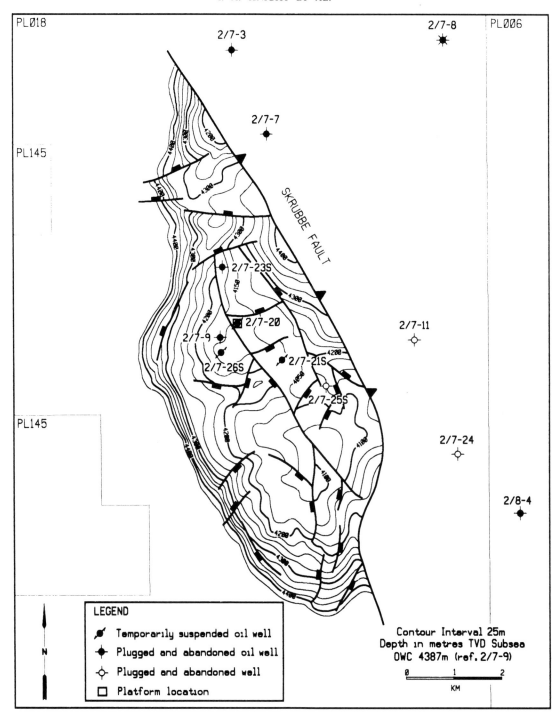

Fig. 5. Embla Field Base Cretaceous Unconformity depth structure map.

The Embla Field is thought to have an areal extent of 19.4 km², as defined by the Base Cretaceous Unconformity and the oil–water contact at 4387 m. A common oil–water contact is currently assumed, but to date has only been reliably observed in the 2/7-09 well. The crest of the structure is at 4033 m, giving a gross oil column of 354 m.

The trap is formed by a combination of structural and stratigraphic elements. Along its margin, the field is dip and fault closed. On the crest of the structure, where reservoir sandstones are truncated by erosion, hydrocarbons are stratigraphically trapped by overlying mudstones of the Upper Jurassic Mandal Formation and/or Lower Cretaceous Cromer Knoll Group.

Core information suggests that the Embla Field is the product of multiple tectonic events. The cores are highly fractured and faulted and demonstrate variable fracture orientations. Intense extensional (both dilational and shear) fracturing dominates. The fractures have moderate to steep dips and occur as both large fault zones and smaller, more pervasive fracture systems. The main fracture orientations are NW–SE (Wells 2/7-20, 2/7-21S, and 2/7-23S), NNW–SSW and NE–SW (Wells 2/7-21S and 2/7-23S).

Seismic data confirm two dominant structural trends at the Base Cretaceous level (Fig. 5). First, a NW–SE fault trend parallel or sub-parallel to the Skrubbe Fault. Faults with this orientation are presumed to be associated with Late Cimmerian tectonics and divide the horst into sub-blocks. A fault between the 2/7-09 and the 2/7-20 wells, with a throw of approximately 120 m at the Base Cretaceous Unconformity level, divides the field into an eastern and western fault block.

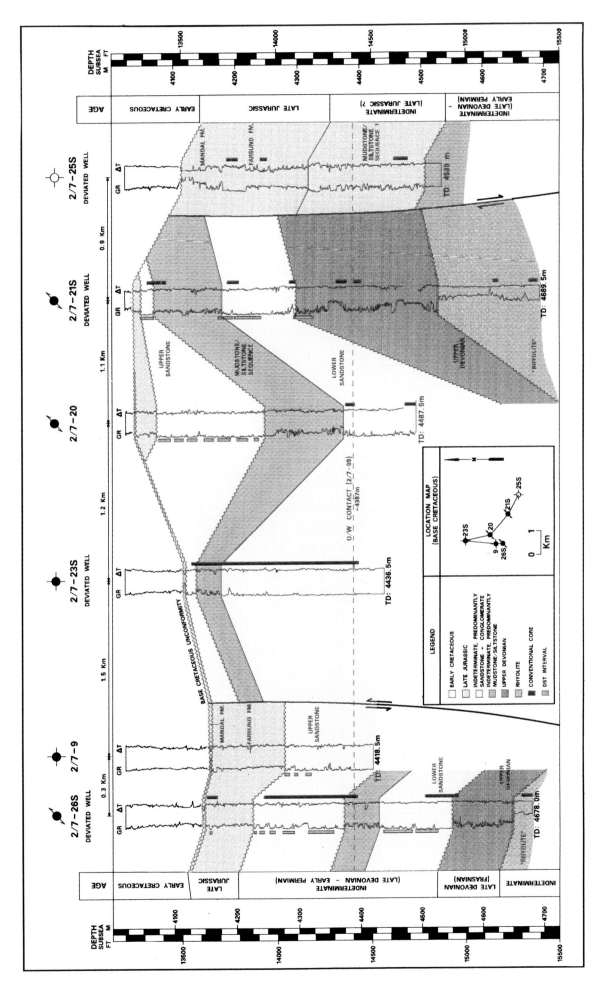

Fig. 6. Structural cross-section.

The second trend strikes NE–SW and is seen in the dip and azimuth attributes (*see above*) of the Blodøks horizon. Many of these faults are traceable into the reservoir, where they generally have less throw than faults of the other trend. The origin of these NE–SW-striking faults is problematic.

A complex fault zone of particular importance to field appraisal was identified after drilling the 2/7-25S well. This feature, located between the 2/7-21S and 2/7-25S wells, appears to mark a limit of the field and juxtaposes the Paleozoic reservoir sandstones against thick Upper Jurassic mudstones (Fig. 6).

Reservoir distribution and sedimentology

The reservoir interval in the Embla Field is composed of two major sandstone units separated by a mudstone/siltstone sequence. Figure 6 shows the pre-Cretaceous interval well correlations. Total thickness of the reservoir sections varies from 322 m (2/7-26S) to more than 415 m (2/7-20).

The upper sandstone is thickest in the 2/7-09, 2/7-26S and 2/7-20 wells, where it varies between 140 and 175 m. To the southeast in the 2/7-21S well the upper sandstone is truncated to form a thin 20 m interval bounded by an unconformity above and a fault cut at its base (seen in core). In the 2/7-23S well to the north, the upper sandstone is thin (15 m), again due to erosional truncation.

The mudstone/siltstone sequence between the upper and lower sandstones varies in thickness between 41 m (2/7-23S) and 126 m (2/7-20).

The lower sandstone thickness varies greatly, from more than 261 m in the north (2/7-23S) to only 120 m in 2/7-26S. However, lateral facies changes and faulting within this interval may render both the correlations and resulting isopach unreliable.

Description of the reservoir facies has been largely dependent on core examination, as the available wireline logging suite cannot adequately discriminate all of the facies types within the productive interval. Well correlations from logs have been variously interpreted, and tend to rely on the identification of the volcanic horizons.

Detailed sedimentological and compositional analyses of cores indicate that the sandstones were deposited in a braided fluvial/alluvial fan to lacustrine environmental setting. Five different facies associations have been recognized:

- A: proximal braided channel/sheetflood sediments;
- B: distal alluvial plain sediments;
- C: flood plain/lacustrine sediments;
- D: braid plain to braid delta sediments; and
- E: volcanic, volcaniclastic and intrusive igneous rocks.

The upper sandstone was almost totally cored in the 2/7-26S well and partially cored in the 2/7-21S and 2/7-23S wells. It consists of fine- to medium-grained sandstones with occasional quartz pebbles. The unit is generally well sorted, with graded bedding in fining-upwards cycles containing thin and discontinuous clay laminae. The interval is interpreted as a distal braided fluvial system of unknown transport direction.

The lower sandstone is totally cored in the 2/7-23S well and partially cored in the 2/7-20, 2/7-21S and 2/7-26S wells. It is significantly coarser than the upper sandstone, and can be divided into two sub-sections: a lower sequence of stacked debris flows; and an overlying sequence of braided pebbly sandstones with thin volcanic intercalations.

Small (<8 m), medium (8–15 m) and large (15–30 m) coarsening- and fining-upward trends are repeated throughout the reservoir section. These trends represent respectively, channel-fill aggradation, downstream transition from pebbly braided

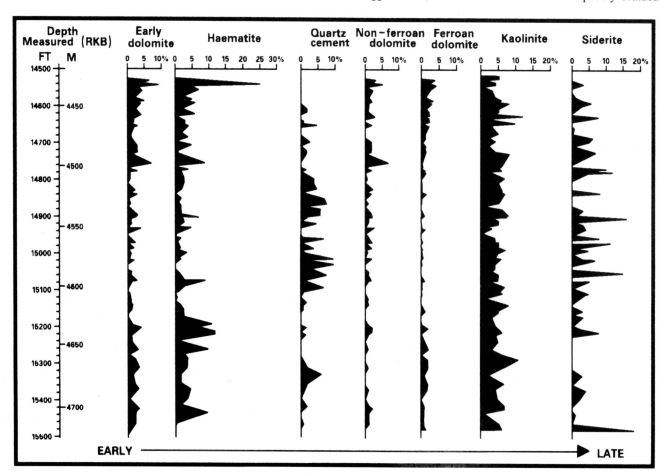

Fig. 7. Temporal and vertical distribution of secondary mineralization, Well 2/7-23S.

streams to more sandy bedload channel systems and finally, the overall stacking or repetition of facies sequences reflecting fault movements or large-scale channel switching.

Compositionally, the sandstones can be classified as lithic and sublithic arenites and lithic wackes (Dott 1964; modified by Pettijohn 1975). Facies-related primary textural parameters have an important control on porosity/permeability, with the coarse, clean facies associations A and D displaying the best characteristics. Both compaction and locally extensive early cementation have resulted in a severe loss of intergranular porosity.

Petrography and diagenesis

Extensive petrographic analysis has been conducted to address the detrital mineralogy, diagenetic history, porosity characterization and fracture fill history. Some of the more relevant observations are summarized below.

1. Quartz dominates the detrital fraction of the matrix. It occurs primarily in mono-crystalline form, but does possess a significant poly-crystalline component in the lower sandstone. Accessory detrital components include micas (mainly muscovite), metamorphic and sedimentary rock fragments and traces of heavy minerals.

2. The secondary (authigenic) mineralization present indicates that the reservoir has experienced a complicated diagenetic history with evidence of repeated porosity loss and enhancement. Repeated phases of silica, carbonate, pyritic and haematitic cements are observed in varying abundance in both matrix and fractures. Kaolinite is the most abundant authigenic clay, occurring in at least two phases. A residual bitumen is also prominent. An example of the secondary mineralogy and the interpreted chronological sequence of mineralization from the 2/7-23S well is shown in Fig. 7. Best reservoir quality in this well is between 4500 and 4600 m. This is attributed to the small yet significant presence of quartz cement forming a rigid pore framework.

3. Most of the effective porosity is secondary, providing the major contribution to reservoir quality. It occurs in the form of intragranular (partial grain dissolution) and mouldic pores (total grain dissolution). Petrographic examination indicates more than one phase of porosity generation. The rare occurrence of feldspar suggests that most of the secondary porosity may have been associated with dissolution of feldspar. A later phase of secondary porosity is indicated by the presence of large, open mouldic pores which clearly occurred after mineral precipitation. Some primary intergranular porosity is recognizable, but is commonly highly altered by both compaction and cementation.

4. Healed fractures demonstrate repeated episodes of opening and re-cementation. The minerals present in the fractures suggest that hydrothermal fluids periodically invaded the reservoir.

Source and migration of hydrocarbons

Oil-source rock correlations indicate that the organic-rich (typically 3–8% TOC) Upper Jurassic Mandal and Farsund formations in the Feda Graben are the source rocks for the mobile oil in the Embla Field. These bituminous mudstones attain a thickness in excess of 1825 m in the Feda Graben to the east.

Visual and petrographic inspection of cores has identified a patchy but pervasive distribution of a bitumen which pre-dates both the mobile oil and some of the authigenic kaolinite within the reservoir. Oil–oil and oil–source correlation studies have been unable to establish definitively either the source or mechanism by which the bitumen was produced.

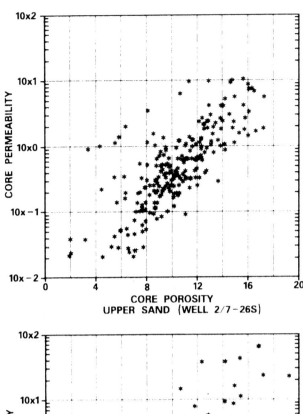

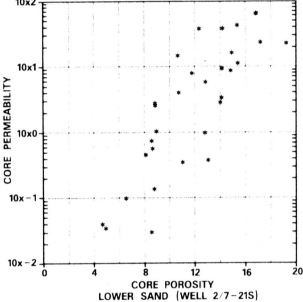

Fig. 8. Comparison of porosity/permeability for upper and lower sandstones.

Rock and fluid properties

The two reservoir sandstones demonstrate contrasting porosity/permeability relationships. Comparison of porosity/log permeability crossplots for the two sandstones shows significantly higher permeabilities in the lower sandstone (Fig. 8). This is mostly attributed to the coarser grain size of the lower sandstone. Porosities of up to 20% have been measured, also reflecting the beneficial effects of overpressuring and hydrocarbon emplacement curtailing chemical diagenesis.

The interpretation of intense faulting from both cores and seismic data indicates a risk of reservoir compartmentalization. Both RFT and DST data have indicated that vertical and horizontal barriers to fluid flow may exist. Interpretation of the DST data has indicated the presence of boundaries representing faults or stratigraphic discontinuities, and slight shifts in the measured RFT values may also reflect barriers to pressure equilibrium (Fig. 9). Barriers to the vertical flow of fluids may take the form of caliche layers, intercalated volcanic horizons

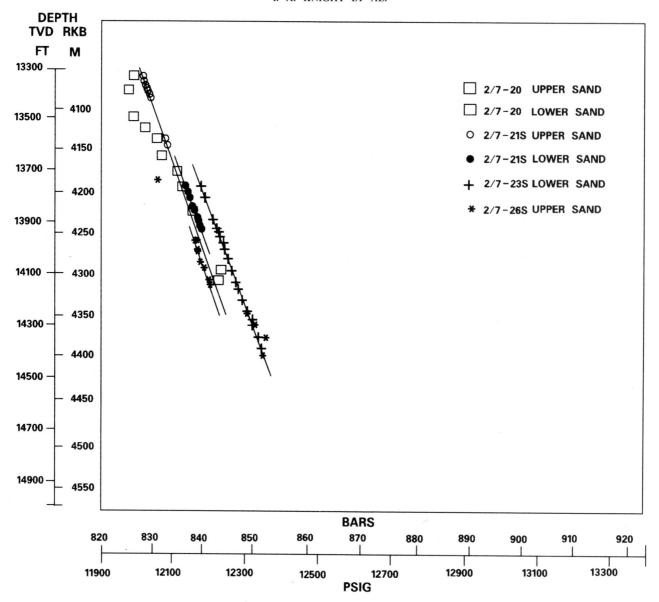

Fig. 9. RFT formation pressure data for all Embla Field wells.

and/or the lacustrine mudstones separating the upper and lower sandstones.

An initial reservoir pressure of approximately 841 bars has been recorded from DSTs, and RFT pressure gradients indicate compositionally similar fluids. However, PVT properties do show some variation within the field. Higher gas–oil ratios have been recorded in the wells in the eastern fault block, suggesting that the more deeply buried source to the east may have contributed a greater gas component. PVT properties for the reservoir oil are presented in the field data summary. Of particular note is the relatively low bubble point compared to the initial reservoir pressure, of importance in assessing how the field development could be managed.

Field development strategy

The encouraging results of the 2/7-20 well in 1988 left the Phillips Group with a challenging dilemma: how best to manage a complicated reservoir in a high pressure/high temperature environment where conventional field delineation methods may prove both costly and unsuited to the geological characteristics. Compounding this were the limitations of the 3D seismic data, which gave a poorly resolved and highly inter-

pretative display of the pre-Cretaceous stratigraphy and reservoir geometry.

The high level of geological uncertainty was a factor in the estimation of the field volumetrics and resulting economic appraisal. Full field estimates based on the results of the 2/7-20 and 2/7-21S wells suggested in-place volume as high as 163×10^6 m^3 (1028 MMBBL). However, the complexity of the field suggested caution, and the initial field economics were based upon only the proven reserves attributable to these two wells. As a result, the in-place volume of 34×10^6 m^3 (215 MMBBL) of oil served as the minimum case for economic assessment.

Despite the high geological risk, the Embla Field offered new possibilities. Close proximity to the infrastructure of the Ekofisk complex provided the means to produce economically what had previously been viewed as a marginal field. Consequently, both subsea and fixed surface facilities were assessed for suitability. Long-term testing from a purpose-built floating facility was also considered, but dismissed because of unavailability. Economic and feasibility studies indicated that, given adequate reserves, a remote-controlled surface facility would be as economic as subsea completion, yet allow the flexibility of further drilling, workover or well monitoring. As a consequence, the three-slot temporary template was installed over the 2/7-20 well and the 2/7-21S well was drilled and deviated to the southeast. The purpose of the 2/7-21S well was to prove the southward extension of the field and the necessary in-place hydrocarbon volumes to justify development economically. The well was successful and the test results exceeded expectations.

Based on the DST and geological data from the 2/7-20 and 2/7-21S wells, a 'Plan for Development and Operation' was approved by the Norwegian Authorities in 1990. The development strategy for the Embla Field had by then evolved into the following:

Phase One

1. Phase One would be to pre-drill up to eight deviated appraisal/production wells from a permanent template of fifteen slots located at the 2/7-20 well using a mobile rig. The drilling radius from the template would enable the Phase One development to cover the northern end of the field (Fig. 10). Each well would be drilled on a step-out basis using 1.3 km² (320 acre) spacing. Extensive coring would be performed to counter the limitations of the seismic and log data, plus providing information for reservoir modelling and possible improved recovery projects. It was believed that the step-out philosophy would maximize the chances of correctly interpreting the reservoir correlations and geometry.
2. Limit or simplify well testing to avoid formation damage and to enable later tie-back to a production facility.
3. Install a remote-controlled production facility from which a period of reservoir monitoring could be conducted. This would enable long-term testing of wells and interference tests to identify more reliably the degree of reservoir compartmentalization, reservoir anisotropy and the need for further drilling. In light of the large pressure difference between initial reservoir pressure and the bubble point, critical performance data would be obtained while the reservoir remained in single-phase flow.
4. Proceed with further drilling as the above results dictated.

Phase Two

5. Phase Two would be to repeat the procedure described above for the southern part of the field lying beyond the drilling radius of the Phase One platform. This area of the field is geologically the most complex and consequently of higher risk. Both further seismic evaluation and delineation drilling would be necessary to assess the best means of developing this area, either through a second platform or subsea development.

Phase Three

6. Phase Three would be implementation of an improved recovery programme, designed from the core data, geological/geophysical description and production data gathered during Phases One and Two.

In the two years that have elapsed since the project commenced the 3D seismic volume has been reprocessed and three further wells have been drilled. The added well control has aided the seismic interpretation, providing a more confident description of the fault pattern. However, the reservoir stratigraphy and Base Cretaceous subcrop geometry remain open to interpretation. This was most evident in the drilling of the 2/7-25S well, which has prompted a significant downgrading of the in-place hydrocarbon volumes and resulted in even higher risk being assigned to the potential of the Phase Two (southern) area.

Delays in the drilling programme will result in only four wells being available for production monitoring. Present plans are to install the Phase One jacket and topsides in late 1992, with tie-back and initial production early 1993.

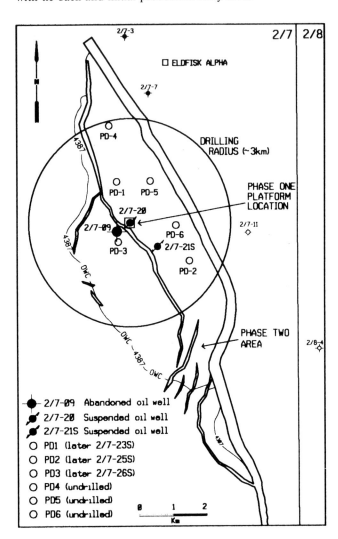

Fig. 10. Phase One development area as proposed in 1990 showing the step-out philosophy of the pre-drilling programme.

The authors wish to thank Fina Exploration Norway Inc., Norsk Agip a.s., Elf Aquitaine Norge a.s., Norsk Hydro a.s., Total Norge a.s., Den Norske stats oljeselskap a.s., Elf Rep Norge a.s., Elf Rex Norge a.s. and Norminol for permission to publish this paper. We also acknowledge the work of Dr A. Bray, Dr R. Kimber and Dr T. Needham of Simon-Robertson Company, Professor R. Steel (University of Bergen) Dr D. G. Fraser (University of Oxford) and Dr J. E. Dixon (University of Edinburgh). Study of this complex field is still in its early stages, and the information presented here is the authors' current interpretation, which does not necessarily reflect the opinions of all of the Licence 018 Group.

References

DALLEY, R. M., GEVERS, E. C. A., STAMPFLI, G. M., DAVIES, D. J., GASTALDI, C. N., RUIJTENBERG, P. A. AND VERMEER, G. J. O. 1989. Dip and azimuth displays for 3-D seismic interpretation. *First Break*, **7**(3), 86–95.

DOTT, R. H., Jr. 1964. Wacke graywacke and matrix—what approach to immature sandstone classification? *Journal of Sedimentary Petrology*, **35**, 625–632.

GOWERS, M. B. AND SÆBØE, A. 1985. On the structural evolution of the Central Trough in the Norwegian and Danish sectors of the North Sea. *Marine and Petroleum Geology* **2**, 298–318.

PETTIJOHN, F. J. 1975. *Sedimentary Rocks*. Harper and Row, London.

Embla Field data summary

	Upper Sandstone (2/7-20)	Lower Sandstone (2/7-235)
Trap		
Type	Combination stratigraphic/structural	
Top reservoir (2/7-21S) (m TVDSS)	4033 (13233 ft)	
Oil–Water Contact (2/7-9) (m TVDSS)	4387 (14394 ft)	
Oil column (m)	354 (1161 ft)	
Area (km²)	19.4 (4791 acres)	
Reservoir		
Age	Indeterminate (Late Devonian–Early Permian)	
Gross thickness (m)	175 (573 ft)	261 (857 ft)
Net/gross sand ratio (using a Vshale cut-off of 30%)	0.80	0.99
Net pay (m) (using cut-offs of porosity 8%; Vshale 30%; Sw 60%)	137 (450 ft)	136 (446 ft)
Average net porosity (%)	14.8	11.3
Hydrocarbon saturation (%)	90	70
Permeability (range) (mD)	0.05–10	0.05–100
Productivity index (BOPD/psi)	Variable 0.5–2	Variable up to 9.0
Reservoir conditions		
Temperature (°C)	160 (320°F)	
Initial pressure (bars)	841 (12200 psia)	
Pressure gradient (KPa m⁻¹)	5.88 (0.26 psi ft⁻¹)	
Drive mechanism	Solution gas	
Hydrocarbons		
Oil gravity (g cm⁻³)	0.816 (42°API)	
Oil type	Undersaturated	
Oil viscosity (at 12000 psi, 320°F) (cp)	0.17	
Initial FVF (m³/m³)	1.89 (1.89 rb/STB)	
Bubble point (bars)	330 (4779 psia)	
Gas/oil ratio (m³/m³)	Eastern Block: 329 (1900 SCF/STB)	
	Western Block: 276 (1550 SCF/STB)	
Formation water		
Salinity (ppm NaCl equiv.)	58000	
Resistivity (ohm m at 151.7°C)	0.031 (2/7-9 SP log)	
Production		
Start-up date	February 1993	
Development scheme	Remotely-operated steel platform	
Production rate (m³/day)	Estimated 6356 (40000 BOPD)	

The evolution of a fractured chalk reservoir: Machar Oilfield, UK North Sea

P. T. FOSTER and P. R. RATTEY

BP Exploration, 301 St Vincent Street, Glasgow G2 5DD, UK

Abstract: Machar is the first of the 'Diapir' fields discovered in the eastern trough of the Central North Sea, in BP sole interest Block 23/26a. It comprises a fractured Cretaceous chalk and Paleocene sand reservoir overlying a steeply dipping salt structure. The oil column is greater than 1100 m and reserve estimates are in the region of 100 million barrels of oil equivalent.

The history of the Machar reservoir represents a complex interplay of salt movement, oil migration and burial diagenesis. Structural studies have shown that the Permian Zechstein salt has been in a state of buoyant equilibrium for most of its geological history, with only one phase of true diapirism identified in the mid-Miocene. The buoyancy resulted in the deposition of a condensed (<300 m) Triassic–Eocene section over the salt high. In the surrounding area, sediment loading and related salt withdrawal caused 'downbuilding' and the deposition of a thick (3000 m) sediment package. A 50–300 m section of chalk was thus deposited over the diapir with a mature Jurassic oil source in the adjacent depocentre. Paleocene sands are thin or absent over the crest of the structure, but thicken to about 100 m downflank.

Regional subsidence and oil migration commenced in the early Oligocene, the trap filling from the top down. Here, shallow burial depth (*c.* 100 m) and early oil emplacement ensured porosity was preserved in the chalk matrix. At the same time, on the flanks of the structure greater depth of burial and diagenesis were destroying chalk porosity from the bottom up. These competing influences resulted in a diagenetic front marking the limit of pay in the chalk. That front lies in the depth range 2000–2500 m.

Passive movement of the chalk during periods of buoyancy, plus regional tectonic compression in the mid-Miocene, created a widespread fracture system. Studies have identified a continuous process of fracture, stylolite and cement formation across a variety of brittle and ductile deformation styles. Average spacing of open fractures in 200 m of chalk core from Well 23/26a-13 is 7.5 cm. About a third of these are open.

The burial history, oil emplacement and fracturing resulted in a reservoir with a dual porosity system. The fine-grained pelagic chalk matrix has primary porosity in the range 12–35% with permeability generally less than 1 mD. Secondary fracture porosity, although difficult to measure, is probably less than 1%. Production test data over fractured intervals, however, show permeability in excess of 1000 mD. The fracture system is the key to developing the chalk reservoir as it provides effective drainage of the tight matrix. In a wider context the recognition of buoyancy, not piercement, as the dominant structural process provides a model for exploration elsewhere.

The Machar Field comprises a combined Paleocene sand and fractured Cretaceous chalk reservoir overlying a salt-cored high. It is situated at the edge of the Central Graben Jurassic depocentre in the Central North Sea (Figs 1 and 2). Field development is currently under consideration.

This paper focuses on the interplay between structural history, oil generation and migration that preserves a trap of highly productive fractured chalk overlying the salt.

Exploration and appraisal history

Machar is located in Block 23/26a, part of first round licence P.057 awarded in 1964. The first exploration well 23/26a-1RD, was drilled in 1976 targeted at Tertiary sands overlying a salt high (Fig. 3). The well encountered oil shows in thin (18 m) Paleocene sands and tested oil at 4870 BOPD from a condensed sequence of fractured chalk underlying the sand. Additional pay was present in a diagenetic unit of celestite in the salt caprock.

The second well was drilled in 1978 as a flank appraisal: 23/26a-2 encountered oil shows in the Paleocene sands, but was abandoned due to drilling problems. The well was side-tracked to a deeper Jurassic target. No chalk pay was encountered in the well.

Interest waned until 1986 when a crestal appraisal well was drilled, 23/26a-5. This encountered no reservoir; Tertiary shales rested directly on the salt. Undeterred, further appraisal went ahead down-dip with 23/26a-6Y in 1987. This found a thick (206 m) fractured chalk reservoir section which flowed at 7700 BOPD. In the same year, a 3D seismic survey was acquired. Two more successful wells followed, 23/26a-10 and -10Z.

Wells were drilled on the northern flank of the field, 23/26a-12Z (1989) and in the northwest 23/26a-13 and -13ZA (1990). The latest well, -16, was designed to investigate the field OWC in the flank sands.

Database

Cores

Wells 10, 12Z, 13, 13ZA have full reservoir core coverage. Combined with smaller intervals from the other wells, total core footage is 845 m.

Wireline logs

The Schlumberger FMS borehole imaging tool was run in Wells 13 and 13ZA. This proved an effective means of identifying and orientating principal fractures.

Production tests

All hydrocarbon-bearing wells were production tested. No tests were carried out in Wells 2 and 5. Acid stimulation yielded a pronounced improvement in reservoir performance in all cases. Running of PLT logs in these wells identified producing zones.

Seismic data

A 3D seismic survey was acquired in 1987. It consists of 4300 km sail line data shot in an area of 9.2 × 9.2 km. The data were acquired on a 25 m sub-surface line spacing with

From *Petroleum Geology of Northwest Europe: Proceedings of the 4th Conference* (edited by J. R. Parker).

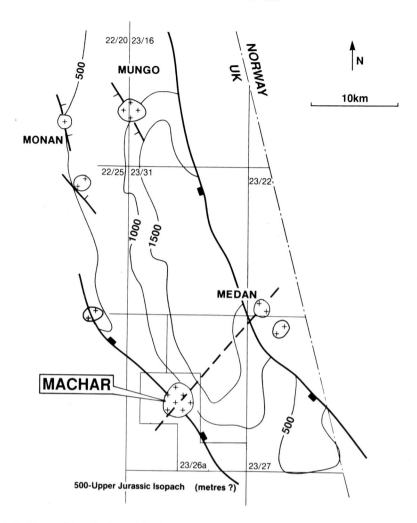

Fig. 1. Location map, 'diapir' fields: Eastern Trough, Central Graben (line of section in Fig. 2 is indicated).

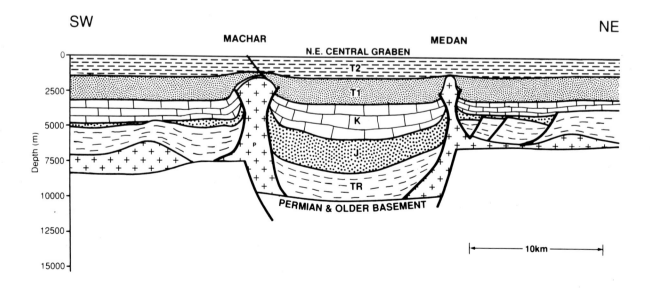

Fig. 2. Regional cross-section.

a 13.33 m (cdp) separation. The data were interpolated in the cross-line direction pre-migration to give a final grid of 12.5 × 13.33 m. In addition to these 3D data, a number of 2D regional lines were used to place the structure in its regional context.

The seismic data are of good quality over the main field area. On the steep flanks of the structure, it has not been possible to image the seismic energy using conventional processing techniques.

Wells 23/26a-3 and -7 were used to calibrate the regional picks.

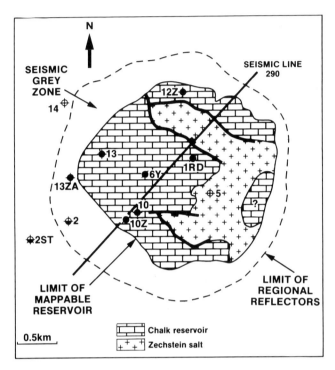

Fig. 3. Well locations, Block 23/26a.

Regional setting

The Machar salt structure is immediately obvious on cross-sections and suggests, at first sight, a structural history of active salt diapirism and piercement. However, drilling has shown that a condensed Mesozoic and early Cenozoic section is preserved over the salt and that the structural history is more complex, being dominated by long periods of buoyant equilibrium of the salt with downbuilding in the surrounding area. These observations have important implications for reservoir distribution and hydrocarbon fill and so warrant further description.

The structural and stratigraphic evolution of the Machar area of the Eastern Trough is summarized in Fig. 4, which is a cross-section based on regional backstripping. Four time intervals which are key to the development of the Machar structure are presented.

Due to the incompressibility of halite, the salt structure was remodelled at each time interval to accommodate the space vacated by decompacting the overburden.

Structural development

Triassic to base Cretaceous

Following deposition of Zechstein evaporites, Late Permian to Early Triassic extension opened up the proto-Eastern Trough. Early fault-controlled deposition triggered salt pillowing along

the graben margins. Subsequent sediment loading continued to drive the pillowing; Triassic and Jurassic sediments were eroded or not deposited over the developing salt highs. The Machar structure was initiated over one of these highs.

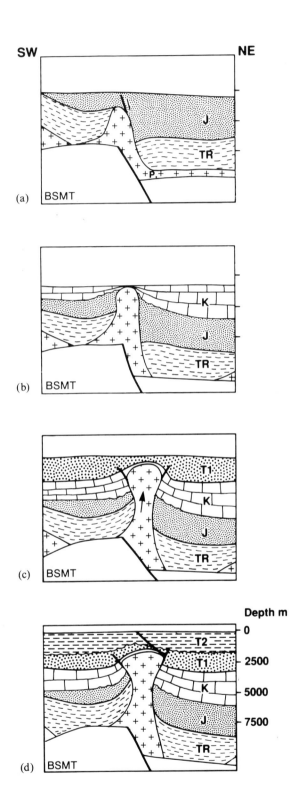

Fig. 4. Machar structural history. (**a**) Triassic–base Cretaceous (TR, J): salt pillowing along graben margin, followed by late Jurassic tectonic event. Basement blocks are Permian–Devonian age. (**b**) Cretaceous–end Paleocene (K): condensed sequence over structural high. Salt in buoyant equilibrium, subsidence in surrounding basinal areas. (**c**) Eocene–mid-Miocene (T1): continued buoyant equilibrium followed by regional tectonic event in mid-Miocene, causing minor diapiric movement. (**d**) Late Miocene–Recent (T2): rapid regional subsidence and overpressuring.

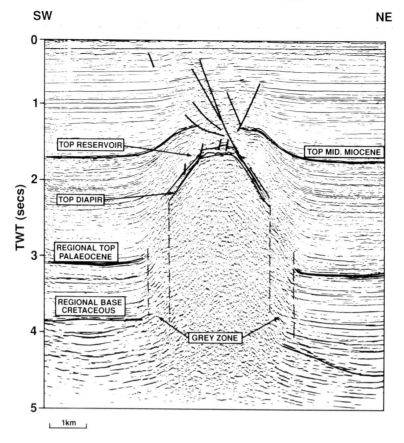

Fig. 5. Seismic section 3D inline 290. (Location of section shown on Fig. 3.)

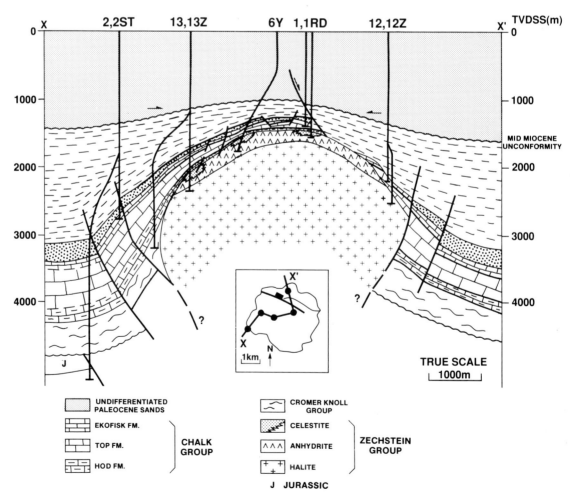

Fig. 6. Structural cross-section.

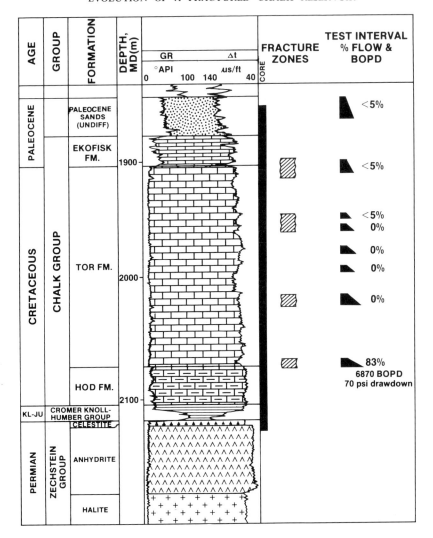

Fig. 7. Type section, Well 23/26a-13.

Cretaceous to end Paleocene

Throughout this period, the salt was in buoyant equilibrium and remained near the seabed. Continued salt withdrawal caused grounding of the Triassic pod near the root of the structure during the Cretaceous, reducing the potential for lateral movement of salt. A condensed chalk sequence less than 300 m thick was deposited over the salt, with 1000 m deposited in the surrounding basinal area. During the Paleocene, the salt may have formed a seabed high and acted as an obstruction to deep marine sand fans flowing from the NW. This has important implications for sand distribution around the structure (Mutti and Normark 1987).

Eocene–mid-Miocene

Buoyant equilibrium continued until a regional Alpine-related inversion in the mid-Miocene. This caused the only true phase of diapiric salt piercement in the history of Machar. Compression and uplift of the salt is manifested by the pronounced unconformity visible on seismic and confirmed by missing section in wells. Thermal modelling of the Kimmeridge Clay Formation source rock of the Central Graben indicates that oil migration began in the Early Oligocene.

Late Miocene–Recent

Rapid regional subsidence during this episode caused over-

pressuring of the Machar reservoir. Machar is currently buried more deeply than at any time in its history. Diagenesis of the diapir caprock probably began in the late Miocene, forming the celestite reservoir. Some residual positive buoyancy remains in the salt; a modest (< 10 m) bathymetric high exists on the seabed.

Seismic definition

The field is identified by a high amplitude seismic reflector at the top of the salt structure (Fig. 5). Top reservoir is not clearly expressed on seismic and may only be picked with confidence near well control.

Figures 5 and 6 clearly show the steeply dipping flanks of the structure and faulting on a variety of scales. A large down-to-the-east normal fault splits the field into its major segments, Machar and Machar North, and small faults mappable at top salt illustrate the fractured nature of the overburden. At the edge of the salt, seismic reflections become incoherent (the 'Grey Zone') due to near-vertical beds and intense faulting. The 'lick-up' of regional top Paleocene and Base Cretaceous reflectors defines the outer limit of the Grey Zone. The fault pattern has been mapped using dip/azimuth displays on a 3D seismic workstation.

Reservoir units

The principal Machar reservoir and focus of this paper is

fractured Paleocene and Cretaceous chalk: the Ekofisk, Tor and Hod formations. Other pay intervals include Paleocene sands and celestite (a diagenetic unit in the salt caprock). Figure 7 shows a type section from 23/26a-13.

Fig. 8. Scanning electron photomicrograph of chalk matrix from 1889.5 m, 23/26a-13, Machar Field. Porosity 29%, permeability 0.22 mD.

Chalk lithology is dominated by clean pelagic bioclastic mudstones of the Tor Formation. The matrix scanning electron photomicrography (Fig. 8) shows the narrow (1–3 μm) pore throats between coccolith fragments. This accounts for the low matrix permeability. Minor packstone and wackestone units are found in redeposited chalk units, probably formed as a result of local erosion during periods of salt movement. These form less than 10% of the lithologies observed in core. This is in contrast to the extensive redeposited chalk units in the Norwegian sector (Kennedy 1980; Watts 1980).

The reservoir has a dual porosity system. Primary, matrix porosity is in the range 12–35% with permeability less than 1 mD.

Secondary fracture porosity, although difficult to measure, is probably less than 1%. However, permeability calculated from production test data exceeds 1000 mD.

The fracture zones provide a means of draining the tight matrix and form a highly conductive path to the well bore. A typical well crosses a few narrow zones of intense fracturing. Production tests imply these tap a widespread interconnected reservoir network. The inflow is dominated by contribution from only one or two of the zones per well.

Fracture style and spacing is extremely variable and an example of the range of fracture types in the chalk is given in Fig. 9. A comprehensive fracture study has been undertaken, but is outside the scope of this paper.

Trap evolution

Seismic interpretation, structural modelling, core and FMS log interpretation have been combined to provide a four-stage trap evolution history (Fig. 10).

(1) *Early Paleocene* Chalk reservoir facies were deposited in a 'wedge' thinning towards the crest of the seabed high. Salt may have protruded at seabed, or the thin chalk was subsequently eroded. Minor eroded and redeposited chalk units testify to syn-sedimentary faulting, probably due to passive jointing and gravity sliding of beds away from the structural high. Faults and fractures formed in this way are likely to have consisted of a component parallel to the

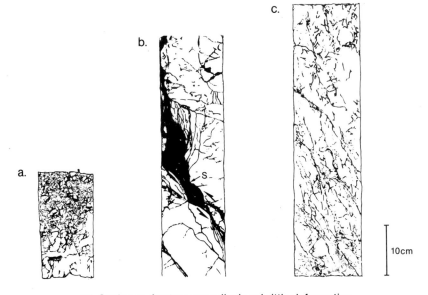

(a) Conjugate fracture zone displays brittle deformation.

(b) Thick stylolite (S) acting as shear zone.

(c) Sub parallel planar fractures.

Fig. 9. Examples of fracturing in chalk.

diapir edge, forming a concentric pattern, and a component radial to the structural high. Paleocene sands may have been preferentially deposited on the NW side of the high.

(2) *Early Oligocene* By this stage the reservoir was sealed by Eocene and Oligocene mudstones. Thermal modelling of the Jurassic source rock shows the earliest oil generation/

migration had begun and the crest of the structure may have been partly filled with oil. Gravity sliding continued in the chalk to the extent that displacement began to segment the field into its main component areas. On the deepest parts of the flanks (500–1000 m), burial diagenesis commenced with stylolite development in the shalier chalk units.

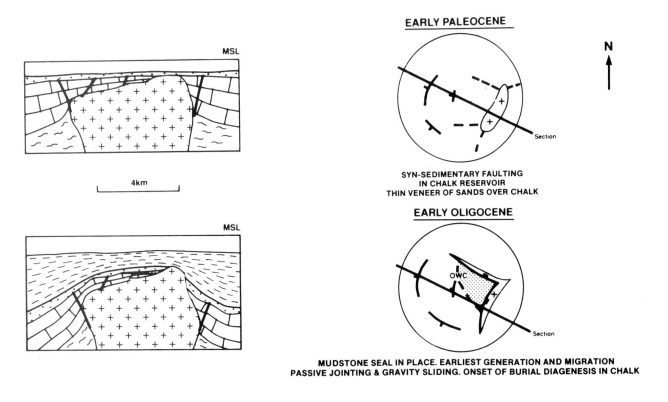

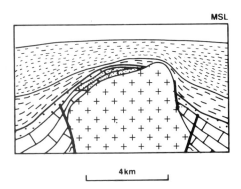

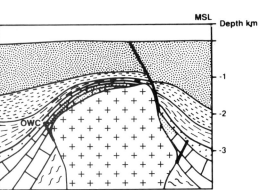

Fig. 10. Trap evolution.

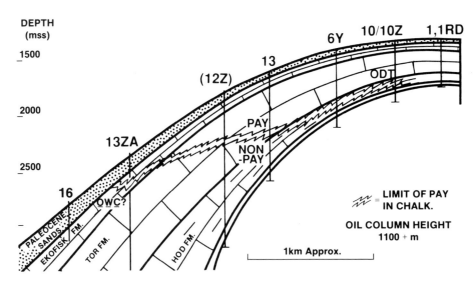

Fig. 11. Pay summary.

(3) *Mid-Miocene* A regional alpine tectonic event in the mid–Late Miocene initiated the only true diapiric movement during the Machar trap evolution. Compression and uplift of the salt caused penetrative tectonic fracturing of the chalk and break-up of the field into its main segments: Machar and Machar North. This coincided with the acme of oil generation and migration. Matrix and fracture pore space in the oil leg are likely to have been preserved. Below the palaeo-oil–water contact, diagenesis continued with fractures becoming cemented in the aqueous pore fluids and stylolite formation particularly well developed in the shaley chalks.

(4) *Present* Rapid subsidence during the late Tertiary to Recent has probably contributed to overpressuring in the reservoir (*c*. 1100 psia at 1700 m ss). This may be exacerbated by continued hydrocarbon charge from the Jurassic. Final trap filling occurs in the deep flank sands, but chalk porosity has been destroyed on the deeper flanks of the structure. Trap leakage is also occurring and shallow gas is a potential hazard.

Impact of evolution on reservoir

The reservoir cross-section (Fig. 11) illustrates the effects of the key evolutionary events on net reservoir. The limit of pay in the chalk is defined by a diagenetic front, where oil migration has been halted by porosity destruction. Chalk pay is progressively lost downflank due to burial diagenesis and late oil emplacement (cf. Taylor and Lapre (1987) for a discussion of this phenomenon in the Norwegian and Danish sectors). The free water level has not yet been encountered but probably lies in the Paleocene sands. Within the area of mappable top salt, total average net/gross in sands and chalk is about 0.75.

Conclusions

The Machar Field consists of a condensed sequence of fractured Paleocene/Cretaceous chalk overlying a salt high. Dia-

piric movement was restricted to one tectonic event in the mid-Miocene, otherwise the structural history is dominated by long periods of buoyant equilibrium. This has preserved a condensed chalk sequence in reservoir condition at shallow depths, surrounded by 'downbuilt' basinal areas which contain mature Jurassic oil source rocks.

Fracturing in the chalk has mostly been due to passive jointing and gravity sliding. Where open fractures are preserved, they form an effective high permeability drain for the tight chalk matrix. Production from wells to date has been concentrated in a few fracture zones.

Net pay in the chalk matrix is progressively destroyed downflank by burial diagenesis. The field OWC has not yet been found, but probably exists in the sand 'skirt' that extends downflank beyond the diapir.

We are grateful to BP Exploration for permission to publish this paper and to the whole Machar team: Alan H. Smith and Bob M. Allan for the seismic interpretation; Kevin Schofield for the sedimentology and help in deriving the geological model; Lauren B. Segal for the petrophysical analysis and Alasdair H. Duncan for the geochemical review.

References

KENNEDY, W. J. 1980. Aspects of Chalk Sedimentation in the Southern Norwegian Offshore. *Proceedings of the Symposium 'The Sedimentation of the North Sea Reservoir Rocks'*. Norwegian Petroleum Society, Geilo.

MUTTI, E. AND NORMARK, W. R. 1987. Comparing Examples of Modern and Ancient Turbidite Systems; Problems and Concepts. *In*: LEGGET, J. K. AND ZUFFA, G. G. (eds) *Marine Clastics Sedimentology*. Graham & Trotman, London, 1–38.

TAYLOR, S. R. AND LAPRE, J. F. 1987. North Sea chalk diagenesis: its effect on reservoir location and properties. *In*: BROOKS, J. AND GLENNIE, K. (eds) Petroleum Geology of NW Europe, Graham & Trotman, London.

WATTS, N. L. *et al.* 1980. Upper Cretaceous and Lower Tertiary Chalks of the Albuskjell Area, North Sea: Deposition in a Slope and Base of Slope Environment. *Geology* **8**, 217–221, May 1990.

The Bruce Field

A. BECKLY,[1] C. DODD[2] and A. LOS[1]

[1] BP Exploration, Farburn Industrial Estate, Dyce, Aberdeen AB2 0PD, UK
[2] C. Dodd Associates, Glenlogie, Pitcaple, Inverurie, UK

Abstract: The Bruce Field is a giant gas condensate accumulation which lies on the western side of the Viking Graben at the northern end of the Beryl Embayment. Although discovered in 1974, the Annex B was not approved until 1990 by which time the field had been shown to extend across three exploration licence blocks (9/8a, 9/9b and 9/9a) and to be extremely complex. There are three main elements to the complexity: structure, reservoir distribution and hydrocarbon charge.

The major structural elements are the Western Flank (a rotated fault block) and the Eastern High with a central low between, the Central Panel; however, relatively poor seismic data quality makes detailed structural interpretation difficult. Each of these structural elements has a different hydrocarbon–water contact, with a proven range of nearly 500 m. The shallowest contact is found on the Eastern High and though the contact in the central low is nearly 350 m deeper, pressure data indicate that the gas leg is in communication with that in the Eastern High. The distribution of the main reservoir, the Middle Jurassic Beryl Group, and the magnitude of the unconformity that underlies it, appear to be controlled by earlier faults which are cross-cut by the present structure. The result of these complexities is that long hydrocarbon columns can be found in the structural lows, and thick reservoir successions can occur on the structural highs.

The Bruce Field is a giant gas condensate accumulation with recoverable reserves (at the time of Annex B approval), estimated at 2.6 TCF of sales gas and 120 MMBBL of 52°API condensate. The main reservoir is the Middle Jurassic Beryl Embayment Group (Knutson and Munro 1991), although subsidiary accumulations have been found locally in the Lower Jurassic Nansen Formation and an Upper Cretaceous limestone unit.

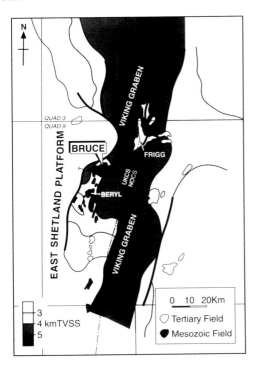

Fig. 1. Location map. Structure on top Middle Jurassic.

The field is located 350 km NE of Aberdeen in a water depth of 120 m, and extends across three exploration licence blocks: 9/8a, 9/9b and 9/9a. It lies at the northern end of the Beryl Embayment (Fig. 1), a complex area of fault blocks between the East Shetland Platform and the Viking Graben. The prolonged appraisal history from initial discovery in 1974 to Annex B approval in 1990 reflects the complexity of the field and its considerable areal extent. The first successful well in each of the licence blocks was effectively the discovery of a new structural segment of the field (Fig. 2), each with a different hydrocarbon contact. This paper is the first published description of the field and, therefore, of necessity describes the structure and stratigraphy. Key aspects that are also discussed are the complexities in reservoir development with respect to the structure and the highly variable, and unusual, fluid distribution.

Field history

Hamilton Brothers discovered the Bruce Field in 1974 with Well 9/8-1, which found wet gas in Middle Jurassic sandstone formations on the dip-slope of a tilted fault block. For the remainder of the 1970s drilling concentrated on the search for a western extension, with disappointing results (Fig. 3). The only well drilled in the east at this time, Well 9/9b-1, was no more successful, but is now thought to have penetrated the fault that separates the eastern and central areas of the field (see Fig. 9).

It was therefore not until 1981 that an eastward extension of the field was proved with a successful well in the Central Panel. This was followed by extensive drilling across the entire area now recognized as the Bruce Field throughout the early 1980s (Fig. 3), and culminated in the discovery and appraisal of the Eastern High between 1984 and 1986. It was the promising results from this area that gave a major impetus to the development of the field. In 1982, 3D seismic was shot over the central and western parts of the field, and over the eastern part in 1985.

The Annex B was approved in 1990 and development drilling started the same year. Three phases are planned in the development of the Bruce Field. Phase I, the pre-drilling of 12 wells in the east of the field through a subsea template, is now nearing completion. This is to be followed by the installation of a drilling platform over the template. Phase II is the drilling of additional wells in the east from this facility. Phase III is the development of the western area.

First sales gas is scheduled for delivery in October 1993. Peak production will be reached in 1994 with sales to British Gas of some 480 MMSCFD and sales to Corby Power Project of some 50 MMSCFD. Liquids production will peak at 80 000 BOPD. The ultimate field life is projected to be approximately 25 years.

From *Petroleum Geology of Northwest Europe: Proceedings of the 4th Conference* (edited by J. R. Parker).
© 1993 Petroleum Geology '86 Ltd. Published by The Geological Society, London, pp. 1453–1463.

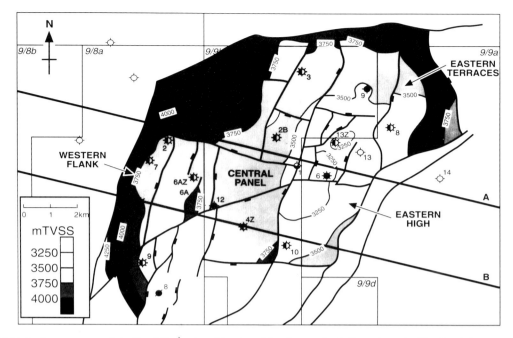

Fig. 2. Bruce Field depth map to top of the Beryl Formation. Line A and B refer to seismic lines shown in Figs 6 and 7.

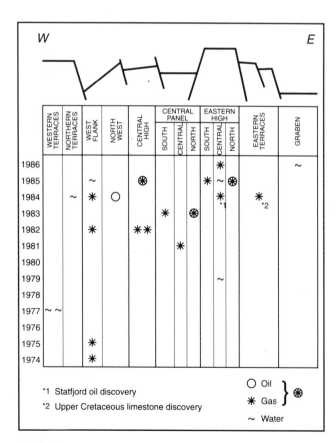

Fig. 3. Schematic representation of the history of appraisal drilling on the Bruce Field.

Stratigraphy

The Upper Triassic and Jurassic stratigraphy of the Bruce Field is shown in Fig. 4, which compares published interpretations with the informal nomenclature in use for the field. Since definition and refinement of lithostratigraphic units is

not a primary aim of this paper, those proposed for the Beryl Field (Knutson and Munro 1991) are adopted. However, for the reservoir interval the informal Bruce nomenclature is retained for two reasons.

1. The subdivisions of the Beryl Formation recognized over the Beryl Field are of differing importance in the Bruce Field.
2. The division between the B and C Sands is difficult over much of the Bruce Field and hence use of the Linnhe Formation for the latter would place undue emphasis on an uncertain boundary.

The wireline response of the reservoir interval is shown in Fig. 5.

The Upper Triassic to Upper Jurassic succession comprises two transgressive megasequences separated by an erosional unconformity (Fig. 4). The main difference between these sequences is that the lower transgressive megasequence has a layer-cake stratigraphy, whereas the upper transgressive megasequence shows abrupt thickness changes in response to contemporary fault activity. The lateral facies changes invoked by Richards (1991) for the Lower Jurassic have not been recognized over the Bruce Field.

Cormorant Group (Lewis Formation)

The top of this group is taken at the top of the Lewis Formation (Fig. 4). Most wells in the Bruce Field do not reach it; however, where penetrated, the Lewis Formation is a thick succession of interbedded sandstone (often pebbly) and red–brown mudstone. Where cored, the sandstones are often cross-bedded. Based on this sparse information the formation is interpreted to have been deposited by braided rivers, although the muddier intervals may represent flood plain or lacustrine intervals. Frostick *et al.* (1992) review the Triassic in the Bruce–Beryl area: the Lewis Formation appears to be equivalent to their Middle and Lower Lunde Formation. However, they describe this interval as lacking coarse detritus and interpret it to have been dominated by a playa lake.

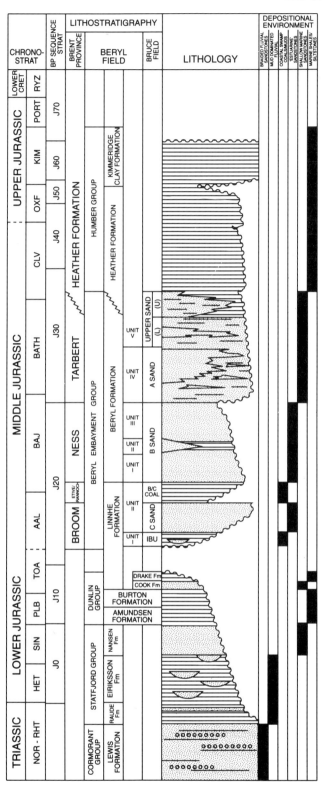

Fig. 4. Schematic representation of Jurassic stratigraphy of the Bruce Field. The lithostratigraphy for the Beryl Field is taken from Knutson and Munro (1991) and the sequence stratigraphy is that of Rattey and Hayward (1993). The Middle Jurassic interval is expanded to approximately double scale compared with the Lower and Upper Jurassic.

Statfjord Group

Raude Formation This formation is typically red–brown and because of this has historically been included in an undifferentiated 'Triassic'. The formation is generally around 200 m thick

and is dominated by red–brown siltstone and mudstone with thin sandstone beds towards the top. Where cored, the formation is interpreted as the deposit of ephemeral lakes. At the top of the formation there is often an interval of 20–30 m characterized by poor hole conditions. This could be the product of weathering, but is seen both when the apparently conformable Eiriksson Formation is present and when the Raude Formation is significantly truncated by the base Beryl Group unconformity. This implies a depositional hiatus between the Eiriksson and Raude formations. The scant biostratigraphic control for the Raude Formation in the Bruce Field is all consistent with a Rhaetian age. This is equivalent to the Upper Lunde Formation of Frostick *et al.* (1992) which they interpreted as a playa lake deposit.

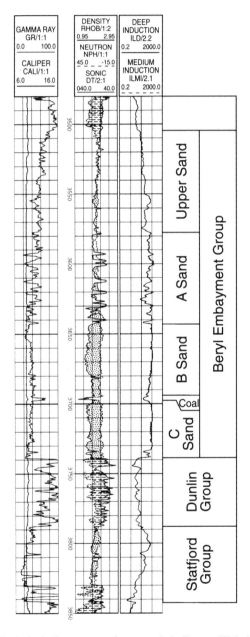

Fig. 5. Typical wireline response for reservoir in Eastern High development well. Note that in this well the Beryl Embayment Group is gas-filled and the Nansen Formation is oil-filled.

Eiriksson Formation The formation comprises interbedded sandstone and mudstone, the latter generally grey and occasionally carbonaceous, interpreted to be the deposits of a fairly high sinuosity fluvial system. Where present, the Eiriksson Formation varies in thickness from 20–80 m. The formation normally occurs to the south of a major SW–NE lineament (see Fig. 10); therefore there may be a significant fault control.

Nansen Formation The Nansen Formation is a very clean, typically highly bioturbated, shallow marine sandstone unit with a blocky log response. Although the total thickness ranges from 16–55 m the formation is normally 40–50 m thick. Two very thin, muddier intervals, which cause characteristic gamma ray spikes, divide the sandstone package into three correlatable units of fairly even thickness (Fig. 5). The Nansen Formation is an oil reservoir in the east of the field where it has an average porosity of 16.4% and flowed on test at 4000 BOPD from Well 9/9a-6.

Dunlin Group

Four units can be recognized within the Dunlin Group over the Bruce Field and since these correspond lithologically to the units defined in the type wells (Deegan and Scull 1977) all four formation names are used. Within the area of the Bruce Field there is no evidence for the lateral facies changes proposed by Richards (1991) and, in contrast, the group has an extremely well-developed layer-cake stratigraphy (particularly well defined by the FDC/CNL logs). The group is dominated by marine siltstone and mudstone, apart from the Cook Formation which is a distinct, cleaning upward marine sandstone unit. The thickness of the Dunlin Group varies from 40–70 m but much of this is accounted for by variation in thickness of the Drake Formation. Whether this is the result of primary deposition or of later truncation is not clear.

Beryl Embayment Group

Basal Unconformity The major unconformity within the reservoir succession is at the base of the Beryl Embayment Group and therefore divides the two megasequences. The lack of any discernible facies or stratigraphic variation within the Dunlin Group does not support a depositional pinch-out of this unit. In the north of the field, the Beryl Embayment Group rests directly on truncated Raude Formation.

Controls on the level of erosion are less certain. Broadly, the Lower Jurassic is preserved to the south of a line trending from northeast to southwest across the field and shown as a fault on Fig. 10. This is similar to the thickening trend observed for the Eiriksson Formation and suggests the existence of a structural lineament in this orientation.

Linnhe Formation: (Interbedded Unit and C Sand): Units I and II of the Linnhe Formation in the Beryl Area (Knutson and Munro 1991) are equivalent to the Interbedded Unit (IBU) and C Sand, respectively, in the informal Bruce nomenclature. In the Bruce area these units appear to onlap the underlying structure. At the extreme onlap edge the Linnhe Formation is almost absent.

Interbedded Unit (IBU) The IBU can be up to 75 m thick, but is only present in areas where the Dunlin Group is present. It is composed of coal, siltstone and rare channel sandstone and is interpreted as the deposit of a coastal swamp lying in structural lows.

C Sand The C sand is a good reservoir over much of the field. It is dominated by cross-bedded sandstone interpreted as an estuarine deposit (Richards *et al.* 1988) with variable dominance of marine and fluvial influence. In the east a conglomeratic base interpreted as a beach deposit is locally developed. The C Sand reaches 100 m in thickness; average porosities range from 14–19% and net sand can exceed 90%.

Beryl Formation The B and C sands are divided in some wells by the B/C Coal. This is a thin coal/siltstone interval that reaches a maximum thickness of 15 m. This is also interpreted as a coastal swamp deposit, but is stratigraphically equivalent to virtually the entire Etive and Rannoch formations of the Brent Province.

B Sand The B Sand is similar in facies and depositional environment to the C Sand, and tends to mirror the thickness variations of the latter, reaching a maximum of almost 150 m. The B Sand is generally a massive, indivisible sand unit but where it is thickest the three units recognized in the Beryl Field (Fig. 4) can be identified. However, when the B/C Coal is absent, distinction between the B and C Sands is almost impossible on lithology alone. The scant biostratigraphic control suggests that in some cases the B Sand may pinch-out completely, but with little evidence for erosion of the underlying C Sand.

The B Sand is the best reservoir in the field with porosity reaching 19% and net to gross ratio generally greater than 90%.

A Sand The base of the A Sand is marked by transgression of the underlying estuarine succession, represented by a thin muddy interval. The A Sand contains intervals with abundant shallow marine burrows, and sheet sands that are unburrowed. The unburrowed sands are interpreted as probable storm deposits. Sporadic carbonate cemented intervals occur through the unit. Rare siltstone deposits also occur. The thickness variations of the unit tend to mirror the present structure with the thickest developments in the structural lows to north and west where it reaches 75 m. Best reservoir quality is in the west of the field where the average porosity locally exceeds 19%.

Upper Sand The Upper Sand is composed of similar facies to the A Sand, but these are finer grained and the burrowed sands are more dominant and argillaceous. This is interpreted as the deposit of a deeper water shelf. The top of the A Sand and base of the Upper Sand is a sharp boundary identified by a particularly argillaceous interval. This is interpreted as the deposit of a transgressive 'flooding event'.

Table 1. Typical Bruce Field reservoir properties

Interval	Thickness range (m)	Typical thickness (m)	Typical net to gross (%)	Maximum net to gross (%)	Typical core helium porosity (%)	Typical core air permeability (mD)	Typical hydrocarbon saturation (%)
Upper Sand	0–200	100	50	90	13.5	5	85
A Sand	0–110	70	75	90	15	25	90
B Sand	0–150	50	95	100	17	300	95
C Sand	0–100	55	80	100	16	100	90
Nansen	0–50	40	95	96	16	150	80

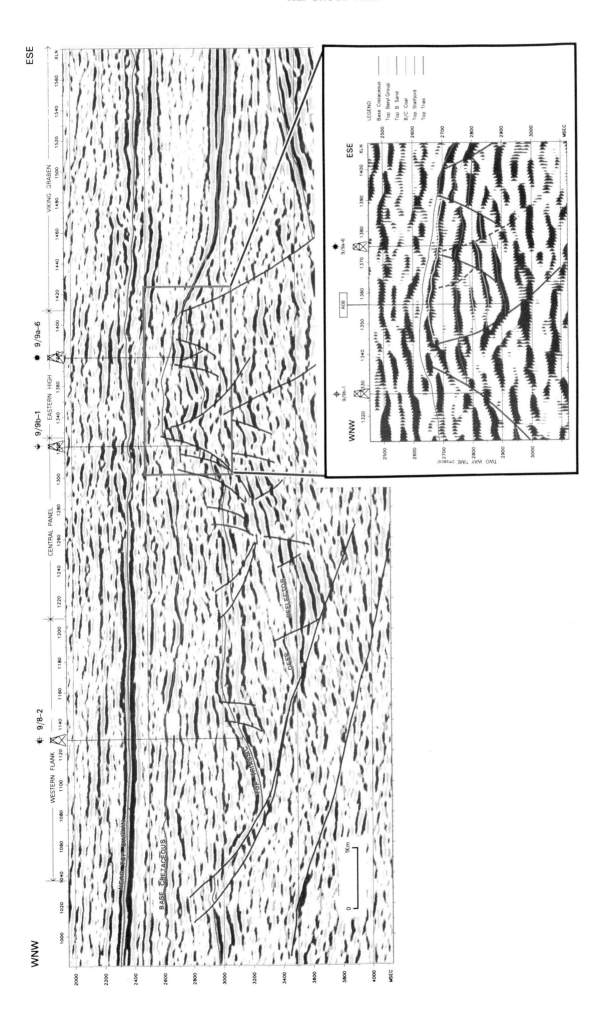

Fig. 6. Interpreted seismic line across the Bruce Field (Line A on Fig. 2), with detail of the Eastern High.

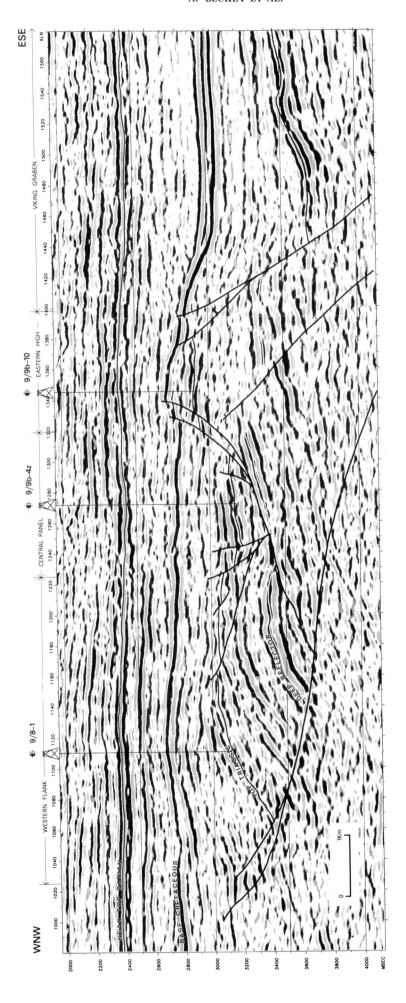

Fig. 7. Interpreted seismic line across the Bruce Field (Line B on Fig. 2).

Above this, the Upper Sand comprises a coarsening-upward unit with an increasing concentration of storm sands, suggesting progressive shallowing on the shelf. These are then capped by a further flooding event and a typically thick interval of very argillaceous and burrowed sand which has very low reservoir quality. The cleaner storm sands in the lower part of the Upper Sand constitute most of the net sand in this unit. The boundary with the Humber Group is diachronous and in some areas the middle flooding event effectively marks the top of the sand.

The Upper Sand can reach considerable thickness, exceeding 150 m in the northwest of the field. As in the A Sand, the best reservoir quality is in the west.

Field mapping

Seismic

Two 3D surveys cover the field. The first, shot in 1982, covers the central and western parts of the field and was acquired using a 75 m line spacing. The second survey was shot in 1985 over the east of the field with a 50 m line spacing. Both surveys were shot in a WNW–ESE direction. To provide a uniform dataset both surveys were reprocessed and merged in 1990.

The original 1982 and 1985 surveys were used in the preparation of maps for the Annex B, but the quality of these datasets is only fair to poor. At the reservoir level this only allowed the mapping of the 'Key Horizon': a reflector at or below the base of the reservoir.

The reprocessing and merging of the two 3D surveys in 1990 considerably improved the quality of the seismic data-set particularly in the east of the field, although in other parts it remained poor. These new data have allowed some fine tuning of the structural interpretation and, in areas of the Eastern High, direct mapping of reservoir distribution.

Field structure

Three major elements dominate the field structure: the Western Flank, the Central Panel and the Eastern High (Figs 6 and 7). The most clearly defined of these is the Western Flank, which is a westward-dipping rotated fault block. The western limit of the field is along the listric fault which creates this feature, and is about 7 km basinward of the East Shetland Platform.

This fault exerts a major control on the field structure, and the three major elements of the field can all be matched to structures identified from sandbox models of listric faults (cf. Ellis and McClay 1988; Fig. 8). However, there are a number of factors which suggest that this comparison, though useful as a means of visualizing the field in cross-section, is too simplistic.

The Bruce area is cut by cross-faults of variable orientation (Fig. 2) possibly reflecting the dual influence of a NE–SW Caledonian trend and a NW–SE Tornquist trend. Such elements, in addition to affecting the internal structure of the field, are also the major influence on field closure to north and south, though this is achieved by dip rather than fault closure. To the south, the controlling NE–SW cross-fault produces an overall eastward offset of the Eastern High of the Bruce Field relative to the Beryl Field. To the north, the Eastern High and Central Panel dip into an ENE–WSW-trending cross-fault which separates the Beryl Embayment from the East Shetland Platform. The slightly different orientation of this fault to that of the Caledonian trend may reflect the presence of a granite batholith in the footwall (Donato and Tully 1982). The deepening of the top reservoir to the north probably reflects oblique-slip movement (*sensu* Gibbs 1987) on this fault.

Within the field the offset on faults at top Triassic are the cumulative effects of more than one phase of faulting. Different faults were active through time, and this may reflect changes in extension direction. This has a marked effect on reservoir distribution.

Reservoir distribution

The appraisal drilling on Bruce showed considerable variation in the thickness and stratigraphy of the Jurassic succession. The seismic data prior to reprocessing were not adequate to map this variation directly. Following reprocessing, it is now possible on the Eastern High to map the Top B Sand directly and distinguish between the top Triassic and Top Statfjord. This has allowed detailed mapping of reservoir distribution which has been confirmed by development drilling.

One of the surprises from appraisal drilling was the presence of a thick reservoir section on the structurally elevated Eastern High, overlying a fully preserved Lower Jurassic succession. Thin and incomplete Middle to Lower Jurassic successions had already been found in structurally lower areas to the west. The majority of the structural differences in depth to top reservoir are infilled by the Humber Group and, therefore, there was early evidence of changing structural controls from Early to Late Jurassic times.

A large fault downthrowing to the west separates the Eastern High from the Central Panel. The throw on this fault is largely taken up by thickness changes in the Humber Group. However, a combination of well evidence and the reprocessed seismic data showed an even larger fault to be present within the Eastern High, causing growth during deposition of the lower part of the Beryl Embayment Group, and preservation of the Lower Jurassic (Figs 6 and 9). At the level of the top Lewis Formation the throw on this fault is about 400 m and yet there is little evidence of any displacement at top reservoir.

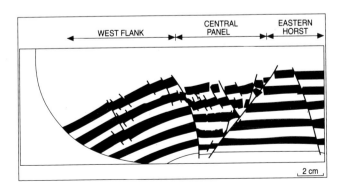

Fig. 8. Line drawing of sandbox model of listric faults, with interpreted comparison with the Bruce structure (after Ellis 1988).

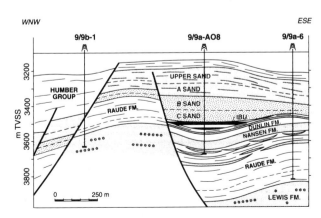

Fig. 9. Geological cross-section over the Eastern High.

NANSEN AND EIRIKSSON FORMATION

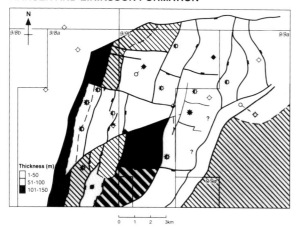

UPPER AND A SAND

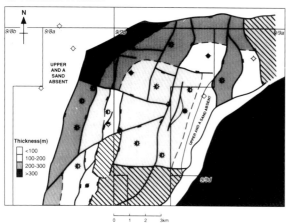

B AND C SAND

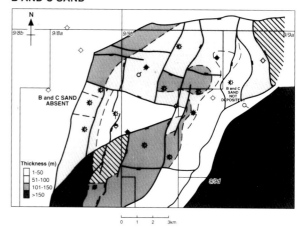

HUMBER GROUP

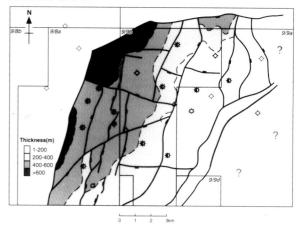

Fig. 10. Reservoir distribution maps. Cross-hatching indicates areas of low confidence.

Though the growth fault on the Eastern High is orientated N–S and hence parallel to the faults which dominate the structure, faults with a more E–W orientation also appear to have a major influence on Lower Jurassic distribution. The Lower Jurassic is not preserved over the northern part of the Eastern High and this observation implies the presence of a cross fault with significant vertical throw, although this is not clearly defined on seismic. Over the field area, well evidence suggests that a NE–SW-trending lineament is the main control on Lower Jurassic preservation (Fig. 10) and this is comparable in orientation to the cross faults interpreted to affect Triassic sedimentation (Frostick *et al.* 1992). The true nature of this lineament is unclear because of the difficulty in mapping it on seismic and it may be more complex than the single fault shown on maps (Fig. 2).

Controls on the distribution of the lower part of the Bruce Group are probably the most difficult to interpret (Fig. 10). There is clearly some influence from the faults that control the Lower Jurassic, particularly in the distribution of the Interbedded Unit, but there are also areas where the lower part of the Beryl Embayment Group is relatively thick but rests directly on the Triassic. One feature that does appear to emerge is an eastward narrowing 'high' trending across the middle of the field, terminated by the growth fault mapped on the Eastern High (Fig. 10). By contrast the upper, more marine, part of the reservoir shows greater similarities in distribution to the Humber Group above (Fig. 10) and hence to the structure which now dominates the field.

Fluid distribution

All the panels of the Bruce Field have a gas condensate leg underlain by a black oil rim of variable thickness. Although there is some variation in the composition of these phases between the panels (Table 2), the much more significant contrast is in the level of the fluid contacts (Fig. 12).

The range of oil–water contacts for the field is over 500 m with the deepest in the northwest and the shallowest on the Eastern High. A cross-plot of pressure v. depth (Fig. 13) shows each of the major panels of the field to have a water leg at different pressure and therefore not to share a common aquifer. By contrast, despite a difference in contact of some 300 m, the gas legs of the Central Panel and the Eastern High lie on the same line (Fig. 13). This suggests that there is communication between these panels within the gas leg and may indicate that gas filling caused perched water contacts to be developed. More subtle variations of the gas–oil contact are seen within panels (15 m on the Eastern High) and may result from similar, small-scale control between segments. A complication caused by the common gas column is that it is impossible to interpret from pressure data the level of the contact associated with wells where the reservoir is gas filled.

Although the majority of the field is normally pressured, overpressure and depletion provide evidence for vertical permeability barriers on both geological and production time scales.

The Eastern Terraces are the only area of the field in which

Table 2. Typical Bruce Field fluid properties

Accumulation ⟍ Properties	East (north) gas	East (south) gas	Central and western areas gas	Field oil rim	Statfjord oil accumulation
Condensate:gas ratio/ Gas:oil ratio	55–80 (STB/MMSCF)	80–105 (STB/MMSCF)	90–110 (STB/MMSCF)	900–1400 (SCF/STB)	900 (SCF/STB)
Liquid density (°API)	50–52	50–52	50–52	36–42	38
Saturation pressure (PSIA)	4900–5450	5050–5350	5250–5850	4150–5050	3550
Reservoir pressure (PSIA)	5600	5600	5700–5900	5650–5950	5700
Reservoir temperature (°F)	205–215	205–215	215–225	210–225	210

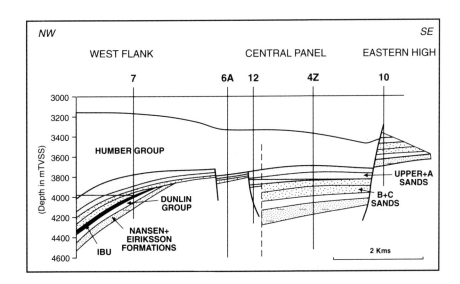

Fig. 11. Schematic geological cross-section over the Bruce Field, showing variation in hydrocarbon contacts. Line of section equivalent to line B on Fig. 2.

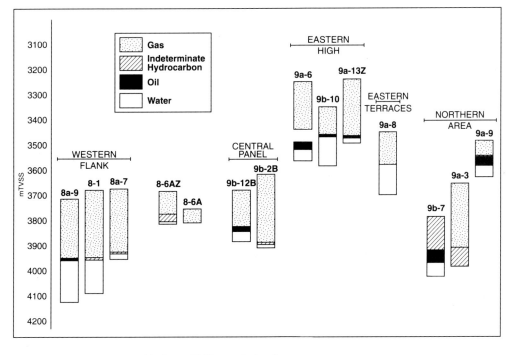

Fig. 12. Reservoir fluid columns as encountered in Bruce Field appraisal wells.

the Beryl Embayment Group is overpressured. However, on the Eastern High, overpressure is encountered in an Upper Cretaceous limestone unit and in the Raude and Lewis formations. This indicates effective top and bottom seal to the Beryl Embayment Group.

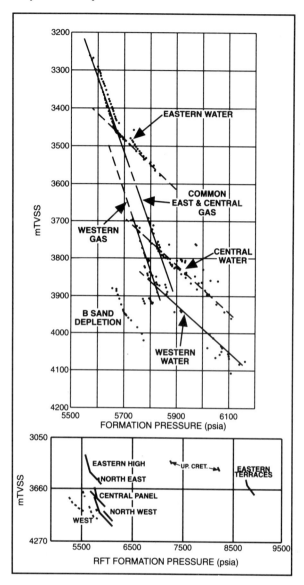

Fig. 13. Summary of pressure data from Repeat Formation Tester for the Bruce Field.

On the Western Flank, pressure depletion due to production from the Beryl Field indicates the presence of vertical permeability barriers within the Beryl Embayment Group. The B Sand in this area is depleted whereas the reservoir units above and below are not. This suggests that the B–C Coal and mudstone at the base of the A Sand both form laterally extensive permeability barriers, whereas the depletion indicates lateral connectivity within the B Sand to Beryl wells 6 km to the south.

The data presently available indicate some of the potential complexity in fluid distribution within the Bruce Field. However, the implications for production will only become known when there is a substantial volume of dynamic data.

Conclusions

The Bruce Field lies in a structurally complex area which was evolving at the time of reservoir sand deposition. This has

resulted in the Middle Jurassic reservoir distribution being controlled by faults now largely obscured by the later (Upper Jurassic) development of the structure. However, such features can be mapped where seismic data allow the interpretation of intra-reservoir reflectors.

In contrast, hydrocarbon distribution appears to be mainly controlled by the faults which were active during the Upper Jurassic and offset the reservoir. This has resulted in hydrocarbon–water contacts varying by over 500 m across the field. However, pressure data show that over 300 m of this variation is between areas with apparent communication in the gas leg. The implications of this for connectivity and field drainage will probably not be answered until there is a considerable production history.

The authors would like to thank Pete Ellis (structural geology), Regis Marion (petrophysics), Brian Cornock (reservoir engineering) and Dave Ewen (biostratigraphy) for their technical contributions to the compilation of this paper; also all those who have worked on Bruce, far too numerous to name, without whose efforts our understanding would be poorer. We would also like to acknowledge our partners in the Bruce Field: Elf UK plc, Hamilton Oil Company, Total Oil Marine plc, and Lasmo North Sea Ltd for permission to publish this paper, while emphasizing that the interpretations presented may not be shared by them. Thanks also to Damien Theaker and the BP Exploration Drawing Office for all their drafting work associated with this paper.

References

DEEGAN, C. E. AND SCULL, B. J. 1977. *A Standard Lithostratigraphic Nomenclature for the Central and Northern North Sea.* Report of the Institute of Geological Sciences 77/25.

DONATO, J. A. AND TULLY, M. C. 1982. A proposed granite batholith along the western flank of the North Sea Viking Graben. *Geophysical Journal of the Royal Astronomical Society,* **69**, 187–196.

ELLIS, P. G. 1988. *Geometry and Kinematic Analysis of Extensional Faulting from Analogue Model Studies.* PhD Thesis, University of London.

—— AND MCCLAY, K. R. 1988. Listric extensional fault systems—results of analogue model experiments. *Basin Research,* **1**, 55–70.

FROSTICK, L. E., LINSEY, T. K. AND REID, I. 1992. Tectonic and climatic control of Triassic sedimentation in the Beryl Basin, northern North Sea. *Journal of the Geological Society, London,* **149**, 13–26.

GIBBS, A. 1987. Development of extension and mixed-mode sedimentary basins. *In:* COWARD, M. P., DEWEY, J. F. AND HANCOCK, P. L. (eds) *Continental Extensional Tectonics.* Geological Society, London, Special Publication, **28**, 19–33.

KNUTSON, C. A. AND MUNRO, I. C. 1991. The Beryl Field, Block 8/13, UK North Sea. *In:* ABBOTTS, I. L. (ed.) *United Kingdom Oil and Gas Fields 25 Years Commemorative Volume.* Geological Society, London, Memoir **14**, 33–42.

RATTEY, R. P. AND HAYWARD, A. B. 1993. The Middle Jurassic to Early Cretaceous basin evolution of the Central and Northern North Sea. *In:* PARKER, J. R. (ed.) *Petroleum Geology of Northwest Europe: Proceedings of the 4th Conference.* Geological Society, London, 215–249.

RICHARDS, P. C. 1991. Evolution of Lower Jurassic coastal plain and fan delta sediments in the Beryl Embayment, North Sea. *Journal of the Geological Society, London,* **148**, 1037–1047.

——, BROWN, S., DEAN, J. M. AND ANDERTON, R. 1988. A new palaeogeographic reconstruction for the Middle Jurassic of the northern North Sea. *Journal of the Geological Society, London,* **145**, 883–886.

Discussion

Question (G. Robertson, Mobil North Sea Ltd):

The seismic line presented through Bruce Field showed the western bounding fault to the field to be a listric fault. Can you explain why this is interpreted as a listric fault and what stratigraphic interval or lithological units is this fault decolling on?

Answer (A. Beckly):

The west flank of the field is thought to be bounded by a listric fault for two main reasons. Firstly, the structural geometry of the field is compatible with movement on such a fault as is illustrated by the similarity to sandbox models of listric faults. Secondly, on some seismic lines a listric fault appears to be imaged directly. The level of the detachment is, however, uncertain. The 'Deep Reflector' (possibly of top Zechstein) appears to be rotated by the fault and therefore the detachment is probably within or below the Paleozoic succession.

Hyde: a proposed field development in the Southern North Sea using horizontal wells

R. P. STEELE, R. M. ALLAN, G. J. ALLINSON and A. J. BOOTH

BP Exploration, 301 St Vincent Street, Glasgow G2 5DD, UK

Abstract: The Hyde Field was discovered in 1982. It is a small gas field with a poor quality Lower Leman Sandstone reservoir, close to existing facilities. It may be typical of many future developments in the Southern North Sea. A development plan prepared for the field in 1989 was withdrawn because of the unacceptable economic risk. Since then, a revised plan has been prepared. It is proposed to produce the gas through three near-horizontal wells drilled close to the crest of the structure. These wells will target reservoir zones of moderate permeability (*c.* 5 mD) formed from the deposits of large aeolian bedforms.

The viability of the project was sensitive to the economics of the downside case. The risk was investigated by a statistical analysis of field performance. This involved assigning a range to each of the parameters controlling field GIIP and deliverability. These were then combined statistically to produce a probability distribution of economic performance which in turn was used in the project evaluation. Following this analysis, funds were obtained for the drilling of an appraisal well to test the major remaining uncertainty: the deliverability of a horizontal well. This well was completed on 20 February 1992. The planning, drilling, results and consequences are described below.

The Hyde Field is a dry gas accumulation located in the Southern North Sea immediately to the north of West Sole and some 55 km to the east of the North Humberside coast (Fig. 1). It was discovered in 1982 with the Well 48/6-25 and lies mostly in Blocks 48/6 and 47/10 (BP 100%) with smaller volumes of gas in Blocks 47/5a and 48/1.

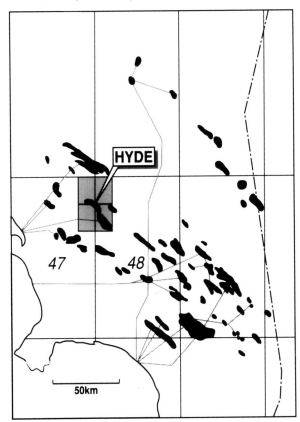

Fig. 1. Location map showing Hyde in relation to existing fields and facilities in the Southern North Sea.

The field is located on a northwest-plunging, faulted anticline (Fig. 2). The dominant fault direction is NW–SE with dips to both the NE and SW. Most of the faults are in net extension. The faults are all reactivated Carboniferous features, which in turn probably reflect an earlier Caledonian fault trend. Cross-faults are uncommon and tend to be transfer faults that are short and have small vertical displacements (20–50 m).

The reservoir is a poor to moderate quality aeolian sandstone in the Lower Leman Sandstone Formation of the Rotliegendes Group.

Four appraisal wells, 48/6-26, -27, 47/5a-2 and -3, were drilled between 1982 and 1985 (Fig. 2). These wells, together with the 3D seismic survey acquired in 1986, effectively appraised the discovery. During this time, estimates of GIIP ranged up to 1.5 TCF, reducing to *c.* 300 BCF following the 47/5a wells.

In 1989 a proposed development of the field using 5 hydraulically fractured deviated wells was rejected by BP management because the economic exposure and risk was judged unacceptable.

Two further wells have been drilled on Hyde within the last two years. Well 48/1-1, drilled by Amerada Hess in 1990, proved the small, low-relief extension of the field into Block 48/1. In February 1992 BP completed Well 48/6-34 and sidetrack 34Z.

A revised development plan for the field envisages three near-horizontal production wells in a core area of the field lying in Block 48/6 (Figs 2 and 3). These wells are designed to improve substantially the initial deliverability in comparison with vertical equivalents. Field drainage and thus economic reserves will also be enhanced. The GIIP in the proposed development area is estimated at around 210 BCF, with reserves of around 150 BCF. The gas will be produced through a minimum facility platform operating in 'not-normally-manned' mode, and exported to shore via BP's West Sole installations.

This paper describes some of the work of the 'sub-surface team' (i.e. geoscientists, reservoir engineers and petrophysicists) that has contributed to the success of the 48/6-34Z appraisal well and the submission of the revised development scheme. A large part of the improvement in the field economics embodied in the revised plan is due to changes in the commercial circumstances of the field and the facilities engineering of the development. This includes changes in the tax position of the field, the deregulation of the gas market, 'cloning' the platform design from Amethyst and, not least, Statoil taking a 45% equity stake in the field following the results of 48/6-34 well. The 'sub-surface' contribution has been to minimize the risk and facilitate a low-cost development.

From *Petroleum Geology of Northwest Europe: Proceedings of the 4th Conference* (edited by J. R. Parker).
© 1993 Petroleum Geology '86 Ltd. Published by The Geological Society, London, pp. 1465–1472.

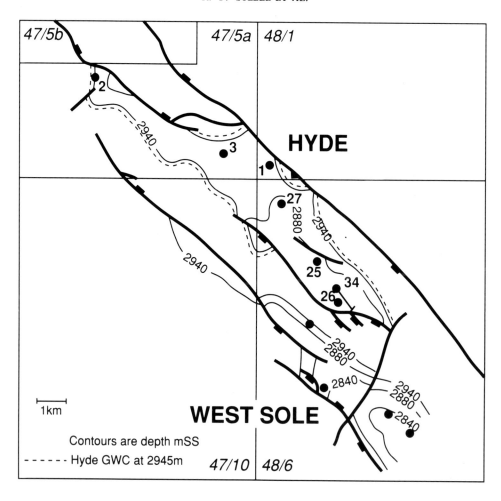

Fig. 2. Depth to top reservoir, with the appraisal wells marked. The proposed core development area lies in Block 48/6. The structural separation of Hyde from West Sole became clear only after a detailed depth conversion of the Zechstein.

The present account covers the reservoir geology, a statistical analysis of field performance that was conducted primarily to assess the confidence in the downside case, and a review of the 48/6-34Z well.

The reservoir

Engineering considerations

A number of features impose important constraints on the development of the field.

1. Almost all of the field has partial gas column. The reservoir is around 110 m thick and the maximum gas column is about 105 m.
2. The permeability of the reservoir ranges from 0.01–40 mD with a modal value of 1 mD (uncorrected permeability to air of air-dried samples). Capillary effects lead to high water saturation. This is nowhere lower than 40%, even in the best quality rock at the crest of the field. The whole reservoir is effectively a transition zone, with relative permeability, and thus gas saturation, a very important control on well performance.
3. The maximum unstimulated flow rate achieved from the field prior to February 1992 was 12 MMSCFD from 48/6-26 (normalized to 600 PSIA WHFP and 5.5 inch tubing).

These facts indicate that wells require some form of stimulation to achieve economic flow rates. Horizontal drilling and

hydraulically fractured deviated wells are the two principal options.

The partial gas column raises the possibility of water production from fractured wells. Fracturing would also necessitate the installation of facilities to handle the back-produced fluid and proppant, at considerable capital cost. These risks can be minimized or eliminated with horizontal wells. The questions that arise are: is the reservoir suitable for exploitation by horizontal wells and what is the likely performance of those wells?

Geology

The five-fold zonation of the reservoir is illustrated in Figs 3 and 4.

- *Zone 1* at the base is thin, fluvial, tight and in the water leg.
- *Zone 2* is 38–46 m thick in the core area. Except for one fluvial horizon in the south it is entirely aeolian in the core area. Its top is arbitrarily defined within an aeolian succession except in the south where it is picked at the base of another fluvial unit.

No further breakdown of the zone has been sought. Its appearance over most of the area as a continuous aeolian succession with no damp horizons is suggestive of a climbing dry erg. However, the presence of fluvial horizons in the south and their possible continuity with similar facies in West Sole may indicate that there are supersurfaces within this succession marking periods of contraction or destruc-

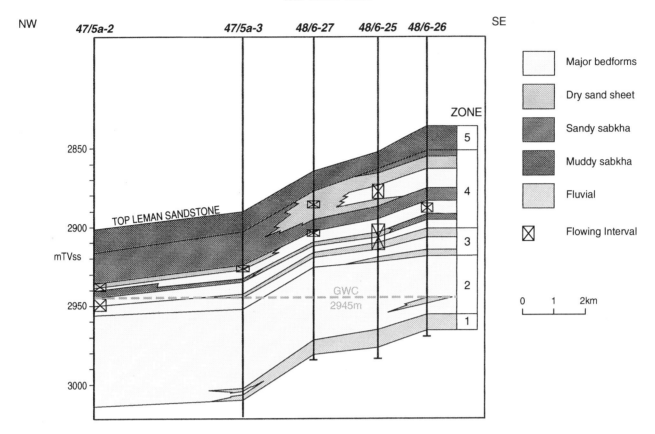

Fig. 3. Cross-section along the axis of the field. The plunge of the structure to the northwest is clear. The diagram also shows the facies change in the upper part of the reservoir that occurs into Block 47/5a. Aeolian layers with reasonable quality rock in Block 48/6 pass laterally into sandy sabkha facies with very low permeability in Block 47/5a. This effectively confines the core development area of the field to the southeast end of the section.

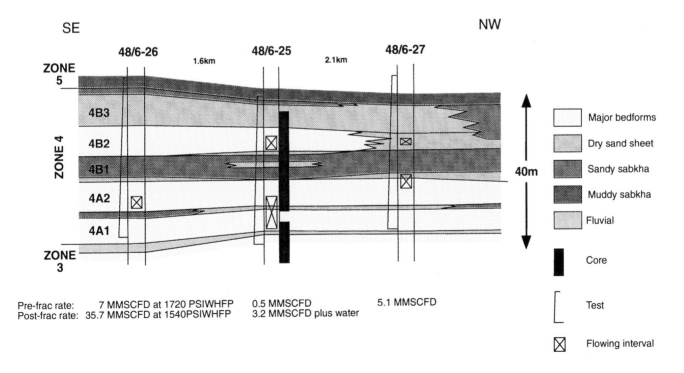

Fig. 4. Facies correlation of the stratigraphic and geographic core area of the field. The layering of the reservoir is of a scale and continuity that can be targeted with horizontal wells. The primary targets of the 48/6-34Z well (which twinned 48/6-26) were zones 4A2 and 4B2.

tion of the erg. Such surfaces are dry north of 48/6-25 and therefore difficult, if not impossible, to distinguish.

- *Zone 3* is 9–19 m thick in the core area. It is aeolian sandstone except for a fluvial unit at the base in the south and a fluvial horizon defining the top of the zone across the core area.
- *Zone 4*, 40–50 m thick in the core area, is a sequence of interbedded sabkha and aeolian facies. These facies are interpreted to be organized into subzones 6–12 m thick that are continuous and correlatable across the core area. This correlation is illustrated in Fig. 4.

The stratigraphy of Zone 4 records the repeated expansion and contraction of an erg. The erg is represented by layers of aeolian sand, the erg margin by layers of sandy sabkha and dry sand sheet facies (dry sand sheet is a distinctive association of small sets of aeolian cross-bedding, much flat or low-angle wind-ripple stratification and a minor content of sandy sabkha). The immediate cause of this layering is changes in the amount or direction of sand supply or in the position of the water table. The relative contribution and the ultimate causes of these may be speculated upon.

The story of sedimentation in Zone 4 is that on four occasions an erg was deflated from the erg-margin inwards to create a near-flat surface: a supersurface. Erg-margin facies accumulated on this surface, accommodated by a relative rise in the water table. A positive sand budget in dry conditions has later resumed and large bedforms (draa) nucleated in the area. These formed a dry erg and climbed to accumulate continuous aeolian sand. Conditions then changed again and the erg deflated to re-start the cycle.

The aeolian sand left behind by these ergs contains a high proportion of grain-flow stratification. This is the most permeable subfacies in the reservoir. Consequently, in these layers the most likely radial permeability to a vertical well is 5 mD, compared to a modal permeability for the whole reservoir of 1 mD. The top 20% of the permeability distribution of the reservoir is concentrated into these laterally extensive, 6–12 m thick layers, labelled 4A1, 4A2 and 4B2 in Fig. 4. Because they have the best absolute permeability and are high in the reservoir, these layers also have the best gas saturations and, therefore, the best relative permeability to gas. These layers also contributed most of the pre-frac flow in the appraisal well tests (Figs 3 and 4). Thus, they are ideal targets for horizontal wells.

A statistical analysis of field performance

Since 1989 significant changes have been made to the Hyde development concept with the introduction of not-normally-manned facilities, horizontal wells and the selection of a core area. This resulted in a significant improvement in the field economics. However, the downside case still did not meet the required performance criteria. With this in mind, early in 1991 a statistical analysis was undertaken to assess the validity of the deterministic downside case. Was it a P1 or a P10 case (i.e. a 1% or 10% chance that the downside volume of gas was this figure or less)? What was the significance of the assumptions that had been made in its derivation?

The statistical analysis consisted of five phases:

1. identify the critical parameters affecting the production potential;
2. estimate the ranges of uncertainty for those parameters, identify the critical assumptions and assign dependencies;
3. statistically combine the input ranges using Parametric or Monte-Carlo techniques;

4. reality check;
5. identify actions to reduce the risk or uncertainty.

The two main inputs to the production potential are GIIP and deliverability. The controls on these, in turn, were identified from single-well and full-field ECLIPSE modelling and are listed in Table 1. There are no surprises in this list.

Table 1. Key factors controlling gas-in-place and deliverability, derived from the ECLIPSE modelling. Each of these parameters was assigned a minimum (P10), median (P50) and maximum (P90) value for the statistical analysis

GIIP	Deliverability
Bulk rock volume (BRV)	Kv/Kh
Porosity	Skin
Gas saturation	Horizontal well effective length
Net-to-gross ratio	Relative permeability
Gas expansion factor	Reservoir thickness
	Absolute permeability

Uncertainty ranges for these parameters were assigned. This process was in some cases very subjective, in others more objective. It relies on the quality and quantity of the appraisal data. The two key assumptions made at this point are:

1. the dataset from the appraisal wells is large enough and of sufficient quality to represent the parent populations of the various parameters;
2. the deterministic geological model is correct.

For Hyde the quality of the appraisal data is fairly good and there is a fairly extensive database, with approximately one appraisal well per 40 BCF of reserves in the core area.

Several geological models were investigated, offering different interpretations of the continuity of permeability within Zone 4. Only one (illustrated in Fig. 4 and described above) was judged to fit the data with a high degree of confidence. This confidence was justified as results from the 48/1-1 well, drilled in 1990, fitted the model.

The ranges of the parameters were then statistically combined using a Monte-Carlo sampling for the GIIP calculation and a Parametric method (Smith and Buckee 1985) for the deliverability. This gave two distributions, one for GIIP and one for field deliverability. These were sampled to give values (with associated probability) for input into the ECLIPSE reservoir model. This in turn generated reserves profiles which were used to evaluate the economic performance of the development.

One result of the statistical analysis was to reduce the *range* of uncertainty from that previously derived deterministically. The new analysis gave more confidence in the downside case. It turned out that the deterministic downside case was more akin to a P1 than the P10 it had been thought to be. This is because the deterministic case involved the minimum values of many parameters combined in a manner that implied their simultaneous occurrence. Even allowing for degrees of dependency between some of the parameters, this is very unlikely.

The 'reality check' process underpinned all the activities during the statistical analysis with the emphasis on seeking alternative views and challenging the many assumptions inherent in the method. The process of recognizing and challenging assumptions enforced by the method was a major benefit, ultimately enhancing the understanding of the reservoir.

The statistical analysis was very useful in quantifying the uncertainties and identifying the critical parameters affecting the economics of the project, but to add value (i.e. to *reduce* the uncertainties), required an action. It was very clear which single action would significantly reduce the uncertainty on the project economics: drill a well to prove the deliverability from a horizontal hole. Success with this well would indicate the eventual deliverability of the development and reduce significantly the uncertainty of some of the key parameters. Failure would severely dampen confidence and enthusiasm in the project.

The idea of drilling an appraisal well is common sense. An additional value of the rigorous statistical approach was as a means of demonstrating confidence in the project sufficient to secure the release of funds for the well. Planning commenced in June 1991.

Well 48/6-34 and 34Z

Well planning

The primary objective of the well was deliverability. Could a near-horizontal hole be drilled to target the high-permeability layers which would give the increased flow rate predicted by the single-well modelling and statistical analysis?

Given the importance of structural elevation in deliverability, the crest of the field, where the most successful appraisal well was drilled (48/6-26), was chosen as the location for the horizontal well. The well was to be drilled from the surface location of the future production facility in the centre of the field and run towards the SE about 150 m to the NE of the 48/6-26 well. It was these criteria, rather than any model of permeability anisotropy in the reservoir, that determined the azimuth of the near-horizontal section.

A threefold increase in deliverability from the 48/6-26 pre-frac flowrate of 12 MMSCFD would give 36 MMSCFD steady state flow, which would equate to a transient rate (initial test rate) of approximately 50 MMSCFD. Such a result would prove the credibility of the technical work and provide sufficient data to confirm robust economics which would lead to approval of the project.

The major uncertainties affecting deliverability are skin (formation damage), effective length (length of high-permeability layers in the well-bore), Kh and Kv/Kh (Table 1). The evaluation requirements for the well were aimed at understanding and decreasing these uncertainties.

It was planned to core the pilot hole in Zone 4. The porosity and permeability data together with the PLTs and the test data from the horizontal section would be expected to improve our understanding of Kh and Kv/Kh. Skin was to be minimized by using a water-based, salt-saturated mud system. The effective length was to be optimized by attempting to steer the wellbore along the high-permeability layers using LWD logs integrated with the interpretation of the seismic data.

Well results

The well was spudded on 20 October 1991 and the pilot hole, 48/6-34, was drilled to a TD of 3449 m BRT (2976.1 m TVDSS), 28 m into the Carboniferous Coal Measures. The well was drilled at approximately 60° deviation through the reservoir. Two cores were taken over Zone 4. LWD logs were acquired over the whole reservoir and wireline logs were also run to permit a detailed comparison of the two systems.

The pilot hole came in different to prognosis in two important ways. Firstly, top reservoir was 12.3 m structurally high. Secondly, two fluvial horizons with an aggregate thickness of about 3 m, which had not been identified in the field previously, were encountered at the top of Zone 4A1. However, the quality and thickness of the two main targets, Zones 4B2 and 4A2, were close to prognosis.

Using these data, the planned trajectory for the horizontal section was revised. The pilot hole was plugged back and sidetracked within the Silverpit Formation at 3132 m BRT.

Soon after penetrating the first high-permeability layer (4B2), the LWD tool became differentially stuck. To counter this the mud was substantially diluted, from 1.26 to 1.15 SG, and no further problems arose. Steering the well using the LWD data to target the high-permeability layers was a success, significantly enhancing the well's effective length. Though the drill bit did not always respond as expected, the near-horizontal section was eventually drilled to a TD of 3913 m BRT (2897.1 m TVDSS).

The structural interpretation shown (Fig. 5) is 12.3 m higher than prognosis at the pilot hole, 5 m high close to the 48/6-26 well and 18 m high at the anticline, 150 m from TD. Because this unexpected elevation was encountered, the major bedforms in Zones 4A1 and 3 were targeted in the latter 300 m of the well.

The log interpretation is shown as Fig. 6. It illustrates the better reservoir quality on the neutron log ('CNPhi' in the header) in red and density log ('SFD' in the header) in yellow. Using these logs qualitatively, the well appears to have approximately 420 m of potentially productive reservoir.

For the testing programme, a 5 inch pre-drilled liner was run on a $9\frac{5}{8}$ inch production packer. This was to permit the well to be suspended with gas across the perforations to minimize the risk of formation damage. The well was then tested through a temporary completion string which was sized with the topsides equipment to allow the well to flow at the expected production conditions. This removed the need to extrapolate the test results to production conditions and hence removed uncertainty.

The well produced at a transient rate of 69 MMSCFD, which translates to a steady-state rate of between 45 and 60 MMSCFD (the remaining uncertainty is due largely to Kv/Kh). This rate lies between the P50 and P90 values from the statistical analysis. From the test interpretation the values for skin and Kh are interpreted as 0.3 and 1000 mD ft, which are very close to expectation. The effective producing length from the PLT temperature profile closely ties in with the LWD logs showing that a significant proportion of the well is productive, as prognosed.

Conclusions

Within two weeks of the test well, Statoil entered the project with a 45% equity stake and BP management approved the field development.

48/6-34Z is now suspended as a future production well. With first gas scheduled for October 1993, the plan now is to spud the second well in October 1992 and target the crestal area between the 48/6-25 and 48/6-27 appraisal wells (Fig. 2). If that second well is as successful as 48/6-34Z, the first year's production might be met from only two wells and the drilling of the third development well could be deferred. These options, and the precise specification of the second well, are under investigation now.

48/6-36Z, the second production well, was successfully tested and suspended in March 1993 after drilling a horizontal section of 2256 m. The well produced at a transient rate of 56 MMSCFD. As a result of the success of 48/6-34Z and -36Z, BP and Statoil are now looking to defer the third development well.

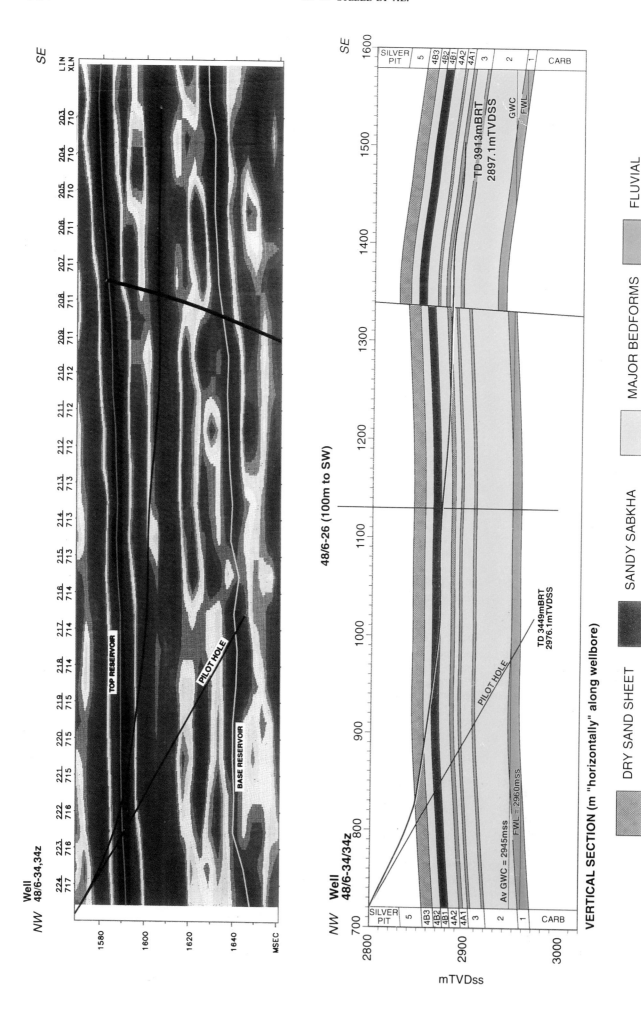

Fig. 5. Seismic section (amplitude intensity display) and interpreted structural cross-section along the 48/6-34 and -34Z well paths through the reservoir. The fault interpretation is on the limit of seismic resolution using dip, azimuth and attribute mapping techniques. Facies colours are as Figs 3 and 4.

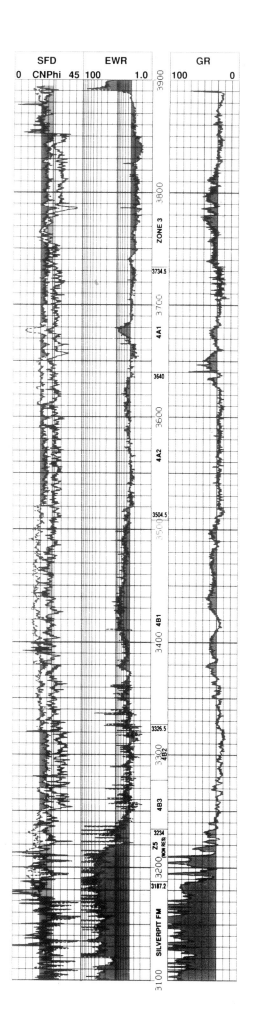
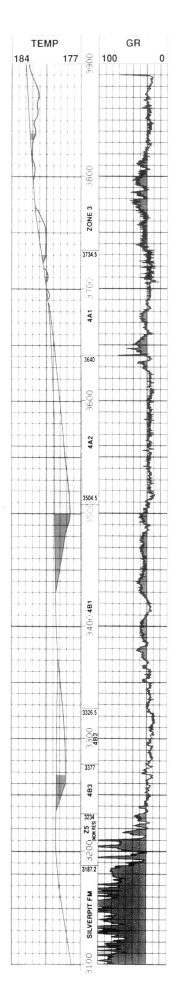

Fig. 6. LWD suite of logs for 48/6-34Z reservoir section (upper). GR: gamma ray; EWR: resistivity; SFD: density; CNPhi: neutron porosity. Shut-in pass temperature log and GR over same interval (lower). The temperature log, run immediately after the 69 MMSCFD flow, highlights in yellow the zones which were flowing. These zones show temperature drops below the background gradient.

Hyde field data summary

Trap

Type	Faulted anticline
Depth to crest (m TVDSS)	2840
Gas–water contact (m TVDSS)	2945
Gross gas column (m)	105
Net area (acres)	2000

Pay zone

Formation	Lower Leman Sandstone
Age	Early Permian
Gross thickness (average) (m)	110
Net-to-gross (average/range) (%)	96/94–98
Porosity (zonal averages/range) (%)	13/11–15%
Water saturation (average) (%)	64
Permeability (range) (mD)	0.01–40

Hydrocarbons

Gas gravity	0.596
Viscosity (cp at 188°F)	0.02
Gas expansion factor (SCF/RCF)	244
Condensate/gas ratio (BBL/MMSCF)	1.6

Formation water

Water gradient (psi/ft)	0.5
Salinity (ppm)	146 000
Resistivity (ohm m at 188°F)	0.0195
Other ions	Sulphate 430 ppm, sodium 70 000 ppm, potassium 1480 ppm, calcium 20 700 ppm

Reservoir conditions

Temperature (average) (°F)	188
Pressure (average) (psia)	4316
Pressure gradient in reservoir (psi/ft)	0.2

Field size

Area (acres)	2000
Recovery factor (%)	70
Recoverable gas (BCF)	150
Drive mechanism	Depletion

Production

Number and type of wells	2 or 3 near-horizontal producers
Start-up date	1 October 1993
Development scheme	Single 'not-normally-manned' platform
Production rate (average)	45 MMSCFD × 1.7 swing in year 1

Reference

SMITH, P. J. AND BUCKEE, J. W. 1985. Calculating in-place and recoverable hydrocarbons: a comparison of alternative methods. SPE 13776, Hydrocarbons and Evaluation Symposium, Dallas, Texas.

Discussion

Question (H. J. Duyverman, Statoil UK, London):

During drilling the horizontal well, was fracturing ever considered, or was BP satisfied with the achieved test rates?

Answer (R. P. Steele):

The achieved test rate was most satisfactory. In the event of a much poorer result the planned response was to perforate selectively and to stimulate clean-up with a diesel wash or base oil (assuming an oil-based mud had been used). Hydraulic fracturing was ruled out because of the need that would then arise to handle back-produced material and because of the risk of water production.

Estimation of recoverable reserves: the geologist's job

A. F. CORRIGAN

Corrigan Associates, 2 The Drove, Ditchling, Sussex BN6 8TR, UK

Abstract: The accurate prediction of recoverable reserves in the North Sea has proved to be a difficult task for many operators. This paper summarizes the performance of a number of fields by tracking the way in which reserves have changed with time, compared to the volume of reserves thought to exist at the time that development approval was given. In many fields reserves have changed by between 20 and 70%, with the areal distribution also exhibiting considerable variation from that in the field development plan. The changes have been both positive and negative. In the latter case, the economic consequences of a significant reduction in reserves are obvious. However, increases in reserves also bring financial penalties in poorly sited and sized facilities, additional well requirements and inefficient use of funds.

Although this work has shown that there are a number of reasons for poor estimation, the underlying cause is usually insufficient recognition of reservoir complexity and thus the use of simple or inappropriate models to predict reserves. It is also clear that there are geological factors that are common to fields that have exhibited large losses or increases in reserves. In the latter category are fields that are structurally simple, often deposited in submarine fan environments, where the volume of movable oil was underestimated. Fields suffering large drops in reserves are often structurally complex, highly compartmented and contain high permeability zones.

It is proposed that geologists should take a much more active role in reserve estimation than has generally been the case in the past. In particular, at the appraisal stage, early recognition of complexity is vital, and understanding of the complexity should become one of the major goals of the appraisal wells. In addition, the geologist needs to become closely involved in the construction and operation of the simulation models that are used by most reservoir engineering groups to predict reserves. This is to ensure that geological features which affect drainage are correctly modelled in the simulators.

The recoverable reserves estimated for a field are probably the single most important number generated by an oil company's technical organization. The reserves figure will control whether or not a field is developed, in what fashion, and also provides one of the principal foundations upon which agencies such as banks, the stock market and governments base their evaluation of the company.

In spite of the undoubted importance of producing as accurate a value as possible for recoverable reserves, there has been little discussion in the literature as to precisely how to do this, and especially as to the role of the geologist. For those companies with stock exchange listing in the USA, the Securities and Exchange Commision regulations require annual technical audits of fields, but these are dominated by a review of the reservoir engineering aspects of recovery and particularly the validity of the simulation model. However, the uncertainty associated with the geological model, and its link to the simulation model, often seem to receive less attention. An earlier paper (Corrigan 1988) outlined some of the problems that have occurred in the North Sea. This paper seeks to demonstrate that the geologist has a critical role in reserves estimation, but that, in order to fulfil this, he or she must understand the mechanism by which reserve estimation is carried out.

In order to provide some perspective, the paper first examines the record of reserve estimation in a number of fields from the UK sector of the North Sea. From this review it is clear that the industry's record on reserve estimation is highly variable, with some fields showing huge changes with time. The record is examined to assess to what degree lessons can be learned and whether there are particular geological features that lead to poor estimation.

The methods used to calculate reserves are examined particularly from the point of view of their reliance on the recognition and inclusion of geological complexity. This discussion largely revolves around the role of the simulation model which, at least in the North Sea, is always the basis for reserve calculation. It is shown that there are some geological situations in which the simulator tends to be overly optimistic in its estimation of reserves unless careful control is exerted by both reservoir engineer and geologist.

Definition of reserves

The precise manner in which reserves are defined has been the subject of a large number of papers (see summaries by Martinez *et al.* 1987; Cronquist 1991). In most companies, and certainly those with a listing on the New York Stock Exchange, the definition of reserves is controlled by the requirements of the US Securities and Exchange Commission. In the UK, the Department of Energy (now the Department of Trade and Industry) has three categories of reserves: proved reserves are defined as 'virtually certain' of being produced; probable reserves have 'above 50% certainty'; and possible reserves have 'less than 50% certainty' (*Development of the Oil and Gas Resources of the United Kingdom*, HMSO, London, annual publication).

Recently there has been some discussion in the reservoir engineering literature as to both the method of classification of reserves and the degree to which probabilistic methods can, or should, be used (Cronquist 1991; Cockcroft 1992; Hagoort 1992). The latter problem is further analysed later in this paper.

Much of the discussion in the papers on reserves revolves around semantics and will not be added to in this paper. Instead, some reasons are discussed for the industry's erratic record on reserve estimation, in spite of the care that has been taken in these calculations both by the operators concerned, and independent auditors working within carefully controlled guidelines.

Reserve estimation in the UK North Sea

Information on recoverable reserves is not generally easily accessible in the public domain. What is more, since different companies use different criteria to judge reserves, comparison

From *Petroleum Geology of Northwest Europe: Proceedings of the 4th Conference* (edited by J. R. Parker).
© 1993 Petroleum Geology '86 Ltd. Published by The Geological Society, London, pp. 1473–1481.

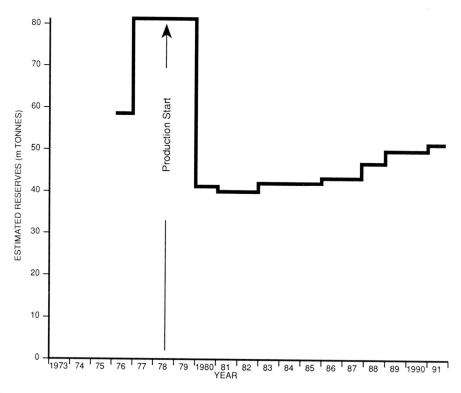

Fig. 1. Dunlin Field.

between companies is difficult and usually not valid. However, the source of data for this paper is the *Development of the Oil and Gas Resources of the United Kingdom*, an annual HMSO publication. Usually known as the 'Brown Book', this publication provides statistics on the production of hydrocarbons from the United Kingdom based on returns to the Department of Energy (now DTI), in which each field operator is required to provide an estimate of a field's recoverable reserves as part of the record.

The precise guidelines used to establish the recoverable reserves figure are not recorded, and these will almost certainly vary somewhat from operator to operator. However, this is not important for the purposes of this study, since it is the variation in the figures from year to year that is of interest, not the absolute values. It is assumed that individual companies are unlikely to have changed their method of calculation of reserves from year to year to the extent that it causes large swings in reserves. Thus, where significant changes in reserves do occur, it is as the result of increased understanding of the field during development. This then poses the question as to whether the factors that cause reserves to be revised can be recognized at the appraisal stage, and whether development programmes can be designed to accelerate the recognition of problems.

Data covering the last 17 years from the 'Brown Book' have been plotted for a number of fields, with the estimated reserves in millions of tonnes shown against the year in which the value was quoted. One million tonnes of oil are equivalent to approximately 7.5 million barrels or 1.19 million cubic metres, assuming an oil density of 37° API. Also plotted is the year in which production from the field commenced.

Fields that exhibit decreasing reserves

A number of North Sea fields show a marked drop in reserves through time. These fields will be reviewed briefly in terms of their reservoir geology and production history. This will be followed by a discussion of any features common to the fields

that may have led to overestimation of the recoverable reserves.

Figure 1 shows the results from the Dunlin Field and illustrates features that are common to most of this group of fields. The reserves defined at the end of the appraisal period were relatively high, but after only a year the number was drastically reduced by about 50%. Following this, reserves have been relatively stable, but, over the last few years, have been increasing by modest amounts. The oil in Dunlin is produced from the Middle Jurassic Brent Sand section. Several other Brent Sand fields exhibit similar reserve profiles; Fig. 2 shows the profile from Thistle while those from Heather, NW Hutton and Hutton are summarized in Fig. 3. All these fields have suffered severe reductions in reserves, some more rapid than others, while the majority show the same slight increases in later years that is seen in Dunlin.

Drastic drops in reserves are not restricted to the Brent Sand fields. Figure 4 shows the Tartan Field record, with a 70% reduction occurring after less than two years of production, followed again by a modest recovery. Other fields, such as Clyde, have recorded smaller decreases.

Obviously, the reductions in reserves, outlined above, represent huge economic setbacks for the fields and their owners. It would thus be of great value to ascertain, for future reference, which particular factors led to the overestimation of reserves. Unfortunately, this process is not straightforward since the operators concerned, perhaps not surprisingly, generally have not published details of the fields in either geological or reservoir engineering journals. However, enough has been published to indicate that a major factor has been the role played by field structure and the initial underestimation of its complexity.

Nadir (1980) discusses the evolution in structural understanding of the Thistle Field. The field development plan envisaged only limited faulting, with the early drilled crestal producers being supported, in part, by aquifer drive. However, the producers suffered rapid pressure decline and it soon became apparent that the field contained a large number of

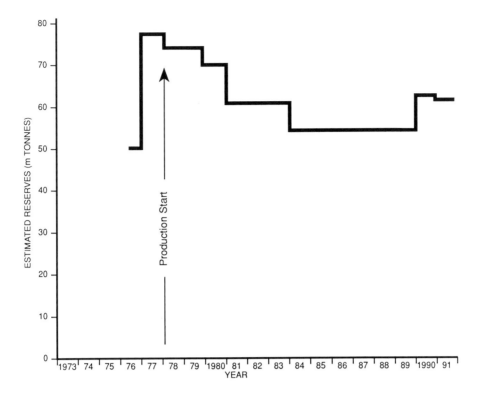

Fig. 2. Thistle Field.

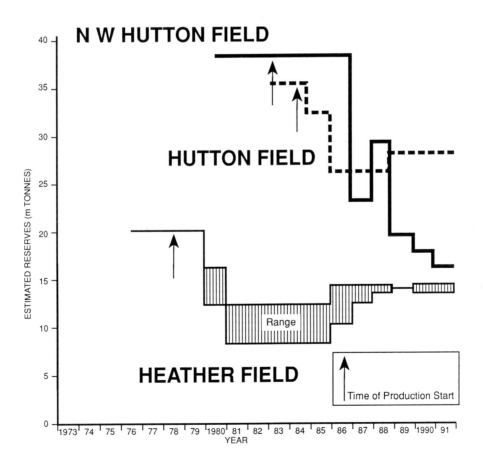

Fig. 3. NW Hutton Field/Hutton Field/Heather Field.

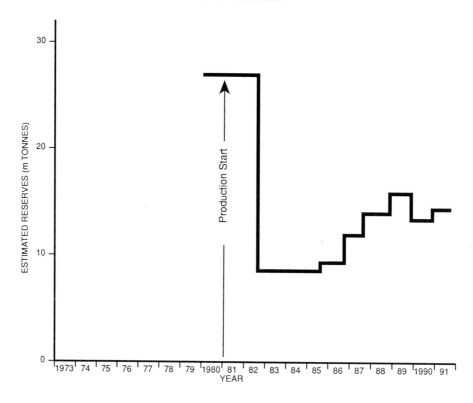

Fig. 4. Tartan Field.

sealing faults which reduced the role of the aquifer and, more importantly, divided the field into isolated compartments. A similar history for the Dunlin Field is recorded by Van Rijswijk *et al.* (1980). Development well drilling and a 3D seismic survey indicated that the volume and distribution of the hydrocarbons differed from that assumed in the development plan and that severe compartmentation was present.

The other major factor that has complicated the realization of reserves is the presence of highly permeable 'thief' zones particularly in the Etive section of these fields. Baumann and O'Cathain (1991) and Williams and Milne (1991) record the influence of these zones in Dunlin and Thistle, respectively. In these, and other Brent Sand fields, the Etive section can cause problems in several ways. Because it contains sands in which the permeability can be two orders of magnitude higher than the surrounding rock, it allows water from injectors to move very rapidly through the reservoir to producers, especially where it is confined by faulting or channelling. As a result, the productivity of the wells declines and, more importantly, some parts of the reservoir may be completely bypassed by the injection water. The sweep efficiency is not only reduced in an areal sense. Where significant reserves are present in the poor quality Rannoch section under the Etive, the presence of the Etive thief zones severely limits the engineering of an effective water flood for the Rannoch.

In the Tartan Field, which produces from sandstones of the Piper Formation, Coward *et al.* (1991) describe a situation of considerable tectonic complexity upon which is superimposed both sedimentological and stratigraphic complexity. The field is divided by one of the bounding faults of the Witch Ground Graben. In the south, the upthrown Piper Sandstone reservoir is 2000 ft shallower than the Piper Sandstone and Volgian turbidite reservoirs to the north. The reservoir is compartmented by faulting and stratigraphic pinch-out. This is so extreme in the downthrown block that, for a while, the Piper Sandstone reservoir had to be produced under depletion drive until sufficient water injection capacity could be made available (Coward *et al.* 1991).

Further evidence for the role of faulting in causing compartmentation and damaging reserve estimation can be found in the Northwest Hutton Field (Johnes and Gauer 1991). As can be seen from the structure map, the field is severely faulted, restricting the chances of an efficient areal sweep. Similarly, Penny (1991) describes the structural complexity of the Heather Field and how the production problems were compounded by water breakthrough along high permeability zones. It seems likely that the small increases in reserves which occur late in the lives of several of the fields illustrated, reflect the improved structural control gained by development drilling and further seismic surveys, which, in turn, enables more efficient targeting of the remaining reserves.

Thus, an important common factor in these fields is not just that they have suffered severely from faulting, but that the faults act as barriers, reducing the drainage volume of wells and profoundly affecting the efficiency of water flooding. The effects of compartmentation are exacerbated by the presence of high permeability thief zones in the Brent Sand fields. Clearly, it is important to define the level and complexity of faulting in a field as early as possible during appraisal, and to assess the degree to which the faults seal. Additionally, the presence and geometry of anomalously high permeability zones must be defined.

Data on the original oil-in-place in the fields are not included in the 'Brown Book', so that it is not certain to what extent these have changed in parallel with the recoverable reserves. However, it appears that while in some of the above fields the oil-in-place was overestimated, in many others it has not changed significantly. Thus, the decline in recoverable reserves was caused by overly optimistic estimates of recovery efficiency.

Fields that exhibit increasing reserves

A number of fields have shown marked, and continuous, increases in reserves through time. Figure 5 shows the Forties Field. The field's reserves have increased by 35% since the start

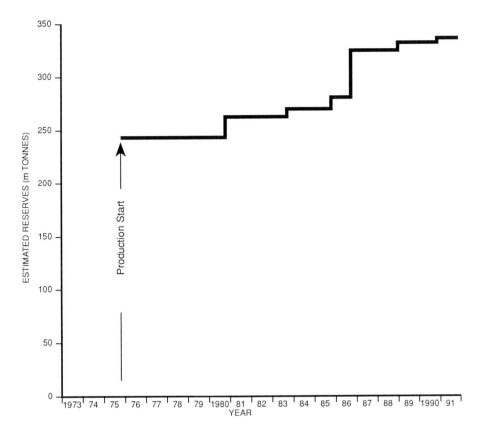

Fig. 5. Forties Field.

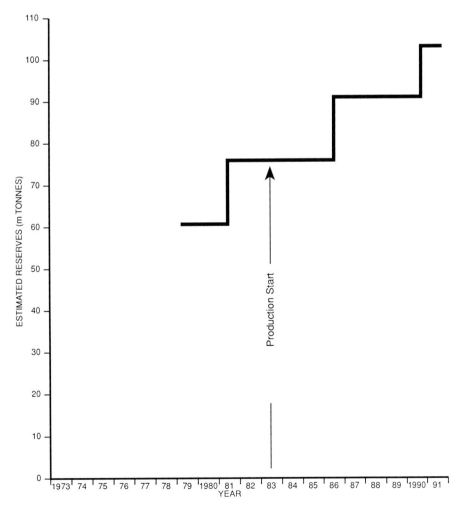

Fig. 6. Magnus Field.

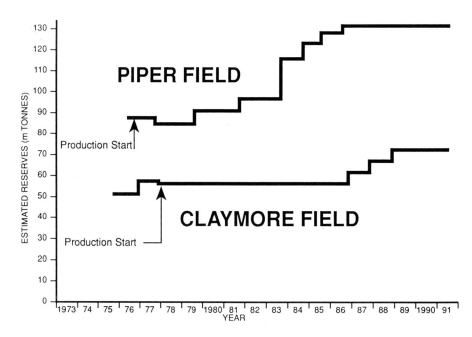

Fig. 7. Piper Field/Claymore Field.

of production, but the area of the field as defined by the oil–water contact has remained approximately the same. The sandstone reservoir was deposited in a Paleocene submarine fan and the trap was formed by a draping of the sands into a four-way dip closure. The structure of the field is simple, with only limited faulting.

The Magnus Field (Fig. 6), which produces from Upper Jurassic submarine fan deposits shows a 70% increase in reserves, while the Maureen Field, another Paleocene submarine fan, has had an increase of 25%. Major increases in reserves have also occurred in the Piper and Claymore fields (Fig. 7). Piper produces from the Piper Sandstone on the northern margin of the Witch Ground Graben only 10 km from Tartan and in an analogous position. The Claymore Field straddles the southern margin of the Witch Ground Graben, 20 km northwest of Tartan, and produces both from Upper Jurassic and Lower Cretaceous turbidite sands.

Thus, even this small sample indicates that fields of a wide range of ages and depositional environments can show large increases in reserves. One common feature that supports the conclusions of the previous section is that all of the above fields are structurally relatively simple. Forties and Maureen post-date the major episodes of faulting in the North Sea, while Magnus occupies a relatively simple dipping fault block. The Piper Field, although close to a graben margin, contains most of its hydrocarbons in the dip-slope away from the complexly faulted margin itself (Maher and Harker 1987), and contains only minor faults. Although Claymore has suffered more faulting in the Upper Jurassic succession to the south, the reservoir is much simpler than that in Tartan. The northern reservoir in Claymore, which appears to straddle the major graben margin faults, largely post-dates the tectonics and remains relatively simple (Harker *et al.* 1991).

Another important common factor in these fields is that the sand units tend to be areally extensive which maximizes the drainage efficiency. In addition, all of the reservoirs under discussion are highly layered, in that the sand units are separated vertically by extensive shales. Rather than hinder recovery, this situation has assisted the increase in reserves since it allows careful monitoring and control of fluid fronts. This is described in Piper by Schmitt and Gordon (1991) where off-take was reduced for a period to improve reservoir management. Another positive feature of the fields is that while good

permeability is common in the reservoirs, zones of very high permeability which reduce sweep efficiency are rare. Both the Piper and Forties fields are produced using a peripheral water-flood scheme, but both have been found to have a better than expected natural aquifer giving improved sweep and greater reservoir energy (Schmitt and Gordon 1991; Wills 1991).

Other factors leading to increases in reserves seem to be specific to individual fields. In Piper, the individual reservoir units of coarsening-up sands have an almost perfect permeability profile to maximize recovery from water-flood. In addition, the very well-sorted nature of the sands causes extremely low residual oil saturations. The combined effect of all the positive factors has led to a quoted recovery factor of 70% (Maher and Harker 1987; Schmitt and Gordon 1991) which would have been beyond the range of expectation when Piper came on stream in 1976.

In Forties, the reservoir performance has been characterized by the excellent vertical sweep obtained in the sands (Wills 1991). Furthermore, additional reserves were found in the southeast of the field which necessitated the addition of a fifth platform. It is interesting to note that a significant pressure barrier exists in the field, separating South East Forties from the Main Sand, which ensured that the potential of South East Forties was not recognized for some time.

A further major factor in the estimation of reserves must be the accuracy of petrophysical parameters used, and in particular the residual oil saturation assumed. The latter is very difficult to measure, and almost certainly has been severely underestimated in some fields while being overestimated in others. When combined with a poor estimate of vertical and areal sweep efficiency, then the large changes in reserves discussed above are easily explicable.

Fields that show little change in reserves

A number of North Sea fields have experienced little change in reserves throughout their history. The Brent Field, for example, has only deviated from the Annex B reserves by a maximum of 5%, while Statfjord has changed by about 7%. In both of these fields it is noticeable that the level of tectonic complexity is much lower than that seen in Thistle, Dunlin and Heather. In Brent and Statfjord only the crestal areas of the fields have suffered from gravitational deformation while the bulk of the reserves occupy the undeformed flanks (Struijk and Green 1991).

In summary, it appears that the influence of tectonic complexity has led to downward revision of reserves in a number of fields, and the effect is compounded by the presence of large permeability contrasts in the reservoir section. The reasons for underestimation of reserves are not so clear, but all the fields examined share a lack of significant faults, and are strongly layered with very effective sweeps of the sands within these layers. Environments of deposition that promote these conditions are sand-rich submarine fans and the high-energy marginal marine systems of the Piper Sandstone.

The economic consequences of overestimation of reserves are obvious, and become more unpleasant when, even to recover the reduced reserves, it becomes necessary to drill more wells, install more water handling facilities, and perhaps to strengthen platforms. However, it is important to realize that underestimation of reserves also carries economic penalties. Extra platforms or subsea facilities may be required because the original facilities are undersized or in the wrong place. At the very least, cash flow from the field will not have been maximized.

The above discussion would suggest that the recognition of particular geological features early in field life may help to define how difficult it will be to recover the hydrocarbons in place. The geologist should, therefore, be encouraged to take an increasing role in both the definition and realization of reserves.

The geologist's role

The geologist has two principal roles in the process of reserve estimation: recognition and measurement of reservoir complexity at the appraisal stage; and an involvement in the simulation modelling that will be used to calculate reserves.

As was discussed above, it is features such as facies type, which exert control on permeability distribution and reservoir geometry, and tectonic complexity which have had major influences on recovery in North Sea fields. It is, therefore, vitally important to establish as much information as possible about these factors during appraisal before committing to a development plan. Facies and permeability analysis require cores to be taken which cover all the major facies types both areally, and in depth. Structural analysis will require high-quality seismic data (usually 3D), calibrated against wells. However, this will not be sufficient to establish factors such as the extent of fault sealing and the degree of compartmentation in the reservoir. In order to do this, well tests, often run over extensive periods, are needed.

Thus, whoever is responsible for the appraisal programme should be fully aware of the likely data requirements for the field development plan and should prepare a comprehensive assessment of the complexity present in the field. Obviously, there will still be areas where data are insufficient. By recognizing these, some uncertainty can be placed on the reserve estimates, and, if necessary, the data gaps can be addressed during development drilling.

Before outlining the geologist's second major role it is necessary to discuss briefly the method by which recoverable reserves are calculated. In the vast majority of cases in the North Sea, a geological model is constructed which forms the basis for a reservoir engineering simulation model. The validity of the simulation is tested by history matching, in which the model's responses are matched against the known performance of the field, usually measured in terms of volumes of fluids produced against pressure decline, water cut and contact movement in individual wells. The principal factors which govern successful simulation are the degree to which the model duplicates the controls on fluid movement in the sub-surface

and the length of production history available for matching. Unfortunately, at the time when the first simulation models on a field are built, which is usually for the Annex B development approval, there has been no production against which to match the model. Thus, the early estimates of reserves are dependent on the simulation model being a true reflection of sub-surface conditions, and it is here that the geologist's input is important.

The full field simulation models used for North Sea fields are composed of thousands of grid blocks. However, because of the areal extent and thickness of the fields, each of the grid blocks is often tens of thousands of square metres in area and millions of cubic metres in volume. For simulation purposes, each block has one set of reservoir parameters which is assumed to exist uniformly throughout the block. As a result of these restrictions it becomes very difficult to accurately reflect rapidly changing rock properties either stratigraphically or areally. This is illustrated diagramatically in Fig. 8. The permeability distribution in the fault block shown in Fig. 8b will be poorly represented by a single value for each grid block in whatever fashion the permeability is averaged. Similarly, the modelling of the faults poses considerable problems. Faults usually have to be verticalized and made to conform to grid cell boundaries, although the grid itself may be distorted to take account of the fault orientation. However, where the faults are dense and/or listric, it then becomes almost impossible to model them correctly. One solution is to use a finer grid, but if the example shown in Fig. 8b is only a small element of a field then the total number of grid blocks required for the full field simulation model soon becomes impractical. Another solution is to use local grid refinement in which a denser grid is used to cover a part of the model, but again there are practical limitations as to the degree to which these can be used while keeping computer usage to reasonable levels.

A similar problem occurs where restrictions are placed on the number of layers available for the simulation model. Commonly, the models in use for many of the large Brent Sand fields have only one layer assigned to each of the Rannoch and Etive units. This leads to the difficulties illustrated in Fig. 8a. Features such as local high permeability zones, alternations of high and low permeability, vertical permeability gradations and sands with channel geometries cutting into sands with sheet-like geometries cannot be correctly handled without using an unacceptably large number of layers.

Thus, simulation models contain a much simplified version both of the field's stratigraphy and its structure. Because the models are simplified and more homogeneous than reality, they tend to allow much greater sweep efficiencies than is possible in the sub-surface and thus are often somewhat optimistic in their predictions.

Reservoir engineers are well aware of the deficiencies of simulation models, but require the assistance of a geologist to help ensure that critical features are retained in the models. For example, only the geologist can decide on the likely extent and geometry of a high permeability zone and thus whether it warrants a separate layer within the model. Similarly, when simplifying a fault system that is too complex to be included in a simulation model, geological judgement is required. The geologist must be able to take an active role in the building and gridding of the model, and should have used the appraisal wells to obtain an appreciation for those factors that will govern flow in the reservoir. Obviously the model based only on appraisal wells will contain large areas of uncertainty, but by understanding the compromises made in simulation the geologist can play a role in ensuring that these are reflected in uncertainty bands placed on reserves predicted by the model. Where the uncertainties are judged to be critical then the drilling of early production wells can be planned to address the problem. It is, therefore, the geologist working very closely

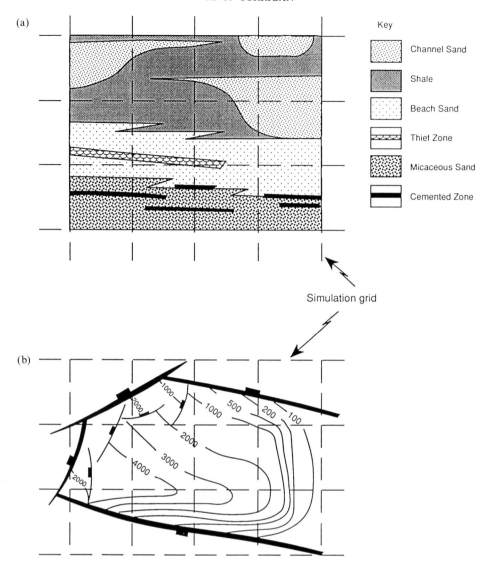

Fig. 8. Examples of geological situations that are difficult to simulate. (**a**) Cross-section; (**b**) permeability in fault block.

with the reservoir engineer that provides the best insurance against poor estimation of reserves. The alternative, in which the reservoir engineer is left alone to modify the geological model for simulation, will at best provide somewhat random results, and the more complex the reservoir, the less the simulation will reflect reality.

Recently, there has been some discussion in the literature regarding the use of probabilistic approaches to reserves estimation (Cronquist 1991; Cockcroft 1992; Hagoort 1992). Most of the reserve estimation for North Sea fields to date has been based on deterministic methods and, as has been described above, these have not been uniformly successful. However, it is not clear that probabilistic methods would have had greater success. Any probabilistic approach relies on the ability of the geologist and reservoir engineer to both identify parameters critical to recovery and to provide realistic probability distributions for those parameters. Clearly, a number of critical parameters, such as the role of sealing faults and high permeability zones, were not recognized early in North Sea developments. Even now, with an understanding of which parameters are critical, it remains a problem as to how to define a probability distribution to handle the possible presence of sealing faults before they have been proven by production. These difficulties are exacerbated by the fact that many parameters are interdependent and thus when their uncertainties are combined a very large composite uncertainty results.

This can be daunting to both management and bankers. The complexity of the problem is illustrated by Ovreberg *et al.* (1992) who describe the importance of not only defining key parameters at an early stage, but also the dependencies between them. Thus, even in probabilistic approaches there is no substitute for high levels of co-operation between geologist and reservoir engineer.

Conclusions

Severe tectonic complexity is a feature that is common to a number of the North Sea fields that have suffered considerable reductions in reserves through time. The tectonic problems have been exacerbated in Brent Sand fields by the presence of very high permeabilities in zones in the Etive Member which drastically reduce the sweep efficiency.

Fields in which reserves have significantly increased do not have such an obvious unifying characteristic. However, they tend to be developed in facies such as the marine-marginal Piper Sand or submarine fans, which are sand rich and strongly layered. Shale barriers separate the layers, but the sands are laterally extensive and disruption by faulting is limited. Using the above criteria it should be possible for the geologist to identify the level of complexity present in a field at the appraisal stage (Fig. 9), and, from the structure and facies present, assess the problems that are likely to occur during simulation. It cannot be overemphasized, however, that these rules are only general and that each field is likely to act in a unique fashion. Thus, it is important to collect data during appraisal that will reduce the uncertainty where possible. This will include extensive coring and perhaps long-term testing if the presence of sealing faults is suspected.

(a)   SIGNIFICANT PROBLEM:

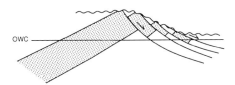

LESS SIGNIFICANT PROBLEM:

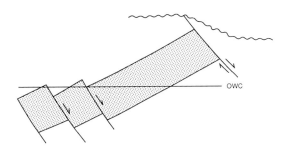

(b)   SIGNIFICANT PROBLEM:

STOIIP Distribution

Tarbert	5%
Ness	30%
Etive	25%
Rannoch	40%

LESS SIGNIFICANT PROBLEM:

STOIIP Distribution

Tarbert	25%
Ness	15%
Etive	40%
Rannoch	20%

Fig. 9. Early recognition of complexity in Brent Sand fields.

It was stated in the introduction that our industry's record in predicting reserves in the past has not been very good. In the future the task is likely to be even more challenging since fields will be smaller, and more complex, and plays more subtle. In the Central Graben, for example, the targets are deep and high pressured and likely to be in complexly faulted environments. Since well costs are high, appraisal drilling will be limited. Taken together, these are exactly the conditions that have led to poor reserve estimation in the past. Perhaps, as geologists, we should therefore ensure that we become as involved in estimating recovery of hydrocarbons as we have been involved in the past in discovering them.

References

BAUMANN, A. AND O'CATHAIN, B. 1991. The Dunlin Field, Blocks 211/23a, 211/24a, UK North Sea. *In*: ABBOTTS, I. L. (ed.) *United Kingdom Oil and Gas Fields 25 Years Commemorative Volume*. The Geological Society, London, Memoir, **14**, 95–102.

COCKROFT, P. 1992. Discussion of Reserves and Probabilities—Synergism or Anachronism? *Journal of Petroleum Technology*, **44**, 366–367.

CORRIGAN, A. F. 1988. 'Factors Controlling Successful Reserve Prediction: A Cautionary Tale from the U.K. North Sea.' Unpublished proceedings of the Norwegian Petroleum Society Conference on Reservoir Management in Field Development and Production.

COWARD, R. N., CLARK, N. M. AND PINNOCK, S. J. 1991. The Tartan Field, Block 15/16, UK North Sea. *In*: ABBOTTS, I. L. (ed.) *United Kingdom Oil and Gas Fields 25 Years Commemorative Volume*. The Geological Society, London, Memoir, **14**, 377–386.

CRONQUIST, C. 1991. Reserves and Probabilities—Synergism or Anachronism? *Journal of Petroleum Technology*, **43**, 1258–1264.

HAGOORT, J. 1992. Discussion of Reserves and Probabilities—Synergism or Anachronism? *Journal of Petroleum Technology*, **44**, 368.

HARKER, S. D., GREEN, S. C. H. AND ROMANI, R. S. 1991. The Claymore Field, Block 14/19, UK North Sea. *In*: ABBOTTS, I. L. (ed.) *United Kingdom Oil and Gas Fields 25 Years Commemorative Volume*. The Geological Society, London, Memoir, **14**, 269–278.

JOHNES, L. H. AND GAUER, M. B. 1991. The Northwest Hutton Field, Block 211/27, UK North Sea. *In*: ABBOTTS, I. L. (ed.) *United Kingdom Oil and Gas Fields 25 Years Commemorative Volume*. The Geological Society, London, Memoir, **14**, 145–152.

MAHER, C. E. AND HARKER, S. D. 1987. Claymore Oil Field. *In*: BROOKS, J. AND GLENNIE, K. W. (eds) *Petroleum Geology of North West Europe*. Graham & Trotman, London, 835–845.

MARTINEZ, A. R., ION, D. C., DESCORCY, G. J., DEKKER, H. AND SMITH, S. 1987. Classification and Nomenclature Systems for Petroleum and Petroleum Reserves 1987 Report. *In*: *Proceedings of the Twelfth World Petroleum Congress, Volume 5*. John Wiley & Sons, Chichester, 259–274.

NADIR, F. T. 1980. Thistle Field Development. *In*: *Proceedings of the European Offshore Petroleum Conference*. Paper EUR 165.

OVREBERG, O., DAMSLETH, E. AND HALDORSEN H. H. 1992. Putting Error Bars on Reservoir Engineering Forecasts. *Journal of Petroleum Technology*, **44**, 732–738.

PENNY, B. 1991. The Heather Field, Block 2/5, UK North Sea. *In*: ABBOTTS, I. L. (ed.) *United Kingdom Oil and Gas Fields 25 Years Commemorative Volume*. The Geological Society, London, Memoir, **14**, 127–134.

SCHMITT, H. R. H. AND GORDON, A. F. 1991. The Piper Field, Block 15/17, UK North Sea. *In*: ABBOTTS, I. L. (ed.) *United Kingdom Oil and Gas Fields 25 Years Commemorative Volume*. The Geological Society, London, Memoir, **14**, 361–368.

STRUIJK, A. P. AND GREEN, R. T. 1991. The Brent Field, Block 211/29, UK North Sea. *In*: ABBOTTS, I. L. (ed.) *United Kingdom Oil and Gas Fields 25 Years Commemorative Volume*. The Geological Society, London, Memoir, **14**, 63–72.

VAN RIJSWIJK, J. J., ROBOTTOM, D. J., SPRAKES, C. W. AND JAMES, D. G. 1980. The Dunlin Field, A Review of Field Development and Reservoir Performance to Date. *In*: *Proceedings of the European Offshore Petroleum Conference*. Paper EUR 168.

WILLIAMS, R. R. AND MILNE, A. D. 1991. The Thistle Field, Blocks 211/18a and 211/19, UK North Sea. *In*: ABBOTTS, I. L. (ed.) *United Kingdom Oil and Gas Fields 25 Years Commemorative Volume*. The Geological Society, London, Memoir, **14**, 199–210.

WILLS, J. M. 1991. The Forties Field, Blocks 21/10, 22/6a, UK North Sea. *In*: ABBOTTS, I. L. (ed.) *United Kingdom Oil and Gas Fields 25 Years Commemorative Volume*. The Geological Society, London, Memoir, **14**, 310–308.

Discussion

Question (C. R. Garland, BP Exploration):

You showed only one company (Union) quoting a range in recoverable reserves. Have other companies adopted this practice, or do they in general persist in stating a single deterministic figure?

Answer (A. F. Corrigan):

In the 'Brown Book' only Unocal UK Ltd is recorded as supplying reserves as a range rather than a single number. Since I do not know what the Department of Energy requires from companies in its request for data for the 'Brown Book' I cannot answer the question more fully. However, since this paper has demonstrated that generally we have not been very accurate in predicting reserves, I would suggest that persisting in using single numbers for reserves rather than a range can be very misleading.

Geological aspects of horizontal drilling in chalks from the Danish sector of the North Sea

S. FINE, M. R. YUSAS and L. N. JØRGENSEN

Mærsk Olie og Gas a.s., Esplanaden 50, 1263 Copenhagen K, Denmark

Abstract: Field development in the Danish chalk fields using horizontal wells has increased productivity in low-permeability reservoirs, maximized fracture intersection with the wellbore and optimized oil rim development.

Horizontal drilling requires extraordinary planning in order to achieve proper well placement. Unfortunately, the available data from logs, shows and field mapping does not always give consistent results and adjustments to the planned well trajectory must be made. A review from Mærsk Oil experience indicates that despite uncertainties related to such factors as structural dip, fluid levels and faults, objectives can nevertheless be satisfactorily attained.

Several techniques incorporating 3D seismic, pilot holes, wellsite biostratigraphy, wellsite geology and logging while drilling are utilized. They have proven crucial for recognizing and implementing rapid changes in the prognosed well trajectory where necessary. Our experience further emphasizes the absolute necessity of ongoing communication between qualified wellsite and office personnel during both the planning and operational phases of drilling a well.

Since the introduction of horizontal drilling in Denmark in 1987, 26 horizontal wells have been completed in five chalk fields (Fig. 1). The length of the horizontal drainhole ($>86°$) has increased from 1136 ft in the first well in the Dan Field to a world record 8204 ft (2500 m) in a recent well from the Tyra Field (Fig. 2). In addition, pay zones with thicknesses of only 20 ft are now being successfully targeted.

The experience of the past five years has continually emphasized the need for multidisciplinary teamwork when planning, evaluating and drilling horizontal wells. Ongoing communication between offshore operations and office-based personnel is critical in order to react quickly to geological surprises while drilling. Misunderstandings must be avoided while the well trajectory is steered to the target.

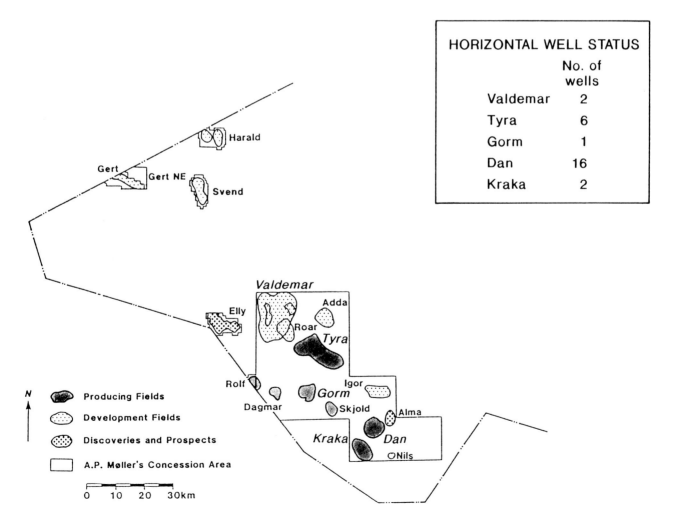

HORIZONTAL WELL STATUS	
	No. of wells
Valdemar	2
Tyra	6
Gorm	1
Dan	16
Kraka	2

Fig. 1. Location map showing fields and number of horizontal wells drilled in the Danish sector (to January 1992).

From *Petroleum Geology of Northwest Europe: Proceedings of the 4th Conference* (edited by J. R. Parker).
© 1993 Petroleum Geology '86 Ltd. Published by The Geological Society, London, pp. 1483–1490.

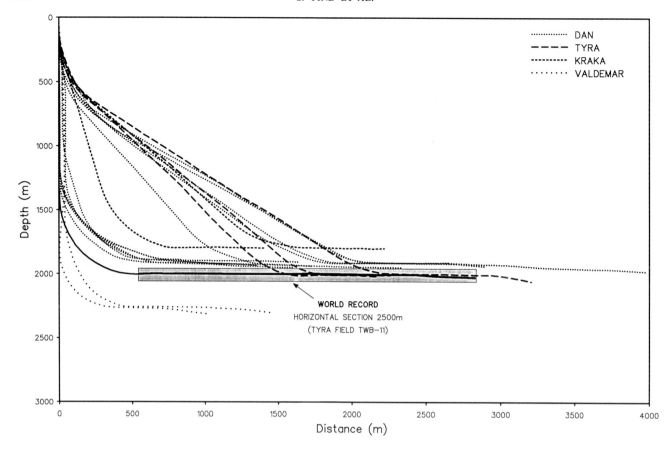

Fig. 2. Length of horizontal wells in the Danish sector.

Horizontal drilling presents unique puzzles in lateral thinking which are sometimes at odds with conventional geological wisdom. Alternative means of analysis are required, demanding both new techniques and problem-solving methodologies. Time is often of the essence. The difficulties, the failures and, fortunately, the solutions all add to the learning curve, optimizing future activities and often allowing for the exploitation of more complex pay zones. Examples from Mærsk Oil experience will be presented here.

Applications of horizontal drilling

In Denmark, as elsewhere, horizontal well development is chosen because of its advantage in addressing field-specific problems (Andersen *et al.* 1988; Andersen and Doyle 1990; Tehrani 1991; West and Bodnar 1991). These include increased productivity on a per-well cost from low permeability reservoirs, maximizing natural fracture intersection with the wellbore and minimizing water or gas coning in oil rim development. In addition, horizontal wells are being designed to appraise reservoir parameters and to avoid shale stability problems in the overburden of extended-reach wells.

Low permeability reservoirs

Some 85% of the oil-in-place in the Danish concession area is found in low-permeability chalks of Lower Tertiary to Lower Cretaceous age (Figs 3 and 4). Thus, optimization of reservoir drainage from a single well was the primary reason for initiating horizontal drilling in Denmark. Subsequent data from the Dan Field (Fig. 5) indicates that gross production from the 16 currently drilled, and hydraulically fractured, horizontal wells have shown an average fourfold increase compared to the 40 conventional producers (Andersen *et al.* 1990; Jørgensen, in press).

Fracture optimization

In the Kraka Field, reservoir simulations indicate that effectively all production comes via natural fractures. The location and orientation of these fractures are therefore crucial for well placement. The first two production wells, both of which are horizontal wells with pilot holes, were therefore designed to maximize appraisal information, in particular information relevant for characterization of the fracture pattern. By interpreting core data, dipmeter, FMS, and PLT logs, it was noted that productivity seemed to be related to wellbore intersection with fracture swarms (Jørgensen and Andersen 1991).

Oil rim development

The Tyra Field was originally developed exclusively as a gas reservoir, but the advent of horizontal drilling technology has since allowed for oil rim development as well. The oil rim, however, is rather thin and due to complexities introduced by tilted fluid levels, has effective target thicknesses varying from 60 ft to less than 20 ft. Selective completion intervals combined with reduced drawdown over the longer horizontal drainhole have served to maximize black oil production and minimize the amount of gas or water coning.

Horizontal well planning

After determining the broader outlines of field development using horizontal drilling, more detailed well trajectories must be defined. This incorporates not only the optimal well spacing as identified by reservoir modelling, but also the operational constraints imposed by geometry, geological uncertainty and drilling technology. In addition, the objectives of the completion programme must be taken into consideration since experience shows that a successful stimulation job, particularly

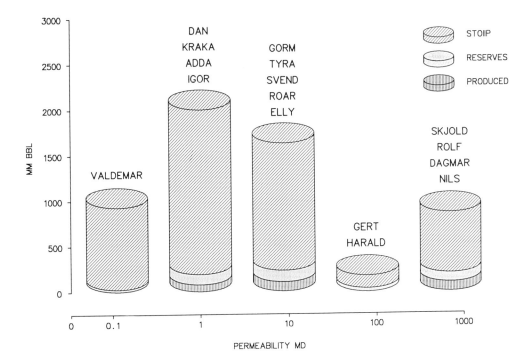

Fig. 3. Most oil-in-place is located in chalk reservoirs with permeabilities less than 10 mD.

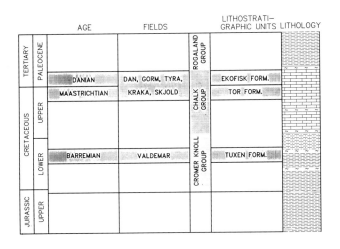

Fig. 4. Danish Central Graben stratigraphy for chalk reservoirs. A hardground is often developed at the Danian–Maastrichtian boundary.

in horizontal drainholes, is not only field specific, but well specific. The following examples will illustrate these points.

Horizontal drilling techniques

Although the objectives of the horizontal wells described above are distinct, the means by which the optimal drainhole is attained can be quite similar. We have found that the drilling of horizontal wells requires one or more of the following techniques:

- 3D seismic, or in its absence, a 2D seismic line for initial well placement;
- a pilot hole to appraise prognosed formation tops and fluid levels, thus allowing for adjustments prior to drilling the horizontal trajectory;

- wellsite biostratigraphy and/or rigorous evaluation of shows in order to optimize the horizontal drainhole while drilling;
- logging while drilling or intermediate logging of the horizontal hole in order to confirm the above data and, if necessary, allow for refinement of the horizontal welltrack.

In order to illustrate the role the above techniques have played in horizontal drilling, specific problems encountered in individual wells will be discussed.

Example 1: structural dip steeper than prognosed

Many of the horizontal wells in the Dan Field (Fig. 6) are located near the top of the Maastrichtian Tor Formation. The basal part of the overlying Danian Ekofisk Formation is of very poor reservoir quality. In two of the early horizontal wells drilled in the Dan Field, the Ekofisk Formation was inadvertently re-entered by the wells in a downflank position. With horizontal wells, a mapping error of only 1° in structural dip is enough to place the well outside the target interval. The net result of this is that a proportion of the drainhole may be outside the main reservoir (Fig. 7).

Recently, 3D seismic has improved the structural mapping considerably, because it is possible to map the top of the Tor Formation directly. With the old 2D seismic, the combined uncertainty of the seismically picked depth and the MWD deviation survey of the well amounted to almost 50 ft. This meant that to make sure the Ekofisk Formation was not re-entered, the wells had to be located more than 50 ft below the top of the Tor Formation. In many instances this would place the well too close to the OWC to allow fracture stimulation of the well.

The solution to this problem lies in the application of onsite biostratigraphy from top reservoir to TD. A wellsite palaeontologist constantly monitors the microfossil assemblage and whenever the trajectory appears to be drifting too high (or too low), the driller is advised and takes the appropriate corrective action. With proper preparation, i.e. analysing the faunal succession in cores from a nearby well, the vertical resolution of the onsite biostratigraphy is very good, often less than 10 ft.

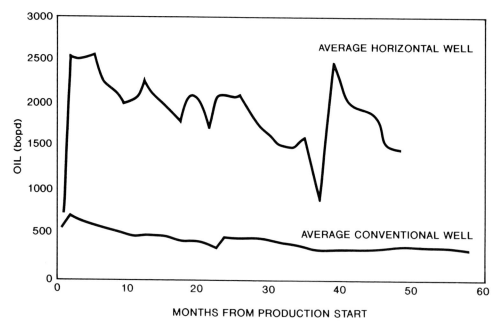

Fig. 5. Comparison of gross productivity in Dan Field.

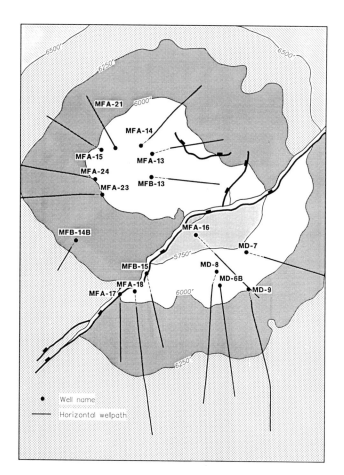

Fig. 6. Dan Field: depth structure map. Note Wells MFA-17 and MD-8 on upthrown B-Block.

The application of wellsite biostratigraphy has been a tremendous success for Mærsk Oil and is now being applied on all wells drilled in the Dan, Kraka and Valdemar fields. Originally, pilot holes were drilled in order to locate a representative reservoir interval prior to drilling the horizontal sidetrack, but with increased experience, it is now often possible to steer the horizontal drainhole exclusively on the basis of

biostratigraphic data collected from the build-up section of the horizontal well. The success of this methodology is not least reflected in the fact that the use of onsite biostratigraphy is also strongly supported by drilling engineers.

Example 2: unexpectedly crossing a fault

Drilling of the first appraisal/production well on the Valdemar Field, the Valdemar-1 well, proved the value of onsite biostratigraphy in successfully placing a horizontal section within a structurally complex reservoir. The target interval was relatively clean chalks of Lower Cretaceous (Barremian) age (Fig. 4) which are interbedded with marls and claystones. A high-resolution biostratigraphic zonation (with a vertical resolution of 5–10 ft) was therefore developed for the field from cores cut in the initial discovery well and from the pilot hole of Valdemar-1. This zonation was based on changes in relative abundances of various microfossil and calcareous nannofossil species which could be used to recognize the anticipated reservoir interval. Prior to drilling, it was hoped that these same biozones could be identified from cuttings while drilling the horizontal section.

Drilling of the horizontal section of Valdemar-1 proved that the high-resolution biostratigraphic zones could be recognized from cuttings and that corrections to the planned trajectory could be made from information available onsite. After drilling approximately one-third of the planned horizontal section, a significant stratigraphic break was noted in the cuttings. Biozones became stratigraphically younger and several zones were missing, representing 30–40 ft of section (Fig. 8). A core was cut in order for the palaeontologist to confirm the break and after reviewing the 3D seismic data, a small, previously unmapped normal fault was interpreted. The well trajectory was then turned down and the remainder of the horizontal section was kept in the reservoir.

Example 3: uncertainties in seismic interpretation

The second appraisal/production well on Valdemar confirmed the importance of onsite biostratigraphy for targeting horizontal wells. The second well, Valdemar-2, was targeted into an area of very large depth uncertainty due to unknown interval seismic velocities. Although depths came in close to prognosis

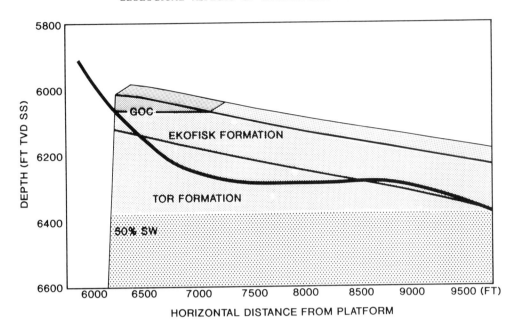

Fig. 7. Dan Field: MFA-17 cross-section showing re-entry into the Ekofisk Formation.

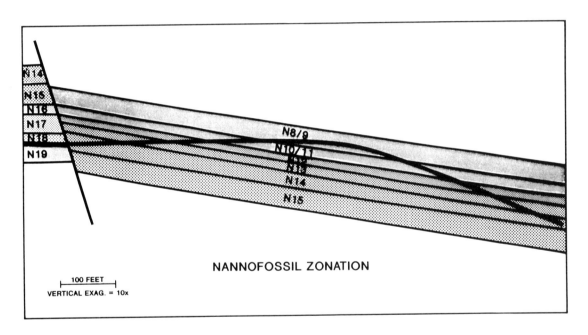

Fig. 8. Biostratigraphic zonation along Valdemar-1 horizontal drainhole. Note jump in four nannofossil zones across fault.

in the pilot hole, while drilling the horizontal section the onsite palaeontologists observed that expected biozone tops were occurring much higher than anticipated from the 3D seismic interpretation. After consultation with drilling engineers, the well trajectory was altered to obtain a higher build rate and avoid drilling the planned horizontal section below the reservoir (Fig. 9). The well was drilled up into the reservoir from below and then turned down slightly to follow the structural dip of the beds, ultimately achieving a horizontal length of over 3000 ft.

Example 4: horizontal well influenced by chert layers

Our experience has shown that chert layers can direct the drillstring such that the wellbore can be kicked up (or down) and out of the reservoir. In addition, it can be difficult to re-

enter through the chert again when drilling horizontally, thus risking the loss of some hundreds of feet of horizontal drainhole. This was the case in MFA-17 (Fig. 7) where hardground development prevented the wellbore from turning back into the Tor Formation after accidentally re-entering the Ekofisk Formation.

If the stratigraphic position of the chert layers can be recognized within the reservoir zonation by wellsite biostratigraphers, then corrective action may be taken before the chert disrupts the wellbore.

Example 5: uncertainties in fluid levels

In the Tyra Field, the limited thickness of the effective oil rim requires extraordinary control for proper well placement and optimal drainage. For the purposes of well planning, the

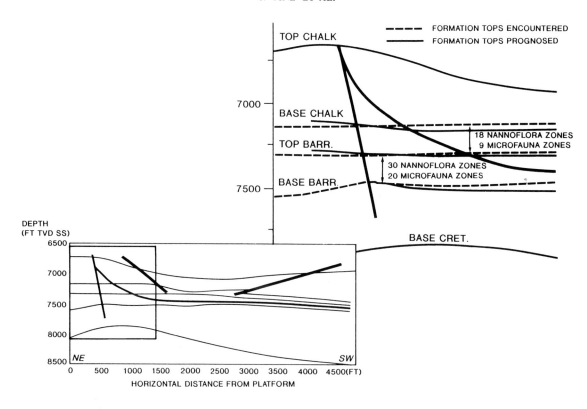

Fig. 9. Valdemar-2 well trajectory. Note how encountered formation tops were high to prognosis.

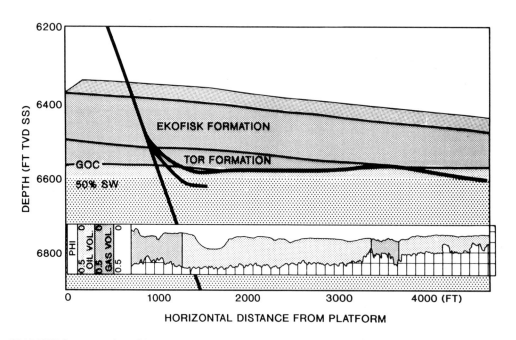

Fig. 10. Tyra Field: TEB-3 cross-section with pilot hole and two horizontal sections. Note re-entry into gas zone around 3500 ft from platform; a result of the bit being kicked upwards by chert.

effective target is defined as the interval between the gas–oil contact (GOC) and the 50% Sw level. Several uncertainties are recognized, however, which complicate the problem.

First, complexities associated with both late-stage structural tilting and regional hydrodynamic gradients suggest that fluid levels cannot be modelled as flat-lying features. Second, differential pressure depletion during field production can distort the original fluid distribution near active producers or injectors. Third, inherent errors in deviation surveys increase with the

horizontal distance drilled and may, in long drainholes, theoretically exceed the thickness of the targeted pay zone.

These problems are recognized in all Danish chalk fields, but, because of relatively thick reservoir sections in most fields, they seldom cause major difficulties. In Tyra, however, their effect may be critical.

In a recent Tyra oil rim pilot hole (Fig. 10), the GOC identified by the pilot hole logs was approximately 15 ft deeper than the GOC observed from shows (in core). The prognosed thickness of the oil rim was only 20 ft thick. The horizontal trajectory was then adjusted downwards from prognosis to adjust for the logged GOC depth. When drilling the horizontal well, however, the wellsite geologist indicated that the shows within the adjusted oil rim were poor and that the wellbore must be too low. The well was plugged back and a second horizontal sidetrack was designed by adjusting the planned trajectory to a depth corresponding to the GOC as recognized from shows in the build-up section of the first horizontal hole. After crossing the new prognosed GOC in this second horizontal sidetrack, an intermediate log was run. In this case, the GOC in the horizontal hole was consistent from both logs and shows, and matched the GOC found in the first horizontal hole. It was shallower, however, than the GOC depths which had been recognized from core and logs in the pilot hole. With this confirmation of fluid levels in the horizontal section, and by continuing close monitoring of shows, the well was then successfully drilled to TD where final drillpipe-conveyed logging indicated that the well had indeed been properly placed.

In another recent Tyra well, the logged GOC and the GOC as recognized from shows coincided in the pilot hole, but was approximately 25 ft high relative to the prognosed GOC depth based on the position of the fluid level in several nearby wells. The wellsite evaluation of shows in the horizontal sidetrack however, seemed to confirm this high GOC and the well was drilled accordingly. TD logging confirmed that nearly 3200 ft of horizontal drainhole was indeed placed within an oil zone having a thickness of only 20–30 ft.

Conclusions

The examples from the Tyra Field, as well as the other examples presented here, clearly show that a critical evaluation of data from logs, shows, cores and regional mapping does not necessarily give consistent results. Horizontal well prognoses, particularly in the Tyra and Valdemar fields, are uncertain and possible adjustments to the planned trajectories must be anticipated, costed, recognized and rapidly implemented. Our experience suggests that correct decisions can be made quickly and that objectives can be satisfactorily attained. Emphasis is placed, however, on the absolute necessity of ongoing communication between qualified wellsite and office personnel.

The financial benefits of choosing the correct package of evaluation techniques are considerable. While a pilot hole can comprise around 15% of the total drilling cost minus completion, a logging while drilling evaluation suite constitutes but 4% and a wellsite geologist or biostratigrapher requires less than 0.2% of the total expenditure. With our increased experience in horizontal drilling, pilot holes are now often replaced by logging while drilling or simply wellsite biostratigraphy/geology in the build-up section of the horizontal hole, which still gives the possibility of a plugback and adjustment should this prove to be necessary. The use of qualified wellsite personnel, as the examples presented here clearly document, is invaluable for optimal reservoir placement.

Our colleagues are thanked for sharing their horizontal drilling experiences with us and helping to describe and discuss the problems, techniques and solutions reviewed in this paper. Management at Mærsk Oil and Gas, Texaco Denmark Inc., and Shell Internationale Petroleum Maatschappij (The Hague) are thanked for their permission to present and publish the information presented here.

References

ANDERSEN, C. AND DOYLE, C. 1990. Review of hydrocarbon exploration and production in Denmark. *First Break*, **8**(5), 155–165.

ANDERSEN, S. A., CONLIN, J. M., FJELDGAARD, K. AND HANSEN, S. A. 1990. Exploring Reservoirs with Horizontal Wells: the Mærsk Experience. *Schlumberger Oilfield Review*, **2**(3), 11–21.

——, HANSEN, S. A. AND FJELDGAARD, K. 1988. *Horizontal Drilling and Completion, Denmark*. Paper SPE 18349.

JØRGENSEN, L. N. in press. *The Dan Field*. AAPG Treatise of Petroleum Geology, Atlas of Oil and Gas Fields. American Association of Petroleum Geologists, Tulsa Oklahoma.

—— AND ANDERSEN, P. M. 1991. *Integrated Study of the Kraka Field*. Paper SPE 23082 presented at the Offshore Europe Conference, Aberdeen, Septemter 3–6.

TEHRANI, A. D. H. 1991. *An Overview of Horizontal Well Targets Recently Drilled in Europe*. Paper SPE 22390 presented at the Second Archie Conference, Houston, November 3–6.

WEST, C. C. AND BODNAR, D. A. 1991. *Review of the Geological Environment for Horizontal Wells*. Paper SPE 23544 presented at the Second Archie Conference, Houston, November 3–6.

Beryl Field: geological evolution and reservoir behaviour

G. ROBERTSON

Mobil North Sea Ltd, Grampian House, Union Row, Aberdeen AB1 1SA, UK

Abstract: The Mobil-operated Beryl Field is located in UK Block 9/13. Hydrocarbons are present in Triassic and Jurassic sandstone reservoirs, within a large fault-bounded structural trap. Interaction of subperpendicular fault trends determined the geological evolution of this structure and controlled the distribution of the sandstone reservoirs.

Hydrocarbon generation and subsequent migration commenced in Paleocene time. Entrapment of these hydrocarbons was influenced by pre-existing geological complexity and reservoir continuity. The distribution of different oil-types and multiple fluid contacts is defined by both structural and stratigraphic compartmentalization. Pre-production, the field reached a state of dynamic equilibrium between influx and leakage of hydrocarbons.

Production of oil commenced in 1976. Gas injection and water injection were initiated to arrest pressure decline. Pressure distribution and fluid movement throughout the field's productive life are strongly influenced by geological controls. Understanding these geological controls has resulted in more effective reservoir management and optimization of the hydrocarbon resource.

The Beryl Field is located within UK Block 9/13, in the Northern North Sea. It is approximately 180 miles northeast of the Scottish mainland and 100 miles south of the Brent oil province in an average water depth of 400 ft (Fig. 1).

Block 9/13 was awarded to a Mobil-operated partnership in the 4th UK Offshore Licensing Round, in 1971. The partnership currently comprises Mobil (45%), Amerada Hess (20%), British Gas (10%), Enterprise Oil (20%) and OMV (5%).

The Beryl Field was discovered in 1972 by the first well drilled on the block, 9/13-1. A total of 2633 MMBBL of oil and 3330 BCF of gas are estimated to have been initially in place in the commercially important Triassic and Jurassic reservoirs (Table 1). Following early appraisal drilling, the Beryl Alpha 40-slot Condeep platform was installed and production commenced in 1976. Beryl Bravo, a 21-slot steel jacket, was added in 1984. To date, 56 Beryl Alpha wells, 30 Beryl Bravo wells and 14 subsea wells have been drilled to develop the field. Oil production has been maintained around 100 000 BOPD since 1979. Gas and water have been injected for pressure support.

Table 1. Oil and gas initially in place in Beryl Field producing units (Fig. 3). STOIIP: stock tank oil initially in place; GIIP: gas initially in place

Producing unit	STOIIP (MMBBL)	GIIP (BCF)
Katrine	96	137
Beryl	1671	2344
Linnhe	226	264
Nansen/Eiriksson	319	344
Lewis	321	241
Total	2633	3330

Exploration and appraisal drilling has defined smaller satellite fields around Beryl: Nevis (9/13-4), Katrine (9/13-18), Ness (9/13b-28A) and Linnhe (9/13c-40). The Ness Field is on production and tied back via subsea facilities to Beryl Bravo. In addition, non-commercial hydrocarbons have been discovered above the field in Paleocene and Eocene reservoirs.

Field description

Trap

The Beryl Field is a large fault-bounded structural trap, 9 miles long and 5 miles wide, on the western edge of the South Viking Graben (Figs 1 and 2). A major NNE–SSW-striking normal fault (F1, Fig. 2) which downthrows to the east defines its eastern boundary. Regional dip is westwards and dip closure defines the western boundary. The field lies in the hangingwall of a second major NNE–SSW normal fault (F3, Fig 2) to the west.

The structure is subdivided into westward-tilted fault blocks by a series of sub-parallel synthetic faults (Fig. 2a). Further compartmentalization is defined by antithetic faulting (Fig. 2b) and cross-cutting NW–SE- and NE–SW-striking faults (Fig. 1). The elongate eastern part of the field can be subdivided into a westward-dipping west flank, a relatively flat central horst and an eastward-dipping east flank (Figs 1 and 2).

Reservoir rocks

Oil and gas are present in late Triassic to late Jurassic stacked sandstone reservoirs. Hydrocarbon shows are also encountered in late Cretaceous limestones. The lithostratigraphic nomenclature used by Mobil to subdivide this sequence is based on Deegan and Scull (1977) with informal modifications (Knutson and Munro 1991). Stratigraphic subdivisions, depositional interpretations and reservoir parameters are summarized in Fig. 3.

Cap rocks

Mid to late Jurassic claystones provide a vertical seal over reservoirs in the west of the field and a lateral seal across the eastern bounding fault (F1, Fig. 2). Early Cretaceous claystones and thick late Cretaceous marls and limestones provide the vertical seal over reservoirs to the east.

Geological evolution

Geological evolution of the Beryl Field is intimately related to regional development of the Viking Graben (Badley *et al.* 1988;

From *Petroleum Geology of Northwest Europe: Proceedings of the 4th Conference* (edited by J. R. Parker).
© 1993 Petroleum Geology '86 Ltd. Published by The Geological Society, London, pp. 1491–1502.

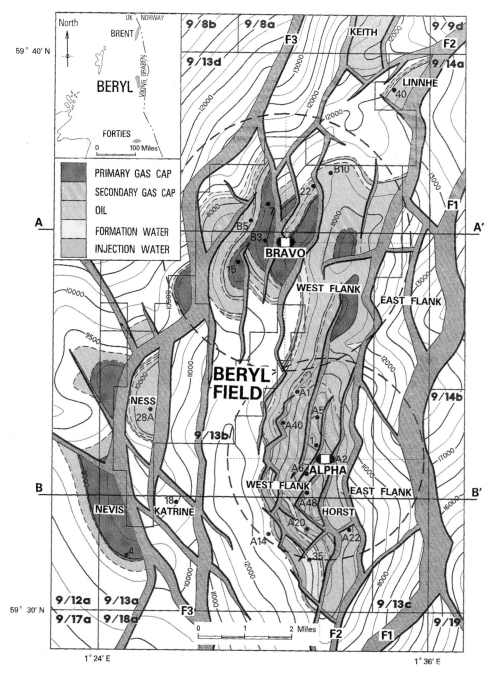

Fig. 1. Beryl Field top mid-Jurassic–Triassic depth map, with 1992 reservoir fluid distribution. Seismic lines A–A′ and B–B′ are presented in Fig. 2. Wells posted are referred to in the text, including Tables 2 and 3. Faults—F1: eastern bounding fault; F2: east flank bounding fault; and F3: western bounding fault.

Ziegler 1990). Triassic to Recent tectonostratigraphic development is subdivided into five well-defined time intervals (cf. Hodgkinson *et al.* 1989):

1. Triassic
2. Latest Triassic to early Jurassic
3. Middle Jurassic
4. Middle to late Jurassic
5. Early Cretaceous to Recent

Triassic (Scythian–Norian)

Sediments of the Cormorant Group were deposited in a semi-arid environment; one formation at the top of this sequence is defined. The Lewis Formation is interpreted as a series of lacustrine and fluvial sandstones and siltstones interbedded

with thick palaeosols. Sedimentation was cyclical and related to climatic variation (Dean 1992).

Early rifting in the Viking Graben commenced in Triassic time and the Triassic sediments in the Beryl area were probably deposited in fault-controlled basins (Frostick *et al.* 1992). O'Donnell (1993) has demonstrated, using well correlation, that by latest Triassic time sedimentation was influenced by faulting. However, seismic definition of these faults is difficult, since Triassic faulting appears to have been overprinted by later Jurassic structures.

Latest Triassic to early Jurassic (Rhaetian–Toarcian)

The Statfjord Group overlies the Cormorant Group conformably, although locally the transition may be markedly unconformable (Tr event, Fig. 3). Strata are subdivided into three well-defined regionally correlatable formations. Raude Forma-

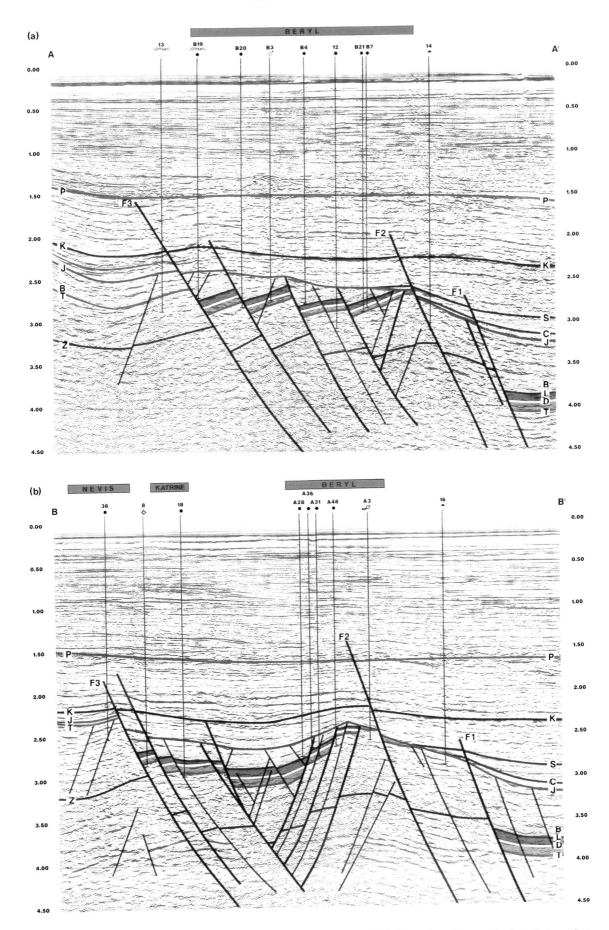

Fig. 2. (a) Seismic line A–A' (Beryl 3D, line 700); **(b)** Seismic line B–B' (Beryl 3D, line 340). Line orientations on Fig. 1. Reflectors—P: top Paleocene; K: top Cretaceous; S: top Turonian limestone; C: top Cromer Knoll Gp; J: Base Cretaceous; B: top Beryl Fm.; L: top Linnhe Fm.; D: top Dunlin Gp; T: top Triassic; and Z: top Zechstein Gp. Faults—F1: eastern bounding fault; F2: east flank bounding fault; and F3: western bounding fault.

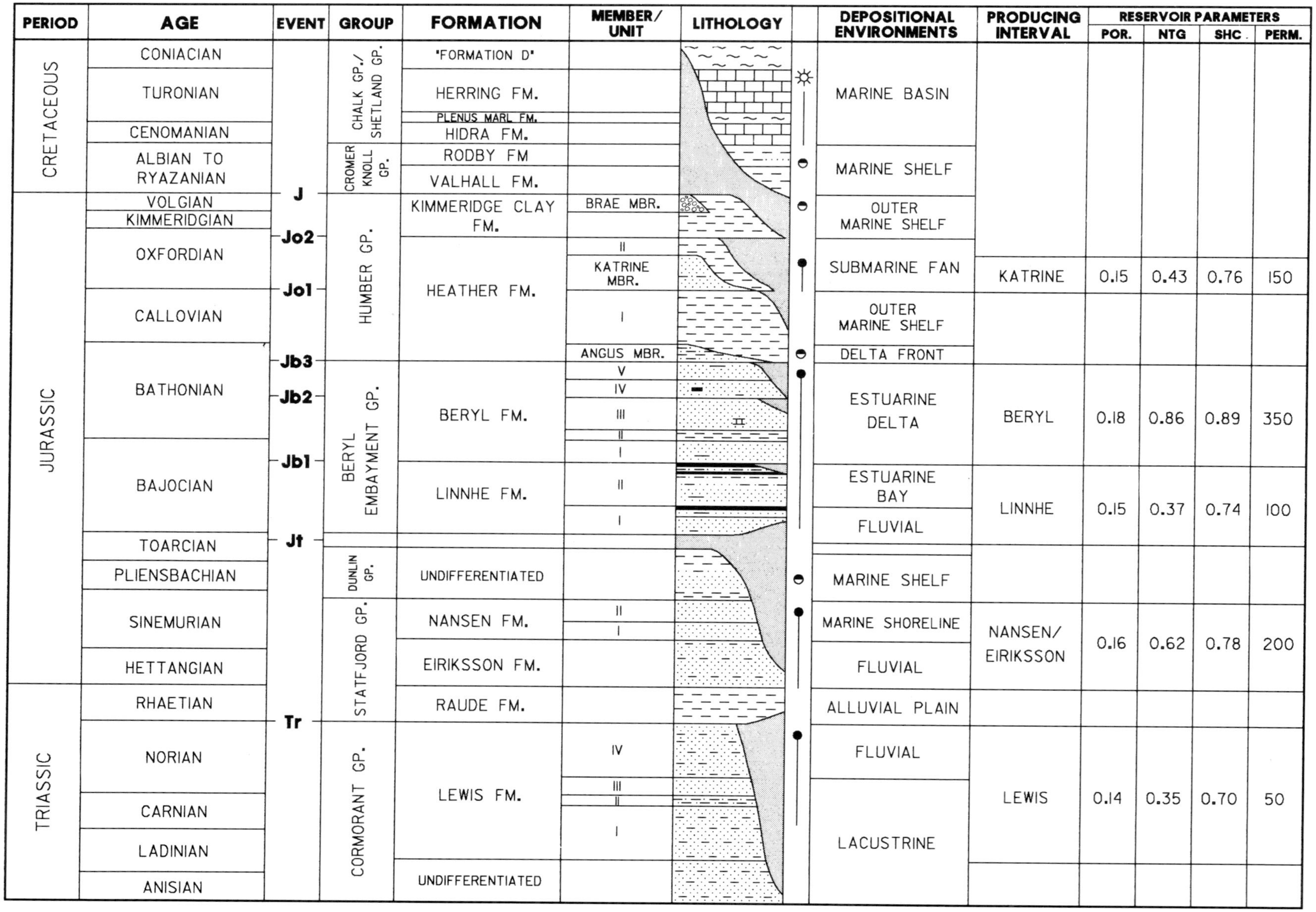

PERIOD	AGE	EVENT	GROUP	FORMATION	MEMBER/UNIT	LITHOLOGY	DEPOSITIONAL ENVIRONMENTS	PRODUCING INTERVAL	RESERVOIR PARAMETERS POR.	NTG	SHC	PERM.
CRETACEOUS	CONIACIAN		CHALK GP./SHETLAND GP.	"FORMATION D"			MARINE BASIN					
	TURONIAN			HERRING FM.			MARINE SHELF					
	CENOMANIAN			PLENUS MARL FM. / HIDRA FM.								
	ALBIAN TO RYAZANIAN			RODBY FM. / VALHALL FM.			OUTER MARINE SHELF					
	VOLGIAN	J	CROMER KNOLL GP.	KIMMERIDGE CLAY FM.	BRAE MBR.		SUBMARINE FAN	KATRINE	0.15	0.43	0.76	150
	KIMMERIDGIAN	Jo2	HUMBER GP.		II / KATRINE MBR.		OUTER MARINE SHELF					
	OXFORDIAN	Jo1		HEATHER FM.			DELTA FRONT					
	CALLOVIAN				ANGUS MBR.		ESTUARINE DELTA	BERYL	0.18	0.86	0.89	350
	BATHONIAN	Jb3	BERYL EMBAYMENT GP.	BERYL FM.	V / IV / III / II / I		ESTUARINE BAY	LINNHE	0.15	0.37	0.74	100
	BAJOCIAN	Jb2		LINNHE FM.	II / I		FLUVIAL					
		Jb1										
	TOARCIAN	Jt	DUNLIN GP.	UNDIFFERENTIATED			MARINE SHELF					
	PLIENSBACHIAN		STATFJORD GP.	NANSEN FM.	II / I		MARINE SHORELINE	NANSEN/EIRIKSSON	0.16	0.62	0.78	200
	SINEMURIAN			EIRIKSSON FM.			FLUVIAL					
	HETTANGIAN			RAUDE FM.			ALLUVIAL PLAIN					
TRIASSIC	RHAETIAN	Tr	CORMORANT GP.	LEWIS FM.	IV / III / II / I		FLUVIAL	LEWIS	0.14	0.35	0.70	50
	NORIAN											
	CARNIAN			UNDIFFERENTIATED			LACUSTRINE					
	LADINIAN											
	ANISIAN											

Fig. 3. Beryl Field stratigraphic column.

tion alluvial plain claystones pass upwards into Eiriksson Formation fluvial channel sandstones and claystones, succeeded by Nansen Formation shallow marine sandstones.

The Dunlin Group conformably overlies the Statfjord Group. Dunlin Group siltstones and sandstones are interpreted as offshore marine sediments.

Latest Triassic to early Jurassic strata represent a transgressive sequence, caused by thermal subsidence and sea-level rise. Evidence for significant fault-controlled deposition (Richards 1991) is not recognized.

At the end of the early Jurassic the South Viking Graben underwent regional uplift. In the Beryl area this uplift is marked by non-deposition of late Toarcian sediments and erosion on structural highs (Jt event, Fig. 3). Nansen Formation and Dunlin Group deposits have been removed by erosion over large areas in the east of the field (Fig. 2).

Middle Jurassic (Bajocian–Bathonian)

The Beryl Embayment Group disconformably overlies the Dunlin Group over the west of the Beryl Field; in the east, the transition is a well-defined angular unconformity. Two distinct formations, which display lateral facies variation, are identified. The Linnhe Formation coals, sandstones and claystones are interpreted as fluvial and estuarine-bay fill deposits. These sediments pass upwards into the sandstone-dominated Beryl Formation, which was deposited in a major estuarine delta.

Active rifting can be recognized in the mid-Jurassic. Thickness variations in the Beryl Embayment Group indicate deposition was structurally controlled. Thick syn-tectonic deposits become gradually thin to absent on structural highs, where localized unconformities are defined (Jb1, Jb2 and Jb3 events, Fig. 3).

Regional extension along NNE–SSW planar basement-controlled faults (e.g. F1 and F3, Fig. 2) created domino-style tilted fault blocks; this fault movement was accompanied by footwall uplift (Yielding 1990). Associated with these major faults, sub-parallel synthetic faults developed. At the same time, NW–SE- and NE–SW-striking transfer faults (*sensu* Roberts *et al.* 1990) interacted with the NNE–SSW faults to define discrete fault-bounded compartments. The orientation of these transfer faults may be related to underlying basement structural trends. As extension continued, antithetic faults, sub-parallel to both NNE–SSW and NW–SE strike directions, developed in response to space accommodation requirements.

Middle to late Jurassic (Bathonian–Volgian)

The Beryl Embayment Group sediments pass conformably upwards into the Humber Group strata (Jb3 event, Fig. 3). Humber Group are subdivided into two well-defined regionally correlatable formations. Heather Formation silty claystones and sandstones are interpreted as offshore marine sediments. Two sandstone units within the Heather are recognized: the basal Angus Member which was deposited in a marginal delta front setting and the upper Katrine Member which was deposited by turbidity currents. Kimmeridge Clay Formation anoxic marine claystones overlie the Heather sequence. This transition is conformable in the west of the field but unconformable in the east (Jo2 event, Fig. 3). Within the Kimmeridge Clay, the Brae Member debris flow breccia-conglomerates are locally developed.

Active rifting continued from the mid-Jurassic into the late Jurassic. Sedimentation of the Humber Group was clearly structurally controlled. Non-deposition of Heather Formation and active erosion of underlying Beryl Embayment Group on structural highs (Jb3, Jo1 and Jo2 events, Fig. 3) confirm

footwall uplift. Early Oxfordian emergence and erosion (Jo1 event, Fig. 3) corresponded with deposition of locally derived Katrine turbidites in structural lows. Rifting proceeded throughout Kimmeridge Clay deposition.

In the latest Jurassic, structural collapse occurred on the footwall of the Beryl eastern bounding fault (F1, Fig. 2). Subsidence of this footwall, migration of the structural crest westwards and displacement along a back fault (F2, Fig. 2) marked the end of the Jurassic rifting episode (cf. Badley *et al.* 1988). Brae breccia-conglomerates were deposited in the hangingwall of the active fault.

Early Cretaceous to Recent (Ryazanian–present day)

The Humber Group is unconformably overlain by the Cromer Knoll Group over the east of Beryl and the Chalk Group/Shetland Group elsewhere. Thick Tertiary sandstones and claystones overlie the Cretaceous.

The Base Cretaceous Unconformity (J event, Fig. 3) is well-defined in the Beryl Field. However, over most of the field, this regional event may simply represent a depositional hiatus. Downcutting Base Cretaceous erosion is confined to the footwall of faults, which were still active in the early Cretaceous (eg. F2, Fig. 2). Transgressive onlap occurred in the collapsed eastern area of the field, in the early Cretaceous. By the late Cretaceous, the entire Beryl structure was drowned.

Throughout the Cretaceous and the Tertiary the Beryl area experienced passive thermal subsidence. Faulting, associated with this thermal subsidence, continued into the early Tertiary. Subsidence in the Viking Graben throughout this time appears to be continuous although minor tectonic (or halokinetic) inversion events may be recognized.

Generation and migration of hydrocarbons

Source rocks

Organic-rich claystones of the Kimmeridge Clay Formation are the primary source of hydrocarbons in the Beryl Field. Secondary sources may include Linnhe Formation coals and Heather Formation claystones.

Hydrocarbon generation and subsequent migration commenced in the Viking Graben to the east of Beryl in Paleocene time and continues to the present day. Volumes of hydrocarbons generated are estimated to be in excess of trapping capacities.

Hydrocarbons

Beryl oil is typically medium gravity (36–38° API), undegraded, low sulphur and naphtheno-paraffinic in composition. Chung (pers. comm.) has subdivided the oil into three source types, based on differences in chemical and isotopic compositions:

1. Type A (highly anoxic marine source)
2. Type B (less anoxic marine source with terrestrial affinities)
3. Type C (mature, highly anoxic marine source).

Hydrocarbon migration

The distribution of Beryl oil-types is geographically defined: Type A oil accumulated in the south of the field; Type B oil in the north; and Type C oil to the east. With this distribution, oil migration routes can be inferred from Viking Graben source kitchens (Chung pers. comm., Fig. 4).

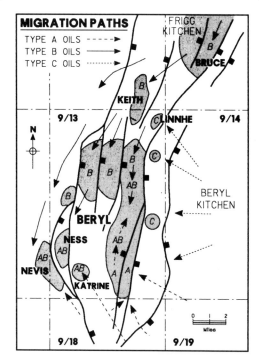

Fig. 4. Beryl Field oil-type distribution and inferred migration routes (Chung pers. comm.).

Entrapment of hydrocarbons

The formation of the Beryl structure pre-dated hydrocarbon generation and migration. Only minor structural movement appears to have occurred after the onset of hydrocarbon generation. Emplacement of hydrocarbons into the trap was, therefore, influenced by pre-existing stratigraphic and structural controls.

Stratigraphic controls

The Beryl Field is sealed by mid- to late Jurassic and Cretaceous cap rocks, which have trapped hydrocarbons by limiting buoyancy-driven migration.

Watts (1987) classifies cap rock seals as either 'membrane seals' (capillary leakage) or 'hydraulic seals' (induced fracture leakage). Membrane seals will trap hydrocarbons until the buoyancy pressure of the hydrocarbon column exceeds the capillary displacement pressure of the seal. Hydraulic seals trap until reservoir pore pressure exceeds seal fracture pressure. Over Beryl, only membrane seals are recognized; reservoir pressures appear to be too low for hydraulic seal failure to occur.

Multiple oil–water contacts are defined in the field. Individual structural compartments are enclosed by cap rock and fault-related seals (see below) and charged with hydrocarbons to different levels. Oil–water contacts are shallowest on the crest of the field and become deeper towards the flanks. Maximum relief of the Beryl structure approaches 2000 ft, but hydrocarbon column heights within structural compartments do not exceed 1000 ft (see Fig. 7). Hydrocarbon levels appear to be controlled by the membrane seal integrity of the overlying cap rocks; compartments were charged to a maximum hydrocarbon column height, at which point the buoyancy force of the hydrocarbons became greater than the membrane seal displacement pressure and the cap rocks leaked.

Faults are defined seismically in the Cretaceous and Tertiary sequences over structural highs (Fig. 2). Reduced seal integrity over these highs is thought likely since faults and associated fractures probably act as low entry-pressure conduits. Fracture

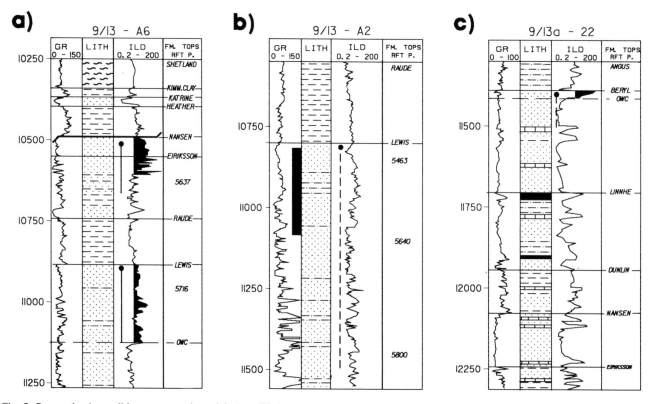

Fig. 5. Pre-production well logs, representing original equilibrium conditions. (**a**) 9/13-A6 (OH log September 1976): stacked reservoir sequence with sandstones charged to different levels and pressures. Katrine and basal Eiriksson sandstones are water bearing, within an oil-bearing sequence. Nansen/Eiriksson reservoir is lower pressure than Lewis reservoir. (**b**) 9/13-A2 (OH log August 1976): thick water-bearing Lewis sandstone sequence with strong oil shows. (**c**) 9/13a-22 (OH log March 1982): thin Beryl oil-column, probably representing migration interval, with strong oil shows below a well-defined oil–water contact. Black shading: oil-bearing reservoir; reservoir pressures are corrected to a −11 200 ft SS datum; OH: open hole.

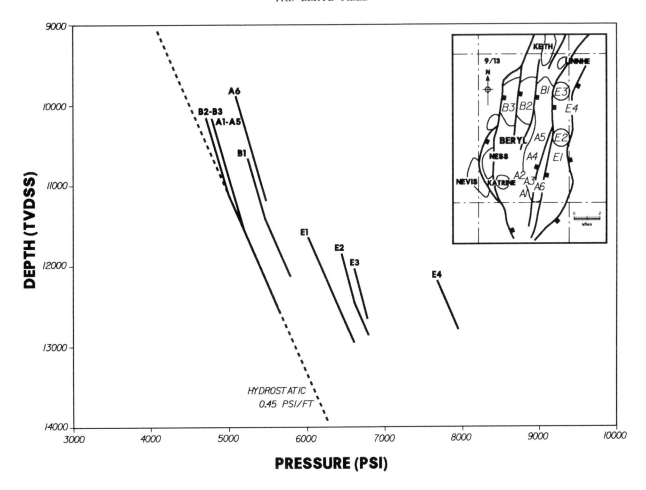

Fig. 6. Beryl Field pre-production reservoir pressure trends. Different fault-bounded pressure regimes are recognized within the field and west to east there is an increase from normal to overpressure conditions. Data compiled by Bala Balakrishnan and Ash Stone.

leakage would also explain how hydrocarbons could migrate vertically to charge Paleocene and Eocene reservoirs.

Lithological heterogeneities, such as claystone beds, coals and cemented layers within reservoir sequences controlled the emplacement of hydrocarbons. Separate reservoirs in the same structural compartment may be charged to different levels and pressures and some reservoirs may even remain uncharged (Fig. 5a). These features can be explained by reservoirs spilling at different levels, variable membrane sealing properties of non-reservoir lithologies and complete structural/stratigraphic isolation of reservoirs.

Structural controls

The geological evolution of the Beryl Field has resulted in a compartmentalized structural trap. In the field, the occurrence of multiple fluid contacts, overpressured reservoirs and differences in oil geochemistry indicate the presence of lateral barriers. These barriers are fault-related seals.

Fault-related seals are classified, by Watts (1987), as either 'sealing faults' (fault plane seal) or 'juxtaposition faults' (reservoir against seal). Sealing faults may result from clay smear, granulation, fault breccia or diagenetic cementation. Faults will trap laterally, in the same way as cap rock membrane seals.

Both sealing and juxtaposition faults are recognized in the Beryl structure. The presence of sealing faults in the field is confirmed by sandstone-to-sandstone contacts across fault planes. Clay smearing is thought to be the most important sealing mechanism (cf. Bouvier *et al.* 1989; Lindsay *et al.* 1993).

In Beryl Field, all the structural compartments from flank to crest are charged with hydrocarbons. Since these compartments are fault-bounded, leakage across faults has occurred during hydrocarbon emplacement.

Geopressure

West to east, across the field, there is a progressive increase in reservoir pressure from normal to overpressure conditions (Fig. 6). Distinct pressure regimes, bounded by fault-related seals, can be identified.

Continuous hydrocarbon generation in adjacent source rocks (Fig. 7), associated with restricted fluid dissipation in the reservoir (cf. Spencer 1987) are thought to be the causes of the overpressure. The degree of overpressuring reflects the relative proximity to the source kitchen.

Oil shows

Hydrocarbons have been identified below petrophysically defined oil–water contacts, from drill-cuttings and cores in Beryl wells drilled prior to production (Figs 5b and c). These oil shows most likely represent hydrocarbon migration in action. Alternatively, the shows may indicate that Beryl was charged to deeper levels in the geological past.

Entrapment model

A hydrocarbon entrapment model is proposed for the Beryl Field (Fig. 7), based on the following observations:

1. source rocks, adjacent to Beryl, are currently generating hydrocarbons;

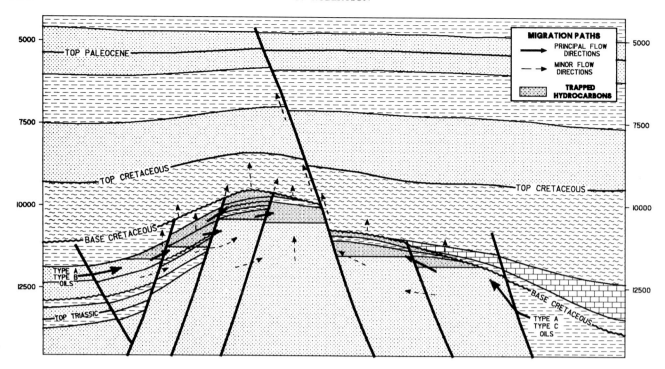

Fig. 7. Beryl Field entrapment model.

2. fast hydrocarbon productivity has exceeded trapping capacity;
3. the Beryl structural trap is compartmentalized;
4. migration of hydrocarbons through the trap is controlled by fault-related seal failure;
5. buoyancy-driven, cap rock seal failure results in hydrocarbon leakage.

This model suggests the Beryl structural trap to be an active system. Pre-production, a state of dynamic equilibrium was reached between influx and leakage of hydrocarbons.

Production of hydrocarbons

Beryl Alpha and Beryl Bravo production and injection histories are summarized in Fig. 8 and Tables 2 and 3. More detailed historical data are presented by Marcum *et al.* (1978), Steele and Adams (1984) and Knutson and Erga (1991).

In production time, Beryl Field reservoir fluids have been extracted and reservoir pressure depleted. To replace voidage and maintain pressures, produced gas and seawater have been injected. As a result of production operations, a continuously changing state of disequilibrium can be described (Fig. 1). Pressure depletion below bubble point and gas injection have resulted in the generation of secondary gas caps. Aquifer expansion and seawater injection have flushed hydrocarbon-bearing reservoirs; seawater breakthrough, from injectors in the aquifer, is headed by formation water fronts.

The changing conditions in the Beryl reservoirs are routinely monitored. Reservoir behaviour, throughout the field's productive life, has been influenced by stratigraphic and structural controls.

Stratigraphic controls

During production, lithological heterogeneities have controlled pressure distribution and fluid movement within the Beryl Field reservoirs. Subtle changes in lithology may have a marked effect on reservoir behaviour.

Table 2. Beryl Alpha production history

Date		Event
December	1971	Block 9/13 awarded to Mobil partnership.
September	1972	9/13-1 discovery well.
July	1975	Beryl Alpha platform on location.
June	1976	First oil production, 9/13-A4.
November	1977	Gas injection initiated in 9/13-A5.
January	1979	Water injection initiated in 9/13a-A14.
September	1980	Peak platform oil production at 122 900 BOPD.
October	1992	Gas production commenced.
November	2017	Production licence expiry.

Table 3. Beryl Bravo production history

Date		Event
April	1975	9/13-7 discovery well.
January	1979	9/13-15 oil production via Beryl Alpha.
June	1983	Beryl Bravo platform on location.
July	1984	First oil production, 9/13a-B5.
December	1984	Gas injection initiated in 9/13a-B3.
July	1986	Water injection initiated in 9/13a-B10.
August	1987	Ness Field oil production via Beryl Bravo.
November	1987	Subsea water injection (BWISS) initiated.
May	1988	Peak platform oil production at 86 700 BOPD.
October	1989	Linnhe Field oil production via Beryl Bravo.
December	1991	Linnhe Field abandoned.
October	1992	Gas production commenced.
November	2017	Production licence expiry.

BWISS: Beryl water injection subsea.

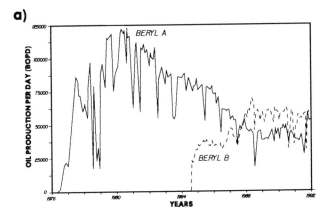

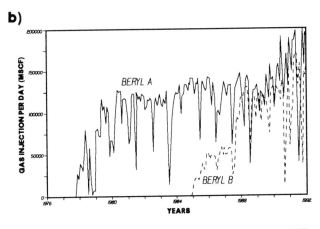

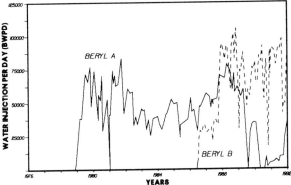

Fig. 8. (a) Beryl Field production history; (b) Beryl Field injection history.

Non-reservoir lithologies act as vertical permeability barriers. Thick, laterally continuous sealing claystones and coals separate the reservoir sandstones into five main producing intervals (Fig. 3 and Table 1). These producing intervals are pressure isolated and behave independently, because of restricted vertical communication.

Claystones, coals and cemented layers can also generate pressure differentials and disrupt fluid movement within reservoir formations (Figs 9a, b, d and e). The effectiveness of these production baffles is determined by their seal integrity and lateral extent.

Variations in sandstone reservoir quality and lateral continuity of sandstone facies can influence reservoir behaviour. Higher permeability sandstones allow better reservoir communication, but may result in channelling (Fig. 9e). Thin, impersistent sandstones in interbedded sequences often display limited reservoir continuity (Figs 9a, b and c).

Understanding stratigraphic controls has influenced Beryl Field operations. Main producing intervals are developed separately to maximize production efficiency and limit cross-flow. Vertical barriers within producing intervals are used to maximize oil production and restrict gas-cap expansion and aquifer encroachment, by selective perforation (Knutson and Erga 1991). Low permeability, laterally discontinuous reservoirs have been fracture-stimulated to improve productivity (Clancey 1990; O'Donnell 1993). The result has been improved well performance and increased oil recovery.

Structural controls

Fault-related seals compartmentalize the Beryl structure. Based on production data, these faults can act as lateral permeability barriers. The effects of faulting are most pronounced in interbedded sequences, since only minor offsets restrict communication.

The Beryl Field can be subdivided into producing compartments, which are bounded by faults (Knutson and Erga 1991). Reservoir communication is good within these compartments; however, pressure distribution and fluid movement may be different in adjacent compartments. Depending on the production/injection histories within producing compartments, pressure history profiles can be significantly different (Fig. 10).

Post-production, pressure-driven fluid movement has replaced buoyancy-driven migration. New communication routes have become established within the field, controlled by pressure gradients and reservoir permeability. This explains how gas and oil have moved down-structure and water has moved up-structure, during production (Figs 1, 9d, e and f).

Sealing faults will leak when the pressure differential across the fault exceeds the capillary displacement pressure of the sealing membrane (Watts 1987). When this occurs, juxtaposed compartments will start to communicate. Faults that sealed during entrapment will leak during production, given suitable pressure conditions (Fig. 9c).

In the Beryl Field, identification of sealing faults is important to maximize production. By-passed oil accumulations in undrilled compartments, are targeted by continued infill drilling. New well locations allow more efficient reservoir drainage and increased oil recovery.

Conclusions

1. The Beryl Field is a large fault-bounded structural trap, with commercial hydrocarbons present in Triassic and Jurassic sandstone reservoirs. Geological evolution of the structure and distribution of sandstone reservoirs was controlled by the interaction of sub-perpendicular fault trends.

2. Hydrocarbon generation and migration commenced in Paleocene time and continues to the present day. The presence of geochemically different hydrocarbons reflects separate source kitchens and migration routes.

3. Emplacement of hydrocarbons was influenced by stratigraphic and structural controls and hydrocarbon column heights appear to be determined by the seal integrity of overlying cap rocks. Prior to production, a state of dynamic equilibrium was reached between influx and leakage of hydrocarbons.

4. In production time, the extraction and injection of different fluids has resulted in a continuously changing state of disequilibrium. Pressure distribution and fluid movement during production are influenced by stratigraphic and structural controls.

5. Understanding the geological controls, which influence reservoir behaviour in Beryl Field, has resulted in more effective reservoir management and optimization of the hydrocarbon resource.

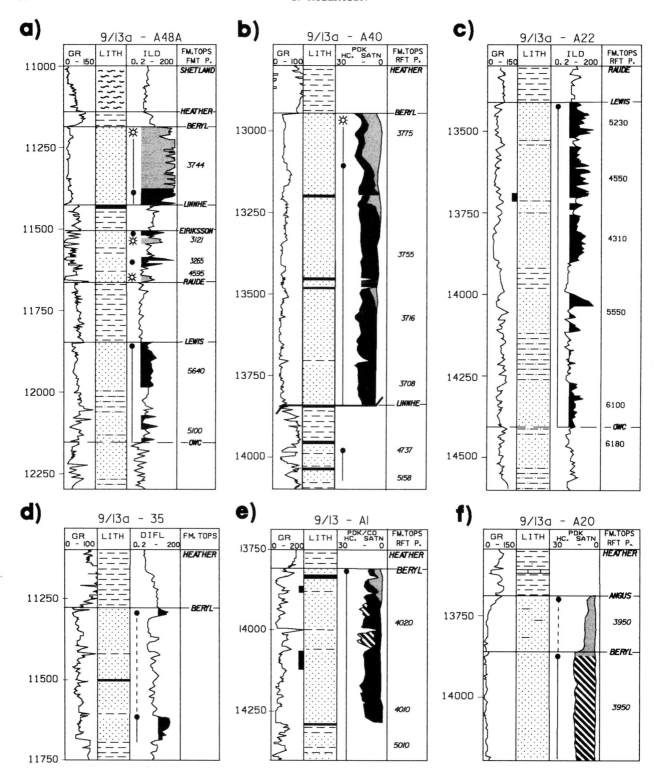

Fig. 9. Post-production well logs, demonstrating disequilibrium conditions following production activities. (**a**) 9/13a-A48A (OH log January 1991): Stacked reservoir sequence with independent producing intervals, separated by regional vertical permeability barriers. Beryl sandstones, although pressure depleted, are located in a structurally isolated fault-bound compartment with a gas–oil contact significantly shallower than the main field gas cap. Eiriksson claystone barriers separate sandstones which are charged with different fluids at different pressures. Lewis sandstones, in part, are undepleted indicating stratigraphic/structural isolation. (**b**) 9/13a-A40 (OH log May 1985 and CH log August 1990): Beryl sandstone sequence with three thin coals, which act as production baffles and control gas movement within the reservoir. Linnhe sandstones are separated by claystone barriers and display differential depletion and limited reservoir continuity. (**c**) 9/13a-A22 (OH log June 1980): Oil-bearing Lewis sandstone sequence located in fault-bound compartment with an independent original oil–water contact. Sandstones are pressure depleted, indicating that faults that sealed during entrapment can leak during production. (**d**) 9/13a-35 (OH log August 1987): Highly water-flushed Beryl sandstone sequence with a thin oil column preserved below a claystone barrier. (**e**) 9/13-A1 (OH log February 1978 and CH log September 1987): Oil-bearing Beryl sandstone sequence displaying gas-cap encroachment and water injection breakthrough. Production baffles control fluid movement. Injection water has broken through in the most permeable sandstone units. (**f**) 9/13a-A20 (OH log September 1979 and CH log August 1989): Gas movement down-structure and injection-water movement up-structure have completely displaced oil from Beryl sandstones. Black shading: oil-bearing reservoir; grey: gas-bearing reservoir; and cross-hatch: injection-water flushed reservoir. Reservoir pressures are corrected to a −11 200 ft SS datum. OH: open hole, CH: cased hole.

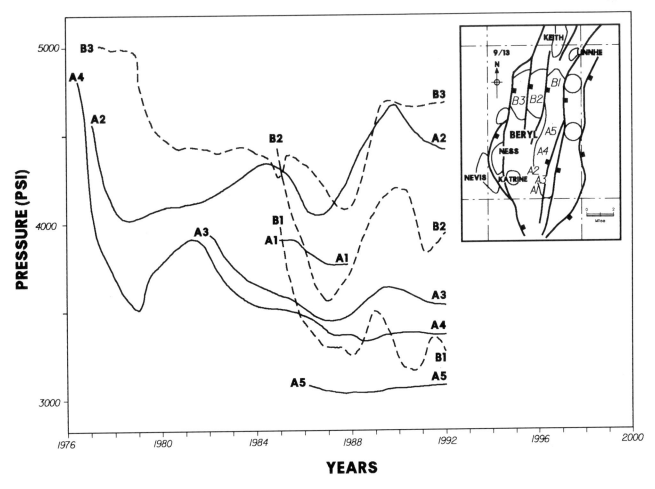

Fig. 10. Beryl Formation pressure history plots for different fault-bound producing compartments.

This study has benefited from the contributions of numerous colleagues. In particular John Allison, Moses Chung, Martin Dru, Bob Dunay, Ragnhild Erga, Ian Munro, Claire Robertson and John Wild provided discussion and helpful comments. Alex Brand, Paul Fleming and Moira McTear prepared the figures and Elison Corstorphan typed the manuscript. Mobil North Sea Ltd and partners permitted publication. Opinions expressed and interpretations made are solely the responsibility of the author.

References

BADLEY, M. E., PRICE, J. D., RAMBECH DAHL, C. AND AGDESTEIN, T. 1988. The structural evolution of the Viking Graben and its bearing upon extensional modes of basin formation. *Journal of the Geological Society, London*, **145**, 455–472.

BOUVIER, J. D., KAARS-SIJPESTEIJN, C. H., KLEUSER, D. F., ONYE-JEKWE, C. C. AND VAN DER PAL, R. C. 1989. Three-dimensional seismic interpretation and fault sealing investigations, Nun River Field, Nigeria. *American Association of Petroleum Geologists Bulletin*, **73**, 1415–1435.

CLANCEY, B. M. 1990. Hydraulic fracture stimulation of permeable North Sea oil wells. *Society of Petroleum Engineers, Europec 1990, The Hague*, 307–314. SPE Paper 20969.

DEAN, K. P. 1992. Sedimentation of the Upper Triassic reservoirs in the Beryl Embayment; lacustrine sequences in a semi-arid environment. *In*: PARKER, J. R. (ed.) *Petroleum Geology of Northwest Europe: Proceedings of the 4th Conference*. Geological Society, London, 581.

DEEGAN, C. E. AND SCULL, B. J. 1977. *A Standard Lithostratigraphic Nomenclature for the Central and Northern North Sea*. Institute of Geological Sciences Report 77/25.

FROSTICK, L. E., LINSEY, T. K. AND REID, I. 1992. Tectonic and climatic control of Triassic sedimentation in the Beryl Basin, northern North Sea. *Journal of the Geological Society, London*, **149**, 13–26.

HODGKINSON, R. J., DUNAY, R. E. AND BARNES, K. R. 1989. Rocks and time: the Jurassic stratigraphy, sedimentation, tectonics and petroleum formation of the 'Beryl Block' (9/13), Viking Graben, UK North Sea. Geological Society Conference on the origin of petroleum: basin evolution and modelling (abstract).

KNUTSON, C. A. AND ERGA, R. E. 1991. Effect of horizontal and vertical permeability restrictions in the Beryl reservoir. *Journal of Petroleum Technology*, **43**, 1502–1509. SPE Paper 19299.

LINDSAY, N. G., MURPHY, F. C., WALSH, J. J. AND WATTERSON, J. 1993. Outcrop studies of shale smears on fault surfaces. *In*: FLINT, S. S. AND BRYANT, I. D. (eds) *The Geological Modelling of Hydrocarbon Reservoirs*. IAS Special Publication, **15**. Blackwells, Oxford, 113–123.

MARCUM, B. L., AL-HUSSAINY, R., ADAMS, G. E., CROFT, M. AND BLOCK, M. L. 1978. Development of the Beryl A Field. *Proceedings of the European Offshore Petroleum Conference*. Society of Petroleum Engineers, London, 319–324.

O'DONNELL, D. 1993. Enhancing the oil potential of secondary Triassic reservoirs in the Beryl A Field, UK North Sea. *In*: SPENCER, A. M. (ed.) *Generation, Accumulation and Production of Europe's Hydrocarbons III*. EAPG Special Publication, **3**. Springer-Verlag, Heidelberg.

RICHARDS, P. C. 1991. Evolution of Lower Jurassic coastal plain and fan delta sediments in the Beryl Embayment, North Sea. *Journal of the Geological Society, London*, **148**, 1037–1047.

ROBERTS, A. M., PRICE, J. D. AND OLSEN, T. S. 1990. Late Jurassic half-graben control on the siting and structure of hydrocarbon accumulations: UK/Norwegian Central Graben. *In*: HARDMAN,

R. F. P. AND BROOKS, J. (eds) *Tectonic Events Responsible for Britain's Oil and Gas Reserves.* Geological Society, London, Special Publication, **55**, 229–257.

SPENCER, C. W. 1987. Hydrocarbon generation as a mechanism for overpressuring in the Rocky Mountain region. *American Association of Petroleum Geologists Bulletin*, **71**, 368–388.

STEELE, L. E. AND ADAMS, G. E. 1984. A review of the northern North Sea's Beryl Field after seven years' production. *Proceedings of the European Petroleum Conference, London*, 51–58. SPE Paper 12960.

WATTS, N. L. 1987. Theoretical aspects of cap-rock and fault seals for single- and two-phase hydrocarbon columns. *Marine and Petroleum Geology*, **4**, 274–307.

YIELDING, G. 1990. Footwall uplift associated with Late Jurassic normal faulting in the northern North Sea. *Journal of the Geological Society, London*, **146**, 455–472.

ZIEGLER, P. A. 1990. Tectonic and palaeogeographic development of the North Sea rift system. *In*: BLUNDELL, D. J. AND GIBBS, A. D. (eds) *Tectonic Evolution of the North Sea Rifts*. International Lithosphere Programme, Publication No. 181, Clarendon Press, Oxford, 1–36.

Clyde: reappraisal of a producing field

P. J. TURNER

BP Exploration Operating Company Ltd, Farburn Industrial Estate, Dyce, Aberdeen, AB2 0BP, UK

Abstract: The complexities inherent in a field are seldom appreciated prior to production and their impact poorly quantified. In order to exploit a reservoir fully, continuous and focused reappraisal is required throughout field life. Evaluation of reservoir performance through the integration of effective surveillance programmes and the analysis of development well results is critical to reduce the uncertainty carried at the planning/early development stage.

Successful development of Clyde has resulted from the application of appropriate techniques within an integrated multidisciplinary team. The timely acquisition of sufficient relevant data also proved an essential part of this process.

The integration of seismic, sedimentological, biostratigraphic and dynamic data from Clyde has provided significant improvements in intra-reservoir resolution, reservoir description and the identification of flow units. This has permitted the discrimination between reservoir models and has reduced uncertainty in field and future infill well performance. Various advanced techniques, such as absolute acoustic impedance and dip and azimuth plots, have been used to enhance our understanding of the Clyde Field. Their application, together with techniques such as parasequence mapping and pressure data analysis, is described. Through the development of a 'reactive' reservoir simulator, improvements in the geoscience model are easily incorporated and the simulator updated. This approach permits rapid quantification of the value of future development wells, recompletions and other reservoir management options. This is part of a continuous process which allows the integrated asset team to optimize the development strategy.

The Clyde Field, which was discovered in 1978, is located on the SW edge of the North Sea Central Graben (Fig. 1). The structure of the field takes the form of a rotated fault block, the Clyde Beta structure, truncated at its crest by a major unconformity (Figs 2 and 3; Gibbs 1984; Smith 1987) and the reservoir is developed in Late Jurassic shallow marine sands of the Fulmar Formation. Estimated STOIIP at Annex B (1982) was 408 MMBBL of oil, of which 154 MMBBL was considered recoverable (Stevens and Wallis 1991). This was revised in 1991 to 380 MMBBL STOIIP and 131.5 MMBBL recoverable.

Production is from a single, centrally located, platform provided with 30 slots. Aquifer support is insufficient to maintain reservoir pressure and sustain a plateau rate of 50 000 BOPD, so a programme of water injection has been implemented.

Appraisal and early development

The results of appraisal drilling were interpreted as a simple 'layered sandbody' model based mainly on Wells 30/17b-2, -3 and -7 (Fig. 3). The original Annex B for Clyde presented two separate producible horizons of uniform properties in a simple 'layer cake' arrangement (Fig. 2a). A redrill of 30/17b-6, 10 years later, (A18) encountered similarly overpressured Late Kimmeridgian sand stringers and was successfully logged and cored. It revealed that this area of the field is not producible. The absence of this data point proved critical in influencing the simplistic reservoir model, used as a basis for the Annex B. Early development drilling rapidly revealed this simple model to be untenable. In particular, whereas Wells A01 and A02 encountered sequences comparable to the crestal 30/17b-2 (discovery) and 30/17b-7 (appraisal) wells, downflank Wells A03 and A04 (Fig. 3) encountered a different section (Fig. 2b). At least three zones were identified, each having different characteristics and recovery factors, with variable reservoir quality.

A much thicker than expected sand interval was also encountered in the upper reservoir in downflank Wells A03,

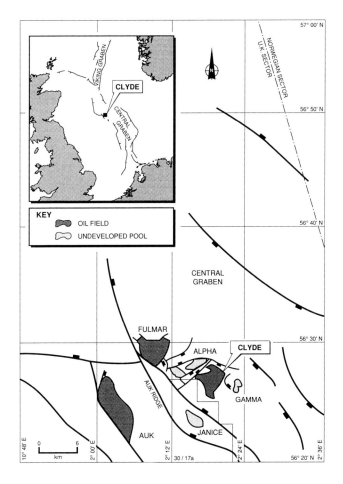

Fig. 1. Clyde Field: location map.

A04 and A06 (Zone A). This sand is of generally poor quality and contains oil of a markedly different composition to the main accumulation. This oil is comparable to that present in

From *Petroleum Geology of Northwest Europe: Proceedings of the 4th Conference* (edited by J. R. Parker).

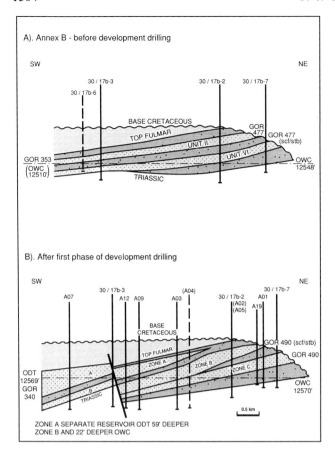

Fig. 2. Clyde Field: schematic cross-sections illustrating broad reservoir zonation.

the topmost sand of the 30/17b-3 well, implying that this sand may form a continuous body between these wells, overlying, but isolated from, the Zone B and Zone C reservoir intervals.

Incorrect correlation of what is now identified as Zone A in 30/17b-3 with Zone B in 30/17b-2, the failure of the 30/17b-6 appraisal well and the resultant optimistic reservoir model, has had a major impact on the development of Clyde and its failure to achieve Annex B performance projections.

A 50 000 BOPD plateau lasting to the end of 1990 was forecast at Annex B. This was achieved by November 1987, following the field's start-up in March 1987, but field production rapidly declined from August 1989. Production during 1990 averaged 36 000 BOPD.

Reservoir problems result primarily from poor reservoir description at the appraisal stage; sand quality is poorer, fault compartmentalization greater, and the reservoir architecture is more complex than originally perceived. This is evident from the poor pressure support to large areas of the field, with wells dying at relatively low water cuts, poor sweep efficiency and early water breakthrough. Consequently, early production profiles were too optimistic.

Most of the first ten wells were drilled in the central part of the field and in crestal Zones B and C (Fig. 2b). Although these encountered poorly prognosed sections, the next eight wells resulted in even greater surprises. The 18 months' drilling results up to January 1990 were disappointing and using prognosed and achieved production figures, a historial 'chance of success' of 1 to 5 was indicated. A success rate of 1 in 2 is required for break-even on most development drilling programmes (i.e. deliver at least 50% of the prognosed 'most likely' hydrocarbon reserves). This made strategic options clear: BP and Co-venturers approved a plan to defer development drilling for up to 18 months, until the fourth quarter of 1991. The geoscience and reservoir engineering team was

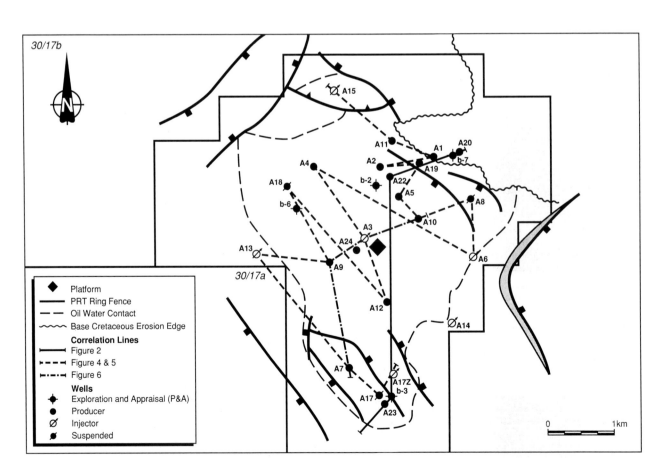

Fig. 3. Clyde Field: well status and correlation location map.

staffed up and complete reservoir re-evaluation was undertaken, aimed at reduction of uncertainty and development plan optimization.

Field reappraisal

The drilling deferral provided the Clyde team with time to re-interpret the seismic data, re-map the reservoir, integrate biostratigraphical, sedimentological and geochemical studies with surveillance data and update the reservoir simulation model. With no further field development, economic estimated recovery would have been less than 85 MMBBL. This study, which cost less than a third of the cost of one new well, has reduced the uncertainty on future infill wells. Non-optimal target locations have been revised and a programme of infill drilling in 1992 and 1993 has been identified. Since drilling recommenced, five consecutive wells have been drilled on prognosis and four wells have been gas lifted. The production profile decline has been arrested and 85% of the field's current production is the result of activities identified in this reappraisal study.

In addition, during the deferral of development drilling, satellite prospectivity has been pursued including successful appraisal of the Clyde Alpha structure leading to expected Annex B approval in 1992. Satellites have impact on the development (i.e. slot utilization and development strategy) and the business plan. An optimized development strategy for all main field zones and the satellites is essential to maximize the value of the whole Clyde catchment area.

A comprehensive data acquisition programme throughout the development phase has proved an essential prerequisite to this study. Seventeen of the 20 development wells were extensively cored. Without core permeabilities and sedimentological analyses, it would not have been possible to evaluate this reservoir adequately. Seven hundred separator well tests, 106 static pressure measurements, formation pressure tests (RFTs) and pressure build-up analyses (PBUs) in every well, and 35 production flowmeter (PLT)/flood front monitoring logs (PDKs) have been taken to date, and significant reduction in uncertainties in the dynamic model has resulted.

Integration of data (e.g. seismic, biostratigraphy, sedimentology and dynamic) has significantly improved intra-reservoir resolution, sand package recognition, reservoir description and potential to identify flow units within the reservoir. This permits discrimination between reservoir models to reduce uncertainty and thus optimize a development strategy.

The approach and key techniques used in the reappraisal are described, then important lessons learned from this field development are summarized.

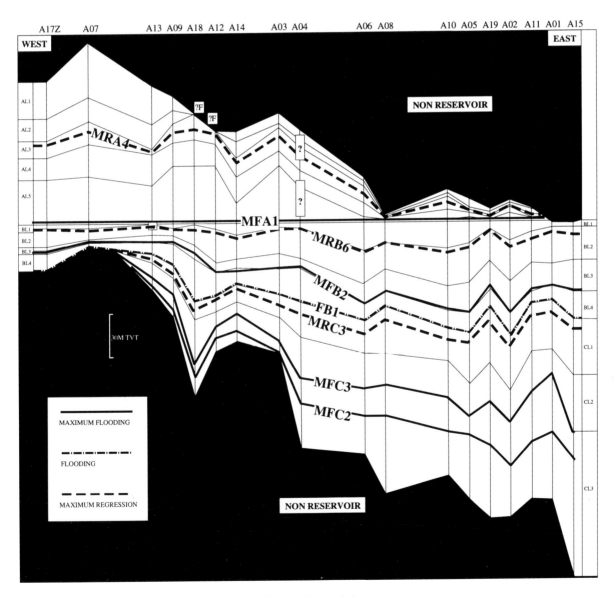

Fig. 4. Clyde Field: relationship between major depositional surfaces and reservoir layers.

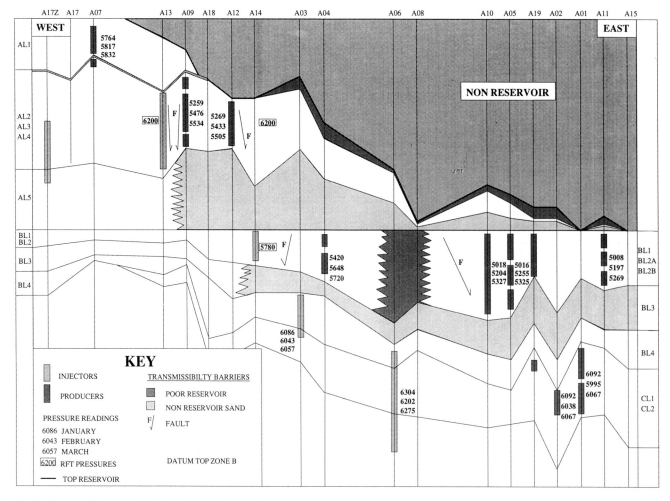

Fig. 5. Clyde Field: pressure barriers within the reservoir.

Approach

There were two stages of well correlation, data integration and reservoir interpretation. The objective of the first stage was to assess the overall zonation. This examined the validity of the then currently interpreted division of the field into the three zones A, B and C. To do this all possible factors were taken into account with no single one providing a definitive solution.

A biostratigraphical study was undertaken to determine the degree of marker reliability. The Clyde Field biostratigraphical database was at the limit of resolution and it was felt that some of the markers may have been extrapolated beyond such limits. It was thus necessary to determine which data were reliable and which were not. The study identified one marker horizon, in particular, with a high degree of confidence. This marker, the lowest down-hole occurrence (LDO) of *P. pannosum*, was situated at the base of the interpreted Zone A and identified the A/B boundary as due to an increase in relative sea level: maximum flooding surface (MF)A1 (Fig. 4). This gave significant confidence to the A and B zone well correlations and provided a suitable datum on which to 'flatten' the reservoir (Figs 4 and 5).

The seismic data were calibrated in a detailed study in the fourth quarter of 1989 and in early 1990, by matching synthetic seismograms to the 3D seismic dataset around the wells. This showed the dataset to be approximately 90° away from zero phase. By applying a bulk phase shift of 90° phase to the entire 3D seismic data, they were converted to approximately zero phase. This allowed the synthetic seismograms, vertical seismic profiles (VSPs) and seismic data to be fully integrated over the entire field. It then became possible to pick the reservoir units

across the field consistently and to define intra-reservoir geometries, incorporating all the well information.

Seismic interpretation recognizes the A/B boundary as the most reliable marker horizon on the field, thus supporting the well log and biostratigraphical interpretation. Interpreted top Zone C is identified as a reliable, if intermittent, marker and reinforces the well log interpretation.

Regional depositional events and the stratigraphic sequence mapping showed a number of field-wide correlatable events (Fig. 4). Following a sedimentological review in 1989, the Clyde reservoir can be subdivided into a series of shallowing and deepening cycles at various scales. These closely correspond to the sub-regional framework, but are more precisely placed through the incorporation of core data (grain-size trends, ichnofacies) with log motif. Correlation of the surfaces has been guided by petrographic variation through the reservoir and by the biostratigraphic review, which has supported the results of the sedimentological and seismic studies. These identify periods of maximum regression and transgression along with their corresponding progradational and retrogradational sequences. The interpretation of these sequences corroborates the A, B and C zonation.

Similarly, pressure data and gas–oil-ratio (GOR) data both appear to indicate that the broad A, B and C zonation (Figs 2b and 5) is correct. The GOR data were derived from fluid analyses at reservoir conditions (PVT analyses) and separator measurements. The large number of pressure readings comprise RFT measurements in new wells, frequent static and pressure fall-off (PFO)/PBU measurements. These proved most valuable in modelling reservoir connectivity and continuity. Particularly important were pressure data fortuitously acquired from all wells simultaneously for 3 months during early 1989, as a result of forced platform shutdown (the Shell/Esso Floating Storage Unit used to transport the oil via Fulmar Field broke from her mooring). One-third of the pressure readings were taken at this stage, to add to the RFT sets and PBUs. These data, integrated with the sedimentological, biostratigraphical and seismic studies, reveal a complex variety of pressure barriers and cross-fault communications operating within the reservoir (summarized in Fig. 5). The most general conclusion to come from the shutdown period was that the pressure in Zone C decreased, whereas in Zones A and B pressure increased. This broadly shows the whole reservoir to be in communication, although the time period over which re-equilibration occurs is considerable. Distinct pressure regimes have been discerned within and between the principal zones since production commenced and careful monitoring of depletion and injection responses has confirmed the presence of significant barriers that impact field development strategy. Future PLT and PDK logs, together with tracer studies and pressure monitoring, should further refine the dynamic model and enhance the development strategy.

Taking all these factors into account, the broad framework of correlation was believed to be correct and the more detailed analysis of individual zones (stage 2) could take place.

A new layering system was created in the geological model which exactly corresponded to the layering to be used in the new simulation model. Correlation of the individual sub-zones across the field was found to be possible. 'Arbitrary percentage splits' of geologically defined units were avoided. For example, a high-permeability layer defined by a percentage split of an interval in the 1989/90 simulation model (Zone B1; Stevens and Wallis 1991 fig. 8) has been individually picked in each of the wells and is defined as layer BL2a (Fig. 5).

Layer BL3 is of good reservoir quality in the west of the field but deteriorates towards the crest (east). It therefore acts as a barrier to vertical communication on the crest so that layer BL4 is in communication with Zone C, but not with layers BL1 and BL2. On the flank, however, Zone C is in good communication with the whole of Zone B.

Zone B shows two distinct petrophysical facies between the flank and the crest of the field. In the 1989/90 model unequivocal cross-field correlation was not considered possible. Subsequent to the study, this change in facies was identified as reservoir quality deterioration within correlatable horizons and the resultant data grid was input into the simulation as a single layer.

Layer AL1 is only of reservoir significance in the southwest of the field, most importantly in the area of the SW horst block. A field-wide pressure barrier was identified at the base of layer AL1. Layers AL2, AL3 and AL4 equate with the most important reservoir section in the A zone. The quality and thickness of these layers rapidly diminishes onto the crest of the field. Layer AL3 has been picked at each well as it is a high (Darcy level) permeability interval between lower permeability (tens to hundreds of millidarcies) intervals. This had previously been done arbitrarily in the simulation model. Layer AL5 is considered non-reservoir and acts as a vertical communication barrier between Zones A and B.

Aspects of the key techniques utilized in reappraising the field are discussed below.

Seismic techniques

Image analysis

New software on seismic workstations provides powerful and accurate automatic tracking programs which were used to pick the horizons consistently at the same phase of the wavelet. This enables the interpreter to incorporate data from every common midpoint in the dataset. Displays of instantaneous dip, instantaneous azimuth and difference were produced using appropriate colour bars to highlight the main features.

For dip lineaments a smooth white–black–white colour bar gives the effect of vertical illumination so that low dips are white, high dips are black. The black lineations parallel the manually interpreted faults but also continue beyond. This indicates where the fault throw is below visual resolution (i.e. less than 4 ms). This indicated that many of the Clyde faults are more extensive and continuous than previously interpreted.

Artificial illumination of the azimuth displays from different directions is achieved by varying where the white appears in a white–black–white colour bar representing 0–360°. Use of this display mode highlights flexures that have been interpreted as locally high ridges. These lines may well prove to be optimum paths for fluid flow.

Difference maps are effective for highlighting subtle faults and misties by displaying the amount of displacement along a fault. A suitable colour bar uses redness for increasing positive throw and blueness for increasing negative throw. The time data from this display were transferred to Zycor and computer contoured at a working scale.

It is possible to produce a 3D map effect, similar to satellite imagery, by combining maps on a light table. If a coloured seismic depth map is overlain by a white–black–white azimuth map and illuminated from below, an apparent 3D image of this field is produced. This visual effect dramatically shows the interplay between the major faulting and the high and low areas of the field.

Absolute acoustic impedance (AAI)

On Clyde there is a direct correlation between acoustic impedance (AI) and reservoir porosity, permitting the prediction of porosity to an accuracy of ±4 porosity units in sections exceeding seismic tuning thickness (about 30 m). The rephased 3D dataset was converted to AAI to predict reservoir porosities away from well control. Absolute porosity values were taken at well locations for each set of AAI time slices and compared with core porosity values for the equivalent sequence in order to 'calibrate' the AAI maps.

Sequence stratigraphy

The sequence stratigraphic approach was adopted during the studies of the Clyde Field in an attempt to resolve the inability of 'conventional' log correlation and biostratigraphy (prior to 1990) to produce a reliable correlation scheme. The approach relies upon the recognition of depositional cycles (parasequences/parasequence sets) which reflect variations in the balance between sediment supply and available accommodation space. These cycles are recognized on the basis of ichnofacies and grain-size trends observed in cores, combined with wireline log correlation.

The technique relies upon the recognition of:

1. maximum flooding (MF) surfaces: these correspond to the finest grain size/'deepest' water facies at the base of coarsening-upwards (top of fining-upwards) cycles;
2. maximum regression (MR) surfaces: these correspond to the coarsest grain size/shallowest water facies at the tops of the coarsening-upwards (base of fining-upwards) cycles;
3. flooding (F) surfaces: these correspond to abrupt upwards facies shifts developed as a result of 'rapid' increase of relative sea-level.

Examination of the cores and logs in the Clyde Field reveals a hierarchy of coarsening- (and more rarely, fining-) upwards sequences from a few metres to approximately 60 m. The large-

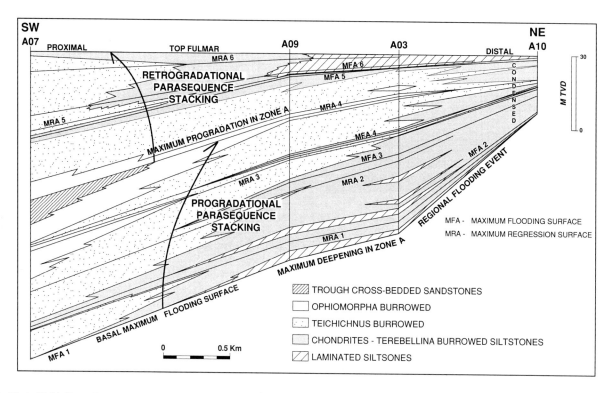

Fig. 6. Clyde Field: Zone A sequence stratigraphy.

scale cycles (parasequence sets) represent major basinal/regional controls (tectonic/eustatic), while the small-scale cycles (parasequences) represent autocyclic sediment switching or minor tectonic controls on relative sea-level.

The initial approach was to identify the large-scale cycles characterized by major facies shifts and correlate them across the field. Once this large-scale framework was established and verified by other techniques, then a more focused approach allowed the recognition and correlation of smaller-scale cycles.

The major surfaces in the Clyde Field are illustrated in Fig. 4. Essentially these surfaces parallel the current major reservoir zone boundaries (A, B and C). The most significant part of this correlation is the relationship between the A and B layers and geometry of the intra-B surfaces. The 'framework' for the correlation of these layers is provided by five surfaces: FB1 (top Zone C); MFB2 (approximately base BL3); MRB6 (within BL2A); MFA1 (approximately base AL5); MRA4 (within AL3). These are briefly described below.

- Top Zone C (FB1) is marked by a major facies shift representing a relative sea-level rise, the incoming of spiculitic sands and the initiation of sedimentation on the southern horst block. This event is readily correlatable across the field.
- Above FB1 the sediments reflect overall relative sea-level rise up to the MFB2 surface which represents the deepest water facies in Zone B and approximates to the base of a fine-grained unit (reservoir layer BL3) that is correlatable across the field. The top of this fine-grained unit is seismically resolvable as an 'intra-B' reflector over much of the field.
- The overall coarsening-upwards from MFB2 reaches a maximum at the MRB6 maximum regression (progradation) surface. This represents the furthest basinward advance of high-energy facies into the basin.

- Base Zone A is marked by a major deepening event (MFA1 surface). Biostratigraphic data show that this interval is characterized by the LDO of *P. pannosum*, which corresponds to the Early Kimmeridgian *Eudoxus* ammonite zone, a period of major deepening across the whole North Sea Basin. As a consequence, there is a high degree of confidence in this correlation and, therefore, that the A zone thins from the SW flank of the field onto the crest in the NE. Base Zone A (top Zone B) is a high confidence seismic reflector that can be recognized across the field further supporting this correlation.
- Zone A coarsens upwards from MFA1 to the MRA4 maximum regression surface. This surface represents the furthest basinwards advance of high-energy facies within the A zone. The distribution of facies, which control reservoir quality, and their relationship to these and smaller-scale cycles within Zone A is illustrated in Fig. 6.

These five major surfaces formed the correlation framework within which the smaller-scale depositional sequences (parasequences) were analysed. Overall, the sequences were deposited by prograding shallow marine shelf (rarely shoreface) systems that were infilling actively developing fault-controlled basins (ultimately controlled by salt withdrawal). The main depocentres shifted position dramatically during Fulmar Sandstone deposition probably as a result of salt withdrawal. This was combined with an overall 'backstepping' of the facies belts with time.

Barriers to vertical fluid movement within the reservoir correspond to fine-grained facies deposited during relative sea-level highstands. High permeability intervals in layers AL3 and BL2 correspond with maximum regression surfaces which mark the furthest advance of high-energy facies into the basin in the A and B zones, respectively.

Because the vertical and lateral permeability variations

within the field are related to depositional controls, facies mapping provides a means of predicting permeability within the reservoir.

Use of analogues

The early to middle Kimmeridgian in the Central Graben contains other shallow marine sandstone reservoirs in, for example, the Fulmar, Ula and Gyda fields, which have been used as analogues. Some comments on their suitability as indicators of controls on reservoir quality, as sources of drilling statistics and of techniques for reservoir description are briefly reviewed below.

The gross depositional architecture of the Upper Jurassic shallow marine reservoirs in the Central Graben is controlled by the development of accommodation space versus the rate of sediment supply. The Clyde Field comprises three reservoir zones, the deposition of which was controlled by the switching of accommodation space due to salt withdrawal. The sandstones of the Clyde Field represent a complex (and varying) mixture of arkosic, spiculitic and argillaceous components. As a consequence, the gamma log, useful in Ula, Gyda and Fulmar (van der Helm *et al.* 1990), cannot be used to correlate depositional sequences or reservoir quality (Stevens and Wallis 1991, Fig. 8). The density/neutron (FDC/CNL) log is used for correlation purposes in combination with core studies, as outlined above.

In the Ula Field (Partington *et al.* 1993; Stewart and Schjerverud 1993), sedimentation was essentially infilling a single developing hole. There are only minor variations in the locus of sedimentation, few lateral facies changes, and as a result the reservoir zonation is virtually layer cake. Variations in reservoir quality are strongly influenced by depth-related diagenesis (mostly quartz overgrowths) rather than lateral facies changes (Gluyas *et al.* 1990).

In the Gyda Field (Partington *et al.* 1993; Stewart and Schjerverud 1993), uplift during sedimentation resulted in crestal stripping and downflank sedimentation. The reservoir zones are thus discontinuous across the field, with large thickness variations. The crestal area of the field was also a site of carbonate hardground development. Lateral reservoir quality variations are strongly controlled by diagenesis–depth related quartz overgrowths and carbonate cement on the crest (Gluyas *et al.* 1993).

Significant reservoir thickness variations, resulting from differential subsidence linked to salt dissolution, have been described from the Fulmar Field (van der Helm *et al.* 1990; Stockbridge and Gray 1991). The former authors also describe significant occlusion of porosity due to dolomite and silica cements.

These differences in reservoir architecture and diagenetic controls on reservoir quality, combined with differences in reservoir fluid properties, result in quite different reservoir performances. Considerable care should be exercised in the use of other fields in a play fairway as analogues.

Reservoir description/simulation interface

The latest Clyde reservoir simulation model has been constructed in a systematic manner. Each step has been selected as being the optimal method of building and running a model in a tight time frame, while still honouring all field data. The simulation model is rapidly responsive to both geoscience and reservoir engineering input, and can be readily understood by each discipline. It is 'user friendly' and allows rapid incorporation of revised petrotechnical data, within days rather than weeks or months. Accordingly, field development decisions can be based on the latest data and thus financial analysis of the resulting production profiles has maximum validity. This model has been named the 'reactive simulator'.

The current 3D simulator model contains the features required to provide a proper tool for optimized field development. The model is designed to reduce the effort in history matching, and to allow rapid incorporation of revised data for development studies.

The important feature here is that the 3D simulator model maps must be exactly conformable with, and exactly overlay, the geoscience maps. The idea is that geoscientists should be able to relate clearly to reservoir parameters such as pressure and flows shown on the simulator maps. Similarly, reservoir engineers should relate to reservoir parameters such as faults and rock properties on the geoscience maps. The map layout, faulting, and well positioning should be identical.

The reservoir simulator grid is designed to parallel exactly the seismic grid. Each reservoir simulation cell is 150 m square (and an average thickness of 20–100 ft), corresponding to exactly six seismic lines by six seismic traces (the spacing being 25 m). The reservoir simulation grid outline is marked on all the depth maps.

To transfer the fault pattern into the simulator, the depth maps were produced at 1:20 000 scale. The transparent simulator grid overlay at the same scale could then be overlain on the maps. The fault pattern was then copied as closely as possible by tracing the fault pattern along the edges of the simulator cells. The depth data from Zycor were directly output as a file of every relevant sixth line and trace coordinate and value. This method of data transfer was both quick and accurate.

In the simulator model, permeability is mapped with maximum weighting on reservoir condition oil permeabilities based on actual well test information. This is critical in Clyde as the core porosity/permeability transforms differ between wells and zones, as does the test/core permeability relationship. This is an important reason to test each layer in turn, then all combined, in appraisal and development well tests.

A key feature is that flow at a 'type' well is exactly modelled, i.e. the known data point is honoured. This key tenet is expanded by grouping wells of similar characteristics to provide specific areas where the known data are honoured. Pseudo-relative permeability curves are based on 'type wells' in areas of common permeability contrast. Grouping is achieved by mapping core and transform-derived permeability profiles on a zonal basis. One well from each area of common characteristics is then selected as the 'type' well and should, therefore, have good core coverage. The fine grid (15 m square and average 2–10 ft thickness) cross-section properties are generated from this well. Each coarse grid block is modelled by 100 fine grid blocks with consistent aspect ratio. Property averaging for the sub-layers is carried out on a PC from the petrophysical foot by foot data derived from the core wherever possible.

The model is most used when matching history to field performance and, subsequently, for running sensitivities for field development decision making. This interrogation is only efficiently carried out if high-quality post-processing is available. Quality maps, with translucent overlays, zonal averaging and tabular output will be generated.

Formal re-mapping is usually the culmination of a long study programme. Interim results are often hand drafted and affect only parts of the field. To incorporate local geoscience model changes, the geoscientist, aware of the 3D grid, will define the area of interest in a box conformable with the reservoir model. To incorporate the changes in this box, the engineer overlays the 3D grid and hand modifies the array inside the box. This is the key to rapid and effective response to new data input, providing profiles by well and field, for optimum decision making.

From a reservoir surveillance perspective, if the pressure match or water break-through/build-up in the model starts to give concern, the fault pattern and fault transmissibility are very easily modified, again by using an overlay to work block face transmissibility arrays in the area of interest. This approach focuses geoscientists and reservoir engineers on specific problems, to provide solutions which honour both of their datasets.

The planning process, whether considering a range of model outcomes, single or multi-development well programme options, is thus utilizing the latest description and understanding of the reservoir.

As a result of the improved reservoir description a more accurate and rapid history match was achieved. The ease with which the simulator model was matched against historical performance indicates that it is more representative than previous models and this increases confidence in the production forecasts it is used to generate. The successful completion of 14 new wells and a sidetrack also reflects the value of this improved model.

Current attention is being directed at allowing the above procedures to be evaluated within a workstation environment.

The multidisciplinary team approach

The Clyde team comprises workers from geoscience, reservoir engineering, production technology, drilling, engineering and commercial disciplines. The team works on an annual 'Life of Field' evaluation of development drilling programme options and facility enhancements which provides the vehicle for approval of overall strategy with Co-venturers. It then focuses through the year on these options. In this process, key uncertainties are identified. These may include drilling hazards, or problems for highly deviated wells (e.g. Alpha 6.4 km measured depth and 5 km step-outs, Fig. 1) or reservoir management and commercial uncertainties. In different zones and areas of the field, STOIIP or fault transmissibilities may carry the critical uncertainties. A range of development drilling programme options are evaluated through the use of the simulator to identify upside, downside and most likely profiles. The economic viability of these options (net present value (NPV), cash flow, payback time) can then be evaluated for different economic scenarios and modifications, such as accelerated production.

Break-even analysis is required to focus studies into potentially productive evaluations. The geoscientist, for example, needs to be aware of the commercial impact of any study and related sensitivities, to ensure that this work is correctly scoped.

Flexibility to respond to changes in field performance is essential. The rapid design and implementation of a low pressure manifold and the introduction of a gaslift programme on Clyde are good examples. Implementation of such actions requires a clear planning and approval structure and an effective reservoir simulator.

The reactive simulator, designed at appraisal for compatibility and refinement during development, should provide a better tool than Monte Carlo techniques in this environment. It is critical to capture the impact of adjacent well performances and complex variables, such as S_w and pseudo-relative permeabilities, and deliver a product (profile) which is easily understood by all disciplines and can be commercially analysed.

Lessons for development

This final section reviews a number of critical lessons based on the experience of appraising and developing the Clyde Field. These personal views have been grouped into seven categories which have widespread application. The first two are principally focused on process: the types of data we collect, how we evaluate them and communicate the results effectively. Lessons three to six focus on technical issues, but also question the way that we process and use data to achieve our objectives. Lesson seven covers the issue of the 'reactive simulator'. We believe such a system, integrated with the multidisciplinary team approach, provides a common language we require for rapid commercial evaluation of the complex variables necessary to maximize asset value in an ever-changing environment.

1. Data acquisition and evaluation for total management purposes

It is important to acquire both sufficient quantity and the appropriate type of data (e.g. dynamic, log and core), during the appraisal and development phases. Equally important is the adequate evaluation of these data to permit the recognition of uncertainty; i.e. define the range of such for different models, and its subsequent reduction through focused data acquisition. This leads to optimization of the development strategy if effectively communicated.

Full core coverage of the reservoir section (including the aquifer) in appraisal wells is important. It provides critical data on reservoir characteristics which may not be acquired subsequently, due to drilling problems, or the prohibitive costs of redrilling a segment. On Clyde, core data are essential for adequate pseudo-relative permeability curve generation, in order to predict the dynamic flow within the reservoir. Core data are also required for sedimentological analysis and reservoir quality evaluation, for correlation and thus reservoir performance prediction.

Reservoir surveillance data have been vital in constructing and history matching the simulator model, particularly data acquired during platform shutdown. Collection of field-wide simultaneous pressure readings should be carefully considered, perhaps in conjunction with a maintenance shutdown, as a periodic requirement on developing fields and as a justifiable cost to reduce uncertainty and improve the dynamic model.

The input of the range of uncertainty and the benefit of reducing uncertainty must be recognized. Late appraisal does not compensate for overspent capital. It is important to capture the range of possible models, and carry these in upside and downside commercial evaluations. The range of possible outcomes is more important than attempting statistical validation of a limited range of models, based on limited data.

2. Develop integrated multidisciplinary teams with continuity from appraisal through development

Communication of uncertainty between all disciplines is critical, and is optimally achieved within a multidisciplinary team. A team needs to comprise petrotechnical, engineering and commercial disciplines for maximum effectiveness. The 'common language' is provided by the profiles generated from the 'reactive' reservoir simulator and the economic scenarios applied to these. The varied lessons from this experience can be considered under three categories.

(a) Effective communication of uncertainty at appraisal and through the development phase   The level of uncertainty analysis must be appropriate for the project size. For small, possibly marginal, developments, identification and communication of the key factors is essential. If the range of uncertainty is wrong, as for the initial development well programme for Clyde, the project will lose money. For larger developments, inadequate evaluation in early phases will result in lost value rather than 'major shocks'. This loss is often hidden until the late stages in the development of an asset.

Many small developments will be unable to afford 'comfort factors', such as drilling wells close to the discovery well, then 'appraising' the field late in development, as happened on Clyde. This can also apply to fields where production quotas drive the project through rapid appraisal, without adequate reduction of uncertainty. Such an approach provides exposure to situations where longer-term commitments may not be achievable, due to catastrophic failure of the 'P50' model.

If a more realistic uncertainty range is carried in the development decision process and the appraisal phase is accelerated, it may be that the range will exceed flexible facility planning options. The downside case should then control the economics of the development, not the 'P50/most likely' outcome.

If a development is small and 'cannot support the appraisal cost' it is particularly important to evaluate the whole asset, including satellite potential before commitment to Annex B. The planning of sidetrack contingencies back into the main field area may be most cost effective at the appraisal stage. The evaluation of satellites once ullage is available is non-optimal.

(b) Continuity and co-location of technical staff and systems are required for optimal total project management   Small fields have such a short plateau period that the whole asset must be evaluated and developed together. Co-location and continuity of technical staff through appraisal and development of the total asset should occur. This team, under one asset manager, should have responsibility for all the area accessible to the facility. This would provide a significant improvement in efficiency, reduce operating costs and allow for a flexible response to changing field performance. As accelerated appraisal and development becomes a reality this will be even more critical.

(c) Flexible response to changes in field performance through integrated planning and approval procedures   The integrated multidisciplinary team, with clear planning, approval structures and an appropriate reservoir simulator is capable of reacting effectively to provide changes in strategy. These changes may need to be developed throughout the year in response to new data and changes in field performance.

Flexibility in staff resourcing is equally important. A vast quantity of data is acquired when drilling back-to-back wells every eight weeks (the Annex B prediction was 15 weeks). The penalty of not keeping up with the data influx can be ill-judged development decisions. If, during development, unexpected complexities arise, then the multidisciplinary team must receive additional resourcing to evaluate the new situation adequately.

The identification of base, planned and upside case commercial outcomes from the team planning allows the value of a study to be quantified and the NPV, for example, of a team of workers to be identified. This can lead to optimization of staffing levels in terms of potential increase in value. The potential value of this Clyde reappraisal study leading to a renewed drilling programme of up to nine wells exceeds 300×10^6.

3. 3D seismic data require adequate processing, integration and interpretation at the earliest possible stage

An important lesson to be learned from our experience of the Clyde Field is that, before interpreting the data, it is necessary to calibrate the 3D seismic data to the well data. Application of a 90° phase shift to the entire 3D dataset was required; this allowed the synthetic seismograms, VSPs and seismic data to be tied over most of the field. It then became possible to define intra-reservoir geometries and heterogeneities and start developing a reasonable reservoir model which could incorporate all the available well information.

4. Appreciate the limitation of analogues as a guide to reservoir complexity and for the identification of suitable techniques for adequate reservoir description

The use of other fields in a play fairway as analogues, whether for drilling success statistics, as an indication of controls on reservoir quality, or as a guide to techniques to be applied in reservoir description, can be inappropriate.

5. Use fields as 'fence posts' for exploration

There is value to be derived from the iteration of field studies to impact on regional exploration and satellite potential. Intra-Fulmar Formation reservoir potential is variable, with different trap locations for each package and reflector recognition can be critical for trap definition in this play fairway. In the Clyde Field, salt movement exerts a major control on reservoir thickness and quality. The Clyde area is clearly a major 'fence post' to the regional work, and regional studies must maximize understanding of the relevant aspects of Clyde as an aid to interpretation of sparser data areas.

6. Integrate tools for intra-reservoir resolution and development strategy optimization

Integration of data (e.g. seismic, biostratigraphy, sedimentology and dynamic) provides intra-reservoir resolution (sand package recognition). This permits discrimination between reservoir models to reduce uncertainty and thus optimize a development strategy.

Geoscience tools are at their limits of resolution in Clyde and an integrated set of tools which work together must be utilized both in field development and regional studies. These have been outlined earlier.

7. Build a 'reactive' reservoir simulator

There is a strong requirement for a 'reactive' reservoir simulator to facilitate simple and rapid interface of geoscience model updates with the simulator and provide sensitivities within days. These are the basis for economic evaluation of uncertainty and the profiles provide the common language within the multidisciplinary team. Such a tool is essential for rapid commercial evaluation of complex variables in an ever-changing environment. The impact of the latest reservoir description updates for a development strategy (e.g. well locations and field performance) is thus maximized.

Conclusions

The Clyde Field is an example of an initial development based upon a limited perception of the impact of variability of both reservoir quality and connectivity. Through a focused reappraisal, timely acquisition of sufficient relevant data, and the application of appropriate techniques within an integrated multidisciplinary team, recovery is now being optimized. This relatively small field highlights important differences between the historically acceptable paradigm for the development of large assets and the consequences of failing to develop a new approach for the future. The importance of production experience, and the ability to learn from reservoir performance and detailed integrated studies of fields, is critical to the development paradigm required to realize asset value.

The contribution made by all individuals to this manuscript, wittingly or otherwise, is gratefully acknowledged. The paper represents the work of an integrated team and I would particularly like to thank J. Ashbridge, T. Gibbons, J. Hunt, A. Leonard, R. McIlroy, C. Milton, N. Morrison, C. Richardson, D. Smith, R. Wallis and A. Wood for their contributions to the study and D. Farquhar and I. Pirie for their help during the preparation of this paper. Figs 4 and 6 are modified from A. Leonard's work. Co-venturer (Shell and Esso) permission to publish is also gratefully acknowledged.

References

GIBBS, A. D. 1984. Clyde field growth fault secondary detachment above basement faults in the North Sea. *American Association of Petroleum Geologists Bulletin*, **68**, 1029–1039.

GLUYAS, J., LEONARD, A. J. AND OXTOBY, N. 1990. Diagenesis and petroleum emplacement: the race for space—Ula trend, North Sea. *13th International Sedimentolgists Conference*. IAP, Utrecht, p. 193.

——, ROBINSON, A., EMERY, D., GRANT, S. AND OXTOBY, N. 1993. The link between petroleum emplacement and sandstone cementation. *In*: PARKER, J. R. (ed.) *Petroleum Geology of Northwest Europe: Proceedings of the 4th Conference*. Geological Society, London, 1395–1402.

VAN DER HELM, A. A., GRAY, D. I., COOK, M. A. AND SCHULTE, A. M. 1990. Fulmar: The development of a large North Sea field. *In*: BULLER, A. T. *ET AL*. (eds) *North Sea Oil and Gas Reservoirs II*. Graham & Trotman, London, 25–45.

PARTINGTON, M. A., MITCHENER, B. C. MILTON, N. J. AND FRASER, A. J. 1993. Genetic sequence stratigraphy for the North Sea Late Jurassic and Early Cretaceous: distribution and prediction of Kimmeridgian–Late Ryazanian reservoirs in the North Sea and adjacent areas. *In*: PARKER, J. R. (ed.) *Petroleum Geology of Northwest Europe: Proceedings of the 4th Conference*. Geological Society, London, 347–370.

SMITH, R. L. 1987. The structural development of the Clyde field. *In*: BROOKS, J. AND GLENNIE, K. D. (eds) *Petroleum Geology of North West Europe*. Graham & Trotman, London, 523–531.

STEVENS, D. A. AND WALLIS, R. J. 1991. The Clyde Field, Block 30/17b, UK North Sea. *In*: ABBOTTS, I. L. (ed.) *United Kingdom Oil and Gas Fields 25 Years Commemorative Volume*. Geological Society, London, Memoir, **14**, 279–285.

STEWART, I. J. 1993. Structural controls on the Late Jurassic age shelf system, Ula trend, Norwegian North Sea. *In*: PARKER, J. R. (ed.) *Petroleum Geology of Northwest Europe: Proceedings of the 4th Conference*. Geological Society, London, 469–483.

STOCKBRIDGE, C. P. AND GRAY, D. I. 1991. The Fulmar Field, Blocks 30/16 & 30/11b, UK North Sea. *In*: ABBOTTS, I. L. (ed.) *United Kingdom Oil and Gas Fields 25 Years Commemorative Volume*. Geological Society, London, Memoir **14**, 309–316.

Wytch Farm oilfield: deterministic reservoir description of the Triassic Sherwood Sandstone

M. B. J. BOWMAN,[1] N. M. McCLURE[2] and D. W. WILKINSON[2]

[1] BP Exploration, 4/5 Long Walk, Stockley Park, Uxbridge, Middlesex UB11 1BP, UK

[2] BP Exploration Operating Company Limited, Blackhill Road, Holton Heath Trading Park, Poole, Dorset BH16 6LS, UK

Abstract: The Triassic Sherwood Sandstone is the principal reservoir in the Wytch Farm Field, located in Dorset, southern England. Composed of predominantly arkosic alluvial sandstones, it has excellent reservoir qualities but includes significant mudstone, calcrete and calcrete conglomerate baffles which affect the vertical passage of fluids. The identification of 'flow units' within this sequence has helped to determine appropriate layers for use in the reservoir simulation model. These flow units were defined using core data from 24 of the 48 wells and by log correlation. The well information was supplemented by outcrop data to help elucidate the geometry of individual layers.

The Wytch Farm Field is the largest onshore oilfield in Western Europe with total reserves estimated at 300 million barrels. BP Exploration is the field operator on behalf of the Dorset Bidding Group (Arco British Ltd, Premier Consolidated Oilfields plc, Clyde Petroleum plc, Goal Petroleum plc and Purbeck Oil plc). The field is located in Dorset, Southern England within onshore Licences PL089 and PL259, extending offshore into P534, and it lies in a structural province defined to the north by the Ryme Intrinseca fault and to the south by the Purbeck Disturbance (Fig. 1). The geological context of the field was described by Colter and Havard (1981). This summary paper outlines how a detailed reservoir description has been integrated directly into a simulation model to optimize the development of this major oilfield.

Reservoir description

The Sherwood Sandstone Group, deposited during the early Triassic period, forms the principal reservoir, containing some 250 million barrels of oil reserves; the remaining reserves occur in the Bridport Sandstone and the 'Frome Clay' limestone, both of Jurassic age (Fig. 2). The structure, which has been defined using 910 km of 2D seismic data, is an intensely faulted anticline, formed by a relatively catastrophic extensional event during the early Cretaceous. Oil migration and fill commenced shortly after the structure developed, sourced from Lower Jurassic shales which were reaching thermal maturity in the down-faulted block south of the Purbeck Disturbance.

The Sherwood reservoir is formed from a complex and heterogeneous arkosic alluvial succession deposited in a semi-arid (red bed) climatic setting. Regional data on thickness and mineralogy show the reservoir in the Wytch Farm area to have developed near the confluence of two major braided alluvial tracts draining northwards from France and westwards from the Brabant Massif (Fig. 3). The most significant elements restricting the passage of fluids within the reservoir are sub-horizontal beds of mudstone, calcrete and calcrete conglomerate. The Mercia Mudstone Group which forms the cap-rock to

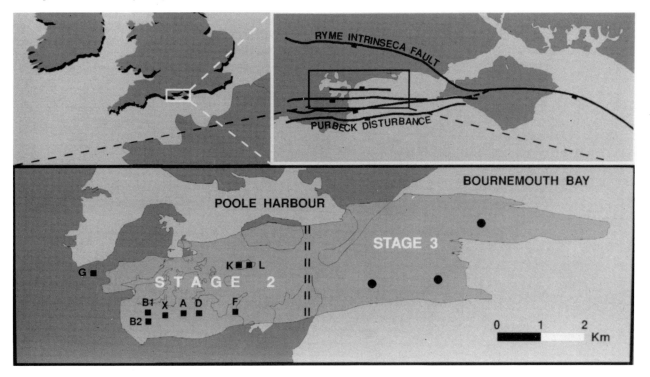

Fig. 1. Wytch Farm Field location map.

From *Petroleum Geology of Northwest Europe: Proceedings of the 4th Conference* (edited by J. R. Parker).

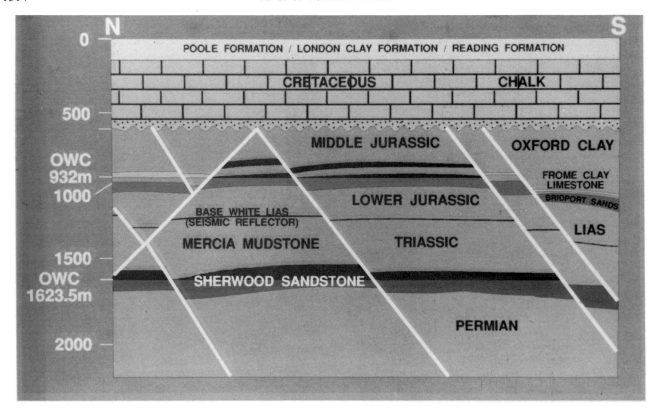

Fig. 2. True-scale structural cross-section.

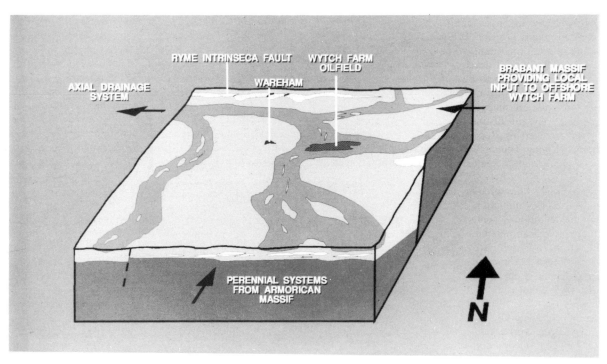

Fig. 3. Sherwood Sandstone palaeogeography.

the reservoir is of Middle and Upper Triassic age (Warrington et al. 1980).

Reservoir development

In Stages I and II of field development, some 67 wells have been drilled from nine well sites. Of these, 44 were targetted at the onshore part of the Sherwood Sandstone reservoir. Cumulative production from the Sherwood reservoir reached 50 million barrels of stabilized crude oil by March 1992. The original reservoir pressure exceeded the bubble point by about 1350 psi. Solution gas is delivered by pipeline into the National Grid and LPG to an adjacent rail terminal.

A reservoir simulator model (SIM 2) has provided a basis for making reservoir management decisions about water injection strategy and the provision of artificial lift (beam pumps and electric submersible pumps). The performance of the SIM 2 model is considerably influenced by the ratio of vertical to horizontal permeability (kv/kh), itself a product of the number and extent of horizontal 'baffles' or layers of reduced permeability which can have varied horizontal extent. It is necessary, therefore, to combine dynamic with static data in a close-knit fabric to achieve a satisfactory performance match in the SIM 2 model. This depends not only on thorough and appro-

Fig. 4. Outcrop panel, Otter Sandstone (Sherwood Sandstone equivalent), south Devon.

priately detailed geological description but also on reservoir engineering insight. Assembling an on-site, fully integrated multidisciplinary team has proved indispensable in ensuring that the reservoir model is geologically realistic.

When funds for field development were approved in 1987, only limited static and dynamic data were available. At that stage, the heterogeneity and effective properties for the Sherwood reservoir were described by stochastic modelling using a coarse and essentially arbitrary layering scheme (Dranfield *et al.* 1987). The additional data from cores, logs and pressure tests provided by Stage II development has now enabled the initially coarse description to be refined and a more detailed subdivision into reservoir 'flow units' to be made.

These flow units are mappable, discrete sandbodies within the reservoir, each with distinct properties, which enable dynamic data provided by Repeat Formation Testers (RFTs), Production Logging Tools (PLTs) and Extended Well Tests (EWTs) to be linked with static data: lithofacies, porosity, permeability, etc. This flow unit scheme forms the basis for reservoir layering in the three-dimensional cellular grid which defines the full field simulation model.

Flow unit definition depends upon being able to correlate individual lithological units between wells drilled on a spacing of about 500 m. In Wytch Farm, 48 Sherwood wells have been drilled and logged. These include 32 producers, 12 water injection wells and 4 offshore appraisal wells. Descriptions of sedimentology and measurements of porosity and permeability are available only from the 24 wells cored in the Sherwood reservoir. Where a laterally extensive lithotype such as a sheet sand is developed, correlation using electric logs is straightforward; however, recognition of a specific channel sand is more difficult. The core study has, therefore, been supplemented by quantitative analysis of the Sherwood equivalent (Otter Sandstone) which outcrops in south Devon, some 80 km west of the field (Fig. 4). This work, which is still ongoing, has enabled the reservoir to be described in terms of a limited number of lithotypes (Table 1) each having a definite range of geometric extent and pore system characteristics. Combining these with the dynamic data from the field (RFTs, PLTs, EWTs) has enabled the definition of the flow units, each with consistent lithological and fluid flow characteristics and hence with consistent effective properties in the simulation model.

The flow units combine to form a distinctly layered reservoir (Fig. 5) with higher permeability and kv/kh at its base where

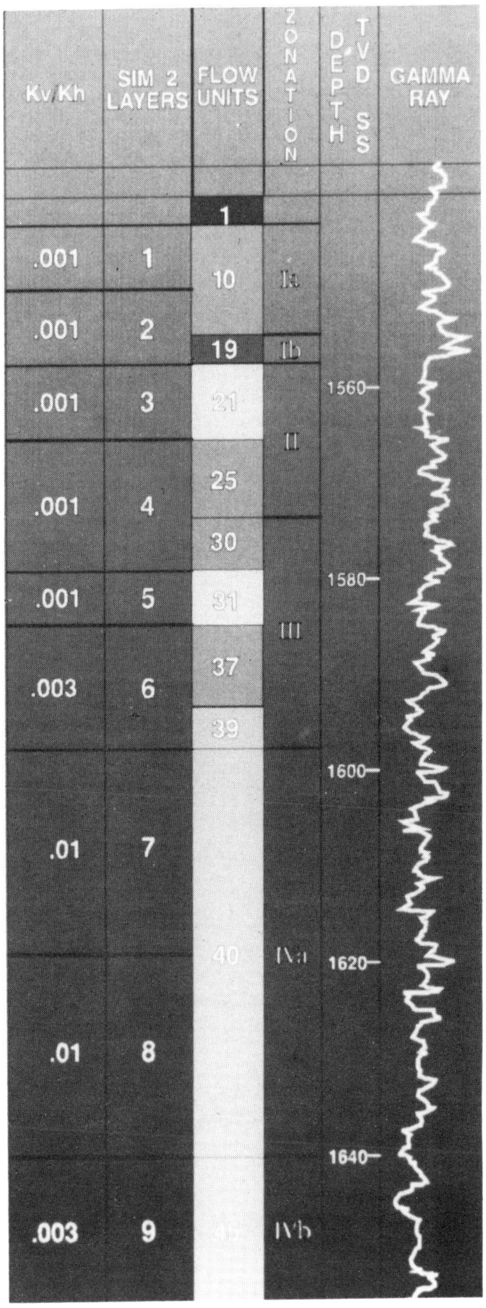

Fig. 5. Well K6SP Sherwood reservoir zonation.

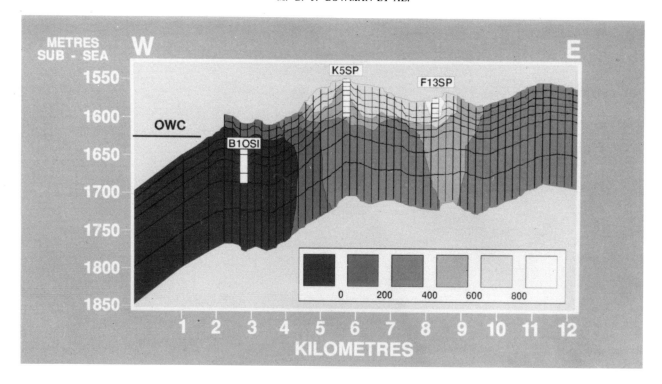

Fig. 6. East–west cross-section through full-field model for Sherwood Sandstone showing flow unit based layering and pressure depletion in psi.

sandbodies are thicker and more continuous. The upper layers are dominated by more discrete and poorly connected sandbodies which consequently have poorer effective fluid communication. Pressure, and consequently production, is supported by water injection below the original oil–water contact utilizing 12 peripheral injector wells. The layered character and variability in effective reservoir properties cause strong vertical and lateral pressure gradients across the field between the injectors and crestal producers (Fig. 6).

The resultant differential pressure depletion, acting together with the vertical variation in reservoir quality, leads to lower productivity from the upper flow units and accelerated water

breakthrough in the more productive lower flow units. There is evidence in one well that en échelon faulting may also contribute to the creation of pathways for early water breakthrough.

Some 100 million barrels of the field's reserves are contained in an offshore extension which is scheduled for Stage III development during 1993–4 (Fig. 1). In view of its location within a designated area of outstanding natural beauty, every effort is being made to minimize the number of drill sites required for Stage III development. Optimizing the number and location of drilling targets with the help of the SIM 2 model will have a crucial impact on the success of the project, particularly as drilling at an angle of up to 86° to reach Sherwood targets located up to 5 km distant from the onshore well sites is planned. In the offshore extension, the net/gross ratio in the upper zones (greater than 0.5) is higher than in the onshore area. This would be expected to have a beneficial effect on the production performance of the offshore extension of the field.

The Sherwood Sandstone Study is an excellent example of detailed reservoir description integrating directly into a simulation model to optimize the development of a major oilfield.

The authors gratefully acknoweldge the earlier contributions of J. P. Hirst and D. McDougall to the geological description of the Sherwood rèservoir. Fig. 4 in this paper derives from the detailed outcrop description by J. A. Lorsong and G. C. Gaynor (ARCO Exploration and Production Technology, Plano, Texas), whose major work on this outcrop will be reported elsewhere. The lithotype nomenclature adopted here derives from the ongoing study by T. McKie (Badley Ashton Ltd) relating core descriptions of Wytch Farm offshore wells with the Otter Sandstone at outcrop. The authors express their gratitude for his helpful advice.

Table 1. *Facies associations and lithofacies, Sherwood Sandstone reservoir*

Facies associations	Lithofacies
1. Channel dominated	
	● Channel sandstone
	● Channel conglomerate
	● Low stage/abandonment
2. Sheetflood dominated	
	● Amalgamated sandstone
	● Thick-bedded sandstone
	● Thin-bedded sandstone
3. Playa/flood plain dominated	
	● Interchannel/flood plain facies
	● Playa margin mudstone
Primary Modifier: Calcrete	

Wytch Farm Sherwood Sandstone Reservoir data summary

Trap

Type	Structural
Depth to crest (m)	1530
Oil–water contact (m)	1625
Gross oil column (m)	95

Pay zone

Formation	Sherwood Sandstone
Age	Triassic
Average gross thickness (m)	160
Net/gross average/range	0.7/0.45–0.95
Porosity average/range	0.18/0.14–0.23
Average water saturation	0.44
Permeability range (mD)	1.0–1500

Hydrocarbons

Oil gravity	38.3° API
Specific gravity	0.835
Oil type	Undersaturated
Gas/oil ratio (SCF/STB)	357
Viscosity (cp)	0.9
Bubble point (psi)	1086
Formation Volume Factor	1.24

Formation water

Specific gravity	1.14
Salinity (mgl^{-1})	209000
Resistivity at 150°F (ohm m)	0.025
Composition (mgl^{-1})	73700 Na, 1290 K, 4410 Ca, 705 mg, 127400 Cl, 1000 sulphate, 280 carbonate

Initial reservoir conditions

Average temperature (°F)	150
Average pressure (psi)	2430

Field size

Area (km^2)	20
Recovery factor (%)	40
Recoverable oil (MMBBL)	240
Drive mechanism	Water drive

Production

Start-up date	June 1990
Development scheme	Onshore drill sites
Number/type of wells	32 production wells, 12 water injection wells
Plateau production rate (BOPD)	60 000
Cumulative production (MMBBL)	50
Secondary recovery methods	Pump/water injection

References

COLTER, V. S. AND HAVARD, D. J. 1981. The Wytch Farm Oil Field, Dorset. *In*: ILLING, L. V. AND HOBSON, G. D. (eds) *Petroleum Geology of the Continental Shelf of Noth-West Europe.* Heyden, London, 494–503.

DRANFIELD, P., BEGG, S. H. AND CARTER, R. R. 1987. Wytch Farm Oilfield: reservoir characterisation of the Triassic Sherwood Sandstone for input into reservoir simulation studies. *In*: BROOKS, J. AND GLENNIE, K. W. (eds) *Petroleum Geology of North West Europe*, Graham & Trotman, London, 149–160.

WARRINGTON, G., AUDLEY-CHARLES, M. G., ELLIOT, R. E., EVANS, W. B., IVIMEY-COOK, H. C., KENT, P. E., ROBINSON, P. A., SHOTTON, F. W. AND TAYLOR, F. M. 1980. *A Correlation of Triassic Rocks in the British Isles.* Geological Society, London, Special Report No. 13.

Meillon–Saint Faust gas field, Aquitaine basin: structural re-evaluation aids understanding of water invasion

D. HALLER and G. HAMON

Elf Aquitaine, CSTJF Avenue Larribau, F-64018 PAU Cedex, France. (Present address: Elf Petroleum Norge A/S, Postboks 168, N-4001 Stavanger, Norway)

Abstract: The case history of the Meillon–Saint Faust gas field is an example of how a structural interpretation based on well data can help in the understanding of the dynamic behaviour of the reservoir.

This field was initially developed with clustered wells at the structural culminations of the trap, relying on good fluid communication in the field. After production of 50–65% of the initial gas-in-place, the productivity had drastically decreased, due to overall water breakthrough at the wells, and the field was apparently approaching the end of its economic life.

Detailed structural analysis and reservoir simulation have now led to an improved understanding of field behaviour. Since the southern border of the field is understood to be a major E–W normal fault, it is apparent that most of the production wells are located in a highly fractured zone and that good fluid circulation is restricted to this southern fringe.

This structural model is taken into account for numerical reservoir simulation. The pressure predictions of these simulations match the pressure history and thus validate the structural model. The main conclusion of the simulation is that water has only invaded the fracture network and that the total water volume entering the reservoirs is low.

The Meillon–Saint Faust gas field is located in southwest France, near the giant Lacq gas field and partly beneath the town of Pau (Fig. 1). The field is on the southern border of the Aquitaine basin and is close to the northern front of the Pyrénées thrust belt. The accumulation is elongated E–W, about 28 km long and 3 km wide. It is a deep (4000–5000 m subsea), hot (150°C) and high pressure field (initial pressure of 48 MPa at 4100 m subsea).

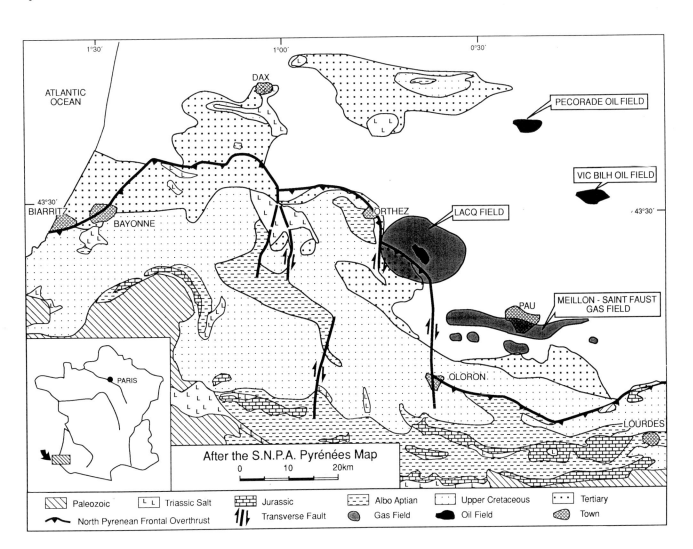

Fig. 1. Location map.

From *Petroleum Geology of Northwest Europe: Proceedings of the 4th Conference* (edited by J. R. Parker).
© 1993 Petroleum Geology '86 Ltd. Published by The Geological Society, London, pp. 1519–1526.

Sour gas containing 6.3% H_2S and 9.6% CO_2 is found in two dolomitic reservoirs of Jurassic age, the Mano dolomite (Portlandian) and the Meillon dolomite (Lower Kimmeridgian to Oxfordian). Both reservoirs have an average thickness of about 200 m. Their petrophysical properties are not encouraging. The porosity ranges between only 1–2% in the Mano dolomite and, in the Meillon dolomite, between 3–5%, up to 8% in vuggy intervals. Matrix permeability is below 1 mD in both reservoirs. The Meillon dolomite is considered as the main reservoir because of its better production potential. It culminates at 4075 m subsea, while its initial gas–water contact was found at 4960 m subsea in well Le Lanot 1. The Mano dolomite is absent or reduced along the southern fringe of the field. Both reservoirs are in pressure communication in the western part of the field, whereas there is no pressure communication in the eastern part, as revealed by the production history.

Initial gas-in-place estimations undertaken after appraisal were fairly inaccurate, mainly due to mapping and petrophysical uncertainties, and ranged from $65–100 \times 10^9$ Sm^3.

Brief history of the field

The Meillon–Saint Faust gas field was discovered in 1965 by the MLN 1 wildcat located in the far east of the field (Fig. 3). The seismic image of the structure at that time was rather unclear and inaccurate.

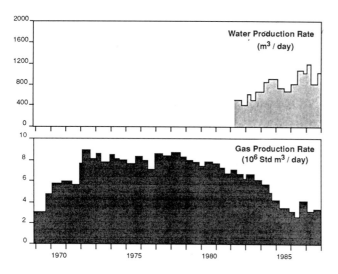

Fig. 2. Gas and water production history.

The first appraisal wells (MLN 2, AST 1, OUS 1; Fig. 3), drilled south of the discovery well, did not encounter gas, the Jurassic being either very deep or not penetrated. After this period of uncertainty, the E–W elongation of the field was interpreted assuming a geometric relation between the deep structure and an overlying anticline, called the Pau anticline. This led to a line of successful appraisal wells in a westerly direction (from east to west: MZS 2, SFT 2, PTS 1, BAY 1; Fig. 3).

Although the two reservoirs showed poor petrophysical properties, the appraisal wells revealed the Meillon reservoir to have a high initial potential, testing with production rates of c. 1×10^6 Sm^3 per day. This was believed to be linked to natural fracturing.

It was initially decided to develop the field with groups of wells drilled in the three culminations along the structure, named (from east to west): Saint Faust, Pont d'As and Baysère. The targets for these production wells were chosen at the crest of the structure. The western part of the field was put on stream in 1968 with three production wells (BAY 1, PTS 1, PTS 2; Fig. 3). After the drilling of additional producers in this western part (BAY 2, PTS 3, PTS 4; Fig. 3) and the start of production in the Saint Faust area with 6 wells (SFT 2, SFT 3, SFT 4bis, SFT 5-5A, SFT 7, SFT 12; Fig. 3), the expected plateau rate of 8×10^6 Sm^3 per day was reached in 1971 (Fig. 2). It was expected that the eastern part of the field (Mazères area), down-dip of the Saint Faust area and partly lying below urban zones, could be drained by the Saint Faust wells, relying upon the inferred good fluid communication in the reservoirs.

For ten years, this plateau rate was maintained without any major problem (Fig. 2). In 1978, after cumulative production of 26×10^9 Sm^3, some wells in which the Meillon reservoir was more than 500 m above the initial gas–water contact started producing water. Water breakthrough at these wells became critical since it drastically decreased gas productivity, or even in some wells led to a cessation of production due to water lifting difficulties. Between 1979 and 1985, the gas rate declined rapidly from $8–3 \times 10^6$ Sm^3 per day (Fig. 2). A new plateau rate of 3×10^6 Sm^3 per day was achieved in the following years (Fig. 2) only by drilling additional wells and, in particular, by putting on stream the eastern part of the field (Mazères–Le Lanot) with five wells (MZS 2, MZS 3, MZS 4, LLT1 D, LLT 2 G; Fig. 3). By 1990, cumulative production reached 47×10^9 Sm^3.

With water production building up in most wells, it became obvious that under its initial development plan the field was approaching the end of its economic life, although 35–50%

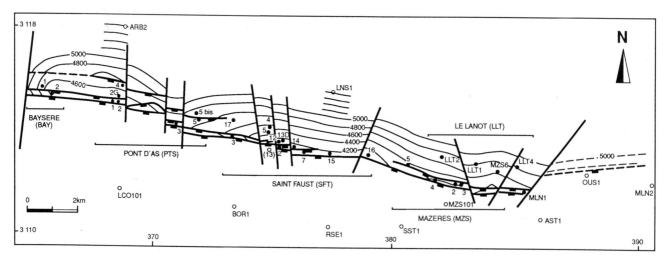

Fig. 3. Structure map at the top of the Meillon reservoir.

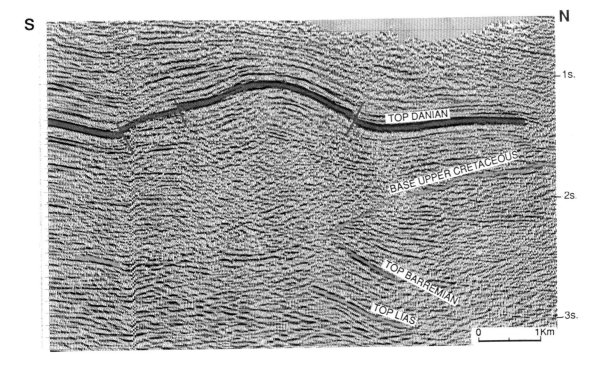

Fig. 4. N–S seismic cross-section (time migrated).

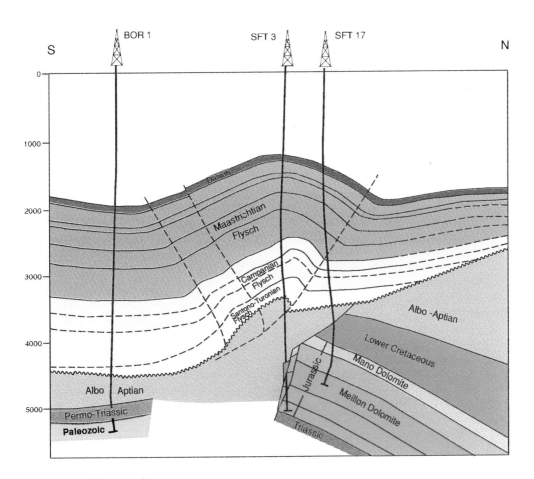

Fig. 5. Corresponding structural cross-section.

of the initial gas-in-place was still unproduced (i.e. 20–40 × 10^9 Sm^3 of gas).

In order to increase the ultimate gas recovery, a complete review of the field was required. Since 1984, Elf Aquitaine has launched several studies, planned as follows:

(1) to build a new structural model of the field;
(2) to build a simulation model which would confirm the structural model by an adequate history match of past field performance; and
(3) based on these results, to determine the optimal method for the final development of the field.

The new structural model

Interpretation of the 2D seismic lines covering the Meillon–Saint Faust field was always considered difficult (Fig. 4). Due to the complex deformation in the overlying series, the trap and in particular its southern closure appears badly imaged by seismic. On the seismic line illustrated (Fig. 4), both reservoirs lie between two deep northward-dipping seismic markers; the picking of the southern extension of these seismic markers appears speculative. As the structural interpretation was poorly constrained by seismic, various tectonic models could be proposed. At the time of discovery and subsequent appraisal, the structure was inferred to be an asymmetric anticline related to backthrusting during the Pyrenean orogeny, i.e. of Eocene age.

The need for tectonic guidelines to aid re-interpretation was obvious. This was achieved by using all the available well data, i.e. lithostratigraphical correlations and dipmeter interpretations, and by setting the field in its regional context. Since the work was performed after field development, a large amount of well data was available. More than 50 deep wells, either wildcats or producers, have been drilled into the accumulation or in its vicinity.

Review of all these interpreted data suggested six key points, which together define the structural configuration. These are summarized below.

(1) The relationship between the overlying Pau anticline and the Meillon–Saint Faust structure

The structure overlying the trap, known as the Pau anticline, is understood to be a pop-up structure induced by a slight thrust movement from the south (Figs 4 and 5). This thrusting is thought to have been initiated by the North Pyrenean Frontal Overthrust, some 10–15 km to the south (Fig. 1). It has propagated towards the north by bedding-plane slip within the Albo-Aptian series and the base of the Upper Cretaceous Flysch. It was deflected upwards over the field, both due to a facies change to the north (the Upper Cretaceous passing from turbidites to massive limestones) and in response to the underlying steep wall of rigid Jurassic carbonates of the Meillon–Saint Faust field, which acted as a 'stop' (Fig. 6). Horizontal motion being impaired, it translated into vertical deformation, resulting in a pop-up structure. The flanks of the pop-up structure are well defined by dipmeter logs recorded in the Upper Cretaceous section, which reveal two very sharp kink-band zones converging at depth, with dip magnitudes up to 50° in both flanks (Fig. 5).

The major implication of this new interpretation is that the origin of the Meillon–Saint Faust structure is no longer necessarily thought to be related to the Pyrenean compression, but appears to be a previously existing structure.

(2) The unconformity at the base of the Upper Cretaceous

The base of the Upper Cretaceous is a major angular unconformity, as can be seen on the seismic lines (Fig. 4). The strong erosion represented by this unconformity was induced by the development of the Upper Cretaceous Flysch trough, which has an E–W elongation.

(3) The nature of the southern limit of the Meillon–Saint Faust structure

The Meillon–Saint Faust trap is defined as a northward-dipping monocline limited to the south by a major E–W

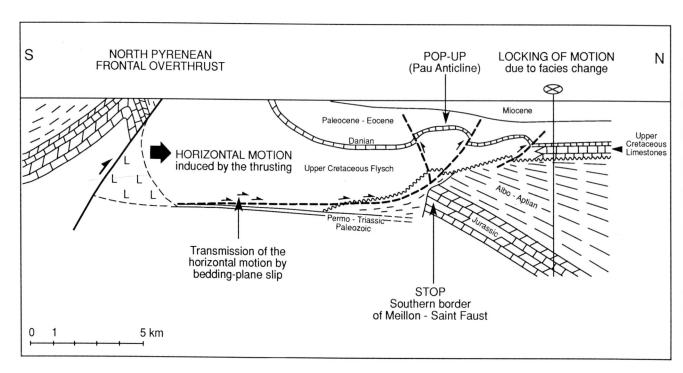

Fig. 6. Tectonic scheme of the Pyrenean deformation.

striking fault. This fault is clearly defined by well data. A cross-section was constructed through the Saint Faust area (Fig. 7), combining all well profiles. The Saint Faust 13 well (SFT 13), drilled in 1979, is highly significant. Located only about 150 m south of the line of the other production wells, it was drilled through a thick Albo-Aptian sequence and failed to penetrate the Meillon reservoir (Fig. 7). Comparing the profiles of SFT 2-A and SFT 13 wells, the only explanation is to infer a very steep southward dipping fault. In the western part of the field, the Pont d'As 2 well was recently side-tracked and has crossed this boundary fault. The dipmeter interpretation, based on a fault plane pick from microresistivity curves, confirms this E–W orientation and defines a fault dip of 70–80° to the south. South of this boundary fault, the Albo-Aptian series thickens significantly, as revealed in the Saint Faust 13 well and also the old appraisal wells MLN 2, AST 1 and OUS 1. These observations support the interpretation of the southern closure of the field as a major E–W normal fault of Albo-Aptian age.

(4) Interpretation of truncation at the crest of the structure

The truncation of the northward-dipping Lower Cretaceous and Upper Jurassic sequence at the crest of the structure is

believed to be directly related to the southern boundary fault. In the Saint Faust area (Fig. 7), this truncation is seen as a plane dipping 30° towards the south, cutting out the Lower Cretaceous and Upper Jurassic formations, which are overlapped by the Albo-Aptian series. Although the feature would appear at first sight to be an erosional surface, it can be shown to be a low angle gravity fault. In fact, dipmeter logs recorded in the Albo-Aptian series above the truncation show general southward dips ranging from 20–40°, and numerous small faults. These southward-dipping series must be considered as rotated and tectonically displaced. Support for this interpretation is given by evidence for tectonic slices along the fault plane. Short Lower Cretaceous sections are 'squeezed' between the Albo-Aptian and the truncated Jurassic sequence, for example in the Saint Faust wells 15 and 16 (Figs 7 and 8, respectively).

(5) Evidence for normal faults within the structure

The structure is cut by several normal faults, revealed by missing sections and associated drag folds on dipmeter logs. Most of the these faults have an E–W orientation and throw down the southern compartment. In the vicinity of the southern boundary fault, it is possible to observe in six wells (MZS 2, MZS 3, MZS 4, MZS 5, SFT 3, PTS 3; Fig. 3) an important rotation of the Jurassic series, which show sub-horizontal to southward dips, related to intense faulting. All these normal faults are obviously of Albo-Aptian age and are probably related to the southern boundary fault (Fig. 7).

(6) The significance of transverse faults

Mapping reveals that the E–W structural trend described above is cut and displaced by transverse faults (Fig. 3). Their strike-slip motions are dated as Pyrenean, i.e. Eocene age. This observation further supports the evidence of northward thrusting. These faults can now be precisely mapped, with the help of a recent 3D seismic survey.

The Meillon–Saint Faust field is thus described as a faulted block related to Albo-Aptian stretching. This Albo-Aptian stretching phase is a well-known episode in the Aquitaine basin and the Pyrenean foreland, which led to the opening of small actively subsiding rhomboidal graben basins. This has been established by structural studies of other fields in the Aquitaine basin, including the Vic Bilh and Pecorade oil fields, where there is no expression of the Pyrenean compression, and also from field studies along the Pyrenean foreland (Canerot 1989).

This structural framework was the basis for accurate mapping of the two reservoirs (Fig. 3), based on geological cross-sections tied at the wells (Fig. 5). This mapping will be up-dated utilizing on-going 3D seismic interpretation.

Structural model and field engineering data

Most of the wells were drilled at the top of the structure and had very good initial productivities. The new structural model shows that they are located very close to the boundary fault, and therefore in a highly fractured area. This is confirmed by comparison between the average matrix permeability, which is about 1 microDarcy at reservoir effective stress, and well permeabilities which range from 1–10 mD (Hamon et al. 1991).

Structural model and fracture permeability anisotropy

Early reservoir simulation indicated that these well permeabilities were apparently not large enough to explain the rapid pressure decline of the well Meillon 1, only a few months after

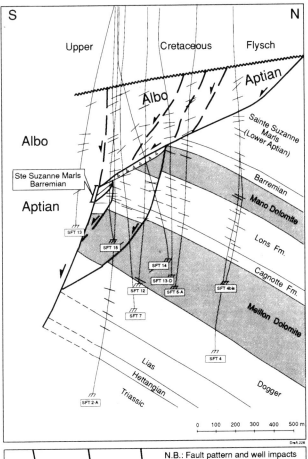

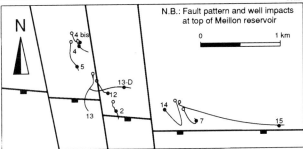

Fig. 7. N–S cross-section through the Saint Faust wells.

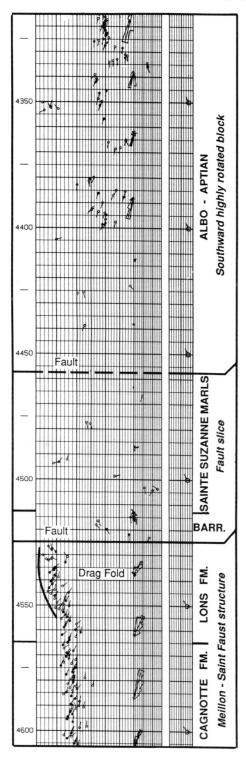

Fig. 8. Dipmeter log of Saint Faust 16 well: an image of the fault system bounding the field to the south.

the geometric mean permeability of an anisotropic system and is calculated as:

$$K_{bu} = \sqrt{(K_{max} \times K_{min})}$$

where K_{max} is the fracture permeability in the east–west direction and K_{min} is the fracture permeability in the north–south direction (Golaz *et al.* 1990).

Despite only moderate overall fracture permeabilities, large fracture permeabilities parallel to the southern boundary can explain the good pressure communication in the east–west direction and the fast pressure decline of the well Meillon 1.

Dynamic validation of the structural model

A full field model was constructed to honour this new structural understanding (Golaz *et al.* 1990). Simple material balance calculations suggested that the overall water influx was rather weak. Consequently, a single phase single porosity simulator was used initially. The agreement between simulated and measured pressures was very good for the first ten years of production and validates the structural model.

Several observations confirmed that water moves preferentially along the fractured border. Firstly, an aquifer influx exists at both ends of the southern border, as seen in wells Baysère 1 and Meillon 1. Located at the west end of the field, Baysère 1 experienced water breakthrough only three years after field start-up and then quickly watered-out. Moreover, simulated well pressure of Meillon 1 declined faster than actual measurements. Periodic monitoring of the gas–water contact of this observation well indicated that the pressure support was due to a significant local water influx. Secondly, the development of the eastern half of the field (which started in 1981) showed that the water encroaching on the reservoir flowed westwards of the Meillon 1. For instance, the wells Mazères 3 and Mazères 2 were drilled 2 km west of Meillon 1 and started producing water after one and three years of production, respectively.

Reservoir characterization study

With water production building up in most wells, gas production decreasing and with significant gas volumes still to be produced, a study was undertaken which incorporated several refinements to previous evaluations (Hamon *et al.* 1991). The characterization of fracture networks was carried out by incorporating several sources of information such as cores, mud logs, open-hole injectivity tests, pressure transient analysis, production logs, borehole imagery (FMS), outcrop studies, structural model and performance history. The merits of each type of data for the determination of fracture parameters are illustrated by Table 1.

(1) Cores and FMS logs Thirteen wells were cored. Natural fractures are common throughout the vertical extent of all wells. Most of them are short, sub-vertical and partially mineralized. Quantitative analysis is not possible due to the abundance of induced fractures. Some Formation MicroScanners logs were run in new wells: abundant fractures are identified; the average vertical spacing between the fractures which cut through the borehole rarely exceeding 3 m. Two dominant conjugate fracture trends were observed.

(2) Mud losses, injectivity tests and production logs Essentially all early wells were cased, cemented and perforated. Because of the thick section the location of mud losses was used to select the perforation intervals. Additional open-hole injectivity tests with flowmeter and temperature surveys were also carried out. Early in the life of the field it was recognized

the field start-up. Meillon 1 is an observation well, located at the far east end of the field, 8 km from the closest gas producer in 1969.

The structural model provides a possible explanation of this behaviour. The southern fringe of the field consists of a series of step faults running west to east. This suggests that fracture permeability is anisotropic, i.e. fracture permeability will be higher parallel to the southern boundary fault than it is perpendicular to the structural elongation. The in-situ permeability K_{bu}, deduced from well pressure build-ups, is in fact

Table 1. Fracture analysis techniques

	Mud losses	Open-hole injectivity test	Well tests	Production logs	Cores borehole imagery (FMS)	Outcrop studies	Structural model
Orientation					*	*	*
Permeability			*				
Porosity			*		*		
Anisotropy			*				
Fracture spacing	*	*	*	*	*	*	
Vertical distribution	*	*		*	*	*	
Areal distribution				*		*	*

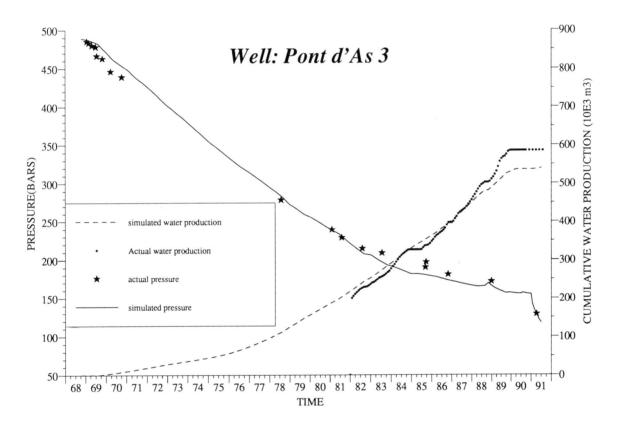

Fig. 9. Pressure and water production history matches for Pont d'As 3 well.

that the gas production often came from a few discrete fractures spaced relatively far apart. The fracture spacing ranges from 15–40 m. It was found in early wells that the productivity after casing and perforation was disappointing even with extensive stimulation. The later wells were completed with a slotted liner and the temperature log data are considered to be representative of the distribution of the natural fractures. The producing fractures cause a cooling effect which correlates well with the location of the drilling fluid losses. This approach provides quantitative estimates of the fracture spacing, which range from 10–60 m throughout the field. Comparison with the fracture spacing deduced from the FMS logs confirms that most of the fractures detected on the FMS logs in the Meillon–Saint Faust field are not really producing and only provide a matrix permeability enhancement.

(3) Pressure transient analysis All available pressure build-up tests have been re-interpreted with modern methods. The primary objective is to assess the correct well permeability thickness (KH), particularly when sealing faults are close to the wells. The KH values range from 40 mD m to 2000 mD m. The area of highest permeabilities lies along the major southern fault. Northern wells display low permeabilities. However, it should be pointed out that the worst well KH values are 50 times higher than the matrix values calculated from the overburden corrected core permeabilities.

Several attempts were made to characterize the matrix-to-fracture contribution. Pressure build-ups were matched with double-porosity, pseudo steady-state analytical models when appropriate. The calculated storativity ratios range from 0.01–0.07 and the interporosity flow coefficients from 3×10^{-6} to 8×10^{-8}. These values suggest that the block sizes are very large and/or the matrix permeability is very low. This is consistent with the results discussed earlier. Moreover, the comparison between FMS and production logs suggest that several fracture networks exist. A high-frequency low-conductivity fracture network (observed on FMS) is nested in a low-frequency high-conductivity fracture set (observed on production logs).

(4) Outcrop studies Outcrop studies were undertaken in the Mailh–Arrouy area to gain insight into the fracture distribution and to try to extrapolate to the undrilled reservoir areas. In fact the selected outcrop (located 20 km south of the Meillon–Saint Faust field) mainly illustrates the fracture pattern resulting from Pyrenean compression. Therefore, these outcrop observations represent only the northern part of the Meillon–Saint Faust field. No outcrop was found representative of the southern fringe of the reservoir. Forty-five measurement stations are distributed over 5 km of east–west outcrop. A wide range of vertical extensions of fractures is observed, from a few centimetres to several tens of metres. A major fracture strike N 10° is confirmed and two minor strikes are also recorded. Some north–south intervals show a high fracture frequency. The east–west distance between these intervals, or between fractures with very high vertical extension, is generally larger than 100 m. Within these highly fractured zones, the frequency of the conjugate fracture sets is low. It is thus concluded that a north–south, widely spaced fracture system may exist in the northern flank of the Meillon–Saint Faust field. Recent analysis of breakouts does not lead to clear regional orientation of the principal horizontal stress and suggests that the horizontal stress anisotropy is weak.

Delineation of fracture characteristics

Sufficient reservoir data have now been gathered to support the use of a dual-porosity model for the full-field simulation. However, an attempt to infer a more detailed and areal distribution of fracture characteristics from the available data was not conclusive. There are two reasons: firstly, some typical features of the Meillon–Saint Faust field and secondly, the gap between the geometric description of a fracture network and its flow properties.

Most wells were drilled along the southern boundary fault whereas only a few were stepped out to the north. However, the origin of the effective fracture system is probably different in the two areas. Along the southern boundary, the Albo-Aptian extension generated east–west fault-related fractures whereas the northern flank of the Meillon–Saint Faust field is mainly affected in a north–south tension fracture system. The lack of accurate seismic data for this deep reservoir also hinders any attempt to extrapolate well observations to undrilled areas.

There is increasing evidence that very few observed fissures really contribute to flow towards the well. This discrepancy between the geometric observations and the flow properties of the fracture network requires a detailed inspection of the evaluation concepts being applied to naturally fractured reservoirs. In the Meillon–Saint Faust field, the fractures do play an important role, but result in only moderate absolute permeabilities. A significant number of fractures may not be connected.

In such a reservoir, it seems that geometric observations (such as core examination, borehole imaging and outcrop studies) provide very useful data about the fracture orientation or lithological control but are not conclusive for the determination of the spacing of the producing fractures. The larger

spacing between producing fractures away from the southern fringe explains the disappointing results of vertical wells which were stepped out to the north, far from the boundary fault. Since they generally did not cross any mega-fracture, their initial productivity was poor. It appears that the delineation of the productive fractures in a moderately fractured reservoir is a very difficult task because they have a large spacing, a limited extent and are below the seismic resolution.

Numerical simulation

Notwithstanding the difficulty of accurately mapping the dynamic characteristics of the fractures, integration of geological and engineering data provided quantitative estimates of the fracture permeability, porosity, anisotropy and spacing. These values were input in a dual-porosity simulator, COMP IV, in order to reproduce the interplay between the geometry of the fracture network and the production mechanisms, i.e. pressure depletion and water imbibition into the matrix.

The numerical simulation yields a satisfactory match of both the pressure decline and the gas and water production histories. We conclude that the aquifer influx is weak. However, the capillary imbibition is not efficient enough to slow down the water encroachment in the fracture network, due to the large fracture spacing and also to the extremely low matrix permeability. Water flows through the mega-fractures, bypasses the gas that is left in the large matrix blocks and wells water-out.

The history matched model is currently used to assess the impact of various development strategies.

Conclusions

The determination of a reliable structural model of the Meillon–Saint Faust field was a very important clue to the understanding of past field performance. Our current structural model substantiates large areal anisotropy of the fracture permeability associated with the southern border fault. The integration of the structural model, other geological information and engineering data leads to a conceptual model which incorporates several scales of fractures. This model is an essential component in the success of the subsequent reservoir simulation of this naturally fractured water-driven gas reservoir.

The authors thank Elf Aquitaine for permission to publish this paper.

References

CANEROT, J. 1989. Rifting éocrétacé et halocinèse sur la marge ibérique des Pyrénées occidentales (France). Conséquences structurales. *Bulletin Technique Exploration-Production Elf Aquitaine*, **13-1**, 87–99.

GOLAZ, P., SITBON, A. J. AND DELISLE, J. G. 1990. Meillon Gas Field: Case History of a Low-Permeability, Low-Porosity Fractured Reservoir With Water Drive. *Journal of Petroleum Technology*, **42**, 1032–1036.

HAMON, G., MAUDUIT, D., BANDIZIOL, D. AND MASSONAT, G. 1991. Recovery Optimization in a Naturally Fractured, Water-Drive Gas Reservoir: Meillon Field. *66th Annual Technical Conference of SPE, Dallas, October 6–9, 1991*, Paper SPE 22915.

The Eakring Dukeswood oil field: an unconventional technique to describe a field's geology

M. W. STOREY[1] and D. F. NASH[2]

[1]Quantock Geological Services, 2 St James Street, Taunton TA1 1JH, UK
[2]BP Exploration Operating Co. Ltd, 4/5 Long Walk, Stockley Park, Uxbridge, Middlesex UB11 1BP, UK

Abstract: The Eakring Dukeswood oil field of Nottinghamshire was last produced in 1971. All wells have since been abandoned and the licence ML 1 expired on 22 April 1992. Cumulative production was small by North Sea standards, but during the Battle of the Atlantic the field produced the bulk of the UK's indigenous crude, peaking at 1600 BOPD in 1941. Cumulative recovery was some 6.5 MMSTB from an estimated 25.6 MMSTB in place in the completed horizons. In 1986 a study was undertaken to evaluate the remaining reserve potential.

The primary tool for such work would normally be a detailed geological model, but since less than 25% of the 197 grid drilled wells have any wireline logs or cores only structural models existed. There is very little seismic. A further complication is that there were nine productive horizons and the majority of well completions were unsegregated over four of them. Only 20% of cumulative production can be assigned unequivocally to any one reservoir.

Reservoir models, and individual reservoir contributions to production, had to be determined using less conventional techniques. Individual well performance under recycled peripheral water flood has proved to be a powerful tool to determine reservoir connectivity. Mapping decline rate reversal and water breakthrough by month indicates the transmissibility fairways and no-flow barriers in the system. It has also proved in some circumstances a useful discriminator of reservoir contribution.

The reservoirs are mainly Westphalian A to Namurian B siliciclastics. All were undersaturated at initial conditions. Porosity is often secondary, rarely above 20% and absolute permeability ranges from less than 1 to several hundred mD. Irreducible water saturation (S_{wi}) from mercury injection data ranges between 30 and 50% and water flood experiments indicated residual oil saturation (S_{or}) at 30%. The oil is waxy, up to 16% by weight, and has a high viscosity with correspondingly unfavourable mobility. API gravities range from 31.9–37.6°. Most production was obtained from the Loxley Edge Rock, the Sub Alton and Crawshaw sandstones—all by depletion drive, and from the Ashover Grit—in part water driven. Post-war water flooding was directed mainly at the Sub Alton and Crawshaw units.

Regional studies and sedimentological examination of two recent adjacent wells established the likely broad reservoir anatomy. This was then refined for the field from the qualitative transmissibility maps constructed from flood front behaviour. Geological models could then be assembled with some confidence and bypassed oil and secondary accumulations were identified. Relative permeability experiments on the modern core indicate an additional field reservoir/accumulation, previously unrecognized.

The Eakring Dukeswood oil field of Nottinghamshire (Fig. 1) was developed during the 1939–1945 war. It provided much of the UK's indigenous crude oil at that time, and was last produced in 1971. Modern reservoir data acquisition was then in its infancy, and less than 25% of the wells have any petrophysical logs. A recent field study has evaluated its remaining production potential from consideration of well production rates and fluid cuts. The main findings are presented here.

It is normal practice to validate static geological models through consideration of dynamic data, reservoir and well performance. Models are constructed using well log correlations, sedimentological studies and seismic interpretations. At Eakring Dukeswood these building blocks are mostly absent and the process has had to be reversed. We have created our static models primarily from dynamic data and in the process have identified the areas of unproduced oil. The applicability of the techniques are not of course confined to evaluating old fields with poor static data: they are essential to reservoir management and can be used to improve conventional reservoir description and to cross-check reservoir simulation studies.

The problems facing us in our review of the field's remaining potential are simply stated:

- less than 25% of the 197 wells have any wireline logs;
- less than 25% of the wells have basic reservoir core data (not necessarily those wells which have been logged);
- there are nine productive horizons, but with unsegregated completions. Only 20% of cumulative production can be assigned unequivocally to any one reservoir unit;
- there is very little seismic coverage.

The conventional geological database is thus restricted to a few gamma ray (GR) logs, drillers' logs (of dubious value), and core plug analyses. Electrical Survey (ES) logs were run in two wells only. Other than structure contour maps, no geological modelling was attempted in the past. However, there are excellent production records for oil and water through the field's life. This fact, combined with the close well spacing, allows the progress of both natural aquifer influx and induced water flood breakthrough to be followed in time so as to produce qualitative permeability maps for the reservoirs. The reservoirs studied were the Loxley Edge Rock (LER), the Sub Alton/Crawshaw sandstones (undifferentiated) (SAC) and the Ashover Grit (AG 5-1), which together constitute 62% of STOIIP and 97% of cumulative production. Minor and unproduced accumulations are also briefly described.

History

The field was discovered in Prospecting Licence A60 granted in April 1937. Its early history is described by Lees and Taitt (1945), and engineering aspects by Dickie and Adcock (1954). Mining Licence 1 (ML1), covering the field (with the exception of the Eakring North accumulation) was granted in April 1942 and expired on 22 April 1992. Production commenced in 1940 and the last sustained production was in 1966. A number of wells were recommissioned in 1968/69 but shut in by early 1971, since when all wells have been plugged and abandoned.

From *Petroleum Geology of Northwest Europe: Proceedings of the 4th Conference* (edited by J. R. Parker).
© 1993 Petroleum Geology '86 Ltd. Published by The Geological Society, London, pp. 1527–1537.

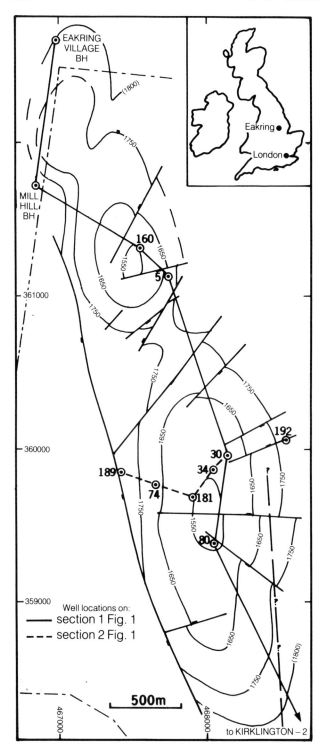

Fig. 1. Eakring Dukeswood oil field showing simplified structure contours (Alton marine band, feet BOD). Contour interval is 100 ft.

Reservoir stratigraphy

A folded Carboniferous sequence unconformably underlies monoclinal Triassic and Permian (Zechstein carbonates) rocks, with a very thin basal Permian clastic unit infilling the erosion surface. Over the structural crest the oldest penetrated sequence is the mid-Westphalian A and this becomes progressively younger downflank. Beneath the uneroded Kilburn Coal, and down to the 'Lower Carboniferous' Limestone, all porous units are oil bearing. The nomenclature used here is derived from BP's regional studies and is compared in Table 1 to that of the earlier field description by Lees and Taitt (1945).

The distribution of these reservoir units is shown in a north to south well correlation, tied to two modern wells (Fig. 2). The datum is the ?*Bilinguites bilingue* marine band, an intra-Namurian B marker, and the panel shows the overall thinning of the sequence and loss of the Rough Rock and Chatsworth Grit reservoirs between Eakring and Dukeswood. The correlation presented is controversial in that the unit identified here as Ashover Grit Unit 5 in the Kirklington well is believed by others to be the Chatsworth Grit (B. Mitchener pers. comm.). The identification of the main productive horizon as Sub Alton/Crawshaw is also debatable since it contains a field-wide intraformational shale that could contain the *G. subcrenatum* marine band. If so, the lower member should indeed be the Rough rock of regional usage.

Structure and fluid distribution

The Eakring Dukeswood Field is contained in three contiguous anticlines on the hangingwall of the Eakring Foston Fault, one of the post-Caledonian rift boundaries of the East Midland Province (Fraser *et al.* 1990). The fault reversed in the Variscan orogeny and the en échelon form suggests an element of lateral displacement as well. Southwards, the individual culminations increase in size (Fig. 1). The Kirklington Field lies on a trend to the south, but outside the ML1 area and is a separate accumulation. The Eakring North Culmination underlies the village of Eakring, and is inferred from well data. It has not been drilled.

The key structure map is on the *G. listeri*–Alton marine band, readily identifiable on the GR logs, and on drillers' logs overlying the 'Green Fireclay' marker (see Strong in Falcon and Kent 1960). Small displacement normal faults within the structure are interpreted to account for interval thickness anomalies and pressure history/production behaviour not directly attributable to reservoir property variations. Residual pressure maps for 1967, two years after production ceased, highlight five hydraulic production compartments for the main reservoir (see below). The western NNW–SSE-trending longitudinal fault is detectable on seismic and GR log correlation. It appears to be the main control on the accumulations' dynamic water levels.

Oil is continuous from Eakring North to Dukeswood South in the WF, LER and SAC reservoirs only.

Reservoir description

Loxley Edge Rock: LER

This reservoir comprises all sandstones found between the Norton and Alton marine bands. In its best development, two sandstones are encountered, and the whole package is interpreted to be a fluviatile sequence of channel sand fills and thin overbank sands. In the Eakring sector gross sand rarely exceeds 10 ft, whereas in the stacked channels of Dukeswood up to 40 ft is encountered, but in places there may be no sand at all. Figure 3 is a west–east well correlation across northern Dukeswood showing the LER in its better development. However, as noted above, these channel sands are generally discontinuous. Mercury injection data are only available for the better channel sands; S_{wi} is *c.* 30% (k 89 mD) and 45% (k 22 mD); in both cases zero water fractional flow level (fw = 0) occurs 50 ft above the free water level (FWL).

Sub Alton/Crawshaw sandstones (undifferentiated): SAC

This grouping is the major reservoir of the field both in terms of STOIIP and recovery. It lies directly below the Alton marine band complex. There are two sandstone developments, an upper dominantly clean unit 15–25 ft thick extending over

Table 1. Reservoir stratigraphy

Regional marker	Reservoir	Identifier	Lees and Taitt	Depth range (feet BOD)
Kilburn Coal	Wingfield Flags	WF	No change	1300–1500
G. amaliae MB (Norton)	Loxley Edge Rock	LER	B Sandstone	1500–1700
G. listeri MB (Alton)	Sub Alton/Crawshaw sandstones (undifferentiated)	SAC	Rough rock*	1550–1750
G. subcrenatum MB (Pot Clay)	Rough Rock	RR	D Sandstone	1650–1850
G. cancellatum MB	Chatsworth Grit	CG	Longshaw Grit	1700–1900
B. superbilingue MB	Ashover Grit unit 5	AG5	Chatsworth Grit	1700–1900
	unit 4–1	AG4		
R. gracile MB	Kinderscout Grit	KG	No change	2000–2150
Base Namurian Clastics	Lower Carboniferous Limestone	LC	No change	2150

* The early misidentification of SAC as 'Rough Rock' was corrected by Howitt, in Brunstrom (1963).
MB: Marine Band
BOD: Below Ordnance Datum

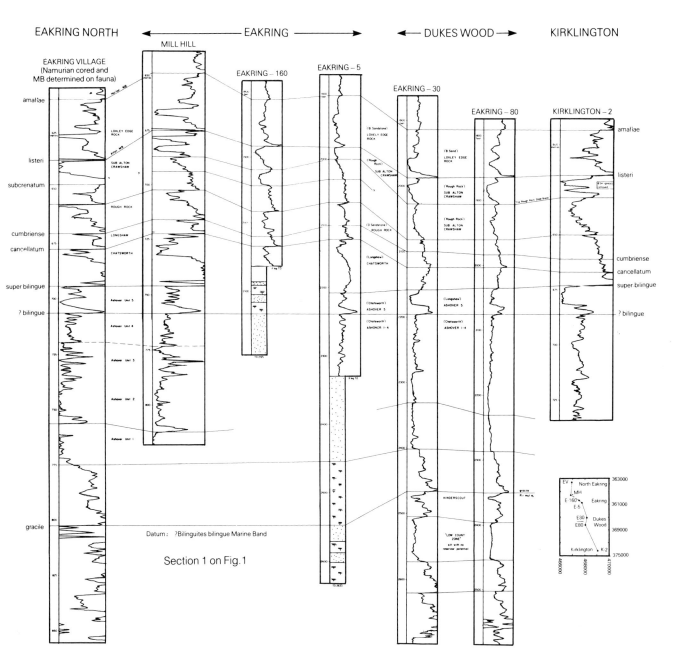

Fig. 2. Eakring North to Kirklington: gamma ray log correlation of Silesian interval (Norton marine band to top Dinantian).

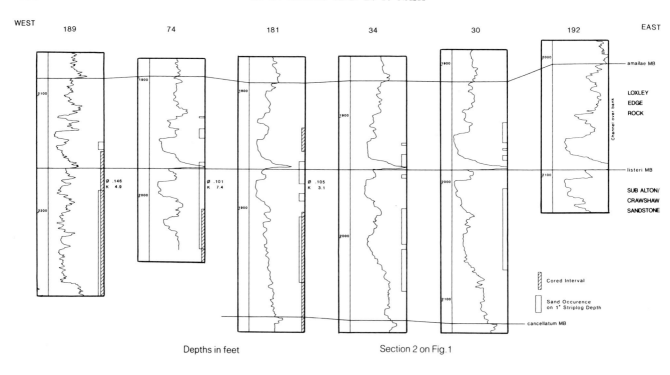

Fig. 3. Loxley Edge Rock and Sub Alton/Crawshaw sandstones: gamma ray log section showing reservoir development.

most of the field, except in Eakring North and east of Eakring. Below a field-wide shale break is a lower coarsening-upward sequence of variable extent and thickness. Compare for example Wells 189, 74 and 181 on Fig. 3: whereas the gross sand interval may reach 80 ft, net sand rarely exceeds 45 ft (using GR log cut-off).

proved impossible to separate the data above and below the shale break, they are combined (Fig. 4). Mercury injection data give S_{wi} 45% (average porosity = 14.6, k = 23.0 mD) and the fw = 0 level is c. 60 ft above FWL.

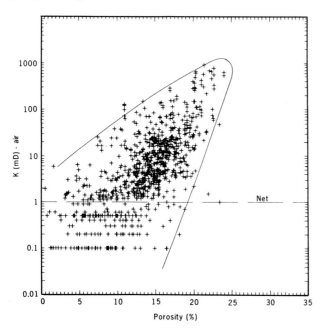

Fig. 4. Sub Alton/Crawshaw reservoir: rock properties (data from 33 wells).

Core data are available from 33 wells for this reservoir. Five of these wells have GR logs, but they are not readily correlatable to the core. Neither, in the unlogged wells, is it clear where the core was actually taken, but the general impression is that the upper unit is the more permeable. Since it has

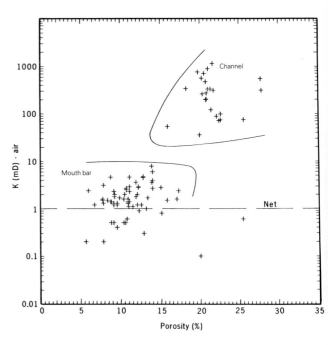

Fig. 5. Ashover Grit, Unit 5 reservoir: rock properties (data from 6 wells).

Ashover Grit, Units 5–1 AG5-1

The Ashover Grit is a sequence of sandstones over 200 ft thick lying below the *B. superbilingue* and above the *R. gracile* marine bands. Its uppermost unit, AG5, is defined by a field-wide 10–15 ft shale break believed to contain the *B. bilingue*

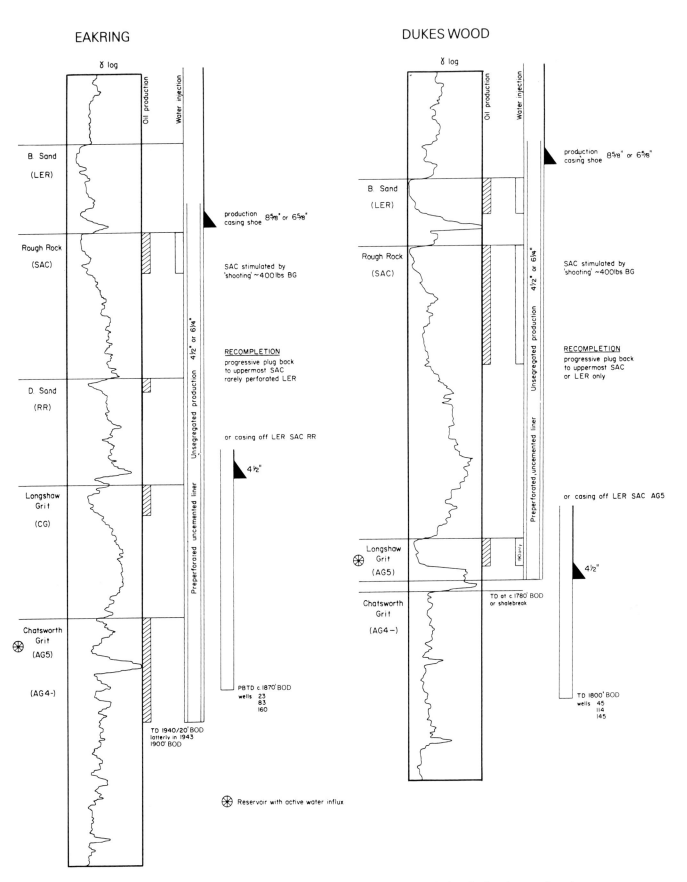

Fig. 6. Well completion and production practice at Eakring and Dukeswood. (Note: new terminology is given in parentheses.)

MB. Well test data show it to be a barrier to vertical flow. The lower units are thought to be in vertical communication, with the exception of Unit 1 in Eakring North (see below).

The AG5 is approximately 20 ft thick. From regional studies and core in the modern wells Eakring Village BH, an NCB exploration borehole deepened by BP, and Kirklington 2, it is interpreted as a distributary channel/mouthbar complex (B. Mitchener pers. comm.). The two facies have very different reservoir properties (Fig. 5), and are readily identified by their production characteristics. The channel sands possess excellent petrophysical properties and S_{wi} is 30% with only a 20 ft capillary transition zone above FWL.

AG Units 4–1 are dominantly coarse-grained sandstones approximately 2000 ft thick with minor intraformational shales. They possess poor to moderate reservoir properties, in general.

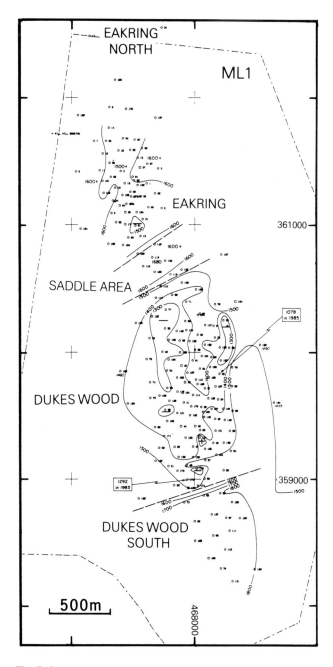

Fig. 7. Isopressure map of Loxley Edge Rock–Sub Alton/Crawshaw reservoir in 1967, two years after field shut-in. Injection and production connection compartments are shown (pressure in psig, 1000 initial).

Sub-surface development

All reservoirs were initially undersaturated, with gas–oil ratios between 40 and 80 SCF/STB. API gravities ranged from 31.9–37.6° and all oils were waxy, up to 16% by weight. The initial field development during 1939–1941 was on a five acre spacing with an 8.625-inch production string set above the main reservoir section. The latter was then drilled with stabilized crude to TD, and a 6.25-inch mixed casing/slotted liner string made up and set against the reservoirs. Production under beam pump lift was unsegregated, with up to four horizons exposed. In the Eakring sector these were the SAC, RR, CG and AG to 1940 ft BOD, and at Dukeswood the LER, SAC and AG to 1780 ft BOD, or base AG5 (Fig. 6). Completion philosophy differed because at Dukeswood the AG5 was recognized as separate from the underlying AG4, and most wells were terminated in the shale break above AG4. At Eakring the distinction was not made, the AG being treated as one continuous unit.

During the 1942–43 infill drilling campaign, the producing area well spacing was reduced to 1.25 acres and completions were 6.625 inches/4.5 inches, again unsegregated, but drilled with mud. TD at Eakring was raised to 1900 ft BOD.

At this time it was clear that the AG had localized but strong aquifer support whereas the LER and SAC were dominantly solution gas driven. The LER/SAC reservoir pressures were as low as 200 psi (c. 1000 psi initial) and well below bubble point since producing GOR was four times the initial GOR. Supplementary recovery was implemented in full in 1948 by peripheral water injection to the LER and SAC. Response was rapid and injection continued, reaching pressure equilibrium at twice initial reservoir pressure. The process evolved into a recycled water flood operation and the production mechanism was effectively by viscous stripping of the SAC oil due to the poor mobility ratio. Capillary and gravity forces were overwhelmed since gross liquid production rates were often as high as ten times initial pre-flood rates. Well stimulation was by 'shooting', exploding up to 200 kg of gelignite across the reservoir in open hole, against LER and SAC only (Waters 1946).

As water cuts increased due either to aquifer influx (AG) or water flood breakthrough (LER/SAC), recompletion policy was either progressively to plug the hole back up for cut off, or less commonly, case off the LER/SAC leaving only the CG, AG open.

Production geology

It is normal for the production geologist to construct a static model for reservoir engineers to simulate under dynamic conditions. The model can then be tested against production history and refined. Unswept areas and bypassed oil can be identified. For Eakring Dukeswood the process has been reversed since the basic building blocks for geological modelling are absent.

The following engineering data were used: reservoir pressure, well gross liquid rate (GLR) production and water cut variation.

In the field several wells were maintained for pressure observation in the LER/SAC. These provide valuable, if localized, records of reservoir depletion and later repressurization of the water flood. For connectivity studies, residual pressures measured at the wellhead for two years after field shut-in in 1965–66 have proved most useful. Converted to reservoir datum, they highlight the field's major no-flow barriers and allow field subdivision into LER/SAC production compartments (Fig. 7). In particular, note the barrier between the Eakring and Dukeswood culminations, the Saddle Area, and the distinctly separate Dukeswood South unit.

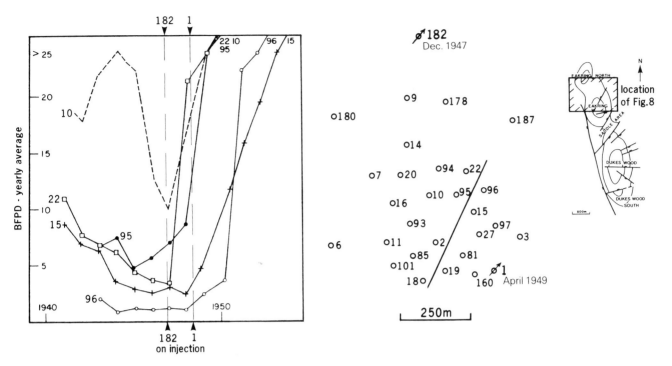

Fig. 8. Eakring area production behaviour in response to water flood, showing barriers exemplified by Wells 10/22/95 and 15/96.

The wells were production tested monthly and the GLR record is excellent. It can be utilized in two ways: firstly by mapping the time of decline rate reversals in response to the water flood; and secondly to study the effects of individual injectors. The first technique points to the transmissibility fairways and relative magnitudes. The second identifies localized flow barriers (e.g. Fig. 8): injector Well 182 was commissioned at the end of 1947 and producing wells to the south all responded during 1948, but only west of Wells 15 and 96. These, and other eastern wells, continued to decline until Well 1 was converted to injection in April 1949.

Water cut changes can be used to point to transmissibility variations both in the flooded horizons and where there was active aquifer influx. Additionally, since there are only two wells with electrical surveys, oil–water levels for STOIIP computations are taken as the level above which dry oil was initially produced. The sparse capillary pressure data indicate transition zones between 20 and 60 ft from FWL to fw = 0, for all reservoirs.

Discussion

Over the structural culminations, wells were completed across both LER/SAC and AG, but towards the periphery, completions were only in the LER/SAC. Recording the date of the onset of more than 10% water cut in the peripheral wells, therefore, indicates natural aquifer ingress to the LER/SAC until the flood commenced. Influx to the LER/SAC system was limited from the west at Eakring and east at Dukeswood (Fig. 9). Injection commenced in earnest in 1947 at Eakring and 1948 at Dukeswood, and the 10% water cut isochrons are seen then to move rapidly, although selectively, up-dip. Eakring was overrun in the SAC within three years but at Dukeswood two areas of low transmissibility are indicated.

In contrast, in the AG5 of Dukeswood (Fig. 10), high water cuts developed from the start in some central wells; that is, those completed on both LER/SAC and AG5. Since they lie outside the LER/SAC 10% contour, the water can only be AG5 aquifer influx. A high transmissibility channel system is thus indicated, open to the regional aquifer to the east and south. Approximately half the Dukeswood AG wells went

'wet' in this manner, the remainder staying 'dry' until the arrival of the LER/SAC flood. The channel boundaries cutting the mouth bar sands of the AG5 are clearly defined by this behaviour.

Most of the high productivity 'wet' wells were recompleted out of the AG5 before the arrival of the LER/SAC flood front to the borehole. In the 'dry' wells the AG5 was still open and its contribution to total production would have been killed by the higher pressure LER/SAC water. Since AG5 'dry' well transmissibilities were probably less than the overlying SAC, and the 'wet' wells had been plugged back, the net effect of the LER/SAC flood was to kill all AG5 production. There was no benefit to the AG5 and for the past 40 years capillary and gravitational forces have acted on this partly-depleted reservoir to create a secondary accumulation in the coincidentally high permeability crestal areas.

Although the AG5/4 separation was not then recognized at Eakring, a similar 'wet' and 'dry' situation occurred. Wet wells were not plugged back but the dump flood from the overrunning SAC injection will have killed AG production in the same manner.

The water flood was targeted at the SAC at Eakring and LER/SAC in Dukeswood. In the former area the LER, being thin (less than 10 ft) was not normally open. Only at Dukeswood was it incorporated into the primary completion interval. There, in the early–middle 1950s, some 30 wells with very high water cuts were plugged back to leave only the LER open. GLRs dropped dramatically, as did water cut, implying little pressure support to the LER. Only in wells with an LER injector close by were high water cuts maintained. For example, the situation in 1959 on the Dukeswood west flank is shown in Fig. 11. In the south, LER recompletions are close to injector 132 and all have high water cuts; whereas to the north, with no adjacent LER injection, water cuts of less than 30% were recorded.

The inevitable conclusion is that LER/SAC injection was dominantly to the continuous SAC unit and that the LER, being apparently discontinuous on logs and certainly in behaviour, was poorly swept, other than by intra-well cross-flow effects. Where there were no LER recompletions, recovery will have been inefficient (Fig. 12).

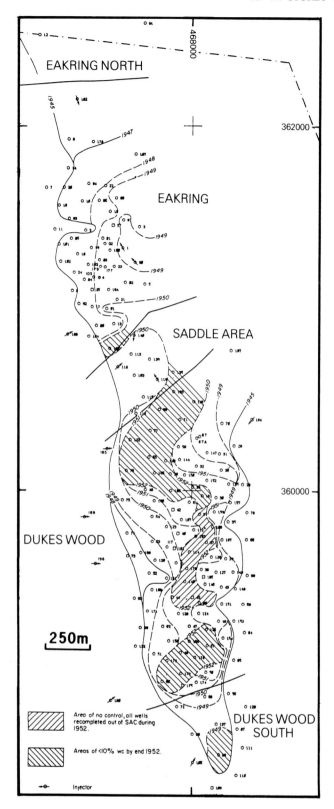

Fig. 9. Sub Alton/Crawshaw reservoir 10% water cut isochronology to the end of 1952, showing negligible natural water influx.

As noted above, only the upper SAC (I) reservoir is field-wide and the lower SAC (II) is variably developed. The flood water would, therefore, have been concentrated into the upper unit. Although there are no rock relative permeability curves for the SAC, the high dead oil viscosity of *c*. 20 cp and low initial gas–oil ratio of 80 SCF/BBL suggests that the mobility ratio will have been unfavourable and much oil will have been bypassed by viscous fingering. Counteracting this effect will

have been viscous stripping due to the large water throughput volumes, of 6.7 times oil production at Eakring and 5.9 times at Dukeswood. Under this regime much of the SAC II oil will have been unswept, and after field shut-in in 1965–66, capillary and gravitational forces will have encouraged crestal migration.

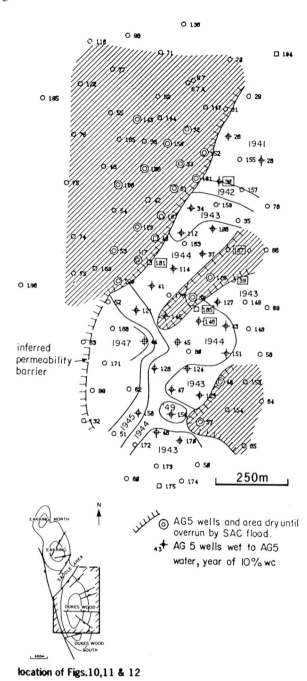

location of Figs.10,11 & 12

Fig. 10. Dukeswood Ashover Grit, Unit 5 reservoir 10% water cut isochronology to the end of 1947. This shows localized water influx controlled by the depositional environment.

The lowermost AG, Unit 1, was never considered to be of importance during the field development, as it lies well below the perceived oil–water level. No GR logs were run over it at Eakring, and in Dukeswood it is not an obviously separate unit. Its potential becomes apparent from the core analysis conducted on the Eakring Village BH. Here, and at nearby Mill Hill BH, it is overlain by a 10 ft thick shale (Fig. 2). Mercury injection and oil–water relative permeability data are summarized for this unit in Fig. 13 with the well's petrophysical evaluation data superimposed. The well appears to have penetrated a mobile transition zone overlying S_{or}, and that a gain of 20 ft in elevation would put the top of the zone above the dry oil-producing level. Since the well is off structure, the implication is that a hitherto unsuspected accumulation exists beneath the undrilled Eakring North culmination possibly extending under the main Eakring structure. In the latter field area there are no GR logs but the shale marker is recorded on the geologists' logs in crestal Wells 177 and 179, although shows in the sands are not.

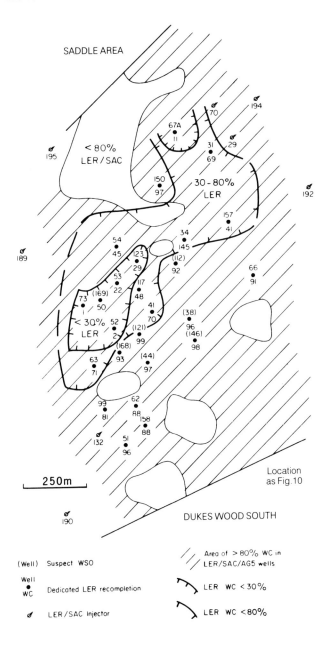

Fig. 11. Dukeswood water cut values in 1959 for dedicated Loxley Edge Rock recompletions, showing the limited effect of injection.

STOIIP and recoveries

Uncertainties in STOIIP are huge. While core data provide useful information on porosity and hydrocarbon saturation, reservoir net and gross thickness in the heterogeneous units remains poorly known since the bulk of the well lithology logs were prepared from the drillers' log books. The latter sand records bear little comparison to the gamma ray logs run a decade or more later (Fig. 3). Nonetheless, reservoir thick-

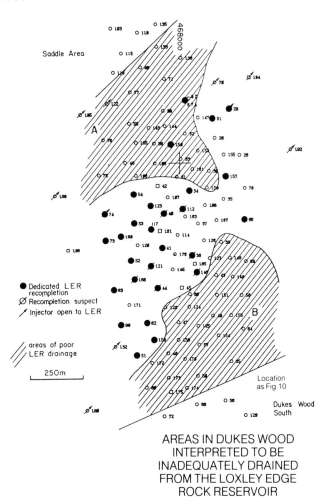

AREAS IN DUKES WOOD INTERPRETED TO BE INADEQUATELY DRAINED FROM THE LOXLEY EDGE ROCK RESERVOIR

Fig. 12. Loxley Edge Rock dedicated completions, Dukeswood main.

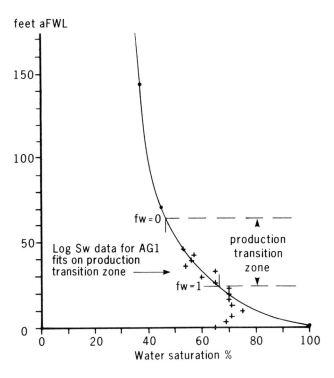

Fig. 13. Mercury injection relative permeability data for Ashover Grit, Unit 1, in the Eakring Village BH.

Table 2. STOIIP and ultimate recoveries

| | Eakring* | | | Dukes Wood† | | |
	STOIIP (MSTB)	Recovery (MSTB)	%	STOIIP (MSTB)	Recovery (MSTB)	%
Wingfield Flags	2758	2	0	9606	12	0
Loxley Edge Rock	813 ⎫			4127 ⎫		
Sub Alton/Crawshaw sst	2677 ⎭	1675	47	7494 ⎭	3830	33
Rough Rock	688	34‡	5	Absent		
Chatsworth Grit	644	32‡	5	Absent		
Ashover Grit 5	1612	277	7	3338	391	12
Ashover Grit 4-	2460			1086	109	10
Kinderscout Grit	244	37	15	279	10	5
L. Carboniferous	0			178	69	38

*Includes Eakring North and Saddle Area
†Includes Dukes Wood South
‡Estimates

nesses recorded in the geological monthly reports have to be used. Reservoir well elevations are not so critical since these are low dip structures (10° maximum) and the areal extent of fluid levels is reasonably well known. Since the objective was to locate undrained oil potential rather than produce an accurate recovery factor, the deterministic Equivalent Oil Column (EOC) method was used.

Assignment of individual reservoir contribution has been through consideration of the production mechanism and completion policy. Production from the WF, KG and LC was from dedicated wells but there were no dedicated RR and CG wells and since both are low permeability units, an arbitrary 5% recovery has been assigned to them.

For AG5 production, it is concluded that in the 'dry' well areas there was little contribution, and the wells' production is assigned solely to the LER/SAC. In the 'wet' well areas, production is assigned 50/50 between LER/SAC and AG5. At Dukeswood AG4 production is known from dedicated recompletions and back-extrapolation of their decline rate curves, but at Eakring no distinction can be made with the AG5.

Production from the LER/SAC has proved impossible to separate, although most will have been from the SAC, and the two are therefore combined. Evaluations are shown in Table 2.

The high recoveries assigned to the LER/SAC are supportable. Water flood experiments on SAC core plugs established S_{or} at 30% with 98% water cut. Since S_{wi} can be reasonably taken at 40%, 50% recovery is possible, with 100% sweep efficiency from the SAC encouraged by the very close well spacing.

Other reservoirs

Wingfield Flags

Low permeability micaceous sands 50–60 ft thick which proved poorly productive despite propped fracture stimulation. The oil has a pour point coincident with reservoir temperature. Seven dedicated completions were made, but the accumulation is effectively unproduced. It extends across the whole field area. Average porosity: 12.1%; permeability: 4.2 mD (63 samples from 4 wells).

Rough rock

Only present in the Eakring North and Eakring sectors. It is rarely thicker than 15 ft and exhibits poor reservoir properties. There were no dedicated completions. Average porosity: 13.9%; permeability: 4.9 mD (17 samples from 3 wells).

Chatsworth Grit

Only present in the Eakring sector but is up to 40 ft in thickness. Reservoir properties are poor and there were no dedicated completions. Average porosity: 12.5%; permeability: 0.9 mD (61 samples from 5 wells).

Kinderscout Grit

Contained small accumulations at both Eakring and Dukeswood. The former exhibited water drive and is effectively exhausted by one dedicated well. There were three wells at Dukeswood where the drive was by solution gas, and reservoir quality much poorer. Average porosity: 13.2%; permeability: 51.9 mD (21 samples from 3 wells).

Lower Carboniferous

Contained oil at Dukeswood only where 69 MSTB was obtained from seven dedicated wells. The reservoir is poorly understood. Traditionally the top of the Lower Carboniferous was taken at the first occurrence of limestones, underlain by erratic sandbodies. These may be channels or karst infill, although whatever their interpretation, production was only obtained when the carbonates were naturally fractured.

Conclusion

The main findings of this work are that the reservoirs are not fully depleted and that the field has remaining production potential. We show that with extensive production history, particularly under pressure maintenance, the lack of modern data is no detriment to the creation of geological models for the evaluation of remaining potential. Water movement through the reservoirs has effectively delineated the sedimentological heterogeneity and tectonic barriers.

The production mechanism for the LER, SAC and AG5 reservoirs as interpreted in this study is summarized in Fig. 14. Initial recoveries from the LER and SAC were by depletion drive, and partly by water drive in the AG. The LER/SAC water flood was concentrated through the upper SAC leaving much of the LER and lower SAC unswept, and effectively killing AG5 production. The low permeability facies of the AG5 are effectively undrained.

Production rates were uncontrolled in the water-driven AG wells as was SAC water injection, both encouraging viscous fingering. Since field shut-in 25 years ago, bypassed and unswept oil will have migrated up-dip to form secondary accumulations.

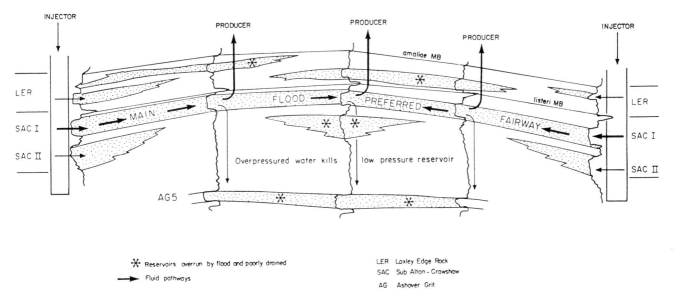

Fig. 14. Schematic field production system.

The Eakring North structural culmination is undrilled and is prospective not only for the established LER/SAC/AG5/4 system, but also for a hitherto unsuspected AG1 unit.

Finally, it should be noted that the Wingfield Flags reservoir accumulation remains essentially unproduced.

We wish to thank BP management for allowing us to make public this work, Bill Walbank and Alan James for critical discussions during preparation of the paper, Pia Walmsaess for BP Norway for the illustrations, and Jennifer Lythell for coping with many text revisions.

References

BRUNSTROM, R. G. W. 1963. Recently discovered oilfields in Britain. *Proceedings of the 6th World Congress*, **1/49** 1–10.

DICKIE, R. K. AND ADCOCK, C. M. 1954. Oil production in Nottinghamshire oilfields. *Journal of the Institute of Petroleum*, **40**, 179–190.

FALCON, N. L. AND KENT, P. E. 1960. Geological Results of Petroleum Exploration in Britain 1945–57. Geological Society of London, Memoir No. 2, 53–54.

FRASER, A. J., NASH, D. F., STEELE, R. P. AND EBDON, C. C. 1990. A regional assessment of the intra Carboniferous play of northern England. *In*: BROOKS, J. (ed.) *Classic Petroleum Provinces*. Geological Society, London, Special Publication, **50**, 417–440.

LEES, G. M. and TAITT, A. H. 1945. The geological results of the search for oil fields in Great Britain. *Quarterly Journal of the Geological Society of London*, Vol. **ci** parts 3 and 4 (for 1945), 225–317.

WATERS, J. F. 1946. Some problems encountered during well-shooting operations in the Nottinghamshire Oilfields. *Journal of the Institute of Petroleum*, **32** (267), 119–126.

Core workshop
and discussion forum

Core workshop and discussion forum

C. D. OAKMAN[1] and J. H. MARTIN[2]

[1] Reservoir Research Ltd, 4.03 Kelvin Campus, West of Scotland Science Park, Glasgow G20 0SP, UK
[2] Department of Geology, Imperial College of Science, Technology and Medicine, Prince Consort Road, London SW7 2BP, UK (Present address: Reservoir Geological Consultant, 150 Croxted Road, London SE21 8NW, UK)

The concept of public core workshops is well established in North America, while in recent years smaller displays in the UK have been arranged by, for example, the British Geological Survey in conjunction with the Petroleum Exploration Society of Great Britain, and the British Sedimentological Research Group. The workshop and discussion session held in connection with this conference was, however, the most ambitious arranged to date in NW Europe.

Originally envisaged as a forum in which contributing authors could illustrate their oral or poster presentations by displaying original material, it expanded to cover representative sections of the major depositional systems within the North Sea. The final outcome was that material submitted by contributing authors (Table 1) was organized and displayed along sequence stratigraphic lines. Around 600 m of core from 58 wells, mainly representing the UK sector, but with shorter sections from the Norwegian and Danish sectors, was displayed. Hand specimens were also included to illustrate field studies in Greenland and NW Scotland (Christiansen et al. 1993; England et al. 1993; Morton 1993).

Table 1. Summary of core segments displayed

WELL	FIELD	GEOGRAPHICAL AREA	AGE
United Kingdom			
9/9b-10	Bruce	Beryl Embayment	Jurassic
9/13a-27 9/13c-33 9/13a-34 9/13a-42 9/13a-47 9/13b-53 9/13a-A02 9/13a-A30 9/13a-A46 9/13a-A50 9/13a-B10 9/13a-B17 9/13a-B21	Beryl	Beryl Embayment	Jurassic and Triassic
9/18b-**	Gryphon	South Viking Graben	Tertiary
11/30-2 11/30a-A4	Beatrice	Inner Moray Firth	Jurassic
12/27-1	—	Inner Moray Firth	Jurassic
12/29-1	—	Inner Moray Firth	Jurassic
13/28-2	Ross	South Halibut Graben	Jurassic
15/20a-5 15/20a-7	—	Fladen Ground Spur	Tertiary
16/7b-20 16/8b-A2	Miller	South Viking Graben	Jurassic
16/17-14 16/17-15	Tiffany	South Viking Graben	Jurassic
16/18-2	—	South Viking Graben	Tertiary
20/2-5	—	Buchan Graben	Jurassic
21/9-4	—	East Buchan Graben	Tertiary
21/9-5A	—	Forties Montrose High	Tertiary
21/18-3	Kittiwake	West Central Graben	Jurassic
22/6a-9	Nelson	Forties Montrose High	Tertiary
22/10a-4	Everest	Central Graben	Tertiary
22/21-5	Gannet	Central Graben	Tertiary
22/25-2	—	Central Graben	Tertiary
30/17b-9	Clyde	Central Graben	Jurassic
44/23-9	Caister	Southern North Sea	Permian and Carboniferous
48/6-34	Hyde	Southern North Sea	Permian
206/8-8	Clair	Rona Ridge	Devonian
211/12-2 211/12-3A 211/12-5 211/12a-M1 211/12a-M12	Magnus	East Shetland Basin	Jurassic
211/23-7	—	East Shetland Basin	Jurassic
Norway			
2/7-26S	Embla	Central Graben	? Devonian
Denmark			
Dan-1 Dan-2	Dan	Central Graben	Cretaceous
Gorm-1	Gorm	Central Graben	Cretaceous
Kraka-1 Kraka-2	Kraka	Central Graben	Cretaceous
Tyra-1	Tyra	Central Graben	Cretaceous
Valdemar-1	Valdemar	Central Graben	Cretaceous
Iris-1	Iris	Central Graben	Jurassic
Gert-1	Gert	Central Graben	Jurassic
Ravn-1	Ravn	Central Graben	Jurassic
West Lulu-2 West Lulu-3	West Lulu	Central Graben	Jurassic

**Well not released

Organization was not without its logistical problems as material was obtained from oil company and service company facilities in Aberdeen, Glasgow, London and SE England, as well as from Stavanger and Copenhagen. Selection and transport of core material required good co-ordination, and considerable time. Resinated slabs were mainly used, although several non-resinated cores required careful transport and handling while on display. During the final days prior to the

From *Petroleum Geology of Northwest Europe: Proceedings of the 4th Conference* (edited by J. R. Parker).
© 1993 Petroleum Geology '86 Ltd. Published by The Geological Society, London, pp. 1541–1542.

conference material was stored in a secure area in London: access to such a facility would be a prerequisite if a similar workshop were held at forthcoming conferences.

All contributors (too numerous to name individually) made a substantial contribution to the success of the conference as a whole; we will mention here only a few of the displays which attracted much attention:

- core from horizontal chalk wells, displaying fine stylolitization as well as unique sedimentary structures, the subject of debate amongst carbonate experts present (Fine *et al.* 1993);
- unconsolidated core from the Gryphon Field, with spectacular deformation structures (Newman *et al.* 1993);
- important thematic displays from the Tertiary (Anderton 1993) and also Jurassic sections from the Beryl Field (Dean 1993; Robertson 1993).

Subject to approval of contributing authors a volume containing 1:5 scale colour core photographs of released material displayed at the workshop will be published as a supplement to these volumes.

During the conference, 1700 individual entries were recorded to the workshop area: certainly we believe that the vast majority of all registered delegates visited the display at least briefly. All visitors felt that the workshop gave a unique opportunity to discuss geological concepts in the light of the rocks themselves. Most authors commented on the high degree of feedback received from delegates: generally higher than that from conventional presentations.

We hope that the success of this inaugural event will encourage the adoption of a similar event in future conferences.

We would like to thank all contributors and their respective companies who provided core, hand specimens and poster displays, and are particularly grateful to those individuals who helped to select material for thematic displays. Jane Hill of Conference Associates Ltd and members of the Conference Committee dealt with ever more extensive administrative requests during the months preceding the conference, without knowing if the outcome could be successful.

CDO thanks Reservoir Research Ltd for use of facilities and time, and JHM acknowledges financial help from the Natural Environment Research Council during the compilation of a North Sea cored well database used in the early stages of selection of core material. Griff Cordey from Shell UK generously provided access to an in-house database which considerably eased the task of compiling material for display. Facilities of the Royal School of Mines, Imperial College, London, were used for core storage.

References

ANDERTON, R. 1993. Sedimentation and basin evolution in the Paleogene of the Northern North Sea and Faroe–Shetland basins. *In*: PARKER, J. R. (ed.) *Petroleum Geology of Northwest Europe: Proceedings of the 4th Conference*. Geological Society, London, 31.

CHRISTIANSEN, F. G., LARSEN, H. C., MARCUSSEN, C., PIASECKI, S. AND STEMMERIK, L. 1993. Late Paleozoic plays in East Greenland. *In*: PARKER, J. R. (ed.) *Petroleum Geology of Northwest Europe: Proceedings of the 4th Conference*. Geological Society, London, 657–666.

DEAN, K. P. 1993. Sedimentation of Upper Triassic reservoirs in the Beryl Embayment: lacustrine sedimentation in a semi-arid environment. *In*: PARKER, J. R. (ed.) *Petroleum Geology of Northwest Europe: Proceedings of the 4th Conference*. Geological Society, London, 581.

ENGLAND, R. W., BUTLER, R. W. H. AND HUTTON, D. H. W. 1993. The role of Paleocene magmatism in the Tertiary evolution of basins on the NW seaboard. *In*: PARKER, J. R. (ed.) *Petroleum Geology of Northwest Europe: Proceedings of the 4th Conference*. Geological Society, London, 97–105.

FINE, S., YUSAS, M. R. AND JØRGENSEN, L. N. 1993. Geological aspects of horizontal drilling in chalks from the Danish sector of the North Sea. *In*: PARKER, J. R. (ed.) *Petroleum Geology of Northwest Europe: Proceedings of the 4th Conference*. Geological Society, London, 1483–1489.

MORTON, N. 1993. Potential reservoir and source rocks in relation to Upper Triassic to Middle Jurassic sequence stratigraphy, Atlantic margin basins of the British Isles. *In*: PARKER, J. R. (ed.) *Petroleum Geology of Northwest Europe: Proceedings of the 4th Conference*. Geological Society, London, 285–297.

NEWMAN, M. St J., REEDER, M. L., WOODRUFF, A. H. W. AND HUTTON, I. R. 1993. The geology of the Gryphon Oil Field. *In*: PARKER, J. R. (ed.) *Petroleum Geology of Northwest Europe: Proceedings of the 4th Conference*. Geological Society, London, 123–133.

ROBERTSON, G. 1993. Beryl Field: geological evolution and reservoir behaviour. *In*: PARKER, J. R. (ed.) *Petroleum Geology of Northwest Europe: Proceedings of the 4th Conference*. Geological Society, London, 1491–1502.

Appendix: *A list of common abbreviations*

°API	Oil gravity, American Petroleum Institute degrees
BBL	Barrel
BCF	Billion cubic feet
BCPD	Barrels of condensate per day
BGS	British Geological Survey
BMSL	Below mean sea-level
BOPD	Barrels of oil per day
BWPD	Barrels of water per day
DEN	UK Department of Energy (now Department of Trade and Industry)
DST	Drill stem test
FIT	Formation interval test
FT SS	Feet, subsea
GIIP	Gas initially in place
GOC	Gas–oil contact
GOR	Gas–oil ratio
GWC	Gas–water contact
Ma	million years
mD	millidarcies
MMBBL	Million barrels
MMSCF(D)	Million standard cubic feet (per day)
ms	milliseconds
NGL	Natural gas liquids
NOCS	Norwegian Continental Shelf
OWC	Oil–water contact
PSI	Pounds per square inch
PVT	Pressure, volume, temperature
RFT	Repeat formation test
RKB	Below kelly bushing
r_o	Resistivity (oil)
R_w	Resistivity (water)
SCF	Standard cubic feet
Sm^3	Standard cubic metres
SS	Subsea
STB	Stock tank barrel
STOIIP	Oil initially in place
TCF	Trillion cubic feet
TD	Total depth
TOC	Total organic carbon
TVD	True vertical depth
TV(D)SS	True vertical depth, subsea
TWT	two-way time
UKCS	United Kingdom Continental Shelf

Index

References in italic are to figures and in bold are to tables.